Intelligent Vehicles

Intelligent Vehicles

Editors

David Fernández-Llorca
Ignacio Parra Alonso
Iván García Daza
Noelia Hernández Parra

MDPI • Basel • Beijing • Wuhan • Barcelona • Belgrade • Manchester • Tokyo • Cluj • Tianjin

MDPI

Editors
David Fernández-Llorca
Universidad de Alcalá
Spain

Ignacio Parra Alonso
Universidad de Alcalá
Spain

Iván García Daza
Universidad de Alcalá
Spain

Noelia Hernández Parra
Universidad de Alcalá
Spain

Editorial Office
MDPI
St. Alban-Anlage 66
4052 Basel, Switzerland

This is a reprint of articles from the Special Issue published online in the open access journal *Sensors* (ISSN 1424-8220) (available at: https://www.mdpi.com/journal/sensors/special_issues/intel_veh).

For citation purposes, cite each article independently as indicated on the article page online and as indicated below:

LastName, A.A.; LastName, B.B.; LastName, C.C. Article Title. *Journal Name* **Year**, *Article Number*, Page Range.

ISBN 978-3-03943-402-2 (Hbk)
ISBN 978-3-03943-403-9 (PDF)

Contents

About the Editors

David Fernández-Llorca received the MSc and PhD degrees in Telecommunications Engineering (2003 and 2008) from the University of Alcalá (UAH) and was the recipient of the Best Master Thesis Award from ADA Lectureship Madrid (2004), the Best Academic Record Award from IVECO (2004) and the Best PhD Thesis UAH (2010). He is Full Professor and Head of the INVETT Research Group. He is the author of more than 120 publications. He was ranked as the #1 most collaborative author in ITS in Europe during the period of 2010–2013. He is co-inventor in 15 patents (10 in exploitation). He has been a recipient of the IEEE ITSS Young Researcher Award (2018, under 40), the Young Research Medal of the Real Academía de Ingeniería (2018, under 40), Consejo Social Award (UAH, 2018), Best National Patent Award (UAH, 2017), the Best Team with Full Automation of GCDC 2016, the 2013 IEEE ITSS Outstanding Application Award, the Best Young Researcher Award from UAH in 2013, the Best Research Award in the domain of Automotive and Vehicle Applications in Spain in 2008 and the 3M Foundation Awards in 2009. He is currently Editor-in-Chief of the *IET Intelligent Transport Systems* journal and he has served as Associate Editor of the *IEEE Transactions on Intelligent Transportation Systems* and the *Journal of Advanced Transportation*.

Ignacio Parra Alonso received the MSc and PhD degrees in Telecommunications Engineering from the University of Alcalá (UAH) in 2005 and 2010, respectively. He currently works as an Associate Professor at the Computer Engineering Department, UAH. His research interests include intelligent transportation systems and computer vision. He received the Master Thesis Award in eSafety from the ADA Lectureship at the Technical University of Madrid, Spain, in 2006 and the 3M Foundation Award under the category of eSafety in 2009.

Iván García Daza received the MSc and PhD degrees in Telecommunications Engineering from the University of Alcalá (UAH), Madrid (Spain), in 2004 and 2011, respectively. His thesis presented a new approach for estimating the drowsiness level of the driver to provide alerts about dangerous situations in the driving task by using artificial intelligence techniques. At present, he is Assistant Professor with the Computer Engineering Department at UAH, and he has been a member of the INVETT research group since 2007. His research interests are mainly focused on intelligent transportation systems and intelligent vehicles, more specifically on topics such as the HD mapping process, modeling the car dynamics and cinematics to improve the behavior of Bayesian filters, sensor data fusion, LiDAR perception and optimization algorithms to improve the odometry systems errors. He is the author of more than 20 referenced papers in international journals and conference proceedings.

Noelia Hernández Parra received the MSc and PhD degrees in Advanced Electronics Systems (Intelligent Systems) from the University of Alcalá (UAH) in 2009 and 2014, respectively. Her thesis presented a new approach for estimating the global position of a mobile device in indoor environments by using WiFi, receiving the Best PhD Award from the UAH in 2014. She currently works as an Assistant Professor at the Computer Engineering Department of the UAH. Her research interest mainly focuses on indoor and outdoor localization, artificial intelligence and intelligent transportation systems.

Preface to "Intelligent Vehicles"

The concept of an intelligent vehicle is quite simple: a "vehicle" endowed with "intelligence". The word "vehicle" can be applied here to different modes of transport, including road vehicles, trains, planes or ships. The word "intelligence", however, can be a topic of discussion that may involve writing an entire book to clarify it. Sometimes the definition of intelligence is simplified to human intelligence (e.g., "driving as humans"), but human intelligence is even more complicated to address and understand than artificial intelligence (e.g., theory of multiple intelligences). Furthermore, limiting the potential of artificial systems to what humans can do is like driving with only low beam headlights. In fact, although intelligent vehicles thus far generally have lower performance than humans, there is no reason not to think that they may far exceed the abilities of humans in the not too distant future.

An intelligent vehicle is a highly complex system, with a large set of technological components and subsystems, such as advanced electronics, mechatronics, communications, sensors and actuators, to deal with many different problems such as perception, sensing and situation awareness, positioning, localization, navigation and planning, low-level vehicle control, vehicular communications, actuation, trajectory planning and prediction. Among all the problems that intelligent vehicles must solve, we venture to establish that tasks such as perception, scene understanding, situation awareness are perhaps the most complex and the most critical in ensuring the safety and efficiency of their operation (and, therefore, their future adoption). The proper execution of such tasks is probably the main bottleneck that the scientific community and industry must solve. One of the main questions is what we mean here when we say "solve", since the variability of possible scenarios in which an intelligent vehicle has to operate is almost infinite. This has been considered as the "never-ending problem", and it is very reasonable to imagine a future in which advances and improvements in perception systems are always in progress.

As of today (mid-2020), we can say that intelligent vehicles are not yet above human driving, and there is still a long way to go. This book provides 32 manuscripts included in the Sensors Special Issue on Intelligent Vehicles as a humble contribution towards the advancement of vehicles endowed with intelligence. Submissions were accepted between March 2019 and May 2020, and the result is a well-representative set of current research and developments related to perception and sensor technologies for intelligent vehicles. More specifically, the topics of the published manuscripts include advanced driver-assistance systems, automatic vehicle operation, vehicle positioning and localization, fault diagnosis, fail-aware and fail-operational positioning systems, object detection, tracking and prediction, road segmentation, lane detection, traffic light recognition, smart regenerative braking systems for electric vehicles, driver behavior modeling, simulation-based approaches and intelligent sensing, among others. We hope the reader will find these manuscripts interesting and useful.

David Fernández-Llorca, Ignacio Parra Alonso, Iván García Daza, Noelia Hernández Parra

Editors

sensors

MDPI

Editorial

Sensors and Sensing for Intelligent Vehicles

David Fernández Llorca *, Iván García Daza, Noelia Hernández Parra and Ignacio Parra Alonso

Computer Engineering Department, University of Alcalá, 28805 Madrid, Spain; ivan.garciad@uah.es (I.G.D.); noelia.hernandez@uah.es (N.H.P.); ignacio.parrra@uah.es (I.P.A.)

* Correspondence: david.fernandezl@uah.es

Received: 2 September 2020; Accepted: 3 September 2020; Published: 8 September 2020

Abstract: Over the past decades, both industry and academy have made enormous advancements in the field of intelligent vehicles, and a considerable number of prototypes are now driving our roads, railways, air and sea autonomously. However, there is still a long way to go before a widespread adoption. Among all the scientific and technical problems to be solved by intelligent vehicles, the ability to perceive, interpret, and fully understand the operational environment, as well as to infer future states and potential hazards, represent the most difficult and complex tasks, being probably the main bottlenecks that the scientific community and industry must solve in the coming years to ensure the safe and efficient operation of the vehicles (and, therefore, their future adoption). The great complexity and the almost infinite variety of possible scenarios in which an intelligent vehicle must operate, raise the problem of perception as an "endless" issue that will always be ongoing. As a humble contribution to the advancement of vehicles endowed with intelligence, we organized the Special Issue on Intelligent Vehicles. This work offers a complete analysis of all the mansucripts published, and presents the main conclusions drawn.

Keywords: intelligent vehicles; sensors; sensing; percepction; scene understanding; object detection and tracking; scene segmentation; vehicle positioning; fail-x systems; driver behavior modelling; automatic operation

1. Introduction

When referring to intelligent vehicles, we must be somewhat precise and establish an appropriate definition. The definition of the first word, vehicles, is straightforward. When talking about "vehicles" we can consider different modes of transport, including road vehicles and trains for land transportation, planes for air transportation and ships for water transportation. However, when dealing with the definition and understanding of the second word, intelligence, things get very complicated. Comparing the intelligence of vehicles with the intelligence of humans is one of the most common steps. Still, human intelligence is even more complicated to address and understand than artificial intelligence (e.g., the theory of multiple intelligences). Furthermore, limiting the potential of artificial systems to what humans can do is like driving with only low beam headlights. In fact, although so far intelligent vehicles, in general, have a lower performance than humans (which is reasonable since all the possible components of transportation have been devised for humans), nothing prevents us thinking that in the not-too distant future, they can far exceed the abilities that humans have to drive them.

Intelligent vehicles are highly complex systems designed with a broad set of technological components and sub-systems such as advanced electronics, mechatronics, communications, sensors and actuators, to deal with many different problems including perception, sensing and situation awareness, positioning, localization, navigation and planning, low-level vehicle control, vehicular communications, actuation, trajectory planning, prediction, etc. Among all the problems that intelligent vehicles must solve, the capability to sense, interpret, and fully understand the operational environment, as well as to infer future states and potential hazards, can be considered as the main challenges and, perhaps, the most complex and the most critical tasks to ensure the safety and efficiency of their operation (and, therefore, their future adoption). Perception and scene understanding are probably the main bottlenecks that the scientific community and industry must solve. One of the main questions is what we mean here when we say "solve", since the variability of possible scenarios in which an intelligent vehicle has to operate is almost infinite. This has been considered as the "never-ending problem", and it is very reasonable to imagine a future in which advances and improvements in perception systems are always in progress.

Still, during the last decades, both industry and academy have made tremendous advancements in this field, and a considerable number of prototypes are now driving our roads, railways, air and sea autonomously. As an example of the advances, we depict Figures 1 and 2, where we can observe the evolution of approaches such as pedestrian detection and road segmentation.

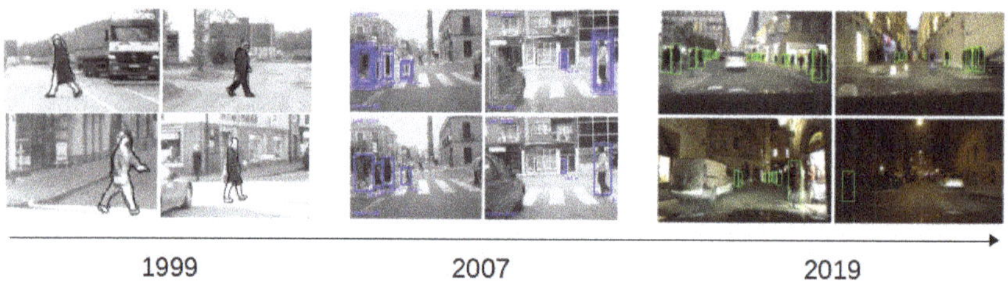

Figure 1. Evolution of pedestrian detection for intelligent vehicles with three examples from [1–3].

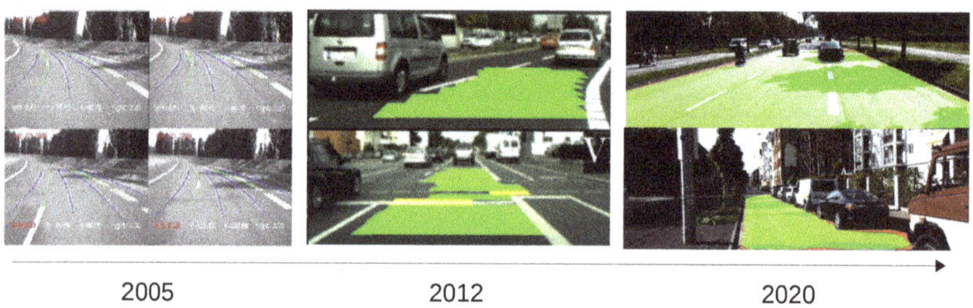

Figure 2. Evolution of road/lane segmentation for intelligent vehicles with three examples from [4–6].

The essential improvements in sensing for intelligent vehicles and intelligent transportation systems have come from the introduction of deep learning methodologies [7]. However, there is also a current trend focused on reducing the complexity of the models, as well as the dependence on data and learning-based approaches, by combining classical detection methods [8] or using prior information from different sources such as accurate digital maps [9].

In any case, as of today (mid-2020) we can say that intelligent vehicles are not yet above human driving and there is still a long way to go. In these difficult times, it is even more pressing that the transport of people and goods is done in the most automated way possible, allowing social distance when necessary [10], including assistive intelligent transportation systems [11] or even taking into account disruptive solutions [12].

As a humble contribution towards the advancement of vehicles endowed with intelligence, we organized the Special Issue on Intelligent Vehicles. In the next sections we provide a generic description of the Special Issue as well as a brief introduction to each of the manuscripts published in it.

2. Special Issue on Intelligent Vehicles

The call for papers of the Special Issue on Intelligent Vehicles was released on February 2019. The main goal of the issue was defined as "to contribute to the state-of-the-art, and to introduce current developments concerning the perception and sensor technologies for intelligent vehicles". Submissions were accepted until May 2020, a total of 32 manuscripts were finally accepted. The published papers are a well-representative set of current research and developments related to perception and sensor technologies for intelligent vehicles. They include topics such as advance driver assistance systems, automatic vehicle operation, vehicle positioning and localization, fault diagnosis, fail-aware and fail-operational positioning systems, object detection, tracking and prediction, road segmentation, lane detection, traffic lights recognition, smart regenerative braking systems for electric vehicles, driver behavior modeling, simulation-based approaches and intelligent sensing, among others.

As an example of the main concepts included in the Special Issue, a word cloud has been elaborated by using the titles and abstracts of the 32 manuscripts, which can be seen in Figure 3.

Figure 3. Word cloud from the titles and abstracts of the 32 publications included in the Special Issue (elaborated using tagcrowd).

To provide a better view of the distribution of sensors used in all the papers published in the Special Issue, we present in Figure 4 an alluvial diagram. We can see the correlation between the different proposals and the primary sensor used for each application. Note that, in most cases, the works adopt a multi-sensor approach, and the list of sensors used includes cameras (monocular and stereo), radar, LiDAR, GPS, IMU, vehicle sensors (using OBD interface to read from the CAN Bus), strain sensors, and smartphones.

Figure 4. Alluvial diagram showing the correlation between the sensors and the manuscripts.

Concerning the different topics and subtopics, we have identified up to seven main categories, and some sub-categories that are presented in the following list (the number of papers per each category/sub-category is enclosed in parentheses):

- Object detection and scene understanding (11)
  - Vehicle detection and tracking (4): [13–16].
  - Scene segmentation and interpretation (7)
    - Road segmentation (2): [17,18].
    - Shadow detection (1): [19].
    - Lane detection (2): [20,21].
    - Traffic lights detection (2): [22,23].
- Driver behavior modeling (5)
  - Lane change modeling (2): [24,25].
  - Driver behavior understanding (2): [26,27].
  - Driving behavior for simulation (1): [28].
- Fail-x systems and fault diagnosis (3)
  - Fail-x vehicle positioning (2): [29,30].
  - Fault diagnosis of vehicle motion sensors (1): [31].
- Vehicle positioning and path planning (5)
  - Simultaneous Localization and Mapping (1): [32].
  - Vehicle localization (3): [33–35].
  - Path planning (1): [36]
- Smart regenerative braking systems for electric vehicles (2): [37,38].
- Physical intelligence in sensors and sensing (3): [39–41]
- Driver assistance systems and automatic vehicle operation (3)
  - Advanced driver assistance systems (1): [42].

- Automatic parking of road vehicles (1): [43].
- Automatic train operation (1): [44].

As can be observed, most of the proposals are focused on object detection, scene understanding and vehicle localization, including fail-aware and fail-operational approaches. In the following sections, we provide a detailed description of the papers published in the Special Issue, following the aforementioned categorization.

3. Object Detection and Scene Understanding

3.1. Vehicle Detection and Tracking

In [13] a monocular-based real-time multiple vehicles tracking system is proposed by using a novel Siamese network with a spatial pyramid pooling (SPP) layer which is applied to calculate pairwise appearance similarity (see Figure 5). The motion model captured from the bounding boxes provides the relative movements of the vehicles. An online-learned policy treats each tracking period as a Markov Decision Process (MDP) to maintain long-term, robust tracking. The approach achieves significant performance in terms of the "Mostly-tracked", "Fragmentation", and "ID switch" variables on the well-known KITTI dataset.

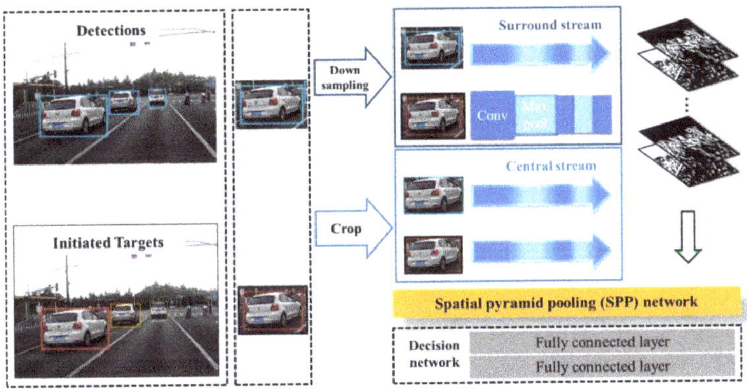

Figure 5. Central-surround two-channel spatial pyramid pooling network (CSTCSPP) based on the Siamese-type architecture (image obtained from [13]).

Hu et al. propose in [14] an improved edge-oriented segmentation-based method to detect the objects from a 3D point cloud by applying three main steps. First, 2D bounding boxes are selected by edge detection and stixel estimation in 2D images from the stereo system. Second, 3D sparse point clouds are computed in the selected areas. Finally, the dense 3D point clouds of objects are segmented by matching the 3D sparse point clouds of objects with the whole scene point cloud, as can be observed in Figure 6. After comparison with existing segmentation methods, it is demonstrated that the proposed approach improves precision.

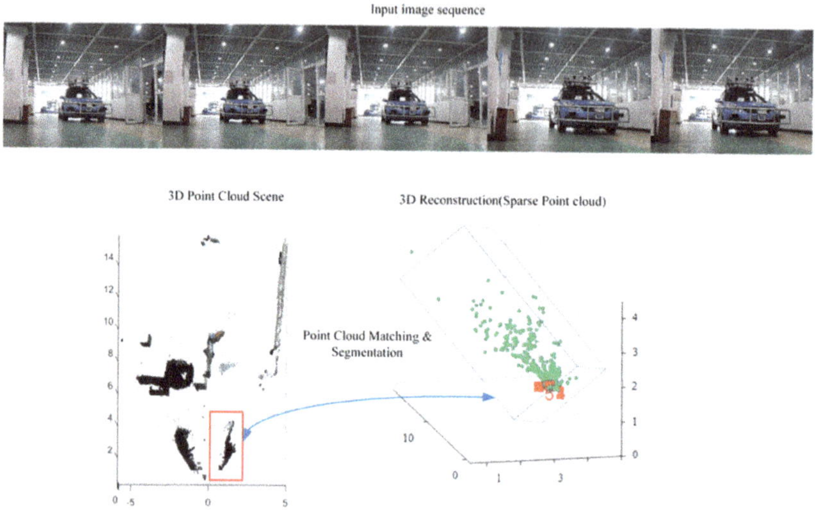

Figure 6. Point cloud matching and segmentation (image obtained from [14]).

Song and Wu present in [15] a method to estimate and identify the motion of target-vehicles (surrounding vehicles) using data from a millimeter-wave radar. Based on the Square-Root Cubature Kalman Filter (SRCKF), the Sage–Husa noise statistic estimator (SH-EKF) and the fading memory exponential weighting method are combined to derive a time-varying noise statistic estimator for non-linear systems. This approach is named the Improved Square-Root Cubature Kalman Filter (ISRCKF), and it improves the filtering accuracy of the longitudinal distance and speed about 50% and 25% respectively, with respect to SH-EKF and SRCKF methods. As depicted in Figure 7 the experimental platform used during the experiments includes one front radar to get raw data, one camera to obtain images from the scenes, and two LIDAR sensors to generate the ground truth. The classification and recognition results of the target-vehicle motion state are consistent with the actual target-vehicle motion state.

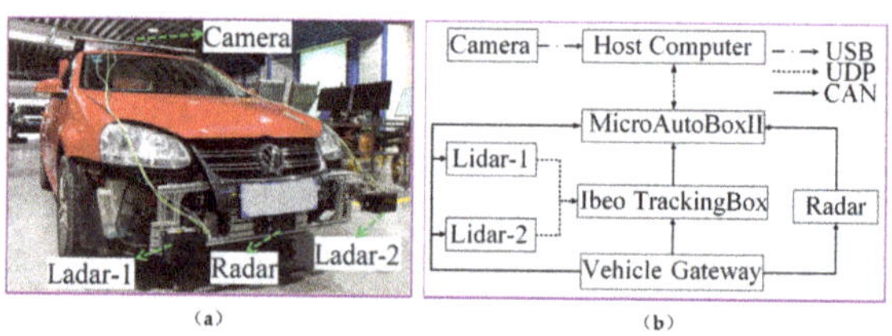

Figure 7. Experimental vehicle. (**a**) Test platform equipment. (**b**) Experimental platform communication (images obtained from [15]).

In [16] a vehicle collision warning system is proposed based on a Kalman filter-based approach for high-level fusion of multiple sensors, including radar, LIDAR, camera and wireless communication. The trajectories of remote targets are predicted, and an appropriate warning to the driver is provided based on the TTC (Time-To-Collision) estimate and the risk assessment. The multi-sensor approach is validated using a virtual driving simulator (see Figure 8) with two different Euro NCAP test scenarios: a vehicle–vehicle and a vehicle–pedestrian collision scenarios.

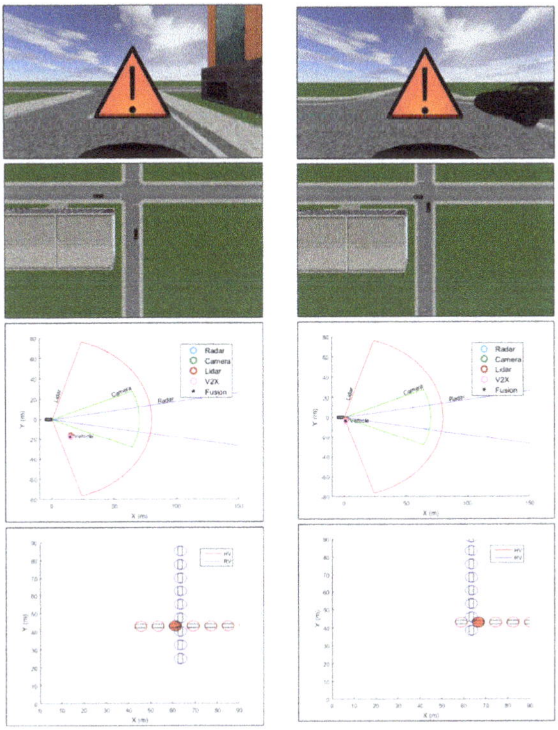

Figure 8. Vehicle–vehicle collision simulation results and snapshots of the experimental environment at two different time points (images obtained from [16]).

3.2. Scene Segmentation and Interpretation

The availability of large training dataset demanded by deep learning algorithms is not always possible. One potential solution is transfer learning from different domains. However, this method may not be a straightforward task considering issues such as original network size or large differences between the source and target domains. In [17], transfer learning is applied for semantic segmentation of off-road driving environments using a lightweight deconvolutional network (depicted in Figure 9) which is half the size of the original DeconvNet architecture. Transfer learning and fine-tuning is applied from the original pre-trained DeconvNet to the lightweight version. In addition, a synthetic dataset is used as an intermediate domain. It is observed that fine-tuning the model trained with the synthetic dataset that simulates the off-road driving environment provides more accurate results for the segmentation of real-world off-road driving environments than transfer learning without using a synthetic dataset does, as long as the synthetic dataset is generated considering real-world variations.

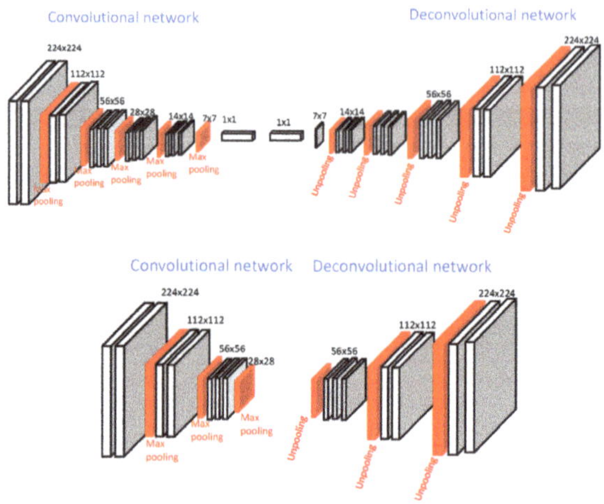

Figure 9. **Top**: Original DeconvNet architecture, **Bottom**: Proposed light-weight network architecture (images obtained from [17]).

Wang et al. address in [18] the problem of performing semantic segmentation for occluded areas (the overall idea is depicted in Figure 10). This is a complex problem that requires a comprehensive understanding of the geometry and the semantics of the visible environment. A specific dataset (KITTI-occlusion-free road segmentation) based on KITTI dataset is created and a specific lightweight fully convolutional neural network, named OFRSNet (occlusion-free road segmentation network), is designed to predict occluded portions of the road in the semantic domain by looking around foreground objects and visible road layout. The global context module is used to build up the down-sampling and joint context up-sampling block in the network and a spatially-weighted cross-entropy loss is used. Extensive experiments on different datasets verify the effectiveness of the proposed methodology, and comparisons with current methods show that the proposed approach outperforms the baseline models by obtaining a better trade-off between accuracy and computational cost, which makes the presented approach appropriate to be applied to intelligent vehicles in real-time.

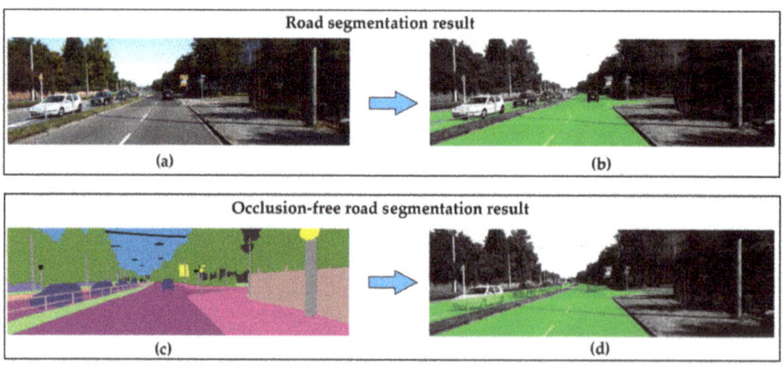

Figure 10. Comparison of road segmentation and proposed occlusion-free road segmentation. (**a**) RGB image; (**b**) standard road segmentation; (**c**) semantic segmentation; (**d**) occlusion-free road segmentation (images obtained from [18]).

One of the most challenging vision-based perception problems is to mitigate the effects of shadows on the road, which increases the difficulty of fundamental tasks such as road segmentation or lane detection. In [19], a new shadow detection method is proposed based on the skylight and sunlight contributions to the road surface chromaticity. Six constraints on shadow and non-shadowed regions are derived from these properties. The chrominance properties and the associated constraints are used as shadow features in an effective shadow detection method intended to be integrated on an onboard road detection system where the identification of cast shadows on the road is a determinant stage. The underlying principle is based on the following assumption: a non-shadowed road region is illuminated by both skylight and sunlight, whereas a shadowed region is illuminated by skylight only and, thus, their chrominance values vary. Shadow edges are detected and classifying by verifying whether the pixel chrominance values of regions on both sides of the edges satisfy up to six different constraints. Experiments on real traffic scenes (some examples can be seen in Figure 11) demonstrated the effectiveness of the proposed shadow detection method, outperforming previous approaches based on physical features.

Figure 11. Examples of the results of the shadow edge detection method (images obtained from [19]).

In [20] a novel approach based on multiple frames is proposed by taking advantage of the fusion of vision and Inertial Measurement Units (IMU). Hough space is employed as a storage medium where lane markings can be stored and visited conveniently. A CNN-based classifier is introduced to measure the confidence probability of each line segment, and transforms the basic Hough space into a probabilistic Hough space, as depicted in Figure 12. Pose information provided by the IMU is applied to align previous probabilistic Hough spaces to the current one and a filtered probabilistic Hough space is acquired by smoothing the primary probabilistic Hough space across frames. The proposed approach is applied experimentally, and the results demonstrate a satisfying performance compared to various existing methods.

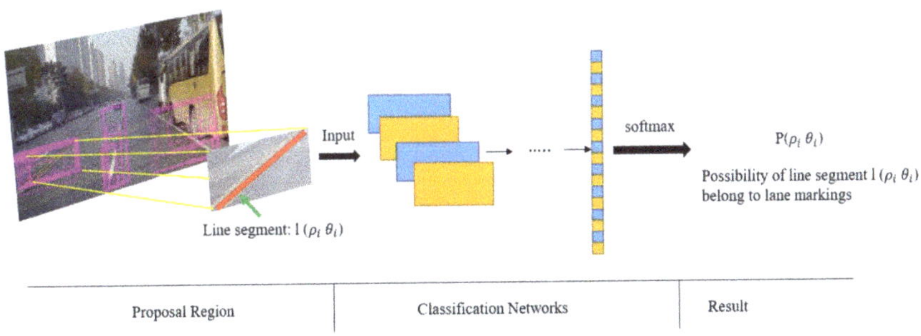

Figure 12. Line segments classified by the proposed network into the probabilistic Hough space which records the the confidence probability of each line segment (image obtained from [20]).

Still, lane detection and tracking in a complex road environment is one of the most important research areas in highly automated driving systems. Studies on lane detection cover a variety of difficulties, such as shadowy situations, dimmed lane painting, and obstacles that prohibit lane feature detection. Jeong et al. have carefully selected typical scenarios in which the extraction of lane candidate features can be easily corrupted by road vehicles and road markers that lead to degradation in the road scene understanding [21]. They introduced a novel framework combining a lane tracker method integrated with a camera and a radar forward vehicle tracker system, which is especially useful in dense traffic situations. An image template occupancy matching method is integrated with the vehicle tracker which allows to avoid extracting irrelevant lane features caused by forward target vehicles and road markers. In addition, a robust multi-lane detection method is presented by tracking adjacent and ego lanes. The proposed approach is comprehensively evaluated using a real dataset comprised of difficult and complex road scenarios recorded with a experimental vehicles with an appropriate sensor setup (see Figure 13). Experimental result shows that the proposed method is reliable for multi-lane detection at the presented difficult situations.

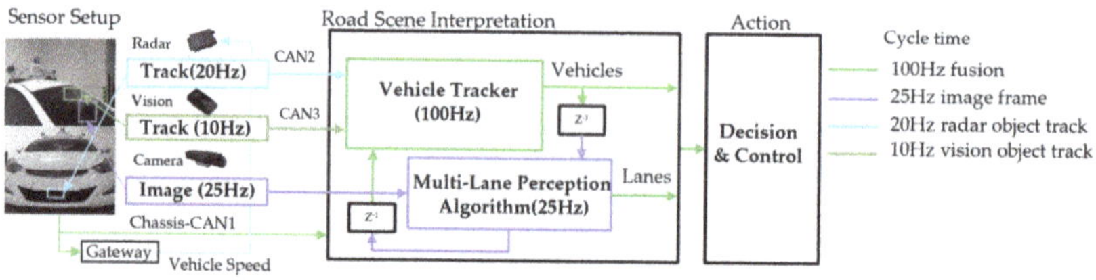

Figure 13. Multi-sensor setup and fusion chart (image obtained from [21]).

Another important type of object that should be identified by intelligent vehicles in urban areas are the traffic lights. In [22] traffic lights and arrow lights are recognized by image processing using digital maps and precise vehicle positioning. The overall approach is shown in Figure 14. The use of a digital map allows the determination of a region-of-interest (ROI) in the image to reduce the computational cost and false alarms. In addition, this study develops an algorithm to recognize arrow lights using relative positions of traffic lights which allows for the recognition of far distant arrow lights that are difficult for humans to see clearly. Quantitative evaluations indicate that the proposed method achieved 91.8% and 56.7% of the average f-value for traffic lights and arrow lights, respectively. The proposed arrow-light detection method can recognize small arrow objects even for sizes smaller than 10 pixels. The experiments indicate that the performance of the proposed method meets the necessary requirements for smooth acceleration/deceleration at intersections in automated driving.

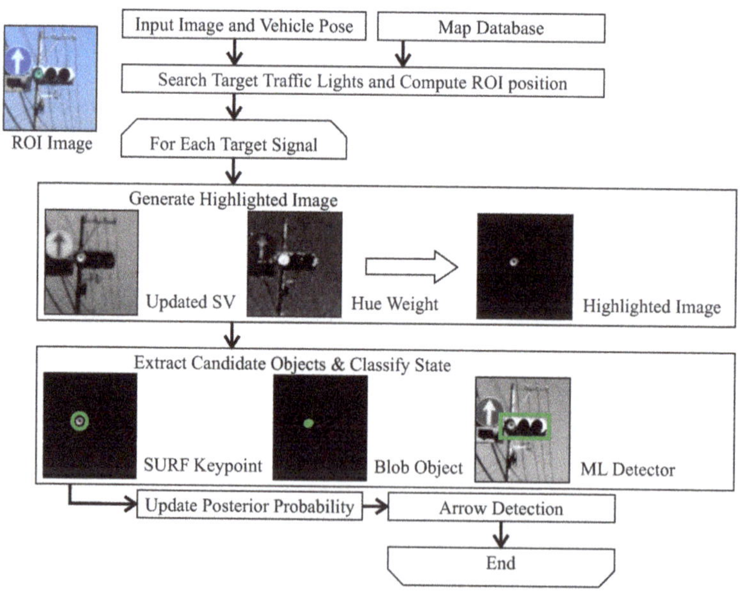

Figure 14. Flowchart of the method proposed in [22].

Yabuuchi et al. address the problem of detecting LED traffic lights which blink at high frequencies using a high-speed camera [23]. The method is composed of six modules, which includes a band-pass filter and a Kalman filter. All modules run simultaneously to achieve real-time processing. Actually, they can run up to 500 FPS for images with a resolution of 800 × 600 pixels. The proposed technique is robust under various illuminations because it can detect traffic lights by extracting information from the blinking pixels at a specific frequency (see Figure 15). An original dataset was created with the high-speed camera including images under different illumination conditions such as a sunset or night scene. The results show that the system can detect traffic lights with a different appearance. The most important benefits of this work are that neither parameter-tuning nor learning from a dataset are needed.

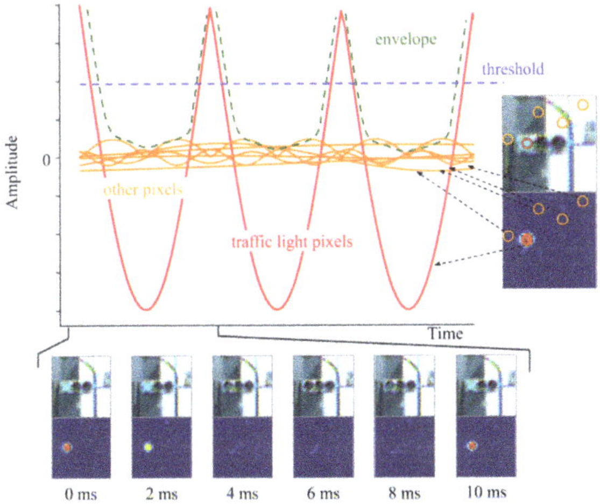

Figure 15. Brightness variation (blinking) for each pixel can be captured by the high-speed camera (images obtained from [23]).

4. Driver Behavior Modeling

Studying and modeling drivers behaviors is a common practice to address motion and action recognition and prediction of surrounding vehicles, as well as to deal with advanced driver assistance systems in the ego vehicle. One of the most risky maneuvers of road vehicles is lane change. Most lane-changing models deal with lane-changing maneuvers solely from the merging driver's standpoint and thus ignore driver interaction. To overcome this shortcoming, in [24] a game-theoretical decision-making model is developed. Validation makes use of empirical merging maneuver data at a freeway on-ramp. Specifically, this paper advances the repeated game model by using updated payoff functions. Validation results using the Next Generation SIMulation (NGSIM) empirical data show that the developed game-theoretical model provides better prediction accuracy compared to previous work, giving correct predictions approximately 86% of the time. The proposed lane change model, which captures the collective decision-making between human drivers, can be applied to develop automated vehicle driving strategies.

Determining an appropriate time to execute a lane change is a critical issue for the development of autonomous vehicles. However, few studies have considered the rear and the front vehicle-driver's risk perception while developing a human-like lane-change decision model. In [25], Wang et al. propose a lane-change decision model by identifying a two level threshold that conforms to a driver's perception of the ability to safely change lanes with a rear vehicle approaching fast. Based on the signal detection theory and extreme moment trials on a real highway, two thresholds of safe lane change are determined with consideration of risk perception of the rear and the subject vehicle drivers, respectively. The rear vehicle's Minimum Safe Deceleration (MSD) during the lane change maneuver of the subject vehicle is selected as the lane change safety indicator, and it is calculated using the human-like lane-change decision model. The results of this paper show that, compared with the driver in the front extreme moment trial, the driver in the rear extreme moment trial is more conservative during the lane change process. To meet the safety expectations of the subject and rear vehicle drivers, the primary and secondary safe thresholds were determined to be 0.85 m/s^2 and 1.76 m/s^2, respectively. A multi-sensor platform was used on an experimental vehicle to record the data and validate the system (see Figure 16). The proposed decision model can be of great help to make intelligent vehicles safer and more polite during lane changes, improving acceptance and safety.

Figure 16. Experimental vehicle used to validate the proposed lane-change decision model in [25].

As mentioned above, understanding the driver's behavior is a key factor for assistance systems, but also for monitoring the state of the driver for automated driving with SAE levels 2 or 3. The use of the phone is of particular interest to prevent drivers from being distracted. As stated in [26], recent studies have shown that 70% of the young and aware drivers are used to texting while driving. There are many different technologies used to control mobile phones while driving, including electronic device control, global positioning system (GPS), on-board diagnostics (OBD)-II-based devices, etc. However, we can even think of mobile phones as a sensor device to be used inside the vehicle to monitor the driver's behavior. These devices acquire vehicle information such as the car speed and use the information to control the driver's phone such as preventing them from making or receiving calls at specific speed limits. The information from the devices is interfaced via Bluetooth and can later be used to control mobile phone applications. The main aim of the work presented in [26] is to design a portable system (the overall structure is depicted in Figure 17) for monitoring the use of a mobile phone while driving and for controlling the driver's mobile phone, if necessary, when the vehicle reaches a specific speed limit (>10 km/h). A paper-based self-reported questionnaire survey was carried out among 600 teenage drivers from different nationalities to see the driving behavior of young drivers in Qatar. A mobile application is presented to monitor the mobile usage of a driver and a OBD-II module-based portable system was designed to acquire data from the vehicle to identify drivers' behavior with respect to phone usage, sudden lane changes, and abrupt breaking/sharp speeding. The presented application, which combines the sensors of the mobile phone and the vehicle, can significantly improve drivers' behavior.

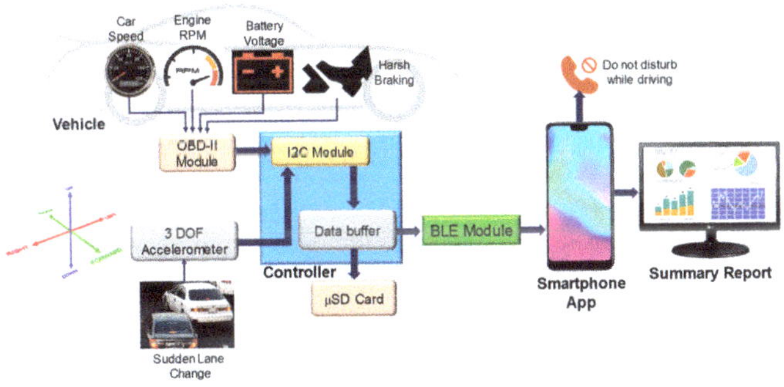

Figure 17. Complete block system diagram proposed in [26].

Assuncao et al. propose in [27] the use of the Statistical Process Control (SPC) and Exponentially Weighted Moving Average methods for the monitoring of drivers using approaches based on the vehicle and the driver's behavior. Different detection methods are independently devised for lane departure, sudden driver behaviors and driver fatigue. All of them consider information from sensors scattered by the vehicle in a multi-sensor fashion. The results show the efficiency of the proposed approach. Lane departure detection obtained results of up to 76.92% (without constant speed) and 84.16% (speed maintained at 60 kmh aprox.). Furthermore, sudden movements detection obtained results of up to 91.66% (steering wheel) and 94.44% (brake). The driver fatigue is detected in up to 94.46% situations.

The use of simulation is also getting more attention in the research community and industry. In [28] an automatic approach of simulating dynamic driving behaviors of vehicles in traffic scene represented by image sequences is proposed (see Figure 18). The spatial topological attributes and appearance attributes of virtual vehicles are computed separately, according to the constraint of geometric consistency of sparse 3D space organized by image sequence. To achieve this goal, three main problems must be solved. First, registration of vehicle in a 3D space of road environment. Second, to generate the vehicle's image observed from corresponding viewpoint in the road scene. Third, to maintain consistency between the the vehicle and the road environment. After the proposed method was embedded in a scene browser, a typical traffic scene including the intersections is chosen for a virtual vehicle to execute the driving tasks of lane change, overtaking, slowing down and stop, right turn, and U-turn. The experimental results show that different driving behaviors of vehicles in typical traffic scenes can be exhibited smoothly and realistically. The proposed method can also be used for generating simulation data of traffic scenes that are difficult to collect in real driving scenarios.

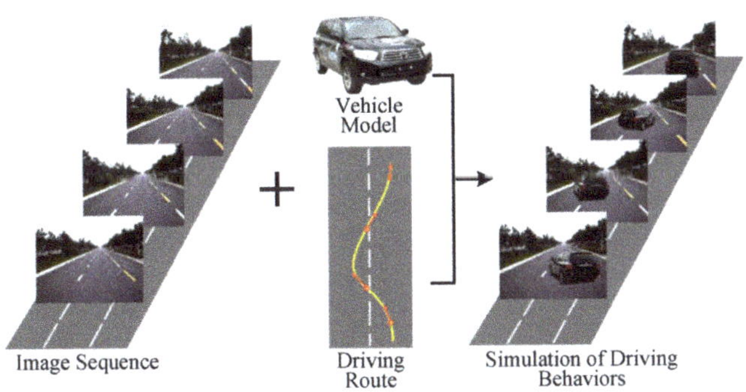

Figure 18. Simulation of driving behaviors with image sequences collected from real road environment (image obtained from [28]).

5. Fail-X Systems and Fault Diagnosis

Robust sensing under different lighting and weather conditions, and considering all the complexity that the vehicle can face in real world scenarios becomes mandatory. To that end, it is fundamental to advance towards fail-aware, fail-safe, and fail operational systems, including fault diagnosis.

As explained in [29], presently, in the event of a failure in automated driving systems, control architectures rely on hardware redundancies over software solutions to assure reliability, or wait for human interaction in takeover requests to achieve a minimal risk condition. As user confidence and final acceptance of autonomous vehicles are strongly related to safety, automated fall-back strategies must be assured as a response to failures while the system is performing a dynamic driving task. In the work presented by Matute-Peaspan et al. [29], a fail-operational control architecture approach and dead-reckoning strategy in case of positioning failures are developed. A fail-operational system is capable of detecting failures in the last available positioning source, warning the decision stage to set up a fall-back strategy and planning a new trajectory in real time. The surrounding objects and road borders are considered during the vehicle motion control after failure, to avoid collisions and lane-keeping purposes. A case study based on a realistic urban scenario (depicted in Figure 19) is simulated for testing and system verification, showing that the proposed approach always bears in mind both the passenger's safety and comfort during the fall-back maneuvering execution.

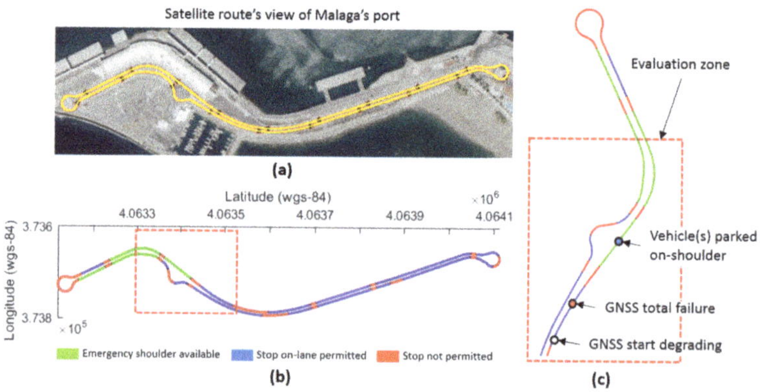

Figure 19. Realistic environment scenario for automated driving system tests on simulation. (**a**) Satellite's view of urban route, (**b**) permitted and non-permitted stops in case of total positioning failure, and (**c**) evaluation zone for test case study. (image obtained from [29]).

García-Daza et al. state in [30] that, currently, even the most advanced architectures require driver intervention when functional system failures or critical sensor operations take place, presenting problems related to driver state, distractions, fatigue, and other factors that prevent safe control. All odometry systems have drift error, making it difficult to use them for localization tasks over extended periods. In this work, a redundant, accurate and robust LiDAR odometry system with specific fail-aware features that can allow other systems to perform a safe stop man oeuvre without driver mediation is presented. A fail-aware indicator is designed which estimates a time window in which the system can manage the localization tasks appropriately. The odometry error is minimized by applying a dynamic 6-DoF model and fusing measures based on the Iterative Closest Points (ICP), environment feature extraction, and Singular Value Decomposition (SVD) methods. A general overview of the proposed system can be seen in Figure 20. The obtained results are promising for two reasons. First, in the KITTI odometry data set, the ranking achieved by the proposed method is twelfth, considering only LiDAR-based methods, where its translation and rotation errors are 1% and 0.0041 deg/m, respectively. Secondly, the encouraging results of the fail-aware indicator demonstrate the safety of the proposed LiDAR odometry system. The results depict that, in order to achieve an accurate odometry system, complex models and measurement fusion techniques must be used to improve its behavior. Furthermore, if an odometry system is to be used for redundant localization features, it must integrate a fail-aware indicator for use in a safe manner.

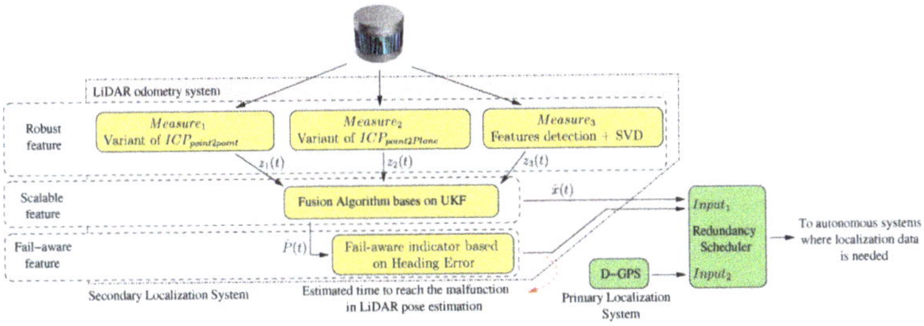

Figure 20. General diagram. The developed blocks are represented in yellow. The horizontal blue strips represent the main features of the odometry system. A framework where the LiDAR odometry system can be integrated within the autonomous driving cars topic is depicted with green blocks, such as a secondary localisation system (image obtained from [30]).

Sensor fault diagnosis is a necessary step to design fail-x systems. In [31] a fault diagnosis logic and signal restoration algorithms for vehicle motion sensors are presented. The primary idea of the proposed fault detection system is the conversion of measured wheel speeds into vehicle central axis information and the selection of a reference central axis speed based on this information. Thus, the obtained results can be employed to estimate the speed for all wheel sides, which are compared with measured values to identify faults and recover the fault signal. For fault diagnosis logic, a conditional expression is derived with only two variables to distinguish between normal and fault. Further, an analytical redundancy structure and a simple diagnostic logic structure are proposed. Finally, an off-line test is conducted using test vehicle information to validate the proposed method. It demonstrates that the proposed fault detection and signal restoration algorithm can satisfy the control performance required for each sensor failure.

6. Vehicle Positioning and Path Planning

One of the key features needed to perform autonomous navigation is to have accurate global locations of the vehicle. Vehicle positioning is a fundamental task for long-term navigation in a digital map, but also to perform short-term path planning in the local environment. In the previous section, we showed two manuscripts related with fail-x vehicle positioning [29,30]. In this section we present five more papers that deal with this important topic.

In [32], an enhanced visual SLAM algorithm based on the sparse direct method is proposed to deal with illumination sensitivity problems of mobile ground equipment. The presented procedure can be described as follows. First, the vignette and response functions of the input sequences are optimized based on the photometric formation of the camera. Second, the Shi–Tomasi corners of the input sequence are tracked, and optimization equations are established using the pixel tracking of sparse direct visual odometry (VO). Third, the Levenberg–Marquardt (L–M) method is applied to solve the joint optimization equation, and the photometric calibration parameters in the VO are updated to realize the real-time dynamic compensation of the exposure of the input sequences, thus reducing the effects of the light variations on accuracy and robustness. Finally, a Shi–Tomasi corner filtered strategy is designed to reduce the computational complexity of the proposed algorithm, and the loop closure detection is realized based on the oriented FAST and rotated BRIEF (ORB) features. The proposed algorithm is tested using TUM, KITTI, EuRoC, and a specific environment. Experimental results show that the positioning and mapping performance of the proposed algorithm is promising.

Lin et al. propose in [33] a sensor fusion approach to reduce typical problems of global positioning system (GPS) such as noisy signal and multi-path routing in urban environments. To localize the vehicle position, a particle-aided unscented Kalman filter (PAUKF) algorithm is proposed. The Unscented Kalman Filter (UKF) updates the vehicle state, which includes the vehicle motion model and non-Gaussian noise affection. The Particle Filter (PF) provides additional updated position measurement information based on an onboard sensor and a high definition (HD) digital map. This methodology is validated in a simulated environment. The obtained results show that this method achieves better precision and comparable stability in localization performance compared to previous approaches.

In [34], the authors propose a method that improves autonomous vehicle positioning using a modified version of the probabilistic laser localization like the Monte Carlo Localization (MCL) algorithm. The weights of the particles are enhanced by adding Kalman filtered Global Navigation Satellite System (GNSS) information. GNSS data are used to improve localization accuracy in places with fewer map features and to prevent the kidnapped robot problems. Besides, laser information improves accuracy in places where the map has more features and GNSS higher covariance, allowing the approach to be used in specifically difficult scenarios for GNSS such as urban canyons. The algorithm is tested using the KITTI odometry dataset proving that it improves localization compared with classic GNSS + Inertial Navigation System (INS) fusion and Adaptive Monte Carlo Localization (AMCL). The presented approach is also tested in the autonomous vehicle platform of the Intelligent Systems Lab (LSI), of the University Carlos III of Madrid (depicted in Figure 21), providing promising qualitative results.

Figure 21. (**Left**) Autonomous vehicle platform used during the experiments. (**Right**) Visualization of the LiDAR point cloud (images obtained from [34]).

Diaz-Arango et al. propose in [36] a multiple-target collision-free path planning based on homotopy continuation capable to calculate a collision-free path in a single execution for complex environments. The method exhibits better performance, both in speed and efficiency, and robustness compared to the original Homotopic Path Planning Method (HPPM). Among the new schemes that improve their performance are the Double Spherical Tracking (DST), the dummy obstacle scheme, and a systematic criterion to a selection of repulsion parameter. The case studies, although focusing on robotics indoor environments, show the efficiency to find a solution path in just a few milliseconds, even if they have narrow corridors and hundreds of obstacles. Additionally, a comparison between the proposed method and sampling-based planning algorithms (SBP) with the best performance is presented. The method exhibits better performance than SBP algorithms for execution time, memory, and, in some cases, path length metrics. To validate the feasibility of the computed paths two simulations using the pure-pursuit controlled and differential drive robot model contained in the Robotics System Toolbox of MATLAB are presented. The proposed approach will be further applied and validated in intelligent vehicle environments.

Autonomous racing provides very similar technological issues than standard autonomous driving while allowing for more extreme conditions in a safe human environment. In [35] a localization architecture for a racing car that does not rely on Global Navigation Satellite Systems (GNSS) is presented. More specifically, two multi-rate Extended Kalman Filters and an extension of a state-of-the-art laser-based Monte Carlo localization approach that exploits some prior knowledge of the environment and context are combined. When driving near the friction limits, localization accuracy is critical as small errors can induce large errors in control due to the nonlinearities of the vehicle's dynamic model. The proposed method is compared with a solution based on a widely employed state-of-the-art implementation, outlining its strengths and limitations within the experimental scenario. The architecture is then tested both in simulation and experimentally on a full-scale autonomous electric racing car during an event of Roborace Season Alpha. The results show its robustness in avoiding the robot kidnapping problem typical of particle filters localization methods, while providing a smooth and high rate pose estimate. The pose error distribution depends on the car velocity, and spans on average from 0.1 m (at 60 km/h) to 1.48 m (at 200 km/h) laterally and from 1.9 m (at 100 km/h) to 4.92 m (at 200 km/h) longitudinally.

7. Smart Regenerative Braking Systems

Two manuscripts from the same research group have been published in the Special Issue which provides classification and prediction approaches to deal with regenerative braking for Electric Vehicles (EV). Smart regenerative braking systems are an autonomous version of one-pedal driving in EVs. To implement them, a deceleration planning algorithm is necessary to generate the deceleration used in automatic regenerative control. To reduce the discomfort from the automatic regeneration, the deceleration should be similar to human driving. In [37], a deceleration planning algorithm based on Multi-Layer Perceptron (MLP) is proposed. The MLP models can mimic human driving behavior by learning the driving data. In addition, the proposed deceleration planning algorithm has a classified structure to improve the planning performance in each deceleration condition. Therefore, the individual MLP models are designed according to three different deceleration conditions: car-following, speed bump, and intersection. The proposed algorithm is validated through driving simulations using a test vehicle equipped with multiple sensors and logging software (see Figure 22). Time to collision (TTC) and similarity to human driving is analyzed. The results show that the minimum TTC was 1.443 s and the velocity root-mean-square error (RMSE) with human driving was 0.302 m/s.

Figure 22. Test vehicle, sensors and logging system used by [37].

The vehicle state prediction on decelerating driving conditions can be applied to automatic regenerative braking in EVs. However, drivers can feel a sense of heterogeneity when regenerative control is performed based on prediction results from a general prediction model. As a result, a deceleration prediction model which represents individual driving characteristics is required to ensure a more comfortable experience with automatic regenerative braking control. Thus, in [38], a deceleration prediction model based on the parametric mathematical equation and explicit model parameters is presented. The model is designed specifically for deceleration prediction by using the parametric equation that describes deceleration characteristics. Furthermore, the explicit model parameters are updated according to individual driver characteristics using the driver's braking data during real driving situations. An overview of the proposed methodology is depicted in Figure 23.

The method is integrated and validated on a real-time embedded system, and it is applied to the model-based regenerative control algorithm as a case study.

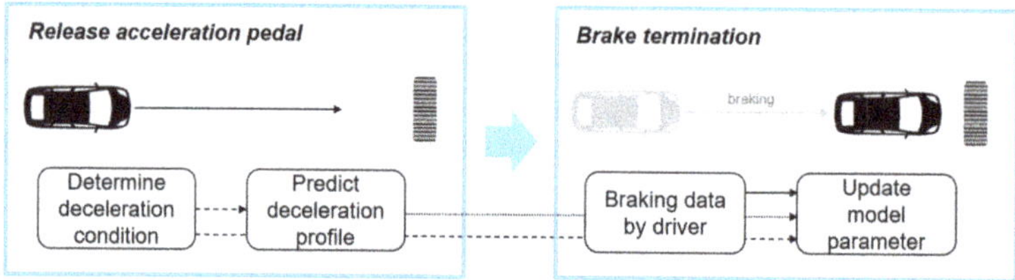

Figure 23. Overview of the methodology presented in [38].

8. Physical Intelligence in Sensors and Sensing

Adding intelligence to some important components and sensors of the vehicles is a fundamental approach to increase the intelligence of the vehicles, providing new measurements and variables very useful for higher-level decision modules. For example, in [39] the possibility of using tires as active sensors is explored using strain sensors. The concept is known as Intelligent or Smart Tires and can provide relevant vehicle dynamics information. The main goal of this work is to estimate all tire forces, based only on deformations measured in the contact patch. Through an indoor test rig data, an algorithm is developed to select the most relevant features of strain data and correlate them with tire parameters. Tire contact patch information is transmitted to a fuzzy logic system to estimate the tire parameters (see Figure 24 for an overview of the proposed system architecture). To evaluate the reliability of the proposed estimator, the well-known simulation software CarSim is used to back up the estimation results. The obtained estimations are checked with the simulation results enabling the behavior of the intelligent tire to be tested for different maneuvers and velocities. The proposed methodology provides key information about the tire parameters directly from the only contact that exists between the vehicle and the road.

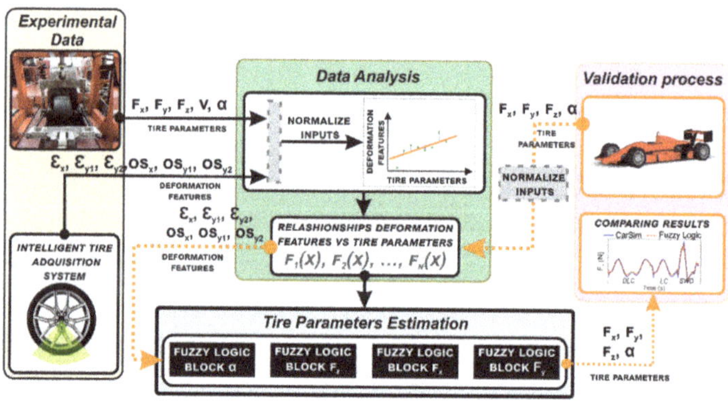

Figure 24. Overview of the methodology presented in [39] to develop the strain-based tire state estimation system.

Gao et al. propose in [40] a nonlinear observer aided by vehicle lateral displacement information for estimating the road friction coefficient, which is a key parameter for autonomous vehicles and vehicle dynamic control. A modified version of the tire brush model is proposed to describe the tire characteristics more precisely in high friction conditions using tire test data. Then, on the basis of vehicle dynamics and a kinematic model, a nonlinear observer is designed, and the self-aligning torque of the wheel, lateral acceleration, and vehicle lateral displacement are used to estimate the road friction coefficient during steering. Slalom and Double Line Change (DLC) tests in high friction conditions are conducted to verify the proposed estimation algorithm. The results show that the proposed method performs well during steering and the estimated road friction coefficient converges to the reference value rapidly.

In the study presented in [41] the virtual testing of intelligent driving is addressed by examining the key problems in modeling and simulating millimeter-wave radar environmental clutter, and proposing a modeling and simulation method for the environmental clutter of millimeter-wave radar in intelligent driving. Based on the attributes of millimeter-wave radar, the classification characteristics of the traffic environment of an intelligent vehicle and the generation mechanism of radar environmental clutter are analyzed. The statistical distribution characteristics of the clutter amplitude, the distribution characteristics of the power spectrum, and the electromagnetic dielectric characteristics are studied. Conditions such as road surface, rainfall, snowfall, and fog are deduced and designed. Experimental comparison results are utilized to validate the proposed model and simulation method.

9. Driver Assistance Systems and Automatic Vehicle Operation

The last three papers listed in the Special Issue are devoted to driver assistance systems and automatic vehicle operation. Thus, in [43] an automatic parking system (APS) based on reinforcement learning is presented. The parking path is planned based on the parking slot detected by the cameras. The path tracking module guides the vehicle to track the planned parking path. However, since the vehicle is a non-linear dynamic, path tracking error inevitably occurs, leading to the inclination and deviation of the parking. Accordingly, in the presented work, a reinforcement learning-based end-to-end parking algorithm is proposed to achieve automatic parking. The vehicle can continuously learn and accumulate experience from numerous parking attempts and then learn the command of the optimal steering wheel angle at different parking slots. Based on this end-to-end parking, errors caused by path tracking can be avoided. Moreover, to ensure that the parking slot can be obtained continuously in the process of learning, a parking slot tracking algorithm is proposed based on the combination of vision and vehicle chassis information (vehicle sensors). Furthermore, given that the learning network output is hard to converge, and it is easy to fall into local optimum during the parking process, several reinforcement learning training methods, in terms of parking conditions, are developed. Lastly, by the real vehicle test, it is proved that using the proposed method can achieve a better parking attitude than using the path planning and path tracking-based methods. An overview of the whole process is shown in Figure 25.

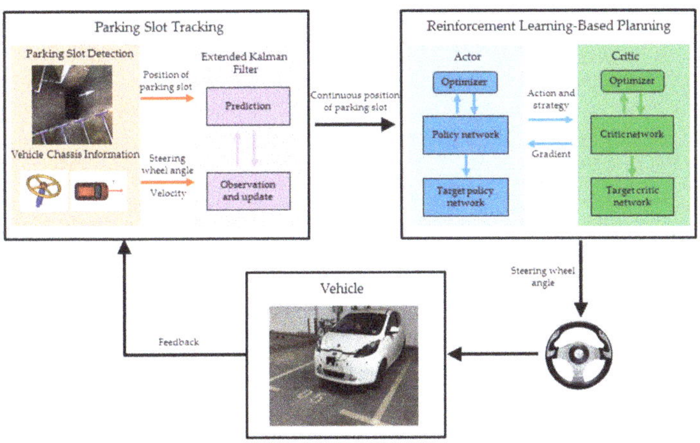

Figure 25. Overview of the reinforcement learning-based end-too-end parking method presented in [43].

Wang et al. present in [44] an improved model predictive controller based on the online obtaining of softness factor and fusion velocity for automatic train operation to enhance the tracking control performance. Specifically, the softness factor of the improved model predictive control algorithm is not a constant, conversely, an improved online adaptive adjusting method for softness factor based on fuzzy satisfaction of system output value and velocity distance trajectory characteristic is adopted, and an improved whale optimization algorithm has been proposed to solve the adjustable parameters (see Figure 26). The system output value for an automatic train operation is not sampled by a normal speed sensor. Instead, an online velocity sampled method for the system output value based on a fusion velocity model and an intelligent digital torque sensor is applied. The proposed strategies show a good performance in tracking precision, are simple and easily conducted, and can ensure the accomplishing of computational tasks in real-time. Finally, to verify the effectiveness of the model predictive controller, the MATLAB/Simulink and hardware-in-the-loop simulation (HILS) are adopted for automatic train operation tracking control, and the results also indicate that the proposed predictive controller has better tracking control effectiveness compared with the existing traditional model predictive controller.

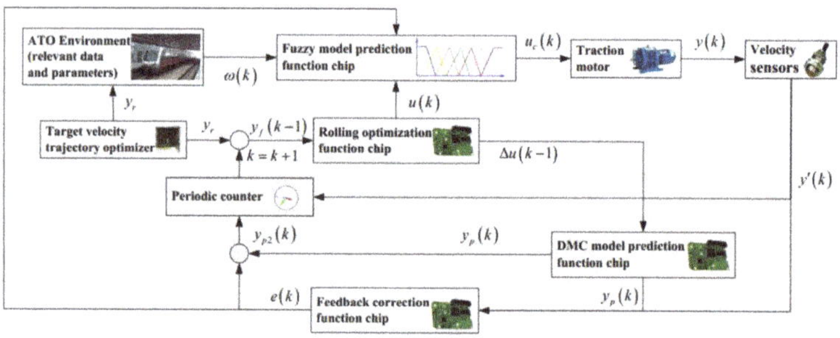

Figure 26. Schematic diagram of the Fuzzy Dynamic Matrix Control (DMC) Model Predictive Controller (MPC) proposed in [44] for automatic train operation.

In [42] the development of an intelligent driving assistant system based on vehicle telemetry and road accident risk map analysis is presented. The goal of the system is to alert the driver in order to avoid risky situations that may cause traffic accidents. In performance evaluations using real cars in real environments, the on-board intelligent assistant reproduced real-time audio-visual alerts. The method is mainly based on fuzzy reasoning and correctly warns the driver in real-time according to the telemetry data, the vehicle environment and the principles of safe driving practices and transportation regulation laws. Experimental results and conclusions emphasizing the advantages of the proposed intelligent driving assistant in the improvement of the driving task are obtained.

10. Conclusions

After analyzing all the manuscripts, some important and exciting conclusions arise. All the papers deal with well-known problems, which have been widely addressed by the research community during the last years. This is a clear example of the complexity of the problems involving sensors and sensing for intelligent vehicles. The issues most addressed in the Special Issue have been scene understanding, including object detection and tracking, and vehicle localization. We believe that it is a faithful representation of the importance of these problems for the advance of intelligent vehicles.

Fail-x systems (including fail-aware, fail-operational, fail-safe and fault diagnosis) are starting to emerge as an essential feature to be considered in future research works. Indeed, the ability to operate safely in the face of various types of failure and to be aware of and diagnose such failures, are fundamental issues in ensuring the safety of intelligent vehicles.

The complexity and variety of scenarios in which intelligent vehicles have to operate, as well as the need to function in the event of failure, are crucial elements that lead to most of the proposed solutions having a multi-sensor approach, including cameras, radar, LiDAR, IMUs, strains, vehicle sensors, and even smartphones as sensors onboard the vehicles. Intelligent vehicle sensing will be an "ever-present" problem, which will attract a lot of attention from both the research community and industry in the coming years. It is identified as the main bottleneck for enabling intelligent vehicles to operate autonomously and safely, and therefore to achieve good public acceptance.

On top of that, the current global situation, in which physical distance is socially necessary, is accelerating the need for intelligent and autonomous transportation, being the intelligent vehicles may be the most relevant topic in the field. We expect an increase in both the number of publications and the impact on this particular topic in the next coming years.

Funding: This research received no external funding.

Acknowledgments: The editors would like to thank all the authors who have submitted their manuscripts to this Special Issue, including the authors of the rejected papers, and all the reviewers for their valuable evaluation. We also acknowledge to the Electronic Component Systems for European Leadership Joint Undertaking through the European Union's Horizon 2020 Research and Innovation Program and Germany, Austria, Spain, Italy, Latvia, Belgium, The Netherlands, Sweden, Finland, Lithuania, Czech Republic, Romania,and Norway, under Grant 737469 (Autodrive Project) which has provided support to topics related with fail-x systems. Finally, we acknowledge grants S2018/EMT-4362SEGVAUTO 4.0 (Community Region of Madrid, Spain) and DPI2017-90035-R (Spanish Ministry of Science and Innovation).

Conflicts of Interest: The authors declare no conflict of interest.

References

1. Gavrila, D.M. Pedestrian Detection from a Moving Vehicle. In Proceedings of the European Conference on Computer Vision, Dublin, Ireland, 26 June–1 July 2000; pp. 37–49.
2. Parra-Alonso, I.; Fernández-Llorca, D.; Sotelo, M.A.; Bergasa, L.M.; Revenga de Toro, P.; Nuevo, J.; Ocana, M.; García-Garrido, M.A. Combination of Feature Extraction Methods for SVM Pedestrian Detection. *IEEE Trans. Intell. Transp. Syst.* **2007**, *8*, 292–307. [CrossRef]
3. Braun, M.; Krebs, S.; Flohr, F.B.; Gavrila, D.M. EuroCity Persons: A Novel Benchmark for Person Detection in Traffic Scenes. *IEEE Trans. Pattern Anal. Mach. Intell.* **2019**, *41*, 1844–1861. [CrossRef]
4. Sotelo, M.A.; Nuevo, J.; Bergasa, L.M.; Ocana, M.; Parra, I.; Fernández, D. Road vehicle recognition in monocular images. In Proceedings of the IEEE International Symposium on Industrial Electronics, Dubrovnik, Croatia, 20–23 June 2005; pp. 1471–1476.
5. Kuhnl, T.; Kummert, F.; Fritsch, J. Spatial ray features for real-time ego-lane extraction. In Proceedings of the 2012 15th International IEEE Conference on Intelligent Transportation Systems, Anchorage, AK, USA, 16–19 September 2012.
6. Hernández, A.; Woo, S.; Corrales, H.; Parra, I.; Kim, E.; Llorca, D.F.; Sotelo, M.A. 3D-DEEP: 3-Dimensional Deep-learning based on elevation patterns for road scene interpretation. *arXiv* **2020**, arXiv:2009.00330.
7. Nguyen, H.; Kieu, L.-M.; Wen, T.; Cai, C. Deep learning methods in transportation domain: A review. *IET Intell. Transp. Syst.* **2018**, *12*, 998–1004. [CrossRef]
8. Fernández, C.; Fernández-Llorca, D.; Sotelo M.A. A Hybrid Vision-Map Method for Urban Road Detection. *J. Adv. Transp.* **2017**, *2017*, 7090549. [CrossRef]
9. Fernández, C.; Munoz-Bulnes, J.; Fernández-Llorca, D.; Parra, I.; García-Daza, I.; Izquierdo, R.; Sotelo M.A. High-Level Interpretation of Urban Road Maps Fusing Deep Learning-Based Pixelwise Scene Segmentation and Digital Navigation Maps. *J. Adv. Transp.* **2018**, *2018*, 2096970. [CrossRef]
10. Fernández-Llorca, D. Future trends of ITS in difficult times: A message from the new Editor-in-Chief of IET Intelligent Transport Systems. *IET Intell. Transp. Syst.* **2020**, *14*, 469–470. [CrossRef]
11. Fernández-Llorca, D.; Quintero-Minguez, R.; Parra-Alonso, I.; Fernández-López, C.; García Daza, I.; Sotelo, M.A.; Alen Cordero, C. Assistive Intelligent Transportation Systems: The Need for User Localization and Anonymous Disability Identification. *IEEE Intell. Transp. Syst. Mag.* **2017**, *9*, 25–40. [CrossRef]

12. Díez-Jiménez, E.; Fernández-Munoz, M.; Oliva-Dominguez, R.; Fernández-Llorca, D.; Sotelo, M.A. Personal Rapid Transport System Compatible With Current Railways and Metros Infrastructure. *IEEE Trans. Intell. Transp. Syst.* **2020**. [CrossRef]
13. Zou, Y.; Zhang, W.; Weng, W.; Meng, Z. Multi-Vehicle Tracking via Real-Time Detection Probes and a Markov Decision Process Policy. *Sensors* **2019**, *19*, 1309. [CrossRef]
14. Hu, F.; Yang, D.; Li, Y. Combined Edge- and Stixel-based Object Detection in 3D Point Cloud. *Sensors* **2019**, *19*, 4423. [CrossRef]
15. Song, S.; Wu, J. Motion State Estimation of Target Vehicle under Unknown Time-Varying Noises Based on Improved Square-Root Cubature Kalman Filter. *Sensors* **2020**, *20*, 2620. [CrossRef]
16. Baek, M.; Jeong, D.; Choi, D.; Lee, S. Vehicle Trajectory Prediction and Collision Warning via Fusion of Multisensors and Wireless Vehicular Communications. *Sensors* **2020**, *20*, 288. [CrossRef]
17. Sharma, S.; Ball, J.E.; Tang, B.; Carruth, D.W.; Doude, M.; Islam, M.A. Semantic Segmentation with Transfer Learning for Off-Road Autonomous Driving. *Sensors* **2019**, *19*, 2577. [CrossRef] [PubMed]
18. Wang, K.; Yan, F.; Zou, B.; Tang, L.; Yuan, Q.; Lv, C. Occlusion-Free Road Segmentation Leveraging Semantics for Autonomous Vehicles. *Sensors* **2019**, *19*, 4711. [CrossRef]
19. Ibarra-Arenado, M.J.; Tjahjadi, T.; Pérez-Oria, J. Shadow Detection in Still Road Images Using Chrominance Properties of Shadows and Spectral Power Distribution of the Illumination. *Sensors* **2020**, *20*, 1012. [CrossRef]
20. Sun, Y.; Li, J.; Sun, Z. Multi-Stage Hough Space Calculation for Lane Markings Detection via IMU and Vision Fusion. *Sensors* **2019**, *19*, 2305. [CrossRef]
21. Jeong, J.; Yoon, Y.H.; Park, J.H. Reliable Road Scene Interpretation Based on ITOM with the Integrated Fusion of Vehicle and Lane Tracker in Dense Traffic Situation. *Sensors* **2020**, *20*, 2457. [CrossRef]
22. Yoneda, K.; Kuramoto, A.; Suganuma, N.; Asaka, T.; Aldibaja, M.; Yanase, R. Robust Traffic Light and Arrow Detection Using Digital Map with Spatial Prior Information for Automated Driving. *Sensors* **2020**, *20*, 1181. [CrossRef]
23. Yabuuchi, K.; Hirano, M.; Senoo, T.; Kishi, N.; Ishikawa, M. Real-Time Traffic Light Detection with Frequency Patterns Using a High-Speed Camera. *Sensors* **2020**, *20*, 4035. [CrossRef]
24. Kang, K.; Rakha, H.A. A Repeated Game Freeway Lane Changing Model. *Sensors* **2020**, *20*, 1554. [CrossRef]
25. Wang, C.; Sun, Q.; Li, Z.; Zhang, H. Human-Like Lane Change Decision Model for Autonomous Vehicles that Considers the Risk Perception of Drivers in Mixed Traffic. *Sensors* **2020**, *20*, 2259. [CrossRef] [PubMed]
26. Khandakar, A.; Chowdhury, M.E.; Ahmed, R.; Dhib, A.; Mohammed, M.; Al-Emadi, N.A.M.A.; Michelson, D. Portable System for Monitoring and Controlling Driver Behavior and the Use of a Mobile Phone While Driving. *Sensors* **2019**, *19*, 1563. [CrossRef] [PubMed]
27. Assuncao, A.N.; Aquino, A.L.L.; Camara de M. Santos, R.C.; Guimaraes, R.L.M.; Oliveira, R.A.R. Vehicle Driver Monitoring through the Statistical Process Control. *Sensors* **2019**, *19*, 3059. [CrossRef] [PubMed]
28. Zhao, D.; Li, Y.; Liu, Y. Simulating Dynamic Driving Behavior in Simulation Test for Unmanned Vehicles via Multi-Sensor Data. *Sensors* **2019**, *19*, 1670. [CrossRef]
29. Matute-Peaspan, J.A.; Perez, J.; Zubizarreta, A. A Fail-Operational Control Architecture Approach and Dead-Reckoning Strategy in Case of Positioning Failures. *Sensors* **2020**, *20*, 442. [CrossRef] [PubMed]
30. García-Daza, I.; Rentero, M.; Salinas Maldonado, C.; Izquierdo Gonzalo, R.; Hernández Parra, N.; Ballardini, A.; Fernandez Llorca, D. Fail-Aware LIDAR-Based Odometry for Autonomous Vehicles. *Sensors* **2020**, *20*, 4097. [CrossRef]
31. Byun, Y.-S.; Kim, B.-H.; Jeong, R.-G. Sensor Fault Detection and Signal Restoration in Intelligent Vehicles. *Sensors* **2019**, *19*, 3306. [CrossRef]
32. Liu, P.; Yuan, X.; Zhang, C.; Song, Y.; Liu, C.; Li, Z. Real-Time Photometric Calibrated Monocular Direct Visual SLAM. *Sensors* **2019**, *19*, 3604. [CrossRef]
33. Lin, M.; Yoon, J.; Kim, B. Self-Driving Car Location Estimation Based on a Particle-Aided Unscented Kalman Filter. *Sensors* **2020**, *20*, 2544. [CrossRef]
34. de Miguel, M.Á.; García, F.; Armingol, J.M. Improved LiDAR Probabilistic Localization for Autonomous Vehicles Using GNSS. *Sensors* **2020**, *20*, 3145. [CrossRef] [PubMed]

35. Massa, F.; Bonamini, L.; Settimi, A.; Pallottino, L.; Caporale, D. LiDAR-Based GNSS Denied Localization for Autonomous Racing Cars. *Sensors* **2020**, *20*, 3992. [CrossRef] [PubMed]
36. Diaz-Arango, G.; Vazquez-Leal, H.; Hernandez-Martinez, L.; Jimenez-Fernandez, V.M.; Heredia-Jimenez, A.; Ambrosio, R.C.; Huerta-Chua, J.; De Cos-Cholula, H.; Hernandez-Mendez, S. Multiple-Target Homotopic Quasi-Complete Path Planning Method for Mobile Robot Using a Piecewise Linear Approach. *Sensors* **2020**, *20*, 3265. [CrossRef] [PubMed]
37. Sim, G.; Min, K.; Ahn, S.; Sunwoo, M.; Jo, K. Deceleration Planning Algorithm Based on Classified Multi-Layer Perceptron Models for Smart Regenerative Braking of EV in Diverse Deceleration Conditions. *Sensors* **2019**, *19*, 4020. [CrossRef] [PubMed]
38. Min, K.; Sim, G.; Ahn, S.; Sunwoo, M.; Jo, K. Vehicle Deceleration Prediction Model to Reflect Individual Driver Characteristics by Online Parameter Learning for Autonomous Regenerative Braking of Electric Vehicles. *Sensors* **2019**, *19*, 4171. [CrossRef]
39. Mendoza-Petit, M.F.; Garcia-Pozuelo, D.; Diaz, V.; Olatunbosun, O. A Strain-Based Method to Estimate Tire Parameters for Intelligent Tires under Complex Maneuvering Operations. *Sensors* **2019**, *19*, 2973. [CrossRef]
40. Gao, L.; Xiong, L.; Lin, X.; Xia, X.; Liu, W.; Lu, Y.; Yu, Z. Multi-sensor Fusion Road Friction Coefficient Estimation During Steering with Lyapunov Method. *Sensors* **2019**, *19*, 3816. [CrossRef]
41. Li, X.; Tao, X.; Zhu, B.; Deng, W. Research on a Simulation Method of the Millimeter Wave Radar Virtual Test Environment for Intelligent Driving. *Sensors* **2020**, *20*, 1929. [CrossRef]
42. Terán, J.; Navarro, L.; Quintero M.C.G.; Pardo, M. Intelligent Driving Assistant Based on Road Accident Risk Map Analysis and Vehicle Telemetry. *Sensors* **2020**, *20*, 1763. [CrossRef]
43. Zhang, P.; Xiong, L.; Yu, Z.; Fang, P.; Yan, S.; Yao, J.; Zhou, Y. Reinforcement Learning-Based End-to-End Parking for Automatic Parking System. *Sensors* **2019**, *19*, 3996. [CrossRef]
44. Wang, L.; Wang, X.; Sheng, Z.; Lu, S. Model Predictive Controller Based on Online Obtaining of Softness Factor and Fusion Velocity for Automatic Train Operation. *Sensors* **2020**, *20*, 1719. [CrossRef] [PubMed]

sensors

MDPI

Article

Multi-Vehicle Tracking via Real-Time Detection Probes and a Markov Decision Process Policy

Yi Zou, Weiwei Zhang *, Wendi Weng and Zhengyun Meng

College of Mechanical and Automotive Engineering, Shanghai University of Engineering Science, Shanghai 201620, China; zoezou9@163.com (Y.Z.); wangxinchensues@163.com (W.W.); mzysues@163.com (Z.M.)
* Correspondence: zwwsues@163.com; Tel.: +86-1560-192-1376

Received: 25 January 2019; Accepted: 9 March 2019; Published: 15 March 2019

Abstract: Online multi-object tracking (MOT) has broad applications in time-critical video analysis scenarios such as advanced driver-assistance systems (ADASs) and autonomous driving. In this paper, the proposed system aims at tracking multiple vehicles in the front view of an onboard monocular camera. The vehicle detection probes are customized to generate high precision detection, which plays a basic role in the following tracking-by-detection method. A novel Siamese network with a spatial pyramid pooling (SPP) layer is applied to calculate pairwise appearance similarity. The motion model captured from the refined bounding box provides the relative movements and aspects. The online-learned policy treats each tracking period as a Markov decision process (MDP) to maintain long-term, robust tracking. The proposed method is validated in a moving vehicle with an onboard NVIDIA Jetson TX2 and returns real-time speeds. Compared with other methods on KITTI and self-collected datasets, our method achieves significant performance in terms of the "Mostly-tracked", "Fragmentation", and "ID switch" variables.

Keywords: tracking-by-detection; multi-vehicle tracking; Siamese network; data association; Markov decision process

1. Introduction

Advanced driver-assistance systems (ADASs) and autonomous driving have consistently been a popular research area. An intelligent vehicle is expected to interact with other vehicles as well as other traffic participants, in which case relative movement tendencies of a multi-vehicle environment is of great concern. An accurate multi-vehicle tracker is necessary for several tasks such as location, navigation, and traffic behavior analysis.

In the research area of single-object tracking (SOT), most state-of-the-art methods tend to learn a discriminative classifier on labeled sample patches within a neighborhood area [1–3]. Especially, when deep neural networks (DNNs) show powerful effectiveness in feature selection, the performance of tracking significantly improves [4–6]. Multi-object tracking (MOT) comes from SOT, and it has wide applications in visual surveillance, traffic monitoring [7–9], sports analysis, ADAS, and autonomous driving. The goal of MOT is to estimate the locations of multiple objects in real-time while maintaining each identity consistently and yielding individual trajectories [10–13]. However, multi-object tracking faces special challenges that can be even more serious with moving camera platforms. Firstly, multiple targets may share a similar appearance in complex scenarios, and appearance may change dramatically at any time. Secondly, observable motion cues are more complicated since new emerging targets and tracked targets always overlap with each other. When it comes to onboard moving camera platforms, these conditions deteriorate, and tracking models need to put more computational overhead on real-time performance. All the above factors contribute to tracking drift and even failure.

Multi-object tracking benefits significantly from advances in object detection in recent years. Tracking-by-detection frameworks [3,11,12,14–16] have achieved extremely reliable test results and have shown great potential in handling object appearance variations and model drifts. Distinguished from the detection-free tracking method that needs to calibrate targets manually first, the tracking-by-detection approach is more feasible in handling new targets at each time step in a dynamic environment. This kind of approach detects objects in each frame and then matches them in the following frames to form complete trajectories. The batch tracking system [12,14,17] utilizes a set of detection results collected by temporal sliding windows of whole frames to generate global trajectories. Although such offline tracking methods perform well in obtaining an optimal, theoretical global solution in partial time snippets, they are not applicable in handling dramatic model changes in online, long-term tracking. Specifically, the real-time tracking application requires online methods [16,18–20] to handle up-to-time observations and sequentially extend existing trajectories with current detections based on frame-by-frame associations.

Date association and matching play a vital role in MOT identity assignment. The Hungarian method [21] is applied to achieve matching of bipartite graphs by finding the minimum point solution of the assignment matrices. The feature of appearance (e.g., color histogram, histogram of oriented gradients (HOG) feature, shapes feature, texture, and optical flow) is usually extracted as a part of a measurement. The rigid characteristics of vehicles benefits this under positive conditions for generating discriminative appearance models in data association. Inspired by multiple neural network architectures [22], the two-channel network is used to learn a richer hierarchical feature of patches and output pairwise similarity. Moreover, combined with spatial pyramid pooling (SPP) layers [23], the network reduces the size limitation, and thus becomes more reasonable in practice.

On the other hand, there are inaccurate detections of occluded and novel objects, so the process of learning to track is a trend that can deal with these ambiguities in data association [15,16,18,19,24–26]. In this study of tracking with a moving camera, scenarios are more complex and unpredictable. ID switch is one of the most common problems in long-term tracking, where the previous methods are less reliable to handle. In order to improve long-term tracking robustness, a Markov decision processes (MDP) is introduced to manage the state of each object and alleviate track drift. Furthermore, reinforcement learning is applied to learn data association policies, which could effectively cope with the appearance/disappearance of each vehicle by state transition.

In this paper, an integrated framework is proposed to track frontal vehicles with an onboard monocular camera, which can assist intelligent vehicles with substantial benefits in high-performance and safe distance maintenance. The main contributions of this paper are threefold:

- An offline-trained vehicle detector is customized to generate robust and fine detections by an onboard monocular camera. Data augmentation benefits the detector to meet various traffic conditions in moving scenes.
- A well-designed association strategy adopts multi-dimensional information to score pairwise similarity. A Siamese convolution network is designed to score pairwise similarity, wherein a dual-resolution in two specific channels could efficiently improve the performance of image matching. Any size of the input patches can still maintain the fixed output dimensionality through the SPP layer. A tracking-by-detection framework is applied to accomplish linear assignments by linking new detections with initial tracks.
- The tracking process is formulated as the Markov decision process. Four states are designed to manage the lifetime of each vehicle, which is more adaptable to the changeable traffic scenes. With reinforcement learning, an updated policy is applied to reduce false positives and improve tracking accuracy.

The rest of this paper is organized as follows. Related work is discussed in Section 2. Section 3 describes the specific methods from three parts in details. Experimental results are analyzed in Section 4, and Section 5 concludes the paper.

2. Related Work

Recently, the tracking-by-detection framework has become the leading paradigm in MOT because of its remarkable processes in object detection. These approaches formulate MOT as a data association problem, in which the main task is linking individual detections to build longer tracklets. Sadeghian et al. [15] followed this paradigm, whereby temporal detections were encoded across appearance, motion, and interactions for tracking multiple targets. In [26], a continuous confidence of detectors was proposed, and then target-specific classifiers were learned to select high-confidence detections and were associated to targets for robustly tracking multiple people in complex scenes. Coifman et al. [7] proposed a video image processing system to realize effective traffic surveillance. They took corner points of vehicles as the relevant feature, which made the system less sensitive to partial occlusions. Bae and Yoon [20] formulate an MOT problem based on tracklet confidence, in which fragmented tracklets were linked up with others, relying on online-provided detections. Sanchez-Matilla et al. [25] associated strong and weak detection responses for tracking, which denoted that high confidence detections could initialize targets while weak confidence detections only supported the propagation of labels. In this work, the tracking task of each vehicle is initialized frame-by-frame according to the latest detections.

The core of multi-object tracking is based on data association, which is to identify correspondence between trajectories and new detections. The key in corresponding is how to compute a matching score that models multiple cues from the past, such as object interactions, appearances, and motions. A tracking method based on the template matching was reported in [8], which can dynamically switch modules to handle various conditions in real sequences. Yoon et al. [16] utilized a structural model to realize the best assignment by minimizing total cost, in which an event aggregation approach was developed to integrate structural constraints in assignment cost. However, it showed limited camera motion performance because a single metric model was used. The association cost in [25] relied only on the position and size, so nearby targets were hard to discriminate. Besides motion information, Wojke et al. [27] integrated an appearance model and a deep association metric, which was trained on a large-scale person re-identification dataset to improve the performance of real-time tracking [28]. In [20], both tracklet confidence and learned-appearance models were designed to support a reliable association for multi-object tracking problems. In such methods above, the Hungarian algorithm [21] helps to solve the bipartite matching problem of possible tracker-detection anchors.

Bromley et al. [29] proposed a two-stream Siamese architecture for signature verification. Similarly, this architecture was introduced for face verification in [30], where two identical convolutional networks were trained to realize similarity metric learning. Inspired by successful progress in the convolutional neural network, deep neural networks are employed in Siamese invariance networks to learn the generic matching function for single object tracking. Tao et al. [31] focused on the learning strategy of matching functions, but they had a large gap in handling specific MOT problems, e.g., occlusion or model update. In this multi-vehicle tracking task, an improved Siamese network with a dual-resolution stream is used to generate similarity between pairs of candidates for data association. Specifically, an SPP layer [23] is embedded to release size constraints by fixed dimensional characteristics. Consequently, the network becomes more variable in managing arbitrary patches in practical tracking scenarios.

Recently, the MDP [32] has been widely used in computer vision to learn policy parameters. Karayev et al. [33] found a dynamic policy of optimizing feature selection and classification strategies by formulating the problem as an (MDP). Kitani et al. [34] incorporated uncertainty and noise observations into the hidden variable MDP (hMDP) model to realize activity understanding and forecasting in computer vision. In [35], in order to balance the cost and accuracy in the study of human–machine collaboration in object annotation, the MDP was used to automatically quantify the best tradeoff. Inspired by previous research, the proposed state transition framework is designed to manage each single object tracker as a separate agent in MDP. Each action is responsible for a specific situation, such as in false alarms and missed detection in cluttered traffic scenes. The potential for ambiguous tracking can be alleviated by correcting detection errors and recovering observations from an occluded period.

3. Methods

The proposed tracking scheme consisted of detecting targets and matching their identities frame by frame, which led to a set of target trajectories over time. The tracking-by-detection method was used to address this problem. Figure 1 shows the overview of the proposed multiple-vehicle tracking framework. The detection probes produced simultaneous current results, and the tracker guaranteed long-term tracking. New detections were linked to the activated tracks at each time step by solving the linear assignment problem. The motion and appearance model were integrated to create a pairwise matching score matrix, where traditional methods and deep learning were both involved. The initialized targets T_t^i and the new detections D_t^j were gathered in a bipartite graph, and the Hungarian algorithm was used to find the optimal assignments that maximized the total matching score. Finally, to realize stable tracking, each object was initialized with its own MDP that could manage lifetime based on real-time state transition. Moreover, it relied on online reinforcement learning to learn a policy for data association between training tracks and ground truth.

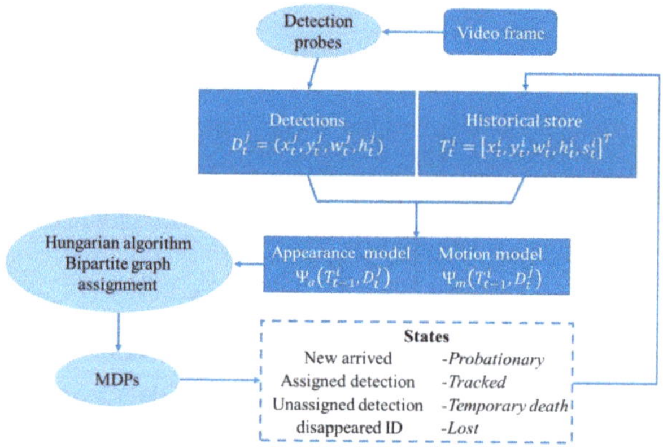

Figure 1. The overview of the proposed multiple vehicle tracking system. Discriminative appearance similarity and motion model are implemented to perform pairwise associations and Markov decision processes (MDPs) to define the real-time state.

3.1. *Vehicle Detection Probes*

Based on the tracking-by-detection framework, the robustness of the real-time tracking system takes advantage of high-precision detection results. The single shot detector YOLOv3 runs significantly faster than other detection methods, which makes it more suitable to be applied in real-time tasks. The proposed vehicle traction probes were trained based on YOLOv3 in rich datasets to improve the precision of vehicle detection.

The vehicle images formed the KITTI Vision Benchmark [33] and a self-collected dataset that were both integrated to increase the diversity of training samples, which involved multi-scale vehicles in different scenes containing occlusions and truncations. Furthermore, facing various appearances of vehicles in dynamic traffic scenes, data augmentation was adopted to improve generalization. Specifically, the brightness, contrast, and saturation of the images were changed to adapt to various light conditions. The straighten angle was rotated to deal with different tracking views. The training dataset contained a total of 18,952 images with 480 $\times$ 640 pixels, which contained various appearances of vehicles in different light conditions. Since the batch size was set to 50, one epoch needed to iterate 18,952/50 = 379 times. The training epochs were set to 60, and thus the number of iterations was

160 × 379 = 60,640. Different vehicle types, such as MPVs, SUVs, sedans, hatchbacks, vans, minibuses, pickups, and other types were trained to annotate as "vehicle". Furthermore, an intersection over union threshold of 0.7 was adopted for evaluation. The precision of the bounding box was highly demanded while the position feature sets were used for calculating matching measurements. In this work, an iterative refinement framework [36,37] was conducted to improve localization accuracy by tight object-bounding boxes.

By comparing tracking performances by switching the detector component, the evaluation result could verify the effectiveness of the proposed detection probes, and it could demonstrate that detection quality plays a significant role in the tracking-by-detection framework for MOT.

3.2. Diversity Feature Extraction

The goal of data association is to identify the correspondence between pre-existing tracks and new detections. A set of linear corresponding constraints between an initialized trajectory T_t^i and a current detection D_t^j is defined to discriminate how well a pair of candidate patches match. Motion and appearance models are integrated into this problem formulation by addressing appropriate metrics.

3.2.1. Motion and Size Models

Small changes in object positions are the critical components of data associations in traffic scenes. The motion model used the Mahalanobis distance to measure relative movements, which defines the distance between the initialized target T_{t-1}^i and the current detection D_t^j. The bounding coordinates of initial and detected scenes are represented as: $T_{t-1}^i = \left(x_{t-1}^i, y_{t-1}^i, w_{t-1}^i, h_{t-1}^i\right)^T$, $D_t^j = \left(x_t^j, y_t^j, w_t^j, h_t^j\right)^T$,

$$d_{(T_{t-1}^i, D_t^j)} = \sqrt{(D-T)^T \Sigma^{-1} (D-T)} \tag{1}$$

$$\Sigma_{T_{t-1}^i D_t^j} = \begin{bmatrix} \mathrm{E}\left[\left(T_{t-1}^i - E\left[T_{t-1}^i\right]\right)\left(T_{t-1}^i - E\left[T_{t-1}^i\right]\right)\right] & \mathrm{E}\left[\left(T_{t-1}^i - E\left[T_{t-1}^i\right]\right)\left(D_t^j - E\left[D_t^j\right]\right)\right] \\ \mathrm{E}\left[\left(D_t^j - E\left[D_t^j\right]\right)\left(T_{t-1}^i - E\left[T_{t-1}^i\right]\right)\right] & \mathrm{E}\left[\left(D_t^j - E\left[D_t^j\right]\right)\left(D_t^j - E\left[D_t^j\right]\right)\right] \end{bmatrix} \tag{2}$$

where j is the number of current detections in frame t, and $\left(x_t^j, y_t^j\right)$ denotes the upper-left corner of the detection bounding box in the image. The width w_t^j and the height h_t^j correspond to the size of the bounding box. As the vehicle is rigid, the area scale and the aspect ratio of the bounding box are also considered. The area scale α and the aspect ratio r of the detection are computed by wh and $\frac{w}{h}$, respectively. Σ represents the covariance matrix in the Mahalanobis distance, where the operator E denotes the expected value of its argument.

Given a pairwise object patch, the similarity score of motion is obtained as follows:

$$\Psi_m\left(T_{t-1}^i, D_t^j\right) = \frac{1}{d_{(T_{t-1}^i, D_t^j)} + \left(r_j - r_i\right)^2 + \left(\alpha_j - \alpha_i\right)^2} \tag{3}$$

3.2.2. Central-Surround Two-Channel Spatial Pyramid Pooling (SPP) Network

In the data association process, the similarity of appearance is definitely a crucial cue in matching score computations. In this section, a Siamese network was designed to compare corresponding targets and to output their pairwise similarities for discriminative appearance models. The framework is presented in Figure 2, and Table 1 details the architecture of each convolutional layer.

The so-called two-stream network was constructed of a central stream and a surrounding stream. It enabled this process in a spatial domain, in which two different resolutions were applied. The inputs of the network were pairs of image patches from the initial identity store and scaled current detection results. Besides the area caught by the tight bounding box, the surrounding environment also mattered to combat any similar appearances. The architecture of the network was inspired by VGG-M, which contained two branches with exactly the same set of weights. Different branches played unique roles in feature extraction functions.

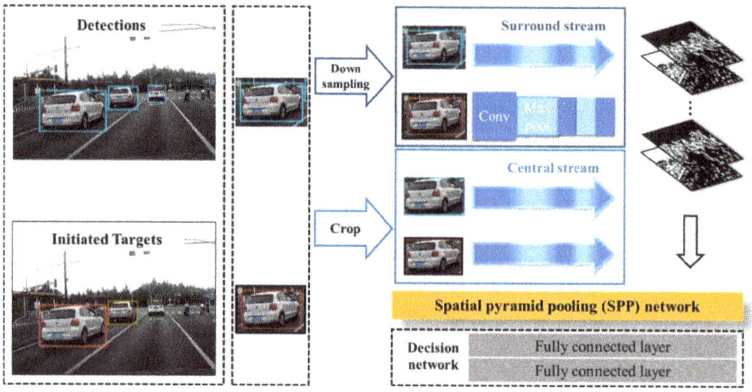

Figure 2. Central-surround two-channel spatial pyramid pooling network (CSTCSPP). This network uses the Siamese-type architecture to extract shallow features with different resolutions and then calculates pairwise similarity. A spatial pyramid pooling layer embedded before the top decision network allows patches to be free of size limitations. All convolution layers are followed by Rectified Linear Units (ReLU), which could increase the nonlinear relation between each layer of the neural network.

Table 1. Details of each branch network.

Layer	Type	Kernel Size	Stride
Input	Raw data		
Conv1	Convolution	7×7	2
Pool1	Max pooling	3×3	2
Conv2	Convolution	5×5	2
Pool2	Max pooling	3×3	2
Conv3	Convolution	3×3	1
Output	FC		

To calculate similarity in the two-channel network, the patches of each target were cropped to $(x - 0.15w,\ y + 0.15h,\ 1.3w,\ 1.3h)$ by experimental experience. Surrounding context features could enhance comparability, and large expansion may not only increase computation but also decrease accuracy. These patches go through down-sampling or cropping processes, and they are then transferred into the surrounding and central steams, respectively. Down-sampled patches in the surrounding low-resolution stream match the surrounding context features when the targets have a similar appearance. High-resolution patches in the central stream supplied more details about vehicle features. Two streams were designed to extract discriminative features, where the pixels of the vehicle and the periphery were all taken into consideration.

The prevalent convolutional neural networks (CNNs) require a fixed input image size due to the definition of the fully-connected layers, which limits both the aspect ratio and the scale of the inputs. In practical tracking scenarios, the detection patches are caught with arbitrary sizes under different distances and angles. With the help of a spatial pyramid pooling (SPP) layer, the network could aggregate features through spatial pooling and then generate a fixed-length representation. The top decision network consisted of two linear, fully connected layers with 512 hidden units. They were separated by the ReLU activation layer, which could increase the non-linearities inside the network and make the decision function more discriminative.

The parameters of the network were trained offline, based on self-collected datasets. In order to improve the efficiency in retrieving patch pairs and storing all the input images in Graphics

Processing Unit (GPU) memory, data augmentation and preprocessing were adopted to train the model. The training data were augmented by flipping both patches horizontally and vertically and operating multi-degree rotation to reduce overfitting problems.

The learning function is calculated based on the L_2-norm regularization and hinge loss:

$$J(\omega) = \min_{\omega} \tfrac{\lambda}{2} \|\omega\|_2 + \sum_{i=1}^{N} \max(0, 1 - y_i \mu_i) \tag{4}$$

where ω is the weights of the neural network, $y_i \in \{-1, 1\}$ is the corresponding label of the patch pairs with -1 and 1 denoting a non-matching and a matching pair, respectively. And $\mu_i \in (-1, 1)$ represents the network output for the i–th training sample. Asynchronous stochastic gradient descent (ASGD) with a constant learning rate 1.0, momentum of 0.9, and weight decay of λ = 0.0005 was used to train the models. Weights were initialized randomly and all models were trained from scratch.

3.2.3. Feature Representation

Constitute a tracklets historical store $T_t = \{T_t^1, T_t^2, \ldots, T_t^i\}$, $T_t^i = \left[x_t^i, y_t^i, w_t^i, h_t^i, s_t^i\right]^T$.

Where i is the number of initialized targets in the last frame t. Specifically, T_t^i corresponded to the historical store of tracked targets in the previous frame, which contained multi-dimensional information about the location $\left[x_t^i, y_t^i\right]^T$, the shape of bounding box $\left[w_t^i, h_t^i\right]^T$, and the latest state $\left[s_t^i\right]^T$ in frame t. Generally, the store was preferable in this application, where facing dynamic situations involved false alarms and missed detections.

The similarity of motion

$$\Psi_m\left(T_{t-1}^i, D_t^j\right) = \frac{1}{\eta d_{\left(T_{t-1}^i, D_t^j\right)} + \delta\left(r_j - r_i\right)^2 + \rho\left(\alpha_j - \alpha_i\right)^2} \tag{5}$$

where η, δ, ρ are the weighing parameter to balance the value of distance, aspect ratio, and area scale, respectively. All parameters were found experimentally and remained unchanged for all datasets.

The similarity of appearance

$$\Psi_a\left(T_{t-1}^i, D_t^j\right) \in (-1, 1) \tag{6}$$

The goal of data association is to find the set of trajectories T_{t-1} that best explains the detections D_t^j. This means we needed to find the best linear assignment to get bipartite graph maximum matching scores. The matching score defined how probable a match was for pairwise objects between the tracked target and the current detection.

$$M_{\left(T_{t-1}^i, D_t^j\right)} = max\left[\lambda \Psi_m\left(T_{t-1}^i, D_t^j\right) + (1 - \lambda)\left(\Psi_a\left(T_{t-1}^i, D_t^j\right) + 1\right)\right] \tag{7}$$

Matching matrix

Consider a scenario where there are m preexisting tracks and n new detections at frame t. A matrix $M_t \in \mathbb{R}^{m \times n}$, which is $M_{\left(T_{t-1}^i, D_t^j\right)} \in M$, represents the matching score of assigning detection j to track i at time t. The Hungarian algorithm was introduced to find the global optimal assignment matrix so that the total matching score was maximized.

3.3. *Markov Decision Processes (MDPs)*

This part focuses specifically on how to maintain robust multi-vehicle tracking, which is a tough challenge in MOT. Four states were utilized to handle false alarms and missed detections occurring in crowded scenes so that the tracker could re-identify the target with the same ID from any short-term occlusion.

3.3.1. Overview of the MDPs

Due to multiple vehicles moving with varying speeds, inter-object occlusion and truncation often occurs in onboard, multi-object tracking tasks. Distinguished from SOT, multiple-object tracking depends on detection that often suffers from track drift when the appearance dramatically changes as a result of frequent inter-object occlusions.

A Markov decision process (MDP) is the Markov reward process with a decision. In this framework, the lifetime of each target is modeled with an MDP that consists of four components $(S, A, T(\cdot), R(\cdot))$. $s \in S$ encodes the status of the target in a particular time, which is determined by its previous action. Action $a \in A$ can be performed to transfer the state in each frame. T represents the transition function, which can be described as $T : S \times A \rightarrow S$, and it describes the effect of each action in each state. $R : S \times A \rightarrow \mathbb{R}$ defines the immediate reward received after executing action a to state s. Each target had its own corresponding MDP to handle the lifetime, and the process of state transition is detailed in Figure 3. Reinforcement learning provided a framework that was concerned with how the agent took action within a given state so as to maximize the cumulative reward.

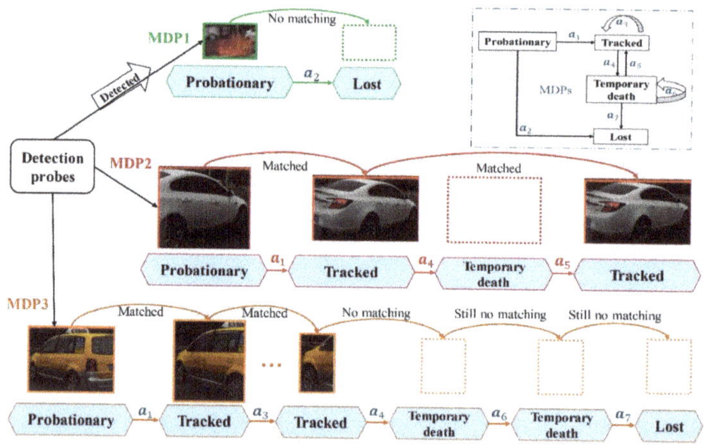

Figure 3. Online multi-vehicle tracking problem formulated as decision-making in MDP. The upper-right framework represents the transition map of four categorized states at each time step. Each target is initialized with a unique MDP to manage their lifetimes, depicted in different colors.

The state space in the target MDP was parted into four subspaces, where each state encoded the global information of the target depending on feature representation, such as location, size, and appearance. Firstly, each object caught by the detector was activated to enter the "probationary" state. Vehicles in this state could transition to the "tracked" state only if they matched in the consecutive frames. Otherwise, the false alarm triggered entry to the "lost" state and removed the historical data. A tracked target could stay "tracked", or transition into "temporary death" if the vehicle was lost due to occlusion by other vehicles, acceleration, or being out of view. Likewise, vehicles in the "temporary death" state had the chance to get back to "tracked" if it could complete successful matching, otherwise it transitioned to the "lost" state forever. Seven possible transitions were designed between the states of a target, which corresponded to seven actions in MDP.

3.3.2. Policy in the Probationary State

Each detection that was unclaimed by any track underwent a probationary period where the target could be consistently detected to accumulate enough evidence. This period made up for the defect of false alarm and avoided an unnecessary increase of ID.

To handle targets in the probationary state, the MDP needed to decide whether it should switch to the "tracked" state or transfer into the "lost" state. If the tracked vehicles were not able to successfully associate any detection responses D_t^i in the next consecutive frame, the MDP recognized the failure of tracking initialization, and transitioned the object to the "lost" state. In the meantime, redundant data was deleted for efficiency. Otherwise, the target finished the preprocessing step of tracking and was transferred to a "tracked" state.

This is equivalent to learning the reward function in the probationary period state:

$$\mathrm{R}_p(\mathrm{s},\mathrm{a}) = \begin{cases} y(a),\ if\ M_{(T_t^i,D_t^j)} \geq m_0 \\ -y(a), \quad otherwise \end{cases}, \tag{8}$$

where $y(a)$ = +1 if action a = a_1, and $y(a) = -1$ if a = a_2.

3.3.3. Policy in the Tracked State

To handle targets in the tracked state, the MDP needed to decide whether to keep tracking or to transfer it to temporary death. If the activated trajectory could associate with the corresponding detection pair, the MDP recognized this target as still under tracking, otherwise transferred it to the "temporary death" state.

The reward function in the tracked state is defined as followed:

$$\mathrm{R}_{tracked}(\mathrm{s},\mathrm{a}) = \begin{cases} y(a),\ if\ M_{(T_t^i,D_t^j)} \geq m_0 \\ -y(a),\ otherwise \end{cases}, \tag{9}$$

where $y(a)$ = +1 if action a = a_3, and $y(a) = -1$ if a = a_4.

3.3.4. Policy in the Temporary Death State

In data association progress, unassociated tracks transitioned to the temporary death period. In addition, their coded feature and current state were historically stored just in case it was re-tracked (the red line in Figure 3). Trajectory terminated if they continued to fail to match with each input of detections, which meant this vehicle accelerated to speed away or was left behind (the yellow line in Figure 3). The linear function $\mathcal{L}\left(T_t^i, D_t^j\right) = W^T\tau\left(T_t^i, D_t^j\right) + b$ was used to make the decision rule. $\tau\left(T_t^i, D_t^j\right)$ is the feature vector which represented the similarity between the initialized target and detection. Moreover, the coding message of the vehicle was deleted after action a7, and thus, this object would be activated with a new ID if it was re-detected.

Consequently, the reward function in the temporary death is defined as:

$$R_{td}(s,a) = y(a)\left(\max_{1 \leq j \leq M^t} \left(W^T\tau\left(T_t^i, D_t^j\right) + b\right)\right), \tag{10}$$

where $y(a)$ = +1 if action a = a_5, and $y(a) = -1$ if a = a_6. j indexes Q candidate detections for data association.

3.3.5. Reinforcement Learning

The tracking drift problem is highlighted in onboard, multi-vehicle tracking tasks. A learned policy was performed to handle the tracking robustness. The binary classifier with enforcement learning was trained offline in public KITTI datasets and self-collected datasets where each sequence was marked with ground truth. In the training process, each MDP took unique action as indicated by the ground truth trajectory. The goal in this part was training an MDP policy that could be used to track all these targets. Reinforcement learning defined a set of actions $a \in A$ that made achieving the maximum reward possible. This policy was updated only when the MDP made a mistake in data association.

To obtain a max-margin classifier for data association, the training function is used as follows:

$$\min_{w,b,\xi} \frac{1}{2}\|W\|^2 + C\sum_{k=1}^{Q}\xi_k \tag{11}$$

$$\text{subject to } y_k\left[W^T\tau\left(T_t^i, D_t^j\right)+b\right] \geq 1-\xi_k, \xi_k \geq 0, k=1,2,\ldots,Q, \tag{12}$$

where ξ_k, k are the slack variables, and C is a regularization parameter. The policy was kept iterated when the classifier was updated until all the visible and correct targets were successfully tracked.

4. Experiments

In this section, dataset and evaluation metrics are presented in the first part. The comprehensive experiments were conducted in three stages. First, the comparison of different components was evaluated in three typical scenes on a self-collected dataset. Second, the motion and appearance models were disabled sequentially to evaluate the contribution of each component. Finally, the proposed method was compared with five state-of-the-art methods on KITTI datasets to assess the contribution of the work in terms of six evaluation metrics. As shown in Figure 4, comprehensive tests and analyses were performed on NVIDIA Jetson TX2 with an on-board camera.

Figure 4. (**a**) NVIDIA Jetson TX2 with 256 GPU cores; (**b**) Comprehensive tests are validated in the moving vehicle in different scenes (e.g., highway).

4.1. Dataset and Evaluation Metrics

Datasets.

To evaluate the performance of the proposed multi-vehicle tracking method, extensive experiments were conducted on the KITTI Vision Benchmark Suite dataset [38], which is the widely used benchmark for multiple vehicle tracking. The training dataset consisted of 21 sequences with 8008 frames, and the testing dataset consisted of 29 sequences with 11,095 frames. Despite the dataset having labeled eight different classes, only the class "car" was considered in our work. Especially, the KITTI dataset provided object detection as well as tracking results in a full-face perspective based on its comprehensive annotations. It was crucial to the research of tracking by detection with a frontal, onboard monotonous camera. In the self-collected datasets, 50 annotated sequences of three typical traffic scenes in various light conditions were acquired from a moving camera with 480×640 pixels. All sequences had a varying number of objects and lengths with unique motion scenarios. The differences of size and orientation, occlusion pattern, and illumination were considered in our datasets.

Evaluation metrics.

For quantitative evaluation, the average precision (AP) was first taken into account to evaluate detection performance. A widely accepted protocol, CLEAR MOT metrics [39], were adopted, which included multiple-object tracking precision (MOTP) and multiple-object tracking accuracy (MOTA). The MOTP measured the ability of the tracker to estimate precise object positions. Furthermore, fragmentation (FRAG), ID switches (IDS), mostly-tracked (MT), and mostly-lost (ML) were also indispensable in valuing the performance in MOT. ID switch happened when a ground-truth trajectory was matched with another wrong identity. The MT and ML represented the percentage of the ground truth trajectories covered by the tracker output for more than 80% in length or less than 20% in length, respectively. Identification F1 score (IDF1) was the ratio of correctly identified detections over the average number of ground-truth and computed detections, which evaluated identification precision.

4.2. Performance Evaluation

The combined multi-vehicle tracking frameworks were evaluated on the self-collected dataset, which contained different motion patterns on campuses, urban roads, and highways. The previous algorithms "SSD" [40] and "YOLOv3" [41] performed well in object detection domains. By switching partial components, Table 2 shows the performance of detection and tracking in three typical traffic scenes. The bold results present relatively better performance.

Table 2. Comparative results under different traffic scenes.

Detector	Evaluation of Detection (AP)			Tracker	Evaluation of Tracking (MOTA)		
	Campus	Urban	Highway		Campus	Urban	Highway
SSD	65.25%	60.16%	68.84%	Proposed	70.64%	72.62%	74.32%
YOLOv3	63.55%	62.99%	70.19%	Proposed	74.65%	**77.22%**	77.98%
Detection probes	**68.84%**	**63.66%**	**72.03%**	Proposed	**75.29%**	76.06%	**78.14%**

The evaluation results note that better detection results led to better scores in tracking. In moving scenes, the size of the target vehicle varied while the distance changed. YOLO was relatively sensitive to the changing scale objects, and the generalization ability of objects with large-scale changes was poor. Detection probes trained in augmented vehicle dataset significantly improved the detection performance (measured as AP) under diverse scenes. The customized detector combined with the proposed tracking scheme could stay competitive in different environments.

In a campus environment, the tracking scenario was relatively simple, where most of the target vehicles were parked on the roadside. But on the urban road, inter-object occlusion and truncation frequently occurred due to cluttered traffic scenes. Facing traffic signals and lane marks, the motion of each vehicle became relatively complicated. In the urban traffic intersection, vehicles show different shapes in our view, The traffic flow became smoother on the highway, in which vehicles kept moving in the same direction with typical highway situations, like cruising, overtaking, following, etc. They were free from other distractions, e.g., pedestrians or bicycles.

The trade-off between accuracy and speed was quite tough in detection and tracking tasks. The offline, pre-trained detector on the portable NVIDIA Jetson TX2 with 256 GPU cores could achieve real-time performance while maintaining competitive tracking accuracy. As the computation speed depended on the number of targets in the video sequence, tests were applied in three typical traffic scenes and returned about 25 frames per second (FPS).

Inspired by the deep-sort method [27], only appearance information was used in the association cost term during the experiments when there was substantial camera motion. The motion model describes the movement of the object while the appearance model focused on the similarities of the surface features. In order to demonstrate the effectiveness of each component, the contribution of each model was investigated under two typical situations. Figure 5a illustrates the tracking performance under different situations in terms of IDF1 and MOTA. IDF1 is a major tracking metric that measures how often objects are correctly identified by the same tracking identity. As expected, significant performance drops happened when the single feature model was taken into account.

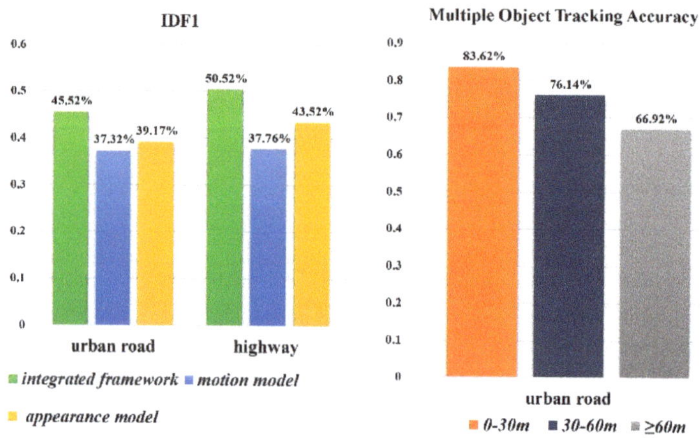

(**a**) (**b**)

Figure 5. Comprehensive analyses of the proposed framework. (**a**) The contribution of each components in two typical scenes respectively; (**b**) The tracking accuracy in different distance and the threshold selection depends on the image size.

More specifically, tracking on the urban roads performs worse than on the highway because of the volume of road traffic facilities and inter-object occlusions. Appearance cues became less discriminative in over-crowed tracking backgrounds. One single cue was not reliable to capture the correlation of pairwise targets. The motion model only figured the relative location change, but it still had a gap in handling false positives near the target. Appearance constraints could significantly reduce this ambiguity. On the other hand, no motion model may contribute to the mishandling of target vehicles sharing the same characteristics. These limitations indicate that only considering both of the factors is sufficient to guarantee the robustness of MOT in dynamic and complex traffic scenes.

In terms of using the track method in the domains of intelligent vehicles to increase safety, the distance between the ego-vehicle and other objects is worth taking into account. Three distance thresholds were observed and analyzed in an urban road environment. The threshold selection depended on the image size in this test phase. As shown in the right histogram of Figure 5b, the multiple object tracking accuracy (MOTA) performed better when the targets were closer, in which they were highly threatened.

The proposed method was evaluated on the KITTI Tracking benchmark and only the "car" class was considered. A quantitative comparison between our method and other state-of-the-art tracking systems [42–46] is given in Table 3. Here, ↑ represents that higher scores indicate better results and ↓ notes lower are better. The bold results present relatively better performance.

Table 3. Comparison of our proposed methods with five state-of-the-art methods on KITTI.

Method	MOTA ↑	MOTP ↑	FRAG ↓	IDS ↓	MT ↑	ML ↓
Proposed	76.53%	81.19%	**349**	11	**82.12%**	9.92%
SSP [39]	57.85%	77.65%	704	**7**	29.38%	24.31%
RMOT [40]	65.83%	75.42%	727	209	40.15%	9.69%
MDP [41]	69.35%	82.10%	387	130	52.15%	13.38%
ExtraCK [42]	79.99%	82.46%	938	342	62.15%	5.54%
MOTBeyondPixels [43]	**84.24%**	**85.73%**	944	468	73.23%	**2.77%**

The proposed method showed strong competition with other multi-object trackers. In particular, the number of "mostly tracked" increased by at least 8.89% while the FRAG, IDS, and other evaluated metrics were still robust.

The high-precision detections can potentially reduce false positives and improve the tracking accuracy (measured as MOTA). The significant score of MT implied that this method could generate a more integrated trajectory. The result of identity switches was 11, which was really close to the best result of 7 the SSP algorithm. The ability to maintain target identity denoted that the tracking scheme could initialize and terminate targets effectively and keep robust trajectories, which was enhanced by the proper policy with reinforcement learning in MDP. The competitive comparison results verified the effectiveness of the multi-vehicle tracking method. The exemplary tracking results on campuses, urban roads, highways, and the KITTI dataset are shown in Figure 6.

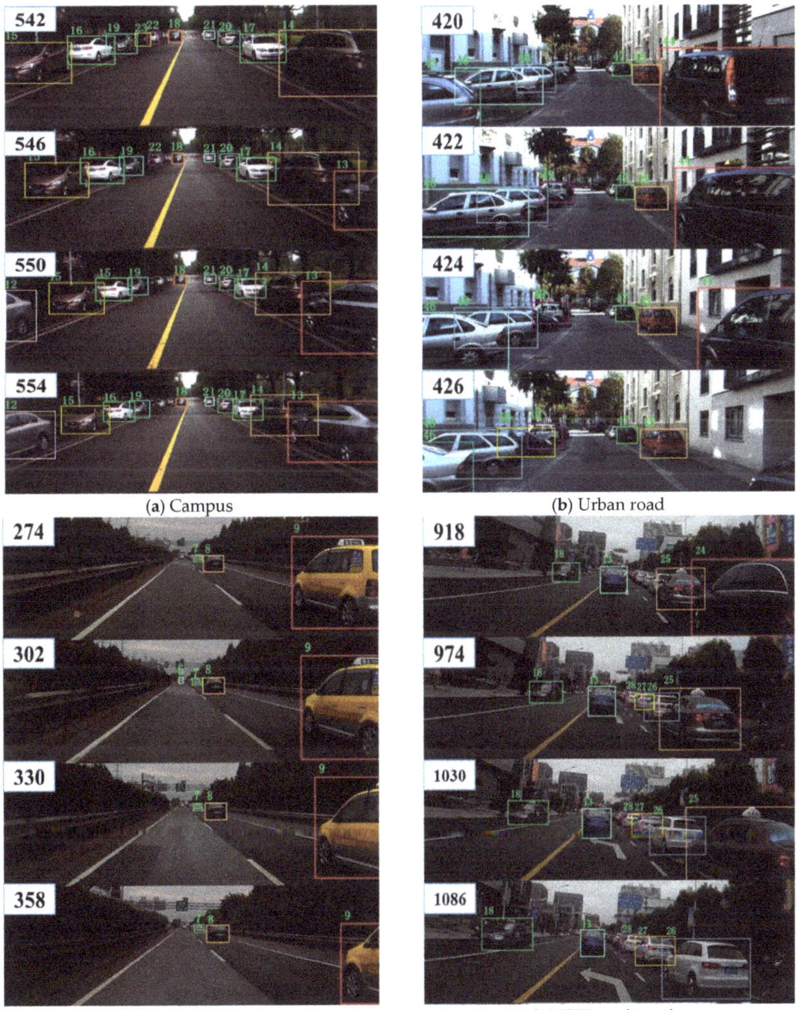

(**a**) Campus (**b**) Urban road (**c**) Highway (**d**) KITTI tracking data

Figure 6. Exemplary output under four typical traffic scenes.

5. Conclusions

In this paper, a novel method was customized to realize robust tracking of multi-vehicles with an onboard monocular camera in dynamic environments. Based on the tracking-by-detection framework, the detection probes were utilized to detect vehicles in real-time. A multi-feature model was designed to generate the matching matrix. The central-surround two-channel SPP (CSTCSPP) network generated discriminative similarity of appearance, while the motion model was used to account for the relative movements. Based on corresponding cues, the Hungarian algorithm helped to generate best matches in the data association process. Furthermore, to alleviate tracking drift, MDPs with reinforcement learning were implemented to transfer the state at each time step. The comparative experiments were conducted in different scenes to evaluate quality. The comprehensive performance analyses showed that our method was effective for real-time, long-term tracking and achieved an efficient improvement in robustness. In the future, we plan on expanding this application by adding more direction perspectives under different light conditions to employ in various scenes. 3D object detection, as well as related applications, will be considered in the next step, and the additional 3D object labels will be added to further improve the tracking performance. In addition, the system is planned to employ other specific kinds of objects, e.g., faces, pedestrians, and animals.

Author Contributions: Conceptualization, Y.Z.; Data curation, Y.Z.; Formal analysis, Y.Z.; Funding acquisition, W.Z.; Investigation, Y.Z.; Methodology, Y.Z. and W.Z.; Resources, Y.Z.; Software, Y.Z.; Supervision, W.Z.; Writing—original draft, Y.Z.; Writing—review & editing, Y.Z., W.Z., W.W. and Z.M.

Funding: This research was funded in part by National Natural Science Foundation of China (No. 51805312), in part by Shanghai Sailing Program (No. 18YF1409400), in part by Training and Funding Program of Shanghai College young teachers (No. ZZGCD15102), in part by Scientific Research Project of Shanghai University of Engineering Science (No. 2016-19), and in part by the Shanghai University of Engineering Science Innovation Fund for Graduate Students (No. 18KY0613).

Conflicts of Interest: The authors declare no conflict of interest.

References

1. Henriques, J.F.; Caseiro, R.; Martins, P.; Batista, J. High-speed tracking with kernelized correlation filters. *IEEE Trans. Pattern Anal. Mach. Intell.* **2015**, *37*, 583–596. [CrossRef] [PubMed]
2. Danelljan, M.; Hager, G.; Khan, F.S.; Felsberg, M. Learning spatially regularized correlation filters for visual tracking. In Proceedings of the IEEE International Conference on Computer Vision, Santiago, Chile, 7–13 December 2015.
3. Babenko, B.; Yang, M.H.; Belongie, S. Robust object tracking with online multiple instance learning. *IEEE Trans. Pattern Anal. Mach. Intell.* **2011**, *33*, 1619–1632. [CrossRef] [PubMed]
4. Wang, L.; Ouyang, W.; Wang, X.; Lu, H. Visual tracking with fully convolutional networks. In Proceedings of the IEEE International Conference on Computer Vision, Santiago, Chile, 7–13 December 2015.
5. Wang, L.; Ouyang, W.; Wang, X.; Lu, H. STCT: Sequentially Training Convolutional Networks for Visual Tracking. In Proceedings of the IEEE Conference on Computer Vision and Pattern Recognition (CVPR), Las Vegas, NV, USA, 27–30 June 2016.
6. Chu, Q.; Ouyang, W.; Li, H.; Wang, X.; Liu, B.; Yu, N. Online Multi-Object Tracking Using CNN-Based Single Object Tracker with Spatial-Temporal Attention Mechanism. In Proceedings of the IEEE International Conference on Computer Vision, Venice, Italy, 22–29 October 2017.
7. Coifman, B.; Beymer, D.; McLauchlan, P.; Malik, J. A real-time computer vision system for vehicle tracking and traffic surveillance. *Transp. Res. Part C Emerg. Technol.* **1998**, *6*, 271–288. [CrossRef]
8. Battiato, S.; Farinella, G.M.; Furnari, A.; Puglisi, G.; Snijders, A.; Spiekstra, J. An integrated system for vehicle tracking and classification. *Expert Syst. Appl.* **2015**, *42*, 7263–7275. [CrossRef]
9. Peña-González, R.H.; Nuño-Maganda, M.A. Computer vision based real-time vehicle tracking and classification system. In Proceedings of the IEEE 57th International Midwest Symposium on Circuits and Systems (MWSCAS), College Station, TX, USA, 3–6 August 2014; pp. 679–682.
10. Ding, R.; Yu, M.; Oh, H.; Chen, W.H. New multiple-target tracking strategy using domain knowledge and optimization. *IEEE Trans. Syst. Man, Cybern. Syst.* **2017**, *47*, 605–616. [CrossRef]

11. Bae, S.H.; Yoon, K.J. Confidence-Based Data Association and Discriminative Deep Appearance Learning for Robust Online Multi-Object Tracking. *IEEE Trans. Pattern Anal. Mach. Intell.* **2018**, *40*, 595–610. [CrossRef] [PubMed]
12. Fagot-Bouquet, L.; Audigier, R.; Dhome, Y.; Lerasle, F. Improving multi-frame data association with sparse representations for robust near-online multi-object tracking. In Proceedings of the 14th European Conference, Amsterdam, The Netherlands, 11–14 October 2016.
13. Berclaz, J.; Fleuret, F.; Türetken, E.; Fua, P. Multiple object tracking using k-shortest paths optimization. *IEEE Trans. Pattern Anal. Mach. Intell.* **2011**, *33*, 1806–1819. [CrossRef] [PubMed]
14. Milan, A.; Roth, S.; Schindler, K. Continuous energy minimization for multitarget tracking. *IEEE Trans. Pattern Anal. Mach. Intell.* **2014**, *36*, 58–72. [CrossRef] [PubMed]
15. Sadeghian, A.; Alahi, A.; Savarese, S. Tracking the Untrackable: Learning to Track Multiple Cues with Long-Term Dependencies. In Proceedings of the IEEE International Conference on Computer Vision, Venice, Italy, 22–29 October 2017.
16. Yoon, J.H.; Lee, C.-R.; Yang, M.-H.; Yoon, K.-J. Online Multi-object Tracking via Structural Constraint Event Aggregation. In Proceedings of the IEEE Conference on Computer Vision and Pattern Recognition (CVPR), Las Vegas, NV, USA, 27–30 June 2016.
17. Leal-Taixé, L.; Fenzi, M.; Kuznetsova, A.; Rosenhahn, B.; Savarese, S. Learning an image-based motion context for multiple people tracking. In Proceedings of the IEEE Computer Society Conference on Computer Vision and Pattern Recognition, Columbus, OH, USA, 23–28 June 2014; 2014.
18. Kim, S.; Kwak, S.; Feyereisl, J.; Han, B. Online multi-target tracking by large margin structured learning. In Proceedings of the 11th Asian Conference on Computer Vision, Daejeon, Korea, 5–9 November 2013.
19. Xiang, Y.; Alahi, A.; Savarese, S. Learning to track: Online multi-object tracking by decision making. In Proceedings of the IEEE International Conference on Computer Vision, Santiago, Chile, 7–13 December 2015.
20. Bae, S.H.; Yoon, K.J. Robust online multi-object tracking based on tracklet confidence and online discriminative appearance learning. In Proceedings of the IEEE Computer Society Conference on Computer Vision and Pattern Recognition, Columbus, OH, USA, 23–28 June 2014.
21. Kuhn, H.W. The Hungarian method for the assignment problem. In *50 Years of Integer Programming 1958–2008: From the Early Years to the State-of-the-Art*; Springer: Berlin/Heidelberg, Germany, 2010; ISBN 9783540682745.
22. Zagoruyko, S.; Komodakis, N. Learning to compare image patches via convolutional neural networks. In Proceedings of the IEEE Computer Society Conference on Computer Vision and Pattern Recognition, Boston, MA, USA, 7–12 June 2015.
23. He, K.; Zhang, X.; Ren, S.; Sun, J. Spatial Pyramid Pooling in Deep Convolutional Networks for Visual Recognition. *IEEE Trans. Pattern Anal. Mach. Intell.* **2015**, *37*, 1904–1916. [CrossRef] [PubMed]
24. Li, Y.; Huang, C.; Nevatia, R. Learning to associate: Hybridboosted multi-target tracker for crowded scene. In Proceedings of the 2009 IEEE Computer Society Conference on Computer Vision and Pattern Recognition Workshops, Miami, FL, USA, 20–25 June 2009.
25. Sanchez-Matilla, R.; Poiesi, F.; Cavallaro, A. Online multi-target tracking with strong and weak detections. In *Proceedings of the European Conference on Computer Vision*; Springer: Berlin, Germany, 2016.
26. Breitenstein, M.D.; Reichlin, F.; Leibe, B.; Koller-Meier, E.; Van Gool, L. Online multiperson tracking-by-detection from a single, uncalibrated camera. *IEEE Trans. Pattern Anal. Mach. Intell.* **2011**, *33*, 1820–1833. [CrossRef] [PubMed]
27. Wojke, N.; Bewley, A.; Paulus, D. Simple online and realtime tracking with a deep association metric. In Proceedings of the International Conference on Image Processing, Beijing, China, 17–20 September 2018.
28. Bewley, A.; Ge, Z.; Ott, L.; Ramos, F.; Upcroft, B. Simple online and realtime tracking. In Proceedings of the International Conference on Image Processing, Phoenix, AZ, USA, 25–28 September 2016.
29. Bromley, J.; Bentz, J.W.; Bottou, L.; Guyon, I.; Lecun, Y.; Moore, C.; Säckinger, E.; Shah, R. Signature Verification using A "siamese" time delay neural network. *Int. J. Pattern Recognit. Artif. Intell.* **1993**. [CrossRef]
30. Chopra, S.; Hadsell, R.; LeCun, Y. Learning a similarity metric discriminatively, with application to face verification. In Proceedings of the 2005 IEEE Computer Society Conference on Computer Vision and Pattern Recognition, CVPR 2005, San Diego, CA, USA, 20–26 June 2005; IEEE: Piscataway, NJ, USA, 2005.

31. Tao, R.; Gavves, E.; Smeulders, A.W.M. Siamese Instance Search for Tracking. In Proceedings of the IEEE Conference on Computer Vision and Pattern Recognition (CVPR), Las Vegas, NV, USA, 26 June–1 July 2016; pp. 1420–1429.
32. Bellman, R. A Markovian Decision Process. *J. Math. Mech.* **1957**, *6*, 679–684. [CrossRef]
33. Karayev, S.; Fritz, M.; Darrell, T. Anytime recognition of objects and scenes. In Proceedings of the IEEE Computer Society Conference on Computer Vision and Pattern Recognition, Columbus, OH, USA, 23–28 June 2014.
34. Kitani, K.M.; Ziebart, B.D.; Bagnell, J.A.; Hebert, M. Activity forecasting. In Proceedings of the 12th European Conference on Computer Vision, Florence, Italy, 7–13 October 2012.
35. Russakovsky, O.; Li, L.J.; Fei-Fei, L. Best of both worlds: Human-machine collaboration for object annotation. In Proceedings of the IEEE Computer Society Conference on Computer Vision and Pattern Recognition, Boston, MA, USA, 7–12 June 2015.
36. Rajaram, R.N.; Ohn-Bar, E.; Trivedi, M.M. RefineNet: Refining Object Detectors for Autonomous Driving. *IEEE Trans. Intell. Veh.* **2016**, *1*, 358–368. [CrossRef]
37. Eshed, O.B.; Rajaram, R.N.; Trivedi, M.M. RefineNet: Iterative refinement for accurate object localization. In Proceedings of the IEEE Conference on Intelligent Transportation Systems, Rio de Janeiro, Brazil, 1–4 November 2016.
38. Geiger, A.; Lenz, P.; Urtasun, R. Are we ready for autonomous driving? the KITTI vision benchmark suite. In Proceedings of the IEEE Computer Society Conference on Computer Vision and Pattern Recognition, Providence, RI, USA, 16–21 June 2012.
39. Bernardin, K.; Stiefelhagen, R. Evaluating multiple object tracking performance: The CLEAR MOT metrics. *J. Image Video Process.* **2008**. [CrossRef]
40. Liu, W.; Anguelov, D.; Erhan, D.; Szegedy, C.; Reed, S.; Fu, C.Y.; Berg, A.C. SSD: Single shot multibox detector. In Proceedings of the 14th European Conference, Amsterdam, The Netherlands, 11–14 October 2016.
41. Redmon, J.; Farhadi, A. YOLOv3: An Incremental Improvement. *arXiv*, 2018; arXiv:1804.02767.
42. Lenz, P.; Geiger, A.; Urtasun, R. FollowMe: Efficient Online Min-Cost Flow Tracking with Bounded Memory and Computation. In Proceedings of the IEEE International Conference on Computer Vision, Santiago, Chile, 7–13 December 2015.
43. Yoon, J.H.; Yang, M.H.; Lim, J.; Yoon, K.J. Bayesian multi-object tracking using motion context from multiple objects. In Proceedings of the IEEE Winter Conference on Applications of Computer Vision, Waikoloa, HI, USA, 5–9 January 2015.
44. Xiang, Y.; Choi, W.; Lin, Y.; Savarese, S. Subcategory-Aware convolutional neural networks for object proposals & detection. In Proceedings of the IEEE Winter Conference on Applications of Computer Vision, Santa Rosa, CA, USA, 24–31 March 2017.
45. Gündüz, G.; Acarman, A.T. A Lightweight Online Multiple Object Vehicle Tracking Method. In Proceedings of the IEEE Intelligent Vehicles Symposium, Changshu, China, 26–30 June 2018.
46. Sharma, S.; Ansari, J.A.; Murthy, J.K.; Krishna, K.M. Beyond Pixels: Leveraging Geometry and Shape Cues for Online Multi-Object Tracking. In Proceedings of the IEEE International Conference on Robotics and Automation (ICRA), Brisbane, QLD, Australia, 21–25 May 2018.

sensors

MDPI

Article

Portable System for Monitoring and Controlling Driver Behavior and the Use of a Mobile Phone While Driving

Amith Khandakar [1,*], Muhammad E.H. Chowdhury [1], Rashid Ahmed [2], Ahmed Dhib [1], Mohammed Mohammed [1], Nasser Ahmed M A Al-Emadi [1] and Dave Michelson [3]

1 Electrical Engineering Department, College of Engineering, Qatar University, Doha-2713, Qatar; mchowdhury@qu.edu.qa (M.E.H.C.); ad1404376@student.qu.edu.qa (A.D.); mm1207599@student.qu.edu.qa (M.M.); alemadin@qu.edu.qa (N.A.A.-E.)

2 Industrial and Mechanical Engineering Department, Qatar University, Doha-2713, Qatar; rashid.ahmed@qu.edu.qa

3 Electrical Engineering Department, University of British Columbia, Vancouver, BC V6T 1Z4, Canada; davem@ece.ubc.ca

* Correspondence: amitk@qu.edu.qa; Tel.: +974-403-4235

Received: 25 February 2019; Accepted: 28 March 2019; Published: 31 March 2019

Abstract: There is an utmost requirement for technology to control a driver's phone while driving, which will prevent the driver from being distracted and thus saving the driver's and passenger's lives. Information from recent studies has shown that 70% of the young and aware drivers are used to texting while driving. There are many different technologies used to control mobile phones while driving, including electronic device control, global positioning system (GPS), on-board diagnostics (OBD)-II-based devices, mobile phone applications or apps, etc. These devices acquire the vehicle information such as the car speed and use the information to control the driver's phone such as preventing them from making or receiving calls at specific speed limits. The information from the devices is interfaced via Bluetooth and can later be used to control mobile phone applications. The main aim of this paper is to propose the design of a portable system for monitoring the use of a mobile phone while driving and for controlling a driver's mobile phone, if necessary, when the vehicle reaches a specific speed limit (>10 km/h). A paper-based self-reported questionnaire survey was carried out among 600 teenage drivers from different nationalities to see the driving behavior of young drivers in Qatar. Finally, a mobile application was developed to monitor the mobile usage of a driver and an OBD-II module-based portable system was designed to acquire data from the vehicle to identify drivers' behavior with respect to phone usage, sudden lane changes, and abrupt breaking/sharp speeding. This information was used in a mobile application to control the driver's mobile usage as well as to report the driving behavior while driving. The application of such a system can significantly improve drivers' behavior all over the world.

Keywords: driving behavior; real-time monitoring; driver distraction; mobile application; portable system

1. Introduction

Over the last few years, road accident incidents have seen a tremendous increase mainly associated with driver distraction caused by mobile phone calls. The U.S. federal government has reported [1,2] that more than 30,000 people are killed on U.S. roads every year in crashes related to distracted driving. Despite knowing the risk of using a mobile phone while driving, 70% of drivers use a mobile phone during driving. There have been recent studies showing the effect of popular social networking games, such as Pokémon GO, leading to many accidents, where users play games while driving [3]. The chances of crashing become higher if mobile phones are used because this reduces driving performance. The control of the sidelong and longitudinal position of the car is one of the basic needs during safe driving; however, frequent use of mobile phones while driving results in poorer lane-keeping, slower response time, and more variable speed [4,5]. In addition, the distance between the driver car and the front car, the fact of wandering out of the driving lane, and a reduced awareness of surroundings are some factors that lead to frequent road accidents and mortalities. In order to control the abrupt use of mobiles phones while driving, we need a smart portable system, which can control and monitor the behavior of a driver whenever the distracted driving exceeds a threshold and the risk of road crashing becomes imminent.

The consequences of using a mobile phone during driving are more lethal than communicating with fellow passenger, and research has shown that drivers busy in discussions over the phone have missed highway exits four times more frequently than those talking with passengers. The drivers talking with travelers did not demonstrate any significant differences with the lone drivers in the simulator environment [6]. The use of cell phones for chatting, messaging, playing media, web browsing, gaming, using the global positioning system (GPS), or working other telephone applications or apps is a dangerous act leading to distracted driving and road crashes [6]. This was evident from a report in 2010 from the U.S. National Highway Traffic Safety Administration (NHTSA) stating that 995 drivers died only because of distraction caused by the use of phones. Similarly, another study in March 2011, carried out by a U.S. insurance agency, State Farm Insurance, declared that 19% of the drivers involved in road accidents were busy talking on a cell phone while driving [7].

The tendency of mobile phone use during driving is more frequent in youth, with one of the studies reporting that more than 90% of college students were involved in initiating, reading, or replying to messages while driving. Messaging while driving, is considered the most harmful among all types of distraction using a mobile phone while driving, and it has been reported that there is a six-fold increase in distraction-related crashes due to text messaging [7]. Mobile phone texting, using MP3s, and other distractions may hinder the capacity of young drivers to control the vehicle and their ability to anticipate and manage hazards [8]. However, collision avoidance systems, electronic stability control, vehicle tracking systems, and intelligent speed adaption may help to reduce the problem even though technology alone cannot make the young driver safe.

A study [9] was conducted that was based on a self-reported questionnaire survey carried out among 242 young drivers in Riyadh, Saudi Arabia to obtain detailed insights into traffic violations committed by young Saudi drivers. The study showed that excessive speeding, which is mainly caused due to running late or testing a car's performance, is the leading cause of traffic accidents and traffic violations. Moreover, driving very close to the front car, which inhibits the driver to stop in an emergency, is another significant factor leading to traffic accidents. A study was carried out at the American University of Beirut (AUB), Lebanon and at George Washington University (GWU), United States of America to investigate the differences in red-light violations and driving behavior of drivers in those two countries. It has been reported that AUB students engage more in dangerous driving behavior than GWU students do, whereas GWU students are prone to violate traffic rules and red-light signals in the simulator [10]. A study was carried out on 83 new license holder young drivers in private cars, where the system was acquiring driving performance including secondary task engagement and driving environment logging. This study showed that teenage drivers are frequently engaged in secondary tasks and tend to regulate themselves poorly based on the surrounding environmental conditions. Moreover, the teenagers are greatly influenced by peers with respect to engaging in secondary tasks [11]. A system was demonstrated in the study [12], which merged the driver's background data and driving data to assess the good/bad driving pattern. The score for driving performance managed through the in-vehicle smart system, which provided feedback to the drivers to improve their driving, was found to be useful for taxi companies. There have also been studies where the authors have tried to detect driver distraction using semi-supervised machine learning without developing a prototype [13].

Various technologies and smartphone applications have been developed in recent years in order to limit the frequent use of mobile phones during driving. One of these is Google Glass, but it is still not safe to use [14]. In this same vein, a three-axis accelerometer of an Android-based smartphone was built up with multiple sensors to improve a driver's awareness to maximize safety [15]. A hardware device that can detect mobile phone use while driving and later block mobile communications could be a very fruitful option for monitoring and controlling road accidents. For example, radio frequency identification (RFID) technology could be used to record the data and send the vehicle's plate number to a control center when the driver uses the mobile phone and a radio frequency (RF) blocker can be used to block the mobile phone. However, the regulatory commission of some countries does not allow an RF blocker or jammer to be implemented in the car. The use of smartphone accelerometer sensors is another important technique to monitor vehicle status that involves the application of a principal component analysis (PCA) algorithm with time, frequency as well as power spectral density features of the sensor data to predict the vehicle status. This mobile sensor proved beneficial in identifying driving behavior using mobile phone applications [16]. However, this requires a high-performance computational capability in the smartphone application, or the application must be implemented in the cloud. The low-speed following mode (LSM) uses millimeter wave radar to identify the speeding up, deceleration, and stopping of the front car to appraise the distance from the front car; in the interim, the driver likewise controls the brake and the fuel systems to keep up the vehicle distance within the safety range. When the front car encounters a strange condition, the system produces an alert sound to warn the driver [17]. However, this assistive technology is implemented in some expensive cars and is not easy to implement in all available vehicles. The Lane Keeping Assist is a useful system to monitor the passing or separating lane by using a camera fixed to the front of the vehicle. The system produces a cautioning signal when the driver crosses or enters the opposite side of the passing line without using the correct light direction indicator [17]. However, this camera-assisted system is prone to making mistakes during rough weather conditions and bad road conditions, and the image processing requires a powerful computer to be installed on board. In a very recent work [18], a content analysis was conducted on 29 English smartphone applications to identify the stopping, preventing, or reducing phone use behavior while driving, detected by the applications. The functionality of these applications was determined based on application–mobile phone interaction, application–driver interaction, and application–context interaction. Most of these applications focused on blocking specific phone functions; however, the applications did not focus on simplifying phone tasks while driving and none of them was designed to study driving behavior. Another recent work by Delgado et al. [19] showed that the strongest perceived benefit of cellphone blocking apps was decreasing distraction (86%). The predominant reason among young drivers for not wanting to use this technology was not wanting parents to monitor their behavior (60%). This work shows the importance of developing driver-friendly applications while controlling a driver's mobile usage. The systems that have been reported so far in the literature did not present a feasible solution that could acquire the driving behavior from the car and use it to control the mobile phone automatically.

In this work, we have proposed a hybrid (hardware and software combined) solution to monitor driving behavior and keep track of a driver's mobile phone usage, and to control the mobile phone call when the car speed reaches a certain threshold. The on-board diagnostics (OBD)-II port of the vehicle was used to get the vehicle's real-time data. It was used to obtain the car speed very accurately, the accelerometer (ACM) sensors were used to identify some aspects of the driving behavior, and a mobile application written in the Android platform was used to monitor, log, and report driving behavior and control mobile phone calls. We decided on several driving behaviors to be studied and monitored in this work, based on an anonymous self-reported survey that was conducted with 600 male and female teenager drivers belonging to different nationalities. The survey was done in order to determine the pattern and frequency of mobile phone use while driving, and to find out about driving behavior, distracted driving due to mobile phones, and drivers' level of recommendations regarding the use of technologies to assist drivers or control mobile phones in Qatar.

2. Experiment Details and Methods

In this section, we provide the details of the pilot study conducted to gather self-reported information regarding teenager driving practice in Qatar. Although this survey was conducted for Qatar, this result should reflect the driving practices of Middle East and North African (MENA) countries very closely. This section also provides a detailed description of the hardware and software design of the prototype system used to monitor and control driving behavior.

2.1. Pilot Study

A self-reported paper-based pilot study was carried out in order to assess the driving behavior and perception of using a mobile phone while driving in Qatar. Both male and female subjects aged 18 to 26 years belonging to different nationalities were selected and data were collected from a population of 600 subjects. An approval was obtained from the Qatar University Human Research Ethics Committee prior to the start of the study. The participants were selected from different undergraduate classes from the universities of Qatar and were given a paper-based questionnaire to choose answers anonymously [13–17,20]. The survey was conducted in both an Arabic and an English version based on the user's choice (the English version of the survey questionnaire is shown in the Appendix Figure A1). Instructions were given to the participants that the study was to determine the perception of teenage drivers in Qatar about the use of mobile phones while driving. The questionnaire was prepared carefully so that there was no repetitive questions. Additionally, participants were asked to fill in the questionnaire honestly by circling or ticking the number that best suited their opinion after going through it carefully. Extreme care was taken to maintain the anonymity of the study and the confidentiality of the responses by preventing the identification of data obtained from the participants. Furthermore, the identification of participants was prevented by analyzing and reporting the data in a cumulative manner. Moreover, the survey was carried out to address the question of whether the driving behavior played any role based on gender, age range, and nationality or the car model being driven (expensive or inexpensive).

2.2. Design of Experiments

A complete block diagram of the prototype is shown in Figure 1. Any vehicle manufactured after 1996 is equipped with an on-board diagnostics (OBD) II system, which allows access to the vehicle's real-time information from its electrical control unit (ECU). The vehicle information from the OBD-II module was sent to a controller unit (microcontroller) and stored in the secure data (SD) card and transmitted via Bluetooth to the mobile phone. The hardware module was powered by the OBD-II module, which took power from the OBD-II port. Therefore, the system only ran while the vehicle was running and it did not drain the vehicle's battery. There was a three-degree-of-freedom (DOF) accelerometer module interfaced with the controller to keep track of the acceleration in the x-, y-, and z-directions. In the mobile phone, an in-house developed smartphone application made decisions based on the information received from the controller. The mobile application was designed to make decisions based on certain set thresholds, which were determined by detailed experiments on different subjects and will be discussed in the next section.

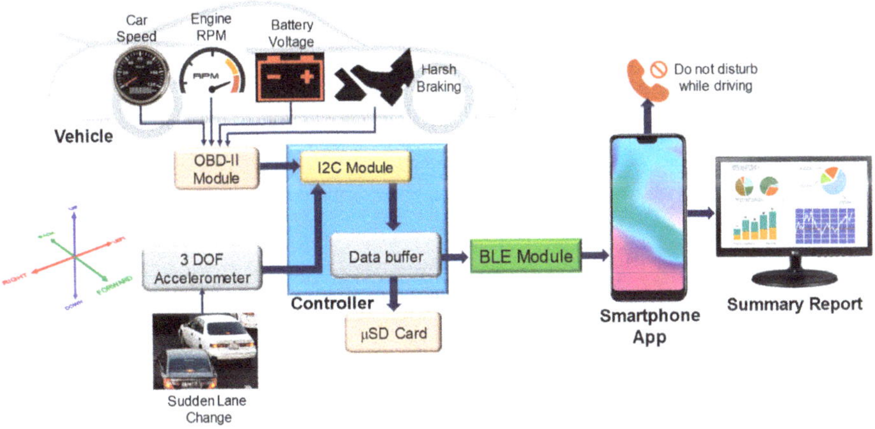

Figure 1. Complete system block diagram.

2.2.1. Hardware Modules

Various hardware modules are discussed in detail below.

OBD-II Module: The OBD-II adapter as displayed in Figure 2 was plugged into the OBD port of the vehicle to access various data from the car (car speed, engine rpm, battery voltage, etc.). The data were merged to measure the frequency of sudden breaking-like driving behavior. The connection of the OBD-II module to the OBD port is shown in Figure 3.

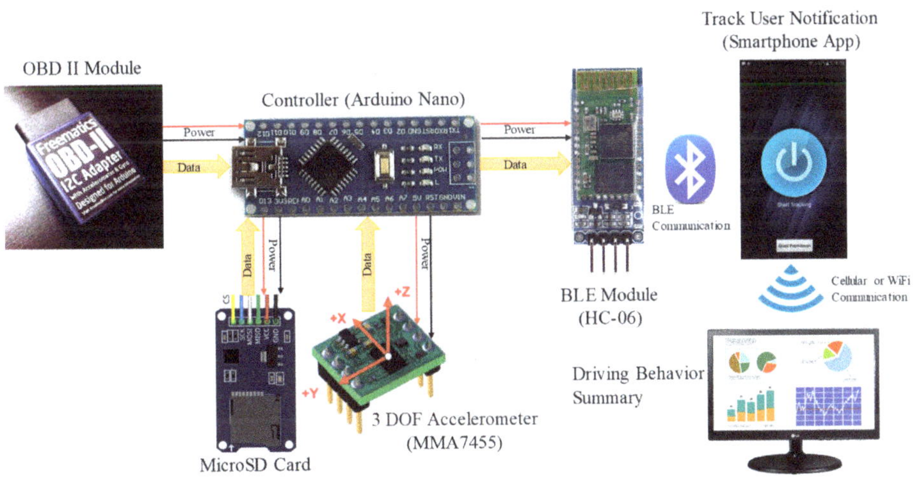

Figure 2. Connection diagram for different modules.

Controller Module: The OBD-II module was interfaced to an Arduino Nano microcontroller to gather the information from it. This information was packaged with 3-DOF accelerometer data and sent to the mobile phone via Bluetooth. The microcontroller and the modules were powered by the car battery through the OBD-II module. The inter-integrated circuit (I^2C) interface was used for connecting the OBD-II and 3-DOF modules to the microcontroller, whereas the serial peripheral interface (SPI) was used for connecting the Micro SD card module to the microcontroller.

3-DOF Accelerometer Module: A 3-DOF accelerometer module (MMA7455) was used to collect x- (forward-backward), y- (left-right) and z-axis (up-down) acceleration of the vehicle. This was used to identify normal left-right turning, sudden right-left turning, and U-turn. The MMA7455 module was connected with the Arduino Nano using the I^2C interface (Figure 4A).

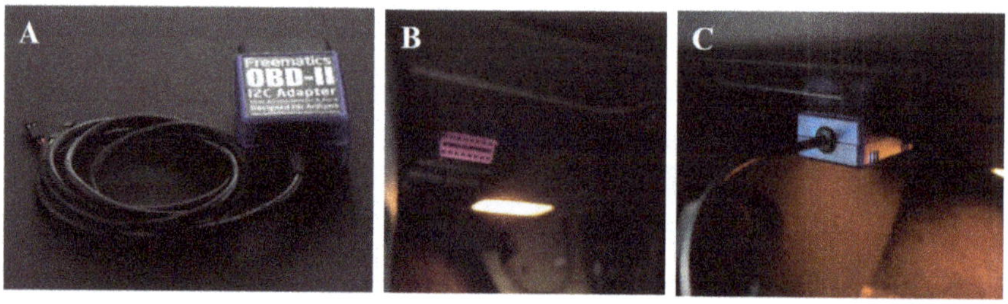

Figure 3. (**A**) On-board diagnostics (OBD)-II module, (**B**) OBD-II port, and (**C**) OBD-II module connected to the car's OBD-II port.

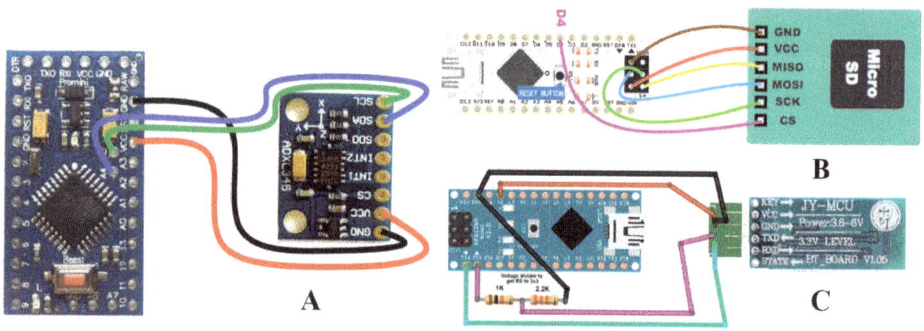

Figure 4. Arduino Nano interfacing wiring with (**A**) MMA7455, (**B**) Micro SD card module, and (**C**) Bluetooth Low Energy (BLE) module.

Micro SD Card Module: The information received from the OBD-II and 3-DOF modules was sent to the mobile phone via Bluetooth and stored in a Micro SD card as a backup. This was to ensure that if the connection with the mobile phone and the controller disconnected for some reason, the controller would not lose any data. As soon as the connection was established, the controller updated the mobile application logger to log the information. The details of the connection between the microcontroller and the Micro SD module are shown in Figure 4B.

Bluetooth Low Energy (BLE) Module: The controller was used to gather useful data (x-, y-, and z-acceleration from the accelerometer and car speed and engine revolution per minute (rpm) from the OBD-II module) continuously every 50 ms with a sampling frequency at 20 Hz, which was sent continuously to the mobile phone. An HC-06 Bluetooth module was used for wireless communication between the controller and the mobile phone. The HC-06 was initially paired with the mobile phone. The microcontroller received the data and combined them in packets and then sent the data to the mobile phone using the HC-06 Bluetooth module. The connection diagram of the microcontroller and the Bluetooth module is shown in Figure 4C.

2.2.2. Android Mobile Application: Track User Notification

A tracking application in the Android platform was developed to read the data from the hardware wirelessly, log the data locally for 24 h, and identify the driver's behavior based on the logged data and pre-set threshold. The application was designed to monitor car speed and control the driver's mobile phone by restricting receiving or generating phone calls and monitoring behavior while driving if the car speed went above 10 km/h. The application logged only the driver's other mobile usage information like texting, browsing, playing, etc. At midnight, the application automatically sent the obtained information about car speed, sudden lane change behavior, call blocking, and the driver's mobile usage to a pre-specified email address. Although the driver was not allowed to generate or receive calls if the speed was above 10 km/h, they could receive a call if the speed dropped below that threshold. Different stages of the mobile application's workflow are shown in Figure 5.

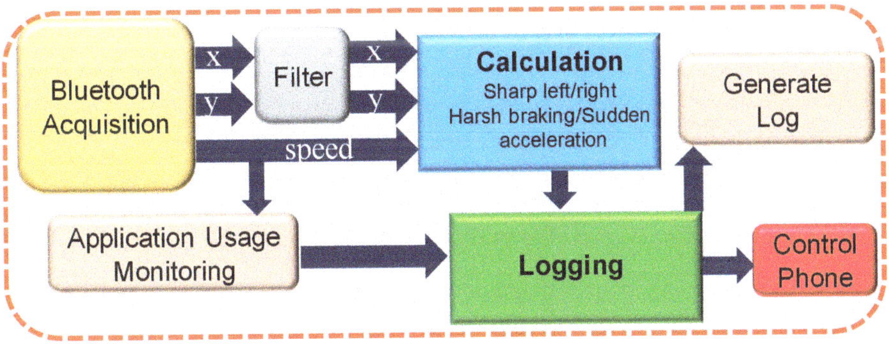

Figure 5. Block diagram to show the stages of the application's operation.

BLE Acquisition: The Arduino Nano in the hardware module received the data and combined them in a packet and sent the data to the mobile phone using the HC-06 Bluetooth module. The data packet template used for the BLE communication is shown in Figure 6. In the packet template, "E" is a header, "," is a data separator, and ";" is a data terminator. As shown in Figure 5, there is a module for Bluetooth data reception, and a call constant function was developed to check whether Bluetooth was on or not.

E	**x**	**,**	**y**	**,**	**z**	**,**	**speed**	**,**	**rpm**	**;**

Figure 6. Data packet template.

Grant Permission: In order to respect the user's privacy, the application first asked the user for permission to access and track by enabling the button during the installation process (Figure 7 (left)). If the permission from the user was enabled and Bluetooth was enabled, the application continued. If it found either one disabled, it exited instead of continuing the process.

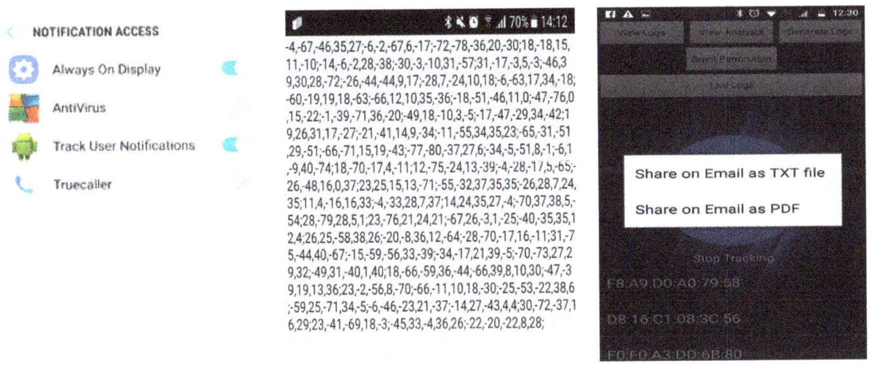

Figure 7. Screen shots of some of the mobile application's features.

Automatic Tracking: The app was designed so that it started automatically to track the driver's mobile phone if the car speed increased above 10 km/h.

Data Logging and Reporting: The user could see the data acquired from the Bluetooth module using the user interface (Figure 7 (center)). The user could also share the day's log by email as a text file or as a portable document format (PDF) file, as shown in Figure 7 (right). However, the log was reported automatically to a pre-specified email address at midnight.

The application home interface had five features (and one button for manually starting the tracking), as shown in Figure 8.

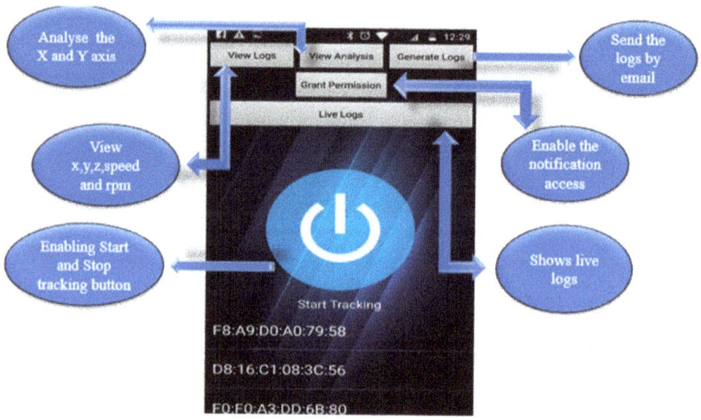

Figure 8. User interface of the Track User Notification mobile application.

A set of tests were designed to study the performance of the prototype system.

2.2.3. Study 1: Hardware Module Evaluation using Emulator

The prototype hardware was initially tested using an emulator before using it on a real car. Freematics OBD-II Emulator MK2 (Freematics, Wahroonga, New South Wales, Australia) (Figure 9A) is an OBD-II emulator with controlled area network (CAN) bus simulation that provides a 16-pin female OBD-II port identical to that of a real car. This device is very useful to get the car's OBD-II facility on the desk to simulate the real car behavior.

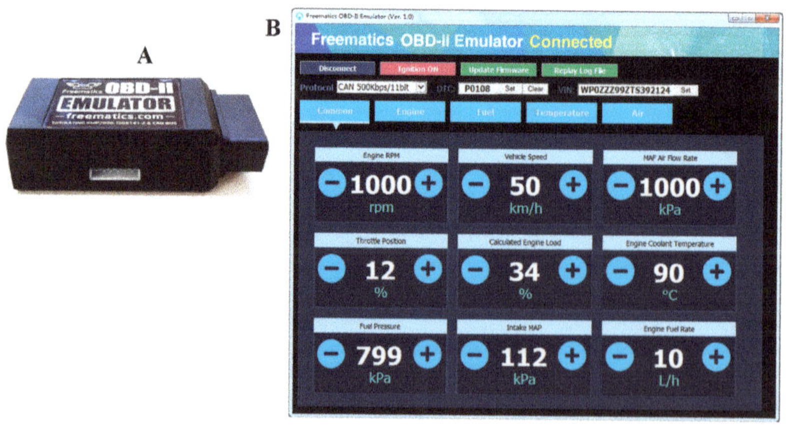

Figure 9. Freematics OBD-II Emulator MK2 (**A**) and Graphical User Interface (GUI) (**B**).

An open-source GUI software is available to see the car parameters on screen and vary them to check the OBD-II device's performance on the desk. A series of tests were carried out using the emulator to check the performance of the complete system before testing it on a real-car environment.

The performance of the Bluetooth communication between the vehicle and the controller was evaluated by displaying the received packet from the emulator to the controller (interfaced to a personal computer over a USB interface), and the performance between the controller and the mobile phone was evaluated by analyzing the received packet in the mobile phone, comparing it with the packets sent from the controller.

2.2.4. Study 2: Hardware Module Evaluation in Vehicle Environment

Experiments were conducted to evaluate the performance of the hardware prototype in reliably acquiring the vehicle information and driving behavior in the real-car environment. The speed of the vehicle reported by the OBD-II module was logged and the speed displayed in the dashboard was recorded synchronously and an off-line comparison was done.

2.2.5. Study 3: Evaluation of Driving Behavior

This study was designed to obtain the thresholds for lane change and sudden acceleration/braking behavior of the driver using the data acquired from the OBD-II module and the accelerometer. These thresholds were used to make decisions about driver behavior, such as sharp left, sharp right, sudden brake, or sharp acceleration. We asked ten teenage drivers to perform a series of driving experiments in the pattern listed in Table 1. The OBD-II and accelerometer data were recorded for the various user trials and the threshold for each case was calculated.

Table 1. Driving pattern for data collection to calculate the thresholds.

Duration	Actions
0 s–10 s	Normal
10 s–20 s	Turn left
20 s–30 s	Turn right
30 s–40 s	Sudden change lane toward right
40 s–50 s	Sudden change lane toward left
50 s–60 s	U-turn(leftward)
60 s–70 s	Sudden acceleration
70 s	Harsh brake

2.2.6. Study 4: Evaluation of the Phone Control, Data Logging, and Reporting

Finally, the application was tested to check if it was working with all the log viewing features and could properly report all the activities through email or not. The initial testing of the application was accomplished by monitoring and tracking all the tasks performed by the driver while data were sent from the controller. We evaluated whether the call controlling feature was activated at 10 km/h or not and whether the user's activity information (incoming, outgoing, duration of call, etc.), all push notifications (SMS, social media apps, third party messaging apps, that is, any conversation, incoming and outgoing, duration of that session of activity) was correctly logged or not. In addition, we checked whether the app could generate a summary of call activities and notifications and send it automatically by email or not.

3. Analysis

In this work, survey data analysis was accomplished in Microsoft Excel 2016 and vehicle data were initially analyzed in MATLAB 2018 and later (after the development of the mobile application) done in the mobile phone. Initial development and testing of the smartphone application were carried out on a Samsung Note 8 mobile phone, which is powered by an Exynos 8895 Octa-core processor, along with 6 GB of RAM and 64 GB internal memory. The operating system installed on the phone was Android 7.1.1 (Nougat) and enabled with Bluetooth Low Energy (BLE) 5.0. However, the smartphone application was tested in several lower-end smartphones in the testing phase.

Survey Analysis: Detailed chi square statistics [21] were performed on two major questions, namely, (i) Is it also a good idea to restrict texting, gaming, or browsing while driving and (ii) Is it a good idea to restrict phone calls to only emergency family contacts while driving?, to check the accuracy of the predictions made by the authors based on the literature review. The first question had the assumption that all participants would state that it was a very good idea. For the second question, the authors assumed that all participants would state that it was not safe.

Preliminary Analysis in MATLAB: Initial accelerometer results were smoothened by the moving average filter in MATLAB and averaged over trials and subjects to calculate the mean of the x-, y-, and *z*-axis data (Figure 10). It was observed that, for this work, *z*-axis data were not useful and therefore not used for further processing. Moreover, engine rpm had an offset value when the vehicle was started and changed from that reference and the variation reflected in the engine rpm was also reflected in the speed and therefore rpm was not used for calculating behavior. The accelerometer and OBD-II module data were analyzed to see the trend of driver behavior in relation to the nature of the x and *y*-axis and speed data.

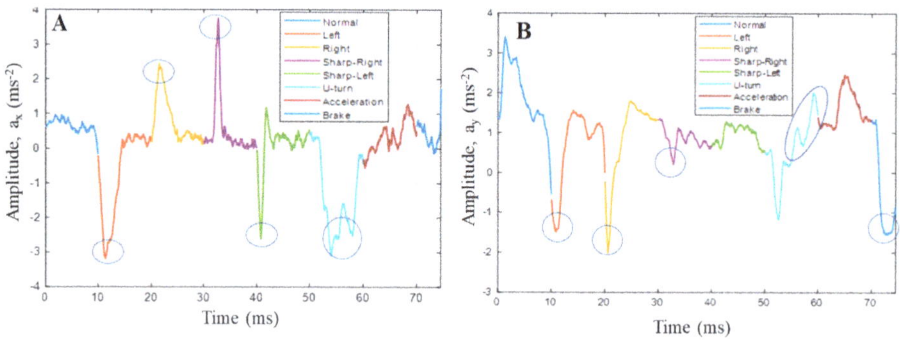

Figure 10. Average x and y data to show (**A**) left/right movements and (**B**) acceleration/braking of the vehicle.

Data Analysis in Mobile Phone: The mobile application used here to collect the vehicle data from the OBD-II device and to store them temporarily in the mobile memory until they were processed to make a decision (i.e., the x, y, and speed data) were buffered in the mobile memory. The buffering duration was kept to 10 s to get enough data to observe the changes while not cluttering the memory of the mobile phone. The algorithm of real-time peak detection is very robust because it constructs a separate moving mean and deviation for the buffered data, such that signals do not corrupt the threshold. Future signals are therefore identified with approximately the same accuracy, regardless of the number of previous signals. The algorithm takes three inputs: lag represents the lag of the moving window, threshold represents the z-score at which the algorithm generates peak, and influence represents the influence (between 0 and 1) of new signals on the mean and standard deviation. For example, a lag of 5 (moving window) will use the last five observations to smooth the data. A threshold of 3.5 (estimated from MATLAB study) will signal if a datapoint is 3.5 standard deviations away from the moving mean. In addition, an influence of 0.5 gives signals half of the influence that normal datapoints have. Likewise, an influence of 0 ignores signals completely for recalculating the new threshold. An influence of 0 is therefore the most robust option; putting the influence option at 1 is least robust.

In the mobile application, there were three subclasses: two for x and y data analysis and the third for the speed data analysis. The subclasses for the x and y data analysis helped to identify the peak of the x and y movements of the vehicle which, in turn, helped to identify normal and abnormal behaviors of the driver, whereas the other subclass sent responses based only on speed data. The filtered speed data were sent to another class, which made a decision based on speed, that is, the vehicle was either in driving mode or stopped, and sent a callback to this class on the current status. Based on the status of the vehicle, the control function started tracking or stopped tracking and the calling function was also deactivated or activated. Figure 11 shows the detailed stages of how the mobile application was designed to log normal or abnormal behavior.

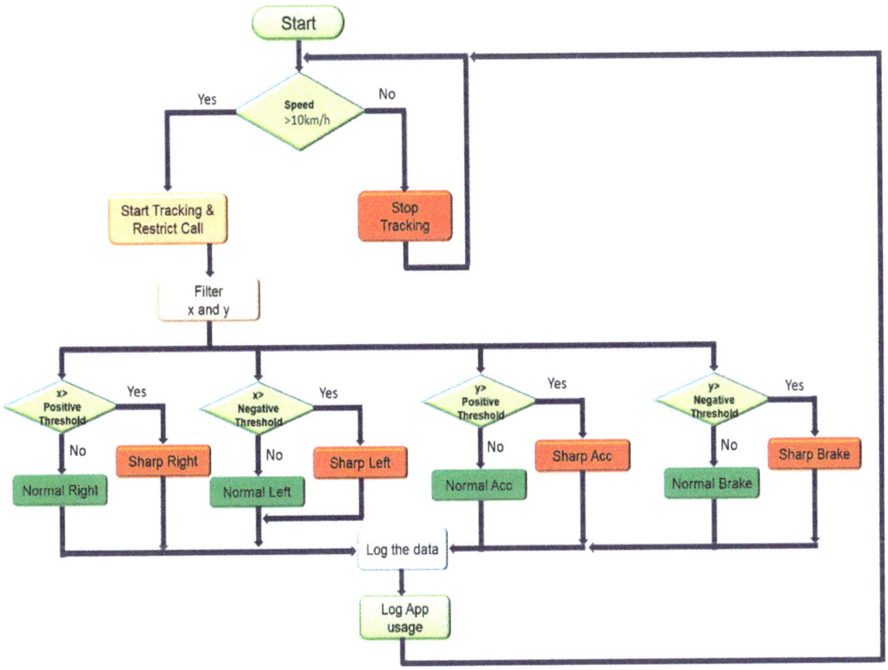

Figure 11. Flowchart of the mobile application's decision stages.

4. Results and Discussion

The outcomes of the different experiments are summarized in this section.

4.1. Summary of Pilot Study

Some important findings from the survey are presented below, and it is interesting to see that 55% of the survey participants had never had any accidents (from Figure 12A) and thus can be considered as careful drivers. Their answers to the survey questions provided motivation for this work and suggestions to improve the cases of accidents due to mobile phone use. Almost half of the teenage drivers who had received their license within three to six years experienced accidents due to their use of a mobile phone. It can be seen from the self-reported survey percentages shown in the bar chart (Figure 12B) that the drivers were occupied with various activities using the mobile phone during driving. This is one of the primary causes of accidents where the user gets distracted. From Figure 12B, it is evident that 83% of the teenagers like to talk, while 66% of them like to text to some extent while driving.

As shown in Figure 13A, the histogram shows that 100% of respondents believe that restricting phone calls except for emergency family calls while driving is a very good idea. However, the restriction of texting, gaming, or browsing was not considered a good idea by all the teenage drivers, although those activities cause more distraction than calling. This clearly highlights the driving behavior of the teenagers. In the same manner, Figure 13B shows that the young drivers mostly do not agree that music can distract the driver, although this point is evident from different research studies conducted in other regions. However, the majority of them agree that using and inputting data to the navigator while driving distract the driver.

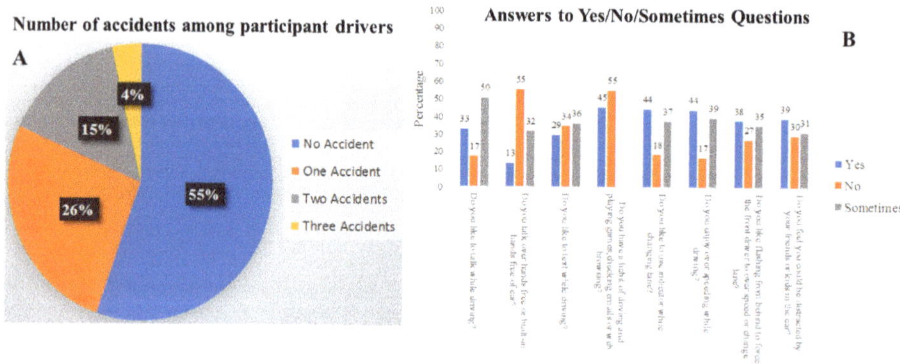

Figure 12. (**A**) Statistics of number of accidents and (**B**) results of some yes/no/sometimes questions among the survey participants.

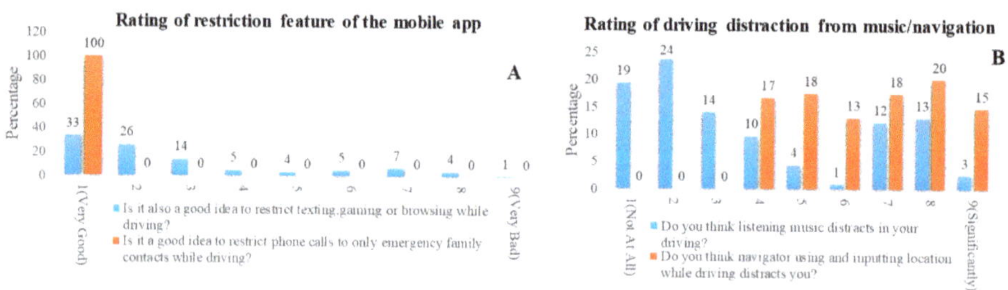

Figure 13. (**A**) Statistics for the questions on rating of restriction feature of the mobile app and (**B**) using mobile phone.

From Figure 14A, it was found that the majority of the survey participants believe that it is not safe to drive while using a mobile phone or when sleepy/drowsy. It can be observed from the self-reported survey percentages shown in the bar chart in Figure 14B that the drivers are occupied with various dangerous activities during driving.

The summary of the chi-squared distribution analysis [19] done on the data is shown in Table 2, where it is observed that such a system is needed for the welfare of drivers and that it is important to raise awareness among teenage drivers in Qatar.

Table 2. Chi-Squared distribution results for two major survey questions.

Question	Prediction	Reject or Accept Depending on Chi-Squared Distribution
Is it a good idea to restrict phone calls to only emergency family contacts while driving?	Almost all of them should say that it is a very good idea	Accept, which is a good motivation to the development of the prototype
How safe or dangerous do you feel using mobile phone while driving?	Almost all of them should say that it is very dangerous	Reject which is a good motivation to increase awareness about it among them.

Comparing the predictions with the actual response and then calculating the chi-squared value of the difference between them and the tabular chi-squared value (for the sample data) for the first question, it was found that the prediction should not be rejected (as the calculated chi-squared value was less than the tabular chi-squared value). However, comparing the predictions with the actual response for question two, it was found that the prediction had to be rejected (as the calculated chi-squared value was more than the tabular chi-squared value).

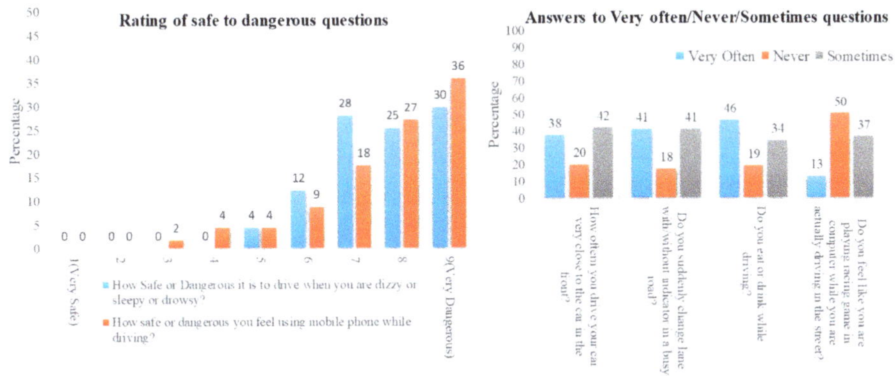

Figure 14. (**A**) Statistics for the rating of safe to dangerous questions and (**B**) results for some very often/never/sometimes questions among the survey participants.

4.2. *Performance Evaluation of the Prototype*

4.2.1. Studies 1 and 2

Initial results from the testing of the prototype module using the car emulator along with the OBD-II module showed that the data packets received in the PC match the data sent from the emulator without missing any packets. Moreover, the data sent from the controller to the mobile application were also tested to evaluate their reliability. Figure 15 illustrates the real-car data (x, y, z, speed, rpm) received reliably on the mobile application using Bluetooth communication.

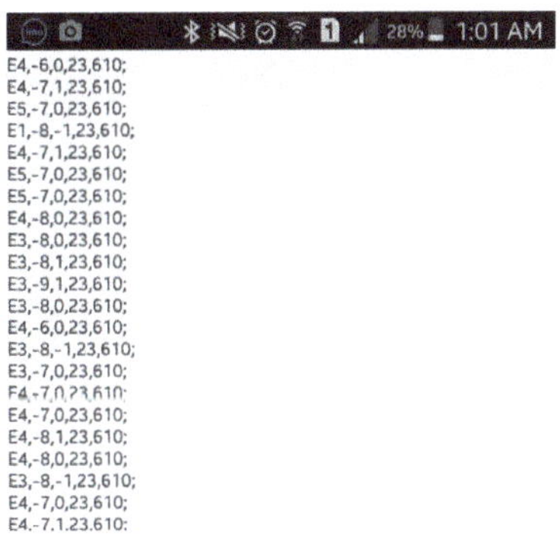

Figure 15. Sample data packets received in the mobile application from real-car testing.

4.2.2. Study 3: Evaluation of the Driving Behavior

After a series of tests with teenage drivers using the hardware, it was found that there was a specific threshold of the x data, y data, and x, y data combined from the accelerometer, which can help in identifying the driving behavior. The application identified the sudden changes of the x data to classify any changes above the pre-specified threshold to detect "Left", "Right", "Sharp Right", "Sharp Left", and "U-Turn". However, only two classes, namely, "Sharp Right" and "Sharp Left", were logged. Moreover, filtered *y*-axis data helped to classify normal and abnormal acceleration and braking action. If the positive change exceeded the positive threshold, it was classified as "Sudden Acceleration", and if there was any negative change which occurred below the negative threshold, it was classified as "Sudden Braking" and was logged.

According to Figure 16, the speed increased over period number 1, whereas for period number 2 there was no significant change in speed and it remained almost stable. However, over period 3 the speed increased rapidly and reached its maximum value (40 km/h). Finally, over period 4 it decreased sharply to its minimum value (zero km/h). Almost similar information was reflected from the rpm data. This can be used to track excessive speeding behavior if the application is pre-loaded with the speed limit of the particular street along with a global positioning system (GPS).

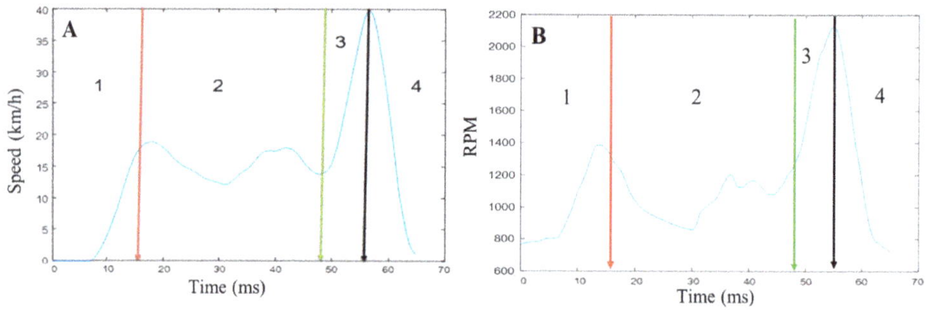

Figure 16. (**A**) Average speed and (**B**) engine rpm for speed monitoring.

4.2.3. Study 4: Evaluation of the Phone Control, Data Logging, and Reporting

Figure 17 shows the Call, Message, Behavior and Application Usage Log Summary while driving for several trials in a day and logged in the application (Tracking User Notification). For example, the Call Log Summary notification shows the call type, Count, and Time of Event. Moreover, the Message Log Summary notification shows the message type (Incoming and Outgoing), the number of total messages, and the time.

The most important implemented feature of the "Track User Application" mobile app is that it checks the speed of the driver's vehicle and, if the car speed goes above 10 km/h, it blocks the mobile phone's call feature. It blocks both the incoming and outgoing mobile calls. This feature was tested on the desk using the OBD-II emulator and also during real driving scenarios. It was observed that in all scenarios, this call blocking feature worked with 100% accuracy. The Call Log Summary in Figure 17 shows that the user attempted to receive and make calls; however, the person was not successful in making/receiving calls because of the enabled blocking feature. However, the mobile application created logs of the incoming/outgoing call attempts in the report. In the report, where the incoming and outgoing call timings were shown, the user was not able to make/receive calls because of the call blocking feature. The Behavior Log Summary shows a summary of some of the driving behavior for that particular day which included sudden left turn, sudden right turn, sudden acceleration, and harsh braking as well as the date and time of the action. This can clearly reflect driving behavior of a particular driver and can easily be modified to monitor accidents as well. It is evident that thresholds help to populate driving behavior from the real-time data. The output from the mobile application was sent to a pre-specified user via email. Moreover, it shows the timing and use of different applications by the mobile user and, more importantly, none of the user privacy data were shared.

Track User Notification

Call Log Summary	Count	Time of Event
Incoming Call	4	14 Dec 2017 20:18:18
	3	14 Dec 2017 19:16:04
	2	14 Dec 2017 19:12:35
	1	14 Dec 2017 18:25:46
Outgoing Call	5	14 Dec 2017 16:35:42
	4	14 Dec 2017 15:47:05
	3	14 Dec 2017 15:46:35
	2	14 Dec 2017 15:34:39
	1	14 Dec 2017 15:26:59
Total	9	15 Dec 2017 00:00:01

Message Log Summary	Count	Time of Event
Incoming Message	3	14 Dec 2017 18:25:26
	2	14 Dec 2017 17:18:05
	1	14 Dec 2017 16:15:40
Outgoing Message	2	14 Dec 2017 20:16:10
	1	14 Dec 2017 15:16:35
Total	5	15 Dec 2017 00:00:01

Behavior Log Summary	Count	Time of Event
Sharp Left	3	14 Dec 2017 20:16:26
	2	14 Dec 2017 17:18:25
	1	14 Dec 2017 15:16:50
Sharp Right	2	14 Dec 2017 17:18:15
	1	14 Dec 2017 15:17:24
Sudden Acceleration	3	14 Dec 2017 20:11:46
	2	14 Dec 2017 17:04:56
	1	14 Dec 2017 15:18:19
Harsh Braking	4	14 Dec 2017 20:10:06
	3	14 Dec 2017 17:03:29
	2	14 Dec 2017 15:55:47
	1	14 Dec 2017 15:20:01
Total	12	15 Dec 2017 00:00:01

Application Usage Log Summary		
Application Name	Count	Time of Event
WhatsApp	8	14 Dec 2017 20:17:23
	7	14 Dec 2017 20:16:01
	6	14 Dec 2017 17:16:22
	5	14 Dec 2017 16:19:05
	4	14 Dec 2017 16:18:59
	3	14 Dec 2017 16:06:03
	2	14 Dec 2017 15:26:49
	1	14 Dec 2017 15:25:35
Browser	2	14 Dec 2017 17:17:15
	1	14 Dec 2017 15:10:24
Viber	3	14 Dec 2017 20:12:46
	2	14 Dec 2017 17:07:06
	1	14 Dec 2017 15:19:19
Waze	4	14 Dec 2017 20:16:56
	3	14 Dec 2017 17:23:19
	2	14 Dec 2017 15:35:49
	1	14 Dec 2017 15:22:11
Total	17	15 Dec 2017 00:00:01

Figure 17. Screenshot of report generated at the end of the day.

5. Conclusions

This work proposed a portable solution to gather vehicle details from the ECU of a vehicle through the OBD-II port and to send data to a mobile phone along with an on-board 3-DOF accelerometer to detect driving behavior. The results from the test subjects show that this can be potentially used to identify drivers' abnormal behaviors. This abnormal driving behavior along with continuous speed monitoring could be used in the mobile application to make decisions on controlling mobile phone activity. The literature reviews and the surveys conducted with a group of teenage drivers in Qatar support the need for such a portable solution. Most of the applications available on the market share the contents of the text, email, or social media message while logging the notifications, whereas our portable solution along with the tracking application does not share any private data in the report, which improves the security of the user. Therefore, the proposed system can be used as a robust system for monitoring the behavior of drivers and controlling them to avoid emergencies. It is clear from the literature and the results of the conducted survey in this work that teenage drivers are willing to stop using their mobile phone while driving. However, incoming messages and calls encourage the user to respond in most of the scenarios and this was observed in the behavior report from the proposed system. The authors therefore consider that significant media awareness through different forms of social media activities along with government and law enforcement involvement (e.g., seat-belt usage enforcement) should be put in place to avoid the life-threatening use of mobile phones while driving. In the future, we plan to add collision detection based on the accelerometer data and the enforcement of Bluetooth for the mobile phone turning-on feature. The system can be modified to allow the user to drive only when Bluetooth is on, thereby enabling the tracking. Further investigation is needed for real-time monitoring of driving behavior, where the system is deployed in several vehicles and monitored for a longer duration to truly benefit from this research.

Author Contributions: Experiments were designed by A.K., M.E.H.C.; Experiments were performed by A.D., M.M.; Results were analyzed by A.K., M.E.H.C., R.D., N.A.A.-E. and D.M.; All authors were involved in interpretation of data and paper writing.

Funding: The publication of this article was funded by the Qatar National Library.

Conflicts of Interest: The authors declare no conflict of interest. The funders had no role in the design of the study; in the collection, analyses, or interpretation of data; in the writing of the manuscript, or in the decision to publish the results.

Appendix A

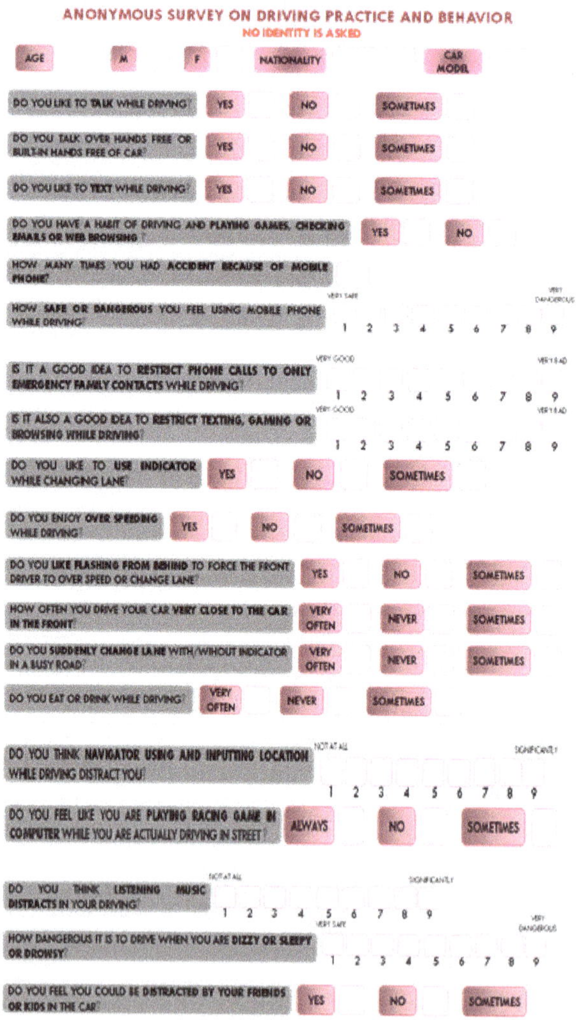

ANONYMOUS SURVEY ON DRIVING PRACTICE AND BEHAVIOR
NO IDENTITY IS ASKED

AGE | M | F | NATIONALITY | CAR MODEL

DO YOU LIKE TO TALK WHILE DRIVING? YES NO SOMETIMES

DO YOU TALK OVER HANDS FREE OR BUILT-IN HANDS FREE OF CAR? YES NO SOMETIMES

DO YOU LIKE TO TEXT WHILE DRIVING? YES NO SOMETIMES

DO YOU HAVE A HABIT OF DRIVING AND PLAYING GAMES, CHECKING EMAILS OR WEB BROWSING? YES NO

HOW MANY TIMES YOU HAD ACCIDENT BECAUSE OF MOBILE PHONE?

HOW SAFE OR DANGEROUS YOU FEEL USING MOBILE PHONE WHILE DRIVING? VERY SAFE 1 2 3 4 5 6 7 8 9 VERY DANGEROUS

IS IT A GOOD IDEA TO RESTRICT PHONE CALLS TO ONLY EMERGENCY FAMILY CONTACTS WHILE DRIVING? VERY GOOD 1 2 3 4 5 6 7 8 9 VERY BAD

IS IT ALSO A GOOD IDEA TO RESTRICT TEXTING, GAMING OR BROWSING WHILE DRIVING? VERY GOOD 1 2 3 4 5 6 7 8 9 VERY BAD

DO YOU LIKE TO USE INDICATOR WHILE CHANGING LANE? YES NO SOMETIMES

DO YOU ENJOY OVER SPEEDING WHILE DRIVING? YES NO SOMETIMES

DO YOU LIKE FLASHING FROM BEHIND TO FORCE THE FRONT DRIVER TO OVER SPEED OR CHANGE LANE? YES NO SOMETIMES

HOW OFTEN YOU DRIVE YOUR CAR VERY CLOSE TO THE CAR IN THE FRONT? VERY OFTEN NEVER SOMETIMES

DO YOU SUDDENLY CHANGE LANE WITH/WIHOUT INDICATOR IN A BUSY ROAD? VERY OFTEN NEVER SOMETIMES

DO YOU EAT OR DRINK WHILE DRIVING? VERY OFTEN NEVER SOMETIMES

DO YOU THINK NAVIGATOR USING AND INPUTTING LOCATION WHILE DRIVING DISTRACT YOU? NOT AT ALL 1 2 3 4 5 6 7 8 9 SIGNIFICANTLY

DO YOU FEEL LIKE YOU ARE PLAYING RACING GAME IN COMPUTER WHILE YOU ARE ACTUALLY DRIVING IN STREET? ALWAYS NO SOMETIMES

DO YOU THINK LISTENING MUSIC DISTRACTS IN YOUR DRIVING? NOT AT ALL 1 2 3 4 5 6 7 8 9 SIGNIFICANTLY

HOW DANGEROUS IT IS TO DRIVE WHEN YOU ARE DIZZY OR SLEEPY OR DROWSY? VERY SAFE 1 2 3 4 5 6 7 8 9 VERY DANGEROUS

DO YOU FEEL YOU COULD BE DISTRACTED BY YOUR FRIENDS OR KIDS IN THE CAR? YES NO SOMETIMES

Figure A1. Survey questionnaire.

References

1. Shabeer, H.A.; Banu, R.W.; Zubar, H.A. Technology to prevent mobile phone accidents. *Int. J. Enterp. Netw. Manag.* **2012**, *5*, 144–155. [CrossRef]
2. Guo, F.; Klauer, S.G.; Fang, Y.; Hankey, J.M.; Antin, J.F.; Perez, M.A.; Lee, S.E.; Dingus, T.A. The effects of age on crash risk associated with driver distraction. *Int. J. Epidemiol.* **2017**, *46*, 258–265. [CrossRef] [PubMed]
3. Ayers, J.W.; Leas, E.C.; Dredze, M.; Allem, J.-P.; Grabowski, J.G.; Hill, L. Pokémon GO—A new distraction for drivers and pedestrians. *JAMA Intern. Med.* **2016**, *176*, 1865–1866. [CrossRef] [PubMed]
4. Burns, P.; Parkes, A.; Burton, S.; Smith, R.; Burch, D. *How Dangerous Is Driving with a Mobile Phone?: Benchmarking the Impairment to Alcohol*; TRL: Wokingham, UK, 2002; Volume 547.
5. Oviedo-Trespalacios, O.; Haque, M.M.; King, M.; Washington, S. Understanding the impacts of mobile phone distraction on driving performance: A systematic review. *Transp. Res. Part C Emerg. Technol.* **2016**, *72*, 360–380. [CrossRef]
6. Coxon, K.; Keay, L. Behind the wheel: Community consultation informs adaptation of safe-transport program for older drivers. *BMC Res. Notes* **2015**, *8*, 764. [CrossRef] [PubMed]
7. Drews, F.A.; Pasupathi, M.; Strayer, D.L. Passenger and cell phone conversations in simulated driving. *J. Exp. Psychol. Appl.* **2008**, *14*, 392. [CrossRef] [PubMed]
8. Lee, J.D. Technology and teen drivers. *J. Saf. Res.* **2007**, *38*, 203–213. [CrossRef] [PubMed]
9. Hassan, H.M. Investigation of the self-reported aberrant driving behavior of young male Saudi drivers: A survey-based study. *J. Transp. Saf. Secur.* **2016**, *8*, 113–128. [CrossRef]
10. Danaf, M.; Hamdar, S.H.; Abou-Zeid, M.; Kaysi, I. Comparative assessment of driving behavior at signalized intersections using driving simulators. *J. Transp. Saf. Secur.* **2018**, *10*, 124–158. [CrossRef]
11. Gershon, P.; Zhu, C.; Klauer, S.G.; Dingus, T.; Simons-Morton, B. Teens' distracted driving behavior: Prevalence and predictors. *J. Saf. Res.* **2017**, *63*, 157–161. [CrossRef] [PubMed]
12. Cen, J.; Wang, Z.; Wang, C.; Liu, F. A System Design for Driving Behavior Analysis and Assessment. In Proceedings of the 2016 IEEE International Conference on Internet of Things (iThings) and IEEE Green Computing and Communications (GreenCom) and IEEE Cyber, Physical and Social Computing (CPSCom) and IEEE Smart Data (SmartData), Chengdu, China, 15–18 December 2016; pp. 882–887.
13. Liu, T.; Yang, Y.; Huang, G.-B.; Yeo, Y.K.; Lin, Z. Driver distraction detection using semi-supervised machine learning. *IEEE Trans. Intell. Transp. Syst.* **2016**, *17*, 1108–1120. [CrossRef]
14. He, J.; Choi, W.; McCarley, J.S.; Chaparro, B.S.; Wang, C. Texting while driving using Google Glass™: Promising but not distraction-free. *Accid. Anal. Prev.* **2015**, *81*, 218–229. [CrossRef] [PubMed]
15. Fazeen, M.; Gozick, B.; Dantu, R.; Bhukhiya, M.; González, M.C. Safe driving using mobile phones. *IEEE Trans. Intell. Transp. Syst.* **2012**, *13*, 1462–1468. [CrossRef]
16. Lu, D.-N.; Nguyen, D.-N.; Nguyen, T.-H.; Nguyen, H.-N. A Novel Mobile Online Vehicle Status Awareness Method Using Smartphone Sensors. In *Information Science and Applications*; Springer: Singapore, 2017; pp. 30–37.
17. Lu, S.-N.; Tseng, H.-W.; Lee, Y.-H.; Jan, Y.-G.; Lee, W.-C. Intelligent safety warning and alert system for car driving. *Tamkang J. Sci. Eng.* **2010**, *13*, 395–404.
18. Oviedo-Trespalacios, O.; King, M.; Vaezipour, A.; Truelove, V. Can our phones keep us safe? A content analysis of smartphone applications to prevent mobile phone distracted driving. *Transp. Res. Part F Traffic Psychol. Behav.* **2019**, *60*, 657–668. [CrossRef]
19. Delgado, M.K.; McDonald, C.C.; Winston, F.K.; Halpern, S.D.; Buttenheim, A.M.; Setubal, C.; Huang, Y.; Saulsgiver, K.A.; Lee, Y.C. Attitudes on technological, social, and behavioral economic strategies to reduce cellphone use among teens while driving. *Traffic Inj. Prev.* **2018**, *19*, 569–576. [CrossRef] [PubMed]
20. Ajzen, I. Constructing a TPB questionnaire: Conceptual and methodological considerations. 2002. Available online: http://chuang.epage.au.edu.tw/ezfiles/168/1168/attach/20/pta_41176_7688352_57138.pdf (accessed on 28 March 2019).
21. Lancaster, H.O.; Seneta, E. Chi-square distribution. *Encycl. Biostzatistics* **2005**, 2. [CrossRef]

sensors

MDPI

Article

Simulating Dynamic Driving Behavior in Simulation Test for Unmanned Vehicles via Multi-Sensor Data

Danchen Zhao [1], Yaochen Li [2] and Yuehu Liu [1,*]

1 Institute of Artificial Intelligence and Robotics, Xi'an Jiaotong University, No. 28 Xianning West Road, Xi'an 710049, Shaanxi, China; danchenzhao@foxmail.com

2 School of Software Engineering, Xi'an Jiaotong University, No. 28 Xianning West Road, Xi'an 710049, Shaanxi, China; yaochenli@mail.xjtu.edu.cn

* Correspondence: liuyh@mail.xjtu.edu.cn; Tel.: +86-029-8266-8802 (ext. 8009)

Received: 25 February 2019; Accepted: 3 April 2019; Published: 8 April 2019

Abstract: Driving behavior is the main basis for evaluating the performance of an unmanned vehicle. In simulation tests of unmanned vehicles, in order for simulation results to be approximated to the actual results as much as possible, model of driving behaviors must be able to exhibit actual motion of unmanned vehicles. We propose an automatic approach of simulating dynamic driving behaviors of vehicles in traffic scene represented by image sequences. The spatial topological attributes and appearance attributes of virtual vehicles are computed separately according to the constraint of geometric consistency of sparse 3D space organized by image sequence. To achieve this goal, we need to solve three main problems: Registration of vehicle in a 3D space of road environment, vehicle's image observed from corresponding viewpoint in the road scene, and consistency of the vehicle and the road environment. After the proposed method was embedded in a scene browser, a typical traffic scene including the intersections was chosen for a virtual vehicle to execute the driving tasks of lane change, overtaking, slowing down and stop, right turn, and U-turn. The experimental results show that different driving behaviors of vehicles in typical traffic scene can be exhibited smoothly and realistically. Our method can also be used for generating simulation data of traffic scenes that are difficult to collect.

Keywords: simulation test; dynamic driving behavior; traffic scene augmentation; corridor model

1. Introduction

Evaluation of the intelligence level and comprehensive performance of unmanned vehicles turns to ontology and phenomenology. According to Turing [1], a system could be said to be intelligent enough for special kind of tasks if, and only if, it could finish all the possible tasks of its kind. Therefore, we can begin to achieve safe and reliable artificial intelligence (AI) systems if, and only if, the tests have clear definitions of tasks and efficient methods to generate abundant data for tests. As a result, appropriate AI testing methods should be task-driven and data-centric [2].

Driving behavior is the main basis for evaluating the performance of unmanned vehicles. Testing the driving behaviors of unmanned vehicles is an important means of giving scientific and just evaluation of the research quality of key technologies such as environment perception, control, and decision [3]. Unmanned vehicles should be tested in a typical traffic environment including static, dynamic, and uncertain factors such as urban roads, highway, and rural roads.

The most authoritative way to test and verify unmanned vehicles is by testing on the real road. For this purpose, competitions are held all over the world, such as DARPA-Urban Challenge in the US and the Future Challenge in China. However, this kind of test method is faced with difficulties such as limitation of test site, unrepeatable test condition, time-consuming, and high-cost procedure. Additionally, the actual testing environment may not be accessible, or may only be accessible at certain times while the simulated environment is always available [4]. Therefore, there is a growing trend in unmanned vehicle research to use a simulated environment and several simulation and test platform (STP) are established. But most of them use virtual data instead of real data collected from actual world which will reduce the reliability of the simulation result. The comparison of three test methods based on different data types is shown in Table 1.

Table 1. Comparison of test methods based on three kinds of data types.

	Field Test	3D CG Simulation	Real Multi-Sensor Data
Time and labor cost	High	Low	Low
Security	Low	High	High
Repeatable test scene	No	Yes	Yes
Repeatable test result	No	Yes	Yes
Environment reality	Real	Simulation	Real
Modifiable environment	Hard	Easy	Easy
Typical Scene			

Faced with these limitations, we try to explore ways to realistically simulate and exhibit typical driving behaviors of vehicles with real data of traffic environment in simulation test of unmanned vehicles, as shown in Figure 1. We use the real multi-sensor data captured from the real traffic environment and augment the traffic scene with vehicles in different driving behaviors.

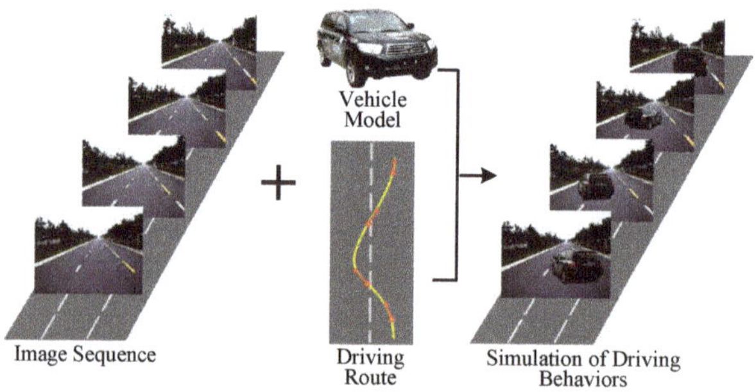

Figure 1. Simulation of driving behaviors with image sequences collected from real road environment.

However, there are quite a few challenges for us to achieve this goal. One is the modeling of virtual vehicles. For "traditional" computer graphics, recent advances in material modeling and global illumination have facilitated the synthesis of realistic and detailed imagery. But it needs painstaking work for the actual visual world, which is very complicated. Each object to be rendered requires lots of work to give surface properties and detailed geometry. Though we can use many image-based approaches to build the model, the data is not acquired straightforward and the methods are not suitable for our relatively large and moving vehicles. Another one is the spatial topology relationship between the virtual vehicles and the road environment. The scene for testing the unmanned vehicles is an image sequence containing spatial topology information. Virtual vehicles have spatial topology attributes, i.e., virtual vehicles should be consistent with reference scene in both appearance and spatial topology relationship.

Encouragingly, we have proposed a simple and effective approach which can exhibit typical dynamic driving behaviors smoothly and realistically.

2. Overview

As part of our work on parallel testing of vehicle intelligence [2], we propose an automatic approach of simulating dynamic driving behaviors of vehicles. The spatial topological attributes and appearance attributes of virtual vehicles are computed separately according to the constraint of geometric consistency of sparse 3D space organized by image sequences.

There are three critical elements for us to solve our task:

Typical Traffic Scene: Unmanned vehicles should be tested in a typical traffic environment including static, dynamic, and uncertain factors such as urban roads, highways, and rural roads. We analyze test tasks and scoring criteria of DARPA-Urban Challenge and the Future Challenge. Then, a typical traffic scene including the intersections is chosen for virtual vehicles to execute the driving tasks of lane change, overtaking, slowing down and stop, right turn, and U-Turn.

3D Road Environment Representation: Road scene simulation and modeling based on vehicle sensors are currently an important research topic in the field of intelligent transportation systems [5–8]. Most current image compositing approaches treat 3D road environment representation as a 2D problem, such as Photoshop. As described by Lalonde et al. [9], we agree with their point of view that any image manipulation must be done in the 3D space of the scene, not in the 2D space of the image. But it is a pity that Lalonde et al. did not precisely define and describe that 3D space of scene. We believe that the three-dimensional road environment is a data field containing image data, spatial topology data, and motion data. Generated novel scene video visualizes these data. As shown in Figure 2, the reference road scene is represented by a sparse ordered image sequence containing data of real filming location of viewpoints. Recent advances in camera calibration, 3D registration, and scene reconstruction have allowed the synthesis of not only images and videos, but also the data of spatial topology. In this paper, we use the GPS coordinates of both viewpoint and virtual vehicles to compute the transformation parameters and open-source software OSM2World to transform a map description exported from OpenStreetMap into a mesh model of road surface. To restrict the motion range of virtual vehicle to the road area and simplify coordinate transformation, we defined a "corridor model" and applied triangle collision detection based on Irrlicht engine [10–12].

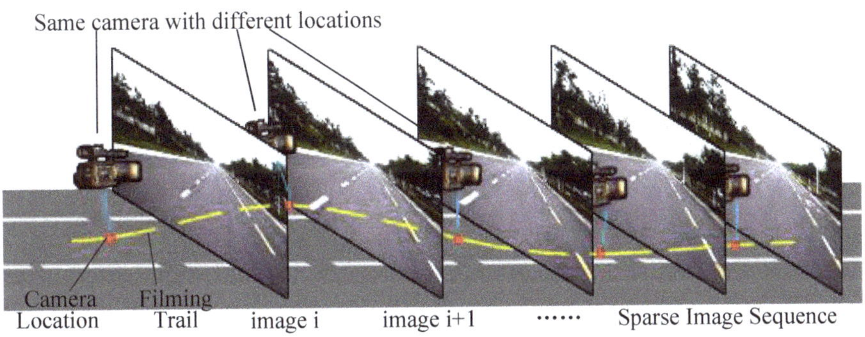

Figure 2. Representation of reference road scene using sparse ordered image sequence containing data of real filming location of viewpoints.

Geometric consistency of vehicle model and road environment: Humans can easily recognize a synthesized object when observing an image. What are the criteria for us? One can judge whether the object should be placed at this location, whether the object should be looked like a side view or a front view, and whether the size of object is fit, too big, or too small. Then, it is easy to understand that the main manifestation of the geometric consistency is that the vehicle should have appropriate position, pose, and size in visualization. In order to solve this problem, we firstly specify the model for virtual vehicle and road environment. The road environment is organized by image sequences. For each image, extrinsic and intrinsic parameters of camera at the corresponding viewpoint are known. For the virtual vehicle, we combine 3D model with multi-viewpoints corresponding to different vehicle images.

3. Related Works

Typical traffic scene: Work of designing and building typical traffic scene for unmanned vehicle tests are being done by researchers. Based road traffic accident data from the years 2000–2010 and from several aspects such as human, vehicle, road environment, and accident form, Zhou et al. [13] selected the greatest impacts of traffic safety to make up the typical dynamic traffic event. In their research, one type of dynamic events in city road scene was selected as a case, which is the conflict between the main vehicle and pedestrians in front of parking bus when pedestrians crossing street on the main road in city. In June, researchers at the University of Michigan announced that they are in the process doing building a fake city center. According to a press release, the fake city center locates on a 13-hectare plot at the school's North Campus just outside Ann Arbor. The faux downtown, to be known as the mobility transformation facility (MTF), will have a four-lane highway, stoplights, intersections, roundabouts, road signs, a railroad crossing, and construction barrels. The facility's designers are also putting up building facades meant to simulate the challenge of transmitting wireless signals inside urban canyons [14].

Prior scene maker: A survey about internet visual media processing [15] showed a number of recent papers demonstrated the work on visual content maker. Lalonde et al. [9] presented the photo clip art system for inserting objects into an image. Users can insert object by specifying a class and a position for the inserted object, which is selected from a clip art database by matching various criteria including camera orientation and lighting conditions. However, their research works on static images rather than continuous dynamic scenario video. Besides, only specific images can be synthesized. For certain objects, not all the possible appearances are available. Eitz et al. [16] proposed a PhotoSketcher system with the goal of synthesizing a new image only given user drawn sketches. The synthesized image is blended from several images, each found using a sketch. However, users must put additionally scribbles on each retrieved image to segment the desired object. The above methods are limited to the synthesis of a single image. To achieve synthesis for image sequence (e.g., scene video as mentioned before), the main additional challenge is to maintain consistency of the same vehicle across successive frames, since candidates for all frames usually cannot be found in the database. Chen et al. [17] proposed the PoseShop system, intended for synthesis of personalized images and multi-frame comic strips. It first builds a large human pose database by collecting and automatically segmenting online human images. Personalized images or comic strips are produced by inputting a 3D human skeleton. Head and clothes swapping techniques are used to overcome the challenges of consistency. However, the PoseShop system did not work very well on dealing with the accurate spatial topology relationship between the background scene and synthesized objects so that geometric consistency criteria can't be matched. Flagg et al. [18] presented a system for capture, analysis, synthesis, and control of video-based crowds. They introduced crowd tubes samples and constraint violations with a conflict graph to avoid collisions. Abdi et al. [19] augmented the traffic signs to provide visual feedbacks to drivers for an enhanced driving experience. They used a virtual 3D model, with a known size, to define a reference coordinate system. They projected the 3D object sign using the corresponding sparse dictionary. Their augmentation is for enhancing driving experience and lacks of reality. The above researches suggest that efforts in this direction are very timely.

The rest of the paper is organized as follows: The geometric consistency of 3D vehicle model and road environment is presented in Section 4. In Section 5, the construction of road scene corridor model and traffic scene augmentation is introduced. Experiments and comparisons are shown in Section 6. Finally, we close this paper with conclusion and future works.

4. Geometric Consistency of 3D Vehicle Model and Road Environment

Vehicle images that match geometric consistency criteria can be obtained by three steps. First, the virtual vehicle should be registered to the 3D road space. Second, the vehicle pose (i.e., vehicle image) corresponding to different viewpoints is obtained. Finally, we compute the scale factor to give the vehicle image an appropriate size.

4.1. Registering 3D Vehicle Model to Road Scene

The position of virtual vehicle and image sequence are not always coincident. So, first of all, we should register the virtual vehicle to a proper position. That is to say, while the real vehicles are at varying distances from the user, the virtual vehicles are all projected to the same distance.

Techniques of augmented reality [20] contribute to this task. By techniques of 3D registration, we can obtain the position of virtual vehicle in the image of road scene. First, we should compute the location of virtual vehicle in road scene. That is, to compute the virtual vehicle's coordinates in virtual 3D space formed by image sequences.

In the real road environment, the road plane is denoted by x-y plane. Each captured image combines with a viewpoint's coordinate $P_1 = (x_1, y_1, z_1)$. The coordinate of the virtual vehicle's center is $P_2 = (x_2, y_2, z_2)$. According to the assumption of x-y plane, z_1 is the vertical height of viewpoint to the road plane while z_2 is the vertical height of virtual vehicle's center to the road plane. $P_v = (x, y, z)$ is the coordinate of virtual vehicle's center in the coordinates of virtual 3D road space. Since the z axes of three coordinates are in parallel with each other, we can get that $z = z_2 - z_1$. Then, we will compute the x and y coordinates of P_v.

We assume that the coordinate origins of real road environment, virtual road scene, and vehicle model are O_w, O_s, and O_v respectively. As shown in Figure 3, the angle between y_s and y_w is θ.

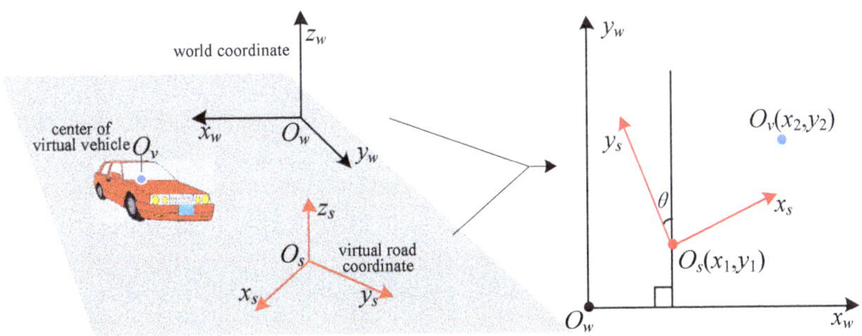

Figure 3. Coordinate transformation between the coordinates of real road environment and the coordinates of virtual road scene.

If we make $P_1^C = [x_1, y_1, 1]^T$, $P_2^C = [x_2, y_2, 1]^T$ and $P^C = [x, y, 1]^T$, then we can get

$$P^C = {}_A^B R P_2^C + P_1^C \tag{1}$$

where ${}_A^B R$ is the rotation matrix for w coordinates to the s coordinates, and

$${}_A^B R = \begin{pmatrix} \cos\theta & \sin\theta & 0 \\ -\sin\theta & \cos\theta & 0 \\ 0 & 0 & 1 \end{pmatrix}. \tag{2}$$

After the 3D coordinate of virtual vehicle's center is obtained, we can use it to get the center's coordinate in each image. If the coordinate of virtual vehicle's center in the image is (u, v), and the projection matrix for 3D to 2D coordinates is $\boldsymbol{P}$, then we have

$$\begin{bmatrix} u \\ v \\ 1 \end{bmatrix} = \boldsymbol{P}_{3\times4} \begin{bmatrix} x \\ y \\ z \\ 1 \end{bmatrix} \tag{3}$$

where $\boldsymbol{P}_{3\times4} = \boldsymbol{A}_{3\times3}\boldsymbol{T}_{3\times4}$. $\boldsymbol{A}_{3\times3}$ is the intrinsic parameters while $\boldsymbol{T}_{3\times4}$ is the extrinsic parameters of the camera. The intrinsic and extrinsic parameters of the camera can be obtained by trilinear method [21].

4.2. *Vehicle Image in Road Scene*

Obtaining the image of virtual vehicle in the road scene consists of two parts, the view image and the scale transformation. The view image shows the virtual vehicle's pose in real time when the vehicle is moving in the road scene. Through the scale transformation, the view image can be synthesized with appropriate size.

4.2.1. The View Image of 3D Vehicle Model

In the 3D vehicle model combined with multi-viewpoints, the distance between the viewpoint and the center of virtual vehicle varies with different viewpoints. For ease of management, we normalize all the viewpoints onto a spherical surface. The center of sphere coincides with the center of virtual vehicle. The spherical radius is R. Figure 4 shows the normalized viewpoint sphere. Each viewpoint lies on the sphere equidistant from the center is denoted by two variables, that is $P_N = (\gamma_1, \gamma_2)$.

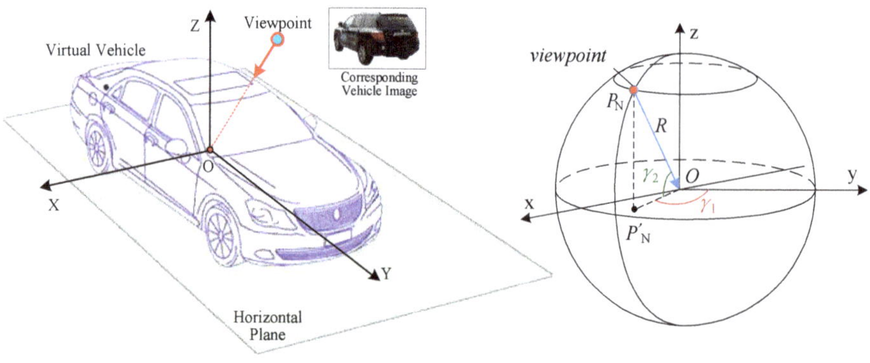

Figure 4. 3D vehicle model combined with multi-viewpoints.

When γ_1 and γ_2 are known, the view image corresponding to the viewpoint can be retrieved. That is to say, all the view images of virtual vehicle can be indexed by two variables. As described above, the coordinates of viewpoint and virtual vehicle are $P_1 = (x_1, y_1, z_1)$ and $P_2 = (x_2, y_2, z_2)$ in real road environment respectively. If the unit vector of the y axis of vehicle model in real traffic environment is (v_i, v_j), we can obtain

$$\gamma_1 = \arctan\frac{y_1 - y_2}{x_2 - x_1} - \arctan\frac{v_i}{v_j} \tag{4}$$

$$\gamma_2 = \arctan\frac{z_1 - z_2}{\sqrt{(x_1 - x_2)^2 + (y_1 - y_2)^2}}. \tag{5}$$

4.2.2. Scale Transformation

After the view image of virtual vehicle is retrieved, we transform its scale to make it fit the scene with appropriate size. The normalized views have the same scale. Based on the model in Figure 4, if P_N is definite, the sight line of viewpoint P_s in virtual road space coincides with that of the viewpoint lies on normalized spherical surface (i.e., OP_N). Based on principle of pin-hole imaging, the sight lines are coincided as that is shown in Figure 5.

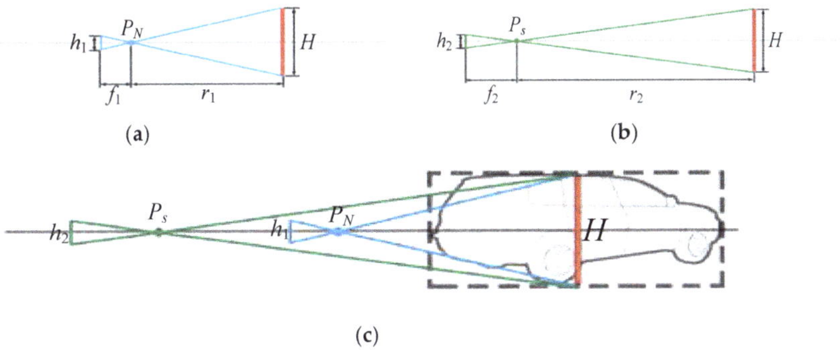

Figure 5. View space of vehicle model computation with scale factor using pin-hole model. (**a**) Pin-hole model at the viewpoint of P_N which lies on normalized spherical surface; (**b**) pin-hole model at the viewpoint of *Ps* which is in the 3D space of road scene; (**c**) overlapping the optical axis in (a) and (b), the scaling relation is clear.

For the viewpoint lying on normalized spherical surface (i.e., P_N), the focal length of camera is f_1, and the size of vehicle image is h_1. The distance from P_N to the center of vehicle is r_1. For the viewpoint (i.e., P_s) in virtual road space, the focal length of camera is f_2, and the size of vehicle image is h_2. The distance from P_s to the center of vehicle is r_2. Then we can easily get

$$h_2 = \frac{r_1 f_2}{r_2 f_1} h_1, \tag{6}$$

where $r = \sqrt{x^2 + y^2 + z^2}$.

So the scale factor is $r_1 f_2 / (r_2 f_1)$.

5. Visual Simulation of Driving Behaviors

In the traffic scene, both static traffic elements that can influence scene semantics such as traffic signs and the subject of traffic flow (vehicle) need explicit geometric description of the road surface to support their structures and motion. On the basis of obtained trail of viewpoint locations corresponding to image sequences, GIS data of the road can be obtained by GIS (geographic information system) or open-source map such as OpenStreetMap [22]. In GIS data, different layers are used to represent the geographical characteristics. A road is represented by a polyline formed by a set of points of geographical locations. In addition, road attributes can be described by parameters such as name, road type, width, and number of lanes.

In practical work, in order to improve efficiency of modeling, we use open-source software OSM2World to transform map description exported from OpenStreetMap into mesh model of road surface, shown in Figure 6.

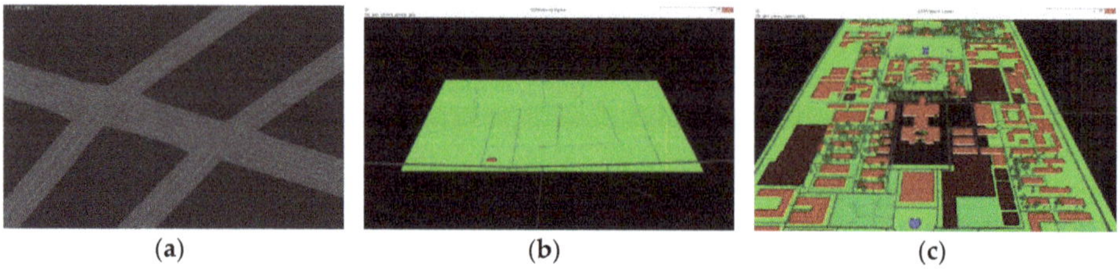

Figure 6. Three kinds of road model built from geographic information system (GIS) data. (**a**) Roads with details such as lanes and intersections; (**b**) roads in a region (data captured from suburb of Xi'an, China); (**c**) model with information besides of road (data captured from Xi'an Jiaotong University, Xi'an, China).

The method in Section 4 and Reference [7] is a pervasive vehicle synthesis method for augmented traffic scene. However, without constraint of road space structure, the synthetic vehicle may appear at illogical location in the road scene, shown in Figure 7. Thus, we propose a logical model named as "corridor model" to restrict the motion range of virtual vehicles.

Figure 7. Augmented traffic scene. Vehicle A and B appear at illogical location (tree lawn).

5.1. Corridor Model Construction

The 3D road space above the road surface is essential for the motion of traffic elements such as vehicles and pedestrians. Leaving out trees, architectures, nature features, and other traffic elements, the real road surface area can be regarded as a ribbon. Boundary control points define the left boundary wall and the right boundary wall. The road ribbon together with left and right boundary walls makes up a 3D space extends infinitely forward. We define this logical model of road geometric space as "corridor model", which is shown in Figure 8.

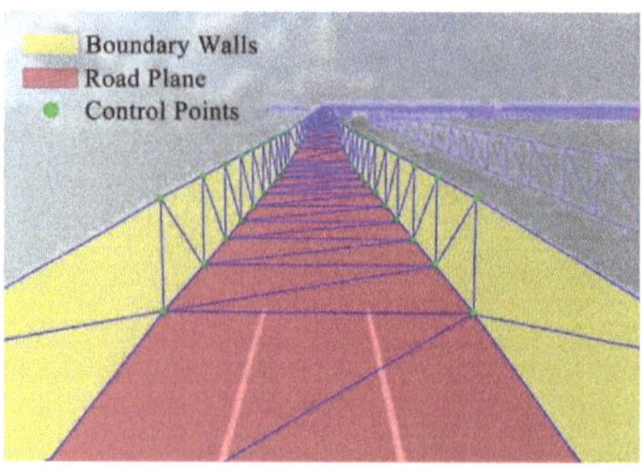

Figure 8. The corridor model and road space.

The corridor model is defined as

$$C = (L_{bottom}, L_{left}, L_{right}), \tag{7}$$

where L_{bottom} is the road plane, L_{left} is the left boundary wall, and L_{right} is the right boundary wall.

The road geometric space based on corridor model (shown in Figure 9) can be implemented by the following steps.

- After the road section is assigned, OSMParser analyses the GIS data to obtain sequence of road center points, lane width, and lane number.
- Sequences of the left boundary and right boundary (LeftVertices[] and RightVertices[]) are figured out through calRoadArea using the data obtained from last step.
- Index of triangle meshes Indices[] is set based on the rendering rules of triangle mesh in computer graphics. Indices[i] records coordinates of three vertexes.
- Sequences of boundary points are connected to adjoint triangle meshes which form the road surface.
- Moving road boundary points through the Y axis, we obtain other two sequences of boundary points. The boundary walls in corridor model can be rendered using these two sequences.

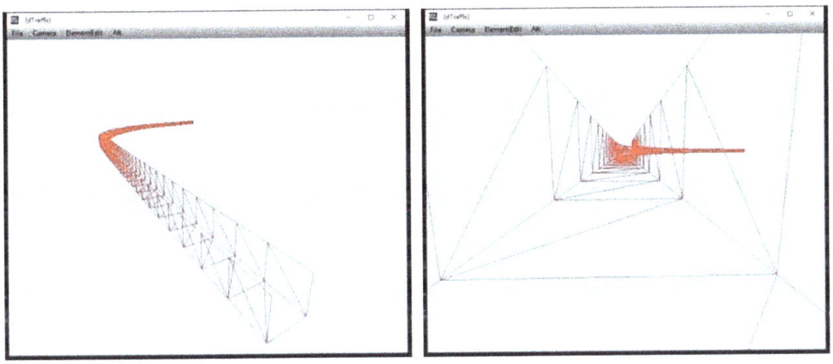

Figure 9. Road geometric space in two different viewpoints.

5.2. Boundary Restraint

The 3D road geometric space is used to restrict the motion range of traffic elements to be in the operating area. Irrlicht engine provides three methods of collision detection respectively based on ellipsoidal bounding box, triangle, and octree. We choose triangle collision detection to achieve our goal. The scene manager builds a triangle picker to judge whether the ray intersects with the plane of triangle mesh. Then, the triangle picker binds to the scene node waiting for collision detection. For example, when the vehicle moving in the road geometric space collides with the boundary, collision response animator controls the vehicle to response to the collision, so that the vehicle keeps moving on the road surface. Figure 10 shows the results of collision detection of the boundary wall and road.

5.3. Registration and Augmentation of Road Scene Data

The perception data of road environment is the data captured by the vehicle sensors, including road scene videos, location information captured by GPS, and pose information from inertial navigation, etc. When existed road scene videos are used for augmentation, no new real-time data is available from the road scene. Thus, we transform the coordinate system of existed perception data to the coordinate system of virtual road space to avoid frequent coordinate transformation between the real world and virtual world. The virtual world uses ENU coordinate system. The transformation relationship can be expressed as

$$C_{ENU} = \boldsymbol{RST}C_e, \tag{8}$$

where C_e is the coordinates in real world, C_{ENU} is the coordinates in ENU coordinate system, $\boldsymbol{R}$ is the rotation matrix, $\boldsymbol{S}$ is the scaling matrix, and $\boldsymbol{T}$ is the translation matrix.

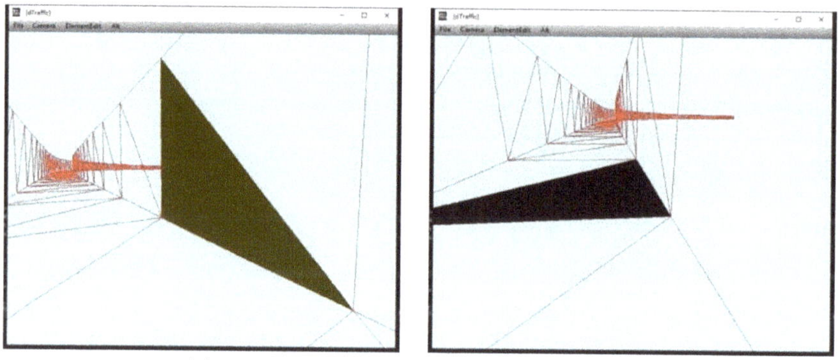

(a) (b)

Figure 10. Road collision and boundary wall collision. (**a**) The dark green area shows the collision with boundary wall; (**b**) the black area shows the collision with road surface.

After the transformation and registration, virtual camera is one-to-one correspondence to real camera. In the virtual road scene, virtual camera moves through the path of the real camera. Each recorded viewpoint has corresponding road scene image. Then, the virtual vehicle is synthesized to the road scene image, shown in Figure 11.

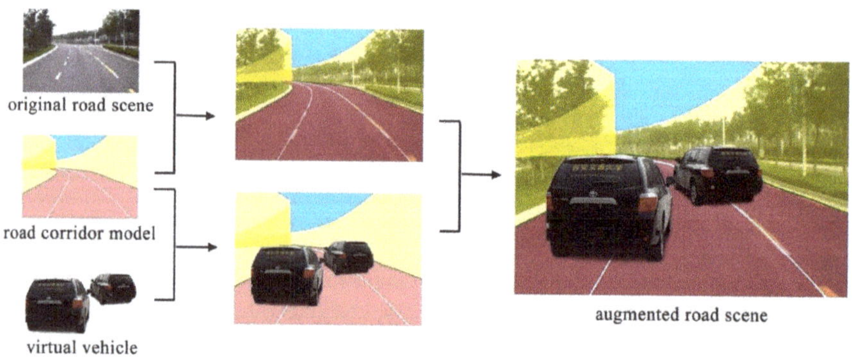

Figure 11. Road scene augmentation based on the corridor model.

6. Experiments

6.1. Dataset of Vehicle Image

Since the acquisition of vehicle image in all viewpoints is unavailable, we discretely capture the image of unmanned vehicle at different viewpoints, as shown in Figure 12. We chose a large square plane without occlusion. The camera moved through a circle centered at the center point of vehicle. The distance between the camera and center of vehicle is 8 m. The height of camera above the ground is 1.8 m, which is the same as the height of camera capturing real road scene videos. We captured one image every other 1°, and obtained a set of 359 images in all. Each image is 2048 × 1536 pixels. Subsequently, we removed the background and rendered shadows manually to obtain a new set of vehicle images in different viewpoints. The new set of vehicle images is indexed by γ_1 and γ_2. The index is organized in the form of hash table to achieve quick load of vehicle images.

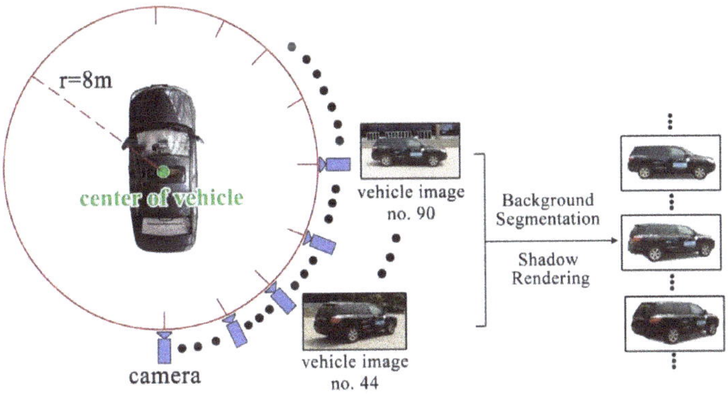

Figure 12. Collection solution of real vehicle images.

However, the images we captured are not enough. When the virtual vehicle moves to different angles, the vehicle image switches abruptly. To solve this problem, we need more vehicle images in other viewpoints. For the vehicle images that have not been captured, we generate them by view interpolation [23] and the result is shown in Figure 13.

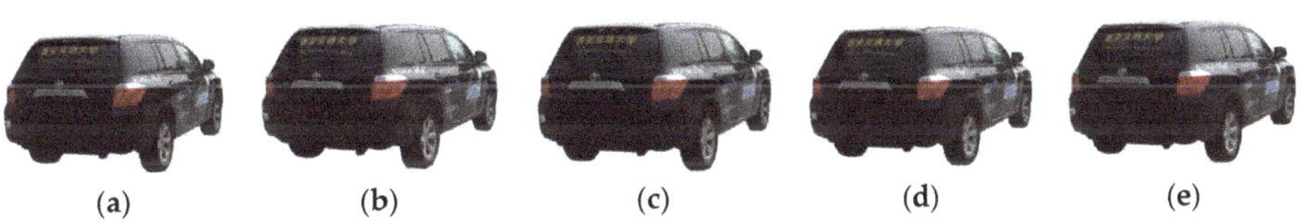

Figure 13. Vehicle image interpolation. (**a**) Image captured at 22°; (**b**) image captured at 24°; (**c**) image interpolated at 22.5°; (**d**) image interpolated at 23°; (**e**) image interpolated at 23.5°.

6.2. Miniature Controllable Environment

We design a miniature scene to verify the reasonability and accuracy of our approach. This is because the advantages of using miniature scene are more controllable than large scene. Besides, most of the current research does not deal with comparisons between real-world images and synthesized scenes. The miniature scene can provide us quantitative data to evaluate the approaches. The coordinates, orientation, and size computed by real location data would be compared with those that are obtained directly from the real image. If these two types of data can match well, then the approach is effective.

As shown in Figure 14, a piece of graph paper in the size of A0 and printed with a chessboard pattern for calibration was laid on a plane to record real locations. The mesh area in the graph paper is 75 cm × 105 cm. Then, the camera was calibrated. It should be noted that the auto-focusing function of the camera should be shut down to keep the intrinsic parameters of camera in constant. The camera was kept still while a car model with model-to-real scale of 1:18 moving step by step. So, the motion area of the miniature environment is equal to the real road area of 13.5 m × 18.9 m. The chessboard on the back of car model is used to obtain the orientation of car model from the original image.

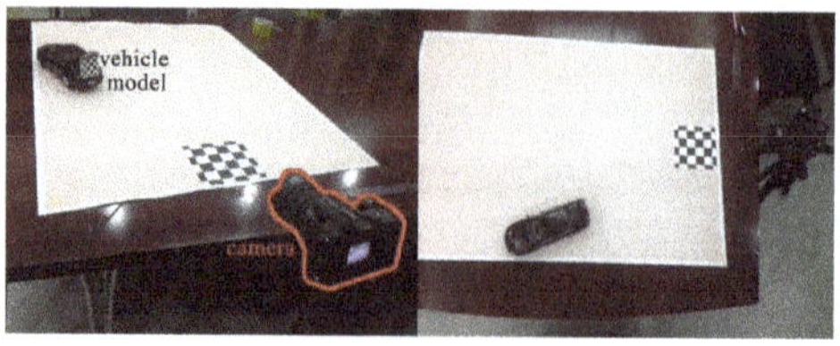

Figure 14. Miniacture controllable experiment environment.

An image, the new location coordinate and orientation of the car model were recorded after the car model had been moved to a new place. Thus, we collect 60 groups of images and location data. The location data is used by our approach to compute relevant parameters. Figures 15 and 16 show the comparison results of our algorithm and original data.

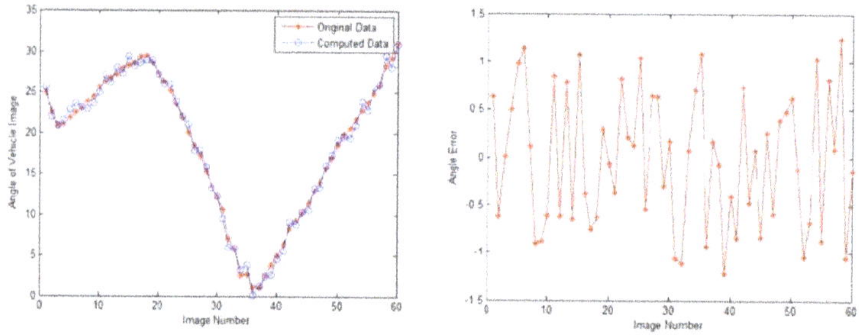

(**a**) (**b**)

Figure 15. Calculation results of angle. (**a**) Comparison result. The blue line is original data and the red line is our result. (**b**) The error of angle is less than 1.2° and the average error is 0.6°.

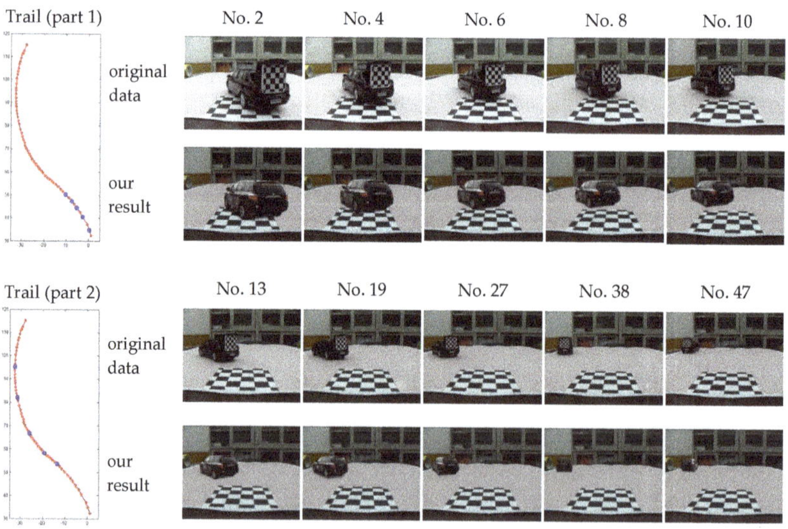

Figure 16. Comparison results of our simulation approach and original image data. The first column shows the trail of vehicle model. The first part of the tail marked with blue dots represent images numbered 2, 4, 6, 8, and 10 from the image sequence, while the second part represent images numbered 13, 19, 27, 38, and 47. In each group of images, the first row shows the original image and the second row shows our simulation results.

6.3. Dynamic Simulation in Scene Browser

For the purpose of further verification of our method, we embed the algorithm in a scene browser and choose a typical traffic scene including the intersections for virtual vehicle to execute the driving tasks of lane change, overtaking, slowing down and stop, right turn, and U-turn.

Our experiments were undertaken on a computer with an Intel i5 processor @3.33 GHz and with 16 GB Memory). The experimental data was mostly taken from the TSD-max dataset [24], which was constructed by the Institute of Artificial Intelligence and Robotics at Xi'an Jiaotong University in China. The dataset is composed of road images captured from urban roads, rural roads, highways, etc.

The traffic image sequences are formed by frames of videos captured from real road environment. The fame rate of the video is 24 fps. When we collect the videos of real road environment, we get the GPS data at the same time. However, GPS data only record the ground and lacks the height information. The height of viewpoint is manually measured. In our experiment, the height of viewpoint is 1.8 m. The height of virtual vehicle's center related to the type of vehicle. Thus, GPS data and height data provide the coordinates of virtual vehicle's center and that of viewpoint corresponding to each image. Applying the algorithm and method mentioned in Sections 4 and 5, we synthesized and augmented the real road scene video, as shown in Figure 17. The computing time is less than 2 ms. The rendering time is about 17 ms. The all process time is about 28 ms. Since the video frame rate is 24 fps, the all process time can meet the real-time requirement. The motion of augmented vehicle is smooth and realistic.

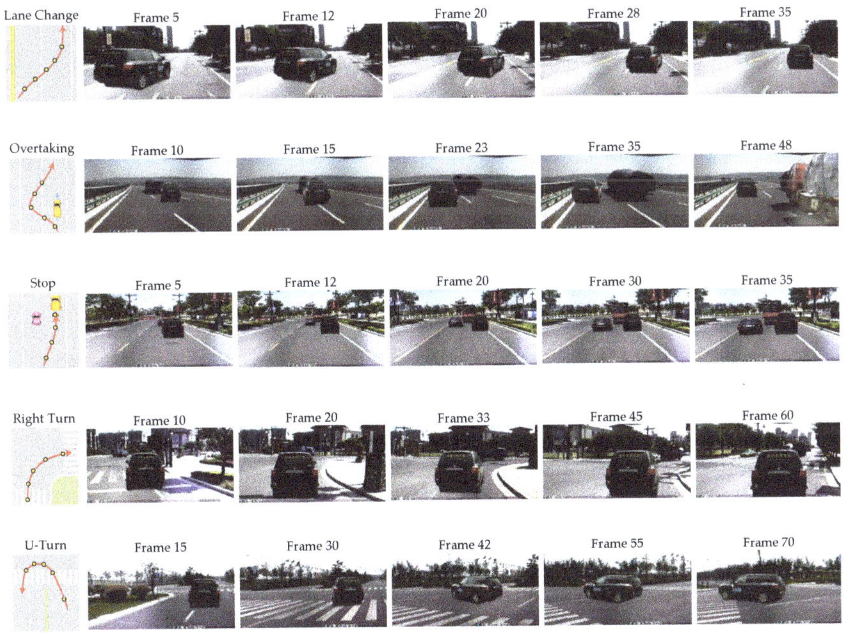

Figure 17. Simulation of typical driving behaviors. The first column shows the driving routes and sampled points of each driving task. Images from the second to sixth columns show the simulation results of sampled points marked in the first column.

7. Conclusions and Future Work

In this paper, we propose a simple and effective approach of simulating dynamic driving behaviors in the traffic scene organized by image sequences collected from real road environment. In order to obtain the geometric consistency of 3D vehicle model and road environment, we use GPS data to accomplish the registration, obtaining vehicle pose and scale transformation. A logical model named as "corridor model" is defined to restrict the motion range of virtual vehicles. The experimental results verify good performance of our method on simulation of dynamic driving behaviors in typical traffic scenes.

For further work, we will build a more complete simulation system with the function of editing traffic scene freely and easy to use. For example, light and weather condition impacts the performance of visual task for unmanned vehicles. Some detectors based on machine learning, such as CNN-based detectors, highly rely on data augmentation techniques to stimulate performance; training detectors with both day and night images are necessary so as to make them more general. In future, we will generate image data in different light and weather condition via generative adversarial networks for varied scenes.

Author Contributions: D.Z. designed the algorithms, wrote the manuscript, and conducted the experiments in 6.1 and 6.2. Y.L. (Yuehu Liu) conducted the experiment in 6.3 and took part in the manuscript revision. Y.L. (Yaochen Li) managed the project.

Funding: This research received no external funding.

Acknowledgments: We give warm thanks to Feng Xu in Lenovo Research, Ye Tian in Huawei Inc., Xiao Huang in CSIC, Chi Zhang, Zhichao Cui and Congcong Hua in Emotion Computing Group, Institute of Artificial Intelligence and Robotics for helpful discussions and valuable advice.

Conflicts of Interest: The authors declare no conflict of interest.

References

1. Turing, A.M. Computing Machinery and Intelligence. *Mind* **1950**, *59*, 433–460. [CrossRef]
2. Li, L.; Wang, X.; Wang, K.; Lin, Y.; Xin, J.; Chen, L.; Xu, L.; Tian, B.; Ai, Y.; Wang, J.; et al. Parallel Testing of Vehicle Intelligence via Virtual-real Interaction. *Sci. Robot.* **2019**, *4*, eaaw4106. [CrossRef]
3. Sun, Y.; Xiong, G.; Chen, H. Evaluation of the Intelligent Behaviors of Unmanned Ground Vehicles Based on Fuzzy-EAHP Scheme. *J. Automot. Eng.* **2014**, *36*, 22–27.
4. Pepper, C.; Balakirsky, S.; Scrapper, C. Robot Simulation Physics Validation. In Proceedings of the 2007 Workshop on Performance Metrics for Intelligent Systems, Washington, DC, USA, 28–30 August 2007; pp. 97–104.
5. Li, Y.; Liu, Y.; Su, Y.; Hua, G.; Zheng, N. Three-dimensional traffic scenes simulation from road image sequences. *IEEE Trans. Intell. Transp. Syst.* **2016**, *17*, 1121–1134. [CrossRef]
6. Hao, J.; Li, C.; Kim, Z.; Xiong, Z. Spatio-temporal traffic scene modeling for object motion detection. *IEEE Trans. Intell. Transp. Syst.* **2014**, *14*, 1662–1668. [CrossRef]
7. Zhao, D.; Liu, Y.; Zhang, C.; Li, Y. Autonomous Driving Simulation for Unmanned Vehicles. In Proceedings of the 2015 IEEE Winter Conference on Applications of Computer Vision, Waikoloa, HI, USA, 5–9 January 2015; pp. 185–190.
8. Li, Y.; Liu, Y.; Zhang, C.; Zhao, D.; Zheng, N. The "Floor-Wall" Traffic Scenes Construction for Unmanned Vehicle Simulation Evaluation. In Proceedings of the 2014 IEEE International Conference on Intelligent Transportation Systems, Qingdao, Shandong, China, 8–11 October 2014; pp. 1726–1731.
9. Lalonde, J.F.; Hoiem, D.; Efros, A.A.; Rother, C.; Winn, J.; Criminisi, A. Photo Clip Art. *ACM Trans. Graph.* **2007**, *26*, 3. [CrossRef]
10. Pokorny, P. Using Irrlicht engine to create a real-time application. *Ann. DAAAM Proc.* **2009**, 1167–1169.
11. Diehl, M. 3-D graphics programming with Irrlicht. *Linux J.* **2009**, *180*, 5.
12. Aung, S.; Johanners, S. *Irrlicht 1.7 Realtime 3D Engine Beginner's Guide*; Packt Publishing Ltd.: Birmingham, UK, 2011.
13. Zhou, Y.; Yan, L.; Wu, Q.; Gao, S.; Wu, C. Design and Implementation of the Typical Dynamic Traffic Event with Virtual Reality Technology. *J. Transp. Inf. Saf.* **2013**, *31*, 128–132.
14. Available online: http://spectrum.ieee.org/cars-that-think/transportation/self-driving/university-of-michigan-to-open-robo-car-test-track-in-the-fall (accessed on 20 June 2016).
15. Hu, S.; Chen, T.; Xu, K.; Cheng, M.; Martin, R.R. Internet Visual Media Processing: A Survey with Graphics and Vision Applications. *J. Visual Comput.* **2013**, *29*, 393–405. [CrossRef]

16. Eitz, M.; Richter, R.; Hildebrand, K.; Boubekeur, T.; Alexa, M. Photosketcher: Interactive Sketch-Based Image Synthesis. *IEEE Trans. Comput. Graphics Appl.* **2011**, *31*, 56–66. [CrossRef] [PubMed]
17. Chen, T.; Tan, P.; Ma, L.; Cheng, M.; Shamir, A.; Hu, S. PoseShop: Human Image Database Construction and Personalized Content Synthesis. *IEEE Trans. Vis. Comput. Graph.* **2013**, *19*, 824–837. [CrossRef] [PubMed]
18. Flagg, M.; Rehg, J.M. Video based Crowd Synthesis. *IEEE Trans. Vis. Comput. Graph.* **2013**, *19*, 1935–1947. [CrossRef] [PubMed]
19. Abdi, L.; Medded, A. Deep Learning Traffic Sign Detection, Recognition and Augmentation. In Proceedings of the Symposium on Applied Computing, Marrakech, Morocco, 3–7 April 2017; pp. 131–136.
20. Krevelen, D.W.F.; Poelman, R. A Survey of Augmented Reality Technologies, Applications and Limitations. *Int. J. Virtual Real.* **2010**, *9*, 1–20.
21. Li, Q.; Zheng, N.; Zhang, X. Calibration of External Parameters of Vehicle-mounted Camera with Trilinear method. *J. Opto-Electron. Eng.* **2004**, *31*, 8.
22. Ramm, F.; Topf, J.; Chilton, S. *OpenStreetMap: Using and Enhancing the Free Map of the World*, 1st ed.; UIT Cambridge: Cambridge, UK, 2010.
23. Chen, S.E.; Williams, L. View Interpolation for Image Synthesis. In Proceedings of the 20th Annual Conference on Computer Graphics and Interactive Techniques (SIGGRAPH 93), Anaheim, CA, USA, 2–6 August 1993; pp. 279–288.
24. Institute of Artificial Intelligence and Robotics at Xi'an Jiaotong University in China. Available online: http://trafficdata.xjtu.edu.cn/index.do (accessed on 3 February 2019).

sensors

MDPI

Article

Multi-Stage Hough Space Calculation for Lane Markings Detection via IMU and Vision Fusion

Yi Sun, Jian Li * and Zhenping Sun

The College of Intelligence Science and Technology, National University of Defense Technology, Changsha 410073, China; 15575191281@163.com (Y.S.); 13974913933@139.com (Z.S.)

* Correspondence: lijian@nudt.edu.cn; Tel.: +86-133-6731-9240

Received: 13 April 2019; Accepted: 14 May 2019; Published: 19 May 2019

Abstract: It is challenging to achieve robust lane detection based on a single frame, particularly when complicated driving scenarios are present. A novel approach based on multiple frames is proposed in this paper by taking advantage of the fusion of vision and Inertial Measurement Units (IMU). Hough space is employed as a storage medium where lane markings can be stored and visited conveniently. The detection of lane markings is achieved by the following steps. Firstly, primary line segments are extracted from a basic Hough space, which is calculated by Hough Transform. Secondly, a CNN-based classifier is introduced to measure the confidence probability of each line segment, and transforms the basic Hough space into a probabilistic Hough space. In the third step, pose information provided by the IMU is applied to align previous probabilistic Hough spaces to the current one and a filtered probabilistic Hough space is acquired by smoothing the primary probabilistic Hough space across frames. Finally, valid line segments with probability higher than 0.7 are extracted from the filtered probabilistic Hough space. The proposed approach is applied experimentally, and the results demonstrate a satisfying performance compared to various existing methods.

Keywords: IMU; vision; classification networks; Hough transform; lane markings detection

---

1. Introduction

With the development of artificial intelligence, intelligent driving technology has made great progress thanks to the advancement of different kinds of sensors and powerful processors. It is a trend where intelligent vehicles play important roles in a safe and efficient transportation environment. Lane detection is an essential research field of intelligent driving, which could be employed to provide lane departure warning in Advanced Driver Assistance System (ADAS) and provide local road navigation for autonomous vehicles, especially when the GPS signal is disturbed.

Many methods are proposed to improve the performance of the lane-marking detection system. Line-segment extraction is a common step to detect lane markings. Well-known methods such as Hough Transform and LSD are very often employed. However, false positive results are given, and a post process is necessary to distinguish whether these line segments belong to lane markings or not. Geometry constraints (e.g., width-based constraints) are always used in this type of classification, but it is difficult to deal with particular kinds of line segments, such as those extracted from fences. Meanwhile, numerous end-to-end networks are proposed to detect lanes in images. Nevertheless, it is of difficulty to merge human logistical knowledge into the networks, and large amounts of labeled images are required.

Due to the disturbance of different kinds of noise, the detection results extracted from a single frame are not reliable for system control. Hence, the integration of sequential information is vital for the development of a robust method. On the other hand, though lane-marking tracking based on sequential information is already frequently employed, the movement information of the vehicle is usually obtained by estimation. As a result, the estimation error will reduce the tracking performance and make it hard to track lane markings at a large time scale. Therefore, obtaining more accurate vehicle information via Inertial Measurement Unit (IMU) is of great necessity.

To solve the problems mentioned above, a novel approach is proposed to extract lane markings by the fusion of vision and IMU. This work aims at obtaining a reliable Hough space which measures each line segment with a probability value. Finally, line segments with high probability values will be extracted from this Hough space. We divide this approach into two steps as follows:

Constructing primary probabilistic Hough space: a primary probabilistic Hough space is extracted from a single frame, which measures each line segment with a probability value. In this section, an efficient Hough Transform with edge gradient constraints [1] is employed for line-segment extraction and a CNN-based classifier is proposed for line-segment classification. The proposed probabilistic Hough space is constructed by the outputs of this classification network and each point in this probabilistic space describes the confidence possibility of the corresponding line segment. A threshold ξ (which is set to 0.7) is used to choose the valid line segments from the probabilistic Hough space. It is necessary to mention that, because Hough space makes it convenient for storing the results across frames, we construct a primary probabilistic Hough space to record the classification results of each frame.

Filtering probabilistic Hough space across frames by IMU and vision data fusion: due to the disturbance of occlusion, vehicle movement, and classification error, the primary probabilistic Hough space extracted from a single frame is not reliable. For example, the change of vehicle pose significantly could affect the classification results of the corresponding line segments. Consequently, the same lane markings might have different values in the probabilistic Hough space. To solve this, sequential information is included, and a Kalman Filter is employed to smooth the probabilistic Hough space across frames. While the vehicle is moving, the line segments extracted from images always have different positions in Hough space at different times, though they lie on the same lane markings. Movement information provided by the IMU makes it possible to align previous and current line segments in the current Hough space, which is essential for the filtering process. The final filtered probabilistic Hough space is used to extract the final line segments. Line segments with low probability value will be eliminated and those with high value will be kept and tracked.

This paper consists of 6 sections. Related works will be introduced in Section 2. Section 3 describes the construction of the primary probabilistic Hough space depending on single frame. In Section 4, the primary probabilistic Hough space is filtered across frames by the fusion of IMU and vision data. The discussion of detailed experiments and conclusions are presented in Sections 5 and 6, respectively. Figure 1 shows the workflow of the proposed method.

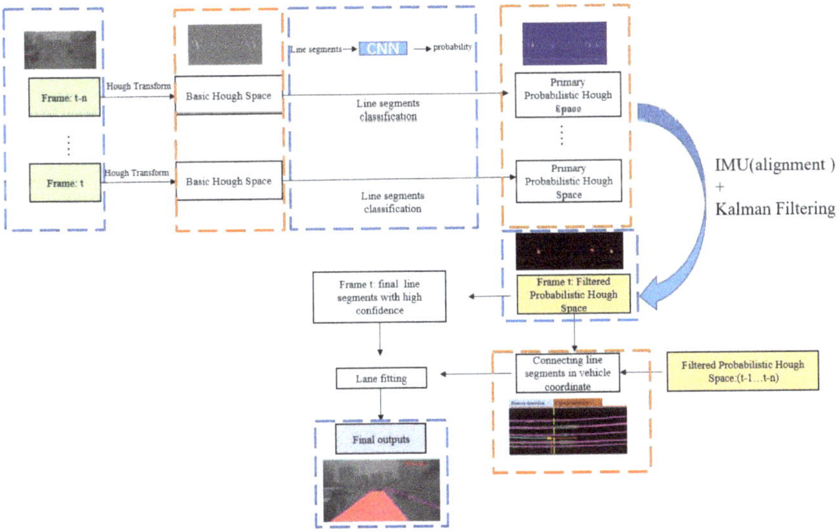

Figure 1. Workflow of the proposed approach: Hough Transform and Classification networks are used to extract the primary probabilistic Hough space. Kalman filter is introduced to smooth the probabilistic Hough space across frames, where sequential information is employed. Movement information provided by IMU is applied to make the previous line segments aligned in the same Hough space. The final filtered probabilistic Hough space is used to extract the final line segments with high probability value. By connecting valid line segments in the vehicle coordinates which are detected at different times, lane fitting could be solved with more sequential information and the final result would be more robust.

2. Related Works

Lane detection plays a fundamental role in current intelligent driving systems such as ADAS or autonomous driver systems. A large amount of vision-based methods has been proposed.

2.1. Conventional Algorithms without CNNs

In conventional lane-detection approaches, edge is a common and important feature for the extraction of lane markings. In [2–4], Canny is used to extract and locate the edge position in image. Many pre-processing algorithms are proposed to strengthen the feature of lane markings. In [2], an LDA model is applied to make it more distinguishable between the lane markings and background in RGB color space. A brightness stretching function named PLSF is proposed in [3] which makes lane markings become clearer than before. Each edge extraction method has its own strength and weakness, so [5] combines different strategies and uses local thresholds to extract edges, which make the edge extraction more robust. Prior information and top-to-bottom constraints are actually useful for eliminating false detection. For example, meaningful edge points are always located in the neighbor of line segments. Thus, in [4], a two-stage feature extraction method is proposed.

Hough Transform is a classical and robust approach to extract line segments from image. To purify these extracted line segments [6], uses SVM to classify these line segments. In [7,8], approaches to estimating the vanishing-point position are proposed and they use the road-tendency information provided by the vanishing point to estimate the optimal parameters of the lane model. A Conditional Random Function (CRF) model is also proposed to extract lane structure in [9].

2.2. Lane Detection with CNNs

Convolutional neural networks free us from designing handicraft features and rules, which have achieved state-of-art performance in many datasets. In [10], a multi-task network named VPG-net is proposed where multi-task training is proved that can improve the network performance. Fully convolutional networks for semantic segmentation are very suitable to solve lane-detection problems, and its encoder-decoder structure has been used in many research works, such as those of [11,12]. In [13], an instance-segmentation network is proposed, which can extract lane markings and divide them into different lane instances. In [14], a Spatial CNN (SCNN) is proposed, which can make the best of the relationship between pixels across rows and columns. Generative adversarial networks (GANs) are also studied in this field; for example, EL-GAN [15] uses GANs and embedding loss to train an end-to-end network.

3. Single Frame: Primary Probabilistic Hough Space via Lane Markings Extraction

In this section, a primary probabilistic Hough space is constructed by the line-segment extraction and classification. Firstly, a combination of Hough Transform and Random Sample Consensus (RANSAC) paradigm algorithm is employed to extract line segments efficiently. Then, the proposed CNN is used to classify these line segments and construct the primary probabilistic Hough space by using the output confidence of each line segment.

3.1. Line Segments Extraction by Hough Transform and RANSAC

Traditional Hough Transform actually leads to extensive computation cost because of its large voting range of direction which usually ranges from 0 to 360 degrees. An efficient Hough Transform [1] is used in this paper via the employment of edge direction to limit the voting range of direction. Defining the edge direction as ϕ and setting $H(\rho,\theta)$ as the Hough space, c represents the column number in image and r represents the row number, θ is limited by the right part of Equation (1). δ is set to 1 (degree) in this paper. This approach can make the extraction of line segments more efficient.

$$\rho = c * \cos(\theta) + r * \sin(\theta) \quad \theta \in [\phi - \delta \ \phi + \delta] \tag{1}$$

However, these line segments extracted by Hough Transform are easily influenced by noisy edge map as shown in Figure 2. A revision process is carried out by RANSAC. These line segments provide RANSAC with numbers of Regions of Interest (ROI) which are extracted from the neighbor of themselves, and RANSAC is then used to extract better line segments in these regions. Detailed information is described by Algorithm 1.

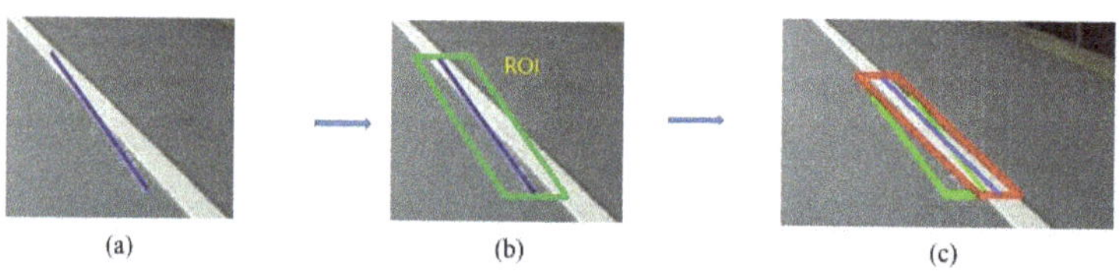

Figure 2. (**a**) A line segment disturbed by edge noise. (**b**) The original ROI which is proposed by the line segment. (**c**) The result of RANSAC. (green ROI: provided by the line segment before revision; red ROI: provided by the line segment after revision).

Algorithm 1 Revising line segments by RANSAC: R represents ROI. (P1,P2) are two edge points randomly extracted from R. Defining l is the original line segment. k represents the slope of l and b is the bias, n is the number of iterations(n=40), lf is the final output

Input: $R, l:(k,b)$
Output: lf
  function REVISE(R, l)
    while n **do**
      $(P1, P2) \leftarrow$ *Get two edge points randomly from R*
      $\hat{l} : (\hat{k}, \hat{b}) \leftarrow$ *Use* $(P1, P2)$ *to fit straight line*
      if $\hat{l}$ *has less outliers than* l **then** $l = \hat{l}:(\hat{k}, \hat{b})$
      end if
      $n = n - 1$
    end while
    $lf = l$
  end function

3.2. Constructing Primary Probabilistic Hough Space by Classification Networks

After extraction of line segments, a post process is necessary to eliminate false detections such as those line segments overlapping fences. To solve this, a CNN-based classification network is proposed to classify line segments, and a probabilistic Hough space is constructed to record the confidence probability of each line segment. Valid line segments extracted from lane markings are labeled with high probability value in this proposed space (Figure 3). Table 1 shows the structure of the networks. The probabilistic Hough space is constructed by the outputs of the classification networks as demonstrated in Figure 4. A threshold ξ (which is set to 0.7) is used to choose the final valid line segments from the probabilistic Hough space.

Why do we need to construct the primary probabilistic Hough space? Indeed, we can choose the valid line segments by the proposed classification networks without constructing this Hough space. However, it is necessary to record the classification results of each frame when integrating sequential information to improve the performance of detection. Hough space is a convenient storage medium of storing the results of each frame.

Table 1. Structure of our classification network.

Layer Index	1	2	3	4	5	6
Layer Name	Data	Conv+Relu	Pooling	Conv+Relu	Interp	Pooling
Output Size	(64, 64, 3)	(62, 62, 40)	(31, 31, 40)	(29, 29, 20)	(28, 28, 20)	(14, 14, 20)
Layer Index	7	8	9	10	11	12
Layer Name	Conv	Pooling	Conv	Inner-Product	Inner-Product	Softmax
Output Size	(10, 10, 20)	(5, 5, 20)	(1, 1, 50)	(1, 1, 500)	(1, 1, 2)	(1, 1, 2)

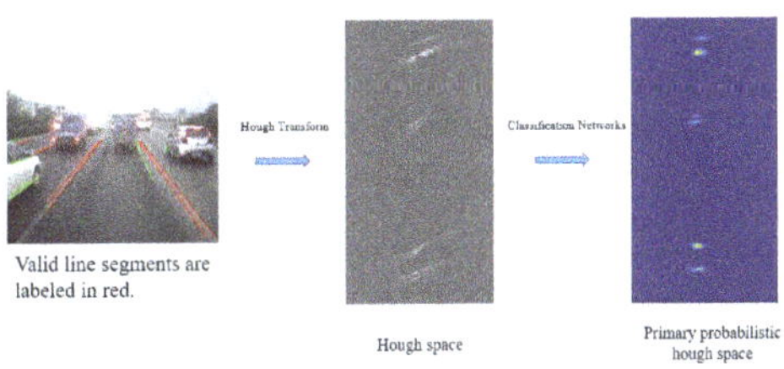

Figure 3. Primary probabilistic Hough space.

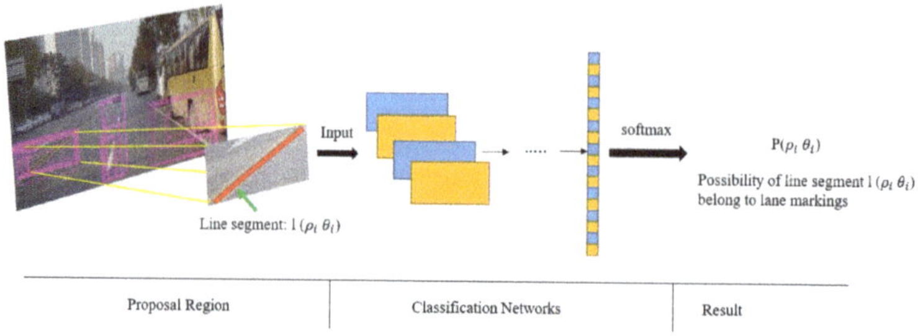

Figure 4. Process of line-segment classification by using the proposed network: the inputs are proposed by line segments and this classification network is used to measure each line segment by the metrics of possibility. The probabilistic Hough space is employed to record the confidence probability of each line segment.

The input of this network is provided by each line segment. The diagonal points of these input images will be calculated according to Equations (2) and (3). Firstly, (x_1, y_1) and (x_2, y_2) are two endpoints of line segment l in vehicle coordinate, and k is the slope of l. W is the max width of traffic lane markings. Two new endpoints $(\widehat{x}_1, \widehat{y}_1)$ and $(\widehat{x}_2, \widehat{y}_2)$ can be obtained according to Equation (3). Finally, these two new diagonal points can be projected into the image plane by Equation (2) and provide us with a reasonable patch as the blue one in Figure 5. In Equation (2), (x, y, z) is a point in the vehicle coordinates and $\boldsymbol{H}$ represents the perspective transformation matrix.

$$s * \begin{pmatrix} c \\ r \\ 1 \end{pmatrix} = \boldsymbol{H} * \begin{pmatrix} x \\ y \\ z \\ 1 \end{pmatrix} \tag{2}$$

$$\begin{aligned} (\widehat{x}_1, \widehat{y}_1) &= (x_1, y_1) + (-\frac{k}{|k|} \times w, -\frac{w}{|k|}) \\ (\widehat{x}_2, \widehat{y}_2) &= (x_2, y_2) + (\frac{k}{|k|} \times w, \frac{w}{|k|}) \end{aligned} \tag{3}$$

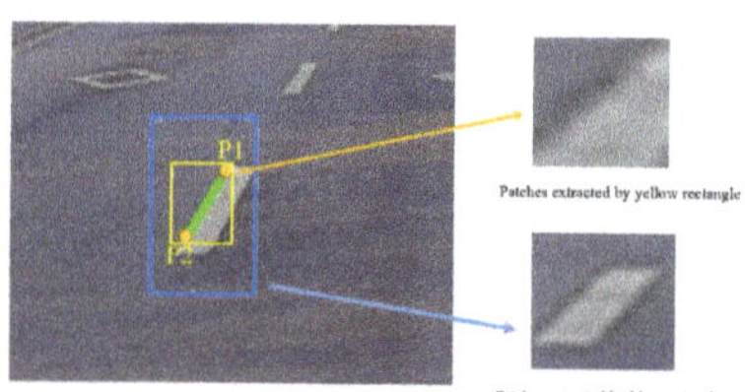

Figure 5. Yellow rectangle is proposed by the two endpoints (P1,P2) of line segments. Blue rectangle is proposed by two new calculated diagonal points by Equation (3).

A dataset was established for training and testing as illustrated in Figure 6. The positive samples are line segments which truly belong to traffic lane markings and negative samples are false line segments. A total of 50,000 pictures were collected.

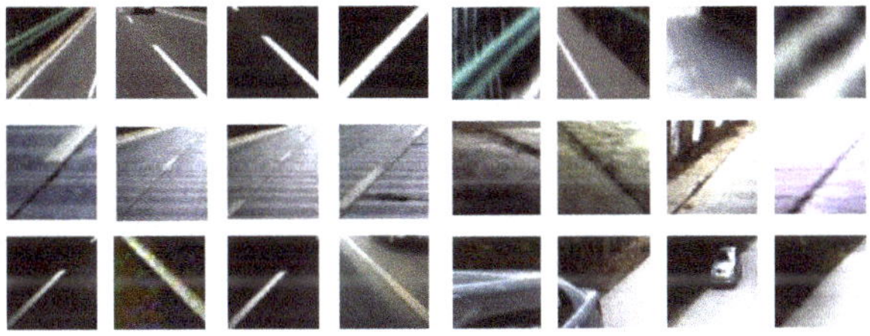

(a) (b)

Figure 6. (**a**) Positive samples. (**b**) Negative samples.

4. Sequential Frames: Filtered Probabilistic Hough Space via IMU and Vision Data

The existence of lane markings is consistent in the sense that they rarely abruptly appear or disappear in the view. Therefore, it is very likely for a line segment with a sudden appearance or disappearance to be false. On the contrary, if valid line segments appear in the same place often, the corresponding positions will keep high probability values for line segments. However, the primary probabilistic Hough space mentioned above is easily disturbed by occlusion, movement of vehicle and classification error (Figure 7). Thus, a Kalman Filter is used to smooth the primary probabilistic Hough space across sequence frames in this section. Movement information provided by IMU is applied to make the line segments extracted at different times aligned in the current Hough space.

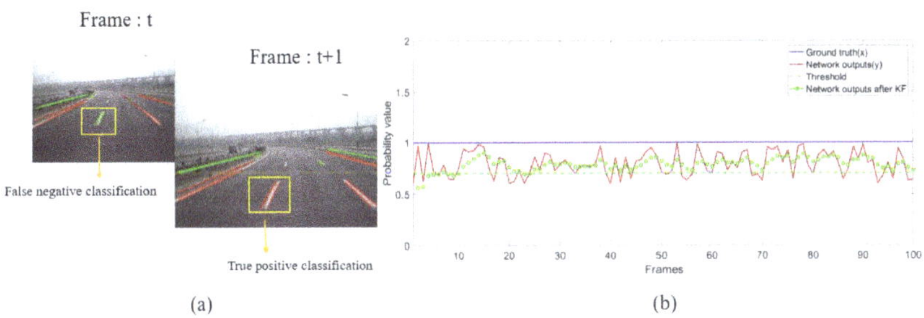

Figure 7. (**a**) due to the vehicle movement and the classification error of the networks, the same line segment has different classification results at time t and $t + 1$. (**b**) plotting the probability values before and after Kalman filtering.

4.1. Filtering Primary Hough Space with Kalman Filter

Kalman filtering makes the filtering process more efficient by using the Markov Assumption. Setting x as the probability value of a line segment l and y as the output confidence of the classification networks. Theoretically, x is equal to 1 if l is valid, otherwise x is equal to 0. The state-transition matrix A and the observation matrix C are both set to be a unit matrix . The noise matrix B is a zero matrix as the attribute of the l should be kept consistent with the previous frames. D is the observation noise caused by vehicle movement and classification error of networks. Equation (4) are the state equation for Kalman filtering.

$$\begin{aligned} x_t &= A * x_{t-1} + B \\ y_t &= C * x_t + D \end{aligned} \tag{4}$$

The filtered probabilistic Hough space describes the probability of whether a line segment belongs to traffic lane markings or not and is more reliable than the primary probabilistic Hough space.

4.2. Aligning Previous Line Segments in the Current Hough Space

As shown in Figure 8, line segment l has different positions at different times because of the movement of the vehicle. Therefore, it is necessary for Kalman filtering to obtain its observed value y from sets of probabilistic Hough spaces which extracted at different times, meaning alignment of $l_{t-1}(\rho_{t-1}, \theta_{t-1})$ and $l_t(\rho_t, \theta_t)$ should be performed in the current Hough space.

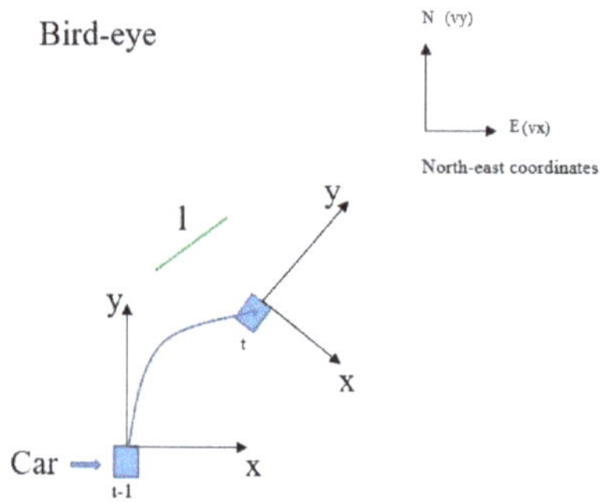

Figure 8. The line segment l has different positions in vehicle coordinate and Hough space at different times. Velocity V and acceleration A are measured in north-east coordinates.

To begin with, $l_{t-1}(\rho_{t-1}, \theta_{t-1})$ from previous vehicle coordinate needs to be projected into the current coordinate based on IMU information including velocity $V=(vx, vy, vz)$, acceleration $A=(ax, ay, az)$ and Euler Angle (α, β, γ). Rotation matrix and transition matrix are calculated by Equations (5) and (6), respectively. Defining $([x^1_{t-1}, y^1_{t-1}, z^1_{t-1}], [x^2_{t-1}, y^2_{t-1}, z^2_{t-1}])$ as the two endpoints of l at time $t-1$ in the vehicle coordinates. Its position at time t can be calculated by Equation (7) (i = 1, 2). Finally, (ρ_t, θ_t) is solved by perspective mapping (Equation (2)) and Hough Transform (Equation (8)).

$$R(\alpha, \beta, \gamma) = \begin{pmatrix} 1 & 0 & 0 \\ 0 & \cos\alpha & \sin\alpha \\ 0 & -\sin\alpha & \cos\alpha \end{pmatrix} * \begin{pmatrix} \cos\beta & 0 & -\sin\beta \\ 0 & 1 & 0 \\ \sin\beta & 0 & \cos\beta \end{pmatrix} * \begin{pmatrix} \cos\gamma & \sin\gamma & 0 \\ -\sin\gamma & \cos\gamma & 0 \\ 0 & 0 & 1 \end{pmatrix} \tag{5}$$

$$\Delta T = \int_{t-1}^{t} V(t) + \frac{1}{2} \times A(t) \times t^2 dt \tag{6}$$

$$\begin{bmatrix} x^i_t \\ y^i_t \\ z^i_t \end{bmatrix} = R(\Delta\alpha, \Delta\beta, \Delta\gamma) * \begin{bmatrix} x^i_{t-1} \\ y^i_{t-1} \\ z^i_{t-1} \end{bmatrix} + R(\alpha_t, \beta_t, \gamma_t) * \Delta T \tag{7}$$

$$\begin{aligned} \theta_t &= \arctan\left(-\frac{r^1 - r^2}{c^1 - c^2}\right) + \frac{\pi}{2} \\ \rho_t &= c^1 * \cos(\theta) + r^1 * \sin(\theta) \end{aligned} \tag{8}$$

Despite the effort described above, precision alignment is hard to achieve due to some factors such as the noise of IMU. So we regard all the $\{(\hat{\theta}, \hat{\rho})\}$(calculated by Equation (9)) as the alignment results of $l_{t-1}(\rho_{t-1}, \theta_{t-1})$ at time t. The alignment error r is set to be 49 in this paper. The final result is demonstrated by Figure 9. The current detections are labeled in red and the previous results (after alignment) are labeled in yellow.

$$(\hat{\theta} - \theta_t)^2 + (\hat{\rho} - \rho_t)^2 <= r \tag{9}$$

Figure 9. The result of alignment during neighbor frames. The current detections are labeled in red and the previous results (after alignment) are labeled in yellow. The bottom part shows the result in the Hough space and the blue circles represent the range of alignment error.

4.3. Final Lane Fitting Using the Result of Sequential Frames

By connecting valid line segments detected across frames as illustrated in Figure 10, the problem of lane fitting can be solved with extensive sequential information. To give the final outputs, a region-growth algorithm is used to divide these foreground points into different lane instances and a parabolic model is used to fit each lane in the current vehicle coordinate. Figure 10 shows the full process mentioned above. To limit the risk of over-fitting, L2 norm is added into the loss function as Equation (10) where α_1 (set to be 0.9) and α_2 (set to be 0.3) are tradeoff coefficients.

$$E = \alpha_1 \sum (ax^2 + bx + c - y)^2 + \alpha_2 ||a||^2 \tag{10}$$

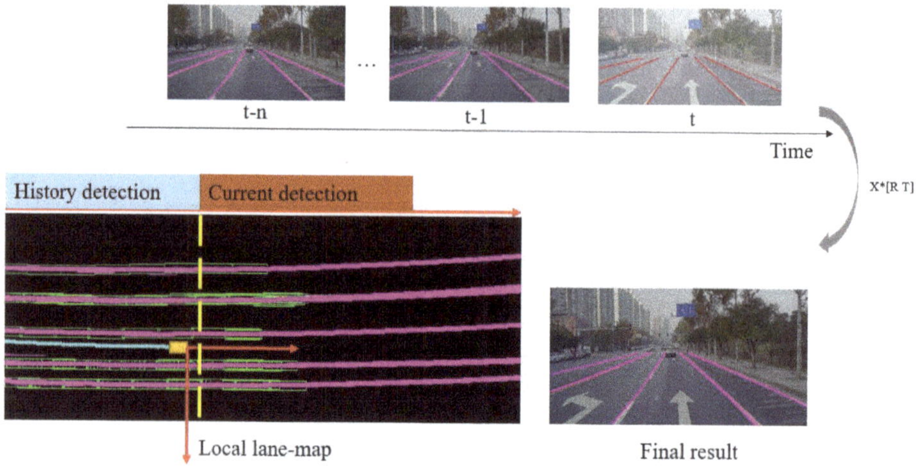

Figure 10. Local lane-map is constructed by connecting those recorded results from *t-n* to *t* in the same vehicle coordinate. It makes the final output more stable by providing useful information for the fitting stage in a larger spatial and time scale than single frame.

5. Results and Discussion

We run the proposed algorithm on an Intel(R) Core(TM) i7-7700HQ 2.80GHz CPU with a NVIDIA GTX1050ti GPU. The average total time cost is 52.3 ms. Processing steps mentioned in Section 3 cost 22 ms (Hough Transform: 10.6 ms, classification networks (one-line segment): 0.3 ms). Processing steps mentioned in Section 4 cost 25 ms (Kalman filtering:2ms, Alignment: 9 ms, lane fitting: 11 ms). One camera and one IMU are employed. The type of the camera is OV10650 and the IMU is Epson G320. Figure 11 shows the vehicle used to carry out the experiments.

Figure 11. The experimental vehicle produced by the Dongfeng Motor Corporation.

This section is divided into two parts. In the first part, detailed analysis of the performance of the classification networks will be introduced. Experiments about the filtered probabilistic Hough space will be discussed in the second part, where the fusion of IMU and vision is employed.

5.1. Performance of the Classification Networks

The performance of the classification networks was tested under Caltech dataset [16]. This dataset contains four video sequences all sampled in urban areas. Easy conditions and challenging scenarios are all included, such as shadows or writing. Please note that only two lines in the current lane were detected in this part. Comparison between the used algorithm and other ones was carried out using this dataset based on the metrics of Accuracy Rate (AR) and False Negative Rate (FNR).

Figure 12 demonstrates the test result of the proposed method with the Caltech dataset. Table 2 shows that the proposed method for line segments extraction and classification can achieve a more satisfying performance compared to Niu's method.

Figure 12. Performance with Caltech datasets.

Table 2. Performance of Different Algorithms with Caltech Dataset.

Clip	Total	Niu's Method [4]		Our Method	
		AR(%)	FN(%)	AR(%)	FN(%)
cordova1	466	92.2	5.4	97.25	2.7
cordova2	472	97.7	1.8	97.05	1.2
washington1	639	96.9	2.5	95.84	3.7
washington2	452	98.5	1.7	95.63	3.1

5.2. Performance of the Filtered Probabilistic Hough Space

To employ sequence information for lane detection, the information provided by vision and IMU needs to be integrated. More specifically, Euler angle and velocity obtained from IMU were used to align history results in the same coordinate. This alignment helps to match the same line segments at different times, which is necessary for Kalman filtering at a later stage. The filtered probabilistic Hough space has a higher reliability compared to the primary probabilistic Hough space.

To evaluate our algorithm, four parts of the road data (Figure 13b) were chosen to test the performance of our method with the measurement metric of accuracy(ACC). Those annotated pictures are labeled in the form of line segments as shown in Figure 13a.

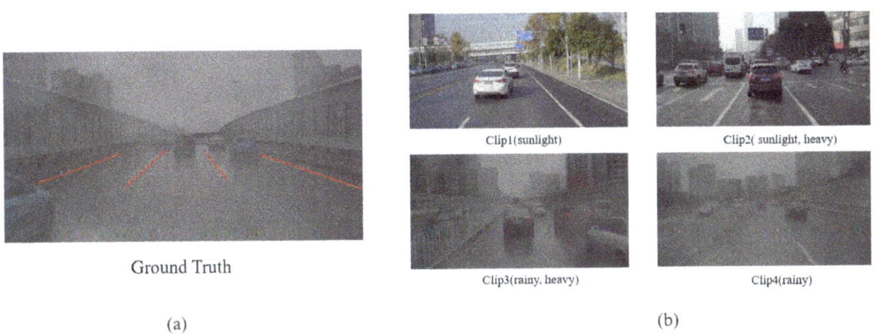

Figure 13. (**a**) Ground truth is labeled in the form of line segments. (**b**) Four parts of data are chosen to test the algorithm: clip1 (sunlight), clip2 (sunlight, heavy), clip3 (rainy, heavy), clip4 (rainy).

A threshold ξ(set to be 0.7) was used to choose the final valid line segments from the filtered probabilistic Hough space (Equation (11)).

$$Attribute = \begin{cases} valid, & p(\rho,\theta) \geq \xi \\ false, & p(\rho,\theta) < \xi. \end{cases} \tag{11}$$

Table 3 lists the accuracy of classification when using the primary probabilistic Hough space and the filtered probabilistic Hough space. It is proven that the filtering process can evidently enhance the accuracy of line-segment classification.

Table 3. Accuracy of the line segments extraction.

Datasets	clip1	clip2	clip3	clip4
Filtered probabilistic Hough space (sequential frames)	0.95	0.93	0.91	0.94
CNNs-based classification (single frame)	0.91	0.89	0.88	0.92

Figure 14 illustrates the result of line-segment detection and tracking. The first and third rows in this figure are the corresponding probabilistic Hough space where the points with high brightness represent the valid line segments.

Figure 14. The first and third rows show the probabilistic Hough space where the points with high brightness represent the possible valid line segments. The second and fourth rows show the corresponding result of line segments extraction where green line segments are the result of detection and red ones are the result of tracking.

Table 4 is the comparison of the performance between the proposed approach in this paper and Neven's method [13], demonstrating that, most of the time, our method outperforms Neven's, especially in terms of false-positive rate due to the use of sequential information.

Table 4. Performance of each algorithm under our own dataset.

Clip	Total	Neven's Method [13]		Our Method	
		TP(%)	FP(%)	TP(%)	FP(%)
part1	927	61.8	6.7	72.2	0.6
part2	174	78.2	38.5	72.9	1.5
part3	647	83.6	6.1	87.3	1.7
part4	713	82.5	5.9	76.5	0.1

By connecting the line segments stored in the past, the problem of lane fitting could be solved with more history information. The results of the proposed approach are showed by Figure 15. Figure 16 describes the final results in the image coordinates and vehicle coordinates which would make it more intuitive to understand the proposed approach.

Figure 15. Detection under different scenarios.

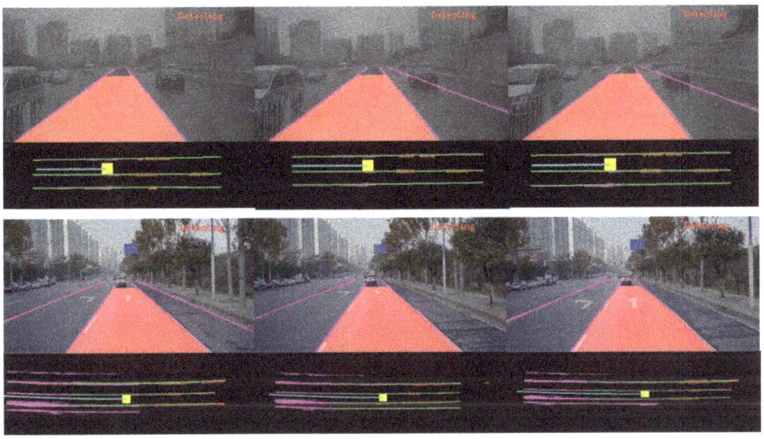

Figure 16. The final results are displayed in the image coordinates and vehicle coordinates. In the second and fourth rows, the yellow rectangle represents the center of the vehicle. Line segments detected in the past are labeled in purple and those extracted from the current frame are labeled in red. The results of lane fitting are labeled in green. The cyan points represent the trace of the vehicle which are calculated by the IMU data.

6. Conclusions

In this paper, a multi-stage Hough space calculation was proposed for a lane-detection task by the fusion of vision and IMU. An efficient Hough Transform and a classification CNNs were introduced to extract and classify line segments from images. By using the outputs of the proposed classification networks, a novel primary probabilistic Hough space was constructed. Kalman filtering was later employed to smooth the probabilistic Hough space across frames for the purpose of eliminating the disturbance from occlusion, movement of vehicle, and classification error. After that movement, information provided by the IMU was applied for aligning the previously detected line segments with the current ones in the current Hough space. The filtered probabilistic Hough space was finally used to clean out line segments with low probability values (threshold was set as 0.7) which were considered false, and to output those with high probability values as the final valid line segments. Though the current method already has a better performance compared to various existing ones mentioned in the paper, more developments are still being sought to further improve the algorithm in the future.

Author Contributions: Y.S.: Conceptualization, software, Methodology, Writing-original draft; J.L.: Methodology, Funding acquisition, Project administration, Supervision, Writing-review; Z.S.: Methodology, Funding acquisition, Project administration, Supervision, Writing-review.

Funding: NSFC Grants 61473303.

Acknowledgments: This work is supported by NSFC Grants 61473303.

Conflicts of Interest: The authors declare no conflict of interest.

References

1. Zhao, Y.; Pan, H.; Du, C.; Zheng, Y. Principal direction-based Hough transform for line detection. *Opt. Rev.* **2015**, *22*, 224–231. [CrossRef]
2. Yoo, H.; Yang, U.; Sohn, K. Gradient-Enhancing Conversion for Illumination-Robust Lane Detection. *IEEE Trans. Intell. Transp. Syst.* **2013**, *14*, 1083–1094. [CrossRef]
3. Gaikwad, V.; Lokhande, S. Lane Departure Identification for Advanced Driver Assistance. *IEEE Trans. Intell. Transp. Syst.* **2014**, *16*, 1–9. [CrossRef]
4. Niu, J.; Lu, J.; Xu, M.; Lv, P.; Zhao, X. Robust Lane Detection Using Two-stage Feature Extraction with Curve Fitting. *Pattern Recognit.* **2016**, *59*, 225–233. [CrossRef]
5. Pollard, E.; Gruyer, D.; Tarel, J.; Leng, S.-S.; Cord, A. Lane marking extraction with combination strategy and comparative evaluation on synthetic and camera images. In Proceedings of the 14th IEEE Intelligent Transportation Systems (ITSC), Washington, DC, USA, 5–7 October 2011; pp. 1741–1746.
6. Zhao, K.; Meuter, M.; Nunn, C.; Müller, D.; Müller-Schneiders, S.; Pauli, J. A novel multi-lane detection and tracking system. In Proceedings of the IEEE Intelligent Vehicles Symposium, Alcala de Henares, Spain, 3–7 June 2012; pp. 1084–1089.
7. Zhou, S.; Jiang, Y.; Xi, J.; Gong, J.; Xiong, G.; Chen, H. A novel lane detection based on geometrical model and gabor filter. In Proceedings of the IEEE Intelligent Vehicles Symposium, La jolla, CA, USA, 21–24 June 2010; pp. 59–64.
8. Ozgunalp, U.; Fan, R.; Ai, X.; Dahnoun, N. Multiple Lane Detection Algorithm Based on Novel Dense Vanishing Point Estimation. *IEEE Trans. Intell. Transp. Syst.* **2016**, *18*, 621–632. [CrossRef]
9. Hur, J.; Kang, S.N.; Seo, S.W. Multi-lane detection in urban driving environments using conditional random fields. In Proceedings of the IEEE Intelligent Vehicles Symposium (IV), Gold Coast City, Australia, 23–26 June 2013; pp. 1297–1302.
10. Lee, S.; Kim, J.; Yoon, J.S.; Shin, S.; Bailo, O.; Kim, N.; Lee, T.-H.; Hong, H.; Han, S.-H.; Kweon, I.S. Vpgnet: Vanishing point guided network for lane and road marking detection and recognition. In Proceedings of the IEEE International Conference on Computer Vision, Venice, Italy, 22–29 October 2017; pp. 1947–1955.
11. Wang, Z.; Ren, W.; Qiang, Q. LaneNet: Real-Time Lane Detection Networks for Autonomous Driving. *arXiv* **2018**, arXiv:1807.01726.
12. Zhang, W.; Mahale, T. End to End Video Segmentation for Driving: Lane Detection For Autonomous Car. *arXiv* **2018**, arXiv:1812.05914.
13. Neven, D.; De Brabandere, B.; Georgoulis, S.; Proesmans, M.; Gool, L.V. Towards end-to-end lane detection: An instance segmentation approach. In Proceedings of the IEEE Intelligent Vehicles Symposium (IV), Changshu, China, 26–30 June 2018; pp. 286–291.
14. Pan, X.; Shi, J.; Luo, P.; Wang, X.; Tang, X. Spatial as deep: Spatial cnn for traffic scene understanding. In Proceedings of the Thirty-Second AAAI Conference on Artificial Intelligence, Hilton New Orleans Riverside, New Orleans, LA, USA, 2–7 February 2018.
15. Ghafoorian, M.; Nugteren, C.; Baka, N.; Booij, O.; Hofmann, M. EL-GAN: Embedding loss driven generative adversarial networks for lane detection. In Proceedings of the European Conference on Computer Vision (ECCV), Munich, Germany, 8–14 September 2018.
16. Aly, M. Caltech. 2014. Available online: http://www.vision.caltech.edu/malaa/datasets/caltech-lanes/ (accessed on 12 June 2018).

sensors

MDPI

Article

Semantic Segmentation with Transfer Learning for Off-Road Autonomous Driving

Suvash Sharma [1], John E. Ball [1], Bo Tang [1,*], Daniel W. Carruth [2], Matthew Doude [2] and Muhammad Aminul Islam [3]

[1] Department of Electrical and Computer Engineering, Mississippi State University, Starkville, MS 39762, USA; ss3795@msstate.edu (S.S.); jeball@ece.msstate.edu (J.E.B.)
[2] Center for Advanced Vehicular Systems, Mississippi State University, Starkville, MS 39762, USA; dwc2@cavs.msstate.edu (D.W.C.); mdoude@cavs.msstate.edu (M.D.)
[3] Department of Electrical Engineering and Computer Science, University of Missouri, Columbia, MO 65211, USA; mig5g@missouri.edu
* Correspondence: tang@ece.msstate.edu

Received: 2 April 2019 ; Accepted: 30 May 2019; Published: 6 June 2019

Abstract: Since the state-of-the-art deep learning algorithms demand a large training dataset, which is often unavailable in some domains, the transfer of knowledge from one domain to another has been a trending technique in the computer vision field. However, this method may not be a straight-forward task considering several issues such as original network size or large differences between the source and target domain. In this paper, we perform transfer learning for semantic segmentation of off-road driving environments using a pre-trained segmentation network called DeconvNet. We explore and verify two important aspects regarding transfer learning. First, since the original network size was very large and did not perform well for our application, we proposed a smaller network, which we call the light-weight network. This light-weight network is half the size to the original DeconvNet architecture. We transferred the knowledge from the pre-trained DeconvNet to our light-weight network and fine-tuned it. Second, we used synthetic datasets as the intermediate domain before training with the real-world off-road driving data. Fine-tuning the model trained with the synthetic dataset that simulates the off-road driving environment provides more accurate results for the segmentation of real-world off-road driving environments than transfer learning without using a synthetic dataset does, as long as the synthetic dataset is generated considering real-world variations. We also explore the issue whereby the use of a too simple and/or too random synthetic dataset results in negative transfer. We consider the Freiburg Forest dataset as a real-world off-road driving dataset.

Keywords: semantic segmentation; transfer learning; autonomous; off-road driving

1. Introduction

Semantic segmentation, a task based on pixel-level image classification, is a fundamental approach in the field of computer vision for scene understanding. Compared to other techniques such as object detection in which no exact shape of object is known, segmentation exhibits pixel-level classification output providing richer information, including the object's shape and boundary. Autonomous driving is one of several fields that needs rich information for scene understanding. As the objects of interest, such as roads, trees, and terrains, are continuous rather than discrete structures, detection algorithms often cannot give detailed information, hindering the performance of autonomous vehicles. However, this is not true of semantic segmentation algorithms, as all the objects of interests are detected on a pixel-by-pixel basis. Nonetheless, to use this technique, one needs careful annotations of each object of interest in the images along with a complex prediction network. Despite these challenges, there has been tremendous work and progress in object segmentation in images and videos.

Convolutional Neural Networks (CNNs) such as Alexnet [1], VGGnet [2], and GoogleNet [3] have been used extensively in several seminal works in the field of semantic segmentation. For semantic segmentation, either existing classification networks are adopted as a baseline or completely new architectures are designed from scratch. For the segmentation task that uses an existing network as a baseline, the learned parameters on that network are used as a priori information. Semantic segmentation can also be considered as a classification task in which each pixel is labeled with the class of the corresponding enclosing object. The segmentation algorithm can either be single-step or multi-step. In a single-step segmentation process, only the classification of pixels is carried out, and the output of the segmentation network is considered to be the final result. When the segmentation is a multi-step process, the network output is subjected to a series of post-processing steps such as conditional random fields (CRFs) and ensemble approaches. CRFs provide a way of statistical modeling for the structured prediction. In semantic segmentation, CRFs help to improve the boundary delineation in the segmented outputs. Ensemble approaches help to pool the strengths of several algorithms. The results of these algorithms are fused using some rules to achieve better performance. However, these techniques increase the computational cost, making them inapplicable to our problem of scene segmentation for autonomous driving. Therefore, the application of these post-processing steps depends upon the type of domain. The performance and usefulness of the segmentation algorithms are evaluated on the basis of parameters such as accuracy over a benchmark dataset, algorithm speed, boundary delineation capability, etc.

As segmentation holds its importance in the identification/classification of objects, investigating the abnormalities, etc., it applies to a number of fields, such as agriculture [4,5], medicine [6,7], and remote sensing [8–10]. A multi-scale CNN and a series of post-processing techniques are applied in [11] to provide a scene labeling on several datasets. The concept of both segmentation and detection is used in [12,13] to classify the images in a pixel-wise manner. Although there has been a lot of work in semantic segmentation, the major improvement was recorded after [14], which demonstrates the superior results on the Pascal Visual Object Classes (VOC) dataset. It performs the end-to-end training and supervised pre-training for segmentation avoiding any post-processing steps. In terms of architecture, it uses the skip layers method to combine the coarse higher-layer information with fine lower-layer information. The methods described in [15,16] are based on an encoder–decoder arrangement of layers that use the max-pooling indices transferred to the decoder part making the network more memory efficient. In both of these works, the mirrored version of the convolutional part acts as the deconvolutional or decoder part. The concept of dilated convolution to avoid information loss due to the pooling layer was used in [17]. A fully connected CRF is used in [18] to enhance the object representation along the boundary. A CRF is used as a post-processing step that improves the segmentation results produced by the network. An enhanced version of [18] is used in [19] which is based on spatial pyramid pooling and the concepts of dilated convolution presented in [17]. A new technique using a pooling called pyramid pooling is introduced in [20] so as to increase the contextual information along with the dilated convolution technique.

All the works mentioned above are evaluated on several benchmark datasets, and one is said to be better than another based on the performance on those datasets. However, in real-life scenarios, there are several areas in which adequate training data are not available. The deep convolutional neural networks require huge amount of training data so that they can generalize well. Lack of enough training data in the domain of interest is one of the main reasons for using Transfer Learning (TL). In TL, the knowledge from a domain, known as the source domain, is transferred to the domain of interest, known as the target domain. In this technique, the deep neural network is first trained in the domain where enough data are available. After this, the useful features are incorporated into the target domain as a priori information. This technique is effective and beneficial when the source and target domain tasks are comparable. The nature of the convolutional neural network to learn general features through lower layers and specific features through higher layers makes the technique of TL effective [21,22]. In particular, in fields such as medicine and remote sensing where datasets with correct annotations are rarely available, the transfer learning technique is a huge benefit. In [23,24], the transfer learning technique is applied for the segmentation of brain structures in brain images from different imaging protocols. Fine-tuning of fast R-CNN [25] for traffic sign detection and classification for autonomous vehicles is performed in [26].

Apart from finding different applications where transfer learning might be used, there has been a constant research effort in effective transfer of knowledge from one domain to another. As it is never the case that all of the knowledge learned from the source task is useful for the target task, deciding what to transfer and how to transfer it holds an important role for the optimum performance of the TL approach. A TL method which automatically learns what and how to transfer from previous experiences is proposed in [27]. A new way of TL for segmentation is devised in [28], which transfers the learned features from a few strong categories, using pixel-level annotations to predict the classes that do not have any annotations (known as weak categories). For a similar transfer scenario, Hong et al. [29] proposes an encoder–decoder architecture combined with an attention model to semantically segment the weak categories. In [30], an ensemble technique, which is a TL approach that trains multiple models one after the other, is demonstrated when the source and target domains have drastic differences.

In our work, we use the TL approach for semantic segmentation specifically for off-road autonomous driving. We use the semantic segmentation network proposed in [16] as a baseline network. This network is trained with the Pascal VOC datasets [31] for segmentation. This domain has a large difference from the one that we are interested in (the off-road driving scene dataset). On the other hand, the off-road driving scene contains fewer classes compared to the Pascal VOC datasets, consisting of 20 classes. Because of this, we propose decreasing the network size, and performing transfer learning on the smaller network. To bridge the difference between the real-world off-road driving scene and Pascal VOC datasets, we use different synthetic datasets as an intermediate domain which might help in performance boosting for the data-deprived domain. Similarly, to correspond to the lower complexity and the latency required for the off-road autonomous driving domain, a smaller network is proposed. Motivated by previous TL approaches in CNN [22,32] and auto-encoder neural networks for classification [33], we transfer the trained weights from the original network to the corresponding layers in the proposed smaller network. However, while most of the state-of-the-art TL methods perform fine-tuning without making any changes to the original architecture (with the exception of the last layer), to the best of our knowledge, this is the first attempt to perform transfer learning from a bigger network to a smaller network, which is helpful to address the two important requirements of autonomous driving. With several experiments using synthetic and real-world datasets, we verify that the network size trained in the source domain may not transfer the best knowledge to the target domain. However, a smaller chunk of the same architecture might work better based on the complexity embedded in the target domain. On the other hand, this work also explores the effect of using various synthetic datasets as an intermediate domain during TL by assessing the performance of the network on a real-world dataset.

The main contributions of this paper are listed as follows:

- We propose a new light-weight network for semantic segmentation. Basically, the DeconvNet architecture is downsampled to half the original size which performs better for the off-road autonomous driving domain;
- We use the TL technique to segment the Freiburg Forest dataset. During this, the light-weight network is initialized with the trained weights from the corresponding layers in the Deconvnet architecture;
- We study the effect of using various synthetic datasets as an intermediate domain to segment the real-world dataset in detail.

The rest of the paper is organized as follows. We briefly review the background and related work in the semantic segmentation of off-road scenes in Section 2. The details of the proposed methods, including Deconvnet segmentation network and our proposed light-weight network, are explained in Section 3. In Section 4, we describe all the experiments and the corresponding results including, the descriptions of the datasets used. Section 5 provides the brief analysis and discussion about the obtained results. The final section of the paper includes our conclusions and notes on future work.

2. Background and Related Work

2.1. Background

2.1.1. Convolutional Neural Networks (CNN)

The simple CNN architecture is composed of five important layers: the input layer, convolutional layer, activation layer, pooling layer, and fully connected layer. For the purpose of classification, a series of these layers can be used on the basis of the complexity of the dataset under consideration. The convolutional layer extracts the structural and spatial relationships from the image. According to [34], in order to improve the learning task, this layer leverages three important ideas: sparse interactions, parameter sharing, and equivalent representations. The convolutional layer is followed by a sub-sampling layer called the pooling layer. This layer is supposed to capture the high-level information of feature maps in compressed form. Thus, it helps to make the features invariant to smaller transitions and translations which results in CNNs being capable of focusing on the useful properties and ignoring the less important features in the feature space. Max-pooling is the famous pooling technique which takes the maximum value of pixels within a defined boundary as its output. The pooling layer may either alternate with convolutional layer or reside sparsely in the network, depending upon the nature of the classification task.

Another important operation within a CNN architecture is activation. This layer, called the activation layer, introduces the non-linearity in input–output relationship, making CNN a universal function approximator. The last layer in most classification-based CNN architecture is the fully connected layer. The fully connected layer takes the flattened data as input, and is responsible for mixing the signals from each dimension so as to introduce the generalization. However, in most segmentation tasks, this layer is not suitable as it increases the computational cost. CNNs are trained in the same way as multilayer perceptrons, which are trained using back propagation algorithm. Back propagation is based on minimizing the cost function with respect to the weight and adjusting those weights based on the gradient as follows:

$$L = \frac{1}{N}\sum_{i}^{N} p(y^i \mid X^i), \tag{1}$$

where N is the total number of images or training samples per batch, X^i represents the ith input sample, and y^i represents the corresponding label. $p(.)$ is the probability of correct classification for corresponding input data. For any layer l, W_l^t is the weight vector at lth layer at time instant t, and U_l^t is the required update in weight. If α_l is momentum, and μ is the learning rate, learning in the network occurs as follows:

$$U_l^{t+1} = \alpha U_l^t - \mu \frac{\partial L}{\partial W_l} \tag{2a}$$

$$W_l^{t+1} = W_l^t + U_l^{t+1}. \tag{2b}$$

2.1.2. Transfer Learning

As specified earlier, TL is a way of utilizing the knowledge of a trained model to solve the problem at hand. In the case of the CNN, the network trained on one domain, called the source domain, might have learned some features that would also be relevant to another domain, called target domain. Therefore, the network with the learned features in the source domain could be a better baseline network to accomplish the task in the target domain. Hence, TL involves the use of an existing trained model, modifying its learned features, called knowledge, into target domain features such that it gives acceptable test performance on the target domain. On the basis of this, several TL techniques are notable. In [35], Pan et al. categorize the TL approaches as inductive transfer learning, transductive transfer learning, and unsupervised transfer learning. However, in deep learning, we can also distinguish them differently as: the fine-tuning approach, the feature extraction approach, multitask learning, and meta learning. In the fine-tuning approach, the nature of the CNN to learn general features through the lower layers and specific features through the higher layers is better utilized. The weights learned by the original trained model in lower layers are frozen as they are related to the general properties of images and have greater similarity with the general features of data in the target domain. Only the few higher layers are modified with the dataset in the target domain. The number of higher layers being trained may vary depending on the data distribution differences between the source and target domains. In the feature extraction approach, only the most important features from the source domain that might better represent the features in the target domain are extracted, and the model is trained with those features mixed with target domain dataset. Multitask learning, on the other hand, trains a model on multiple source tasks so as to increase the generalization capability of the network and is finally fine-tuned with the target domain. Meta learning in TL helps the model to learn about what to learn so that the knowledge will be best fitted for the target domain. In this work, we are dealing with a fine-tuning approach.

2.2. Related Work

With the advent of powerful GPU technology, CNN-based deep learning techniques have been receiving much attention. Semantic segmentation is one of the fields benefiting from this change. Equally, the interest in intelligent autonomous vehicles has been growing and there has been a large amount of research over recent years. The segmentation of road scenes holds a major role in the functionality of such systems. There have been many works directed at city road environment segmentation. However, there have only been a few works for off-road driving scene segmentation. Daniel et al. perform the semantic mapping for off-road navigation using custom convolutional networks in [36]. In [37], a deep neural network is applied in order to classify the off-road scene as trail and non-trail parts using image patches. It successively applies the dynamic programming to delineate the light-weight trail from sub-light-weight network output. In [38], the TL approach is used to semantically segment the off-road scene using the network trained with on-road scenery. Our work is different in the sense that [16] is trained with Pascal VOC images and we transfer the knowledge to the target, which has very different data distributions. Furthermore, we change the original network size, proposing a smaller network that transfers the optimum knowledge considering the real-time issues required by the autonomous vehicle.

3. Proposed Methods

3.1. Segmentation Network Structure

The first part of this work aims at finding the light-weight network structure that suits the target domain. This process is largely dependent upon the complexity of the target domain and upon the extent of the source and target domain differences. While designing the autonomous driving systems, two aspects come into play: the safety and processing speed of the autonomous system software. Safety can be seen from a much wider point of view, which is mostly the function of vehicle hardware design and decisions made by the system software. As a result of the nature of autonomous vehicles, a fast processing speed is required for scene understanding and inferencing which ultimately gives robust control over decision making of the vehicle. We consider this requirement to be very important in this work, thus we aim for the smallest possible network size with the highest possible accuracy. In addition to this, transferring all the weights from large pre-trained networks provided sub-optimal results for our synthetic and real-world dataset as the target domains are simpler than the source domain. Therefore, to use the best size of convolutional network (which achieves a suitable processing speed) as well as having an acceptable accuracy level, we propose a smaller convolutional network, called the light-weight network, taking [16] as a base model. Our proposed network, which better suits our application, is half the size of the original Deconvnet architecture. Figure 1 shows the structure of our light-weight network architecture.

The DeconvNet [16] is learned on top of the VGG-16 network [2] and takes 2D images 224 × 224 pixels in size. The deconvolutional part is a mirrored version of the convolutional part and contains 13 layers on both the convolutional and deconvolutional side. The convolutional part is converged into two fully connected layers augmented at the end to impose class-specific projections. It is trained using a two-stage training procedure in which the first step involves training with easy examples. The second stage involves fine-tuning of the network learned in first stage with more challenging images. Our light-weight network consists of seven convolutional layers and three pooling layers towards the convolutional side. The deconvolutional network is the mirrored version of the convolutional network. The major modification in architecture [16] is the removal of some intermediate layers, including fully connected layers, which improves the computational complexity of the network. Both the architectures, DeconvNet and light-weight, are called encoder–decoder-based architectures, in which the convolutional part downsamples and the deconvolutional part upsamples the feature maps. Such architectures allow the use of max-pooling indices during upsampling which helps to obtain better segmentation maps with preserved global context information. However, the use of max-pooling indices slightly increases the computational cost. The original DeconvNet architecture and proposed light-weight network architectures are shown in Figure 1. The details, including each layer's output and the kernel size of our light-weight network architecture, are shown in Table 1.

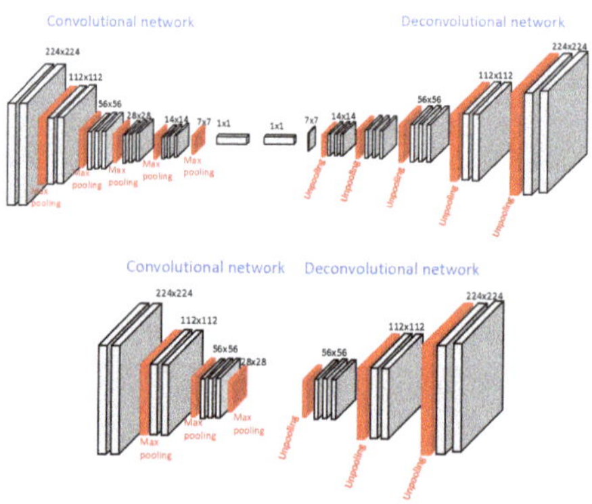

Figure 1. Top: Original DeconvNet architecture, **Bottom**: Proposed light-weight network architecture.

In the following two sections, we do a comparative study of the original and proposed network in terms of computational complexity and latency.

3.1.1. Computational Complexity

For any CNN, the total computational complexity of the convolutional layer can be expressed as follows [39]:

$$O\left(\sum_{l=1}^{d} n_{l-1} s_l^2 n_l m_l^2\right). \tag{3}$$

In Equation (3), l represents the corresponding layer; n_{l-1} represents the number of filters in the $(l-1)$th layer; s_l represents the spatial size (length) of filter in the lth layer; and m_l is the spatial size of the output feature map. DeconvNet consists of 13 convolutional layers and 13 deconvolutional layers, whereas the proposed light-weight network consists of seven convolutional and seven deconvolutional layers. Incorporating the fact that the convolution and deconvolution operations are the same in terms of computation, the overall computational complexity for both networks is shown in Table 2. The proposed light-weight network has a complexity 1.56 times lower compared to that of the original network. This reduction in complexity is in favor of the low latency requirement of autonomous driving.

Table 1. Detailed structure of proposed light-weight network architecture. Note that C is the number of classes.

Layer's Name	Kernel Size	Stride	Pad	Output Size
input	-	-	-	224 × 224 × 3
conv1-1	3 × 3	1	1	224 × 224 × 64
conv1-2	3 × 3	1	1	224 × 224 × 64
pool1	2 × 2	2	0	112 × 112 × 64
conv2-1	3 × 3	1	1	112 × 112 × 128
conv2-2	3 × 3	1	1	112 × 112 × 128
pool2	2 × 2	2	0	56 × 56 × 128
conv3-1	3 × 3	1	1	56 × 56 × 256
conv3-2	3 × 3	1	1	56 × 56 × 256
conv3-3	3 × 3	1	1	56 × 56 × 256
pool3	2 × 2	2	0	28 × 28 × 256
unpool3	2 × 2	2	0	56 × 56 × 256
deconv3-1	3 × 3	1	1	56 × 56 × 256
deconv3-2	3 × 3	1	1	56 × 56 × 256
deconv3-3	3 × 3	1	1	56 × 56 × 128
unpool2	2 × 2	2	0	112 × 112 × 128
deconv2-1	3 × 3	1	1	112 × 112 × 128
deconv2-2	3 × 3	1	1	112 × 112 × 64
unpool1	2 × 2	2	0	224 × 224 × 64
deconv1-1	3 × 3	1	1	224 × 224 × 64
deconv1-2	3 × 3	1	1	224 × 224 × 64
output	1 × 1	1	1	224 × 224 × C

3.1.2. Frame Rate

The scene segmentation algorithms for autonomous driving require a frame rate as high as possible. In this work, we aimed to find a network architecture that provides a better frame rate without compromising the accuracy. We performed this test on a Nvidia Quadro GP100 GPU with 16G memory. In this setup, while maintaining the comparable accuracy, our proposed light-weight network has a frame rate of 21 Frames Per Second (fps), which is better than that of the original network (17.7 fps).

Table 2. Complexity comparison of the two networks.

Network	Complexity	Ratio
DeconvNet	$O\ (2.914 \times 10^{10})$	O (1.56)
Light-weight	$O\ (1.867 \times 10^{10})$	

3.2. Training

The second part of this work is about actual learning and fine-tuning the network with synthetic and real-world datasets. We fine-tuned our proposed light-weight network with synthetic datasets as well as with a real-world dataset and report the result. Here, we explore the advantages and disadvantages of using a synthetic dataset. We used the synthetic dataset as the intermediate domain and the real-world dataset as the final domain. In the first training method, we performed transfer learning using only the real-world data and observed the results. In the second training technique, we trained the light-weight network using the synthetic dataset as an intermediate domain. In this work, we are interested in seeing the effectiveness of our segmentation results in a real-world scenario by fine-tuning the light-weight network trained with synthetic dataset. To do so, we fine-tuned the original model with the synthetic dataset as a first step, and transferred this knowledge for the real-world dataset as a final step. As we are interested in the off-road autonomous driving scenario, we focused on how the transfer learning works in order to segment the real-world dataset with and without using synthetic dataset.

In this work, we used the softmax loss as an optimization function available in Caffe framework [40]. This loss function is basically a multinomial logistic loss that uses softmax of the output in the final layer of the network. The softmax function is the most common function used in the output of CNNs for classification. It is used as a layer in CNN architecture that takes an N-dimensional feature vector and produces the probabilistic values as output in the range (0, 1). Considering $[x_1, x_2, x_3, ..., x_N]$ as the input to the softmax layer and $[o_1, o_2, o_3, ..., o_N]$ as its output, the input–output mapping occurs as in Equation (4).

$$o_i = \frac{e^{x_i}}{\sum_{y=1}^{N} e^{x_y}} \ \forall i \in 1...N. \tag{4}$$

Therefore, in the classification or segmentation of input images, the softmax layer produces the probabilistic values for all possible classes. On the basis of these probabilities, any test data (or pixel in the case of segmentation) is assigned to the class with the maximum probabilistic value. Consider $(x_1, y_1), (x_2, y_2), ..., (x_n, y_n)$ to be any n number of training points ,where x denotes the training data and y denotes the corresponding label. In Caffe [40], the softmax loss is defined in a composite form by applying multinomial logistic loss to the softmax layer's output. In Equation (5), the softmax loss is defined as a cost function to be optimized.

$$J(\theta) = -\frac{1}{n}\left[\sum_{i=1}^{n}\sum_{j=1}^{c} 1\left\{y_i = j\right\} log \frac{e^{\theta_j^T x_i}}{\sum_{k=1}^{c} e^{\theta_k^T x_i}}\right], \tag{5}$$

where c represents the total number of classes. The parameter θ^T represents the transpose of the weight matrix of the network at that instant in time. With this loss function, the training was performed using the stochastic gradient descent method with a learning rate of 0.001, a momentum of 0.9, and a weight decay of 0.0005 in Nvidia Quadro GP100 GPU with 16G memory.

4. Experiments and Results

4.1. Dataset Description

We performed experiments with four different datasets in which three are simulated datasets and one is a real-world off-road dataset. The simulated datasets range from simple two-class datasets to more complex four-class datasets. These datasets were generated considering different real-world aspects such as surface reflectivity of tree trunks or the ground, the shadowing effect, time of the day, etc. The real-world dataset is the off-road autonomous vehicle dataset called Freiburg Forest dataset [41]. In the section below, we describe each of them briefly.

4.1.1. The Synthetic Dataset

Three sets of synthetic data were used which were generated using a specially designed simulator enabled by the MSU Autonomous Vehicle Simulator (MAVS) [42,43]. This simulator is a physics-based sensor simulator for ground vehicle robotics that includes high-fidelity simulations of LiDAR , cameras, and several other sensors. In this work, these datasets are considered to assess the performance of segmentation, transferring the knowledge from the pre-trained convolutional network to the simulated dataset. In addition, we assess the segmentation performance by transferring the knowledge from the synthetic dataset to a real-world off-road driving scenario. As the off-road vehicle domain has very little data to use for training and it is a domain requiring the highest possible level of accuracy, a larger volume of annotated datasets are required. In order to fulfill this requirement, the use of a synthetic dataset can be a help.

The Two-Class Synthetic Dataset

This dataset is the simplest synthetic dataset containing two classes: Ground and Tree. This dataset does not strongly incorporate the characteristics of real-world scenes such as time of the day, shadowing, reflectivity, etc. However, it considers the properties of tree trunks, leaves, and the ground mostly in terms of color and structure. The dataset consists of 5674 images of size 640×480 pixels. We separated 80 percent into the training set and 20 percent into the validation set with no overlap. Some samples of this dataset are shown in Figure 2.

Figure 2. Sample images from two-class synthetic dataset. Best viewed in color.

The Four-Class High-Definition Dataset

This dataset is more complex than the previous two-class dataset. It considers a more complex environment including vegetative structure as well as more realistic forest scenes. The increased complexity of this dataset is mostly due to fine vegetative structures sparsely distributed on the ground. Additionally, we consider the flowering as well as non-flowering vegetation and trees, making this dataset both realistic and complex at the same time. It simulates sky, trees, vegetation, and the ground as four different classes. In total, we have 1700 high-definition images of size 1620×1080 pixels; we separated 80 percent into the training set and 20 percent into the validation set with no overlap. Some typical images from this synthetic dataset are shown in Figure 3.

Figure 3. Three sample images from four-class high-definition dataset. best viewed in color.

The Four-Class Random Synthetic Dataset

Compared to the other two synthetic datasets, this dataset is more natural as more real-world variations are considered. This also includes sky, trees, vegetation, and the ground as the classes for off-road driving scene. We use total 10,726 images of 224 × 224 pixels. In the MAVS simulator, the use of randomized scenes with physics-based simulation of cameras and environments allows for the use of a wide variety of training data. MAVS considers features such as different terrain structure, different time of the day, and haziness of the atmosphere quantified by turbidity [43]. As mentioned in [43], five different times of the day and five different turbidity values are considered, producing 25 unique lighting scenarios in the images. On the other hand, the random dataset includes images from three different environments: an American Southeast forest ecosystem, an American Southeast meadow ecosystem, and an American Southwest desert ecosystem. Because of this set up, this dataset has much more variance than the previous two synthetic datasets. Some sample images from this dataset are shown in Figure 4.

Figure 4. Sample images from four-class random synthetic dataset. Best viewed in color.

4.1.2. The Real-World Dataset

We use Freiburg Forest dataset [41] as real-world dataset. These were collected at 20 Hz with a resolution of 1024 × 768 pixels on three different days to acquire the variability in data caused by lighting conditions. However, in our experiments, we pre-processed the dataset as per our requirement. Before feeding them into our proposed light-weight network, the images were cropped into 224 × 224 size as a pre-processing step to make them compatible with the input layer as well as to acquire simple data augmentation. In [40], cropping can be performed randomly to extract an image patch of a desired dimension. The images in the dataset are in different formats such as RGB, NIR , depth images. For this work, we use the RGB image format only. The dataset includes six different classes: Obstacle, Trail/Road, Sky, Grass, Tree, and Vegetation. While experimenting, we considered the tree and vegetation as a single class as suggested in [41]. Therefore, in terms of training, it is only a five-class dataset. Some sample images from this dataset pool are shown in Figure 5.

Figure 5. Sample images from Freiburg Forest dataset. Best viewed in color.

4.2. Segmentation of the Real-World Dataset with Transfer Learning

In this experiment, we train our light-weight network with the pre-trained weights from DeconvNet architecture. The DeconvNet architecture was originally trained with the Pascal VOC dataset (as a benchmark dataset for segmentation). To transfer the knowledge from this architecture, we initialize our proposed network with the pre-trained weights from DeconvNet corresponding to the existing layers in the light-weight network while ignoring the weights of the remaining layers. We apply fine-tuning by learning up to two layers completely from scratch towards the deconvolutional side of our light-weight network.

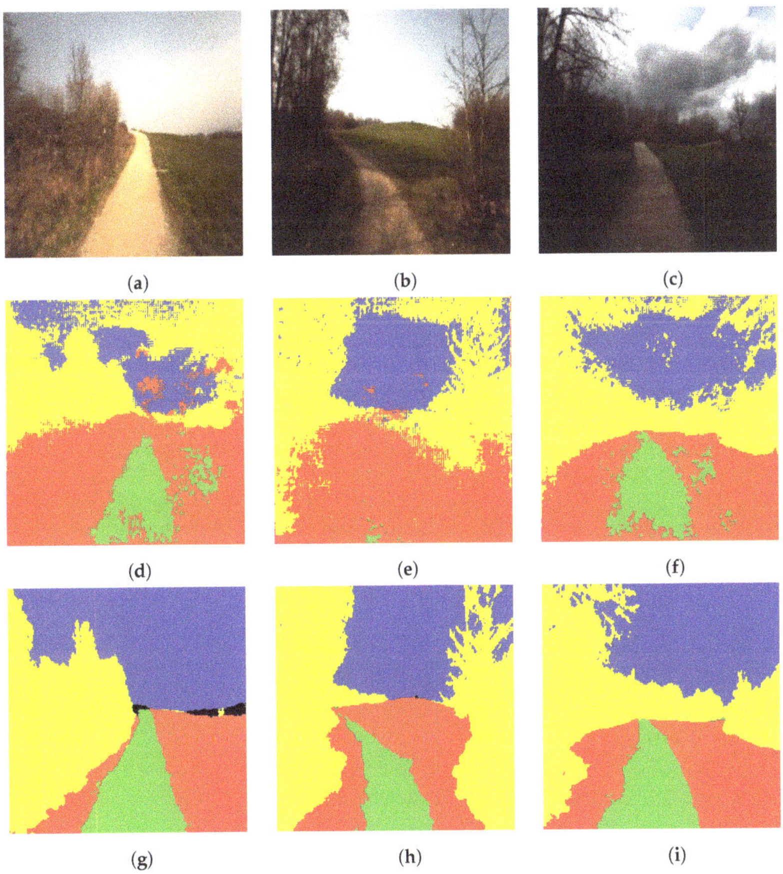

Figure 6. Segmentation of the Freiburg Forest dataset (**a–c**): test images, (**d–f**): corresponding segmented images using DeconvNet, (**g–i**): corresponding segmented images using the proposed light-weight network. Note that the color code for classes is: yellow: tree, green: road, blue: sky, red: ground, black: obstacle. Best viewed in color.

The proposed algorithm achieved 93.1 percent overall accuracy with only the outermost layer learning from scratch and 94.43 percent accuracy with the two outermost layers learning from scratch with a learning rate of 0.01. All the other layers are slowly modified with a learning rate of 0.001. This way of fine-tuning typically means adopting the DeconvNet to the new domain, where the general properties are slowly modified/learned and specific properties are quickly modified/learned. The concern about which layers are to be learned from scratch is an open-ended question and is mostly the function of diversity between the source and target domain. The results produced by the model with the best accuracy (the one that is trained with the two outermost layers learned from scratch) are shown in Figure 6.

4.3. Utilizing the Synthetic Dataset

In this experiment, we use TL approach somewhat differently. This training approach is based on training the network multiple times with multiple domains in order to slowly learn the target domain. We use three different synthetic datasets as the intermediate domain and observe the performance of fine-tuning for the real-world dataset. The obtained overall accuracy of segmentation for all three sets of synthetic datasets are shown in Table 3. The accuracy of the four-class high-definition dataset is lower compared to the other two synthetic datasets. As specified in Section 4.1.1, this dataset has the fine vegetative structures sparsely distributed on the ground which makes them difficult to detect. In addition, the vegetation and the trees are with and without flowers, which makes this dataset realistic and complex at the same time. This complexity inherent to the four-class synthetic dataset resulted into the lower accuracy.

Table 3. Segmentation accuracy on the synthetic dataset. TL—Transfer Learning.

Data	Method	DeconvNet	Light-Weight
Synthetic	TL on two-class	97.62 (%)	99.15 (%)
	TL on four-class high-definition	65.61 (%)	75.71 (%)
	TL on four-class random	73.23 (%)	91.00 (%)

4.3.1. The Two-Class Synthetic Dataset

In this experiment, we first trained our proposed light-weight network with the pre-trained DeconvNet weights using the two-class synthetic dataset. As we specified in the earlier section, this dataset contains trees and ground as two classes and is a simple dataset. This dataset just considers the autonomous driving scenario in terms of color. The tree class is represented with a gray and green color, and the ground with a yellowish color, as shown in Figure 2. Structurally, the trees have minor variations and the ground is uniform. After training and testing with our proposed light-weight network, we obtained 99.15 percent overall pixel-wise accuracy with this synthetic dataset. We used the learning rate of 0.01 for the two outermost layers and 0.001 for all the other layers. We show some segmented results of this synthetic dataset in Figure 7.

The model trained with this two-class synthetic dataset is again fine-tuned with the real-world Freiburg dataset. This time, only the outermost layer of the light-weight network was learned from scratch with a learning rate of 0.01 and all the other layers with 0.0001. As shown in Table 4, we obtained 94.06 percent overall accuracy, which is somewhat below the accuracy given by the previous experiment.

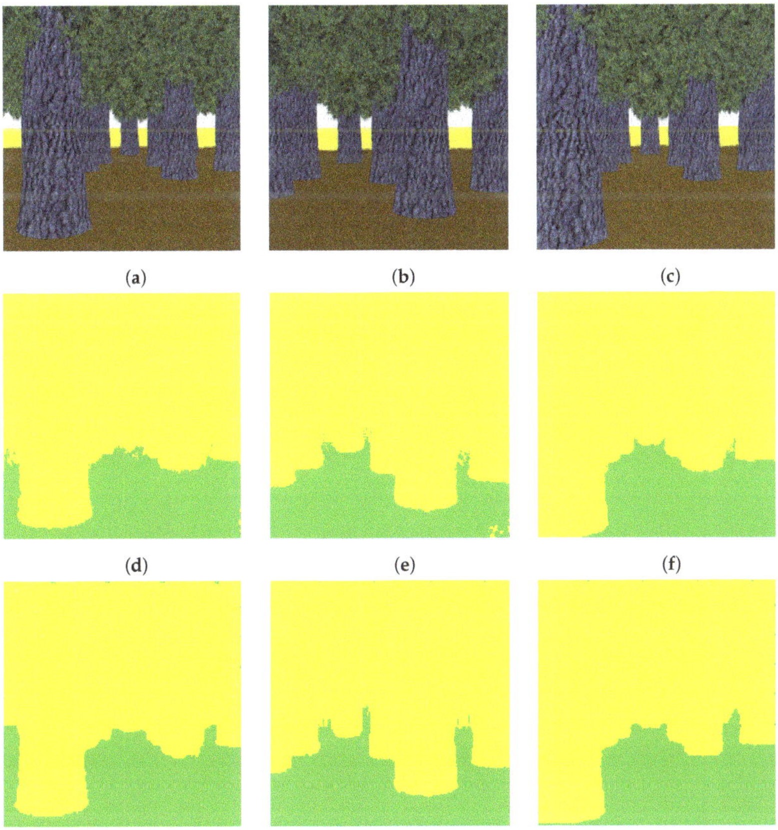

(a) (b) (c) (d) (e) (f) (g) (h) (i)

Figure 7. Segmentation of the two-class synthetic dataset (**a–c**): test images, (**d–f**): corresponding segmented images using DeconvNet, (**g–i**): corresponding segmented images using the proposed light-weight network. Note that the color code for classes is: yellow: tree, green: ground. Best viewed in color.

4.3.2. The Four-Class High-Definition Dataset

In this experiment, we trained our light-weight network with the pre-trained DeconvNet weights using the four-class synthetic dataset. This dataset is more complex than the two-class synthetic dataset in terms of the number of classes and their structure. The four classes in this dataset include ground, vegetation, tree, and sky. The vegetation includes smaller grass and/or bush like structures and contains variations such as flowers or no flower within it. After training and testing our proposed light-weight network with this dataset using the same learning rate setup as in the two-class dataset, we obtained 75.71 percent overall test accuracy. Some results of segmentation of the synthetic dataset are shown in Figure 8.

As in the previous experiment, the model trained with the four-class high-definition dataset is fine-tuned using the Freiburg Forest dataset. Only the outermost layer of the light-weight network was learned from scratch with a learning rate of 0.01 and all the other layers with 0.0001. We obtained the improved segmentation performance when compared with the results that did not use the synthetic dataset as well as with that of the two-class synthetic dataset. This improvement is obvious as the four-class high-definition dataset considers more realistic properties of the real-world environment in terms of number of classes and intra-class variability.

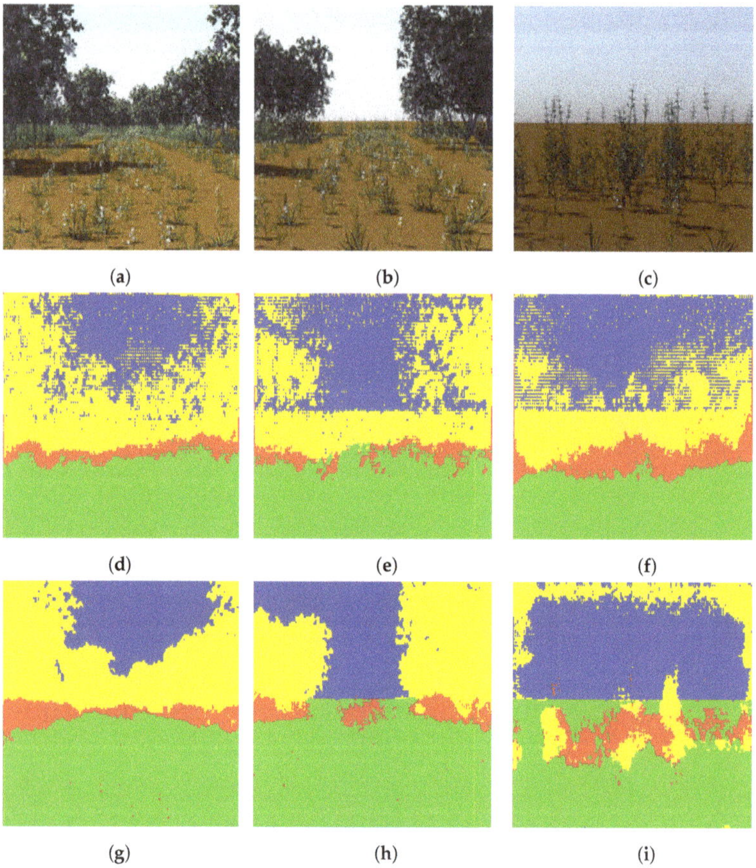

Figure 8. Segmentation of the four-class high-definition dataset (**a**–**c**): test images, (**d**–**f**): segmented images using DeconvNet, (**g**–**i**): segmented images using proposed light-weight network. Note that the color code for classes is: green: ground, red: vegetation, yellow: tree, blue: sky. Best viewed in color.

4.3.3. The Four-class random synthetic dataset

In this experiment, we trained our proposed light-weight network with the pre-trained DeconvNet weights using the four-class random synthetic dataset. We used the same learning rates as in the experiment with the two-class synthetic dataset. As specified earlier, this dataset is complex in the sense that it considers different factors to make it more realistic. Some factors considered are different time of the day, different terrain surface, varying tree structure, etc. As in the dataset used in previous experiment, it also contains four classes including ground, vegetation, tree, and sky. After training and testing our proposed light-weight network with this dataset, we obtained 91 percent overall accuracy. Some of the results of the segmentation of the synthetic dataset are shown in Figure 9.

Again, as with the four-class high-definition dataset, the model trained with the four-class random synthetic dataset is fine-tuned using the real-world Freiburg dataset. In this part, only the outermost layer of the light-weight network was learned from scratch with a learning rate of 0.01 and all of the other layers with 0.0001. As shown in Table 4, the performance of the light-weight network for transfer learning with this dataset decreased somewhat compared to the previous three experiments.

As stated above, consideration of the real-world properties for the forest environment is increased in this dataset. However, the reduction in the overall accuracy could be due to increased variation among the dataset that caused the network to learn the features that are less correlated to the target domain. This phenomenon is sometimes called negative transfer. In Figure 10, we show the comparative results for all the experiments.

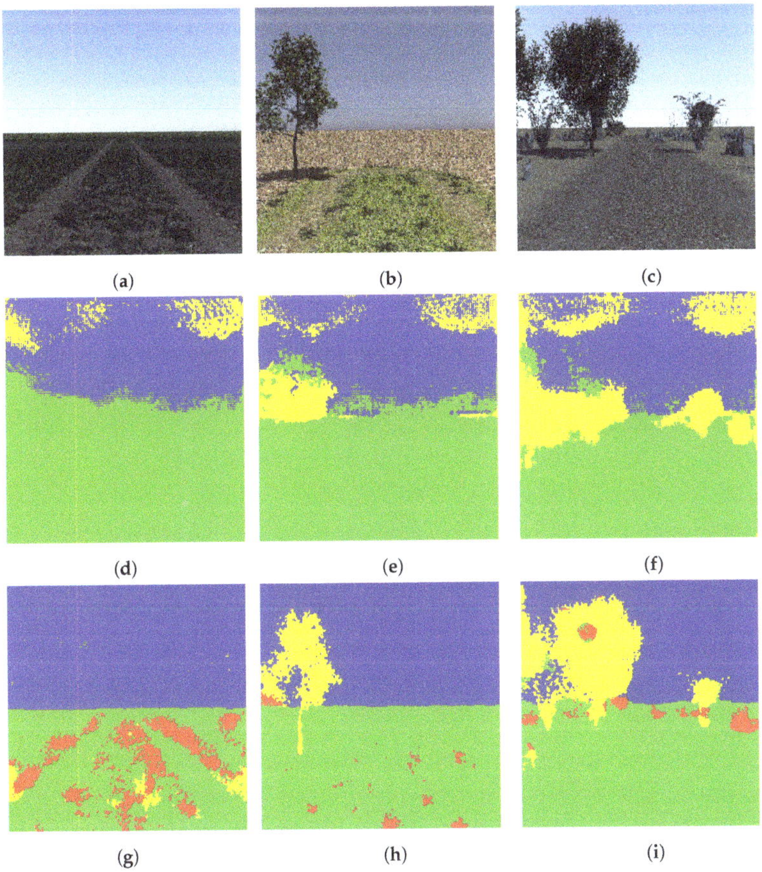

Figure 9. Segmentation of the four-class random dataset (a–c): test images, (d–f): segmented images using DeconvNet, (g–i): segmented images using proposed light-weight network. Note that the color code for classes is: green: ground, red: vegetation, yellow: tree, blue: sky. Best viewed in color.

5. Result Analysis and Discussion

Table 4 shows the comparative results of the proposed method including the baseline [16] method in terms of overall accuracy. We can analyze these results in terms of two aspects: network and TL method. Our proposed light-weight network gives much better results compared to the DeconvNet for all the four experiments. Most surprisingly, the results obtained are much better after stripping down the network to a half of its original size. This result favors the requirement of autonomous driving which needs higher accuracy with reduced latency. On the other hand, if we analyze the table in terms of TL method, we can see mixed results. For the TL with DeconvNet, the use of the synthetic dataset as intermediate domain led to a negative performance. Whereas, with our proposed light-weight network, we achieved an increased performance after using the four-class high-definition datasets compared to that which did not use the synthetic dataset. For both the datasets, the two-class synthetic and the four-class random synthetic, the performance decreased slightly. The two-class synthetic dataset is a simpler dataset which does not take into account the real-world environmental effects in terms of both the number of classes and their properties. This dataset just increased the volume with no helpful information learned before moving into the target domain causing negative transfer performance. On the other hand, the random dataset includes images from different environments. It includes the data from three different environments: an American Southeast forest ecosystem, an American Southeast meadow ecosystem, and an American Southwest desert ecosystem with their various lighting conditions. These different environments caused a high level of randomness and a lower correlation to the target domain; this dataset also added no helpful knowledge while doing the transfer learning. However, it also caused the negative transfer. The four-class high-definition dataset gave the positive TL performance with the accuracy of 94.59% on the Freiburg test set. Different from the two other datasets, this dataset has higher correlation with the target domain. Additionally, the huge randomness caused by the various ecosystems in the four-class random dataset is not available in the four-class high-definition dataset. The forest and ground structure have comparatively more similarity with that of the target domain which causes the improved performance while training with the Freiburg dataset.

Table 4. Quantitative results produced by DeconvNet and the proposed network for various TL experiments. Shading indicates the improvement of one method over another.

Data	Method	DeconvNet	Light-Weight
Freiburg	W/O synthetic data	73.65(%)	94.43(%)
	After using two-class synthetic	66.62(%)	94.06(%)
	After using four-class high-definition	68.7(%)	94.59(%)
	After using four-class random synthetic	68.14(%)	93.89(%)

We show the confusion matrices for each experiment performed with the proposed light-weight network in Table 5. Each entry is the percentage measurement of either the correctly or falsely classified number of pixels in all test images. We can see the obstacle class having the lowest accuracy and the sky class having the highest accuracy in each TL experiment. The cause for the low accuracy regarding obstacles is that the pixels belonging to this class are very limited in the training datasets compared to the other classes. In addition, the obstacle in the training images have less structural uniformity. This results in the network learning less about the obstacle class causing a biased prediction in favor of classes having a higher number of pixels.

Figure 10. Examples to show that the light-weight network produces better results than DeconvNet for each of the experiments. Note that each pair of rows (**a**,**b**) represents the results produced by DeconvNet and the proposed light-weight network, respectively. Best viewed in color.

Table 5. Confusion matrices of the test results produced by the proposed network for different TL experiments on the Freiburg Forest dataset. Note that each entry is an overall percentage. (Top left: without using synthetic dataset, top right: using two-class synthetic dataset, bottom left: using four-class high-definition synthetic dataset, bottom right: using four-class random synthetic dataset).

Top left (True class across; Predicted class down):

Class	Obstacle	Grass	Road	Tree	Sky
Obstacle	60.84	0.11	0.00	0.11	0.03
Grass	12.99	90.83	13.60	3.18	0.09
Road	5.58	3.11	83.09	0.98	0.17
Tree	19.28	5.62	2.44	93.58	4.62
Sky	1.28	0.30	0.84	2.12	95.07

Top right (True class across; Predicted class down):

Class	Obstacle	Grass	Road	Tree	Sky
Obstacle	55.69	0.10	0.01	0.09	0.04
Grass	14.92	91.07	12.89	3.33	0.15
Road	4.76	3.01	84.28	1.02	0.23
Tree	23.06	5.35	2.23	93.32	4.31
Sky	1.54	0.43	0.577	2.21	95.23

Bottom left (True class across; Predicted class down):

Class	Obstacle	Grass	Road	Tree	Sky
Obstacle	59.43	0.10	0.01	0.10	0.03
Grass	16.43	89.89	11.31	2.99	0.05
Road	4.85	3.47	86.72	1.07	0.21
Tree	18.16	6.17	1.43	93.41	4.50
Sky	1.10	0.35	0.50	2.40	95.18

Bottom right (True class across; Predicted class down):

Class	Obstacle	Grass	Road	Tree	Sky
Obstacle	59.05	0.11	0.01	0.11	0.03
Grass	15.22	90.42	11.75	3.21	0.25
Road	4.16	3.38	85.57	1.03	0.36
Tree	19.20	5.81	2.24	93.36	4.49
Sky	2.34	0.25	0.39	2.27	94.85

6. Conclusions and Future Work

In this paper, we explored the transfer learning from the perspective of network size and training techniques with and without the use of synthetic data. We conclude that it is important to find out the size of the network that performs best for the target domain rather than using the original architecture as a whole. In doing so, we proposed a new light-weight network; a network well suited for use in autonomous driving applications due to its low latency, which is initialized with the pre-trained DeconvNet weights from the corresponding layers. Furthermore, we explored the effects of using different synthetic datasets as the intermediate domain. As TL techniques are used for these domains where training datasets are insufficiently available, generating and using synthetic datasets is a good approach, which can help boost performance. While doing so, considering the target domain characteristics as much as possible when generating the synthetic dataset will increase the TL performance. We also conclude that an oversimple and/or too random dataset, as was the case for the two-class synthetic and the four-class random synthetic dataset herein, can cause negative transfer.

The intermediate layers and their weights of DeconvNet are absent in the proposed light-weight network. In order to understand the relationship among the layers and correspondence between layers from source to target network, a detailed theoretical study is needed focusing the semantic meaning, i.e., mapping between features across layers of the target and source domain. While there exists some work to understand what the features means in different layers—e.g., initial layers extract lower level features—for classification task, there is no such study for encoder–decoder architecture targeted for segmentation task. In the future, we plan to study the detailed theoretical underlying regarding those aspects for encoder–decoder-based networks. This would also shed light on how the proposed way of transfer learning leads to better adaptability and performance. Furthermore, we plan to incorporate our road segmentation model into the real off-road autonomous vehicle and study the creation of occupancy grid with the segmentation results to support decisions of path planning.

Author Contributions: S.S. provided conceptualization, implementation and writing; J.E.B. and B.T. gave a detailed revision and suggestions about experimental design; D.W.C., M.D. and M.A.I. provided the data and important suggestions for writing.

Funding: Portions of this work were performed in association with the Halo Project, a research and development project with the Center for Advanced Vehicular Systems (CAVS) at Mississippi State University. Learn more at http://www.cavs.msstate.edu.

Conflicts of Interest: The authors declare no conflict of interest.

References

1. Krizhevsky, A.; Sutskever, I.; Hinton, G.E. Imagenet classification with deep convolutional neural networks. In Proceedings of the NIPS'12 the 25th International Conference on Neural Information Processing Systems, Lake Tahoe, Nevada, 3–6 December 2012; pp. 1097–1105.
2. Simonyan, K.; Zisserman, A. Very deep convolutional networks for large-scale image recognition. *arXiv* **2014**, arXiv:1409.1556.
3. Szegedy, C.; Liu, W.; Jia, Y.; Sermanet, P.; Reed, S.; Anguelov, D.; Erhan, D.; Vanhoucke, V.; Rabinovich, A. Going deeper with convolutions. In Proceedings of the 2015 IEEE Conference on Computer Vision and Pattern Recognition (CVPR), Boston, MA, USA, 7–12 June 2015; pp. 1–9.
4. Lin, K.; Gong, L.; Huang, Y.; Liu, C.; Pan, J. Deep learning-based segmentation and quantification of cucumber Powdery Mildew using convolutional neural network. *Front. Plant Sci.* **2019**, *10*, 155. [CrossRef] [PubMed]
5. Bargoti, S.; Underwood, J.P. Image segmentation for fruit detection and yield estimation in apple orchards. *J. Field Robot.* **2017**, *34*, 1039–1060. [CrossRef]
6. Ciresan, D.; Giusti, A.; Gambardella, L.M.; Schmidhuber, J. Deep neural networks segment neuronal membranes in electron microscopy images. In Proceedings of the NIPS'12 the 25th International Conference on Neural Information Processing Systems, Lake Tahoe, Nevada, 3–6 December 2012; pp. 2843–2851.
7. Kolařík, M.; Burget, R.; Uher, V.; Říha, K.; Dutta, M.K. Optimized High Resolution 3D Dense-U-Net Network for Brain and Spine Segmentation. *Appl. Sci.* **2019**, *9*, 404. [CrossRef]
8. Liu, Y.; Ren, Q.; Geng, J.; Ding, M.; Li, J. Efficient Patch-Wise Semantic Segmentation for Large-Scale Remote Sensing Images. *Sensors* **2018**, *18*, 3232. [CrossRef] [PubMed]
9. Pan, X.; Gao, L.; Zhang, B.; Yang, F.; Liao, W. High-Resolution Aerial Imagery Semantic Labeling with Dense Pyramid Network. *Sensors* **2018**, *18*, 3774. [CrossRef] [PubMed]
10. Papadomanolaki, M.; Vakalopoulou, M.; Karantzalos, K. A Novel Object-Based Deep Learning Framework for Semantic Segmentation of Very High-Resolution Remote Sensing Data: Comparison with Convolutional and Fully Convolutional Networks. *Remote Sens.* **2019**, *11*, 684. [CrossRef]
11. Farabet, C.; Couprie, C.; Najman, L.; LeCun, Y. Learning hierarchical features for scene labeling. *IEEE Trans. Pattern Anal. Mach. Intell.* **2013**, *35*, 1915–1929. [CrossRef] [PubMed]
12. Gupta, S.; Girshick, R.; Arbeláez, P.; Malik, J. Learning rich features from RGB-D images for object detection and segmentation. In *European Conference on Computer Vision*; Springer: Cham, Switzerland, 2014; pp. 345–360.
13. Hariharan, B.; Arbeláez, P.; Girshick, R.; Malik, J. Simultaneous detection and segmentation. In *European Conference on Computer Vision*; Springer: Berlin/Heidelberg, Germany, 2014; pp. 297–312.
14. Long, J.; Shelhamer, E.; Darrell, T. Fully convolutional networks for semantic segmentation. In Proceedings of the 2015 IEEE Conference on Computer Vision and Pattern Recognition (CVPR), Boston, MA, USA, 7–12 June 2015; pp. 3431–3440.
15. Badrinarayanan, V.; Kendall, A.; Cipolla, R. Segnet: A deep convolutional encoder-decoder architecture for image segmentation. *arXiv* **2015**, arXiv:1511.00561.
16. Noh, H.; Hong, S.; Han, B. Learning deconvolution network for semantic segmentation. In Proceedings of the 2015 IEEE International Conference on Computer Vision (ICCV), Santiago, Chile, 7–13 December 2015; pp. 1520–1528.
17. Yu, F.; Koltun, V. Multi-scale context aggregation by dilated convolutions. *arXiv* **2015**, arXiv:1511.07122.
18. Chen, L.C.; Papandreou, G.; Kokkinos, I.; Murphy, K.; Yuille, A.L. Semantic image segmentation with deep convolutional nets and fully connected crfs. *arXiv* **2014**, arXiv:1412.7062.

19. Chen, L.C.; Papandreou, G.; Kokkinos, I.; Murphy, K.; Yuille, A.L. Deeplab: Semantic image segmentation with deep convolutional nets, atrous convolution, and fully connected crfs. *IEEE Trans. Pattern Anal. Mach. Intell.* **2018**, *40*, 834–848. [CrossRef] [PubMed]
20. Zhao, H.; Shi, J.; Qi, X.; Wang, X.; Jia, J. Pyramid scene parsing network. In Proceedings of the IEEE Conference on Computer Vision and Pattern Recognition, Honolulu, HI, USA, 21–26 July 2017; pp. 2881–2890.
21. Long, M.; Cao, Y.; Wang, J.; Jordan, M.I. Learning transferable features with deep adaptation networks. *arXiv* **2015**, arXiv:1502.02791.
22. Yosinski, J.; Clune, J.; Bengio, Y.; Lipson, H. How transferable are features in deep neural networks? In *Advances in Neural Information Processing Systems 27*; Ghahramani, Z., Welling, M., Cortes, C., Lawrence, N.D., Weinberger, K.Q., Eds.; Curran Associates, Inc.: New York, NY, USA, 2014; pp. 3320–3328.
23. Van Opbroek, A.; Ikram, M.A.; Vernooij, M.W.; de Bruijne, M. Supervised image segmentation across scanner protocols: A transfer learning approach. In Proceedings of the International Workshop on Machine Learning in Medical Imaging, Nice, France, 1 October 2012; pp. 160–167.
24. Van Opbroek, A.; Ikram, M.A.; Vernooij, M.W.; De Bruijne, M. Transfer learning improves supervised image segmentation across imaging protocols. *IEEE Trans. Med. Imaging* **2015**, *34*, 1018–1030. [CrossRef] [PubMed]
25. Girshick, R. Fast r-cnn. In Proceedings of the IEEE International Conference on Computer Vision, Santiago, Chile, 13–16 December 2015; pp. 1440–1448. [CrossRef]
26. Wei, L.; Runge, L.; Xiaolei, L. Traffic sign detection and recognition via transfer learning. In Proceedings of the 2018 Chinese Control And Decision Conference (CCDC), Shenyang, China, 9–11 June 2018; pp. 5884–5887.
27. Ying, W.; Zhang, Y.; Huang, J.; Yang, Q. Transfer learning via learning to transfer. In Proceedings of the 35th International Conference on Machine Learning, Stockholm, Sweden, 10–15 July 2018; pp. 5072–5081.
28. Xiao, H.; Wei, Y.; Liu, Y.; Zhang, M.; Feng, J. Transferable Semi-supervised Semantic Segmentation. *arXiv* **2017**, arXiv:1711.06828.
29. Hong, S.; Oh, J.; Lee, H.; Han, B. Learning transferrable knowledge for semantic segmentation with deep convolutional neural network. In Proceedings of the IEEE Conference on Computer Vision and Pattern Recognition, Las Vegas, NV, USA, 27–30 June 2016; pp. 3204–3212.
30. Nigam, I.; Huang, C.; Ramanan, D. Ensemble Knowledge Transfer for Semantic Segmentation. In Proceedings of the 2018 IEEE Winter Conference on Applications of Computer Vision (WACV), Lake Tahoe, NV, USA, 12–15 March 2018; pp. 1499–1508.
31. Everingham, M.; Van Gool, L.; Williams, C.K.; Winn, J.; Zisserman, A. The pascal visual object classes (voc) challenge. *Int. J. Comput. Vis.* **2010**, *88*, 303–338. [CrossRef]
32. Bengio, Y. Deep learning of representations for unsupervised and transfer learning. In Proceedings of the UTLW'11 the 2011 International Conference on Unsupervised and Transfer Learning Workshop, Washington, DC, USA, 2 July 2011; pp. 17–36.
33. Baldi, P. Autoencoders, unsupervised learning, and deep architectures. In Proceedings of the ICML Workshop on Unsupervised and Transfer Learning, Edinburgh, Scotland, 27 June 2012; pp. 37–49.
34. Goodfellow, I.; Bengio, Y.; Courville, A.; Bengio, Y. *Deep Learning*; MIT Press: Cambridge, MI, USA, 2016; Volume 1.
35. Pan, S.J.; Yang, Q. A survey on transfer learning. *IEEE Trans. Knowl. Data Eng.* **2010**, *22*, 1345–1359. [CrossRef]
36. Maturana, D.; Chou, P.W.; Uenoyama, M.; Scherer, S. Real-time semantic mapping for autonomous off-road navigation. In *Field and Service Robotics*; Springer: Berlin/Heidelberg, Germany, 2018; pp. 335–350.
37. Adhikari, S.P.; Yang, C.; Slot, K.; Kim, H. Accurate Natural Trail Detection Using a Combination of a Deep Neural Network and Dynamic Programming. *Sensors* **2018**, *18*, 178. [CrossRef] [PubMed]
38. Holder, C.J.; Breckon, T.P.; Wei, X. From on-road to off: transfer learning within a deep convolutional neural network for segmentation and classification of off-road scenes. In Proceedings of the European Conference on Computer Vision, Amsterdam, The Netherlands, 8–16 October 2016; pp. 149–162.
39. He, K.; Sun, J. Convolutional neural networks at constrained time cost. In Proceedings of the IEEE Conference on Computer Vision and Pattern Recognition, Boston, MA, USA, 7–12 June 2015; pp. 5353–5360.
40. Jia, Y.; Shelhamer, E.; Donahue, J.; Karayev, S.; Long, J.; Girshick, R.; Guadarrama, S.; Darrell, T. Caffe: Convolutional architecture for fast feature embedding. In Proceedings of the 22nd ACM international conference on Multimedia, Orlando, FL, USA, 3–7 November 2014; pp. 675–678.

41. Valada, A.; Oliveira, G.; Brox, T.; Burgard, W. Deep Multispectral Semantic Scene Understanding of Forested Environments using Multimodal Fusion. In Proceedings of the 2016 International Symposium on Experimental Robotics (ISER 2016), Tokyo, Japan, 3–6 October 2016.
42. Hudson, C.R.; Goodin, C.; Doude, M.; Carruth, D.W. Analysis of Dual LIDAR Placement for Off-Road Autonomy Using MAVS. In Proceedings of the 2018 World Symposium on Digital Intelligence for Systems and Machines (DISA), Kosice, Slovakia, 23–25 August 2018; pp. 137–142.
43. Goodin, C.; Sharma, S.; Doude, M.; Carruth, D.; Dabbiru, L.; Hudson, C. *Training of Neural Networks with Automated Labeling of Simulated Sensor Data*; SAE Technical Paper; Society of Automotive Engineers: Troy, MI, USA, April, 2019.

sensors

MDPI

Article

A Strain-Based Method to Estimate Tire Parameters for Intelligent Tires under Complex Maneuvering Operations

Mª Fernanda Mendoza-Petit [1], Daniel Garcia-Pozuelo [1], Vicente Diaz [1] and Oluremi Olatunbosun [2,*]

1 Mechanical Engineering Department, Universidad Carlos III de Madrid, Avd. De la Universidad, 28911 Madrid, Spain

2 School of Mechanical Engineering, University of Birmingham, Edgbaston B15 2TT, UK

* Correspondence: mmendoza@ing.uc3m.es; Tel.: +34-91-624-9912

Received: 30 April 2019; Accepted: 2 July 2019; Published: 5 July 2019

Abstract: The possibility of using tires as active sensors opens the door to a huge number of different ways to accomplish this goal. In this case, based on a tire equipped with strain sensors, also known as an Intelligent Tire, relevant vehicle dynamics information can be provided. The purpose of this research is to improve the strain-based methodology for Intelligent Tires to estimate all tire forces, based only on deformations measured in the contact patch. Firstly, through an indoor test rig data, an algorithm has been developed to pick out the relevant features of strain data and correlate them with tire parameters. This information of the tire contact patch is then transmitted to a fuzzy logic system to estimate the tire parameters. To evaluate the reliability of the proposed estimator, the well-known simulation software CarSim has been used to back up the estimation results. The software CarSim has been used to provide the vehicle parameters in complex maneuvers. Finally, the estimations have been checked with the simulation results. This approach has enabled the behaviour of the intelligent tire to be tested for different maneuvers and velocities, providing key information about the tire parameters directly from the only contact that exists between the vehicle and the road.

Keywords: tire-road forces estimation; slip angle estimation; gauge sensors; fuzzy logic system; load transfer estimation; simulation results; normalization; lateral force empirical model

1. Introduction

Nowadays, the tire is largely regarded as a passive element in the automotive field, even though many studies have pointed out the importance of the tire in the dynamic behaviour of the vehicle. In spite of the interesting developments in intelligent tires along the last years, one of the greatest achievements of the intelligent tire systems commercialized to date is the Tire Pressure Monitoring System (TPMS). This is a proof of the difficulty for a more ambitious intelligent tire concept to meet all the requirements in commercial tires. Overall, the number of sensors in a vehicle continues to increase. Many of these sensors provide vital information such as longitudinal and lateral acceleration, yaw rate, engine torque, etc. Nonetheless, very few sensors are able to provide accurate information related to tire-road interaction parameters.

Safety is the foremost reason to develop an intelligent tire as an active sensor able to provide useful information which is otherwise hard to measure, e.g., load transfer, tangential forces, tire conditions, road conditions or friction coefficients. Such information would enhance the functionality of different control systems such as Anti-Lock Braking Systems (ABS), Traction Control Systems (TCS), Electronic Stability Control (ESC), and Suspension Control Systems (SCS). Matsuzaki and Todoroki [1] suggested the possibility of developing an optimized braking control and road condition warning system using a strain-based system, where the road condition warning would be actuated if the recorded friction coefficient at certain slip ratio is lower than a safe reference value. Using Intelligent Tires to complement the current control systems may help prevent losing control of the vehicle in adverse road conditions.

There is a large number of interesting published works about tire parameter estimation. An example is, the neural model based on recursive lazy learning to predict the tire characteristics. However, adjustment of the model is required each time new data is presented [2]. Doumiati et al. [3,4], implemented virtual sensors to estimate the lateral tire force and sideslip angle, from a simplified four-wheel vehicle model. Despite the fact the experimental results show the potential of the estimation method, it needs to measure the data from the on-board vehicle sensors to estimate the tire parameters. According to Lee et al. [5], observers estimation provides, under the severe driving condition accompanied with large combined slips and abrupt wheel load changes, unreliable values and also, considerable uncertainties can be accumulated during the estimation process, which can be due to the use of a simple vehicle model. One of the most used is the Magic Formula [6–8], nonetheless, according to Rajamani [9], the results from analytical elastic foundation models (the brush models) can match the data very well for cases of pure lateral or pure longitudinal force generation. However, the analytical models do not always lead to quantitatively accurate results. Differences from experimental data are observed, especially at large slip and at combined slip.

Other instrumented equipment, e.g. dynamometric hubs and dynamometric plates, allow measuring tire forces, but they are usually expensive and not easy to use in conventional vehicles.

Due to the influence of tire forces and the slip angle on the dynamic control of the vehicle, it is important to estimate reliable results. Following different approaches and transducers many prototypes of instrumented tires have been produced to develop the concept, complementing the current vehicle control systems. The results of these studies in terms of the achieved reliability are diverse, but the main goal of the sensors used into the intelligent tires is to characterize the contact patch in all cases. For this reason, these sensors are usually located as near as possible to the contact patch and present advantages and disadvantages as it is explained in next analysis.

Accelerometers based on micromechanical systems (MEMS) are widely used. This type of sensor assures signal linearity and stability over time and insensitivity to temperature change, however, it is very sensitive to the noise (high frequency vibration) generated when the tire rolls on the road. Hence, it is difficult to extract the characteristics of interest without advanced signal processing [5]. On the other hand, the high vibration levels inside a tire have the potential to generate electrical power using vibration based energy harvesting techniques. As result, it has been proposed a piezoelectric energy harvesting system with energy storage as intelligent tire sensor based on piezoelectric transducers (accelerometers). The onboard vibration energy harvesting unit has been adapted to the tire vibration spectra and the superimposed acceleration signal [10,11]. To optimize the harvester performance with changing dominant tire vibration frequencies, there were used an artificial neural network (ANN) trained at different tire operating conditions to ensure broadband operation of the harvester [12].

Other types of sensors used in the intelligent tire field are based on the deformation of the tire as results of the contact surface interaction are introduced. Xiong and Tuononnen [13], explained that the deformation of parts of a tire is the direct result of tire–road interactions. In this study, a flexible ring tire model and a laser sensor were used to analyze the in-plane deformations of the tire carcass. It was found that the radial deformation of the tire carcass provides information on both the longitudinal and vertical forces acting on the tire, being mainly attributed to the coupled effect between the in-plane tire deformations caused by the inextensibility of the tire carcass in a modern radial tire.

Roveri et al. [14] proposed a device consisting of Fiber Bragg Grating (FBG) sensors and a light spectrum analyzer (optical strain measurements in rolling tires). This device allows the acquisition of the tire strain along an array of measurement points. Based on experimental and theoretical methodologies, it can set the pillars for a new system for real-time identification of the tire stress during rolling and the residual grip estimation.

Different studies have been conducted in intelligent tires based on strain gauges. One of the first issues with this sort of sensors is their location. Cheli et al [15] studied the influence of the strain gauge position on the wheel rim on a measurement system. It was concluded that the results were apparently independent of the position of the strain gauge. Some other researches have successfully developed intelligent tires based on strain gauge measurements, achieving advanced systems to estimate tire characteristics [16–20]. Matsuzaki and Todoroki [1] investigated the relationship between variation in strain on the inner surface of pneumatic tires and tire mechanical parameters by carrying out finite element analysis and simulating the strain sensor signal when a tire rotates. In general, it has been found that the strain sensors measurements can be directly related to tire deflection and operational conditions [5] with a little margin for error under dynamic conditions [19,20]. In most cases, they are attached to the inner surface of the tire to contribute to a lower probability of damage.

Other studies with strain gauges as sensors have been fitted for real-time conditions. This is the case of Garcia-Pozuelo et al. [21,22]. In this document a real-time physical model suitable describe the dynamics of intelligent tires based on measurements of strains grounded on a flexible ring on a viscoelastic foundation. It can reproduce the tire longitudinal dynamics for both concentrated and distributed forces by a discrete approach. The solution of the model dynamics has been obtained in the closed form and the model parameters have been identified from experimental data. This model could be used for predicting the dynamical behavior of both the strain-based intelligent tire sensor and the laser based one. It also may be used in the design of real-time observers for tire condition based on tire strain measurements.

Yunta et al. [23] prove the camber angle influence on the strain signal. In an indoor tire test rig experimental tests were carried out to measure the tire tread deformation by means of strain gauges under different working conditions, such as vertical load, slip angle, camber angle. This study shows in a very clear way the influence of the strain gauges location on the results.

These studies evidence that the strain sensors meet the requirements to estimate the tire parameters regarding the working condition and tire properties; as well as the reliability, repeatability and easily installed [16,17,19,20] without adding weight to the measured surface (which can represent adding stress to the surface). The achievement of an intelligent tire system is a concept that is being currently developed and presents many future ways and possibilities. In this manuscript some improvements over the strain-based intelligent tire are proposed, providing new information for driving behavior and warnings about slippery road by measuring the potential friction and the road condition where the tires' rolling on.

In this study experimental data has been used to develop an estimator of the tire-road contact surface parameters based on strain data measurement. Firstly, the experimental data has been analyzed to derive relationships between the strain gauge measurements and the tire working conditions. To accomplish this analysis, herein has developed an algorithm in MATLAB able to make a selection of the key points of the strain measurements. The implementation of this algorithm allows a more accurate and standardized data selection reducing the margin of error in the relationships between the strain gauge and tire working conditions. Following on from the relationships derived from the analysis, a fuzzy logic system has been implemented to estimate the following tire parameters; slip angle, vertical and tangential forces in the contact patch. These results complement the conclusions of previous studies [16–23] to find the capacity of strain sensors to measure all the wheel forces.

As validation process, herein has been tested the outputs of the fuzzy logic estimator within complex maneuvers. The simulation software CarSim has been used to provide the data needed to accomplish this goal. Also, it has been used a semi-empirical model (Pacejka's model) based on experimental data of the lateral force to compare the output of the fuzzy logic system at the same maneuvers.

One of the biggest features of this work has been to show the operational behaviour of intelligent tires as active sensors forming part of a vehicle. Further, it illustrates the information that may be provided by these sensors to ensure better dynamic performance of vehicles.

2. Strain-Based Intelligent Tire Systems

This section is about the process of strain data acquisition and collection of principal features. It will give the reader an overview of the experimental steps needed to develop this research, as well as, the data processing realized to pick out the main features.

2.1. Experimental Strain-Based Intelligent Tire System

Many research studies have chosen the strain sensor as the active element to provide information about the forces at the tire-road contact patch. Indeed, as explained in previous section, the tires as sensors can provide a strong relation with driving condition parameters, thereupon, through the deformation, it is possible to estimate tire parameters based on the strain [16–24]. This study is grounded on collected data that had been carried out in an indoor tri-axial tire test rig with strain-based intelligent tire system in the University of Birmingham Vehicle Dynamics Laboratory (see Figure 1). The indoor tire test rig allows variation in the speed, vertical load and slip angle. It also allows to measure the values of the longitudinal force for traction and braking rolling test and the lateral force for tire steady-state cornering rolling test [16–24].

(a) (b)

Figure 1. (**a**) Indoor tire test rig; (**b**) Acquisition system SoMat2000®.

The drum's curved surface has a large diameter (2.44 m) so, its curvature has an insignificant effect on the results [20,25]. The indoor tire test rig is equipped with a drum speed controller and a signal conditioner to control the displacement and load.

The intelligent tire prototype system is comprised of a test tire, the strain sensors, a SoMat2000®data acquisition system and a SoMat 2000 field computer. This data acquisition system is particularly suited for portable data collection.

The selected tire used was a DUNLOP SP SPORT 175/505 R13 (tubeless) slick radial tire. The test tire has been chosen taking into account the tire test rig limitations and the possibility to be used in an "FSAE" prototype as test car. Further stages of this study will be done in this "FSAE" prototype in order to extend the conclusions in real conditions, out of a laboratory tire test rig.

The experimental operational range of parameters implemented are defined according to the tire test rig limitation and the tested tire. The experimental operational range of parameters implemented are as follows:

- Tire inflation pressure: 0.8 bar–1.4 bar, step size: 0.2 bars.
- Tire preload: 250 N–1000 N, step size: 250 N.
- Tire speed: 10 km/h–50 km/h, step size: 10 km/h.
- Tire slip angle: 0°–10°, step size: 2°.
- Tire camber angle: 0° (due to the limitation in the test rig).

Three multiaxial strain gages (2 mm length—120 Ω gauge resistance) were set up, placed symmetrically at different points on the inner liner of the tire tread (see Figure 2). Two of them were placed in the same cross-section and the third one separated by 123.75 degrees of angular rotation. Figure 2 shows the strain gauges' location scheme. The distance "d" and "l" are about 0.040 m and 0.515 m, respectively.

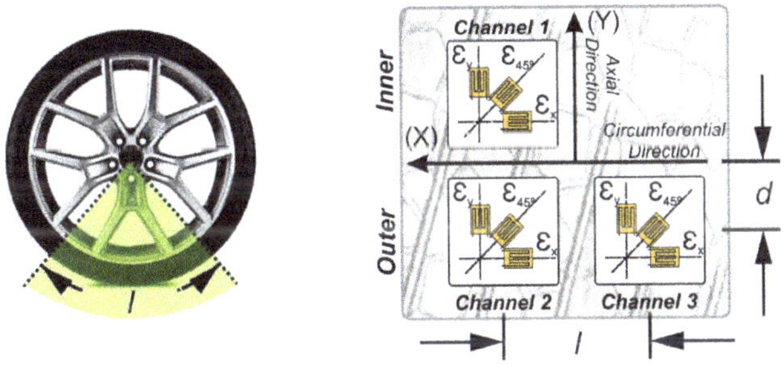

Figure 2. Multiaxial strain gauge disposition.

To measure the experimental data three channels were turned on at a sampling frequency of 1000 Hz, two of them for the axial strain measurement ($\mu\varepsilon_{y1}$, $\mu\varepsilon_{y2}$—channels 1 and channel 3, respectively) located symmetrically with respect to tread center line and the other one to measure the circumferential strain ($\mu\varepsilon_{x}$— channel 2). The channel 2 and 3, are located in the same cross section to measure the strains in the circumferential and axial directions when the tire is deflected by the influence of the lateral force. Additionally, this location provides interesting information due to the nonsymmetrical lateral behavior of the tire and certain external conditions such as camber angle [23].

The experimental measurements were collected by a set of data carried out at steady-state conditions for each tire operational condition. The measurement system was calibrated and zero error were checked before start the measurements of the system.

2.2. *Selection of Characteristics for Measured Tire Strain*

The data collected from experimental strain-based intelligent tire system is organized by the sets of data according to the slip angle, vertical force and tire rolling speed. Each set is formed by a time history of the strain measured at steady state condition. This order help to detail the effect of varying the tire operational conditions over the strain measurement.

In Park et al. [26] it was pointed out that the distribution of contact stresses suffers variabilities due to changes of vertical load and inflation pressure. Under a constant tire load, the distribution of tire contact stresses are changed from higher contact stresses at the edge to higher stresses at the middle of the contact patch as tire inflation pressure increases. At the constant tire inflation pressure, the tire contact stresses at the edge area become higher as tire load increases. This information is of interest for the tire deformations measurements because they are carried out in three different points of the contact patch (see Figure 2, the strain multiaxial gauge disposition). Hence, in this work constant pressure under load variabilities has been considered to assure that the distribution of contact stresses is equal along the contact patch. Therefore, this work has considered the tire normal inflation pressure of 0.8 bar [24].

Figure 4a illustrates the set of data measured for channel 1 at the tire operational condition of 0.8 bar, 50 km/h and preloaded at 500 N. It can be noticed the repeatability in all the time history. Hence, an objective of this section is to automate the features extraction from the tire strain time history, being needed first to define the points of interest. According to Garcia-Pozuelo et al. [16,19,20], the most influential features to estimate tire parameters (e.g., vertical load, speed, and slip angle) are the tensile strain peaks and the offsets on the curve of strain in the time history, not considering the pressure effects despite the existence of a direct relationship between the pressure and the maximum compressive strain values, due to the increment in the stiffness of the tire [20]. For this purpose, this study focuses on the study of the tensile strain peaks in axial (lateral) and circumferential (longitudinal) directions, and the offset at each channel.

In Figure 3 the most influential features (tensile strain peaks and the offsets) according to García-Pozuelo et al. [20] have been pointed out. These features are represented over the strain curves of the tire at channel 1, channel 2 and channel 3. The notation used for deformation features is as following, the axial strain offset, OS_x, the axial strain offset, OS_{y2}, the front/rear tensile axial strain peaks, $\varepsilon_{y1f}/\varepsilon_{y1r}$, the tensile circumferential strain peak, ε_x, and the front/rear tensile axial strain peaks, $\varepsilon_{y2f}/\varepsilon_{y2r}$. Finally, it is developed an algorithm capable of detecting the strain features into the strain curve at each operational condition. It automates the collection of the maximum tensile values for each channel and analyzes its variation for the different working conditions (speed, slip angle, and vertical load).

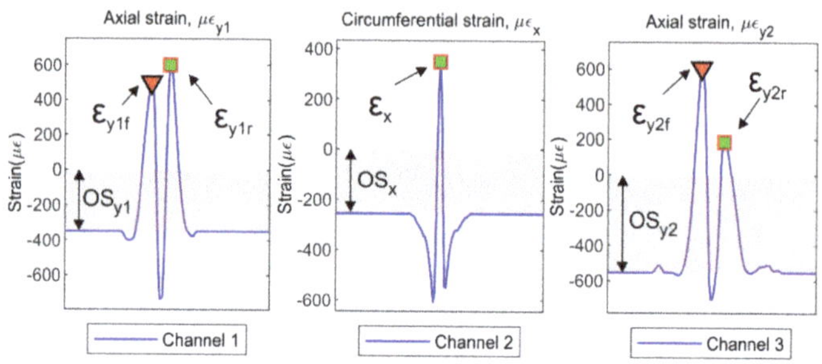

Figure 3. Features of the tire strain curve (0°, 0.8 bar, 500 N, 50 km/h).

Figure 4b illustrates the implementation of the algorithm for the tire strain time history at circumferential and axial directions. This algorithm is capable of detect the front and rear tensile peaks for the axial direction curves. Axial peaks are marked with the red triangle as front peak, and the green square as rear peak (see Figure 3). In order to analyze the collected data and achieve general conclusions the mean value for the set of representative peaks has been calculated (see Figure 4). The variation of these mean values for different tire conditions (vertical load, slip angle and rolling velocity) has been studied to propose robust estimation systems.

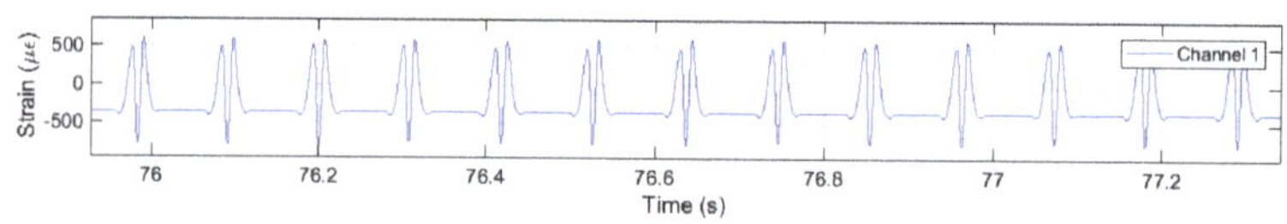

(**a**)

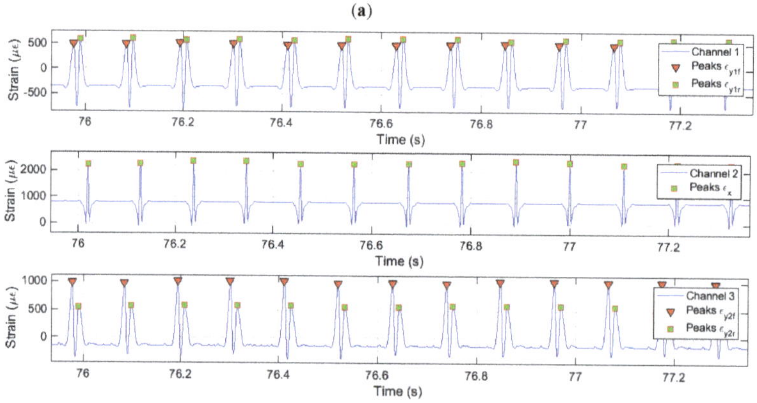

(**b**)

Figure 4. Intelligent Tire Strain time history at 0°, 0.8 bar, 500 N, 50 km/h. (**a**) Collected data of channel 1; (**b**) Detection peaks algorithm implemented.

3. Strain-Based Method

In this work, the process of developing the strain-based method to detect tire parameters for Intelligent Tires is divided into four steps. The first and second steps are related to the Intelligent Tire Data Acquisition System and the data analysis, as previously described.

In the third step, the Tire's Parameter Estimation started with the fuzzy logic computational method, where, the available knowledge about the variations of the strain features is included in the fuzzy estimator by the formulation of the rules. Hence, the surface and curve fitting over the deformation features are the inputs whereas the values of the slip angle, radial load, and the tangential forces describe the outputs. Figure 5 shows the steps taken to develop the strain-based method proposed.

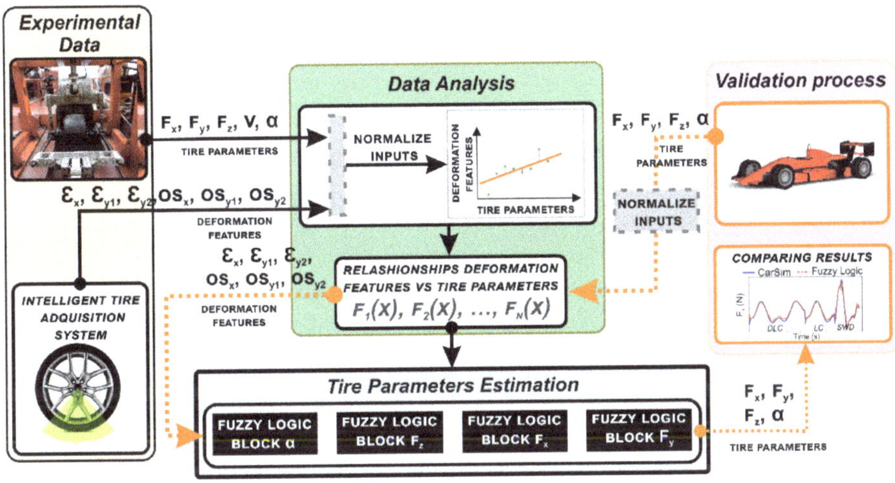

Figure 5. Working scheme used to develop the strain-based method.

Once the proposed method is capable of detecting the tire parameters' values, this work seeks to confirm the results to demonstrate the feasibility of the strain-based method. Therefore, a validation process is the fourth step (this process is indicated in Figure 5 with the dot arrows). It starts with the values of the tire parameters (lateral force, longitudinal force, vertical load and slip angle) described during a simulation maneuver.

In this case, the tire parameters were obtained from CarSim [27]. CarSim is a simulation software widely used in the automotive industry based on parametric modeling. The main advantage of using simulations software is the capacity to perform diverse types of vehicle maneuvers. The Formula 3 vehicle configuration (F3) was set up using experimental data. The maneuvers used to compare the results are Double Lane Change (DLC), Lane Change (LC), and Sine With Dwell (SWD) with speed of 30 km/h and 80 km/h on dry pavement.

The simulation parameters needs to be normalized to set up them into a range of [−1, 1]. According to the strain features relationships in the data analysis module, their inputs need to be into that bounds to then being turned them into deformation features (as it is depicted in Figure 5).

The next step is applying the Tire's Parameter module (see Figure 5), where the resultant deformations are the inputs, and the outputs are the longitudinal, lateral and vertical forces and, the slip angle. Then a parallel display of the fuzzy logic estimations and the CarSim simulations is made to compare the results.

In order to back up the graphical evidence, provided by the validation process, about the effectiveness of the fuzzy logic estimation, the errors for different maneuvers, have been computed, picking as target output the simulations. According to Vargas-Meléndez et al, and Boada et al. [28,29], the normalized error as a function of time may be calculated by Equations (1)–(3), where, y_l and $\widetilde{y_l}$, are the simulation data and the fuzzy estimation results during a time period expressed as T, and the mean value μ_l of the target output:

$$E_t = \frac{\varepsilon_t}{\sigma_t} \tag{1}$$

$$\varepsilon_t{}^2 = \int_0^T (y_l - \widetilde{y_l})^2 \, dt \tag{2}$$

$$\sigma_t{}^2 = \int_0^T (y_l - \mu_l)^2 \, dt \tag{3}$$

Furthermore, the experimental data has been used to develop an empirical model based on the so-called Magic Formula, chosen for its ability to produce characteristics that closely match measured curves for the side force [6–9]. For this case, it matches with the lateral force as a function of their respective wheel slip angle and vertical load. Afterward, this empirical model for describing lateral force behaviour has been used to compare with the proposed estimation.

3.1. Analysis of Operational Tire Parameters by Means of Experimental Tire Strains

Preliminary measurements of tire dynamic strain have been used to build relationships for developing the proposed estimation methodology. The tire strain features are regarded as the response variables and the tire working conditions as the input variables. Figures 6 and 7 show the experimental data fitted by curves and by surfaces, where the trend of the strain features for different working conditions can be observed. In Figure 6 the blue dots show the experimental strain data at every specific condition and their trends are depicted by the red lines. In Figure 7 the fitted surfaces enable us to assess the strain features with two experimental conditions simultaneously.

After identifying the pattern between the variables, it is possible to estimate a strain value at tire working conditions different from the experimental ones.

On the other hand, during the indoor experimental test the values of the lateral force, f_y, normal force, f_z, slip angle, α, rolling speed, V, and tire pressure, P, were collected, whereas, the longitudinal force value, f_x, was estimated according to Yang et al. [18]. There is established a relation between the tensile strain peaks ratio and longitudinal force values. Thus, the longitudinal force, f_x, for the experimental tests may be roughly equivalent to the results yielded by the relations claimed by Yang et al. [18,24].

As discussed above, this research is grounded on the Intelligent Tire experimental tests. It relates the strain features (OS_{y2}, ε_x, ε_{y1f}, ε_{y1r}, ε_{y2f}, ε_{y2r}, OS_x, OS_{y1}) to the tire working conditions, such as the slip angle, α, vertical force, f_z, lateral force, f_y, and longitudinal force, f_x. In this work the experimental data have been normalized in order to facilitate the data treatment for a deep analysis (see Figure 5). A range of data included between −1 and 1 can be analyzed in a more efficient way, providing robust relations not conditioned by its magnitudes. The same normalization method has been applied to the strain measurements and to the working conditions. First and second order polynomial functions were used for fitting the data.

Figure 6 displays the relationship between the strain features and tire parameters, specifically the slip angle, α, and rolling speed, V. Figure 6a, shows that the speed influences the measurements in the axial direction (specifically in the outside part of the tire) and slightly in the circumferential direction. When the velocity increases (from 10 km/h to 50 km/h) it can be observed that the strain measurements at channel 3, tends to decreases; i.e., the offset, OS_{y2}, and strain peaks, ε_{y2f} and ε_{y2r}. Moreover, in Figure 6b the strain measurement at channel 2 shows that the tensile values decrease and converge to a constant value as the slip angle increases from 0° to 10°, with step sizes of 2°.

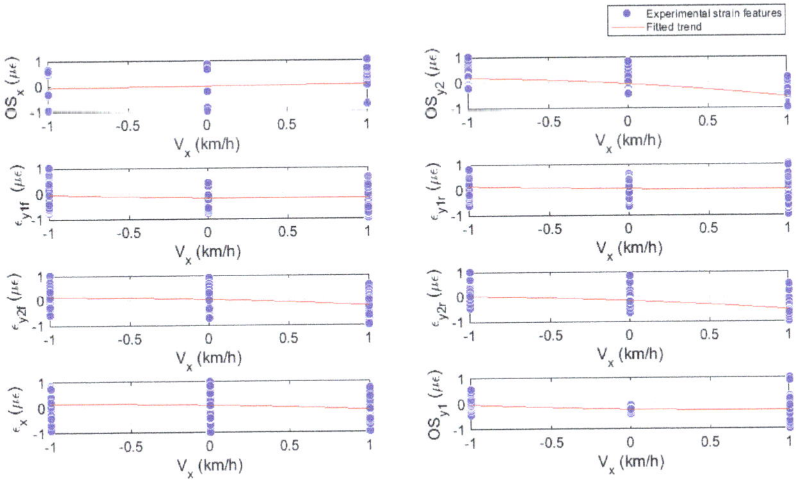

(**a**)

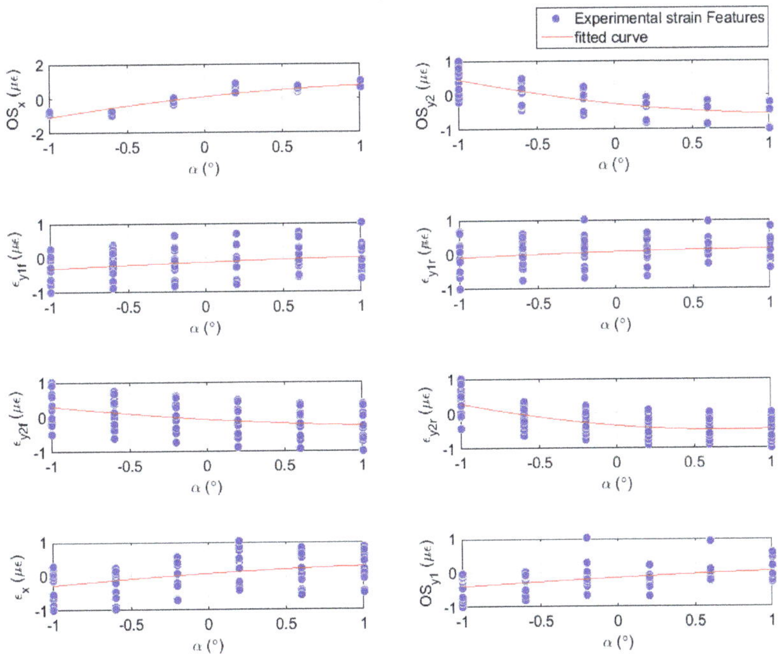

(**b**)

Figure 6. Simple curve fitting, variation of the strains features as function of potential parameters. (**a**) Strains feature as function of rolling speed. (**b**) Strains feature as function of slip angle.

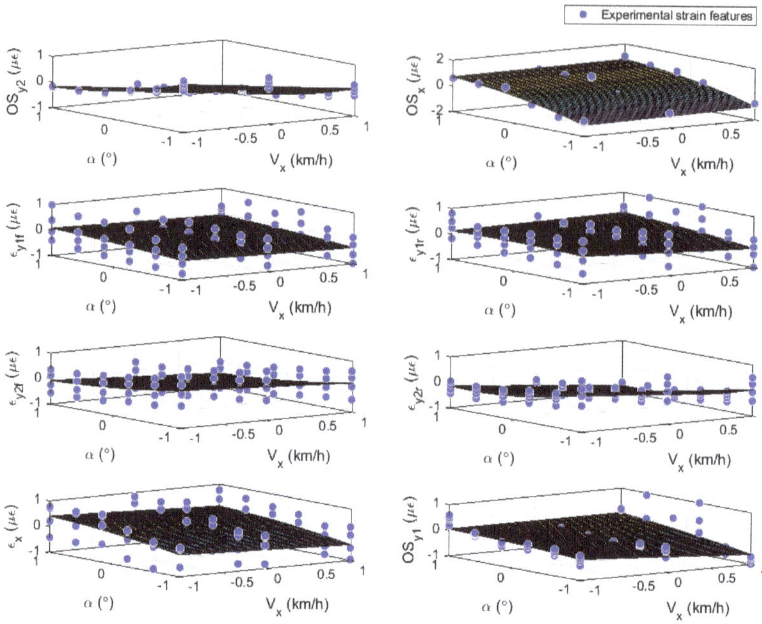

(a)

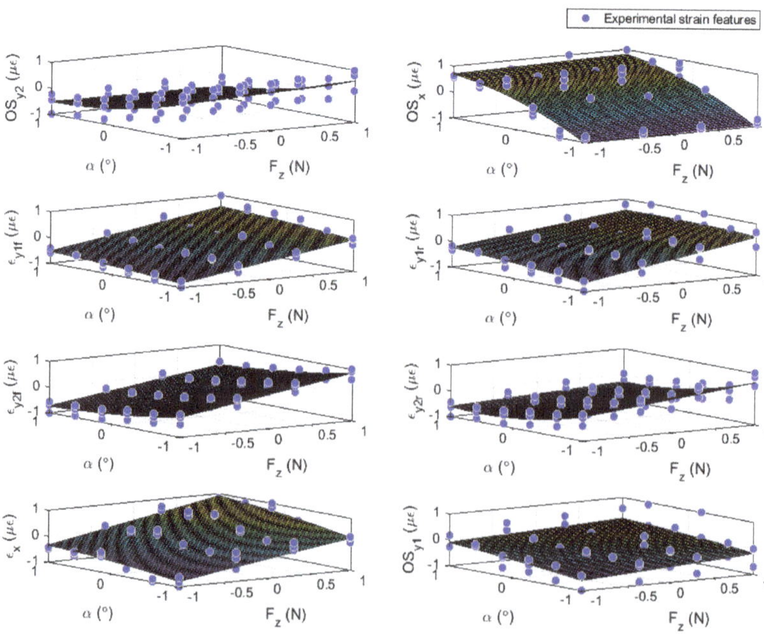

(b)

Figure 7. Curve surface fitting, variation of the strains features as function of potential parameters. (**a**) Strain features as function of slip angle and rolling speed. (**b**) Strain features as function of slip angle and vertical load.

Furthermore, the influence of the slip angle in the channel 2 (measurement in the circumferential direction) can be observed. The offset, OS_x, converge to a constant value and its low deviation makes it adequate to be used for slip angle estimation.

Besides, comparing the trend of the offset of channel 2, OS_x, and the offset of channel 3, OS_{y2}, both strain gauges located in the outer part of the contact patch (see Figure 2), when the slip angle increases, the offset OS_x also increases and the offset OS_{y2} decreases. The variation in the channels located on the outer part are higher than the variation at the channel on the inner part. This fact makes the outer channels more sensitive and adequate to analyze the influence of the slip angle and the velocity changes at the contact tread. The fact that the tensile strain in the half of the contact patch of the tread band decreases agrees with the results of Yang et al. [17], "where the shape of the contact patch and the pressure distribution become asymmetric, such as, the lateral shear stress drastically varies on one side of the tire contact patch rather than the other side". Also, the decreasing and increasing convergence of the axial and circumferential measurement might be attributed to inextensibility of the tire carcass, providing a real information about the tire parameters, e.g. lateral force, slip angle.

Similar plots have been produced for longitudinal force, lateral force and vertical load to note their influence over the strain features measurements. Owing to the high deviation that seems to be evident in most of the curve fitting, it has been considered to assess the variation of the deformation with more than one independent variable.

Therefore, the use of surfaces is contemplated to provide a full picture from an overall perspective. To do this, two independents variables are considered with variables X and Y, as tire working conditions and a variable Z, as the mean of the strain parameters, where X, Y, and Z all fall onto a plane.

In Figures 6 and 7 the experimental points of the strain features for different tire operational conditions have been plotted. Further, the curve fitting and the surface fitting were plotted, depending on the type of plot. Into the figures can be noted the trend of the strain features while the tire operational condition changes, independently of the specific value of this condition. In the surfaces' representations, most of the strain features are structured in layers. The layers show the effect of the gap between the experimental tire operational conditions. As example, in Figure 7a the structured layers show the vertical load influence over the strain features graphic and in Figure 7b the velocity influence is shown. The information provided by the curves and surfaces, evidence the simultaneous influence of the operational conditions over the contact patch measurements on most of the strain features.

In Figure 7 the influence of the tire working condition on the strain features is clearer. The vertical load seems not to modify the offsets at any channel; nonetheless, the slip angle and the velocity influence them, more clearly at those channels in the outer part of the tread. The offset in the circumferential direction (OS_x) is slightly affected by the velocity, but the magnitude of the offset in the axial direction (at channel 2) decreases with the increment of the tire velocity (see Figure 7a,b). Examining the tensile strain features, the tensile strain values at channel 1 show a linear increasing trend due to the rise of vertical load independent of the slip angle, whereas, at channel 2, the increment is also linear but dependent on the slip angle and its magnitude decreased with increment in the velocity; in the circumferential direction the vertical load influenced the tensile strain magnitude (ε_x).

Figure 7a shows an example of surface fitting where the strain features OS_{y2} and OS_x, are clearly functions of the tire operating parameters velocity and slip angle with a slight influence of the vertical load. It is evident that for these features (OS_{y2}, OS_x, and ε_{y2r}) which exhibit low deviation from the surface, the slip angle has a greater influence. However, contrasting Figures 6a and 7a, the offset of the channel 2, OS_{y2}, is the only one which is notably affected by the velocity.

This information suggests that the velocity has more effect on the outer part of the tire tread. In Figure 7b the tensile strain peaks of the channel 1 can be compared with the tensile strain peaks for channel 3. The plots for ε_{y2f}, ε_{y2r}, exhibit higher deviation from the surface than the peaks for channel 1. The slip angle clearly affects the offset at the outer part of the tire whereas the vertical load act upon the tensile features throughout the contact patch.

The results shown in Figure 6b indicate a low dependence of the tensile peaks at channel 1 (axial strain) on the slip angle, whereas, the surface fitting (Figure 7b) clearly shows the goodness of fit of other strain features with the tire parameters. Further surface fittings showed similar results with most of the strain fits showing great precision. These findings are important for tire strain systems in order to standardize the variation of the strain features. The relationship between strain data and tire parameters (forces and slip angle) are expressed by the matrix [A]. This matrix includes a set of coefficients which characterize that relationship and it was built by using the study of the variations (see Equation (4)):

$$\begin{bmatrix} OS_{y2} \\ \varepsilon_{y2r} \\ \vdots \\ \varepsilon_{y1r} \end{bmatrix} = [A] \cdot \begin{bmatrix} \alpha \\ f_z \\ f_x \\ f_y \end{bmatrix}, \tag{4}$$

where the dependent variables can be the strain features ($OS_{y2}, \varepsilon_x, \varepsilon_{y1f}, \varepsilon_{y1r}, \varepsilon_{y2f}, \varepsilon_{y2r}, OS_x, OS_{y1}$) and, the working condition (f_x, f_y, f_z, α), can be defined as the independent variables.

First of all, it is interesting to highlight that the apparent variability in the sets of data is due to different operational conditions are shown in each of them and don't implies any repeatability issue. For instance, in Figure 7a the strain data are shown explicitly for different slip angles and velocities, but includes different vertical loads implicitly. This shows a certain stratification in the data for each condition.

The information provided by these curves illustrate the influence of the rolling speed, the vertical load and the slip angle in the tire-road contact surface (two variables are shown explicitly and one of them implicitly). From the measurement can be deduced: the channels located in the outer part of the tire show more sensitivity to slip angle (OS_{2y}, E_{2y} and OS_x) and the speed (OS_{y2}) variations. The vertical load has influenced both sides of the contact patch in a similar way. It should be taken into account that the experimental tests were carried out for steady state cornering conditions. In this condition, the tire rolling under the influence of the vertical load and the slip angle shows a lateral deflection distorting the shape of the contact patch and the strain distribution in it.

In previous works was shown the complexity of studying the tire through the variability found in the strain measurements under the simultaneous influence of operational conditions [1,13,16,19–23,30,31]. Herein has been simplified the analysis of the correlations of the experimental data to demonstrate that tire parameters and working conditions (lateral force, longitudinal force, vertical load and slip angle) can be estimated from the deformations measured in the contact surface with the strain-based Intelligent Tire system, extending the results from previous studies. Performing a deeper study of the experimental data to expand the analysis made and improve the estimator effectiveness is not ruled out.

3.2. *Tire Parameters Estimation*

This research started with investigating the variation of the strain values against the experimental conditions. In this phase, the possibility of fitting curves and surfaces of deformation variations has taken shape, being of great utility for applying a fuzzy logic estimator for longitudinal and lateral forces, vertical load, and slip angle.

3.2.1. Fuzzy Logic System Applied

As discussed above, the Tire Parameter Estimation calls for the variations in the tire strain to characterize the tire deformation under certain operating conditions. The influence of tire operating parameters on the strain features is such that none of the single relationships gives by itself a reliable estimate of a tire parameter. Therefore, the use of fuzzy logic is adopted to correlate the features of the strain to give an accurate estimation of the tire parameter.

Fuzzy Logic is a methodology that makes possible to estimate output values based on initial conditions known as inputs. A fuzzy logic system may divide into three steps as shown in Figure 8. According to Kiencke and Nielsen [32], the first step is fuzzification, where the inputs are converted into linguistic variables with memberships functions between 0 and 1. Afterwards, the linguistic inputs are then evaluated with an inference engine based on fuzzy rules and formed into fuzzy logic outputs. Lastly, the resulting fuzzy output is mapped to a crisp output using the membership functions in the step of defuzzification.

Figure 8. Processing steps of a fuzzy system.

The defuzzification was carried out using the centroid method:

$$y_{Def} = \frac{\int_{-\infty}^{+\infty} y \cdot \mu_{res}(y) \cdot dy}{\int_{-\infty}^{+\infty} \mu_{res}(y) \cdot dy}, \tag{5}$$

where $\mu_{res}(y)$ is the aggregated membership function and y_{Def} is the output variable. This work has been carried out using the Mamdani's fuzzy inference and the triangular membership functions.

Based on previous studies of strain-based methods for Intelligent Tires [16,17,19,20], this study has enhanced the methodology used to estimate tire parameters. To improve the strain-based method requires using the slip and vertical load estimator to estimate the influence of tire operating parameters on tire strain features. Hence, the fuzzy logic system inputs work with normalized values of the fitted curves and surfaces. In Figure 5 the working scheme is shown, where the input and outputs of the fuzzy logic systems which are implemented into the Tire Parameters Estimation module can be noted. The normalization of data is also pointed out there. The normalization process has been done by mapping the minimum and maximum values of each feature to $[y_{min}, y_{max}]$ based on the algorithm shown in Equation (6):

$$\mathrm{y} = \frac{(y_{max} - y_{min}) \cdot (x - x_{min})}{(x_{max} - x_{min}) + y_{min}}, \tag{6}$$

The architecture of the fuzzy logic system applied to develop the proposed strain-based method is displayed in Figure 9. It is composed of four fuzzy logic blocks, each of them to estimate one tire parameter. Their inputs were chosen according to the impact of the tire working conditions on the tire deformation feature. The Tire Parameters Estimation phase in which the Fuzzy Logic Architecture detailed in Figure 9 is contained can be seen in Figure 5. The architecture of the fuzzy logic system starts by determining the slip angle values, followed by the vertical load, the lateral force and finally, the longitudinal force. Then the estimated slip angle values are fed into the vertical load and lateral force blocks (blue dot lines). Similarly, the vertical load value is one of the inputs for the longitudinal force block (purple dash lines). The continuous lines indicate deformation values. The fuzzy logic block to estimate the slip angle is defined by 21 rules and marking out as inputs: the offset of channel 3 ($\mathrm{OSy_2}$), the tensile strain front peak of channel 3 ($\varepsilon_{\mathrm{y2f}}$), and the offset of channel 2 (OS_x). These inputs are backed up by the curve fitting of the deformation features as a function of the slip angle. These strain features are shown to be more influenced by slip variation in Figure 6a,b. The selected inputs ($\mathrm{OSy_2}$, $\varepsilon_{\mathrm{y2f}}$, OS_x) are the same inputs previously used by Garcia-Pozuelo et al. [19] without using the rolling speed of the tire.

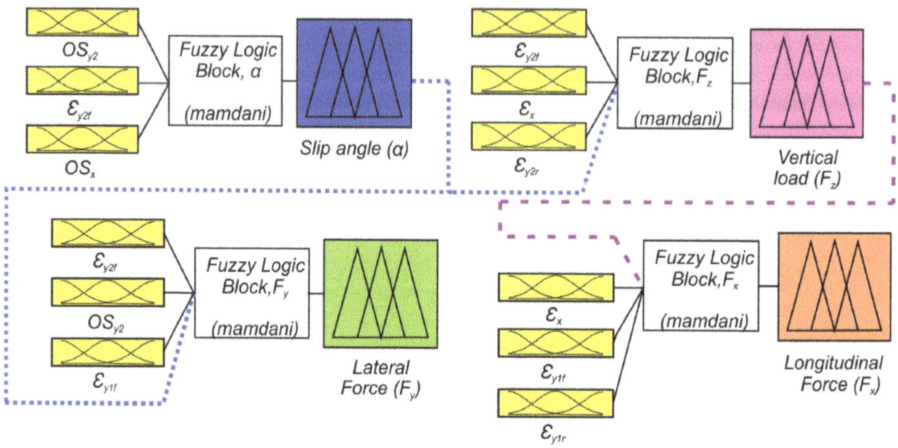

Figure 9. Architecture of the fuzzy logic system applied to develop the strain-based method.

The fuzzy logic block for vertical load estimation has four inputs and 217 rules. The tensile peaks of channel 3 (ε_{y2f} and ε_{y2r}) and channel 2 (ε_x), as well as, the slip angle estimated values are considered as inputs. Emphasizing the influence of the vertical load on the tensile strain peaks, the inputs of this block contemplate use of surfaces fitting as a function of the vertical load and slip angle to define the values of the selected strain features.

Equally, fuzzy logic block to estimate the lateral force takes as inputs the values of tensile peaks channels 1 and 3 (ε_{y1f} and ε_{y2r}) in addition to the offset values of channel 3 (OS_{y2}) and the slip angle. To generate deformation values, relationships of surfaces of ε_{y2f}, OS_{y2} (both as a function of lateral force and normal load) and ε_{y1f} (as a function of lateral force and slip angle) are used. The final input is the estimated slip angle. This block uses a total of four inputs and 460 rules.

The final block is to estimate the longitudinal force with four inputs and a total of 145 rules. The inputs are the values of the tensile strain peaks of channel 2 (ε_x) and channel 1 (ε_{y1f} and ε_{y1r}), in addition to the estimated vertical load values to provide a good estimation. Estimates of the deformation values are provided by a surface fit of the tensile strain peak (ε_x) as a function of the longitudinal force and vertical load, and curve fits of tensile strain peak values as a function of the longitudinal force.

3.2.2. Validating Fuzzy Logic System

In this work a strain-based methodology has been establishing to estimate tire parameters for Intelligent Tires. To check the fuzzy logic estimation system a validation process has been developed. This process has been identified by the orange dashed lines in Figure 5. It starts in the module of validation process where the values of the longitudinal force, lateral force, vertical force and slip angle are extracted from CarSim at severe maneuvers. These data are normalized and turned into deformation by the relationships of deformation features as function of tire parameters. Next, the strain features values are used to tests the fuzzy logic system in order to compare the results with CarSim. The maneuvers consisted of Double Lane Change (DLC), Lane Change (LC) and Sine With Dwell (SWD) as illustrated in Figure 10. Nonetheless, emphasizing the bounded range in the tests, the CarSim simulations at different maneuvers provide values out of the range, therefore, using the relationships of the normalized data the estimator is able to deliver the normalized outputs, and those outputs are compared with the original data from the maneuvers set in CarSim.

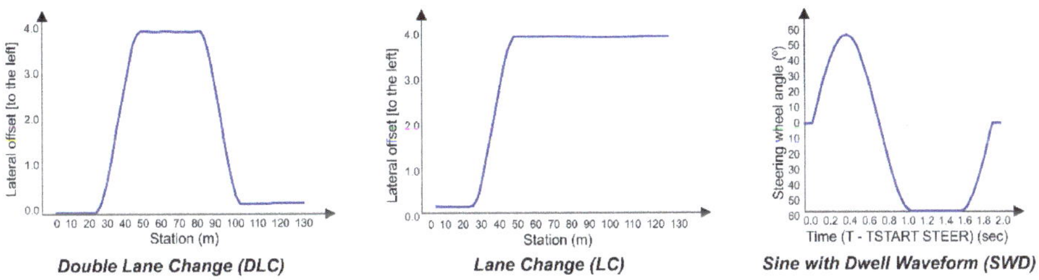

Figure 10. Lateral maneuvers.

Figures 11 and 12 illustrate the estimated results of the tangential forces at the contact patch (F_x and F_y), vertical load (F_z) and slip angle (α) on the left side of the car (at left front wheel, L1, and left rear wheel, L2) for a combination of the three maneuvers. While the simulations were carried out at various velocities, the results displayed in Figures 11 and 12 are for only 30 km/h and 80 km/h, respectively. The results obtained from tests carried out using the proposed estimator in these difficult lateral maneuvers show the good fidelity and sensitivity of the proposed approach. As a further illustration, the normalised errors of the simulations are presented in Table 1 for each wheel and velocity for three different velocities (30, 50 and 80 km/h).

An examination of Figures 11 and 12 reveals that the proposed methodology to estimate tire parameter is highly accurate due to the excellent agreement between the simulation and the proposed estimation results. The vehicle configuration is rear traction wheel, L2. The proposed method is able to estimate the tire parameters at pure lateral condition or at combined slip situation, based solely on the information provided by the Intelligent Tire. Indeed, the relationships established from the strain gauge measurements enable us to describe the intelligent tire parameters for different operational conditions.

The simulation results show that the proposed estimator is capable of predicting, with low margin of error, the target output in three severe maneuvers at 30, 50 and 80 km/h. This is also supported by the average errors indicated in Table 1.

Three velocities have been shown in Table 1: one in the middle of the experimental test values, other on the boundary of the test values and the last one, outside the bounds of the real test in order to illustrate one of the advantages found when using the normalization to feed the fuzzy logic algorithm. The curves for different operational conditions (vertical load and rolling speed) in the surface's representations seem to be morphologically similar.

These figures demonstrate that the most difficult maneuver for accurately estimating the tangential forces is the Sine With Dwell (SWD) maneuver, whereas for slip angle and vertical load the prediction shows good results. Despite this, in the majority of cases the proposed estimator accurately predicts the target output.

The differences between the estimated and the simulated longitudinal force may be attributed to inaccuracies in the experimental longitudinal force values which are estimated rather than measured as indicated above. It is also evident from their error values, which are higher than the rest. Despite this, it has been shown that they are not unduly affected by the assumptions used in their estimation.

It can be observed that the values of normalized errors calculated for the proposed estimator are all less than one. It means that the estimation results are acceptable since the root mean square error between the estimated results and the reference output (CarSim simulations) are within the deviation allowed for the reference output.

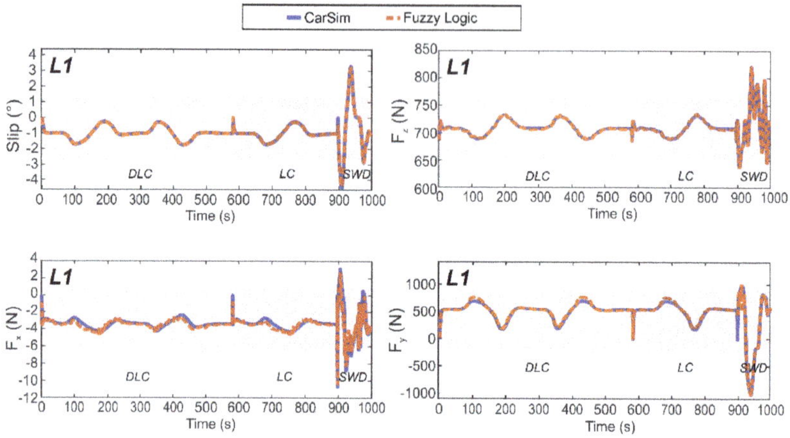

(a)

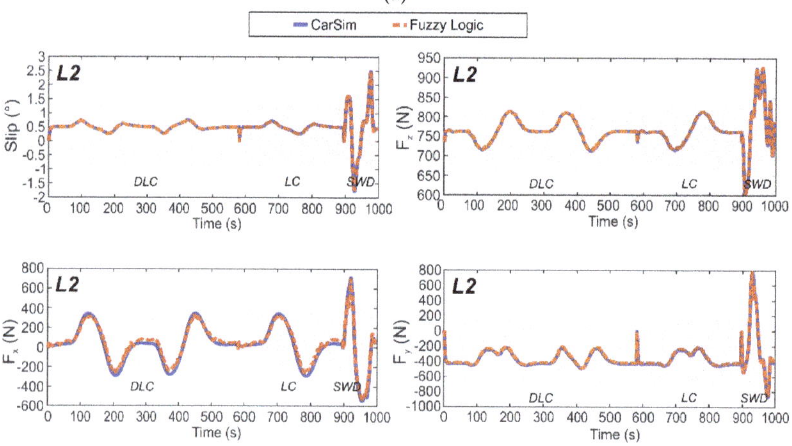

(b)

Figure 11. Simulation results for tire's maneuverings at 30 km/h. (**a**) Steered axle, wheel L1 (**b**) Powered axle, wheel L2.

Table 1. Normalized error of tire parameters simulation results.

Velocity (km/h)	Tire	α	F_z (–)	F_x (–)	F_y (–)
30	L1	0.049	0.063	0.346	0.156
	R1	0.023	0.039	0.419	0.117
	L2	0.023	0.039	0.177	0.121
	R2	0.027	0.037	0.233	0.149
50	L1	0.022	0.036	0.361	0.141
	R1	0.021	0.036	0.337	0.151
	L2	0.024	0.026	0.152	0.138
	R2	0.025	0.025	0.229	0.146
80	L1	0.041	0.026	0.274	0.119
	R1	0.026	0.032	0.256	0.155
	L2	0.026	0.025	0.182	0.142
	R2	0.024	0.025	0.201	0.112

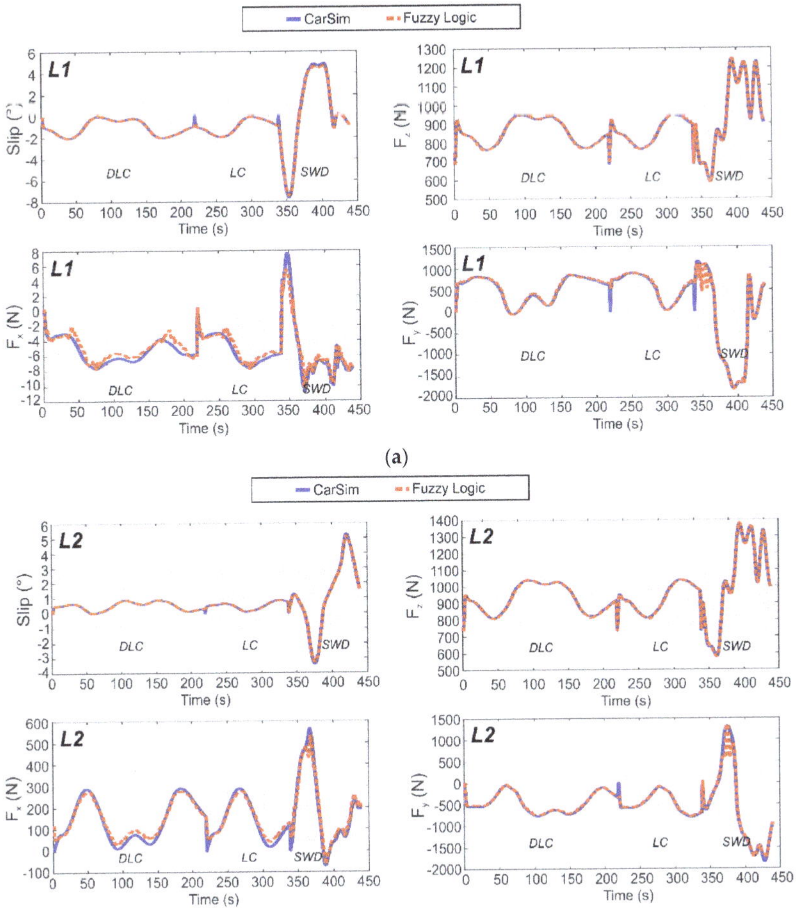

Figure 12. Simulation results for tire's maneuverings at 80 km/h. (**a**) Steered axle, wheel L1 (**b**) Powered axle, wheel L2.

3.2.3. Modeling of the Lateral Force

To describe the interaction between tires and road surface is of vital importance for the study of lateral vehicle motion due to the influence of the tire-road contact forces on vehicle dynamics behaviour. This section focuses on describing the lateral force in a combined slip situation with a mathematical model to test the estimation results with the experimental data.

Additionally, the achieved results by means of the fuzzy logic estimator have been compared to a semi-empirical model proposed by Rajamani, Bakker et al. [9] and Pacejka et al. [6–8]. The lateral force value collected at tire test rig for different tire operational condition has been used to fit the semi-empirical model.

The general form of the Magic Formula, derived from experimental data, is shown in Equations (7) and (8), where the variable Y is the output variable F_y, and x is the slip angle as an input variable. The coefficients B, C, D, E, S_v, S_h, take account of the camber angle, the cornering stiffness and the load variations [9]. To determine their values, MATLAB toolbox of curve fitting is applied to measured data to obtain the best fit. The Magic Formula parameters obtained from the lateral force experimental data are indicated in Table 2. The experimental data was obtained for various values of the vertical load for a zero camber angle, picking it as another input variable:

$$y = D \cdot \sin\left[C \cdot \tan^{-1}\left\{Bx - E\left(Bx - \tan^{-1} Bx\right)\right\}\right] \tag{7}$$

with:

$$\begin{aligned} Y(x) &= y(x) + S_v \\ x &= X + S_h \end{aligned} \tag{8}$$

Table 2. Parameters for empirical model of lateral force fitting and goodness of fit.

	Coefficients	Confidence Bounds at 95%	Goodness of Fit
B	0.142	(−4.771; 5.056)	SSE: 7.699e+04 R-square: 0.990 Adjusted R-square: 0.989 RMSE: 34.416
E	0.121	(−4.258; 4.499)	
C	0.093	(−66.100; 66.290)	
a	7.435e-04	(−0.524; 0.526)	
b	−8.444	(−5973; 5956)	
c	0.099	(0.045; 0.154)	
d	−47.922	(−82.330; −13.510)	

As is shown in Figure 13, the experimental data produces a curve that passes through the origin of the axes of the lateral force and slip angle. It may be observed that the lateral force reaches a maximum value and subsequently tends to a horizontal asymptote through variation of slip angle. Bringing the influence of the vertical force out, the lateral force increases with the increment of vertical load. Consequently the vertical load has been integrated into the model by the coefficients D that represents the peak value and by the vertical shift, S_v. The coefficients, D and S_v, are replaced by the relation (see Equations (10) and (11)) suggested by Rajamani et al. [9] and Pacejka et al. [6–8]. Figure 13 highlights the shift due to the vertical force, which might be due to the ply steer, conicity and rolling resistance explained by Bakker et al. [6]. Although the stiffness factor B and curvature factor E are functions of the vertical load, they can be directly defined by the surface fitted data. Equation (9) shows an adaptation of Equation (7) to include vertical force variation as another input value where a, b, c, d, B, C, E, are constant coefficients for the fitted data:

$$F_y = (F_z \cdot (a \cdot F_z) + b) \cdot \sin\left[C \cdot \tan^{-1}\left\{B \cdot x - E\left(B \cdot x - \tan^{-1}(B \cdot x)\right)\right\}\right] + c \cdot F_z + d, \tag{9}$$

$$\mathrm{D} = F_z \cdot (a \cdot F_z) + b, \tag{10}$$

$$S_v = c \cdot F_z + d \tag{11}$$

Further, Figure 13 illustrates the influence of velocity by splitting the experimental points as the vertical load and slip angle are increased. The comparison has been made at 30 km/h.

In the following, the model has been tested and compared with the simulation result for the same vehicle maneuvers and compared with the proposed estimator results. Table 3 displays the errors estimated for the experimental model and the fuzzy logic estimation in comparison with the simulated values. The complete dataset made up of the linear and non-linear dynamic conditions (severe maneuvers) shows that the proposed estimation model, Pacejka's model, presents certain differences with fuzzy logic (see Figure 14). The most important result is that under these conditions, it can be seen that the peaks of both curves show similar values and these are the most representative points in the range of severe maneuvers. It can also be seen that the curves show better similarities for the front tires, with higher slip angles, than for the rear tires. The differences in the values away from these peaks are less significant than in these key points. These differences might be due to inaccurate coefficients in the model or due to the limitations of Pacejka's model when it is used to predict combined high demanding conditions. This will be analyzed more deeply in further studies.

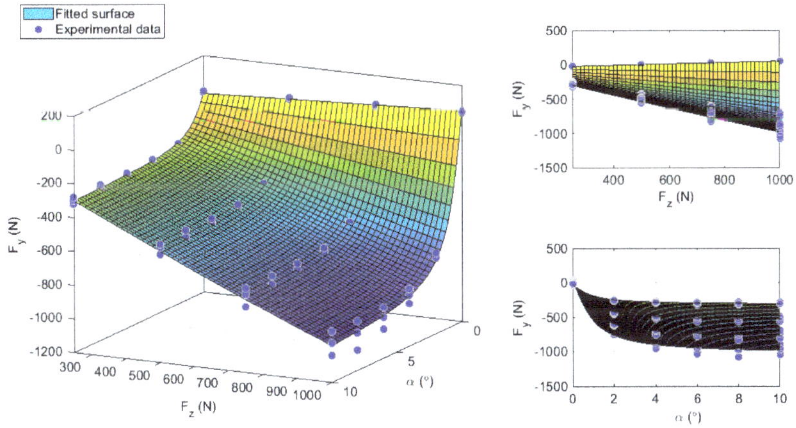

Figure 13. Experimental lateral force fitted surface based on Pacejka model at 30 km/h.

Table 3. Contrasting normalised error of tire lateral force.

Velocity (km/h)	Tire	F_y Fuzzy Logic (-)	F_y Pacejka (-)
30	L1	0.156	0.429
	R1	0.117	0.646
	L2	0.121	0.816
	R2	0.149	0.600
50	L1	0.141	0.316
	R1	0.151	0.471
	L2	0.138	0.567
	R2	0.146	0.402
80	L1	0.119	0.318
	R1	0.155	0.405
	L2	0.142	0.489
	R2	0.112	0.323

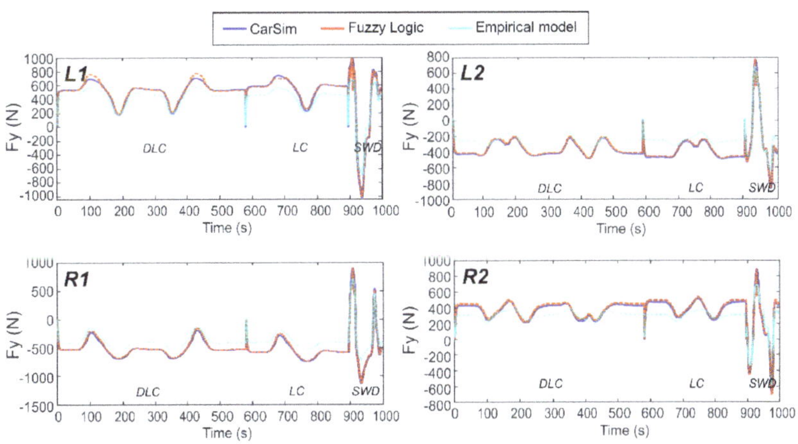

Figure 14. Comparison of the lateral force estimated by the experimental models, estimated by Fuzzy Logic and CarSim simulation at 30 km/h.

To support the graphical evidence, the normalised error is supplied in all cases (see Table 3), in comparison with the target output of the simulation. Figure 14 shows that the proposed estimation (fuzzy logic) is able to estimate the lateral force more accurately. Parallels between the proposed empirical model, the estimation provided by fuzzy logic and the CarSim result can be noted. The parallels of these results is also demonstrated with the computed normalised error values, for the proposed estimation it varies between 0.112–0.156 and for the empirical model it varies between 0.316–0.816, depending on the tire position.

It is evident that the fuzzy logic estimation accurately represented the lateral force at each tire during slip angle variations (see Figure 11a). Since the empirical model is considered less accurate than the experimental results, it is assumed that some improvement of the proposed empirical model is needed, but this is not the aim of this study. As discussed above, the present work is not focused on the use of semi-empirical Pacejka's model. At the contrary, the main aim is to develop a methodology to estimate the tire parameters by means of fuzzy logic algorithms. Pacejka's model is just used in order to compare the results of the estimations with other methodology (a consolidated tire model fitted from experimental data).

The resulting normalised errors are displayed in Table 3. In all cases, the proposed estimator shows a better performance and with considerably smaller errors than the empirical model. It is important to note the influence of the slip angle over the empirical model, i.e. that at large values of slip angle its prediction is better.

4. Discussion

Figure 7a,b shows the influence of the tire parameters on the strain features. Surfaces were formed to display the variation of the strain features resulting from varying of vertical load and rolling velocity. In some cases, the surface fit seems to describe the majority of points (see also Figure A1 in the Appendix A), although for further studies implementation of a deeper analysis of the experimental data is proposed. The use of normalized correlations between the operational condition and the strain gauges measurements allows us to work directly with the trends (avoiding the use of specific experimental points for every estimation). Therefore, the fuzzy logic system is managed by the integration of these trends yielding accurate results. Further, the previous normalization of the data has it made possible to estimate a normalized output and, therefore, estimate tire conditions out of the bounds of the experimental range. As result, the strains collected by the intelligent tire in the contact patch area are used to determine the values of tire parameters for severe maneuvers conditions through the methodology proposed. Figures 11 and 12 show the results yielded by the proposed methodology, where a good fidelity at both the 30 and 80 km/h results is noted.

In Section 3.2.2, the ability of the proposed empirical model to predict the target output is better for a pure slip condition than for combined situations, where the empirical model seems more affected. According to Rajamani [9], analytical models do not always lead to quantitatively accurate results. Differences from experimental data were observed, especially at large slip and at combined slip. This may be happening at steered wheel, L1, and at powered wheel, L2, where their values of vertical force are similar, nonetheless, results for the traction wheel were less accurate using the empirical model (see Figure 11a,b and Figure 14). Something similar happened for the right side of the vehicle (R1 and R2), where the estimation of a combined situation was less accurate.

5. Conclusions

This paper has investigated the detection of the lateral force, longitudinal force, vertical load and the slip angle through strain-based Intelligent Tires using fuzzy logic estimation. The proposed method has made use of curve and surface fittings of the deformation features and normalized the data to correlate the variations between them, thus yielding acceptable estimation for the slip angle and the tire-road contact forces, founded on graphical and numerical evidences. The knowledge about the variations of the tire deformation in the contact patch has been helpful for accomplishing the estimation of all forces. Moreover, it is necessary to highlight the behavior of the tire's strains experimental data at different positions in the tread band (inner and outer parts of the contact patch), being possible to identify that the influence of the velocity and the slip angle are more clear and significant on the outer part of the tread band.

One of the limitations of this kind of study is being able to measure the behaviour of an Intelligent Tire estimator for real car maneuvers. Nonetheless, in this work, it has been possible to assess the behaviour of the proposed method to estimate the tire-road contact forces in complex maneuvering by means of simulation tools. An FSAE prototype is being adapted as test car for further stages of this study in order to extend the conclusions in real conditions, out of a laboratory tire test rig. Looking at the estimations for inputs outside the range of explicit experimental points, it is observed that the process of normalization provides to the fuzzy logic system the capacity to estimate suitable results.

It also demonstrates that deformation of the tire contact patch is highly correlated with the tire parameters emphasizing the essential role of the tire as a vehicle sensor. This combination also is a good approach in order to reduce the number of tests needed.

Furthermore, the proposed method is able to estimate the tire parameters under pure lateral conditions or for combined slip situations, based only on the information provided by the Intelligent Tire, while the empirical tire model has limitations, as has been shown.

The values of normalised error achieved by the proposed estimator for tire road parameters demonstrates the effectiveness of strain-based Method for Intelligent Tires. Further, the ability of the strain-based Intelligent Tire to estimate the tire parameters of each tire through contact patch may be significantly useful for vehicle dynamic behaviour offering the possibility to predict the load transfer or friction coefficient. A whole set of working condition combinations has been taken into account in the design of experiment in order to carry out the test [19,20]. However, a deeper study of the experimental data could yield a more precise estimator. Additionally, in this research only the experimental data of a tire slick radial tire DUNLOP SP SPORT 175/505 R13 (tubeless) has been used to develop a strain-based Intelligent Tire. It would be interesting to implement this study with other types of tire to compare the differences that might exist between them. With this aim an Avon 175/53R13 slick tire has been instrumented to complement these conclusions and the results will be reported in due course.

Author Contributions: D.G.-P. defined the research topic and performed the experiments. M.F.M.-P., D.G.-P., V.D. and O.O. analysed the data and conceived and designed the simulations. O.O. provided direction for experimental methods; and M.F.M.-P. developed the MATLAB implementation about the mathematical algorithms and fuzzy logic implementation. All authors have contributed to the production of the paper and have approved the manuscript.

Funding: This research received no external funding.

Acknowledgments: We acknowledge the University of Birmingham for the facilities and the Universidad Carlos III de Madrid for the financial support that covered the costs to publish in open access.

Conflicts of Interest: The authors declare no conflict of interest.

Appendix A

This appendix contains the rest of the relevant surfaces to show the reader how almost all the experimental points may approximate a surface fit. This research was based on the use of this type of fit to approximate deformation values.

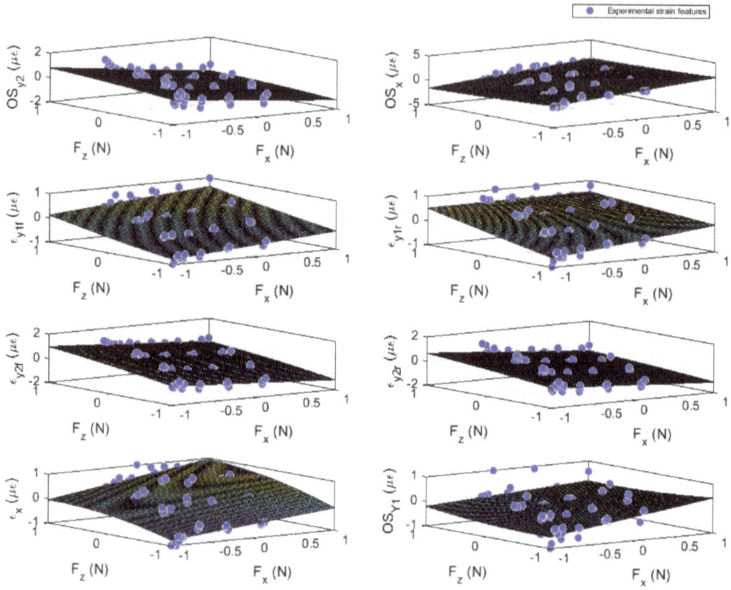

(a)

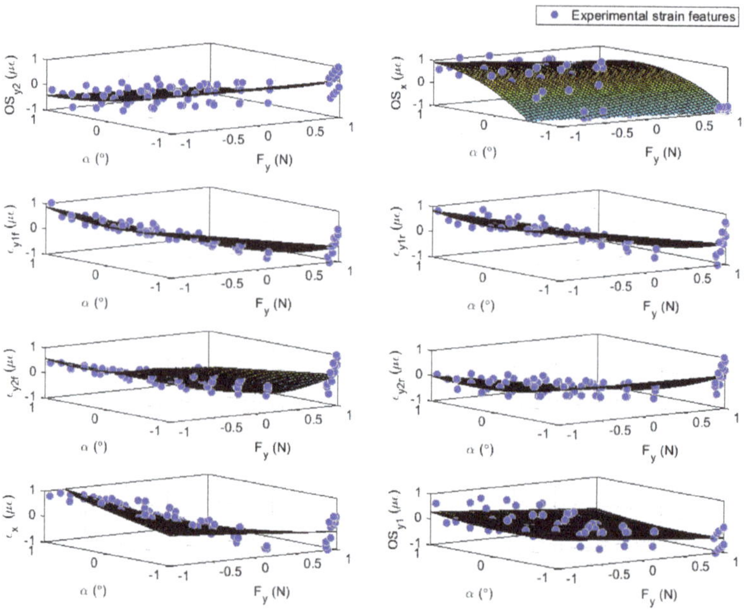

(b)

Figure A1. *Cont.*

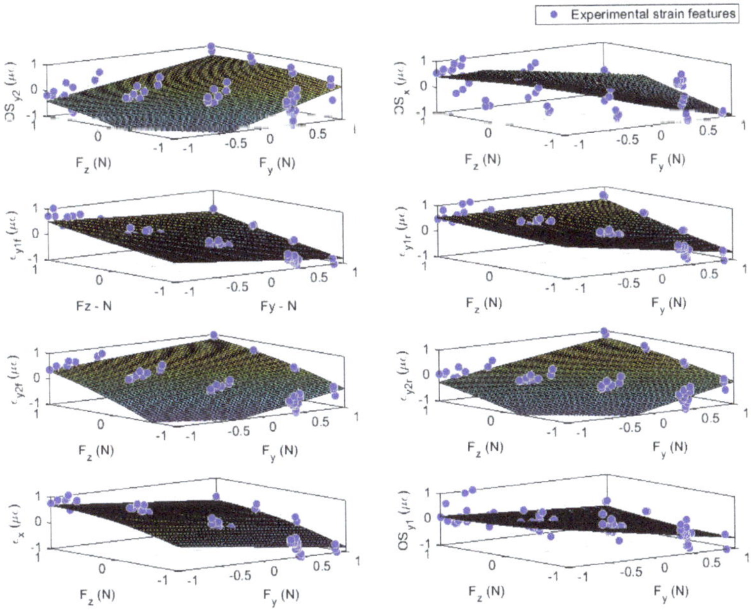

(c)

Figure A1. Curve surface fitting, variation of the strains features as function of potential parameters. (**a**) Strain features as function of vertical load and longitudinal force. (**b**) Strain features as function of slip angle and lateral force. (**c**) Strain features as function of vertical load and lateral force.

References

1. Matsuzaki, R.; Todoroki, A. Intelligent Tires Based on Measurement of Tire Deformation. *J. Solid Mech. Mater. Eng.* **2008**, *2*, 269–280. [CrossRef]
2. Boada, L.M.J.; Boada, L.B.; Garcia-Pozuelo, D.; Diaz, V. Application of Neural Networks for Estimation of Tyre/Road Forces. *Mech. Syst. Control Parts A B* **2009**, *10*, 427–433.
3. Doumiati, M.; Victorino, A.; Charara, A.; Lechner, D. Unscented Kalman filter for real-time vehicle lateral tire forces and sideslip angle estimation. In Proceedings of the 2009 IEEE Intelligent Vehicles Symposium, Xi'an, China, 3–5 June 2009; pp. 901–906.
4. Doumiati, M. *Vehicle Dynamics Estimation Using Kalman Filtering: Experimental Validation*; ISTE: London, UK, 2013.
5. Lee, H.; Taheri, S. Intelligent Tires? A Review of Tire Characterization Literature. *IEEE Intell. Transp. Syst. Mag.* **2017**, *9*, 114–135. [CrossRef]
6. Bakker, E.; Nyborg, L.; Pacejka, H.B. Tyre Modelling for Use in Vehicle Dynamics Studies. *SAE Transact.* **1987**, *96*, 190–204.
7. Pacejka, H.B.; Besselink, I.J.M. Magic Formula Tyre Model with Transient Properties. *Veh. Syst. Dyn.* **1997**, *27*, 234–249. [CrossRef]
8. Pacejka, H.B.; Bakker, E. The magic formula tyre model. *Veh. Syst. Dyn.* **1992**, *21*, 1–18. [CrossRef]
9. Rajamani, R. *Vehicle Dynamics and Control*; Frederick, F., Ling, Eds.; Mechanical Engineering Series; Springer: New York, NY, USA, 2006.
10. Jousimaa, O.J.; Xiong, Y.; Niskanen, A.J.; Tuononen, A.J. Energy harvesting system for intelligent tyre sensors. In Proceedings of the 2016 IEEE Intelligent Vehicles Symposium (IV), Gothenburg, Sweden, 19–22 June 2016; pp. 578–583.
11. Lee, J.; Choi, B. Development of a piezoelectric energy harvesting system for implementing wireless sensors on the tires. *Energy Convers. Manag.* **2014**, *78*, 32–38. [CrossRef]

12. Singh, K.B.; Bedekar, V.; Taheri, S.; Priya, S. Piezoelectric vibration energy harvesting system with an adaptive frequency tuning mechanism for intelligent tires. *Mechatronics* **2012**, *22*, 970–988. [CrossRef]
13. Xiong, Y.; Tuononen, A. The in-plane deformation of a tire carcass: Analysis and measurement. *Case Stud. Mech. Syst. Signal Process.* **2015**, *2*, 12–18. [CrossRef]
14. Roveri, N.; Pepe, G.; Carcaterra, A. OPTYRE—A new technology for tire monitoring: Evidence of contact patch phenomena. *Mech. Syst. Signal Process.* **2016**, *66–67*, 793–810. [CrossRef]
15. Cheli, F.; Braghin, F.; Brusarosco, M.; Mancosu, F.; Sabbioni, E. Design and testing of an innovative measurement device for tyre–road contact forces. *Mech. Syst. Signal Process.* **2011**, *25*, 1956–1972. [CrossRef]
16. Yunta, J.; Garcia-Pozuelo, D.; Diaz, V.; Olatunbosun, O. A strain-based method to detect tires' loss of grip and estimate lateral friction coefficient from experimental data by fuzzy logic for intelligent tire development. *Sensors* **2018**, *18*, 490. [CrossRef] [PubMed]
17. Yang, X.; Olatunbosun, O.; Garcia-Pozuelo Ramos, D.; Bolarinwa, E. Experimental Investigation of Tire Dynamic Strain Characteristics for Developing Strain-Based Intelligent Tire System. *SAE Int. J. Passeng. Cars Mech. Syst.* **2013**, *6*, 97–108. [CrossRef]
18. Yang, X.; Olatunbosun, O.; Garcia-Pozuelo, D.; Bolarinwa, E. FE-Based Tire Loading Estimation for Developing Strain-Based Intelligent Tire System. *SAE Tech. Pap.* **2015**. [CrossRef]
19. Garcia-Pozuelo, D.; Yunta, J.; Olatunbosun, O.; Yang, X.; Diaz, V. A Strain-Based Method to Estimate Slip Angle and Tire Working Conditions for Intelligent Tires Using Fuzzy Logic. *Sensors* **2017**, *17*, 874. [CrossRef] [PubMed]
20. Garcia-Pozuelo, D.; Olatunbosun, O.; Yunta, J.; Yang, X.; Diaz, V. A novel strain-based method to estimate tire conditions using fuzzy logic for intelligent tires. *Sensors* **2017**, *17*, 350. [CrossRef] [PubMed]
21. Garcia-Pozuelo, D.; Olatunbosun, O.A.; Romano, L.; Strano, S.; Terzo, M.; Tuononen, A.J.; Xiong, Y. Development and experimental validation of a real-time analytical model for different intelligent tyre concepts. *Veh. Syst. Dyn.* **2019**. [CrossRef]
22. Garcia-Pozuelo, D.; Olatunbosun, O.; Strano, S.; Terzo, M. A real-time physical model for strain-based intelligent tires. *Sens. Actuators A Phys.* **2019**, *288*, 1–9. [CrossRef]
23. Yunta, J.; Garcia-Pozuelo, D.; Diaz, V.; Olatunbosun, O. Influence of camber angle on tire tread behavior by an on-board strain-based system for intelligent tires. *Measurement* **2019**, *145*, 631–639. [CrossRef]
24. Yang, X. *Finite Element Analysis and Experimental Investigation of Tyre Characteristics for Developing Strain-Based Intelligent Tyre System*; University of Birmingham: Birmingham, UK, 2011.
25. Aguilar-Martínez, J.; Alvarez-Icaza, L. Analysis of tire-road contact area in a control oriented test bed for dynamic friction models. *J. Appl. Res. Technol.* **2015**, *13*, 461–471. [CrossRef]
26. Park, D.W.; Martin, A.E.; Jeong, J.H.; Lee, S.T. Effects of Tire Inflation Pressure and Load on Predicted Pavement Strains. *Balt. J. Road Bridge Eng.* **2008**, *3*, 181–186. [CrossRef]
27. Simulation Software CarSim. Available online: www.carsim.com/products/carsim/ (accessed on 4 July 2019).
28. Vargas-Meléndez, L.; Boada, B.L.; Boada, M.J.L.; Gauchía, A.; Díaz, V. A sensor fusion method based on an integrated neural network and Kalman Filter for vehicle roll angle estimation. *Sensors* **2016**, *16*, 1400. [CrossRef] [PubMed]
29. Boada, M.L.J.; Boada, B.L.; Diaz, V. A novel frequency dependent model based on trigonometric functions for a magnetorheological damper. *Meccanica* **2017**, *11–15*, 2567–2581. [CrossRef]
30. Kim, S.J.; Kim, K.S.; Yoon, S.Y. Development of a tire model based on an analysis of tire strain obtained by an intelligent tire system. *Int. J. Automot. Technol.* **2015**, *16*, 865–875. [CrossRef]
31. Morinaga, H.; Wakao, Y.; Hanatsuka, Y.; Kobayakawa, A.; Bridgestone Corporation, J. The possibility of intelligent tire (Technology of contact area information sensing). In Proceedings of the FISITA world Automotive Congress, Yokohama, Japan, 22–27 October 2006.
32. Kiencke, U.; Nielsen, L. *Automotive Control Systems: For Engine, Driveline, and Vehicle*; Springer: Berlin, Germany, 2005.

sensors

Article

Vehicle Driver Monitoring through the Statistical Process Control

Arthur N. Assuncao [1,2], Andre L. L. Aquino [3,*], Ricardo C. Câmara de M. Santos [2], Rodolfo L. M. Guimaraes [2] and Ricardo A. R. Oliveira [2]

1 Instituto Federal de Educação, Ciência e Tecnologia do Sudeste de Minas Gerais, Santos Dumont, MG 36240-000, Brazil
2 Departamento de Computação, Universidade Federal de Ouro Preto, Ouro Preto, MG 35400-000, Brazil
3 Instituto de Computação, Universidade Federal de Alagoas, Maceió, AL 57072-970, Brazil
* Correspondence: alla@laccan.ufal.br

Received: 23 April 2019; Accepted: 9 July 2019; Published: 11 July 2019

Abstract: This paper proposes the use of the Statistical Process Control (SPC), more specifically, the Exponentially Weighted Moving Average method, for the monitoring of drivers using approaches based on the vehicle and the driver's behavior. Based on the SPC, we propose a method for the lane departure detection; a method for detecting sudden driver movements; and a method combined with computer vision to detect driver fatigue. All methods consider information from sensors scattered by the vehicle. The results showed the efficiency of the methods in the identification and detection of unwanted driver actions, such as sudden movements, lane departure, and driver fatigue. Lane departure detection obtained results of up to 76.92% (without constant speed) and 84.16% (speed maintained at $\approx$60). Furthermore, sudden movements detection obtained results of up to 91.66% (steering wheel) and 94.44% (brake). The driver fatigue has been detected in up to 94.46% situations.

Keywords: driver monitor; lane departure; statistical process control

1. Introduction

There is a consensus regarding the high rate of traffic accidents caused by lane deviation due to momentary driver fatigue [1]. This high rate of road accidents, involving driver failures, such as tiredness and drowsiness, impose a significant burden on society, constituting a severe public health problem. Nowadays, the vehicle technologies mitigate the above problems by introducing into the vehicle several devices composed of both hardware for sensing and software for processing and decision making. Based on generated data, it is possible to provide different applications, on board or in a cellphone, capable of monitoring and interacting with both vehicle and driver behavior.

The Advanced Driver Assistance Systems (ADAS) [2] are those that exploit the knowledge of the environment based on advanced sensors. Galvani [3] presents the different levels of automated driving defined by Society of Automotive Engineers (SAE) (https://www.sae.org/): (i) *No automation*, the driver performs all driving tasks; (ii) *Driver Assistance*, the driver controls the vehicle, but the system includes some driving assist features in the vehicle design; (iii) *Partial automation*, the vehicle has combined automated functions like acceleration and steering, but the driver must remain engaged with the driving task and monitor the environment at all times; (iv) *Conditional automation*, the driver is a necessity but does not monitor the environment. The driver must be ready to control the vehicle at all times with notice; (v) *High automation*, the vehicle is capable of performing all driving functions under certain conditions. The driver has the option to control the vehicle; (vi) *Full automation*, the vehicle is capable of performing all driving functions under all conditions. The driver also has the option to control the vehicle. In particular, in this paper, we investigate the *driver assistance* level, i.e., the

ADAS still leaves the authority to the driver, but we take care of specific driving functions based on conventional sensors.

Our study uses the vehicle motion information, coming from inertial sensors (accelerometer and gyroscope) in the steering wheel and brake pedal, and driver behavior based on the camera. The actions monitored are the lane department, sudden movements detection, and driver fatigue detection. The lane department is concerned with warning the driver when the vehicle begins to move out of its lane. The sudden movements detection is also to detect when the driver is getting drowsy but using sudden movements in the steering wheel and the brake pedal. Finally, the driver fatigue detection corresponds to detecting when the driver is getting drowsy through the eyes monitoring.

To perform this monitoring, we propose the use of the Statistical Process Control (SPC) [4]. In particular, we applied Exponentially Weighted Moving Average control graphs (EWMA) [5] from the SPC to identify the quality of safe driving. In an initial proposal, to lane departure detection, we use the EWMA in the analysis of the data from the X and Y axes of only one accelerometer and the combination of an accelerometer and a gyroscope positioned to the center of the steering wheel. After that, we evaluate the previous proposal considering sudden movements detection, by using the SPC applied to the accelerometer on the steering wheel and potentiometers on the pedals. Finally, for driver fatigue detection, we use a camera positioned in front of the driver, an analysis is performed based on the closed eyes percentage with the use of SPC, and the driver is alerted in case of fatigue.

In the evaluation, we considered different scenarios with several vehicles in a predefined route simulator, where the driver committed incorrect movements when the test evaluator requested. These movements do not differ from involuntary movements, because our method detects anomalies in the steering process, so any movement out of control may be detected. The experiments showed that the proposed methods and systems are viable alternatives to driver monitoring. In the initial experiment, we evaluate the lane departure detection with results of up to 76.92% (without constant speed) and 84.16% (speed maintained at $\approx$60). In the second experiment, we evaluate together sudden movements detection and driver fatigue, achieving, to sudden movements detection, results up to 91.66% (steering wheel) and 94.44% (brake); and, to driver fatigue, results up to 94.46% (blinks). The results reveal that we can apply our method to lane departure, fatigue, and sudden movement detection with acceptable accuracy. Among several contributions of this work, the introduction of SPC to driver monitoring is the main one. Additionally, the solution is low cost, allows the continuous monitoring, without human intervention, and is transparent to the driver. Besides that, the solution executes in a limited embedded system, thus must have a low time complexity cost. This restriction turns our solution more competitive against the traditional ones.

We organized the rest of the paper as follows: Section 2 shows the related works. Section 3 discusses the concepts related to Statistical Process Control and its EWMA graph. Section 4 shows the methods developed for monitoring drivers: Lane departure detection, driver fatigue detection, and sudden movements detection. Section 5 presents the evaluation of each method presented. Finally, Section 6 shows the conclusions and future work.

2. Related Work

Here we present works related to general ADAS systems [2,6], lane departure problem [7], driver fatigue, and sudden movements detections [8]. For ADASs, we analyze the number of solutions, the types of detection performed, and the issuance of alerts. For other ones, we analyze the work considering the approach used (computational vision, vehicle-based or physiological signals), the method cost, the use of computational resources, sensitivity to light, and other interferences.

The literature presents some well-established ADASs proposals, such as iCar [9] which is a system aimed at reducing accidents and assisting the driver in various aspects of driving, such as lane detection, pedestrian detection, car detection, driver fatigue detection and rearview assistance for parking. There is also the ADAS proposed by Wang et al. [10] which presents a system with self-learning, aiming to help the driver in the task of maintaining safety concerning the car-to-front, reducing their workload and reducing accidents and allowing control of cruise speed and frontal collision warning. Finally, the ADAS proposed by Chien et al. [11], driver assistance, and pedestrian safety system that detects lanes, vehicles, and pedestrians in front of the vehicle. These systems generally use a combination of various techniques, such as computational vision, feature extraction, machine learning, object recognition, and human-computer interaction.

For the lane departure problem, Sandström et al. [12] created a method with signals from the steering wheel, avoiding loss in weather conditions and bad roads. Satzoda et al. [13] provided a low-cost video-based system designed to assist the driver by issuing warnings during lane drifting and lane changes implemented on the Snapdragon^{TM} embedded computing processor setup with the camera on top of the windshield. Son et al. [14] show a system with a low-efficiency loss under the most varied lighting conditions. With the use of computer vision, it works even under adverse atmospheric conditions and at night. Thinking about low cost and low resource consumption, Jung et al. [15] developed a system with a camera located in the center of the vehicle facing the road and a computer with low processing power. Unlike our solution that uses a vehicle-based approach, these works use computer vision, which is the standard in the literature.

Sensitivity to light is a relevant factor in the lane departure detection and methods based on computer vision usually suffer from light interference. At this point, different works [12,16,17] do not suffer interference. Our proposal does not suffer interference, but it presents some disadvantages, like the dependence on the sensitivity of the steering wheel of the vehicle for the calibration, needing to calibrate the steering wheel a first time. Thus, it requires specific calibration for the coupled steering wheel. The financial and resource cost is also relevant. In this sense, in addition to our method, only the work of Jung, MinKim [15] aims to be low cost and consumes few computational resources.

Regarding driver fatigue detection, Abulkhair et al. [18] propose a system for detecting driver fatigue to take advantage of the use of the driver's smartphones and its mobility and to avoid the use of larger computers for driver fatigue detection. Patel et al. [19] use a video camera to monitor eye states. Jung et al. [20] developed a method with the use of electrocardiogram analyzing the variability of the heart rate to bypass problems of illumination, common in systems of computer vision. Aiming to perform lane departure detection related to driver fatigue, McDonald et al. [21] developed a vehicle-based approach by applying a Random Forest classifier to the steering wheel angle data. Mehta et al. [22] proposed an approach to detect driver drowsiness using SVM to capture the drivers' face frames and calculate the EAR Eye Aspect Ratio (EAR), then from a threshold value ($EAR = 0.25$) to infer that the Driver is sleepy. They tested the method with different classifiers. We highlight SVM and Random Forest that obtained an accuracy of 80% and 84%, respectively. Finally, Pauly and Sankar [23] presented a method of detecting drowsiness based on the Viola and Jones method with the use of images from a web camera. His detection used SVM as a classifier for eye blinking. Finally, the PERCLOS uses a limit value of six seconds. Pauly and Sankar showed that 91.6% match with the judgments of that of a human rater.

Finally, the most accepted method of drowsiness analysis is the PERCLOS [24], which uses the proportion that the eyes are 80% to 100% closed in a time interval. However, some authors, such as Kong et al. [25], have used a PERCLOS simplified that uses the percentage of the duration of entirely closed eye state at a specific time interval (1 min or 30 s). PERCLOS based systems have a good acceptance and good correlation with drowsiness. However, among their disadvantages, Stanton et al. [26] mention that the use of a system to identify slow eyelid closure generally requires a restricted field of vision. Thus, the head movements of the driver may require more than one camera. Besides, it may have low efficiency in low humidity environments, as users may be prone to close their eyes slowly and keep them closed for a while, thus mistaking fatigue with humidification, resulting in false positive. It is essential to highlight that we do not found any research about the use of SPC methods in ADAS. Thus we agree that this is our main contribution.

3. The Statistical Process Control with EWMA

Statistical Process Control (SPC) is a quality engineering technique that can be used to control and, where possible, make improvements in the production process. The objective of statistical control is to monitor the process to identify sources of variability and, if necessary, to take some corrective action to eliminate the type of event that caused it. We monitor the current state of data distribution accuracy to control this process using variable data. For this, the target value, which corresponds to the desired value for a particular characteristic of a product, is compared to limits indicating its conformity with characteristics of good quality. We perform all this monitoring with the help of control graphs [27].

The control graphs are the main components of the SPC—to allow the identification of the behavior of the process over time or the number of samples and the detection of the incidence of particular causes, we perform this through a history of data. From the identification and detection, it is possible to take actions to prevent and avoid recurrence of the event. This whole process can be performed and controlled in real time. Besides, according to Borror et al. [28], the control graph has the advantage of its operational simplicity and effectiveness in the detection of problems in the process.

The purpose of using statistical process control to monitor the driver is to identify moments of erratic driving, such as actions that can cause accidents. Thus driving should be safe, within certain limits, to maintain the quality of driving. For this, we use a control graph for online monitoring by quickly detecting the occurrence of causes attributable to some event in process changes so that we can take some corrective action before the problem occurs [4].

The control graph represents a quality characteristic concerning the number of samples or time. It has three lines, a call of Central Line (CL) that represents the average value of the quality characteristic and corresponds to the state under control. Two other horizontal lines, called Upper Control Limit (UCL) and Lower Control Limit (LCL), are used to control the range of data variability.

According to Montgomery [4], the control graph works comparing the average of sampled values $\bar{x}$, with the two control limits (UCL and LCL),

$$LCL \leq \bar{x} \leq UCL.$$

If the measure is within limits, the process is under control, and we do not perform any action. Otherwise, when we identify an out-of-control situation, we perform some correction or identify the cause.

The statistical measure chosen to analyze and monitor driving data was the Exponentially Weighted Moving Average (EWMA) [5]. We chose the EWMA because it is the faster and more usual SPC method used in industrial control scenarios. We use an EWMA graph when rapid detection of out-of-control situations is required by calculating the time series of measures [29]. In the first step of the EWMA calculation, the measures of the processes are sampled at specific periods and grouped into subgroups of predefined size. We calculate the average and the standard deviation of each subgroup. Then, the EWMA statistic, z_i at time i, is recursively calculated from the average of the values of the subgroups sampled. We calculate the first value of the EWMA series as the average of the first subgroup. EWMA [5] is given by

$$z_i = \lambda\, x_i + (1 - \lambda)\, z_{i-1},$$

where $i \in \mathbb{N}$, λ is a constant called decay factor, $0 < \lambda \leq 1$. z_{i-1} is the previous value, so that $z_0 = \mu_0$ and x_i indicates the i-th sampling. μ_0 is the mean of initial samples when the process is in-control. However, since the distribution may not be known, the average of some preliminary data is used as the initial EWMA value, $z_0 = \bar{x}$, where $\bar{x}$ is the average of initial samples. Figure 1 illustrates an approximation of the value of the weights for several lambda values, calculated by $(1 - \lambda)\, \lambda^{(i-1)}$.

This factor of decay (λ) allows to adjust the weight of the samples considered in the EWMA and, therefore, to consider more recent samples and to disregard older samples. For example, the graph from Figure 1 demonstrates that higher values of λ, as 0.9, allow a soft decay, thus considering more samples, while low values of λ as 0.2, have a marked decline, where very recent samples are of much higher relevance than the others.

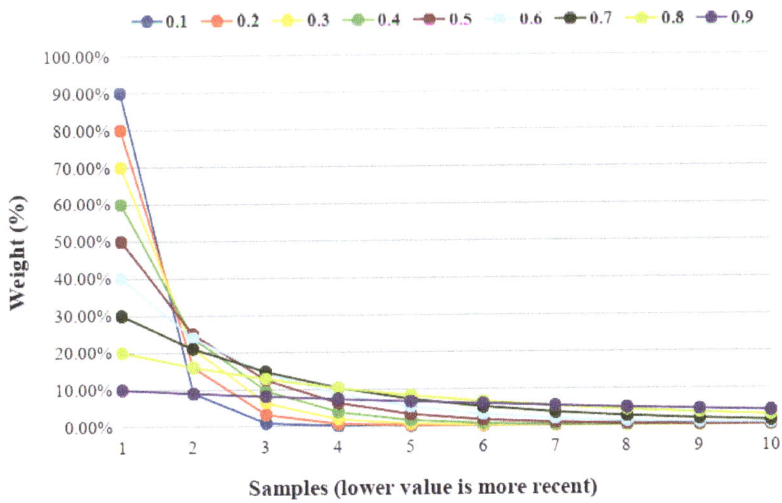

Figure 1. Graphic of the decay of the samples weight from lambda values (λ). Values between 0 and 1.

Thus, for our proposal, a smoother decay is more indicated, so the value 0.9 is ideal. However, a fine-tuning with details of the solution is necessary for this. The main detail is the number of samples that are relevant, for a proposal focused on safety in driving, times greater than 1 s can be very large. Therefore, we fixed that samples in a range of 1 s should have a good weight, noting that the most recent should have a slightly more substantial weight.

From this, since our solutions have an update rate of 10 samples per second, we consider that 10 samples make up the optimal set of our solution. We calculate the EWMA with values close to 0.9, as shown in Figure 2. We can see that in the tenth reading, the value 0.9 is the one with the highest weight $(0.87 = 3.71\%, 0.88 = 3.80\%, 0.89 = 3.85\%, 0.9 = 3.87\%, 0.91 = 3.85\%, 0.92 = 3.78\%, 0.93 = 3.64\%)$. Despite this, the most indicated lambda value is 0.93, because it is softer than other ones. The graph curve shows that the weight of its most recent sample is quite low, only twice as large as the tenth sample. So 0.9 is the best value for our solution. Nevertheless, we can use close values since the weight of each sample is very close and would generate a few differences.

In this way, the control graph can be defined with the following limits

$$UCL = \mu_0 + L\sigma\sqrt{\frac{\lambda}{2-\lambda}} \tag{1}$$

$$CL = \mu_0 \tag{2}$$

$$LCL = \mu_0 - L\sigma\sqrt{\frac{\lambda}{2-\lambda}} \tag{3}$$

where μ_0 is the mean of initial samples when the process is in-control, L determines the width of the control limits and σ is the process standard deviation, $\frac{\lambda}{2-\lambda}$ is the standard deviation component of EWMA statistics and L is a factor that allows a greater opening of the limits, usually 2 or 3. Thus, the L allows the control of the limits.

The EWMA calculation allows incorporating information from all subgroups of previous measures, with weights that increase the relevance of the last calculated sub-group. Thus a control decision is made based on the information from the previous subgroups and the current one.

EWMA is considered to be a quasi-non-parametric procedure, free of distribution, as Borror et al. [30] demonstrates. Moreover, Hunter [5] shows that the EWMA allows later samples to have larger weights. This feature is unusual for the proposal since it allows to consider subgroups with weights. Due to these characteristics, EWMA allows the sample to be more recent and the most relevant for identification.

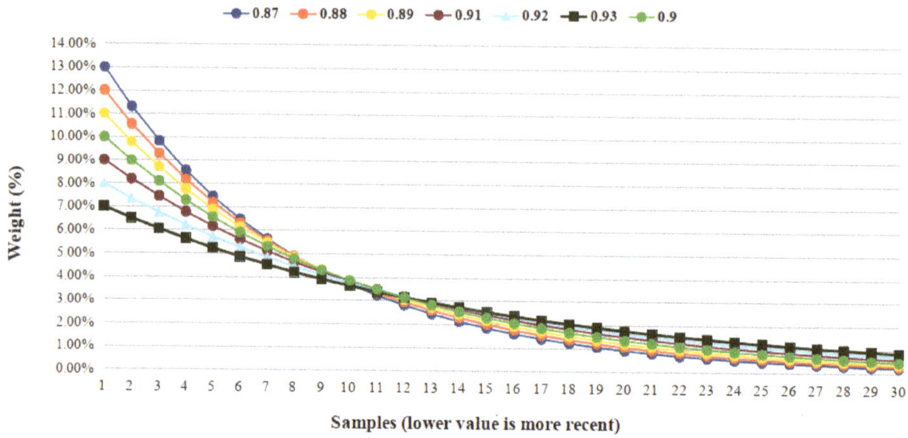

Figure 2. Graphic of the decay of the samples weight from lambda (λ) values close to 0.9. Values between 0.87 and 0.93.

4. Vehicle Drivers Monitoring through EWMA

The use of SPC in driver assistance systems is intended to facilitate the detection of errors and thus assist in the process of monitoring driving through EWMA applied to drive data. In this section, we present the different methods proposed for the monitoring of conductors, using the EWMA. We apply these methods to the following problems: i. Lane departure detection; ii. Sudden driver movement detection. iii. Driver fatigue detection;

4.1. Lane Departure Detection

We develop a method considering wheel data collected through accelerometer and gyroscope variables. The gyroscope decreases the rate of false positives because this sensor allows obtaining the acceleration at a specific moment without any calculations by the system. Thus, the acceleration of motion also becomes better evaluated, allowing better detection at high speeds. Tests involving only the accelerometer analyze more the amplitude of the movement than its acceleration, making this sensor necessary to increase the range of incorrect situations detected.

In this way, the identification of the dangerous movement of lane departure necessitated obtaining the data of the axes X and Y of the accelerometer (acceleration in m/s^2) and of the gyroscope (angular speed in $^{\circ}/s$) connected to the steering wheel, data processing and then the application of the EWMA control graph with $\lambda = 0.9$ and factor $L = 3$ in the data for the identification of events. The application of the EWMA graph first required a specification of steering wheel values, which we call calibration, to identify dangerous movements in lane departure. The values of the specification depend, as mentioned, mainly, on the speed of the vehicle. When a move that does not meet the specification occurs, the EWMA quickly detects it. 30-s specifications were generated for each direction of rotation of the steering wheel, clockwise and counterclockwise, each to detect events in its direction when we apply the EWMA to the X axis and both directions when we apply the control graph to the Y axis. We generate eight control graphs, one for each of the two calibrations (calibration counterclockwise and clockwise) for each axis (X and Y) and each of the two sensors (accelerometer and gyroscope).

The final result of the detections is the combination of the results of the two sensors generated by each of the calibrations. This combination of sensors is given by

$$(accX \cup accY) \cap (girX \cup girY)$$

so that the sets of points $accX$ and $accY$ detected as lane departure along the X and Y axis, respectively; and the sets of points $girX$ and $girY$ detected by the X and Y axes of the gyroscope, respectively. In this way, the control graph can identify out-of-control conditions when points are out of control limits, LCL e UCL.

We can obtain the angular position data through CAN bus of the car, in case of vehicles with electric assist direction. However, the application of EWMA will be similar. How we identify sudden movements is that the solution locates the lanes on the road without any environment perception equipment, like a camera. However, the results (Section 5) present a high number of false positives and negatives. This result motivates the proposal of sudden movements detection (Section 4.3), based on EWMA, and the driver fatigue detection (Section 4.4), based on image processing. Used together, they presented better results than the previous one. However, in order to provide a more complete solution we also consider the lane departure detection by monitoring the road with a camera (Section 4.2), because using only acceleration and yaw rate data would not be sufficient to detect lane departure due to the robustness issue to the disturbance and the unknown relative lane position.

4.2. Lane Departure Detection by Monitoring the Road with a Camera

This lane detection technique is intended to analyze the position of the vehicle concerning the lanes and thus indicate that the vehicle is moving out of lane or even if there was lane departure. For this purpose, we use several image processing techniques in order to be able to design efficient detection. The detection process used is illustrated in the flowchart of Figure 3.

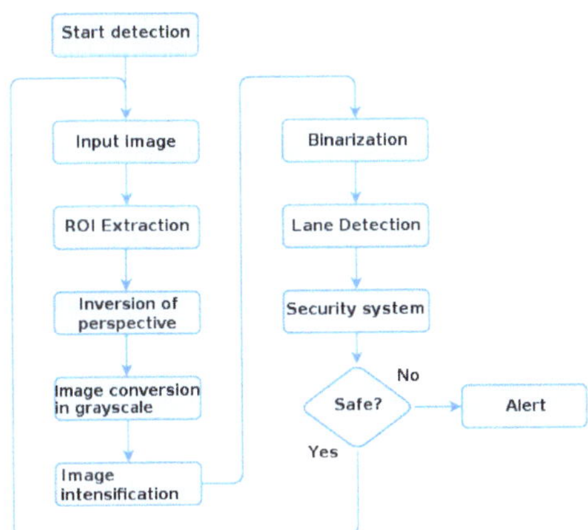

Figure 3. Lane detection flowchart.

The flowchart, for lane detection, consists of the following steps:

1. We extract the region of predefined interest (ROI) of the input image, on this image we apply the inversion of perspective by the technique *bird's-eye image* [31] creating an image with parallel lanes. This step is necessary since the lane identification technique requires that they be parallel to estimate their positioning;
2. A filter is used to intensify the marking of the lane [32] through a precomputed pixel width of the grayscale image, in the image line. The filter intensifies the pixel value of the image lane to another intensity.

3. With the intensified image, a threshold is adopted in order to make the binary image.
4. The pixels that are above the threshold receive the value 1 (white color) represented by the lane and 0 (black color) the rest of the image. The distance transformation formula is applied using as a metric the Euclidean distance in the binarized image [31]. This distance transformation is used to find the distance from the current pixel to the nearest white pixel. From this distance, we generate a gradient image where the darker regions are the lanes.
5. With the lane detected, the road is divided into two sectors, the left lane and right lane. The median of these lanes is the center of the road. From the center of the image (center of the vehicle), we can obtain the variation of the position of the vehicle on the road. With each limit, left and right, as 100%, we consider a position higher than 60% as invading another lane. We detect a zigzag movement by analysis in a time interval.

The application of the detection process used can be seen in Figure 4.

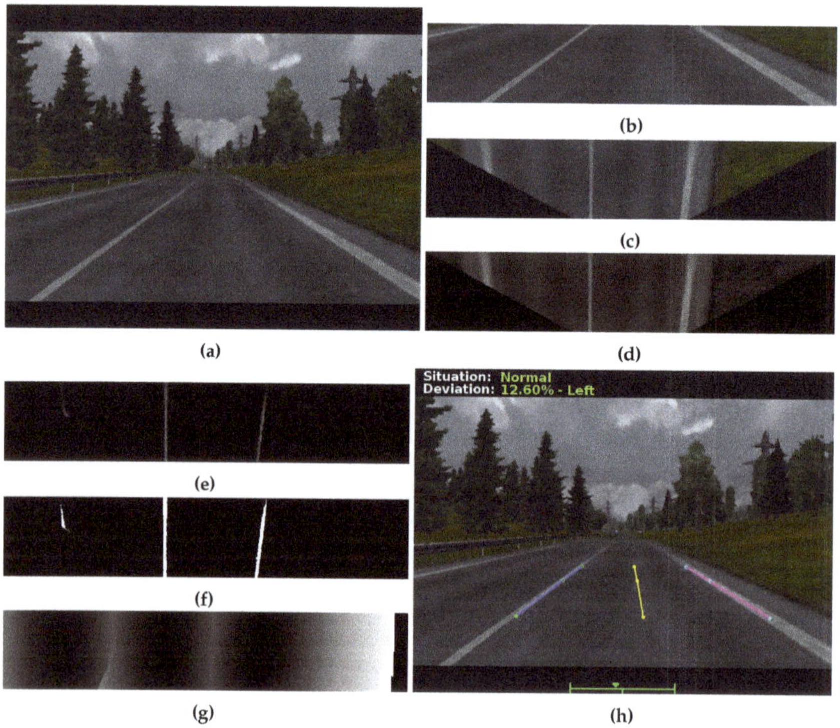

Figure 4. Lane detection on one of the *frames* experiments. (**a**) the input image; (**b**) the predefined Region of Interest (ROI); (**c**) ROI with the technical application *bird'eye* [31]; (**d**) ROI in grayscale; (**e**) the intensified image; (**f**) the binarized image; (**g**) the transformation of distance; and (**h**) the end result of lane detection.

It is worth noting that this method of lane detection has some limitations; as an example, we can list: i. vehicles within the region of interest may impair the binarization phase of the image; ii. although the method has efficiency at night, vehicle headlights in the opposite direction can generate false positives; and iii. if one of the road lanes is erased or covered up, such as by land, detection is impaired. It is essential to highlight that, once this strategy does not use EWMA, we do not use it to evaluate our system (Section 5). The system could use the lane information generated by this strategy to improve the robustness of the system. However, we propose to avoid the frontal camera to keep the system cheap and with a viable time complexity to execute in an embedded system.

4.3. Sudden Movements Detection

The detection of errors caused by sudden movements aims to identify two types of event: i. sudden movements in the steering wheel that cause lane deviation; and ii. sudden movements on the brake pedal. These events are dangerous because they can cause some accidents and even indicate that the driver is drowsy.

From the EWMA control limits, it is possible to identify that there were sudden movements. For sudden movements in the steering wheel that generate lane deviations, we generate specifications of accepted values, which we call calibration. This specification depends on the speed of the vehicle, and the turns that go counterclockwise.

When a move that does not meet the specification occurs, the EWMA quickly detects it. Detection occurs through the application of the control graph to accelerating data ($^{m}/_{s^2}$) of the X axis of the accelerometer. The pedal used is the brake pedal. We calibrate it, and the data analyzed represent the distance between the pedal and its support, in centimeters, so that it is 7 cm when it is at rest.

Due to problems in the accelerometer and potentiometer signals, such as noise and inaccuracy, we apply the Low Pass Filter with Moving Average:

$$new_i = \frac{(sample_i + sample_{i-1} + sample_{i-n-1})}{n},$$

where $i \geq n$ is the index of the new sample, starting at the n. This filter is the average of the last signal samples n, as n being the size of the window. This moving average must be adjusted to each sensor, thus not having an ideal value to obtain a better signal. Tests should be made to verify the noises' remotion without generating significant delays in the signal. Each sensor used a window size that best suited the signal type. We define this window from the analysis of the signal with values that ranged from 7 to 2, in this way, we chose the sizes: 2 for the accelerometer and 3 for the potentiometer.

We obtain the values used in the moving average for the accelerometer, and the potentiometer through tests. We verify the best signal generated for each sensor. The graphics in Figures 5 and 6 present some readings with different window sizes. We can see in Figure 5, that the curve without moving average presents some noise, being a little soft, whereas the curve with the size two window has this problem reduced. Larger window sizes, while showing smoother curves, add a considerable delay, making them less attractive to use. Therefore, size two was considered better for the accelerometer. For the potentiometer (Figure 6), the curve with window size 2 softens the signal. However, the signal needs to be even smoother. Thus, window size three was chosen, even adding a small delay because it has a smoother curve. Larger window sizes generate a much more significant delay with few benefits in the curve.

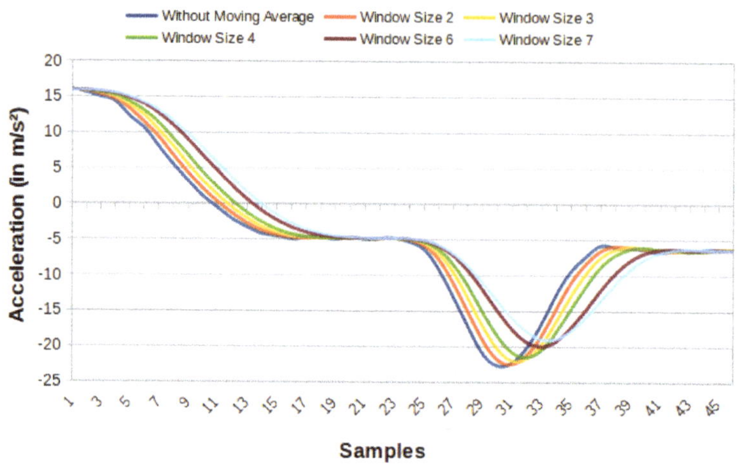

Figure 5. Comparison graph of the accelerometer reading on the steering wheel without moving average and moving average with windows of sizes 2, 3, 4, 6 and 7.

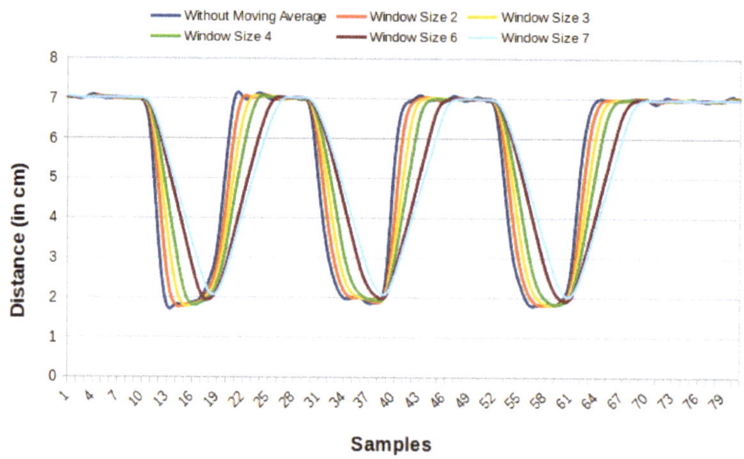

Figure 6. Graph reading comparison of the potentiometer reading on the brake pedal without moving average and moving average with windows of sizes 2, 3, 4, 6 and 7.

4.4. Driver Fatigue Detection

We detect driver fatigue through driver characteristics that are extracted using image processing, pattern recognition, and SPC techniques. This method has as input only the images acquired by the camera facing the driver's face. Camera image processing provides driver related information. This information is the face direction (front, left or right) and eye condition (open or closed) in instant data.

We divide the driver fatigue detector into three stages: i. face detection; ii. eyes classification; and iii. application of the EWMA control graph to the closed eye percentage of the last two seconds. The steps for detecting driver fatigue are illustrated in Figure 7.

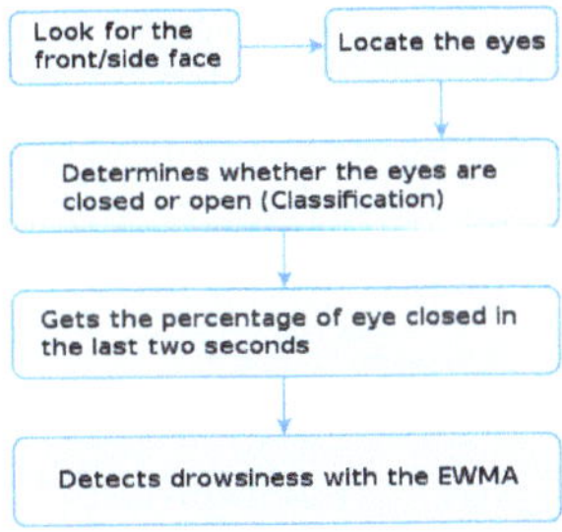

Figure 7. Driver fatigue detection diagram.

Face detection is performed using the method proposed by Viola and Jones [33] with the increases proposed by Lienhart and Maydt [34]. In the first, Viola and Jones use machine learning to detect objects in images, dividing the base into two parts, one containing the object and the other not, to extract features that differ from the two parts. The proposal of Lienhart and Maydt uses more characteristics in the learning frame employing rotations in 45 ° in the classifier training.

In our proposal, we use two classifiers, one trained to detect frontal faces and another trained to detect lateral faces. We train the side face detection classifier with right-sided profile faces, so to detect left-facing faces the image is mirrored before the search process. The first step of the algorithm is to search for the front face, if found, the algorithm passes to the next step, the classification of the eyes. If the front face is not detected, the algorithm searches the side face (left and right) and then it processes the next image.

We perform the classification of the eyes as follows: i. Locate the eyes: during software initialization, the location is made using a trained classifier to locate eyes using techniques used for face detection; ii. The location of the eye is done through template matching, using a template saved at startup. iii. The classification determines if the driver's eyes are open or closed at that moment. We classify partially open eyes as open. The classification is made using three Support Vector Machine (SVM) classifiers, trained with a base 4000 open eyes and 4000 closed eyes, these classifiers receive as input data taken from the following image descriptors: Local Binary Pattern (LBP) [35], Local Ternary Pattern (LTP) [36] and Histograms of Oriented Gradient (HOG) [37]. By classifying the eyes in a sequence of images, it is possible to detect the driver's blinks and also to measure the duration of each of them. The driver fatigue detector extracts the percentage of closed-eye time per reading at each 2 s. We use this percentage as input in our EWMA control graph. In this case, we have only one graph.

For monitoring the driver's attention, we use the EWMA control graph with the eye closed by reading using $\lambda = 0.9$ and factor $L = 3$, indicating moments of driver fatigue. An observer validates fatigue moments. In addition, an audible alert may be issued to indicate to the driver that he is drowsy and that he must resume attention.

5. Evaluations and Results

In this section, we present the results of the evaluations performed for the proposed methods. The objective of the evaluations is to identify the efficiency of the methods and the prototype in a simulated environment.

Because it is not safe to conduct tests related to driving errors on the road, we use a vehicle emulator environment to conduct experiments. Some of the main advantages of its use are experimental control, low cost, efficiency, security, and ease in data collection. The emulator used consists of a computer, a set of monitors, a realistic cockpit, a steering wheel, manual gearbox and clutch pedals, brake and accelerator [38] (Figure 8).

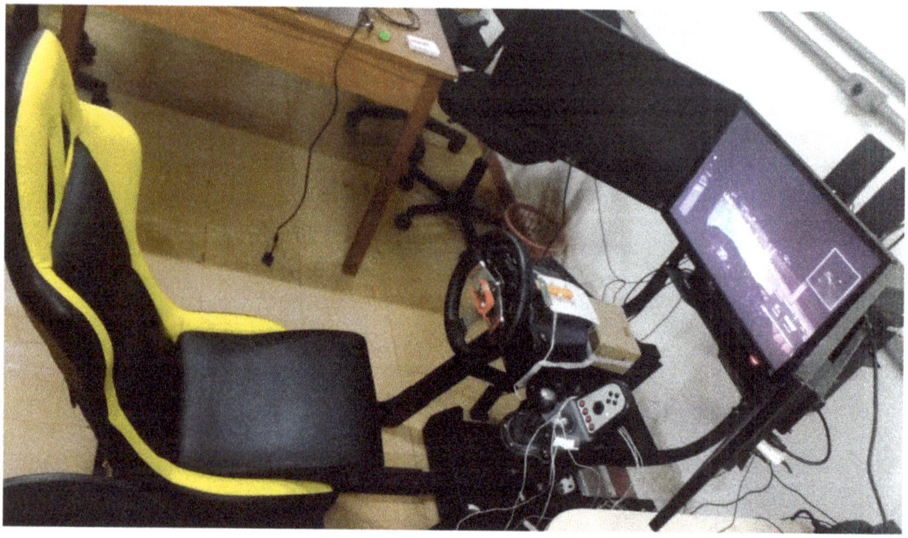

Figure 8. Driving simulator.

Some researchers such as Auberlet et al. [39] and Mayhew et al. [40] perform the validation of the use of emulators to create a real-world environment because this environment seems to include limitations, as the driver does not realize the real risks about real driving. However, Mayhew et al. [40] state that "(...) collectively, the results of the concurrent and discriminant validity studies support the use of the simulator as a valid measure of driving performance for research purposes". Thus, the tests were done using the Euro Truck Simulator 2 (Euro Truck Simulator 2, Last Access, April 2019: http://eurotrucksimulator2.com/) which according to Lee et al. [41], realistically imitates the driving process, and according to the parameters of each scenario. Besides, an automatic transmission was used in the experiments to avoid noise caused by the change of gears in the obtained data.

5.1. Lane Departure Detection

In this section, we present our initial evaluation of the lane departure detection methods with two sensors. The objective is to show that by using two simple sensors, we can detect the drivers' behaviors.

5.1.1. Experiment Setup

To evaluate the lane departure detection technique, we have developed a functional prototype for lane departure detection. The first version (used in the first scenario evaluated) considers a tablet, the GT-P4500, which has 1 GB of LPDDR2 memory, 32 GB of internal memory and a 1 Ghz Dual-Core ARM Cortex-A9 processor, was coupled to the tablet with the function of obtaining the values of the three axes, X, Y and Z, of the accelerometer in m/s^2 and of the gyroscope in °/s. This device has a capacitive type accelerometer, the KXTF9 manufactured by Kionix, and a gyroscope MPL Gyro.

The second and more sophisticated version (used in the second scenario evaluated) obtains the data using an Inertial Measurement Unit (IMU) [42] with an accelerometer, gyroscope, and magnetometer connected to a microcontroller ESP8266 [43] with Wi-Fi interface. We install the microcontroller with the IMU in the center of the steering wheel. Figure 9 illustrates the wiring diagram of the microcontroller to the IMU unit and the battery. It shows that the connection between the microcontroller and the IMU takes place through the respective pairs of pins: 3V3/VIN, GND/GND, SDA(2)/SDA, SCL(14)/SCL. The connection between the microcontroller and the battery is via the JST connector. In order to send the code to the module, we use an FTDI Serial USB RS232 Converter, where the pins of the microcontroller were connected to the converter as follows: GND to GND, NC to

CTS, 3V3 to VCC, RXI to TXO, TXO to RXI and DTR to DTR. This connection was not included in the figure since it is a connection used only for recording the source code of the prototype in the module.

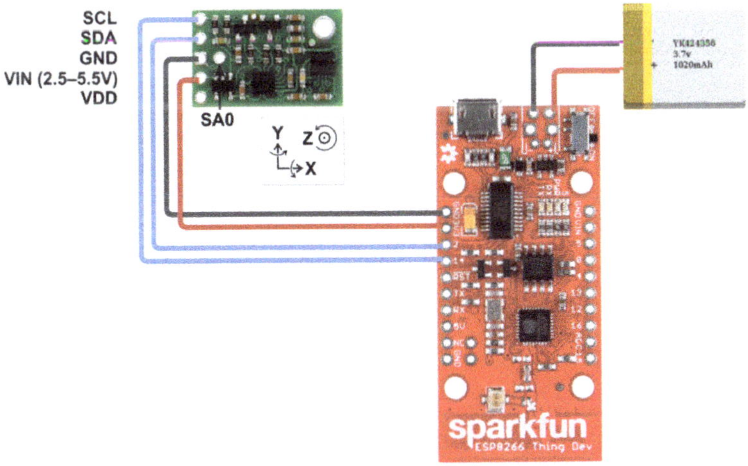

Figure 9. Connection diagram of the prototype components (module ESP8266, IMU unit and battery).

The developed prototype uses a web service on the Wi-Fi module and an Android application to obtain and analyze this data using EWMA graphics. It was developed to take advantage of the processing of mobile devices that are common to users and thus to avoid increasing the cost of the prototype to end consumers. In addition, it has a reduced size, about 22 mm × 66 mm × 42 mm. The application is connected to the module's Wi-Fi network and calibrated by the user from the steering wheel rotations to the two sides for 30 s. The angle to the rotation for each side varies according to the automotive system and its sensitivity, being an angle smaller than one that could generate a lane departure by sudden movement.

We perform the lane departure monitored by the prototype with EWMA under the data of the X and Y axes of the accelerometer and gyroscope. First, the application begins to consume the data from the web service and performs a calibration where the conductor rotates the steering wheel to both sides in $\approx 30^\circ$. We use this value because in the tests carried out. It proved to be the "limit" angle that did not generate speed range between ≈ 40 and ≈ 60 Km/h. We store this calibration in the application for future routes, and we can recalculate if necessary, this prevents the calibration of each route. For the calculation of EWMA control limits we use factor $L = 3$ and $\lambda = 0.9$. We chose these values because they guarantee a higher weight to the most recent readings, but with a soft decay, as presented in Section 3. However, to ensure a more robust system, these values can be reconfigured. After calibration, the system begins to monitor the data it obtains from the web service and, in case of an out-of-control point, as caused by an incorrect movement, the prototype issues an audible alert.

Finally, for both the accelerometer and the gyroscope, the sensor can add Brownian noise [44], obtaining an offset in the expected reading. Besides, reading the sensor data is converted into an electronic signal subject to electronic noise, generating an unexpected departure. For this reason, it is necessary to apply filters and calibrate the data so that it is possible to obtain a data set with greater precision. We used an Average Resting Calibration and the Low Pass Filter with Cumulative Average to mitigate error in sensor readings. For the average resting, we used samples obtained in the range of 30 s, thus, with the steering wheel and sensor at rest, samples were obtained that were used for an average considered as zero reference value. The low-pass filtering consisted of reducing the set of measures by a cumulative mean. For this task, we use a regular number of samples, since using many samples could lead to data loss, while a few samples could remove the noise. This accumulated average was calculated every three samples, because for a proper analysis, some readings per second are necessary, and we were able to register, without errors, 10 readings per second. For the beginning of the tests, an alignment was made to establish the reference attitude. We perform this alignment with the technique "gyro-compassing" [45].

5.1.2. First Scenarios Evaluated

In order to validate the proposed methods, we evaluated the rate and number of identified events, false positives, and false negatives. These experiments were carried out in a controlled manner in a path of the driving emulator presented at the beginning of this section. In the scenarios used, the driver made deliberate errors at some points along the way. The movements were analyzed and validated by the observer for verification with the control graph. We evaluate these movements through the movement of the steering wheel correlated with the position of the vehicle in the lane from the observation of this position.

The test route used (Figure 10) was chosen to include several driving situations, such as straight road, corners, and curves, stretches of double lanes and lanes with opposite directions. An observer evaluated all the experiments in order to validate lane departures for comparison with the graph. The experiments were performed by adult drivers under normal conditions and with a driver's license and repeated 25 times. The drivers or tests have the following specifications:

- Each driving lasted about 5 min.
- The drivings followed a predefined route;
- The driver kept the vehicle speed around 40 km/h.
- The drivers are informed that they must make some mistakes at certain times.
- Sudden movements were performed each 40 s, approximately.
- The first four movements were counterclockwise, and the other four were clockwise, totaling eight movements.
- The lane deviations were observed, only large deviations were accepted where the entire vehicle crossed the lane limits, invading the other lane, characterizing a lane departure.

The drivers were without signs of drowsiness or any other abnormal condition because it was not our goal to assess the causes of lane departures.

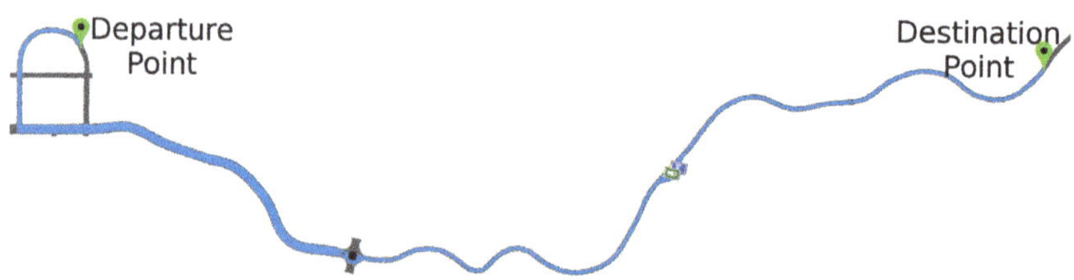

Figure 10. Test route in driving simulator.

In order to apply the EWMA control graph to the samples, specifications were generated using the steering wheel movements within a standard pattern (movements that do not generate large lane deviations—lane departures) for the speed defined in the experiments, 40 km/h. This specification will be called the EWMA calibration. Two types of the specification were generated, with counterclockwise rotations and clockwise rotations, each with data of 30 s of duration. We use the data obtained in the conduction tests in the control graph with the calibration data, thus generating two graphs for each group (X axis and Y axis data) of samples. Each graph is referring to calibration and thus to a direction of rotation (counterclockwise and clockwise). Each graph has as objective to identify errors related to turns in the direction of rotation of the calibration. However, the Y axis allows the identification in two directions.

As seen in Section 3, the EWMA definition has a constant λ and the limits of the control graph have a factor L, as defined in EWMA equations. We chose factor $L = 3$ for the experiments so that the limits were slightly away from the center line, avoiding a high false positive rate. We chose $\lambda = 0.9$ so that several recent readings to weight the results. A value below λ would make only the very recent readings have weight, as presented in Figure 1. For example, with a $\lambda = 0.9$ the first reading would have a 0.1 weight and the tenth 0.03 while with $\lambda = 0.2$, the first reading would have 0.8 weight and the 10th 8.192×10^{-8}, an extremely insignificant value. The vehicle speed was maintained around 40 km/h because in Brazil this is the speed in collecting ways (streets that allow access to and exit of arterial roads, usually with traffic lights and allowing movement within a region of the city). Finally, the refresh rate was 100 ms, since it was the rate at which the application gets data without loss of performance or accuracy. Table 1 shows a summary of the parameters used in the experiments.

Table 1. Parameters used in Exponentially Weighted Moving Average (EWMA) experiments.

Parameter	Value
Factor L (the multiple of σ)	3
Factor of decay (λ)	0.9
Data update rate	100 ms
Vehicle speed	40 km/h

Figure 11 shows the number of events detected in absolute numbers with a combination of the EWMA application results in the accelerometer and gyroscope. The results present the average values with a confidence interval of 95%. The number of errors that could be detected was 8. The combination of the data from the two sensor axes had a good part of the eight errors detected with low false negatives ≈5.76 and only ≈0.11 false positives.

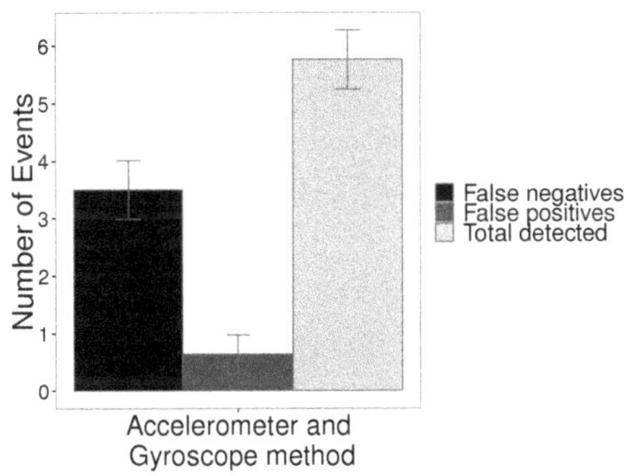

Figure 11. Number of detected EWMA application events on accelerometer and gyroscope.

Figure 12 shows the results of detection rate and false positives with the combination of the results of the axes X and Y. The detection rate was 72% with a low false-positive rate of only 8.07%.

Table 2 illustrates the application of confusion matrix on test data. We can observe that the ability to correctly predict what it values is 62% (sensitivity), showing to be able to recognize essential cases in 98% of the cases (precision) and thus the precision of the test is the F1 Score which was 76%.

Table 2. Metrics from the confusion matrix on test data with Accelerometer and Gyroscope.

Metrics	Value (%)
Sensitivity	62
Accuracy	98
F1 *Score* or *F-measure*	76

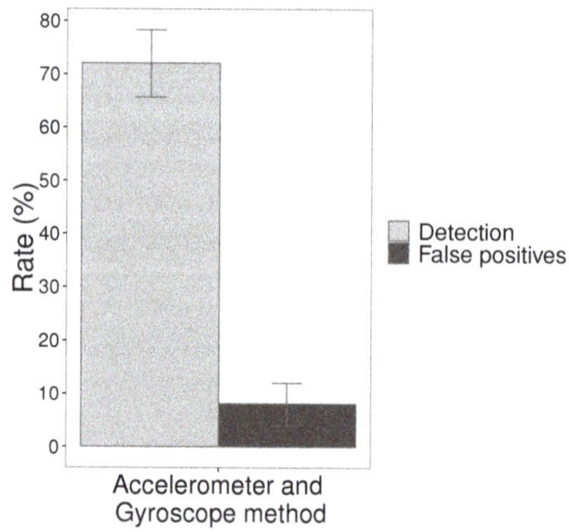

Figure 12. Detection rate and false positive rate with the combination of the results of the two axes, *X* and *Y*.

This efficiency is a feature of the EWMA control chart that allows efficient and rapid detection of out-of-control events in a process. Testing in a controlled environment is one factor that may have contributed to high efficiency.

Finally, the Figures 13 and 14 show the EWMA graphs of one of the test conductions. On the figure, each "Reading" is a sample and then the abscissa values are the samples. The graphs with the axes *X* and *Y* of the accelerometer, Figure 13a,b show all eight errors being detected, gray dots, with few false positives in the limit *UCL* of the graph of the *X* axis. While the EWMA with the axes *X* and *Y* of gyroscope, Figure 13c,d did not identify all the errors, so that the *Y* axis detected only two of the errors. However, the combination of the sensors allowed the detection of most errors.

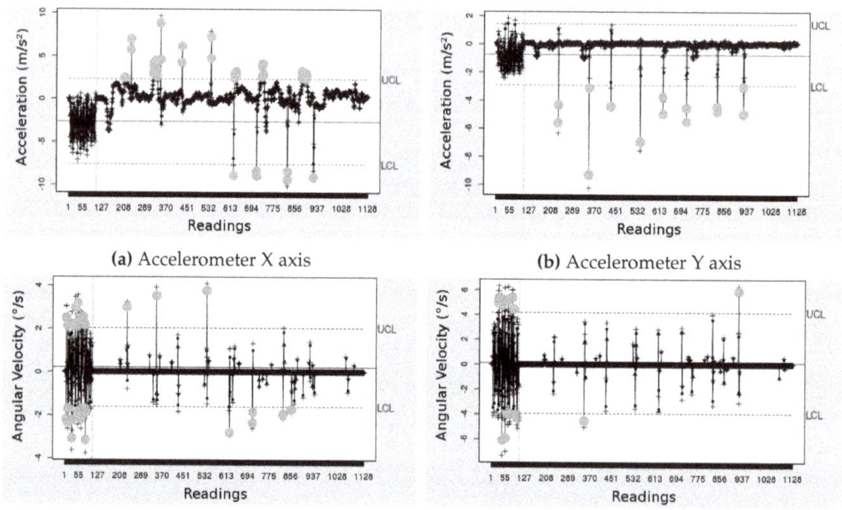

(a) Accelerometer X axis **(b)** Accelerometer Y axis

(c) Gyroscope X axis **(d)** Gyroscope Y axis

Figure 13. EWMA control graphs applied to the accelerometer data, in m/s^2, and the gyroscope, in $^\circ/s$. Calibration with clockwise rotation data.

The graph with the X axis of the accelerometer, Figure 14a shows all eight errors being detected (gray dots), with few false positives in the limit LCL, however, the graph of Figure 14b with the Y axis detected only two errors. While the EWMA with the axes X and Y of the gyroscope, Figure 14c,d did not identify all the errors, so that the Y axis detected only two of the errors. The combination of sensors enabled detection of most errors.

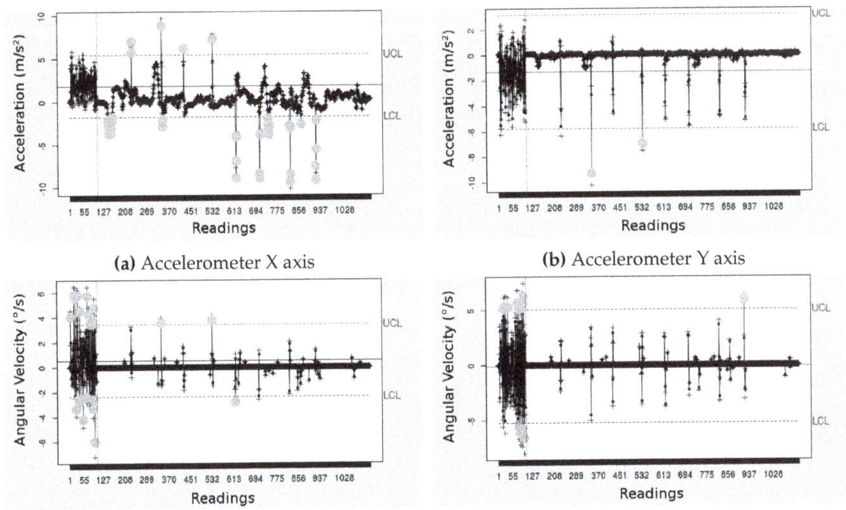

(a) Accelerometer X axis

(b) Accelerometer Y axis

(c) Gyroscope X axis

(d) Gyroscope Y axis

Figure 14. EWMA control graphs applied to the accelerometer data, in m/s^2, and the of the gyroscope, in $^{\circ}/s$. Calibration with rotation data counterclockwise.

Our proposal used the union of the X axis and the Y axis of each sensor, since they are complementary, and the intersection of the two sensors so that the detection was more faithful to the reality when being detected by the two sensors. However, because the sensors intersect, some events can only be detected by the accelerometer (amplitude of movement) and not by the angular velocity (gyroscope) or otherwise, thus reducing its efficiency concerning the gyroscope only. A different combination of the detections of each of the sensors can result in higher precision.

5.1.3. Second Scenario Evaluated

Figure 15 shows the tested route. The experiments consisted of a scenario where the driver committed some lane departures during the course. We analyze the driver by an observer and by the system, in this way, it was possible to analyze when the prototype emitted a correct alarm. We set the period for each record to 30 min. The experiments were performed by adult drivers under normal conditions and with a driver's license and repeated 13 times without constant speed and six times with maintained speed close to 60 km/h and the results presented average values with a confidence interval of 95%.

Figure 15. Route for testing with the prototype in the driving emulator.

Thus, we performed experiments in two ways. In the first one, the drivers had to keep the vehicle, as far as possible, 60 km/h and in the second one, we did not define the speed. Thus, they obeyed the following items:

- 30 min of driving.
- Performing 20 lane departure movements as defined in the proposal.
- The movements were to the two sides, left and right, being thus from turns in the steering wheel in two directions: anti-clockwise and time.

The graph of Figure 16 shows the complete number of results of detections with constant speed and without constant speed. It is possible to observe that at constant speed the number of events detected was slightly higher than without a constant speed, ≈16.83 against ≈15.38. The number of false positives was very close, ≈3 with speed maintained close to 60 km/h and ≈2.69 with speed varying to any value. Finally, the false negatives were a bit lower with constant speed, getting slightly above 3, ≈3.16, different from the other approach that reached ≈4.61.

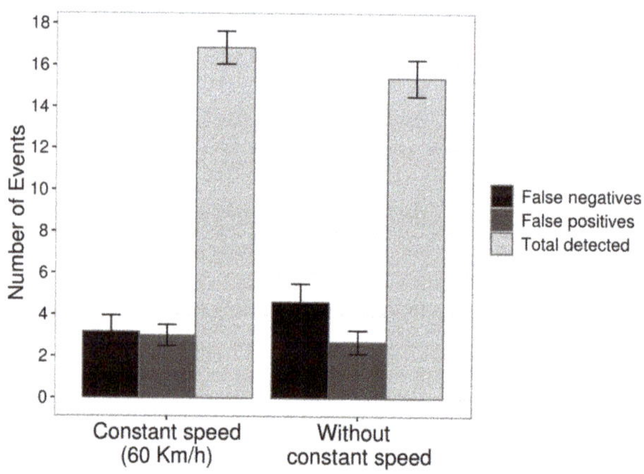

Figure 16. Number of events detected in absolute number by prototype with constant speed and without constant speed.

We divide the percent results into two parts, first the results without the control of the vehicle speed and then with speed maintained close to the 60 km/h. The graph of Figure 17 presents the detection and false positive rates of the proposed system with a constant speed close to 60 km/h and without speed control, the driver can drive the vehicle at any speed, normally between 40 km/h and 80 km/h. The application of the EWMA control chart on the data obtained by the microcontroller located on the steering wheel ensured a detection rate of 76.92% with 11.70% and false positives without constant speed. We calculate the percentage of false positives from the sum of the total expected events during driving plus total false positives.

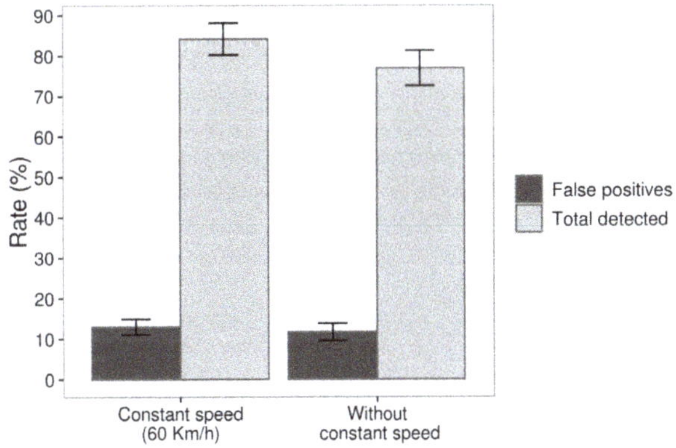

Figure 17. Detection rate and false positive rate of prototype with constant speed ($\approx$ 60 km/h) and without constant speed.

Besides, Figure 17 illustrates the detection rates and false positives of the prototype when drivers have maintained their speed, as close as possible to 60 km/h. We can see that the detection rate reached a value, 84.16%, and 12.98% of false positives. We calculate this percentage of false positives as the previous one.

The comparison between the two forms, constant speed, and no constant speed, indicates that the method has a significant gain when we consider the speed of the vehicle for calibration allowing to control it during driving. Despite this gain, the system had a reasonable detection rate for "more real" drives, where the speed does not remain constant. Besides, a part of the false positives occurred in moments where the vehicle made sharp curves with low speed, like when leaving the garage. However, we consider these false positives because it is a system directed to real environments.

Table 3 illustrates the application of the confusion matrix on test data with constant speed and without constant speed. We can see that the system can correctly predict the condition evaluated in 84% of the cases against 76% tests without a constant speed. Recognizing relevant cases in 86% in both cases, the precision of the tests with constant speed and without constant speed is 85% and 81%, respectively.

Table 3. Metrics from the confusion matrix on the prototype data.

Metrics	Constant Speed (60 km/h)	Non-Constant Speed
Sensitivity	0.84	0.76
Accuracy	0.86	0.86
F1 *Score* or *F-measure*	0.85	0.81

5.2. Driver Fatigue and Sudden Movements Detection

After the first evaluation, we compose a complete evaluation scenario by considering different sensors and elements in order to identify the driver behaviors. Thus, we execute together with the driver fatigue detection and sudden movements detection with sensors in the steering wheel and a brake pedal. To perform this evaluation, we developed a functional prototype for comprehensive driver monitoring. The prototype consists of a video camera positioned in front of the driver connected to a Quad Wandboard board [46] to capture information from the driver, through his face. In the experiments, we used a 15-megapixel camera, Full HD 1080p, model Logitech C920. We also consider a capacitive accelerometer, model MPU6050 of IMU GY-88, in the steering wheel connected to a Mega 2560 Arduino to capture the movements of the steering wheel; and a rotary potentiometer 10 KΩ on the brake pedal connected to an Arduino to capture the pedal data. Figure 18 illustrates how the positioning of the hardware inside a vehicle could be. In the proposed positioning, we can only vary the location of the processing units. We cannot change the position of the onboard computer and sensors and cameras. The onboard computer needs to be close to the driver so that the driver can hear and visualize the warning signs clearly and have easy access to him to perform other activities of his interest. The road facing camera is a lane departure detection method based on image processing. This technique goes beyond the scope of the article, thus it is not evaluated.

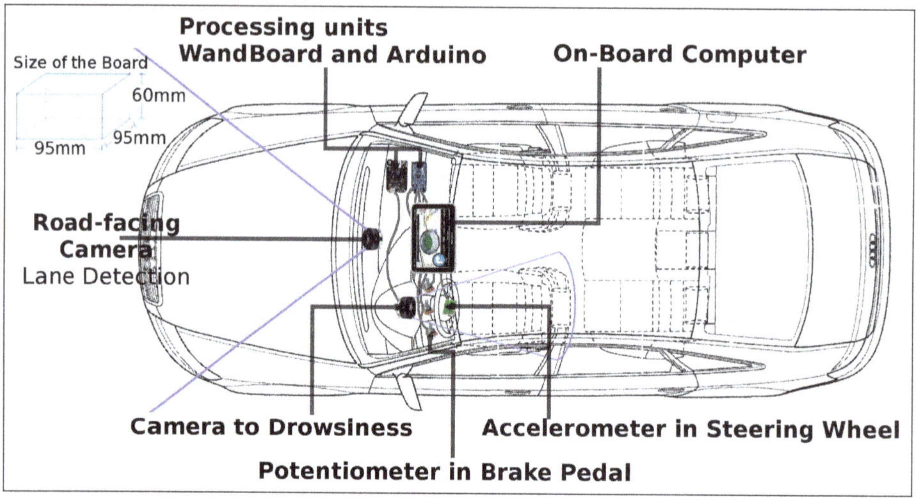

Figure 18. Positioning conceptual the components of hardware of the prototype inside a vehicle.

Figure 19 illustrates the connection diagram of the Arduino with the connection between the Arduino and the IMU unit (with accelerometer) located on the steering wheel and the potentiometer located on the brake pedal.

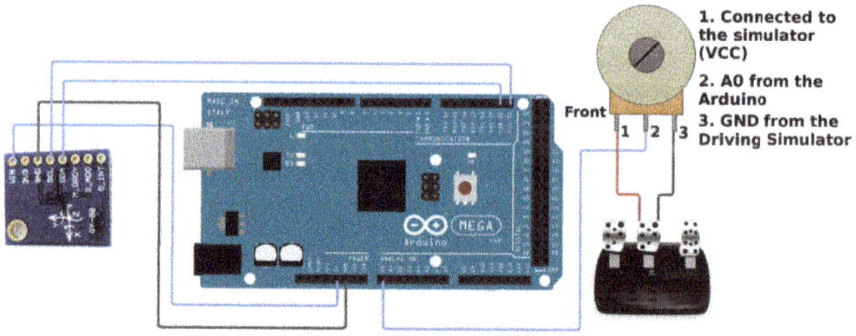

Figure 19. Connection diagram of the Arduino to the accelerometer and potentiometer.

The system meets certain requirements mentioned in [47], acceptable performance when run in real time, ≈20 FPS with a Wandboard and ≈10 readings per second with the Arduino; low cost, around US$30,000; with a reduced size, approximately 95 mm × 95 mm × 60 mm; and low power consumption and flexibility.

We evaluated the driver fatigue detection and the detection of sudden driving movements together. Separated monitoring leads to a non-satisfactory answer as to the inference about errors made by the drivers. For this evaluation, a driving simulator highway, previously presented, was used. The results presented values with a confidence interval of 95%. The test route considered is illustrated in Figure 20. During the tests, the driver started the vehicle and followed the defined path.

Figure 20. Test route in the driving simulator.

The experiments were carried out by six adult drivers, all males aged 20 to 27 years (mean $\pm$ standard deviation (SD): 24 $\pm$ 2.8 years), under normal conditions and with a driver's license in a controlled environment and with the following specifications:

- The driver was asked to keep the vehicle speed around 80 km/h.
- The driving lasted 10 min.
- In the first 5 min, the driver was instructed not to make mistakes and to pay attention.
- After 5 min of driving, the driver was asked to simulate driver fatigue, such as blinking slower and longer. An observer assessed this driver fatigue.
- The driver performs an event at each $\approx$30 s. We use three types of events:

  1. driver fatigue followed by sudden movements on the steering wheel counterclockwise.
  2. driver fatigue followed by sudden movements on the steering wheel clockwise.
  3. driver fatigue followed by sudden braking, simulating a moment where the vehicle approaches some object and needs to brake to avoid a collision.

Drivers were in normal condition and simulated drowsiness, for example with slower actions and lack of attention, such as closing their eyes for a few milliseconds. Our team evaluated these actions and validated by results.

Due to problems in the accelerometer and potentiometer signals, such as noise and inaccuracy, a low-pass filter of the moving average type is applied. Each sensor used a window size that best suited the signal type. We define this window from the analysis of the signal with values that ranged from 7 to 2. In this way, we chose the sizes: 2 for the accelerometer and 3 for the potentiometer.

The Wandboard microcontroller processes and obtains data from the camera positioned in front of the driver and provides:

- The total number of blinks.
- The number of blinks in the last 20 s.
- The percentage of time the closed eyes at each reading (2 s).
- The percentage of time that the closed eyes in the last 50 s.
- The face position (front, right, left, or distracted).

Arduino microcontroller, on the other hand, retrieves and processes data from:

- Accelerometer (X axis) on the steering wheel, in m/s^2.
- Potentiometer on the brake pedal, in cm. Possibly varying from $\approx$7 (resting) to $\approx$1 (pressed).

We apply the EWMA control graph in the samples of each method used as follows:

- Steering wheel monitoring: We generate a specification (calibration) with movements that are not incorrect for the speed of 80 km/h. When a move that does not meet the specification occurs, the EWMA quickly detects it. Detection occurs by applying the graph to the X axis acceleration data of the accelerometer located on the steering wheel.
- Brake monitoring: Uses a specification as made for the steering wheel. The data represent the distance between the pedal and the pedal holder, in centimeters, so it is 7 cm when resting. In case of incorrect movement, EWMA detects incorrect moments.
- Monitoring of driver fatigue: The EWMA was applied to the percentage of closed-eye time per reading (2 s) to monitor the driver fatigue. According to the conduction samples, the EWMA generates the limits that allow identifying the driver fatigue.

For the parameters used, we chose factor $L = 3$ so that the boundaries were slightly away from the center line, avoiding a high false positive rate. In order for several recent readings to weight the results, we chose $\lambda = 0.9$. The vehicle speed was maintained around the 80 kmh because in Brazil this is speed in fast traffic routes (roads with different lanes, without traffic lights, without pedestrian traffic and with great extension) and stretches of highways (paved roads). In this way, the accidents are more severe and are better for testing the various functions of the system. Finally, we define the update rates by the performance of the boards with our algorithms: $\approx$20 FPS with the Wandboard and $\approx$10 readings per second with the Arduino.

Figure 21 shows the detection and false-positive rates of the system components and the complete system. The results present the average values with a confidence interval of 95%. The application of the EWMA graph, in the steering wheel data, guaranteed a rate of detection of sudden movements of 91.66% with only 5.55% of false positives. The brake movement results show detection of 94.44% of sudden braking at a high rate of 24.72% of false positives. This proper detection is a feature of the EWMA control graph that allows efficient and fast detection of events out of control in a process.

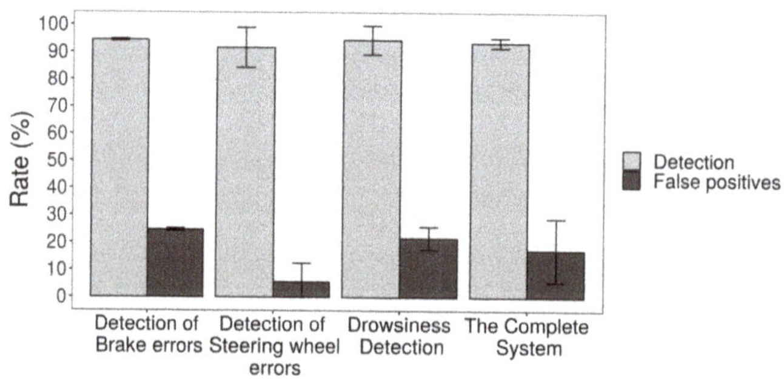

Figure 21. Detection rate of components and complete system.

An observer assesses the driver fatigue, and we compared to the EWMA graphs applied to closed-eye time per reading, Figure 21 shows the graph with the results. Despite the high detection rate (94.46%), driver fatigue detection had a considerable rate of false positives, 21.68%. This false positive occurs because of the way the EWMA works. It is possible that with better refinement, the false positives decrease without a considerable decrease of the detection rate.

In addition, Figure 21 shows that the complete system has a high detection rate (93.52%) and acceptable false positive rates (17.32%), we generate these values from the average detection of the components that had the detected and false positive rates analyzed. Table 4 presents the application of a confusion matrix on test data—it demonstrates that the system's ability to correctly predict the assessed condition is 94%, 91% and 94% for the detection of brake errors, steering wheel errors, and driver fatigue, respectively. The ability to recognize essential cases (accuracy) is 79%, 94% and 81%

and the accuracy of the test which is the average of the precision and sensitivity is 86%, 92% and 87%. These results corroborate even more with the conclusions previously presented.

Table 4. Metrics from the confusion matrix on the evaluated data.

Metric	Brake Pedal Detection (%)	Steering Wheel Detection (%)	Fatigue Detection (%)
Sensitivity	94	91	94
Accuracy	79	94	81
F1 *Score* or *F-measure*	86	92	87

Once we detect some sudden movement, the system triggers an alert to the driver. Such alerts occurred in two situations: i. visual and audible alert indicating that a sudden movement occurred on the steering wheel or brake pedal, and ii. an audible alert when the driver kept his eyes closed for 1 s. The simulator tests showed that the alerts were satisfactory for results above 90% on detection in all methods.

Finally, we present the control graphs of a test run, Figure 22. In all the graphs, very close points were considered referring to the same or false positive. Gray dots indicate out of control points, detections, or false positives. We can see in Figure 22a that all six errors were detected, three sudden movements on the steering wheel in each direction (counterclockwise and clockwise). We detect some errors more than once, due to the rapid detection of the EWMA.

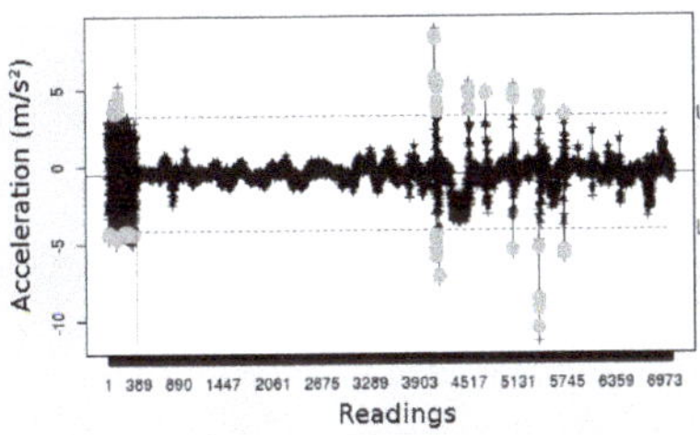

(a) Steering Wheel Detection

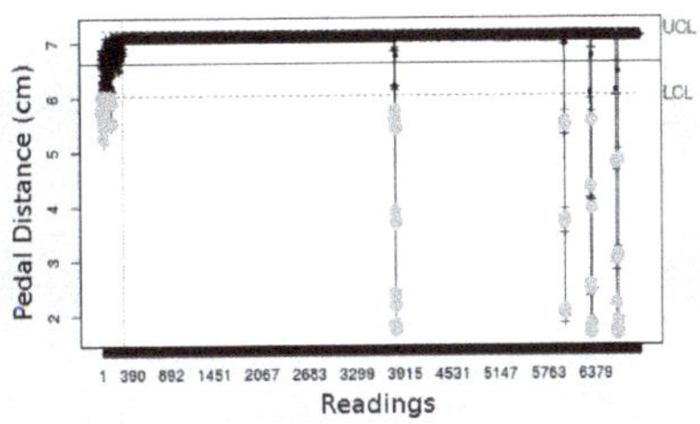

(b) Brake pedal detection

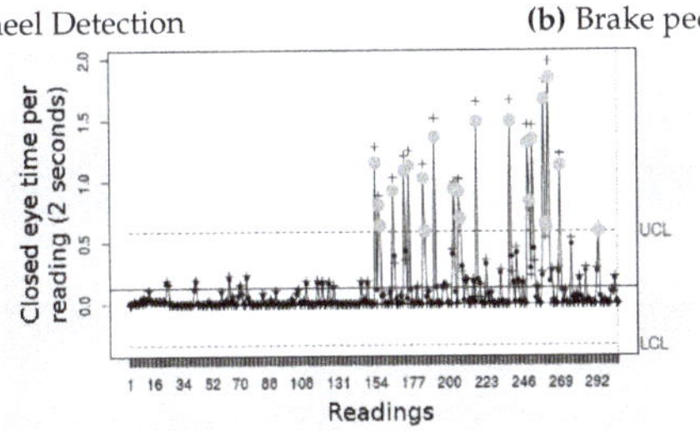

(c) Driver fatigue detection

Figure 22. EWMA graphics for steering wheel, brake pedal and driver fatigue detection. (**a**) Detection on the steering wheel applied to the X axis of the accelerometer, (**b**) detection on the brake pedal and (**c**) driver fatigue detection from closed eye time per reading.

Figure 22b shows the EWMA graph with the brake pedal data. The three vertical lines with gray dots to the right are the simulated errors that were detected. The control plot was able to detect the three errors, several times (near gray dots), with one false positive. This false positive was a moment where the driver had to brake but without risks.

Figure 22c shows the control chart with driver fatigue data. We can see that out of control points were detected from the middle of the graph—this is because we simulate driver fatigue from half the driving. We observe the driver fatigue moments and compare them to the graph—the gray dots show moments of driver fatigue, but they indicate some points at incorrect moments, for that reason the false positive rate was not low for this detection.

5.3. Qualitative Evaluation

For the qualitative evaluation, initially, we analyze the data with PERCLOS using the methodology of Wierwille et al. [24], where we have two distinct categories of wakefulness (awake and drowsy), and three distinct categories (awake, questionable and drowsy), the "questionable" category is at the center line between "awake" and "drowsy". In Figure 23 the first ≈300 points on the abscissa correspond to observation or sample of test 1 (named as T1), the next ≈300 points correspond to samples of test 2 (named as T2), and so on. Each value is an observation, a sample, inside corresponding test. We use the same proportion values of Wierwille et al. where "questionable" is between 0.075 and 0.15 and above 0.15 is "drowsy". The calculation performed by Kong et al. [25]:

$$PERCLOS = \frac{t}{30} \times 100\%,$$

where t is the duration of closed eyes.

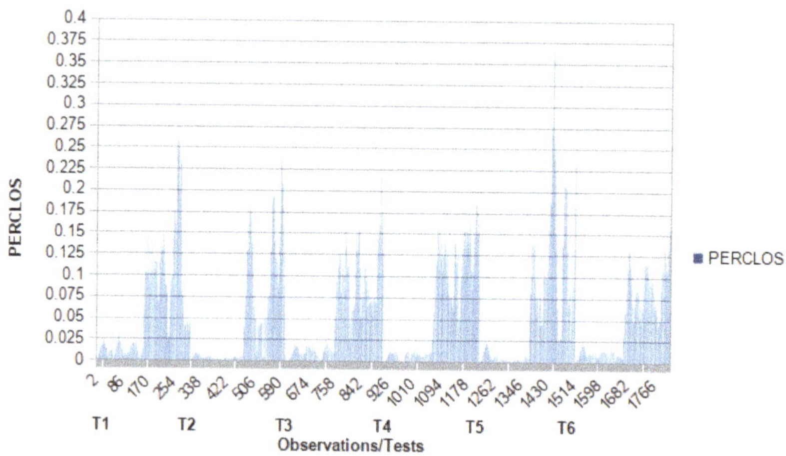

Figure 23. PERCLOS data with upper and lower criterion lines for three categories and single criterion for two categories.

We can observe that during the moments where the driver performed simulations of drowsiness (from half of each test) the PERCLOS considered these as "questionable" or in "drowsiness", which evidences its effectiveness. However, the method does not point out specific circumstances where the driver had some more severe drowsiness signal, as our method provides from the EWMA.

Besides that, we perform a qualitative comparison of two works: Mehta et al. [22] and Pauly and Sankar [23] with our proposal using three analyzes: i. the types of analyses of the works and its limitations; ii. The proposal to processing the features of the driver's face; iii. The complexity of the proposed algorithms.

The main difference between the analyses is that we tested our proposal with the use of a driving environment, a driving simulator, which makes our analysis more relevant for the detection of driver drowsiness. Besides, our detection rate was 94.46%, better than the other proposals.

Fernández et al. [48] presents the Figure 24 and shows the typical steps in most distraction monitoring systems and the most accepted and used in the literature, such as methods based on Viola and Jones [33] to perform face detection and SVM to detect facial features, such as an eye condition.

However, some additional processing is required to detect drowsiness, so this last step is significant in determining the efficiency and innovation of a method.

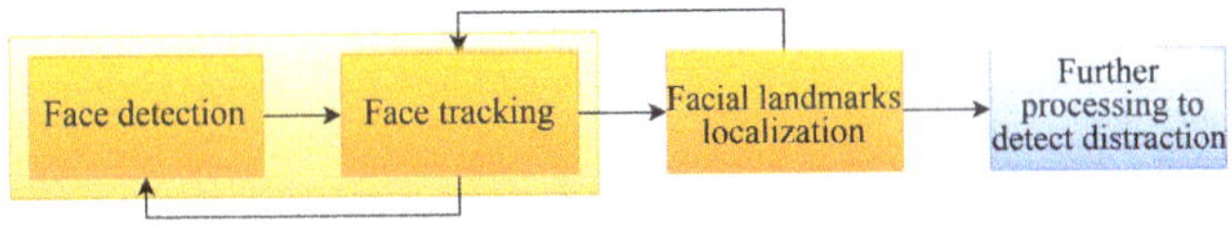

Figure 24. Common steps in most distraction monitoring systems according to Fernández et al. [48].

From this, Mehta et al. use a threshold value of EAR for the detection of sleepiness while Pauly and Sankar use the PERCLOS with a limit of 6 s, our solution is very efficient for using the EWMA to perform such analysis with results as good as PERCLOS.

A part of the computational complexity principle is the time analysis that describes the resources used computationally. For this, we can use the Big O notation to calculate the complexity of any algorithm, describing the upper bound of the increase of the running time as Rauber et al. [49] presents:

$$O(g(n)) = \{f(n) \mid \exists\, c > 0, \exists\, n_0, n_0 \in \mathbb{N}, \forall\, n \geq n_0 : 0 \leq f(n) \leq cg(n)\}$$

Therefore, we compare the complexity (in terms of running time) of these monitoring vehicle drivers methods, using the Big-O notation for each of them.

Lane departure detection: Ahmed et al. [50] explain that the Big-O for EWMA is $O(2)$ for each estimation value. Thus EWMA is a constant complexity algorithm. Therefore, we can say that our lane departure solutions have a constant complexity for each sample, since obtaining the data from the sensors does not depend on N.

Driver fatigue and sudden movements detection: For our sudden movements detection, we use a process similar to Lane departure detection, so its complexity is also constant. However, driver fatigue requires the driver's eye data before performing the calculation with the EWMA. We obtain this data with the use of SVM. According to Abdiansah et al. [51], the SVM has complexity $O(n^3)$. However, we must emphasize that almost every method of computational vision based fatigue analysis needs to obtain the data of the driver's eye with considerable complexity, as it is presented in the related works. However, our method has low complexity after obtaining this data, making the process much less costly.

6. Conclusions

The work brought several contributions to the area of driving monitoring. The main one was the introduction of the use of Statistical Process Control in the development of methods to monitor driving, something not seen in the literature, thus being innovative, opening space for the development of new methods of driver analysis. From the tests performed it was possible to see that this tool can be used in the construction of methods to monitor driving data obtaining results of up to 76.92% (without constant speed) and 84.16% (speed maintained at ≈60) in lane departure detection. Our driver fatigue detection obtained results up to 94.46% and the detection of sudden movements of up to 91.66% (steering wheel) and 94.44% (brakes). The experiments performed in our scenarios presented satisfactory results, considering the average and confidence interval values of 95%. However, they cannot be generalized. We guarantee, statistically, these results since we use the same variables. Eventual other scenarios may involve new variables and, thus, consider other evaluations.

Due to the new use of EWMA from Statistical Process Control in a driving environment, this article also opens the possibility for the use of the SPC to monitor other characteristics of the driving—characteristics that were not addressed by our solutions. Finally, for developers and researchers wishing to implement solutions for driving environments, the information contained in this paper can serve as a starting point for generating new ideas and products that can meet the expectations of drivers and companies in the area of vehicles and development of solutions for monitoring and aiding driving.

For future work, we intend not only to perform lane detection, driver fatigue, and sudden movements, but also to detect pedestrians, signs, and objects on the road. We plan to combine methods such as lane detection with our method by computer vision and vehicle-based approach, verifying the conditions where systems complement each other. We can avoid accidents due to driver failures and pedestrian failures. Thus, we can carry out comprehensive monitoring of the vehicular environment. Besides, we can try to have access to a real, controlled environment to avoid the risks of testing on real highways. In order to improve the representativeness of the results, we intend to increase the number of drivers in each experiment. Finally, we intend to evaluate other non-parametric controls or Support Vector Data Description in order to render our solution robust.

Author Contributions: A.N.A. implemented all software solutions and wrote the manuscript. R.C.C.d.M.S. and R.L.M.G. implemented the software on hardware and conducted, with A.N.A., all experiments. A.L.L.A. and R.A.R.R. formulated the model concepts, structured and organized the manuscript, and supervised the review writing.

Funding: This research was funded by CNPq (grant: 404895/2016-6), Fapemig, FAPEAL (grant: 60030 000346/2017), FAPESP (grant: 2015/24544-5), and SEVA.

Conflicts of Interest: The authors declare no conflict of interest.

References

1. Kaur, G.; Kumar, D. Lane Detection Techniques: A Review. *Int. J. Comput. Appl.* **2015**, *112*, 4–8.
2. Butakov, V.A.; Ioannou, P. Personalized Driver/Vehicle Lane Change Models for ADAS. *IEEE Trans. Veh. Technol.* **2015**, *64*, 4422–4431. [CrossRef]
3. Galvani, M. History and Future of Driver Assistance. *IEEE Instrum. Meas. Mag.* **2019**, *22*, 11–16. [CrossRef]
4. Montgomery, D.C. *Introduction to Statistical Quality Control*, 7th ed.; Wiley: New York, NY, USA. 2009.
5. Hunter, J.S. The exponentially weighted moving average. *J. Qual. Technol.* **1986**, *18*, 203–210. [CrossRef]
6. McCall, J.; Trivedi, M.M. Video Based Lane Estimation and Tracking for Driver Assistance: Survey, System, and Evaluation. *IEEE Trans. Intell. Transp. Syst.* **2006**, *7*, 20–37. [CrossRef]
7. Hsiao, P.Y.; Yeh, C.W.; Huang, S.S.; Fu, L.C. A Portable Vision-Based Real-Time Lane Departure Warning System: Day and Night. *IEEE Trans. Veh. Technol.* **2009**, *58*, 2089–2094. [CrossRef]
8. Ji, Q.; Zhu, Z.; Lan, P. Real-time nonintrusive monitoring and prediction of driver fatigue. *IEEE Trans. Veh. Technol.* **2004**, *53*, 1052–1068. [CrossRef]
9. Kumar, H.; Ahmed, Z.; Shetty, A.; Bangera, N.; Bangera, V. i-Car: An Intelligent and Interactive Interface for Driver Assistance System. *Sci. Technol. Arts Res. J.* **2014**, *3*, 197–200. [CrossRef]
10. Wang, J.; Zhang, L.; Zhang, D.; Li, K. An adaptive longitudinal driving assistance system based on driver characteristics. *IEEE Trans. Intell. Transp. Syst.* **2013**, *14*, 1–12. [CrossRef]
11. Chien, J.C.; Lee, J.D.; Chen, C.M.; Fan, M.W.; Chen, Y.H.; Liu, L.C. An integrated driver warning system for driver and pedestrian safety. *Appl. Soft Comput.* **2013**, *13*, 4413–4427. [CrossRef]
12. Sandström, M.; Lampsijärvi, E.; Holmström, A.; Maconi, G.; Ahmadzai, S.; Meriläinen, A.; Hæggström, E.; Forsman, P. Detecting lane departures from steering wheel signal. *Accid. Anal. Prev.* **2017**, *99*, 272–278. [CrossRef] [PubMed]
13. Satzoda, R.K.; Lee, S.; Lu, F.; Trivedi, M.M. Snap-DAS: A vision-based driver assistance system on a Snapdragon TM embedded platform. In Proceedings of the IEEE Intelligent Vehicles Symposium, Seoul, Korea, 28 June–1 July 2015.
14. Son, J.; Yoo, H.; Kim, S.; Sohn, K. Real-time illumination invariant lane detection for lane departure warning system. *Expert Syst. Appl.* **2015**, *42*, 1816–1824. [CrossRef]
15. Jung, H.; Min, J.; Kim, J. An efficient lane detection algorithm for lane departure detection. In Proceedings of the 4th IEEE Intelligent Vehicles Symposium, Gold Coast, Australia, 23–26 June 2013.
16. Ahmed, H.; Muhammad, T. Accurate attitude estimation of a moving land vehicle using low-cost MEMS IMU sensors. *IEEE Trans. Intell. Transp. Syst.* **2017**, *18*, 1723–1739. [CrossRef]

17. Clanton, J.M.; Bevly, D.M.; Hodel, A.S. A low-cost solution for an integrated multisensor lane departure warning system. *IEEE Trans. Intell. Transp. Syst.* **2009**, *10*, 47–59. [CrossRef]
18. Abulkhair, M.; Alsahli, A.H.; Taleb, K.M.; Bahran, A.M.; Alzahrani, F.M.; Alzahrani, H.A.; Ibrahim, L.F. Mobile Platform Detect and Alerts System for Driver Fatigue. *Procedia Comput. Sci.* **2015**, *62*, 555–564. [CrossRef]
19. Patel, S.P.; Patel, B.P.; Sharma, M.; Shukla, N.; Patel, H.M. Detection of Drowsiness and Fatigue level of Driver. *Int. J. Innov. Res. Sci. Technol.* **2015**, *1*, 133–138.
20. Jung, S.J.; Shin, H.S.; Chung, W.Y. Driver fatigue and drowsiness monitoring system with embedded electrocardiogram sensor on steering wheel. *IET Intell. Transp. Syst.* **2014**, *8*, 43–50. [CrossRef]
21. McDonald, A.D.; Lee, J.D.; Schwarz, C.; Brown, T.L. Steering in a Random Forest Ensemble Learning for Detecting Drowsiness-Related Lane Departures. *Hum. Factors J. Hum. Factors Ergon. Soc.* **2014**, *56*, 986–998. [CrossRef]
22. Mehta, S.; Dadhich, S.; Gumber, S.; Jadhav Bhatt, A. Real-Time Driver Drowsiness Detection System Using Eye Aspect Ratio and Eye Closure Ratio. In Proceedings of the International Conference on Sustainable Computing in Science, Technology and Management (SUSCOM), Jaipur, India, 26–28 February 2019.
23. Pauly, L.; Sankar, D. Detection of drowsiness based on HOG features and SVM classifiers. In Proceedings of the 2015 IEEE International Conference on Research in Computational Intelligence and Communication Networks (ICRCICN), Kolkata, India, 20–22 November 2015; pp. 181–186.
24. Wierwille, W.W.; Wreggit, S.; Kirn, C.; Ellsworth, L.; Fairbanks, R. *Research on Vehicle-Based Driver Status/Performance Monitoring; Development, Validation, and Refinement of Algorithms for Detection of Driver Drowsiness*; Technical Report; U.S. Department of Transportation: Virginia, VA, USA. 1994.
25. Kong, W.; Zhou, L.; Wang, Y.; Zhang, J.; Liu, J.; Gao, S. A system of driving fatigue detection based on machine vision and its application on smart device. *J. Sens.* **2015**, *2015*, 548602. [CrossRef]
26. Stanton, N.; Hedge, A.; Brookhuis, K.; Salas, E.; Hendrick, H. *Manual de Fatores Humanos e Métodos Ergonômicos*; Phorte Editora: São Paulo, Brazil. 2015.
27. Oakland, J.S. *Statistical Process Control*; Butterworth-Heinemann: Burlington, MA, USA. 2008.
28. Montgomery, D.C.; Woodall, W. Research issues and ideas in statistical process control. *J. Qual. Technol.* **1999**, *31*, 376–387.
29. Aradhye, H.B.; Bakshi, B.R.; Strauss, R.A.; Davis, J.F. Multiscale SPC using wavelets: Theoretical analysis and properties. *Am. Inst. Chem. Eng. J.* **2003**, *49*, 939–958. [CrossRef]
30. Borror, C.M.; Montgomery, D.C.; Runger, G.C. Robustness of the EWMA control chart to non-normality. *J. Qual. Technol.* **1999**, *31*, 309–316. [CrossRef]
31. Ruyi, J.; Reinhard, K.; Tobi, V.; Shigang, W. Lane detection and tracking using a new lane model and distance transform. *Mach. Vis. Appl.* **2011**, *22*, 721–737. [CrossRef]
32. Seo, Y.W.; Rajkumar, R.R. Utilizing instantaneous driving direction for enhancing lane-marking detection. In Proceedings of the IEEE Intelligent Vehicles Symposium Proceedings, Dearborn, MI, USA, 8–11 June 2014.
33. Viola, P.; Jones, M. Rapid object detection using a boosted cascade of simple features. In Proceedings of the IEEE Conference on Computer Vision and Pattern Recognition (CVPR'01), Kauai, HI, USA, 8–14 December 2001.
34. Lienhart, R.; Maydt, J. An extended set of haar-like features for rapid object detection. In Proceedings of the International Conference on Image Processing, Rochester, NY, USA, 22–25 September 2002.
35. Zhou, L.; Wang, H. Open/closed eye recognition by local binary increasing intensity patterns. In Proceedings of the IEEE Conference on Robotics, Automation and Mechatronics (RAM'11), Qingdao, China, 17–19 September 2011.
36. Tan, X.; Triggs, B. Enhanced local texture feature sets for face recognition under difficult lighting conditions. *IEEE Trans. Image Process.* **2010**, *6*, 1635–1650.
37. Dalal, N.; Triggs, B. Histograms of oriented gradients for human detection. In Proceedings of the IEEE Conference on Computer Vision and Pattern Recognition (CVPR'05), San Diego, CA, USA, 20–25 June 2005.
38. Kaptein, N.A.; Theeuwes, J.; Van Der Horst, R. Driving simulator validity: Some considerations. *Transp. Res. Rec. J. Transp. Res. Board* **1996**, *1550*, 30–36. [CrossRef]
39. Auberlet, J.M.; Rosey, F.; Anceaux, F.; Aubin, S.; Briand, P.; Pacaux, M.P.; Plainchault, P. The impact of perceptual treatments on driver's behavior: From driving simulator studies to field tests—First results. *Accid. Anal. Prev.* **2012**, *45*, 91–98. [CrossRef]

40. Mayhew, D.R.; Simpson, H.M.; Wood, K.M.; Lonero, L.; Clinton, K.M.; Johnson, A.G. On-road and simulated driving: Concurrent and discriminant validation. *J. Saf. Res.* **2011**, *42*, 267–275. [CrossRef]
41. Lee, B.L.; Lee, B.G.; Chung, W.Y. Standalone wearable driver drowsiness detection system in a smartwatch. *IEEE Sens. J.* **2016**, *16*, 5444–5451. [CrossRef]
42. Fux, S. Development of a Planar Low Cost Inertial Measurement Unit for UAVs and MAVs. Master's Thesis, Swiss Federal Institute of Technology, Zurich, Switzerland, 2008.
43. Espressif Systems IOT Team. *ESP8266EX Datasheet, Version 4.3*; Technical Report; Espressif Systems: Shanghai, China. 2015.
44. Jiang, X.; Wang, F.; Kraft, M.; Boser, B.E. An integrated surface micromachined capacitive lateral accelerometer with $2\mu G/\sqrt{Hz}$ resolution. In Proceedings of the Solid-State Sensor, Actuator and Microsystems Workshop, Hilton Head Island, SC, USA, 2–6 June 2002.
45. Cannon, R. Alignment of inertial guidance systems by gyrocompassing-linear theory. *J. Aerosp. Sci.* **1961**, *28*, 885–895.
46. Freescale. *Wandboard User Guide Revision B1*; Technical Report; Wandboard.org: New Taipei City. 2013.
47. Velez, G.; Otaegui, O. Embedded Platforms for Computer Vision-based Advanced Driver Assistance Systems: A Survey. In Proceedings of the 22nd Intelligent Transport Systems World Congress (ITSWC'15), Bordeaux, France, 5–9 October 2015.
48. Fernández, A.; Usamentiaga, R.; Carús, J.; Casado, R. Driver distraction using visual-based sensors and algorithms. *Sensors* **2016**, *16*, 1805. [CrossRef]
49. Rauber, T.; Rünger, G. Performance analysis of parallel programs. In *Parallel Programming*; Springer: Berlin, Germany, 2013; pp. 169–226.[CrossRef]
50. Ahmed, F.; Tamberg, G.; Le Moullec, Y.; Annus, P. Dual-Source Linear Energy Prediction (LINE-P) Model in the Context of WSNs. *Sensors* **2017**, *17*, 1666. [CrossRef] [PubMed]
51. Abdiansah, A.; Wardoyo, R. Time complexity analysis of support vector machines (SVM) in LibSVM. *Int. J. Comput. Appl.* **2015**. [CrossRef]

sensors

MDPI

Article

Sensor Fault Detection and Signal Restoration in Intelligent Vehicles

Yeun-Sub Byun *, Baek-Hyun Kim and Rag-Gyo Jeong

Korea Railroad Research Institute, 176, Cheoldobangmulgwan-ro, Uiwang, Gyeonggi-do 16105, Korea
* Correspondence: ysbyun@krri.re.kr; Tel.: +82-31-460-5437

Received: 5 June 2019; Accepted: 24 July 2019; Published: 27 July 2019

Abstract: This paper presents fault diagnosis logic and signal restoration algorithms for vehicle motion sensors. Because various sensors are equipped to realize automatic operation of the vehicle, defects in these sensors lead to severe safety issues. Therefore, an effective and reliable fault detection and recovery system should be developed. The primary idea of the proposed fault detection system is the conversion of measured wheel speeds into vehicle central axis information and the selection of a reference central axis speed based on this information. Thus, the obtained results are employed to estimate the speed for all wheel sides, which are compared with measured values to identify fault and recover the fault signal. For fault diagnosis logic, a conditional expression is derived with only two variables to distinguish between normal and fault; further, an analytical redundancy structure and a simple diagnostic logic structure are presented. Finally, an off-line test is conducted using test vehicle information to validate the proposed method; it demonstrates that the proposed fault detection and signal restoration algorithm can satisfy the control performance required for each sensor failure.

Keywords: fault detection; sensor fault; signal restoration; intelligent vehicle; autonomous vehicle; kinematic model

1. Introduction

The desire for convenient and safe passenger transportation has increased the need for automated and intelligent automobiles. Consequently, the function of autonomous navigation has been introduced for passenger safety and convenience, and various in-vehicle devices such as ultrasonic sensors, radars, cameras, and actuators have been installed to detect the surrounding environment, recognize information, and control the motions of the vehicle. In addition, speed sensors, steering angle sensors, gyroscopes, and acceleration sensors can be installed to measure the vehicle operation state and movement and to use the gathered information for control. If faults occur in these sensors or actuators during automatic running, the vehicle may deflect from its route or fail to conduct the precise operation required for control, which may lead to an accident. To prevent accidents caused by these faults, technologies applying the soft computing method in fault detection (FDI) and fault tolerant control (FTC) of vehicles are garnering attention in academia and industry. In conjunction with these developments, various ideas and techniques for FDI/FTC methods, including neural network and fuzzy approaches, are presented [1–3]. To this end, the main purpose is to prevent or mitigate deterioration of the control performance of the system caused by a failure. In the field of fault diagnosis, various studies have been conducted to compensate or detect the fault sensor information by combining it with information of various sensors [4–6]. In a system in which a vehicle is driven by an electric motor, a frequency domain analysis technique may be applied in the fault diagnosis of the electric motor, whereas a frequency component analysis usually deals with the diagnosis of physical faults in rotating machinery [7].

Although studies on fault detection are particularly important for the aircraft sector [8–13], as even minor aircraft faults can lead to serious accidents, the increasing interest in autonomous vehicles has inspired numerous investigations of fault detection in the automotive sector [14–23]. Na et al. [24] applied residual sensitivity as a threshold for predicting the occurrence of vehicle sensor failure, while Emirler et al. [25] employed a virtual sensor featuring a velocity-scheduled Kalman filter to characterize vehicle kinematics and estimate the yaw rate. Huang and Su [26] devised a model-based fault detection and isolation scheme considering disturbance and noise to diagnose single sensor faults in intelligent navigation systems, while analytical redundancy and nonlinear transformation were also used to generate residual values used to detect embedded sensors in intelligent vehicles [27].

In these fault-related studies, when a fault is detected, the fault signal has been largely recovered using either direct or analytical redundancy [28]. In practical environments, analytical redundancy methods are studied and used because of the low cost or installation space required. This analytical fault detection method can be divided into a data-based method and signal model, a model-based method, and knowledge-based method [13]. The data-based method and signal model are applied to compare and analyze the characteristics of data [29]. The model-based method is based on the mathematical model of the target system [30]. Knowledge-based methods are implemented using expert systems or fuzzy logic [31]. Among these methods, the model-based method that is acknowledged in this paper is classified into parity equations [32,33], parameter estimation methods [34], and observer-based methods [35,36]. In addition, this study focuses on fault detection and signal restoration for sensors detecting vehicle motion (e.g., speed, steering angle, and rotational angular velocity sensors), assuming that only one sensor can fail at a given instant, such that the sensor fault detection and restoration algorithm can be applied. In this regard, real-time fault diagnosis and signal estimation of major sensors have been extensively researched [37–46]; major methods and research trends for fault detection have been introduced and investigated by Miljkovic [47]. When attempting to diagnose the case where several sensors are mounted on a vehicle, residuals are generated for the diagnosis from each sensor and threshold is applied to each diagnosis; here, the number of thresholds is equal to the number of sensors to be diagnosed [18,24,26]. Therefore, each of the corresponding threshold values must be carefully set for the appropriate fault diagnosis of each sensor; otherwise, the result may affect the fault identification of other sensors. In the proposed method, it is possible to identify six sensor faults with only two threshold values and conditional expressions. As a result, the possibility of diagnostic error due to the threshold setting can be considerably reduced.

To prepare for unexpected sensor failure, this study develops a method for signal duplication using the analytical redundancy method based on the information provided by sensors installed in the vehicle and a mathematical vehicle model. Moreover, the analytical redundancy method is employed to realize fault detection and signal restoration.

It is difficult to apply existing research results to other vehicles because existing study have a complex logic structure with different types and numbers of sensors applied to the target vehicle. In addition, existing studies have difficulty in practical application because they have a multi-variable structure in which the judgment logic of failure is complex. These results do not demonstrate the effect of the actual running on the failure of each sensor. Furthermore, their research results do not demonstrate the effectiveness of their performance by applying a restored signal for each fault.

This study aims to detect main sensor faults in real time to guarantee occupants' safety in a vehicle automatically following a designated route without a driver. A vehicle model-based fault detection and signal restoration method is proposed, and the information provided by the failed sensor is restored to create a time margin in which the vehicle control system can conduct normal safety control procedures. To judge the fault of each vehicle sensor, one first needs to identify a suitable comparison standard. Therefore, the velocity and direction components in the central vehicle axis are defined as the central information on vehicle motion. In the absence of abnormalities in each sensor, the central axis speeds estimated by each vehicle sensor should have the same value. Moreover, if the center speed is correct, it can be used to estimate individual wheel speeds, and the calculated values should match actual wheel speeds if each sensor is normal. A kinematic model is designed to convert the speed of each wheel into the central axis speed. In this process, relationships for estimating the central axis speed using a steering angle or a gyroscope are defined, and fault classification conditions obtained by applying these are derived.

To validate the proposed method, defects are simulated independently for each normal sensor signal. The signal of the normal sensor is compared with the restored signal, and the resulting error is presented. Additionally, it is confirmed that when the fault signal is restored and applied to the position estimation of the autonomous vehicle, the results of path tracking error are within the valid range.

In summary, the contribution of this study is as follows:

- By conducting studies on fault diagnosis and restoration based on the major motion sensors that are installed in most vehicles, the possibility that this study can be used in several vehicles has increased.
- A vehicle kinematic model based on the central axis of the vehicle, which is used to detect faults and restore fault signals using a structure in which the failure of each sensor does not affect each other, is proposed.
- In the logic for fault detection, finally, a simple diagnostic logic structure is presented; this structure helps discriminate between normal and fault and distinguishes a specific fault using a conditional expression of only two variables.

2. Fault Detection and Recovery

2.1. Configuration of the Autonomous Vehicle

The target vehicle has four speed sensors, one steering angle sensor, and one gyroscope for vehicle control (Figure 1) [48]. It is assumed that two or more sensors are not defective at the same time, i.e., only one fault is assumed to occur at any moment.

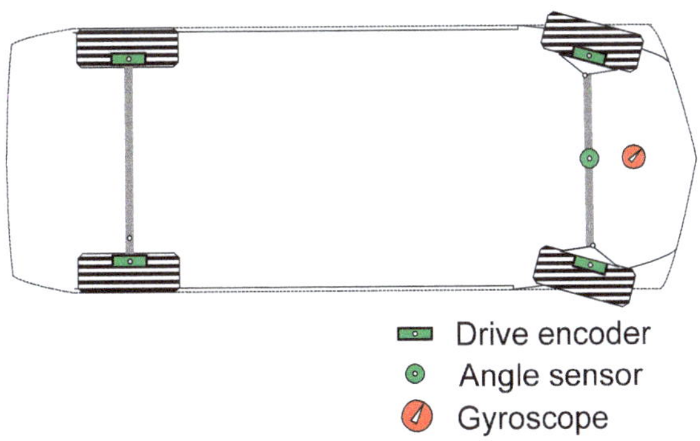

Figure 1. Vehicle sensor configuration.

2.2. Influence of Sensor Fault

The sensor mounted on the vehicle strongly influences the safety of automatic operation. First, the vehicle controller determines the vehicle position during driving in automatic mode and controls the speed and steering angle to reach the destination based on the fusion of various sensors. The positional information of the vehicle is important for driving control and can be estimated by fusing landmark information, GPS information, and vehicle motion sensor information. The vehicle control system selects a certain point in the vehicle and controls the speed and position of this point to match the desired reference. For example, when the central axis of rear wheels is used as the control reference point in a straight-running vehicle, the calculated speed becomes half of the actual speed if the simple average of the left and right rear wheel speeds is used and one of the corresponding sensors fails. Moreover, when this information is used for position estimation, the calculated position deviates from the actual position. If the gyroscope fails in a curved path, and the information provided by this device is used for position estimation, the vehicle running direction is not calculated correctly, and the resulting position error causes a deviation in route guidance control. Finally, depending on the failure situation, a fault of the steering angle sensor may cause a path-following error or deviation from the traveling path.

2.3. Architecture of Fault Detection

The main idea behind sensor fault detection and signal restoration process is the conversion of the speed of each wheel to the central axis speed of the vehicle. If all sensors are normal, all calculated central axis speeds should be identical, whereas different values should be obtained in the case when any sensor fails. Thus, this difference can be used as a sensor fault indicator. Moreover, once a fault is found, the speed of each wheel can be estimated from the center axis speed determined as normal, and the faulty signal is recovered by replacing the speed of the defective wheel with the estimated speed. In this case, steering angle information is used to convert the speed of each wheel to the central axis speed. If there is an error in the steering angle sensor, the calculated value is also erroneous, which highlights the importance of knowing whether the steering angle sensor is normal or not. Therefore, two methods are used to calculate the central axis speed. Specifically, the input variables are divided into the cases of steering angle usage and usage of gyroscope-provided information on rotational angular speed. This strategy allows one to detect speed sensor, steering angle sensor, and gyroscope faults and to restore the affected signals even if either the steering angle sensor or the gyro sensor is abnormal. The employed procedure is illustrated in Figure 2.

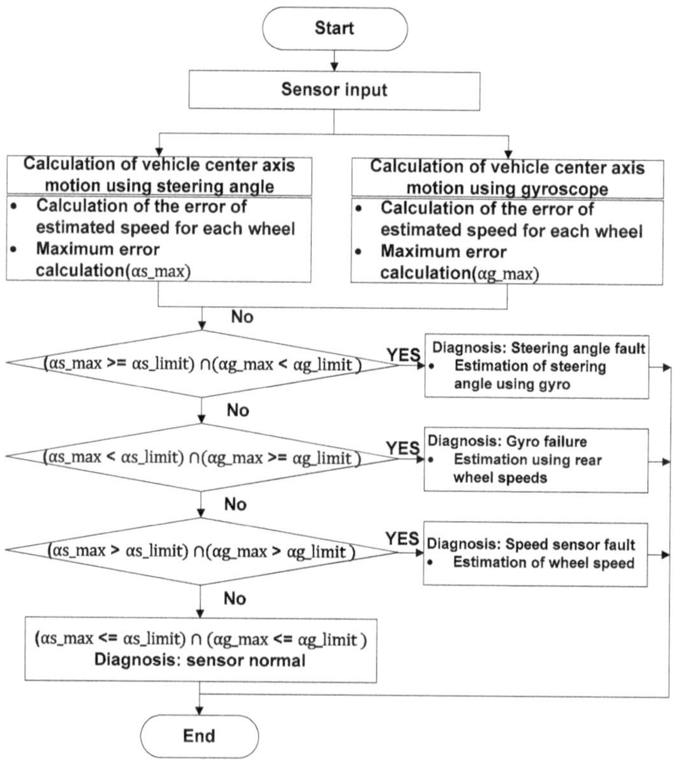

Figure 2. Diagnostic sequence flowchart.

2.4. *Vehicle Geometry for Kinematic Estimation*

Figure 3 depicts the situation in which the central axis of the vehicle moves around the center of rotation (o) to define the kinematic motion of the vehicle and associated parameters.

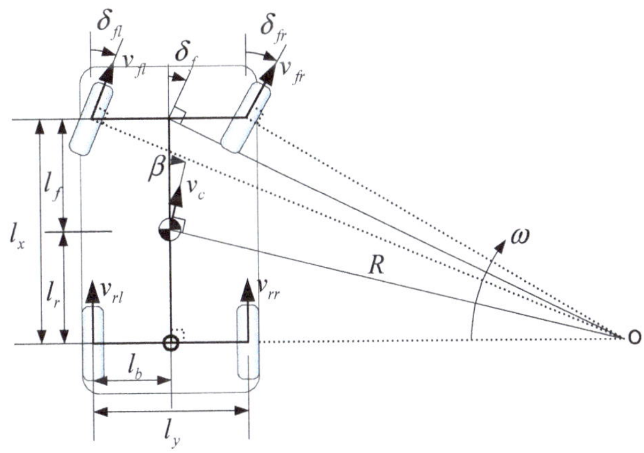

Figure 3. Vehicle geometry for kinematic estimation.

In the above figure, β is the side slip angle, δ_f is the front wheel steering angle, δ_{fl} is the front left wheel steering angle, δ_{fr} is the front right wheel steering angle, l_f is the distance from the front axle to the vehicle central axis, l_r is distance from the rear axle to the central axis, l_x is the distance from the front axle to the rear axle, l_y is front or rear axle width, l_b is the half-width of l_y, v_c is the speed at the central axis, ω is the yaw rate, R is the instant radius at central axis, v_{fl}, v_{fr}, v_{rl}, and v_{rr} are speeds of the front left, the front right, the rear left, and the rear right wheels, respectively.

2.5. *Virtual Redundancy of Sensors and Errors*

Formulas to convert the vehicle wheel speed to the central axis speed are proposed. The velocity and direction of the central axis are obtained based on the steering angle or gyroscope information.

2.5.1. Central Axis Speed Estimation Based on Steering Angle

When the steering angle information and gyroscope information are fused together in the fault detection formula, it becomes difficult to distinguish between each of the corresponding faults. Therefore, to differentiate the steering angle fault from the gyroscope fault, the vehicle central axis speed is calculated from the wheel speed, the steering angle, and vehicle parameters excluding the gyroscope information. First, the side slip angle and curvature are obtained from Equations (1) and (2), respectively, using the front wheel steering angle and vehicle parameters.

$$\beta = \tan^{-1}\left(\frac{l_r \tan(\delta_f)}{l_f + l_r}\right), \tag{1}$$

$$C = \frac{1}{R} = \frac{\cos(\beta)\tan(\delta_f)}{l_f + l_r}. \tag{2}$$

The central axis speed can be obtained from Equation (3) to Equation (6) using individual wheel speeds and Equations (1) and (2).

$$v_{c,fl} = \frac{v_{fl}}{\sqrt{\left(\frac{\cos(\beta)}{\cos(\delta_f)}\right)^2 + l_b{}^2C^2 - 2l_bC\cos(\beta)}}, \tag{3}$$

$$v_{c,fr} = \frac{v_{fr}}{\sqrt{\left(\frac{\cos(\beta)}{\cos(\delta_f)}\right)^2 + l_b{}^2C^2 + 2l_bC\cos(\beta)}}, \tag{4}$$

$$v_{c,rl} = \frac{v_{rl}}{\sqrt{(\cos(\beta))^2 + l_b{}^2C^2 - 2l_bC\cos(\beta)}}, \tag{5}$$

$$v_{c,rr} = \frac{v_{rr}}{\sqrt{(\cos(\beta))^2 + l_b{}^2C^2 + 2l_bC\cos(\beta)}}, \tag{6}$$

where C is the curvature at the central axis, and $v_{c,ij}$ is the central axis speed calculated based on individual vehicle wheel data. In a straight section, the speed of each wheel is the same, whereas in a curved section, the speed inside the curvature radius is lower than that on the outside, as shown in Figure 4.

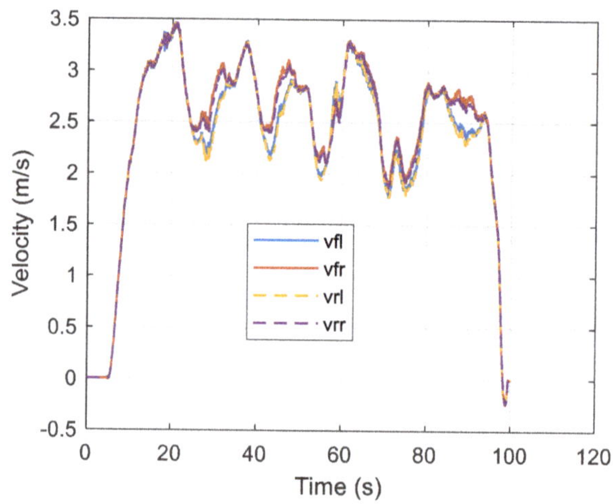

Figure 4. Speeds of the four wheels as functions of time.

The central axis speed satisfies the condition of Equation (7) when all wheel speed and steering angle sensors are normal and hence afford almost identical speed values, as shown in Figure 5. In practice, small differences may occur depending on road conditions and vehicle characteristics, as indicated by the maximum error in Figure 5. Here, maximum error refers to the maximum value of Equation (9). In the vehicle test under normal conditions, the maximum error was measured to be within 0.1 m/s.

$$v_{c,fl} = v_{c,fr} = v_{c,rl} = v_{c,rr}. \tag{7}$$

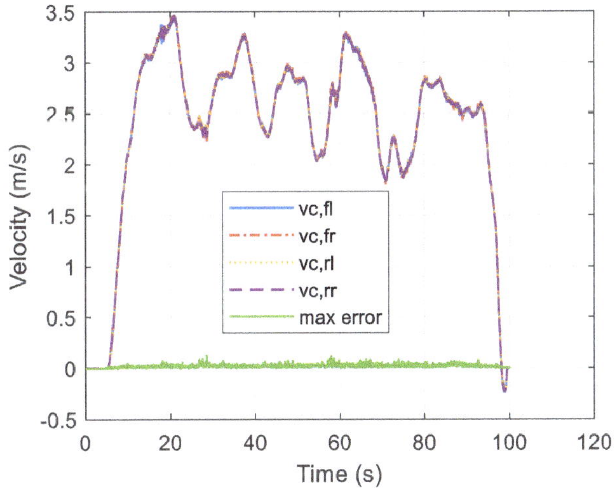

Figure 5. Central axis speeds as functions of time.

Therefore, if the central axis speed error exceeds a predetermined threshold value, the sensor or another part are considered to be abnormal, i.e., such an error may be indicative of an abnormal pressure difference between wheel tires, wheel slippage, wheel encoder abnormality, steering angle error, or communication abnormality.

The steering angle information is included in Equation (1) to Equation (6), i.e., if the steering angle information is defective, all calculated central axis speeds may be erroneous. Therefore, if the steering angle information is assumed to be normal and there is a defect in the speed sensor information, the following procedure is followed to select two judged-to-be-normal speeds from the four calculated central axis speeds. The two most closely matched central axis speeds are selected using Equations (8) and (9).

$$(v_{ci,\min}, v_{cj,\min}) = \min(E_v), \quad i \in \{fl, fr, rl\}, j \in \{fr, rl, rr\}, i \neq j, \tag{8}$$

with

$$\begin{gathered} E_v = [e_{v12}\, e_{v13}\, e_{v14}\, e_{v23}\, e_{v24}\, e_{v34}], \\ e_{v12} = |v_{c,fl} - v_{c,fr}|,\ e_{v13} = |v_{c,fl} - v_{c,rl}|,\ e_{v14} = |v_{c,fl} - v_{c,rr}|, \\ e_{v23} = |v_{c,fr} - v_{c,rl}|,\ e_{v24} = |v_{c,fr} - v_{c,rr}|,\ e_{v34} = |v_{c,rl} - v_{c,rr}|. \end{gathered} \tag{9}$$

The selected two central axis speeds are averaged (Equation (10)) as

$$v_{cr} = \left(v_{ci,\min} + v_{cj,\min}\right)/2. \tag{10}$$

Next, the obtained average is used to estimate the speed of each wheel (Equation (11) to Equation (14)):

$$\hat{v}_{fls} = v_{cr}\sqrt{\left(\frac{\cos(\beta)}{\cos(\delta_f)}\right)^2 + l_b{}^2C^2 - 2l_bC\cos(\beta)}, \tag{11}$$

$$\hat{v}_{frs} = v_{cr}\sqrt{\left(\frac{\cos(\beta)}{\cos(\delta_f)}\right)^2 + l_b{}^2C^2 + 2l_bC\cos(\beta)}, \tag{12}$$

$$\hat{v}_{rls} = v_{cr}\sqrt{(\cos(\beta))^2 + l_b{}^2C^2 - 2l_bC\cos(\beta)}, \tag{13}$$

$$\hat{v}_{rrs} = v_{cr}\sqrt{(\cos(\beta))^2 + l_b{}^2C^2 + 2l_bC\cos(\beta)}. \tag{14}$$

The thus obtained estimated speed of each wheel should match the measured speed of the wheel if there is no fault in each sensor. For the purpose of fault detection, the error between the estimated and the measured wheel speed is calculated as in Equation (16), and the largest among these errors (defined as in Equation (15)) is used as a condition variable value for detecting sensor defects.

$$as_{\max} = \max(E_{vs}), \tag{15}$$

with

$$\begin{aligned} E_{vs} = \begin{bmatrix} e_{vfl} & e_{vfr} & e_{vrl} & e_{vrr} \end{bmatrix}, \\ e_{vfl} = \left|\hat{v}_{fls} - v_{fl}\right| \\ e_{vfr} = \left|\hat{v}_{frs} - v_{fr}\right| \\ e_{vrl} = \left|\hat{v}_{rls} - v_{rl}\right| \\ e_{vrr} = \left|\hat{v}_{rrs} - v_{rr}\right| \end{aligned} \tag{16}$$

2.5.2. Central Axis Speed Estimation Using the Gyroscope

To distinguish between steering angle and gyro faults, gyro information is used instead of steering angle information in the calculation of the central axis speed. For this purpose, the steering angle is estimated from the gyroscope information, wheel speed, and vehicle parameters, and the central steering angle is estimated using Equation (17) to Equation (20).

$$\hat{\delta}_{f1} = \cot^{-1}\left(\cot(\hat{\delta}_{fl1}) + \frac{l_b}{l_x}\right), \ with\ \hat{\delta}_{fl1} = \sin^{-1}\left(\frac{\omega l_x}{v_{fl}}\right), \tag{17}$$

$$\hat{\delta}_{f2} = \cot^{-1}\left(\cot(\hat{\delta}_{fr1}) - \frac{l_b}{l_x}\right), \ with\ \hat{\delta}_{fr1} = \sin^{-1}\left(\frac{\omega l_x}{v_{fr}}\right), \tag{18}$$

$$\hat{\delta}_{f3} = \cot^{-1}\left(\cot(\hat{\delta}_{fl2}) + \frac{l_b}{l_x}\right), \ with\ \hat{\delta}_{fl2} = \tan^{-1}\left(\frac{\omega l_x}{v_{rl}}\right), \tag{19}$$

$$\hat{\delta}_{f4} = \cot^{-1}\left(\cot(\hat{\delta}_{fr2}) - \frac{l_b}{l_x}\right), \ with\ \hat{\delta}_{fr2} = \tan^{-1}\left(\frac{\omega l_x}{v_{rr}}\right), \tag{20}$$

where ω is the rotational angular velocity measured by the gyroscope. The estimated front-center steering angles should be equal to each other as shown in Equation (21) if there are no faults in the speed sensor and gyroscope.

$$\hat{\delta}_{f1} = \hat{\delta}_{f2} = \hat{\delta}_{f3} = \hat{\delta}_{f4} = \delta_f. \tag{21}$$

If the gyroscope is fault-free, a fault of the speed sensor should result in a difference between some estimated steering values. Therefore, to select the steering angle with the smallest error due to the defect, two estimated steering angles with the smallest error among the estimated steering angles ($\hat{\delta}_{f1}$, $\hat{\delta}_{f2}$, $\hat{\delta}_{f3}$, $\hat{\delta}_{f4}$) are determined using Equations (22) and (23).

$$\left(\hat{\delta}_{fi,\min}, \hat{\delta}_{fj,\min}\right) = \min(E_{\delta k}),\ i \in \{1, 2, 3\},\ j \in \{2, 3, 4\},\ i \neq j,\ i \langle j, \tag{22}$$

with

$$\begin{aligned} E_{\delta k} = [e_{\delta 12}\, e_{\delta 13}\, e_{\delta 14}\, e_{\delta 23}\, e_{\delta 24}\, e_{\delta 34}], \\ e_{\delta 12} = \left|\hat{\delta}_{f1} - \hat{\delta}_{f2}\right|,\ e_{\delta 13} = \left|\hat{\delta}_{f1} - \hat{\delta}_{f3}\right|,\ e_{\delta 14} = \left|\hat{\delta}_{f1} - \hat{\delta}_{f4}\right| \\ e_{\delta 23} = \left|\hat{\delta}_{f2} - \hat{\delta}_{f3}\right|,\ e_{\delta 24} = \left|\hat{\delta}_{f2} - \hat{\delta}_{f4}\right|,\ e_{\delta 34} = \left|\hat{\delta}_{f3} - \hat{\delta}_{f4}\right| \end{aligned} \tag{23}$$

The selected estimated steering angles are averaged (Equation (24)), and the obtained value is used as the estimated front wheel steering angle.

$$\hat{\delta}_f = \left(\hat{\delta}_{fi,\min} + \hat{\delta}_{fj,\min}\right)/2. \tag{24}$$

If the front steering angle is estimated using the gyroscope information, a process identical to that used for central axis speed estimation, as described in the previous section, is applied. The speed at each vehicle center is calculated using the speed of each wheel and the estimated steering angle. Next, the side slip angle and curvature are obtained as in Equations (25) and (26), respectively, using the estimated steering angle and vehicle parameters.

$$\beta = \tan^{-1}\left(\frac{l_r \tan(\hat{\delta}_f)}{l_f + l_r}\right), \tag{25}$$

$$C = \frac{1}{R} = \frac{\cos(\beta)\tan(\hat{\delta}_f)}{l_f + l_r}. \tag{26}$$

The central axis speed can be obtained using Equation (27) to Equation (30) by considering individual wheel speeds, Equations (25) and (26).

$$v_{c,fl} = \frac{v_{fl}}{\sqrt{\left(\frac{\cos(\beta)}{\cos(\hat{\delta}_f)}\right)^2 + l_b{}^2C^2 - 2l_bC\cos(\beta)}}, \tag{27}$$

$$v_{c,fr} = \frac{v_{fr}}{\sqrt{\left(\frac{\cos(\beta)}{\cos(\hat{\delta}_f)}\right)^2 + l_b{}^2C^2 + 2l_bC\cos(\beta)}}, \tag{28}$$

$$v_{c,rl} = \frac{v_{rl}}{\sqrt{(\cos(\beta))^2 + l_b{}^2C^2 - 2l_bC\cos(\beta)}}, \tag{29}$$

$$v_{c,rr} = \frac{v_{rr}}{\sqrt{(\cos(\beta))^2 + l_b{}^2C^2 + 2l_bC\cos(\beta)}}. \tag{30}$$

If each wheel speed and the gyroscope are normal, the calculated central axis speeds should be equal to each other as shown in Equation (31) (Figure 6).

$$v_{c,fl} = v_{c,fr} = v_{c,rl} = v_{c,rr}. \tag{31}$$

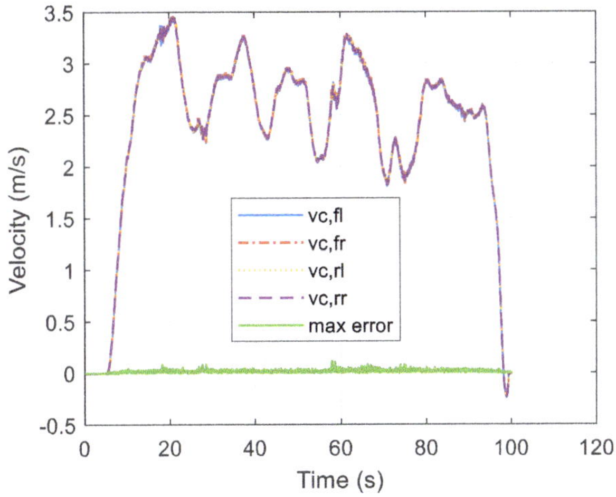

Figure 6. Time-dependent central axis speeds.

As in the previous case, two speeds with the smallest error selected based on the error between the central axis speed of the two combinations are averaged (Equation (32)).

$$v_{cg} = \left(v_{ci,\min} + v_{cj,\min}\right)/2. \tag{32}$$

The thus obtained value is used to estimate the speed of each wheel using Equation (33) to Equation (36).

$$\hat{v}_{flg} = v_{cg}\sqrt{\left(\frac{\cos(\beta)}{\cos(\hat{\delta}_f)}\right)^2 + l_b{}^2C^2 - 2l_bC\cos(\beta)}, \tag{33}$$

$$\hat{v}_{frg} = v_{cg}\sqrt{\left(\frac{\cos(\beta)}{\cos(\hat{\delta}_f)}\right)^2 + l_b{}^2C^2 + 2l_bC\cos(\beta)}, \tag{34}$$

$$\hat{v}_{rlg} = v_{cg}\sqrt{(\cos(\beta))^2 + l_b{}^2C^2 - 2l_bC\cos(\beta)}, \tag{35}$$

$$\hat{v}_{rrg} = v_{cg}\sqrt{(\cos(\beta))^2 + l_b{}^2C^2 + 2l_bC\cos(\beta)}. \tag{36}$$

The thus estimated speeds of each wheel should match the measured values if there is no defect in each sensor. For the purpose of fault detection, the error between each estimated wheel speed and the measured wheel speed is obtained as in Equation (38), and the largest among these errors (defined as in Equation (37)) is used as a condition variable value for detecting sensor faults.

$$ag_{\max} = \max\left(E_{vg}\right), \tag{37}$$

with

$$\begin{aligned} E_{vg} &= \begin{bmatrix} e_{vfl} & e_{vfr} & e_{vrl} & e_{vrr} \end{bmatrix}, \\ e_{vfl} &= \left|\hat{v}_{flg} - v_{fl}\right| \\ e_{vfr} &= \left|\hat{v}_{frg} - v_{fr}\right| \\ e_{vrl} &= \left|\hat{v}_{rlg} - v_{rl}\right| \\ e_{vrr} &= \left|\hat{v}_{rrg} - v_{rr}\right|. \end{aligned} \tag{38}$$

2.6. *Fault Detection, Identification, and Signal Recovery*

To identify sensor faults, Equations (15) and (37), which describe the calculation of maximum estimated speed and the measured speed error, are used for judgment. The maximum values of each error ($as_{\max}$, $ag_{\max}$) and the corresponding limit values (as_{limit}, ag_{limit}) are used to determine whether the sensor is faulty.

2.6.1. Fault-Free Sensor Judgment Condition

If the maximum error ($as_{\max}$, $ag_{\max}$) between the estimated speed and the measured speed is small, all sensors applied to the relational expression can be viewed as not defective. The maximum allowable error limit (as_{limit}, ag_{limit}), which is the defect judgment boundary, is determined based on the observed results under the condition that all sensors are normal. Herein, limit values of as_{limit} = ag_{limit} = 0.025 were selected for the running test of the test vehicle. Therefore, if the maximum error

($as_{\max}$, $ag_{\max}$) is smaller than the above values, all sensors are normal, and if not, a fault is concluded to be present. This condition is expressed in Equation (39).

$$\begin{array}{l} if\ (as_{\max} \leq as_{\text{limit}}) \cap (ag_{\max} \leq ag_{\text{limit}}) \\ \quad TRUE : Normal \\ else \\ \quad FALSE : Other\ fault \\ end \end{array} \tag{39}$$

2.6.2. Fault Detection and Signal Restoration for Steering Angle Information

If the maximum value ($ag_{\max}$) of the error calculated using gyroscope information is smaller than the limit value (ag_{limit}), the gyroscope and speed sensor are both viewed as normal. If the maximum value ($as_{\max}$) of the error calculated using steering angle information is larger than the limit value (as_{limit}), the steering angle sensor is viewed as defective, as there is no defect of the speed sensor under the preceding condition. This condition is expressed in Equation (40).

$$\begin{array}{l} if\ (as_{\max} \geq as_{\text{limit}}) \cap (ag_{\max} \langle ag_{\text{limit}}) \\ \quad TRUE : Steering\ sensor\ fault \\ end \end{array} \tag{40}$$

If a steering angle fault is identified according to the condition of Equation (40), the measured steering angle sensor information is replaced with the estimated steering angle information calculated using Equation (24).

2.6.3. Gyroscope Fault Detection and Signal Restoration

If the maximum value ($as_{\max}$) of the error calculated using steering angle information is smaller than the limit value (as_{limit}), both the steering angle sensor and the speed sensor are viewed as normal. In this case, if the maximum value ($ag_{\max}$) of the error calculated using the gyroscope information is larger than the limit value (ag_{limit}), one can judge that only the gyroscope is defective, as it follows from the above that the speed sensor is not defective.

$$\begin{array}{l} if (as_{\max} \langle as_{\text{limit}}) \cap (ag_{\max}\ \geq\ ag_{\text{limit}}) \\ \quad TRUE : Gyroscope\ fault \\ end \end{array} \tag{41}$$

If a gyroscope fault is identified according to the condition of Equation (41), rotational angular velocity is estimated from Equation (42) by applying Equations (13) and (14), which describe the estimation of rear wheel speeds based on steering angle information.

$$\hat{\omega} = \frac{\hat{v}_{rrs} - \hat{v}_{rls}}{2l_b}. \tag{42}$$

2.6.4. Fault Detection of the Speed Sensor and Signal Restoration

Under the condition that the gyroscope and the steering angle sensor do not fail simultaneously, the speed sensor is viewed as defective if (i) the maximum value ($as_{\max}$) of the error calculated based on steering angle information is greater than the limit value (as_{limit}), and (ii) the maximum value ($ag_{\max}$) of the error calculated based on the gyroscope information is greater than the limit value (ag_{limit}) [the error of the speed sensor affects both sides]. In this case, when the speed error is larger than the limit value, fault identification of each speed sensor can be performed as follows.

$$\begin{array}{l} if(as_{\max}\rangle\, as_{\text{limit}}) \cap (ag_{\max}\rangle\, ag_{\text{limit}}) \\ \quad TRUE : Speed\,encoder\,fault \\ \quad if\ (e_{vfl}\rangle as_{\text{limit}}) \\ \quad\quad TRUE : Front\,left\,speed\,encoder\,fault \\ \quad end \\ \quad if\ (e_{vfr}\rangle as_{\text{limit}}) \\ \quad\quad TRUE : Front\,right\,speed\,encoder\,fault \\ \quad end \\ \quad if\ (e_{vrl}\rangle as_{\text{limit}}) \\ \quad\quad TRUE : Rear\,left\,speed\,encoder\,fault \\ \quad end \\ \quad if\ (e_{vrr}\rangle as_{\text{limit}}) \\ \quad\quad TRUE : Rear\,right\,speed\,encoder\,fault \\ \quad end \\ end \end{array} \tag{43}$$

If a speed sensor fault is identified according to the condition of Equation (43), the value provided by the defective sensor is replaced with the value estimated using Equation (11) to Equation (14).

3. Results and Discussion

To verify the validity of the proposed algorithm, we performed an off-line test using the sensor data collected from the test vehicle [48]. The vehicle was set up to run in loop guided 235-m test tracks in automatic path guided mode, and sensor information was collected during driving. The vehicle control system collected the signal of magnetic markers embedded in the road and combined it with vehicle motion sensor information to determine real-time vehicle location/orientation and perform automatic guidance control. Faults, failures, and malfunctions observed during driving (e.g., those of the speed sensor, the steering angle sensor, and the rotational angular velocity sensor) may result in erroneous normal position estimation, guidance control errors, or deviation from the suggested path. The results of the off-line test showed how each sensor fault affects vehicle control. In addition, the validity of the proposed method was verified by identifying the defective sensor and replacing the corresponding signal with the estimated value to show whether one can maintain the normal traveling orbit within the effective error range.

3.1. Speed Sensor Fault Test

One of the four wheels simulated the fault of one rear right wheel speed sensor from 20 to 80 s. A zero was injected during this time instead of the normal signal to simulate a failure. Therefore, the actual speed was known. The faulted signal, the normal signal, and the signal estimated using Equation (13) are shown in Figure 7, which compares the estimated signal and the normal signal to show that the sensor signal can be well estimated even if it changes to zero because of failure simulation, i.e., the normal signal is well restored by the estimator. In this case, the error between the normal value and the estimated value was within 0.05 m/s. For simulated faults, the rear right wheel speed signal decreased to 70% of value of the normal signal during the same time interval. Figure 8 compares the estimated signal with the normal signal for the case in which the measured signal decreased from normal to 70% when tire pressure loss or puncture was assumed.

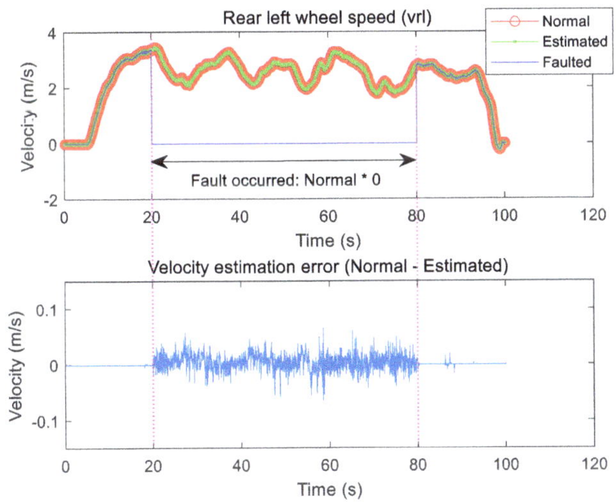

Figure 7. Type 1: Rear left wheel speed sensor fault and estimation.

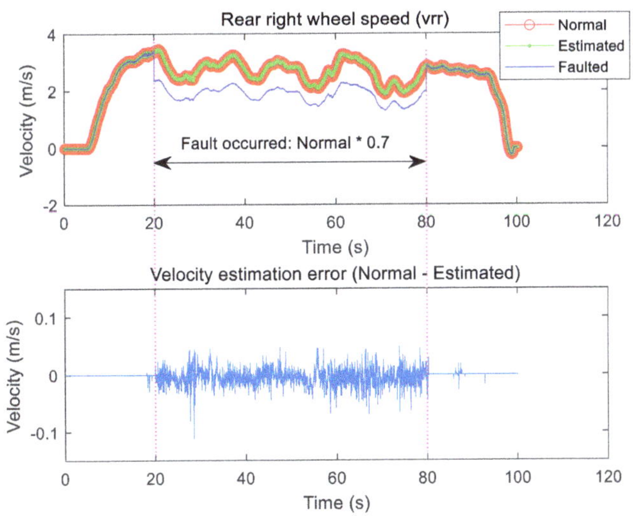

Figure 8. Type 2: Rear right wheel speed sensor fault and estimation.

3.2. Steering Angle Sensor Fault

Only the case where the steering sensor is faulty was considered, and the fault was chosen to occur between 20 and 80 s. During this period, the steering angle information was changed to zero, and the signal was estimated using the proposed algorithm and compared with the normal sensor signal (Figure 9). Because of the fault, the steering angle sensor value was fixed at a constant value for the same time period. The estimated steering angle information was found to be in good agreement with the normal steering angle information (Figure 10).

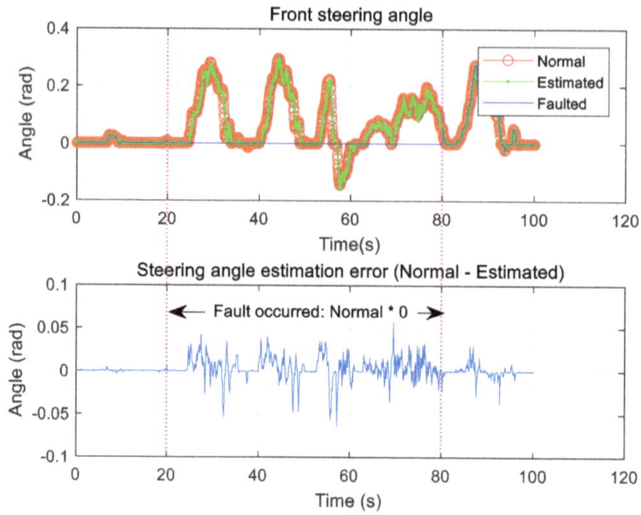

Figure 9. Type 1: Front steering angle sensor fault and estimation.

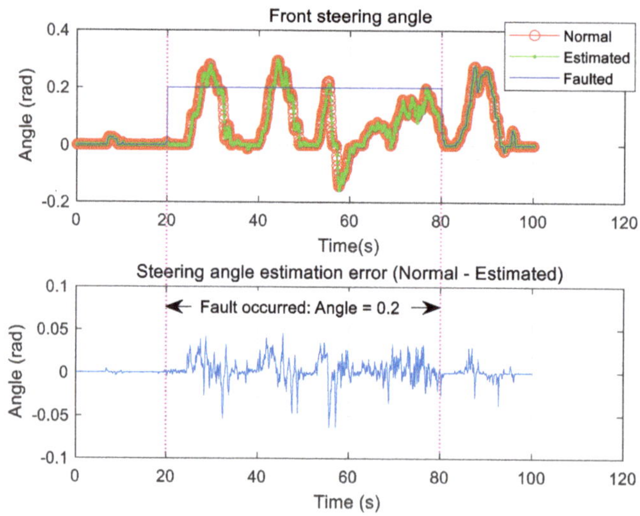

Figure 10. Type 2: Front steering angle sensor fault and estimation.

3.3. Gyroscope Fault

Only the case where the gyroscope is faulty was considered, and the fault was chosen to occur between 20 and 80 s. During this period, the signal was estimated using the proposed algorithm and compared with the normal sensor signal. Notably, the proposed method allowed for good recovery of the normal signal. Figure 11 shows the result of estimation based on Equation (42). We also tested the scenario of a fault when the signal decreased to 50% of the normal signal over the same time interval. In this case, good signal estimation results were also observed (Figure 12).

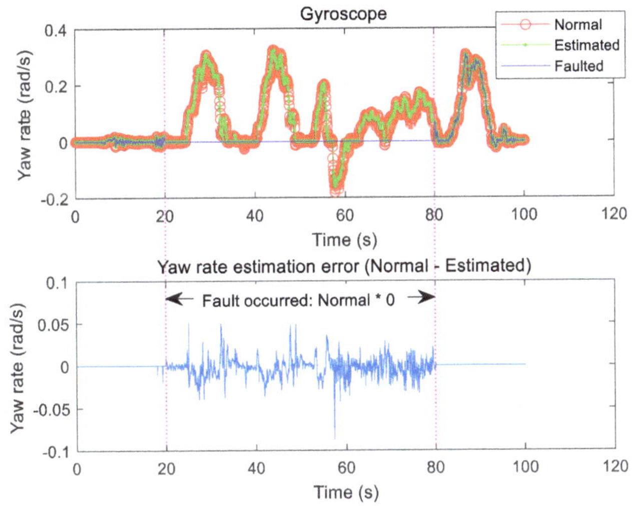

Figure 11. Type 1: Gyroscope fault and estimation.

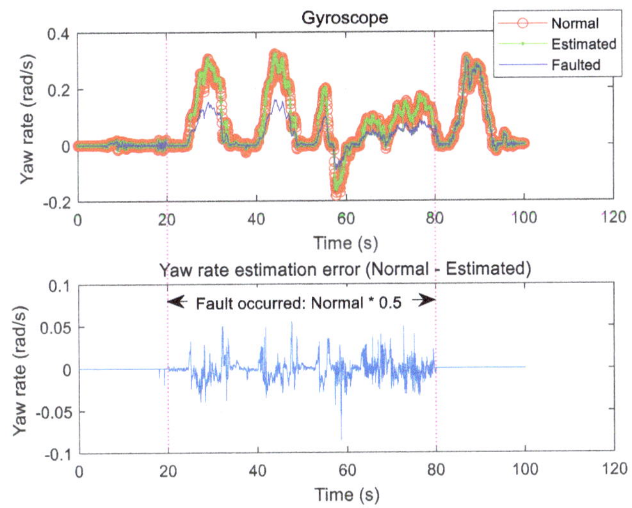

Figure 12. Type 2: Gyroscope fault and estimation.

3.4. *Influence of Defects in Automatic Running*

This test aimed to check the effect of sensor faults on automatic running and to determine whether the proposed fault detection and signal restoration method is valid. In the case of a running vehicle, the sensor was configured so that only one fault occurs at the same time.

3.4.1. Fault-Free Driving

The estimated position and tracking control state of the vehicle were checked in the case where all sensors were normal, and the observed performance was compared to that in the case of sensor fault. Figure 13 shows the driving trajectory of a vehicle that automatically ran the designed path with that of the fault-free vehicle.

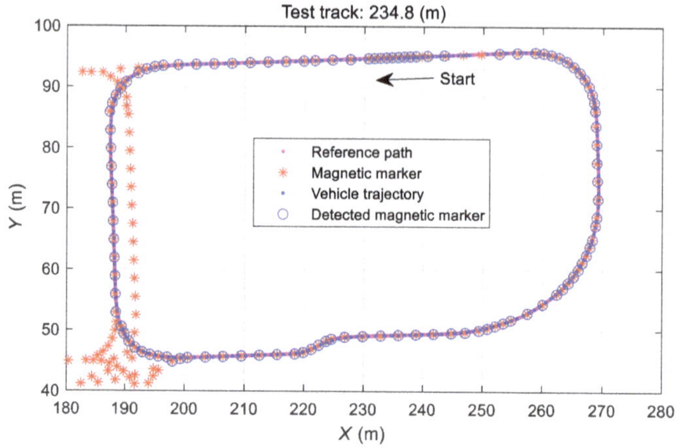

Figure 13. Normal automatic path tracking.

3.4.2. Effect of Rear Left-Wheel Speed Sensor Defect during Automatic Driving

The case when only the rear left wheel speed sensor of the vehicle is defective was considered. When the average of the left- and right wheel speed is used as the center speed, the calculated travel distance is half of the actual travel distance when one of the speed sensors fails during vehicle operation, which results in erroneous position calculation and deviation from the traveling path (Figure 14). As shown in Figure 14, when the speed sensor error occurred before the curve was entered, the vehicle trajectory gradually deviated toward the inside of the reference path.

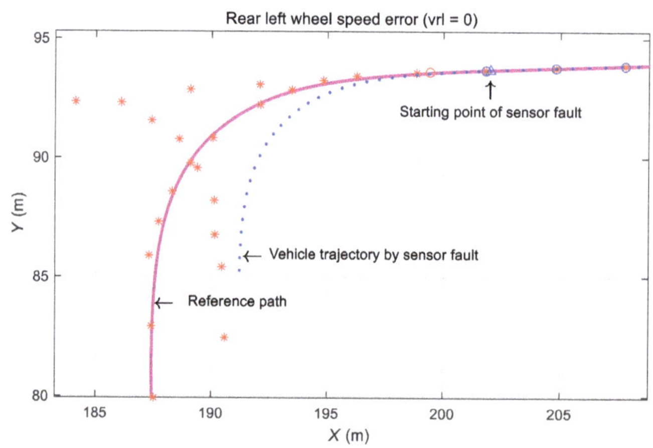

Figure 14. Test result for a wheel speed sensor fault.

3.4.3. Effect of Gyroscope Defect during Automatic Driving

When a gyroscope fault occurs during vehicle operation, an error is generated in the calculation of the running direction, which increases the error in the calculation of vehicle position. As a result, the vehicle deviates from the traveling path. As shown in Figure 15, the vehicle was not able to follow the travel route because the state of the rotational angular velocity sensor was "zero" before the curve was entered, and a straight line trajectory was generated.

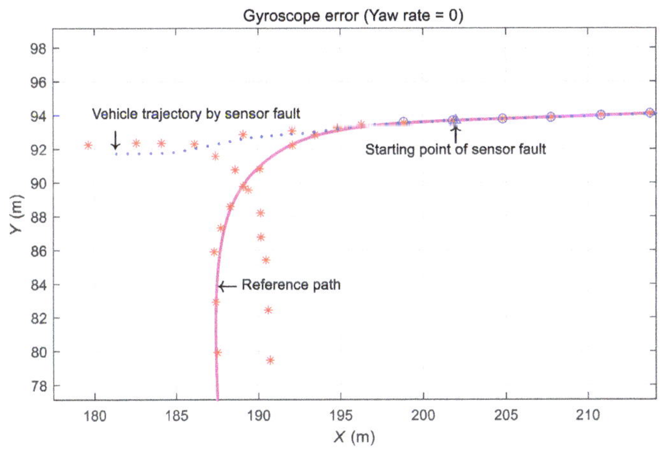

Figure 15. Test result for a gyroscope fault.

Figure 16 shows path-following errors of the vehicle under the fault condition of each sensor. The upper and lower thick solid lines in this figure are the maximum allowable travel error boundaries on normal driving, equaling 15 cm on the straight line portion and increasing along the curved portion. Until 40 s before entering the curved section, the vehicle followed the trajectory with an error within 15 cm, but after 40 s, it deviated from the set error margin according to the fault of each sensor.

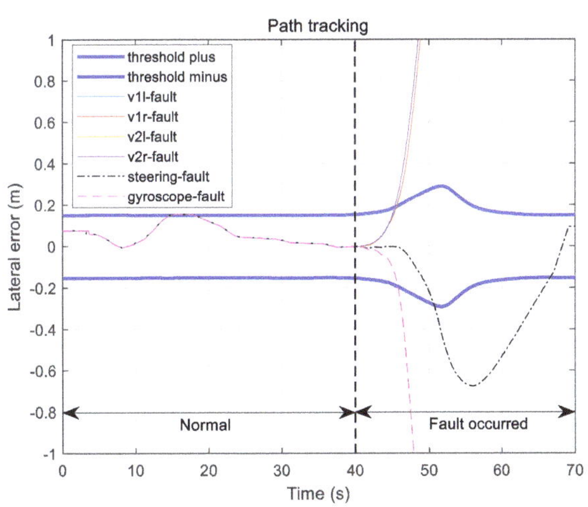

Figure 16. Path tracking errors for each sensor fault.

Figure 17 shows the travel path follow-up error for fault signal restoration under each fault condition. Fault detection and signal restoration were performed 40 s after each sensor fault to afford a running result within the allowable limit error for the normal running of the vehicle. Consequently, the effectiveness of fault detection and the recovery method was successfully verified.

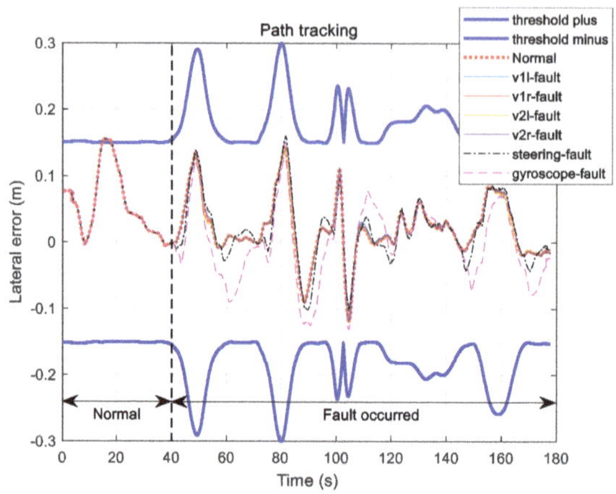

Figure 17. Path tracking errors for each sensor fault recovery.

4. Conclusions

In this paper, fault diagnosis logic and signal restoration algorithms for vehicle motion sensors are presented. To this end, a central axis kinematic model is applied to each wheel speed of the vehicle. Both steering angle and gyroscope information are considered to distinguish failure effects. For fault diagnosis logic, we derive a conditional expression with only two variables to distinguish between normal and fault, and an analytical redundancy structure and a simple diagnostic logic structure are presented to distinguish specific faults. This study further assumes that only one sensor can fail at any given instant, which may limit the current scope of application of the proposed fault diagnosis scheme.

To verify the validity of this method, vehicle sensor data are collected under normal driving conditions, and the algorithm used in the actual vehicle is applied to estimate the vehicle position and orientation in an off-line test. The risk of automatic driving according to each failure is examined through the addition of faults to normal sensor information. It is shown that the autonomous vehicle can satisfy the valid normal driving conditions when the proposed fault detection and signal restoration method is applied under fault conditions. To improve vehicle safety, the authors plan to investigate diagnostic methods for multiple sensor faults.

Author Contributions: Conceptualization, Y.B.; Methodology, Y.B.; Software, Y.B.; Validation, Y.B., B.K. and R.J.; Writing—Original Draft Preparation, Y.B.; Writing—Review and Editing, B.K.; Supervision, R.J.

Funding: Financial support for this research was provided by a grant from the R&D Program of the Korea Railroad Research Institute (KRRI) of the Republic of Korea.

Conflicts of Interest: The authors declare no conflict of interest.

References

1. Karimi, H.; Chadli, M.; Shi, P. Fault detection, isolation, and tolerant control of vehicles using soft computing methods. *IET Control Theory Appl.* **2014**, *8*, 655–657. [CrossRef]
2. Aouaouda, S.; Chadli, M.; Boukhnifer, M. Speed sensor fault tolerant controller design for induction motor drive in EV. *Neurocomputing* **2016**, *214*, 32–43. [CrossRef]
3. Oudghiri, M.; Chadli, M.; El Hajjaji, A. Robust observer-based fault tolerant control for vehicle lateral dynamics. *IJVD* **2008**, *48*, 173–189. [CrossRef]
4. Realpe, M.; Vintimilla, B.; Vlacic, L. Sensor fault detection and diagnosis for autonomous vehicles. *Proc. MATEC Web Conf.* **2015**, *30*, 04003. [CrossRef]
5. Abid, A.; Tahir-khan, M.; De-silva, C.W. Fault detection in mobile robots using sensor fusion. In Proceedings of the 10th ICCSE, Cambridge, UK, 22–24 July 2015; pp. 8–13.

6. Agogino, A.; Alag, S.; Goebel, K. A framework for intelligent sensor validation, sensor fusion, and supervisory control of automated vehicles in IVHS, Intelligent Transportation: Serving the User through Deployment. In Proceedings of the 1995 Annual Meeting of ITS America, Washington, DC, USA, 15–17 March 1995; pp. 77–87.
7. Nembhard, A.D.; Sinha, J.K.; Yunusa-Kaltungo, A. Development of a generic rotating machinery fault diagnosis approach insensitive to machine speed and support type. *JSV* **2015**, *337*, 321–341. [CrossRef]
8. Napolitano, M.; An, Y.; Seanor, B.; Pispistos, S.; Martinelli, D. Application of a neural sensor validation scheme to actual Boeing B737 flight data. In Proceedings of the '99 AIAA Guidance, Navigation and Control Conference, Portland, OR, USA, 9–11 August 1999.
9. Napolitano, M.; An, Y.; Seanor, B. A fault tolerant flight control system for sensor and actuator failures using neural networks. *Aircr. Des.* **2000**, *3*, 103–128. [CrossRef]
10. Napolitano, M.; Windon, D.; Casanova, J.; Innocenti, M.; Silvestri, G. Kalman filters and neural-network schemes for sensor validation in flight control systems. *IEEE Trans. Control Syst. Technol.* **1998**, *6*, 596–611. [CrossRef]
11. Rago, C.; Prasanth, R.; Mehra, R.K.; Fortenbaugh, R. Failure Detection and Identification and Fault Tolerant Control Using the IMMKF with Applications to the Eagle-Eye UAV. In Proceedings of the 37th Conference on Decision and Control, Tampa, FL, USA, 18 December 1998.
12. Heredia, G.; Remuß, V.; Ollero, A.; Mahtani, R.; Musial, M. Actuator Fault Detection in Autonomous Helicopters. In Proceedings of the 5th IFAC/EURON Symposium on Intelligent Autonomous Vehicles, Lisbon, Portugal, 5–7 July 2004.
13. Heredia, G.; Ollero, A.; Mahtani, R.; Béjar, M.; Remuss, V.; Musial, M. Detection of Sensor Faults in Autonomous Helicopters. In Proceedings of the IEEE International Conference on Robotics and Automation, Barcelona, Spain, 18–22 April 2005; pp. 2229–2234.
14. Garg, V. Fault Detection in Nonlinear Systems: An Application to Automated Highway Systems. Ph.D. Thesis, University of California, Berkeley, CA, USA, July 1995.
15. Agogino, A.; Chao, S.; Goebel, K.; Alag, S.; Cammon, B.; Wang, J. *Intelligent Diagnosis Based on Validated and Fused Data for Reliability and Safety Enhancement of Automated Vehicles in an IVHS*; PATH Research Report UCB-ITS-P RR-98-17; University of California: Berkeley, CA, USA, 1998.
16. Isermann, R. Diagnosis methods for electronic controlled vehicles. *Veh. Syst. Dyn.* **2001**, *36*, 77–117. [CrossRef]
17. Yi, J.; Howell, A.; Horowitz, R.; Hedrick, K.; Alvarez, L. *Fault Detection and Handling for Longitudinal Control*; California PATH Research Report UCB-ITS-P RR-2001-21; University of California: Berkeley, CA, USA, 2001.
18. Rajamani, R.; Howell, A.S.; Chen, C.; Hedrick, J.K.; Tomizuka, M. A complete fault diagnostic system for automated vehicles operating in a platoon. *IEEE Trans. Control Syst. Technol.* **2001**, *9*, 553–564. [CrossRef]
19. Howell, A. Nonlinear Observer Design and Fault Diagnostics for Automated Longitudinal Vehicle Control. Ph.D. Thesis, University of California, Berkeley, CA, USA, July 2002.
20. Chen, R.H.; Ng, H.K.; Speyer, J.L.; Mingori, D.L. *Testing and Evaluation of Robust Fault Detection and Identification for a Fault Tolerant Automated Highway System*; UC Berkeley, California Partners for Advanced Transportation Technology: Berkeley, CA, USA, 2002. Available online: https://escholarship.org/uc/item/2bw4f9fw (accessed on 25 July 2019).
21. Okatan, A.; Hajiyev, C.; Hajiyeva, U. Fault detection in sensor information fusion Kalman filter. *AEU Int. J. Electron. Commun.* **2009**, *63*, 762–768. [CrossRef]
22. Rodger, J.A. Toward reducing failure risk in an integrated vehicle health maintenance system: A fuzzy multi-sensor data fusion Kalman filter approach for IVHMS. *Expert Syst. Appl.* **2012**, *39*, 9821–9836. [CrossRef]
23. Arogeti, S.A.; Wang, D.; Low, C.B.; Yu, M. Fault detection isolation and estimation in a vehicle steering system. *IEEE Trans. Ind. Electron.* **2012**, *59*, 4810–4820. [CrossRef]
24. Na, W.; Park, C.; Lee, S.; Yu, S.; Lee, H. Sensitivity-based fault detection and isolation algorithm for road vehicle chassis sensors. *Sensors* **2018**, *18*, 2720. [CrossRef] [PubMed]
25. Emirler, M.T.; Kahraman, K.; Sentürk, M.; Güvenç, B.A.; Güvenç, L.; Efendioglu, B. Vehicle yaw rate estimation using a virtual sensor. *Int. J. Veh. Technol.* **2013**, *2013*, 582691. [CrossRef]
26. Huang, W.; Su, X. Design of a fault detection and isolation system for intelligent vehicle navigation system. *IJNO* **2015**, *2015*, 279086. [CrossRef]

27. Pous, N.; Gingras, D.; Gruyer, D. Intelligent vehicle embedded sensors fault detection and isolation using analytical redundancy and nonlinear transformations. *J. Control Sci. Eng.* **2017**, *2017*, 1763934. [CrossRef]
28. Aldridge, H.A. Robot position sensor fault tolerance. Ph.D. Thesis, Carnegie Mellon University, Pennsylvania, PA, USA, June 1996.
29. Klancar, G. Fault detection and isolation by means of principal component analysis. In Proceedings of the Cybernetics & Informatics Eurodays Workshop, Prague, Czech Republic, 26–30 September 2000; pp. 26–30.
30. Isermann, R. Model-based fault-detection and diagnosis status and applications. *Annu. Rev. Control.* **2005**, *29*, 71–85. [CrossRef]
31. Duan, F.; Wang, H.; Zhang, L. Study on fault-tolerant filter algorithm for integrated navigation system. In Proceedings of the 2007 International Conference on Mechatronics and Automation, Harbin, China, 5–8 August 2007; pp. 2419–2423.
32. Chan, C.W.; Hua, S.; Hong, Y.Z. Application of fully decoupled parity equation in fault detection and identification of DC motors. *IEEE Trans. Ind. Electron.* **2006**, *53*, 1277–1284. [CrossRef]
33. Muenchhof, M. Comparison of change detection methods for a residual of a hydraulic servo-axis. *FAC Proc. Vol.* **2005**, *38*, 317–322. [CrossRef]
34. Escobet, T.; Trave-Massuyes, L. Parameter estimation methods for fault detection and isolation. *Bridge Workshop Notes* **2001**, *40*, 1–11.
35. Hilbert, M.; Kuch, C.M.; Nienhaus, K. Observer based condition monitoring of the generator temperature integrated in the wind turbine controller. In *EWEA 2013 Scientific Proceedings;* EWEA: Vienna, Austria, 2013; pp. 189–193.
36. Heredia, G.; Ollero, A. Sensor fault detection in small autonomous helicopters using observer/Kalman filter identification. In Proceedings of the 2009 IEEE International Conference on Mechatronics, Malaga, Spain, 14–17 April 2009; pp. 1–6.
37. Garcia, E.A.; Frank, P.M. Deterministic nonlinear observer-based approaches to fault diagnosis: A survey. *Control Eng. Pract.* **1997**, *5*, 663–670. [CrossRef]
38. Patton, R.J.; Chen, J. Observer-based fault detection and isolation: Robustness and applications. *Control Eng. Pract.* **1997**, *5*, 671–682. [CrossRef]
39. Zhou, D.H.; Frank, P.M. Fault diagnostics and fault tolerant control. *IEEE Trans. Aerosp. Electron. Syst.* **1998**, *34*, 420–427. [CrossRef]
40. Frank, P.M. Fault diagnosis in dynamic systems using analytical and knowledge-based redundancy. A survey and some new results. *Automatica* **1990**, *26*, 459–474. [CrossRef]
41. Jayaram, S. A new fast converging Kalman filter for sensor fault detection and isolation. *Sens. Rev.* **2010**, *30*, 219–224. [CrossRef]
42. Hwang, I.; Kim, S.; Kim, Y.; Seah, C.E. A survey of fault detection, isolation, and reconfiguration methods. *IEEE Trans. Control Syst. Technol.* **2010**, *18*, 636–653. [CrossRef]
43. Gertler, J. Fault detection and isolation using parity relations. *Control Eng. Pract.* **1997**, *5*, 653–661. [CrossRef]
44. Isermann, R. Process fault detection based on modelling and estimation methods-a survey. *Automatica* **1984**, *20*, 387–404. [CrossRef]
45. Patton, R.J.; Uppal, F.J.; Lopez-Toribio, C.J. Soft Computing Approaches to Fault Diagnosis for Dynamic Systems: A Survey. In Proceedings of the 4th IFAC Symposium on Fault Detection Supervision and Safety for Technical Processes, Budapest, Hungary, 14–16 June 2000; pp. 298–311.
46. Isermann, R.; Ballé, P. Trends in the application of model based fault detection and diagnosis of technical processes. *Control Eng. Pract.* **1997**, *5*, 709–719. [CrossRef]
47. Miljkovic, D. Fault Detection Methods: A Literature Survey. In Proceedings of the 34th International Convention on Information and Communication Technology Electronics and Microelectronics, Opatija, Croatia, 23–27 May 2011; pp. 750–755.
48. Byun, Y.S.; Jeong, R.G.; Kang, S.W. Vehicle position estimation based on magnetic markers: Enhanced accuracy by compensation of time delays. *Sensors* **2015**, *15*, 28807–28825. [CrossRef]

sensors

MDPI

Article

Real-Time Photometric Calibrated Monocular Direct Visual SLAM

Peixin Liu, Xianfeng Yuan *, Chengjin Zhang, Yong Song, Chuanzheng Liu and Ziyan Li

School of Mechanical Electrical and Information Engineering, Shandong University, Weihai 264209, China
* Correspondence: yuanxianfeng@sdu.edu.cn

Received: 28 June 2019; Accepted: 16 August 2019; Published: 19 August 2019

Abstract: To solve the illumination sensitivity problems of mobile ground equipment, an enhanced visual SLAM algorithm based on the sparse direct method was proposed in this paper. Firstly, the vignette and response functions of the input sequences were optimized based on the photometric formation of the camera. Secondly, the Shi–Tomasi corners of the input sequence were tracked, and optimization equations were established using the pixel tracking of sparse direct visual odometry (VO). Thirdly, the Levenberg–Marquardt (L–M) method was applied to solve the joint optimization equation, and the photometric calibration parameters in the VO were updated to realize the real-time dynamic compensation of the exposure of the input sequences, which reduced the effects of the light variations on SLAM's (simultaneous localization and mapping) accuracy and robustness. Finally, a Shi–Tomasi corner filtered strategy was designed to reduce the computational complexity of the proposed algorithm, and the loop closure detection was realized based on the oriented FAST and rotated BRIEF (ORB) features. The proposed algorithm was tested using TUM, KITTI, EuRoC, and an actual environment, and the experimental results show that the positioning and mapping performance of the proposed algorithm is promising.

Keywords: visual SLAM; sparse direct method; photometric calibration; corner detection and filtering; loop closure detection

1. Introduction

Recently, many visual simultaneous localization and mapping (SLAM) systems have been proposed, since they are fundamental building blocks for many emerging technologies, such as autonomous cars, virtual reality, and augmented reality [1]. Mobile ground equipment estimates its own position and reconstructs a three-dimensional map in real time using specific sensors without any prior environmental information [2].

At present, the SLAM system based on vision sensors has gained popularity in the field [3]. According to its algorithmic principle, the visual SLAM system can be divided into the direct formulation and the indirect formulation [4]. Compared with the indirect visual SLAM, the direct formulation can establish dense, semi-dense, sparse 3D reconstructions that are valuable for the navigation of ground mobile equipment [5]. In addition, research has shown that the mapping performance of the direct approach was more robust than the indirect one for the low-texture-features environment [6].

The direct and semi-direct formulations optimize the photometric error based on the grayscale invariant assumption to estimate the camera motion, since the sensors provide the photometric measurements [7]. J. Engel et al. [8] proposed the LSD-SLAM (large-scale direct monocular SLAM) with indirect loop closure detection based on the angular relationship between the pixel gradient and the polar line in dense reconstruction. The LSD-SLAM easily loses the tracked visual features as the camera moves quickly, since it is sensitive to the camera's internal parameters and exposure conditions. C. Forster et al. [9] proposed SVO (semi-direct visual odometry), which is a visual odometry

(VO) without back-end optimization, loop closure detection and re-localization. SVO tracks the features from accelerated segment test (FAST) feature points and surrounding pixels by minimizing the photometric error to estimate the camera motion. The speed rate of SVO can reach 100 frames per second and up to 400 frames per second in SVO2.0 [10]. To improve the robustness of the system, P. Kim et al. [11] proposed patch-based VO in 2015 using linear illumination models to compensate for the local brightness variations. Patch-based VO enhances the robustness of sudden illumination changes but has a high dependency on the scene's textural features. J. Engel et al. [12] proposed DSO (direct sparse odometry), which is a direct pixel-tracking model with photometric parameters that calculates the residual of the pixel projection from the dominant frame to the current frame. When DSO is tracking the pixels, the system retains several key frames using a sliding window to establish the minimized energy function to obtain the pose and the inverse depth of the current camera status as the back-end [13]. In order to enhance the performance of direct visual odometry, P. Bergmann et al. [14] proposed an online photometric calibration, which dynamically estimates the photometric parameters by solving the least squares equation of the feature tracker and adjusts the exposure situation of the input sequence. It is a milestone in improving the positioning and mapping accuracy for direct formulation. Stereo DSO, which was proposed by M. Schwörer et al. [15] and improved by N. Yang et al. [16], further enhances the precise depth estimation. X. Gao et al. [17] proposed LDSO (direct sparse odometry with loop closure)—which is a SLAM system with indirect formulation loop closure detection—and evaluated it on multiple sets of datasets but not in an actual environment. In the field of direct SLAM, the computer vision group at the Technical University of Munich has made a major contribution.

In the DSO series and LDSO, the photometric parameters were introduced to compensate for the vignetting and response function of input images as constants. However, the compensation, based on pre-trained photometric calibration files, could not update photometric parameters for dynamic illumination in real time. Inspired by [14] and [17], in order to further improve the robustness of the direct formulation visual SLAM system in positioning and mapping, we reinforced the LDSO algorithm by introducing real-time photometric calibration to update the exposure condition of the input sequence. In addition, a Shi–Tomasi corner filtering mechanism was designed to reduce the computational complexity of loop closure detection. The flow chart of the proposed SLAM system is shown in Figure 1. Firstly, a photometric parameter model was introduced to compensate the input sequence according to the photometric formation of the automatic exposure camera. Secondly, we utilized the robust Kanade-Lucas-Tomasi tracking method (KLT) tracker to obtain the continuous feature points between the input sequence to establish an optimization equation which integrated the KLT tracker with the direct-tracked pixels in VO. Then, the photometric calibration parameters of the visual odometry were updated in real time to optimize the exposure situation of the input sequence. Finally, a Shi–Tomasi corner filtering mechanism was introduced in the indirect back-end to realize the relocation and loop closure detection based on the ORB feature. On the generic dataset, we demonstrated that the drift error of the proposed algorithm was significantly reduced with respect to LDSO and the performance on KITTI was similar to mono-ORBSLAM. In addition, the proposed algorithm was evaluated in the actual illumination challenge environment and the experimental results indicate that the mapping performance of the proposed method is promising.

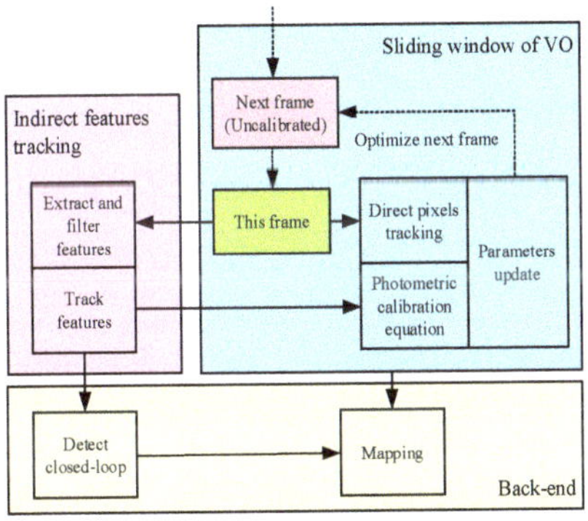

Figure 1. The flow chart of the proposed system. We divided the approach into three elements: The indirect feature tracker, the sliding windowed photometric compensation and the back-end optimization. The green one is the last frame that joins in the localization, mapping, and photometric parameters' calculation. Then, the parameters were utilized to compensate the red frame based on the photometric parameters.

2. Photometric Calibration Model

According to the photometric parameters of the auto exposure camera, an optimization equation was established based on the corner point tracker. Then, the vignetting factors and the response function were dynamically updated to compensate for the input sequence illumination condition to realize the photometric calibration in real time.

2.1. Vignetting and Response Function

A scene point is illuminated by a light source and reflects the energy back into space [18]. The global light intensity, which is called the radiance of the scene point, is independent of the viewing angle of the observer.

When the vision sensor captures the scene as an image, the radiance of the scene's points are converted into irradiance B by a lens. For the formation of the image each time, the total energy that is received by the sensor depends on the irradiance that passes through the camera shutter during the exposure time t. Finally, the energy turns into the pixel intensity according to the response function G [19]. The flow chart of the photometric image formation process is shown in Figure 2.

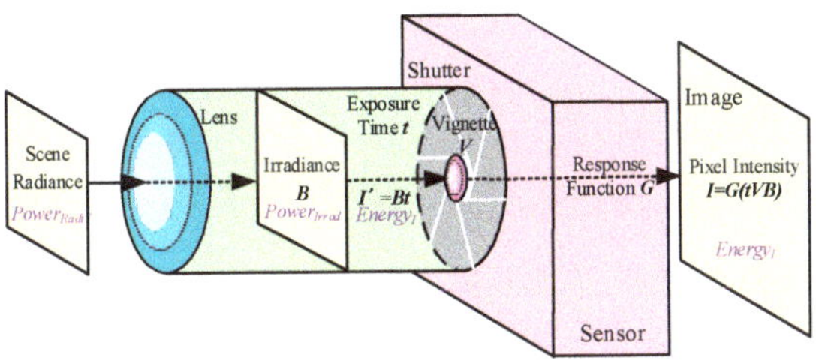

Figure 2. The flow chart of photometric image formation. The original energy that is emitted from the scene, which is called the radiance, is affected by the vignetting effort of the lens and the exposure time of the shutter.

The effective incident off-axis light of the front lens was changed according to the size of the aperture and the exposure time, which caused the pixel intensity of the image to gradually weaken from the center out. In the auto-exposure mode, the adaptive exposure time was determined by the different scenes. The response function is a process through which the received photon is nonlinearly converted into a brightness value.

The imaging model of the photometric image formation in Figure 2 can be defined by Equations (1) and (2).

$$I_i(x) = G_i(t_i V_i(x) B_i(x)) \tag{1}$$

$$I'_i(x) = t_i B_i(x) = \frac{G_i^{-1}(I_i(x))}{V_i(x)} \tag{2}$$

where I_i is the pixel intensity observed in frame i, B_i is the power of irradiance, I'_i is the received energy of irradiance during once exposure time, t_i is the exposure time, V is the lens attenuation (vignetting), and G is the response function.

Vignette $V{:}\Omega \to [0,1]$. Assuming that the pixel intensity attenuation factors are symmetric around the center of the image, vignetting is defined as follows [20].

$$V(x) = 1 + \sum_{n=1}^{3} v_n R(x)^{2n} \tag{3}$$

where $R(x)$ is the normalized radius of pixel x with respect to the center of the image.

Response function $G{:}\,\mathbb{R} \to [0,\ 255]$. When the frames are underexposed and overexposed, their brightness values are 0 and 255, respectively. Linearization is applied to G.

$$G(x) = g_0(x) + \sum_{k=1}^{\mathrm{n}} c_k h_k(x) \tag{4}$$

The main response function g_0 and the basis function h_k were obtained by PCA (principal component analysis). When the coefficient vector $\boldsymbol{c_k}$ and v_n were iteratively calculated using the photometric calibration equation, the adaptive vignetting and response function compensation of the input sequence were realized.

2.2. Photometric Calibration Equation

After constructing the model of the vignetting factors V and the response function G, the feature points of the images were extracted to track the input sequences to establish an optical flow equation. The equation, including the residual of the last M frames, were minimized to update the vignetting factors V and the response function G [14]. The flow of the photometric parameters' optimization is described Figure 1.

The Shi–Tomasi corners are generally utilized as the global features to represent an image, owing to their good affine invariance. Those corners are tracked by the Kanade–Lucas (LK) optical flow on the image pyramid, which is called the KLT tracker, to construct an optical flow energy equation.

We segmented an image into 32 × 32 regions and defined a constant of tracked candidate points to obtain a good effect. When the tracked feature was lost, a new candidate was extracted from the high-gradient region that contained fewer points. If the max gradient of a region was lower than the threshold, the region was filtered [21].

For a set of tracked points $\mathcal{P}$ in one frame, the proposed approach designs the function of the energy residual using its co-visual frames. The tracked pixel intensity was restored to the irradiance estimation energy during one exposure, written as $I'^{p'}_i$, based on the photometric formation. Then the received energy from irradiance during once exposure, written as I'^{p}_j, was utilized to calculate the residual between the energy of the co-visual frames. According to Equation (2), the Huber norm of the pixel residual energy function is defined as Equations (5) and (6).

$$E_{I'} = \sum_{p \in \mathcal{P}} \sum_{i \in \mathcal{F}_p} w_i{}^p \| I'^{p}_{j} - I'^{p'}_{i} \|_h \tag{5}$$

$$r_i = t_i B_i - \frac{G_i^{-1}\left(I_i^p\right)}{V_i^p} \tag{6}$$

$\mathcal{F}_p$ is the set of frames that can observe the points $\mathcal{P}$. The photometric parameters were dynamically chosen by minimizing Equation (5) using the L–M approach. The optimization process of the photometric parameters is described in detail in Section 3.

The estimation of the vignetting and response functions requires multiple images, which are difficult to collect in time during one calibration. Therefore, the states of the current vignetting and response function were maintained after the current estimation to compensate the last frame. Then, the compensated frame was used for localization, mapping, and evaluating the photometric parameters in the next frame, as shown in Figure 1. The adaptive photometric compensation results of the dataset were randomly selected, as shown in Figure 3.

(**a**) Tracked frame of KITTI (**b**) Compensated frame of KITTI

Figure 3. The partial photometric calibration results of KITTI sequence 00. The subfigures (**a**) are the tracked original frames, and the subfigures (**b**) are the compensated frames. It can be seen in subfigure (**b**) that the global exposure was enhanced, especially at the edge of the image. In addition, the brightness values of subfigure (**b**) remained continuous.

3. The Combination of Photometric Calibration and Direct SLAM

In the direct formulation, the minimized photometric error was utilized to achieve the camera pose based on the grayscale invariance assumption. DSO integrates the photometric parameters to simulate the vignetting effect and the gamma attenuation to enhance the robustness of VO. However, the scene radiance of VO was calculated using the pretrained photometric calibration files that were proposed in [1].

To further enhance the performance of the direct formulation, the photometric calibration should adapt to a continuous pixels' brightness value to respond to the illumination challenge. When the last frame enters the sliding window, it is compensated based on the previous frames and then applied to the front-end of the SLAM system.

3.1. Direct Sparse Model

After the frame is compensated, it enters the VO model to join the localization and mapping. The covisual pixels in all frames of the sliding window are projected onto the current frame to build the photometric error equation, which is given by Equation (7).

$$minE_{photo} = min \sum_{i\in\mathcal{F}_w} \sum_{p\in\mathcal{P}_i} \sum_{j\in obj(p)} E_{p_j} \tag{7}$$

where $\mathcal{F}_w$ is the set of all frames in the sliding window, $\mathcal{P}_i$ is the set of all observed pixels on the host frame, and $obj(p)$ is the set of co-visual frames that can observe the pixel p. The estimation of the inverse depth and camera pose were achieved by minimizing Equation (7), which can be shown as the flowing factor graph [12].

There are at most N_f active key frames in each sliding window. When a new frame enters the sliding window, it is tracked to determine whether to create a new key frame. After obtaining enough key frames, the redundant key frames are deleted according to the marginalized strategy to reduce the calculation costs [7,12].

In Section 2.2 of Chapter 2, the input image was divided into 32 × 32 regions. The pixel tracking only happens where the maximum pixel gradient is greater than the threshold. If the value of the threshold of each region equals the average gradient, then one a constant must be added.

3.2. Parameters Update

The tracking of the indirect feature, which was utilized to update the response function and the vignetting factor, was simultaneously performed with the estimations of the inverse depth and the camera pose in the sliding window. Thus, the tracking equations in the sliding window are rewritten as Equations (8) and (9).

$$r_i = t_j B_j - e^{-a_i} \frac{G^{-1}\left(I_i^{p'} - b_i\right)}{V_i} \tag{8}$$

$$E_{p_j} = \sum_{p\in N_{p_k}} w_p \|r_i\|_h \tag{9}$$

where N_{p_k} epresents the neighboring pixels of pixel p_k; t_i and t_j are the exposure times of images I_i and I_j, respectively; a and b are the affine brightness transform parameters [22]; and p' is a reprojection pixel of p on I_j. Combined with Equations (3) and (4), we set $\boldsymbol{x}$ as the total number of variable parameters to optimize Equation (9).

$$\boldsymbol{x} = [\xi, a, b, \boldsymbol{c}, \boldsymbol{v}]^T \tag{10}$$

$\xi \in \mathbb{R}^6$ the camera state, $\boldsymbol{c} = (c_1, c_2, c_3, c_4)$ is the coefficient vector of the response function G, and $\boldsymbol{v} = (v_1, v_2, v_3)$ is the vignetting coefficient vector. Considering that the exposure time t can be estimated by two consecutive frames, the proposed approach suggests decoupling the exposure time t estimation from the other parameters [14]. According to Equations (8) and (9), the visual odometry with the adaptive exposure compensation equation is introduced as Equation (11).

$$minE_{CalibVo} = min \sum_{i\in\mathcal{F}_w} \sum_{p\in\mathcal{P}_i} \sum_{j\in obj(p)} E_{p_j} \tag{11}$$

The L–M algorithm is applied to calculate the Jacobian matrix of the residual r_i as

$$J_{x_i} = \frac{\partial r_i}{\delta \boldsymbol{x}_i} = \left(\frac{\partial r_i}{\delta \xi_i}, \frac{\partial r_i}{\delta a_i}, \frac{\partial r_i}{\delta b_i}, \frac{\partial r_i}{\delta \boldsymbol{c}_i}, \frac{\partial r_i}{\delta \boldsymbol{v}_i}\right) \tag{12}$$

$$\boldsymbol{H}\delta\boldsymbol{x}_i = -\boldsymbol{b} \tag{13}$$

$$H = J_{x_i}{}^T W_{x_i} J_{x_i} + \lambda I \tag{14}$$

$$b = J_{x_i}{}^T W_{x_i} r_i \tag{15}$$

Equations (12)–(14) describe the iterative solution process of Equation (11) based on the L–M approach. Here, the weight matrix W_{x_i}. was inversely proportional to the image gradient of x_i [12].

By solving $\delta x = (\delta\xi, \delta a, \delta b, \delta c, \delta v)$, the vignetting V and the response function G were updated to realize the real-time photometric calibration of the input sequence. Figure 4 shows that when a new frame I_i arrived, its response function and vignetting, which were calculated based on the last $\mathcal{F}_w$ frames, were removed to restore the scene's radiance. The system maintains the previous vignetting and response function state to restore the current frame I_j, and simultaneously updates the previous vignetting state and response function. Then, the incoming frames are calibrated based on the last response function and the vignetting estimation.

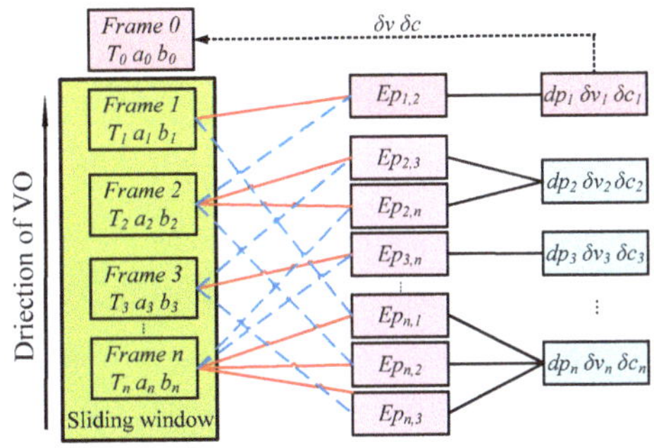

Figure 4. The factor graph of the direct formulation. The tracked pixel of the host frame is represented by the solid red line, which is linked to co-visual frames by the dotted blue line. For each term of the tracked pixel, an energy function of the residual was established to calculate the inverse depth and photometric parameters, which are shown by the black line. Then, the parameters were utilized to compensate the next frame.

3.3. Window Optimization

When a new key frame is inserted, the current sliding window is optimized using the bundle adjustment (BA) [7] like the local loop closure of ORBSLAM [23,24] to reduce the drift localization error, as shown in Figure 5.

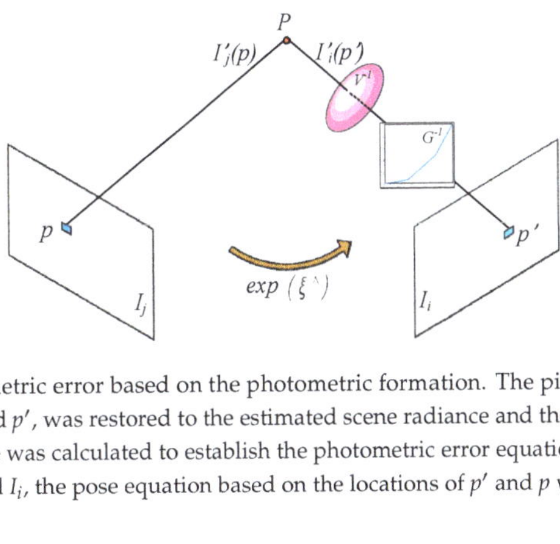

Figure 5. The photometric error based on the photometric formation. The pixel intensity of tracked point p, which is called p', was restored to the estimated scene radiance and then the residual with the current scene radiance was calculated to establish the photometric error equation. For the camera pose change between I_j and I_i, the pose equation based on the locations of p' and p was utilized to calculate the $\mathfrak{se}(3)$.

In Figure 5, $\xi \in \mathbb{R}^6$ includes the element of $\mathfrak{se}(3)$, and $\hat{\xi}$ is the anti-symmetric matrix of ξ. For all keyframes $\mathcal{F}_{wkey}$ in the sliding window, the camera pose optimization equation is established as follows.

$$minE_{wkey} = min \sum_{i\in\mathcal{F}_{wkey}} \sum_{p\in\mathcal{P}_i} \|p' - \frac{1}{s_i} Kexp(\hat{\xi})p\|_2^2 \tag{16}$$

The least squares problem represented by Equation (16) can be iteratively solved using the L–M algorithm, and then the best current camera pose $\hat{\xi}$ can be obtained. The graph optimization was based on g2o library [25].

3.4. Loop Closure Detection

In LDSO, the loop closure detection was realized by calculating the ORB descriptor of the tracked Shi–Tomasi corner. However, the tracked candidate points may be extracted in the low gradient region, which increases the calculation burden and affects the performance of the loop closure detection [26]. In the proposed approach, a corner extraction and screening strategy was designed so that the 32 × 32 regions were segmented in each image and the low-gradient regions were screened, as in Section 2.2 of Chapter 2. When the tracked features are lost, new candidate points will be extracted from the smaller region and the total number of points is constant. This strategy improved the localization performance of the system while enhancing the effective calculation capability. The results of the robust Shi–Tomasi corner detection on the EuRoC [27] V1_03_difficult dataset are shown in Figure 6.

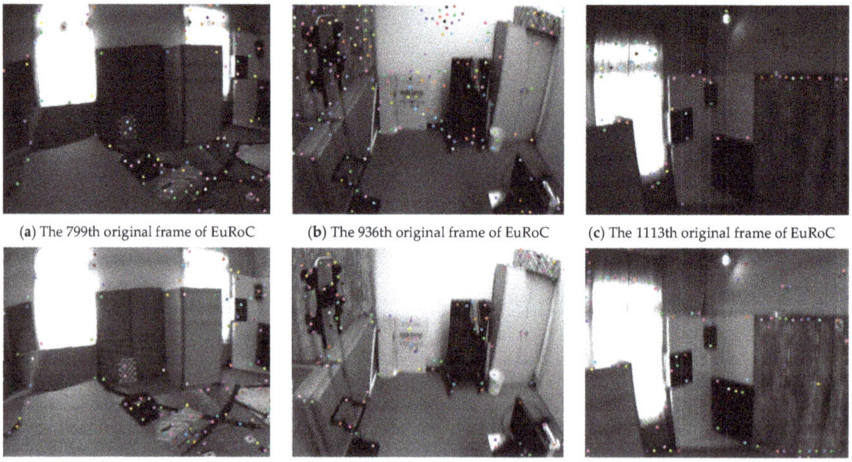

(**a**) The 799th original frame of EuRoC (**b**) The 936th original frame of EuRoC (**c**) The 1113th original frame of EuRoC

(**d**) The 799th calibrated frame of EuRoC (**e**) The 936th calibrated frame of EuRoC (**f**) The 1113th calibrated frame of EuRoC

Figure 6. The experimental results on the EuRoC V1_03_difficult dataset. Subfigures (**d**), (**e**) and (**f**) are modified from subfigures (**a**), (**b**) and (**c**) respectively.

As can be seen from Figure 6, the number, the distribution and the area texture of the detected features are obviously adjusted. With the help of photometric calibration and the Shi–Tomasi corner filtering mechanism, the features in the low texture regions (e.g., the wall in Figure 6b,e) are filtered and the features in the high texture regions (e.g., the cabinet in Figure 6c,f) are increased. In the loop closure detection process, DBoW3 was used to build the database of the bag of words (BoW) model to achieve loop closure detection [28]. The corresponding descriptors are re-coded based on the pixels' intensity around the features. It can be concluded that the stable quantity and reasonable distribution of the tracked features and their descriptors can make the score—which is calculated by the BoW model— become more reliable. Then the tree node of the marginalized key frames in the BoW model are selected to realize the loop closure detection. Thereafter, the loop closure detection performance of the proposed system can be facilitated. The loop closure detection experiment of the proposed system and the LDSO can be seen in Section 4.1.

4. Experiments

The proposed algorithm was operated on a laptop with an Intel i7-8750H CPU, 16G of memory, and Ubuntu16.04. GPU acceleration was not adopted during the experiments. The simulation and the actual experiments were designed to evaluate the localization accuracy, loop closure performance, and point cloud map of the proposed algorithm.

The experimental designs were grouped according to the dataset in this paper. Firstly, the localization accuracy, the photometric parameters 'calibration and the pixels' tracking performance were tested on the TUM dataset [29]. Secondly, the localization accuracy, the timing cost and the loop closure detection performance were evaluated on the KITTI [30] dataset. Thirdly, the feature detection of loop closure and point cloud map were shown on the EuRoC challenging illumination dataset. Finally, the proposed method was tested in a custom environment where the illumination in the room was being changed under control.

4.1. Experiments Based on Different Datasets

The proposed algorithm was evaluated using the TUM-Mono and KITTI Odometry datasets in a monocular setting. The TUM dataset [29] is a scenic dataset of the Technical University of Munich, including 50 laboratory and outdoor sequences. The proposed algorithm was tested using the TUM-Mono dataset sequences 04 and 31, and compared the localization accuracy with the original DSO [12], the original LDSO [17], and the enhanced DSO facilitated by the online photometric calibration [14], respectively. The experimental results are illustrated in Figure 7.

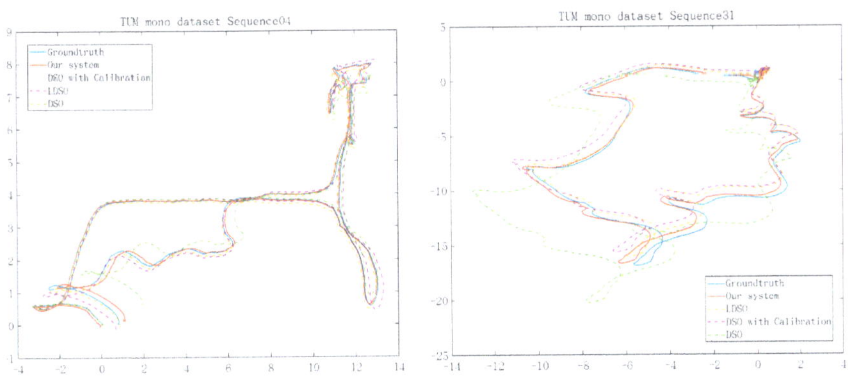

(**a**) The experimental results on TUM-Mono sequence 04 (**b**) The experimental results on TUM-Mono sequence 31

Figure 7. The experimental results on the TUM-Mono dataset. Subfigures (**a**) and (**b**) show the trajectories of sequences 04 and 31, respectively, along the *x*-axis and *z*-axis on our system. Down are direct sparse odometry (DSO); DSO with loop closure (LDSO); and enhanced DSO, which was integrated with the algorithms proposed in [12] and [14].

Figure 7 shows that the trajectories obtained by the online photometric calibration-enhanced DSO [14] are evidently better than that of the original DSO [12]. However, the performance was limited due to the lack of loop closure detection. As Figure 7 shows, our approach corrected the partially distorted segments of the LDSO and obtained the best overall performance among DSO, LDSO, and the enhanced DSO [14], due to the loop closure detection which was improved by the adaptively compensated pixel intensity.

As shown in Figure 8, we calculated the errors of the translation and the Euler angle transformation with respect to ground truth. In the subfigures, the residuals were controlled within a reasonable range.

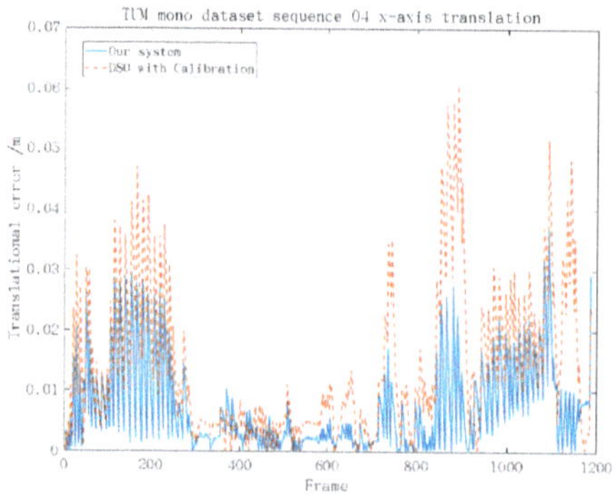

(**a**) The *x*-axis error of TUM 04

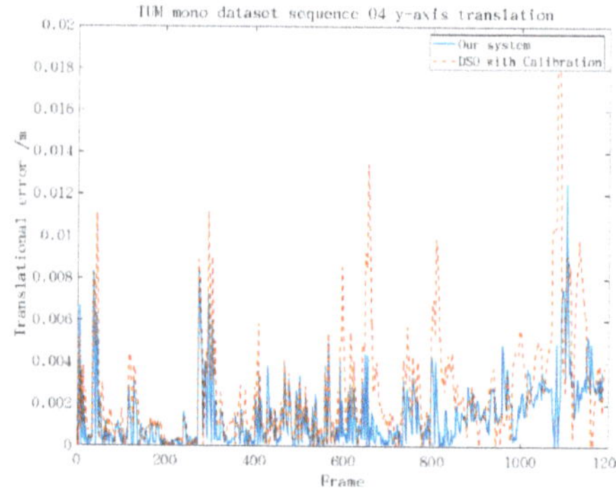

(**b**) The *y*-axis error of TUM 04

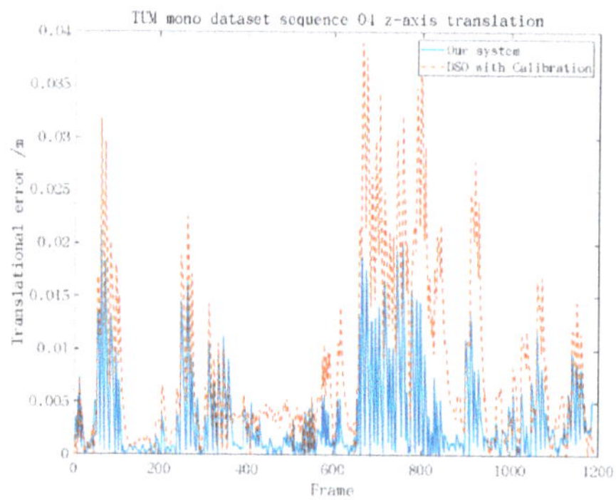

(**c**) The *z*-axis error of TUM 04

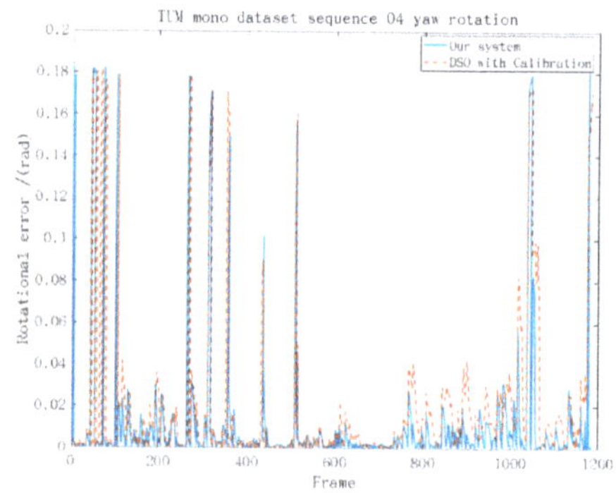

(**d**) The yaw error of TUM 04

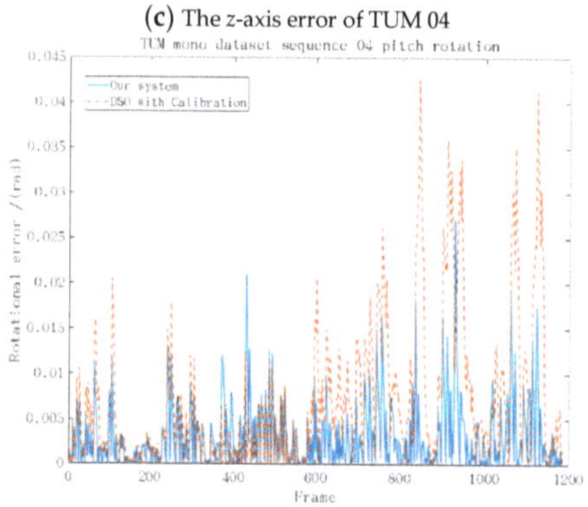

(**e**) The pitch error of TUM 04

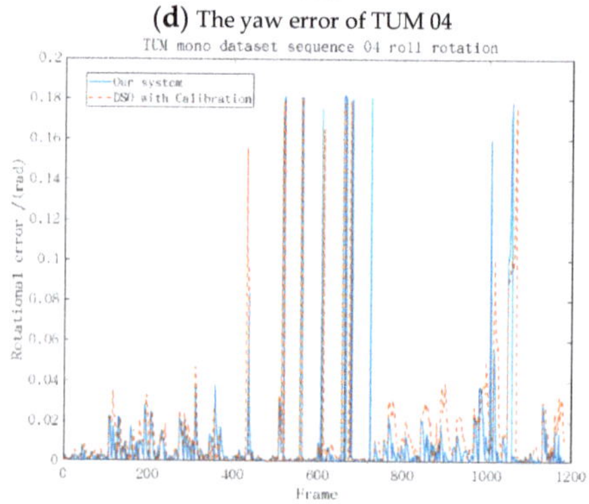

(**f**) The roll error of TUM 04

Figure 8. The error of the 6-degree of freedom (6-DoF) on TUM-Mono sequence 04 with respect to the ground truth between our system and the enhanced DSO [12] and [14].

Figure 8a–c demonstrates that the proposed system had smaller errors along the *x*-axis, *y*-axis, and *z*-axis compared with the enhanced DSO [14] and the errors of our method stayed within a reasonable range on TUM sequence 04. However, Figure 8 d–f indicates that the proposed algorithm had a similar performance in the Euler angle transformation of the enhanced DSO [14].

The photometric parameters of randomly selected frames were calculated and are shown in Figure 9. The vignette and response function were dynamically estimated when the irradiance function had accumulated to a reasonable range while the exposure time could be estimated frame by frame. The results show that the estimated exposure times were closer and closer to the ground truth as the frame number increased, and the response function and vignette were dynamically adjusted around the ground truth to fit the different exposure conditions.

(**a**) The 1274th frame of TUM 04

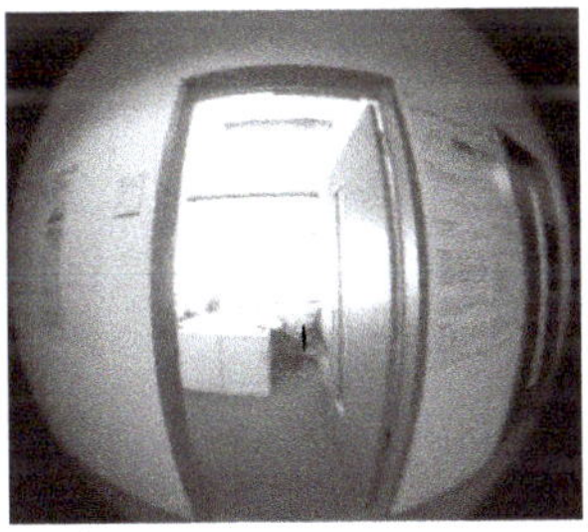

(**b**) The 4385th frame of TUM 04

(**c**) The 5484th frame of TUM 04

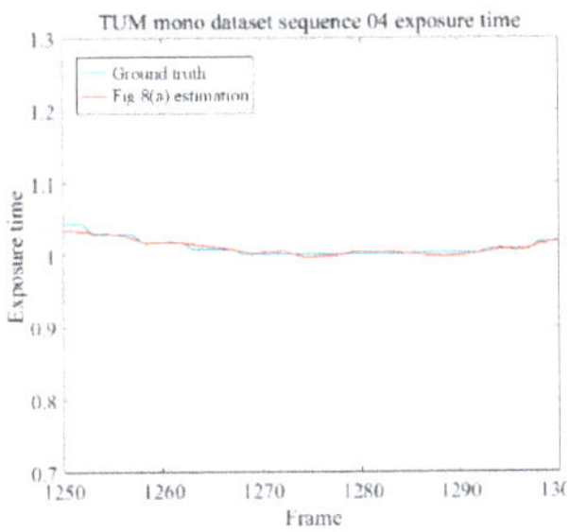

(**d**) The exposure time of subfigure (**a**)

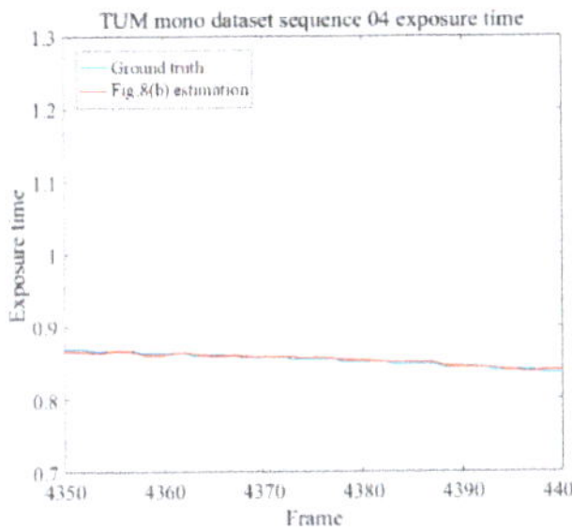

(**e**) The exposure time of subfigure (**b**)

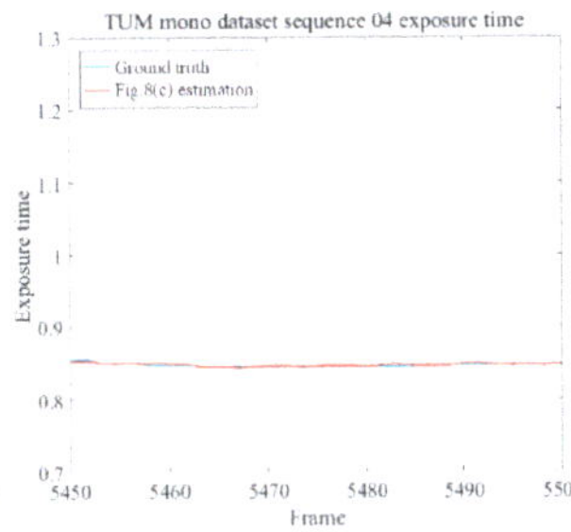

(**f**) The exposure time of subfigure (**c**)

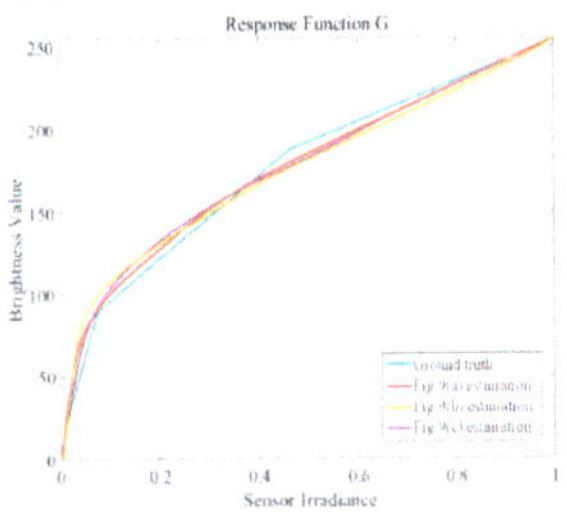

(**g**) The response function of the frames

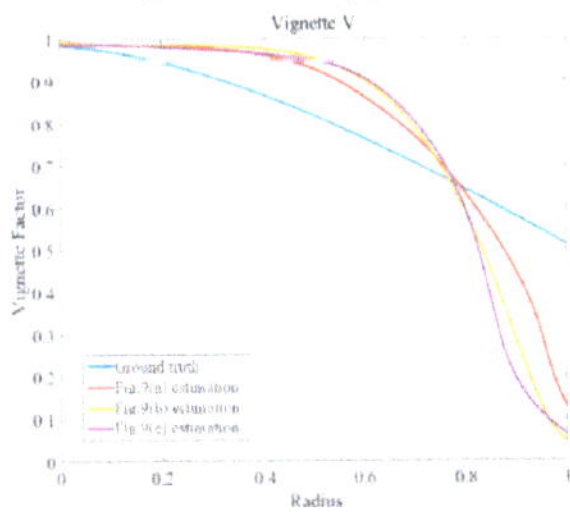

(**h**) The vignette of the frames

Figure 9. The photometric parameters of randomly selected frames. It can be seen that the estimated exposure times were very close to the ground truth. However, the parameters of response function and vignette were acutely adjusted to fit the pixels' intensity of the different scenes.

As can be seen in Figure 9, the vignette and response function were dynamically estimated when the irradiance function had accumulated to a reasonable range, while the exposure time was estimated frame by frame. The results show that the estimated exposure times were closer and closer to the ground truth in the frames across time, and the response function and vignette were dynamically adjusted around the ground truth to fit the different exposure conditions.

The last frame of TUM mono dataset sequence 04 was captured to compare the condition of exposure and tracking performance after photometric calibration and the experimental results are shown in Figure 10.

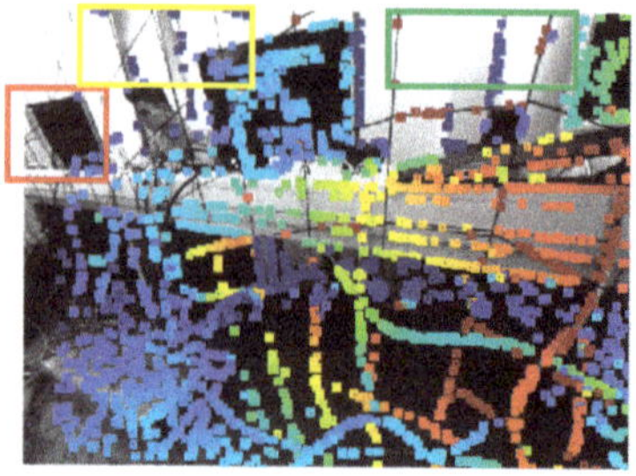

(a) The tracking condition of the last frame

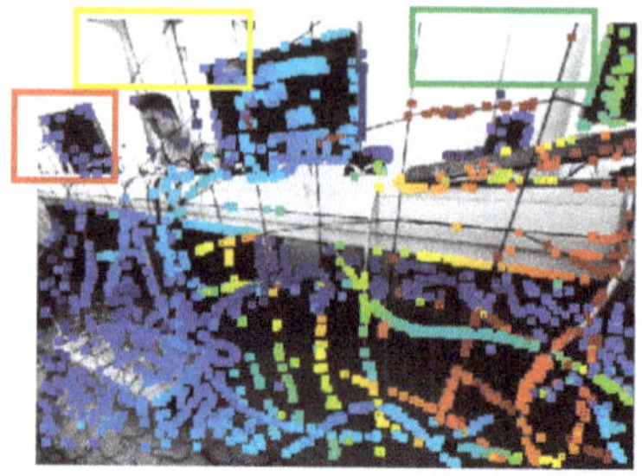

(b) The tracking condition of the last frame after calibration

Figure 10. The exposure condition and pixel tracking of TUM mono dataset sequence 04.

Figure 10a is the original frame and Figure 10b is the frame after photometric calibration. As can be seen, there are two main differences. Firstly, the exposure of Figure 10b was enhanced. The global brightness and pixel contrast were obviously improved by the adaptive response function and vignette. Secondly, the tracked pixels in the low texture region were filtered like the window at the upper part of the image (regions in yellow and green boxes) and the pixels in the high texture region were increased, like the computer at the middle part of the image (region in red box). The enhancement was set to improve the robustness of the tracking, which would then promote the depth estimation accuracy of visual odometry.

To further verify the performance, the KITTI dataset was utilized to test the localization accuracy, the loop closure detection performance, and the timing cost of the proposed algorithm. The KITTI dataset [30] was jointly produced by the Karlsruhe Institute of Technology in Germany and the Toyota Institute of Technology in the United States. It is currently the largest computer vision algorithm evaluation dataset in the world for autonomous driving scenarios.

In the evaluation of [15], monocular VO was considered to be unusable for such a large-scaled dataset, which was overcome by introducing loop closure detection in [17]. Therefore, the 00-10 sequences of the KITTI dataset were applied to test the proposed algorithm. As shown in Figure 11, the experimental results were compared with DSO [12], LDSO, the enhanced DSO [14] and mono-ORBSLAM [24].

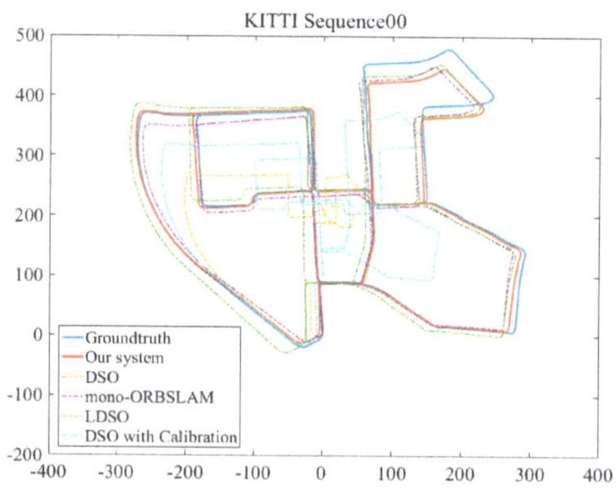

Figure 11. The experimental trajectories results of our system, DSO, LDSO and the enhanced DSO which was integrated with the algorithms proposed in [12] and [14] along the x-axis and z-axis of the KITTI dataset sequences 00.

The trajectories along the x-axis and z-axis were recorded in Figure 10. The experimental results of the proposed method for KITTI sequences 00-10 were better than those of DSO, LDSO and had a similar performance to mono-ORBSLAM. In addition, we also evaluated the performance of the enhanced DSO [14] on KITTI sequence 00. Evidently, the trajectory of the enhanced DSO was distinctly corrected with respect to the DSO's result, and it could almost close the loop. However, the correction was limited due to the lack of an excellent loop closure detection strategy.

At present, the SLAM algorithm's performance indicators mainly include the ATE (absolute trajectory error) and RPE (relative position error). The ATE was utilized to compare the localization accuracy of the SLAM algorithm due to its directness. We calculated the ATEs of DSO, LDSO, and the proposed algorithm on KITTI and collected them in Table 1.

Table 1. The absolute trajectory errors (ATE) of our system, DSO, LDSO and ORB-SLAM on KITTI.

Sequence	DSO [17]	LDSO [17]	ORBSLAM [17]	Our System
00	126.7	9.322	8.27	**7.48**
01	165.03	**11.68**	-	20.15
02	138.7	31.98	26.86	**12.14**
03	4.77	2.85	**1.21**	2.04
04	1.08	1.22	0.77	**0.13**
05	49.85	5.1	7.91	**5.09**
06	113.57	13.552	12.54	**11.08**
07	27.99	2.96	3.44	**0.56**
08	120.17	129.02	**46.81**	105.4
09	74.29	**21.64**	76.54	26.90
10	16.32	17.36	**6.61**	17.45

The ATE of mono-ORBSLAM could not be obtained because of the tracking failure around the 585th frame of the KITTI dataset sequence 01 as Figure 12.

Figure 12. The lost ORB features during tracking on the sequence 01 of KITTI dataset.

The experimental results in Table 1 show that the proposed algorithm had a better positioning performance on most KTITI sequences and had similar performance to that of mono-ORBSLAM [24]. However, the loop closure detection performance of LDSO in sequence 09 declined because of the repeated frames around the loop closure that were too small to detect. When the previous bright frames participated in compensating the frames around the loop closure based on the photometric formation, the repeated images became much brighter than before. However, those frames were dark at the initial stage of the sequence. Therefore, after the compensation, the global brightness difference around the loop closure reduced the positioning performance.

The 6-DoF rigid body motion error was calculated, as shown in Figure 13, to demonstrate the performance along the frames related to LDSO. The unit of the translational error is meters, and the unit of the rotational error is radians.

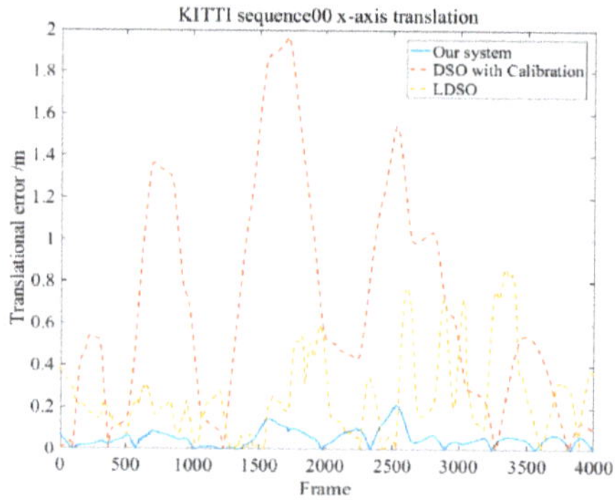

(**a**) The *x*-axis error of KITTI sequence 00

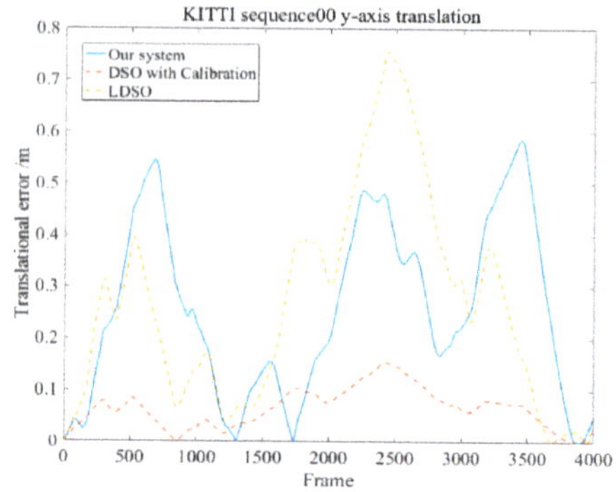

(**b**) The *y*-axis error of KITTI sequence 00

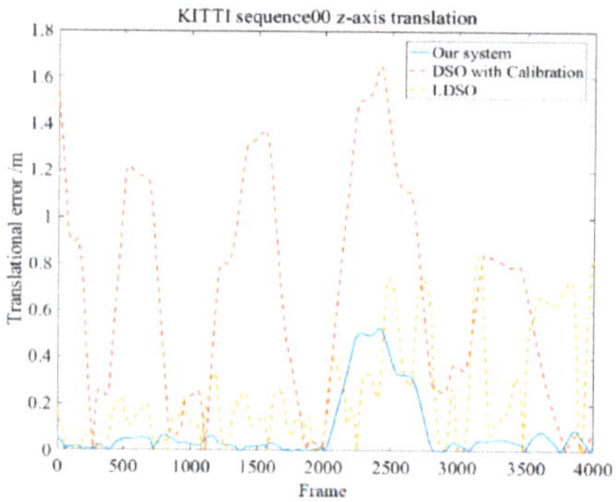

(**c**) The *z*-axis error of KITTI sequence 00

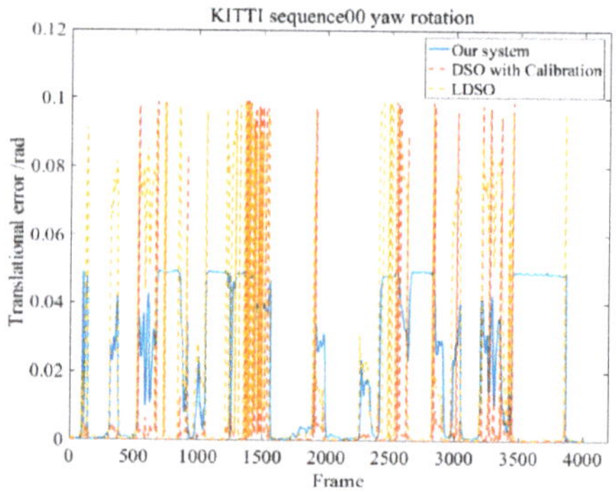

(**d**) The yaw error of KITTI sequence 00

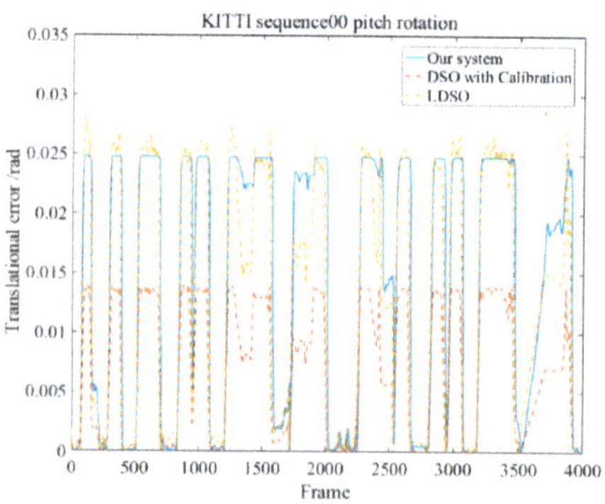

(**e**) The pitch error of KITTI sequence 00

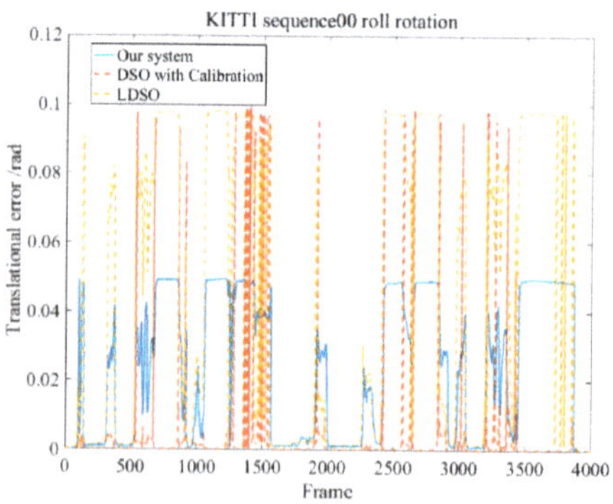

(**f**) The roll error of KITTI sequence 00

Figure 13. The residuals of the 6-DoF on KITTI sequence 00 including the translations and Euler angle of rotations. The residuals of subfigure (**b**) were obviously larger in the both proposed system and LDSO because of the introduction of loop closure detection. However, the error tendencies on *x*-axis and *z*-axis were primarily lower than LDSO and the enhanced DSO [14]. In general, the translation error and rotational error of the proposed system were stably maintained as reasonable values along all frames of KITTI sequence 00.

Figure 13 shows that the error of the proposed system with respect to ground truth was in a reasonable range and better than the enhanced DSO [14]. The max translational residuals along the *x*-axis and the *z*-axis were 0.2153 m and 0.5259 m, respectively, and the rotational error of the camera pose was less than 0.05 rad, which further proves that the proposed system can better cope with the illumination changes in the KITTI dataset sequence 00.

The precision–recall (PR) curve is widely applied to evaluate the performance of loop closure detection. The percentage of loop closures, which were correctly detected in all the detections, was represented as the precision ratio. The percentage of loop closures, which were correctly detected in all real loop closures, was defined as the recall ratio. To compare the loop closure detection performance of the proposed system and LDSO, the PR curve of the loop closure detection is illustrated in Figure 14, from which we can see that when the recall ratio equals 0.5, the precision ratio of our system is larger than that of LDSO. In addition, the proposed system has a higher max recall ratio when the precision ratio equals 1. Figure 14 indicates that the proposed system has better loop closure detection performance compared with LDSO.

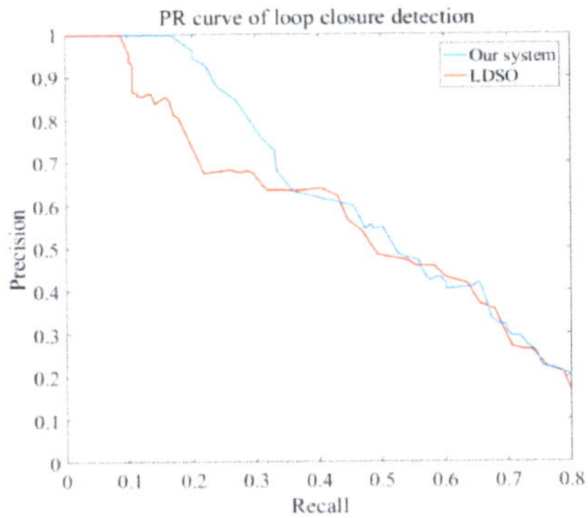

Figure 14. The contrast of precision-recall ratios between our system and LDSO on KITTI dataset.

The research in this paper was mainly based on the improvement of LDSO [17] and online photometric calibration [14]. With the introduction of real-time photometric calibration, the average processing costs of a single-frame image for KITTI sequences 00-10 are provided in Table 2.

Table 2. Timing results of our system and LDSO.

System	Sections	Time
LDSO	Total	894.43ms
Our system	Filtering and tracking feature	40.21 ms
	Exposure time estimation	3.24 ms
	Parameters v and c update	193.15 ms
	Input frame I optimization	135.23 ms
	Back-end	564.51 ms
	Total	936.34 ms

The experimental results show that the real-time photometric calibration direct SLAM system can obtain a 19.7% higher accuracy performance and 4.7% bigger timing costs than LDSO. Due to the performance improvement of the direct SLAM, the extra time burden is acceptable.

The EuRoC micro aerial vehicle datasets [27] were produced by the Swiss Federal Institute of Technology Zurich, including stereo images and inertial measurement unit (IMU) data. In order to analyze the limits of the proposed system, the point cloud maps of our system and LDSO on the EuRoC dataset, V1_03_difficult challenging illumination sequence, are segmentally shown in Figure 15.

(**a**) The segment of EuRoC V1_03 dataset (**b**) The segmental map of LDSO (**c**) The segmental map of the proposed system

Figure 15. The segmental experimental results of LDSO and proposed system on the EuRoC dataset, V1_03_difficult sequence.

As can be seen from Figure 15c, the point cloud map has less noise than Figure 15b. The online photometric calibration can improve the performance of mapping by compensating the exposure condition; however, the effect is not very satisfying. The exposure condition of the image becomes unstable due to the overdue response function and vignetting. The instability may be attributed to the KLT tracker's intrinsic sensibility to illumination change and the blurry images, which were captured in the fast-changing scenes by a violent shaking micro aerial vehicle. To further improve the robustness of tracking, the indirect feature matching, the more robust descriptors, and the deblurring strategy, can be tested in online photometric calibration for future work.

4.2. Actual Experiment

We emphatically evaluated the proposed algorithm in an actual environment to prove the enhancement related to LDSO. To collect the real-time images, the Basler acA1920-155uc camera was selected, which is a global shutter complementary metal oxide semiconductor (CMOS) industrial camera produced by Basler, Germany. It has a 1920 × 1200 maximum resolution and a 164 fps maximum rate.

The camera and notebook are equipped on a TurtleBot2, which is a robot operation system (ROS)-based mobile research platform that was produced by YUJIN, Korea. The experimental platform and the camera calibration are shown in Figure 16.

Figure 16. The camera and notebook were installed on the mobile ground equipment. Then, the camera was calibrated using a checkerboard to eliminate radial distortion.

To achieve better real-time performance, the adopted resolution was 640 × 480 during the experiment, which can be modified in the calibration file of the camera.

After the preparation of the experiment, the ROS (Robot Operation System, ROS) was utilized to control the ground equipment to acquire the scene's visual information and perform positioning and mapping. The process of the experiment was recorded as Figure 17.

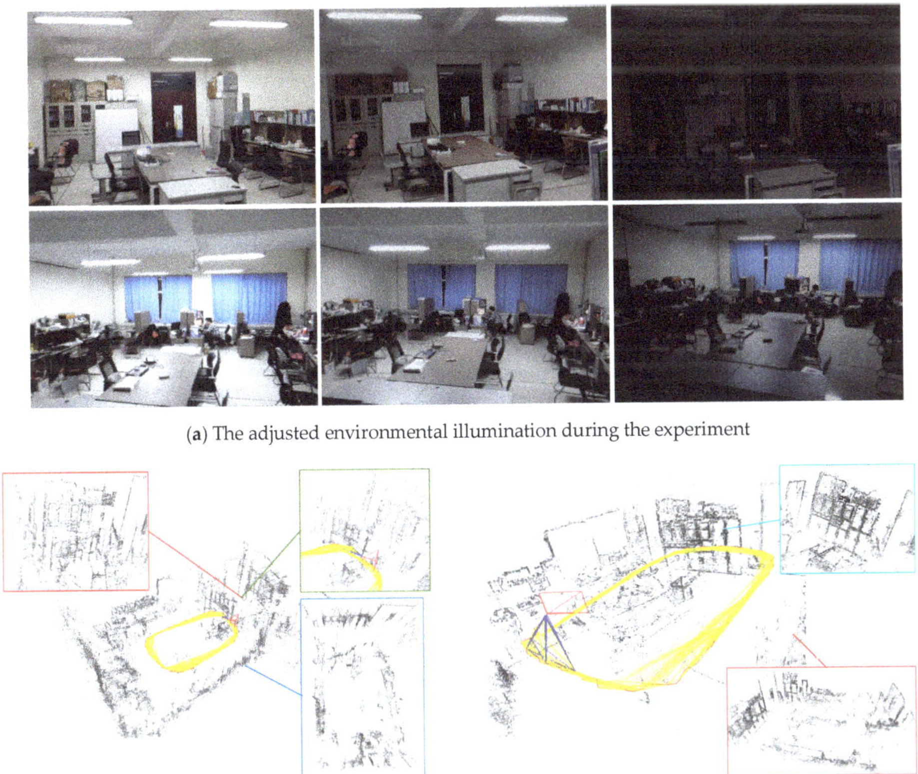

(a) The adjusted environmental illumination during the experiment

(b) The experimental result of LDSO

(c) The experimental result of the proposed algorithm

Figure 17. The actual environmental experiment of LDSO and the proposed algorithm. Subfigure (**a**) shows the adjusted environmental illumination during the experiment. Then, we tested the LDSO and proposed algorithm with respect to subfigure (**a**).

In Figure 17a, after the initialization of the camera, the light source of the experimental scene was successively adjusted. We gradually reduced the brightness of the laboratory during the equipment by moving around. The localization and mapping effects of LDSO and the proposed algorithm are shown in subfigures Figure 17b,c, respectively. In Figure 17b, the point cloud of the cabinet was repeated as the green part, and the scene splicing was distorted as the blue part. In Figure 17c, the distorted construction of Figure 17b was calibrated. The comparison between Figure 17b and 17c shows that the construction of the point cloud map was greatly affected by the exposure of the scene. When the adaptive exposure compensation parameters are not introduced in the direct visual SLAM, the brightness of the input sequence is discontinuous, which causes deviations in localization and mapping. Therefore, the proposed visual SLAM overcame the discontinuous brightness of the scene using adaptive compensation to calculate a more robust point cloud map and camera pose estimation.

5. Conclusions

In this paper, a real-time photometric calibrated monocular direct visual SLAM was proposed to dynamically compensate for the input sequence's exposure. It solved the problem that the LDSO had—poor positioning and mapping robustness, due to illumination challenges. The enhanced sparse direct visual SLAM formulation was more suitable for the research and application of navigation and positioning in mobile ground equipment. Firstly, the vignetting and response function according to the photometric formation were introduced into the front-end. Secondly, the Shi–Tomasi corners of the input sequence were filtered and added to the tracking optimization equation using the tracked pixel in visual odometry. Then, the L–M approach was utilized to iteratively calculate the photometric parameters in the sliding window to compensate for the exposure condition of the input sequence. Finally, the tracked Shi–Tomasi corners in the adaptive photometric calibration and their ORB features were applied to achieve loop closure detection. The results of multiple simulations and experiments show that the proposed method had a better positioning and mapping performance than DSO and LDSO. In particular, the DSO which was integrated with the online photometric calibration, was also complementally tested to prove the generalization performance of the algorithm [14] to a certain extent, and further illustrated the advantages of our algorithm. The positioning accuracy and the point cloud map's clearness from the proposed system, in most sequences of the KITTI and TUM-Mono datasets, were better; and the performance of our system on KITTI was similar to mono-ORBSLAM. In the actual experiment, the proposed approach was evaluated using an artificial dynamic illumination environment. As in the simulation experiment, we still obtained better positioning and mapping effects on both the TUM and the KITTI datasets than LDSO.

In future work, we will consider enhancing the adaptability of the online photometric calibration to further calibrate the visual odometry. The insertion mechanism of the key frames and the loop closure strategy will be adjusted to improve the calculation efficiency. In addition, we will seek to introduce semantic information into the direct SLAM to achieve a better loop closure detection.

Author Contributions: Conceptualization, P.L.; methodology, P.L., X.Y., and C.L.; software, P.L.; validation, P.L., X.Y., and C.L.; investigation, X.Y.; resources, P.L. and C.L.; writing—original draft preparation, P.L. and Z.L.; writing—review and editing, X.Y., C.Z., and Y.S.; visualization, P.L. and Z.L.

Funding: This work is supported by the National Natural Science Foundation of China (under Grants 61803227, 61603214, 61673245, and 61573213), the National Key Research and Development Plan of China (under Grant 2017YFB1300205), the Shandong Province Key Research and Development Plan (under Grants 2016ZDJS02A07 and 2018GGX101039), Shandong Provincial Natural Science Foundation (under Grant ZR2017PEE022), the China Postdoctoral Science Foundation (under Grant 2018M630778), and the Independent Innovation Foundation of Shandong University (under Grant 2082018ZQXM005).

Acknowledgments: We would like to convey our deep appreciation to the editors and the anonymous reviewers for their insightful comments and constructive suggestions, which were very helpful in the improvement of our manuscript.

Conflicts of Interest: The authors declare no conflict of interest. The funders had no role in the design of the study; in the collection, analyses, or interpretation of data; in the writing of the manuscript; or in the decision to publish the results.

References

1. Engel, J.; Usenko, V.; Cremers, D. A photometrically calibrated benchmark for monocular visual odometry. *arXiv* **2016**, arXiv:1607.02555.
2. Gostar, A.K.; Fu, C.; Chuah, W.; Hossain, M.I.; Tennakoon, R.; Bab-Hadiashar, A.; Hoseinnezhad, R. State Transition for Statistical SLAM Using Planar Features in 3D Point Clouds. *Sensors* **2019**, *19*, 1614. [CrossRef] [PubMed]
3. Wang, L.; Wu, Z. RGB-D SLAM with Manhattan Frame Estimation Using Orientation Relevance. *Sensors* **2019**, *19*, 1050. [CrossRef] [PubMed]
4. Yu, N.B.; Wang, S.R.; Xu, C. Monocular semidirect visual odometry for large-scale outdoor localization. *IEEE Access* **2019**, *7*, 57927–57942.

5. Jiang, L.; Zhao, P.; Dong, W.; Li, J.; Ai, M.; Wu, X.; Hu, Q. An Eight-Direction Scanning Detection Algorithm for the Mapping Robot Pathfinding in Unknown Indoor Environment. *Sensors* **2018**, *18*, 4254. [CrossRef] [PubMed]
6. Lee, S.H.; Civera, J. Loosely-Coupled Semi-Direct Monocular SLAM. *IEEE Robot. Autom. Lett.* **2019**, *4*, 399–406. [CrossRef]
7. Gao, X.; Zhang, T.; Liu, Y.; Yan, Q.R. *Visual SLAM XIV: From Theory to Practice*; Electronic Industry Press: Beijing, China, 2017; pp. 184–188, 203–204, 245–255.
8. Engel, J.; Schöps, T.; Cremers, D. LSD-SLAM: Large-Scale Direct Monocular SLAM. In Proceedings of the European Conference on Computer Vision (ECCV), Zurich, Switzerland, 6–12 September 2014; pp. 834–849.
9. Forster, C.; Pizzoli, M.; Scaramuzza, D. SVO: Fast semi-direct monocular visual odometry. In Proceedings of the 2014 IEEE International Conference on Robotics and Automation (ICRA), Hong Kong, China, 31 May–7 June 2014; pp. 15–22.
10. Forster, C.; Zhang, Z.; Gassner, M.; Werlberger, M.; Scaramuzza, D. SVO: Semidirect Visual Odometry for Monocular and Multicamera Systems. *IEEE Trans. Robot.* **2017**, *33*, 249–265. [CrossRef]
11. Kim, P.; Lim, H.; Jin, K.H. Robust visual odometry to irregular illumination changes with RGB-D camera. In Proceedings of the 2015 IEEE/RSJ International Conference on Intelligent Robots and Systems (IROS), Hamburg, Germany, 28 September–2 October 2015; pp. 3688–3694.
12. Engel, J.; Koltun, V.; Cremers, D. Direct sparse odometry. *IEEE Trans. Pattern Anal. Mach. Intell.* **2018**, *40*, 611–625. [CrossRef]
13. Kim, C.; Kim, P.; Lee, S.; Kim, H.J. Edge-Based Robust RGB-D Visual Odometry Using 2-D Edge Divergence Minimization. In Proceedings of the 2018 IEEE/RSJ International Conference on Intelligent Robots and Systems (IROS), Madrid, Spain, 1–5 October 2018; pp. 1–9.
14. Bergmann, P.; Cremers, D.; Wang, R. Online Photometric Calibration of Auto Exposure Video for Realtime Visual Odometry and SLAM. *IEEE Robot. Autom. Lett.* **2018**, *3*, 627–634. [CrossRef]
15. Schwörer, M.; Cremers, D.; Wang, R. Stereo DSO: Large-Scale Direct Sparse Visual Odometry with Stereo Cameras. In Proceedings of the 2017 IEEE International Conference on Computer Vision (ICCV), Venice, Italy, 22–29 October 2017; pp. 1–6.
16. Yang, N.; Wang, R.; Stückler, J.; Cremers, D. Deep Virtual Stereo Odometry: Leveraging Deep Depth Prediction for Monocular Direct Sparse Odometry. In Proceedings of the European Conference on Computer Vision (ECCV), Munich, Germany, 8–14 September 2018; pp. 835–852.
17. Gao, X.; Wang, R.; Demmel, N.; Cremers, D. LDSO: Direct Sparse Odometry with Loop Closure. In Proceedings of the IEEE/RSJ International Conference on Intelligent Robots and Systems (IROS), Madrid, Spain, 1–5 October 2018; pp. 2198–2204.
18. Sloan, P.; Kautz, J.; Snyder, J. Precomputed radiance transfer for real-time rendering in dynamic, low-frequency lighting environments. *ACM Trans. Graphy.* **2017**, *21*, 527–536.
19. Huo, Y.; Zhang, X. Single image-based HDR image generation with camera response function estimation. *IET Image Process.* **2017**, *11*, 1317–1324. [CrossRef]
20. Dan, B.G.; Chen, J.H. Vignette and exposure calibration and compensation. *IEEE Trans. Pattern Anal. Mach. Intell.* **2010**, *32*, 2276–2288.
21. Kalal, Z.; Mikolajczyk, K.; Matas, J. Forward-Backward Error: Automatic Detection of Tracking Failures. In Proceedings of the 2010 20th International Conference on Pattern Recognition, Istanbul, Turkey, 23–26 August 2010; pp. 1–4.
22. Park, S.; Pollefeys, M.; Schops, T. Illumination change robustness in direct visual SLAM. In Proceedings of the 2017 IEEE International Conference on Robotics and Automation (ICRA), Singapore, 29 May–3 June 2017; pp. 4523–4530.
23. Mur-Artal, R.; Montiel, J.M.; Tardós, J.D. ORB-SLAM: A versatile and accurate monocular SLAM system. *IEEE Trans. Robot.* **2017**, *31*, 1147–1163. [CrossRef]
24. Mur-Artal, R.; Tardos, J.D. ORB-SLAM2: An Open-Source SLAM System for Monocular, Stereo, and RGB-D Cameras. *IEEE Trans. Robot.* **2017**, *33*, 1255–1262. [CrossRef]
25. Wu, Y.X.; Wang, C.; Xian, Y.X. SLAM based on sparse direct method and graph optimization for mobile robot. *Chin. J. Sci. Instrum.* **2018**, *39*, 257–263.
26. Zhou, S.-C.; Yan, R.; Li, J.-X.; Chen, Y.-K.; Tang, H. A brain-inspired SLAM system based on ORB features. *Int. J. Autom. Comput.* **2017**, *14*, 564–575. [CrossRef]

27. Burri, M.; Nikolic, J.; Gohl, P.; Schneider, T.; Rehder, J.; Omari, S.; Achtelik, M.W.; Siegwart, R. The EuRoC micro aerial vehicle datasets. *Int. J. Robot. Res.* **2016**, *35*, 1157–1163. [CrossRef]
28. Kim, H.K.; Kim, H.; Cho, S. Bag-of-concepts: Comprehending document representation through clustering words in distributed representation. *Neurocomputing* **2017**, *266*, 336–352. [CrossRef]
29. Sturm, J.; Engelhard, N.; Endres, F.; Burgard, W.; Cremers, D. A benchmark for the evaluation of RGB-D SLAM systems. In Proceedings of the 2012 IEEE/RSJ International Conference on Intelligent Robots and Systems, Vilamoura, Portugal, 7–12 October 2012; pp. 573–580.
30. Geiger, A.; Lenz, P.; Urtasun, R. Are we ready for autonomous driving? The KITTI vision benchmark suite. In Proceedings of the 2012 IEEE Conference on Computer Vision and Pattern Recognition, Providence, RI, USA, 16–21 June 2012; pp. 1–8.

sensors

MDPI

Article

Multi-sensor Fusion Road Friction Coefficient Estimation During Steering with Lyapunov Method

Letian Gao [1,2], Lu Xiong [1,2,*], Xuefeng Lin [1,2], Xin Xia [1,2], Wei Liu [1,2], Yishi Lu [1,2] and Zhuoping Yu [1,2]

1 School of Automotive Studies, Tongji University, Shanghai 201804, China; letiangao_515@126.com (L.G.)
2 Clean Energy Automotive Engineering Center, Tongji University, Shanghai 201804, China
* Correspondence: xiong_lu@tongji.edu.cn

Received: 2 July 2019; Accepted: 3 September 2019; Published: 4 September 2019

Abstract: The road friction coefficient is a key parameter for autonomous vehicles and vehicle dynamic control. With the development of autonomous vehicles, increasingly, more environmental perception sensors are being installed on vehicles, which means that more information can be used to estimate the road friction coefficient. In this paper, a nonlinear observer aided by vehicle lateral displacement information for estimating the road friction coefficient is proposed. First, the tire brush model is modified to describe the tire characteristics more precisely in high friction conditions using tire test data. Then, on the basis of vehicle dynamics and a kinematic model, a nonlinear observer is designed, and the self-aligning torque of the wheel, lateral acceleration, and vehicle lateral displacement are used to estimate the road friction coefficient during steering. Finally, slalom tests and DLC (Double Line Change) tests in high friction conditions are conducted to verify the proposed estimation algorithm. Test results showed that the proposed method performs well during steering and the estimated road friction coefficient converges to the reference value rapidly.

Keywords: road friction coefficient; tire model; nonlinear observer; self-aligning torque; lateral displacement; Lyapunov method

1. Introduction

Vehicle safety-related state estimation [1–3] and control systems [4–6] have received much attention in recent decades. Vehicle dynamic control systems, such as the ASR (Anti-slip Regulation System), ESC (Electronic Stability Control), and AEB (Autonomous Emergency Brake), are realized by controlling the driving forces or braking forces so that the forces exerted by the road on the tires can be changed to maintain the stability of the wheels and the vehicle. The road friction coefficient is a key parameter for vehicle dynamic control systems [7,8], because it can reflect the dynamic motion limitations to a certain extent [9]. For human-operated vehicles, drivers can estimate the motion limitation of the vehicle and adapt their driving style using experience to prevent the vehicle from driving into critical conditions. However, with the development of intelligent vehicles, progressively more ADAS (Advanced Driving Assistant System) functions are being implemented by automated systems, which means that driving and braking forces and steering angles need to be calculated and controlled by control units, such as ACC (Adaptive Cruise Control) and LKA (Lane Keep Assistant). Therefore, an accurate road friction coefficient provides the automated system with the current motion limitation of the vehicle. Furthermore, for highly automated vehicles, the road friction coefficient is critical for decision-making, trajectory planning, and trajectory tracking.

Road friction coefficient estimation methods can be divided into two general types: cause-based methods and effect-based methods. The state-of-the-art methods of road friction estimation have been reviewed [10]. The principle of cause-based methods is the direct determination of road surface characteristics by special sensors, such as cameras, laser scanners, optical sensors, and so on. Alonso, J. et al. [11] proposed an asphalt status classification system based on real-time acoustic analysis of the tire-road interaction, but only wet and dry asphalt states were covered. Roychowdhury, S. et al. [12] proposed a two-stage method based on images captured by the front camera. A convolutional neural network model was applied to learn the road characteristics, and then the road states were divided into three types according to a rule-based strategy. Although cause-based methods can accurately characterize road states with special sensors, only a rough road friction coefficient is estimated, and the interval may vary within a very large range. The road friction coefficient describes the interaction effects between the road and tires [13], which means it cannot be estimated accurately by cause-based methods without considering the vehicle and tire characteristics. Effect-based methods estimate the road friction coefficient from the dynamic or kinematic motion responses [14] of the vehicle or wheels due to the tire forces caused by the interaction between the tires and the road. Normally, effect-based methods estimate the road friction coefficient by using models, and the estimation results are more accurate than those of cause-based methods. Effect-based methods fall into two categories: methods based on longitudinal dynamics and methods based on lateral dynamics. In order to obtain the road friction coefficient in all conditions, Ahn, C. et al. [15] divided driving conditions according to the slip ratio, the sideslip angle, and lateral acceleration. Then, different estimation methods were applied to estimate road friction coefficients under different conditions. Using longitudinal dynamics, Castillo Aguilar, J.J. et al. [16] applied a fuzzy logic algorithm to estimate the road friction according to the variation curve of the relationship between the road friction coefficient and the longitudinal slip ratio, and the algorithm was utilized in the hydraulic pressure control system of the EHB (Electric Hydraulic Brake) [17]. Enisz, K. et al. [18] designed an augmented vehicle model with the road friction coefficient and slip ratio, and the vehicle speed, wheel rotation speed, slip ratio, and road friction coefficient were simultaneously estimated by the EKF (Extended Kalman Filter). Taking advantage of the fact that the wheel torque of distributed drive electric vehicles is available and can be controlled precisely, Xia, X. et al. [19] proposed a road friction coefficient estimation algorithm under driving conditions using a nonlinearobserver. The observer performed well with strong longitudinal excitation, and its stability was proved. Moreover, many studies have focused on road friction coefficient estimation methods based on lateral dynamic characteristics. Using the relationship between lateral force and the sideslip angle, Wang, R. et al. [20] proposed a road friction coefficient estimation method that is effective when the vehicle has enough lateral excitation. Qi, Z. et al. [21] designed a Kalman Filter to estimate the longitudinal and lateral forces of each tire, as well as the derivatives of the two forces, using a 4 DOF(Degree of Freedom) vehicle model, and then the road friction coefficient was estimated according to the estimated tire lateral force. Compared with lateral force, self-aligning torque enters the nonlinear region earlier, so less lateral excitation is needed to estimate the road friction coefficient. Therefore, the relationship between the sideslip angle and self-aligning torque has been applied in many studies to obtain an accurate road friction coefficient. Luque, P. et al. [22] used a Kalman Filter to estimate longitudinal and lateral tire forces, and the self-aligning torque of the tire was calculated by a pretrained neural network. Then, the road friction coefficient was obtained by the relation curves between self-aligning torque and the sideslip angle in different road states. Matsuda, T. et al. [23] considered the road friction coefficient as a state and designed an EKF using a nonlinear 2 DOF single-track vehicle model, and self-aligning torque was measured to update the states. With varying road friction, Hsu, Y.-H.J. et al. [24] estimated the road friction coefficient from the relationships between (i) self-aligning torque and the tire trail and (ii) the tire trail and the sideslip angle. Ahn, C et al. [25] used a Kalman Filter to estimate the self-aligning torque of tires on the basis of the steering system and designed a nonlinear observer to estimate the road friction coefficient using self-aligning torque and a nonlinear vehicle model, and the stability and robustness of the nonlinear observer were proved. Shao, L. and Jin, C. [26] adopted a novel strategy to estimate the front axle lateral force. Then, combined with an indirect measurement based on total aligning torque estimation, a nonlinearadaptive observer was designed to estimate the road friction coefficient with guaranteed stability. The self-aligning torque of the front axle is coupled with the sideslip angle; so, to precisely calculate self-aligning torque, we must know the current sideslip angles of each tire. Therefore, an accurate sideslip angle contributes to improvement in the estimation accuracy of the road friction coefficient using self-aligning torque-based estimation methods. With the development of intelligent vehicles, besides conventional onboard sensors, such as steering wheel angle sensors, wheel speed sensors, and inertial measurement units, information from new sensors equipped on intelligent vehicles can also be used to estimate the vehicle state. Yoon, J. and Peng, H. [27] used velocity measurements from two GPS receivers to estimate the sideslip angle. To reduce the cost, they took advantage of the direction measurement using a magnetometer and proposed a sideslip angle estimation method that integrated a magnetometer with a GPS [28]. Wang, Y. et al. [29] proposed a combined vehicle and vision model to increase the robustness of the body-slip-angle estimation to uncertain vehicle parameters, and multi-rate and time-delay issues were explained. Furthermore, camera-aided estimation of the lateral state for the integrated control of automated vehicles was discussed in Reference [30].

Since new sensors equipped on intelligent vehicles facilitate the estimation of vehicle states, they could be useful for improving the accuracy of the results of road friction coefficient estimation. In this paper, we introduced vehicle lateral displacement, which is based on the relationship between road friction and the self-aligning torque of the front axle, to the framework of road friction coefficient estimation. On the one hand, lateral displacement information contributes to improvement in the estimation accuracy of the vehicle's sideslip angle so that tire forces can be estimated more precisely. For intelligent vehicles, vehicle lateral displacement information can be obtained from cameras, GNSS (Global Navigation Satellite System) and maps, or V2X (Vehicle to Everything) systems. We acquired this information from a high-accuracy GNSS and a pre-established lane line map. On the other hand, compared with methods based on the relationship between road friction and the longitudinal or lateral forces of the tires, the self-aligning torque-based method requires fewer excitations, so the road friction coefficient can be estimated before the vehicle drives into critical conditions. We adjusted the tire brush model to fit the tire test data more accurately. Then, by integrating lateral displacement information, self-aligning torque measurement, lateral acceleration measurement, the tire model, and the vehicle model, we developed a nonlinear observer for road friction coefficient estimation. The stability of the observer was proved, and the observer's robustness was analyzed.

The main contributions of this paper are summarized as follows:

- A novel modified tire brush model based on tire test data is proposed. Compared with the traditional tire brush model, new mapping relationships between lateral tire force and the sideslip angle and between self-aligning torque and the sideslip angle are established, which can model tire forces and self-aligning torque more precisely. Further, the simple expression form of the modified tire model functions facilitates the proof of the non-linear observer's stability.
- Lateral displacement information is introduced into the estimation system. Lateral displacement information can be obtained from new sensors equipped on intelligent vehicles, and it can be useful for accurate sideslip angle estimation, so that the road friction coefficient can be calculated more precisely.
- A non-linear observer for the road friction coefficient is proposed. The stability of the nonlinear observer is proved thorough the Lyapunov method, and the robustness is analyzed.

The remainder of this paper is organized as follows. In Section 2, the vehicle model and modified tire model are introduced. In Section 3, the nonlinear observer for road friction coefficient estimation is proposed, and its stability and robustness are analyzed. Section 4 presents experiments that were conducted to prove the proposed estimation method, and the experimental results are discussed. Finally, the conclusions are summarized in Section 5.

2. Vehicle and Tire Model

2.1. Vehicle Model

A nonlinear 2 DOF vehicle model is introduced to express the vehicle lateral dynamics, as shown in Figure 1. Both longitudinal and lateral load transfer are considered, and the dynamic model is expressed as:

$$\dot{\omega} = \frac{l_f}{I_z}(F_{yfl}\cos\delta_{fl} + F_{yfr}\cos\delta_{fr}) - \frac{l_r}{I_z}(F_{yrl} + F_{yrr}) \quad (1)$$

$$\dot{\beta} = \frac{a_y}{v_x} - \omega \quad (2)$$

where ω is the yaw rate; l_f and l_r are the distance between the COG (Center of Gravity) and the front and rear axles, respectively; I_z is the vehicle yaw moment of inertia; F_{yfl}, F_{yfr}, F_{yrf}, and F_{yrr} are the lateral forces of the front left tire, front right tire, rear left tire, and rear right tire, respectively; δ_{fl} and δ_{fr} are the steering angles of the front left wheel and front right wheel, respectively; β is the sideslip angle; a_y is the lateral acceleration of the vehicle; v_x is the longitudinal speed of the vehicle.

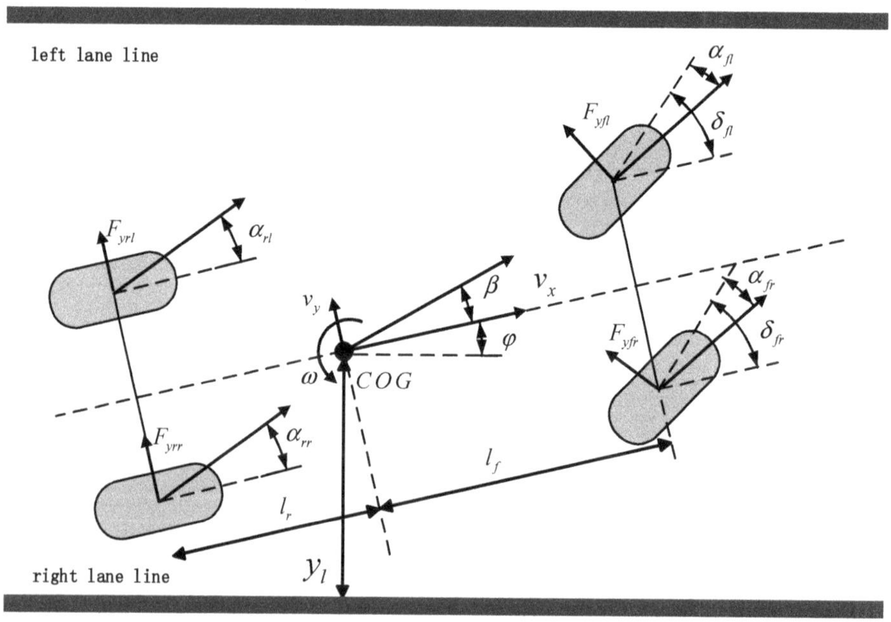

Figure 1. Vehicle model.

Lateral force F_y is calculated by the tire model, and it corresponds to the road friction coefficient μ and sideslip angle α of each tire and vertical load F_z. The sideslip angles of tires are:

$$\begin{cases} \alpha_{fl} = \frac{v_y + l_f\omega}{v_x - \frac{b}{2}\omega} - \delta_{fl} & \alpha_{rl} = \frac{v_y - l_r\omega}{v_x - \frac{b}{2}\omega} \\ \alpha_{fr} = \frac{v_y + l_f\omega}{v_x + \frac{b}{2}\omega} - \delta_{fr} & \alpha_{rr} = \frac{v_y - l_r\omega}{v_x + \frac{b}{2}\omega} \end{cases} \quad (3)$$

where v_y is the lateral speed of the vehicle, and b is track base. Given the load transfer, the vertical load of each tire is:

$$\begin{cases} F_{zfl} = mg\frac{l_r}{2(l_f+l_r)} - ma_x\frac{h}{2(l_f+l_r)} - ma_y\frac{hl_r}{b(l_f+l_r)} \\ F_{zfr} = mg\frac{l_r}{2(l_f+l_r)} - ma_x\frac{h}{2(l_f+l_r)} + ma_y\frac{hl_r}{b(l_f+l_r)} \\ F_{zfl} = mg\frac{l_r}{2(l_f+l_r)} + ma_x\frac{h}{2(l_f+l_r)} + ma_y\frac{hl_f}{b(l_f+l_r)} \\ F_{zfl} = mg\frac{l_r}{2(l_f+l_r)} + ma_x\frac{h}{2(l_f+l_r)} - ma_y\frac{hl_f}{b(l_f+l_r)} \end{cases} \quad (4)$$

where F_{zfl}, F_{zfr}, F_{zrf}, and F_{zrr} are the vertical forces of the front left tire, front right tire, rear left tire, and rear right tire, respectively; m is the vehicle mass; a_x is longitudinal acceleration; h is the height of the COG.

2.2. Tire Model

A novel modified tire brush model is applied to describe lateral force F_y and the self-aligning torque of the tire M_z. Compared with the traditional tire brush model, the proposed modified tire model has a simpler form and fits the tire test data better, so it is more convenient for road friction coefficient estimation during normal steering conditions. In the proposed modified tire model, the relationship between $\frac{\alpha}{F_z^{0.15}}$ and $\frac{F_y}{F_z^{0.81}}$ and the relationship between $\frac{\alpha}{F_z^{0.45}}$ and $\frac{M_z}{F_z^{1.85}}$ are established, as shown in Equations (5) and (6), respectively. With the new mapping relationships, the tire model can describe the variation in lateral force and self-alignment with different vertical loads more precisely.

$$\frac{F_y}{F_z^{0.81}} = -\text{sign}(\alpha)\bullet 3d_1\mu\frac{c_1}{3\mu}\left(\frac{|\alpha|}{F_z^{0.15}}\right)\left\{1 - \frac{c_1}{3\mu}\left(\frac{|\alpha|}{F_z^{0.15}}\right) + \frac{1}{3}\left[\frac{c_1}{3\mu}\left(\frac{|\alpha|}{F_z^{0.15}}\right)\right]^2\right\} \quad (5)$$

$$\frac{M_z}{F_z^{1.85}} = \text{sign}(\alpha)\bullet\frac{d_2}{2}\mu\frac{c_2}{3\mu}\left(\frac{|\alpha|}{F_z^{0.45}}\right)\left[1 - \frac{c_2}{3\mu}\left(\frac{|\alpha|}{F_z^{0.45}}\right)\right]^3 \quad (6)$$

where d_1, d_2, c_1, and c_2 are parameters, and μ is the road friction coefficient.

Tire tests were conducted on a tire test bench to verify the proposed tire model. The raw test data are shown in Figure 2. If we normalize the lateral tire force with the vertical load using the traditional tire brush model, the curves for different tire loads do not line up very well, as shown in Figure 3, which means that F_y and M_z are not directly correlative with F_z. With the new relationship proposed in the modified tire brush model, the test data are normalized, as shown in Figure 4. For lateral tire force, Figure 4a reveals that the relationships between $\frac{\alpha}{F_z^{0.15}}$ and $\frac{F_y}{F_z^{0.81}}$ calculated by the modified tire model are nearly the same for different tire loads. Similarly, the test results in Figure 4b show that the relationships between $\frac{\alpha}{F_z^{0.45}}$ and $\frac{M_z}{F_z^{1.85}}$ are the same for different vertical loads. Therefore, the proposed modified tire brush model can calculate the lateral tire force more precisely with tire load variation.

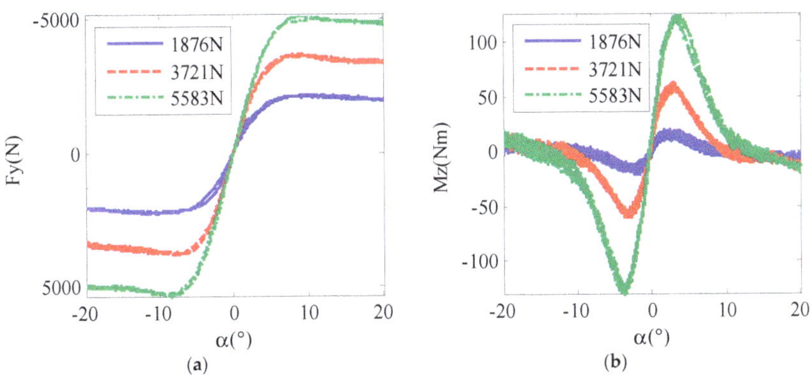

Figure 2. Row tire test data: (**a**) lateral force; (**b**) self-aligning torque.

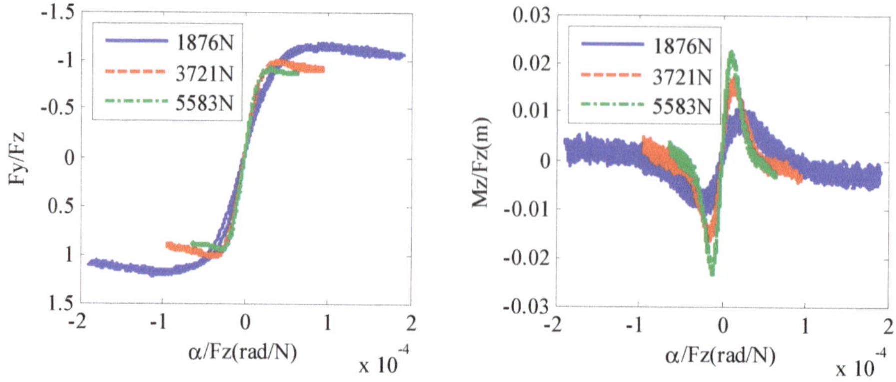

(**a**) (**b**)

Figure 3. Normalization of tire test data with the original tire brush model: (**a**) lateral force; (**b**) self-aligning torque.

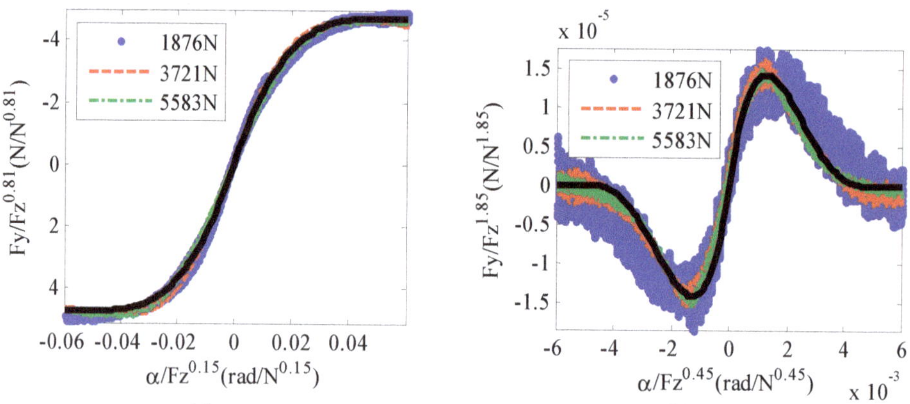

(**a**) (**b**)

Figure 4. Normalization of tire test data with the modified tire brush model: (**a**) lateral force; (**b**) self-aligning torque.

The tire test data in Figure 4 prove the function form of the modified tire model, so the next step is to use the test data normalized by the vertical load to fit the tire model function. The fitting results are shown as the black line in Figure 4, and the proposed modified tire model can be expressed as:

$$F_y(\alpha, F_z, \mu) = -\text{sign}(\alpha) \bullet 3d_1 \mu F_z^{0.81} \frac{c_1}{3\mu}\left(\frac{|\alpha|}{F_z^{0.15}}\right)\left\{1 - \frac{c_1}{3\mu}\left(\frac{|\alpha|}{F_z^{0.15}}\right) + \frac{1}{3}\left[\frac{c_1}{3\mu}\left(\frac{|\alpha|}{F_z^{0.15}}\right)\right]^2\right\} \tag{7}$$

$$M_z(\alpha, F_z, \mu) = \text{sign}(\alpha) \bullet \frac{d_2}{2} \mu F_z^{1.85} \frac{c_2}{3\mu}\left(\frac{|\alpha|}{F_z^{0.45}}\right)\left[1 - \frac{c_2}{3\mu}\left(\frac{|\alpha|}{F_z^{0.45}}\right)\right]^3 \tag{8}$$

3. Nonlinear Observer Design for Road Friction Coefficient Estimation

3.1. NonlinearObserver Design

Assuming that the road friction coefficient μ is piecewise constant and using the vehicle dynamic model (2) we have:

$$\dot{\mu} = 0 \tag{9}$$

$$\dot{v}_y = a_y\left(\mu, v_y, F_z\right) - \omega v_x \tag{10}$$

where $a_y\left(\mu, v_y, F_z\right)$ is lateral acceleration, which is:

$$\begin{aligned} a_y\left(\mu, v_y, F_z\right) &= \frac{1}{m}\left(F_{yfl}\cos\delta_{fl} + F_{yfr}\cos\delta_{fr} + F_{yrl} + F_{yrr}\right) \\ &= \frac{1}{m}\left[F_y\left(\mu, \alpha_{fl}, F_{zfl}\right)\cos\delta_{fl} + F_y\left(\mu, \alpha_{fr}, F_{zfr}\right)\cos\delta_{fr} + F_y(\mu, \alpha_{rl}, F_{zrl}) + F_y(\mu, \alpha_{rr}, F_{zrr})\right] \end{aligned} \tag{11}$$

Since the lane line map information is available, we can measure (i) the distance between the COG of the vehicle and the lane line y_l and (ii) the angle between the lane line and vehicle heading φ. y_l and φ can either be obtained by a camera installed on the vehicle or calculated through the location information from a GNSS receiver and lane map, which is a priori knowledge. The lateral displacement can also be obtained through V2X technology; for example, through the UWB (Ultra-Wideband) localization technique, the distance between the vehicle and infrastructure along the road can be calculated. According to the kinematic relationships of vehicle motion, the dynamics of the distance between the COG and the lane line can be expressed as:

$$\dot{y}_l = v_x \sin\varphi + v_y \cos\varphi \tag{12}$$

From the system defined by Equations (10)–(12), the corresponding nonlinear observer isdesigned as:

$$\dot{\hat{\mu}} = k_1 \mathrm{sign}(f_1)[\mathrm{M}_k - f_k(\hat{\mu}, \hat{\alpha}, F_z)] + k_2 \mathrm{sign}(f_2)\left[a_y - a_y(\hat{\mu}, \hat{\alpha}, F_z)\right] \tag{13}$$

$$\dot{\hat{y}}_l = v_x \sin\varphi + \hat{v}_y \cos\varphi + k_3(y_l - \hat{y}_l) \tag{14}$$

$$\dot{\hat{v}}_y = a_y\left(\hat{\mu}, \hat{v}_y, F_Z\right) - r v_x + k_4(y_l - \hat{y}_l) + k_5\left[a_y - a_y\left(\hat{\mu}, \hat{v}_y, F_Z\right)\right] + k_6 \int_0^t \left[a_y - a_y\left(\hat{\mu}, \hat{v}_y, F_Z\right)\right] dt \tag{15}$$

where the superscript ^ denotes an estimated value; k_1, k_2, k_3, k_4, k_5, and k_6 are parameters; k_1, k_2, k_3 and k_6 are positive, $k_4 = \cos\varphi$. From Equation (13) f_1 is defined as:

$$\begin{cases} f_1 > 0 & \alpha_f > 0, \hat{\alpha}_f > 0 \\ f_1 < 0 & \alpha_f < 0, \hat{\alpha}_f < 0 \\ f_1 = 0 & else \end{cases} \tag{16}$$

and f_2 is defined as:

$$\begin{cases} f_2 < 0 & \alpha_f > 0, \hat{\alpha}_f > 0, \alpha_r > 0, \hat{\alpha}_r > 0 \\ f_2 > 0 & \alpha_f < 0, \hat{\alpha}_f < 0, \alpha_r < 0, \hat{\alpha}_r < 0 \\ f_2 = 0 & else \end{cases} \tag{17}$$

M_k is the self-aligning torque at the kingpin, and it can be calculated as:

$$M_k = F_{kl} L_l(\delta_{fl}) - F_{kr} L_r(\delta_{fr}) \tag{18}$$

where F_{kl} and F_{kr} are tie rod forces on the left and right sides, respectively. $L_l(\delta_{fl})$ and $L_r(\delta_{lr})$ are the distances from the steering rods to the kingpin on the left and right sides, respectively. It has to be mentioned that, in our approach, F_{kl} and F_{kr} are measured by tension and compression force sensors installed at the left and right tie rods. This measurement can be provided by the steer-by-wire system of intelligent vehicles. It can also be estimated by the steering system if the algorithm of the EPS (Electric Power Steering) system is available. $f_k\left(\hat{\mu}, \hat{v}_y, F_z\right)$ is the self-aligning torque at the kingpin estimated by the wheel steering model and tire model, which can be expressed as:

$$\begin{aligned} f_k(\mu, v_y, F_z) &= M_{zfl} + M_{zfr} + l_{ml}F_{yfl} + l_{mr}F_{yfr} \\ &= M_z(\mu, \alpha_{fl}, F_{zfl}) + M_z(\mu, \alpha_{fr}, F_{zfr}) + l_{ml}F_y(\mu, \alpha_{fl}, F_{zfl}) + l_{mr}F_y(\mu, \alpha_{fr}, F_{zfr}) \end{aligned} \quad (19)$$

where l_{ml} and l_{mr} are the mechanical trails of the left and right front tires, respectively.

3.2. Stability Analysis

According to the system dynamics in (9), (10), (12) and nonlinear observer in (13), (14), (15), the corresponding error dynamics is:

$$\dot{\tilde{\mu}} = -k_1\text{sign}(f_1)\left[f_k(\mu, v_y, F_z) - f_k(\hat{\mu}, \hat{v}_y, F_z)\right] - k_2\text{sign}(f_2)\left[a_y(\mu, v_y, F_z) - a_y(\hat{\mu}, \hat{v}_y, F_z)\right] \quad (20)$$

$$\dot{\tilde{y}}_l = -k_3\tilde{y}_l + \tilde{v}_y\cos\varphi \quad (21)$$

$$\begin{aligned} \dot{\tilde{v}}_y &= (1-k_5)\left[a_y(\mu, v_y, F_z) - a_y(\hat{\mu}, \hat{v}_y, F_z)\right] - k_6\int_0^t\left\{\left[a_y(\mu, v_y, F_z) - rv_x\right] - \left[\hat{a}_y(\hat{\mu}, \hat{v}_y, F_z) - rv_x\right]\right\}dt - k_4\tilde{y}_l \\ &= -k_6\tilde{v}_y - (k_5-1)\left[a_y(\mu, v_y, F_z) - a_y(\hat{\mu}, \hat{v}_y, F_z)\right] - k_4\tilde{y}_l \end{aligned} \quad (22)$$

where the superscript ~denotes the error between the estimated value and the real value.

The Lyapunov function is defined as:

$$V = \frac{1}{2}\tilde{\mu}^2 + \frac{1}{2}\tilde{y}_l^2 + \frac{1}{2}\tilde{v}_y^2 \quad (23)$$

Then, we have:

$$\begin{aligned} \dot{V} = &\tilde{\mu}\left\{-k_1\text{sign}(f_1)\left[f_k(\mu, v_y, F_z) - f_k(\hat{\mu}, \hat{v}_y, F_z)\right] - k_2\text{sign}(f_2)\left[a_y(\mu, v_y, F_z) - a_y(\hat{\mu}, \hat{v}_y, F_z)\right]\right\} \\ &+\tilde{y}_l\left\{-k_3\tilde{y}_l + (\cos\varphi)\tilde{v}_y\right\} + \tilde{v}_y\left\{-k_6\tilde{v}_y - (k_5-1)\left[a_y(\mu, v_y, F_z) - a_y(\hat{\mu}, \hat{v}_y, F_z)\right] - k_4\tilde{y}_l\right\} \end{aligned} \quad (24)$$

Using the mean value theorem, we have:

$$f_k(\mu, v_y, F_z) - f_k(\hat{\mu}, \hat{v}_y, F_z) = \frac{\partial \overline{f}_k}{\partial \mu}\tilde{\mu} + \frac{\partial \overline{f}_k}{\partial v_y}\tilde{v}_y \quad (25)$$

$$a_y(\mu, v_y, F_z) - a_y(\hat{\mu}, \hat{v}_y, F_z) = \frac{\partial \overline{a}_y}{\partial \mu}\tilde{\mu} + \frac{\partial \overline{a}_y}{\partial v_y}\tilde{v}_y \quad (26)$$

where

$$\begin{cases} \frac{\partial \overline{f}_k}{\partial \mu} = \frac{\partial f_k}{\partial \mu}(\overline{\mu}, \overline{v}_y) \\ \frac{\partial \overline{f}_k}{\partial v_y} = \frac{\partial f_k}{\partial v_y}(\overline{\mu}, \overline{v}_y) \\ \frac{\partial \overline{a}_y}{\partial \mu} = \frac{\partial a_y}{\partial \mu}(\overline{\mu}, \overline{v}_y) \\ \frac{\partial \overline{a}_y}{\partial v_y} = \frac{\partial a_y}{\partial v_y}(\overline{\mu}, \overline{v}_y) \end{cases} \quad (27)$$

In Equation (27) $\overline{\mu}$ is a median between μ and $\hat{\mu}$; $\overline{v}_y$ is a median between v_y and $\hat{v}_y$. Substituting (25), (26) and (27) into (24) we have:

$$\begin{aligned} \dot{V} = &-\left[k_1\text{sign}(f_1)\frac{\partial \overline{f}_k}{\partial \mu} + k_2\text{sign}(f_2)\frac{\partial \overline{a}_y}{\partial \mu}\right]\tilde{\mu}^2 - k_3\tilde{y}_l^2 - \left[(k_5-1)\frac{\partial \overline{a}_y}{\partial v_y} + k_6\right]\tilde{v}_y^2 \\ &-\left[k_1\text{sign}(f_1)\frac{\partial \overline{f}_k}{\partial v_y} + k_2\text{sign}(f_2)\frac{\partial \overline{a}_y}{\partial v_y} + (k_5-1)\frac{\partial \overline{a}_y}{\partial \mu}\right]\tilde{\mu}\tilde{v}_y + (-k_4+\cos\varphi)\tilde{y}_l\tilde{v}_y \\ = &-\begin{bmatrix} \tilde{\mu} & \tilde{y}_l & \tilde{v}_y \end{bmatrix} A \begin{bmatrix} \tilde{\mu} \\ \tilde{y}_l \\ \tilde{v}_y \end{bmatrix} \end{aligned} \quad (28)$$

where

$$A = \begin{bmatrix} A_{11} & 0 & A_{13} \\ 0 & k_3 & A_{23} \\ A_{31} & A_{32} & A_{33} \end{bmatrix} \tag{29}$$

$$A_{11} = k_1 \mathrm{sign}(f_1)\frac{\partial \overline{f}_k}{\partial \mu} + k_2 \mathrm{sign}(f_2)\frac{\partial \overline{a}_y}{\partial \mu} \tag{30}$$

$$A_{13} = A_{31} = \frac{1}{2}k_1 \mathrm{sign}(f_1)\frac{\partial \overline{f}_k}{\partial v_y} + \frac{1}{2}k_2 \mathrm{sign}(f_2)\frac{\partial \overline{a}_y}{\partial v_y} + \frac{1}{2}(k_5-1)\frac{\partial \overline{a}_y}{\partial \mu} \tag{31}$$

$$A_{23} = A_{32} = -\frac{1}{2}k_4 + \frac{1}{2}\cos\varphi \tag{32}$$

$$A_{33} = (k_5-1)\frac{\partial \overline{a}_y}{\partial v_y} + k_6 \tag{33}$$

If A is a positive definite matrix, then $\dot{V} < 0$ holds, which means that the estimation system is stable, and the estimation error will converge to zero as time $t \to \infty$. To ensure that A is a positive definite matrix, all sequential principal minors of A need to be positive. According to the modified tire model in (7) and (8) and the symbolic function defined in (16) and (17) it can be deduced that:

$$\mathrm{sign}(f_1)\frac{\partial \overline{f}_k}{\partial \mu} \geq 0 \tag{34}$$

$$\mathrm{sign}(f_2)\frac{\partial \overline{a}_y}{\partial \mu} \geq 0 \tag{35}$$

Therefore, if we choose k_1, k_2, and k_3 as positive constants, the first-order and second-order sequential principal minors of A are positive. If we choose $k_4 = \cos\varphi$, then the third-order sequential principal minor of A is:

$$\begin{aligned} |A| = \; & k_3\left[k_1 \mathrm{sign}(f_1)\tfrac{\partial \overline{f}_k}{\partial \mu} + k_2 \mathrm{sign}(f_2)\tfrac{\partial \overline{a}_y}{\partial \mu}\right]\left[(k_5-1)\tfrac{\partial \overline{a}_y}{\partial v_y} + k_6\right] \\ & -\tfrac{k_3}{4}\left[k_1 \mathrm{sign}(f_1)\tfrac{\partial \overline{f}_k}{\partial v_y} + k_2 \mathrm{sign}(f_2)\tfrac{\partial \overline{a}_y}{\partial v_y} + (k_5-1)\tfrac{\partial \overline{a}_y}{\partial \mu}\right]^2 \end{aligned} \tag{36}$$

Since $\frac{\partial \overline{f}_k}{\partial v_y}$, $\frac{\partial \overline{a}_y}{\partial v_y}$, and $\frac{\partial \overline{a}_y}{\partial \mu}$ are bounded according to the modified tire model, there exists a parameter k_5 such that:

$$\left[k_1 \mathrm{sign}(f_1)\frac{\partial \overline{f}_k}{\partial v_y} + k_2 \mathrm{sign}(f_2)\frac{\partial \overline{a}_y}{\partial v_y} + (k_5-1)\frac{\partial \overline{a}_y}{\partial \mu}\right]^2 = \left[A_{11} + (k_5-1)\frac{\partial \overline{a}_y}{\partial \mu}\right]^2 = 0 \tag{37}$$

If the chosen value of k_6 is large enough, then $|A| > 0$ holds. Therefore, $\dot{V} < 0$ holds, and it can be deduced that:

$$\begin{bmatrix} \tilde{\mu} \\ \tilde{y}_l \\ \tilde{v}_y \end{bmatrix} \to \begin{bmatrix} 0 \\ 0 \\ 0 \end{bmatrix} \text{ as } t \to \infty \tag{38}$$

and the estimation error can converge to zero exponentially.

3.3. Robustness Analysis

The uncertainties of the tire and vehicle models or measurements from sensors introduce perturbance to the system. It is necessary to analyze the performance of the estimator with bounded external excitation. According to the error dynamics of the system in (20), (21) and (22) without external inputs, the error dynamics of the system with inputs can be expressed as

$$\dot{\tilde{\mu}} = -k_1 \text{sign}(f_1)\left(\frac{\partial \bar{f}_k}{\partial \mu}\tilde{\mu} + \frac{\partial \bar{f}_k}{\partial v_y}\tilde{v}_y\right) - k_2 \text{sign}(f_2)\left(\frac{\partial \bar{a}_y}{\partial \mu}\tilde{\mu} + \frac{\partial \bar{a}_y}{\partial v_y}\tilde{v}_y\right) + u_1 \tag{39}$$

$$\dot{\tilde{y}}_l = -k_3\tilde{y}_l + \tilde{v}_y \cos\varphi + u_2 \tag{40}$$

$$\dot{\tilde{v}}_y = -k_6\tilde{v}_y - (k_5 - 1)\left(\frac{\partial \bar{a}_y}{\partial \mu}\tilde{\mu} + \frac{\partial \bar{a}_y}{\partial v_y}\tilde{v}_y\right) - k_4\tilde{y}_l + u_3 \tag{41}$$

where u_1, u_2, and u_3 are external bounded inputs. u_1 includes the uncertainties of the tire model and the uncertainties in the estimation results of v_y. u_2 includes the uncertainties in the measurements of lateral distance between the COG and the lane line and uncertainties in the estimation results of v_y.u_3 includes the uncertainties in lateral distance measurements and uncertainties in the estimation results of μ.

The Lyapunov function is chosen and defined in Equation (23) thus, according to (39), (40), and (41) we have:

$$\begin{aligned}\dot{V} = {} & \tilde{\mu}\left\{-k_1\text{sign}(f_1)\left(\tfrac{\partial \bar{f}_k}{\partial \mu}\tilde{\mu} + \tfrac{\partial \bar{f}_k}{\partial v_y}\tilde{v}_y\right) - k_2\text{sign}(f_2)\left(\tfrac{\partial \bar{a}_y}{\partial \mu}\tilde{\mu} + \tfrac{\partial \bar{a}_y}{\partial v_y}\tilde{v}_y\right) + u_1\right\} \\ & + \tilde{y}_l\left\{-k_3\tilde{y}_l + (\cos\varphi)\tilde{v}_y + u_2\right\} + \tilde{v}_y\left\{-k_6\tilde{v}_y - (k_5-1)\left(\tfrac{\partial \bar{a}_y}{\partial \mu}\tilde{\mu} + \tfrac{\partial \bar{a}_y}{\partial v_y}\tilde{v}_y\right) - k_4\tilde{y}_l + u_3\right\} \\ = {} & -A_{11}\tilde{\mu}^2 - k_3\tilde{y}_l^2 - A_{33}\tilde{v}_y^2 - \left[A_{11} + (k_5-1)\tfrac{\partial \bar{a}_y}{\partial \mu}\right]\tilde{\mu}\tilde{v}_y + (\cos\varphi - k_4)\tilde{v}_y\tilde{y}_l + \tilde{\mu}u_1 + \tilde{y}_l u_2 + \tilde{v}_y u_3\end{aligned} \tag{42}$$

Since $k_4 = \cos\varphi$, substituting (37) into (42) we have

$$\begin{aligned}\dot{V} = {} & -A_{11}\tilde{\mu}^2 - k_3\tilde{y}_l^2 - A_{33}\tilde{v}_y^2 + \tilde{\mu}u_1 + \tilde{y}_l u_2 + \tilde{v}_y u_3 \\ \leq {} & -A_{11}|\tilde{\mu}|^2 - k_3|\tilde{y}_l|^2 - A_{33}|\tilde{v}_y|^2 + |\tilde{\mu}||u_1| + |\tilde{y}_l||u_2| + |\tilde{v}_y||u_3| \\ = {} & \left[-A_{11}(1-\theta_1)|\tilde{\mu}|^2 - A_{11}\theta_1|\tilde{\mu}|^2 + |\tilde{\mu}||u_1|\right] + \left[-k_3(1-\theta_2)|\tilde{y}_l|^2 - k_3\theta_2|\tilde{y}_l|^2 + |\tilde{y}_l||u_2|\right] \\ & + \left[-A_{33}(1-\theta_3)|\tilde{v}_y|^2 - A_{11}\theta_1|\tilde{v}_y|^2 + |\tilde{v}_y||u_3|\right] \\ = {} & -A_{11}(1-\theta_1)|\tilde{\mu}|^2 - k_3(1-\theta_2)|\tilde{y}_l|^2 - A_{33}(1-\theta_3)|\tilde{v}_y|^2 \\ & -\left(A_{11}\theta_1|\tilde{\mu}|^2 - |\tilde{\mu}||u_1|\right) - \left(k_3\theta_2|\tilde{y}_l|^2 - |\tilde{y}_l||u_2|\right) - \left(A_{11}\theta_1|\tilde{v}_y|^2 - |\tilde{v}_y||u_3|\right)\end{aligned} \tag{43}$$

where $0 < \theta_1 < 1$, $0 < \theta_2 < 1$, and $0 < \theta_3 < 1$. Therefore, if the bounded inputs satisfy:

$$\begin{cases} |u_1| < A_{11}\theta_1 \\ |u_2| < k_3\theta_2 \\ |u_3| < A_{33}\theta_3 \end{cases} \tag{44}$$

then we have:

$$\dot{V} \leq -A_{11}(1-\theta_1)|\tilde{\mu}|^2 - k_3(1-\theta_2)|\tilde{y}_l|^2 - A_{33}(1-\theta_3)|\tilde{v}_y|^2 \tag{45}$$

If we define:

$$A_{11}(1-\theta_1)|\tilde{\mu}|^2 + k_3(1-\theta_2)|\tilde{y}_l|^2 + A_{33}(1-\theta_3)|\tilde{v}_y|^2 = W(x) \tag{46}$$

where $x = \begin{bmatrix} \tilde{\mu} & \tilde{y}_l & \tilde{v}_y \end{bmatrix}^T$, then inequalities (47) (48) hold:

$$\frac{1}{4}\|x\|^2 \leq V \leq \|x\|^2 \tag{47}$$

$$\frac{\partial V}{\partial t} + \frac{\partial V}{\partial x}\dot{x} \leq -W(x). \tag{48}$$

By applying Theorem 4.19 from [31], we can reason that the system expressed in (39), (40), and (41) is input-state stable; thus, if the estimation system is interfered by bounded inputs, then the system will still stay stable.

4. Experimental Validation

Tests based on an electric vehicle were conducted to verify the proposed road friction coefficient estimation algorithm. The experimental setup and results are discussed in this section.

4.1. Experimental Setup

4.1.1. Test Vehicle

The test vehicle is an electric vehicle and shown in Figure 5a, and the vehicle parameters are listed in Table 1. Information about the wheel speed and steering wheel angle was obtained through CAN-Bus. The GNSS receiver is Novatel 718D, which provides the absolute position and heading angle of the vehicle. It is necessary to point out that the lane lines were modeled in advance in the navigation coordinates so that the distance between the vehicle and lane line could be calculated in real time. ADIS16495 is the IMU (Inertial Measurement Unit), which measures the acceleration and angular velocities. The angle between the vehicle heading and lane line can be calculated by integrating the yaw rate of the vehicle. The steering tie rods are cut off and two Kistler tension and compression force sensors 9321B are installed on the left and right tie rods, respectively, as shown in Figure 5b. The tension and compression force sensors measure the force at the tie rods so that the self-aligning torque of the wheel can be measured indirectly.

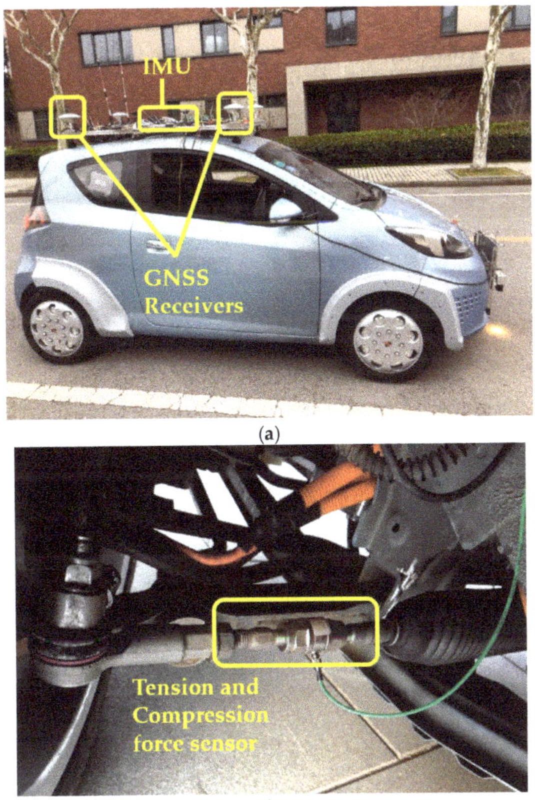

(a)

(b)

Figure 5. Test vehicle implementation: (**a**) test vehicle; (**b**) steering tie rod with a tension and compression force sensor.

Table 1. Vehicle parameters.

Parameters	Value
m/(kg)	1343.8
b/(m)	1.356
l_f/(m)	1.112
l_r/(m)	1.193
I_z/(kg·m^2)	1785

4.1.2. Test Road

To verify the proposed road friction coefficient estimation algorithm, slalom tests and DLC (Double Line Change) tests were conducted on the road shown in Figure 6. The white lane lines shown in Figure 6 were mapped in advance. According to a large number of emergency braking experiments, the real road friction coefficient is considered to be around 0.8.

Figure 6. Test road.

4.2. Experimental Results and Analysis

4.2.1. Slalom Test

Slalom test results are shown in Figure 7. Figure 7a shows the vehicle speed measured by the GNSS receiver. The blue line and green line show the steering wheel angle and yaw rate during the test, respectively. Figure 7c shows the accelerations recorded by the IMU, and we can see that the maximum lateral acceleration reaches 8 ms^{-2}, which means that the vehicle has enough lateral excitation. Figure 7d shows the estimated self-aligning torque at the kingpin according to the tension and compression force sensors installed at the steering tie rods. Figure 7e shows the estimated lateral distance between the COG of the vehicle and the lane line. The reference is calculated by the absolute position and the lane line map. The road friction coefficient estimation result is shown in Figure 7f. Since the road friction coefficient is related not only to the road surface but also to the tires, an absolutely precise road friction coefficient could not be obtained. From braking tests, we know that the real friction coefficient is about 0.8; therefore, we set 0.75–0.85 as the reference region. The initial road friction coefficient is set at 1. From Figure 8f, we can see that at around 9 s, the estimated road coefficient converges to the reference region, and the convergence time was about 3 s with continuous lateral excitation. After 9 s, the estimated value remains within the reference region, although there is a slight fluctuation, which means that the nonlinear estimator performs well. If the reference value of the road friction is considered as 0.8, then the estimation accuracy was about 97.2%. From 16 s onward, the vehicle drives straightly, and the road friction coefficient estimation algorithm stops without lateral excitation.

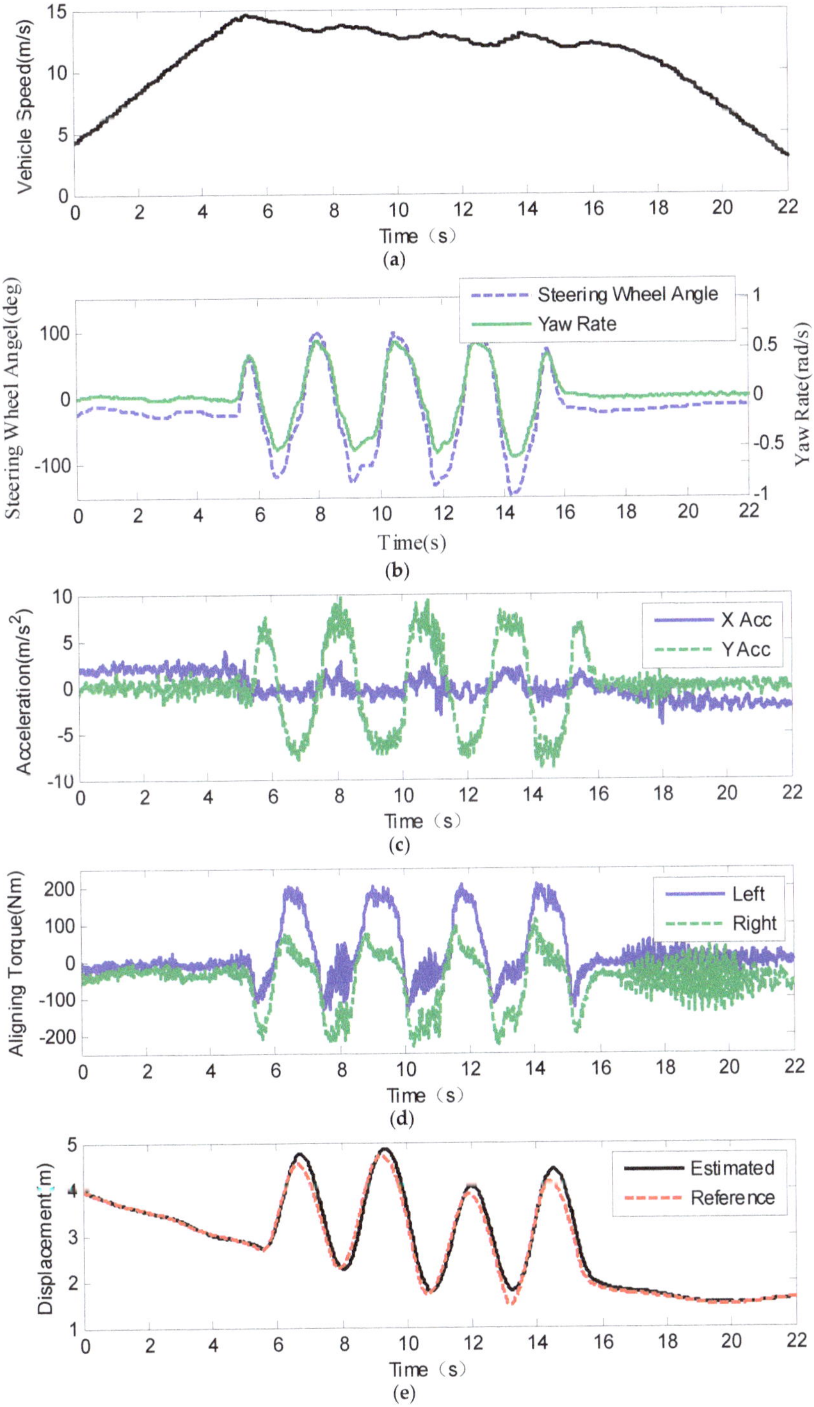

Figure 7. *Cont.*

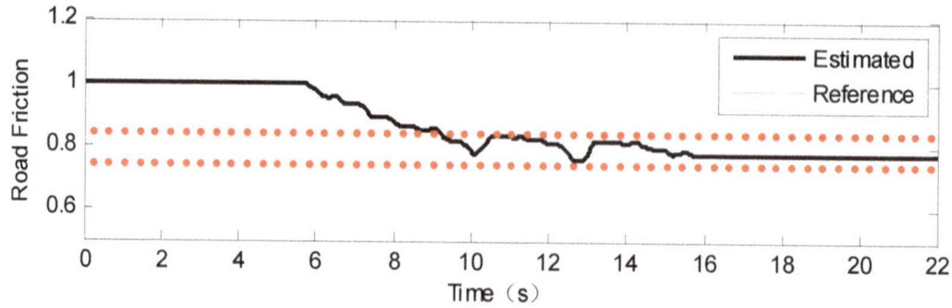

(f)

Figure 7. Results of the slalom test: (**a**) vehicle speed; (**b**) steering wheel angle and yaw rate; (**c**) longitudinal and lateral acceleration; (**d**) self-aligning torque at the kingpin; (**e**) distance between the vehicle and the left lane line; (**f**) road friction coefficient.

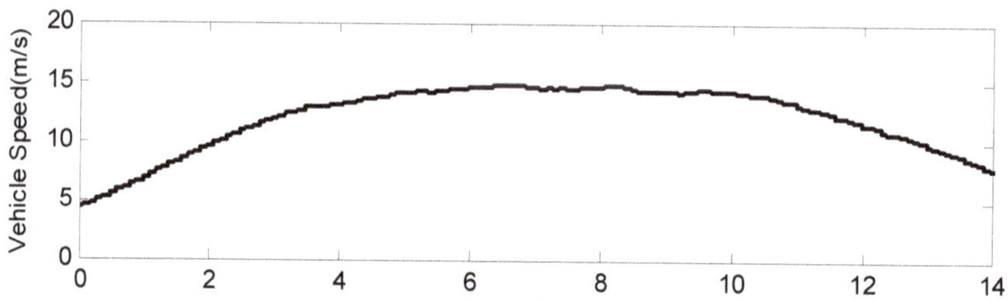

(a)

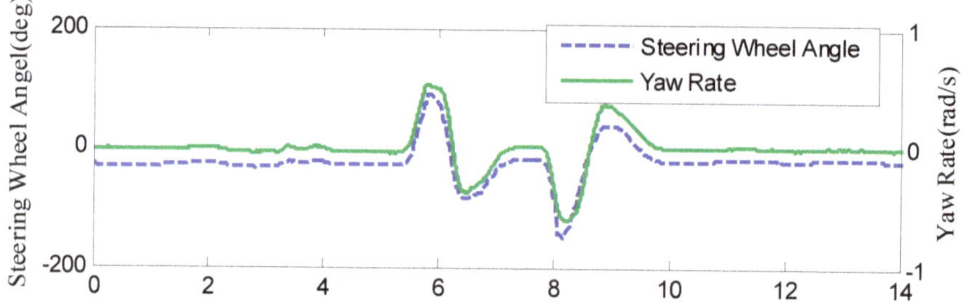

(b)

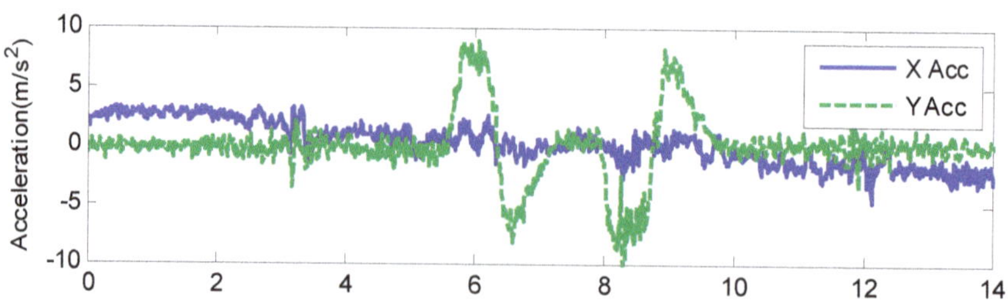

(c)

Figure 8. *Cont.*

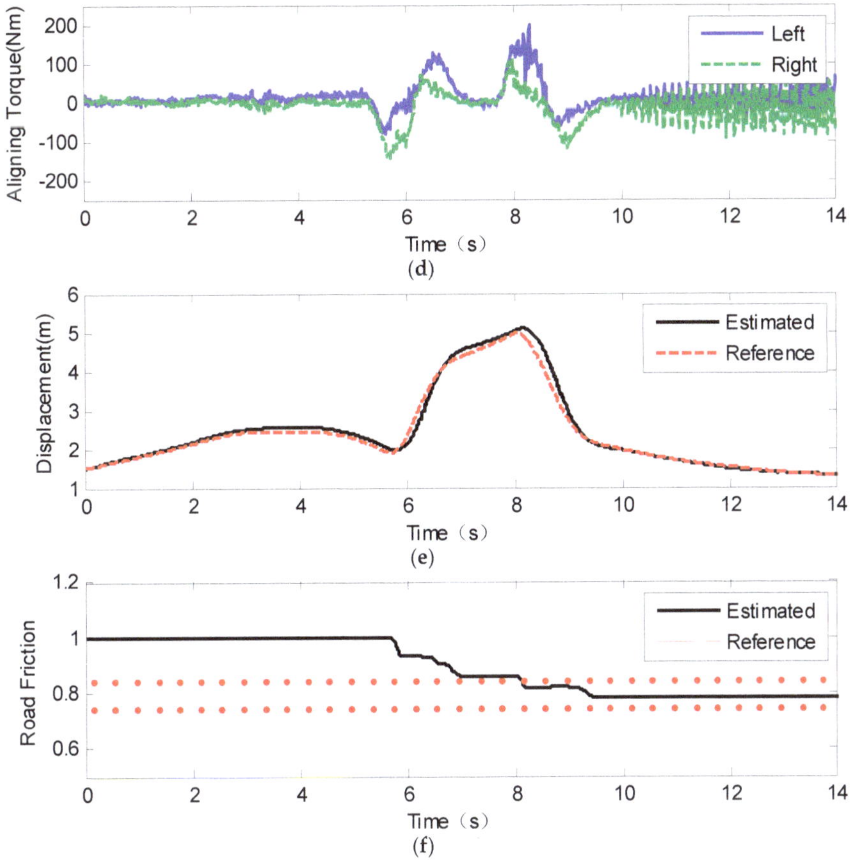

Figure 8. Results of the DLC (Double Line Change) test: (**a**) vehicle speed; (**b**) steering wheel angle and yaw rate; (**c**) longitudinal and lateral acceleration; (**d**) self-aligning torque at the kingpin; (**e**) distance between the vehicle and the left lane line; (**f**) road friction coefficient.

4.2.2. DLC Test

Figure 8 shows the experimental results of DLC maneuvering. Figure 8a shows the vehicle speed during the test, and in Figure 8b represents the steering wheel angle and yaw rate of the vehicle by the blue line and green line, respectively. Figure 8c shows the longitudinal and lateral accelerations of the vehicle, and the maximum lateral acceleration is between 8 and 9 ms^{-2}. The estimated aligning torques of the left and right kingpins are shown in Figure 8d. Figure 8e shows the estimated lateral distance between the vehicle and the lane line, and the estimated value tracks the reference value with little error. The road friction coefficient estimation results are shown in Figure 8f. If the reference value of the road friction is considered as 0.8, then the estimation accuracy was about 97.8%. Since the nonlinear estimator only works during steering, the estimation holds if the vehicle's lateral acceleration is relatively small during DLC, for example, from 7 to 8 s. Compared with the slalom test results, the road friction estimation results dose not fluctuate because the lateral excitation is not continuous. From the experimental results, we can see that the road friction coefficient rapidly converges to the reference value.

5. Conclusions

In this paper, a nonlinear observer for the road friction coefficient during steering based on the self-aligning torque characteristics of the tires aided by vehicle lateral displacement information was proposed. A modified tire brush model was established according to the tire test data, and the model describes the tire characteristics more precisely than the original model. A nonlinear observer using vehicle lateral displacement information was designed, and the stability and robustness were analyzed. Experiments were conducted to verify the proposed road friction coefficient estimation algorithm. The test results demonstrate that the proposed method performs well during vehicle steering, and the estimated road friction coefficient converges to the reference value very rapidly.

6. Future Work

We have modified the tire brush model according to the tire test data, and the results show that the modified model describes the tire characteristics properly. However, the tire tests were done with only one type tire in high friction condition, which is not sufficient to verify that the modified tire model is suitable for other tires with different sizes or types. Tire tests with more tires in different friction conditions should be conducted to verify the modified tire brush model.

The experiments were conducted only in high friction condition due to the test condition limitation. The algorithm should also be validated in low friction condition and high to low or low to high friction transition conditions in the future.

Author Contributions: methodology, L.G., L.X. and X.X.; software, L.G. and X.L.; validation, X.X., W.L. and Y.L.; resources, L.X. and Z.Y.; data curation, L.G. and X.X.; writing—original draft preparation, L.G.; writing—review and editing, W.L., X.X. and Y.L.; supervision, L.X. and Z.Y.; funding acquisition, L.X. and Z.Y.

Funding: This research is supported by the National Key Research and Development Program of China (grant no. 2016YFB0100901) and Shanghai Scientific Research Project (grant no. 16DZ1100700).

Conflicts of Interest: The authors declare no conflict of interest.

References

1. Cheng, S.; Li, L.; Chen, J. Fusion Algorithm Design Based on Adaptive SCKF and Integral Correction for Side-Slip Angle Observation. *IEEE Trans. Ind. Electron.* **2018**, *65*, 5754–5763. [CrossRef]
2. Chen, J.; Song, J.; Li, L.; Jia, G.; Ran, X.; Yang, C. UKF-based adaptive variable structure observer for vehicle sideslip with dynamic correction. *IET Control Theory and Appl.* **2016**, *10*, 1641–1652. [CrossRef]
3. Lv, C.; Xing, Y.; Zhang, J.; Na, X.; Li, Y.; Liu, T.; Cao, D.; Wang, F.Y. Levenberg-Marquadt Backpropagation Training of Multilayer Neural Networks for State Estimation of a Safety-Critical Cyber-Physical System. *IEEE Trans. Ind. Inf.* **2018**, *14*, 3436–3446. [CrossRef]
4. Li, L.; Jia, G.; Chen, J.; Zhu, H.; Cao, D.; Song, J. A novel vehicle dynamics stability control algorithm based on the hierarchical strategy with constrain of nonlinear tyre forces. *Veh. Syst. Dyn.* **2015**, *53*, 1093–1116. [CrossRef]
5. Lv, C.; Zhang, J.; Li, Y.; Yuan, Y. Directional-stability-aware brake blending control synthesis for over-actuated electric vehicles during straight-line deceleration. *Mechatronics* **2016**, *38*, 121–131. [CrossRef]
6. Lv, C.; Zhang, J.; Li, Y.; Yuan, Y. Novel control algorithm of braking energy regeneration system for an electric vehicle during safety-critical driving maneuvers. *Energy Convers. Manage.* **2015**, *106*, 520–529. [CrossRef]
7. Li, L.; Song, J.; Li, H.Z.; Shan, D.S.; Kong, I.; Yang, C.C. Comprehensive prediction method of road friction for vehicle dynamics control. *J. Automobile Eng.* **2009**, *223*, 987–1002. [CrossRef]
8. Li, L.; Yang, K.; Jia, G.; Ran, X.; Song, J.; Han, Z.Q. Comprehensive tire-road friction coefficient estimation based on signal fusion method under complex maneuvering operations. *Mech. Syst. Sig. Processing* **2015**, *56–57*, 259–276. [CrossRef]
9. Cabrera, J.A.; Castillo, J.J.; Perez, J.; Velasco, J.M.; Guerra, A.J.; Hernandez, P. A Procedure for Determining Tire-Road Friction Characteristics Using a Modification of the Magic Formula Based on Experimental Results. *Sensors* **2018**, *18*, 896. [CrossRef]

10. Khaleghian, S.; Emami, A.; Taheri, S. A technical survey on tire-road friction estimation. *Friction* **2017**, *5*, 123–146. [CrossRef]
11. Alonso, J.; Lopez, J.M.; Pavon, I.; Recuero, M.; Asensio, C.; Arcas, G.; Bravo, A. On-board wet road surface identification using tyre/road noise and Support Vector Machines. *Appl. Acoust.* **2014**, *76*, 407–415. [CrossRef]
12. Roychowdhury, S.; Zhao, M.; Wallin, A.; Ohlsson, N.; Jonasson, M. Machine Learning Models for Road Surface and Friction Estimation using Front-Camera Images. In Proceedings of the 2018 International Joint Conference on Neural Networks, Rio de Janeiro, Brazil, 8–13 July 2018; pp. 1–8.
13. Yunta, J.; Garcia-Pozuelo, D.; Diaz, V.; Olatunbosun, O. A Strain-Based Method to Detect Tires' Loss of Grip and Estimate Lateral Friction Coefficient from Experimental Data by Fuzzy Logic for Intelligent Tire Development. *Sensors* **2018**, *18*, 490. [CrossRef] [PubMed]
14. Ambroz, M.; Hudomalj, U.; Marinsek, A.; Kamnik, R. Raspberry Pi-Based Low-Cost Connected Device for Assessing Road Surface Friction. *Electronics* **2019**, *8*, 341. [CrossRef]
15. Ahn, C.; Peng, H.; Tseng, H.E. Robust estimation of road friction coefficient using lateral and longitudinal vehicle dynamics. *Veh. Syst. Dyn.* **2012**, *50*, 961–985. [CrossRef]
16. Castillo Aguilar, J.J.; Cabrera Carrillo, J.A.; Guerra Fernandez, A.J.; Carabias Acosta, E. Robust Road Condition Detection System Using In-Vehicle Standard Sensors. *Sensors* **2015**, *15*, 32056–32078. [CrossRef] [PubMed]
17. Castillo, J.J.; Cabrera, J.A.; Guerra, A.J.; Simon, A. A Novel Electrohydraulic Brake System With Tire-Road Friction Estimation and Continuous Brake Pressure Control. *IEEE Trans. Ind. Electron.* **2016**, *63*, 1863–1875. [CrossRef]
18. Enisz, K.; Szalay, I.; Kohlrusz, G.; Fodor, D. Tyre-road friction coefficient estimation based on the discrete-time extended Kalman filter. *J. Automobile Eng.* **2015**, *229*, 1158–1168. [CrossRef]
19. Xia, X.; Xiong, L.; Sun, K.; Yu, Z.P. Estimation of maximum road friction coefficient based on Lyapunov method. *Int. J. Automot. Technol.* **2016**, *17*, 991–1002. [CrossRef]
20. Wang, R.; Wang, J. Tire-road friction coefficient and tire cornering stiffness estimation based on longitudinal tire force difference generation. *Control Eng. Pract.* **2013**, *21*, 65–75. [CrossRef]
21. Qi, Z.; Taheri, S.; Wang, B.; Yu, H. Estimation of the tyre-road maximum friction coefficient and slip slope based on a novel tyre model. *Veh. Syst. Dyn.* **2015**, *53*, 506–525. [CrossRef]
22. Luque, P.; Mantaras, D.A.; Fidalgo, E.; Alvarez, J.; Riva, P.; Giron, P.; Compadre, D.; Ferran, J. Tyre-road grip coefficient assessment—Part II: Online estimation using instrumented vehicle, extended Kalman filter, and neural network. *Veh. Syst. Dyn.* **2013**, *51*, 1872–1893. [CrossRef]
23. Matsuda, T.; Jo, S.I.; Nishira, H.; Deguchi, Y. Instantaneous Estimation of Road Friction based on Front Tire SAT using Kalman Filter. *SAE Int. J. Passenger Cars Mech. Syst.* **2013**, *6*, 147–153. [CrossRef]
24. Hsu, J.; Laws, S.M.; Gerdes, J.C. Estimation of Tire Slip Angle and Friction Limits Using Steering Torque. *IEEE Trans. Control Syst. Technol.* **2010**, *18*, 896–907. [CrossRef]
25. Ahn, C.; Peng, H.; Tseng, H.E. Robust Estimation of Road Frictional Coefficient. *IEEE Trans. Control Syst. Technol.* **2013**, *21*, 1–13. [CrossRef]
26. Shao, L.; Jin, C.; Lex, C.; Eichberger, A. Robust road friction estimation during vehicle steering. *Veh. Syst. Dyn.* **2019**, *57*, 493–519. [CrossRef]
27. Yoon, J.H.; Peng, H. A Cost-Effective Sideslip Estimation Method Using Velocity Measurements From Two GPS Receivers. *IEEE Trans. Veh. Technol.* **2014**, *63*, 2589–2599. [CrossRef]
28. Yoon, J.H.; Peng, H. Robust Vehicle Sideslip Angle Estimation Through a Disturbance Rejection Filter That Integrates a Magnetometer With GPS. *IEEE Trans. Intell. Trans. Syst.* **2014**, *15*, 191–204. [CrossRef]
29. Wang, Y.; Binh Minh, N.; Fujimoto, H.; Hori, Y. Multirate Estimation and Control of Body Slip Angle for Electric Vehicles Based on Onboard Vision System. *IEEE Trans. Ind. Electron.* **2014**, *61*, 1133–1143. [CrossRef]
30. Wang, Y.; Liu, Y.; Fujimoto, H.; Hori, Y. Vision-Based Lateral State Estimation for Integrated Control of Automated Vehicles Considering Multirate and Unevenly Delayed Measurements. *ASME Trans. Mechatron.* **2018**, *23*, 2619–2627. [CrossRef]
31. Khalil, H.K. *Nonlinear Systems*, 3rd ed.; Prentice Hall: Upper Saddle River, NJ, USA, 2002.

sensors

MDPI

Article

Reinforcement Learning-Based End-to-End Parking for Automatic Parking System

Peizhi Zhang [1], Lu Xiong [1,*], Zhuoping Yu [1], Peiyuan Fang [1], Senwei Yan [2], Jie Yao [2] and Yi Zhou [2]

1 School of Automotive Studies, Tongji University, Shanghai 201804, China; zhangpeizhi@tongji.edu.cn (P.Z.); yuzhuoping@tongji.edu.cn (Z.Y.); funpayyuan@foxmail.com (P.F.)
2 SAIC Motor Corporation Limited, Shanghai 201800, China; adam_yan@foxmail.com (S.Y.); yaojie@saicmotor.com (J.Y.); zhouyi02@saicmotor.com (Y.Z.)
* Correspondence: xiong_lu@tongji.edu.cn

Received: 31 July 2019; Accepted: 12 September 2019; Published: 16 September 2019

Abstract: According to the existing mainstream automatic parking system (APS), a parking path is first planned based on the parking slot detected by the sensors. Subsequently, the path tracking module guides the vehicle to track the planned parking path. However, since the vehicle is non-linear dynamic, path tracking error inevitably occurs, leading to inclination and deviation of the parking. Accordingly, in this paper, a reinforcement learning-based end-to-end parking algorithm is proposed to achieve automatic parking. The vehicle can continuously learn and accumulate experience from numerous parking attempts and then learn the command of the optimal steering wheel angle at different parking slots. Based on this end-to-end parking, errors caused by path tracking can be avoided. Moreover, to ensure that the parking slot can be obtained continuously in the process of learning, a parking slot tracking algorithm is proposed based on the combination of vision and vehicle chassis information. Furthermore, given that the learning network output is hard to converge, and it is easy to fall into local optimum during the parking process, several reinforcement learning training methods in terms of parking conditions are developed. Lastly, by the real vehicle test, it is proved that using the proposed method can achieve a better parking attitude than using the path planning and path tracking-based method.

Keywords: automatic parking system (APS); end-to-end parking; reinforcement learning; parking slot tracking

1. Introduction

The average proportion of cars and parking slots in big cities is about 1:0.8, and that in small and medium-sized cities is nearly 1:0.5, according to the data released by the National Development and Reform Commission of China. The lack of parking space makes the designed parking slot increasingly narrower. Accordingly, parking environment is becoming complex progressively, and the increasingly higher requirement of the parking operation accuracy is raised, bringing great troubles to many drivers. Automatic parking system (APS) can increase parking safety and utilization rate of parking slot, so it has wide market application prospects.

However, the smaller size of the parking slot requires very high parking accuracy for APS. Take the perpendicular parking slot as an example; it raises a higher demand of parking attitude for its narrow width. The BS ISO 16787-2016 [1] stipulates that the perpendicular parking inclination angle of APS should be confined within ±3°, imposing huge challenges to the performance of APS.

The current mainstream APS architecture is a path planning and path tracking-based method. To be specific, a parking path is first planned based on the parking slot detected by the sensors (e.g., camera and ultrasonic radar), and then the path tracking module controls the vehicle to track the planned parking path. However, since the vehicle is nonlinear dynamic, the control error of path tracking is inevitable during the path tracking, leading to inappropriate parking attitude.

How can we avoid the path tracking error to ensure the ideal parking attitude? Let us think about how humans park their cars. Actually, we directly turn the steering wheel according to the position of the parking slot, which is an end-to-end parking mode. Moreover, as the number of parking increases, we gain more experience, and parking is increasingly accurate. In fact, reinforcement learning is an algorithm in which the agent gets the greatest reward in the process of interactive learning with the environment, thus learning the optimal mapping from environment to action. Therefore, we try to apply reinforcement learning to APS to improve the parking attitude.

1.1. Related Work

1.1.1. Mainstream APS

The mainstream APS first plans a parking path according to the parking slot detected by sensors. Path planning can be split into geometric method, sampling method, and numerical optimization method. The geometric method adopts Reeds–Shepp (RS) curve [2], B-spline curve [3], η^3-splines [4], and arcs optimized by cyclotron curve [5,6] to plan parking path based on the non-holonomic constraints of the vehicle. The sampling method aims to spread the points evenly in the sampling space, filter the points by a certain method, and connect the selected points into the required path, covering Rapidly-exploring Random Tree (RRT) [7] and target bias RRT [8]. The numerical optimization method is to consider the parking process as a dynamic system, and the length [9,10] or curvature [11,12] of the parking path is the optimization goal of this dynamic system. The constraints of this dynamic system include the non-holonomic constraints of the vehicle, the starting point, and the target location of the parking.

After completing the path planning, the path tracking module of APS controls the vehicle to track the planned parking path. Path tracking can be divided into Ackerman steering model-based open loop control method and vehicle dynamics model-based closed loop control method. The Ackerman steering model-based open loop control method considers that there is no tire sideslip, and vehicle satisfies the non-holonomic constraints. The most typical one is the pure tracking control algorithm [13]. The vehicle dynamics model-based closed loop control method considers tire sideslip. Feedforward control is designed using the two-degree-of-freedom vehicle dynamics model, and closed-loop feedback control is implemented by proportional-integral-differential (PID) algorithm [14,15] or sliding mode control (SMC) algorithm [16,17]. In fact, no matter which control method is used, the control error of the path tracking is inevitable since the vehicle is nonlinear dynamic [18,19], which makes the vehicle unlikely to completely track the planned parking path. Though there have been studies to reduce the path tracking error [20,21], the path tracking error cannot be eliminated.

1.1.2. Reinforcement Learning

As mentioned above, path planning and path tracking-based method may result in poor parking attitude due to the inevitable control error (the experimental results in Section 3 also confirmed this). Accordingly, we take the "human-like" parking mode based on reinforcement learning, which cannot only avoid the error caused by path tracking through the end-to-end method of environment-to-action but also continuously learn and accumulate experience from considerable parking attempts, as well as learning the optimal steering wheel angle command at different parking slots relative to vehicle. How to choose a suitable reinforcement learning method for APS? To answer this question, different reinforcement learning methods are first reviewed.

Reinforcement learning mainly includes value-based method, policy-based method, and Actor-Critic method.

The value-based method evaluates the cumulative expectation reward by the value function after taking action and then chooses the action with the largest cumulative reward expectation [22]. Deep Q network (DQN) [23,24] is a typical value-based method. It is based on Q-learning and replaces Q-table with deep neural network (DNN) to solve the problem that Q-learning is prone to dimension disasters when state space is high-dimensional. However, value-based method makes value function continuous and chooses action based on the value function of each action, so it is not suitable for continuous action spaces (e.g., continuous steering wheel angle command for APS).

Compared with the value-based method, the policy-based method directly optimizes the policy based on the sampling method and constantly calculates the gradient of the policy expectation reward about the parameters of the policy network during the training process [25]. Though the policy-based method is applicable to high-dimensional continuous action spaces, each iterative step should sample a batch sequence to update the parameters, resulting in a large variance of the policy gradient estimation and making it easy to fall into local optimum.

The Actor-Critic method, which combines the value-based method and the policy-based method, adopts policy-based method to update the policy, and adopts the value function as the evaluation method of the policy [26–28]. By introducing the value function as the evaluation criterion in the policy search, the loss of sequential difference about the reward can be minimized, so that the variance of the policy gradient estimation can be reduced effectively. Although Actor-Critic method can realize the learning of continuous action space and can reduce the variance of the strategy gradient estimation, it only has one actor network and one critic network, which easily leads to unstable training. Deep Deterministic Policy Gradient (DDPG) algorithm [29,30] has made some improvements on the basis of Actor-Critic method. On one hand, it creates target networks for actor network and critic network, respectively, significantly enhancing the stability of learning. On the other hand, it uses experience pool-based replay caching technology to cut off the data correlation.

As mentioned above, we believe that DDPG is applicable to APS for the following reasons: first, Actor-Critic architecture can realize the learning of continuous action space (since the steering wheel angle for APS is a continuous action). Second, introducing the value function as the evaluation criterion in the policy search can reduce the variance of the policy gradient estimation, which is more efficient. Lastly, DDPG creates target networks for actor network and critic network, respectively, which makes it closer to the supervised learning and significantly enhance the stability of learning.

1.2. Objectives and Contributions

In brief, the current path planning and path tracking-based method cannot easily ensure the ideal parking attitude, especially the perpendicular parking slot, for its narrow width, which requires a higher demand for parking. To solve this problem, a reinforcement learning-based end-to-end parking algorithm is proposed in this paper for perpendicular parking. The main contributions are as follows:

- We innovatively apply DDPG to perpendicular parking so that the vehicle can continuously learn and accumulate experience from considerable parking attempts, learn the optimal steering wheel angle command at different parking slots relative to vehicle, as well as achieve the real "human-like" intelligent parking. Moreover, because it realizes the end-to-end control from the parking slot to the steering wheel angle command, the control errors caused by path tracking are fundamentally avoided;
- Since the parking slot needs to be continuously obtained in the course of learning, we propose a parking slot tracking algorithm, which uses extended Kalman filter (EKF) to fuse the parking slot information with vehicle chassis information to achieve continuous tracking of parking slot;
- Given that the learning network output is hard to converge and it is easy to fall into local optimum in the parking process, several reinforcement learning training methods in terms of parking conditions, e.g., manual guided exploration for accumulating initial experience sequence, control cycle phased setting, and training condition phased setting, are designed. Besides, the well-trained network in the simulation environment is migrated to the real vehicle training.

1.3. Paper Outline

The rest of this paper is organized as follows. In Section 2, our reinforcement learning-based end-to-end parking method is introduced. In Section 3, the experimental results are showed. In Section 4, some discussions are contained. Lastly, this paper is concluded in Section 5.

2. Method

The overview of the proposed method is shown in Figure 1. It primarily includes two modules, parking slot tracking and reinforcement learning-based planning. Parking slot tracking is used to provide continuous position of parking slot for reinforcement learning, and reinforcement learning is adopted to achieve end-to-end planning from the parking slot to steering wheel angle.

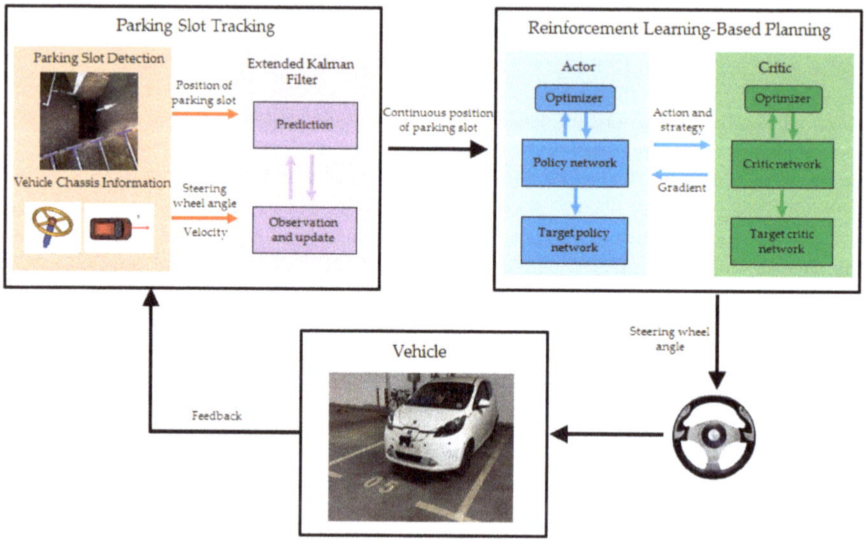

Figure 1. Overview of the reinforcement learning-based end-to-end parking method.

2.1. Parking Slot Tracking

In this section, the parking slot detection is first introduced, followed by the EKF-based parking slot tracking.

2.1.1. Parking Slot Detection

The sensors of surround view parking slot detection system are outfitted with four fisheye cameras in the front, rear, left, and right positions of the vehicle, respectively, with 180° of FOV horizontally and 140° of FOV vertically, as shown in Figure 2a.

The parking slot detection consists of two steps: one is to yield a surround view based on the images taken by the four fisheye cameras; the other is to detect the corner points of parking slots using the surround view. The flow chart is shown in Figure 2b.

For the generation of surround view, first, the distortion parameters of fisheye camera are calculated by Zhang Zhengyou's calibration method [31], and the mapping table T_{UF} from undistorted image coordinate system (CS) to fisheye image CS is yielded. Subsequently, based on the checkerboard calibration site, the homography matrix M_{VU} from vehicle CS to undistorted image CS is calculated using the least square method. Lastly, after confirming the scope and the image size of the surround view, the similarity transformation matrix M_{SV} of surround view CS to vehicle CS is calculated. Four fisheye perspectives are joined into one surround view by the comprehensive mapping table T_{BF} constructed above. The whole process is illustrated in Figure 3.

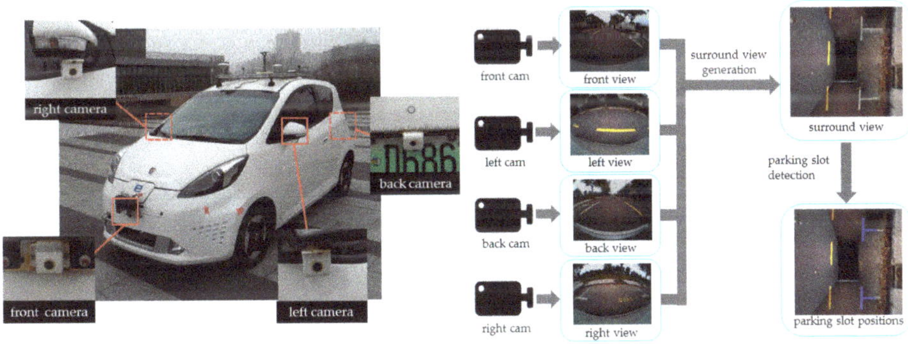

(**a**) (**b**)

Figure 2. Surround view parking slot detection system. (**a**) Test vehicle and camera installation location. (**b**) Surround view generation and parking slot detection.

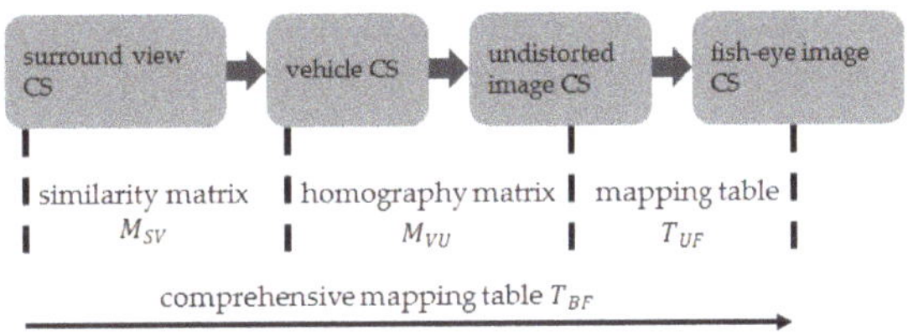

Figure 3. Surround view generation process.

The method proposed by Li et al. [32] is adopted to detect parking slot, which is an AdaBoost-based slot detection method that detects parking slots from surround view. This method is primarily used to detect common "L" and "T" corner points, as shown in Figure 4. The basic principle is to use Adaboost algorithm and decision tree to design a binary classifier to detect corner point patterns. The input of the classifier refers to an image patch, and the output is a Boolean value, indicating whether the input local block is a corner pattern. Because of the limited FOV of surround view, the length of the parking slot can be inferred following some prior rules after the detection of the corner points.

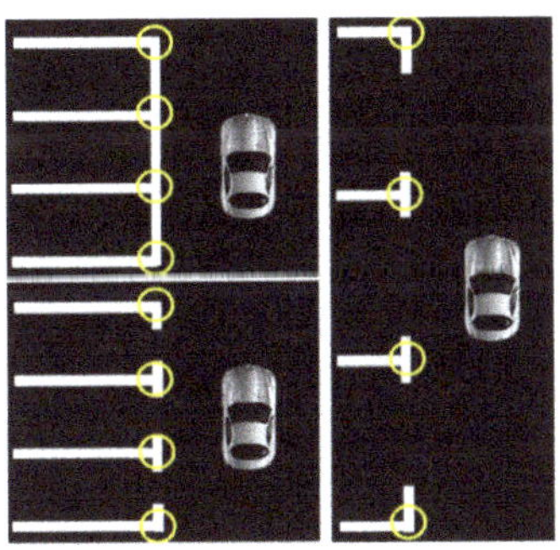

Figure 4. The "L" and "T" corner points.

The above analysis reveals that when detecting the corner points, the parking slot relative to the vehicle can be calculated through coordinate transformation. However, we find that the parking slot is difficult to continuously identify by relying solely on the vision. For instance, the corner points of the parking slot are sometimes not detected during parking due to image distortion, illumination change, occlusion, as well as the limited FOV, as shown in Figure 5.

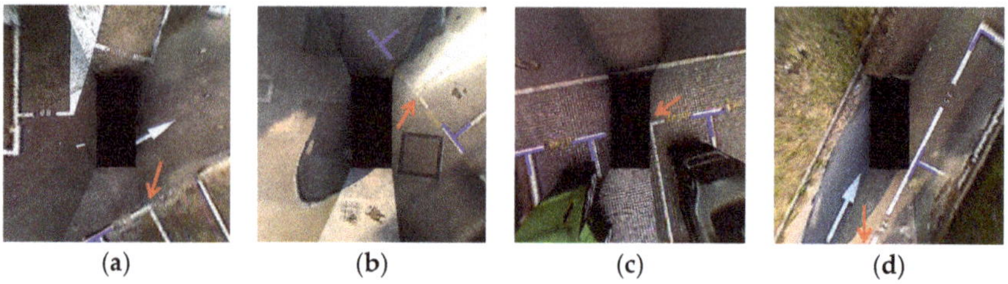

Figure 5. Missed parking slot detections (indicated by red arrows). (**a**) Image distortion. (**b**) Illumination change. (**c**) Occlusion. (**d**) Limited FOV.

2.1.2. EKF-Based Parking Slot Tracking

To track parking slot continuously and accurately, EKF is employed to achieve the fusion of vision and vehicle chassis information. First, we take the position of the corner point of parking slot relative to vehicle as the EKF's observation, building the constraint relationship between the position of vehicle and parking slot in the reference CS, i.e., the EKF's observation model. Second, the vehicle kinematics model is taken as the EKF's motion model, and the steering wheel angle and velocity obtained from vehicle chassis act as the EKF's control input. Lastly, based on EKF's "prediction" and "update" process, the fusion is completed to achieve the maximum posterior estimation of parking slot in the presence of noise.

The definition of CS and parameters is shown in Figure 6, where (x, y), φ and (x_i, y_i) are the vehicle coordinates, vehicle heading angle and the i th corner points (i = 1 and 2 represent the left and right corner point of parking slot, respectively) coordinates in the reference CS, respectively. (x_{vi}, y_{vi}) obtained by parking slot detection system in Section 2.1.1 denotes the coordinates of the i th corner point in the vehicle CS. Its distance from the center of the rear axle of the vehicle is r_i, and its angle relative to the axis of the vehicle CS is θ_i. r_i and θ_i can be derived from the Equation (1).

$$\begin{aligned} \theta_i(k) &= \arctan\left(\frac{y_{vi}(k)}{x_{vi}(k)}\right) \\ r_i(k) &= \sqrt{x_{vi}(k)^2 + y_{vi}(k)^2} \end{aligned} \tag{1}$$

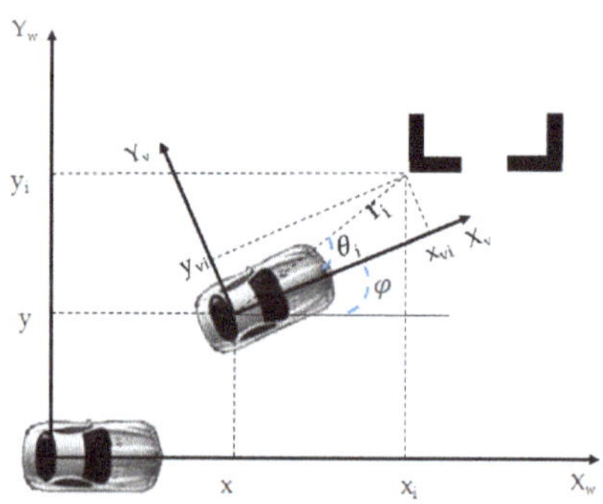

Figure 6. CS and parameter definition.

The state variable of EKF is $X = (x, y, \varphi, x_i, y_i)^T$, and its covariance matrix of the error is denoted as P. The observed variable of the system is expressed as $Z = (r_i, \theta_i)^T$.

The discrete observation model is expressed as Equation (2),

$$Z(k) = h(X(k)) + \vartheta(k)$$
$$\begin{bmatrix} \sqrt{(x_i(k) - x(k))^2 + (y_i(k) - y(k))^2} \\ \arctan\left(\frac{y_i(k)-y(k)}{x_i(k)-x(k)}\right) - \varphi(k) \end{bmatrix} + \vartheta(k) \quad (2)$$

where $\vartheta(k)$ denotes the noise of parking slot detection system, assuming a Gaussian distribution; its covariance matrix is R.

According to the Ackerman steering model of the vehicle, the discrete motion model can be expressed as Equation (3),

$$X(k) = f(X(k-1), U(k)) + w(k)$$
$$= \begin{bmatrix} x(k-1) + Tv(k)cos\varphi(k) \\ y(k-1) + Tv(k)sin\varphi(k) \\ \varphi(k-1) + \frac{Tv(k)\tan(\delta(k)/i_0)}{L} \\ x_i(k-1) \\ y_i(k-1) \end{bmatrix} + w(k) \quad (3)$$

where δ denotes the steer wheel angle; v the velocity; δ and v can be obtained directly from the vehicle chassis; i_0 the steering gear ratio; L the wheelbase; T the period; $w(k)$ the noise of the motion model, assumed to be Gaussian noise, and its covariance matrix is expressed as Q.

EKF can be split into two steps (prediction and update). First, the system state and its error covariance matrix at the k th iteration time are predicted, as expressed in Equation (4),

$$\begin{aligned} \hat{X}(k)^- &= f\left(\hat{X}(k-1), U(k)\right) \\ P(k)^- &= F(k)P(k-1)F(k)^T + Q \end{aligned} \quad (4)$$

where $F(k)$ denotes the Jacobian of function $f(X(k), U(k))$ with respect to $X(k)$.

Subsequently, it is the update process. First, the Kalman gain $K(k)$ is calculated, which is the key to the maximum posteriori estimation of $X(k)$ in the presence of noise, as shown in Equation (5). Second, $X(k)$ and $P(k)$ are updated by $K(k)$, as expressed in Equation (6),

$$K(k) = P(k)^- H(k)^T \left(H(k)P(k)^- H(k)^T + R\right)^{-1} \quad (5)$$

$$\begin{aligned} \hat{X}(k) &= \hat{X}(k)^- + K(k)\left[Z(k) - H(k)\hat{X}(k)^-\right] \\ P(k) &= (I - K(k)H(k))P(k)^- \end{aligned} \quad (6)$$

where $H(k)$ is the Jacobian of function $h(X(k))$ with respect to $X(k)$.

Equation (6) reveals that $K(k)$ helps fuse the vision and vehicle chassis information, and the updated $\hat{X}(k)$ is calculated by $K(k)$ in the presence of noise, satisfying the maximum posterior estimate of $X(k)$. Accordingly, EKF-based fusion is more accurate than relying solely on vision detection. Moreover, $X(k)$ is continuous because the vehicle chassis information continues to be inputted. Thus, the continuous and accurate position of the parking slot relative to the vehicle can be derived from the above.

2.2. Reinforcement Learning-Based Planning

In this section, reinforcement learning is adopted to achieve end-to-end planning from the parking slot to steering wheel angle. We first introduce the appropriate reinforcement learning model for APS, followed by the system settings and training process of DDPG, and finally the improved training measures applied in parking.

2.2.1. Appropriate Reinforcement Learning Model for APS

The basic process of reinforcement learning is a Markov decision-making process, which can be expressed by the quaternion $\{S, A, P, R\}$ composed of state S, action A, state transition probability P and reward R.

When a policy π is executed at time t, cumulative reward G_t can be calculated:

$$G_t = R_t + \gamma R_{t+1} + \gamma^2 R_{t+2} + \cdots = \sum_{k=0} \gamma^k R_{t+k+1} \tag{7}$$

where γ is the discount factor, which is used to reduce the reward weight corresponding to the long-term decision.

The action value function $Q_\pi(s,a)$ is the expectation of the cumulative reward G_t after taking action a at the current state s, as expressed in Equation (8).

$$Q_\pi(s,a) = E_\pi[G_t | S_t = s, A_t = a] \tag{8}$$

The action valued function $Q_\pi(s,a)$ satisfies the Bellman equation (Equation (9)), which transforms the solution of $Q_\pi(s,a)$ into an iterative process of dynamic programming.

$$Q_\pi(s,a) = E_\pi[R_{t+1} + \gamma q_\pi(S_{t+1}, A_{t+1}) | S_t = s, A_t = a] \tag{9}$$

The goal of reinforcement learning is to find an optimal policy for obtaining the maximum $Q_*(s,a)$, as shown in the following equation.

$$Q_*(s,a) = \max_\pi Q_\pi(s,a) \tag{10}$$

According to the different optimization objects, reinforcement learning methods can be divided into value-based method, policy-based method, and Actor-Critic method.

- Value-based method

Q-learning is a basic value-based method. Q-learning first chooses action a according to Q value at the current state s in each step of the cycle (e.g., using $\varepsilon - greedy$ method: $1 - \varepsilon$ probability of selecting action $\underset{a}{\mathrm{argmax}} Q(s,a)$, ε probability of randomly selecting action). After the selected action is taken, the immediate reward R and the next state s' are observed, and then $Q(s,a)$ is updated, as expressed in Equation (11). Repeat the process until the final state is reached,

$$Q(s,a) \leftarrow Q(s,a) + \alpha\left[R + \gamma \max_a Q(s', a) - Q(s,a)\right] \tag{11}$$

where α is the learning rate.

DQN replaces Q-table with DNN with parameter w (Equation (12)) to solve the problem that Q-learning is prone to dimension disasters when state space is high-dimensional.

$$Q_w(s,a) \approx Q(s,a) \tag{12}$$

The updating objective of $Q_w(s,a)$ is to minimize the mean square deviation of the objective value $Q_\pi(s,a)$ and the actual value $Q_w(s,a)$, as shown in Equation (13). If gradient descent method is used, the gradient $\nabla_w J(w)$ of the objective function $J(w)$ relative to the parameter w is first calculated, and then the parameter w changes along the opposite direction of the gradient $\nabla_w J(w)$, as shown in Equation (14),

$$J(w) = E_\pi\Big[(Q_\pi(s,a) - Q_w(s,a))^2\Big] \tag{13}$$

$$w \leftarrow w + \alpha \nabla_w J(w) = w + \alpha[R + \gamma Q_w(s',a') - Q_w(s,a)]\nabla Q_w(s,a) \tag{14}$$

where $Q_\pi(s,a)$ can be solved by temporal-difference method, i.e., $Q_\pi(s,a) = R + \gamma Q_w(s',a')$.

Since the action space of the value-based method is discrete, it is not suitable for the continuous action space of parking control. Though the continuous action space can be discretized, too large discrete spacing will lead to the algorithm not getting the optimal action, and too small discrete spacing will lead to dimension disaster.

- Policy-based method

The policy-based method directly optimizes the policy based on the sampling method, and constantly calculates the gradient $\nabla_\theta J(\theta)$ of the policy expectation reward $J(\theta)$ about the policy parameter θ during the training process, as expressed in Equations (15) and (16),

$$J(\theta) = E_{\pi_\theta}[R] = \sum_{s\in S} d^\pi(s) \sum_{a\in A} \pi_\theta(s,a) R_{s,a} \tag{15}$$

$$\nabla_\theta J(\theta) = \sum_{s\in S} d^\pi(s) \sum_{a\in A} \pi_\theta(s,a) \nabla_\theta log\pi_\theta(s,a) R_{s,a} = E_{\pi_\theta}[\nabla_\theta log\pi_\theta(s,a) R_{s,a}] \tag{16}$$

where $\pi_\theta(s,a)$ is the policy of action selection, which represents the probability of choosing action a under the state s; $d^\pi(s)$ is the static distribution of the state s under the policy π.

If the Monte Carlo policy gradient algorithm is used, the iteration equation of the policy parameter θ is as follows:

$$\theta \leftarrow \theta + \alpha \nabla_\theta \; log\pi_\theta(s,a) v \tag{17}$$

where v is equal to the cumulative reward G_t of the current step minus the average cumulative reward $(1/T)\sum_{t=1}^{T} G_t$, i.e., if the action gets better evaluation than before and v is positive, it will increase the probability of this action being selected.

Though the policy-based method is applicable to high-dimensional continuous action spaces, each iterative step should sample a batch sequence to update the parameters, resulting in a large variance of the policy gradient estimation and making it easy to fall into local optimum.

- Actor-Critic method

The Actor-Critic method, which combines the value-based method and the policy-based method, consists of two updating processes: The critic network is responsible for updating the network parameters of the action value function, observing the action and reward, and evaluating the policy. The actor network is responsible for updating the actor network parameters according to the guidance of the critic networks. By introducing the value function as the evaluation criterion in the policy search, the loss of sequential difference about the reward can be minimized so that the variance of the policy gradient estimation can be reduced effectively.

Although Actor-Critic method can realize the learning of continuous action space and can reduce the variance of the policy gradient estimation, it only has one policy network and one critic network, which easily leads to unstable training.

In order to solve this problem, DDPG constructs the target network with parameter θ', which is used to calculate the target value. The target network is adopted to track the actor network and critic network slowly to update the parameter θ', as expressed in Equation (18). This means that the target value is limited to slow change, which greatly improves the stability of learning. This improvement brings the reinforcement learning closer to supervised learning.

$$\theta' \leftarrow \tau\theta + (1-\tau)\theta', \tau \leq 1 \tag{18}$$

In addition, a challenge in reinforcement learning using neural networks is that most optimization algorithms assume that the samples are independent and identically distributed. Obviously, this assumption is no longer valid when the samples are sequentially explored in the environment. DDPG uses a finite size experience pool to cut off the data correlation. The experience sequence is sampled from the environment according to the exploratory strategy, and the tuples are stored in the experience pool. When the experience pool is full, discard the oldest sample. At each time step, the actor network and critic network are updated by uniformly sampling small batches from the experience pool.

As mentioned above, we believe that DDPG is applicable to APS for the following reasons: First, Actor-Critic architecture can realize the learning of continuous action space (since the steering wheel angle for APS is a continuous action). Second, introducing the value function as the evaluation criterion in the policy search can reduce the variance of the policy gradient estimation, which is more efficient. Third, DDPG creates target networks for actor network and critic network, respectively, which makes it closer to the supervised learning and significantly enhances the stability of learning. Lastly, the experience pool is adopted to cut off the data correlation.

2.2.2. System Settings of DDPG

In this section, we mainly introduce the system settings of DDPG, including input and output, reward and network.

- Input and output

The input state s of DDPG refers to the parking slot relative to the vehicle, i.e., the coordinates of the four corner points in the vehicle CS. The output action a of DDPG is the steering wheel angle, capable of controlling the vehicle backing into the parking slot.

- Reward

At present, the reward of reinforcement learning mainly depends on expert experience. The goal of proposed algorithm is to make the vehicle parked in the middle of the parking slot, avoiding inclination, deviation, and line-pressing. We take these factors into consideration and through a large number of simulation training get a better reward setting as shown in Equations (19) to (24).

The total reward R consists of three parts, as expressed in Equation (19). The first part R_{cp} considers the reward that the vehicle tends to the center of the parking slot, and vehicle longitudinal axis parallels to the parking slot. The second part P_l and the last part P_d consider the punishment of line-pressing and the punishment of the vehicle's deviation to one side of the parking slot, respectively.

$$R = R_{cp} + P_l + P_d \tag{19}$$

As shown in Figure 7a, when the rear axle center of the vehicle is outside the outer line of the parking slot, P_{cp} is defined as:

$$\begin{aligned} P_{cp} &= P_c + P_p \\ &= \left(5 - 5\left(\tfrac{1}{2}abs\left(Y_{p_0} + Y_{p_1}\right) + \tfrac{1}{2}abs\left(Y_{p_2} + Y_{p_3}\right)\right)\right) + \left(5 - 5abs\left(\frac{Y_{p0} - Y_{p3}}{X_{p0} - X_{p3}}\right)\right) \end{aligned} \tag{20}$$

where P_c denotes the reward for the vehicle to be close to the center of the parking slot; P_p the reward for the vehicle's longitudinal axis parallel to the parking slot; (X, Y) the coordinates of the corner points ($P_0 - P_3$ in Figure 7) of the parking slot in the vehicle CS.

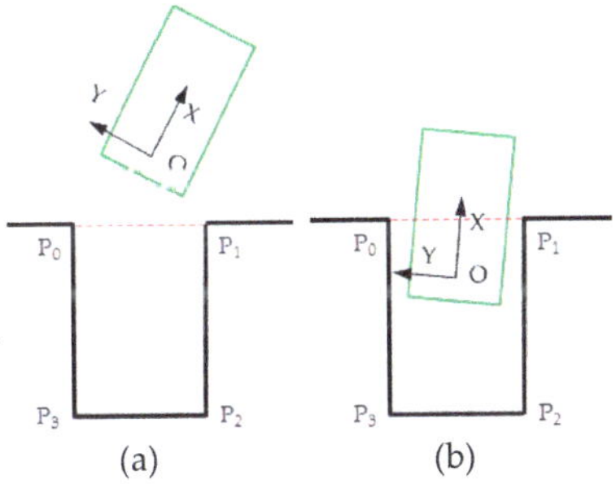

Figure 7. Different parking stages. (**a**)The rear axle center of the vehicle is outside the outer line of the parking slot. (**b**) The rear axle center of the vehicle crosses the outer line of the parking slot.

As shown in Figure 7b, when the rear axle center of the vehicle crosses the outer line of the parking slot, R_{cp} is defined as:

$$R_{cp} = min\{R_c, R_p\} + \frac{1}{2}max\{R_c, R_p\} + R_n \quad (21)$$

The reason why the larger values in R_c and R_p are reduced is to prevent falling into one better and conceal the worse performance of the other, so more attention is paid to the worse performance of the two. Besides, when the vehicle enters the parking slot, we pay more attention to the parallelism between the vehicle and the parking slot. Accordingly, a reward R_n is set, which is expressed in Equation (22). It is limited to a value of less than 10 to avoid covering other rewards.

$$R_n = abs(\frac{1}{10}min\left\{1/abs\left(\frac{Y_{p_0} - Y_{p_3}}{X_{p_0} - X_{p_3}}\right) + eps\right), 100\}))\quad (22)$$

If any outer contour boundary of the vehicle intersects with the parking slot lines, it is considered a line-pressing, P_l is defined as:

$$P_l = -10 \quad (23)$$

If the vehicle is biased towards one side of the parking slot, P_d is defined as:

$$P_d = -10 \quad (24)$$

- Network

The input of our network is not the image but the result of the parking slot detection. Thus, the deep neural networks are not necessarily required to be used. For actor network and critic network of DDPG, we just use back propagation neural network in this paper. Besides, we build the target network with an identical structure but different parameters for actor network and critic network and the relationship between network parameter θ and its target network parameters θ' is $\theta' \leftarrow \tau\theta + (1-\tau)\theta', \tau \ll 1$, significantly enhancing the stability of learning.

The structure of actor network and target actor network is illustrated in Figure 8. The number of nodes for the coordinates of four corner points is 8, and the number of nodes in hidden layer la_1 and hidden layer la_2 is 100 and 200, respectively. Then, the number of nodes for steering wheel angle is 1. All activation functions are Rectified Linear Unit (ReLU).

The structure of critic network and target critic network is shown in Figure 9. The number of nodes for the coordinates of four corner points is 8, and the number of nodes in hidden layer ls_1 and hidden layer ls_2 are both 100. The number of nodes for steering wheel angle is 1, and the number of nodes in hidden layer lc_1 is 200. Subsequently, the number of nodes in hidden layer l_1 and hidden layer l_2 is 300 and 200, respectively. Lastly, the number of nodes for reward is l. All activation functions are also ReLU.

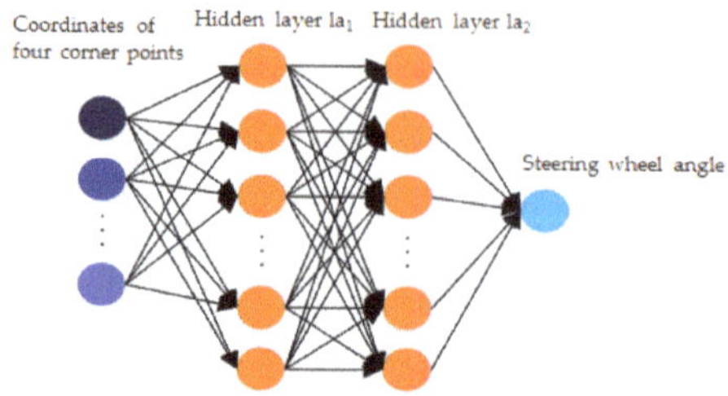

Figure 8. The structure of actor network and target actor network.

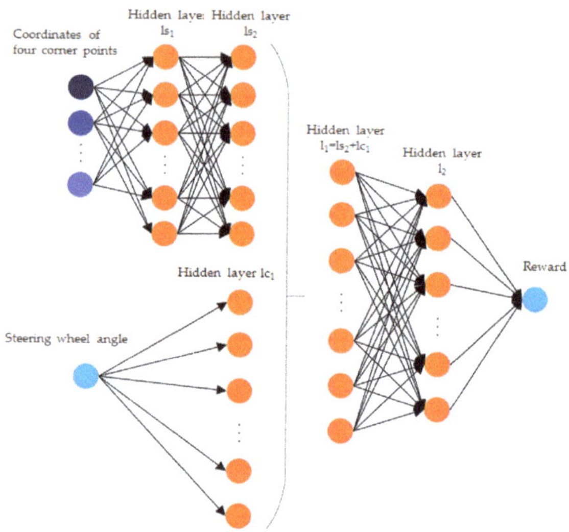

Figure 9. The structure of critic network and target critic network.

2.2.3. Training Process of DDPG

First, the training is conducted in the simulation environment. The simulation platform is shown in Figure 10. We use PreScan, MATLAB/Simulink, and Python in sequence to build the parking environment, build the vehicle model, and then run our algorithm, respectively. After the simulation training, the well-trained network migrates to the real vehicle training. In Section 3, the real vehicle platform will be introduced.

The training architecture of DDPG is shown in Figure 11, and the corresponding training process is shown in Algorithm 1.

Algorithm 1: DDPG Algorithm

Randomly initialize critic network $Q(s, a/\theta^Q)$ and actor network $\mu(s/\theta^\mu)$ with parameters θ^Q and θ^μ
Initialize target critic network Q' and target actor network μ' with parameters $\theta^{Q'}$ and $\theta^{\mu'}$
Set up a replay memory buffer (experience pool) for the sampling experience sequence with the total number of buffers M
for each episode:
Initialize a random process for action exploration
Receive initial state s_1
for $t = 1, T$:

1. Select action a_t according to the current policy and exploration noise:

$$a_t = \mu(s_t/\theta^\mu) + N_t$$

   where N_t denotes Gaussian noise.
2. Execute action a_t and obtain the reward r_t and the next state s_{t+1}
3. Store the transition (s_t, a_t, r_t, s_{t+1}) in the experience pool
4. Randomly sample N experience sequences from experience pool as a mini-batch training data for the critic network and actor network
5. This step is adopted to update the parameters of the critic network. With a method similar to supervised learning, loss is defined as:

$$L = \frac{1}{N}\sum_i \left(y_i - Q(s_i, a_i/\theta^Q)\right)^2$$

   where y_i is calculated based on μ' and Q':

$$y_i = R_i + \gamma Q'\left(s_{i+1}, \mu'\left(s_{i+1}/\theta^{\mu'}\right)/\theta^{Q'}\right)$$

   Calculate the gradient $\nabla_{\theta^Q} L$, and then update θ^Q with gradient descent method:

$$\theta^Q = \theta^Q + \alpha\, \nabla_{\theta^Q} L$$

   where α is the learning rate.
6. After the critic network is updated, the actor network is updated using the policy gradient method:

$$\nabla_{\theta^\mu} J \approx \frac{1}{N}\sum_i \nabla_a Q(s, a/\theta^Q)/_{s=s_i, a=\mu(s_i)} \nabla_{\theta^\mu} \mu(s/\theta^\mu)/_{s_i}$$

   Update θ^μ with $\nabla_{\theta^\mu} J$ based on gradient descent method:

$$\theta^\mu = \theta^\mu + \alpha \nabla_{\theta^\mu} J$$

7. Update the target networks:

$$\theta^{Q'} \leftarrow \tau\theta^Q + (1-\tau)\theta^{Q'}$$
$$\theta^{\mu'} \leftarrow \tau\theta^\mu + (1-\tau)\theta^{\mu'}$$

end for
end for

2.2.4. Improved Training Measures Applied in Parking

Given that the learning network output is hard to converge and it is easy to fall into local optimum in the parking process, several reinforcement learning training methods in terms of parking conditions are designed.

- Manual guided exploration for accumulating initial experience sequence

Before the training of network, exploration should be conducted to gain the initial experience sequence database. In the initialization stage, instead of random exploration, we conduct manual guidance on exploration, which is realized by setting a series of control commands for the initial parking slot relative to the vehicle (the driver's control sequence is collected in the simulation or real vehicle test). Based on the manual control commands, the appropriate noise is added to give the model a better space for policy exploration and trial-and-error. In such a way, compared with random exploration, considerable experience sequences will receive higher rewards, which can make the training converge to excellent policy faster. The reward can converge eventually with manual guided exploration, as shown in Figure 12.

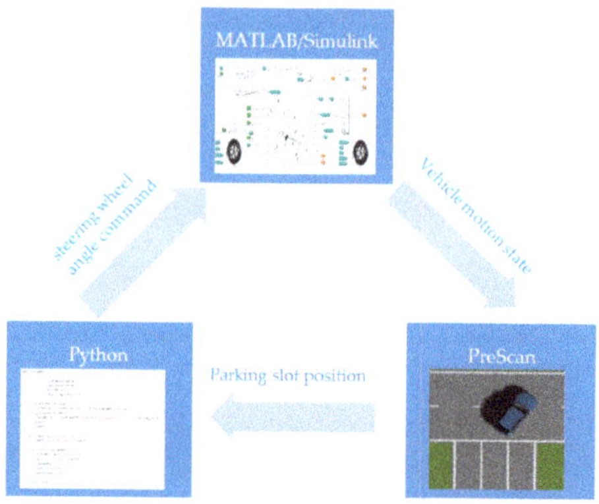

Figure 10. Simulation platform.

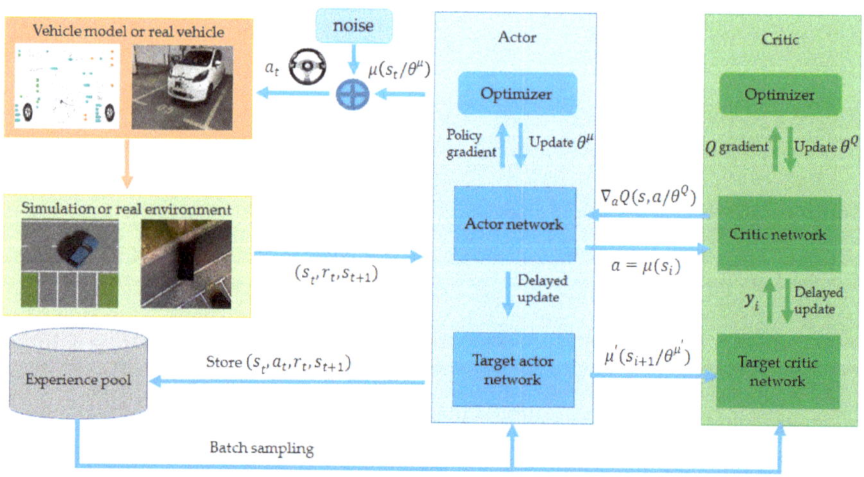

Figure 11. The training architecture of DDPG.

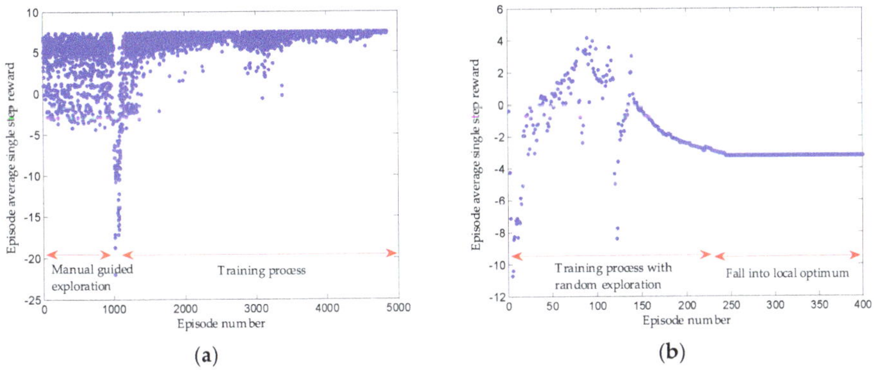

Figure 12. (**a**) Training process by manual guided exploration. (**b**) Training process by random exploration.

- Control cycle phased setting

Given that the vehicle model has inertia delay characteristics, it is found that if the period of steering wheel angle change is too small, it will cause the loss of Markov characteristics of some collected experience sequences. The state of the current cycle of the vehicle depends on both the state of the previous cycle and the action taken. To weaken this adverse effect, the first round of training sets the control cycle to 1000 ms. In such a way, the actions executed in the current cycle will retain sufficient execution time, which will be the major factor affecting the state of the next cycle and can be approximated to Markov decision-making process. When the network converges to the optimum, the training control cycle of the following training can be reduced, which can make the control cycle closer to the actual situation and achieve better results. Figure 13a shows that the 1000 ms control cycle is first trained, then the 100 ms control cycle is trained, and lastly the reward lastly converges. Figure 13b suggests that the reward does not converge if we start with a 100 ms control cycle training directly.

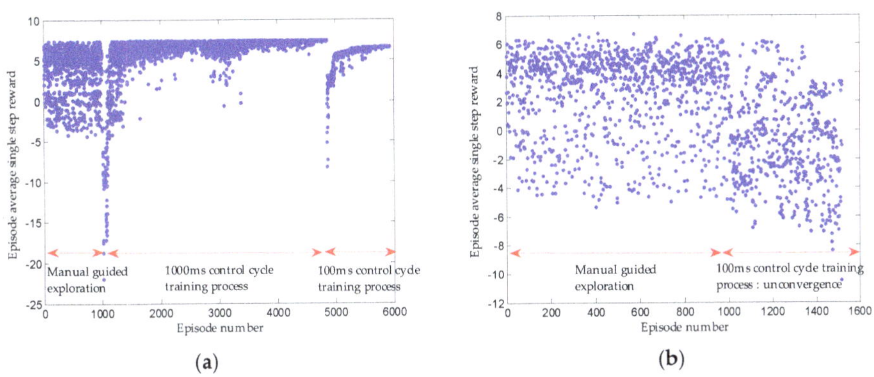

Figure 13. (**a**) Training process first by 1000 ms control cycle and followed by 100 ms control cycle. (**b**) Training process only by 100 ms control cycle.

- Training condition phased setting

Usually the perpendicular parking can be split into two steps, as shown in Figure 14. Just like human parking, the "step two" plays a major role in the final parking attitude. Thus, we currently primarily apply reinforcement learning to the "step two".

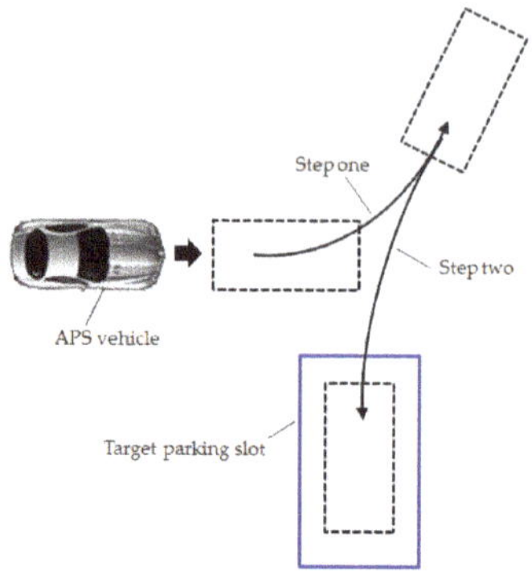

Figure 14. Common perpendicular parking process.

Since there will be different initial angles of "step two" between the vehicle and the parking slot, we first train at 30° and then expand the initial angle to 0° to 90° to continue training. Figure 15 suggests that based on 30° well-trained network, the networks between 0° to 90° can converge quickly, i.e., the 30° well-trained network has ideal generalization ability. Since the initial angle of the vehicle relative to the parking slot is different in different episodes, the sequence of states experienced in each episode is different, so the average single step reward is also different.

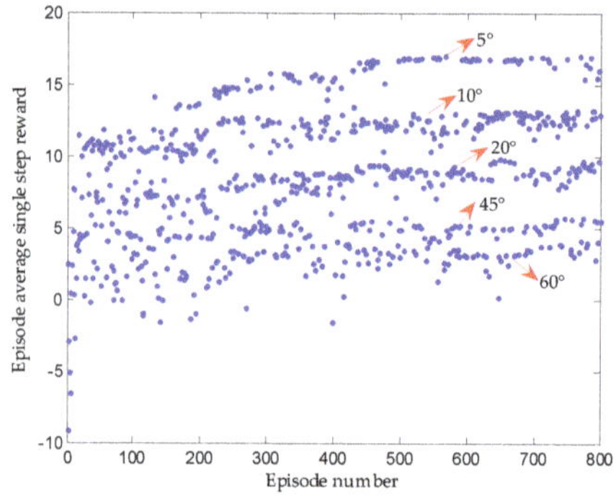

Figure 15. Extended training of other initial angles based on 30° well-trained network.

- Real vehicle training migration

Because the real vehicle training takes a lot of manpower, time and resources, it is better to train in the simulation environment and then transfer it to the real vehicle. Since the sensor model and vehicle model used in simulation will differ from the real vehicle, the same control command may produce different observation results. Accordingly, the real vehicle should be continuously trained based on well-trained network in simulation. Figure 16 reveals that the result of real vehicle migration training is ideal.

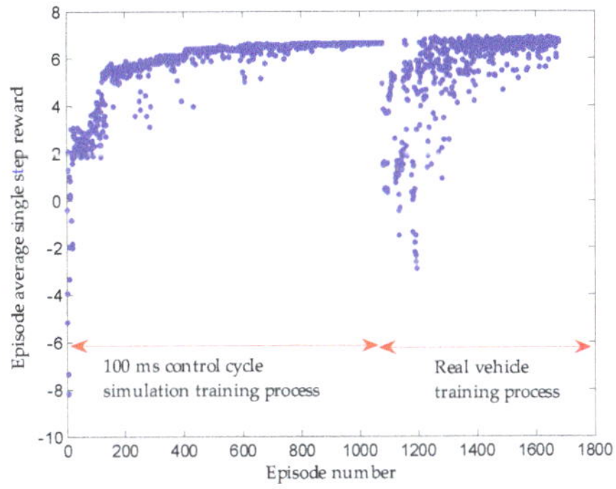

Figure 16. Real vehicle training migration.

3. Experimental Results

After the above training, we can ascertain the performance of the trained algorithm. This section shows the experimental platform, experimental scenes, and results.

3.1. Experimental Platform

The experimental platform is refitted from Rongwei E50 pure electric vehicle (Figure 17). Four fisheye cameras act as sensors for parking slot detection. The algorithm running platform is an industrial computer (i5 processor, 8G memory, 128G solid-state hard disk). Chassis control and information exchange is performed in the vehicle control unit, i.e., the controller of vehicle chassis. The RT3000 navigation system is employed to acquire the position information of the vehicle. During the test, notebook computer is employed to record data.

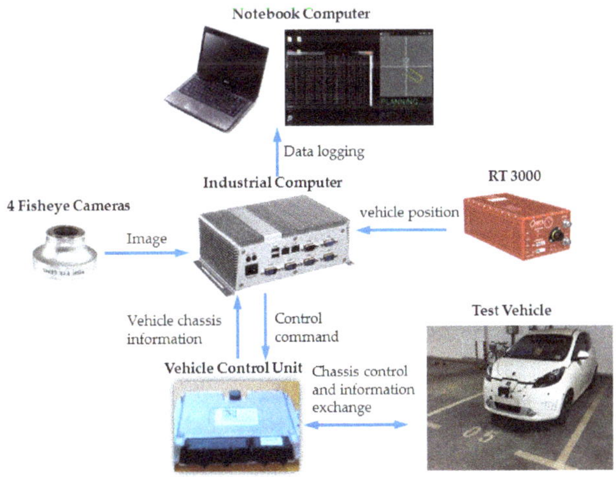

Figure 17. Experimental platform.

3.2. Experimental Scenes

To ascertain the performance of the proposed algorithm in the "step two" perpendicular parking, we choose three parking scenes with initial angles of 60°, 45°, and 30° between the vehicle and the parking slot, which are common "step two" scenes. Figure 18 illustrates the experimental scenes expressed in the surround view. The blue marking points represent the target parking slots; the width of these parking slots ranges from 2.4 m to 2.44 m and the length is between 5.6 m and 5.8 m, which basically meets the test requirements of BS ISO 16787-2016. As described in Section 2.1.1, since the FOV of surround view cannot cover the entire parking slot, only the nearby corner points can be detected, i.e., the width of the parking slot can be detected, and the length can only be inferred from priori rules.

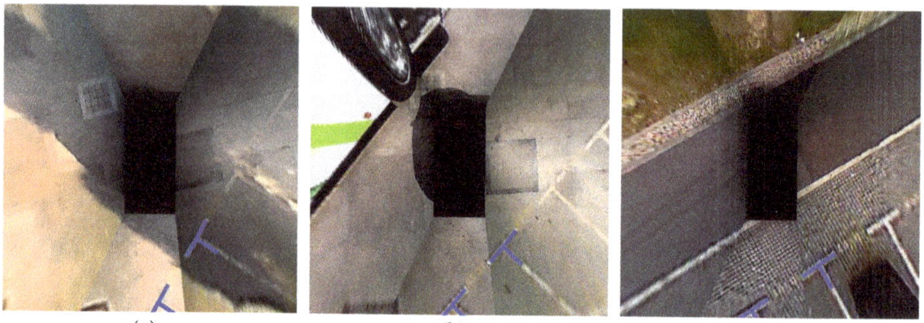

(a) (b) (c)

Figure 18. Experimental scenes. (**a**), (**b**) and (**c**) present parking scenes with initial angles of 60°, 45°, and 30°, respectively.

Three parking methods are adopted in the experiment: geometric method-based path planning with PID-based path tracking, geometric method-based path planning with SMC-based path tracking, and reinforcement learning-based end-to-end parking. The first two represent the current mainstream parking methods, and the last one represents the method used in this paper. Each parking method is at the same starting point and reversed at the same speed (4km/h).

Subsequently, the parking performances of different parking methods are compared. According to BS ISO 16787-2016, the inclination angle of the vehicle with respect to the parking slot, the deviation between the four tire contact points of the vehicle and the parking slot, and the deviation between the rear of the vehicle and the parking slot are measured. The measurement parameters are presented in Figure 19.

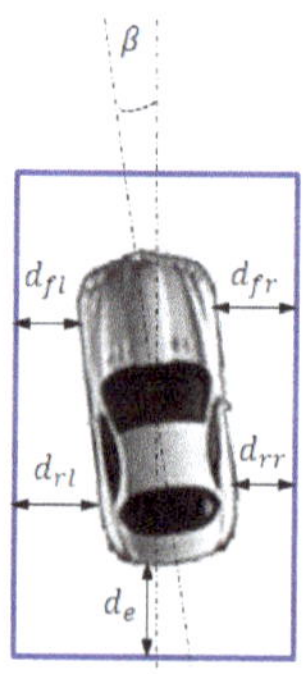

Figure 19. Measurement of parking attitude.

3.3. Results

The experimental results of 60° perpendicular parking are presented in Figure 20. Figure 20b shows that only planned path can ensure the ideal parking attitude, whereas the two path tracking methods (PID and SMC) cannot completely track the planned parking path. The existing control error causes the final vehicle to deviate from the ideal parking attitude, as shown in Figure 20b,c. Figure 20b also shows that the parking performance of reinforcement learning is better than those of the other two methods. Besides, the changes of the parking slot in the vehicle CS are recorded in the case of only visual detection and parking slot tracking in the experiment of reinforcement learning, as shown in Figure 20d. It is suggested that visual detection has missed detection, and it cannot provide continuous parking slot for reinforcement learning. Thus, the test cannot be performed normally. However, the parking slot tracking has not missed detection.

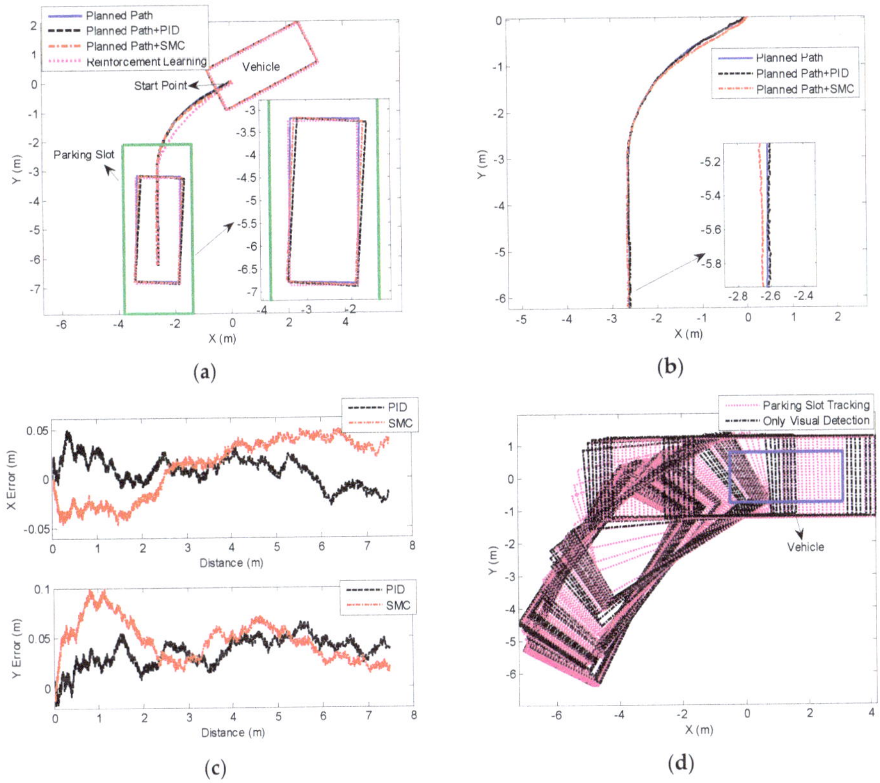

Figure 20. Experimental results of 60° perpendicular parking. (**a**) Parking performance of different methods; (**b**) and (**c**) present the control effect and error of different path tracking methods, respectively. (**d**) Parking slot detection and tracking in the parking process.

The experimental data corresponding to Figure 20 is listed in Table 1. This table suggests that reinforcement learning can achieve an inclination angle of –0.747°, satisfying the requirements of the BS ISO 16787-2016 (≤±3°). Moreover, these deviations are relatively uniform, satisfying the requirements of the BS ISO 16787-2016 (>0.1 m). As mentioned above, only planned path can ensure that the ideal parking and inclination angle and deviation meet the requirements of the standard. However, when path tracking is practically performed, these deviations of path planning with PID and path planning with SMC are not uniform, and the inclination angles are –3.638° and –3.126°, respectively, which do not satisfy the requirements. These two path tracking methods have errors of more than 0.02 m in both X and Y directions. Lastly, it is suggested that the loss rate of visual detection is 37.35%, and that of the parking slot tracking reaches 0%.

Table 1. Experimental data of 60° perpendicular parking.

	Planned Path	Planned Path+PID	Planned Path+SMC	Reinforcement Learning
Inclination angle β (°)	−1.051	−3.638	−3.126	−0.747
Deviation d_{fr} (m)	0.423	0.304	0.379	0.457
Deviation d_{fl} (m)	0.468	0.589	0.514	0.463
Deviation d_{rr} (m)	0.465	0.45	0.504	0.487
Deviation d_{rl} (m)	0.425	0.443	0.388	0.434
Deviation d_e (m)	1.087	1.016	1.047	1.053
X average error (m)	\	0.021	0.028	\
Y average error (m)	\	0.033	0.048	\
Loss rate of visual detection (%)	\	\	\	37.35
Loss rate of parking slot tracking (%)	\	\	\	0

The experimental results of perpendicular parking at 45° and 30° initial angle are shown in Figures 21 and 22. On the whole, the results are consistent with the 60° test. The parking performance of reinforcement learning is obviously superior over those of the other two methods, suggesting that our algorithm can adapt to parking scenario with different initial angles. Likewise, both PID and SMC have control errors, making it unlikely for the vehicle to track the parking path accurately, and eventually the vehicle has inclination angle and uniform deviation.

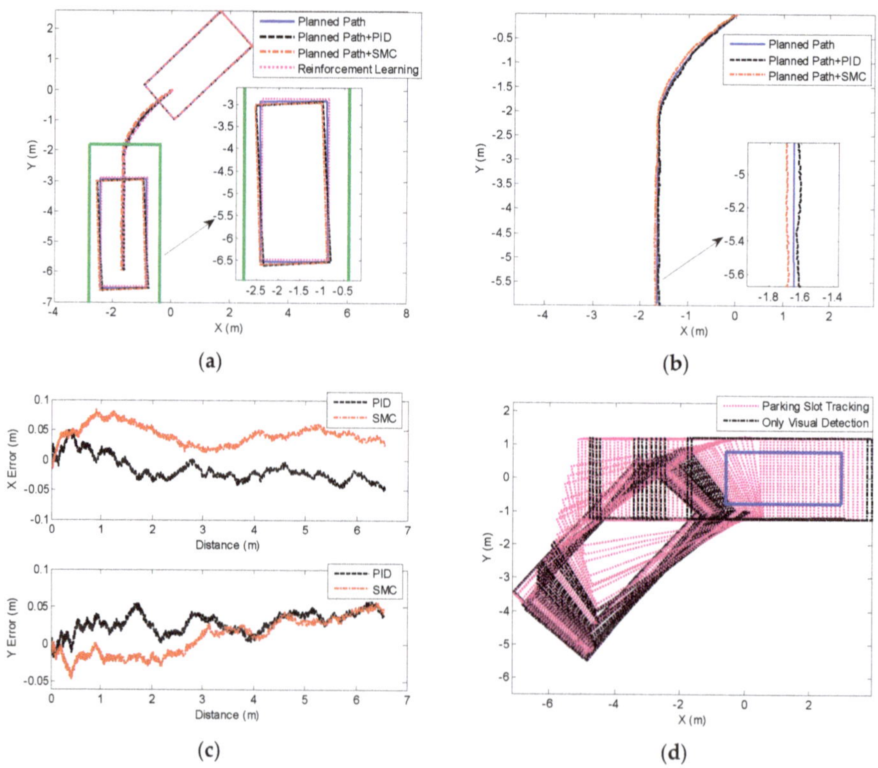

Figure 21. Experimental results of 45° perpendicular parking. (**a**) Parking performance of different methods; (**b**) and (**c**) represent the control effect and error of different path tracking methods, respectively. (**d**) Parking slot detection and tracking in the parking process.

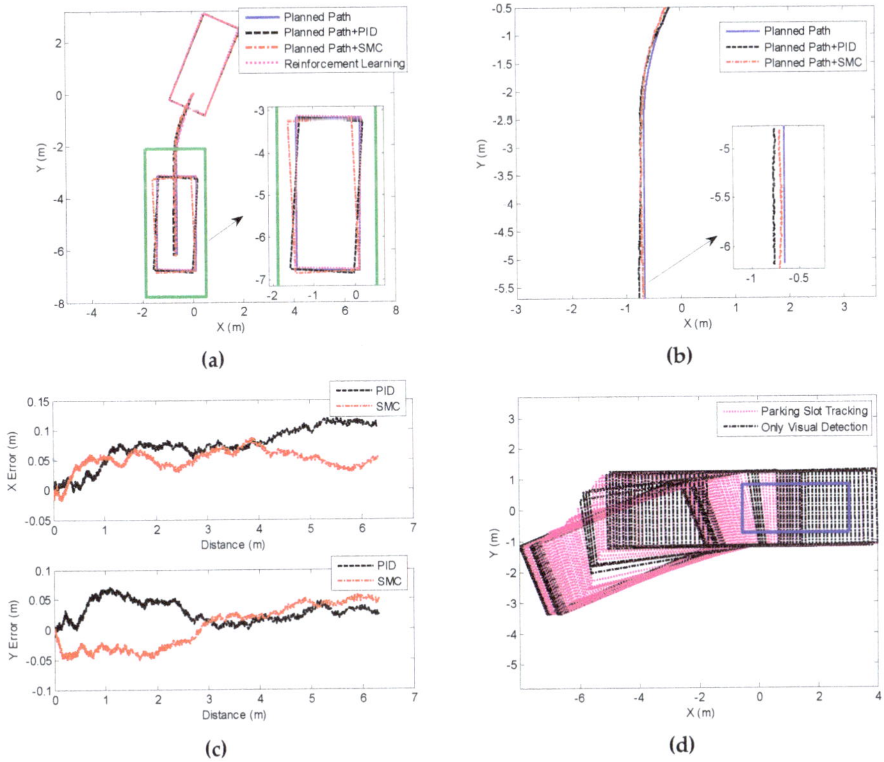

Figure 22. Experimental results of 30° perpendicular parking. (**a**) Parking performance of different methods; (**b**) and (**c**) present the control effect and error of different path tracking methods, respectively. (**d**) Parking slot detection and tracking in the parking process.

Tables 2 and 3 suggest that reinforcement learning can achieve an inclination angle below 1°, satisfying the standard requirements. The other two methods have large inclination angle due to control error, especially the PID exceeding 3°. The path tracking errors of these two methods in X and Y directions are basically above 0.02 m. Besides, the loss rate of visual detection in these two scenarios reaches over 30%, and the parking slot tracking ensures that the position of the target parking slot can be continuously achieved.

Table 2. Experimental data of 45° perpendicular parking.

	Planned Path	Planned Path+PID	Planned Path+SMC	Reinforcement Learning
Inclination angle β (°)	0.313	3.088	2.011	−0.573
Deviation d_{fr} (m)	0.493	0.557	0.59	0.438
Deviation d_{fl} (m)	0.377	0.315	0.281	0.436
Deviation d_{rr} (m)	0.48	0.433	0.509	0.461
Deviation d_{rl} (m)	0.391	0.439	0.361	0.413
Deviation d_e (m)	0.872	0.788	0.795	0.918
X average error (m)	\	0.032	0.042	\
Y average error (m)	\	0.024	0.019	\
Loss rate of visual detection (%)	\	\	\	43.68
Loss rate of parking slot tracking (%)	\	\	\	0

Table 3. Experimental data of 30° perpendicular parking.

	Planned Path	Planned Path+PID	Planned Path+SMC	Reinforcement Learning
Inclination angle β (°)	−0.223	−3.782	2.416	−1.02
Deviation d_{fr} (m)	0.394	0.363	0.552	0.376
Deviation d_{fl} (m)	0.456	0.49	0.299	0.474
Deviation d_{rr} (m)	0.403	0.515	0.455	0.417
Deviation d_{rl} (m)	0.447	0.339	0.396	0.434
Deviation d_e (m)	1.031	0.952	0.948	1.056
X average error (m)	\	0.056	0.043	\
Y average error (m)	\	0.028	0.031	\
Loss rate of visual detection (%)	\	\	\	31.48
Loss rate of parking slot tracking (%)	\	\	\	0

4. Discussion

The above experimental results reveal that the existing mainstream parking methods of path planning with path tracking can basically park the vehicle into the parking slot, whereas the final inclination angle of the vehicle does not meet the strict requirements of the standard. This method is feasible for some wide parking slots. However, with the increasing number of vehicles, the design of parking slot will become narrower and narrower. Thus, the accuracy of parking should be enhanced. Besides, we can also see that it is not difficult to plan an ideal parking path according to the parking slot. However, due to the nonlinear dynamic characteristics of the vehicle, path tracking will inevitably produce control errors that cause the vehicle to deviate from the planned path, thereby resulting in inclination angle and uniform deviation of the parking attitude.

The reinforcement learning-based end-to-end planning method can not only achieve the end-to-end parking from parking slot to steering wheel angle, avoiding errors caused by path tracking, but also learn the best steering wheel angle through a lot of training. Thus, the reinforcement learning-based end-to-end planning can achieve better parking attitude. Besides, because we have fused the vision and vehicle chassis information, we can continuously get the position of parking slot to ensure the normal training and testing of reinforcement learning.

However, future research can still make some improvements: (1) The reward setting of this article is obtained by artificial setting and experimental adjustment. Though the final effect converges to an ideal level, it cannot be proved that it is the optimal reward setting. Accordingly, we will consider the method of inverse reinforcement learning [33,34] to optimize the reward. (2) In this paper, the reinforcement learning-based parking only has the function of reversing (e.g., "step two" in Figure 14), and it cannot automatically adjust the gear forward and backward. If the vehicle needs to judge the gear, we will consider selecting the Long Short-Term Memory (LSTM) network [35].

5. Conclusions

In this study, we innovatively adopt reinforcement learning to perpendicular parking so that the vehicle can continuously learn and accumulate experience from considerable parking attempts, learn the command of the optimal steering wheel angle at different parking slots relative to vehicle, as well as achieve real, "human-like" intelligent parking. Moreover, such end-to-end planning can avoid errors caused by path tracking. Besides, to ensure that the parking slot can be obtained continuously in the course of learning, a parking slot tracking algorithm is proposed based on fusion of vision and vehicle chassis information. Besides, since the learning network output is hard to converge and it is easy to fall into local optimum in the parking process, several reinforcement learning training methods in terms of parking conditions are designed (e.g., manual guided exploration for accumulating initial experience sequence, control cycle phased setting, and training condition phased setting). Lastly, the well-trained network in the simulation environment is migrated to the real vehicle training.

In the subsequent study, on one hand, inverse reinforcement learning will be used to set rewards to ensure optimal reward settings; on the other hand, the LSTM network will be used to achieve gear adjustment in the parking process.

Author Contributions: Conceptualization, L.X., P.Z. and Z.Y.; funding acquisition, Z.Y.; investigation, P.Z. and P.F.; methodology, P.Z. and L.X.; software, P.Z., S.Y.; validation, P.Z., P.F., S.Y., J.Y. and Y.Z.; resources, L.X. and Z.Y.; supervision, X.L. and Z.Y.; writing—original draft, P.Z.; writing—review and editing, P.Z. and L.X.

Funding: This research is supported by the National Key Research and Development Program of China (grant no.2016YFB0100901) and Shanghai Scientific Research Project (grant no. 17DZ1100202).

Conflicts of Interest: The authors declare no conflict of interest.

References

1. British Standards Institution. BS ISO 16787:2016. Intelligent transport systems-Assisted Parking System (APS)-Performance requirements and test procedures. BSI Standards Limited: UK, 2016. Available online: https://www.iso.org/standard/63626.html (accessed on 10 September 2019).
2. Fraichard, T.; Scheuer, A. From reeds and shepp's to continuous-curvature paths. *IEEE Trans. Robot.* **2004**, *6*, 1025–1035. [CrossRef]
3. Gómez-Bravo, F.; Cuesta, F.; Ollero, A. Parallel and diagonal parking in nonholonomic autonomous vehicles. *Eng. Appl. Artif. Intell.* **2001**, *14*, 419–434. [CrossRef]
4. Lini, G.; Piazzi, A.; Consolini, L. Multi-optimization of η3-splines for autonomous parking. In Proceedings of the 50th IEEE Conference of Decision and Control/European Control Conference, Orlando, FL, USA, 12–15 December 2011; pp. 6367–6372.
5. Vorobieva, H.; Minoiu-Enache, N.; Glaser, S.; Mammar, S. Geometric continuous-curvature path planning for automatic parallel parking. In Proceedings of the 2013 IEEE 10th International Conference on Networking, Sensing and Control, Evry, France, 10–12 April 2013; pp. 418–423.
6. Vorobieva, H.; Glaser, S.; Minoiu-Enache, N.; Mammar, S. Automatic parallel parking in tiny spots: Path planning and control. *IEEE Trans. Intell. Transp. Syst.* **2015**, *16*, 396–410. [CrossRef]
7. Han, L.; Do, Q.H.; Mita, S. Unified path planner for parking an autonomous vehicle based on RRT. In Proceedings of the 2011 IEEE International Conference on Robotics and Automation, Shanghai, China, 9–13 May 2011; pp. 5622–5627.
8. Zheng, K.; Liu, S. RRT based path planning for autonomous parking of vehicle. In Proceedings of the 2018 IEEE 7th Data Driven Control and Learning Systems Conference, Enshi, China, 25–27 May 2018; pp. 627–632.
9. Li, B.; Shao, Z. A unified motion planning method for parking an autonomous vehicle in the presence of irregularly placed obstacles. *Knowledge-Based Syst.* **2015**, *86*, 11–20. [CrossRef]
10. Takei, R.; Tsai, R. Optimal trajectories of curvature constrained motion in the Hamilton-jacobi formulation. *J. Sci. Comput.* **2013**, *54*, 622–644. [CrossRef]
11. Micelli, P.; Consolini, L.; Locatelli, M. Path planning with limited numbers of maneuvers for automatic guided vehicles: An optimization-based approach. In Proceedings of the 2017 25th Mediterranean Conference on Control and Automation, Valletta, Malta, 3–6 July 2017; pp. 204–209.
12. Roald, A.L. Path planning for vehicle motion control using numerical optimization methods. Master's Thesis, Norwegian University of Science and Technology, Trondheim, Norway, 2015.
13. Thrun, S.; Montemerlo, M.; Dahlkamp, H.; Stavens, D. Stanley: The robot that won the DARPA Grand Challenge. *J. Field Robot.* **2006**, *23*, 661–692. [CrossRef]
14. Park, H.; Ahn, K.; Park, M.; Lee, S. Study on robust lateral controller for differential GPS-based autonomous vehicles. *Int. J. Precis. Eng. Manuf.* **2018**, *19*, 367–376. [CrossRef]
15. Mvemba, P.; Lay-Ekuakille, A.; Kidiamboko, S.; Rahman, M. An embedded beamformer for a PID-based trajectory sensing for an autonomous vehicle. *Metrol. Meas. Syst.* **2018**, *25*, 561–575.
16. He, X.; Liu, Y.; Lv, C.; Ji, X.; Liu, Y. Emergency steering control of autonomous vehicle for collision avoidance and stabilization. *Veh. Syst. Dyn.* **2019**, *57*, 1163–1187. [CrossRef]
17. Wu, Y.; Wang, L.; Zhang, J.; Li, F. Path following control of autonomous ground vehicle based on nonsingular terminal sliding mode and active disturbance rejection control. *IEEE Trans. Veh. Technol.* **2019**, *68*, 6379–6390. [CrossRef]

18. Samuel, M.; Maziah, M.; Hussien, M.; Godi, N. Control of autonomous vehicle using path tracking: A review. In Proceedings of the International Conference on Science, Engineering, Management and Social Sciences, Johor Bahru, Malaysia, 6–8 October 2016; pp. 3877–3879.
19. Cui, Q.; Ding, R.; Zhou, B.; Wu, X. Path-tracking of an autonomous vehicle via model predictive control and nonlinear filtering. *Proc. Inst. Mech. Eng. Part D-J. Automob. Eng.* **2018**, *232*, 1237–1252. [CrossRef]
20. Yu, Z.; Zhang, R.; Xiong, L.; Fu, Z. Robust hierarchical controller with conditional integrator based on small gain theorem for reference trajectory tracking of autonomous vehicles. *Veh. Syst. Dyn.* **2019**, *57*, 1143–1162. [CrossRef]
21. Lidberg, M.; Muller, S. Special issue on motion control for automated driving and autonomous functions on road vehicles. *Veh. Syst. Dyn.* **2019**, *57*, 1087–1089. [CrossRef]
22. Sun, R.; Silver, D.; Tesauro, G.; Huang, G. Introduction to the special issue on deep reinforcement learning: An editorial. *Neural Netw.* **2018**, *107*, 1–2. [CrossRef] [PubMed]
23. Van, H.; Guez, A.; Silver, D. Deep reinforcement learning with double Q-Learning. In Proceedings of the 30th AAAI Conference on Artificial Intelligence, Phoenix, AZ, USA, 12–17 February 2016; pp. 2094–2100.
24. Sharifzadeh, S.; Chiotellis, I.; Triebel, R.; Cremers, D. Learning to drive using inverse reinforcement learning and Deep Q-Networks. *arXiv* **2016**, arXiv:1612.03653.
25. Gu, S.; Lillicrap, T.; Ghahramani, Z.; Turner, R.; Levine, S. Q-Prop: Sample-efficient policy gradient with an off-policy critic. *arXiv* **2016**, arXiv:1611.02247.
26. Jagodnik, K.; Thomas, P.; Branicky, M.; Kirsch, R. Training an Actor-Critic reinforcement learning controller for arm movement using human-generated rewards. *IEEE Trans. Neural Syst. Rehabil. Eng.* **2017**, *25*, 1892–1905. [CrossRef] [PubMed]
27. Fan, Q.; Yang, G.; Ye, D. Quantization-Based Adaptive Actor-Critic Tracking Control With Tracking Error Constraints. *IEEE Trans. Neural Netw. Learn. Syst.* **2018**, *29*, 970–980. [CrossRef] [PubMed]
28. Skach, J.; Kiumarsi, B.; Lewis, F.; Straka, O. Actor-Critic off-policy learning for optimal control of multiple-model discrete-time systems. *IEEE T. Cybern.* **2018**, *48*, 29–40. [CrossRef]
29. Lillicrap, T.; Hunt, J.; Pritzel, A.; Heess, N.; Erez, T. Continuous control with deep reinforcement learning. *arXiv* **2015**, arXiv:1509.02971.
30. Xu, J.; Hou, Z.; Wang, W.; Xu, B.; Zhang, K.; Chen, K. Feedback deep deterministic policy gradient with fuzzy reward for robotic multiple peg-in-hole assembly tasks. *IEEE Trans. Ind. Inform.* **2019**, *15*, 1658–1667. [CrossRef]
31. Zhang, Z. A flexible new technique for camera calibration. *IEEE Trans. Pattern Anal. Mach. Intell.* **2000**, *22*, 1330–1334. [CrossRef]
32. Li, L.; Zhang, L.; Li, X.; Liu, X.; Shen, Y.; Xiong, L. Vision-based parking-slot detection: A benchmark and a learning-based approach. In Proceedings of the 2017 IEEE International Conference on Multimedia and Expo, Hong Kong, China, 10–14 July 2017; pp. 649–654.
33. Imani, M.; Braga-Neto, U. Control of gene regulatory networks using bayesian inverse reinforcement learning. *IEEE-ACM Trans. Comput. Biol. Bioinform.* **2019**, *16*, 1250–1261. [CrossRef] [PubMed]
34. Pan, W.; Qu, R.; Hwang, K.; Lin, H. An ensemble fuzzy approach for inverse reinforcement learning. *Int. J. Fuzzy Syst.* **2019**, *21*, 95–103. [CrossRef]
35. Gao, C.; Yan, J.; Zhou, S.; Chen, B.; Liu, H. Long short-term memory-based recurrent neural networks for nonlinear target tracking. *Signal Process.* **2019**, *164*, 67–73. [CrossRef]

sensors

MDPI

Article

Deceleration Planning Algorithm Based on Classified Multi-Layer Perceptron Models for Smart Regenerative Braking of EV in Diverse Deceleration Conditions

Gyubin Sim [1], Kyunghan Min [2], Seongju Ahn [2], Myoungho Sunwoo [2] and Kichun Jo [3,*]

1 Department of Automotive Electronics and Controls, Hanyang University, Seoul 04763, Korea; gbcompany27@gmail.com
2 Department of Automotive Engineering, Hanyang University, Seoul 04763, Korea; kyunghah.min@gmail.com (K.M.); bingju159@gmail.com (S.A.); msunwoo@hanyang.ac.kr (M.S.)
3 Department of Smart Vehicle Engineering, Konkuk University, Seoul 05030, Korea
* Correspondence: kichun.jo@gmail.com

Received: 12 August 2019; Accepted: 16 September 2019; Published: 18 September 2019

Abstract: The smart regenerative braking system (SRS) is an autonomous version of one-pedal driving in electric vehicles. To implement SRS, a deceleration planning algorithm is necessary to generate the deceleration used in automatic regenerative control. To reduce the discomfort from the automatic regeneration, the deceleration should be similar to human driving. In this paper, a deceleration planning algorithm based on multi-layer perceptron (MLP) is proposed. The MLP models can mimic the human driving behavior by learning the driving data. In addition, the proposed deceleration planning algorithm has a classified structure to improve the planning performance in each deceleration condition. Therefore, the individual MLP models were designed according to three different deceleration conditions: car-following, speed bump, and intersection. The proposed algorithm was validated through driving simulations. Then, time to collision and similarity to human driving were analyzed. The results show that the minimum time to collision was 1.443 s and the velocity root-mean-square error (RMSE) with human driving was 0.302 m/s. Through the driving simulation, it was validated that the vehicle moves safely with desirable velocity when SRS is in operation, based on the proposed algorithm. Furthermore, the classified structure has more advantages than the integrated structure in terms of planning performance.

Keywords: deceleration planning; multi-layer perceptron; smart regenerative braking; driving behavior; electric vehicles

---

1. Introduction

"One-pedal driving" is one of the many remarkable changes as the paradigm of vehicle platforms switches from an internal combustion engine to electric vehicles (EVs) [1–3]. It makes driving with only the accelerator pedal possible by generating regenerative braking torque when the accelerator pedal is released. This can increase driver convenience as there are fewer pedals to shift.

The smart regenerative braking system (SRS) is one of the "one-pedal driving" technologies, which can take advantage of both the improvement of driver convenience and energy efficiency [4]. It is a type of advanced driver assistance system (ADAS) in that it supports driving using a radar sensor. Unlike general one-pedal driving, it does not always generate regenerative torque when the accelerator pedal is released. However, it does generate adequate regenerative torque according to car-following situations. The amount of regenerative torque is appropriately determined by relative distance, relative velocity, and speed of the ego vehicle.

SRS is an intelligent braking system which controls the friction braking or regenerative braking to meet certain goals such as safety, energy efficiency, and braking performance. There were various studies regarding the intelligent braking system. In Reference [5], an ADAS system regarding brake assistance based on the driver behavior and situation was proposed. Lin et al. suggested active control of regenerative braking to improve the braking performance of an electric vehicle [6]. In Reference [7], a brake-by-wire actuator was designed to shorten the braking distance and time. Similar to Reference [5], SRS improves the safety of the vehicle by intelligent braking. In addition, the system can enhance driver convenience.

Since SRS requires the deceleration to be used as a reference value in the automatic regenerative torque control, a deceleration planning algorithm is necessary. There are some points to consider when designing a deceleration planning algorithm for SRS. The first essential point is the harmony between the acceleration by the human driver and the deceleration by the SRS. It can be achieved by generating a deceleration profile that is similar to human driving [8,9]. A second crucial point is the applicability of SRS to diverse deceleration conditions to maximize the advantages of SRS, such as driver convenience and energy efficiency. A third important point is the usability in vehicles online.

In order to mimic human driving, a machine learning technique is an appropriate method because it can learn the characteristics of human driving. Various researches were conducted to predict acceleration and velocity using an artificial neural network (ANN) [10–16]. In particular, Reference [17] proposed an ANN model to predict acceleration with four inputs: relative distance, relative velocity, velocity, and desired speed. In Reference [18], a fuzzy rule-based neural network which used relative distance, relative speed, vehicle speed, and actions of the previous time step was designed to capture the vehicle motions. Khodayari et al. proposed a neural network model focusing on human driving behavior in Reference [16]. Unlike other researches, it used estimated instantaneous reaction delay to capture the realistic driver behavior of moving the foot between the accelerator and the brake pedal. These researches showed an acceptable performance of acceleration prediction. However, they lacked harmonization with the acceleration by human driver when applied to SRS because they could not accurately represent the driver behavior, especially in deceleration scenarios. In addition, these algorithms could only be applied in car-following conditions.

To maximize the advantages of SRS, the deceleration planning algorithm should generate deceleration in not only car-following condition, but other deceleration conditions as well. Intelligent transportation system (ITS) information is actively used in ADAS and energy management systems in various driving situations [19,20]. The deceleration planning algorithm can be applied in diverse deceleration situations using ITS information.

A third important point is the usability in vehicles. The deceleration planning algorithm should operate in vehicles online. Accordingly, only the information which is acquired in vehicles on-board can be used as an input of the algorithm.

Some researches related to deceleration planning and speed prediction were conducted. Yeon et al. developed a recurrent neural network (RNN) model to predict vehicle speed with a 10-s prediction horizon [21]. The proposed algorithm showed better prediction performance than other algorithms that were suggested. However, it had a limitation in that it could only be applied in the specific route used to train the RNN model because the algorithm used the position in the route as an input of the model. Min et al. proposed an RNN model to generate a deceleration profile at braking conditions. The accuracy of the model was improved by using a physical constraint to stop at the specific location. However, the model could only be used in braking conditions at a traffic light.

To overcome these limitations of previous research, this paper proposes a deceleration planning algorithm using classified multi-layer perceptron (MLP) models. The noticeable feature of the proposed algorithm is the classified structure. To improve the planning performance of the model, the deceleration models were developed individually in three different deceleration conditions: car-following, speed bump, and intersection. Each model was trained with the driving data acquired through vehicle experiments. Unlike the previous studies on driver models, the suggested MLP model considers the human reaction delay in deceleration, which results in the generated deceleration mimicking the coasting behavior. In particular, learning of the coasting behavior was adequately performed by appropriately processing the target data. In addition, the reference value of acceleration was used as input. As a result, the vehicle reached the required velocity successfully with the suggested algorithm in the three deceleration conditions. Moreover, the model was applicable in more diverse situations by using ITS information.

Because there are three models which are specialized in the three deceleration conditions, the planning algorithm should select the MLP model to be used. For this, a state recognition algorithm was designed. Using data such as accelerator pedal position, velocity, relative distance, and distance to speed bump, it recognizes the necessity of deceleration and the cause of it. Using the cause of deceleration, the adequate model which has specialized inputs in each deceleration condition is selected and used in planning. This results in acceptable performance in using the suggested algorithm in the SRS.

The proposed algorithm was validated through driving simulations using driving data. Then, the safety of the proposed algorithm was evaluated. In addition, the similarity to human driving was analyzed. Moreover, the planning results of the proposed algorithm were compared to the results with integrated structure.

The rest of the paper is organized as follows: Section 2 illustrates an overview of the entire algorithm. Section 3 describes the state recognition algorithm which determines the driving state and the cause of deceleration, called the "deceleration condition". The set of input and hidden layers of the MLP model, and the hyper-parameter optimization methods are described in Section 4. In Section 5, the training of the MLP models and the data used in the training are described. In Section 6, the simulation results of the suggested algorithm are shown and compared to the algorithm of integrated structure. Section 7 discusses the results and concludes the paper.

2. System Overview

2.1. Description of Deceleration Conditions

To minimize the number of braking instances by a driver, the deceleration planning algorithm should be able to generate deceleration in diverse situations. In this research, three deceleration situations were selected: car-following, speed bump, and intersection. Car-following was chosen because car-following is most common in both urban and highway driving. Speed bump and intersection were selected because they are major deceleration causes in urban driving. With these three deceleration conditions, the algorithm can be applied in most deceleration situations. In urban driving, traffic lights represent a major deceleration situation. However, its signal cannot be acquired in the target vehicle, which was used in the vehicle experiments on-board; thus, the condition is excluded in this research because the proposed algorithm was designed considering its usability in vehicles online. Details of each deceleration condition are given below.

The car-following condition is a driving situation where the leading car gets close to the ego vehicle. In this situation, the relative velocity is negative and the relative distance decreases for a few seconds.

The second situation is a speed bump, where the human pushes the brake pedal to pass through it smoothly. The location of the speed bump and ego vehicle in the route are acquired from the navigation device in the vehicle used for the experiment in this research. Distance to speed bump was calculated using the locations and used to check the reason for deceleration.

The third condition is an intersection. At an intersection, there are three options of driving according to the path: right turn, left turn, and straight. In right-driving countries, the movement of the vehicle is decided by the traffic signal for a left turn and going straight. To design the deceleration model in the two situations, the signal of the traffic light is required. However, it cannot be acquired from the current navigation device in the vehicle. Therefore, only the right turn was included in the intersection condition.

2.2. Algorithm Overview

As mentioned in Section 1, the suggested algorithm has a classified structure. Compared to the integrated structure, the number of inputs in each MLP model is reduced, and the size of the model decreases. In addition, better planning performance can be anticipated, because each model is trained with the deceleration profile which is matched to each deceleration situations.

Because of the classified structure, the model used in the deceleration planning should be selected. In addition, the vehicle motion should be monitored because automatic regeneration can operate only when the driver does not push the pedals. To monitor the vehicle motion and determine the current cause of deceleration, a state recognition algorithm was designed.

Figure 1 shows the overall structure of the entire algorithm. Firstly, the state recognition algorithm determines the driving state which refers to the vehicle motion based on the driver's behavior of pushing the pedal. Then, it determines the cause of deceleration, called the "deceleration condition". After the deceleration condition is selected, the deceleration model generate the deceleration of the next time step using the trained MLP model suitable for the deceleration condition.

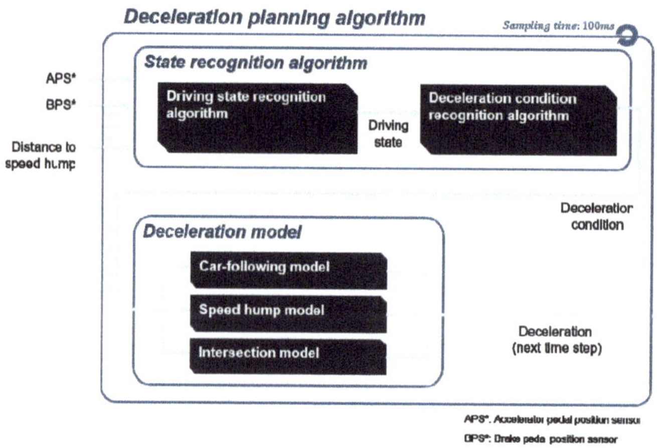

Figure 1. Overall structure of the deceleration planning algorithm.

3. State Recognition Algorithm

3.1. Driving State Recognition Algorithm

The SRS operates only in situations where deceleration is needed. Specifically, the SRS should not work when the driver pushes the accelerator pedal. To determine whether SRS can operate or not, the "driving state" has to be defined. It refers to the state of vehicle movement based on the behavior of drivers pushing the accelerator or brake pedal. There are four driving states categorized by the driving state recognition algorithm: driving, coasting, deceleration, and stopping. The meaning of each driving state is as follows:

- Driving: driver pushes the accelerator pedal.
- Coasting: driver pushes neither the accelerator nor the brake pedal.
- Deceleration: driver pushes the brake pedal and the velocity is not zero.
- Stop: vehicle speed is zero, which means that the vehicle does not move at all.

The driving state recognition algorithm was designed as a simple state flow chart, as shown in Figure 2. State transition occurs from one state to another using on-board sensor data. Data such as accelerator pedal position (AP), brake pedal position (BP), and velocity are used as conditions for state transitions. In Figure 2, "AP on" and "BP on" refer to pushing each pedal. Likewise, "AP off" and "BP off" refer to releasing each pedal.

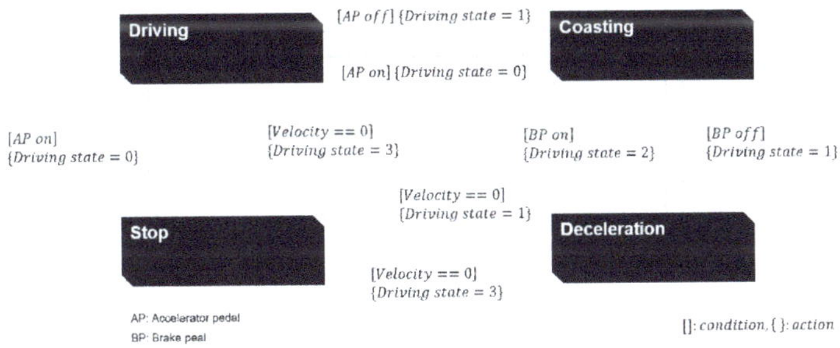

Figure 2. State flow chart of driving state recognition algorithm.

3.2. Deceleration Condition Recognition Algorithm

"Deceleration condition" means the cause of deceleration. There are three deceleration factors as described in Section 2.1: car-following, speed bump, and intersection. The deceleration condition recognition algorithm compares the effect of each deceleration factor and chooses one of the three factors as the deceleration condition.

When there are more than two deceleration factors at the same time, the algorithm should determine which deceleration factor has a greater impact on the necessity of deceleration. A constant acceleration (CA) model is used to evaluate the significance of each deceleration factor. It is a parametric equation which calculates the acceleration to reach the required velocity at a specific location. It consists of current vehicle speed, required velocity, and the distance to a specific location, as shown in Equation (1) and Figure 3.

$$a_{CA} = \frac{v_2^2 - v_1^2}{2d}. \quad (1)$$

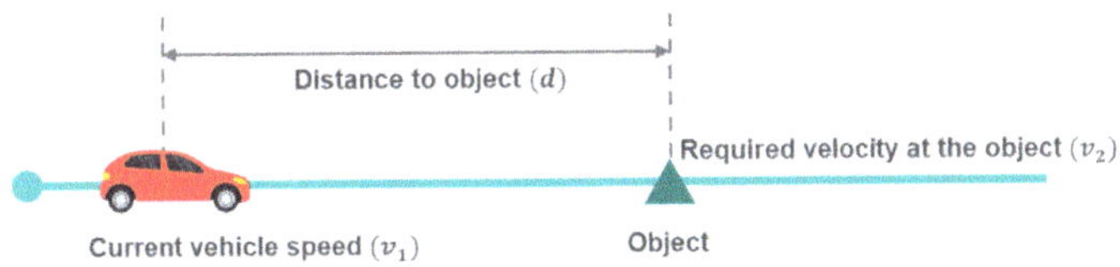

Figure 3. Driving situation for constant acceleration model.

Three parameters in the CA model (current vehicle speed (v_1), required velocity (v_2), and the distance (d) to the object) should be determined to calculate the acceleration. In the three parameters, the required velocity and the distance to the object depend on the deceleration factors. Their categorization according to deceleration factors is shown in Table 1. The values of minimum velocity for speed bump and intersection were confirmed as fixed values in each deceleration condition based on experimental data.

After the values of acceleration were calculated by the CA model in the three deceleration factors, they were compared to each other to determine the most influential deceleration factor. A large absolute value of acceleration in a deceleration factor means that the deceleration factor is dominant. However, the method of selecting the biggest value can cause a problem of frequent switching when two deceleration factor values are the same level. To prevent this type of problem, hysteresis was applied in determining the deceleration condition. Thus, the deceleration condition changes from one deceleration factor to another when the difference in absolute value calculated by the CA model is bigger than 0.2 m/s^2.

Table 1. Meanings of constant acceleration (CA) model parameters depending on deceleration conditions.

Deceleration Factor	Required Velocity	Distance to Object
Car-following	Preceding vehicle speed	Relative distance
Speed bump	Minimum velocity (30 km/h)	Distance to speed bump
Intersection	Minimum velocity (15 km/h)	Distance to intersection

4. Deceleration Model Based on Multi-Layer Perceptron

The suggested algorithm uses an MLP model to generate the deceleration used in automatic regeneration. MLP is one of the simplest types of ANN. Although it has a simpler structure compared to other types of ANN such as RNN or convolutional neural network (CNN), it can capture the nonlinear characteristics between multiple inputs and outputs, which is adequate when considering the human driving behavior. In addition, it is more suitable for application in vehicles because it usually has a smaller size than other ANN structures.

In this section, the input layer of the MLP model was designed based on the analysis of the human driving behavior. Then, the grid search algorithm was used to optimize the structure of the hidden layer. In the grid search, the data acquired from the vehicle experiments were used to train the model.

4.1. *Design of the Input Layer*

4.1.1. Driver Behavior during Deceleration

Prior to designing the deceleration model, deceleration profiles in each deceleration situation were analyzed to select the proper input set of the MLP model. In the data acquired from the vehicle experiments, the deceleration part in the three deceleration conditions were sliced and analyzed.

The deceleration profiles in each deceleration situation are shown in Figure 4. They are sliced deceleration profiles from when the accelerator pedal was released to when the brake pedal was released. The deceleration when both the accelerator and brake pedal were released was processed as zero. Additionally, the deceleration profiles of a human driver were compared to the deceleration profiles calculated by the CA model, which was modified from the original version.

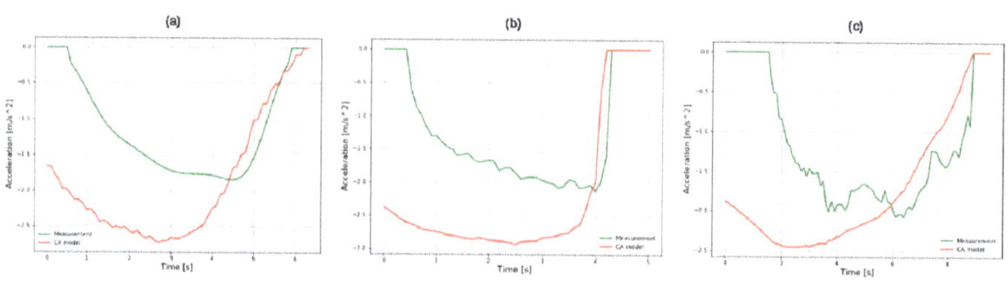

Figure 4. Deceleration profiles in (**a**) car-following; (**b**) speed hump; and (**c**) intersection.

The deceleration profiles had two common features regardless of the deceleration conditions. Firstly, drivers constantly released both the accelerator and the brake pedal. This driving behavior is usually called "coasting". Although the duration of coasting behavior is different depending on the driving situation, and although there is uncertainty in that behavior, human drivers always display the coasting behavior.

Secondly, the deceleration gradually finished as the acceleration by a CA model operated in accordance with the deceleration condition. Therefore, the shape of the deceleration profiles in each deceleration condition were highly similar to the acceleration profiles generated from the modified constant acceleration model. The modification depending on the deceleration condition is described in the next few paragraphs.

In a car-following condition, drivers usually control the speed of the vehicle to follow the speed of the preceding vehicle. However, the following vehicle's velocity can often range between 0.5 m/s slower and 0.5 m/s faster than the preceding vehicle. Therefore, Equation (1) was modified to Equation (2) to consider the driver intention in each sliced deceleration profile, where v_1 and v_2 in the equation are the speed of the ego and leading vehicle, $v_{1,final}$ and $v_{2,final}$ are the specific parameters of the ego and leading vehicle's speed at the end of the each sliced deceleration profile, and d is relative distance.

$$a_{CA} = \frac{\left(v_2 + \left(v_{1,final} - v_{2,final}\right)\right)^2 - v_1^2}{2d}. \tag{2}$$

In speed bump and intersection conditions, the minimum velocity is considered in calculating constant acceleration instead of the velocity of the leading vehicle. However, the minimum velocity is different in each deceleration profile because the deceleration aim of the driver differs. Likewise, the minimum velocity depends on the driver's intention at intersections. Therefore, the original constant acceleration model was modified to Equation (3), to consider the driver's intention in the sliced deceleration profiles.

$$a_{CA} = \frac{v_{1,final}^2 - v_1^2}{2d}. \tag{3}$$

4.1.2. Selection of the Input Set

The selection of the input set for the ANN model is important because the model performance highly depends on the appropriate set of inputs [22]. There were various researches to represent the microscopic motion of vehicles in car-following conditions. Regardless of the type of methods used in car-following models, there are three common elements in their input sets: ego vehicle speed, relative distance, and relative velocity, which shows that they are highly correlated with the movement of the vehicle in car-following conditions.

The three inputs mentioned in the previous paragraph were also used in this research. However, there were no relative distance and relative velocity data in the speed bump and intersection conditions. Therefore, these two inputs were replaced with other information that can be acquired by the navigation device. The relative distance was replaced with distance to speed bump and intersection. The relative velocity was replaced with difference between the speed of ego vehicle and minimum velocity. The minimum velocity changes depending on each deceleration profile and deceleration condition. Therefore, it is defined as the velocity at the end of each deceleration profile.

The deceleration models use two more inputs in addition to these three inputs. One is "coasting time" to simulate the coasting behavior of human driving and the other one is "reference acceleration" to deal with the end condition in each deceleration condition. One of the important points in designing the deceleration model for the SRS is the harmony between the acceleration of the human driver and the deceleration by the SRS. The stability is increased by mimicking the coasting behavior because drivers usually feel discomfort when the regenerative torque is generated right after the accelerator pedal is released. In addition, to guarantee safety and reach adequate velocity, the performance of the model should be improved. The reference acceleration helps the training of the MLP models and improves the model performance.

Costing time means the time taken after the accelerator pedal is released. The learning of coasting behavior with only the velocity, relative distance, and relative velocity is possible to some extent, but it cannot take the delay from the human reaction into account. When the coasting time is added to the input set of models, the algorithm can consider the nonlinear characteristic of coasting behavior based on the reaction delay and driving situations. The coasting time is calculated by counting the tick every 100 ms after the value of the accelerator pedal position sensor (APS) becomes zero.

Reference acceleration refers to the acceleration calculated by the modified CA model shown in Equations (2) and (3). When the modified constant acceleration model is applied, the deceleration profile follows the profile generated from the modified CA model in the final part of the profile in each deceleration condition as shown in Figure 4. Therefore, the acceleration calculated from the modified CA model is used as an input for the MLP models to help training by providing the standard value in the final part of the deceleration.

4.1.3. Normalization of Inputs

In addition to selecting an adequate set of inputs, normalization of inputs is crucial in designing the input layer. The selected inputs have different ranges of values depending on their type. The different scales of values can slow down the training process, and produce a poor performance result. Therefore, normalization of each input was conducted by min–max normalization method as shown in Equation (4), which makes the value of each parameter between 0 and 1. The overall structure of the MLP model based on the selected input set and normalization is shown in Figure 5.

$$x_{nrom} = \frac{x_{in} - \min(x_{in})}{\max(x_{in}) - \min(x_{in})}. \quad (4)$$

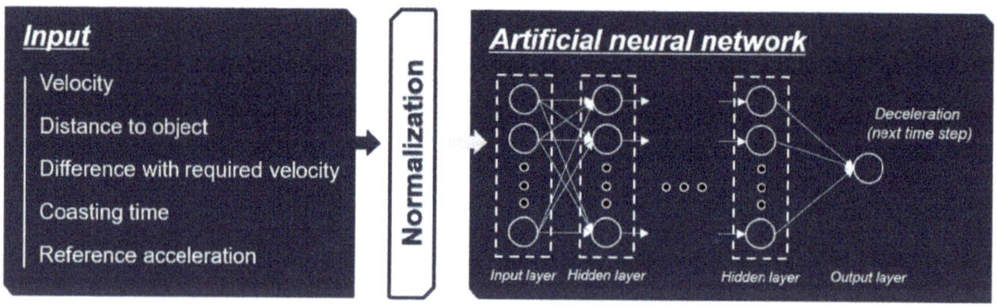

Figure 5. Overall structure of the multi-layer perceptron (MLP) model.

4.2. Design of the Hidden Layer

To design the hidden layer of the MLP model, various hyper-parameters should be selected such as the number of hidden layers and hidden nodes, the activation function, the optimizer, and the training iteration. Before optimizing the hyper-parameter set, hyper-parameters in an adequate range were found by hand-tuning as shown in Table 2. The number of hidden layers was fixed as two considering the small number of inputs. The model with only two hidden layers showed acceptable performance. As candidates of optimizers, four types of optimizer were used in optimization: Stochastic gradient descent (SGD), Adaptive subgradient (ADAGRAD), nesterov-accelerated ADAM (NADAM), and'RMSprop which was proposed by Geoff Hinton in his lecture.

Table 2. List of hyper-parameters.

Hyper-Parameter	Specifications
Number of hidden nodes in first hidden layer	20–40 (5 units)
Number of hidden nodes in second hidden layer	20–40 (5 units)
Activation function in each hidden layer	Relu, Sigmoid, Tanh, elu
Optimizer	SGD, ADAGRAD, NADAM, RMSprop
Dropout	0.1–0.3 (0.1 units)
Iteration of training	200, 400

Because the size of the model with determined range of the hyper-parameter set was small, the set was optimized based on a grid-search algorithm. The grid-search algorithm is the optimization method which tries all candidates and selects the best set. Due to the small model size, training time of the model was short, which made it possible to use the grid-search algorithm.

5. Experiments

In this section, the experimental environments in each deceleration condition are described. The experimental data were divided into three different parts to prevent overfitting. Then, the MLP models were trained with the training dataset, and the optimum set of hyper-parameters was selected by the grid-search algorithm.

5.1. Experiment Environments

5.1.1. Test Vehicle Configuration

Vehicle experiments were conducted to acquire the data used to train the MLP models. The vehicle used in the experiment and the data acquisition system are shown in Figure 6. The specifications of the radar sensor are in Table 3. We obtained information related to the leading car such as the relative distance velocity through the radar sensor. Moreover, the ITS information, such as location of speed bumps and intersections, was collected using the navigation device in the test vehicle.

Figure 6. Test vehicle and logging environment.

Table 3. Specifications of the radar sensor.

Index	Value
Maximum range	150 m
FOV (field of view)	±10° over 60 m ±45° under 60 m
Update rate	50 ms

5.1.2. Test Route

The vehicle experiment was conducted in Incheon, Korea in the routes shown in Figures 7–9. A test driver drove the vehicle; thus, the different driving characteristics depending on the driver were excluded. In addition, in each experiment, the cause of the deceleration was excluded. For example, there was no preceding vehicle in the experiment for the speed bump situations.

Experiments were conducted in the route shown in Figure 7. The ego vehicle decelerated multiple times depending on the velocity change of the preceding vehicle. To maintain a safe distance, the ego vehicle was decelerated by the test driver. The deceleration started from 10 to 20 m/s and finished at various speeds, including zero.

The experiments for speed bumps and intersections were conducted in the routes shown in Figures 8 and 9, and the locations of speed bumps and intersections are represented in each figure. To exclude the effect of the preceding vehicle, the experiment was conducted in situations without a leading car.

Figure 7. Test route for car-following situations.

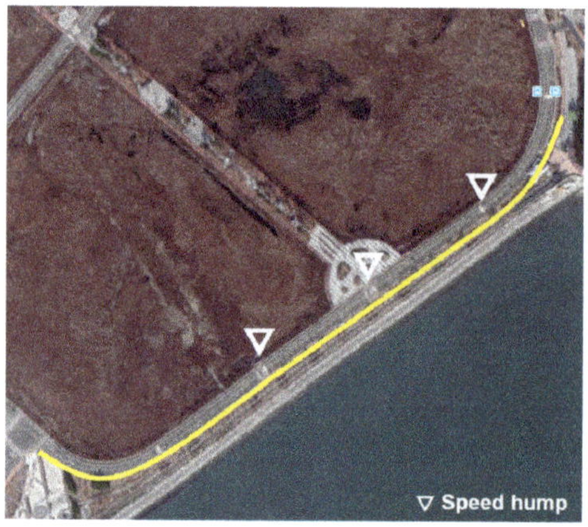

Figure 8. Test route for speed bump situations.

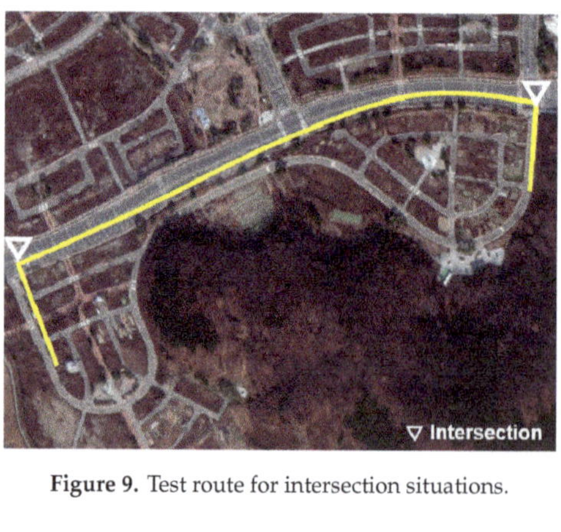

Figure 9. Test route for intersection situations.

5.2. Training

5.2.1. Input and Target Data

From the experimental data acquired through the vehicle experiments illustrated in Section 5.1, the input and target datasets were generated. The five types of inputs in each deceleration condition were extracted, and they were normalized as described at Section 4.1.3. The target data were the measured acceleration of the next time step because the model should generate the deceleration of the next time step. Therefore, there was a gap in the time step between the input and target data. The target data were processed as zero when both the accelerator and brake pedals were not pushed to assist the model learning the coasting behavior. The time step of the input and target data was set as 100 ms.

5.2.2. Dataset Splitting

If the hyper-parameters are not suitable to estimate the target value, the model's ability to estimate with new data is aggravated, which is called overfitting. Cross-validation is used to check overfitting. For cross-validation, there are three types of dataset: training, validation, and test datasets. The training dataset is a group of data used in training the model, which usually takes 70% of the entire data. The validation dataset is unseen data in training to check the overfitting. After the model performance is measured using the validation dataset, hyper-parameters are tuned to improve the prediction performance. Although the validation dataset is not used in the training directly, it is used to improve the model performance because hyper-parameters are selected by validation results. Therefore, a new dataset is needed to check the performance of the model, which is called the test dataset.

The dataset in this research was split in the way mentioned in the previous paragraph. The proportions of training, validation, and test datasets were 70%, 20%, and 10%, respectively. As for the evaluation index of validation, root-mean-square error (RMSE) was used, which is a representative method to compare two types of values, as shown in Equation (5), where n is the number of data, $a_{meas,i}$ is the i-th measured acceleration from human driving, and $a_{pred,i}$ is the i-th predicted acceleration by MLP models.

$$RMSE = \sqrt{\frac{1}{n}\sum_{i=1}^{n}\left(a_{meas,i} - a_{pred,i}\right)^2}. \tag{5}$$

5.2.3. Training and Hyper-Parameter Optimization

Each model which has one of the combinations of hyper-parameters described in Table 2 was trained with training dataset. Then, the best model which had the smallest RMSE in the validation dataset was selected through the grid-search algorithm. To train the models, Keras, one of the representative libraries for deep learning, was used based on Anaconda, which is the software platform for data science. The optimized sets of hyper-parameters depending on each deceleration condition are shown in Table 4. The number of hidden nodes in each hidden layer was different according to deceleration conditions, but they commonly used Relu as the activation function.

Table 4. Optimized hyper-parameter set in each deceleration condition.

Hyper-Parameter	Car-Following	Speed Bump	Intersection
Number of hidden nodes in first hidden layer	25	30	25
Number of hidden nodes in second hidden layer	25	30	30
Activation function in each hidden layer	Relu	Relu	Relu
Optimizer	RMSprop	NADAM	NADAM
Dropout	0.1	0.1	0.1
Iteration of training	400	400	400

The results of training, validation, and testing with the MLP models with optimized hyper-parameters are shown in Figures 10–12. They compare the target value, which is the measured acceleration from human driving, and the predicted value by trained MLP models in each deceleration condition. They are expressed as the absolute value for visibility. In addition, the results of RMSE and the number of points in training, validation, and test are also provided.

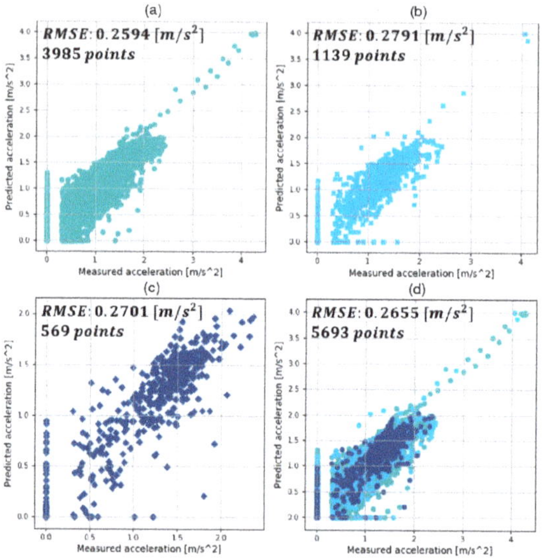

Figure 10. Results of (**a**) training; (**b**) validation; (**c**) test; (**d**) and total using the best model in car-following conditions.

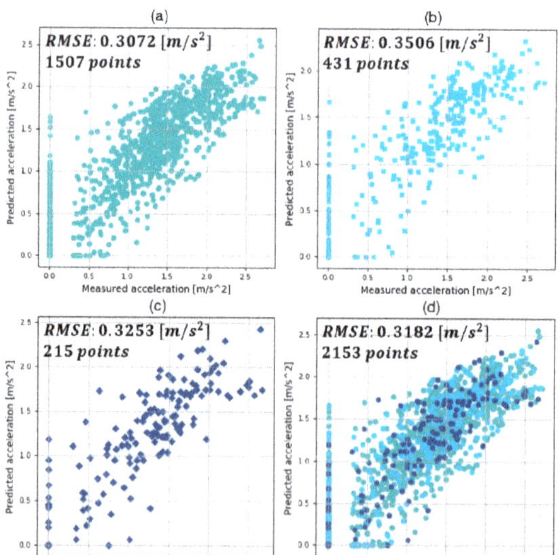

Figure 11. Results of (**a**) training; (**b**) validation; (**c**) test; (**d**) and total using the best model in speed hump conditions.

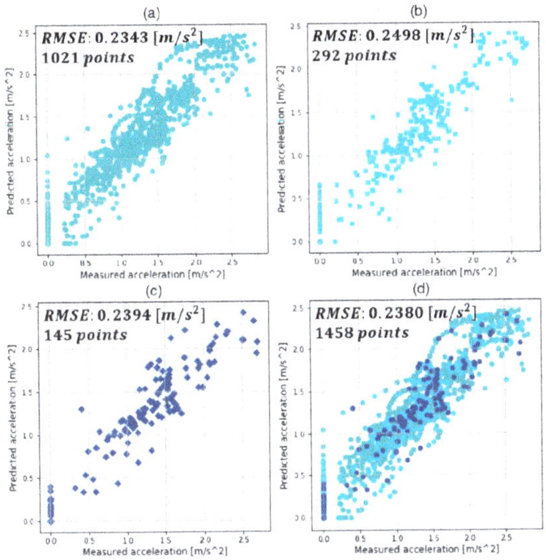

Figure 12. Results of (**a**) training; (**b**) validation; (**c**) test; (**d**) and total using the best model in intersection conditions.

In the comparison graphs shown in Figures 10–12, there are some points in the x-axis and y-axis generated by inaccurate prediction of deceleration start timing. In human driving, there is a high level of uncertainty in the coasting behavior, which makes the prediction of coasting duration difficult. The total numbers of data points were 5693, 2153, and 1458 in each deceleration condition, reflecting 569.3, 215.3, and 145.8 s of data. The three datasets were from 90, 44, and 15 deceleration scenarios in the three deceleration conditions.

6. Validation through Driving Simulation

Since the proposed deceleration planning algorithm is for use in the SRS of EVs, the motion of the vehicle when the algorithm is applied should be validated. Thus, a driving simulation was conducted for algorithm validation using driving data. The driving data were acquired in the vehicle tests including car-following, speed bump, and intersection situations. The motion of the vehicle was simulated using the generated deceleration by the algorithm with the assumption that the vehicle moves in accordance to the generated value. The results of driving simulation were compared to the human driving data, and they were analyzed in terms of satisfaction of required velocity and safety. In addition, the proposed algorithm using the classified MLP models was compared to a deceleration planning algorithm with an integrated structure in terms of model performance.

6.1. Driving Simulation Process

In the driving simulation, the deceleration planning algorithm shown in Figure 1 operated every 100 ms, which was the sampling time. However, there was one more step not represented in Figure 1 called "vehicle motion simulation". In the driving simulation, the vehicle moves as per the generated deceleration by the algorithm. Therefore, the states such as velocity, relative distance, and distance to speed bump and intersection are calculated by the generated deceleration. Then, the planning algorithm generates the deceleration again with the calculated states during the vehicle motion simulation.

Reference acceleration was used as an input of the MLP models, which refers to the acceleration calculated from the modified constant acceleration model mentioned in Section 4.1.2. However, unlike when the models were trained, the values of $v_{1,final} - v_{2,final}$ in car-following conditions and $v_{1,final}$ in speed bump and intersection conditions were not defined from the given driving data when the driving simulation was in progress. Therefore, these values were predetermined as constant values. The value of $v_{1,final} - v_{2,final}$ in car-following conditions refers to the speed difference at the end of deceleration. It was fixed at −0.5, which means that the model had the intention to decelerate to achieve a speed 0.5 m/s slower than the leading vehicle, considering the safety of the vehicle. The value of $v_{1,final}$ refers to the minimum velocity in speed bump and intersection conditions. These values were determined as 30 and 15 km/h in speed bump and intersection conditions, which were average values calculated from the experimental data.

6.2. Data for Driving Simulation

New data for the driving simulation were acquired from vehicle experiments to validate the model performance in the three deceleration conditions. The vehicle test was conducted in the routes shown in Figure 13 with the data acquisition system shown in Figure 7. In Figure 13, the locations of speed bumps and intersections are expressed with different marks. Although the route for the driving simulation was the same as the route for the intersection test shown in Figure 10, a leading vehicle drove in front of the ego vehicle throughout the experiment, which is different from the previous experiment described in Section 5.1. Therefore, the newly acquired data included car-following, speed bump, and intersection situations.

Figure 13. Driving route for driving simulation data acquisition.

The data acquired from the experiment are shown in Figure 14. It shows various information related to the vehicle states, the preceding vehicle, and the ITS. Some descriptions of the data are given below.

- Relative distance: The data of relative distance were acquired from the radar sensor described in Table 1. Sometimes, the distance value decreased to zero, which means that there were no vehicles in front of the ego vehicle.
- Distance to a speed bump: Distance to a speed bump was calculated using the locations of the ego vehicle and speed bumps acquired from the navigation device. When the distance to the speed bump was more than 60 m or when there were no speed bumps in front of the ego vehicle, the value was 60 m.
- Distance to an intersection: Similar to the distance to a speed bump, the distance to an intersection was calculated using the locations of the ego vehicle and intersection from the navigation device. When the distance to the intersection was more than 150 m or when there was no intersection in front of the ego vehicle, the value remained at 150 m.

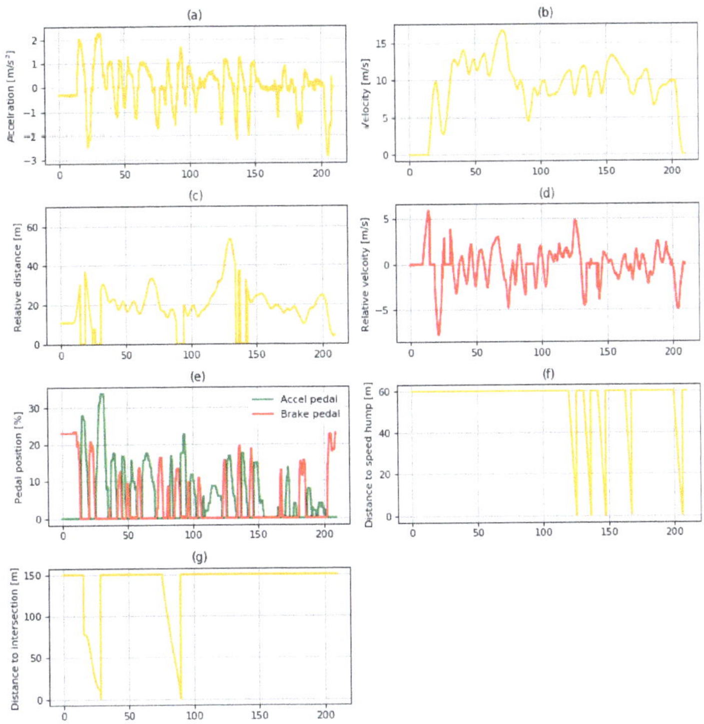

Figure 14. Data of (**a**) acceleration; (**b**) velocity; (**c**) relative distance; (**d**) relative velocity; (**e**) pedal position; (**f**) distance to speed hump; (**g**) and distance to intersection used in driving simulation.

6.3. Results of the Driving Simulation

6.3.1. Overall Description of the Driving Simulation Results

The driving simulation was conducted using the data shown in Figure 14. The entire simulation results are shown in Figure 15. It shows various data such as the results of the state recognition algorithm, the vehicle states, and the ITS information.

The driving conditions and deceleration conditions are shown in Figure 15e,f. As mentioned in Section 3, the SRS does not operate when the driving state is "acceleration" or "stopping". When the driving state is "coasting" or "deceleration", the deceleration is generated depending on the recognized deceleration condition. When the deceleration condition is "none", the generated value of deceleration is 0. When the deceleration condition is one of the other values, the deceleration is generated using the deceleration model adequate for the deceleration condition.

In Figure 15a–c,g,h, the red lines signify the simulated results and the yellow lines denote the measured data. The red line only appears when the SRS is in operation. When the driving state changes from acceleration to coasting, the simulation of the SRS starts. At that time, the values of simulation are initialized as the same value of the measured value. Then, the simulation is conducted while the driving state is kept as coasting or deceleration. The details of simulation results are analyzed for each deceleration condition in the next section.

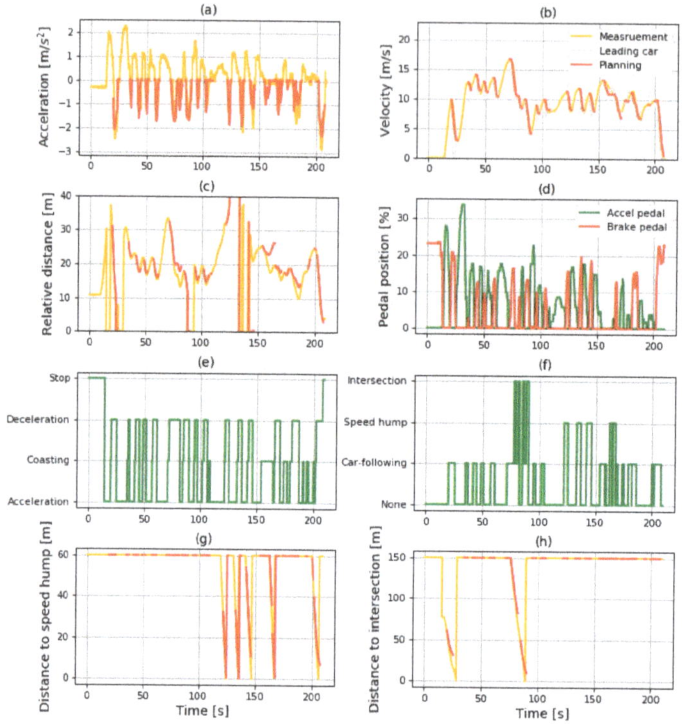

Figure 15. Result of (**a**) acceleration; (**b**) velocity; (**c**) relative distance; (**d**) pedal position; (**e**) driving state; (**f**) deceleration condition; (**g**) distance to speed hump; (**h**) and distance to intersection in driving simulation.

6.3.2. Results in Car-Following Condition

The sliced results in the car-following condition are shown in Figure 16. The driver released the accelerator pedal at about 57 s, and the driving state transited from driving to coasting at the same time. Then, the driving simulation started. The deceleration generated by the car-following model is shown in Figure 16a. Although the duration of coasting was not exactly the same as with the human driver because of the uncertainty of driver behavior, the result of deceleration planning initially shows the coasting behavior. Therefore, the deceleration profile generated by the model has a similar shape to the measured data from human driving.

In addition to checking the coasting behavior and similarity to human driving of the simulated results, the time to collision (TTC) was measured to check the safety in car-following conditions. TTC is a safety indicator to guarantee safety in car-following conditions that was widely used in various researches, calculated by dividing the relative distance by the velocity. The distribution of TTC from the planning results is shown in Figure 17. Furthermore, the minimum value of TTC was 1.443 s, which means that it would take 1.443 s to collide with the leading vehicle if the velocity was maintained at the same level.

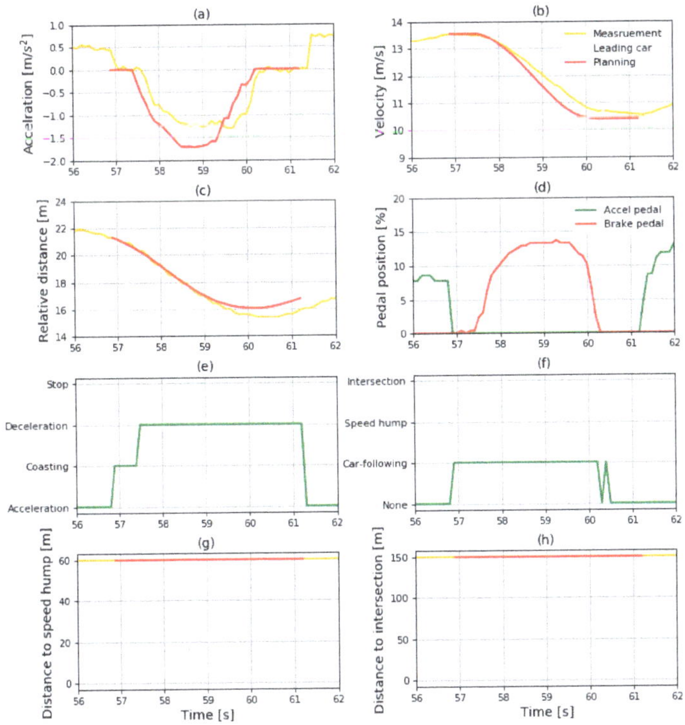

Figure 16. Sliced results of (**a**) acceleration; (**b**) velocity; (**c**) relative distance; (**d**) pedal position; (**e**) driving state; (**f**) deceleration condition; (**g**) distance to speed hump; (**h**) and distance to intersection in car-following condition.

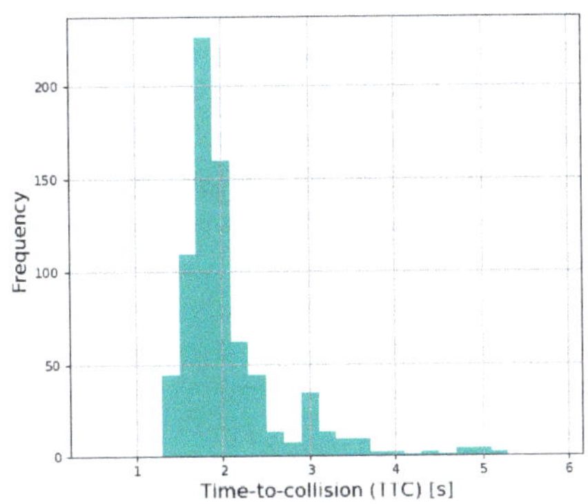

Figure 17. Distribution of time to collision (TTC) in car-following condition.

6.3.3. Results from the Speed Bump Condition

The sliced results where the vehicle passes through the speed bump are shown in Figure 18. Because the deceleration was generated by the speed bump model, the vehicle speed at the end of the profile should meet the criterion of the minimum velocity. As mentioned in Section 6.1, the minimum velocity was set to 30 km/h in the speed bump situation. This requirement seems to be satisfied because the velocity at the end was 8.24 m/s, which is almost the same as 30 km/h.

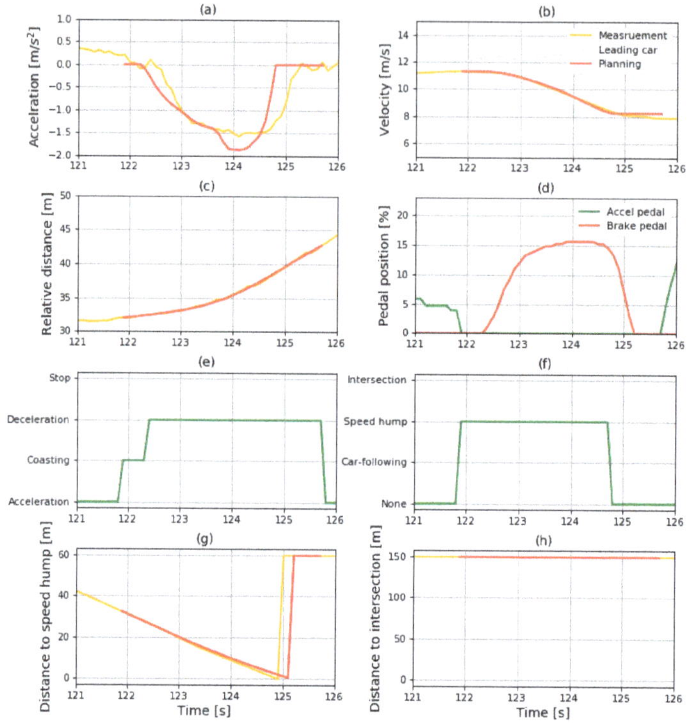

Figure 18. Sliced results of (**a**) acceleration; (**b**) velocity; (**c**) relative distance; (**d**) pedal position; (**e**) driving state; (**f**) deceleration condition; (**g**) distance to speed hump; (**h**) and distance to intersection in speed bump condition.

6.3.4. Results in Intersection Condition

The sliced results from where the vehicle turns right at the intersection are shown in Figure 19. Unlike the previous two sliced results shown in Figures 16 and 18, there were two deceleration factors in Figure 19: car-following and intersection. Therefore, the deceleration condition was determined by comparing the acceleration from the CA model as shown in Figure 19h. Therefore, the deceleration condition was the intersection initially, before changing to car-following in the middle, and transiting to intersection again toward the end of the deceleration.

Because the deceleration condition was intersection toward the end of the deceleration, the vehicle speed at the end of the profile should meet the minimum velocity. As mentioned in Section 6.1, the minimum velocity was set to 15 km/h in the intersection situation. This requirement seems to be satisfied because the velocity at the end was 4.16 m/s, which is almost the same as 15 km/h.

Because the objective of this research was to design a deceleration planning algorithm, and not to predict the acceleration, the motion of the vehicle when the algorithm was applied was compared to human driving using the values of velocity. Table 5 shows the RMSE of planning results and driving data in each sliced deceleration profile. An RMSE of 0.088 m/s was found in the car-following condition, while it was 0.015 m/s in the speed bump condition and 0.27 m/s in the intersection condition.

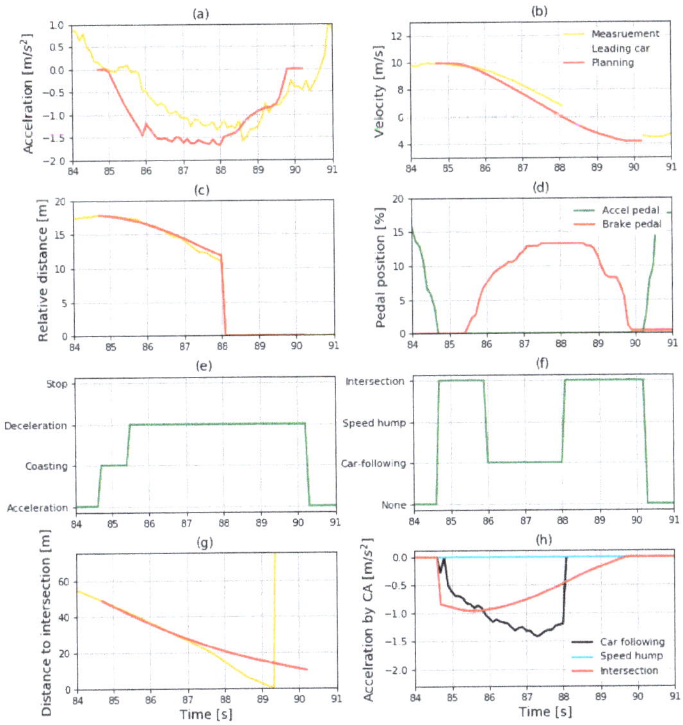

Figure 19. Sliced results of (**a**) acceleration; (**b**) velocity; (**c**) relative distance; (**d**) pedal position; (**e**) driving state; (**f**) deceleration condition; (**g**) distance to intersection; (**h**) and acceleration by CA model in intersection condition.

Table 5. Root-mean-square error (RMSE) of velocity in each sliced result.

Deceleration Condition	RMSE (m/s)
Car-following	0.088
Speed bump	0.015
Intersection	0.27

6.4. Comparison with the Integrated ANN Model

As mentioned above, a noticeable feature of the proposed algorithm is the classified MLP model, where it is expected that the performance of the model is improved. Therefore, the proposed algorithm was compared with the integrated MLP model-based deceleration planning algorithm in terms of the similarity to human driving.

The integrated MLP model has more inputs to generate deceleration in the three deceleration conditions. In addition, it does not require a deceleration condition recognition algorithm because there is no need to select which MLP model to use. The structure of the integrated MLP model is shown in Figure 20. Hyper-parameters are also optimized in the integrated model. The number of hidden layers is two, and the number of hidden nodes in each hidden layer is 40 in the integrated model. In addition, it uses Relu as an activation function.

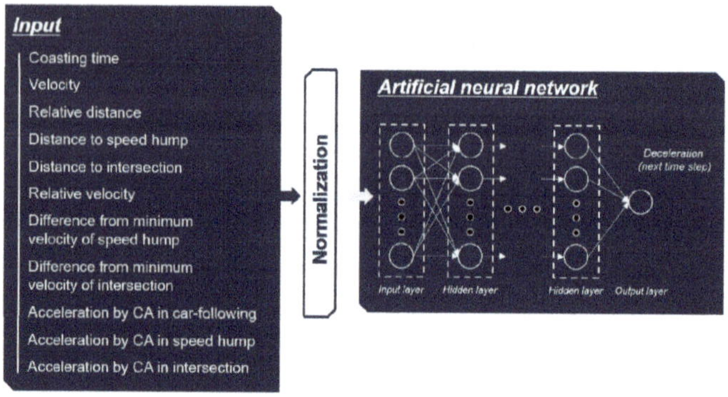

Figure 20. Model structure of integrated MLP model.

The comparison results of velocity RMSE are shown in Table 6. The velocity RMSE was calculated using the entire simulation results shown in Figure 15. The proposed algorithm showed a velocity RMSE of 0.312 m/s, and the planning algorithm with integrated structure showed a velocity RMSE of 0.890 m/s, which indicates that the classified structure improved the planning performance of MLP models.

Table 6. Comparison of classified and integrated structures.

	Classified Structure	Integrated Structure
RMSE of velocity (m/s)	0.312	0.901

According to the comparison results of the two different structures, the classified structure showed better performance in planning the deceleration for automatic regeneration. The classified structure has three different MLP models which have specialized inputs in each deceleration condition. The models have a reference acceleration as the common input, which is described in Section 4.1.2. This value provides a standard to reach the desirable velocity and keep a safe distance in each deceleration condition. Although the integrated structure uses three types of reference acceleration, as shown in Figure 20, it does not have the structure shown in Figure 1, which has a state recognition algorithm. As a result, the model seems to have difficulty in determining the most influential element without recognizing the deceleration condition. This results in weaker performance of the integrated structure compared to the classified structure.

7. Conclusions

This study suggested a deceleration planning algorithm which consisted of classified MLP models. The MLP models were trained with the human driving data acquired from vehicle experiments. In addition, diverse hyper-parameters were selected by hand-tuning and a grid-search algorithm. The validation results in each deceleration condition are summarized as follows:

- The best model in car-following showed a validation RMSE of 0.2791 m/s^2 and a total RMSE of 0.2655 m/s^2;
- The best model in speed bump showed a validation RMSE of 0.3506 m/s^2 and a total RMSE of 0.3182 m/s^2;
- The best model in intersection showed a validation RMSE of 0.2498 m/s^2 and a total RMSE of 0.2380 m/s^2.

The driving simulation was conducted using trained models with the best hyper-parameters. Through the driving simulation, it was validated that the vehicle motion with SRS based on the proposed algorithm is safe in car-following conditions and satisfies the minimum velocity at speed bumps and intersections. In addition, the deceleration planning results showed the coasting behavior in the initial part of the deceleration by using the "coasting time" and the "reference acceleration" as inputs for the MLP models. This can reduce the driver discomfort in using the SRS and improve the acceptability of the SRS. The proposed algorithm was compared with other deceleration planning algorithms with an integrated MLP model. The results showed that the planning algorithm with a classified structure has more similarity with human driving than the integrated structure.

In the future, this research will be applied to the SRS in EVs via integration with a regenerative torque controller. Furthermore, the deceleration conditions will be extended by using more types of ITS information such as curvature, traffic lights, and speed limits.

Author Contributions: Conceptualization, G.S. and K.M.; investigation, G.S.; methodology, G.S.; validation, G.S. and S.A.; visualization, K.M. and S.A.; writing—review and editing, M.S. and K.J.

Funding: This work was financially supported by the BK21 plus program (22A20130000045) under the Ministry of Education, Republic of Korea, the Industrial Strategy Technology Development Program (No. 10039673, 10060068, 10079961), the International Collaborative Research and Development Program (N0001992) under the Ministry of Trade, Industry and Energy (MOTIE Korea), and National Research Foundation of Korea (NRF) grant funded by the Korean government (MEST) (No. 2011-0017495).

Acknowledgments: The experiment vehicle was supported by the Hyundai motor company.

Conflicts of Interest: The authors declare no conflicts of interest.

References

1. Van Boekel, J.J.P.; Besselink, I.J.M.; Nijmeijer, H. Design and realization of a One-Pedal-Driving algorithm for the TU/e Lupo EL. *World Electr. Veh. J.* **2015**, *7*, 226–237. [CrossRef]
2. Jung, M.; Kessler, F.; Müller, P.; Wahl, S. Vehicle Integration and Driving Characteristics of the BMW Active E. *ATZ Wordwide* **2012**, *114*, 52–56. [CrossRef]
3. Helmbrecht, M.; Olaverri-Monreal, C.; Bengler, K.; Vilimek, R.; Keinath, A. How electric vehicles affect driving behavioral patterns. *IEEE Intell. Transp. Syst. Mag.* **2014**, *6*, 22–32. [CrossRef]
4. Lee, J.; Kim, H. Control Apparatus and Method for Regenerative Braking of Eco-Friendly Vehicle. K.R. Patent 101558772, 21 June 2014.
5. McCall, J.C.; Trivedi, M.M. Driver Behavior and Situation-aware Brake Assistance for Intelligent Vehicles. *Proc. IEEE* **2007**, *95*, 374–387. [CrossRef]
6. Lin, C.L.; Hung, H.C.; Li, J.C. Active control of regenerative brake for electric vehicles. *High. Throughput* **2018**, *7*, 84. [CrossRef]
7. Qian, G.; Wang, G. Research on the Anti-Disturbance Control Method of Brake-by-Wire Unit for Electric Vehicles. *World Electr. Veh. J.* **2019**, *10*, 44.
8. Hasenjager, M.; Wersing, H. Personalization in advanced driver assistance systems and autonomous vehicles: A review. *IEEE Conf. Intell. Transp. Syst. Proc. ITSC* **2018**, 1–7. [CrossRef]
9. Moon, S.; Yi, K. Human driving data-based design of a vehicle adaptive cruise control algorithm. *Veh. Syst. Dyn.* **2008**, *46*, 661–690. [CrossRef]
10. Morton, J.; Wheeler, T.A.; Kochenderfer, M.J. Analysis of Recurrent Neural Networks for Probabilistic Modeling of Driver Behavior. *IEEE Trans. Intell. Transp. Syst.* **2017**, *18*, 1289–1298. [CrossRef]
11. Colombaroni, C.; Fusco, G. Artificial neural network models for car following: Experimental analysis and calibration issues. *J. Intell. Transp. Syst. Technol. Plan. Oper.* **2014**, *18*, 5–16. [CrossRef]
12. Zheng, J.; Suzuki, K.; Fujita, M. Car-following behavior with instantaneous driver-vehicle reaction delay: A neural-network-based methodology. *Transp. Res. Part. C Emerg. Technol.* **2013**, *36*, 339–351. [CrossRef]
13. Zhou, M.; Qu, X.; Li, X. A recurrent neural network based microscopic car following model to predict traffic oscillation. *Transp. Res. Part. C Emerg. Technol.* **2017**, *84*, 245–264. [CrossRef]
14. Chong, L.; Abbas, M.M.; Medina, A. Simulation of Driver Behavior with Agent-Based Back-Propagation Neural Network. *Transp. Res. Rec. J. Transp. Res. Board* **2011**, *2249*, 44–51. [CrossRef]

15. Cheng, Z.; Chow, M.-Y.; Jung, D.; Jeon, J. A big data based deep learning approach for vehicle speed prediction. In Proceedings of the IEEE 26th International Symposium on Industrial Electronics (ISIE), Edinburgh, UK, 19–21 June 2017; pp. 389–394.
16. Khodayari, A.; Ghaffari, A.; Kazemi, R.; Braunstingl, R. A Modified Car-Following Model Based on a Neural Network Model of the Human Driver Effects. *IEEE Trans. Syst. Man Cybern. Part A Syst. Hum.* **2012**, *42*, 1440–1449. [CrossRef]
17. Jia, H.; Juan, Z.; Ni, A. Develop a car-following model using data collected by five-wheel system. In Proceedings of the 2003 IEEE International Conference on Intelligent Transportation Systems, Shanghai, China, 12–15 October 2003; Volume 1, pp. 346–351.
18. Chong, L.; Abbas, M.M.; Medina Flintsch, A.; Higgs, B. A rule-based neural network approach to model driver naturalistic behavior in traffic. *Transp. Res. Part C Emerg. Technol.* **2013**, *32*, 207–223. [CrossRef]
19. Lang, D.; Schmied, R.; Del Re, L. Prediction of Preceding Driver Behavior for Fuel Efficient Cooperative Adaptive Cruise Control. *SAE Int. J. Engin.* **2014**, *7*, 14–20. [CrossRef]
20. Li, Y.; Wang, H.; Wang, W.; Liu, S.; Xiang, Y. Reducing the risk of rear-end collisions with infrastructure-to-vehicle (I2V) integration of variable speed limit control and adaptive cruise control system. *Traffic Inj. Prev.* **2016**, *17*, 597–603. [CrossRef] [PubMed]
21. Yeon, K.; Min, K.; Shin, J.; Myoungho, S.; Han, M. Ego-vehicle speed prediction using a long short-term memeory based recurrent neural network. *Int. J. Automot. Technol.* **2019**, *20*, 713–722. [CrossRef]
22. Drezga, I.; Rahman, S. Input variable selection for ann-based short-term load forecasting. *IEEE Trans. Power Syst.* **1998**, *13*, 1238–1244. [CrossRef]

sensors

MDPI

Article

Vehicle Deceleration Prediction Model to Reflect Individual Driver Characteristics by Online Parameter Learning for Autonomous Regenerative Braking of Electric Vehicles

Kyunghan Min [1], Gyubin Sim [2], Seongju Ahn [1], Myoungho Sunwoo [1] and Kichun Jo [3,*]

1 Department of Automotive Engineering, Hanyang University, Seoul 04763, Korea; kyunghah.min@gmail.com (K.M.); bingju159@gmail.com (S.A.); msunwoo@hanyang.ac.kr (M.S.)
2 Department of Electronics and Controls, Hanyang University, Seoul 04763, Korea; gbcompany27@gmail.com
3 Department of Smart Vehicle Engineering, Konkuk University, Seoul 05030, Korea
* Correspondence: kichun.jo@gmail.com

Received: 12 August 2019; Accepted: 22 September 2019; Published: 26 September 2019

Abstract: The connected powertrain control, which uses intelligent transportation system information, has been widely researched to improve driver convenience and energy efficiency. The vehicle state prediction on decelerating driving conditions can be applied to automatic regenerative braking in electric vehicles. However, drivers can feel a sense of heterogeneity when regenerative control is performed based on prediction results from a general prediction model. As a result, a deceleration prediction model which represents individual driving characteristics is required to ensure a more comfortable experience with an automatic regenerative braking control. Thus, in this paper, we proposed a deceleration prediction model based on the parametric mathematical equation and explicit model parameters. The model is designed specifically for deceleration prediction by using the parametric equation that describes deceleration characteristics. Furthermore, the explicit model parameters are updated according to individual driver characteristics using the driver's braking data during real driving situations. The proposed algorithm was integrated and validated on a real-time embedded system, and then, it was applied to the model-based regenerative control algorithm as a case study.

Keywords: vehicle speed prediction; driver behavior modeling; electric vehicle control; driver characteristics online learning

1. Introduction

The prediction of vehicle states based on Intelligent Transportation System (ITS) information can be widely used to improve energy efficiency and driver convenience [1–4]. For electric vehicles, a smart regenerative control algorithm is an advanced driver assistance system which uses the prediction information of a vehicle decelerating from ITS information. This system recognizes the driving conditions that vehicle should decelerate by radar sensor. Then, it controls the regenerative torque of the electric motor automatically. This system provides the one-pedal driving technology, since the driver does not need to step on the brake pedal while accelerating using the acceleration pedal [5–8]. Thus, this autonomous braking system can serve driver convenience by reducing the driver's pedal shifting. Furthermore, braking by using regeneration torque without the hydraulic braking can improve efficient energy management. However, the driver feels the heterogeneity due to this autonomous braking because each driver has different driving characteristics according to diverse driving situations. Thus, an appropriate prediction algorithm for vehicle states is required to generate the deceleration trajectory as the set point of this autonomous braking control on diverse deceleration conditions. Also, this algorithm should consider individual driver characteristics to apply this automatic braking system more practically [4,9–14].

In order to predict vehicle states, a number of studies on two methods have been typically introduced. The first method is a parametric model-based prediction. The intelligent driver model is a representative parametric model for vehicle-state prediction according to driver characteristics [14–19]. This model consists of a mathematical equation based on the physical behavior for car-following situations with some explicit parameters. The explicit model parameters are configurable variables, and these represent driver characteristics.

The other modeling approach is the data-driven method using the driving data of each driver. The deep-learning [20–22] and hidden Markov models [23–25] are representative prediction algorithms of data-driven methods. These algorithms can efficiently cover the modeling uncertainties and can provide accurate predictions. Due to the deep-learning algorithm's advantages, much research has been conducted based on this algorithm.

The introduced research can predict the vehicle states with driver characteristics well. However, to apply the prediction algorithm for regenerative braking control, the algorithm should recognize diverse deceleration conditions such as stopping for a traffic light, slowing down for a speed bump, adhering to speed limits, or braking for a preceding vehicle. The algorithm should be specific for each deceleration condition. Furthermore, to reduce the sense of heterogeneity, the algorithm has to update the model parameters which represent the individual driver characteristics by online learning during real driving conditions. Most introduced data-driven algorithms require a premeasured data set to obtain the model and model parameters. Thus, typical data-driven methods are not appropriate for the online update algorithm.

In order to solve these problems, we proposed a parametric deceleration model and online learning algorithm for its parameters in an earlier work. The proposed model predicts the vehicle deceleration profile when the vehicle is braking in front of traffic lights. This model consists of mathematical equations; those equations are converted depending on the physical situation and explicit model parameters that represent individual driver characteristics. The online learning algorithm updates explicit model parameters during real driving conditions because the algorithm was designed to require less calculation time and memory for the real-time embedded system. However, the previous model was designed for only the deceleration condition for traffic light stops and the algorithm runs the prediction algorithm when determining deceleration start conditions.

This paper proposes the extended parametric deceleration model and Deceleration Condition Recognition Algorithm (DCRA) to cover diverse deceleration conditions. The operation process of the proposed algorithm is described at Figure 1. When the driver releases the acceleration pedal, the DCRA determines whether the deceleration will occur in the future and the cause of the deceleration if deceleration will occur. According to the cause of the deceleration, the parametric deceleration model predicts the deceleration profile similarly to our previous research while the driver pushes the brake pedal. The online learning algorithm updates the model parameters using the prediction error between the predicted deceleration profile and the measured deceleration profile by the real driver when the braking ends. The updated model parameters are also managed according to the cause of the deceleration from the DCRA since the deceleration characteristics can vary depending on the cause of the deceleration as well as the driver.

To sum contributions of the paper up, the algorithm can predict the deceleration-specified profile based on the proposed parametric model structure and can represent the individual driver characteristics to the predicted deceleration profile for diverse deceleration conditions. The proposed algorithm was validated through processes in the loop simulation using real-driving data. Subsequently, we analyzed and classified individual driver characteristics using the explicit model parameters of validation results. Furthermore, based on the proposed algorithm, the regenerative control was conducted as a case study in the car-following deceleration conditions.

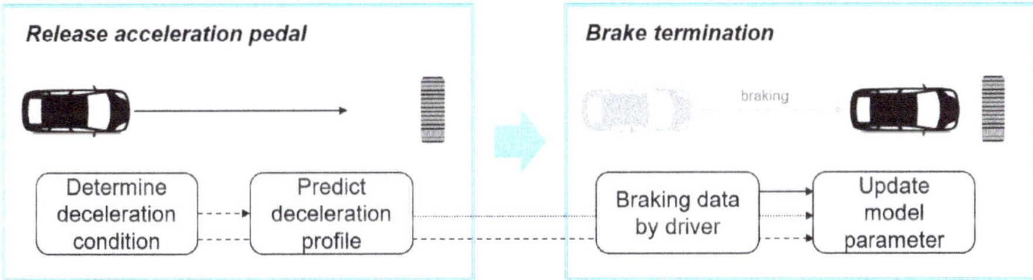

Figure 1. Algorithm operation process.

The rest of this paper is arranged as follows. Section 2 briefly introduces the system overview and the well-known intelligent driver model, which is applied as a frame structure of the proposed model. Section 3 presents the Deceleration Condition Recognition Algorithm (DCRA). According to the recognized deceleration condition, the prediction process of the model will be described in Section 4, and the online learning will be described in Section 5. Section 6 shows validation results through a vehicle experiment with regenerative control. The final section, Section 7, of this paper provides a conclusion.

2. System Architecture

2.1. Vehicle Configuration

In order to acquire driving data for designing the proposed algorithm, a Hyundai KONA Electric Vehicle (EV) was used. The experimental vehicle was equipped with additional sensors and equipment to acquire ITS information, as the proposed algorithm requires precise information of the driving environment. To recognize the deceleration conditions at the car-following situation, the vehicle measures the preceding vehicle velocity and relative distance between the two vehicles using the equipped radar sensor, as described in Table 1. In addition, the algorithm should calculate precise positions of the ego-vehicle and other objectives to determine the deceleration condition by the road objectives in the driving environment. The low-cost Global Positioning System (GPS), high-definition (HD) map, and traffic-light recognition system are additionally installed in the vehicle to calculate the ego-vehicle position and to recognize road objective positions. Figure 2 shows the experimental apparatus of the vehicle. The proposed prediction model and learning algorithm are integrated on the embedded system, which is 32-bit microprocessor MPC5674 manufactured by NXP. Using the Vector VX1000, we have measured and calibrated the proposed algorithm through the universal measurement and calibration protocol (XCP). The additional personal computer calculates the relative distance between the vehicle and other objects. The calculated distance is transported to the embedded system through the Controller Area Network (CAN). Radar information and vehicle states which are measured by the vehicle-equipped sensor are also acquired through the CAN using a Vector VN1640 tool.

Table 1. Sensor specification.

Sensor	Specification
Radar	Maximum range: 150 m FOV: +/− 10 degrees over 60 m, +/− 45 degrees under 60 m Update rate: 50 ms
GPS	Accuracy (Root mean square): 2.5 m Update rate: 20 ms

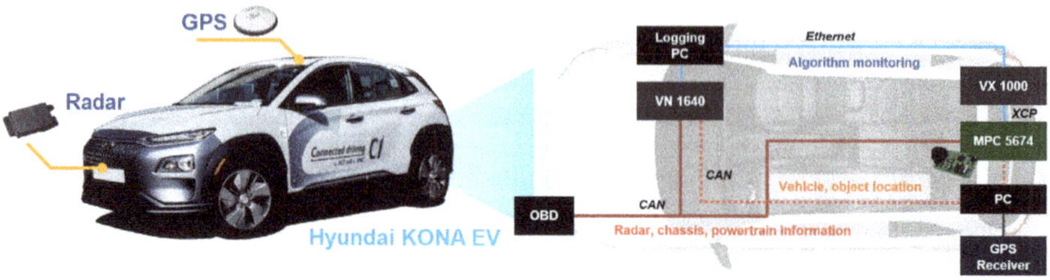

Figure 2. Vehicle configuration.

2.2. *ITS Information Fusion*

As mentioned in previous section, the precise positioning of the ego-vehicle and the static objectives are important to determine the deceleration conditions and, hence, to apply the prediction algorithm. To calculate the precise vehicle position, a localization algorithm which is introduced in References [26,27] is applied. The applied localization algorithm calculates the ego-vehicle position based on the GPS measurement position and in-vehicle motion sensor. On the other hand, the HD map provides the precise position of static objective and road curvature. Thus, the relative distance between the ego-vehicle and the static objective is calculated. Finally, the traffic-light recognition system classifies the traffic-light status based on the deep-learning algorithm that is introduced in Reference [28]. Figure 3 shows the algorithm architecture for ITS fusion.

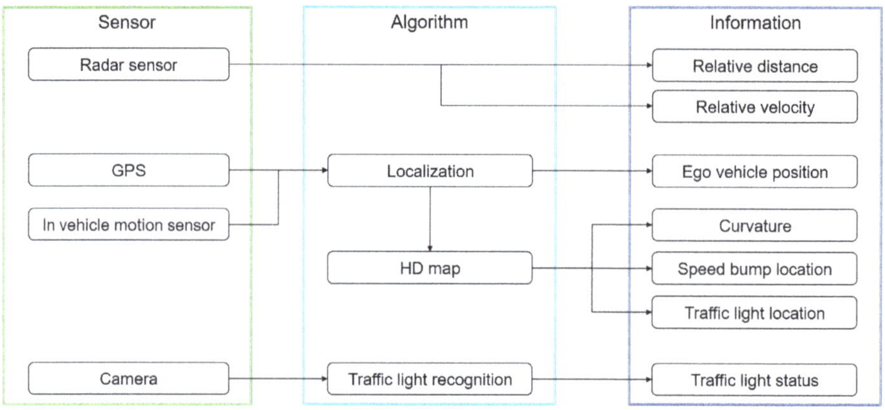

Figure 3. Intelligent Transportation System (ITS) fusion algorithm.

2.3. *Overall Structure of the Proposed Algorithm*

The proposed algorithm consists of three parts, which can be seen in Figure 4. The Deceleration Condition Recognition Algorithm (DCRA) decides the driver intention and deceleration conditions depending on the cause of decelerating, which is measured by ITS information. According to the determined deceleration condition, a deceleration prediction model predicts the deceleration profile using the parametric equation with regard to the brake section and model parameters. After the prediction process ends, an online learning algorithm updates the model parameters using the measured deceleration data by the driver.

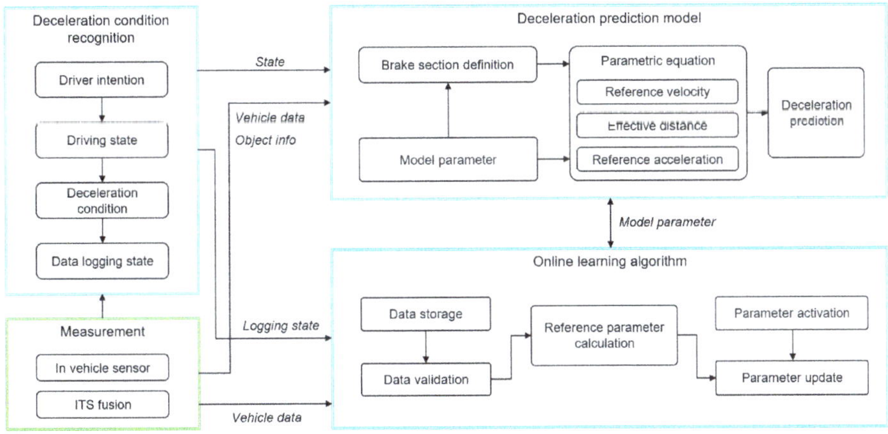

Figure 4. Overall algorithm structure.

3. Deceleration Condition Recognition Algorithm

In the urban driving condition, a vehicle faces diverse driving states [24,29]. To apply the proposed deceleration prediction model and learning algorithm appropriately, the future driving state is clearly defined. If the future driving state defines that the vehicle will decelerate, the cause of the deceleration is also identified to select the corresponding model parameters. Thus, the DCRA determines the driving state using the driver's intention and the deceleration condition according to the cause of deceleration. The proposed condition recognition algorithm consists of four state machines as shown in Figure 5.

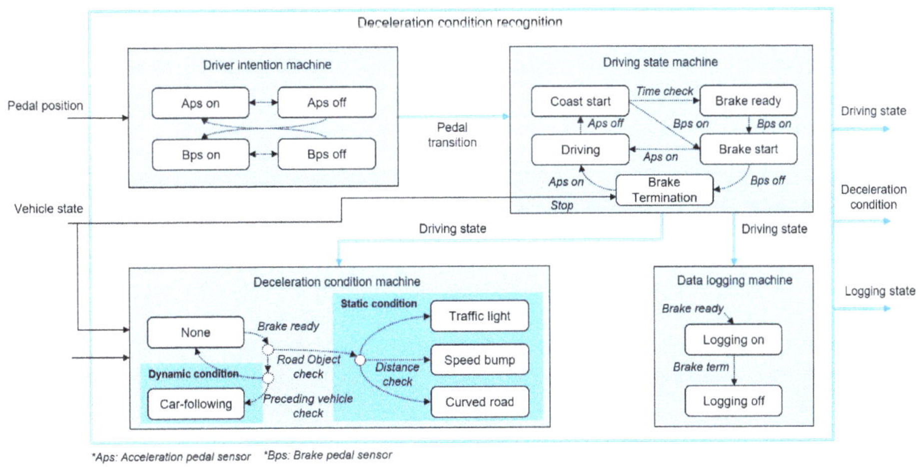

Figure 5. Diagram of the deceleration condition recognition algorithm.

At first, the driver intention machine checks the driver's pedal operation. When the driver presses or releases the accelerator pedal, the driver intention machine determines the pedal transition state as an accelerator pedal transition state. It also gives the brake pedal transition state for the brake pedal operation, and both the accelerator and brake pedal positions are measured by the in-vehicle network.

According to the pedal transition state of the driver intention machine, the driving state machine determines the driving state to affirm that the braking situation will occur. When the accelerator pedal is in the "off" state, the driving state machine sets the "coast start" state and checks the pedals' off time.

If the driver does not step on the accelerator pedal again during a specific period of time, the driving state machine will judge that the driving state is "brake ready". This checking for pedal off-time can prevent the frequent state transition due to frequent acceleration. When the brake pedal is in the "on" state, the driving state machine gives a "brake start" state. The "brake termination" state then turns on when the driver releases their foot from the brake pedal. Finally, the state is determined as the "driving" state when the driver steps on the accelerator pedal.

The deceleration condition machine classifies the deceleration condition if the driving state is determined as a brake-ready state. The deceleration condition is classified according to the deceleration source, which causes the braking situation in a driving environment. In this study, we determine the deceleration conditions to have two categories: dynamic deceleration and static deceleration. The dynamic deceleration condition is a condition where a vehicle follows the preceding vehicle. This state is determined using dynamic information such as the relative velocity and the relative distance. Dynamic information comes from vehicle-equipped sensors such as the acceleration sensor, the wheel speed sensor, and the radar sensor. If the preceding vehicle decelerates so that the relative velocity is negative, the current deceleration condition is in a car-following deceleration state.

The static deceleration condition means the condition where the driver decelerates due to road objects on the driving route. The proposed algorithm selects three road objects: traffic lights, speed bumps, and road curvature. The algorithm can know the location of the traffic lights and speed bumps using ITS information like an HD map and GPS. Using this information, the distance to the road objects on the driving route can be calculated. Thus, when the driving state is brake ready and the relative distance to road objects are closer than configurable road-object distance criterion, the deceleration condition decides the vehicle will brake by the traffic light or the speed bump, respectively. The road curvature state is also another road object where a driver decelerates to reduce velocity to negotiate the curve without heterogeneity [30,31]. Similar to other road objects, the decision of the curve state is made using ITS information.

Finally, the data logging state is determined as the data logging judgment machine for the learning algorithm. When braking starts, this state machine gives that the data logging state is on. During the data logging on state, the vehicle data and ITS information are stored. Then, the learning algorithm updates the model parameter using the stored data after braking is terminated.

4. Deceleration Prediction Model

4.1. Prediction Model Description

The deceleration prediction will be conducted based on the well-known Intelligent Driver Model (IDM) [19]. This intelligent driver model can describe the microscopic car-following behavior using Equation (1). The vehicle acceleration is calculated according to several parameters such as the maximum acceleration a_m, the reference velocity v_{ref}, and the effective distance d_{eff}. In addition, the current vehicle velocity and relative distance of the preceding vehicle are also used in the model.

$$a = a_m(1 - (v/v_{ref})^{\delta} - (d_{eff}/d_{rel})^2) \tag{1}$$

In this paper, we proposed parametric equations for the reference velocity and effective distance to predict deceleration situations more specifically. Figure 6 shows the prediction process of the deceleration profile using the determined parametric equations. As shown in the figure, the parametric equations are determined depending on the braking sections and model parameters. Then, using this parametric equation, the deceleration profile $\hat{a}(k)$ is calculated at each time-step iteratively. The braking section is a time-range parameter which is determined using common patterns in a braking situation. The model parameters represent the individual driver's characteristics explicitly and determine the parametric equation. Since parametric equations are defined differently according to each section and model parameters, the predicted deceleration profile that is a calculation result based on the intelligent driver model can minutely represent the deceleration characteristics depending on each section and driver characteristics. The detailed braking section and model parameters will be explained in the next section.

Furthermore, we proposed one more parametric equation that guarantees the end states for each of the deceleration conditions. That is a reference acceleration profile of Equation (2) which is calculated based on the constant acceleration model for location and velocity of the end of braking.

$$a_{ref}(k) = 0.5(v_{min}^2 - v(k)^2)/d_{rel}(k) \tag{2}$$

where a_{ref} is a reference acceleration, v_{min} is the minimum velocity for each deceleration state, v is ego-vehicle velocity, and d_{rel} is the relative distance to object. The minimum velocity is determined according to the deceleration conditions. In regard to the static deceleration condition, the minimum velocity and termination condition are predefined as a model parameter. In the traffic-light stop condition, the minimum velocity is determined as the zero value, while in the curve and speed bump conditions, the minimum velocity parameters are defined according to the road conditions. In the dynamic deceleration condition, the minimum velocity is the velocity of the preceding vehicle.

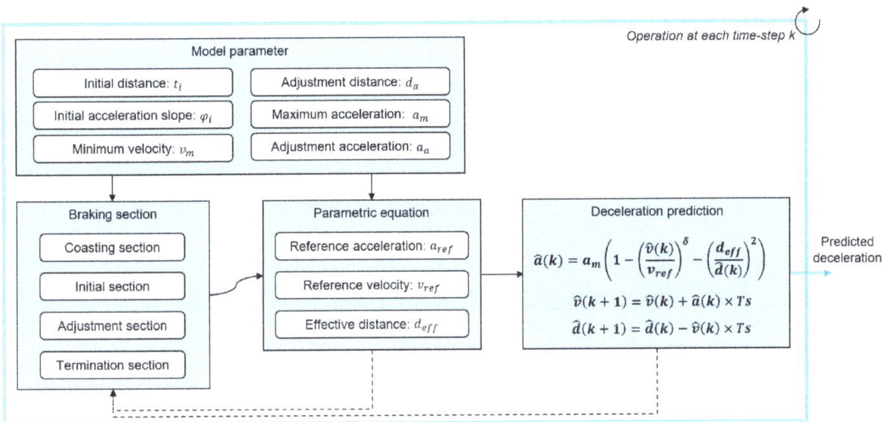

Figure 6. Diagram of the acceleration profile prediction algorithm.

4.2. Braking Section Analysis

The deceleration data is analyzed to predict the deceleration profile based on drivers' characteristics in each deceleration condition. Although the deceleration profile differs according to the driver's driving style and deceleration conditions, the common feature in the declaration pattern is discovered. Therefore, we split the deceleration profile into four braking sections and its start points as shown in Figure 7. Details of each braking section are described as follows:

1. Coasting section: The coasting section is the time range from when a driver releases the accelerator pedal to when the driver pushes the braking pedal.
2. Initial section: In the initial section, the driver applies the brake pedal to reach the deceleration of the adjustment point. The driver decelerates with the jerk in this section. If the initial section is short and the jerk is large, the driver feels the rapid change of deceleration.
3. Adjustment section: In the adjustment section, the driver applies a specific braking pedal force to maintain the deceleration without abrupt changes. The pattern of deceleration in this section depends on the driver's characteristic and the deceleration state.
4. Termination section: The termination section is the last part of the braking sections. Its definition also depends on driving conditions. In the static condition, the driver controls the braking pedal to satisfy the stop condition or minimum velocity in front of the static objectives. In the dynamic condition, the driver begins to remove the driver's foot from the brake pedal.

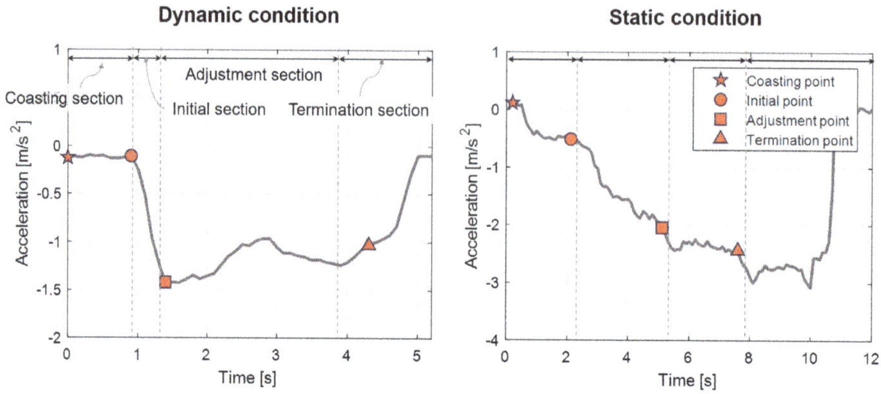

Figure 7. Description of the braking section about various deceleration conditions.

4.3. Model Parameter Description

In order to represent the deceleration characteristics of the individual driver according to each determined braking section, explicit model parameters are determined. As mentioned in the previous section, these model parameters determine the time range of each braking section. The parameters also determine the parametric equations which are reference velocity and effective distance. Consequently, the vehicle acceleration in deceleration condition is predicted depending on the driver's characteristics by model parameters affecting the braking section and parametric equations.

Figure 8 shows the model parameters in the deceleration. Transition-timing parameters are the transition timings of braking sections. According to the section transition, the parametric equations are also adapted. With regards to the acceleration value, the coasting acceleration and the adjustment acceleration parameters are the acceleration value when each section starts, and the minimum acceleration parameter is the negative maximum value during braking. This parameter is used as a maximum acceleration parameter at the classic IDM equation in Equation (1).

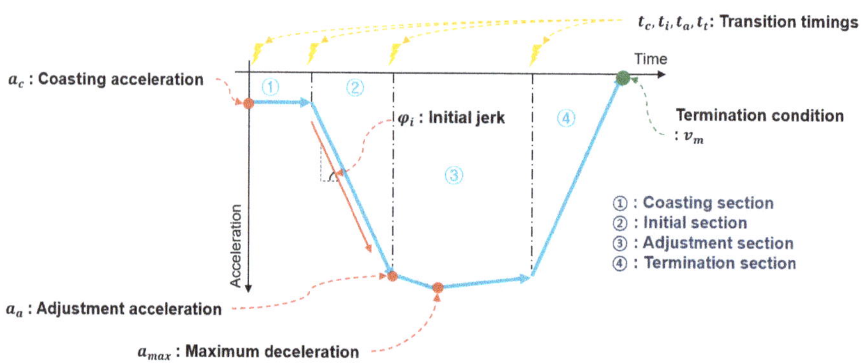

Figure 8. Model parameter description.

The initial acceleration slope parameter represents the jerk state when braking starts. Since the jerk can cause uncomfortable sensations, this parameter should be treated as an individual driver parameter. The minimum velocity determines the braking termination condition. About the static deceleration condition, the minimum velocity is defined as a model parameter according to each object. The minimum velocity for traffic-light conditions is zero because the vehicle should stop in front of the traffic light. The minimum velocity for curve and speed-bump conditions are determined depending on the driving conditions and the driver's driving characteristics. On the other hand, the minimum velocity of the car-following condition is defined depending on the preceding vehicle velocity.

In addition, gain parameters determine the convergence of the predicted acceleration profile. When the braking progress is applied to the adjustment section, the acceleration profile converges to the reference acceleration because the driver follows the stop condition. The relative distance parameters affect the transition time from the initial section to the adjustment section. The detailed process of the braking section transition and transformation of parametric equations is described in the next section.

4.4. Calculation Process of the Parametric Model

As mentioned above, the deceleration prediction model calculates the deceleration profile using parametric equations of the reference velocity v_{ref} and effective distance d_{eff}. These equations are determined using the model parameters according to the braking section as shown in Figure 9. As shown in the figure, parametric equations are determined differently to represent the specified deceleration characteristic of each braking section. At this time, the model parameters are applied to determine these equations. If the transition condition is satisfied, the prediction model shifts the braking section and the parametric equations are also changed. Model parameters also have an effect on the braking section transition as well as on determination of the parametric equations. The deceleration profile is calculated based on the modified intelligent driver model, and other predicted vehicle states about velocity and distance are calculated using this deceleration profile. Then, those predicted vehicle states are used to determine the parametric equations of the next step. This prediction process is operated iteratively while the prediction ends.

The descriptions of parametric equations for each braking section are explained as follows. In the coasting section, the model predicts the coasting deceleration with value a_c. The measured data shows that the deceleration values are around zero for all deceleration states. To continue this value, the value of the effective distance is determined as zero. Then, the reference velocity is updated according to the deceleration value as coasting starts.

In the initial section, the model predicts the jerk characteristics when braking starts. In this section, the effective distance is also set to zero, likewise in the coasting section. The reference velocity is calculated according to the acceleration slope φ of the model parameters. The parameters for each deceleration state determine the end time of the initial section. On the static deceleration condition, the drivers have a deceleration tendency to maintain the initial jerk until the deceleration reaches a specific deceleration value. Thus, the initial section is finished when the acceleration value reaches a specific acceleration value, which is an adjustment acceleration a_a. On the dynamic deceleration condition, the initial jerk is determined depending on the preceding vehicle state. Thus, the initial section of dynamic deceleration is finished when the relative distance is smaller than the adjustment relative distance d_a.

When the adjustment section starts, the driver normally adjusts the brake pedal to converge to the reference acceleration profile. To represent this convergence, the effective distance is determined using the acceleration error value between the predicted acceleration and the calculated reference acceleration profile. Since the acceleration is calculated by the ratio between the effective distance and relative distance, the initial value of effective distance is set to the adjustment relative distance. Then, the effective distance is adapted depending on the adjustment gain. This adjustment gain is defined differently depending on the respective deceleration conditions. The reference velocity is also calculated to keep the velocity ratio to the estimated velocity when the adjustment section starts. Finally, the adjustment section is finished when the acceleration reaches the reference acceleration.

When the braking is almost over, the driver controls the brake pedal to satisfy specific safety conditions. In the static deceleration conditions, the driver controls the acceleration to trace the reference acceleration to satisfy the minimum velocity at the front of the static object. On the other hand, the driver tends to keep a relatively safe distance according to the current vehicle speed of the car-following deceleration state. To guarantee these specific safety conditions, the model updates the effective distance similarly to the adjustment section using the termination gain parameter. When the vehicle velocity is slower than the minimum velocity or preceding vehicle velocity, the termination section is finished.

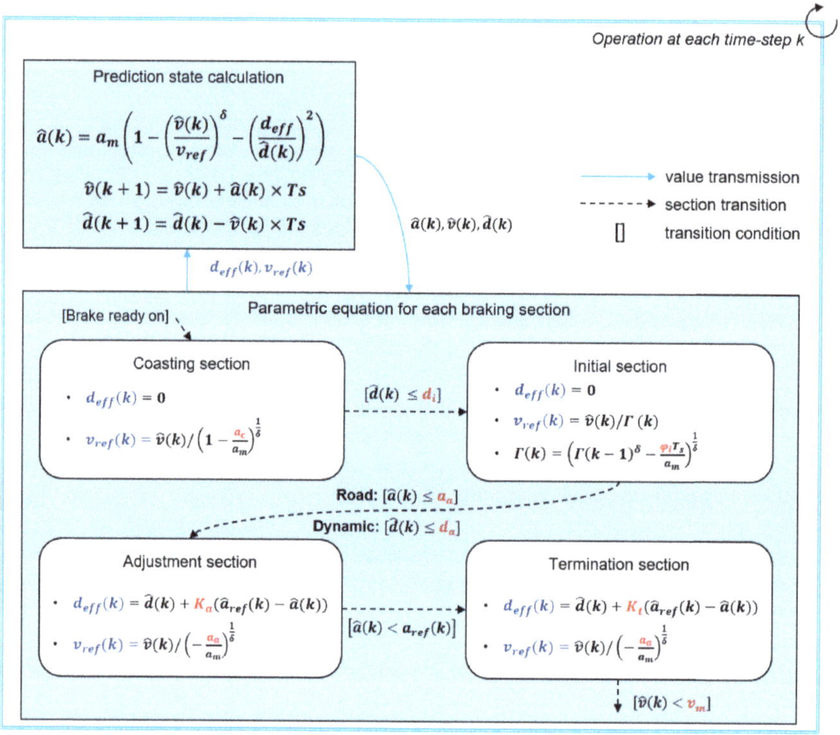

Figure 9. Detailed process for calculation of parametric model equations to predict deceleration profile.

5. Online Learning Algorithm

5.1. Overview of Online Learning Algorithm

Model parameters are used to determine parametric equations and braking-section transition of the deceleration-prediction algorithm. Thus, the deceleration-prediction algorithm predicts the various deceleration profiles depending on the parameter values. The proposed online learning updates these model parameters to reflect individual driver characteristics for various deceleration conditions. This algorithm consists of two parts; parameter activation and parameter update as shown in Figure 10.

The parameter-activation algorithm selects the parameter values of each model parameter using their learned vector arrays when deceleration starts according to the deceleration states. Then, the parameter-update algorithm updates the vector array value of each parameter using the reference parameter. The reference parameter is determined using the driver's braking data after deceleration is finished.

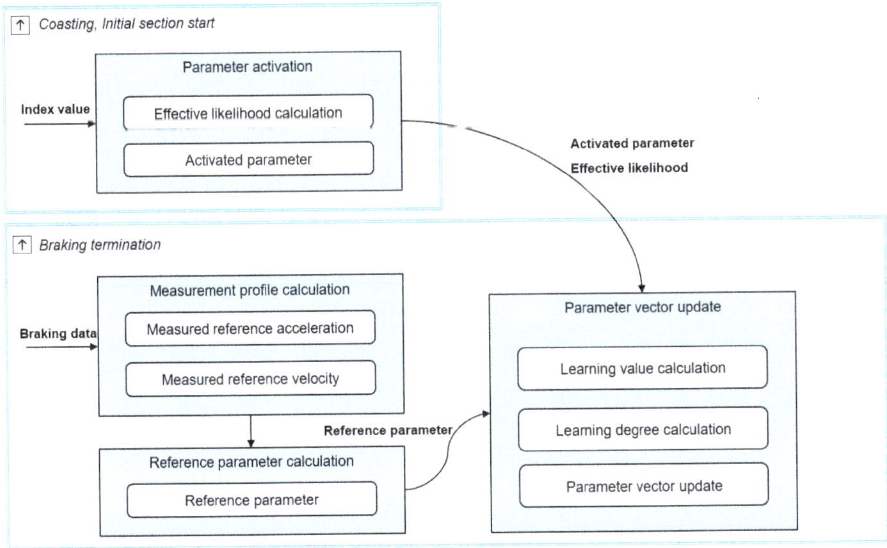

Figure 10. Process of the online learning algorithm.

5.2. Model Parameter Management

In order to manage parameters according to the deceleration states, the parameter values are defined as the vector array value depending on each vector index. The deceleration state which affects the selection of the parameter value is defined as the vector index of the vector array such as the initial deceleration condition, relative distance to the object, or ego-vehicle velocity. This vector index for each parameter is determined as a correlated deceleration state to its model parameter value. Thus, the parameter activation algorithm can select appropriate parameter values depending on the deceleration states. For example, a maximum acceleration parameter is highly correlated to the coast acceleration indicator for all drivers, which can be seen in Figure 11a. The coast acceleration indicator is determined as the difference between the reference acceleration value and the vehicle acceleration value when the coast section starts and is given in Equation (3). This parameter represents the deceleration start states due to the larger value of this parameter, which suggests that the driver should brake more strongly to satisfy the deceleration criterion. Therefore, the parameter activation algorithm should select the large maximum acceleration parameter value to predict strong deceleration if the deceleration state is a large coast acceleration indicator. By determining the vector array value and its index for the maximum acceleration parameter as shown in Figure 11b, the parameter activation algorithm can select the appropriate parameter value of the maximum acceleration according to the deceleration start state. Then, the online learning algorithm updates the parameter vector array value using the measured braking data of the individual driver when braking terminates. At this time, the updated value of the parameter vector is also related to the index indicating the deceleration states of braking data. As a result, the proposed algorithm learns and reflects on the individual driver's driving style according to various deceleration states, which can be seen in Figure 11c.

$$a_{ind,cst} = a(t_c) - a_{ref}(t_c) \tag{3}$$

Figure 12 shows the model parameters and their indices for static deceleration conditions. As mentioned in the explicit model parameters section, model parameters which are related to the value of acceleration represent the driver's characteristics explicitly. Thus, we determined two model parameters about the acceleration value: the maximum acceleration parameter and the adjustment acceleration parameter. As mentioned above, the maximum acceleration parameter is correlated to the coast acceleration indicator, which represents the start condition of deceleration. Similarly, the adjustment acceleration parameter is correlated to the initial acceleration indicator, which represents the deceleration state when the driver starts to step on the brake pedal.

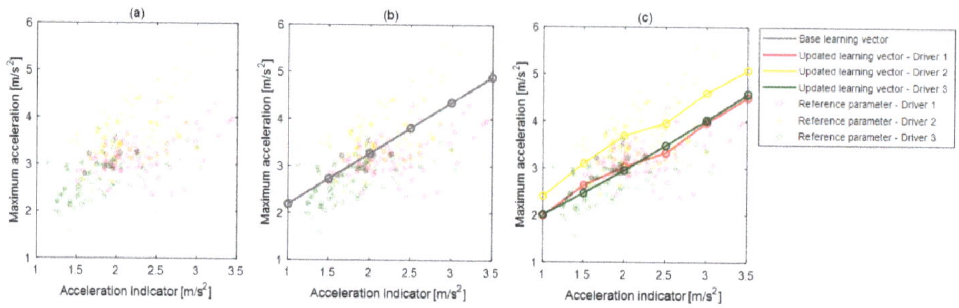

Figure 11. Maximum acceleration parameter and its updated vector array by the learning algorithm. (**a**: reference parameter value, **b**: based learning vector, **c**: updated learning vector.)

The initial jerk parameter, which is an acceleration differential value at the initial section, is also defined as a model parameter with the index of the initial acceleration indicator. As shown in Figure 12, the driver tends to decelerate strongly with larger adjustment acceleration and the initial jerk when the acceleration indicator is large like the maximum acceleration parameter. The last parameter is a distance difference from the coasting section to the initial section. This parameter is correlated with the time-to-collision value of coasting start. It represents that the driver's pedal-shift time decreases as the value of time to collision decreases.

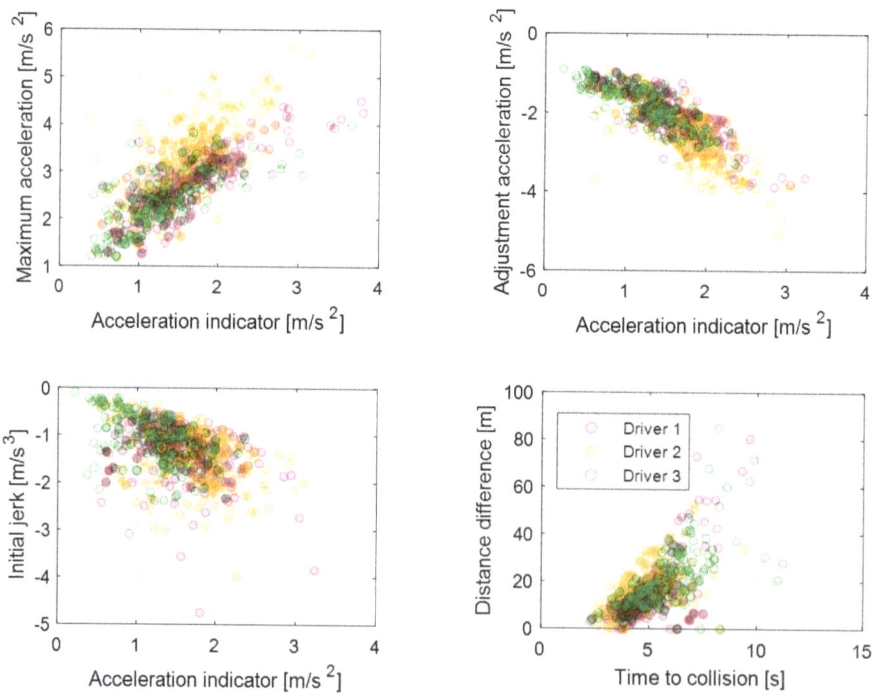

Figure 12. Model parameters and parameter indices for static deceleration condition.

In the dynamic deceleration condition, a driver's driving style is affected based on the preceding vehicle behavior. Therefore, the model parameters are determined as relative states from the preceding vehicle. The relative distance of the initial section is determined by the relative distance of the coast section as shown in Figure 13. This denotes that the driver tends to keep a relative distance during a car-following situation. Subsequently, the vector index of each relative distance parameter is determined by the relative distance of the previous braking section. The initial jerk parameter is also important for the dynamic deceleration condition. However, the correlation with the acceleration index is lower than the correlation of the static dynamic condition because the acceleration index be affected by preceding vehicle behavior. The minimum velocity difference is a velocity difference between the ego-vehicle's minimum velocity and the preceding vehicle's minimum velocity during the deceleration. This parameter does not have a large correlation with the acceleration index, but the average value can represent drivers' characteristics about the deceleration termination condition.

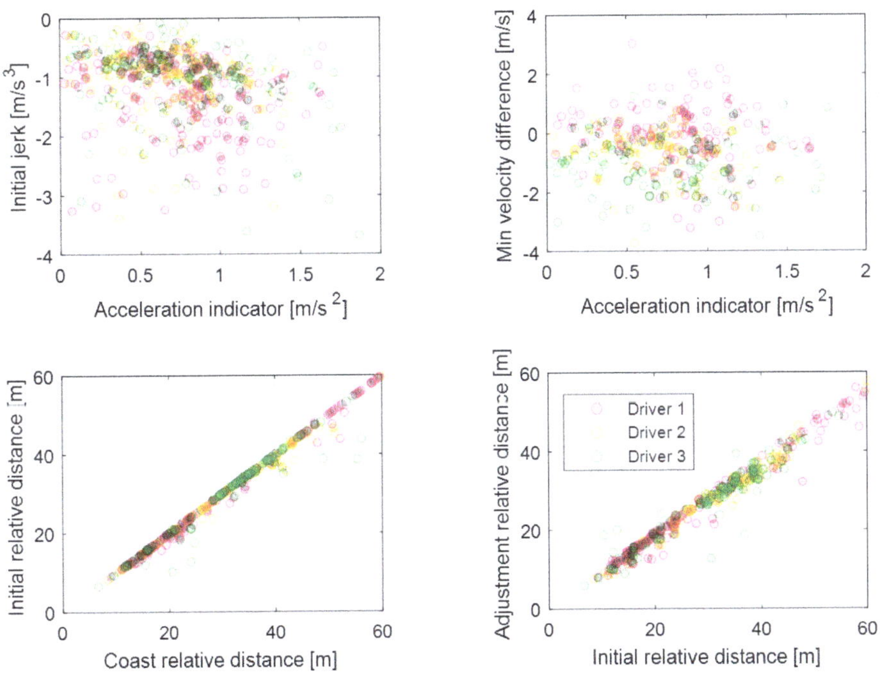

Figure 13. Model parameters and the parameter indices for dynamic deceleration condition.

5.3. Parameter Activation

To use the model parameters for the prediction model, the parameter activation algorithm selects the parameter value of each parameter from the its parameter vector array value. The selected parameter value should be related according to the relevant deceleration states. In addition, the parameter value is treated as stochastic because the parameter vector is updated based on the data-driven method [32].

As mentioned above, the vector index is correlated with the parameter vector and represents deceleration states. Thus, the parameter activation algorithm selects the parameter value by selecting the highly correlated vector index to the current deceleration states. To obtain the degree of relation between the vector index and deceleration states, an effective likelihood is determined by normalizing the Gaussian distribution for each index value as can be seen in Equation (4).

$$P(i) = Norm\left(\frac{1}{\sigma\sqrt{2\pi}}\exp^{-\frac{1}{2}\frac{(i-i_{val})}{\sigma}^2}\right) \tag{4}$$

where $P(i)$ is the effective likelihood for each vector index i, i_{val} is the index value for the current driving state, and σ is the standard deviation of the index value.

For example, when the acceleration indicator is calculated as a 2.1 value according to the deceleration states as shown in Figure 14a, the effective likelihood of the vector index is also determined based on the Gaussian distribution as shown in Figure 14b. The effective likelihood of acceleration indicator value 2 and 2.5 are larger values than the others because the acceleration indicator of the current deceleration state is a 2.1 value. Then, the parameter value is determined by inner product of the effective likelihood and parameter vector value set as shown in Equation (5). Through this process, the parameter-activation algorithm can select the parameter value that is correlated with current deceleration states to use for the deceleration prediction model.

$$\theta_{act} = P(i) \bullet V_\theta(i) \tag{5}$$

where θ_{act} is an activated parameter and $V_\theta(i)$ is a vector array value for parameter θ.

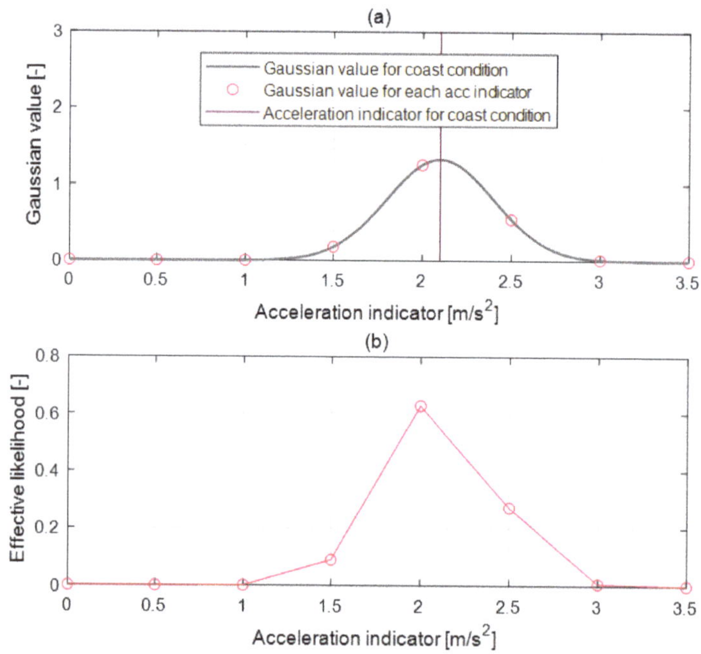

Figure 14. Gaussian value and effective likelihood according to acceleration indicator value. (**a**: Gaussian value, **b**: Effective likelihood).

5.4. Reference Parameter Calculation

In order to update the vector value of each parameter to reflect individual driver characteristics, the online learning algorithm calculates the reference values of each parameter using the braking data of the driver. At first, the transition points of each braking section are determined because the proposed model parameters are determined related to the braking section transition. For example, the adjustment acceleration is determined as the acceleration value when the adjustment section starts. The initial jerk is also determined using the acceleration values between the start point of the initial section and start point of the adjustment section.

The transition times for the coasting section and the initial section are determined easily using the pedal transition information of the driver. When the driver releases the acceleration pedal, the coasting section starts. Then, the initial section starts when the driver pushes the brake pedal. To determine the transition times for the adjustment section and termination section, the reference acceleration and reference velocity are calculated using the measured braking data. These two profiles are described in the prediction model section. The reference velocity is calculated based on the IDM equation to find the adjustment point. By setting the effective distance as a zero value and by using the minimum velocity parameter, the reference velocity profile $v_{ref,measure}$ is defined as Equation (6). Then, the velocity difference is calculated between the reference velocity and the measured velocity, as shown in Figure 15. As shown in the figure, the adjustment point can be determined as the maximum velocity difference point. The timing point is determined as the termination point when the vehicle acceleration is converged to the measured reference acceleration.

$$v_{ref,measure}(k) = (v(k) - v_{min}) / \left(1 - \frac{a(k)}{a_m}^{\frac{1}{\sigma}}\right) \tag{6}$$

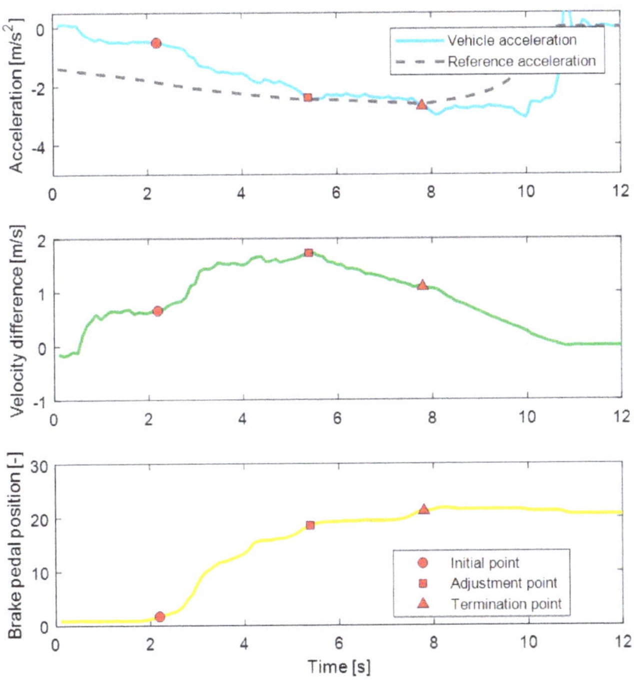

Figure 15. Detailed process for calculation of parametric model equations to predict deceleration profile.

Reference variables of relative distance parameters are defined by using the measured relative distance from the radar sensor at each section transition point. The reference value of the minimum velocity parameter is defined using the measured vehicle velocity during the braking conditions. Similarly, the reference value of maximum acceleration is also defined using the measured vehicle acceleration.

5.5. Parameter Vector Update

After calculation of the reference value using the braking data, the online learning algorithm updates the vector array value. The error value is calculated using the activated parameter value

θ_{act} from the previous vector array value and the reference value θ_{ref} from the braking data. Then, the error value defines an update target value δ_θ with a learning rate α as Equation (7). Through the simple value incremental learning process [33], the update algorithm adds the update values to each vector value as Equation (8).

$$\delta_\theta = \alpha(\theta_{ref} - \theta_{act}) \tag{7}$$

$$V_\theta(i)^+ = V_\theta(i)^- + \psi(i)\delta_\theta \tag{8}$$

In this process, the update algorithm applies the learning degree $\psi(i)$, which is a weight value to determine the updating weights for each vector value. Similar to the parameter activation, the parameter vector value, which is highly correlated to the current driving status, should be updated with a large update weight. Thus, the learning degree is determined based on the effective likelihood to consider the current deceleration states for model parameter as displayed in Equations (9) and (10). Equation (9) is derived from Equations (5) and (8). It means that the activated parameter value using the updated vector array value $P(i) \bullet V_\theta(i)^+$ should be the same as the reference parameter value θ_{ref} if the deceleration states are the same. Using these equations, the learning degree resolves the value of parameter vector to be updated correctly. In addition, the learning degree leads the larger update value for the larger effective probability.

$$\psi(i) = \left(\sum P(\sim i) \times \sum \left(\frac{P(i)}{\sum P(\sim i)}\right)\right)^{-1} \tag{9}$$

$$\sum P(\sim 1) = P(2) + P(3) + \cdots + P(8) \tag{10}$$

6. Vehicle Experimental Results

6.1. Experimental Condition

The proposed algorithm was validated through the vehicle experiment. To validate the static deceleration condition, the algorithm has to know the exact position of the objectives and its status. Since the predefined HD map can provide the position of objectives, the validation experiment for the static deceleration condition was conducted at a specific test cite. The proving ground is Yeongjongdo in Incheon, South Korea, whereby this site contains a speed bump, traffic light, and curvature road conditions to successfully carry out experiments. The algorithm for the dynamic deceleration condition in the car-following situation was validated in various driving conditions for urban driving and highway-driving environments. To specify the driving characteristics for the individual driver, three drivers conducted the vehicle experiments according to the case listed above. The purpose of the proposed research is to present a predictive model for various deceleration conditions and to explicitly update the driving characteristics of individual drivers and not to macroscopically classify them using machine learning methods. Therefore, even if a model is defined and validated using a small number of driver's data, the model is valid if it can clearly show the individual driver's characteristics for various deceleration conditions.

6.2. Deceleration Prediction Results Under Various Driving Conditions

Figures 16–18 show the deceleration prediction results of the model in static deceleration conditions. The static deceleration condition contains the traffic-light stop condition, curved road condition, and speed bump condition. About the static deceleration condition, the model parameters were respectively updated according to individual driver for each deceleration condition. Therefore, the proposed model can predict the deceleration profiles depending on the deceleration conditions, as shown in figures. The prediction results about the traffic-light stop condition are more accurate than the curved road condition and speed bump condition. Since the termination velocity parameter affects the prediction accuracy, the traffic-light stop condition which determines the termination velocity parameter as the fixed zero-velocity value shows a more accurate prediction than other conditions. About curved road conditions, the termination velocity is determined according to the maximum road curvature and its position from the ego-vehicle. However, since the termination velocity is not constant, even though the max curvature and driver are same, it is difficult to determine the exact termination velocity. Therefore, the prediction error that occurred in some deceleration cases of curve condition by the inaccurate activation of the termination velocity parameter, such as the deceleration case of about 50 s for driver 1 or of about 60 s for driver 3. The deceleration profiles about the speed bump condition show similarities to the stop case. According to the tests, driver 2 tends to decelerate more aggressively. Especially for the bump case, the maximum deceleration values of driver 2 are almost near 5 m/s^2. The prediction model cannot represent this strong braking at an early braking state. The model does not work at deceleration cases around 100 s of bump condition because the deceleration occurs due to other reasons and not just due to the speed bump.

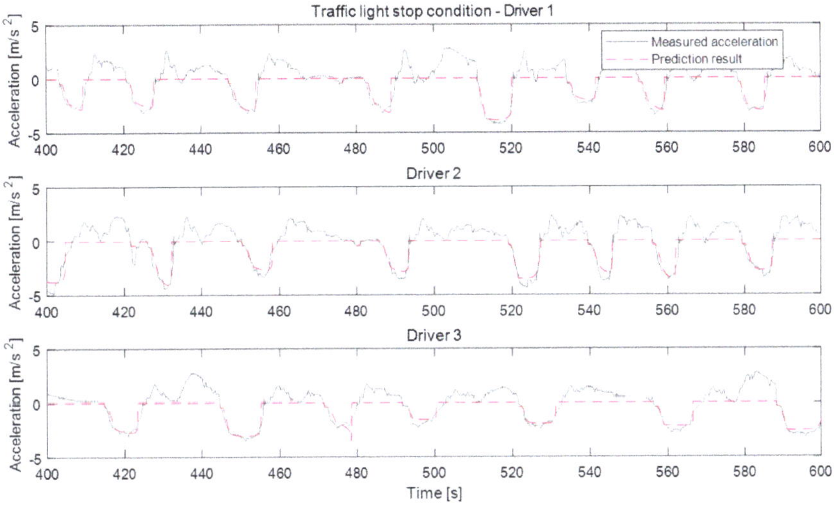

Figure 16. Deceleration prediction results on traffic-light stop condition.

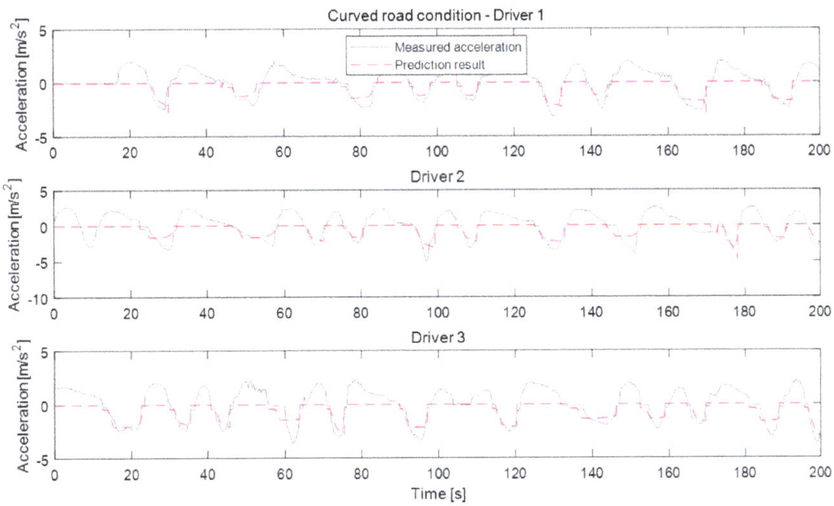

Figure 17. Deceleration prediction results on curved road condition.

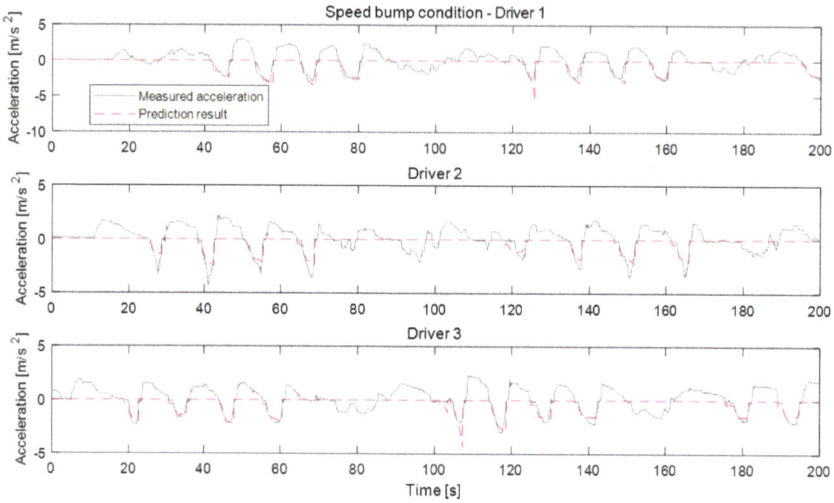

Figure 18. Deceleration prediction results on speed bump condition.

The deceleration model about dynamic conditions was validated through proving ground driving and real road driving. Real road driving contains various urban and highway driving environments. In the proving ground results, the vehicle deceleration occurred by deceleration of a preceding vehicle. The results shown in Figure 19 are consistent in the stable deceleration as the preceding vehicle decelerated intentionally. The prediction results show also that the model predicts the deceleration profile at each deceleration case though the preceding vehicle affects deceleration as well. The model did not predict the deceleration on some deceleration cases because the vehicle deceleration was not caused by the preceding vehicle. The prediction about deceleration cases for around 280 s for driver 1 or 290 s for driver 3 are terminated even before the deceleration ends. This prediction termination occurred because the preceding vehicle deviated from the driving course.

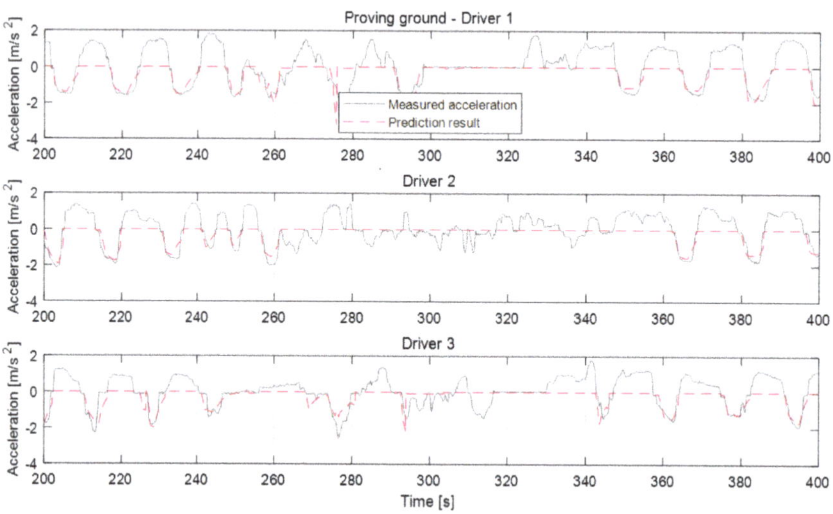

Figure 19. Deceleration prediction results in the dynamic deceleration condition on proving ground.

6.3. Parameter Learning Results

As mentioned in the parameter management section, the model parameter values are managed as vector values according to the highly correlated index. At first, using the vehicle experiment data for all drivers, we determined the base parameter vector to apply the proposed learning algorithm. As shown in the algorithm overview in Figure 4, the proposed algorithm updates the parameter vector when each deceleration is terminated. This process was concurrent with the verification of the prediction algorithm through the various experiment data that are described in the previous prediction validation section.

Parameter vectors of the static condition deceleration model are updated according to each deceleration condition for the individual drivers. Figures 20–23 depict the learning results of the static model parameters: maximum acceleration, adjustment acceleration, initial jerk, and distance difference. Those updated results for the model parameters can represent the driver characteristics on each deceleration condition. As shown in Figure 20, the updated learning vectors about the maximum acceleration parameter shows that, generally, driver 2 decelerates with larger deceleration values than the other drivers in any deceleration conditions. Similar driver characteristics are also described as the adjustment acceleration parameter and initial jerk parameter in Figures 21 and 22. The learning results of these parameters on driver 2 have larger negative values than the learning results of the other drivers. These results could mean that the driving characteristic of driver 2 is more aggressive than the other drivers. The initial jerk and the distance difference parameters represent the different characteristics according to not only the driver but also the deceleration condition. The initial jerk parameter which affects the deceleration feeling in early deceleration is larger for the traffic-light stop condition than for other deceleration conditions.

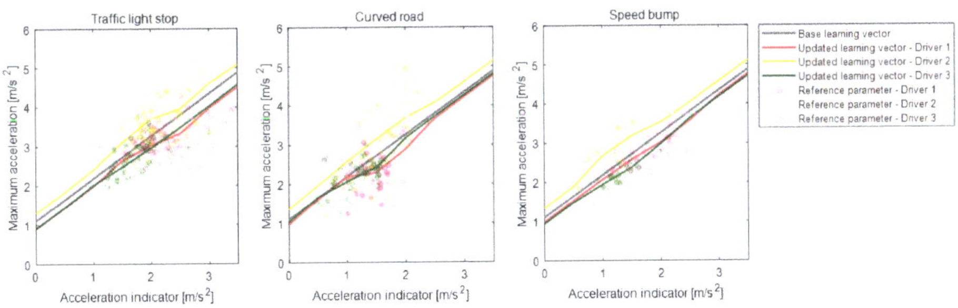

Figure 20. Learning results for parameter maximum acceleration for each road object.

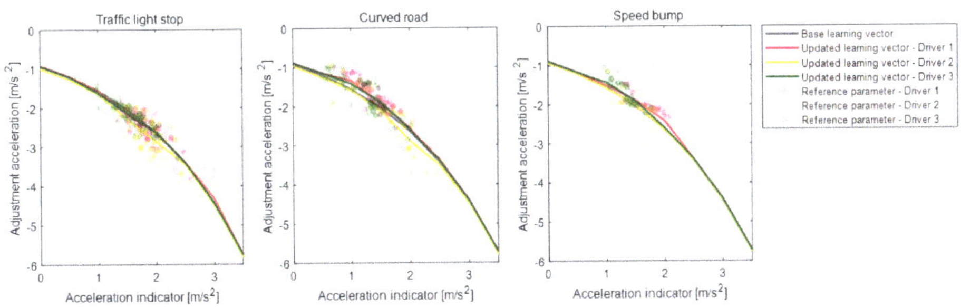

Figure 21. Learning results for parameter adjustment acceleration for each road object.

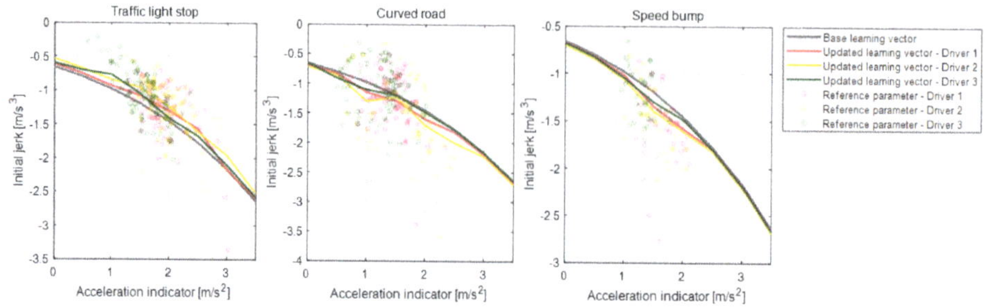

Figure 22. Learning results for parameter initial jerk for each road object.

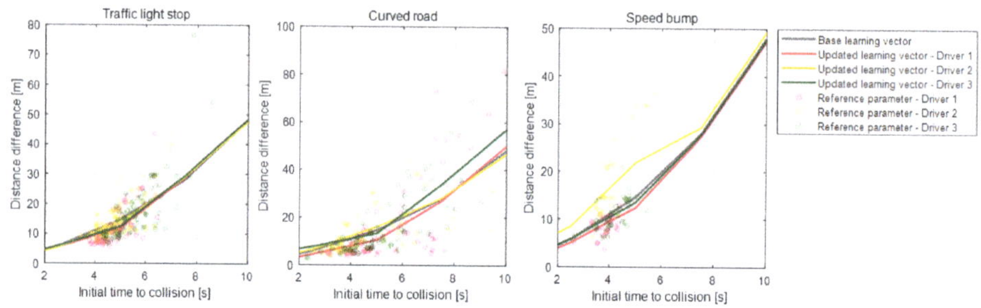

Figure 23. Learning results for parameter distance difference for each road object.

Figure 24 shows the updated results of the parameter vectors about the dynamic deceleration condition. According to the drivers' results, driver 1 shows deceleration characteristics with a larger initial jerk than the others. These results show that driver 1 pushes the brake pedal with more strength when braking starts. The updated learning vector on the minimum velocity difference parameter of driver 1 is larger than of the other drivers. It means that driver 1 is slower than other drivers when the driver terminates deceleration. The updated learning vector on the initial relative distance parameter of driver 1 is smaller than of the other drivers. This suggests that driver 1 uses more time to pedal shift from the accelerator pedal to the brake pedal. Learning results about the adjustment relative distance show a similar tendency with the initial relative distance parameter for each driver.

6.4. Case Study to Prediction Model-Based Regenerative Control

The proposed algorithm was applied to the smart regenerative control system of an electric vehicle as a case study. The controller structure is shown in Figure 25. The deceleration condition recognition algorithm determines the vehicle deceleration when the driver releases their foot from the acceleration pedal. Afterwards, the motor torque control algorithm generates the regenerative torque if the driver does not push the brake pedal. At this time the acceleration profile from the prediction model is used as the acceleration set point of the motor torque control algorithm. On the other hand, the motor torque control algorithm does not operate if the driver pushes the brake pedal. In this case, the prediction model just predicts the acceleration profile; then, the learning algorithm updates the parameter vectors using the braking data.

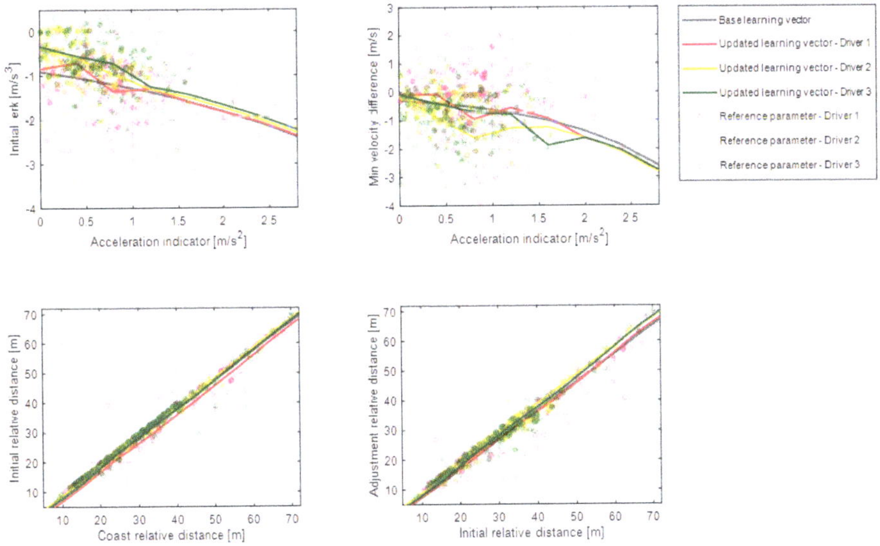

Figure 24. Learning results for model parameters on the dynamic deceleration condition.

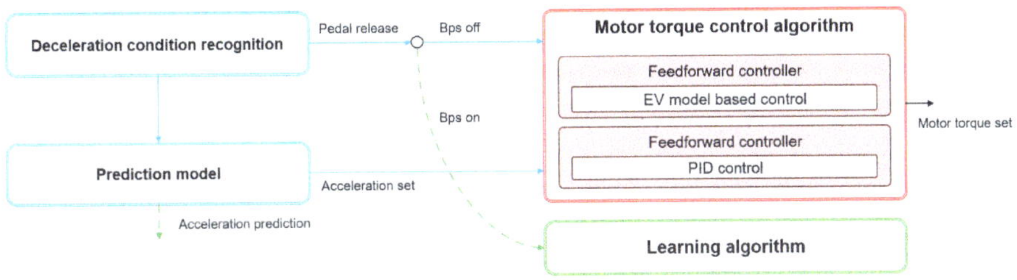

Figure 25. Structure of the regenerative controller based on the proposed driver model.

The torque control algorithm consists of two controllers: the feedforward controller and the feedback controller. The feedforward controller determines the motor torque based on the electric vehicle model. Using the electric vehicle model, the feedforward controller calculates the desired regenerative motor torque to generate the acceleration set point. The feedback controller is a well-known Proportional-Integral-Derivative controller (PID controller) to trace the acceleration set point from the prediction model.

The Vehicle Control Unit (VCU) controls the motor torque according to the generated regenerative torque from the control algorithm. Consequently, the vehicle decelerates without the driver's braking pedaling action. Thus, it provides driving convenience by excluding pedaling of the vehicle's brake. Figure 26 shows the deceleration control results using the proposed driver model and the torque controller; the red dashed line in the top graph is the actual vehicle acceleration, and the gray solid line is the predicted acceleration profile from the model. As shown in the figure, vehicle acceleration traces the acceleration set point from the model from 40 s to 170 s and around 230 s. In contrast, the model only predicts the acceleration profile when the driver pushes the brake pedal around 200 s. At this time, the vehicle is decelerated according to the driver's braking action and the learning algorithm updates the model parameter vectors.

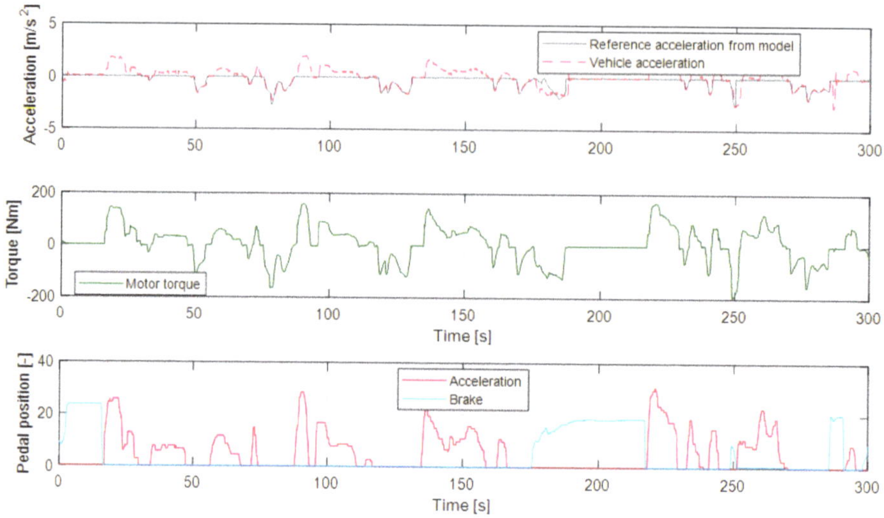

Figure 26. Deceleration prediction model-based regenerative control results.

7. Conclusions

In this paper, we proposed a deceleration prediction model based on individual driver characteristics. The proposed prediction model is designed based on the mathematical intelligent driver model, and it is modified especially to the deceleration prediction with some explicit model parameters. First, we defined the braking section that describes the deceleration characteristics to specifically predict the deceleration state. Then, the parametric equations were calculated according to the braking section with explicit model parameters to predict the deceleration profile. The model parameters represent the individual driver characteristics by updating the vector array values online. These vectors are also updated according to the various deceleration conditions about the various road objects or preceding vehicle. Thus, the proposed model can be used for various deceleration conditions by considering the individual driver's characteristics. The proposed algorithm was validated through vehicle experiments and applied to smart regenerative control. Since the model can execute various deceleration conditions and driver characteristics, the smart regenerative control based on the proposed model does not cause the sense of heterogeneity for an individual driver. In future research, the proposed algorithm will be applied to smart regenerative systems on real driving situations. To achieve this, the deceleration condition recognition algorithm will be modified more practically. The algorithm can substitute the navigation device for the HD map and GPS to determine information of road objectives. Furthermore, the dynamic deceleration condition also is extended to consider traffic jam or cut-in situations.

Author Contributions: Conceptualization, K.M. and G.S.; funding acquisition, M.S.; methodology, K.M.; project administration, M.S.; software, K.M., G.S. and S.A.; supervision, K.J.; writing—original draft, K.M.; writing—review and editing, G.S., S.A. and K.J.

Funding: This work was financially supported by the BK21 plus program (22A20130000045) under the Ministry of Education, Republic of Korea; by the Industrial Strategy Technology Development Program (No. 10039673, 10060068, and 10079961); by the International Collaborative Research and Development Program (N0001992) under the Ministry of Trade, Industry, and Energy (MOTIE Korea); and by the National Research Foundation of Korea (NRF) grant funded by the Korean government (MEST) (No. 2011-0017495).

Conflicts of Interest: The authors declare no conflict of interest.

References

1. Cheung, E.; Bera, A.; Manocha, D. Efficient and safe vehicle navigation based on driver behavior classification. In Proceedings of the IEEE Computer Society Conference on Computer Vision and Pattern Recognition Workshops, Salt Lake City, UT, USA, 18–22 June 2018; pp. 1137–1144. [CrossRef]
2. Gong, Q.; Li, Y.; Peng, Z.; Gong, Q.; Li, Y.; Peng, Z.R. Trip-Based Optimal Power Management of Plug-in Hybrid Electric Vehicles. *IEEE Trans. Veh. Technol.* **2008**, *57*, 3393–3401. [CrossRef]
3. Dib, W.; Serrao, L.; Sciarretta, A. Optimal control to minimize trip time and energy consumption in electric vehicles. In Proceedings of the 2011 IEEE Vehicle Power and Propulsion Conference, Chicago, IL, USA, 6–9 September 2011; pp. 1–8. [CrossRef]
4. Wan, J.; Wu, C. The Effects of Driver Speed Prediction-Based Battery Management System on Li-ion Battery Performance for Electric Vehicles. *Proc. Hum. Factors Ergon. Soc. Annu. Meet.* **2014**, *58*, 515–519. [CrossRef]
5. Qi, Y.; Xiang, C.; Wang, W.; Wen, B.; Ding, F. Model Predictive Coordinated Control for Dual-Mode Power-Split Hybrid Electric Vehicle. *Int. J. Automot. Technol.* **2018**, *19*, 345–358. [CrossRef]
6. Balasubramanian, B.; Huzefa, A.C. Development of regeneration braking model for electric vehicle range improvement. In Proceedings of the 2017 IEEE Transportation Electrification Conference (ITEC-India), Pune, India, 13–15 December 2017; pp. 1–5. [CrossRef]
7. Xu, G.; Li, W.; Xu, K.; Song, Z.; Xu, G.; Li, W.; Xu, K.; Song, Z. An Intelligent Regenerative Braking Strategy for Electric Vehicles. *Energies* **2011**, *4*, 1461–1477. [CrossRef]
8. Dou, J.; Cui, G.; Li, S.; Zhao, X.; Lu, X.; Yu, Z. MPC-based cooperative braking control for rear-wheel-drive electric vehicle. In Proceedings of the 2017 Chinese Automation Congress (CAC), Jinan, China, 20–22 October 2017; pp. 4419–4424. [CrossRef]
9. Wang, J.; Zhang, L.; Zhang, D.; Li, K. An adaptive longitudinal driving assistance system based on driver characteristics. *IEEE Trans. Intell. Transp. Syst.* **2013**, *14*, 1–12. [CrossRef]
10. Smiley, A. Behavioral Adaptation, Safety, and Intelligent Transportation Systems. *Transp. Res. Rec. J. Transp. Res. Board* **2000**, *1724*, 47–51. [CrossRef]
11. Lefèvre, S.; Carvalho, A.; Gao, Y.; Tseng, H.E.; Borrelli, F. Driver models for personalised driving assistance. *Veh. Syst. Dyn.* **2015**, *53*, 1705–1720. [CrossRef]
12. McCall, J.C.; Trivedi, M.M. Driver Behavior and Situation-aware Brake Assistance for Intelligent Vehicles. *Proc. IEEE* **2007**, *95*, 374–387. [CrossRef]
13. Martinez, C.M.; Heucke, M.; Wang, F.Y.; Gao, B.; Cao, D. Driving Style Recognition for Intelligent Vehicle Control and Advanced Driver Assistance: A Survey. *IEEE Trans. Intell. Transp. Syst.* **2017**, *19*, 1–11. [CrossRef]
14. Pariota, L.; Galante, F.; Bifulco, G.N. Heterogeneity of Driving Behaviors in Different Car-Following Conditions. *Period. Polytech. Transp. Eng.* **2016**, *44*, 105–114. [CrossRef]
15. Mitra, P.; Eric, B.; Florian, Fürer; Björn, L.; Klaas, B. Calibration and Evaluation of Car Following Models Using Real-World Driving Data. In Proceedings of the 2017 IEEE 20th International Conference on Intelligent Transportation Systems (ITSC), Jinan, China, 20–22 October 2017; pp. 1453–1458. [CrossRef]
16. Liebner, M.; Klanner, F.; Baumann, M.; Ruhhammer, C.; Stiller, C. Velocity-based driver intent inference at urban intersections in the presence of preceding vehicles. *IEEE Intell. Transp. Syst. Mag.* **2013**, *5*, 10–21. [CrossRef]
17. Malinauskas, R. The Intelligent Driver Model: Analysis and Application to Adaptive Cruise Control. Master Thesis, Clemson University, Clemson, SC, USA, 2014.
18. Zheng, L.J.; Tian, C.; Sun, D.H.; Liu, W.N. A new car-following model with consideration of anticipation driving behavior. *Nonlinear Dyn.* **2012**, *70*, 1205–1211. [CrossRef]
19. Treiber, M.; Hennecke, A.; Helbing, D. Congested traffic states in empirical observations and microsopic simulations.pdf. *Phys. Rev. E* **2000**, *62*, 1805–1824. [CrossRef] [PubMed]
20. Huang, X.; Sun, J.; Sun, J. A car-following model considering asymmetric driving behavior based on long short-term memory neural networks. *Transp. Res. Part C Emerg. Technol.* **2018**, *95*, 346–362. [CrossRef]
21. Yu, H.; Wu, Z.; Wang, S.; Wang, Y.; Ma, X. Spatiotemporal recurrent convolutional networks for traffic prediction in transportation networks. *Sensors* **2017**, *17*, 1501. [CrossRef] [PubMed]
22. Morton, J.; Wheeler, T.A.; Kochenderfer, M.J. Analysis of Recurrent Neural Networks for Probabilistic Modeling of Driver Behavior. *IEEE Trans. Intell. Transp. Syst.* **2017**, *18*, 1289–1298. [CrossRef]

23. Oliver, N.; Pentland, A.P. Driver behavior recognition and prediction in a SmartCar. *Proc. SPIE Int. Soc. Opt. Eng.* **2000**, *4023*, 280–290.
24. Kuge, N.; Yamamura, T.; Shimoyama, O.; Liu, A. A Driver Behavior Recognition Method Based on a Driver Model Framework. *SAE Tech. Pap. Ser.* **2010**, *1*. [CrossRef]
25. Wang, W.; Xi, J.; Zhao, D. Learning and Inferring a Driver's Braking Action in Car-Following Scenarios. *IEEE Trans. Veh. Technol.* **2018**. [CrossRef]
26. Jo, K.; Chu, K.; Sunwoo, M. Interacting Multiple Model Filter-Based Sensor Fusion of GPS With In-Vehicle Sensors for Real-Time Vehicle Positioning. *IEEE Trans. Intell. Transp. Syst.* **2012**, *13*, 329–343. [CrossRef]
27. Jo, K.; Chu, K.; Sunwoo, M. GPS-bias correction for precise localization of autonomous vehicles. In Proceedings of the 2013 IEEE Intelligent Vehicles Symposium (IV), Gold Coast, QLD, Australia, 23–26 June 2013; pp. 636–641.
28. Jang, C.; Cho, S.; Jeong, S.; Suhr, J.K.; Jung, H.G.; Sunwoo, M. Traffic light recognition exploiting map and localization at every stage. *Expert Syst. Appl.* **2017**, *88*, 290–304. [CrossRef]
29. Demir, M.; Çavus, A. A new driver behavior model to create realistic urban traffic environment. *Transp. Res. Part F* **2012**. [CrossRef]
30. Deng, Z.; Chu, D.; Wu, C.; He, Y.; Cui, J. Curve safe speed model considering driving style based on driver behaviour questionnaire. *Transp. Res. Part F* **2018**. [CrossRef]
31. Reymond, G.; Kemeny, A.; Droulez, J.; Berthoz, A. Role of Lateral Acceleration in Curve Driving: Driver Model and Experiments on a Real Vehicle and a Driving Simulator. *Hum. Factors* **2001**, *43*, 483–495. [CrossRef] [PubMed]
32. Solomatine, D.; See, L.; Abrahart, R. Data-Driven Modelling: Concepts, Approaches and Experiences. In *Practical Hydroinformatics*; Springer: Berlin/Heidelberg, Germany, 2009; pp. 17–30._2. [CrossRef]
33. Sutton, R.S. Learning to predict by the methods of temporal differences. *Mach. Learn.* **1988**, *3*, 9–44. [CrossRef]

sensors

MDPI

Article

Combined Edge- and Stixel-based Object Detection in 3D Point Cloud

Fangchao Hu [1,*], Dong Yang [2] and Yinguo Li [2]

1 Department of Computer Science and Technology, Chongqing University of Posts and Telecommunications, Chongqing 400065, China
2 Department of Automation, Chongqing University of Posts and Telecommunications, Chongqing 400065, China; s170331069@stu.cqupt.edu.cn (D.Y.); liyg@cqupt.edu.cn (Y.L.)
* Correspondence: fangchaohu1211@126.com; Tel.: +86-0236-248-7312

Received: 13 August 2019; Accepted: 10 October 2019; Published: 12 October 2019

Abstract: Environment perception is critical for feasible path planning and safe driving for autonomous vehicles. Perception devices, such as camera, LiDAR (Light Detection and Ranging), IMU (Inertial Measurement Unit), etc., only provide raw sensing data with no identification of vital objects, which is insufficient for autonomous vehicles to perform safe and efficient self-driving operations. This study proposes an improved edge-oriented segmentation-based method to detect the objects from the sensed three-dimensional (3D) point cloud. The improved edge-oriented segmentation-based method consists of three main steps: First, the bounding areas of objects are identified by edge detection and stixel estimation in corresponding two-dimensional (2D) images taken by a stereo camera. Second, 3D sparse point clouds of objects are reconstructed in bounding areas. Finally, the dense point clouds of objects are segmented by matching the 3D sparse point clouds of objects with the whole scene point cloud. After comparison with the existing methods of segmentation, the experimental results demonstrate that the proposed edge-oriented segmentation method improves the precision of 3D point cloud segmentation, and that the objects can be segmented accurately. Meanwhile, the visualization of output data in advanced driving assistance systems (ADAS) can be greatly facilitated due to the decrease in computational time and the decrease in the number of points in the object's point cloud.

Keywords: autonomous vehicle; objects' edge detection; stixel histograms accumulate; point cloud segmentation

1. Introduction

To securely and efficiently drive in increasingly complex traffic, the drivers must have a distinct and correct understanding of the environment. Nevertheless, obtaining and processing driving-related information is a great challenge due to the complexity of the environment and evolving traffic dynamics. For instance, complex architecture (e.g., the flyovers, switchback, abrupt slope), moving pedestrians and vehicles in an urban area, dim illumination in underground parking lots, feeble GPS signal, and inaccurate positioning increases the difficulty of driving [1,2]. Such challenges exist not only for human drivers, but also for autonomous vehicles whose safety heavily relies on knowledge of the surrounding environment.

To address these issues, many vehicle manufacturers focus on developing the advanced driving assistance system (ADAS) to assist drivers with decision making [3]. It plays a more and more important role to ensure safety nowadays with advanced technology. The most common facilities for perception on autonomous vehicles are radars, LiDAR (Light Detection and Ranging), cameras, GPS (Global Positioning System), and INS (Inertial Measurement Unit) [4]. LiDAR, which is equipped in Google self-driving cars to identify objects and obstacles during driving, can be utilized to sense the driving environment extremely efficiently and precisely. However, LiDAR cannot be widely employed in ordinary vehicles due to its stiff price. While LiDAR can directly provide important three-dimensional (3D) information, the required data lack rich appearance. This is where a camera-based system bears high potential [5]. By comparison, the stereo camera is capable of achieving similar functions while maintaining low operational costs. The stereo camera has several advantages in practice. First, it is able to obtain more real-time environmental data for driving, due to the short work cycle. Second, the number of 3D points that is produced by LiDAR is more redundant than that of the stereo camera. Therefore, this paper exploits the stereo camera-based method, which reduces the computation burden of image processing to achieve real-time driving assistance.

This paper attempts to identify obstacles in the driving environment, which is the most essential thing for driving. Numerous algorithms have been proposed for obstacle segmenting in different applications. Yu et al. [6] proposed a graph matching-based scene parsing framework to segment 3D point cloud. The graph matching approach can effectively interpret the street scene, but it is computationally expensive as it works on a graph processing at the voxel-wise level. Liu et al. [7] developed a four-step method to label 3D point cloud. This method enhances the inference accuracy by transferring the reliable two-dimensional (2D) labeling results into 3D. However, it demands a huge amount of points that are obtained from a 3D scanner, and significantly adds to the computational burden. Xiao et al. [8] utilized mobile laser scanning (MLS) data to detect street-side vehicles and classify the type of vehicles. This method improves the recognition rate and the localization precision. The laser scanner produces many points to process, and it costs a lot of time and computational resources. Zhang et al. [9] deployed 2D convolutions to segment 3D point cloud. It was faster and required less memory, but it requires pre-training. Narksri et al. [10] utilized the geometric characteristics of point cloud to primarily segment point cloud. Barea et al. [11] deployed the RGB-based CNN and projected into LiDAR point cloud to segment point cloud. Pan et al. [12] presented a top-down method for segmenting main bridge components, combined with rule-based classification, to produce a labeled 3D model from point cloud. Huang et al. [13] represented a multi-scale feature to classify the point cloud into a more effective and discriminative performance. Pan et al. [14] proposed a graph-cut-based method to segment point clouds automatically from multi-view images. This method does not require manual labeling of points and can be automatically run. Moreover, some object proposal methods are proposed to improve the performance of segmentation. For example, Sun et al. [15] proposed a multi-scale approach to detect a vehicle's edges using three different image resolutions. Betke [16] suggested a coarse-to-fine search technique, in which the coarse search identified groups of prominent edges and a finer search performed on these regions detected rectangular-shaped objects. However, with the stereo camera-based method, the 3D points are generated from images with an indirect method, and the precision of localization of geometry extraction and 3D modeling are still challenging tasks which remain unresolved.

There are lots of researches on 3D scene segmentation, many of which deal with 3D point cloud directly, rather than the abovementioned indirect method. Because 3D point clouds contain essential information of shape and spatial, which can characterize the contextual and spatial relationships between different objects, these characteristics provide the cues to segment the objects in 3D point cloud [17]. On the other hand, compared with the objects in indoor environments, the characteristics of objects in outdoor environments are dissimilar, e.g., the objects are apparently larger and the influences of illumination are stronger [18]. Besides, the speed of motion of objects outdoors are faster. These characteristics result in objects that are difficult to recognize and segment.

In this paper, we focus on edge detecting-based object segmentation in 3D point cloud scenes. Specifically, we aim to address the problem of object segmentation in 3D scenes with unsupervised learning by using the stixel estimation and edges of objects extracted from sequential 2D corresponding images. The unsupervised learning method has lower computation complexity than the supervised learning method, especially the deep learning method which cannot run on a vehicle-mounted computing element. Besides, the unsupervised learning method can avoid system error, which is caused by a lack of classifying. It should be noted that the proposed method works in the scenario where a whole scene point cloud is merged from multi-view point clouds. The steps of the proposed method in this paper are stated as follows. First, edges of the objects in corresponding images, which are acquired by the stereo camera, are detected respectively. Second, the feature points in images are detected in the detected area of the edges. Third, the feature points are projected into the 3D scene and match with the whole scene point cloud. Finally, we segment the matched points from the whole scene point cloud. The proposed method utilizes few feature points in the detected area of edges to match and segment the whole scene point cloud, rather than segmenting the whole scene point cloud directly. The unsupervised learning method can be easily implemented in autonomous vehicles; the double threshold includes edge-based and stixel-based estimations which are more accurate in object proposal. Furthermore, the local reconstructed point clouds of objects are efficient to register and match with whole scene point clouds. The points that need to be matched are reduced. We partitioned the object point cloud from whole scene point cloud instead of segmenting the point cloud point-wisely. It implements real-time 3D object detection for autonomous vehicles.

This paper is organized as follows. Section 2 reviews the related work in segmentation and scene analysis. The proposed method is explicitly presented in Section 3. Sections 4 and 5 discuss the experimental results and comparisons with other existing methods. Finally, we draw a conclusion of this paper in Section 6.

2. Related Work

A stereo camera is a cost-effective device and can efficiently perceive the driving environment. Thereby, it is extensively selected to mount on the autonomous vehicle by many vehicle manufacturers and research institutions [19,20]. The interested objects can be segmented and tracked effectively by the depth information of the stereo camera or the fusion information with monocular cues. Depth information enables robust feature tracking over a long distance. Recently, there has been an increasing research interest in semantic segmentation of 3D point cloud [21]. It is applicable for both indoor and outdoor environments and has corresponding tasks, such as the segmentation of tables, desks, trees, cars, roads, and pedestrians. Hitherto, most robotic scene understanding works have focused on 3D indoor scenes, and the technologies have become ripened with the development over the decades. Related applications such as mapping and autonomous driving can be deployed in the well-studied method of indoor environments. In this paper, the task is to segment point clouds from whole scene point clouds into a few dense object point clouds [22]. Outdoor scenes have more challenges, because they are covered by more extensive objects. A major challenge of this task has arisen from the fact that urban transport scenes of great majority are disordered and obstructed. Although significant progress has been made in the existing study of object segmentation of large architectural elements (e.g., walls, road edges), the performance is still far from satisfactory for on-road objects such as vehicles and pedestrians.

With the evolution of 3D reconstruction techniques such as parallel tracking and mapping (PTAM) [23], dense tracking and mapping (DTAM) [24], and Kinect fusion [25], a potential solution is to exploit the merged point cloud, which can achieve high performance for indoor scenes where laser data are classified and modeled into a few rough geometric feasibilities to label the point cloud of indoor scenes. The other solutions implement segmentation and detection on individual frames, respectively, and the outputs are merged into a point cloud and undergo a joint optimization procedure. Most of these approaches use features that describe local shapes and appearances to segment individual 3D laser points or voxels, and through the inference of graphical model, the joint optimization is typically accomplished by spatial and temporal smoothing.

The most related work to ours is presented in [14]. The authors utilized a weighted graph whose nodes represent points and edges that connect each point to its k-nearest neighbors. Next, after refining the initial segmentation, GMM (Gaussian Mixture Model) are created from the color and density features of points in object and background classes, respectively. The downside of their graph-cut segmentation approach is the spoil of object boundaries, leading to the misclassification of objects. In this paper, in order to maintain accurate object boundaries, we present a detection-based scheme, in combination with edge detection of 2D image pairs, to segment the 3D whole scene point clouds. First, we combine the edge of objects and stixel histograms in corresponding images to obtain a boundary region of the objects as the initial distribution. Second, we reconstruct the 3D sparse point cloud of objects based on the boundary region. Third, reconstructed points are projected to 3D coordinate fit with the whole scene point cloud. Finally, we segment the point cloud to obtain the objects which are represented in 3D dense point cloud.

The contribution of the proposed method is twofold. First, we utilize the edge of objects and stixel histograms to find the object proposal area, reduce the computation time, and find the objects without training or classification. Second, the proposed method simplifies the procedure of object location, and improves the positioning accuracy. At the stage of sparse point cloud matching with dense point cloud, less points will be matched with the current point cloud. It not only improves the accuracy of partition, but also reduces the time cost of 3D point cloud partition. Moreover, it facilitates visualization and storage of the points of objects in future applications.

3. Edge Detection and Stixel Estimation-Based Object Segmentation

Edge detection includes manifold methods of mathematics that study to identify points in a digital image, at which the image brightness varies sharply or discontinuities [26]. The contours which are typically organized by a set of curved line segments of varied brightness can be termed as edges. The application of edge detection algorithms to images can significantly reduce the amount of data to be processed and can therefore filter out information that may be regarded as less relevant, while preserving the important structural properties of the image. If the edge detection step is successful, the subsequent task of interpreting the objects in the original image may therefore be substantially simplified. In this paper, search-based methods are applied to detect edges, followed by stereo reconstruction in detected edge areas. Then, the sparse point clouds of objects are obtained and dense point clouds are matched. Finally, the point clouds of objects are partitioned to represent the 3D objects in driving environments. Figure 1 illustrates the proposed workflow of the edge detection and stixel estimation-based segmentation method in 3D point clouds.

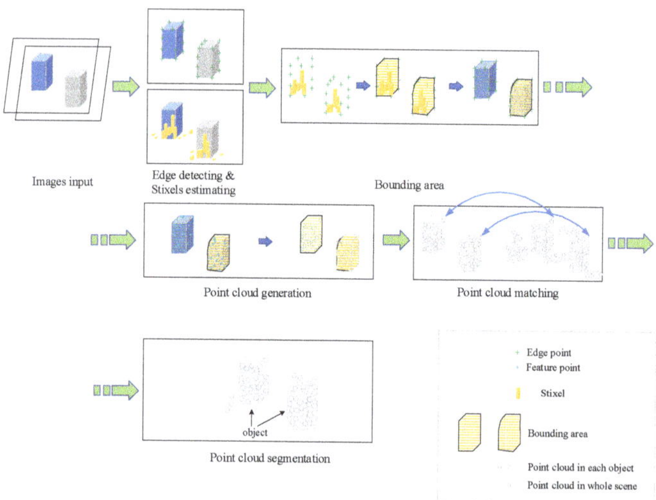

Figure 1. Workflow of the proposed method.

To satisfy the real-time and accuracy in this paper, the edge detection and stixel estimation-based method is proposed to implement efficient segmentation of 3D point clouds. The performance of the proposed method depends on edge detection, and fast edge detection methods are acknowledged as wide-used methods with good performance. Therefore, this paper takes advantage of a fast edge detector [27] to detect the object contour. Finally, we partition the 3D point cloud of the specific object according to the detected contour. The steps of the proposed method are as shown in Algorithm 1.

Algorithm 1 Edge- and Stixel-Oriented 3D Point Cloud Segmentation

1. Capture the corresponding images $image_{left}$, $image_{right}$ from the stereo camera.
2. Detect the edges $\left\{e_l^1, e_l^2, \ldots, e_l^n, e_r^1, e_r^2, \ldots, e_r^m \middle| n, m \in N\right\}$ using the structured forests-based edge detector.
3. Generate the edge candidate points $\left\{c_l^1, c_l^2, \ldots, c_l^p, c_r^1, c_r^2, \ldots, c_r^q \middle| p, q \in N\right\}$ according to the rules given in the following Section 3.1.
4. Identify the utmost 4 bounding areas $\left\{b_l^1, b_l^2, b_l^3, b_l^4, b_r^1, b_r^2, b_r^3, b_r^4\right\}$ in each image based on the edge candidate points.
5. Generate the utmost 4 sparse point clouds (SPCs) $\{SPC_1, SPC_2, SPC_3, SPC_4\}$ of objects in the bounding areas.
6. Match the sparse point cloud SPC_i with the dense point cloud DPC.
7. Segment the dense point cloud DPC_j of objects from the sensed dense point cloud DPC.

3.1. Edge Detection, Stixel Estimation, and Boundary Area Selection

In this work, the goal of edge detection is to determine whether an object exists in an image. Instead of searching for an object at every image location and scale, a set of object bounding box proposals is proposed with the goal of reducing the set of positions that need to be further analyzed. Edge detection works as an object proposal generator to improve the recall and efficiency.

The structured forest-based edge detector is an outstanding performer to extract useful structural information from different visual objects. It dramatically reduces the amount of data to be processed, because edge detection can select meaningful regions rather than the whole image. Among the edge detection methods developed so far, the structured forest-based edge detection algorithm is one of the most strictly defined methods that provides good and reliable detection. Most vehicles' rear-view shows horizontal and vertical edges, and it may be useful for bounding area generation. A group of horizontal and vertical edges that form a rectangular shape with an aspect ratio between 0.4 and 1.6 are good candidates for potential vehicles [28]. The clue has been employed to locate the position of a vehicle after initial ROI (Region of Interesting) was found based on the cue using the shadow underneath the vehicle [29]. In this work, the object proposals are given by edge box and stixel estimation, as shown in Figure 2. Similar to the v-disparity image, the frequency histograms of the edge detector are detected from the horizontal and vertical directions, respectively. In the binary image, we can find that there are more contour lines over the objects, and the values of frequency in the histograms are higher in the corresponding parts. Combined with stixel estimation, the double thresholds are deployed for object proposal, which allows the location of the object to be more accurately determined.

The process of the edge detection-based method deployed the structured forest based-detector to find the edge in this paper, and then constituted a closed region for 3D segmentation to form a semantic segmentation in the 3D point cloud. As the Algorithm 2 shown, the edge detection-based algorithm can be divided into six steps.

The edge points are detected in the corresponding images, and the edge points of objects are more than the background. The region where the value of edged histograms is higher than the other regions can be regarded as the area that objects are located. We found the candidate point of bounding area near the area of high histogram value.

Algorithm 2 Edge Detection and Stixel Estimation-Based Segmentation

1. Use $y \in Y = \mathbb{Z}^{d \times d}$ to denote the segmentation mask, use $y' \in Y' = \{0,1\}^{d \times d}$ for a binary edge map.
2. Find the intensity gradients of the image. Generate candidate features $x(i,j,k)$.
3. Define a mapping $\Pi : Y \rightarrow Z$.
4. Run the structured edge detector on the original, half, and double resolution version of I and average the result of the three edge maps after resizing to the original image dimensions.
5. Track the edges by hysteresis thresholding: Finalize the detection of the edges by suppressing all the other edges that are weak and not connected to strong edges.

$$H_{ij} = \frac{1}{2\pi\sigma^2} \exp\left(-\frac{(i-(k+1))^2+(j-(k+1))^2}{2\sigma^2}\right); \\ 1 \le i,j \le (2k+1)$$

6. Generate bounding area of candidate points.

$$C_{horizon} = \sum_{i}^{x} \frac{1}{2\pi\sigma^2} \exp\left(-\frac{(i-(k+1))^2+(j-(k+1))^2}{2\sigma^2}\right) * G_x; 1 \le i,j \le (2k+1)$$

$$C_{vertical} = \sum_{j}^{y} \frac{1}{2\pi\sigma^2} \exp\left(-\frac{(i-(k+1))^2+(j-(k+1))^2}{2\sigma^2}\right) * G_y; 1 \le i,j \le (2k+1)$$

where x, y are the coordinates of pixel, σ is the standard deviation of the Gaussian distribution, I is the matrix of the image, and $C_{horizon}$, $C_{vertical}$ are the horizontal and vertical direction values of the histograms in Figure 2.

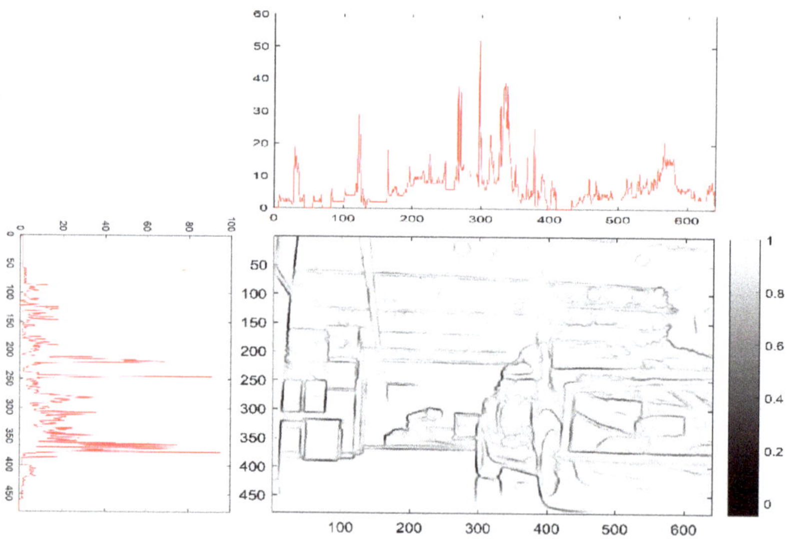

Figure 2. Frequency histograms of edge binary image in horizontal and vertical directions.

Candidate edge points are then verified by comparing the distance of the candidate edge points with other points around the candidate edge points. According to the Lampert assumption [30], compared with the background, there are more edge points on the objects. As Figure 3 shows, if the five adjacent candidate edge points of shortest distance exceed the threshold, this candidate point will be ignored, which means the point cannot be the edge point. If the five adjacent candidate edge points of shortest distance do not exceed the threshold, then they are judged with the following rules. As shown in Figure 4, if the accumulated angle summation of the five points around the candidate edge point does not exceed $3\pi/2$, the candidate point can be determined as the edge point according to the non-maximum suppression principle of the canny edge detector. The algorithms of edge point determination are stated in Algorithm 3.

Algorithm 3 Determining the Edge Points from Candidate Edge Points

Input: Candidate Edge Points $P_n(0 < n < N)$
Output: Edge Points P_m $(0 < m < M)$
While$(P_n \neq \phi)$
  If $P_n(0 < n < N) \in$ **Candidate Edge Points**
    If $distance(P_n) < threshold$
      $acc\_arc = \sum_{i=1}^{5} arc(P_n \cdot P_i)$, P_i is the 5 nearest candidate edge points.
        If $acc\_arc < threshold$
          $P_m = P_n$
          $m = m + 1, n = n + 1$
        end if
      end if
    end if
end while

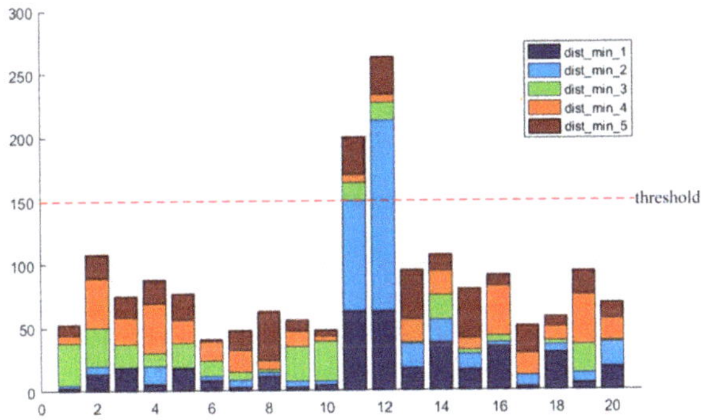

Figure 3. The five shortest distances from 20 candidate edge points.

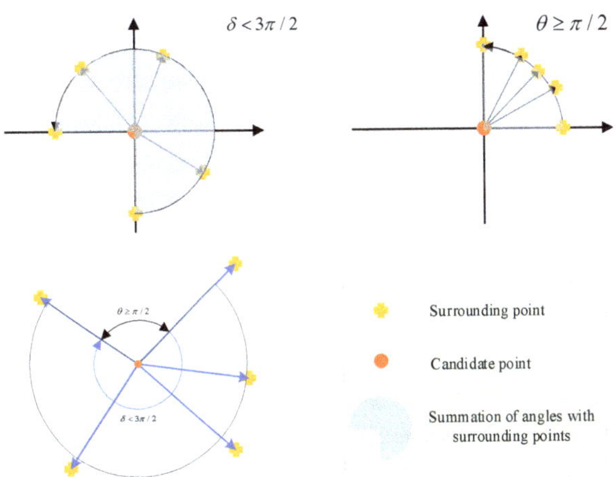

Figure 4. Summation of the angles of the five surrounding points.

Then, we deployed a stixel estimation method to locate the objects as the double threshold with edge proposal. As shown in Algorithm 4, the stixel estimation method can be divided into four steps:

Algorithm 4 Algorithm of Stixels Estimation

1. A pixel-wise cost volume $C_{ijk} = \|p_{ij} - p_{i(j-k+2)}\|_1$ is computed from the input rectified stereo images;
2. Generating the v-disparity image and detecting the ground plane on it;
3. Generating the cost matrices $c_o = \sum_{v(k)} c(i,j,k)$, $c_g = \sum_{|V|} c\left(i,j,f_{ground}(k)\right)$, $v(k) = f^{-1}_{ground}(k)$;
4. The estimated stixel disparities are used to estimate the stixel heights $c_h = \sum_{v(k)} |c_o - 1| + \sum_{v(k)} |c_o + 1|$.

As shown in Figure 5, the objects in a country road are surrounded by the estimated stixels.

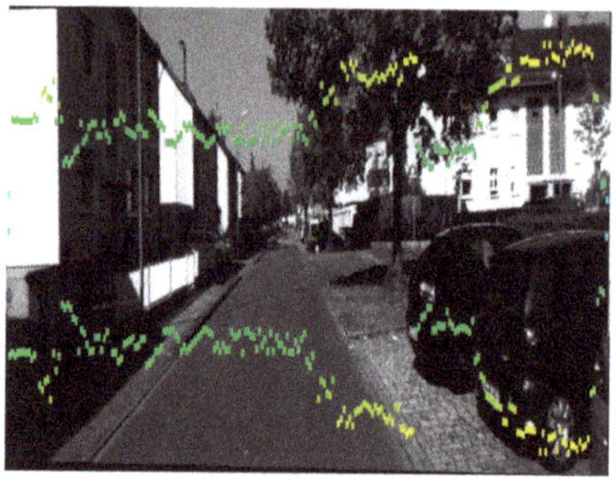

Figure 5. Stixel estimation in a country road.

The location of edge points can be ascertained by the edge detection and stixel estimation, and the uncertainty of the edge points can be reflected by covariance estimation, and measured by the inverse of Hessian matrix:

$$\Sigma^i = \begin{bmatrix} D_{xx}(\mathbf{p},\sigma_i) & D_{xy}(\mathbf{p},\sigma_i) \\ D_{xy}(\mathbf{p},\sigma_i) & D_{yy}(\mathbf{p},\sigma_i) \end{bmatrix}^{-1} \tag{1}$$

where D_{xx}, D_{xy}, and D_{yy} are the second derivative of $D(\mathbf{p},\sigma_i)$, and $\mathbf{p} = (x,y)^{\mathrm{T}}$ is the location of the edge point on σ_i. Besides, we estimated the transform matrix T_i, T_j from continuous images by tracking the edge points and estimated stixels, respectively. Moreover, we deployed bundle adjustment to minimize the reprojection error to determine the location of the objects.

$$E = \sum_i \left(\mathbf{P}_{ci} - (\mathbf{r}_i\mathbf{P}_{pi} + \mathbf{t}_i)\right) + \sum_j \left(\mathbf{P}_{cj} - (\mathbf{r}_j\mathbf{P}_{pj} + \mathbf{t}_j)\right) \tag{2}$$

where $\mathbf{P}_{ci}$, $\mathbf{P}_{cj}$ are the edge points and stixel points in the current frame, respectively, and $\mathbf{P}_{pi}$, $\mathbf{P}_{pj}$ are the projection points of edge points and stixel points in next frame. The transformation matrix $T_i = \begin{bmatrix} \mathbf{r}_i & \mathbf{t}_i \end{bmatrix}$, $T_j = \begin{bmatrix} \mathbf{r}_j & \mathbf{t}_j \end{bmatrix}$. We used the spatial cue, stixel estimation and temporal cue, and optimized transformation matrix to locate the objects.

As shown in Figure 6, the edge points were tracked by using the nearest ORB (Oriented FAST and Rotated BRIEF) feature points, due to the edge points without feature descriptors. Therefore, we deployed the ORB feature points to track the edge points from $frame_n$ to $frame_{n+1}$, and optimize the transformation matrix by using bundle adjustment. The results of the combined method optimized by bundle adjustment are shown in Figure 7.

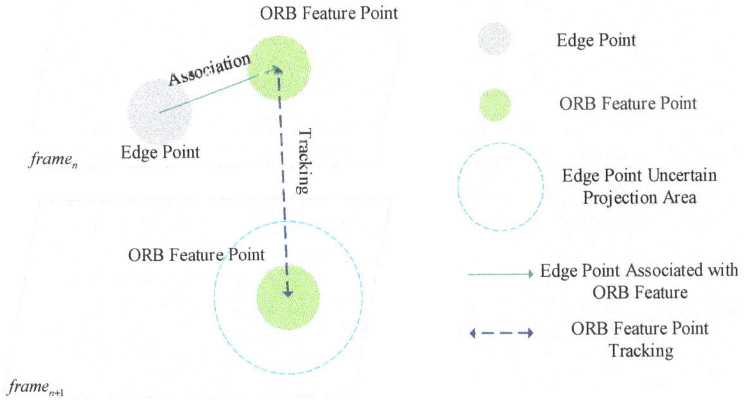

Figure 6. Edge point tracking and optimizing strategy.

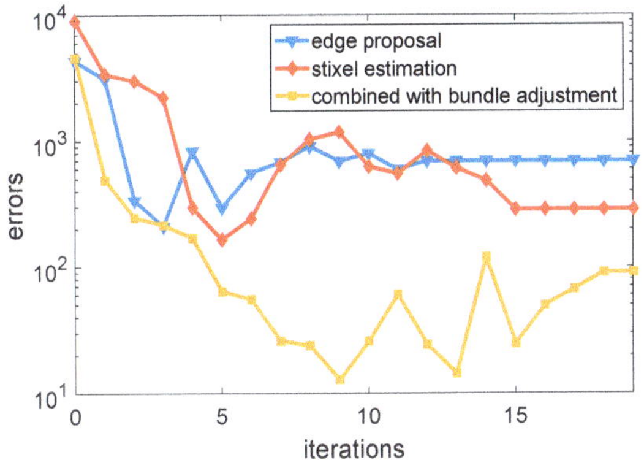

Figure 7. Edge point tracking without/with optimizing.

After calculating the angle of the surrounding points, the candidate edge point can be verified as the edge point. The remaining edge points are connected to form a polygonal bounding area. The object is deemed to be covered by the bounding area. As shown in Figure 8, Figure 8a–d shows the scenarios with a single object, of which Figure 8d is the scenario of an indoor parking lot and Figure 8a,b shows the scenarios of an outdoor parking lot, and Figure 8c is the scenario that objects are in weak light. Furthermore, Figure 8e,f shows the scenarios with two objects, and Figure 8c,g shows the scenarios on a country road from KITTI. The green and blue polygon regions are the bounding areas.

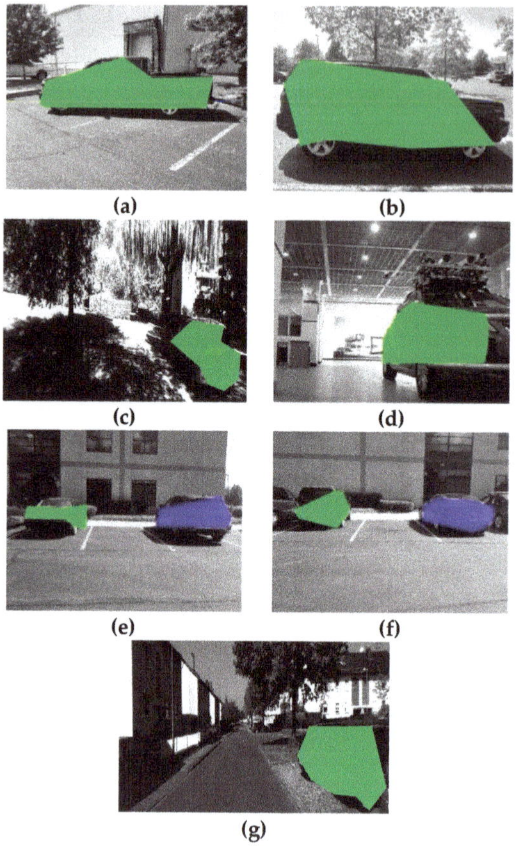

Figure 8. Bounding area in 2D images based on edge points in different scenarios. (**a**) outdoor parking lot, single object; (**b**) outdoor parking lot, single object; (**c**) weak light, single object; (**d**) indoor parking lot, single object; (**e**) outdoor parking lot, multi objects; (**f**) outdoor parking lot, multi objects; (**g**) roadside, single object.

3.2. 3D Object Reconstruction with 2D Image Pairs

The detected edge and estimated stixels provide a bounding area for feature detection in 2D image pairs. The image features are detected in the bounding area. In the bounding area, we take advantage of the good performance of the Speed Up Robust Features (SURF) operator in image transformations to detect the feature points. The feature points in the corresponding image pairs are stereo matched to reconstruct these objectives in a 3D coordinate system. Based on the previous section, we consider the bounding areas covered on the objects. Then, the objects are reconstructed with corresponding images in these bounding areas, and we use the 3D point cloud of these objects to match the wide-angle point cloud. Finally, the objects in dense point cloud are segmented. The next section describes the matching and segmenting of objects in detail.

As Figure 9 shows, Figure 9a is the matched points in the bounding area based on edge detection in 2D corresponding images. The blue shadow area in Figure 9b is the bounding area according to the edge points. Figure 9c is the superimposed area of Figure 9a,b. This step helps to reconstruct the 3D point cloud of an object in an interesting area.

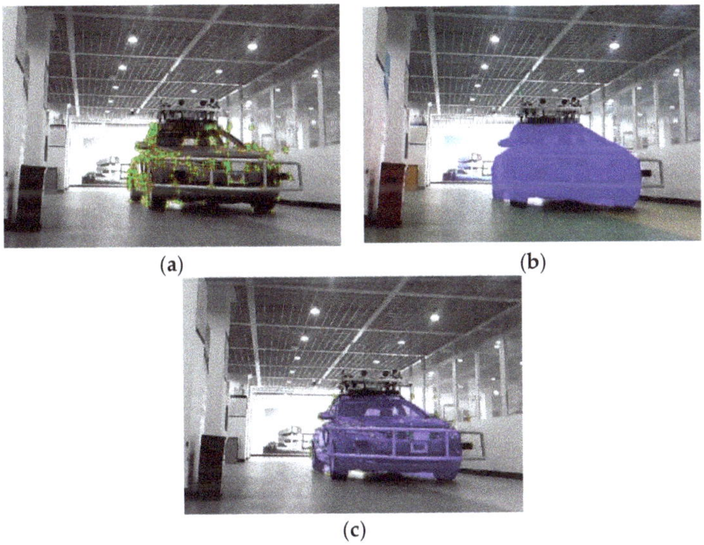

Figure 9. Feature detection in the object area. (**a**) matched points in bounding area; (**b**) bounding area; (c) superimposed area of (**a**,**c**).

3.3. *3D Point Cloud Matching and Segmentation*

In this paper, our aim is to segment objects accurately and rapidly from the whole scene point cloud. We only match the 3D point cloud of objects with the whole scene point cloud. In this step, the norms of the point cloud of an object are matched with norms of the whole scene point cloud. It signifies that the reconstructed point cloud of an object can match the exact location directly in the whole scene point cloud. Therefore, the real location of objects can be found in the real running environment veritably. The matched point clouds in the whole scene point cloud remain, and the rest of the points are removed. The remaining point clouds represent the objects we want to segment. The details of 3D point cloud matching and segmentation will be given in the following subsections.

3.3.1. 3D Point Cloud Matching

We get the reconstructed objects in the bounding areas from Section 3.2 to obtain the norms $\{n_1, n_2, \ldots n_O, O \in N\}$ of the 3D point cloud of this reconstructed object. The 3D points in the sparse point cloud are reconstructed by 2D image feature points. According to the method of perspective n points, the location of the reconstructed points of an object in the whole scene point cloud can be calculated according the number of frames and the transformation matrix. The point cloud of an object can be located in the whole scene point cloud according to the frame number, after registering the spare point cloud in the whole scene point cloud. The norms are matched with the norms $\{n_{w1}, n_{w2}, \ldots n_{wP}, P \in N\}$ of the whole scene point cloud by calculating the summation of distance, $distance_{sum}$. The area of objects in the point cloud of the whole scene are ascertained according to the $distance_{sum}$ of norms. If the $distance_{sum}$ is less than the threshold $\varepsilon < 1000$, the point cloud of objects is matched, otherwise the objects are not regarded in the scene.

As Figure 10 shows, the norms of objects in the bounding areas are matched with the point cloud of the whole scene. Then, the 3D point cloud will be segmented as per the following statement.

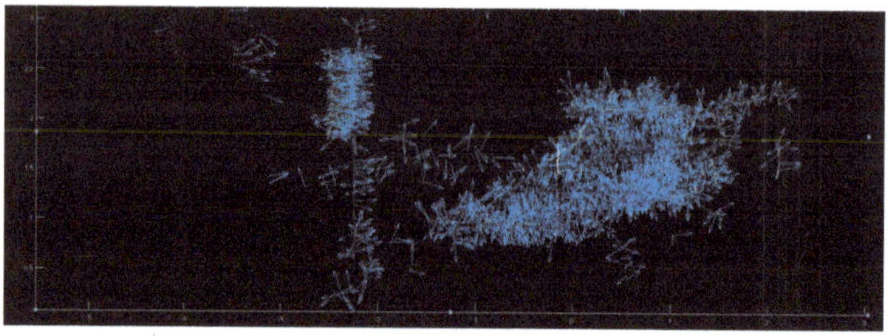

Figure 10. Estimated norms of the point cloud without segmentation.

3.3.2. 3D Point Cloud Partition

First the center point of the sparse point cloud is calculated, then the points are projected to the $x-o-y$ plane, and the minimum rectangle enclosure of the objects in the 2D plane are generated. The heights of objects are from the sparse point cloud of objects. Thus, the 3D bounding boxes of objects are generated. If the $\sum_i distance^i_{sum}$ of the location between the point cloud of objects and the whole scene point cloud are lower than the threshold $\Delta < 1000$, the 3D bounding box are drawn in the whole scene point cloud according to the location of center point. The points in the 3D bounding box located in whole scene can remain. Then, obtain the limit points in the x, y, z axes, respectively. We partition the object point clouds along the z axis into several layers at every 10 points, thus the contour profile of objects is ascertained by compounding the limit points and the edge points. The rest of points, which are regarded as the background, are deleted. We get the partitioned point cloud of objects from the point cloud of the whole scene.

We can find out from Figure 11 that the input is image sequence. The sparse point cloud of objects according to the algorithms which are stated in this section are gained, and the sparse point cloud is matched with the dense point cloud of the whole scene. We set the maximum values of the sparse point cloud in the x, y, z axes, respectively, and then, the values form a bounding box in 3D point cloud. At last, the points in the bounding box remain and the rest of the points are removed. Thus, the result of segmenting the objects from the whole scene 3D point cloud has been implemented. As shown in Figure 12, the point cloud of the object is segmented. Figure 12a is the front view and Figure 12b is the top view. According to the coordinate values in Figures 11 and 12, the volume of the point cloud is reduced and the points are also reduced. The objects can be detected and segmented accurately and quickly.

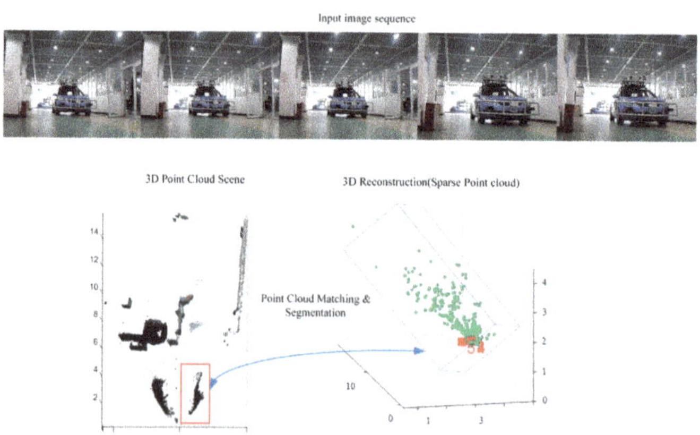

Figure 11. Point cloud matching and segmentation.

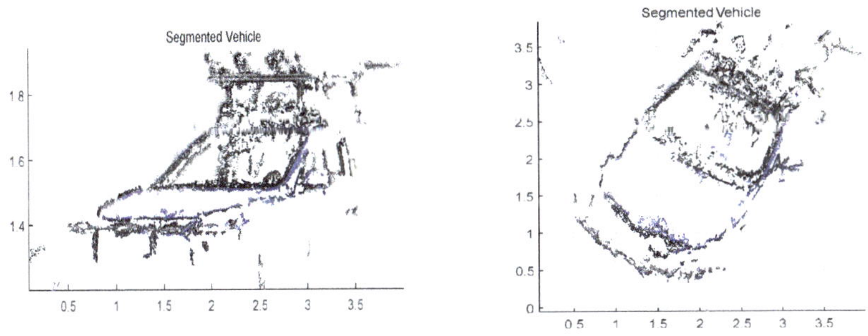

(**a**) (**b**)

Figure 12. Segmented vehicle in a 3D point cloud with front and top view. (**a**) 3D point cloud of segmented vehicle from front view; (**b**) 3D point cloud of segmented vehicle from top view.

4. Experimental Result

We tested the proposed algorithm on two types of datasets, one of which was the dataset of the point cloud that was generated from stereo camera mounted on our experiment vehicle. The stereo camera was built up by two monocular cameras (Basler ace-acA2500-60uc), and the baseline was 60 cm. The number of image pairs was 385, and the size of the dataset was greater than 1 GB. The image size was 1920 × 1080 (this dataset can be download from https://pan.baidu.com/s/18JNburRaGKSVlrOq9ARy1A#list/path=%2F, password: 66ib). We acquired these images in the campus scene and country road. The other was downloaded from the website of KITTI (a project of Karlsruhe Institute of Technology and Toyota Technological Institute at Chicago, IL, USA) [31]. The number of pairs was 200. The size of the dataset was 320 MB, and the image size was 1242 × 375. The downloaded dataset only provided stereo images for training and testing of segmentation. The dataset from KITTI did not provide the stereo-generated point cloud, we just tested the segmentation in 2D images. In this paper, we tested the performance of segmentation with different algorithms, and we compared several performance criterions, including F-measure with different scale of data, F-measure with different algorithms, accuracy and the number of points after segmentation. We deployed the semantic image as the ground truth of segmentation. The segmented pixels were projected in one class of semantics and if the IoU (Intersection over Union) was greater than 0.5, we regarded them as true positives. There were four scenarios in our own datasets: The first scenario only contained one vehicle; the second scenario contained one vehicle and one pedestrian; the third scenario contained two vehicles; and the last scenario contained two vehicles and one pedestrian. In these test scenarios, the boxes represented the fixed objects, and the pedestrian represented the moving object.

To present a quantitative evaluation of the proposed method, we employed three criteria [32]:

$$\text{Precision} = \frac{|GT \cap DR|}{|DR|}$$

$$\text{Recall} = \frac{|GT \cap DR|}{|GT|}$$

$$\text{F-measure} = \frac{2 \times \text{Recall} \times \text{Precision}}{\text{Recall} + \text{Precision}}$$

where GT represents the set of pixels that are classified to a specific category by the proposed method, and DR represents the set of pixels that are manually labeled to a specific category (i.e., ground truth). F-measure is the weighted harmonic mean of Precision and Recall, which is used to quantify the overall performance of the segmentation.

Figure 13a illustrates the performance of four different methods with different data scales. It shows that the proposed method maintains exceptionally high performance when the data scales is close to 4000. As compared with other three methods, the average of the proposed method is 4.48% higher than that of the other three methods. Figure 13b shows the performance of four different methods. It indicates that all methods can most accurately segment the objects in the simplest scenario where there is only one box. The performance of these methods decreases with the increase of the scenario's complexity. Moreover, results of the proposed algorithm are 3.2%, 1.5%, and 1.1% higher than the other three methods, respectively. Tables 1 and 2 compares the precision and recall of the four different methods, respectively.

In the stage of point cloud segmentation, Figure 13c shows the number of points after the objects are segmented from the point cloud of the whole scene. It shows that the proposed method is able to represent the same objects with a fewer number of points. The average number of points in the proposed method is 37.3% less than the other three methods. Figure 13d explains the computation time of the four methods with four experimental scenarios. The proposed method consumes the least time to segment the same objects, which is helpful for real-time visualization and path planning. From the figures and tables, we can find that the accuracy and real-time satisfy the requirement of autonomous vehicles.

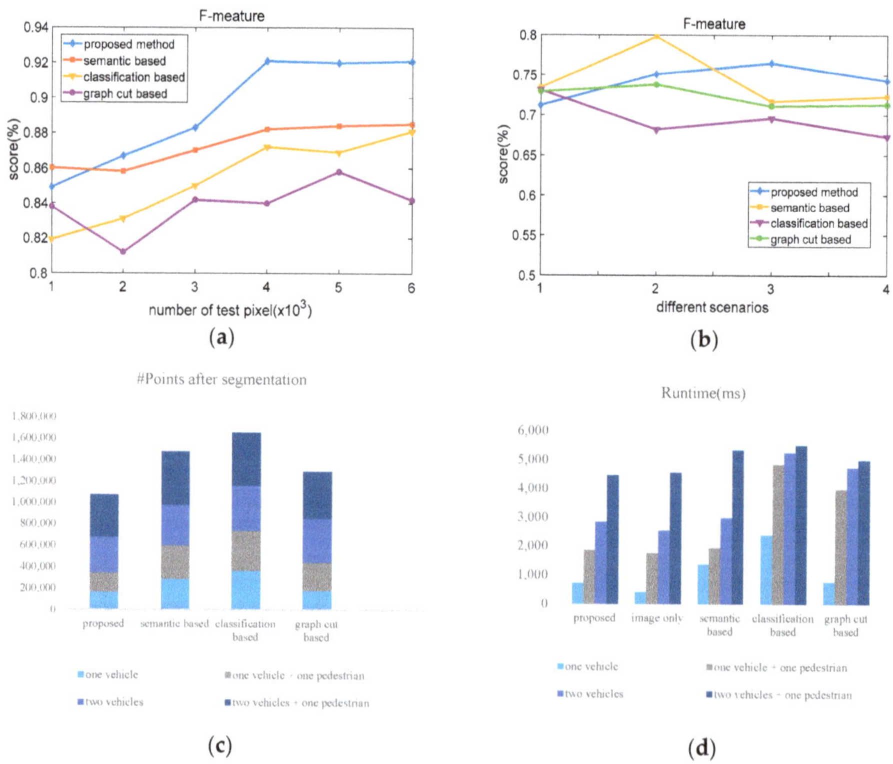

Figure 13. Experiment results with different algorithms. (**a**) F-measure of four algorithms with different data scales; (**b**) F-measure of four different algorithms in four scenarios; (**c**) number of points after segmenting with four different algorithms in four scenarios; (**d**) runtime of four different algorithms in four scenarios.

Table 1. Precision of four different algorithms in four scenarios.

	One Vehicle	One Vehicle + One Pedestrian	Two Vehicles	Two Vehicles + One Pedestrian
Graph-cut-based	0.807	0.823	0.801	0.886
Semantic-based	0.813	0.731	0.735	0.775
Classification-based	0.819	0.788	0.725	0.823
Proposed	0.824	0.848	0.801	0.869

Table 2. Recall of four different algorithms in four scenarios.

	One Vehicle	One Vehicle + One Pedestrian	Two Vehicles	Two Vehicles + One Pedestrian
Graph-cut-based	0.781	0.796	0.686	0.723
Semantic-based	0.787	0.801	0.673	0.642
Classification-based	0.8	0.834	0.782	0.761
Proposed	0.806	0.741	0.763	0.713

As shown in Figure 14, taking a scene in KITTI as the example, the objects are segmented according to the determined bounding area from the corresponding image pair, followed by Figure 15, in which the objects are reconstructed locally without other elements. Therefore, the objects can be located in 3D world coordinates. The availability of the proposed method is verified for the application in 3D object detection for autonomous vehicles.

(a) (b)

(c) (d)

Figure 14. Object segmentation in corresponding image pairs. (**a**) Bounding area of objects determined in left view image; (**b**) bounding area of objects determined in right view image; (**c**) object segmentation in the left view image; (**d**) object segmentation in the right view image.

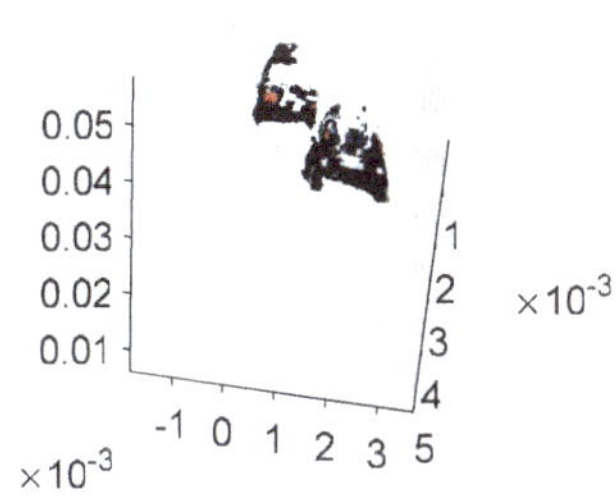

Figure 15. Object reconstruction according to the bounding area.

As shown in Figure 16, the ground truth semantic segmentation images are covered by the deep blue and light green bounding area. In Figure 16a, the bounding areas cover two objects as the light green and deep blue polygons, in the center of left view image. Similarly, in Figure 16b, the two objects are covered in the right view image. The performance of the proposed method can be compared directly. Moreover, the accuracy of the proposed method can be calculated in Table 3.

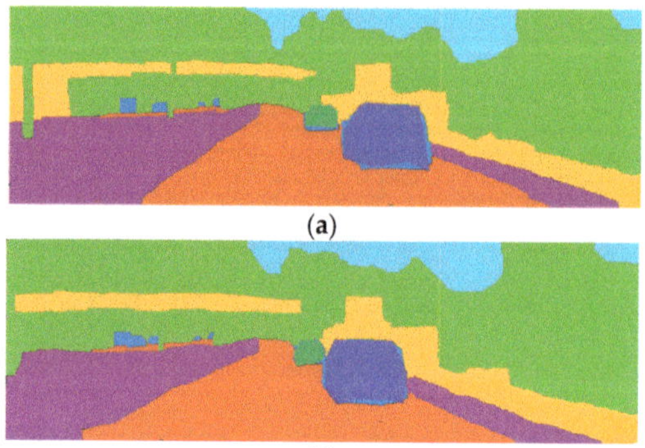

Figure 16. Ground truth semantic segmentation and bounding area. (**a**) Ground truth semantic segmentation-covered bounding area of objects in the left view image; (**b**) Ground truth semantic segmentation-covered bounding area of objects in the right view image.

Table 3. Accuracy measures and runtime for three Intersection over Union (IoU) metrics with four algorithms.

	IoU = 0.5		IoU = 0.6		IoU = 0.7		Runtime (ms)
	Accuracy	Recall	Accuracy	Recall	Accuracy	Recall	
Semantic-based [9]	64%	86%	36%	55%	48%	44%	1658
Classification-based [11]	58%	79%	45%	76%	62%	78%	2594
Graph-cut-based [14]	38%	69%	78%	66%	75%	65%	1583
Proposed	75%	78%	91%	84%	74%	82%	836

As shown in Table 3, the performance of the proposed method increases 31% compared to the other three methods when the IoU is 0.6, and the accuracy is always higher than the other three. The runtime of the proposed method is approaching quasi real-time. Despite the changing of IoU, the accuracy is maintained at a high level.

Figure 17 illustrates that the point clouds of objects are segmented from raw data point clouds from Lidar, according to the 3D coordinates of objects from Figure 15. As shown in Figure 17a, the red points almost belong to the objects in the raw data point cloud. Figure 17b shows the detected objects with the 3D bounding boxes. Figure 17c shows the segmented local point cloud of objects. The accurate point clouds of objects are segmented rapidly from whole scene point cloud.

Figure 18 explains that the location of the segmented 3D point cloud of objects in 3D world coordinates compared with the location of the ground truth point cloud which is generated by Lidar; the error is 0.49 m and the effective range is 50 m, so the error rate is 0.98%. The accuracy can satisfy the requirement of 3D object detection for autonomous vehicles.

(a)

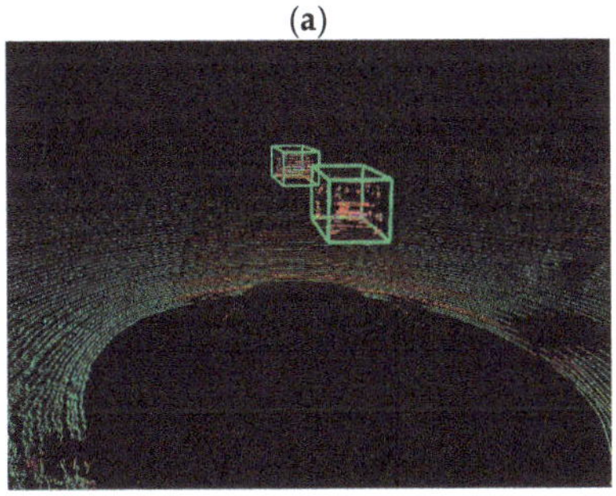

(b)

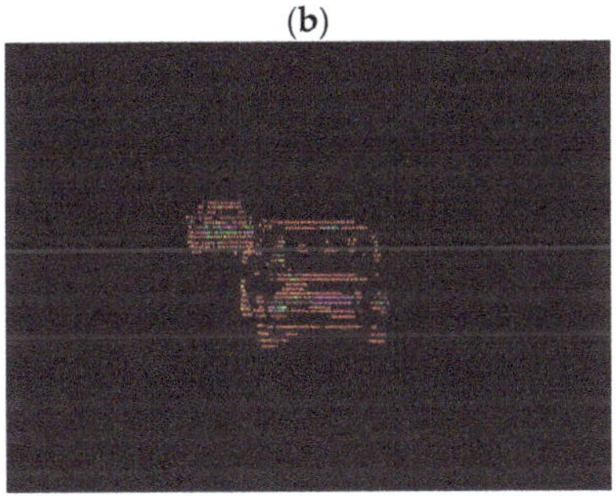

(c)

Figure 17. Raw data point cloud and segmented point cloud. (**a**) Raw data point cloud acquired from Lidar; (**b**) object detection in the 3D box; (**c**) object segmentation in the point cloud.

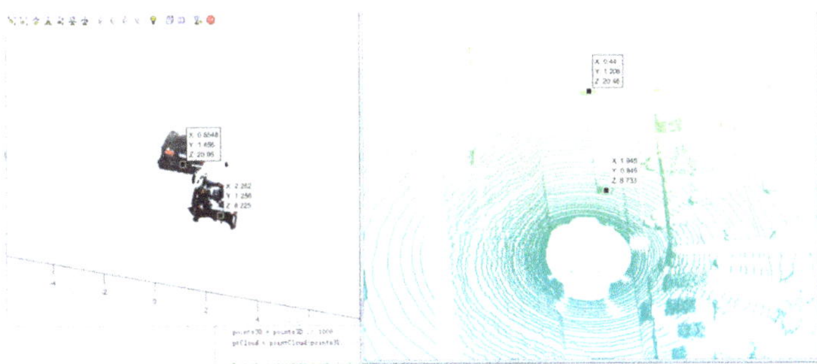

Figure 18. The comparison of 3D coordinates between the image 3D object point cloud and the Lidar point cloud.

5. Discussion

The proposed method averts the difficulty of identifying the objects directly by finding the dramatic change with edge effect and surface effect of objects to reduce computational costs and improve accuracy of objects detection. Compared with the semantic-based method, the proposed method is more time-efficient, despite the time of manual labeling. Besides, the scene of traffic is complex, and semantic-based methods are insufficient for autonomous vehicles. Similarly, the performance of classification-based methods depends on the types we gave, and classification-based methods are sensitive to scene change at a fast speed. Graph-cut-based methods easily destroy the objects, which is fatal for autonomous vehicles. After comparison with the existing methods of segmentation, the experimental results demonstrate that the proposed edge-oriented segmentation method improves the precision of 3D point cloud segmentation, and the objects can be segmented accurately. Besides, compared with the Lidar point cloud, the partitioned point cloud has a satisfied distance accuracy, which is important for autonomous vehicles, and the partitioned point cloud contains the information of color and semantics, which is helpful for object tracking. Meanwhile, the visualization of output data in ADAS (advance driving assistance system) can be greatly facilitated due to the decrease in computational time and the decrease in number of points in the object point cloud. Furthermore, the proposed method can also be available for path planning and obstacle avoidance for autonomous vehicles.

6. Conclusions

We propose a method for the segmentation of point clouds from a multi-view system based on edge detection. By using edge detection in 2D image pairs to initialize the bounding area of segmentation, the objects can be identified directly without hard constraints and artificial labeling. The proposed method was achieved through two main tasks. Firstly, the precision of segmentation was improved by using the bounding area in 2D image pairs, which can decrease the region in the process of 3D reconstruction. Moreover, it provides the location of objects in 3D coordinates, which helps the autonomous vehicle implement path planning. Secondly, after segmenting the dense point cloud, only the point clouds of objects are displayed and the background points are removed. It improves the performance of visualization that focuses on the location of the objects. The number of points in the point cloud is reduced due to the segmentation, which is helpful to display the point cloud of objects efficiently.

Author Contributions: Conceptualization, F.H. and Y.L.; methodology, F.H.; software, D.Y.; validation, F.H. and Y.L.; formal analysis, F.H.; investigation, F.H.; resources, F.H.; data curation, F.H.; writing—original draft preparation, F.H.; writing—review and editing, F.H.; visualization, D.Y.; supervision, Y.L.; project administration, Y.L.; funding acquisition, F.H.

Acknowledgments: This work was supported by the Chongqing Science and Technology Commission (cstc2017rgzn-zdyfX0039), the Ph.D.'s high-end talent program (Grant no. BYJS2016001), and the China Scholarship Council (Grant no. 201608500103).

Conflicts of Interest: The authors declare no conflicts of interest.

References

1. Levinson, J.; Askeland, J.; Becker, J.; Dolson, J.; Held, D.; Sokolsky, M. Towards Fully Autonomous Driving: Systems and Algorithms. In Proceedings of the IEEE Intelligent Vehicles Symposium, Baden-Baden, Germany, 5–9 June 2011; pp. 163–168.
2. Wei, Y.; He, Y. Shadow verification–based waterline detection for unmanned surface vehicles deployed in complicated natural environment. *Int. J. Adv. Robot. Syst.* **2018**, *15*, 1–12. [CrossRef]
3. Michalke, T.P.; Glaeser, C.; Buerkle, L.; Niewels, F. The Narrow Road Assistant—Next Generation Advanced Driver Assistance in Inner-City. In Proceedings of the International IEEE Conference on Intelligent Transportation Systems, The Hague, The Netherlands, 6–9 October 2013; pp. 2173–2180.
4. Mukhtar, A.; Xia, L.; Tang, T.B. Vehicle Detection Techniques for Collision Avoidance Systems: A Review. *IEEE Trans. Intel. Transp. Syst.* **2015**, *16*, 2318–2338. [CrossRef]

5. Ji, X.; Zhang, G.; Chen, X.; Guo, Q. Multi-perspective tracking for intelligent vehicle. *IEEE Trans. Intel. Transp. Syst.* **2018**, *19*, 518–529. [CrossRef]
6. Yu, T.; Wang, R. Scene parsing using graph matching on street-view data. *Comput. Vision Image Underst.* **2016**, *145*, 70–80. [CrossRef]
7. Luo, H.; Wang, C.; Wen, C.; Cai, Z.; Chen, Z.; Wang, H.; Li, J. Patch-based semantic labeling of road scene using colorized mobile LiDAR point clouds. *IEEE Trans. Intel. Transp. Syst.* **2015**, *17*, 1286–1297. [CrossRef]
8. Xiao, W.; Vallet, B.; Schindler, K.; Paparoditis, N. Street-side vehicle detection, classification and change detection using mobile laser scanning data. *ISPRS J. Photogramm. Remote Sens.* **2016**, *114*, 166–178. [CrossRef]
9. Zhang, C.; Luo, W.; Urtasun, R. Efficient convolutions for real-time semantic segmentation of 3d point clouds. In Proceedings of the 2018 International Conference on 3D Vision (3DV), Verona, Italy, 5–8 September 2018.
10. Narksri, P.; Takeuchi, E.; Ninomiya, Y.; Morales, Y.; Akai, N.; Kawaguchi, N. A slope-robust cascaded ground segmentation in 3D point cloud for autonomous vehicles. In Proceedings of the 2018 21st International Conference on Intelligent Transportation Systems (ITSC), Maui, HI, USA, 4–7 November 2018.
11. Barea, R.; Pérez, C.; Bergasa, L.M.; López-Guillén, E.; Romera, E.; Molinos, E.; Ocaña, M.; López, J. Vehicle detection and localization using 3d lidar point cloud and image semantic segmentation. In Proceedings of the 2018 21st International Conference on Intelligent Transportation Systems (ITSC), Maui, HI, USA, 4–7 November 2018.
12. Pan, Y.; Dong, Y.; Wang, D.; Chen, A.; Ye, Z. Three-dimensional reconstruction of structural surface model of heritage bridges using UAV-based photogrammetric point clouds. *Remote Sens.* **2019**, *11*, 1204. [CrossRef]
13. Huang, R.; Hong, D.; Xu, Y.; Yao, W.; Stilla, U. Multi-Scale Local Context Embedding for LiDAR Point Cloud Classification. *IEEE Geosci. Remote Sens. Lett.* **2019**, 1024. [CrossRef]
14. Pan, R.; Taubin, G. Automatic segmentation of point clouds from multi-view reconstruction using graph-cut. *Vis. Comput.* **2016**, *32*, 601–609. [CrossRef]
15. Sun, Z.; Miller, R.; Bebis, G.; DiMeo, D. A real-time pre-crash vehicle detection system. In Proceedings of the Sixth IEEE Workshop on Applications of Computer Vision, 2002. (WACV 2002), Orlando, FL, USA, 4 December 2002; pp. 171–176.
16. Betke, M.; Haritaoglu, E.; Davis, L.S. Real-time multiple vehicle detection and tracking from a moving vehicle. *Mach. Vision Appl.* **2000**, *12*, 69–83. [CrossRef]
17. Derrouz, H.; Elbouziady, A.; Abdelali, H.A.; Thami RO, H.; El Fkihi, S.; Bourzeix, F. Moroccan Video Intelligent Transport System: Vehicle Type Classification Based on Three-Dimensional and Two-Dimensional Features. *IEEE Access* **2019**, *7*, 72528–72537. [CrossRef]
18. Song, W.; Yang, Y.; Fu, M.; Qiu, F.; Wang, M. Real-time obstacles detection and status classification for collision warning in a vehicle active safety system. *IEEE Trans. Intel. Transp. Syst.* **2017**, *19*, 758–773. [CrossRef]
19. Winkler, S.; Min, D. Stereo/multiview picture quality: Overview and recent advances. *Signal Process. Image Commun.* **2013**, *28*, 1358–1373. [CrossRef]
20. Takemura, K.; Takahashi, K.; Takamatsu, J.; Ogasawara, T. Estimating 3-D point-of-regard in a real environment using a head-mounted eye-tracking system. *IEEE Trans. Hum.-Mach. Syst.* **2014**, *44*, 531–536. [CrossRef]
21. Long, J.; Shelhamer, E.; Darrell, T. Fully convolutional networks for semantic segmentation. In Proceedings of the IEEE Conference on Computer Vision and Pattern Recognition, Boston, MA, USA, 7–12 June 2015; pp. 3431–3440.
22. Hackel, T.; Wegner, J.D.; Schindler, K. Fast semantic segmentation of 3D point clouds with strongly varying varying density. *ISPRS Ann. Photogramm. Remote Sens. Spat. Inf. Sci.* **2016**, *3*, 177–184. [CrossRef]
23. Belter, D.; Skrzypczynski, P. Precise self-localization of a walking robot on rough terrain using parallel tracking and mapping. *Ind. Robot Int. J.* **2013**, *40*, 229–237. [CrossRef]
24. Newcombe, R.A.; Lovegrove, S.J.; Davison, A.J. DTAM: Dense tracking and mapping in real-time. In Proceedings of the 2011 IEEE International Conference on Computer Vision (ICCV), Barcelona, Spain, 6–13 November 2011; pp. 2320–2327.
25. Newcombe, R.A.; Izadi, S.; Hilliges, O.; Molyneaux, D.; Kim, D.; Davison, A.J.; Fitzgibbon, A. KinectFusion: Real-time dense surface mapping and tracking. In Proceedings of the 2011 10th IEEE International Symposium on Mixed and Augmented Reality, Basel, Switzerland, 26–29 October 2011; pp. 127–136.

26. Kaur, B.; Garg, A. Mathematical morphological edge detection for remote sensing images. In Proceedings of the 2011 3rd International Conference on Electronics Computer Technology, Kanyakumari, India, 8–10 April 2011; Volume 5, pp. 324–327.
27. Dollár, P.; Zitnick, C.L. Structured forests for fast edge detection. In Proceedings of the IEEE International Conference on Computer Vision, Sydney, Australia, 1–8 December 2013.
28. Teoh, S.S. Development of a Robust Monocular-Based Vehicle Detection and Tracking System. Ph.D. Thesis, University of Western Australia, Perth, Australia, 2011.
29. Li-Sheng, J.; Bai-yuan, G.; Rong-ben, W.; Yi-bing, Z.; Lin-hui, L. Preceding vehicle detection based on multi-characteristics fusion. In Proceedings of the 2006 IEEE International Conference on Vehicular Electronics and Safety, Shanghai, China, 13–15 December 2006; pp. 356–360.
30. Schomaker, L.; Bulacu, M. Automatic writer identification using connected-component contours and edge-based features of uppercase western script. *IEEE Trans. Pattern Anal. Mach. Intel.* **2004**, *26*, 787–798. [CrossRef] [PubMed]
31. Arbelaez, P.; Maire, M.; Fowlkes, C.; Malik, J. Contour detection and hierarchical image segmentation. *IEEE Trans. Pattern Anal. Mach. Intel.* **2011**, *33*, 898–916. [CrossRef] [PubMed]
32. Zhang, C.; Wang, L.; Yang, R. Semantic segmentation of urban scenes using dense depth maps. In *European Conference on Computer Vision. ECCV 2010*; Springer: Berlin/Heidelberg, Germany, 2010; pp. 708–721.

sensors

MDPI

Article

Occlusion-Free Road Segmentation Leveraging Semantics for Autonomous Vehicles

Kewei Wang [1,2,3], Fuwu Yan [1,2,3], Bin Zou [1,2,3,*], Luqi Tang [1,2,3], Quan Yuan [1,2,3] and Chen Lv [4]

1 Hubei Key Laboratory of Advanced Technology for Automotive Components, Wuhan University of Technology, Wuhan 430070, China; wkw199q@whut.edu.cn (K.W.); yanfuwu@vip.sina.com (F.Y.); tlqqidong@163.com (L.T.); 231943@whut.edu.cn (Q.Y.)

2 Hubei Collaborative Innovation Center for Automotive Components Technology, Wuhan University of Technology, Wuhan 430070, China

3 Hubei Research Center for New Energy & Intelligent Connected Vehicle, Wuhan 430070, China

4 School of Mechanical and Aerospace Engineering, Nanyang Technological University, 639798, Singapore; lyuchen@ntu.edu.sg

* Correspondence: zoubin@whut.edu.cn; Tel.: +86-138-7115-3253

Received: 3 September 2019; Accepted: 24 October 2019; Published: 30 October 2019

Abstract: The deep convolutional neural network has led the trend of vision-based road detection, however, obtaining a full road area despite the occlusion from monocular vision remains challenging due to the dynamic scenes in autonomous driving. Inferring the occluded road area requires a comprehensive understanding of the geometry and the semantics of the visible scene. To this end, we create a small but effective dataset based on the KITTI dataset named KITTI-OFRS (KITTI-occlusion-free road segmentation) dataset and propose a lightweight and efficient, fully convolutional neural network called OFRSNet (occlusion-free road segmentation network) that learns to predict occluded portions of the road in the semantic domain by looking around foreground objects and visible road layout. In particular, the global context module is used to build up the down-sampling and joint context up-sampling block in our network, which promotes the performance of the network. Moreover, a spatially-weighted cross-entropy loss is designed to significantly increases the accuracy of this task. Extensive experiments on different datasets verify the effectiveness of the proposed approach, and comparisons with current excellent methods show that the proposed method outperforms the baseline models by obtaining a better trade-off between accuracy and runtime, which makes our approach is able to be applied to autonomous vehicles in real-time.

Keywords: autonomous vehicles; scene understanding; occlusion reasoning; road detection

---

1. Introduction

Reliable perception of the surrounding environment plays a crucial role in autonomous driving vehicles, in which robust road detection is one of the key tasks. Many types of road detection methods have been proposed in the literature based on monocular camera, stereo vision, or LiDAR (Light Detector and Ranging) sensors. With the rapid progress in deep learning techniques, significant achievements in segmentation techniques have significantly promoted road detection in monocular images [1–5]. Generally, these algorithms label each and every pixel in the image with one of the object classes by color and textual features. However, the road is often occluded by dynamic traffic participants as well as static transport infrastructures when measured with on-board cameras, which makes it hard to directly obtain a full road area. When performing decision-making in extremely challenging scenarios, such as dynamic urban scenes, a comprehensive understanding of the environment needs to carefully tackle the occlusion problem. As to the road detection task, road segmentation of the visible area is not sufficient for path planning and decision-making. It is necessary to get the whole structure and layout of the local road with an occlusion reasoning process in complex driving scenarios where clutter and occlusion occur with high frequency.

Inspired by the fact that human beings are capable of completing the road structure in their minds by understanding the on-road objects and the visible road area, we believe that a powerful convolution network could learn to infer the occluded road area as human beings do. Intuitively, to the occlusion reasoning task, the color and texture features are of relatively low importance, what matters is the semantic and spatial features of the elements in the environment. As far as we know, semantic segmentation [6–8] is one of the most complete forms of visual scene understanding, where the goal is to label each pixel with the corresponding semantic label (e.g., tree, pedestrian, car, etc.). So, instead of an RGB image, we performed the occlusion reasoning road segmentation using semantic representation as input, which could be obtained by popular deep learning methods in real applications or human-annotated ground truth in the training phase. As shown in Figure 1, traditional road segmentation takes RGB image as input and labels road only in the visible area. As a comparison, our proposed occlusion-free road segmentation (OFRS) intends to leverage the semantic representation to infer the occluded road area in the driving scene. Note that the semantic input in the figure is just a visualization of the semantic representation, the actual input is the one-hot type of semantic label.

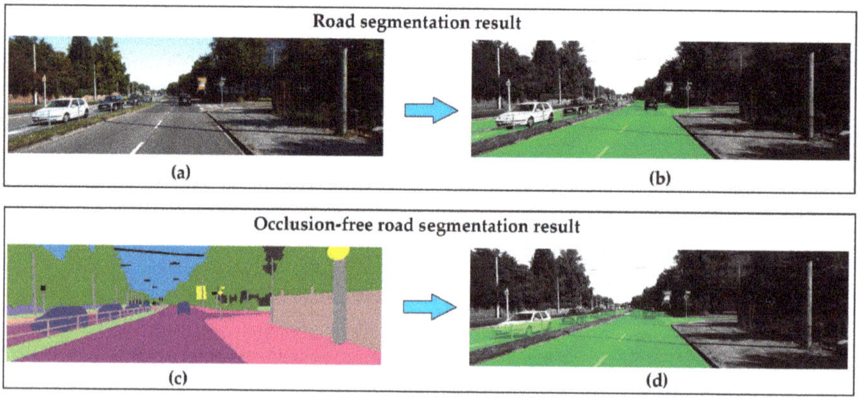

Figure 1. Comparison of road segmentation and proposed occlusion-free road segmentation. (**a**) RGB image; (**b**) visualization of the results of road segmentation; (**c**) visualization of the semantic representation of the scene, which could be obtained by semantic segmentation algorithms in real applications or human annotation in training phase; (**d**) visualization of the results of the proposed occlusion-free road segmentation. Green refers to the road area in (**b**) and (**d**).

In this paper, we aim to infer the occluded road area utilizing the semantic features of visible scenes and name this new task as occlusion-free road segmentation. First, a suitable dataset is created based on the popular KITTI dataset, which is referred to as the KITTI-OFRS dataset in the following. Second, an end-to-end lightweight and efficient fully convolutional neural networks for the new task is proposed to learn the ability of occlusion reasoning. Moreover, a spatially-dependent weight is applied to the cross-entropy loss to increase the performance of our network. We evaluate our model on different datasets and compare it with some other excellent algorithms which pursue the trade-off between accuracy and runtime in the semantic segmentation task.

The main contributions of this paper are as follows:

- We analyze the occlusion problem in road detection and propose the novel task of occlusion-free road segmentation in the semantic domain, which infers the occluded road area using semantic features of the dynamic scenes.

- To complete this task, we create a small but efficient dataset based on the popular KITTI dataset named the KITTI-OFRS dataset, design a lightweight and efficient encoder–decoder fully convolution network referred to as OFRSNet and optimize the cross-entropy loss for the task by adding a spatially-dependent weight that could significantly increase the accuracy.
- We elaborately design the architecture of OFRSNet to obtain a good trade-off between accuracy and runtime. The down-sampling block and joint context up-sampling block in the network are designed to effectively capture the contextual features that are essential for the occlusion reasoning process and increase the generalization ability of the model.

The remainder of this paper is organized as follows: First, the related works in road detection are briefly introduced in Section 2. Section 3 introduces the methodology in detail, and Section 4 shows the experimental results. Finally, we draw conclusions in Section 5.

2. Related Works

Road detection in autonomous driving has benefited from the development of the deep convolutional neural networks in recent years. Generally, the road is represented by its boundaries [9,10] or regions [1,2,11]. Moreover, road lane [12–14] and drivable area [15,16] detection also attract much attention from researchers, which concern the ego lane and the obstacle-free region of the road, respectively. The learning-based methods usually outperform the model-based methods due to the developed segmentation techniques. The model-based methods identify the road structure and road areas by shape [17,18] or appearance models [19]. The learning-based methods [3,6,7,16,20,21] classify the pixels in images as road and non-road, or road boundaries and non-road boundaries.

However, the presence of foreground objects makes it hard to obtain full road despite the occlusion. To infer the road boundaries despite the occlusion, Suleymanov et al. [22] presented a convolutional neural network that contained intra-layer convolutions and produced outputs in a hybrid discrete-continuous form. Becattini et al. [23] proposed a GAN-based (Generative Adversarial Network) semantic segmentation inpainting model to remove all dynamic objects from the scene and focus on understanding its static components (such as streets, sidewalks, and buildings) to get a comprehension of the static road scene. In contrast to the above solutions, we conduct occlusion-free road segmentation to infer the occluded road area as a pixel-wise classification task.

Even though the deep-learning methods have achieved remarkable performance in the pixel-wise classification task, to achieve the best trade-off between accuracy and efficiency is still a challenging problem. Vijay et al. [20] presented a novel and practical deep fully convolutional neural network architecture for semantic pixel-wise segmentation termed SegNet, which follows encoder–decoder architecture that is designed to be efficient both in memory and computational time in inference phase. Adam et al. [24] proposed a fast and compact encoder–decoder architecture named ENet that significantly has fewer parameters, and provides similar or better accuracy to SegNet. Romera et al. [25] proposed a novel layer design that leverages skip connections and convolutions with 1D kernels, which highly reduces the compute cost and increase the accuracy. Inspired by these networks, we follow the encoder–decoder architecture and enhance the down-sampling and up-sampling blocks with contextual extraction operations [26–28], which are proved to be helpful for segmentation-related tasks. This contextual information is even more essential and effective for our occlusion reasoning task, which needs a comprehensive understanding of the driving scenes.

3. Occlusion-Free Road Segmentation

3.1. Task Definition

The occlusion-free road segmentation task is defined as a pixel-level classification as the traditional road segmentation but with occlusion reasoning process to obtain a full representation of the road area. The input is fed to the model as a one-hot encoded tensor of the semantic segmentation labels or predicted semantic segmentation probabilities' tensor $I \in [0,1]^{W \times H \times C}$, where W is the width of the image, H its height, and C the number of classes. In the same way, we trained the network to output a new tensor $O \in [0,1]^{W \times H \times 2}$ with the same width and height but containing only two categories belonging to road and non-road.

3.2. Network Architecture

The proposed model is illustrated in Table 1 and visualized in Figure 2, and was designed to get the best possible trade-off between accuracy and runtime. We followed the current trend of using convolutions with residual connections [29] as the core elements of our architecture, to leverage their success in classification and segmentation problems. Inspired by SegNet and ENet, an encoder–decoder architecture was adopted for the whole network architecture. The residual bottleneck blocks of different types were used as the basic blocks in the encoder and decoder. Dilated convolution was applied in the blocks to enlarge the respective field of the encoder. What is more, the context module was combined with regular convolution to obtain a global understanding of the environment, which is really essential to infer the occluded road area. In the decoder, we proposed a joint context up-sampling block to leverage the features of different resolutions to obtain richer and global information.

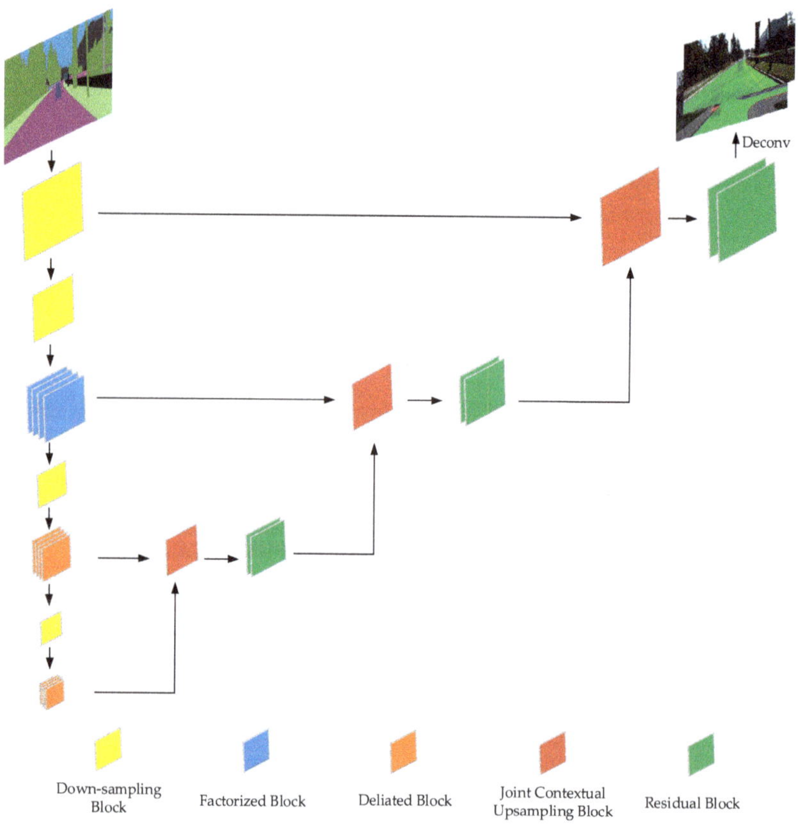

Figure 2. The proposed occlusion-free road segmentation network architecture.

Table 1. Our network architecture in detail. Size refers to output feature maps size for an input size of 384 × 1248.

Stage	Block Type	Size
Encoder	Context Down-sampling	192 × 624 × 16
	Context Down-sampling	96 × 312 × 32
	Factorized blocks	96 × 312 × 32
	Context Down-sampling	48 × 156 × 64
	Dilated blocks	48 × 156 × 64
	Context down-sampling	24 × 78 × 128
	Dilated blocks	24 × 78 × 128
Decoder	Joint Context Up-sampling	48 × 156 × 64
	Bottleneck Blocks	48 × 156 × 64
	Joint Context Up-sampling	96 × 312 × 32
	Bottleneck Blocks	96 × 312 × 32
	Joint Context Up-sampling	192 × 624 × 16
	Bottleneck Blocks	192 × 624 × 16
	Deconv	384 × 1248 × 2

Context Convolution Block Recent works have shown that contextual information is helpful for models to predict high-quality segmentation results. Modules which could enlarge the receptive field, such as ASPP [21], DenseASPP [30], and CRFasRNN [31], have been proposed in the past years. Most of these works explore context information in the decoder phase and ignore the surrounding context when encoding the features in the early stage. On the other hand, the attention mechanism has been widely used for increasing model capability. Inspired by the non-local block [27] and SE block [26], we proposed the context convolution, as shown in Figure 3. A context branch from [28] was added, bypassing the main branch of the convolution operation. As can be seen in Equation (1), the context branch first adopted a 1 × 1 convolution W_k and softmax function to obtain the attention weights, and then performed the attention pooling to obtain the global context features; then the global context features were transformed via a 1 × 1 convolution W_v and was added to the features of the main convolution branch.

$$z_i = x_i + W_v \sum_{j=1}^{N_p} \frac{\exp\left(W_k x_j\right)}{\sum_{m=1}^{N_p} \exp(W_k x_m)} x_j, \tag{1}$$

where W_k and W_v denote linear transformation matrices.

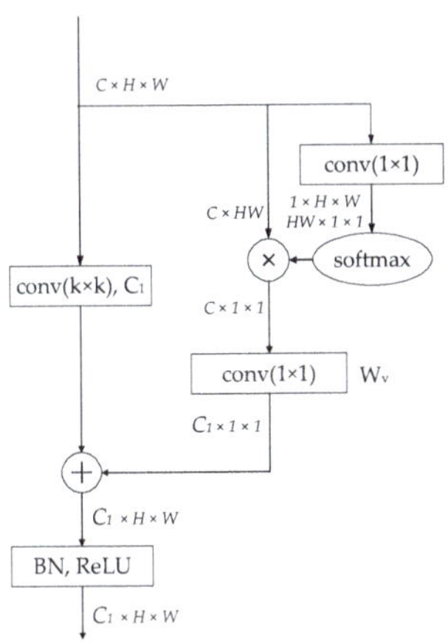

Figure 3. The context convolution block.

Down-Sampling Block In our work, the down-sampling block performed down-sampling by using a 3×3 convolution with stride 2 in the main branch of a context convolution block, as stated above. The context branch extracted the global context information to obtain a global understanding of features. Down-sampling lets the deeper layers gather more context (to improve classification) and helps to reduce computation. And we used two down-sampling blocks at the start of the network to reduce the feature size and make the network works efficiently for large input.

Joint Context Up-Sampling Block In the decoder, we proposed a joint context up-sampling block, which takes two feature maps from different stages in the encoder, as shown in Figure 4. The feature map from the earlier stage with bigger resolution and fewer channels carry sufficient details in spatial, and the feature map from the later stage with a smaller resolution and more channels contain the necessary facts in context. The joint context up-sampling block combines these two feature maps gently and efficiently using a context convolution block and bilinear up-sampling. The two branches of the two feature maps were concatenated along the channels, and a context convolution block was applied to the concatenated feature map. As shown in Figure 2, the joint context up-sampling blocks follow a sequential architecture, the current block utilized the former results and the corresponding decoder features, which made the up-sampling operation more effective.

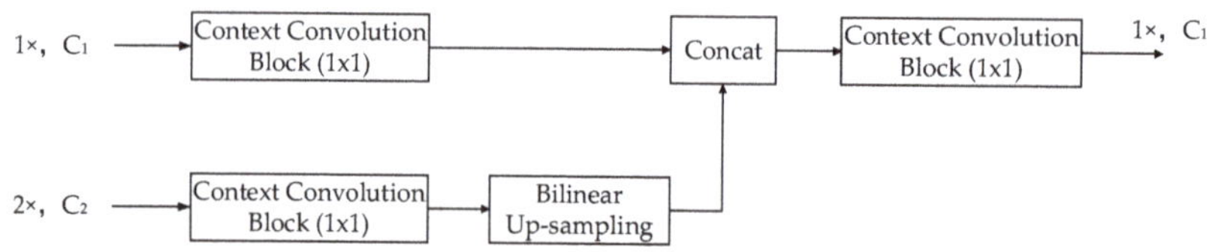

Figure 4. The joint context up-sampling block.

Residual Bottleneck Blocks Between the down-sampling and up-sampling blocks, some residual blocks were inserted to perform the encoding and decoding. In the early stage of the encoder, we applied factorized residual blocks to extract dense features. As shown in Figure 5b, a 3×3 convolution was replaced by a 3×1 convolution and a 1×3 convolution in the residual branch to reduce parameters and computation. In the later stage of the encoder, we stacked dilated convolution blocks with different rates to obtain a larger receptive field and obtain more contextual information. The dilated convolution block applied a dilated convolution on the 3×3 convolution in the residual branch compared to the regular residual block, as shown in Figure 5c. The dilate rates in the stacked dilated residual blocks were 1, 2, 5, and 9, which were carefully chosen to avoid the gridding problem when inappropriate dilation rate is used. One dilated residual block consisted of two groups of stacked dilated residual blocks in our network. In the decoder phase, two continuous regular residual blocks were inserted between the joint context up-sampling blocks.

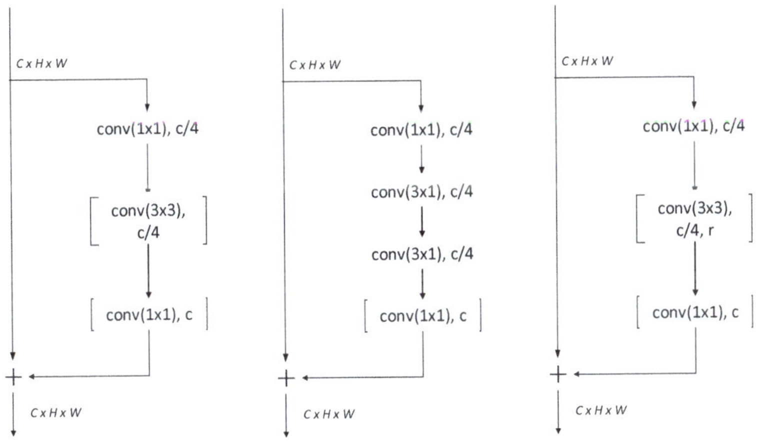

(a) Regular Residual Block (b) Factorized Residual Block (c) Dilated Residual Block

Figure 5. Residual blocks in our network.

3.3. Loss Function

As to the classification tasks, the cross-entropy loss has proved very effective. However, in our task, the road edge area needs more attention paid to it when performing the inference process, and the faraway road in the image took fewer pixels. We proposed a spatially-dependent weight to handle this problem to enhance the loss on the road edge region and faraway road area. The road edge region (ER) was defined as a set of the pixels around the road edge pixels E, which was obtained from the ground truth label image using the Canny algorithm [32], as shown in Figure 6. The Manhattan distance was adopted to calculate the distance between other pixels and edge pixels, and $T_W \in R$ was used to control the region size. Then the weight is defined as Equation (3), which takes into account the road edge region and the faraway distance factor. The loss function with spatial weight is shown in Equation (4), which is referred to as CE-SW, and the traditional cross-entropy loss is referred to as CE in our paper. The experiment showed that the CE-SW could significantly improve the performance of the models on the occlusion-free road segmentation task.

$$\mathrm{ER} = \left\{ \mathrm{v}(\mathrm{i}',\mathrm{j}') \middle| \left|\mathrm{i}-\mathrm{i}'\right| + \left|\mathrm{j}-\mathrm{j}'\right| < \mathrm{T_W},\ \mathrm{e}(\mathrm{i},\mathrm{j}) \in \mathrm{E},\ \mathrm{v}(\mathrm{i}',\mathrm{j}') \in \mathrm{Img} \right\}, \quad (2)$$

$$\mathrm{w}(\mathrm{i},\mathrm{j}) = \begin{cases} 1, & if\ p(i,j) \in \overline{ER} \\ \frac{k*|i-i_0|+|j-j_0|}{k*h+w/2} * 2 + 2, & if\ p(i,j) \in ER \end{cases}, \quad (3)$$

where w and h are the width and height of the image, $k=h/w$ is the rate to balance the height and width of the image, i and j are the pixel index, i_0 and j_0 the bottom center pixel index.

$$\mathrm{Loss}(\mathrm{y},\mathrm{p}) = \sum_i^H \sum_j^W -w(i,j)\left[\mathrm{y}_{\mathrm{i,j}} \log p_{i,j} + \left(1-y_{i,j}\right)\log\left(1-p_{i,j}\right)\right], \quad (4)$$

where y is the ground truth, p is the predict logits, i and j are the pixel index in the image.

(a) (b) (c)

Figure 6. Visualization of the road edge region. (**a**) The road segmentation label; (**b**) road edge obtained from (a) by the Canny algorithm; (**c**) road edge region with a width of 10 pixels.

4. Experiments

In this section, we provide qualitative and quantitative results for experiments carried out to test the performance of our approach. There are numerous approaches in semantic segmentation; we mainly compare our method to those pursuing a good tradeoff between high quality and computation, such as SegNet, ENet, and ERFNet. Moreover, to compare [22], we verified the model of inferring occluded road boundaries by replacing the decoder part of the model with a new one that is suitable for our task. The verified model is referred to as ORBNet in our work, which retained the encoder and employed a decoder similar to that in the DeepLabv3+ algorithm [6]. We present quantitative results based on evaluations with our manually annotated dataset based on the KITTI dataset named KITTI-OFRS dataset. The presented results appear all to be based on the manual dataset annotations except the qualitative results on Cityscapes dataset using predicted semantics as input. We first trained the models on the proposed KITTI-OFRS dataset, and the experimental results demonstrate that the proposed approach spends less time on inference and obtains better performance. Then, we compared the performance of those models when trained with traditional cross-entropy loss function and the proposed spatially-weighted cross-entropy loss function. Moreover, we tested the generalization performance of the models on the Cityscapes dataset. Finally, the performance of the models based on automatically inferred semantics was visualized to show that our network works well in the real system.

4.1. Datasets

There were no available datasets for the proposed occlusion-free road segmentation task, so we built our own datasets. We built a real-world dataset named KITTI-OFRS based on the public KITTI semantic segmentation benchmark, which is used for training and evaluation. Moreover, we qualitatively tested our well-trained model on the Cityscape dataset [33] for a view of its generalization ability.

KITTI-OFRS Dataset The real-world dataset was built on the public KITTI semantic segmentation benchmark, which is part of the KITTI dataset [34]. The KITTI dataset is the largest data collection for computer vision algorithms in the world's largest autopilot scenario. The dataset is used to evaluate the performance of computer vision technologies and contains real-world image data collected from scenes such as urban, rural, and highways, with up to 15 vehicles and 30 pedestrians per image, as well as varying degrees of occlusion. The KITTI semantic segmentation benchmark consists of 200 semantically annotated train as well as 200 test samples corresponding to the KITTI Stereo and Flow Benchmark 2015. We only annotated the available 200 semantically annotated training samples for our task and randomly split them into two parts, one contained 160 samples for training, and the other contained 40 samples for evaluation. We named this real-world dataset as KITTI-OFRS dataset. One sample in this dataset contained the RGB image, normal semantic labels, and occlusion-free road segmentation labels, as demonstrated in Figure 7.

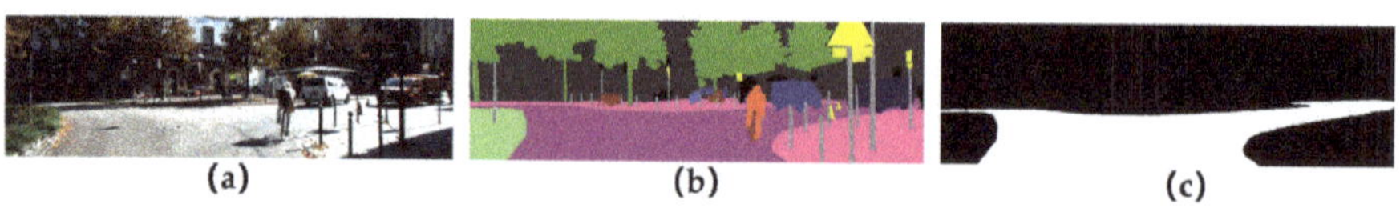

Figure 7. An example of the KITTI-occlusion-free road segmentation (KITTI-OFRS) dataset sample. (**a**) the RGB image; (**b**) annotation of semantic segmentation; (**c**) annotation of full road area, white denotes road.

Cityscapes Dataset The Cityscapes dataset contains 5000 images collected in street scenes from 50 different cities. The dataset is divided into three subsets, including 2975 images in the training set, 500 images in the validation set, and 1525 images in the testing set. High-quality pixel-level annotations of 19 semantic classes are provided in this dataset. We only used this dataset for the generalization ability test.

Classes Transformation The occlusion-free road segmentation network was designed to apply in the semantic domain. However, different semantic segmentation datasets may have different categories, and one category may have a different class labels in different datasets. It is obvious that some categories are not involved in occluding the road, such as sky, and some categories could be aggregated to one category to get a more compact representation, for example, car, truck, bus, train, motorcycle, and bicycle could be aggregated to vehicle. Therefore, a classes transformation layer is proposed to transform different semantic representations to a unify form before being fed to the occlusion-free road segmentation network.

The classes transformation layer is a matrix multiplication operation, taking one-hot liked encoded semantic representation of variable categories $R_{in} \in [0,1]^{W \times H \times C}$ as input and output one-hot representation of a unify categories $R_{out} \in [0,1]^{W \times H \times C_u}$.

$$R_{out} = R_{in} * T, \tag{5}$$

$$T(i,j) = \begin{cases} 1, & if\ C(i) \to C_u(j) \\ 0, & otherwise \end{cases}, \tag{6}$$

where $T \in \{0,1\}^{C \times C_u}$ is the transformation matrix, C is the set of original class labels and C_u the set of target class labels. $C(i) \to C_u(j)$ refers to that the i-th label in C should be set to the j-th label in C_u.

The classes transformation layer could aggregate and unify labels of different semantic segmentation representations from different datasets or different semantic segmentation algorithms. In our work, the unified semantic representation contained 11 classes, namely road, sidewalk, building, wall, fence, pole, traffic sign, vegetation, person, vehicle, and unlabeled.

Data Augmentation In the training phase, the training data was augmented with random cropping and padding, flipping left to right. Moreover, to tackle the uncertainty of the semantic labels due to annotation errors, we augmented the training data by the technique of label smoothing, which is firstly proposed in InceptionV2 [35] to reduce over-fitting and increase the adaptive ability of the model. We used this method to add noise to the semantic one-hot, which could make our model more adaptive to annotation errors and prediction errors from other semantic segmentation methods. Unlike the original usage that takes α a constant value for all the samples, we choose α as a random value between 0.1 and 0.2 following uniform distribution, which was independent of each pixel in a training batch.

$$y_k^{LS} = y_k + (1 - \alpha) + \alpha / K. \tag{7}$$

4.2. Evaluation Metrics

For quantitative evaluation, precision (PRE), recall (REC), F1 score, average precision (ACC), and intersection-over-union (IoU) were used as the metrics within a region around the road edges within 4 pixels. The metrics acting on such a region are more powerful to test the network performance than on the whole pixels taking into account the primary task of occlusion reasoning. The metrics are calculated as in Equations (8)–(12), where TP, TN, FP, FN are, respectively, the number of true positives, true negatives, false positives, and false negatives at the pixel level. Our experiments considered an assessment that demonstrates the effectiveness of our approach for inferring occluded road in the semantic domain.

$$PRE = \frac{TP}{TP + FP}, \tag{8}$$

$$REC = \frac{TP}{TP + FN}, \tag{9}$$

$$F1 = \frac{2PRE \cdot REC}{PRE + REC}, \tag{10}$$

$$ACC = \frac{TP + TN}{TP + FP + TN + FN}, \tag{11}$$

$$IoU = \frac{TP}{TP + FP + FN}. \tag{12}$$

4.3. Implementation Details

In the experiments, we implemented our architectures in PyTorch version 1.2 [36] (FaceBook, State of California, USA) with CUDA 10.0 and cuDNN back-ends. All experiments were run on a single NVIDIA GTX-1080Ti GPU. Due to GPU memory limitations, we had a maximum batch size of 4. During optimization, we used the SGD optimizer [37] with a weight decay of 0.0001 and a momentum of 0.9. The learning rate was set using the poly strategy with a start value of 0.01 and a power of 0.9. The edge region width T_w was set to 10 in the training phase and 4 in the evaluation phase.

4.4. Results and Analysis

To evaluate the effectiveness of our method on the occlusion-free road segmentation task, we trained the proposed model on the KITTI-OFRS dataset, as well as some other lightweight baseline models, such as ENet, SegNet, ERFNet, and ORBNet. The samples were resized to 384 × 1248 when training and testing. The quantitative and qualitative results are shown in Table 2 and Figure 8, respectively. As shown in Table 2 and Figure 8, both models achieved comparable results on the proposed task, and our method was superior to the baseline models in both accuracy and runtime. In Figure 8, red denotes false negatives; blue areas correspond to false positives, and green represents true positives. The models both performed well in the semantic domain containing more compact information of the driving environment, which indicates that the semantic and spatial information were more essential for occlusion reasoning than color and textural features. As can be seen from Figure 8, the models obtained significant results on both simple straight roads and complex intersection areas. Variable occlusion situations could be handled well, even though there were some heavy occlusion scenes. Based on the results of the proposed task, the whole road structure could be obtained and could be easily transformed into 3D world representations by an inverse perspective transformation without the affectation of the on-road objects. Empirically, higher road detection precision may lead to a better road model for better path planning.

Comparison of accuracy and computation complexity Our model achieved a significant trade-off between accuracy and efficiency, which conclusion is drawn by comparing with other models. To compare the computation complexity, we employed several parameters, GFLOPs, and frames per second (FPS) as the evaluation metrics. FPS was measured on an Nvidia GTX1080Ti GPU with an input size of 384 × 1248 and was averaged among 100 runs. As can be seen from Table 2, our model outperformed ENet by 1.5% in the F1 score and 2.6% in the IoU while runs were only a little slower than it. Our model ran almost two times faster than ERFNet and improved 1.0% in the F1 score and 1.7% in the IoU. Compared to SegNet and ORBNet, our model got a little improvement in accuracy but achieved three times faster in the inference phase. In conclusion, our model achieved a better trade-off between accuracy and efficiency.

Table 2. Evaluation results of models trained with spatially-weighted cross-entropy loss (CE-SW).

Model	Parameters	GFLOPs	FPS	ACC	PRE	REC	F1	IoU
ENet	0.37M	3.83	**52**	91.8%	92.1%	89.3%	90.7%	82.9%
ERFNet	2.06M	24.43	25	92.3%	92.6%	89.7%	91.2%	83.8%
SegNet	29.46M	286.03	16	92.9%	93.6%	90.2%	91.8%	84.9%
ORBNet	1.91M	48.48	11.5	92.7%	93.4%	89.9%	91.6%	84.5%
OFRSNet	0.39M	2.99	46	**93.2%**	**94.2%**	**90.3%**	**92.2%**	**85.5%**

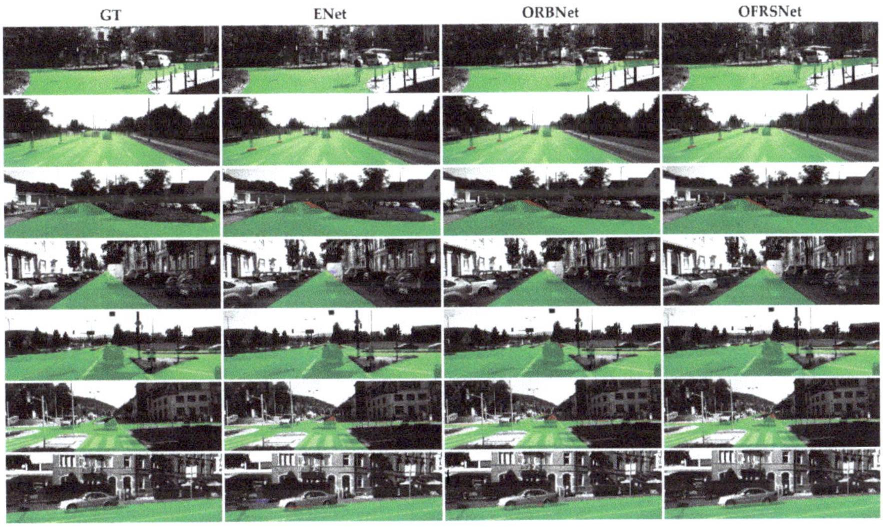

Figure 8. Qualitative results on the KITTI-OFRS dataset. The columns from left to right are the results of GT, ENet, ORBNet, and OFRSNet, respectively. Red denotes false negatives; blue areas correspond to false positives, and green represents true positives.

Comparison of loss function To evaluate the effectiveness of the proposed spatially-weighted cross-entropy loss, we trained the models both with traditional cross-entropy loss (CE) and the spatially-weighted cross-entropy loss (CE-SW), and the evaluation results of the CE and metrics degradation are shown in Table 3. When trained with CE, the models saw obvious metrics degradation compared to CE-SW. The values in parentheses are the metrics degradation compared to that when models were trained with CE-SW, which shows that the spatially-weighted cross-entropy loss was very beneficial for increasing accuracy. Intuitively, the spatially-weighted cross-entropy loss forced the models to take care of the road edge region where the occlusion occurs mostly.

Table 3. Evaluation results of models trained with cross-entropy loss (CE). The values in parentheses are the metrics degradation compared to that when models were trained with spatially-weighted cross-entropy loss (CE-SW).

Model	ACC	PRE	REC	F1	IoU
ENet	90.4%(−1.4%)	90.5%(−1.6%)	87.6%(−1.7%)	89.0%(−1.7%)	80.2%(−2.7%)
ERFNet	90.5%(−1.8%)	90.9%(−1.7%)	87.3%(−2.4%)	89.1%(−2.1%)	80.3%(−3.5%)
SegNet	92.1%(−0.8%)	92.6%(−1.0%)	89.4%(−0.8%)	91.0%(−0.8%)	83.5%(−1.4%)
ORBNet	91.5% (−1.2%)	92.2% (−1.2%)	88.4% (−1.5%)	90.2% (−1.4%)	82.2% (−2.3%)
OFRSNet	91.7%(−1.5%)	92.4%(−1.8%)	88.6%(−1.7%)	90.5%(−1.7%)	82.6%(−2.9%)

Comparison of convolution with and without context To evaluate the benefits of the context convolution block, we replaced the context convolution block with regular convolution operation in the down-sampling and up-sampling blocks. As shown in Table 4, the model with context information outperformed the model without that by 0.6% in the F1 score and 1.0% in the IoU, which demonstrates that the context information is desirable for the proposed approach.

Table 4. Performance comparison of the model with and without context.

Model	Context	Parameters	GFLOPs	ACC	PRE	REC	F1	IoU
OFRSNet	w/o	0.34M	2.96	92.7%	92.8%	90.4%	91.6%	84.5%
OFRSNet	w/	0.39M	2.99	**93.2%**	**94.2%**	**90.3%**	**92.2%**	**85.5%**

Generalization on Cityscape Dataset To further test the generalization ability of our model, we conducted qualitative test experiments on the Cityscape dataset with the model trained only on the KITTI-OFRS dataset. As can be seen from Figure 9, the well-trained model performed well on the complex real-world Cityscapes dataset, which indicates that our model obtained quite a good generalization ability on the occlusion-free road segmentation task. The generalization ability of our model benefited from inferring the occluded road in the semantic domain, which made the model focus on learning the occlusion mechanism in the driving scenes without the affectation of sensing noise. In the scenes, the color and textual features may differ very much in the same position due to different camera configurations and lighting conditions while the semantic features share a similar distribution. The occlusion situations were able to understand that the occluded road area was correctly inferred in variable occlusion scenes by the proposed method according to the results. As shown in Figure 9, the detection results obtained the overall structure of the road and accurate segmentation despite occlusion. Moreover, it is applicable to combine our method with other semantic segmentation algorithms in the real system due to its lightweight and efficiency. As shown in Figure 10, when taking the predicted semantics obtained by the DeepLabv3+ algorithm as input, the proposed OFRSNet still works well to predict the occluded road areas and outperforms ENet and ORBNet in terms of accuracy and robustness.

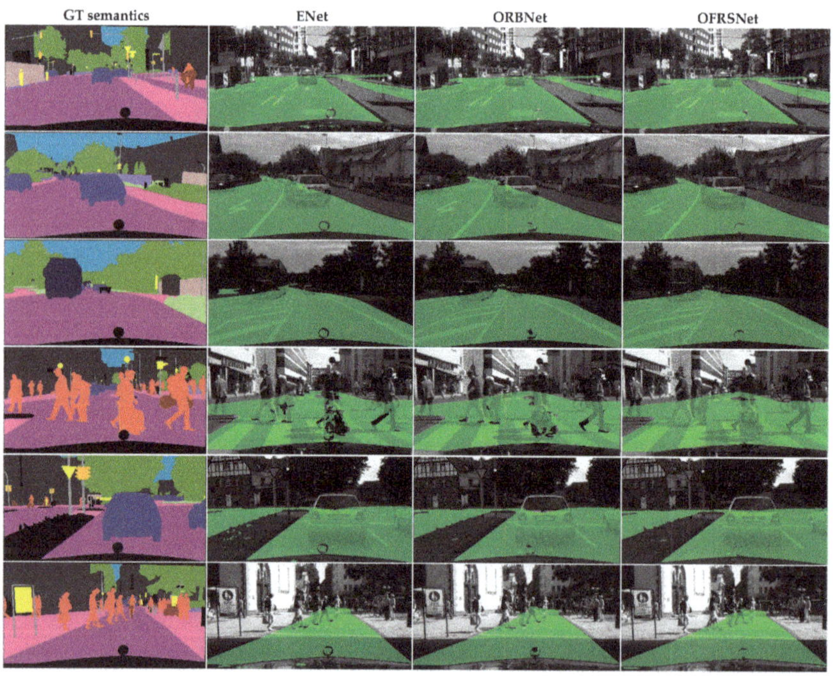

Figure 9. Qualitative results on the Cityscapes dataset using ground truth semantics as input. Green represents the detected full road area.

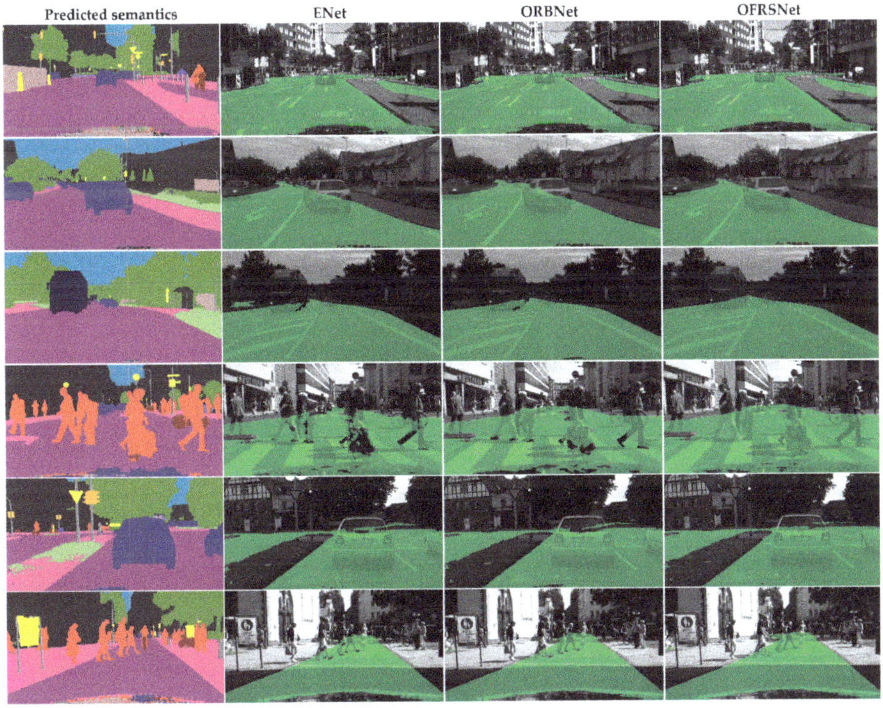

Figure 10. Qualitative results on the Cityscapes dataset using predicted semantics as input, which were obtained by the DeepLabv3+ algorithm. Green represents the detected full road area.

5. Conclusions

In this paper, we presented an occlusion-free road segmentation network to infer the occluded road area of an urban driving scenario from monocular vision. The model we presented is a lightweight and efficient encoder–decoder fully convolutional architecture that contains down-sampling and up-sampling blocks combined with global contextual operations. Meanwhile, a spatially-weighted cross-entropy loss was proposed to induce the network to pay more attention to the road edge region in the training phase. We showed the effectiveness of the model on the self-built small but efficient KITTI-OFRS dataset. Compared to other recent lightweight semantic segmentation algorithms, our network obtained a better trade-off between accuracy and runtime. The comparisons of the models trained with different loss functions highlighted the benefits of the proposed spatially-weighted cross-entropy loss for the occlusion reasoning road segmentation task. The generalization ability of our model was further qualitatively tested on the Cityscape datasets, and the results clearly demonstrated our model's inferring ability of the occluded road even in complex scenes. Moreover, the proposed OFRSNet could be efficiently combined with other semantic segmentation algorithms due to its small size and minimal runtime. We believe that being able to infer occluded road regions in autonomous driving systems is a key component to achieve a full comprehension of the scene and will allow better planning of the ego-vehicle trajectories.

Author Contributions: Conceptualization, K.W., F.Y., and B.Z.; Data curation, K.W. and Q.Y.; Formal analysis, K.W., B.Z., L.T., and C.L.; Funding acquisition, F.Y.; Investigation, K.W., L.T., Q.Y., and C.L.; Methodology, K.W.; Project administration, F.Y. and B.Z.; Resources, B.Z.; Software, K.W. and L.T.; Supervision, K.W.; Validation, K.W., B.Z., L.T., and C.L.; Visualization, K.W. and Q.Y.; Writing—original draft, K.W. and B.Z.; Writing—review and editing, K.W. and B.Z.

Funding: This research was funded by the National Natural Science Foundation of China (Grant No. 51975434), the Overseas Expertise Introduction Project for Discipline Innovation (Grant No. B17034), the Innovative Research Team in University of Ministry of Education of China (Grant No. IRT_17R83) and the Department of Science and Technology, the Hubei Provincial People's Government (Grant No. 2017BEC196).

Acknowledgments: Thanks for the help of reviewers and editors.

Conflicts of Interest: The authors declare no conflict of interest.

References

1. Oliveira, G.L.; Burgard, W.; Brox, T. Efficient deep models for monocular road segmentation. In Proceedings of the 2016 IEEE/RSJ International Conference on Intelligent Robots and Systems (IROS), Daejeon, Korea, 9–14 October 2016; pp. 4885–4891.
2. Mendes, C.C.T.; Fremont, V.; Wolf, D.F. Exploiting Fully Convolutional Neural Networks for Fast Road Detection. In Proceedings of the 2016 IEEE International Conference on Robotics and Automation, Stockholm, Sweden, 16–21 May 2016; Okamura, A., Menciassi, A., Ude, A., Burschka, D., Lee, D., Arrichiello, F., Liu, H., Eds.; IEEE: New York, NY, USA, 2016; pp. 3174–3179.
3. Zhang, X.; Chen, Z.; Wu, Q.M.J.; Cai, L.; Lu, D.; Li, X. Fast Semantic Segmentation for Scene Perception. *IEEE Trans. Ind. Inform.* **2019**, *15*, 1183–1192. [CrossRef]
4. Wang, B.; Fremont, V.; Rodriguez, S.A. Color-based Road Detection and its Evaluation on the KITTI Road Benchmark. In Proceedings of the 2014 IEEE Intelligent Vehicles Symposium Proceedings, Dearborn, MI, USA, 8–11 June 2014; pp. 31–36.
5. Song, X.; Rui, T.; Zhang, S.; Fei, J.; Wang, X. A road segmentation method based on the deep auto-encoder with supervised learning. *Comput. Electr. Eng.* **2018**, *68*, 381–388. [CrossRef]
6. Chen, L.C.; Zhu, Y.; Papandreou, G.; Schroff, F.; Adam, H. Encoder-decoder with atrous separable convolution for semantic image segmentation. In Proceedings of the European Conference on Computer Vision (ECCV), Munich, Germany, 8–14 September 2018; pp. 801–818.
7. Long, J.; Shelhamer, E.; Darrell, T. Fully Convolutional Networks for Semantic Segmentation. In Proceedings of the 2015 Ieee Conference on Computer Vision and Pattern Recognition, Boston, MA, USA, 7–12 June 2015; pp. 3431–3440.
8. Chen, L.; Papandreou, G.; Kokkinos, I.; Murphy, K.; Yuille, A.L. DeepLab: Semantic Image Segmentation with Deep Convolutional Nets, Atrous Convolution and Fully Connected CRFs. *IEEE Trans. Pattern Anal. Mach. Intell.* **2018**, *40*, 834–848. [CrossRef] [PubMed]
9. Mano, K.; Masuzawa, H.; Miura, J.; Ardiyanto, I. *Road Boundary Estimation for Mobile Robot Using Deep Learning and Particle Filter*; IEEE: New York, NY, USA, 2018; pp. 1545–1550.
10. Li, K.; Shao, J.; Guo, D. A Multi-Feature Search Window Method for Road Boundary Detection Based on LIDAR Data. *Sensors* **2019**, *19*, 1551. [CrossRef]
11. Khalilullah, K.M.I.; Jindai, M.; Ota, S.; Yasuda, T. *Fast Road Detection Methods on a Large Scale Dataset for assisting robot navigation Using Kernel Principal Component Analysis and Deep Learning*; IEEE: New York, NY, USA, 2018; pp. 798–803.
12. Son, J.; Yoo, H.; Kim, S.; Sohn, K. Real-time illumination invariant lane detection for lane departure warning system. *Expert Syst. Appl.* **2015**, *42*, 1816–1824. [CrossRef]
13. Li, Q.; Zhou, J.; Li, B.; Guo, Y.; Xiao, J. Robust Lane-Detection Method for Low-Speed Environments. *Sensors* **2018**, *18*, 4274. [CrossRef] [PubMed]
14. Cao, J.; Song, C.; Song, S.; Xiao, F.; Peng, S. Lane Detection Algorithm for Intelligent Vehicles in Complex Road Conditions and Dynamic Environments. *Sensors* **2019**, *19*, 3166. [CrossRef] [PubMed]
15. Liu, X.; Deng, Z. Segmentation of Drivable Road Using Deep Fully Convolutional Residual Network with Pyramid Pooling. *Cogn. Comput.* **2017**, *10*, 272–281. [CrossRef]
16. Cai, Y.; Li, D.; Zhou, X.; Mou, X. Robust Drivable Road Region Detection for Fixed-Route Autonomous Vehicles Using Map-Fusion Images. *Sensors* **2018**, *18*, 4158. [CrossRef] [PubMed]
17. Aly, M. Real time Detection of Lane Markers in Urban Streets. In Proceedings of the Intelligent Vehicles Symposium, Eindhoven, The Netherlands, 4–6 June 2008.
18. Laddha, A.; Kocamaz, M.K.; Navarro-Serment, L.E.; Hebert, M. Map-supervised road detection. In Proceedings of the Intelligent Vehicles Symposium, Gothenburg, Sweden, 19–22 June 2016; pp. 118–123.

19. Alvarez, J.M.; Salzmann, M.; Barnes, N. Learning Appearance Models for Road Detection. In Proceedings of the Intelligent Vehicles Symposium, Gold Coast, QLD, Australia, 23–26 June 2013.
20. Badrinarayanan, V.; Handa, A.; Cipolla, R. SegNet: A Deep Convolutional Encoder-Decoder Architecture for Robust Semantic Pixel-Wise Labelling. *arXiv* **2015**, arXiv:1505.07293.
21. Chen, L.-C.; Papandreou, G.; Schroff, F.; Adam, H. Rethinking atrous convolution for semantic image segmentation. *arXiv* **2017**, arXiv:1706.05587.
22. Suleymanov, T.; Amayo, P.; Newman, P. Inferring Road Boundaries Through and Despite Traffic. In Proceedings of the 2018 21st International Conference on Intelligent Transportation Systems, Maui, HI, USA, 4–7 November 2018; pp. 409–416.
23. Becattini, F.; Berlincioni, L.; Galteri, L.; Seidenari, L.; Del Bimbo, A. Semantic Road Layout Understanding by Generative Adversarial Inpainting. *arXiv* **2018**, arXiv:1805.11746.
24. Paszke, A.; Chaurasia, A.; Kim, S.; Culurciello, E. ENet: A Deep Neural Network Architecture for Real-Time Semantic Segmentation. *arXiv* **2016**, arXiv:1606.02147.
25. Romera, E.; Álvarez, J.M.; Bergasa, L.M.; Arroyo, R. ERFNet: Efficient Residual Factorized ConvNet for Real-Time Semantic Segmentation. *IEEE Trans. Intell. Transp. Syst.* **2017**, *19*, 263–272. [CrossRef]
26. Hu, J.; Shen, L.; Sun, G. Squeeze-and-Excitation Networks. In Proceedings of the 2018 IEEE/CVF Conference on Computer Vision and Pattern Recognition, Salt Lake City, UT, USA, 18–23 June 2018; pp. 7132–7141.
27. Wang, X.; Girshick, R.; Gupta, A.; He, K. Non-local Neural Networks. In Proceedings of the 2018 IEEE/CVF Conference on Computer Vision and Pattern Recognition, Salt Lake City, UT, USA, 18–23 June 2018; pp. 7794–7803.
28. Cao, Y.; Xu, J.; Lin, S.; Wei, F.; Hu, H. GCNet: Non-local Networks Meet Squeeze-Excitation Networks and Beyond. *arXiv* **2019**, arXiv:1904.11492.
29. He, K.; Zhang, X.; Ren, S.; Sun, J. Deep Residual Learning for Image Recognition. In Proceedings of the 2016 IEEE Conference on Computer Vision and Pattern Recognition (CVPR), Las Vegas, NV, USA, 27–30 June 2016; pp. 770–778.
30. Yang, M.; Yu, K.; Zhang, C.; Li, Z.; Yang, K. DenseASPP for Semantic Segmentation in Street Scenes. In Proceedings of the 2018 IEEE/CVF Conference on Computer Vision and Pattern Recognition, Salt Lake City, UT, USA, 18–23 June 2018; pp. 3684–3692.
31. Zheng, S.; Jayasumana, S.; Romera-Paredes, B.; Vineet, V.; Su, Z.; Du, D.; Huang, C.; Torr, P.H.S. Conditional Random Fields as Recurrent Neural Networks. In Proceedings of the 2015 IEEE International Conference on Computer Vision (ICCV), Santiago, Chile, 7–13 December 2015; pp. 1529–1537.
32. Canny, J. A Computational Approach to Edge Detection. *IEEE Trans.Pattern Anal. Mach. Intell.* **1986**, *8*, 679–698. [CrossRef] [PubMed]
33. Cordts, M.; Omran, M.; Ramos, S.; Rehfeld, T.; Enzweiler, M.; Benenson, R.; Franke, U.; Roth, S.; Schiele, B. The cityscapes dataset for semantic urban scene understanding. In Proceedings of the IEEE Conference on Computer Vision and Pattern Recognition, Las Vegas, NV, USA, 27–30 June 2016; pp. 3213–3223.
34. Geiger, A.; Lenz, P.; Stiller, C.; Urtasun, R. Vision meets robotics: The KITTI dataset. *Int. J. Robot. Res.* **2013**, *32*, 1231–1237. [CrossRef]
35. Szegedy, C.; Vanhoucke, V.; Ioffe, S.; Shlens, J.; Wojna, Z. Rethinking the Inception Architecture for Computer Vision. In Proceedings of the 2016 IEEE Conference on Computer Vision and Pattern Recognition (CVPR), Las Vegas, NV, USA, 27–30 June 2016; pp. 2818–2826.
36. PyTorch. Available online: http://pytorch.org/ (accessed on 1 September 2019).
37. Bottou, L. *Large-Scale Machine Learning with Stochastic Gradient Descent*; Physica-Verlag HD: Heidelberg, Germany, 2010; pp. 177–186.

sensors

Article

Vehicle Trajectory Prediction and Collision Warning via Fusion of Multisensors and Wireless Vehicular Communications

Minjin Baek, Donggi Jeong, Dongho Choi and Sangsun Lee *

Department of Electronics and Computer Engineering, Hanyang University, Seoul 04763, Korea; mjbaek@hanyang.ac.kr (M.B.); motive5aa9@hanyang.ac.kr (D.J.); cdh5375@hanyang.ac.kr (D.C.)
* Correspondence: ssnlee@hanyang.ac.kr; Tel.: +82-2-2299-0372

Received: 18 November 2019; Accepted: 1 January 2020; Published: 4 January 2020

Abstract: Driver inattention is one of the leading causes of traffic crashes worldwide. Providing the driver with an early warning prior to a potential collision can significantly reduce the fatalities and level of injuries associated with vehicle collisions. In order to monitor the vehicle surroundings and predict collisions, on-board sensors such as radar, lidar, and cameras are often used. However, the driving environment perception based on these sensors can be adversely affected by a number of factors such as weather and solar irradiance. In addition, potential dangers cannot be detected if the target is located outside the limited field-of-view of the sensors, or if the line of sight to the target is occluded. In this paper, we propose an approach for designing a vehicle collision warning system based on fusion of multisensors and wireless vehicular communications. A high-level fusion of radar, lidar, camera, and wireless vehicular communication data was performed to predict the trajectories of remote targets and generate an appropriate warning to the driver prior to a possible collision. We implemented and evaluated the proposed vehicle collision system in virtual driving environments, which consisted of a vehicle–vehicle collision scenario and a vehicle–pedestrian collision scenario.

Keywords: advanced driver assistance system; trajectory prediction; risk assessment; collision warning; connected vehicles; vehicular communications; vulnerable road users

1. Introduction

The incidence of road traffic crashes is one of the leading causes of death worldwide, and the reduction of the number of traffic-related crashes has become a major social and public health challenge, considering the ever-increasing number of vehicles on the road. One of the most common causes of vehicle crashes is driver inattention. One study conducted by the National Highway Traffic Safety Administration (NHTSA) reported that approximately 80 percent of vehicle crashes and 65 percent of near-crashes involved driver inattention within three seconds prior to the incident [1]. Taking into account that human life expectancy is continuously getting longer, it has become crucial that we assist those who are older and those who are physically impaired in driving and achieve higher road safety measures through research and development of advanced driver assistance systems (ADAS) technology.

The safety functions of ADAS require accurate information on the environment surrounding the vehicle. A popular approach in recent years to obtain the information on the vehicle surroundings involves fusing the data generated by multiple types of sensors (e.g., radar, lidar, and cameras) equipped on the vehicle [2–7]. This way, it is possible to overcome the functional and environmental limitations of each type of sensor and generate the estimate of the state of each surrounding object with higher accuracy. However, this sensor fusion approach has its limits on the reliability and data collection range. The sensor accuracy of driving environment information is affected by a number of factors such as weather and solar irradiance. In addition, no data can be acquired when the target is outside the field of view of the sensors or when the line of sight to the target is obstructed. In order to further enhance road safety, it is therefore critical to improve the reliability and the detection range of the perception system and also find a way to obtain information on objects in non-line-of-sight (NLOS) regions.

A wireless vehicular communication system can be viewed as a new type of automotive sensor that allows engineers to design the next generation of ADAS, enabling drivers to exchange information on their own vehicles as well as the environment surrounding them. Whereas on-board sensor data obtained with radar, lidar, and cameras enable the estimation of target vehicle information such as relative position, speed, and heading, vehicular communication data additionally provide us with the best possible measurements on vital vehicle data including speed, yaw rate, and steering angle, which are obtained directly from the remote vehicle bus. This communication network can further extend its reach when vehicles, roadside infrastructures, and vulnerable road users (e.g., pedestrians, cyclists, and motorcyclists) are equipped with wireless communication devices. Wireless vehicular communications, often referred to as vehicle-to-everything (V2X) communications, can be classified into different types including vehicle-to-vehicle (V2V), vehicle-to-infrastructure (V2I), and vehicle-to-pedestrian (V2P) communications. While V2V communications involves two or more vehicles exchanging data with each other, V2I communications allows data exchange between vehicles and roadside units. Furthermore, V2P communications involves vehicles exchanging data with pedestrians. Studies have shown that combining V2V and V2I technologies can help address about 80 percent of all vehicle crashes [8].

Such significant advantages of V2X communications in road safety can become even more augmented when combined with the on-board sensor measurements via data fusion. Figure 1 summarizes the positives and the negatives of perception through V2X communications and those of remote sensing with on-board sensors such as radar, lidar, and cameras. The two groups of data complement each other, resulting in a more accurate, robust, and complete perception of the vehicle surroundings. As mentioned earlier, the implementation of V2X communications greatly enhances the perception capability, as it enables detection of targets in NLOS regions and extends the detection range up to 1 km [9], while the longest detection range that can be achieved with on-board sensors is 200–250 m (through radar systems). Exchanging V2X communication data is possible regardless of weather conditions, whereas the accuracy and reliability of on-board sensors can be significantly reduced by adverse weather conditions such as rain, snow, and fog [10]. Furthermore, safety applications of camera systems such as collision warning and pedestrian detection are often inactive in a dark environment or during night time. V2X communication data also include accurate target dimension information (width, length, and height), but the dimension information obtained with on-board sensors are often inaccurate or even unavailable due to the effects of occlusion and the limitations of the sensor field of view (FOV). On the other hand, there are some negative aspects to perception based solely on V2X communications. Transmitted V2X communication data can be delayed or even lost in an adverse radio frequency propagation environment (e.g., blockage and multipath) and/or a high communication channel load scenario (e.g., heavily congested urban intersections). In addition, V2X safety messages such as the basic safety message (BSM) are transmitted at a period of 100 ms, whereas on-board sensor measurements can be collected with a period of about 50 ms or even at a faster rate depending on the sensor model. Locating targets through V2X communications is also limited in that vehicles must be equipped with vehicular communication devices to participate in the exchange of the safety messages, and that the accuracy and reliability of positioning are largely dependent on the quality and availability of the global navigation satellite system (GNSS) signals. In an environment where GNSS signals are not available (e.g., inside a tunnel and under an overpass), vehicles can no longer transmit the safety messages, which results in a discontinuous acquisition of data on surrounding vehicles.

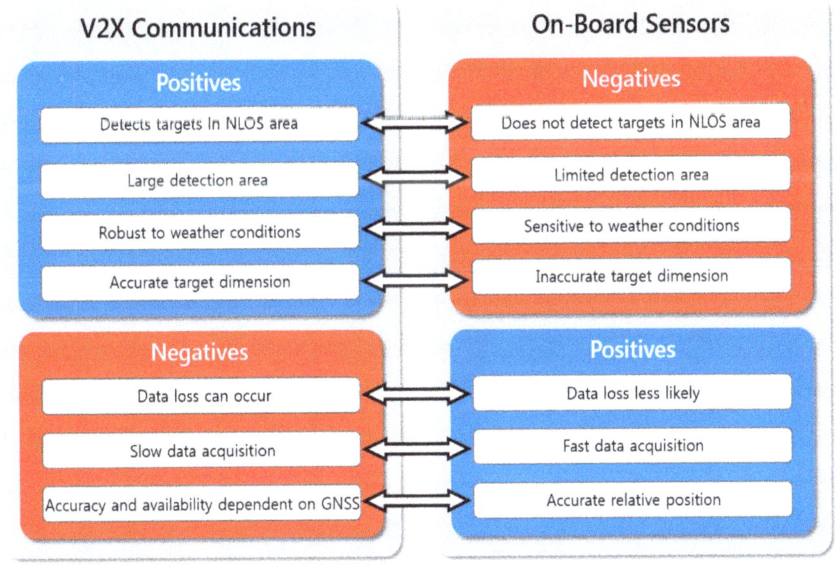

Figure 1. Positive and negative characteristics of perception using vehicle-to-everything (V2X) communications and on-board automotive sensors for remote sensing.

In this paper, we propose a method for vehicle trajectory prediction and collision warning through fusion of multisensors and V2X communications. In order to enhance the perception capabilities and reliability of traditional on-board sensors, we employ a Kalman filter-based approach for a high-level fusion of radar, lidar, camera, and V2X communication data. To verify the performance of the proposed method, we constructed co-simulation environments using MATLAB/Simulink and PreScan [11], which is designed for simulation of ADAS and active safety systems. In addition to radar, lidar, and camera sensor systems, the host vehicle is equipped with a dedicated short-range communications (DSRC) transceiver, which enables the collection of information on the surrounding vehicles and vulnerable road users (VRUs) equipped with DSRC devices through exchanging safety messages. The performance of the proposed vehicle collision warning system is evaluated in a vehicle–vehicle collision scenario and a vehicle–pedestrian collision scenario.

The rest of the paper is organized as follows. Section 2 introduces related research work. In Section 3, we describe the architecture of the proposed system and discuss background information about automotive sensors for remote sensing and V2X communications. The proposed method for vehicle collision warning is presented in Section 4, and the experimental results are given in Section 5. Finally, Section 6 concludes the paper by summarizing the main points and addressing future work.

2. Related Work

Vehicle collision warning systems have been studied by many researchers. Typical vehicle collision warning systems are based on sensor measurements from radar and camera sensors. Vehicle collision warning and automatic partial braking systems based on radar sensors that have been implemented in commercially available Mercedes-Benz cars are described in [12]. A vehicle collision warning system with a single Mobileye camera is presented in [13], where rear-end collision scenarios are considered and the warning is generated based on the time-to-collision (TTC) calculation. More recently, there have been efforts to develop cooperative collision warning systems that utilize vehicular communications. In [14], a crossroad scenario with two vehicles equipped with GPS receivers and vehicular communication devices is considered, where the trajectory prediction is performed with a

Kalman filter and TTC is used for the collision risk indicator. A rear-end collision warning model based on a neural network approach is presented in [15], where participating vehicles are equipped with GPS receivers and vehicular communication devices and are assumed to be moving in the same lane.

Despite the advantages of vehicular communications, the cooperative sensing approach based on vehicular communications and on-board sensor fusion has not been examined extensively yet by researchers. Inter-vehicle object association using point matching algorithms is proposed in [16] to determine the relative position and orientation offsets between measurements taken by different vehicles. In [17], a vision-based multiobject tracking system is presented to check the plausibility of the data received via V2V communications. Radar and V2V communication fusion approach is suggested in [18] for a longer perception range and lower position and velocity errors. In the case of maritime navigation, the automatic radar plotting aids (ARPA) and the automatic identification system (AIS) technologies are widely implemented to identify and track vessels and to prevent collisions between vessels based on radar measurements as well as static and dynamic information (e.g., vessel name, call sign, position, course, and speed) of other AIS-equipped vessels exchanged over the marine VHF radio channels [19,20]. Although these papers present promising applications, the potential of the fusion of on-board sensor data and V2X communication data in the context of ADAS applications, such as vehicle collision prevention, has not been extensively investigated.

3. System Overview

As each type of sensors has its advantages and disadvantages, combining data from multiple types of sensors is necessary in order to maximize detection and tracking capability. In this work, a high-level fusion of radar, lidar, cameras, and V2X communication data was performed to predict the trajectories of the nearby targets and generate an appropriate warning to the driver prior to a possible collision. In an effort to perform simulations under close-to-real conditions, the characteristics of local environment perception sensors that have been widely considered for ADAS functions in commercially available vehicles were employed.

3.1. Architecture of the Proposed System

The framework of the proposed vehicle collision warning system is illustrated in Figure 2. The first step of the proposed system involves perception. For the purpose of estimating the relative position of the target in the surrounding space with respect to the host vehicle, the host vehicle obtains the relative range and azimuth from the radar and the lidar, the relative lateral and longitudinal position from the camera, and the GNSS measurements of the remote target as well as its dynamic information such as speed and yaw rate via the DSRC transceiver. The measurements from each sensor are processed with a Kalman filter algorithm, which reduces the measurement noise and outputs the state and error covariance at each time step. Note that, in the case of computing the relative target position and orientation from V2X communication data, it is necessary to consider the heading and GNSS measurements of the host vehicle as well. A high-level fusion is performed using the estimated quality scores for sensor data, which are based on the error covariance computed through the prediction and update steps of the Kalman filter. Trajectory prediction for the targets detected in the perception stage is performed by employing the constant turn rate and velocity (CTRV) motion model. In risk assessment steps, possible vehicle collisions are detected based on the results from the previous trajectory prediction step. A preliminary assessment that requires significantly less computation load is first carried out to detect possible collisions, and if collisions are expected, a more detailed assessment is performed to estimate precise TTC. Finally, appropriate visual and audible warnings are generated to the driver based on the TTC estimate, where the warning information is provided through the human–machine interface (HMI) in four different threat levels.

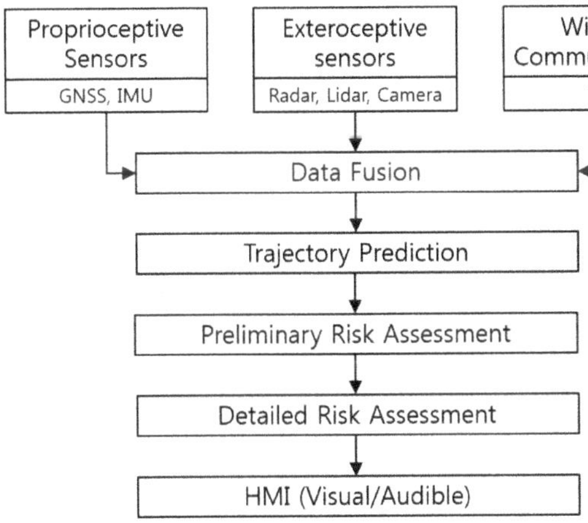

Figure 2. Block diagram summarizing the steps for collision warning generation.

3.2. Automotive Sensors for Remote Sensing

We selected on-board sensors that have already been adopted in production vehicles such that by adding V2X communication devices we can evaluate the benefits of introducing V2X communications to today's vehicles in terms of road safety. The types of sensors installed on vehicles produced in recent years include radar, cameras, and also lidar, which enable ADAS features such as forward collision warning (FCW), automatic emergency braking (AEB), adaptive cruise control (ACC), and lane keeping assist system (LKAS).

Automotive radar, which is an active ranging sensor designed for detecting and tracking remote targets in the surrounding environment, is one of the most used ranging sensors for ADAS functions these days. The most widely found long-range radar sensors on production vehicles include Delphi ESR, Bosch LRR, and Continental ARS series, of which characteristics are shown in Table 1. The specification values are from the respective manufacturer's specification sheet. In this work, the technical data of Delphi ESR were employed to model the radar in the experimental environment.

Table 1. Automotive radar specifications.

Type	Delphi ESR	Bosch LRR3	Continental ARS 30X
Frequency band	76.5 GHz	76–77 GHz	76–77 GHz
Range	174 m	250 m	200 m
Range accuracy	0.5 m	0.1 m	0.25 m
Angular accuracy	0.5 deg	n/a [1]	0.1 deg
Horizontal FOV	20 deg	30 deg	17 deg
Data update	50 ms	80 ms	66 ms

[1] Information not provided in the specification.

Lidar is an active ranging sensor that operates in a similar fashion to radar except that it utilizes light rather than radio waves. Most automotive lidars currently use near-infrared light with a wavelength of 905 nm. Lidar became a popular choice for automated driving technology research since it was used by a large number of teams who participated in the DARPA Grand Challenges. Lidar offers more accurate ranging performance compared with radar and cameras, but despite its advantage, most automakers are yet to adopt lidar mainly due to its tremendous cost. However, it appears that automakers will gradually consider using lidar in the near future because low-cost lidar sensors are becoming more available. Audi became the first automaker to adopt lidar in the production vehicle when they recently started shipping their flagship sedan equipped with an on-board lidar sensor [21]. The performance of the Ibeo Scala sensor is summarized in Table 2.

Table 2. Automotive lidar specifications.

Type	Ibeo Scala B3.0
Laser wavelength	905 nm
Range	80 m
Range accuracy	0.1 m
Horizontal resolution	0.25 deg
Horizontal FOV	145 deg
Data update	40 ms

Contrary to other ranging sensors, vision sensors do not directly provide range information. Instead, range information is often estimated using the road geometry and the point of contact of the vehicle and the road [22], optical flow velocity vectors [23], bird's-eye view [24], and object knowledge [24]. Considering that the detection and tracking performance of a vision-based system may largely vary depending on the algorithm used, the technical data of the Mobileye vehicle detection system, as reported in [22], were employed to model the vision sensor. Table 3 shows the performance characteristics of the Mobileye system.

Table 3. Automotive vision sensor specifications.

Type	Mobileye Camera
Frame size	640 × 480 pixels
Range	70 m (detection) 100 m (tracking)
Accuracy	5% error at 45 m 10% error at 90 m
Horizontal FOV	47 deg

3.3. V2X Communications

The IEEE 802.11p and the IEEE 1609 family of standards are collectively called wireless access in vehicular environments (WAVE) standards. The IEEE has developed the IEEE 802.11p as an amendment to the IEEE 802.11 to include vehicular environments [25]. This amendment was required to support wireless communications among vehicles and infrastructure. The IEEE 1609 protocol suite is a higher-layer standard based on the IEEE 802.11p. In the case of V2V communications, on-board units (OBUs) are installed in vehicles to enable wireless communication. These devices operate independently and exchange data using the 5.9 GHz DSRC frequency band, which is divided into seven 10-MHz channels. One of them is the control channel (CCH), which is used for safety and control messages, while other six are the service channels (SSHs), which are used for data transfer [26]. The characteristics of the WAVE standards are summarized in Table 4.

Table 4. Vehicular wireless communications characteristics.

Type	WAVE Standards
Frequency	5.850–5.925 GHz
Channel	1 CCH, 6 SCH
Bandwidth	10 MHz
Data rate	3–27 Mbps
Maximum range	1000 m

For the purpose of V2X communications, the host vehicle in this work is equipped with a DSRC antenna in addition to the sensors described in the previous section. This makes it possible for the host vehicle to gather information on the remote vehicles in the surrounding area (up to a distance of 1000 m) by exchanging BSMs, which are sent over the CCH channel with a period of 100 ms. The BSM, which is defined in the SAE J2735 message set dictionary [27], contains safety data regarding the vehicle state such as the GNSS position, speed, heading, and yaw rate of the vehicle, as well as the vehicle size. A BSM consists of two parts: Part I and Part II. The BSM Part I contains the core data that must be included in every BSM, whereas the BSM Part II content is optional. Table 5 describes the data contained in a BSM.

Table 5. Basic safety message (BSM) format.

Message	Content
BSM Part I	Message count Temporary ID Time Position (latitude, longitude, elevation) Position accuracy Transmission state Speed Heading Steering wheel angle Acceleration Yaw rate Brake system status Vehicle size (width, length)
BSM Part II	Event flags Path history Path prediction RTCM package

Similar to the BSM, the personal safety message (PSM) contains important kinematic state information on VRUs, such as pedestrians, bicyclists, and road workers. It is possible to detect VRUs located within the DSRC coverage area by collecting the PSMs transmitted from the VRU communication devices. The PSM, which is also defined in the SAE J2735 message set dictionary [27], is currently under development, but the core data elements that must be included in a PSM are specified in advance, as shown in Table 6.

Table 6. Personal safety message (PSM) format.

Message	Content
PSM	Personal device user type Time Message count Temporary ID Position (latitude, longitude, elevation) Position accuracy Speed Heading

The accuracy of the BSM and the PSM information we assumed in the implementation of the proposed vehicle collision system is presented in Table 7. For the BSM, typical measurement noise characteristics of a relatively simple differential GPS (DGPS) receiver, as well as those of a wheel speed sensor and a yaw rate sensor are considered. It is important that the position data included in the BSM meet a lane-level accuracy, which is described in the United States Department of Transportation

(USDOT) report on vehicular safety communications [28] as a minimum relative positioning requirement for collision warning applications. With regard to the PSM, the parameter settings for the VRU safety as reported in the SAE J2945/9 VRU safety message performance requirements [29] are employed in this work for V2P communications.

Table 7. Information accuracy for the BSM and the PSM.

Message	Type	Accuracy
BSM	Position	0.5 m
	Heading	0.3 deg
	Speed	0.3 m/s
	Yaw rate	0.5 deg/s
PSM	Position	1.5 m
	Heading	5 deg
	Speed	0.56 m/s

4. Implementation

4.1. Kalman Filtering

A Kalman filter-based approach was employed in this work for high-level fusion of V2X communications and on-board automotive sensors for remote sensing. Kalman filtering [30–32] is a recursive algorithm that keeps track of the state estimate as well as the uncertainty of the estimate, given the prior knowledge of the state and the measurements collected at the present time. Kalman filtering enables to reduce the measurement noise and obtain the errors associated with each estimated state element. In order to detect the current locations of the remote targets and predict their future trajectories, we utilized position, speed, heading, yaw rate, and size information from V2X communications; range and azimuth information from both radar and lidar; and relative longitudinal and lateral distance information from the camera. In addition, the position and heading measurements from the host vehicle were used to compute the relative position and heading to the target with respect to the host vehicle.

The motion equations of remote targets are typically presented in Cartesian coordinates. However, automotive ranging sensors such as radar and lidar provide measurements in polar coordinates, so transformation to Cartesian coordinates is necessary. Polar-to-Cartesian transformation is a nonlinear process, for which an extended Kalman filter (EKF) is often used. EKF is obtained via a linear approximation of a nonlinear system, and this is consistent only for small errors [33]. A converted measurement Kalman filter performs the coordinate transformation without bias and computes the correct covariance for the converted measurements. This filter is nearly optimal and achieves higher accuracy compared with EKF [34]. The unbiased converted measurement Kalman filter algorithm as presented in [31,35] was employed in this work.

The state vector at time step k is defined by

$$x_k = [X_k \ Y_k \ v_{x,k} \ v_{y,k}]^T \tag{1}$$

where X_k and Y_k describe the position of the target, and $v_{x,k}$ and $v_{y,k}$ describe the target relative velocity in longitudinal and lateral directions, respectively. The measured range and azimuth are

$$r_m = r + \omega_r \tag{2}$$

$$\theta_m = \theta + \omega_\theta \tag{3}$$

where r and θ are the true range and azimuth values. The range and azimuth measurement noises are denoted by ω_r and ω_θ, respectively, of which error standard deviations are σ_r and σ_θ. The unbiased converted measurements are

$$x_m = b_1^{-1} r_m \cos\theta_m \tag{4}$$

$$y_m = b_1^{-1} r_m \sin\theta_m \tag{5}$$

where $b_1 = E[\cos\omega_\theta] = e^{-\sigma_\theta^2/2}$. The unbiased converted measurement vector z_k is

$$z_k = [x_m \;\; y_m]^T \tag{6}$$

and the state $\hat{x}_{k|k-1}$ and error covariance $\hat{P}_{k|k-1}$ are predicted from time step $k-1$ to time step k by

$$\hat{x}_{k|k-1} = A\hat{x}_{k-1|k-1} \tag{7}$$

$$\hat{P}_{k|k-1} = A\hat{P}_{k-1|k-1}A^T \tag{8}$$

where the state transition matrix A is defined as

$$A = \begin{bmatrix} 1 & 0 & \Delta t & 0 \\ 0 & 1 & 0 & \Delta t \\ 0 & 0 & 1 & 0 \\ 0 & 0 & 0 & 1 \end{bmatrix}. \tag{9}$$

The elements of the measurement error covariance R_k obtained from the unbiased conversion are given by

$$R_{11,\,k} = \mathrm{var}(x_m) = \left(b_1^{-2} - 2\right) r_m^2 \cos^2\theta_m + \frac{1}{2}\left(r_m^2 + \sigma_r^2\right)(1 + b_2\cos 2\theta_m) \tag{10}$$

$$R_{22,\,k} = \mathrm{var}(y_m) = \left(b_1^{-2} - 2\right) r_m^2 \sin^2\theta_m + \frac{1}{2}\left(r_m^2 + \sigma_r^2\right)(1 - b_2\cos 2\theta_m) \tag{11}$$

$$R_{12,\,k} = \mathrm{cov}(x_m, y_m) = \left(\frac{1}{2}b_1^{-2}r_m^2 + \frac{1}{2}\left(r_m^2 + \sigma_r^2\right)b_2 - r_m^2\right)\sin 2\theta_m \tag{12}$$

where $b_2 = E[\cos 2\omega_\theta] = e^{-2\sigma_\theta^2}$. Prior to updating the state and the error covariance, the Kalman gain K_k is computed by

$$K_k = \hat{P}_{k|k-1}H^T\left(H\hat{P}_{k|k-1}H^T + R_k\right) \tag{13}$$

where the measurement function matrix H is defined as

$$H = \begin{bmatrix} 1 & 0 & 0 & 0 \\ 0 & 1 & 0 & 0 \end{bmatrix}. \tag{14}$$

Then the state $\hat{x}_{k|k}$ and the error covariance $\hat{P}_{k|k}$ are updated as

$$\hat{x}_{k|k} = \hat{x}_{k|k-1} + K_k\left(z_k - H\hat{x}_{k|k-1}\right) \tag{15}$$

$$\hat{P}_{k|k} = (I - K_k H)\hat{P}_{k|k-1}. \tag{16}$$

For filtering data from the vision sensor and V2X communications, we utilized a linear Kalman filter [30–32] because a polar-to-Cartesian conversion was not necessary for the data we obtained from the two sources. A linear Kalman filter is similar to the filtering process described above, but without the steps for the unbiased conversion. For the purpose of combining the filtered information from multiple sources, we estimated their quality scores based on the error covariance matrices. The quality score matrix $W_{j,\,k|k}$ at time step k for the state obtained from the jth source is given by

$$W_{j,k|k} = \left[\sum_{i=1}^{n} \hat{P}_{i,k|k}{}^{-1} \right]^{-1} \hat{P}_{j,k|k}{}^{-1} \tag{17}$$

$$\sum_{j=1}^{n} W_{j,k|k} = I \tag{18}$$

where $\hat{P}_{j,k|k}$ is the updated error covariance for the jth sensor and I is an identity matrix. Finally, the weight average state $\overline{x}_{k|k}$ for time step k is

$$\overline{x}_{k|k} = \sum_{j=1}^{n} W_{j,k|k}\, \hat{x}_{j,k|k} \tag{19}$$

where $\hat{x}_{j,k|k}$ is the updated state based on the information collected from the jth source.

4.2. Trajectory Prediction and Risk Assessment

Trajectory prediction for each detected remote target is performed by employing a CTRV model. The CTRV state space is constructed with the fused target state estimate as well as the heading and yaw rate information, which was obtained with V2X communications and then filtered with a Kalman filter. Note that the yaw rate of the target was set to zero if the safety message was transmitted from a VRU, considering that yaw rate is not included in the PSM core data. The CTRV state space at time step k is defined as

$$x_k = [X_k \ Y_k \ v_k \ \vartheta_k \ \omega_k]^T \tag{20}$$

where X_k and Y_k describe the relative distance to the target in longitudinal and lateral directions, respectively; v_k is the target velocity; ϑ_k is the relative heading of the target; and ω_k is the target yaw rate. The state transition equation for calculating the state at time step $k+1$ can be written as

$$x_{k+1} = x_k + \begin{bmatrix} \frac{v_k}{\omega_k}(\sin(\vartheta_k + \omega_k \Delta t) - \sin(\vartheta_k)) \\ \frac{v_k}{\omega_k}(-\cos(\vartheta_k + \omega_k \Delta t) + \cos(\vartheta_k)) \\ 0 \\ \omega_k \Delta t \\ 0 \end{bmatrix}. \tag{21}$$

The estimated trajectory of each remote target is then compared with the estimated trajectory of the host vehicle in order to determine whether or not the host vehicle will collide with the remote target. The possibility of a collision is determined by applying a circle model as shown in Figure 3, which illustrates an example for a vehicle–vehicle collision.

The radius of the host vehicle R_{HV} and the radius of the remote vehicle R_{RV} are defined as

$$R_{HV} = \frac{\sqrt{W_{HV}{}^2 + L_{HV}{}^2}}{2} \tag{22}$$

$$R_{RV} = \frac{\sqrt{W_{RV}{}^2 + L_{RV}{}^2}}{2} \tag{23}$$

where W_{HV} and L_{HV} are the width and the length of the host vehicle; and W_{RV} and L_{RV} are the width and length of the remote vehicle. A possible collision is detected if the inequality

$$\sqrt{(X_{HV} - X_{RV})^2} + \sqrt{(Y_{HV} - Y_{RV})^2} \leq R_{HV} + R_{RV} \tag{24}$$

is true. In the case of finding a vehicle–pedestrian collision, the size of the bounding box of a VRU was set according to the dimensions stated in the European New Car Assessment Programme (Euro NCAP) test protocol for AEB VRU systems [36], which are 0.5 m and 0.6 m for an adult pedestrian and 0.5 m and 0.711 m for a child pedestrian.

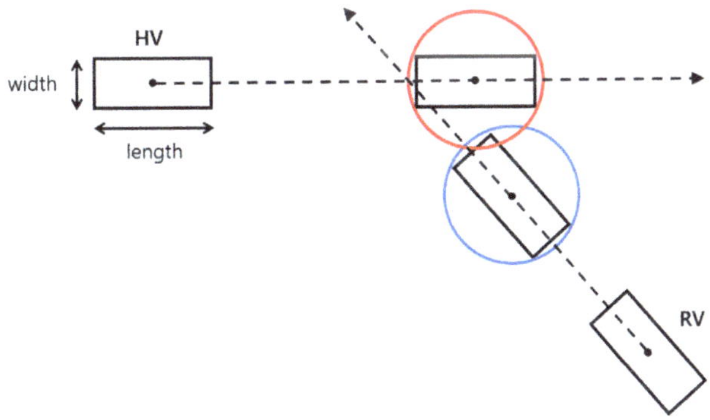

Figure 3. Illustration for finding a possible collision event using the predicted trajectories of the host vehicle and the remote vehicle.

The risk assessment process consists of two stages: the preliminary assessment and the detailed assessment. Figure 4 presents an example of collision event detection and TTC estimation through the two assessment stages. In the preliminary assessment stage, the future positions of the vehicles are computed using a coarse time step, which is computed as

$$\Delta t_{coarse} = \frac{\sqrt{(R_{HV}+R_{RV})^2+(R_{HV}+R_{RV})^2}}{\max(v_{HV},v_{RV})} = \frac{\sqrt{2}(R_{HV}+R_{RV})}{\max(v_{HV},v_{RV})} \tag{25}$$

where v_{HV} and v_{RV} are the speed of the host vehicle and the remote vehicle, respectively. This can be considered as a maximum time step for the preliminary risk assessment, for the collision detection algorithm can fail if longer time steps are used. When the target speed is similar to or lower than the host vehicle speed, longer Δt_{coarse} is used for running the risk assessment for a possible collision with a large remote target such as a bus, while shorter Δt_{coarse} is used for running the assessment for a possible collision with a small target such as a pedestrian. If a collision is detected in the preliminary stage, the future positions of the vehicles are computed using a fine time step in the detailed assessment stage, so that the TTC output is at a resolution of 0.01 s, which corresponds to a distance of a few tens of centimeters in the case of driving on a highway (about 33 cm for a vehicle traveling at 120 km/h).

If a collision is detected in the risk assessment process, an appropriate collision warning is generated to the host vehicle through the HMI according to the estimated TTC. In the case of the detection of multiple collision events, collision warning is generated for the collision associated with the shortest TTC estimation. Table 8 describes the warning generation conditions used in this work, which are similar to those of Daimler PRE-SAFE [12] and Mobileye FCW [37]. Following the suggestions made in the USDOT report on vehicular safety communications [28], we consider four collision warning stages, which include "no threat" in gray, "threat detected" in green, "inform driver" in yellow, and "warn driver" in red. In addition to visual warning, audible warning is generated for the yellow and the red warning level.

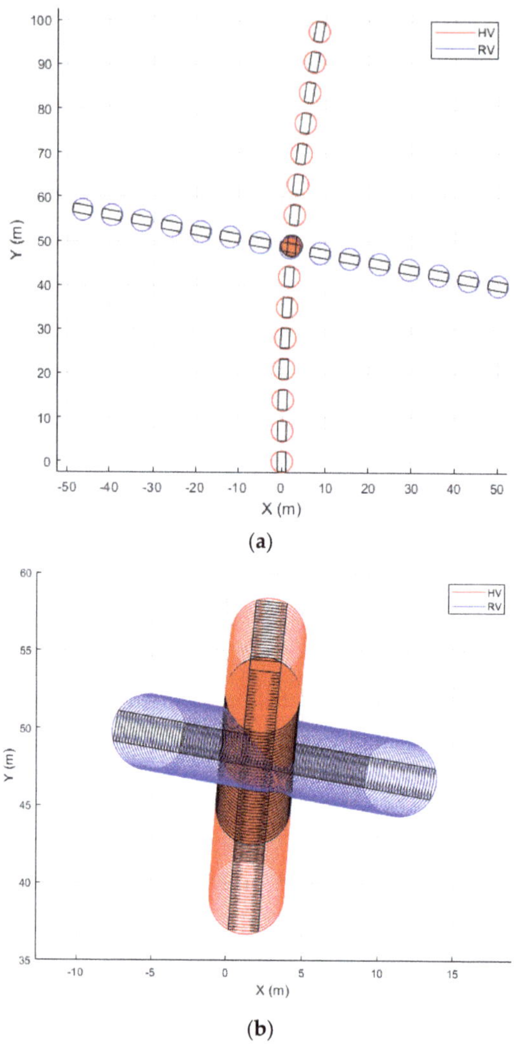

Figure 4. Collision event detection (circles filled in red) and time-to-collision (TTC) estimation using the predicted future trajectories of the host vehicle (HV) and the remote vehicle (RV). (**a**) Preliminary risk assessment step for collision detection; (**b**) detailed risk assessment step for TTC estimation.

Table 8. Conditions for the vehicle collision warning stages.

Condition	Stage	Warning Type	Color
No collision detected	No threat (Level 0)	Visual	Gray
TTC > 2.6	Threat detected (Level 1)	Visual	Green
1.6 < TTC ≤ 2.6	Inform driver (Level 2)	Visual and audible	Yellow
TTC ≤ 1.6	Warn driver (Level 3)	Visual and audible	Red

5. Experiments

The performance of the proposed collision warning system was evaluated experimentally in a simulation environment. Performing tests on vehicular safety systems using a driving simulator is a safer, faster, and cheaper way for system performance evaluation and validation compared with conducting driving tests with real vehicles. In this work, we utilized MATLAB/Simulink and PreScan for designing and evaluating our vehicle collision warning system in virtual driving environments. The simulation was performed in two different types of vehicle collision scenarios: a vehicle–vehicle collision scenario and a vehicle–pedestrian collision scenario.

5.1. Experimental Environment

5.1.1. Vehicle Configuration

According to the specifications described in Section 3, we equipped the host vehicle with remote sensing sensors including a radar, a lidar, and a camera, as well as a DSRC transceiver for V2X communications in the simulation environment. Figure 5 shows the sensor installation illustrations and a bird's-eye view of the vehicle setup within the PreScan model. One long-range radar and one scanning lidar were mounted on the front bumper of the vehicle, and one Mobileye camera was installed on the front windshield. For the sake of simplicity, a GNSS antenna was installed on the center of the bounding box of both the vehicles and the VRUs in our experiments such that the GNSS measurements, obtained from the host vehicle as well as from the remote targets via V2X communications, represent the center position of the two-dimensional bounding box. Some notable dimensions of the vehicle (used for both the host vehicle and the remote vehicle) shown in Figure 5a–d are as follows: length = 5.208 m; width = 2.029 m; and height = 1.447 m. The range and the FOV of each type of sensor installed on the host vehicle are presented in different colors in Figure 5e.

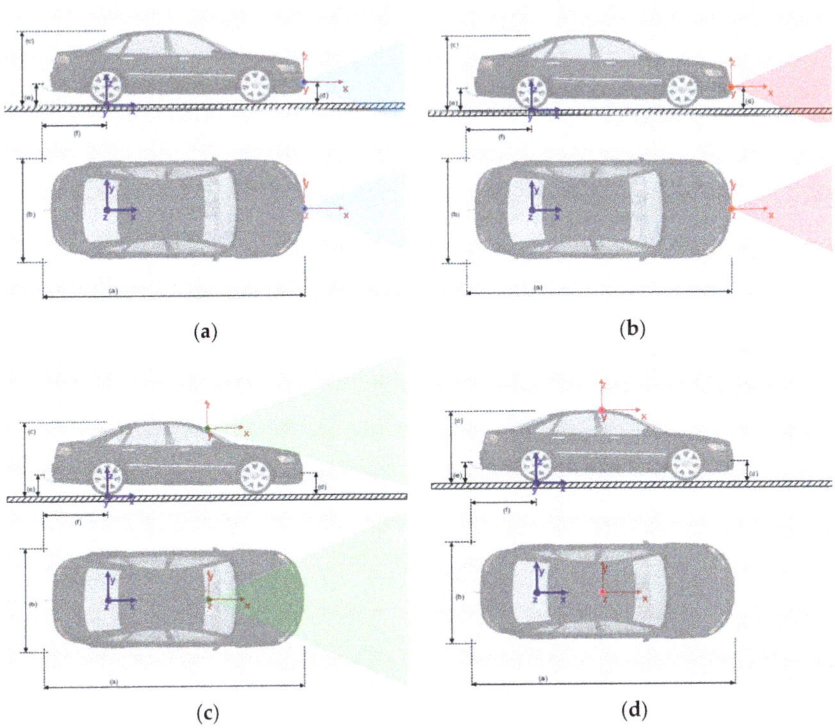

Figure 5. *Cont.*

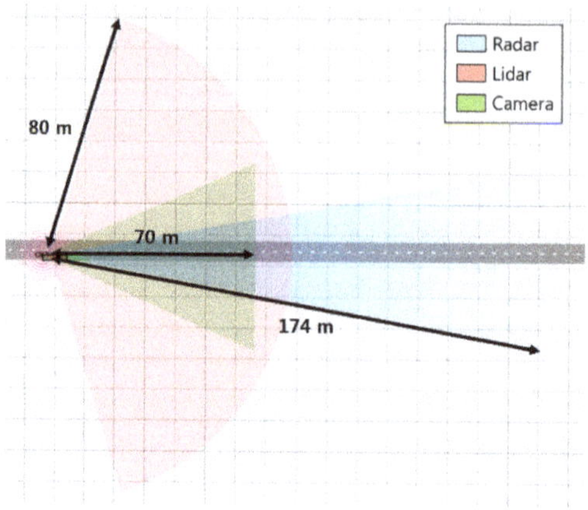

(e)

Figure 5. Locations of the sensors installed on the host vehicle and the sensor coverage. (**a**) Radar; (**b**) lidar; (**c**) camera; (**d**) GNSS antenna; (**e**) sensor range and FOV.

Figure 6 shows the Simulink blocks and subsystems constructed for the proposed vehicle collision warning system. At each time step, measurements from radar, lidar, and camera, as well as safety messages generated from remote targets were collected from the PreScan simulation environment and processed as explained in the previous section in order to estimate the target trajectory and provide the driver with an appropriate warning when a potential collision is detected.

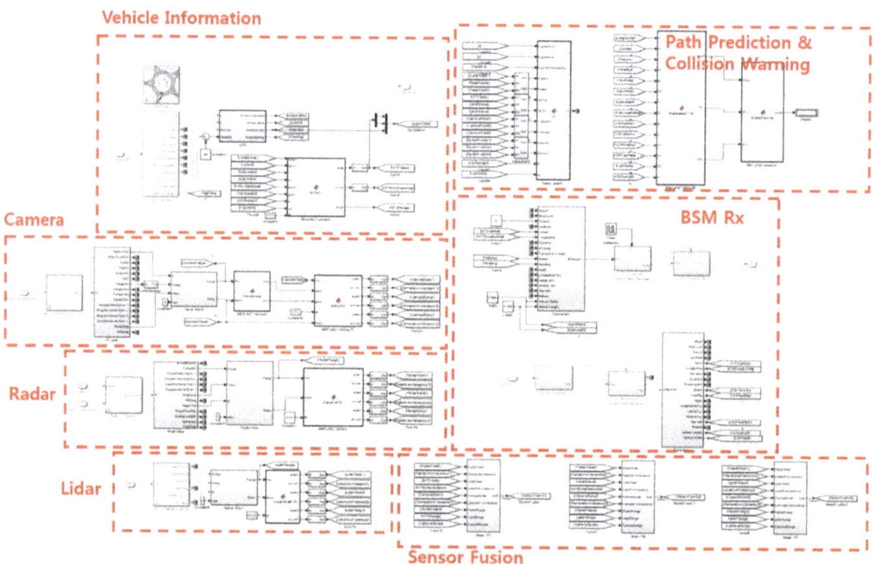

Figure 6. Simulink blocks and subsystems designed for the proposed vehicle collision warning system.

5.1.2. Vehicle–Vehicle Collision Scenario

The vehicle–vehicle collision simulation environment considered in this work is a straight-crossing-paths (SCP) scenario. The SCP scenario at non-signalized junctions ranked the highest among all crashes involving two vehicles in terms of functional years lost [38]. Furthermore, compared with other crossing path collision scenarios at intersections, the SCP scenario is the most frequent collision type when combining the number of crashes at intersections controlled with traffic light signals and stop signs as well as the intersections with no control [39].

A simulation environment for the SCP scenario including two vehicles—a host vehicle and a remote vehicle—was built using PreScan, as shown in Figure 7, to evaluate the performance of the proposed vehicle collision warning system in urban environments. In order to test the proposed system in a challenging yet frequently-occurring scenario, the traveling speed for both vehicles was set to 60 km/h, which corresponds to an upper boundary of average vehicle speed on urban roads with low junction density [40]. The host vehicle traveled from west to east, whereas the remote vehicle traveled from south to north. The two vehicles collided at the end of the simulation where $t = 3.9$ s. An office building was placed in the southwest corner of the intersection to simulate perception in urban driving environments. The width of the sidewalk was set to 1.5 m, and the building was placed 3 m away from the road.

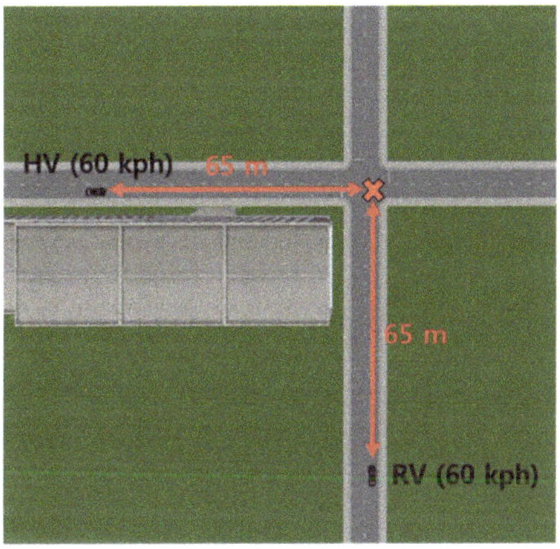

(**a**)

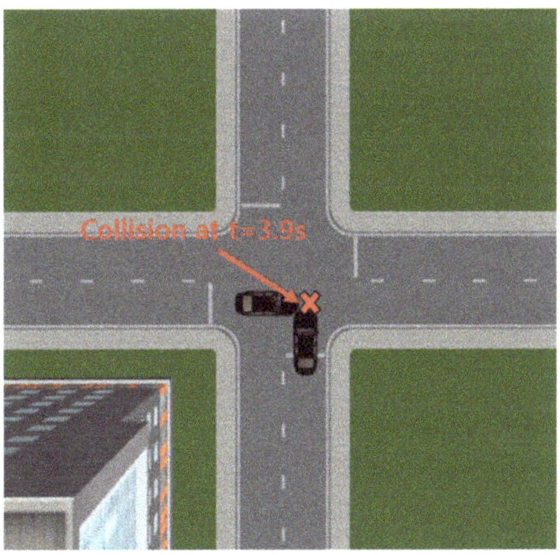

(**b**)

Figure 7. The simulation environment for the vehicle–vehicle collision scenario. (**a**) Experiment setup at the start of the simulation; (**b**) collision between the host and the remote vehicle at the end of the simulation.

5.1.3. Vehicle–Pedestrian Collision Scenario

The vehicle–pedestrian collision simulation environment considered in this work is a scenario where a pedestrian is crossing the road while a vehicle is approaching. According to the USDOT report on vehicle–pedestrian crashes [41], the top four vehicle–pedestrian pre-crash scenarios ranked based on the functional years lost are the following:

1. Pedestrian crossing the road while vehicle going straight.
2. Pedestrian crossing the road while vehicle turning right.
3. Pedestrian crossing the road while vehicle turning left.
4. Pedestrian traveling along/against traffic while vehicle going straight.

Among these four, the first scenario, which is considered for the vehicle–pedestrian collision simulation in this paper, is the most frequent vehicle–pedestrian collision type and accounts for 85 percent of functional years lost for all vehicle–pedestrian pre-crash scenarios.

A simulation environment for this vehicle–pedestrian collision scenario was designed with PreScan in conformity with the Car-to-Pedestrian Nearside Child (CPNC-50) scenario as defined in the Euro NCAP test protocol for AEB VRU systems [36]. As illustrated in Figure 8, the CPNC-50 is a collision where the center of the front side of a vehicle (i.e., 50 percent of the vehicle width) traveling straight strikes a child pedestrian who appears from the nearside, behind obstruction vehicles, and crosses the road. The test protocol also specifies that the vehicle speed should be 20–60 km/h and the pedestrian speed should be 5 km/h. In order to test the performance of the proposed system in the most challenging case, the traveling speeds for the host vehicle and the pedestrian were set to 60 km/h and 5 km/h, respectively. The host vehicle traveled from west to east, while the pedestrian traveled from south to north. At the end of the simulation, the host vehicle and the pedestrian collided at $t = 2.9$ s. The two cars parked roadside were separated by 1 m, and their left side was positioned 1 m away along the lateral direction from the right side of the host vehicle.

5.2. *Performance Evaluation and Analysis*

5.2.1. Vehicle–Vehicle Collision Scenario

The simulation results from the vehicle–vehicle collision scenario along with snapshots of the experimental environment at four different time instances are presented in Figure 9. A set of images shown for each simulation time point includes the forward-looking view from the perspective of the host vehicle, the top-view of the road scene, the sensor fusion result along with filtered measurements from different sources, and finally the collision detection result from the trajectory prediction and preliminary risk assessment algorithms. In the center of the forward-looking view images, an appropriate visual collision warning to the host vehicle is shown as a result of potential collision detection. The color of the visual warning represents the corresponding warning level as explained in Table 8. Throughout the simulation time, the proposed system performed well in providing proper collision warning to the host vehicle. Figure 9a,b correspond to the results for $t = 1$ s and $t = 2$ s, respectively, where, despite the lack of on-board sensor measurements, the results demonstrate successful collision warning based on the BSM data obtained through V2V communications. After $t = 3$ s, the line of sight to the remote vehicle was no longer blocked by the building near the intersection and thus collision detection was carried out with measurements from the lidar in addition to the BSM, as shown in Figure 9c,d.

Figure 10 illustrates the level of the collision warning generated throughout the simulation period from one sequence of the vehicle–vehicle collision simulation. In order to investigate the effectiveness of the implementation of vehicular communications in the SCP collision scenario considered in this paper, the collision warning results provided by the proposed system and those by the identical system with vehicular communications turned off were compared. The proposed system successfully detected a potential collision at the start of the simulation and generated a level-1 warning at $t = 0.1$ s. A level-2 warning and a level-3 warning were subsequently provided to the host vehicle at $t = 1.3$ s and $t = 2.3$ s, respectively, which would give the driver sufficient time to react and slow down the vehicle speed. On the other hand, without vehicular communications the collision warning system failed to provide any warning until only 0.9 s before the collision, which is insufficient for a driver to avoid or mitigate the collision, considering the typical human reaction time of 1.5 s to apply brakes upon the occurrence of unexpected events [42].

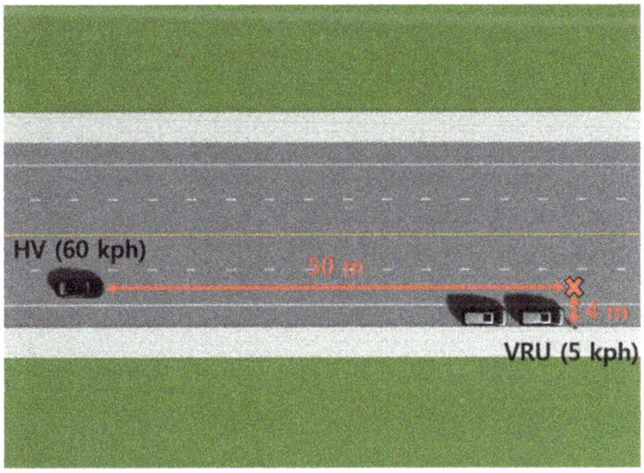

(**a**)

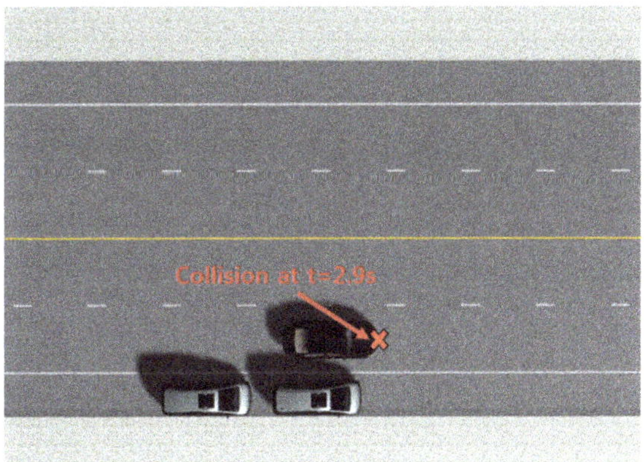

(**b**)

Figure 8. Simulation environment for the vehicle–pedestrian collision scenario: (**a**) Experiment setup at the start of the simulation; (**b**) collision between the host vehicle and the pedestrian at the end of the simulation.

In order to analyze the simulation result in a quantitative manner, we collected the TTC estimates from 10 separate experiments of the vehicle–vehicle collision scenario and grouped them into 1-s bins as presented in Table 9. The mean and the standard deviation of the error in the TTC estimates were computed for each bin. In this analysis, we observe that the accuracy of the TTC estimates becomes significantly better as the actual TTC becomes smaller. The average error and the standard deviation in the TTC estimates for $TTC_{Actual} \leq 1$ are smaller than those for $3 < TTC_{Actual} \leq 4$ by a factor of 20 and 5, respectively. The results confirm that the proposed system is well capable of providing the driver with accurate warning messages in the vehicle–vehicle collision scenario considered in this work.

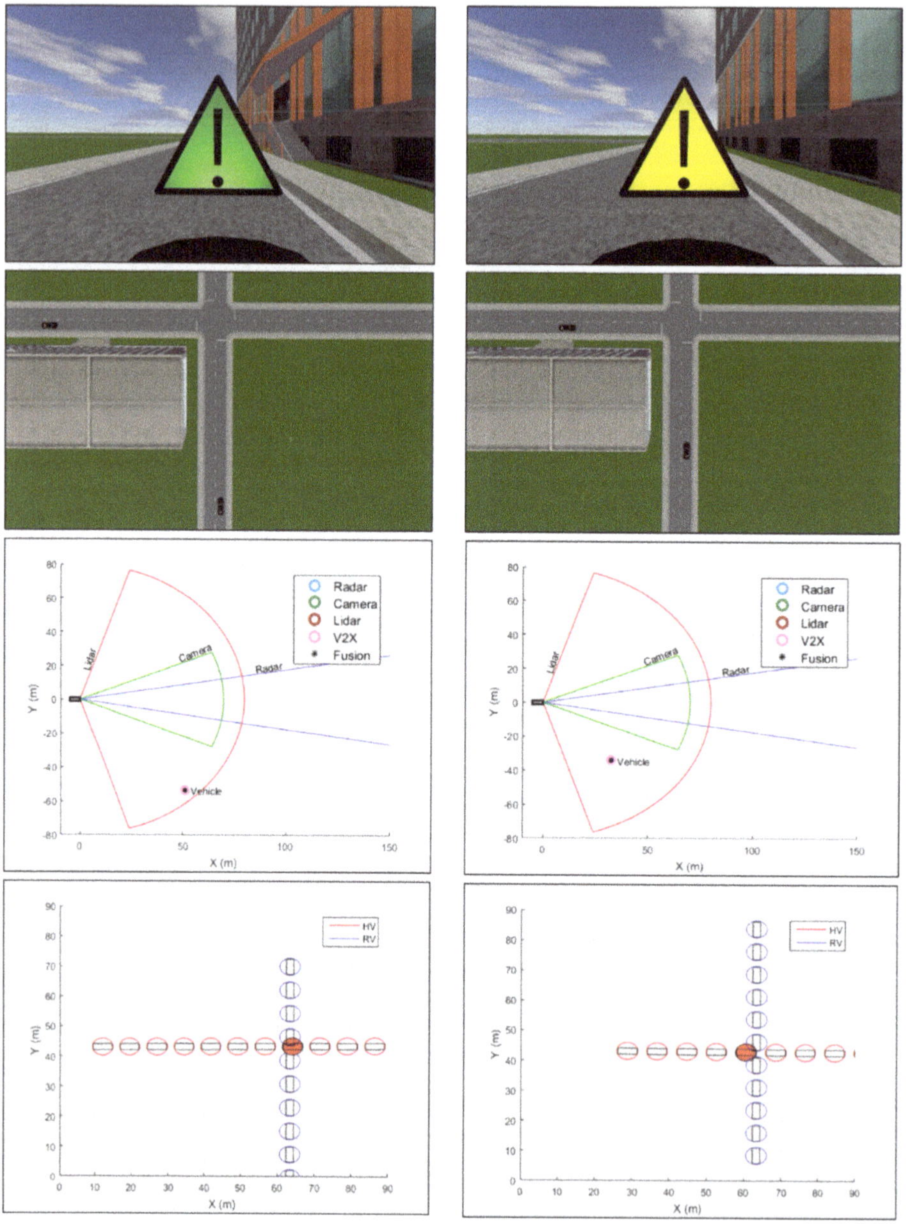

(a) (b)

Figure 9. *Cont.*

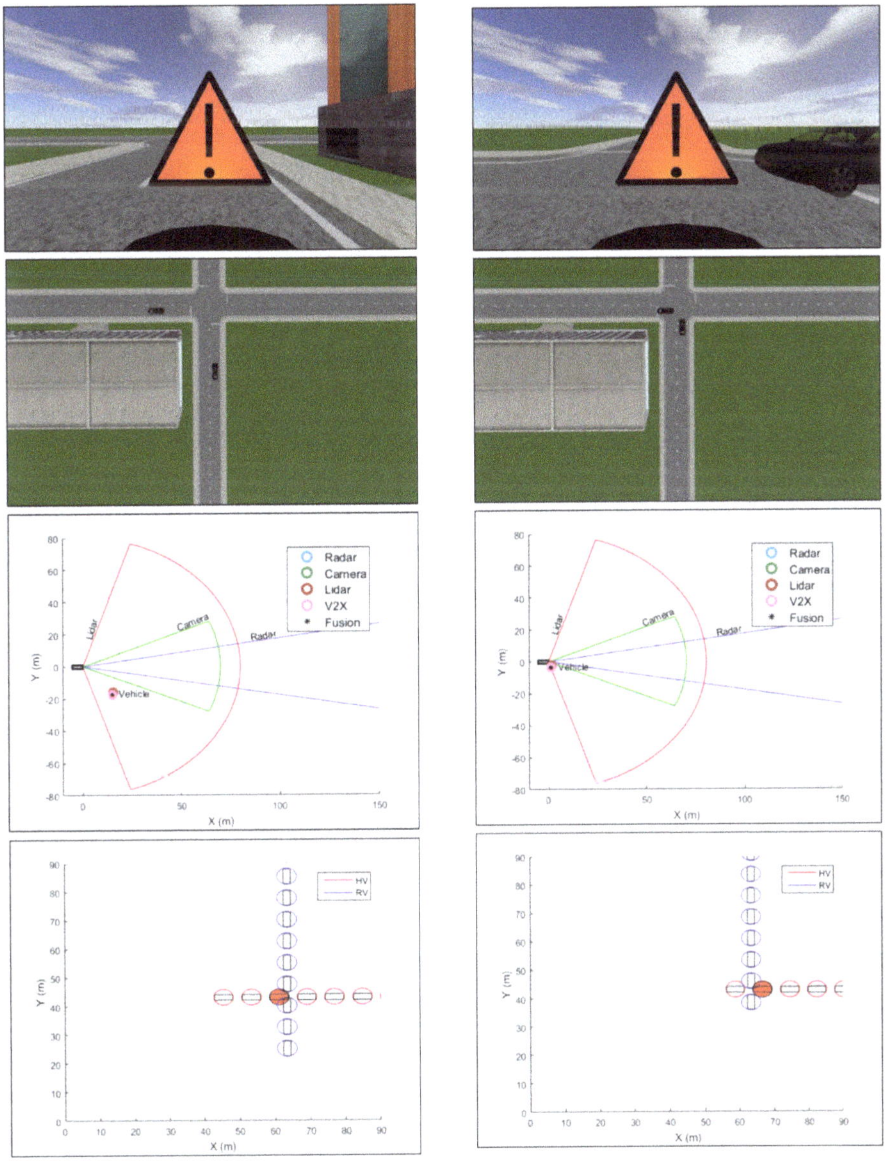

(**c**) (**d**)

Figure 9. Vehicle–vehicle collision simulation results and snapshots of the experimental environment at different time points. Shown in the center of the forward-looking view image is the visual collision warning generated to the host vehicle. The bird's-eye-view image of the road scene shows the locations of the vehicles at the corresponding time instance. The sensor fusion image shows the filtered measurements from various sensors as well as the fusion result. Trajectory prediction and risk assessment enable detection of potential collision location, which is represented by a circle colored in red. (**a**) Results for $t = 1$ s; (**b**) results for $t = 2$ s; (**c**) results for $t = 3$ s; (**d**) results for the time point just before the collision.

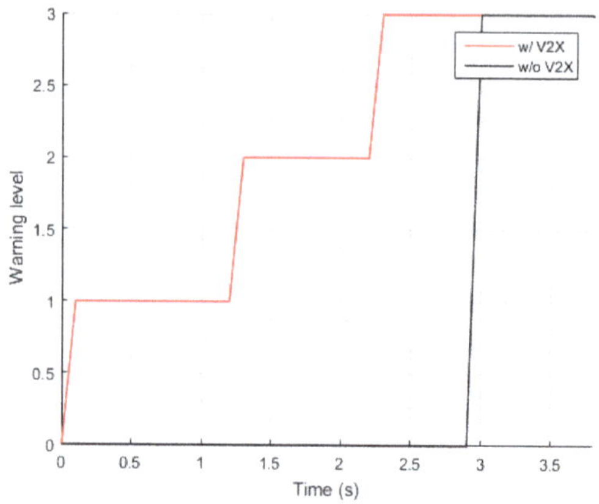

Figure 10. Collision warning generated over time in the vehicle–vehicle collision scenario.

Table 9. Errors in the estimated TTC for the vehicle–vehicle collision scenario.

Data Range	Mean (s)	SD (s)
$3 < TTC_{Actual} \leq 4$	0.08	0.05
$2 < TTC_{Actual} \leq 3$	0.05	0.05
$1 < TTC_{Actual} \leq 2$	0.03	0.02
$TTC_{Actual} \leq 1$	0.004	0.01

5.2.2. Vehicle–Pedestrian Collision Scenario

The simulation results from the vehicle–pedestrian collision scenario along with snapshots of the experimental environment at four different time instances are presented in Figure 11. Four sets of images are presented for four different time instances. For each corresponding time point, the forward-looking view from the perspective of the host vehicle shows the visual collision warning given to the host vehicle, whereas the bird's-eye-view image of the road scene displays where the host vehicle and the pedestrian are located. In the sensor fusion images, we present the positioning results at the corresponding time as well as the results obtained with each sensor. Finally, the collision detection results from the trajectory prediction and preliminary risk assessment algorithms are shown in the images on the bottom. The different colors of the visual warning indicate different warning levels, which are previously defined in Table 8. Throughout the simulation time, we observe that the proposed collision warning system successfully generated appropriate warnings to the host vehicle. Figure 11a corresponds to the results for $t = 0.7$ s, where potential collision with the pedestrian is detected solely based on the PSM data obtained with V2P communications. After the simulation time reached $t = 1.4$ s, the line of sight to the pedestrian was no longer blocked by the two cars parked roadside and thus the collision detection results were based on the measurements collected from the radar, the lidar, the camera, and the PSM collected from the pedestrian, as shown in Figure 11b–d.

The different levels of the collision warning generated from a single sequence of the vehicle–pedestrian collision simulation are shown in Figure 12. The collision warning results provided by the proposed system and those by the identical system with vehicular communications turned off were plotted together to compare the performance of the two systems in the vehicle–pedestrian collision scenario we considered in this work. A potential collision was successfully detected with the proposed system at the start of the simulation and generated a level-1 warning at $t = 0.1$ s. The level of collision warning was soon raised to level 2 at $t = 0.4$ s, which corresponds to 2.5 s before the collision. Although in this particular sequence the level-2 warning was activated 0.1 s later than it was expected, a warning offset of 0.1 s is entirely acceptable in the case of the vehicle–pedestrian scenario we previously defined, considering that the remaining time before the collision is longer than 2 s. A level-3 collision warning was correctly generated to the host vehicle 1.6 s prior to the collision. In the case of the collision warning system without vehicular communications, a warning was not generated until 1.5 s before the collision because the line of sight to the pedestrian had been occluded by the cars parked on the side of the road. When taking into account the typical reaction time of 1.5 s to apply brakes in case of unexpected events [42], this warning may appear to give an attentive driver just enough time to react and slow down; however, it would still be difficult to avoid the collision when considering the vehicle braking distance.

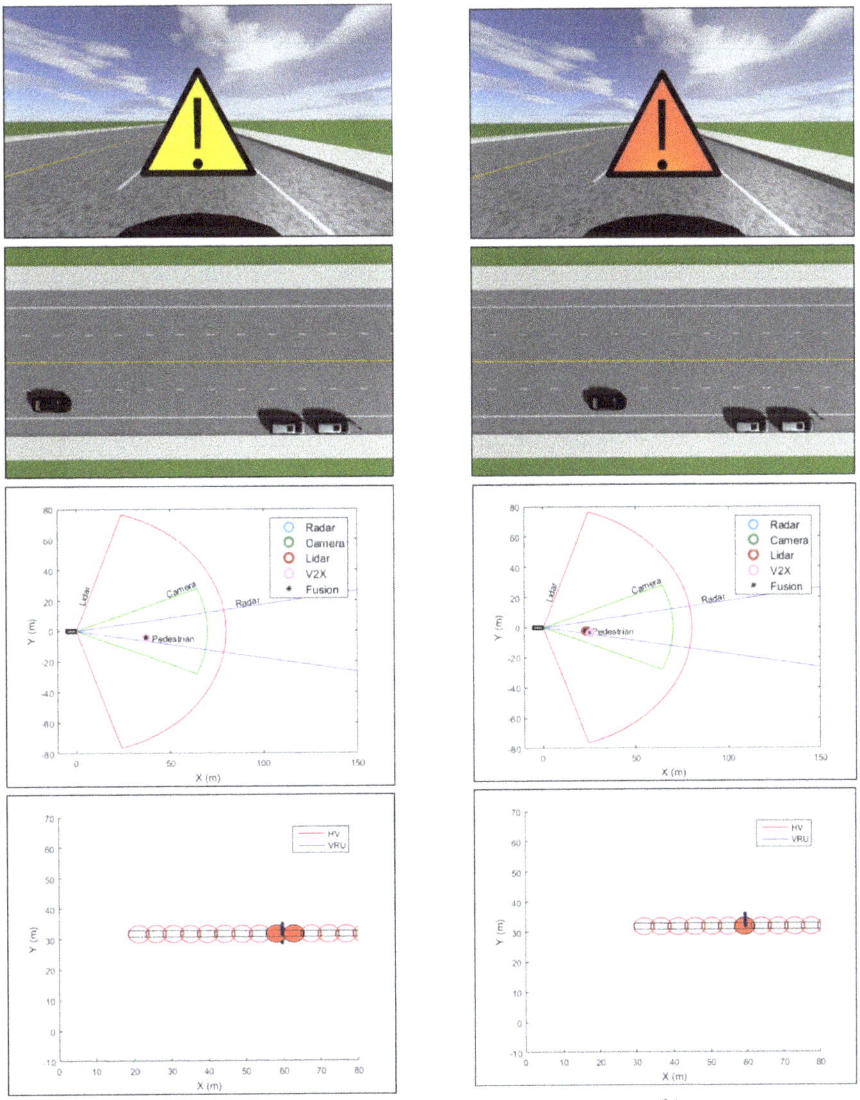

(**a**) (**b**)

Figure 11. *Cont.*

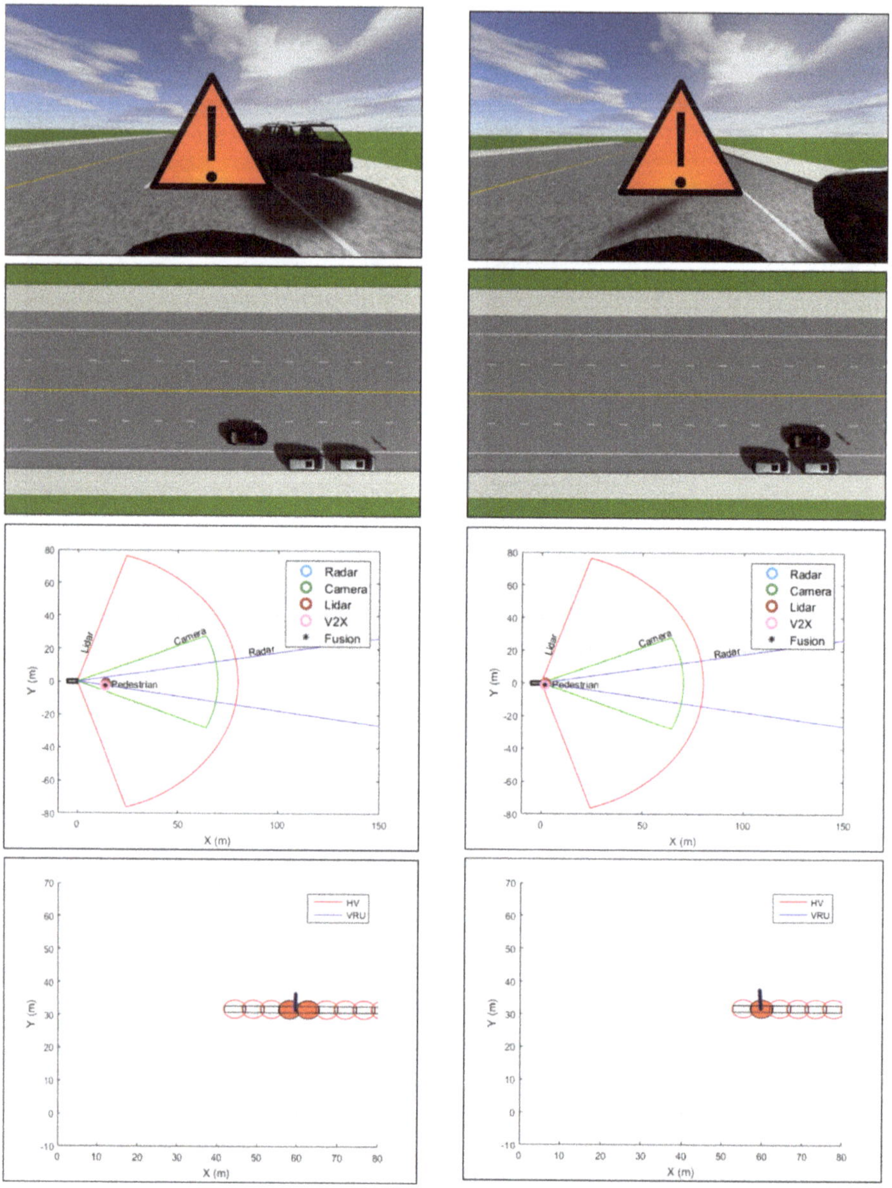

(c) (d)

Figure 11. Vehicle–pedestrian collision simulation results and snapshots of the experimental environment at different time points. Shown in the center of the forward-looking view image is the visual collision warning generated to the host vehicle. The bird's-eye-view image of the road scene shows the locations of the host vehicle and the pedestrian at the corresponding time instance. The sensor fusion image shows the filtered measurements from various sensors as well as the fusion result. Trajectory prediction and risk assessment enable detection of potential collision location, which is represented by a circle colored in red. (**a**) Results for $t = 0.7$ s; (**b**) results for $t = 1.4$ s; (**c**) results for $t = 2.1$ s; (**d**) results for the time point just before the collision.

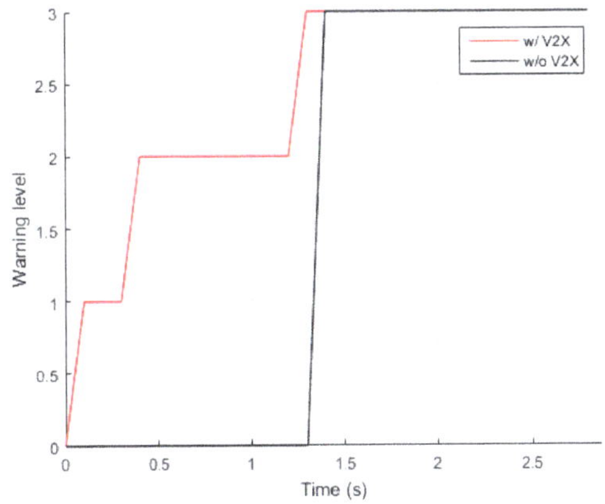

Figure 12. Collision warning generated over time in the vehicle–pedestrian collision scenario.

Table 10 presents the errors in the TTC estimates collected from 10 individual sequences of the vehicle–pedestrian collision simulation. We grouped the TTC estimates into 1-s bins in order to quantitatively investigate how the performance of the proposed system depends on the actual time remaining before the collision. For each 1-s bin, we computed the mean and the standard deviation of the error in the TTC estimates. The results clearly show that the accuracy of the TTC estimates becomes significantly higher as the vehicle nears the collision location. The average error and the standard deviation of the TTC estimates for $TTC_{Actual} \leq 1$ are smaller than those for $2 < TTC_{Actual} \leq 3$ by a factor of 10 and 4, respectively, which shows similar improvement compared to the two sample groups from the results of the vehicle–vehicle collision simulation. The analysis confirms that the proposed system successfully generates timely warnings to the host vehicle in the vehicle–pedestrian collision scenario considered in this paper.

Table 10. Errors in the estimated TTC for the vehicle–pedestrian collision scenario.

Data Range	Mean (s)	SD (s)
$2 < TTC_{Actual} \leq 3$	0.01	0.04
$1 < TTC_{Actual} \leq 2$	0.007	0.03
$TTC_{Actual} \leq 1$	0.001	0.01

6. Conclusions

In this paper, we present the development of a vehicle collision warning system based on multisensors and V2X communications. On-board sensors including radar, lidar, and camera systems that have already been adopted in production vehicles are chosen for this work such that by adding V2X communication devices to the vehicle, we can evaluate the benefits of introducing V2X communications to today's vehicles in terms of road safety. The proposed design employs a Kalman filter-based approach for high-level fusion of V2X communications and on-board automotive sensors for remote sensing. Based on the TTC estimate result from the trajectory prediction and the risk assessment steps, an appropriate visual and audible warning is provided to the driver prior to the collision. The performance of the proposed system is evaluated in virtual driving environments, where two types of vehicle collision scenarios are considered: a vehicle–vehicle collision in an SCP scenario and a vehicle–pedestrian collision in the Euro NCAP test scenario. The results from the proof-of-concept test demonstrate that the proposed system enables higher driver and pedestrian safety through improved perception performance and proper collision warning, even in situations where collision mitigation is difficult with existing safety systems. For future work, we plan to implement the proposed vehicle collision warning method in an in-vehicle prototyping system and evaluate the performance in various driving conditions. In order to ensure the collision warning application reliability, we also aim to investigate the effects of various factors (e.g., distance between vehicles and transmission power) that could adversely affect the reliability of V2X communications.

Author Contributions: Conceptualization, M.B. and S.L.; methodology, M.B.; software, M.B. and D.J.; formal analysis, M.B.; investigation, M.B., D.J. and D.C.; data curation, M.B. and D.J.; writing—original draft preparation, M.B.; writing—review and editing, M.B.; visualization, M.B. and D.J.; supervision, S.L.; project administration, M.B. and D.C.; funding acquisition, S.L. All authors have read and agreed to the published version of the manuscript.

Funding: This research was supported by the Technology Innovation Program (10062375, Development of Core Technologies Based on V2X and In-Vehicle Sensors for Path Prediction of the Surrounding Objects (Vehicle, Pedestrian, Motorcycle)) funded by the Ministry of Trade, Industry and Energy (MOTIE, Korea).

Conflicts of Interest: The authors declare no conflict of interest.

Abbreviations

ADAS	Advanced Driver Assistance Systems
AEB	Automatic Emergency Braking
BSM	Basic Safety Message
CTRV	Constant Turn Rate and Velocity
DSRC	Dedicated Short-Range Communications
FOV	Field of View
FCW	Forward Collision Warning
GNSS	Global Navigation Satellite System
HMI	Human-Machine Interface
NCAP	New Car Assessment Program
NLOS	Non-Line of Sight
OBU	On-Board Unit
PSM	Personal Safety Message
SCP	Straight Crossing Paths
TTC	Time-to-Collision
V2X	Vehicle-to-Everything
V2I	Vehicle-to-Infrastructure
V2P	Vehicle-to-Pedestrian
V2V	Vehicle-to-Vehicle
VRU	Vulnerable Road User
WAVE	Wireless Access in Vehicular Environments

References

1. Dingus, T.A.; Klauer, S.G.; Neale, V.L.; Petersen, A.; Lee, S.E.; Sudweeks, J.; Perez, M.A.; Hankey, J.; Ramsey, D.; Gupta, S.; et al. *The 100-Car Naturalistic Driving Study, Phase II—Results of the 100-Car Field Experiment*; Rep. DOT HS 210 593; National Highway Traffic Safety Administration: Washington, DC, USA, 2006.
2. Urmson, C.; Anhalt, J.; Bagnell, D.; Baker, C.; Bittner, R.; Clark, M.N.; Dolan, J.; Duggins, D.; Galatali, T.; Geyer, C.; et al. Autonomous driving in urban environments: Boss and the Urban Challenge. In *The DARPA Urban Challenge—Autonomous Vehicles in City Traffic*; Buehler, M., Iagnemma, K., Singh, S., Eds.; Springer: Berlin, Germany, 2009; ISBN 978-3-642-03990-4.
3. Montemerlo, M.; Becker, J.; Bhat, S.; Dahlkamp, H.; Dolgov, D.; Ettinger, S.; Haehnel, D.; Hilden, T.; Hoffmann, G.; Huhnke, B.; et al. Junior: The Stanford entry in the Urban Challenge. In *The DARPA Urban Challenge—Autonomous Vehicles in City Traffic*; Buehler, M., Iagnemma, K., Singh, S., Eds.; Springer: Berlin, Germany, 2009; ISBN 978-3-642-03990-4.
4. Wille, J.M.; Saust, F.; Maurer, M. Stadtpilot: Driving Autonomously on Braunschweig's Inner Ring Road. In Proceedings of the IEEE Intelligent Vehicles Symposium, San Diego, CA, USA, 21–24 June 2010; pp. 506–511.
5. Guizzo, E. How Google's Self-Driving Car Works. Available online: http://spectrum.ieee.org/automaton/robotics/artificial intelligence/how-google-self-driving-car-works (accessed on 27 November 2016).

6. Ziegler, J.; Bender, P.; Schreiber, M.; Lategahn, H.; Strauss, T.; Stiller, C.; Dang, T.; Franke, U.; Appenrodt, N.; Keller, C.G.; et al. Making Bertha drive—An autonomous journey on a historic route. *IEEE Intell. Transp. Syst. Mag.* **2014**, *6*, 8–20. [CrossRef]
7. Broggi, A.; Cerri, P.; Debattisti, S.; Laghi, M.C.; Medici, P.; Molinari, D.; Panciroli, M.; Prioletti, A. PROUD—Public road urban driverless-car test. *IEEE Trans. Intell. Transp. Syst.* **2015**, *16*, 3508–3519. [CrossRef]
8. Najm, W.G.; Koopmann, J.; Smith, J.D.; Brewer, J. *Frequency of Target Crashes for IntelliDrive Safety Systems*; Rep. DOT HS 811 381; National Highway Traffic Safety Administration: Washington, DC, USA, 2010.
9. Schmidt, R.K.; Kloiber, B.; Schüttler, F.; Strang, T. Degradation of Communication Range in VANETs Caused by Interference 2.0—Real-World Experiment. In *Communication Technologies for Vehicles*; Strang, T., Festag, A., Vinel, A., Mehmood, R., Garcia, C.R., Röckl, M., Eds.; Springer: Berlin, Germany, 2011; ISBN 978-3-642-19785-7.
10. Zang, S.; Ding, M.; Smith, D.; Tyler, P.; Rakotoarivelo, T.; Kaafar, M.A. The impact of adverse weather conditions on autonomous vehicles. *IEEE Veh. Technol. Mag.* **2019**, *14*, 103–111. [CrossRef]
11. TASS International. PreScan—Simulation of ADAS and Active Safety. Available online: http://www.tassinternational.com/prescan (accessed on 30 September 2019).
12. Bloecher, H.L.; Dickmann, J.; Andres, M. Automotive Active Safety and Comfort Functions Using Radar. In Proceedings of the IEEE International Conference on Ultra-Wideband, Vancouver, BC, Canada, 9–11 September 2009; pp. 490–494.
13. Dagan, E.; Mano, O.; Stein, G.P.; Shashua, A. Forward Collision Warning with a Single Camera. In Proceedings of the IEEE Intelligent Vehicles Symposium, Parma, Italy, 14–17 June 2004; pp. 37–42.
14. Ammoun, S.; Nashashibi, F. Real Time Trajectory Prediction for Collision Risk Estimation between Vehicles. In Proceedings of the IEEE International Conference on Intelligent Computer Communication and Processing, Cluj-Napoca, Romania, 27–29 August 2009; pp. 417–422.
15. Xiang, X.; Qin, W.; Xiang, B. Research on a DSRC-based rear-end collision warning model. *IEEE Trans. Intell. Transp. Syst.* **2014**, *15*, 1054–1065. [CrossRef]
16. Rauch, A.; Maier, S.; Klanner, F.; Dietmayer, K. Inter-Vehicle Object Association for Cooperative Perception Systems. In Proceedings of the IEEE Conference on Intelligent Transportation Systems, The Hague, The Netherlands, 6–9 October 2013; pp. 893–898.
17. Obst, M.; Hobert, L.; Reisdorf, P. Multi-Sensor Data Fusion for Checking Plausibility of V2V Communications by Vision-Based Multiple-Object Tracking. In Proceedings of the IEEE Vehicular Networking Conference, Paderborn, Germany, 3–5 December 2014; pp. 143–150.
18. De Ponte Müller, F.; Diaz, E.M.; Rashdan, I. Cooperative Positioning and Radar Sensor Fusion for Relative Localization of Vehicles. In Proceedings of the IEEE Intelligent Vehicles Symposium, Gothenburg, Sweden, 19–22 June 2016; pp. 1060–1065.
19. Lin, B.; Huang, C.H. Comparison between ARPA radar and AIS characteristics for vessel traffic services. *J. Mar. Sci. Technol.* **2006**, *14*, 182–189.
20. Harati-Mokhtari, A.; Wall, A.; Brooks, P.; Wang, J. Automatic Identification System (AIS): Data reliability and human error implications. *J. Navig.* **2007**, *60*, 373–389. [CrossRef]
21. Ross, P.E. The Audi A8: The World's First Production Car to Achieve Level 3 Autonomy. Available online: https://spectrum.ieee.org/cars-that-think/transportation/self-driving/the-audi-a8-the-worlds-first-production-car-to-achieve-level-3-autonomy (accessed on 25 October 2019).
22. Stein, G.P.; Mano, O.; Shashua, A. Vision-Based ACC with a Single Camera: Bounds on Range and Range Rate Accuracy. In Proceedings of the IEEE Intelligent Vehicles Symposium, Columbus, OH, USA, 9–11 June 2003; pp. 120–125.
23. Shibata, M.; Makino, T.; Ito, M. Target Distance Measurement Based on Camera Moving Direction Estimated with Optical Flow. In Proceedings of the 10th IEEE International Workshop on Advanced Motion Control, Trento, Italy, 26–28 March 2008; pp. 62–67.
24. Fritsch, J.; Michalke, T.; Gepperth, A.; Bone, S.; Waibel, F.; Kleinehagenbrock, M.; Gayko, J.; Goerick, C. Towards a Human-Like Vision System for Driver Assistance. In Proceedings of the IEEE Intelligent Vehicles Symposium, Eindhoven, The Netherlands, 4–6 June 2008; pp. 275–282.
25. *Part 11: Wireless LAN Medium Access Control (MAC) and Physical Layer (PHY) Specifications—Amendment 6: Wireless Access in Vehicular Environments*; IEEE Std. 802.11p; IEEE: New York, NY, USA, 2010.

26. *IEEE Standard for Wireless Access in Vehicular Environments (WAVE)—Multi-Channel Operation*; IEEE Std. 1609.4; IEEE: New York, NY, USA, 2016.
27. *Dedicated Short Range Communications (DSRC) Message Set Dictionary*; SAE J2735; SAE International: Warrendale, PA, USA, 2016.
28. Ahmed-Zaid, F.; Bai, F.; Bai, S.; Basnayake, C.; Bellur, B.; Brovold, S.; Brown, G.; Caminiti, L.; Cunningham, D.; Elzein, H.; et al. *Vehicle Safety Communications—Applications (VSC-A) Final Report*; Rep. DOT HS 811 492A; National Highway Traffic Safety Administration: Washington, DC, USA, 2011.
29. *Vulnerable Road User Safety Message Minimum Performance Requirements*; SAE J2945/9; SAE International: Warrendale, PA, USA, 2017.
30. Kalman, R.E. A new approach to linear filtering and prediction problems. *Trans. ASME J. Basic Eng.* **1960**, *82*, 35–45. [CrossRef]
31. Bar-Shalom, Y.; Li, X.R.; Kirubarajan, T. *Estimation with Applications to Tracking and Navigation: Theory Algorithms and Software*; John Wiley and Sons: New York, NY, USA, 2001; ISBN 0-471-41655-X.
32. Welch, G.; Bishop, G. An Introduction to the Kalman Filter. In Proceedings of the SIGGRAPH, Los Angeles, CA, USA, 12–17 August 2001. Course 8.
33. Julier, S.J.; Uhlmann, J.K. New Extension of the Kalman Filter to Nonlinear Systems. In *Signal Processing, Sensor Fusion, and Target Recognition VI, Proceedings of AeroSense: The 11th International Symposium on Aerospace/Defense Sensing, Simulation, and Controls, Orlando, FL, United States, 21–25 April 1997*; Kadar, I., Ed.; SPIE: Bellingham, WA, USA, 1997; pp. 182–193.
34. Lerro, D.; Bar-Shalom, Y. Tracking with debiased consistent converted measurements vs. EKF. *IEEE Trans. Aerosp. Electron. Syst.* **1993**, *29*, 1015–1022. [CrossRef]
35. Mo, L.; Song, X.; Zhou, Y.; Sun, Z.; Bar-Shalom, Y. Unbiased converted measurements for tracking. *IEEE Trans. Aerosp. Electron. Syst.* **1998**, *34*, 1023–1027.
36. Euro NCAP. European New Car Assessment Programme (Euro NCAP) Test Protocol—AEB VRU Systems, Version 2.0.2. Available online: http://www.euroncap.com/en/for-engineers/protocols/pedestrian-protection/ (accessed on 30 November 2017).
37. Mobileye. Forward Collision Warning (FCW). Available online: https://www.mobileye.com/au/fleets/technology/forward-collision-warning/ (accessed on 10 November 2019).
38. Najm, W.G.; Smith, J.D.; Yanagisawa, M. *Pre-Crash Scenario Typology for Crash Avoidance Research*; Rep. DOT HS 810 767; National Highway Traffic Safety Administration: Washington, DC, USA, 2007.
39. Najm, W.G.; Smith, J.D.; Smith, D.L. *Analysis of Crossing Path Crashes*; Rep. DOT HS 809 423; National Highway Traffic Safety Administration: Washington, DC, USA, 2001.
40. André, M.; Hammarström, U. Driving speeds in Europe for pollutant emission estimation. *Transp. Res. Part D Transp. Environ.* **2000**, *5*, 321–335. [CrossRef]
41. Yanagisawa, M.; Swanson, E.; Najm, W.G. *Target Crashes and Safety Benefits Estimation Methodology for Pedestrian Crash Avoidance/Mitigation Systems*; Rep. DOT HS 811 998; National Highway Traffic Safety Administration: Washington, DC, USA, 2014.
42. Green, M. "How long does it take to stop?" Methodological analysis of driver perception-brake times. *Transp. Hum. Factors* **2000**, *2*, 195–216. [CrossRef]

sensors

MDPI

Article

A Fail-Operational Control Architecture Approach and Dead-Reckoning Strategy in Case of Positioning Failures

Jose Angel Matute-Peaspan [1,2,*], Joshue Perez [1] and Asier Zubizarreta [2]

1 Tecnalia Research & Innovation, 48160 Derio, Spain; joshue.perez@tecnalia.com
2 Department of Automatic Control and Systems Engineering, University of the Basque Country (UPV/EHU), 48013 Bilbao, Spain; asier.zubizarreta@ehu.eus
* Correspondence: joseangel.matute@tecnalia.com

Received: 8 December 2019; Accepted: 10 January 2020; Published: 13 January 2020

Abstract: Presently, in the event of a failure in Automated Driving Systems, control architectures rely on hardware redundancies over software solutions to assure reliability or wait for human interaction in takeover requests to achieve a minimal risk condition. As user confidence and final acceptance of this novel technology are strongly related to enabling safe states, automated fall-back strategies must be assured as a response to failures while the system is performing a dynamic driving task. In this work, a fail-operational control architecture approach and dead-reckoning strategy in case of positioning failures are developed and presented. A fail-operational system is capable of detecting failures in the last available positioning source, warning the decision stage to set up a fall-back strategy and planning a new trajectory in real time. The surrounding objects and road borders are considered during the vehicle motion control after failure, to avoid collisions and lane-keeping purposes. A case study based on a realistic urban scenario is simulated for testing and system verification. It shows that the proposed approach always bears in mind both the passenger's safety and comfort during the fall-back maneuvering execution.

Keywords: fail-operational systems; fall-back strategy; automated driving

1. Introduction

In the last decade, Automated Driving Systems (ADS) has shown significant advances, mainly from the acquisition, perception, control and actuation point of view [1]. Several important developments have been achieved and mentioned in the latest European Commission reports [2], where the challenges on communication technologies and cyber-security, on-board sensors capacities, infrastructure requirements, mobility concepts, and city contexts are playing an active role for sustainable urban transportation developments.

ADS obtain information about the surroundings from different sensors, such as cameras, differential global positioning systems (GPS), Light Detection and Ranging (LiDAR), and Radio Detection and Ranging (RaDAR) systems [3]. Perception tasks are critical for increasing the level of automation of ADS developments, as environment recognition in any scenario, including lighting and weather conditions, should be assured. Moreover, their fail-operational operation during autonomous mode is crucial to ensure passenger safety, as sensor and perception errors can be easily propagated to decision and control stages in different maneuvers, causing fatal accidents [4].

Some authors have considered sensor data fusion for more robust performance on different contexts: obstacles detection [5], perception of the environment [6], localization [7], and Traffic Sign Detection and Recognition [8]. A detailed description of the most popular methods and techniques for performing data fusion is presented in [9], where the author concludes that the appropriate technique to be implemented depends on the type of problem. In the automotive field, the Bayesian approach, extended and unscented Kalman filters (UKF) are mostly used [10–12]. However, these techniques depend mainly on the information directly from on-board sensors without any fall-back strategy.

Currently, the most widely used global localization approaches involve Global Navigation Satellite Systems (GNSS) such as GPS and Galileo [13]. The implemented devices even can implement differential GPS approaches, which have become affordable in recent years, or Inertial Measurement Units (IMU) [7], which may be fused with GNSS data to provide more reliable data. Although this approach works properly in open scenarios such as highways, in urban environments, their localization accuracy is not guaranteed for ADS [13]. Hence, a fail-operational positioning system, which also uses the dynamic model of the vehicle, is required to increase the accuracy. Moreover, the implemented ADS need to have a Dynamic Driving Task (DDT) fall-back strategy approach to be executed when the positioning system fails [14]. Among the different proposed strategies Table 1 summarizes the most important DDT fall-back strategies proposed.

Table 1. DDT fall-back strategies due automated driving system failures.

System	Functionality	Fall-Back Strategy
Perception	Object detection	Create ghost vehicles to replace the hidden ones due high curvatures in highways [15].
		Create ghost objects due sensor failure and perform lane-changing maneuver to emergency shoulder [16].
Decision	Lane centering	switch to differential braking control if electrical power steering fails [17].
	Trajectory planning	Emergency maneuver bring vehicle to stop if collision free trajectories fails [18].
		Emergency trajectory to stop at the slowest lane [19].
Control	Speed profile	Use a future velocity if communication of control messages or the propulsion controller fails [20].
	Collision avoidance	Brake if reception of data packets or inter-vehicle distance from lead vehicle fails [21].
Actuation	Drive-by-wire	Various forms of monitoring and redundancy are considered in failure cases [22].

However, a better assessment of fail-operational strategies for the functions of the dynamic driving task (DDT) is needed to achieve higher levels of automation on ADS (SAE J3016 [23]). This work is focused on this area, and its main contribution of this work is a fail-operational strategy approach considering positioning failures in a last available device, implemented within a general control architecture for automated vehicles. In brief, the improvements presented in this work are:

1. A fail-operational positioning system that comprises a UKF, a virtual sensor, and a monitor system, capable of remaining operative from degraded to total failure of the position reception and warns for fall-back triggering.
2. A real-time trajectory planner that defines the lateral and longitudinal references for the DDT fall-back in degraded mode to achieve a minimal risk condition avoiding rear-end collisions.
3. A vehicle motion control that executes the planned trajectory, including a lateral reference constraint avoiding undesirable lane departures.
4. A case study resembles a real urban scenario demanding a DDT fall-back strategy due to a major positioning failure, working with minimum sensor interface bringing the vehicle to a safe place.

The rest of the paper is organized as follows. In Section 2, the fail-operational control architecture is detailed, explaining each module. An overview of the fail-operational positioning system is presented in Section 3, where the vehicle model, cornering stiffness estimation and the adaptive UKF are the main contributions. Section 4 describes the DDT fall-back strategy proposed for the decision stage, considering a real-time trajectory generation, rear-end collision avoidance and vehicle motion control. In Section 5, a description of the scenario, the test platform and the parameter of configuration for the UKF and MPC are presented. Results and discussion are explained in Section 6. Finally, some remarks and conclusions are presented in Section 7.

2. Fail-Operational Control Architecture

The proposed work has been developed in the framework of the AutoDrive Project [24]. This program targets the development of SAE Level 4 [23] automated driving capabilities. More precisely, a highly automated driving bus to carry passengers in an urban scenario with mixed traffic conditions.

Please note that the required automation level must include, moreover the lateral and longitudinal motion control, a complete Object and Event Detection and Response (OEDR) system, and the capability to be robust enough to support fail-operational operation [25], which is the focus of this work.

The control architecture proposed to achieve this goal is depicted in Figure 1, covering seven stages of those suggested by [1] required for ADS developments (Database, Acquisition, Perception, Supervisor, Decision, Control, and Actuation). This architecture allows the verification of the DDT fall-back strategy after the occurrence of a positioning system failure.

In the next subsections, the different stages that compose the proposed architecture are detailed.

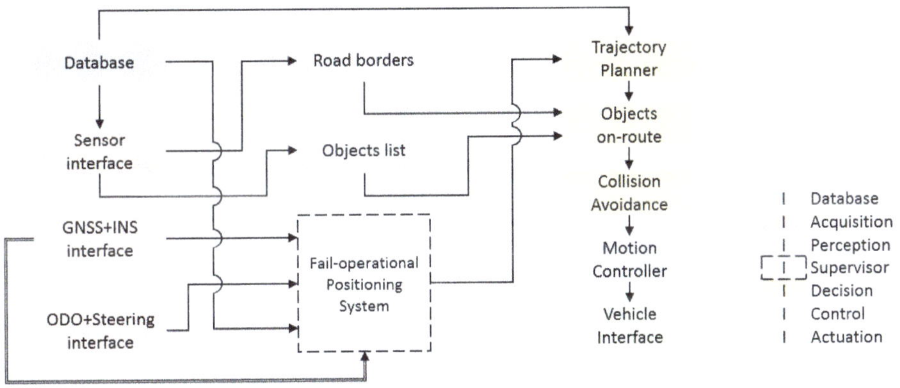

Figure 1. Control Architecture.

2.1. Database

The database is composed of a list of waypoints that contain relevant information of the fixed route, such as global axis coordinates (X, Y), orientation angles (ψ) and velocity limits (v_x). Additionally, information related to safe-parking places is included, considering three different cases: stop not permitted, stop on-lane permitted and stop on-shoulder available. These waypoints will be illustrated in the case study in Section 5.

2.2. Acquisition

The acquisition system interface provides two features: the vehicle surrounding recognition, and the vehicle global position on the route.

For the first feature, a sensor interface is used, which provides relevant information about the surroundings. This will be processed by the perception stage simulating information from road borders and objects. Please note that in the present work's framework the sensor interface will be idealized to provide information from 360 degrees around the vehicle, with a maximum detection radius. Elements outer the range will not be detected.

To define the position of the vehicle in the scenario, a global navigation satellite system integrated with an inertial navigation system (GNSS-INS), odometer and steering sensors are considered. These devices provide the vehicle position (X, Y), front-wheel angle (δ) and inertial parameters as orientation (ψ), acceleration (a_x) and velocity (v_x).

Please note that commercial GNSS-INS interfaces present noise and signal quality reductions. Hence, the failure of this sensor must be handled by the proposed fail-operational positioning system and fall-back strategy.

2.3. Perception

The information provided by the acquisition blocks is used to detect the road borders and objects within the sensor range. These are estimated in coordinates relative to the vehicle. This way, a 4-m road width is considered, so that the road borders are placed at 2 m of lateral distance from the center-lane (X, Y). The lateral distances are estimated from the vehicle's position to the left (e_L) and right (e_R) borders. On the other hand, an object list is provided considering relative distances and velocities.

2.4. Supervisor. Fail-Operational Positioning System

The supervisory system is continuously monitoring the positioning accuracy and the status of the positioning sensor devices (GNSS-INS) of the acquisition blocks.

When the vehicle performance is highly compromised or a relevant sensor failure is detected, a fail-operational strategy is activated. Using the data from the positioning sensors (GNSS-INS, odometer and steering sensors), a virtual positioning sensor is switched on and employed to perform the fall-back strategy that leads the test platform to a safe state. This one of the contributions of this work, and its development is broadly detailed in Section 3.

2.5. Decision

The decision module creates the trajectories to be followed by the automated vehicle and is integrated by the trajectory planner and a collision-avoidance system.

The trajectory planner, in the case of normal operation, will generate optimum trajectories for a specified scenario. However, in the case of system failure, such as the case study analyzed in Section 5, the trajectory planner will adjust the original route. Using the information of the safe-parking places from the database, the planner will modify the route to achieve the nearest safe state. This is an important contribution of this work and will be covered in Section 4.1.

On the other hand, the collision-avoidance system is focused on avoiding rear-end collisions with previous vehicles or objects. Please note that the objects detected by the perception system are first analyzed to evaluate if they are within the trajectory to be followed. Therefore, objects located outside the road borders do not represent a collision risk and adjustments are not necessary over the original trajectory, while objects within the trajectory will require adjustment of the trajectory by maintaining a safe distance to the objects ahead. The proposed approach, which will be detailed in Section 4.2, can work not only in normal operation, but also when there exists a degraded operation due to failure in the positioning sensors.

2.6. Control

The trajectory references estimated in the decision stage are followed by the vehicle motion control, providing reliable inputs to the vehicle interface. The velocity (v_x), acceleration (a_x) and jerk (j_x) are usually considered to be the main state parameters to control the longitudinal vehicle motion behavior. The position in global coordinates (X, Y) and the yaw angle of the vehicle (ψ) are commonly used for lateral and angular vehicle motion control, respectively. Additional relevant state parameters can be included for control improvement as the lateral error concerning the route's center-lane (e_y). A detailed explanation of this stage is presented in Section 4.3.

2.7. Actuation

The actuation module receives information from the control stage. Its task is to move the actuator of the automated vehicle, which means, steering wheel and pedals. Characterization of the actuators is the proposed architecture, as presented in [26].

3. Fail-Operational Positioning System

The proposed approach uses a sensor fusion strategy that combines the information of the GNSS-INS, odometer and steering sensor to provide positioning data even in degraded circumstances. Moreover, it integrates a quality monitor that allows detecting the failure of the sensor. The overall architecture of the proposed fail-operational positioning system is depicted in Figure 2.

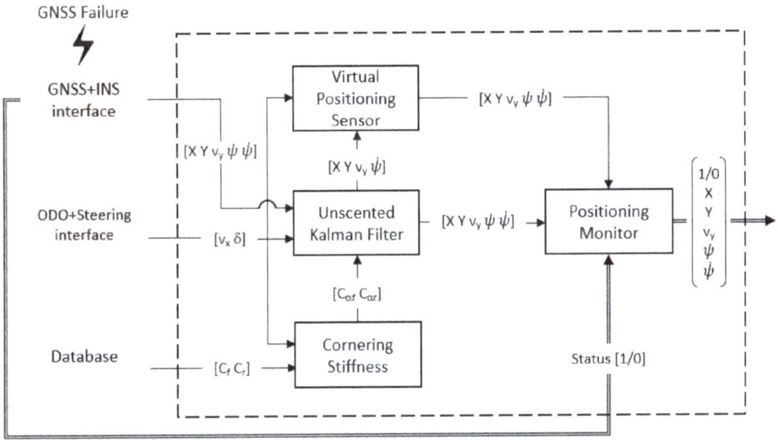

Figure 2. Flowchart on fail-operational Positioning System.

This strategy first uses the current values of velocity (v_x) and front-wheel angle (δ) to interpolate the front (C_{α_f}) and rear (C_{α_r}) cornering stiffness from database values. This data will be used in a second step, where an adaptive UKF is employed to attenuate the accuracy lacking on the GNSS-INS positioning measurement. This is a problem regularly faced in urban environments due to obstructions in the line-of-sight to the satellites [12].

When a relevant performance failure is detected, a virtual positioning sensor is activated. It makes use of the last position acquired by the UKF system, the estimation of the cornering stiffness, and the data provided by the odometer sensors and steering wheel sensor to estimate positioning to perform dead reckoning and perform a safe state on the route.

The detection of a degraded condition and critical failure is performed by a positioning monitor, which selects the source of the positioning data to provide a fail-operational response, and informs the decision stage about the failure.

3.1. Vehicle Model and Cornering Stiffness Estimation

The proposed fail-operational positioning system requires a vehicle model to implement both the lateral vehicle motion control and UKF. In this section, the model used for the development of the UKF is briefly detailed.

3.1.1. The Kinematic and Dynamic Vehicle Models

The lateral motion of a vehicle can be estimated as well as controlled employing simplified vehicle models, being this a technique that reduces the computational effort for real-time implementations while providing enough accuracy for control purposes [27]. For velocities at less than 3 m/s the lateral forces on the tires can be neglected, and the vehicle motion can be calculated entirely on geometric relationships of X, Y and ψ [28,29]. Above 3 m/s, the assumption of no lateral forces on the tires begins to be compromised, as the lateral vehicle motion is affected by its dynamics being necessary to take into consideration a more complex model to improve results [30]. In these cases, a mix of simplified bicycle models for lateral vehicle dynamics (Figure 3) provides a good accuracy vs complexity relationship.

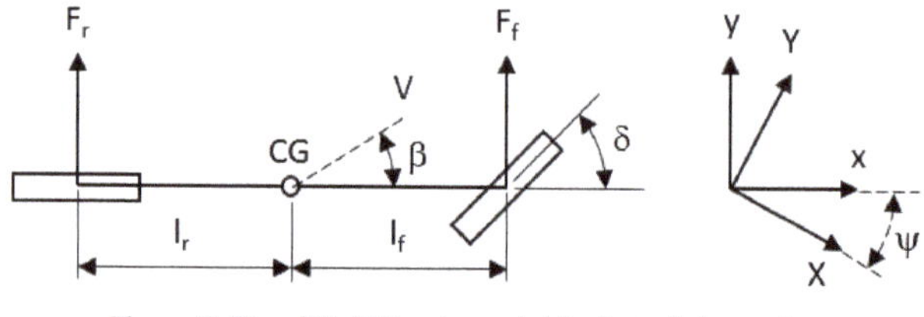

Figure 3. Simplified bicycle model for lateral dynamics.

In Figure 3, V represents the velocity at the center of gravity (CG) and β is the vehicle slip-angle concerning the longitudinal axis of the vehicle (x). The lateral tire forces on the front and rear wheels (F_f, F_r) are strongly affected by the cornering stiffness of each tire (C_{α_f}, C_{α_r}), the slip-angle of each tire (θ_f, θ_r) and the δ of the front wheel. The lateral translation motion and the yaw dynamics for the model can be written as,

$$m(\dot{v}_y + \dot{\psi} v_x) = 2C_{\alpha_f}(\delta - \theta_f) - 2C_{\alpha_r}\theta_r \quad (1a)$$

$$I_z\ddot{\psi} = 2l_f C_{\alpha_f}(\delta - \theta_f) + 2l_r C_{\alpha_r}\theta_r \quad (1b)$$

where m is the mass, $\dot{v}_y$ is the lateral acceleration, $\dot{\psi}$ is the yaw rate, v_x is the longitudinal velocity, I_z is the moment balance around the z axis of the vehicle, $\ddot{\psi}$ is the yaw acceleration and, l_f and l_r are the front and rear wheel distance from the CG. The slip angles of each tire (θ_f and θ_r) can be calculated as,

$$\theta_f = \text{atan}\,\frac{v_y + l_f\dot{\psi}}{v_x} \quad (2a)$$

$$\theta_r = \text{atan}\,\frac{v_y - l_r\dot{\psi}}{v_x} \quad (2b)$$

In this work, a dynamic bicycle model approach is implemented for filtering and positioning. On the other hand, a kinematic bicycle model approach is employed for vehicle motion control.

3.1.2. Cornering Stiffness Estimation

Using the dynamic bicycle model defined in Equation (1) it is possible to estimate the cornering stiffness coefficients C_{α_f} and C_{α_r}. Please note that most parameters are easy to obtain in real applications [31]. This way, if a state-space representation is used,

$$\begin{bmatrix} C_{\alpha_f} \\ C_{\alpha_r} \end{bmatrix} = \begin{bmatrix} 2(\delta - \theta_f) & -2\theta_r \\ 2l_f(\delta - \theta_f) & 2l_r\theta_r \end{bmatrix}^{-1} \begin{bmatrix} m(\dot{v}_y + \dot{\psi} v_x) \\ I_z\ddot{\psi} \end{bmatrix} \quad (3)$$

An open-loop test method for determining the steady-state circular driving behavior described in [32] (ISO 4138) is employed for the cornering stiffness estimation at constant steering wheel angles and velocities. Using this procedure, a cornering stiffness map can be generated and integrated into the control architecture, to implement a cornering stiffness estimator as input to the UKF (Figure 2), so that for any v_x and δ the coefficients C_{α_f} and C_{α_r} can be derived. Results of these tests for the case study analyzed in this work are presented in Section 5.3.

Please note that the use of an off-line cornering stiffness estimator implies that the open-loop tests must cover the whole range of v_x and δ to be performed by the test vehicle within the ODD. Although this can be implemented for on-line estimations, this procedure helps to fix values when necessary avoiding some singularities in circumstances, as no lateral accelerations [33].

3.2. Adaptive Unscented Kalman Filter

An adaptive Unscented Kalman Filter (UKF) is used to attenuate the errors introduced in the GNSS-INS positioning measurement when the satellite signal quality is reduced. In contrast to other Kalman filtering techniques, the UKF frequently provides a lower estimation error and is preferable for implementations in automated driving applications [10]. In this sense, a UKF-based approach capable of adapting the measurement noise covariance matrix is presented here, this is an adaptive UKF.

The development of the UKF requires the space-state transition model of the vehicle detailed in Section 3.1, which can be defined as,

$$\begin{bmatrix} \dot{X} \\ \dot{Y} \\ \dot{y} \\ \dot{v}_y \\ \dot{\psi} \\ \ddot{\psi} \end{bmatrix} = \begin{bmatrix} 0 & 0 & 0 & -\sin\psi & \frac{v_x \cos\psi}{\psi} & 0 \\ 0 & 0 & 0 & \cos\psi & \frac{v_x \sin\psi}{\psi} & 0 \\ 0 & 0 & 0 & 1 & 0 & 0 \\ 0 & 0 & 0 & -\frac{2C_{\alpha_f}+2C_{\alpha_r}}{m v_x} & 0 & -v_x - \frac{2C_{\alpha_f} l_f - 2C_{\alpha_r} l_r}{m v_x} \\ 0 & 0 & 0 & 0 & 0 & 1 \\ 0 & 0 & 0 & -\frac{2l_f C_{\alpha_f} - 2l_r C_{\alpha_r}}{I_z v_x} & 0 & -\frac{2{l_f}^2 C_{\alpha_f} + 2{l_r}^2 C_{\alpha_r}}{I_z v_x} \end{bmatrix} \begin{bmatrix} X \\ Y \\ y \\ v_y \\ \psi \\ \dot{\psi} \end{bmatrix} + \begin{bmatrix} 0 \\ 0 \\ 0 \\ \frac{2C_{\alpha_f}}{m} \\ 0 \\ \frac{2l_f C_{\alpha_f}}{I_z} \end{bmatrix} \delta \tag{4}$$

where X, Y, v_y, ψ and $\dot{\psi}$ are parameters obtained from the GNSS-INS interface. The v_x and δ are parameters received from the odometer and steering angle sensor interface. A linear relationship between the steering angle and the front-wheel angle is employed to obtain the current value of δ. The stiffness coefficients C_{α_f} and C_{α_r} are obtained through the procedure described in Section 3.1.2.

The process noise covariance matrix (Q_n) in a vehicle model is suggested to be calculated as the propagation of each value per time step [11], in this sense, gathering the standard deviation of parameters from the test vehicle circulating in normal conditions helps to determine Q_n.

The measurement noise covariance matrix (R_n) is mainly associated with the accuracy of the acquisition devices. These can be extracted from commercial GNSS devices data-sheet.

3.3. Virtual Positioning Sensor

When a GNSS failure event occurs (lower signal quality or total disconnection), a virtual positioning sensor is used to provide an indirect position measurement by combining information from the remaining physical sensors.

The velocity from odometer (v_x^{odo}), the lateral velocity from the filter (v_y^{ukf}), and the yaw angle obtained due a discrete integration from the filter yaw rate measure (ψ^{int}), are considered to estimate the vehicle velocities in global coordinates ($\dot{X}$,$\dot{Y}$). A state-space representation is described as,

$$\begin{bmatrix} \dot{X} \\ \dot{Y} \end{bmatrix} = \begin{bmatrix} \cos\psi^{int} & -\sin\psi^{int} \\ \sin\psi^{int} & \cos\psi^{int} \end{bmatrix} \begin{bmatrix} {v_x}^{odo} \\ {v_y}^{ukf} \end{bmatrix} \tag{5}$$

The obtained velocities are consequently integrated to obtain X and Y. The last available values before the failure for X, Y and ψ are considered to be the initial values for the newly integrated parameters. The remaining available parameters as v_y and $\dot{\psi}$ are combined with the indirect estimations to maintain the same structure information sent by the UKF.

3.4. Positioning Monitor

The monitor role is to continuously evaluate the positioning quality of the GNSS. In case of a very poor positioning accuracy (quality below 2 in Table 2) or a catastrophic failure (e.g., power supply unavailable), the monitor will instantly switch the information received from the UKF to the one received from the virtual sensor. The output state parameters are combined with a failure tag (1/0) to inform this status to the decision stage so that a degraded condition and proper action taken.

4. Fall-Back Strategies Implementation in the Decision Stage

The fail-operational positioning system provides information on the vehicle position and the existence of a failure to the decision stage of the control architecture depicted in Figure 1. In this section, the fall-back strategy, which includes both the trajectory planner and the collision-avoidance subsystems, will be detailed.

4.1. Real-Time Trajectory Planner

In normal operation, a fixed route is planned off-line based on Bezier and feasible curvatures generation procedure [34]. The velocities are limited considering the curvatures along the route [26] defining bounds for lateral and longitudinal accelerations bearing in mind the passenger comfort [35].

In case of failure, a DDT fall-back strategy starts, and the trajectory planned is modified to achieve a degraded driving mode. The velocity is instantly reduced to a degraded value, to avoid lateral displacements in vehicle motion control while dead-reckoning is performed and maintained until the vehicle is located over a safe-parking spot, where the vehicle stops. The path is not modified until a safe-parking space is available.

The strategy for a degraded velocity (v_x^{degr}) is depicted in Figure 4a. After the failure, a start distance (d_{start}) is defined to reduce the speed at degraded deceleration (a_x^{degr}), as a sudden reduction could affect negatively on the motion controller, producing undesirable and uncomfortable responses. The same procedure is repeated to stop once the vehicle is located over the emergency shoulder.

The strategy for a degraded path is presented in Figure 4b. After the failure, the planned path is maintained until a safe-parking place becomes available ($[X, Y]^{start}$), at this point, the planned route is moved perpendicularly based on a predefined lateral velocity (v_{ey}) to a proper distance in the emergency shoulder (d_{ey}).

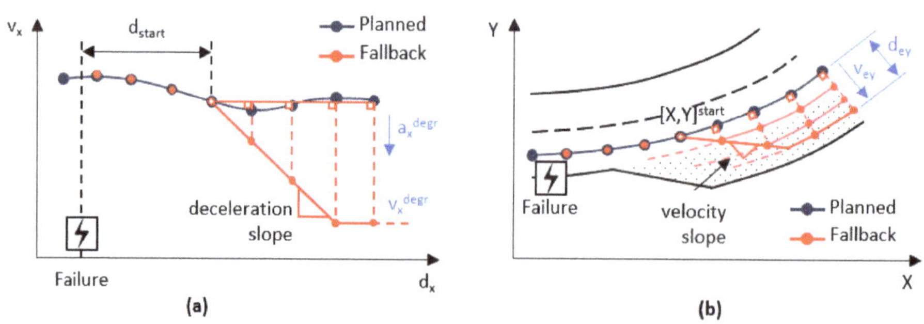

Figure 4. Real-time trajectory planning for (**a**) velocity and (**b**) path.

The degraded path reference is estimated to displace laterally from the original route faster than the vehicle's capabilities, therefore the absolute value of the lateral error increases and decreases during the lane-change maneuver. The d_{ey} magnitude helps to predict when the vehicle goes out the main route ($d_{ey} > 0.64$ m) and afterward is located enough on the emergency shoulder ($d_{ey} < 0.16$ m), finally permitting reduction of the degraded velocity to zero. A flowchart of the real-time trajectory planning is depicted in Figure 5. The practicability of this methodology is discussed in Section 6.1.

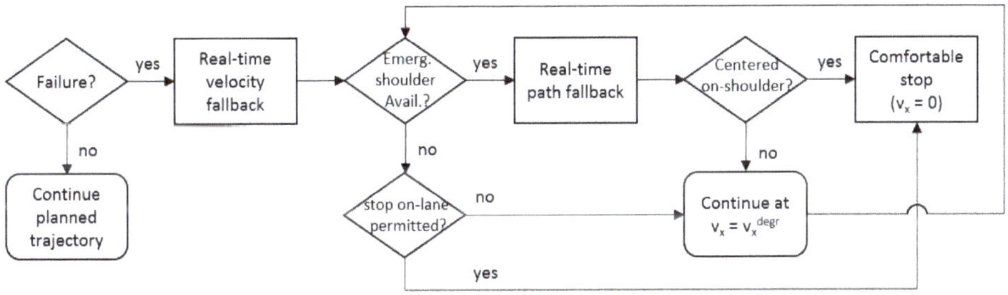

Figure 5. Flowchart on real-time trajectory planning.

4.2. Rear-End Collision Avoidance

In both normal and degraded operation, the automated vehicle implements a rear-end collision-avoidance system using the data provided by the fail-operational positioning system and a detection system of the objects ahead.

For that purpose, a Model Predictive Control (MPC) approach has been implemented, generating the velocity references to be followed by a low-level control, attempting to maintain a safe relative distance (d_r) and velocity (v_r) from objects ahead on-route. A one-dimensional kinematic model is considered to model the longitudinal vehicle motion,

$$\begin{bmatrix} \dot{v_x} \\ \dot{a_x} \\ \dot{d_r} \\ \dot{v_r} \end{bmatrix} = \begin{bmatrix} a_x \\ j_x \\ v_t - v_x \\ a_t - a_x \end{bmatrix} \tag{6}$$

where j_x is the longitudinal jerk, v_t is the target object velocity and a_t is the target object acceleration.

The state parameters $\eta = [v_x \ d_r \ v_r]$ are optimized in the entire prediction horizon (H). The references for v_x are defined by the planned velocity discussed in Section 4.1. The reference values for d_r and v_r are defined as,

$$d_r^{ref} = d_r^{min} + v_x t_{hw} \tag{7a}$$

$$v_r^{ref} = 0 \tag{7b}$$

where d_r^{min} is the minimum safety distance at 5 m, and t_{hw} is a headway time equals to 1 s.

The state parameters weighting matrix Q_w changes according to a d_r-v_r diagram [28], which determines the operation mode to perform velocity or headway control in case an object detection.

The maximum deceleration permitted (a_x^{ref}) changes also with the operation mode, being this an important value to properly perform a longitudinal vehicle motion control. Lower and upper bounds are considered to maintain properly a safe distance from objects ahead as 5 m $< d_r <$ 50 m.

As current detection sensors are mostly incapable of giving a reliable measurement of objects accelerations [28], the target object acceleration a_t is neglected at any time. The d_r and v_r are calculated from a pair of projection points over the tracked route, this functionality is supposed available in system failure condition.

4.3. Vehicle Motion Control

The velocity references provided by the rear-end collision-avoidance system are considered when objects are detected ahead instead of the trajectory references from the real-time planner. The vehicle motion control follows the references based on an MPC strategy. To perform this task, the model implemented by the MPC is a kinematic bicycle model (Section 3.1) with additional equations for j_x and lateral error distance (e_y) to constraint it and assure an accurate lane-keeping,

$$\begin{bmatrix} \dot{X} \\ \dot{Y} \\ \dot{\psi} \\ \dot{\delta} \\ \dot{v_x} \\ \dot{a_x} \\ \dot{e_y} \end{bmatrix} = \begin{bmatrix} v_x \cos(\psi + \beta) \\ v_y \sin(\psi + \beta) \\ v_x \cos(\beta) \tan(\delta)/L \\ \Delta\delta \\ a_x \\ j_x \\ v_x \sin(\psi + \beta - \psi^{ref}) \end{bmatrix} \tag{8}$$

where the ψ^{ref} is the yaw angle reference for a current position of the vehicle over the route.

The state parameters $\eta = [X\ Y\ \psi\ v_x]$ and the control inputs $u = [\Delta\delta\ j_x]$ are optimized for the whole H. The velocity reference is defined by the rear-end collision system when an object ahead is present, in other cases this reference comes from the planned trajectory as well as those for lateral vehicle motion control. The control inputs are after integrated to reproduce the steering and pedal position as actuation signals for the vehicle interface.

5. Case Study

In this section, the case study to evaluate the fail-operational approach is presented. First, the test scenario is defined based on a route in a real urban scenario. Secondly, the test platform to perform the DDT fall-back strategy is detailed. Finally, the parameters considered from the database and for the decision and control are mentioned.

5.1. Realistic Scenario

The realistic scenario considered to validate the proposed approach is a highly automated driving bus to carry passengers at the port of Malaga city (Spain), as depicted in Figure 6.

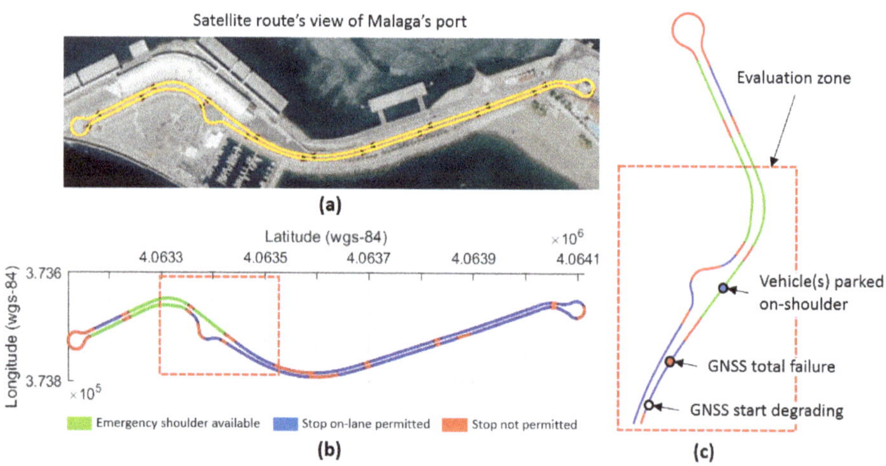

Figure 6. Realistic environment scenario for automated driving system tests on simulation. (**a**) Satellite's view of urban route, (**b**) permitted and non-permitted stops in case of total positioning failure, and (**c**) evaluation zone for test case study.

The selected test route covers a challenging environment with static objects in addition to difficult vehicle motion maneuvers as roundabouts, merging streets and intersections, as seen in Figure 6a.

In case of system failures, the ADS must respond without driver intervention to achieve a minimal risk condition bringing the vehicle to a safe state. In this sense, permitted and non-permitted stops are considered to be portrayed in Figure 6b avoiding to instantly stop.

The test case analyzed in this work is delimited to the evaluation zone depicted in Figure 6c. The failure to be studied is the possible malfunction of the GNSS position receiver (which is a vital part of the ADS), which starts degrading up to total failure and a dynamic driving task (DDT) fall-back strategy must be activated by the ADS.

As an additional issue, the case is considered in which the emergency shoulder cannot be used, as another vehicle is already parked, requiring driving of a longer distance to the next permitted stop while performing dead reckoning. Three different failure scenarios are analyzed, as shown in Figure 7.

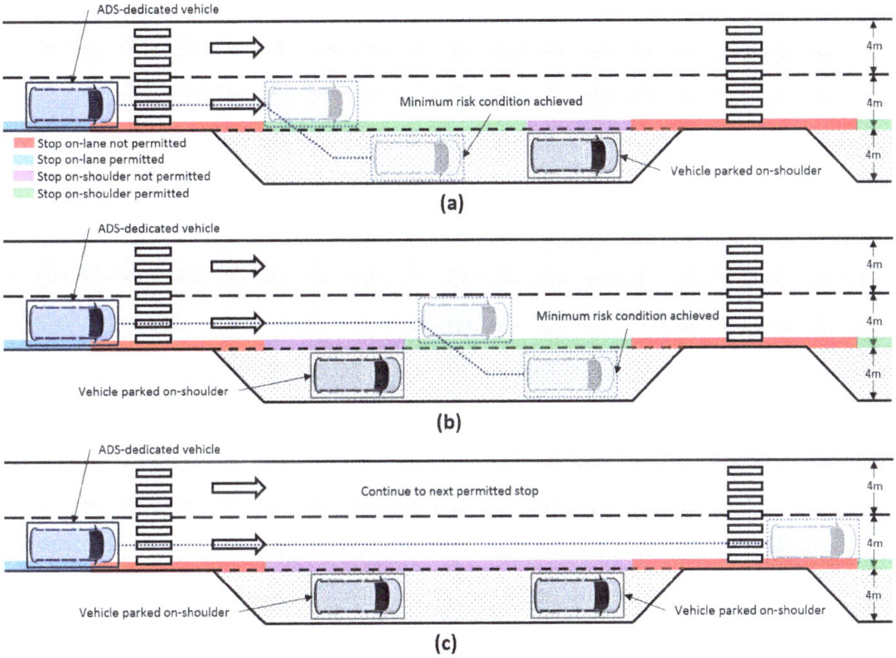

Figure 7. DDT fall-back strategy response under three different scenarios. A minimum risk condition is achieved (**a**) before and (**b**) after an object parked on the emergency shoulder. A next permitted stop necessary due (**c**) no space available on current emergency shoulder.

When a failure of the GNSS occurs, the distance required (d_r^{req}) to achieve a safe-parking is calculated constantly before to initiate the maneuver as presented in the Equation (9). If the d_r^{req} is lower than d_r from and object and the available emergency shoulder longitude, then the lane-change maneuver initiates to achieve a minimal risk condition, parking the vehicle on the emergency shoulder. Moreover, the first one of the two terms in the right-hand side of Equation (9) can be employed to estimate a stop on-lane if permitted. On the contrary, the vehicle continues to the next permitted stop.

$$d_r^{req} = v_x\left(\frac{v_x}{a_x^{degr}} + t^{delay} + t^{timeout}\right) + \frac{d_{ey} - e_y}{v_{ey}} \tag{9}$$

where t^{delay} and $t^{timeout}$ are additional times considered to complete the lane-change maneuver being conservative. The t^{delay} is stated as 0.5 s and related to actuation devices and vehicle's inertia that retard the final stopping time. The $t^{timeout}$ is defined as 1 s considering a required time for the vehicle to be located enough on the emergency shoulder before totally stop.

It should be noted that the case study has been implemented in a simulation environment. This allows introduction of the degrading behavior in the perception system and evaluating the proposed fall-back strategies with minimal risk before future implementation.

5.2. Test Platform

A standard electric bus has been selected as the test platform for the case study scenario. This bus weights 16,000 kg and has a dimension of approximately 12.16 × 3.30 × 2.55 m, with a wheelbase of 5.77 m, a minimum turning radius of 7.2 m and a maximum front-wheel angle of 0.68 rad.

This way, the test platform has been modeled in Dynacar simulator [36], which uses a multi-body formulation to link the chassis with a steering knuckle suspension at the front axle, and a rigid axle suspension type at the rear. The suspensions are also linked to the two wheels at the front axle and four wheels at the rear axle, based on a standard Pacejka tire model defined in [37].

Moreover, the sensors have been simulated from the data obtained from the Dynacar model, introducing measurement errors to simulate degraded scenarios. The exteroceptive sensors can cover 360° around the vehicle, reaching a maximum radius of 60 m for object detection. In the case of the GNSS sensor, which is the focus of the work, a random Gaussian noise associated with the quality signal of the GNSS-INS interface is added around the nominal state parameter obtained from the simulated test platform. The random noise values are introduced considering the quality of the signal to be simulated, as in real commercial devices (Table 2). Future implementation will need some of these exteroceptive sensors, the instrumentation necessary for real vehicles is detailed in [29].

5.3. Parameters

To test the proposed approaches, the following parameter values have been applied.

The noise covariances for the UKF have been calculated as suggested by [11], the process noise covariance matrix (Q_n) is defined assuming the standard deviation of parameters from the test vehicle circulating in normal conditions helps to determine Q_n. The measurement noise covariance matrix (R_n) is selected by taking into account the accuracy of commercially available acquisition devices. The Q_n and R_n are depicted in the Table 2.

Table 2. Process and measurement covariances in UKF.

Position Covariances					Inertial Covariances			
Parameter	Quality	Q_n	R_n	Unit	Parameter	Q_n	R_n	Unit
	5		0.0141	m	σ_y^2	1.26×10^{-2}	2.78×10^{-2}	m
	4		0.2828	m	$\sigma_{v_y}^2$	1.26×10^{-4}	2.78×10^{-4}	$\frac{m}{s}$
σ_{XY}^2	3	1×10^{-3}	0.4243	m	σ_{ψ}^2	2.7×10^{-1}	1.7×10^{-1}	$\frac{rad}{s}$
	2		1.1314	m	$\sigma_{\dot{\psi}}^2$	2.7×10^{-3}	1.7×10^{-3}	$\frac{rad}{s^2}$

The fail-operational positioning system requires the estimation of the Cornering Stiffness coefficients. As detailed in Section 3.1.2, a set of open-loop tests is carried out to determine the steady-state circular driving behavior described in [32], obtaining a set of data that can be used to create a cornering stiffness map.

For that purpose, the open-loop tests must cover the whole range of v_x and δ for the test platform detailed in Section 5. Hence, the steering angle has been modified from −0.5 to 0.5 rad, in 0.1 rad steps, while the longitudinal speeds have taken the values of 0.5, 1, 2, 3, 4, 5 m/s.

The resulting cornering stiffness map is shown in Figure 8. From this map, intermediate values required by the fail-operational positioning system are estimated using interpolation.

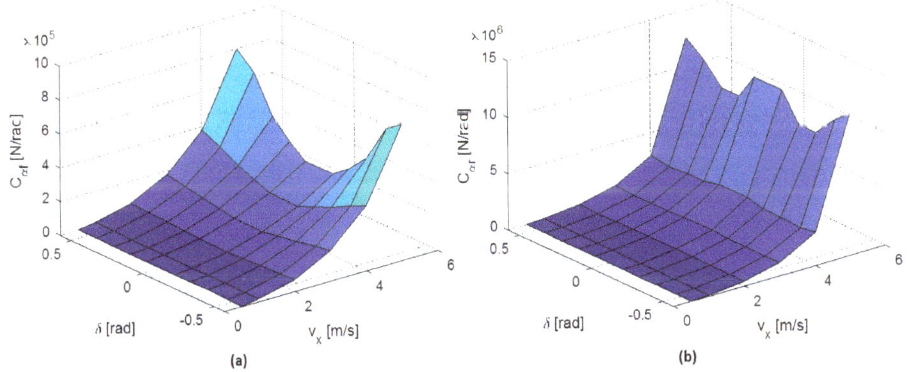

Figure 8. (**a**) Front and (**b**) rear cornering stiffness estimation

In the case of the Real-Time trajectory planner, the start distance d_{start}, longitudinal velocity and deceleration in degraded mode are fixed to 5 m, 1.5 m/s and 0.2 m/s^2, respectively, while (v_{ey} and d_{ey}) are fixed for the case study proposed to 0.2 m/s and 4 m, respectively.

To implement the vehicle motion MPC controller, the following parameters have been used. A prediction horizon of $H = 10$ is defined with a constant time step of 0.5 s. The states parameters and control input weights for optimization are intuitively defined as Q_w = diag([1 1 25 1]) and R_w = diag([10 10]), respectively, giving more importance to the vehicle orientation over the route. The physical constraints for state parameters and control inputs are summarized in Table 3.

Table 3. Constraints in the low-level control.

Parameter	Lower	Upper	Unit
$\Delta\delta$	1	1	$\frac{rad}{s}$
j_x	1	1	$\frac{m}{s^3}$
δ	−0.68	0.68	rad
v_x	0	v_x^{ref}	$\frac{m}{s}$
a_x	$-a_x^{ref}$	0.2	$\frac{m}{s^2}$
e_y	e_{yL}^{ref}	e_{yR}^{ref}	m

where v_x^{ref} are the velocity references, a_x^{ref} depends to the longitudinal operation mode defined in Section 4.2, and the e_{yL}^{ref} and e_{yR}^{ref} are the left and right lateral error distances to the borders, respectively, according the current position of the vehicle on-lane.

Finally, it should be noted that the open-source ACADO Toolkit is employed to solve the optimal control problem both in the rear-end collision-avoidance system and vehicle motion control. A continuous output Implicit Runge–Kutta integrator of second-order simulates the system 1 integration step in both cases. The H is parametrized to obtain 10 elements with a constant time step of 0.5 s.

6. Results and Discussions

In this section, the most relevant results associated with the proposed fail-operational positioning system and the defined fall-back strategies are analyzed. Moreover, the effect of the proposed approach in the comfort of the passengers is also analyzed.

6.1. Evaluation of the Fail-Operational Positioning System

The robustness of the control architecture is evaluated here performing complete laps on the test circuit. The Figure 6a,b shows the route defined for the evaluation of the fail-operational positioning system based on UKF. This trajectory is executed using the control architecture proposed in Section 2. Four different scenarios are proposed, with different GNSS positioning qualities (from 2 to 5).

In each simulation, the positioning data gave by the raw GNSS-INS sensor (with Gaussian noise), the output of the UKF filter and the real position of the vehicle are measured, and compared with the trajectory reference, to calculate the lateral positioning error e_y (m).

Results for the four signal quality scenarios are shown in Figure 9, where the statistic distribution of the lateral positioning error e_y (m) is calculated considering the raw GNSS-INS sensor data (raw), the UKF filter output (UKF) and the real position of the vehicle (real).

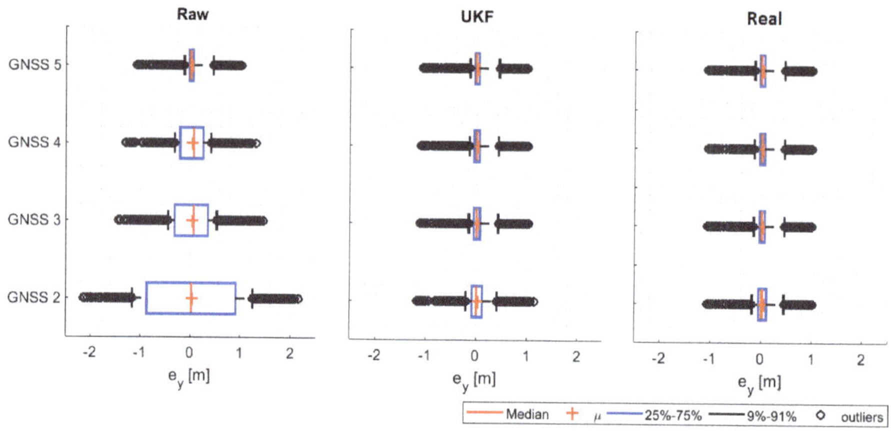

Figure 9. Lateral error under different GNSS positioning quality.

From the results, the decrease in the quality of the GNSS signal increases significantly the lateral error if the raw data is used (raw case). It could be fatal in an automated vehicle operation such as the one analyzed. Moreover, the errors could introduce instability in the controllers, depending on the nature of the noise. This emphasizes the need for providing robust solutions to positioning measurements in automated vehicles.

Results also demonstrate the positive performance of the proposed UKF approach (UKF case), which can reduce in more than 90% the error associated with e_y in the poorest quality condition (GNSS 2). This demonstrates the validity of the proposed approach. In addition, the level of performance that can be achieved using the UKF is demonstrated in the real position of the vehicle (real case).

6.2. Evaluation of the Dynamic Driving Task Fall-back Strategy

In this section, the proposed fall-back strategy performance is evaluated in the three different scenarios depicted in Figure 7: stopping before a parked vehicle, after a parked vehicle and continuing to a next permitted stop due to no space availability.

In all three scenarios, the same GNSS failure sequence will be evaluated, as depicted in Figure 6c. At the beginning of the test, the GNSS system has a higher signal quality, sequentially reducing it until a total failure exists. At that point, the fall-back strategy will have to take on the control of the automated bus and lead it to a minimum risk position using the data provided by the virtual positioning controller.

Figure 10 indicates, for each scenario, the fall-back sequence carried out. In the first row, the point in which the failure occurs (the same in the three cases) is depicted. In the second one, the activation of the degraded condition is shown, in which the speed of the vehicle is reduced. Then, when a free parking spot is activated, the lane-changing maneuver is activated, to finally brake and stop. Please note that the black lines represent the road borders, the green dotted line is the central line of the road, while the red and violet lines are the executed and calculated trajectories.

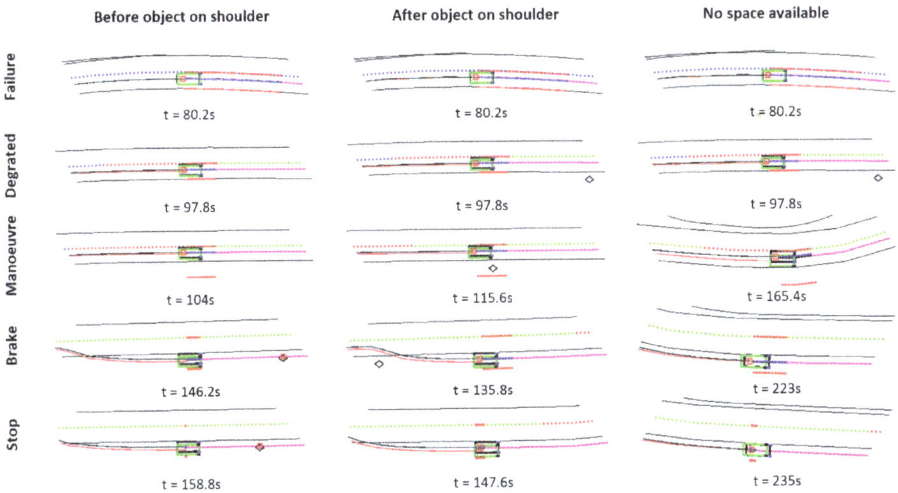

Figure 10. DDT fall-back strategy under different use cases.

The performance data in the three scenarios are shown in Figure 11. In this graph, the vertical dashed lines define the starting points of the failure, degraded, maneuver and brake phases (stop is considered the end of the graph). Moreover, four main performance indicators are analyzed for each scenario. In the first row (v_x) the longitudinal speed reference given by the trajectory planner (*Reference*) and the real speed of the vehicle (*Ego-vehicle*) is depicted. In the second one (d_x), the longitudinal distance to the nearest object (parked vehicle) (*ObjectDistance*) and to the next emergency shoulder (*SpaceAvailable*) is shown. These distances are calculated with the position of these items in the planned trajectory. Also, the longitudinal distance required for performing the lane-change maneuver is shown (*SpaceRequired*). This calculation is detailed in the Equation (9). In the third row, the time evolution of the lateral error e_y for the planned trajectory is shown, considering the raw data provided by the GNSS system (which fails) (*raw*), the output of the UKF (*UKF*) and the real position of the vehicle (*real*). Finally, the computational cost of the high and low-level controllers is shown.

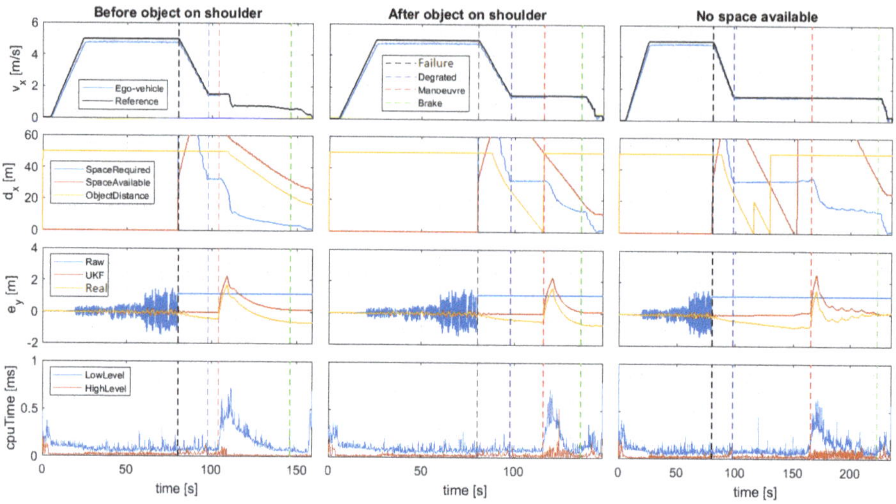

Figure 11. DDT fall-back response due GNSS total failure after continuously degrading position.

From these graphs, several conclusions can be drawn. First, the robustness of the proposed UKF-based position estimator is demonstrated in all scenarios. If e_y is analyzed, it can be noted that the effect of GNSS quality degradation directly affects the noise of the positioning system, which causes important e_y errors (up to 1 m). However, as previously analyzed, the use of the proposed UKF-based estimator reduces the effect substantially.

Second, the proposed fail-operational Positioning System proves an effective approach in a total failure case. When total failure happens (black vertical dashed line), the data provided by the GNSS remains constant and no longer can be used to estimate the position. At this point, the Positioning Monitor of the fail-operational positioning system switches to the Virtual Positioning Controller, entering degraded mode while making use of the odometer and the steering wheel to estimate the position of the vehicle. Please note that due to the nature of the selected sensors, estimation errors in e_y graphs will accumulate in time (see *real*), creating a drift. This effect is better seen in the third scenario, in which the nearest emergency shoulder is not available (is full) and therefore, the vehicle needs to move to the next one, operating more time in degraded mode. Hence, the degraded mode is intended to be used in emergencies for limited amounts of time or small distances, which is a valid assumption in urban environments such as the ones analyzed in the case study.

Third, the longitudinal speed (v_x) and distance (d_x) shows that the proposed fall-back strategy performs properly using the data provided by the fail-operational positioning system. When the failure occurs (black vertical dashed line), the vehicle reduced its speed to 1.5 m/s in all cases, entering a degraded state (blue vertical dashed line) once constant speed is achieved.

In this state, the trajectory planner searches for available spaces on the next emergency shoulder. For that purpose, the planner calculates the required space for the emergency parking maneuver (*SpaceRequired*) which depends on the current velocity and compares it with the distance to the nearest object/vehicle parked (*ObjectDistance*) and the available emergency shoulder distance (*SpaceAvailable*). Only if both are higher than the required distance to maneuver, the trajectory planner modifies the original route to start the lane-change maneuver. Please note that the object detection distance limit is 50 m and that the emergency shoulder-distance limit detection is 60 m, hence higher distances are limited to the maximum value.

The first scenario (parking before an object/vehicle in a shoulder), is the simplest one. It can be seen that when the degraded state is activated (97 s), the required space is less than the available shoulder distance, and the distance to the next vehicle, activating the lane-change maneuver (which implies a peak in e_y due to the lateral reference change), and moving through the shoulder until the maneuver has been completed. In the second scenario (parking after an object/vehicle in a shoulder), the degraded state is activated at the same time, but in this case, the emergency shoulder is still available, but a vehicle is already parked and the relative distance to it is too low to maneuver. Hence, the vehicle continues moving until the parked vehicle is surpassed (115 s). At this point, 50 m of emergency shoulder remains, which is more than the space required for the maneuver. In the third scenario (no space), there are two vehicles parked in the emergency shoulder. Hence, when the degraded state is activated, the distance to the first vehicle, and then, to the second, is detected (the vehicle change is shown as a peak at time 115 s). When the second vehicle is surpassed, however, the remaining shoulder distance is not enough to maneuver safely, and the vehicle continues moving to the next emergency shoulder.

Finally, if the computational cost graphs are considered, it can be seen that the proposed approach is computationally efficient, requiring less than 1ms to execute in an Intel Core i7-6600u CPU, 2.60GHz and 16GB RAM. This demonstrates that the approach can be implemented in real time.

Evaluation of Passenger Comfort

Comfort is a key issue when considering automated driving solutions. Traditionally, comfort has been related to the magnitude of the lateral and longitudinal accelerations, being higher ones less comfortable for passengers.

In Figure 12 the lateral and longitudinal accelerations associated with the three scenarios analyzed in the previous section are depicted. Two situations are considered, the first row depicts the acceleration results when a degraded GNSS quality (level 2) exists when the failure happens. The second situation considers the case in which an optimal quality (level 5).

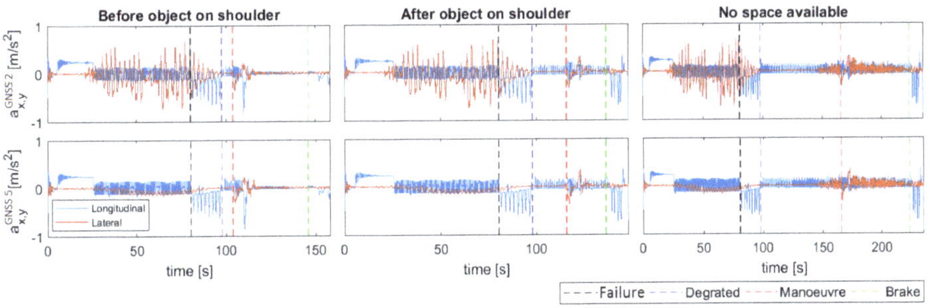

Figure 12. Longitudinal and lateral accelerations in DDT fall-back strategy.

As can be seen, even before the failure, the differences in the lateral acceleration are important due to the noise that the GNSS presents in lower qualities. Lateral accelerations are an order of magnitude higher in these cases, resulting in more uncomfortable driving. Therefore, sensor quality directly can affect passenger comfort.

Please note that the longitudinal acceleration is not affected in this case, due to the odometry is used to estimate it. When failure occurs, similar behavior is achieved. However, the vehicle speed and accelerations are reduced when the positioning monitor switches to the virtual positioning sensor.

7. Conclusions and Future Works

Although the research and development in automated driving has considerably helped the implementation of higher SAE automation levels, current control architectures rely on the driver as a backup in case of system failures. Moreover, hardware redundancy is the usual action plan to ensure fail-operational systems, as few software solutions exist in the literature.

The present work targets the issue of the vehicle bringing itself to a safe state in degraded mode after a major failure in the position receiver. Instead of a progressive deceleration on the current lane, the system focuses on seeking a permitted space on the route, performing a lane-change maneuver to the emergency shoulder, and then executes a safe stop.

The fail-operational control architecture and systems proposed here are explained in depth. They include basic ADS features to achieve a minimal risk condition along a route, according to a realistic case study presented for bus shuttling services as: fail-operational positioning system, real-time trajectory planner, collision-avoidance system and vehicle motion controller.

The fail-operational positioning system comprises a UKF to improve the vehicle location due to lack of quality in the position receiver, an issue very common in urban scenarios where the satellite line-of-sight would be constantly obstructed. A virtual sensor switches on by a positioning monitor in case of total failure in the position sensor is detected, then a DDT fall-back strategy is possible performing dead reckoning with database information. A previous cornering stiffness estimation through open-loop tests provides useful information for the vehicle model employed.

The real-time trajectory planner is capable of comfortably slow-down the velocity reference after the failure, expecting an available and permitted space to perform a lane-change maneuver and safely locating the vehicle on the emergency shoulder. The benefits of having an object parked in advance are considered, hence the available space to initiate the parking maneuver is contrasted constantly with a required space calculation. A rear-end collision-avoidance system is activated at all times adapting the velocity reference to remain a safe distance to objects ahead.

Both the collision-avoidance system and the vehicle motion controller are based on MPC. It is possible to optimize the trajectory bearing in mind safety and comfort in maneuvers. The vehicle motion controller includes a lateral position restriction aiming to improve the lane-keeping performance, being possible to enhance it on one side when lane-change maneuvers are required avoiding that the vehicle goes out the road boundaries.

As this paper was focused on presenting a fail-operational control architecture approach in case of positioning failures, future works will consider in depth the maximum time–distance travel capacity in dead-reckoning circumstances under the influence of real instrumentation and the urban scenario presented in this article.

Author Contributions: Conceptualization, J.A.M.-P. and J.P.; methodology, J.A.M.-P.; software, J.A.M.-P.; formal analysis, J.A.M.-P. and A.Z., investigation J.A.M.-P. and J.P.; writing–original draft preparation, J.A.M.-P.; writing–review and editing, J.A.M.-P. and A.Z. and J.P.; supervision, A.Z. and J.P. All authors have read and agreed to the published version of the manuscript.

Funding: This research was funded by AutoDrive within the Electronic Components and Systems for European Leadership Joint Undertaking (ECSEL JU) in collaboration with the European Union's H2020 Framework Programme (H2020/2014-2020) and National Authorities, under grant agreement number 737469.

Conflicts of Interest: The authors declare no conflict of interest.

References

1. González, D.; Pérez, J.; Milanés, V.; Nashashibi, F. A Review of Motion Planning Techniques for Automated Vehicles. *IEEE Trans. Intell. Transp. Syst.* **2016**, *17*, 1135–1145. [CrossRef]
2. Alonso Raposo, M.; Ciuffo, B.; Ardente, F.; Aurambout, J.; Baldini, G.; Braun, R.; Vandecasteele, I. *The Future of Road Transport—Implications of Automated, Connected, Low-Carbon and Shared Mobility*; Publications Office of the European Union: Rue Mercier, Luxembourg, 2019.
3. Marti, E.; de Miguel, M.A.; Garcia, F.; Perez, J. A Review of Sensor Technologies for Perception in Automated Driving. *IEEE Intell. Transp. Syst. Mag.* **2019**, *11*, 94–108. [CrossRef]
4. Llorca, D.F.; Milanés, V.; Alonso, I.P.; Gavilán, M.; Daza, I.G.; Pérez, J.; Sotelo, M.Á. Autonomous pedestrian collision avoidance using a fuzzy steering controller. *IEEE Trans. Intell. Transp. Syst.* **2011**, *12*, 390–401. [CrossRef]

5. Bernini, N.; Bertozzi, M.; Castangia, L.; Patander, M.; Sabbatelli, M. Real-time obstacle detection using stereo vision for autonomous ground vehicles: A survey. In Proceedings of the 17th International IEEE Conference on Intelligent Transportation Systems (ITSC), Qingdao, China, 8–11 October 2014; pp. 873–878.
6. Hillel, A.B.; Lerner, R.; Levi, D.; Raz, G. Recent progress in road and lane detection: a survey. *Mach. Vision Appl.* **2014**, *25*, 727–745. [CrossRef]
7. Milanés, V.; Naranjo, J.E.; González, C.; Alonso, J.; de Pedro, T. Autonomous vehicle based in cooperative GPS and inertial systems. *Robotica* **2008**, *26*, 627–633. [CrossRef]
8. Guan, H.; Yan, W.; Yu, Y.; Zhong, L.; Li, D. Robust traffic-sign detection and classification using mobile LiDAR data with digital images. *IEEE J. Sel. Top. Appl. Earth Obs. Remote Sens.* **2018**, *11*, 1715–1724. [CrossRef]
9. Castanedo, F. A review of data fusion techniques. *Sci. World J.* **2013**, *2013*, 1–19. [CrossRef] [PubMed]
10. Thrun, S.; Montemerlo, M.; Dahlkamp, H.; Stavens, D.; Aron, A.; Diebel, J.; Fong, P.; Gale, J.; Halpenny, M.; Hoffmann, G.; et al. Stanley: The robot that won the DARPA Grand Challenge. *J. Field Rob.* **2006**, *23*, 661–692. [CrossRef]
11. Balzer, P.; Trautmann, T.; Michler, O. Epe and speed adaptive extended kalman filter for vehicle position and attitude estimation with low cost gnss and imu sensors. In Proceedings of the 2014 11th International Conference on Informatics in Control, Automation and Robotics (ICINCO), Vienna, Austria, 1–3 September 2014; pp. 649–656.
12. De Ponte Müller, F. Survey on Ranging Sensors and Cooperative Techniques for Relative Positioning of Vehicles. *Sensors* **2017**, *17*, 271. [CrossRef] [PubMed]
13. Shladover, S.; Bishop, R. Road transport automation as a Public-Private Enterprise. *White Pap.* **2015**, *1*, 14–15.
14. Thorn, E.; Kimmel, S.C.; Chaka, M.; Hamilton, B.A. *A Framework for Automated Driving System Testable Cases and Scenarios*; Technical Report; United States. Department of Transportation. National Highway Traffic Safety Administration: Washington, DC, USA, 2018.
15. Emzivat, Y.; Ibanez-Guzman, J.; Martinet, P.; Roux, O.H. Dynamic driving task fallback for an automated driving system whose ability to monitor the driving environment has been compromised. In Proceedings of the 2017 IEEE Intelligent Vehicles Symposium (IV), Los Angeles, CA, USA, 11–14 June 2017; pp. 1841–1847.
16. Xue, W.; Yang, B.; Kaizuka, T.; Nakano, K. A Fallback Approach for an Automated Vehicle Encountering Sensor Failure in Monitoring Environment. In Proceedings of the 2018 IEEE Intelligent Vehicles Symposium (IV), Changshu, China, 26–30 June 2018; pp. 1807–1812.
17. Lee, J.W.; Moshchuk, N.K.; Chen, S.K. Lane Centering Fail-Safe Control Using Differential Braking. U.S. Patent 8,670,903, 11 March 2014.
18. Magdici, S.; Althoff, M. Fail-safe motion planning of autonomous vehicles. In Proceedings of the 2016 IEEE 19th International Conference on Intelligent Transportation Systems (ITSC), Rio de Janeiro, Brazil, 1–4 November 2016; pp. 452–458.
19. Ruf, M.; Ziehn, J.R.; Willersinn, D.; Rosenhahny, B.; Beyerer, J.; Gotzig, H. Global trajectory optimization on multilane roads. In Proceedings of the 2015 IEEE 18th International Conference on Intelligent Transportation Systems, Las Palmas, Spain, 15–18 September 2015; pp. 1908–1914.
20. Mudalige, U.P. Fail-Safe Speed Profiles for Cooperative Autonomous Vehicles. U.S. Patent 8,676,466, 18 March 2014.
21. An, N.; Mittag, J.; Hartenstein, H. Designing fail-safe and traffic efficient 802.11 p-based rear-end collision avoidance. In Proceedings of the 2014 IEEE Vehicular Networking Conference (VNC), Paderborn, Germany, 3–5 December 2014; pp. 9–16.
22. Isermann, R.; Schwarz, R.; Stolzl, S. Fault-tolerant drive-by-wire systems. *IEEE Ctrl. Syst. Mag.* **2002**, *22*, 64–81.
23. SAE International. *Taxonomy and Definitions for Terms Related to Driving Automation Systems for On-Road Motor Vehicles*. SAE International: Detroit, MI, USA 2018.
24. ECSEL-JU. AutoDrive. Available online: https://autodrive-project.eu/ (accessed on 28 November 2019).
25. Council, E.R.T.R.A. *Connected Automated Driving Roadmap*, Version 8; ERTRAC Working Group "Connectivity and Automated Driving": Brussels, Belgium, 2019.
26. Matute, J.A.; Marcano, M.; Zubizarreta, A.; Perez, J. Longitudinal model predictive control with comfortable speed planner. In Proceedings of the 2018 IEEE International Conference on Autonomous Robot Systems and Competitions (ICARSC), Torres Vedras, Portugal, 25–27 April 2018; pp. 60–64.

27. Kong, J.; Pfeiffer, M.; Schildbach, G.; Borrelli, F. Kinematic and dynamic vehicle models for autonomous driving control design. In Proceedings of the 2015 IEEE Intelligent Vehicles Symposium (IV), Seoul, Korea, 28 June–1 July 2015; pp. 1094–1099.
28. Rajamani, R. *Vehicle Dynamics and Control*; Springer Science & Business Media: New York, NY, USA, 2012.
29. Matute, J.A.; Marcano, M.; Diaz, S.; Perez, J. Experimental Validation of a Kinematic Bicycle Model Predictive Control with Lateral Acceleration Consideration. *IFAC-PapersOnLine* **2019**, *52*, 289–294. [CrossRef]
30. Kabzan, J.; Valls, M.d.l.I.; Reijgwart, V.; Hendrikx, H.F.C.; Ehmke, C.; Prajapat, M.; Bühler, A.; Gosala, N.; Gupta, M.; Sivanesan, R.; et al. AMZ Driverless: The Full Autonomous Racing System. *arXiv* **2019**, arXiv:1905.05150.
31. Sierra, C.; Tseng, E.; Jain, A.; Peng, H. Cornering stiffness estimation based on vehicle lateral dynamics. *Veh. Syst. Dyn.* **2006**, *44*, 24–38. [CrossRef]
32. ISO. *ISO 4138: Passenger Cars–Steady-State Circular Driving Behaviour–Open-Loop Test Methods*; ISO:Geneva, Switzerland, 2012.
33. Hsu, L.Y.; Chen, T.L. Vehicle dynamic prediction systems with on-line identification of vehicle parameters and road conditions. *Sensors* **2012**, *12*, 15778–15800. [CrossRef] [PubMed]
34. Lattarulo, R.; González, L.; Martí, E.; Matute, J.; Marcano, M.; Pérez, J. Urban motion planning framework based on n-bézier curves considering comfort and safety. *J. Adv. Transp.* **2018**, *2018*, 1–13. [CrossRef]
35. Villagra, J.; Milanés, V.; Pérez, J.; Godoy, J. Smooth path and speed planning for an automated public transport vehicle. *Rob. Autom. Syst.* **2012**, *60*, 252–265. [CrossRef]
36. Pena, A.; Iglesias, I.; Valera, J.; Martin, A. Development and validation of Dynacar RT software, a new integrated solution for design of electric and hybrid vehicles. *EVS26 Los Angeles* **2012**, *26*, 1–7.
37. Pacejka, H. *Tire and Vehicle Dynamics*; Butterworth-Heinemann: Oxford, UK, 2012.

sensors

MDPI

Article

Shadow Detection in Still Road Images Using Chrominance Properties of Shadows and Spectral Power Distribution of the Illumination

Manuel José Ibarra-Arenado [1,*], Tardi Tjahjadi [2] and Juan Pérez-Oria [3]

1 Department of Electrical and Energy Engineering, University of Cantabria, Avda. Los Castros s/n, 39005 Santander, Spain

2 School of Engineering, University of Warwick, Gibbet Hill Road, Coventry CV4 7AL, UK; t.tjahjadi@warwick.ac.uk

3 Department of Electronic Technology and Automatic Systems, University of Cantabria, Avda. Los Castros s/n, 39005 Santander, Spain; juan.perezoria@unican.es

* Correspondence: ibarramj@unican.es; Tel.: +34-942-201-360; Fax: +34-942-201-385

Received: 30 December 2019; Accepted: 10 February 2020; Published: 13 February 2020

Abstract: A well-known challenge in vision-based driver assistance systems is cast shadows on the road, which makes fundamental tasks such as road and lane detections difficult. In as much as shadow detection relies on shadow features, in this paper, we propose a set of new chrominance properties of shadows based on the skylight and sunlight contributions to the road surface chromaticity. Six constraints on shadow and non-shadowed regions are derived from these properties. The chrominance properties and the associated constraints are used as shadow features in an effective shadow detection method intended to be integrated on an onboard road detection system where the identification of cast shadows on the road is a determinant stage. Onboard systems deal with still outdoor images; thus, the approach focuses on distinguishing shadow boundaries from material changes by considering two illumination sources: sky and sun. A non-shadowed road region is illuminated by both skylight and sunlight, whereas a shadowed one is illuminated by skylight only; thus, their chromaticity varies. The shadow edge detection strategy consists of the identification of image edges separating shadowed and non-shadowed road regions. The classification is achieved by verifying whether the pixel chrominance values of regions on both sides of the image edges satisfy the six constraints. Experiments on real traffic scenes demonstrated the effectiveness of our shadow detection system in detecting shadow edges on the road and material-change edges, outperforming previous shadow detection methods based on physical features, and showing the high potential of the new chrominance properties.

Keywords: advanced driving assistance systems; illumination; shadow detection; shadow edge; road detection

1. Introduction

Increasingly powerful computers and advances in the fields of image processing and computer vision make vision-based systems one of the fastest growing segments in advanced driver assistance systems (ADAS). There are several factors that make onboard systems based on computer vision challenging. Changing scenarios, cluttered backgrounds, variable illumination, and the presence of objects of different class in the scene contribute to making the design of driver assistance tasks such as the detection of roads [1,2] and lanes [3,4] difficult. One of the most challenging factors encountered by a vision-based ADAS system is cast shadows [1,5] (see Figure 1). Shadows on a road may cause apparent merging of objects in road scenes captured by a video camera, as well as alterations in the shape and color of objects and road, which result in poor region segmentation. As a consequence, shadowed road regions can easily be misclassified as objects instead of road, which may lead to system error. Motivated by the undesirable effect of shadows, this paper presents a set of new physical properties to better characterize shadows on the road so as to minimize the possible misclassification of non-shadowed road regions, and objects as shadows. We use the new properties to design a shadow edge detection method intended to integrate a complete onboard road detection system which mainly consists of the classification of image pixels as belonging or not to the road surface [1].

Figure 1. The presence of shadows entails a difficult challenge in vision-based road detection systems.

The identification of cast shadows is not only important in vision-based ADAS systems but in general applications; thus, it is extensively studied [6–9]. Existing shadow detection approaches can be classified into two main categories [6]: model-based and property-based methods. The former is highly dependent on the environment, taking into account a priori scene information such as light source direction and geometry of the objects [10]. They are, thus, not applicable for onboard systems where no assumptions of the scene can be made. Property-based methods on the other hand, are more suitable for general applications. They are based on comparing the pixel properties of a candidate shadow region and those of a non-shadowed reference region of the same material surface. In static background applications consisting of a video sequence captured by a fixed camera [11–16], moving shadows are detected using background subtraction techniques [17–19] and comparing properties of pixels in the current frame of the sequence to background pixels in the reference frame devoid of shadows. However, such a technique is not effective for ADAS, since the road scene is continuously changing. Instead, two alternative strategies are applicable to onboard systems. One strategy focuses on comparing pixel properties between the candidate shadow region and a selected region located at the bottom of the image, which is assumed a free road region in front of the ego-vehicle [1,4,20]. However, depending on the distance to the camera, the varying reflection angles of the illumination may cause color variation of the road; thus, even a well-laid asphalted road can show zones in the image where the pixel properties are significantly different. This fact may lead to shadow misclassification when the candidate shadow region is far from the bottom of the image. The second strategy exploits locality and focuses on the comparison of pixel properties of regions across image edges [6,21,22], where the region on the darker side of an edge is a candidate shadow region and the region on the brighter side is assumed the non-shadowed reference region. As a result, an image edge is classified as an edge due to a shadow boundary or a material change. Once a shadow edge is identified, the image region on the darker side of the edge is assumed the shadow.

In order to compare the pixel properties, property-based methods rely on shadow features such as texture [23], gradients [24], histograms [25], and spectral composition [6,11,13], including luminance and chrominance. The use of spectral composition is mainly based on the assumption that shadows reduce the surface brightness without significantly modifying its chromaticity. This observation is effective for applications where the spectral power distribution of the illumination (SPD) is similar for both shadowed and non-shadowed regions; thus, the surface color components vary linearly. Approaches based on this consideration are known as color-invariant methods and they are widely used in ADAS applications exploiting different color spaces such as red–green–blue (RGB) [26,27], normalized RGB [28], Hue–Saturation–Intensity (HSI) [29,30], Hue–Saturation–Value (HSV) [11,16,31], Improved–Hue–Saturation–Luminance (IHSL) [32], YUV [33–35], c1c2c3 [6], and l1l2l3 [36]. However, in outdoor scenes, the illumination is composed of sunlight and skylight, which have different SPDs.

Non-shadowed regions are illuminated by daylight (i.e., skylight and sunlight), whereas shadowed regions are only illuminated by skylight; thus, their chromaticity varies. In order to address this issue, methods based on physical properties consider the SPD of the illumination and the surface reflectivity properties.

Early physics-based approaches were based on the observation that the intensity of each red, green, and blue (RGB) component of a surface decreases across a shadow edge [37]. This shadow feature is used by practically all methods. The bluish effect of shadows was also widely exploited. The fact that shadows are only illuminated by skylight (predominantly blue) makes the normalized blue component of a shadowed region greater than in a non-shadowed one [38]. In References [29,39], it was assumed that the blue component of shadows was dominant over the red and green. However, this assumption is not always true, since surfaces with a strong dominance of red or green may maintain their dominance when shadowed. In Reference [40], it was also observed that the red component of the sunlight was dominant; thus, the normalized red component of a shadowed region decreased. This observation is generally satisfied in the umbra of shadow but not in the penumbra, owing to it being illuminated by some sunlight. More recently, the method in Reference [12] exploited the fact that the intensity change across a shadow edge was greater in the red and green components than in the blue, whereas the method in Reference [41] presented a set of relationships between the attenuation of each RGB channel across a shadow edge and the RGB values of the non-shadowed region. The former is applicable only on low-saturated surfaces whereas the latter assumes the SPDs of skylight and sunlight as constant, which is not true, since they vary significantly during the day. A well-known shadow feature was presented in Reference [42] where shadows were identified using color ratios across image edges. It assumed that the color ratios across boundaries of shadows cast onto the different surfaces were similar since they were due to the same illumination change. Although these ratios may fail in complex real images [42], several approaches were built upon them [20,22,43,44].

A different physics-based approach to detect shadows is based on illumination invariance. In Reference [45], shadow boundaries were detected by comparing edges in the input RGB image to edges found in the one dimensional illumination-invariant shadow-free image obtained by the color-constancy method in Reference [46]. Despite the fact that this method is not reliable in images where shadow edges are not well defined [45], it was widely exploited [1,47–50]. However, most of these illumination-invariant methods require user intervention, as well as high-quality images with wide dynamic range and calibrated sensors, failing severely with consumer-quality images [44].

Generally, in order to improve robustness, most physics-based methods combine more than one shadow feature [6,12,15,44,51,52]. Currently, some of them address the shadow detection by learning techniques [13,16,21,44,51,53–56]. In Reference [21], support vector machines (SVM) were trained using color ratios to identify shadow edges in typical images. In Reference [51], an SVM classifier was trained using intensity ratio, chromatic alignment, and both color and texture histograms. A conditional random field classifier trained using color ratios and texture information was proposed in Reference [44]. This method focused on detecting shadows on the ground in consumer-quality photographs. In Reference [13], the Gaussian mixture model (GMM) was used to learn the properties of shadowed background surfaces to detect moving cast shadows. In Reference [56], two convolutional neural networks (CNNs) were combined to learn features of regions inside the shadow (umbra) and regions adjacent to the shadow boundaries (penumbra), since both shadowed regions presented different types of features. However, although learning-based methods demonstrated high robustness in specific scenarios, they are likely to fail in images slightly different to those used for their training [22].

Despite the numerous methods, shadow detection remains a very challenging task, since shadow features may be shared by objects whose chrominance features are unpredictable. Therefore, new properties are important to better characterize shadows, minimizing the misclassification of objects and non-shadowed regions.

The contributions of this paper are manifold. We firstly derive and validate the following set of new chrominance properties of shadows based on the Planckian illumination and Lambertian surface model, as well as the SPD of the illumination, to effectively characterize shadows on road:

Property 1. The relationship between the red and green surface reflectances due to sunlight is higher or equal to that due to skylight.

Property 2. The red component of the road reflectance due to sunlight is dominant, being higher than the blue, and higher or equal to the green one. In the same way, the green component of the road reflectance is higher than the blue one.

Property 3. The change in the red–green proportion of the road reflectance due to skylight and sunlight is smaller than the change in the red–blue one. This observation is also valid when comparing the changes of the surface relationships green–red and green–blue.

Associated with Property 1, we propose one constraint between the red and green surface reflectances due to sunlight and skylight. Associated with Property 2, we propose three constraints to consider the effect of sunlight on neutral surfaces. Associated with Property 3, we propose two constraints to take into account both the similarity of the red and green components of the illumination and the large variation of the blue component. These chrominance properties and constraints are utilized as shadow features in a shadow edge detection algorithm as a preprocessing stage intended to integrate a complete onboard road detection system for driver assistance. Since onboard systems deal with still images, there is not a known non-shadowed reference road region in the image to compare the pixel properties. Thus, the method focuses on detecting shadow boundaries by comparing pixel properties across image edges. No prior knowledge of the scene, camera calibration, or spatio-temporal restrictions are required, and static background applications can also be addressed. The method identifies image edges delimiting shadows and non-shadowed road regions by verifying whether the pixel values of regions on both sides of the edge under analysis satisfy the new constraints imposed.

The remainder of this paper is organized as follows: Section 2 presents the reflection model, as well as discusses the SPDs of skylight and sunlight, including their effects on shadowed and non-shadowed asphalt road surface. Section 3 presents the new chrominance properties of shadows, and Section 4 describes the proposed shadow edge detection method. Experimental results are shown and discussed in Section 5. Finally, Section 6 concludes the paper.

2. Physics Basis: Reflection Model and SPD of the Illumination

2.1. *Reflection Model*

Assuming a Planckian illumination and Lambertian surface model as in References [6,22,45,57], the light reflected off a point p on a surface is the product of the SPD of the incident illumination $E(\lambda, p)$ and the surface reflectance $S(\lambda, p)$. Thus, for some illumination and viewing geometry, the response of a digital camera sensor C_i at a given image pixel (x, y), which corresponds to a surface point p of the scene, can be expressed as in References [6,21,22,45,46,57],

$$C_i(x,y) = \int_w E(\lambda,x,y) \times S(\lambda,x,y) \times Q_i(\lambda) \times d\lambda, \tag{1}$$

where $C_i \in \{R, G, B\}$ are the red, green, and blue sensor responses, λ is the wavelength, w is the visible spectrum range, and $Q_i(\lambda) \in \{Q_R(\lambda), Q_G(\lambda), Q_B(\lambda)\}$ are the spectral sensitivities of the three color camera sensors. Assuming camera filters of infinitely narrow bandwidth as in References [11,21,22,40,42,45,58], it is possible to represent them by impulse functions which are centered on the filter's characteristics (e.g., Dirac delta function: $Q_i(\lambda) = q_i \cdot \delta(\lambda - \lambda_i)$ [21,22,45]). With this approximation, Equation (1) becomes

$$C_i(x,y) = E(\lambda_i, x, y) \times S(\lambda_i, x, y) \times q_i, \tag{2}$$

where λ_i is the center frequency of the i-th channel filter, and $q_i \epsilon \{q_R, q_G, q_B\}$ are spectral sensitivity factors of the three color camera sensors. Equation (2) represents the three color components of the reflected light due to a single illumination source. However, in outdoor scenes, the illumination is due to the contribution of two illumination sources, sunlight ($E_{sun}(\lambda_i)$) and skylight ($E_{sky}(\lambda_i)$), with different SPDs. In line with References [6,12,41,47], the inter-reflections due to nearby objects can be disregarded, since the energy of inter-reflection decays exponentially for each reflection [12]. Therefore, the sensor measurement for an image pixel (x, y) corresponding to a non-shadowed surface point of the scene is

$$C_i(x,y)_{non-sha} = \left[E_{sky}(\lambda_i, x, y) + E_{sun}(\lambda_i, x, y)\right] \times S(\lambda_i, x, y) \times q_i, \tag{3}$$

giving a three-dimensional (3D) color vector $C_i(x, y)_{non\text{-}sha} = [R_{non\text{-}sha}, G_{non\text{-}sha}, B_{non\text{-}sha}]$. The response for a pixel in the shade is obtained from Equation (3) making $E_{sun}(\lambda_i, x, y) = 0$, i.e.,

$$C_i(x,y)_{sha} = E_{sky}(\lambda_i, x, y) \times S(\lambda_i, x, y) \times q_i, \tag{4}$$

giving a color vector $C_i(x, y)_{sha} = [R_{sha}, G_{sha}, B_{sha}]$.

2.2. *SPD of the Illumination in Outdoor Scenes*

The sun emits white light, which penetrates the atmosphere and is scattered in all directions by gas molecules in the air. However, due to the small size of the molecules, the scattering (Rayleigh scattering) is more effective at short wavelengths which correspond to blue, thus giving the sky a bluish color. On the other hand, most of the light comprising the remaining wavelengths (from green to red) passes through the atmosphere and reaches the earth surface. The mixture of red and green produces a yellowish sunlight, which may attain a reddish tone at certain hours of the day.

2.2.1. SPD of Skylight

A shadow on a road appears when an object occludes the sunlight; thus, only the bluish skylight illuminates the road. Although the intensity of the skylight can vary depending on the time of day and atmospheric conditions, during most parts of the day, the red and green components ($E_{sky}(\lambda_R)$, $E_{sky}(\lambda_G)$) present similar values, combining to give a higher blue component ($E_{sky}(\lambda_B)$). However, as the sun gets lower in the sky, the sunlight passes through more of the atmosphere, which produces an increase in the scattering of its green wavelength. Hence, the green component of the skylight increases relative to red. This intensity difference between the red and green components is generally small and does not cause a significant change in the appearance of the sky (a greenish skylight is not usual). Thus, depending on the time of day when skylight illuminates a road, the green component of the light reflected from the road surface can be considered to be affected by a similar or a slightly higher quantity than the red, whereas the blue is affected by a larger quantity, i.e.,

$$\begin{aligned} E_{sky}(\lambda_R) &\leq E_{sky}(\lambda_G), \\ E_{sky}(\lambda_R) &< E_{sky}(\lambda_B), \\ E_{sky}(\lambda_G) &< E_{sky}(\lambda_B), \end{aligned} \tag{5}$$

2.2.2. SPD of Sunlight

A non-shadowed road region is illuminated by both skylight and sunlight. During most part of the day, the sunlight is yellowish since its red and green components ($E_{sun}(\lambda_R)$, $E_{sun}(\lambda_G)$) remain very similar with respect to each other, combining to give a smaller blue component ($E_{sun}(\lambda_B)$).

The red–green equilibrium is attained at noon. As the sun gets lower in the sky, the green component of the sunlight decreases relative to red, making the sunlight orangish until sunset when it may attain reddish. From sunrise to noon the process is similar but reversed, from reddish to yellowish. Thus, depending on the time of day when sunlight illuminates a road, the red component of the light reflected from the road surface can be considered to be affected by a similar or a higher quantity of light than the green, whereas the blue intensity of sunlight is the lowest, i.e.,

$$\begin{aligned} E_{sun}(\lambda_R) &\geq E_{sun}(\lambda_G), \\ E_{sun}(\lambda_R) &> E_{sun}(\lambda_B), \\ E_{sun}(\lambda_G) &> E_{sun}(\lambda_B), \end{aligned} \tag{6}$$

Based on the reflection model and both skylight and sunlight contributions to the surface chromaticity, we propose the above set of three chrominance properties of shadows.

3. New Shadow Features

According to the reflection model, when comparing a pixel in the shadowed region (x_{sha}, y_{sha}) to a non-shadowed one $(x_{non\text{-}sha}, y_{non\text{-}sha})$ of the same material surface in an image, the contribution of sunlight on $(x_{non\text{-}sha}, y_{non\text{-}sha})$ according to the reflection model is given by

$$C_i(x_{non-sha}, y_{non-sha})_{sun} = C_i(x_{non-sha}, y_{non-sha}) - C_i(x_{sha}, y_{sha}), \tag{7}$$

giving a color vector $C_i(x_{non\text{-}sha}, y_{non\text{-}sha})sun = [R_{sun}, G_{sun}, B_{sun}]$.

Property 1. Considering the red and green components of skylight and sunlight, a relationship between a shadowed region and a non-shadowed one of the same material surface is proposed by taking into account the different components dominating the illumination. From Equation (4), the red and green components of a pixel in a shadowed region are respectively

$$\begin{aligned} R_{sha} &= E_{sky}(\lambda_R) \times S(\lambda_R) \times q_R, \\ G_{sha} &= E_{sky}(\lambda_G) \times S(\lambda_G) \times q_G, \end{aligned} \tag{8}$$

By taking a ratio of the two components, the red and green surface reflectances are related by

$$\frac{R_{sha}}{G_{sha}} = \frac{E_{sky}(\lambda_R) \times S(\lambda_R) \times q_R}{E_{sky}(\lambda_G) \times S(\lambda_G) \times q_G,} \Rightarrow \frac{S(\lambda_R) \times q_R}{S(\lambda_G) \times q_G} = \frac{E_{sky}(\lambda_G)}{E_{sky}(\lambda_R)} \times \frac{R_{sha}}{G_{sha}}. \tag{9}$$

Similarly, the contribution of sunlight on the red and green components of the non-shadowed surface is obtained from Equation (3) by making $E_{sky}(\lambda_i, x, y) = 0$, i.e.,

$$\begin{aligned} R_{sun} &= E_{sun}(\lambda_R) \times S(\lambda_R) \times q_R, \\ G_{sun} &= E_{sun}(\lambda_G) \times S(\lambda_G) \times q_G. \end{aligned} \tag{10}$$

Taking a ratio of the two components gives

$$\frac{R_{sun}}{G_{sun}} = \frac{E_{sun}(\lambda_R) \times S(\lambda_R) \times q_R}{E_{sun}(\lambda_G) \times S(\lambda_G) \times q_G} \Rightarrow \frac{S(\lambda_R) \times q_R}{S(\lambda_G) \times q_G} = \frac{E_{sun}(\lambda_G)}{E_{sun}(\lambda_R)} \times \frac{R_{sun}}{G_{sun}}. \tag{11}$$

Equating Equations (9) and (11) yields

$$\frac{E_{sky}(\lambda_G)}{E_{sky}(\lambda_R)} \times \frac{R_{sha}}{G_{sha}} = \frac{E_{sun}(\lambda_G)}{E_{sun}(\lambda_R)} \times \frac{R_{sun}}{G_{sun}} \Rightarrow \frac{E_{sky}(\lambda_G)}{E_{sky}(\lambda_R)} \times \frac{E_{sun}(\lambda_R)}{E_{sun}(\lambda_G)} = \frac{G_{sha}}{R_{sha}} \times \frac{R_{sun}}{G_{sun}}. \tag{12}$$

According to Equation (5), the green component of the skylight is generally equal to or slightly higher than the red one, whereas, according to Equation (6), the red component of the sunlight is generally equal to or higher than the green; thus, the left-hand side of Equation (12) satisfies

$$\frac{E_{sky}(\lambda_G)}{E_{sky}(\lambda_R)} \times \frac{E_{sun}(\lambda_R)}{E_{sun}(\lambda_G)} \geq 1. \tag{13}$$

According to Equations (12) and (13), when comparing a pixel in the shadow to a non-shadowed pixel of the same material surface, the following constraint is satisfied:

$$\frac{G_{sha}}{R_{sha}} \times \frac{R_{sun}}{G_{sun}} \geq 1, \tag{14}$$

where $R_{sun} = R(x_{non\text{-}sha}, y_{non\text{-}sha}) - R(x_{sha}, y_{sha})$ and $G_{sun} = G(x_{non\text{-}sha}, y_{non\text{-}sha}) - G(x_{sha}, y_{sha})$. This equation shows that the relationship between the red and green surface reflectances due to sunlight is equal to or higher than the red–green one due to skylight.

The first row of Figure 2 illustrates two representative traffic scenes of our dataset, where a region of interest (ROI) focusing on the road is overlaid onto the images. The second row of Figure 2 shows the edges (in green) of the ROI image, as well as the two regions across them. The darker region (in blue) of each edge is candidate to be a shadow, whereas the brighter one (in red) is assumed the non-shadowed reference region. The method to determine the edges, the darker and brighter regions, and the surface reflectance values of the regions is described in detail in Section 4. Figure 3 illustrates the detected shadow obtained using the constraint of Property 1, i.e., Equation (14). Those edges where the surface reflectances corresponding to the dark ($C_i(x_{sha}, y_{sha}) = [R_{sha}, G_{sha}, B_{sha}]$) and bright ($C_i(x_{non\text{-}sha}, y_{non\text{-}sha}) = [R_{non\text{-}sha}, G_{non\text{-}sha}, B_{non\text{-}sha}$) regions satisfy Equation (14) are classified as shadow edges (in red). Otherwise, the edges are classified as edges due to a material change and are removed.

Figure 2. (**First row**) Bounding box containing the region of interest (ROI) of two input red–green–blue (RGB) images of our dataset. (**Second row**) Darker (in blue) and brighter (in red) regions on both sides of each edge (in green) of the ROI image.

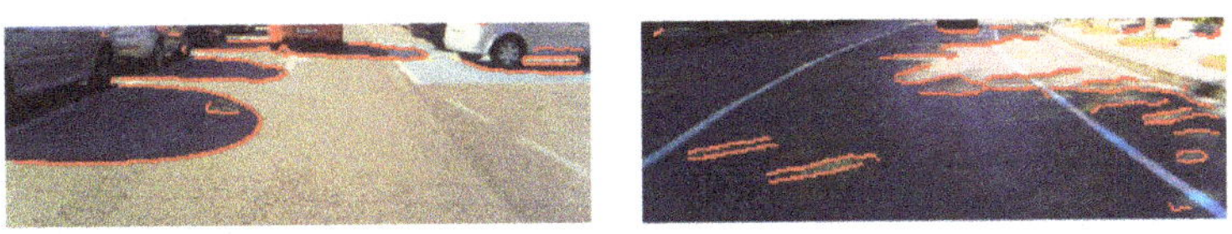

Figure 3. Shadow edge detection of images in Figure 2 obtained using Property 1, i.e., Equation (14).

Property 2. Three constraints are introduced to consider the effect of sunlight on neutral surfaces. Asphalt roads are generally neutral surfaces, which present similar reflectance for each component ($S(\lambda_R) \approx S(\lambda_B) \approx S(\lambda_G)$); thus, their RGB distribution is practically proportional to the SPD of the incident illumination. This implies that the reflectance components of a non-shadowed neutral surface due to the sunlight contribution are proportional to the red, green, and blue components of sunlight. The RGB reflectance components of a non-shadowed pixel due to sunlight are obtained from Equation (3) by making $E_{sky}(\lambda_i, x, y) = 0$. Taking the ratios of two components gives

$$\begin{array}{l} \frac{R_{sun}}{G_{sun}} = \frac{E_{sun}(\lambda_R)\times S(\lambda_R)\times q_R}{E_{sun}(\lambda_G)\times S(\lambda_G)\times q_G}, \\ \frac{R_{sun}}{B_{sun}} = \frac{E_{sun}(\lambda_R)\times S(\lambda_R)\times q_R}{E_{sun}(\lambda_B)\times S(\lambda_B)\times q_B}, \\ \frac{G_{sun}}{B_{sun}} = \frac{E_{sun}(\lambda_G)\times S(\lambda_G)\times q_G}{E_{sun}(\lambda_B)\times S(\lambda_B)\times q_B}, \end{array} \tag{15}$$

Assuming $S(\lambda_R) = S(\lambda_B) = S(\lambda_G)$ and considering, in line with References [6,12,41,47], similar spectral sensitivity constants of the three color camera sensors (i.e., $q_R = q_G = q_B$), Equation (15) becomes

$$\frac{R_{sun}}{G_{sun}} = \frac{E_{sun}(\lambda_R)}{E_{sun}(\lambda_G)}, \ \frac{R_{sun}}{B_{sun}} = \frac{E_{sun}(\lambda_R)}{E_{sun}(\lambda_B)}, \ \frac{G_{sun}}{B_{sun}} = \frac{E_{sun}(\lambda_G)}{E_{sun}(\lambda_B)} \tag{16}$$

Thus, taking into account the SPD of the yellow sunlight which presents a red component higher or equal to the green (depending on the time of day) and higher than the blue, as well as a green component higher than the blue (i.e., Equation (6)), the RGB reflectances of the non-shadowed surface due to sunlight contribution satisfy the three constraints

$$\frac{R_{sun}}{G_{sun}} \geq 1, \ \frac{R_{sun}}{B_{sun}} > 1, \ \frac{G_{sun}}{B_{sun}} > 1\,. \tag{17}$$

The effect of illumination on neutral surfaces was first considered in Reference [39], where the focus was on the skylight contribution to the shadowed road regions. In that case, the blue component of shadows was dominant over the red and green ones; thus, the constraints $R_{sha} > G_{sha}$ and $R_{sha} > B_{sha}$ were satisfied in the umbra of shadows, which was illuminated only by skylight. However, the dominance of the blue light in penumbras was not so strong because they were also illuminated by some amount of sunlight. Thus, even low-saturated surfaces such as asphalt road surface may maintain their dominant color component when softly shadowed. However, the constraints in Equation (17) are also satisfied when comparing penumbras to non-shadowed road, since they do not focus on the skylight contribution to the shadowed road surface but on the sunlight contribution to the non-shadowed one. Figure 4 illustrates the shadow detection of the images in Figure 2 after applying the three constraints associated with Property 2, i.e., Equation (17).

Figure 4. Shadow edge detection obtained by the three constraints in Property 2.

Property 3. A set of two relationships is introduced to take into account both the similarity of the red and green components of the illumination and the large variation of the blue component. For both skylight and sunlight, the red and green intensities are close to each other. Thus, the relationship between the red and green components reflected off the surface illuminated by skylight is close to that due to sunlight. However, the red and blue intensities are very different in magnitude and sign (i.e., $E_{sky}(\lambda_R) < E_{sky}(\lambda_B)$, $E_{sun}(\lambda_R) \geq E_{sun}(\lambda_G)$). Thus, the relationship red–blue of the surface illuminated by skylight is significantly different from that due to sunlight.

Focusing on neutral surfaces and assuming $S(\lambda_R) = S(\lambda_B) = S(\lambda_G)$, as well as $q_R = q_G = q_B$ as in Property 2, the red component proportion of the surface related to the green, i.e., rg, and to the blue, i.e., rb, can be expressed as

$$\begin{aligned} rg &= \frac{R}{R+G} = \frac{E(\lambda_R)\times S(\lambda_R)\times q_R}{E(\lambda_R)\times S(\lambda_R)\times q_R + E(\lambda_G)\times S(\lambda_G)\times q_G} = \frac{E(\lambda_R)}{E(\lambda_R)+E(\lambda_G)}, \\ rb &= \frac{R}{R+B} = \frac{E(\lambda_R)\times S(\lambda_R)\times q_R}{E(\lambda_R)\times S(\lambda_R)\times q_R + E(\lambda_B)\times S(\lambda_B)\times q_B} = \frac{E(\lambda_R)}{E(\lambda_R)+E(\lambda_B)}. \end{aligned} \tag{18}$$

When comparing a shadowed region to a non-shadowed one of the same material surface, the difference in the red–green and red–blue proportions due to skylight and sunlight are respectively

$$\begin{aligned} \left|rg_{sha} - rg_{sun}\right| &= \left|\frac{R_{sky}}{R_{sky}+G_{sky}} - \frac{R_{sun}}{R_{sun}+G_{sun}}\right| = \left|\frac{E_{sky}(\lambda_R)}{E_{sky}(\lambda_R)+E_{sky}(\lambda_G)} - \frac{E_{sun}(\lambda_R)}{E_{sun}(\lambda_R)+E_{sun}(\lambda_G)}\right|, \\ \left|rb_{sha} - rb_{sun}\right| &= \left|\frac{R_{sky}}{R_{sky}+B_{sky}} - \frac{R_{sun}}{R_{sun}+B_{sun}}\right| = \left|\frac{E_{sky}(\lambda_R)}{E_{sky}(\lambda_R)+E_{sky}(\lambda_B)} - \frac{E_{sun}(\lambda_R)}{E_{sun}(\lambda_R)+E_{sun}(\lambda_B)}\right|. \end{aligned} \tag{19}$$

According to Equation (5), the green component of the skylight is smaller than the blue one (i.e., $E_{sky}(\lambda_G) < E_{sky}(\lambda_B)$), whereas, from Equation (6), the green component of the sunlight is higher than the blue (i.e., $E_{sun}(\lambda_G) > E_{sun}(\lambda_B)$); thus,

$$\begin{aligned} \frac{E_{sky}(\lambda_R)}{E_{sky}(\lambda_R)+E_{sky}(\lambda_G)} &> \frac{E_{sky}(\lambda_R)}{E_{sky}(\lambda_R)+E_{sky}(\lambda_B)} \Rightarrow rg_{sha} > rb_{sha}, \\ \frac{E_{sun}(\lambda_R)}{E_{sun}(\lambda_R)+E_{sun}(\lambda_B)} &< \frac{E_{sun}(\lambda_R)}{E_{sun}(\lambda_R)+E_{sun}(\lambda_B)} \Rightarrow rg_{sun} < rb_{sun}. \end{aligned} \tag{20}$$

This implies that the change in the red–green proportion due to skylight and sunlight is smaller than the change in the red–blue one, i.e., the following first constraint of Property 3:

$$\left|rg_{sha} - rg_{sun}\right| < \left|rb_{sha} - rb_{sun}\right| \Rightarrow \frac{\left|rg_{sha} - rg_{sun}\right|}{\left|rb_{sha} - rb_{sun}\right|} < 1. \tag{21}$$

This reasoning is also valid when comparing the changes of the surface relationships green–red and green–blue. From Equation (5), the red component of the skylight is smaller than the blue (i.e., $E_{sky}(\lambda_R) < E_{sky}(\lambda_B)$), whereas, from Equation (6), the red component of the sunlight is higher than the blue (i.e., $E_{sun}(\lambda_R) > E_{sun}(\lambda_B)$); thus, $gr_{sha} > gb_{sha}$ and $gr_{sun} < gb_{sun}$. Therefore, the change in the green–red proportion due to skylight and sunlight is smaller than the change in the green–blue one, i.e., the following second constraint of Property 3:

$$\left|gr_{sha} - gr_{sun}\right| < \left|gb_{sha} - gb_{sun}\right| \Rightarrow \frac{\left|gr_{sha} - gr_{sun}\right|}{\left|gb_{sha} - gb_{sun}\right|} < 1. \tag{22}$$

For simplicity, Equations (21) and (22) are obtained considering neutral surface conditions (i.e., $S(\lambda_R) = S(\lambda_B) = S(\lambda_G)$); thus, they are especially applicable to asphalt roads which are generally colorless surfaces with similar reflectance for each component. Figure 5 illustrates the shadow edge detection for both scenes in Figure 2 using the constraint associated with Property 3, which relates the red–green and red–blue proportions of the surface, i.e., Equation (21). Figure 6 shows, on the other hand, the shadow edge detection using the constraint associated with Property 3 relating the green–red and green–blue proportions, i.e., Equation (22).

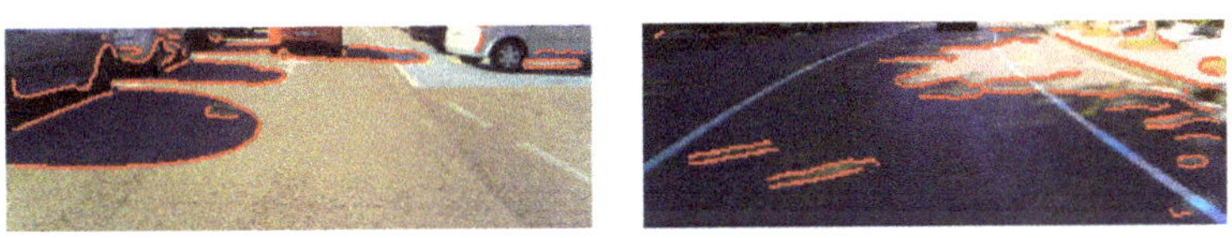

Figure 5. Shadow edge detection obtained by the constraint which relates the red–green and the red–blue proportions of Property 3, i.e., Equation (21).

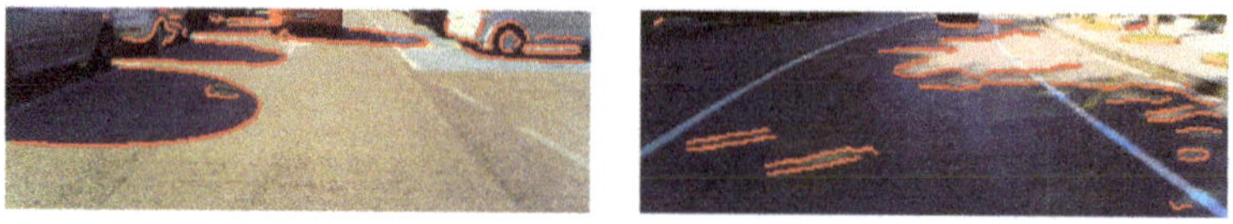

Figure 6. Shadow edge detection obtained by the constraint which relates the green–red and the green–blue proportions of Property 3, i.e., Equation (22).

Figure 7 illustrates the shadow edge detection after applying the proposed three properties, i.e., Equations (14), (17), (21) and (22). As can be observed, the accumulation of shadow feature constraints contributes to a better characterization of shadows since the possibility of errors in the classification of edges due to a material change decreases.

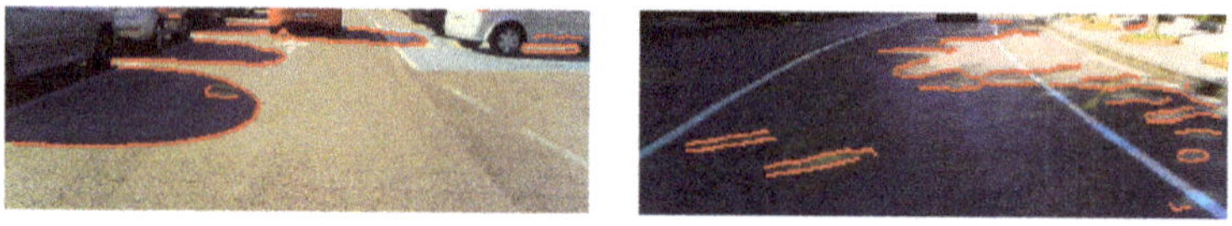

Figure 7. Shadow edge detection results obtained after applying the three properties all together.

4. Shadow Edge Detection Method

On the basis of the new chrominance properties of shadows, a shadow detection method for onboard road detection is proposed. As long as onboard systems deal with still images, there is not a known non-shadowed reference road region in the incoming image to compare the pixel properties. Thus, the shadow detection method focuses on comparing pixel properties across image edges, where the darker region of an image edge is the candidate shadow region and the brighter one is assumed to be the non-shadowed reference region. The method comprises four main stages: extraction of the image edges, selection of the bright and dark regions across the edges, extraction of the strong edges, and shadow edge classification.

In the image edge extraction stage, we break T- and X-junctions that connect different edges, thus obtaining an edge map consisting of individual edges. In order to achieve robustness in the edge classification, we exploit regions across each edge instead of single pixels. Thus, in line with Reference [22], we use two regions of pixels along both sides of the edges to compute the chrominance properties of the surface. Prior to edge classification, an intensity filtering is applied to eliminate noisy edges on the asphalt, thus retaining only the strong ones in the image. Finally, edge classification is carried out by verifying whether the regions on both sides of the strong edges satisfy the six constraints associated with the proposed three chrominance properties of shadow, thus classifying each image edge as a shadow edge or a material-change edge.

Since the method addresses the detection of shadow edges on the road, in order to simplify a captured road scene, as well as reduce the number of false positive detections outside the road surface, an ROI in the incoming color images is defined on the road by using knowledge of the scene perspective and assuming flat road surface as in References [4,59]. The camera is installed beside the rear-view mirror of the ego-vehicle, and the ROI is a rectangular area covering the road region ahead, excluding most of the image areas which do not correspond with the ground (see Figure 8).

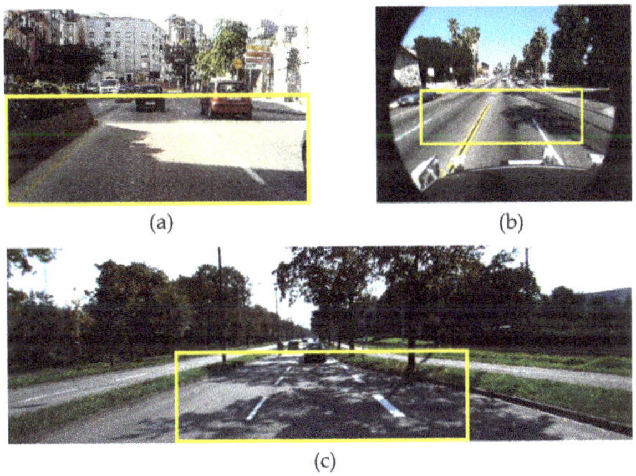

Figure 8. Bounding box containing the ROI in the incoming color images. (**a**) For our 240 × 320 camera, the ROI covers 110 × 320 pixels. (**b**) For the Caltech Lane dataset containing 480 × 640 images, the ROI considered covers 130 × 410 pixels. (**c**) For the Kitti dataset containing 375 × 1242 images, the ROI considered covers 170 × 574 pixels.

4.1. Extraction of the Image Edges

After an averaging low-pass filtering to reduce image noise, the edges in the ROI of the incoming RGB image are extracted by applying the Canny operator [60] owing to its robustness. The resulting edge map consists of edges due to both shadow boundaries and material changes (see Figure 9a). However, a troublesome effect of the edge extraction is the generation of T- and X-junctions, which affect the shadow edge classification since they connect different edges (see Figure 9b). The edge classification requires separating edges into two regions only; thus, individual edges are generated by removing X- and T-junctions. To this end, the edge map is scanned bottom-up and a 3 × 3 kernel centered on each edge pixel is matched with a total of 18 T- and X-masks, as shown in Figure 10. For a positive match, a junction is broken by removing from the edge map the pixels involved in the junction (see Figure 9c). The result is an edge map consisting of individual edges that only separate two regions of the image (see Figure 9d, where the edges are in green).

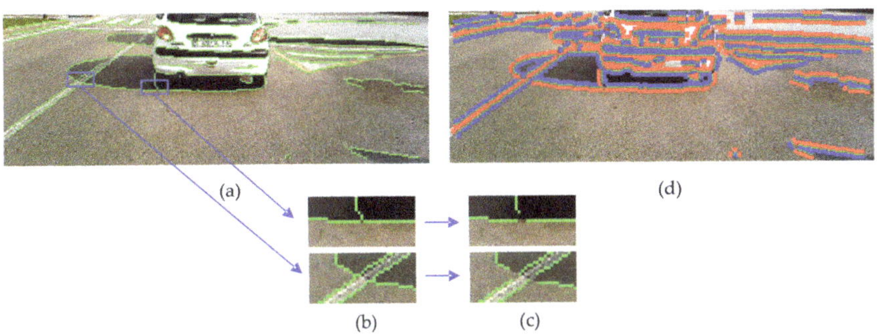

Figure 9. (**a**) Canny edge map (in green) overlaid on the ROI of an incoming image of our dataset. (**b**) Two examples of T-junction connecting different edges. (**c**) Individual edges after removing T-junctions. (**d**) Brighter (in red) and darker (in blue) regions across edges (in green) after T- and X-junction removal.

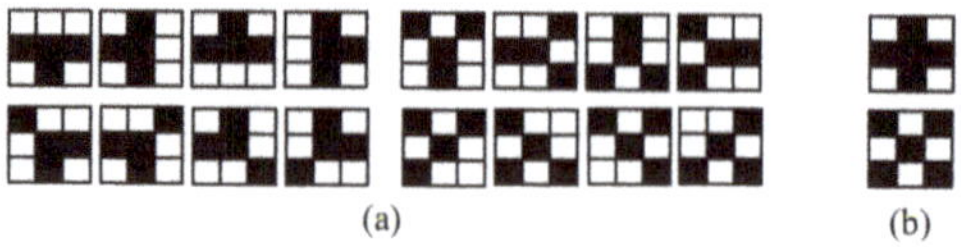

Figure 10. Masks used for the detection of (a) T-junction and (b) X- junction.

4.2. Extraction of the Bright and Dark Regions across the Edges

Since pixel-based methods that use information of a pixel or a small neighborhood around a pixel are prone to noise [44], we impose spatial consistency which employs a higher-level neighborhood on both sides of each edge. To this end, we compute the gradient orientation of each pixel of the edge and consider three pixels along this orientation and on both sides. This operation is performed for all the pixels of the edge, thus obtaining two different regions across each edge. The reflectance components of the darker (R_{sha}, G_{sha}, B_{sha}) and brighter ($R_{non\text{-}sha}$, $G_{non\text{-}sha}$, $B_{non\text{-}sha}$) regions are computed as the mean pixel values of each region as in Reference [22]. The region with the larger mean pixel values (the brighter) is assumed the non-shadowed reference region of the edge, whereas the region with the smaller values (the darker) is assumed the candidate shadow region. Figure 9d illustrates both regions across each edge of the image, where the blue areas represent the darker regions and the red areas represent the brighter ones. If a pixel of a region is also an edge pixel of a different boundary of the image, it is not included in the mean value computation. This may happen with edges three or fewer pixels away.

4.3. Extraction of Strong Edges

Asphalt roads are generally textured surfaces that usually generate noisy edges in the image. The regions on both sides of a noisy edge fall onto the road; thus, the intensity difference between them is generally small. Moreover, depending on the illumination and type of asphalt, the intensity difference between a shadowed road region and a non-shadowed one vary. However, their difference is generally significant. In order to discard a noisy edge on the road, a filtering strategy based on the intensity difference between both sides of the edge is proposed. An image edge is removed from the edge map if the intensity difference between the regions on both sides of the edge I_{sun} is smaller than the 20% of the intensity of the darker region I_{sha}, i.e.,

$$I_{sun} < 0.2 \times I_{sha}, \tag{23}$$

where $I_{sun} = I_{non\text{-}sha} - I_{sha}$, $I_{non\text{-}sha} = (R_{non\text{-}sha} + G_{non\text{-}sha} + B_{non\text{-}sha})/3$ and $I_{sha} = (R_{sha} + G_{sha} + B_{sha})/3$. The choice of 20% of the shadowed region intensity is conservative because the aim of the filter is not to identify shadow edges but discard those whose small intensity difference does not clearly correspond to the intensity difference across a shadow edge (see Figure 11).

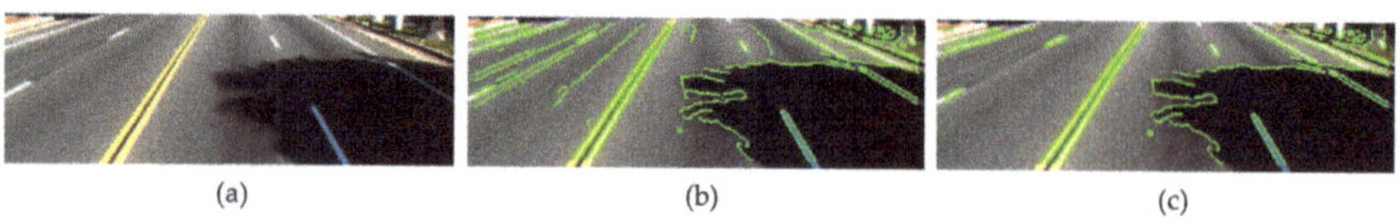

Figure 11. (a) ROI of an input image of the Caltech Lane dataset. (b) ROI overlaid with Canny edge map. (c) Enhanced edge map D after applying the intensity filter, i.e., Equation (23).

4.4. Shadow Edge Classification

Edge classification is the final stage of the method, where each individual edge D_k is classified as a shadow edge if the reflectance components of its darker (R_{sha}, G_{sha}, B_{sha}) and brighter ($R_{non\text{-}sha}$, $G_{non\text{-}sha}$, $B_{non\text{-}sha}$) regions satisfy the six constraints associated with the three chrominance properties of shadow, i.e., the six chrominance constraints in Equations (14), (17), (21) and (22). Otherwise, if one of the constraints is not satisfied, then the edge is classified as an edge due to a material change, i.e.,

$$D_k = \begin{cases} \text{if} & \begin{matrix} property\ 1 \\ \text{and } property\ 2 \\ \text{and } property\ 3, \end{matrix} & \text{Shadow Edge,} \\ \text{otherwise,} & & \text{Material Change Edge.} \end{cases} \tag{24}$$

5. Experimental Results

We firstly discuss the individual performance of each of the proposed three chrominance properties and then the complete shadow edge detection method. In addition, we compare the proposed method with four state-of-the-art shadow detection methods. In our experiments, we used image sequences acquired using an onboard camera which provided 240 × 320 color image frames with an 8-bit pixel depth. A total of 6600 road images in 22 sets of 300 frames were acquired in real traffic. The data consist of urban traffic scenes in the presence of a variety of cast shadows on the road and scenes which do not contain shadows. We also used the publicly available Caltech Lane dataset [61] for driving assistance systems, which consists of 1225 PNG road images of 480 × 640, and the Kitti Road dataset [62] for road and lane detection, which includes 579 PNG images of 175 × 1242 pixels captured under different illumination.

5.1. Individual Performance of the Proposed Shadow Properties

5.1.1. Qualitative Results

Figure 12 illustrates two representative traffic scenes of the Caltech Lane dataset (left and middle images) and one scene of the Kitti Road dataset (right image), which cover a selection of different types of shadows. The image on the left contains a very dark shadow caused by the traffic light which generates well-defined shadow edges on the road. The image in the middle contains shadows caused by the branches of the palm tree which generate soft shadow boundaries. In the image on the right, the trees cause shadows with both well-defined and soft edges on the road.

Figure 12. (**Row 1**) input images; (**Row 2**) ROI of the input images; (**Row 3**) ROI overlaid with brighter (red) and darker (blue) regions across strong edges (green).

Property 1. Figure 13 shows the shadow edge detection results using the constraint associated with Property 1, i.e., Equation (14). As can be observed in the three images, both the well-defined and the soft shadow boundaries are successfully identified as shadow edges, whereas most of the edges caused by material changes such as those caused by lane markings and curbs are also correctly classified and removed from shadow edge map. In the three images, the number of misclassified edges due to material changes is small, and they occur in image regions outside the road. In the right side of the middle image, the material-change edge caused by the grass and the sidewalk is misclassified as shadow edge and thus retained in the shadow edge map. Note that the Canny edge detector provides one-pixel-thick edges; however, the thickness of the edges in the images was increased for a better visualization.

Figure 13. Shadow edge detection results of images in Figure 12 obtained by Property 1.

Property 2. Figure 14 shows the shadow edge detection results using the three constraints associated with Property 2, i.e., Equation (17). As can be observed in the three images, Property 2 also demonstrates effectiveness in the detection of both well-defined and soft shadow edges. However, despite of the fact that the number of misclassified boundaries due to material changes is also small, there are some errors in the classification of strong noise edges on the asphalt road that remained after the intensity filtering. In the middle image, two noisy edges on the asphalt are falsely classified as shadow edges. In fact, they are not material change edges since they do not separate two different materials but two regions with different reflectance of a same surface. On the other hand, in the middle image, it can be observed that the material-change edge caused by the grass and the sidewalk is successfully identified as a material-change edge.

Figure 14. Shadow edge detection results of images in Figure 12 obtained by Property 2.

Property 3. Figure 15 shows the shadow edge detection results using both constraints associated with Property 3, i.e., Equations (21) and (22). The ability to detect both well-defined and soft shadow edges is clearly shown. However, the effectiveness in classifying material-change edges decreases when compared with using Properties 1 and 2. Some edges on the road region due to lane markings and curbs, as well as material-change edges outside the road, are misclassified as shadow edges.

Figure 15. Shadow edge detection results of images in Figure 12 obtained by Property 3.

5.1.2. Quantitative Results

In order to quantitatively evaluate the performance of the proposed three chrominance properties, we compute the commonly used metrics of precision, recall, and F-measure, i.e.,

$$\text{Precision} = \frac{TP}{TP + FN}, \quad \text{Recall} = \frac{TP}{TP + FP}, \quad \text{F} - \text{measure} = 2 \times \frac{\text{Precision} \times \text{Recall}}{\text{Precision} + \text{Recall}}, \tag{25}$$

where *TP* (true positive) is the number of pixels correctly detected as shadow edges, *FP* (false positive) is the number of pixels due to material changes misclassified as shadow edges, and *FN* (false negative) is the number of pixels due to shadow edges misclassified as material-change edges. Higher values of precision, recall, and F-measure denote better results. The evaluation was performed on 300 images consisting of 100 images of the Caltech Lane dataset, 100 images of the Kitti Road dataset, and 100 images of our dataset. The set of images for evaluation includes road scenes captured under different illumination in the presence of a variety of cast shadows on the road, as well as scenes which do not contain shadows. *TP*, *FP*, and *FN* are determined by a pixel-wise comparison between the resulting shadow edge map obtained using the property under evaluation and the ground-truth shadow edge map manually extracted (see Figure 16).

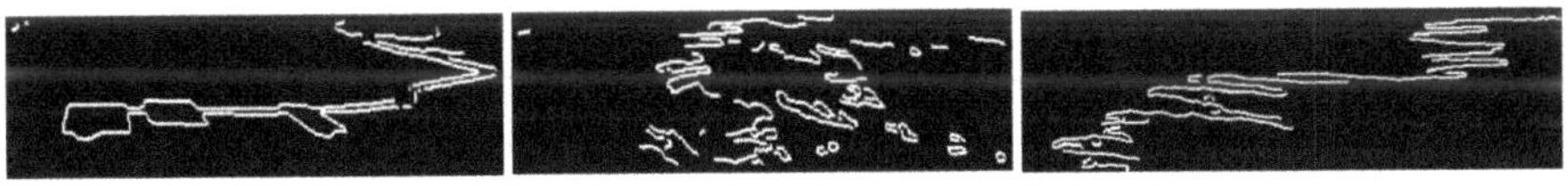

Figure 16. Ground-truth shadow edge maps of images in Figure 12.

Table 1 shows the metrics of each of the properties evaluated on each of the 300 images. The table shows that the precision of the three properties on each dataset is high, achieving values of 0.950, 0.959, and 0.932, respectively. This indicates high effectiveness of each property in the classification of shadow edges, which in turn implies a high number of true positives together with a low number of false negatives.

Table 1. Precision (*P*), recall (*R*), and F-measure (*F-m*) indicators.

	Caltech Dataset (100 Images)			Kitti Dataset (100 Images)			Our Dataset (100 Images)			All Datasets (300 Images)		
	P	*R*	*F-m*	*P*	*R*	*F-m*	*P*	*R*	*F-m*	*P*	*R*	*F-m*
Property 1	0.964	0.791	0.869	0.960	0.727	0.827	0.913	0.672	0.774	0.950	0.730	0.826
Property 2	0.968	0.795	0.873	0.963	0.815	0.883	0.931	0.518	0.666	0.959	0.737	0.833
Property 3	0.935	0.675	0.784	0.939	0.737	0.826	0.918	0.688	0.787	0.932	0.706	0.804
Method	0.901	0.869	0.884	0.926	0.895	0.910	0.888	0.750	0.813	0.905	0.884	0.894

Table 1 shows for each property that the recall values are lower than the precision ones, achieving 0.730, 0.737, and 0.706, respectively. The recall is an indicator of the effectiveness in the classification of material-change edges. A higher recall suggests fewer misclassified material-change edges (false positives). Thus, lower recall values indicate that, in addition to shadow edges, the proposed properties can also be satisfied by some material changes. The recall values decrease the F-measure, achieving values of 0.826, 0.833, and 0.804, respectively.

Table 1 also shows that Property 2 achieves the highest precision (0.959), recall (0.737), and F-measure (0.833), which indicates that it is the most robust property. The precision due to Property 3 is also high (0.932) but the recall (0.706) is the lowest, which makes Property 3 effective in shadow edge detection but least reliable in the classification of edges due to a material change.

Table 1 shows that the three chrominance properties achieve better results on the Caltech Lane and Kitti Road datasets, but the lowest values are obtained on ours. This is because the Caltech Lane and Kitti Road datasets comprise better-quality images with higher definition.

From the qualitative and quantitative results, we can extract three main conclusions:

1. The three chrominance properties demonstrate their effectiveness in identifying shadow edges, validating the considerations made in the reflectance model, the SPD of the illumination, and the properties.
2. Each property demonstrates its effectiveness in the detection of both well-defined and soft shadow edges (penumbras). This is because each property focuses on the sunlight contribution to the non-shadowed road, which occurs whether the latter is compared with the umbra or penumbra road regions (external shadow contours) or when the umbra is compared with the penumbra (internal shadow contours in the middle image of Figures 13–15).

3. The three chrominance properties can be satisfied by some edges separating material changes, i.e., false positives; thus, the individual application of each of them is not sufficient to unequivocally classify an image edge.

5.2. *Performance of the Proposed Shadow Edge Detection Method*

5.2.1. Qualitative Results

Figure 17 shows the shadow edge detection results of the images in Figure 12 after applying the proposed method, which incorporates the six constraints associated with Properties 1, 2, and 3, i.e., Equation (24). The results show that both the well-defined and the soft shadow edges are successfully identified as shadow edges. The majority of boundaries caused by material changes are also correctly classified and removed from shadow edge maps.

Figure 17. Shadow edge detection results of images in Figure 12 obtained by the proposed method.

In the proposed method, an edge has to satisfy the six constraints to be classified as shadow edge; otherwise, the edge is classified as material change. On the one hand, the accumulation of properties makes the effectiveness of the method in the shadow edge classification decrease in relation to the effectiveness of each property. Each individual property provides shadow edges (i.e., true positives); however, they do not have to be the same for the three properties. Thus, the number of shadow edges correctly classified by the method is lower than that provided by each individual property. However, as the three properties demonstrate high effectiveness in the shadow edge detection, the effectiveness of the method is consequently high. In addition, the fact that a shadow edge has to satisfy the six constraints makes the shadow edge detection very reliable.

On the other hand, the accumulation of constraints makes the material-change edge classification more effective than each individual property. An edge is classified as a material-change edge if just one of the six constraints associated with the three properties is not satisfied. Each individual property provides false detections; however, the false detections do not have to be the same for each property (e.g., in the middle image of Figure 12, the material-change edge caused by the grass and the pavement is misclassified as a shadow edge by Properties 1 and 3; however, it is correctly classified as material change by Property 2). Thus, if just one constraint correctly classifies a material change, the method accepts it regardless of the results due to the other five constraints. Figure 17 shows the reduction of misclassified material-change edges when compared to Figures 13–15. To better show the performance of the method, Figure 18 shows some example results in challenging road scenes of the three datasets.

(a)

(b)

(c)

Figure 18. Example results of the proposed shadow edge detection method on (**a**) images of our dataset, (**b**) images of the Caltech Lane dataset, and (**c**) images of the Kitti Road dataset.

5.2.2. Quantitative Results

The bottom row of Table 1 shows the results of the complete shadow edge detection method on each of the three datasets. It shows that the precision, recall, and F-measure on each dataset are high, achieving values of 0.905, 0.884, and 0.870, respectively, on the 300 images. Furthermore, the precision value achieved by the method (0.905) is lower than that of each individual property, whereas the recall value (0.884) is higher. The F-measure value achieved by the method (0.870) indicates effectiveness not only in the classification of shadow edge classification but also material change.

From the qualitative and quantitative results of the proposed method, we can extract two main conclusions:

1. The proposed shadow edge detection method demonstrates effectiveness in identifying shadow edges, achieving a precision value of 0.905. In addition, the accumulation of chrominance properties makes the shadow edge detection more reliable since a shadow edge has to satisfy the six constraints.
2. The accumulation of shadow properties improves the effectiveness in the classification of material-change edges, making the method achieve recall and F-measure values of 0.884 and 0.870, respectively.

5.2.3. Results on Images without Shadows

An effective shadow detection method does not only detect shadows but also does not provide false detections in scenes that do not contain shadows. Figure 19 illustrates two cluttered traffic scenes of our dataset (left and middle images) and one scene of the Kitti Road dataset (right image), which cover two different types of illumination that do not cause shadows on the road. The images on the left and in the middle were captured under cloudy conditions, whereas, in the image on the right, the ego-vehicle is traveling along a street in the shade. As can be observed in the three images, most of the edges caused by material changes are correctly classified and removed from shadow edge map. The number of false detections is small, and they occur in image regions outside the road surface. In the left image, the material-change edges caused by the pedestrians, cyclists, a motorcyclist, and manhole covers on the road are successfully classified as material changes. In the middle image, the material-change edges due to the pedestrian crossing, pedestrians, and curbs are also correctly identified. In the right image, most of the material-change edges caused by the road boundaries, asphalt noise, and vehicles are correctly detected, except for some noisy edges caused by some reflection on vehicles. However, it can be observed that the shadows underneath both vehicles on the right are misclassified as material change and removed from the image. Overall, Figure 19 shows the effectiveness of the proposed method in identifying of material-change edges in shadow-free images.

Figure 19. (**Row 1**) input images; (**Row 2**) ROI overlaid with brighter (red) and darker (blue) regions across strong edges (green); (**Row 3**) results of the proposed shadow edge detection method.

5.2.4. Limitations

The experiments demonstrate effectiveness of the proposed method in the classification of both shadow edges and material changes. However, there are three limitations.

Limitation 1. Overexposed image regions may lead to shadow edge misclassification. Sunlight may cause oversaturated road regions in the image, whose RGB components saturate to a gray level of 255; thus, the surface chrominance is undermined. Figure 20a shows the over-exposure problem where the RGB components of the non-shadowed road are saturated leading to false negatives, i.e., misclassification of shadow edges. This problem is inherent to any physics-based method and could be mitigated using cameras of higher dynamic range.

(a) (b) (c)

Figure 20. Detection errors: (**a**) oversaturated road region; (**b**) shadow underneath a vehicle; (**c**) yellow marking edges.

Limitation 2. Owing to the lack of light underneath a vehicle, the shadow underneath is generally a very dark road region whose RGB reflectance components have very low values. Unlike over-exposed road regions, the shadow underneath a vehicle may be saturated to the minimum value of the range, i.e., 0, making the surface chrominance feature unstable [59]. Some examples of Figures 18 and 19c, and the middle image of Figure 20 show this problem, resulting in misclassification of edges due to the shadow underneath. However, for some applications such as road detection and front vehicle detection, this misclassification is in fact a positive outcome as the edge of the shadow underneath a vehicle is the bottom part of its contour [59].

Limitation 3. In some cases, edges due to yellow markings on the road may satisfy the proposed shadow constraints, leading to an edge misclassification. The middle image of Figure 17 and some examples of Figure 18a,b show correct classification of yellow marking edges. However, the right image of Figure 20 is an example where, even in a same image, some edges due to yellow markings are correctly classified as material change, whereas others are misclassified as shadow edges. This error could be addressed by assuming that the darker region of a yellow marking edge (i.e., the road) is bright enough to not be considered as a shadowed region.

5.2.5. Comparison with Previous Works

The performance of the proposed method is compared with the following five state-of-the-art shadow detection methods:

Method 1. The physics-based method in Reference [29] for shadowed road detection exploits intensity thresholding and normalized RGB color space to compute the bluish effect of shadowed road. The normalized blue component b and the intensity I of shadowed pixels have to satisfy $b \geq 1/3$ and $I \leq I_{road,avg} - 2\sigma_{road}$, respectively, where we compute the mean $I_{road,avg}$ and variance σ_{road} of the non-shadowed road from the brighter region of the edge under evaluation.

Method 2. The physics-based method in Reference [12] exploits the fact that the intensity change between shadowed and non-shadowed surfaces is higher in the red and green components than in the blue. This method uses shadow pixel intensity reduction and albedo ratio test. We adapt the latter to still images using neighboring pixels of regions across the edge under evaluation.

Method 3. Using the HSV color space, the color invariance method in Reference [11] assumes that shadow reduces the luminance v and saturation s components of the surface, whereas the hue h varies within a range. We set the threshold values as in Reference [11], i.e., α, β, τ_s, and τ_h equal to 0.4, 0.6, 0.1, and 0.5, respectively.

Method 4. The method in Reference [45] is based on illuminant invariance where an edge is classified as shadow edge if, at a given location, the original image has a strong edge but the illuminant-invariant image has a weak one, or if both images have strong edges but their orientation is different. As in Reference [45], the thresholds τ_1, τ_2, and τ_3 are set to 0.4, 0.1, and $\pi/4$, respectively, and the image characteristic direction θ is computed using the minimum entropy method in Reference [1], obtaining 47°, 54°, and 73° for the Caltech Lane, Kitti Road, and our dataset, respectively.

Method 5. The method in Reference [50] is also based on illuminant invariance which determines the illumination spectral direction (ISD) by means of shadowed and non-shadowed road regions. To this end, the method generates two maps of potential shadow pixels and potential lit pixels. Potential shadow pixels are defined as having a low-percent variance (<2%) in each color band, and a color which is roughly neutral but not more blue than a neutral surface is the ISD of sunset. Potential lit pixels are defined as having a low-percent variance (<2%) in each color band, and no color bands are 45% brighter than the other. In addition, the difference between the shadow and lit maps must be at least 0.3 in log(RGB) space in all channels. The ISD computed from the shadow and lit maps must be within a Euclidean distance of 0.1 of the arc on the unit sphere defined by a neutral ISD = (0.577, 0.577, 0.577) and sunset ISD = (0.789, 0.547, 0.299).

Method 1 compares the pixel intensity between candidate shadow pixels and a sample region on the non-shadowed road, whereas Methods 2 and 3 are for image sequences where an initial frame void of shadows is required for comparing the surface properties. Method 5 finds potential shadow and lit maps from a trapezoidal ROI which corresponds approximately to the road surface. Since the aim is to compare the shadow properties exploited by each method, the methods are evaluated under the same conditions for our shadow edge detection strategy. Thus, after extracting the strong edges of the ROI, the darker regions are candidate shadow regions for the five methods, whereas their respective brighter regions are assumed to be the reference non-shadowed regions. For Method 4, we use the strong edges obtained by our classification strategy as edges of the original image.

Figure 21 shows the shadow edge detection results obtained by the five methods and ours in four challenging traffic scenes of the Kitti Road dataset. The scenes in Figure 21b,c contain dark shadows caused by the building and parked vehicle, respectively, which generate well-defined shadow edges on the road. The scenes in Figure 21d,e show weak shadows caused by the branches of the trees which generate soft shadow boundaries. The results and comparisons are summarized as follows:

1. Figure 21b,c show that the six methods successfully detect the well-defined shadow edges caused by the building and parked vehicle. The right side of Figure 21b shows that the contour of the thin and weak shadow (penumbra) on the road is also correctly detected by Methods 2 and 4, as well as ours, but not Methods 1, 3, and 5. Similarly, the scene of Figure 21c presents weak shadows on the bottom-left and upper-center regions of the image, whose edges are correctly detected by Methods 1, 2, and 4, as well as ours, but not Methods 3 and 5.

2. Figure 21d,e demonstrate the effectiveness of Methods 1 and 2,as well as ours, in the detection of soft edges separating umbra from penumbra (internal shadow contours) and penumbra from non-shadowed road (external shadow contours), whereas Methods 3–5 miss most of them.

3. Regarding the classification of material-change edges, the four scenes show that the effectiveness of Methods 1–5 is lower than that of our method. Figure 21b shows that Methods 1, 2, and 4 fail to classify the edges due to the asphalt change. Furthermore, they and Methods 3 and 5 misclassify some edges due to material changes in the background of the images of Figure 21b,e. Figure 21c,d show that Methods 1–3 and 5 incorrectly detect bright and material changes on the parked vehicles, whereas Method 4 correctly classifies them in Figure 21c but fails in Figure 21d. An important drawback of Methods 1 and 3–5 is that they easily misclassify edges due to white lane markings on the road (see Figure 21e), which makes them unreliable for lane detection applications. In contrast, the effectiveness of our method in identifying material-change edges can be observed in the correct detection of edges corresponding to asphalt patches, background objects, parked vehicles, and white

lane markings. To better show the performance of Methods 1–5 and ours, Figure 22 shows some example results on images of the three datasets. The quantitative results of Methods 1–5 and ours on each of the three datasets are summarized in Table 2. The precision achieved by Methods 1 and 2, as well as ours, is high, with Method 2 being the most effective with a value of 0.911, closely followed by ours with 0.905. The precision of Methods 3–5 is lower, achieving 0.420, 0.637, and 0.594, respectively. With regard to recall, our method achieves 0.884, which indicates high effectiveness in the classification of material-change edges, whereas the recall of Methods 1–5 is lower, achieving values of 0.511, 0.583, 0.442, 0.476, and 0.545, respectively. Finally, our method achieves the highest F-measure (0.894), followed by Method 2 with a value of 0.711. Since the F-measure indicates the global performance of the methods, encompassing both precision and recall, the F-measure achieved by our method demonstrates effectiveness and robustness in classifying both shadow and material-change edges.

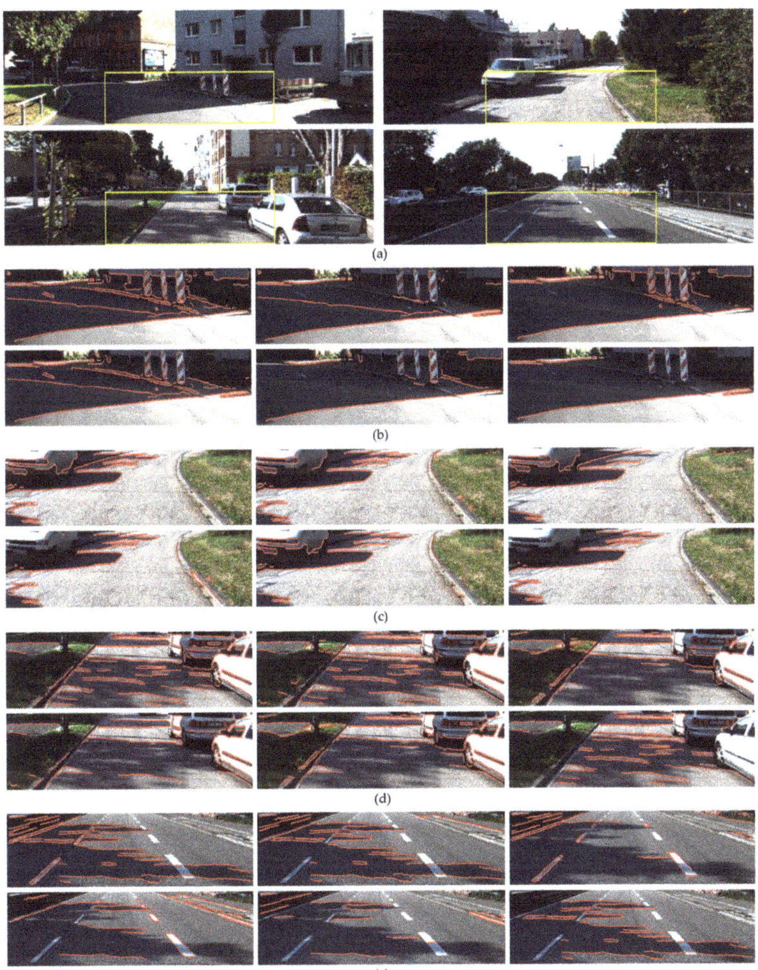

Figure 21. Shadow edge detection results obtained using the five methods. (**a**) ROI overlaid onto incoming images. (**b–e**) *Top-left* results of Method 1; *top-center* results of Method 2; *top-right* results of Method 3; *bottom-left* results of Method 4; *bottom-center* results of Method 5; *bottom-right* results of our method.

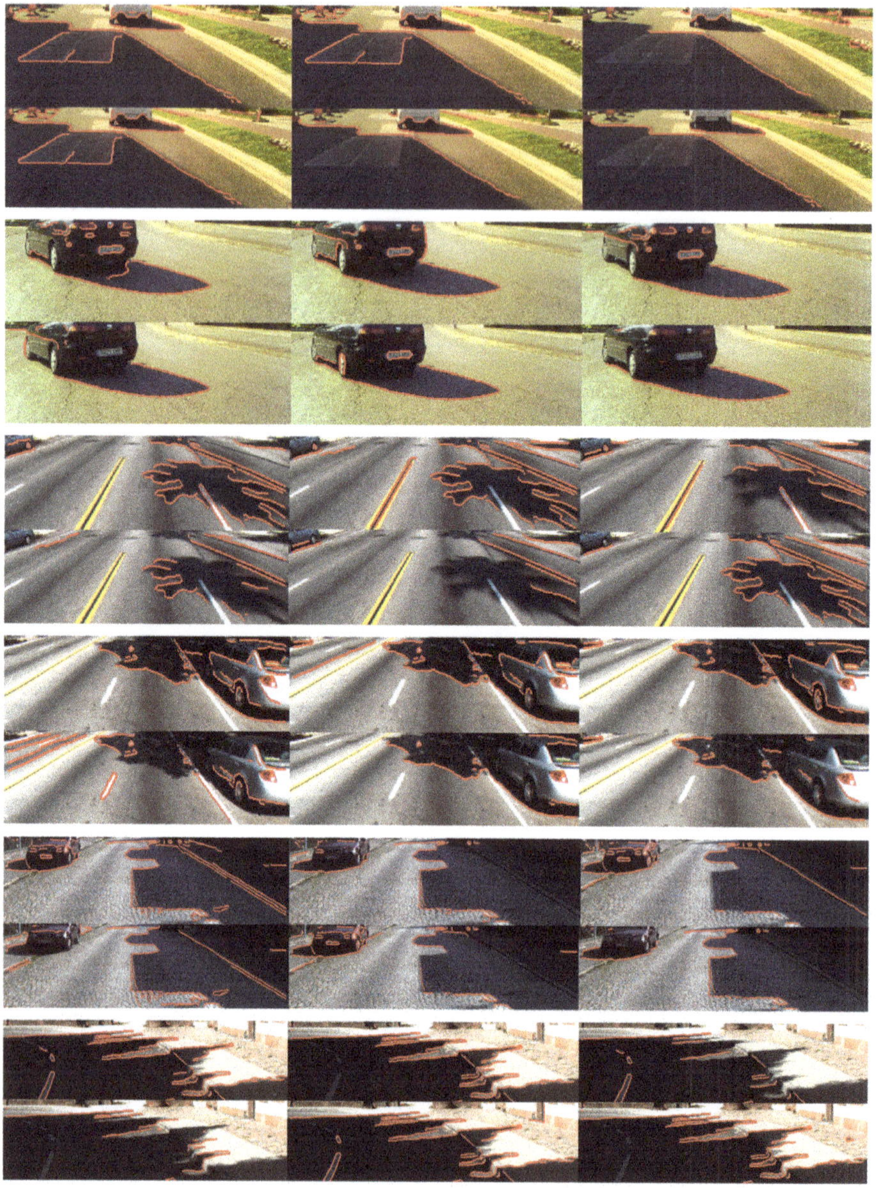

Figure 22. Shadow edge detection results obtained using the five methods. (**First and second scenes**) ROI of images of our dataset; (**third and fourth scenes**) ROI of images of the Caltech Lane dataset; (**fifth and sixth scenes**) ROI of images of the Kitti Road dataset. For each scene: *Top-left* results of Method 1; *top-center* results of Method 2; *top-right* results of Method 3; *bottom-left* results of Method 4; *bottom-center* results of Method 5; *bottom-right* results of our method.

It must be said that Methods 1–3 do not focus on detecting shadow edges but on shadow segmentation, coping with umbras where their indicators are significantly higher. On the other hand, the performance of Methods 4 and 5 is highly dependent of the image quality. A higher image quality results in a more accurate illuminant invariant image and better shadow detection results. This is shown in Table 2, as Methods 4 and 5 achieve better results on the Caltech Lane and Kitti roads datasets. It must be said that both Methods 4 and 5 would achieve much more better results using high-quality images with a wide dynamic range captured by calibrated sensors.

Table 2. Precision (*P*), recall (*R*), and F-measure (*F-m*) indicators of the four methods.

	Caltech Dataset (100 Images)			Kitti Dataset (100 Images)			Our Dataset (100 Images)			All Datasets (300 Images)		
	P	*R*	*F-m*	*P*	*R*	*F-m*	*P*	*R*	*F-m*	*P*	*R*	*F-m*
Method 1	0.827	0.492	0.617	0.870	0.468	0.609	0.822	0.575	0.677	0.839	0.511	0.634
Method 2	0.919	0.596	0.723	0.938	0.566	0.706	0.876	0.588	0.704	0.911	0.583	0.711
Method 3	0.350	0.426	0.384	0.466	0.440	0.453	0.445	0.461	0.453	0.420	0.442	0.430
Method 4	0.646	0.447	0.528	0.724	0.578	0.643	0.541	0.405	0.418	0.637	0.476	0.544
Method 5	0.608	0.565	0.585	0.698	0.630	0.662	0.477	0.442	0.458	0.594	0.545	0.568
Our Method	0.901	0.869	0.884	0.926	0.895	0.910	0.888	0.750	0.813	0.905	0.884	0.894

6. Conclusions

Vision-based driving assistance methods are significantly affected by shadows on the road, which hinder important tasks such as road and lane detections. Additionally, the identification of shadows is not easy since shadow properties may be shared by objects in the scene. The aim of this work was, on the one hand, to find new physical properties to better characterize shadows on the road so as to minimize possible misclassification of objects and non-shadowed image regions as shadows, and, on the other hand, to use the new chrominance properties to design an effective shadow detection method for integration in an onboard road detection system for driver assistance.

We discussed the illumination in outdoor scenes under sunny conditions, which comprise two light sources with different SPDs, i.e., skylight and sunlight, as well as their effect on the road surface. Unlike other methods, when comparing shadowed and non-shadowed regions of the same material surface, we observed the importance of the sunlight contribution to the non-shadowed surface chromaticity, and then derived three new chrominance properties of shadows. Based on the six constraints associated with these properties, we proposed a shadow edge detection method for onboard systems. In as much as onboard systems deal with still images, our method focuses on distinguishing shadow boundaries from material changes by comparing properties of regions across image edges. However, as no prior knowledge of the scene, camera calibration, or spatio-temporal restrictions are required, static background applications can also be addressed.

Tests carried out on different datasets demonstrated the effectiveness of the proposed method in identifying both well-defined and soft shadow edges, achieving precision of 0.905. However, the most remarkable feature of our method is its ability in identifying material-change edges. This demonstrates that the accumulation of shadow feature constraints contributes to a better characterization of shadows by minimizing the possibility of errors in the classification of objects and non-shadowed regions. The proposed method achieved recall and F-measure of 0.884 and 0.894, respectively, showing better performance compared with five state-of-the-art methods.

Although the experiments demonstrated the effectiveness and reliability of our method, there are three limitations, including misclassification of edges due to yellow markings on the road, and missing shadow edges in overexposed and underexposed image regions. The former can be addressed by taking into account the intensity of the candidate shadow region, and the latter could be minimized by using cameras of higher dynamic range.

As future work, we will address the limitations of our method, as well as develop a road detection system for driver assistance.

Author Contributions: Conceptualization, investigation, validation, writing—original draft, and writing—review and editing, M.J.I.-A., T.T., and J.P.-O. All authors have read and agreed to the published version of the manuscript.

Funding: This research received no external funding.

Conflicts of Interest: The authors declare no conflicts of interest.

References

1. Alvarez, J.M.; Lopez, A.M. Road Detection Based on Illuminant Invariance. *IEEE Trans. Intell. Transp. Syst.* **2011**, *12*, 184–193. [CrossRef]
2. Song, Y.; Ju, Y.; Du, K.; Liu, W.; Song, J. Online Road Detection under a Shadowy Traffic Image Using a Learning-Based Illumination-Independent Image. *Symmetry* **2018**, *10*, 707. [CrossRef]
3. Yoo, J.H.; Lee, S.G.; Park, S.K.; Kim, D.H. A Robust Lane Detection Method Based on Vanishing Point Estimation Using the Relevance of Line Segments. *IEEE Trans. Intell. Transp. Syst.* **2017**, *18*, 3254–3266. [CrossRef]
4. Hoang, T.M.; Baek, N.R.; Cho, S.W.; Kim, K.W.; Park, K.R. Road Lane Detection Robust to Shadows Based on a Fuzzy System Using a Visible Light Camera Sensor. *Sensors* **2017**, *17*, 2475. [CrossRef] [PubMed]
5. Bertozzi, M.; Broggi, A.; Fascioli, A. Vision-based intelligent vehicles: State of the art and perspectives. *Robot. Autonom. Syst.* **2000**, *32*, 1–16. [CrossRef]
6. Salvador, E.; Cavallaro, A.; Ebrahimi, T. Cast shadow segmentation using invariant color features. *Comput. Vis. Image Underst.* **2004**, *95*, 238–259. [CrossRef]
7. Sanin, A.; Sanderson, C.; Lovell, B.C. Shadow detection: A survey and comparative evaluation of recent methods. *Pattern Recognit.* **2012**, *45*, 1684–1689. [CrossRef]
8. Prati, A.; Mikic, I.; Trivedi, M. Detecting Moving Shadows: Algorithms and Evaluation. *IEEE Trans. Pattern Anal. Mach. Intell.* **2003**, *25*, 918–923. [CrossRef]
9. Russel, M.; Zou, J.; Fang, G. An evaluation of moving shadow detection techniques. *Comput. Vis. Media* **2017**, *2*, 195–217. [CrossRef]
10. Yoneyama, A.; Yeh, C.H.; Kuo, C.C.J. Moving cast shadow elimination for robust vehicle extraction based on 2d joint vehicle/shadow models. In Proceedings of the IEEE Conference on Advanced Video and Signal Based Surveillance, Miami, FL, USA, 21–22 July 2003; pp. 229–236.
11. Cucchiara, R.; Grana, C.; Piccardi, M.; Prati, A. Detecting moving objects, ghosts, and shadows in video streams. *IEEE Trans. Pattern Anal. Mach. Intell.* **2003**, *25*, 1337–1342. [CrossRef]
12. Nadimi, S.; Bhanu, B. Physical models for moving shadow and object detection in video. *IEEE Trans. Pattern Anal. Mach. Intell.* **2004**, *26*, 1079–1087. [CrossRef]
13. Martel-Brisson, M.; Zaccarin, A. Learning and removing cast shadows through a multidistribution approach. *IEEE Trans. Pattern Anal. Mach. Intell.* **2007**, *29*, 1133–1146. [CrossRef] [PubMed]
14. Huang, J.B.; Chen, C.S. Moving cast shadow detection using physics-based features. In Proceedings of the IEEE Conference on Computer Vision and Pattern Recognition, Miami, FL, USA, 20–25 June 2009; pp. 2310–2317.
15. Gomes, V.; Barcellos, P.; Scharcanski, J. Stochastic shadow detection using a hypergraph partitioning approach. *Pattern Recognit.* **2017**, *63*, 30–44. [CrossRef]
16. Kim, D.S.; Arsalan, M.; Park, K.R. Convolutional Neural Network-Based Shadow Detection in Images Using Visible Light Camera Sensor. *Sensors* **2018**, *18*, 960. [CrossRef] [PubMed]
17. Piccardi, M. Background subtraction techniques: A review. In Proceedings of the IEEE International Conference on Systems, Man and Cybernetics, The Hague, The Netherlands, 10–13 October 2004; pp. 3099–3104.
18. Maddalena, L.A.; Petrosino, A. Exploiting color and depth for background subtraction. In *New Trends in Image Analysis and Processing—ICIAP 2017. Lecture Notes in Computer Science*; Battiato, S., Farinella, G.M., Leo, M., Gallo, G., Eds.; Springer: Cham, Switzerland, 2017; Volume 10590, pp. 254–265. ISBN 978-3-319-70742-6.
19. Babaee, M.; Dinh, D.T.; Rigoll, G. A deep convolutional neural network for video sequence background subtraction. *Pattern Recognit.* **2018**, *76*, 635–649. [CrossRef]
20. Park, S.; Lim, S. Fast Shadow Detection for Urban Autonomous Driving Applications. In Proceedings of the IEEE/RSJ International Conference on Intelligent Robots and Systems, St. Louis, MO, USA, 11–15 October 2009; pp. 1717–1722.
21. Levine, M.D.; Bhattacharyya, J. Removing shadows. *Pattern Recognit. Lett.* **2005**, *26*, 251–265. [CrossRef]
22. Tian, J.; Qi, X.; Qu, L.; Tang, Y. New spectrum ratio properties and features for shadow detection. *Pattern Recognit.* **2016**, *51*, 85–96. [CrossRef]

23. Leone, A.; Distante, C. Shadow detection for moving objects based on texture analysis. *Pattern Recognit.* **2007**, *40*, 1222–1233. [CrossRef]
24. Mohan, A.; Tumblin, J.; Choudhury, P. Editing soft shadows in a digital photograph. *IEEE Comput. Graph. Appl.* **2007**, *27*, 23–31. [CrossRef]
25. Graham, F.; Steven, H.; Gerald, F.; Tian, G.Y. Illuminant and device invariant colour using histogram equalization. *Pattern Recognit.* **2005**, *38*, 179–190.
26. Horprasert, T.; Harwood, D.; Davis, L.S. A Statistical Approach for Real-time Robust Background Subtraction and Shadow Detection. In Proceedings of the IEEE Frame Rate Workshop, Kerkyra, Greece, 20–27 September 1999; pp. 1–19.
27. Dong, X.; Wang, K.; Jia, G. Moving Object and Shadow Detection Based on RGB Color Space and Edge Ratio. In Proceedings of the 2nd International Congress on Image and Signal Processing, Tianjin, China, 17–19 October 2009; pp. 1–5.
28. Cavallaro, A.; Salvador, E.; Ebrahimi, T. Shadow-aware Object-based Video Processing. *IEE Proc. Vis. Image Signal Process.* **2005**, *152*, 398–406. [CrossRef]
29. Sotelo, M.A.; Rodriguez, F.J.; Magdalena, L.; Bergasa, L.M.; Boquete, L. A Color Vision-Based Lane Tracking System for Autonomous Driving on Unmarked Roads. *Autonom. Robots* **2004**, *16*, 95–116. [CrossRef]
30. Rotaru, C.; Graf, T.; Zhang, J. Color image segmentation in HSI space for automotive applications. *J. Real Time Image Process.* **2008**, *3*, 1164–1173. [CrossRef]
31. Zhang, H.; Hernandez, D.E.; Su, Z.; Su, B. A Low Cost Vision-Based Road-Following System for Mobile Robots. *Appl. Sci.* **2018**, *8*, 1635. [CrossRef]
32. Kampel, M.; Wildenaue, H.; Blauensteiner, P.; Hanbury, A. Improved motion segmentation based on shadow detection. *Electron. Lett. Comput. Vis. Image Anal.* **2007**, *6*, 1–12.
33. Schreer, O.; Feldmann, I.; Goelz, U.; Kauff, P. Fast and robust shadow detection in videoconference applications. In Proceedings of the International Symposium on VIPromCom Video/Image Processing and Multimedia Communications, Zadar, Croatia, 16–19 June 2002; pp. 371–375.
34. Chen, C.T.; Su, C.Y.; Kao, W.C. An enhanced segmentation on vision-based shadow removal for vehicle detection. In Proceedings of the 2010 International Conference on Green Circuits and Systems, Shanghai, China, 21–23 June 2010; pp. 679–682.
35. Lee, S.; Hong, H. Use of Gradient-Based Shadow Detection for Estimating Environmental Illumination Distribution. *Appl. Sci.* **2018**, *8*, 2255. [CrossRef]
36. Gevers, T.; Smeulders, A.M.W. Color-based object recognition. *Pattern Recognit.* **1999**, *32*, 453–464. [CrossRef]
37. Rubin, J.M.; Richards, W.A. Color vision and image intensities: When are changes material? *Biol. Cybern.* **1982**, *45*, 215–226. [CrossRef]
38. Pormerleu, D.A. *Neural Network Perception for Mobile Robot Guidance*; Kluwer Academic Publishers: Boston, MA, USA, 1993.
39. Wallace, R.; Matsuzaki, K.; Goto, Y.; Crisman, J.; Webb, J.; Kanade, T. Progress in Robot Road-Following. In Proceedings of the IEEE Conference on Robotics Automation, San Francisco, CA, USA, 7–10 April 1986; pp. 1615–1621.
40. Mikic, I.; Cosman, P.C.; Kogut, G.T.; Trivedi, M.M. Moving shadow and object detection in traffic scenes. In Proceedings of the IEEE International Conference on Pattern Recognition (ICPR), Barcelona, Catalunya, Spain, 3–7 September 2000; pp. 321–324.
41. Tian, J.; Sun, J.; Tang, Y. Tricolor Attenuation Model for Shadow Detection. *IEEE Trans. Image Process.* **2009**, *10*, 2355–2363. [CrossRef]
42. Barnard, K.; Finlayson, G. Shadow identification using colour ratios. In Proceedings of the IS&T/SID 8th Color Imaging Conference on Color Science, Science, Systems and Application, Scottsdale, AZ, USA, 7–10 November 2000; pp. 97–101.
43. Tappen, M.F.; Freeman, W.T.; Adelson, E.H. Recovering intrinsic images from a single image. *IEEE Trans. Pattern Anal. Mach. Intell.* **2005**, *27*, 1459–1472. [CrossRef]
44. Lalonde, J.F.; Efros, A.A.; Narasimhan, S.G. Detecting ground shadows in outdoor consumer photographs. In Proceedings of the 11th European Conference on Computer Vision, Crete, Greece, 5–11 September 2010; pp. 322–335.
45. Finlayson, G.D.; Hordley, S.D.; Cheng, L.; Drew, M.S. On the Removal of Shadows From Images. *IEEE Trans. Pattern Anal. Mach. Intell.* **2006**, *1*, 59–68. [CrossRef] [PubMed]

46. Finlayson, G.D.; Hordley, S.D. Color constancy at a pixel. *J. Opt. Soc. Am. A* **2001**, *18*, 253–264. [CrossRef] [PubMed]
47. Nielsen, M.; Madsen, C.B. Segmentation of Soft Shadows based on a Daylight and Penumbra Model. *Lect. Notes Comput. Sci.* **2007**, *4418*, 341–352.
48. McFeely, R.; Glavin, M.; Jones, E. Shadow identification for digital imagery using colour and texture cues. *IET Image Proc.* **2012**, *6*, 148–159. [CrossRef]
49. Fernández, C.; Fernández-Llorca, D.; Sotelo, M.A. A hybrid vision-map method for urban road detection. *J. Adv. Transp.* **2017**, *2017*, 1–21. [CrossRef]
50. Maxwell, B.A.; Smith, C.A.; Qraitem, M.; Messing, R.; Whitt, S.; Thien, N.; Friedhohh, R.M. Real-Time Physics-Based Removal of Shadows and Shading From Road Surfaces. In Proceedings of the IEEE Conference on Computer Vision and Pattern Recognition (CVPR), Long Beach, CA, USA, 15–21 June 2019; pp. 88–96.
51. Guo, R.; Qieyun, D.; Derek, H. Paired regions for shadow detection and removal. *IEEE Trans. Pattern Anal. Mach.* **2013**, *35*, 2956–2967. [CrossRef] [PubMed]
52. Khan, S.H.; Bennamoun, M.; Sohel, F.; Togneri, R. Automatic shadow detection and removal from a single image. *IEEE Trans. Pattern Anal. Mach. Intell.* **2016**, *38*, 431–446. [CrossRef]
53. Vicente, T.F.Y.; Hoai, M.; Samaras, D. Leave-One-Out Kernel Optimization for Shadow Detection and Removal. *IEEE Trans. Pattern Anal. Mach. Intell.* **2017**, *40*, 682–695. [CrossRef]
54. Qu, L.; Tian, J.; He, S.; Tang, Y.; Lau, R.W. DeshadowNet: A multicontext embedding deep network for shadow removal. In Proceedings of the IEEE Conference on Computer Vision and Pattern Recognition (CVPR), Honolulu, HI, USA, 21–26 July 2017; pp. 4067–4075.
55. Hu, X.; Zhu, L.; Fu, C.W.; Qin, J.; Heng, P.A. Direction-aware Spatial Context Features for Shadow Detection. *arXiv* **2018**, arXiv:1712.04142.
56. Khan, S.; Bennamoun, M.; Sohel, F.; Togneri, R. Automatic feature learning for robust shadow detection. In IEEE Conference on Computer Vision and Pattern Recognition, Columbus, OH, USA, 24–27 June 2014; pp. 4321–4328.
57. Finlayson, G.D.; Drew, M.S.; Funt, B.V. Spectral sharpening: Sensor transformations for improved color constancy. *J. Opt. Soc. Am. A* **1994**, *11*, 1553–1562. [CrossRef]
58. Stauder, J.; Mech, R.; Ostermann, J. Detection of moving cast shadows for object segmentation. *IEEE Trans. Multimedia* **1999**, *1*, 65–76. [CrossRef]
59. Ibarra-Arenado, M.; Tjahjadi, T.; Pérez-Oria, J.; Robla-Gómez, S.; Jiménez-Avello, A. Shadow-Based Vehicle Detection in Urban Traffic. *Sensors* **2017**, *17*, 975. [CrossRef] [PubMed]
60. Canny, J. A Computational Approach to Edge Detection. *IEEE Trans. Pattern Anal. Mach. Intell.* **1986**, *8*, 679–698. [CrossRef] [PubMed]
61. Aly, M. Real time Detection of Lane Markers in Urban Streets. In Proceedings of the IEEE Intelligent Vehicles Symposium, Eindhoven, The Netherlands, 4–6 June 2008; pp. 7–12.
62. Fritsch, J.; Kühnl, T.; Geiger, A. A new performance measure and evaluation benchmark for road detection algorithms. In Proceedings of the 16th International IEEE Conference on Intelligent Transportation Systems (ITSC 2013), The Hague, The Netherlands, 6–9 October 2013; pp. 1693–1700.

sensors

MDPI

Article

Robust Traffic Light and Arrow Detection Using Digital Map with Spatial Prior Information for Automated Driving

Keisuke Yoneda [1,*,†], Akisuke Kuramoto [2,†], Naoki Suganuma [1], Toru Asaka [1], Mohammad Aldibaja [1] and Ryo Yanase [1]

1 Institute for Frontier Science Initiative, Kanazawa University, Kanazawa, Ishikawa 920-1192, Japan; suganuma@staff.kanazawa-u.ac.jp (N.S.); asaka.toru@stu.kanazawa-u.ac.jp (T.A.); amroaldibaja@staff.kanazawa-u.ac.jp (M.A.); ryanase@staff.kanazawa-u.ac.jp (R.Y.)

2 Department of Mechanical Systems Engineering, Tokyo Metropolitan University, Hino, Tokyo 191-0065, Japan; kuramoto.a.aa@tmu.ac.jp

* Correspondence: k.yoneda@staff.kanazawa-u.ac.jp

† The authors have contributed equally to this work.

Received: 25 December 2019; Accepted: 17 February 2020; Published: 21 February 2020

Abstract: Traffic light recognition is an indispensable elemental technology for automated driving in urban areas. In this study, we propose an algorithm that recognizes traffic lights and arrow lights by image processing using the digital map and precise vehicle pose which is estimated by a localization module. The use of a digital map allows the determination of a region-of-interest in an image to reduce the computational cost and false detection. In addition, this study develops an algorithm to recognize arrow lights using relative positions of traffic lights, and the arrow light is used as prior spatial information. This allows for the recognition of distant arrow lights that are difficult for humans to see clearly. Experiments were conducted to evaluate the recognition performance of the proposed method and to verify if it matches the performance required for automated driving. Quantitative evaluations indicate that the proposed method achieved 91.8% and 56.7% of the average f-value for traffic lights and arrow lights, respectively. It was confirmed that the arrow-light detection could recognize small arrow objects even if their size was smaller than 10 pixels. The verification experiments indicate that the performance of the proposed method meets the necessary requirements for smooth acceleration or deceleration at intersections in automated driving.

Keywords: image processing; traffic light detection; intelligent transportation system

1. Introduction

Automated vehicle technologies are considered to be the next generation transportation system. Many companies and research organizations are involved in the research and development of such technologies. Recent automated vehicle technologies focus more on urban driving, which is a mixed transportation environment with human drivers. Public road demonstration experiments have been carried out in the U.S., European, and Asian countries since the mid-2000s [1–3]. Automated driving in mixed environments human driven vehicles, pedestrians, and cyclists requires recognition of surrounding objects autonomously and decision making according to traffic rules. The following sensors are mainly mounted on automated vehicles for surrounding recognition.

- Ranging sensors: light detection and ranging (LiDAR), millimeter-wave radar (MWR)
- Imaging sensor: camera
- Positioning sensor: global navigation satellite system and inertial navigation system (GNSS/INS).

The above mentioned sensors make it possible to observe surrounding transportation environments. In addition, recent automated driving technologies rely on high-definition maps (HD maps) which include the precise position of static road features such as lane boundaries, lane centerlines, traffic signs, and traffic lights. By referring to the predefined road features, it is possible to reduce false surrounding recognitions and implement accurate decision making, considering road structures. The following functions must be implemented in order to achieve automated driving on public roads.

1. Self-localization: Estimating the position of the vehicle on the HD map in the accuracy of decimeter-level
2. Surrounding perception: Recognizing static/dynamic objects including traffic participants and road objects (e.g., lane marking, traffic signals)
3. Motion planning: Designing the optimal collision-free trajectories following the traffic rules
4. Motion control: Determining adequate control signals such as steering, acceleration and braking.

This study focuses on traffic light (TL) detection using HD maps. There are many studies on TL detection by using vision-sensors in intelligent transportation system. TL is one of the most important road features in order to decide an approaching the intersection. Although HD maps have the information of TL positions, the vehicle must recognize the current state of TLs in real time because it changes dynamically. For safety deceleration, it is necessary to recognize the current state of TLs at distances over 100 m. The required recognition distance can be estimated by calculating the braking distance from the vehicle to the stop line after smoothly recognizing the TL state. In studies on vehicle control [4,5], the deceleration without discomfort to passengers is approximately 0.1 G ($=0.098$ m/s^2). For example, when the vehicle decelerates by 0.1 G while traveling at a velocity of 50 km/h, the braking distance is approximately 98 m. Furthermore, the recognition distance may increase further when considering the case where the TL is located at a position away from the stop line. Recognizing TLs in the ranges exceeding 100 m is required to make a natural intersection approach in automated driving. In order to implement a practical method of TL recognition, it is necessary to discuss the effectiveness of the methods, considering the trade-off between the required performance and the hardware specification. For example, installing a high resolution camera or a telephoto lens is an easy solution to increase the recognition distance. However, increasing the resolution may increase the processing time. In addition, the field of view is narrowed and TLs may be left out. From the point of view of implementing a recognition method, it is important how to recognize small pixel objects.

On the other hand, in automated driving using HD maps, the self-localization module precisely estimates the vehicle pose by map-matching using a range sensor or image sensor [6–8]. Generally, position accuracy of approximately 0.1 to 0.2 m is considered to be necessary for decision-making and path planning in automated driving. Assuming that the precise vehicle position on the digital map is estimated, a region-of-interest (ROI) location for the TLs can be calculated using the registered TL position and current vehicle pose. Extracting the ROI makes it possible to reduce the search region of TLs. It is then possible to reduce false detections such as false-positive and false-negative detections, and computational costs [9–12]. In addition to improving recognition performance, associating TLs registered on a map with TLs in an image is an important aspect of the map-based recognition. In the decision making using the HD maps, it is necessary to grasp the state of the relevant TLs, in order to make an approach decision at the intersection.

The purpose of this study is to achieve small object recognition for both TLs and arrow TLs. Different types of candidate detectors are introduced to realize robust lighting area detection. In addition, in order to implement robust arrow recognition, the proposed method uses spatial information of the positional relationship of lights in TLs as prior information obtained from the HD map. The proposed method is specialized in the recognition of small-sized objects, and evaluate recognition performance for actual driving data. Moreover, verification experiments for automated driving were carried out by introducing the proposed method to investigate the validity of the method.

The rest of this paper is composed as follows. Section 2 describes related works. Section 3 introduces the proposed TL detection method. Section 4 presents evaluations for the proposed method with some discussions. Finally, Section 5 concludes the paper with some remarks.

2. Related Works

Table 1 summarizes the reported results of major research cases on the TL recognition. Based on the state-of-the-art researches, the recognition procedure can be described as follows:

1. Determine the search region: A region-of-interest (ROI) is extracted from the captured image by using the predefined map.
2. Extract candidate objects: Circular lighting areas or rectangular objects are extracted from the search region as candidate TLs.
3. Classify the state of the candidates: Simple color filtering or machine learning algorithms identify lighting colors and arrow light directions.

Table 1. Summary of relative studies for traffic light recognition.

Paper	Method	Object	Resolution	Time	Distance / Min. Pixel	Accuracy
[13], 2009	Blob candidate; Adaptive Template Matcher	TL	640 × 480	37.4 ms (CPU)	6 px	95.38% (Prec.) 98.41% (Recall)
[9], 2011	Blob candidate; Mapping	TL; Arrow	2040 × 1080	4 Hz (CPU)	200 m	99% (Prec.) 62% (Recall)
[10], 2011	Probabilistic Template Matching; Mapping	TL	1.3 megapixel	15 Hz (CPU)	140 m	91.7%
[11], 2014	Blob candidate; CNN; Prior Map	TL	—	300 ms (CPU)	100 m	99.4%
[14], 2016	Blob candidate; PCANet [15]; Multi-Object Tracking	TL; Arrow	1920 × 1080	3 Hz (CPU)	13 px	95.7% (Prec.) 95.7% (Recall)
[16], 2019	SSD [17]; Prior Map	TL	1368 × 1096	17.9 ms (GPU)	150 m	86.9%
[18], 2017	YOLO [19]; DNN Classifier; Tracking	TL; Arrow	1280 × 720	15 Hz (GPU)	4 px	—
Ours	Circle, Blob & Shape candidate; AdaBoost; Prior Map	TL; Arrow	1280 × 960	64 ms (CPU)	150 m 2 px	91.8% (TL) 56.7% (Arrow)

Although different approaches have been developed, most of the methods involve extracting candidates according to their specific color spaces [14,20] and circular shape [10,21], and identifying them as TLs or arrow TLs. Regardless of the country, the TLs mainly consists of circle and arrow shaped lights. In the case of the ROI-based recognition, detection of circular objects is one of the effective approaches to recognize lighting areas because the search region is limited in it. According to Table 1, many methods adopted a blob detector which extracted candidate objects by binarizing the image and segmenting pixels [11,13,14]. It can detect circular objects even if their size is a few pixels. Then, the recognition of the whole shape of the TLs are implemented using specific shape matching and machine learning. Moreover, the effect of introducing object tracking to stabilize the recognition result has been reported [14,22]. In recent years, there have been reports of cases in which performance is improved upon using deep neural network (DNN) [11,14,16,18,23]. In order to detect arrow TLs, machine learning-based detector is a key solution. As shown in Table 1, it has been reported that these methods can recognize TLs at distances exceeding 100 m, with a recognition rate of approximately 90%. However, it is difficult to directly compare each performance, because the specifications of the camera (image sensor, resolution, and field of view), the driving scene, and the quality of data are different. Herein, we discuss algorithm limitations by comparing the pixel size of recognizable objects.

On the other hand, assuming that our algorithm will be introduced into the automated vehicles, real-time proceeding is important for decision making. In addition, in order to reduce the delay in recognition, it is necessary to recognize TLs in an appropriate processing time in accordance with the velocity of the vehicle. For example, when traveling at a velocity of 50 km/h, a vehicle moves about 14 m per second. Then, it is important to estimate the required time in consideration of the responsive deceleration for practical development.

In our previous study [24], the TL recognition method was proposed using sped up robust features (SURF) [25]-based circular object detector. It can detect circular objects like a blob detector without binarization, therefore the robust candidate extraction is expected. In addition, the proposed method estimates the existence probability of lighting objects in an image. It has the advantage of reducing false positive detections which caused by surrounding lighting objects. In this work, we improve our method by integrate the typical candidate detectors such as a circular detector using SURF, a blob detector, and the TL shape detector. In particular, we investigate the performance and limitation of the arrow detection method by introducing prior information in order to robustly detect arrow TLs. Moreover, we verify that the performance requirements for the recognition distance are satisfied, by performing automated driving on an urban road. The followings are the contributions of this paper.

- The performance evaluation is performed in challenging dataset including small objects of TLs and arrow TLs in the day and night.
- The effectiveness of prior information is evaluated with respect to the performance in recognition of distant TLs.
- Verification results are presented by introducing the proposed method in automated driving.

3. Proposed Traffic Light Detection

3.1. *Traffic Light Detection*

Before describing the proposed algorithm, the problem of the TL recognition addressed in this study is explained. In the TL recognition, the task is to recognize the state of the TLs in the image that corresponds to the HD map. As shown in Figure 1a,b, it is necessary to properly recognize the lighting status of TLs both in the day and the night. On the HD map, the position information of the TLs is recorded individually, and then the TL positions in the camera image can be calculated from the position of the TL and the vehicle. Figure 1c indicates the typical ROI image which is extracted by the coordinate transform using the HD map for a driving image. As in the enlarged image in Figure 1c, the extracted ROI image may include TLs other than the ones, and background lighting objects. In implementing automated driving at intersections, the purpose is to recognize the TL associated with the ROI. Therefore, if a different TL is recognized in the specific ROI, it will be a false-positive detection.

Figure 2 shows the TL patterns to be recognized by the proposed method. This study focuses on the recognizing TLs in the Japanese traffic environment. We deal with the TL patterns that include three basic types of lights (green, yellow, red) and three types of arrow lights (left, straight, and right) that exist depending on the road environment in Japan. In addition, because there are horizontal and vertical TLs depending on the area in Japan, the proposed method recognizes these patterns as well. As a special case, there are cases where arrow lights in different positions and arrow lights in different directions as shown in Figure 2, are installed in the actual environment. Evaluation of recognition performance for such special situations has not been performed in this work, but it can be easily extended by using the digital map information described herein, as a prior information. Although the proposed method will be evaluated for Japanese traffic images in this work, the proposed method is able to apply to general traffic lights which consist of circular lights and arrow lights. In the proposed method, the recognition distance can be improved if the arrangement pattern of the signal light and the arrow light is known for the target TLs as shown in Figure 2.

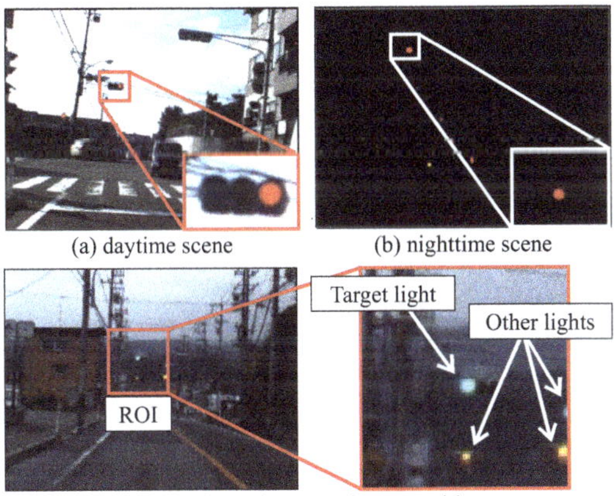

Figure 1. Typical traffic light image at different brightness and region of interest (ROI) image.

Figure 2. Traffic light patterns that can be recognized in the proposed method.

3.2. Predefined Digital Map

A highly-precise digital map is maintained by a growing number of professional mapping companies. Accurate positioning systems, combined with cameras and LiDAR sensors can be possible to generate a precise 3-D maps which contains latitude, longitude and altitude reflectivity. For the purpose of the TL detection, location data of each TL was recorded into the map information. The method described in this paper uses the following information as prior map information:

- the 2-D TL positions (latitude, longitude, and heading direction)
- attribute information of the TL directions (horizontal or vertical)
- attribute information of the type of the TL patterns (see Figure 2a).

Although the exact altitude of the TLs is not used as a prior information, they are installed at a height of approximately 5.0 m above the ground surface on the Japanese road environment. Therefore, the recognition process is performed by considering a height of 5.0 m as a reference height, and providing a margin that assumes a road gradient. Although the standard height of TLs is specified in Japan, the standard height should be different in other countries. It is necessary to set an appropriate height according to the target country or to set a wider recognition area in the image when the height information is unknown.

3.3. Method

Figure 3 illustrates a flowchart of the proposed method. It mainly consists of the following five procedures:

1. Search target TLs and compute ROI
2. Generate a highlighted image as a feature image which emphasizes the lights of TLs in the ROI image

3. Extract candidates for TL lights in the generated highlighted image using three types of different methods
4. Compute the probability of existence area containing TLs using a time-series processing
5. Recognize the arrow light, if the target TL has attribute information of arrow light.

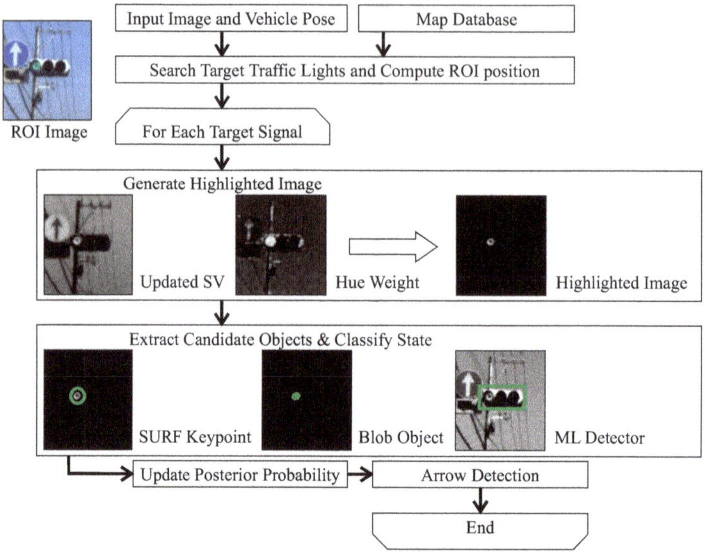

Figure 3. Flowchart of the proposed method.

As described in Section 2, most of the existing recognition methods mainly use individual features such as circular objects, blob regions, and overall shapes using a machine learning (ML) detector to recognize the TL candidates. The proposed method combines them to detect candidates for robust implementation. By using detection methods that focuses on the lighting area of the TLs, it is possible to recognize them even if the overall shape of the TL is not visible, such as during occlusion or at night time. Figure 4 shows typical driving images in an occluded and a dark scenes. Occlusion of TLs is caused by surrounding other vehicles such as a preceding vehicle, a bus, and a tuck. As shown in Figure 4a, there is a situation that the occluded situation where it is difficult to see the whole shape of the TL. However, it is necessary to recognize the TL state only from the lighting area. The situation where such an overall shape cannot be visually recognized is the same even at night as shown in Figure 4b. Section 4 evaluates the contribution of each method by comparing the recognition performances.

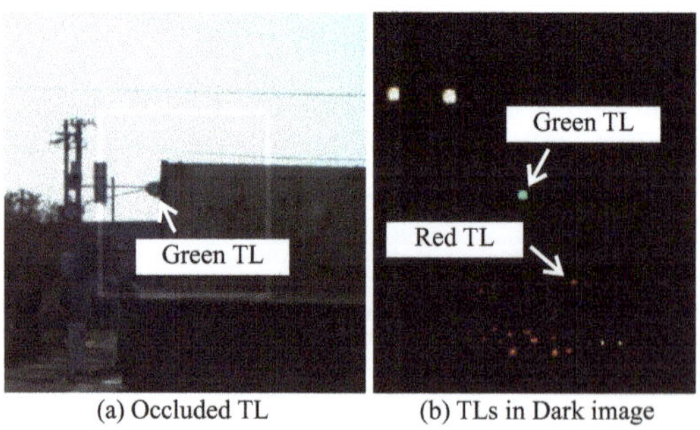

Figure 4. Typical traffic light (TL) images in occluded and dark scenes.

Arrow detection requires recognition of directions, which is affected by unclear images. In order to improve such distant arrow recognition, we propose an arrow recognition method using prior information of the HD map. In addition, this study verifies the effects of using prior information in the arrow light recognition. The algorithms are described in detail in the following section.

3.4. Coordinate System and Traffic Light Selection

Figure 5 illustrates the coordinate systems considered in this work. The latitude and longitude values are converted to 2-D $x_g - y_g$ space in the universal transverse mercator (UTM) coordinate system. The world coordinate system is defined as x_g, y_g and z_g(=altitude). The vehicle coordinate system is centered at the rear wheels. The front direction is x_v, the left direction is y_v and the upper direction is z_v. In a similar way, sensor coordinates ($x_s - y_s - z_s$) and image coordinates ($u - v$) are defined as shown in Figure 5a. The world, vehicle and sensor coordinates can be transformed by using rotation and translation matrices. Sensor and image coordinates are transformed by using intrinsic parameters.

Among the TLs that appear in the frontal camera image, the TLs that are within a certain distance d_T, and whose heading angle difference is within a certain degree θ_T, are extracted as target TLs to be recognized. The target TLs are extracted based on the distance parameter d_T m from faced traffic signals as shown in Figure 5b. In Figure 5b, the red TLs are the extracted target TLs.

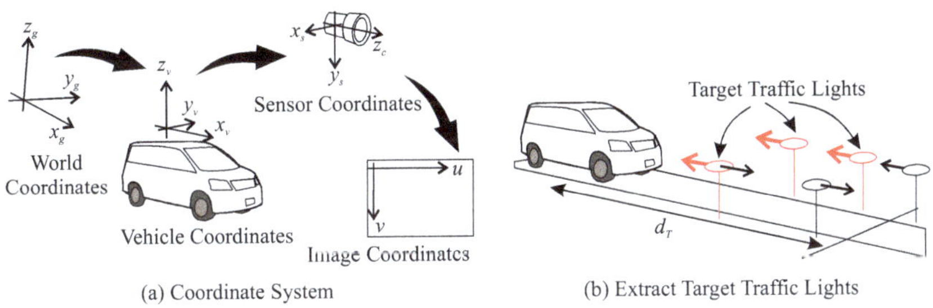

Figure 5. Coordinate systems and traffic light selection.

3.5. ROI Clipping

The ROI image is clipped for each target TL. The location of the ROI can be calculated based on the current pose and the map database. A global TL position $\boldsymbol{x_w} = [x_w, y_w, z_w, 1]$ is converted to $\boldsymbol{x_v} = [x_v, y_v, z_v, 1]$ and $\boldsymbol{x_s} = [x_s, y_s, z_s, 1]$ by the vehicle pose and extrinsic parameters of the camera.

$$\boldsymbol{x_v} = R_{wv}\boldsymbol{x_w} \tag{1}$$

$$\boldsymbol{x_s} = R_{vs}\boldsymbol{x_v}, \tag{2}$$

where R_{wv} and R_{vs} are 4×4 homogeneous transformation matrices for converting world-to-vehicle coordinates and vehicle-to-sensor coordinates, respectively. As described above, if there is no information on the absolute height of the TL, the general TL height is used to compute z_v, assuming a flat surface. In this case, the height z_v is calculated by the following equation:

$$z_v = z_0 - x_v \tan\phi, \tag{3}$$

where z_0 is the general TL height from the road surface (e.g., $z_0 = 5.0$m), and ϕ is the pitch angle of the vehicle. A pixel position u, v of the signal is then calculated based on the intrinsic parameters and the following set of equations:

$$x'' = x_s/z_s, \quad y'' = y_s/z_s, \quad r^2 = x'^2 + y'^2 \tag{4}$$
$$x'' = x'(1 + k_1 r^2 + k_2 r^4) + 2p_1 x' y' + p_2(r^2 + 2x'^2) \tag{5}$$
$$y'' = y'(1 + k_1 r^2 + k_2 r^4) + p_1(r^2 + 2y'^2) + 2p_2 x' y' \tag{6}$$
$$u = f_x x'' + c_x, \quad v = f_y y'' + c_y, \tag{7}$$

where $f_x, f_y, c_x, c_y, k_1, k_2$ are intrinsic parameters of the camera. The ROI is defined as a rectangle with a width w_{roi}, and a height h_{roi} centered at the pixel u, v.

$$w_{roi} = k_{roi} \frac{f_x s_s}{z_s} \tag{8}$$
$$h_{roi} = k_{roi} \frac{f_y s_s}{z_s}, \tag{9}$$

where s_s is the size of a TL and k_{roi} is a constant parameter to determine a scale of the ROI.

3.6. Highlighted Image Generation

The lighting areas of the TLs have higher brightness and saturation compared to other objects. Therefore, the highlighted image can be generated by multiplying the saturation image by the brightness image. In order to extract lighting areas, RGB images are converted into HSVcolor space. The lighting area of the TL gets highlighted as shown in Figure 6a. The highlighted images have higher brightness and saturation to emphasize the lighting areas of TLs. However, in some cases, the highlighted image cannot emphasize the lighting area sufficiently, especially for the lamp-type TLs. In addition, in recognition of distant TLs where the image is unclear, there is a possibility that false-positive detections may occur under the influence of background noise. In order to solve these problems, we have previously reported a method that can reduce false-detection, such as false-positive and false-negative, in distant places, by correcting and weighting the highlighted images [26]. The following processes are suggested to emphasize TLs:

- Normalize the brightness value to emphasize the lighting.
- Update the saturation value to eliminate background noise.
- Weighting with respect to hue value, close to the lighting color of traffic signals.

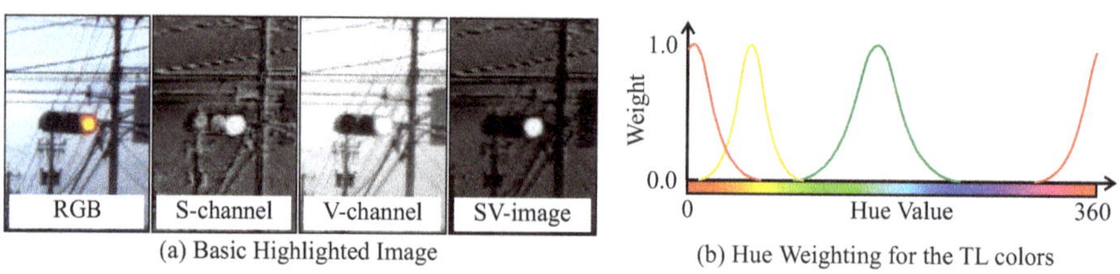

Figure 6. Highlighted image generation.

The first operation updates the image brightness. The brightness value is normalized using the following equation:

$$V_m(u, v) = k_v \bar{V} + \frac{\sigma_v}{\sigma}(V(u, v) - \bar{V}), \tag{10}$$

where $V(u, v)$ and $V_m(u, v)$ are the original brightness from the HSV image and modified brightness value at pixel (u, v), respectively. $\bar{V}$ is the average brightness of the original brightness image $\boldsymbol{V}$. σ is the standard deviation for $\boldsymbol{V}$, and σ_v is the modified standard deviation for the updated image $\boldsymbol{V_m}$. k_v is a constant parameter that increases the average brightness.

The second operation updates the saturation values. The lighting area of the traffic signal generally has saturation values above a certain value. The pixels with saturations lower than this value are reduced using the following equation:

$$S_m(u,v) = S(u,v) \cdot \frac{1}{(1+\exp(-a_s(S(u,v)-b_s))}, \tag{11}$$

where $S(u,v)$ and $S_m(u,v)$ are the original saturation values from the HSV and modified saturation value, respectively. a_s and b_s are constant parameters of the sigmoid function to reduce the saturation value. $\boldsymbol{S_m}$ and $\boldsymbol{V_m}$ are used to generate the highlighted image instead of the SV-image.

The third operation multiplies the pixel values of the highlight image with weight, with respect to the hue values. Figure 6b shows the definition of weighting value. It means that the hue value closest to the lighting color of the traffic signal has a higher weight value.

$$W_G(u,v) = \exp\frac{-(H(u,v)-\mu_G)^2}{2\sigma_G^2} \tag{12}$$

$$W_Y(u,v) = \exp\frac{-(H(u,v)-\mu_Y)^2}{2\sigma_Y^2} \tag{13}$$

$$W_R(u,v) = \exp\frac{-(H(u,v)-\mu_R)^2}{2\sigma_R^2} \tag{14}$$

$$H_w(u,v) = \max(W_G(u,v), W_Y(u,v), W_R(u,v)), \tag{15}$$

where $H(u,v)$ is the original hue value from the HSV. $W_*(u,v)$ is the obtained weight value, μ_* is a mean value and σ_* is the standard deviation for weighting for the corresponding colors. μ_* and σ_* should be determined according to the color balance of the camera.

3.7. Candidate Extraction and State Classification

After the generation of the highlighted image, a lighting area detection is applied to the obtained image. As mentioned in Figure 3, the proposed method introduces three types of methods to extract candidate lighting objects.

The first method is the circle detector. The shape of the lighting area is generally circular shape in the image. A method based on the Hough transform has been adopted to extract circular candidates [21]. However, because a clear circular area cannot be obtained in the image for a distant TL, a blob detector described later was adopted in many works. SURF keypoints have a Hessian matrix H, the types of the edges can be categorized by using $\det(H)$ as shown in Figure 7a. Candidate circle areas can be extracted as keypoints with $\det(H)$ higher than the threshold value H_{min}. This approach can extract circular objects robustly, because SURF is a robust keypoint for illumination change and scale change.

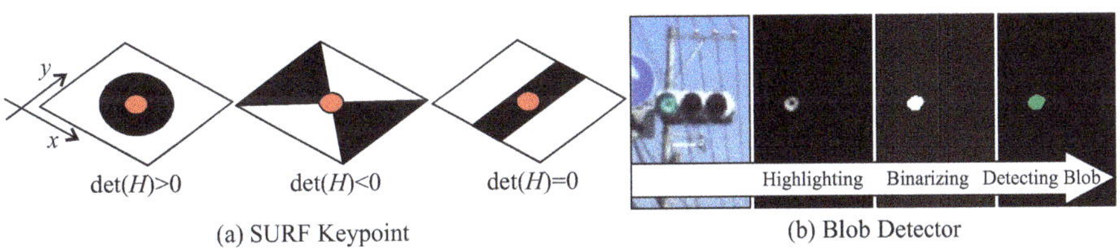

Figure 7. Candidate extraction.

The second method is the blob detector. In the feature image, areas with higher pixel values are distributed near the lighting areas. These areas can be extracted by binarizing and segmenting the image as shown in Figure 7b. Such a method of extracting candidate objects by binarization according to the brightness, saturation and color characteristics has also adopted in many related works [11,13,14]. Results are expected to be similar to circle detection, but the blob detector is expected to be better than the circle detector, when a part of the lighting part is saturated and cannot be visually recognized as a circle shape in the feature image. However, the blob detector is sensitive to threshold adjustment for binarization. If there is a signboard with a color close to the lighting color in the background, it may be detected as false postives. In addition, when the brightness of the lighting area in the image changes due to the influence of the surrounding light, it may be false negatives.

The third method is the ML-detector. Because the detection is performed using camera images, the brightness of the whole image may be affected by the influence of the surroundings, such as sunlight and environmental light. In such cases, the lighting area of the TLs cannot be sufficiently emphasized in the highlighted image. In order to recognize such TLs with lower brightness, it is effective to focus on the whole shape of the TLs together. Generally, machine learning is a common approach to detect such shape of the TL. In the proposed method, the OpenCV cascade detector trained by AdaBoost [27] is used as one of the CPU-based detectors for the TL recognition. In recent years, DNN-based detectors have shown high recognition performance in general object recognition [17,28]. DNN models such as SSD [17] and YOLOv3 [28] are known as typical networks for detecting objects in real-time. However, because the DNN model requires GPU processing, it is necessary to select an appropriate ML-Detector in consideration of the trade-off between computational resources and recognition performance.

The lighting states are classified in the detected objects by these methods, and determined as final candidates. In order to eliminate false-positive detections, objects with comparatively smaller and larger radius are deleted based on the pre-calculated radius $r_l = 0.5 f_x s_l / z_s$ from the HD map. Here, s_l is the diameter of the lamp of the TL. The accepted candidates are extracted according to the minimum radius $k_{min} r_l$, and the maximum radius $k_{max} w_l$, based on the parameters k_{min} and k_{max}. In addition, in order to reduce the processing time, the ML-detector is used only when the circle and blob detectors have not detected any objects.

The lighting color of the candidate object needs to be classified from the hue and brightness distribution of the lighting area. In the proposed method, histograms of the highlighted image and hue image are created for the detected lighting area, and then the AdaBoost classifier is trained using the normalized histograms at the maximum frequency as a feature vector. The generated classifier recognizes the lighting state via four classes, namely Green, Yellow, Red, and Background.

3.8. Probability Updating

The candidates detected in the ROI are potential objects of the target TL. In order to output a likely object from the obtained candidates, a time-series tracking process is performed by computing existence probability. In [14], multi-object tracking is implemented to improve recognition accuracy. In the proposed method, the whole shape of the TL is not always recognized. Therefore, the probability is estimated by calculating the existence probability of the object in the 2-D image space, instead of general tracking using the target as a mass point. The existence probability is computed using a binary Bayes filter (BBF). The following equation shows a relationship between the log-odds l and the probability p for the i-th target signal at time t.

$$p_i(u, v | z_{1:t}, x_{1:t}) = \frac{1}{1 + \exp(-l_{t,i}(u, v))}, \tag{16}$$

where $z_{1:t}$ and $x_{1:t}$ are the observation and the vehicle state until time t, respectively. (u, v) is the pixel location in the image and $l_{t,i}(u, v)$ is a log-odds value at the pixel (u, v) for the i-th TL. The log-odds can be updated by additional computation in BBF.

$$l_{t,i}^{Prior} = \alpha\, l_{t-1,i}^{Post} + l_{t,i}^{Obs}, \tag{17}$$

where $l_{t-1,i}^{Post}$ is the posterior log-odds at the previous time for the i-th target signal. The initial value of $l_{t,i}^{Prior}$ is set to 0. α is the decay rate for the previous posterior probability. $l_{t,i}^{Obs}$ is calculated based on the position and size of the obtained candidates. Figure 8 shows a typical example of the probability updating. For each detected candidate, the rectangular area where the TL may exist is calculated, and the observation distribution $l_{t,i}^{Obs}$ is determined based on the Gaussian distribution around the rectangular areas. Then, it is possible to estimate a likely region by performing time-series processing.

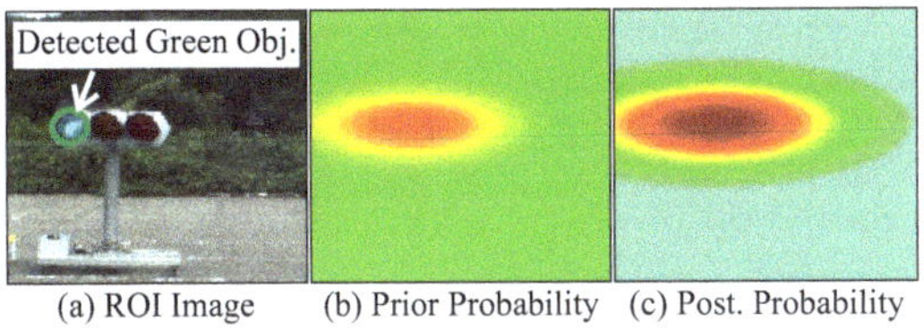

Figure 8. Probability updating.

3.9. Arrow Signal Recognition

In addition to the TL detection, an arrow signal is recognized when the TL has the attribute of arrow lights in the HD map. In Japanese traffic environment, arrow lights are generally lit at red and yellow TLs. After detecting a yellow or red signal, an arrow detection ROI is determined as shown in Figure 9. In the recognition process, the right-arrow detector is trained in advance using AdaBoost (cascade detector in OpenCV), and then it is applied to the extracted ROI. In order to detect left/straight arrows, the ROI image is rotated and the same detector is used to search objects.

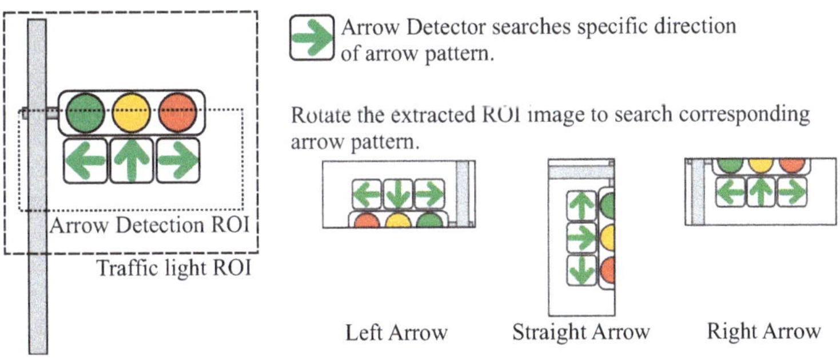

Figure 9. Arrow light recognition.

3.10. Prior Information using Digital Map

By using the proposed method described above, the TL recognition is realized by detecting the candidate objects, classifying the lighting color, and computing the confidence using the existence probability. This work further improves the recognition performance, especially for distant arrow lights, by utilizing the prior information given in the digital map.

In the TL recognition, when there are multiple candidate objects, it is possible to weight candidates according to the distance of the TLs in the probability updating procedure. It is expected to reduce false-positive detections in background.

On the other hand, in arrow recognition, recognition can be improved by providing the pattern of the target TL from Figure 2 as prior information. For example, Figure 10 illustrates the typical arrow recognition scene. In the recognition of a distant arrow light, if it is difficult to visually recognize the direction of the lighting arrow, it may cause false-positives or false-negatives. In Figure 10, it can be seen that some arrow lights are lit in the ROI image, but it is difficult to distinguish the directions. In this case, because the lighting parts of the arrow light is crushed, a candidate point may be detected at the arrow TL as well as the candidate TL. Normally, this detected candidate can be a false-positive detection of a green signal. However, if information on the relative positional relationship of the arrow lights at the TL is provided as a prior information, it is possible to distinguish the direction of the arrow lights. Our work evaluates how this prior information contributes to the recognition of TLs and arrow lights by the proposed method.

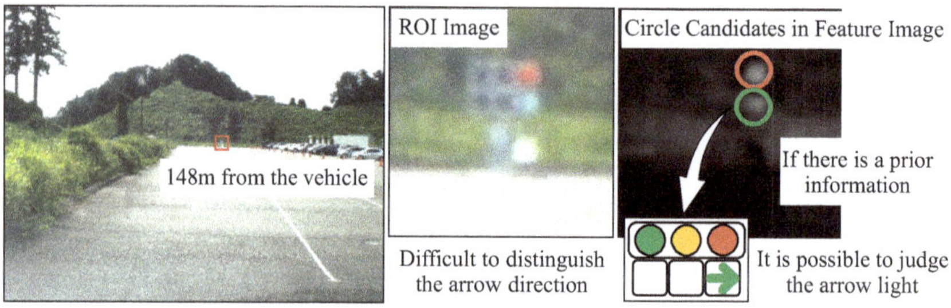

Figure 10. Arrow light recognition using prior information.

4. Evaluations

4.1. Condition

Experiments have been carried out to evaluate the effectiveness of the proposed method with actual driving data. Some driving data have been collected using the automated vehicle owned by our group. Figure 11 shows our automated vehicle. This automated vehicle was equipped with some sensors such as LiDAR, MWR, GNSS/INS and camera to observe its surroundings. A 3-D LiDAR Velodyne HDL-64E S2 with 64 separate beams was mounted on the vehicle to take measurements of the environment. It measured the 3-D omnidirectional distance at a frequency of 10 Hz. An Applanix POS/LV220 coupled GNSS and INS was mounted on the vehicle. It provided an accurate position (latitude, longitude and altitude) and orientation (pitch, yaw, roll) at 100 Hz. In addition, in order to observe the traffic lights, the vehicle was equipped with a mono-camera Pointgrey Flea2, which provided a 1280×960 pixel resolution at 7.5 Hz. There were lamp type and LED type TLs to be recognized. Because the LED type TLs blink at high speeds, the TL may have been turned off when shooting with the camera, depending on the shutter speed. To avoid this problem, the shutter speed was limited from 10 ms to 20 ms for the auto-exposure function. As a result, the image of a dark scene such as the evening was almost dark as shown in Figure 1b and the shape of the traffic light could not be seen.

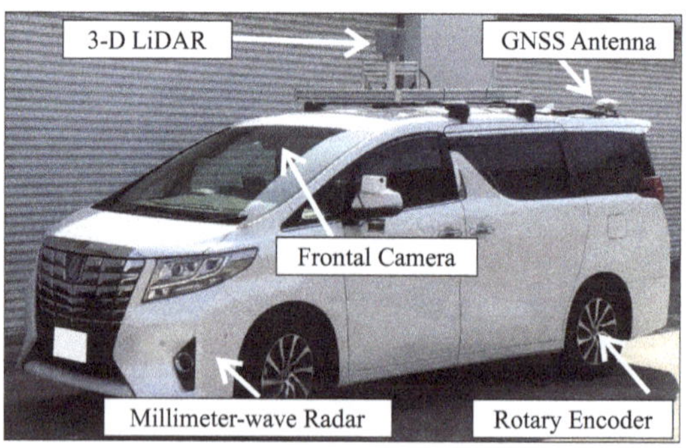

Figure 11. Experimental vehicle.

This vehicle had various functions necessary to enable automated driving in an urban area, and has actually been running in Japan for several years. In previous works, real-time localization algorithms have been developed using different types of sensors such as 3-D LiDARs, cameras or MWRs [7,29,30]. Therefore, it is assumed that the precise vehicle pose has been estimated using such localization algorithms in this evaluation.

Table 2 shows the number of images in the train and test dataset. These datasets were recorded at Kanazawa-city and Suzu-city in Ishikawa, Japan. As described in Section 3.1, the TL recognition aims to recognize a total of six states, three states of TLs (green, yellow, and red) and three states of arrow TLs (left, straight, and right). The training data were used to train the overall shape of the traffic light and the arrow light detector using machine learning. These data consisted of images measured during the daytime as visible data of the overall shape of the TL. On the other hand, the test dataset consisted of not only the daytime scene, but also the dark scene which was images measured in the early morning and evening hours. In addition, Figure 12 shows the frequency distribution of test data for distances from 30 m to 150 m. Although the ratio of the arrow TL data was small overall, the test data were distributed almost uniformly from a short distance to a long distance.

Table 2. Experimental conditions: number of data.

Train/Test	Scene	Green	Yellow	Red	Left	Straight	Right
Train	Daytime	10,211	422	5277	129	293	194
Test	Daytime	30,506	1867	17,721	662	219	890
Test	Dark	9206	646	5005	49	153	247
Test	Total	39,712	2513	22,726	711	372	1,137

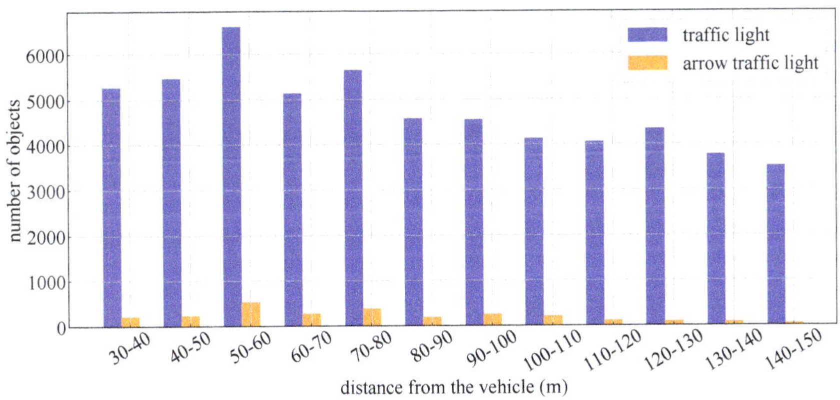

Figure 12. Histogram of number of test data in different distances from the TL.

In evaluating the performance of the proposed method, it was important to objectively compare to the existing methods. However, it was difficult to directly compare the reported performance of other works due to different types of cameras, and sensors, and different experimental conditions such as driving area, and weather conditions. Therefore, in addition to the evaluation based on the recognition distance, the recognition performance for objects with the similar pixel size as other works was also evaluated. Table 3 summarizes the characteristics of the pixel size of bounding boxes in the existing data set and the test data. The table indicates that the test data in this work were challenging data that included objects with large and small pixel sizes for traffic lights and arrow lights, even compared to existing public data sets.

Table 3. Bounding box pixel size in test dataset and other open dataset.

Dataset	Classes	Num. of Objects	W/H	Minimum	Average	Median	Maximum
Our test data (Day & Night)	6 (3 lights & 3 arrows)	67,171	Width Height	2.82 2.22	34.75 15.22	26.72 11.94	213.02 174.98
LARA [13] (Only Datytime)	4 (3 lights & ambiguous)	9,168	Width Height	6.00 14.00	11.45 27.24	11.00 28.00	37.00 93.00
WPI [14] (Only Datytime)	6 (2 lights & 4 arrows)	4,207	Width Height	13.00 13.00	28.03 27.30	28.00 28.00	47.00 48.00
Bosch [18] (Only Datytime)	4 (3 lights & off)	13,493	Width Height	1.88 3.25	9.43 26.75	8.50 24.50	48.38 104.50

The evaluations carried out in this work are summarize bellow:

- Analysis of the contribution to the recognition rate by using spatial prior information in the proposed method
- Comparison between the proposed method and a general object detection algorithm using DNN (YOLOv3 [28])
- Performance comparison of each candidate object detector (SURF, blob, AdaBoost) in the proposed method
- Comparison of processing time in recognition.

YOLOv3 was adopted as one of the state-of-the-art methods because it is a widely used DNN in object detection. The ML detector and the arrow detector (AdaBoost) of the proposed method and YOLOv3 were generated using the training data in Table 2. As described above, AdaBoost detectors were divided into two types (the TL detector and the arrow TL detector). The YOLOv3 model was generated as a model that recognizes six classes of TLs and arrow TLs. In the evaluation of the recognition rate, the data set was divided into intervals of 10 m, and precisions, recalls, and f-values for the data in each interval was used as an evaluation measure. The evaluation metrics were calculated by the following equations.

$$Precision = \frac{TP}{TP + FP} \tag{18}$$

$$Recall = \frac{TP}{TP + FN} \tag{19}$$

$$F\text{-}value = \frac{2Recall \cdot Precision}{Recall + Precision}, \tag{20}$$

where *TP* is the number of true-positives, *FP* is the number of false-positives, and *FN* is the number of false-negatives. The *TL* and the arrow *TL* were evaluated separately due to the difficulty in recognizing and the different number of data.

The relevant parameters of the proposed method were set as follows. $d_T = 150$ m, $\theta_T = 30$ deg, $k_{roi} = 3.0, s_s = 1.2$ m, $s_l = 0.4$ m, $H_{min} = 20000$, $k_{max} = 0.5$, $k_{min} = 2.5$, $\alpha = 0.8$. The computer used for the evaluation was a Windows 10 desktop PC, the CPU was Intel Xeon CPU E5-1620v4 (3.50 GHz), the memory was 16 GB, and the GPU was NVIDIA GeForce GTX TITAN X. The processing on the CPU was operated in a single thread.

4.2. Results

Figure 13 and Table 4 show the experimental results using the proposed method and YOLOv3 for the whole test data. Figure 13 indicates the recognition rate for each interval of the TLs and arrow TL, and Table 4 indicates the average of precision, recall and F-values obtained in each interval. In the TL recognition, the recognition rate of the proposed method was more than 90% even at approximately 100 m. Although the recognition rate decreased as the distance increased, it was confirmed that 80% or more could be maintained even 150 m away. Comparing the effects with and without spatial prior information, the precision, recall, and F-values were slightly improved by using spatial prior information. Therefore, a slight improvement of false-positive and false-negative detections was confirmed using spatial prior information. On the other hand, in YOLOv3, it was confirmed that a similar level of recognition rate was obtained at short distances of less than 40 m. However, the reduction of the recognition rate became larger as the distance increased compared to the proposed method. Comparing the recognition rates of the arrow TLs, it was confirmed that there was a large difference between the proposed method and other methods. Here, Figure 14 shows the average number of pixels of the lighting area in the TLs at each distance. In the object detection using machine learning, it was confirmed that the recognition rate was extremely low for arrow TLs smaller than 10 pixels. However, in the proposed method, it was found that the recognition rate at around 100 m could be greatly improved by suppressing the performance degradation. Therefore, it was shown that the proposed method, which uses the relative position between the TL and the arrow TL as prior information, can improve the recognition distance by 10–20m.

Table 4. Experimental results with or without the prior information.

Object	Method	Mean Precision	Mean Recall	Mean F-Value
traffic light	proposed	0.938	0.899	**0.918**
	without spatial prior	0.937	0.894	0.915
	YOLOv3	0.972	0.598	0.722
arrow light	proposed	0.771	0.517	**0.567**
	without spatial prior	0.542	0.386	0.429
	YOLOv3	0.611	0.352	0.417

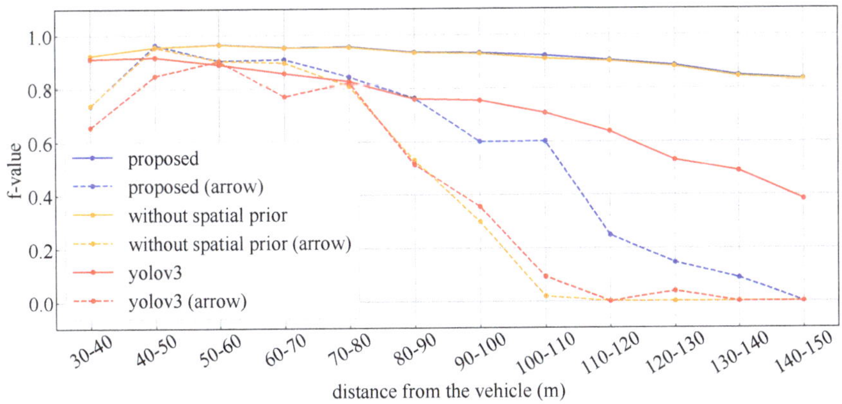

Figure 13. F values for whole data: with or without the prior information.

Next, we objectively evaluate the recognition results of this study against the performance reported in existing works. de Charatte reported a precision of 95.38% and a recall of 98.41% (37.4 ms per frame) as a recognition performance of LARA, a French dataset [13]. In addition, Chen reported a recognition rate of 95.75% (3 Hz per frame) for WPI, a USA dataset [14]. According to Table 3 and Figure 14, in our test data, the data within the range of 120 m (6 pixels or more) correspond to the difficulty of LARA, and the data within 60 m (11 pixels or more) correspond to the difficulty of WPI. Table 5 summarizes the recognition rates of the proposed method for each range of test data. The evaluation results for our test data indicates that the precision value was 97.1% for data within 60 m and 95.7% for data within 120 m. Although there were differences due to the different driving environment, it was shown that approximately the similar precision was obtained. In particular, it has been reported that the processing time of 3 Hz is required for the method [14] that can recognize both TLs and arrow

TLs using the PCANet [15] which is a compact DNN model. On the other hand, in the proposed method, it was possible to simplify the arrow detector model by using prior information, and then the average processing time was 67 ms. Therefore, the proposed method achieved a recognition rate similar to SoAin a compact processing.

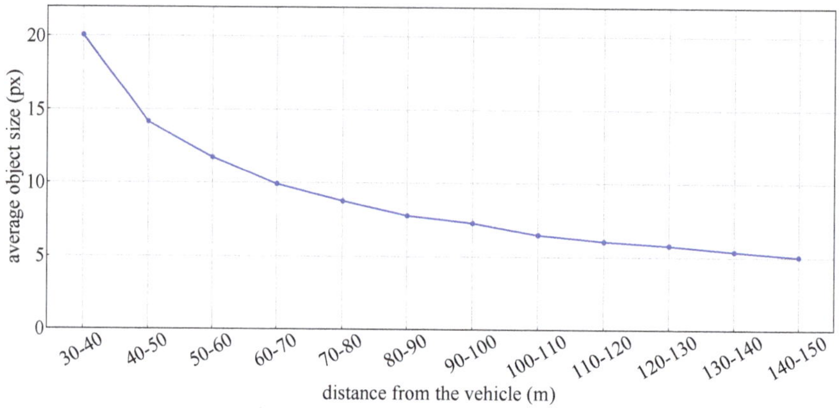

Figure 14. Average pixel size of the lighting area at different distances.

Table 5. Experimental results of the proposed method in different distance ranges.

Object	Distance		Mean Precision	Mean Recall	Mean F-Value
traffic light & arrow light	30–150 m	(ave. pixel size > 5.0)	0.935	0.897	0.910
traffic light & arrow light	30–120 m	(ave. pixel size > 6.0)	0.957	0.907	0.932
traffic light & arrow light	30–60 m	(ave. pixel size > 11.0)	0.971	0.920	0.945

In order to detect candidate objects, the proposed method combined three kinds of methods: circular features by SURF, blob features by binarized images, and features of traffic signal shape by AdaBoost detector. To evaluate the contribution of each method, the obtained recognition rates were compared with the results obtained when each detection method was used alone. Figures 15–17 show the obtained recognition rates for each detection method. These graphs indicate the recognition rate for all test data, the daytime scene data, and the dark scene data, respectively. Table 6 summarizes the average precision, recall, and F-value for the recognition result under each condition. From these results, it was confirmed that the method of circular extraction by SURF showed almost the same level of performance as the proposed method. Introducing the proposed method improved the average recognition rate by approximately 0.4%.

Although the recognition rate of the proposed method and SURF were almost identical, a detailed analysis was performed to verify the superiority of the proposed method. Figure 18 shows a graph summarizing the different precision and recall values between both methods. In this figure, it can be confirmed that the proposed method is superior for short-ranges recall and long-ranges precision values. This means that false-negative rate at short-range and false-positive rate at long-range have been improved. Circular extraction by SURF enables appropriate detection by obtaining a feature image in which the lighting area is sufficiently enhanced. However, in the case of a lamp-type traffic light with weak brightness, the emphasis in the feature image may not be sufficient, and a false-negative detection occurs. In such a case, if the overall shape of the TL is visible, improvement of the false-negative detection is expected by recognizing the TL with the ML-detector. On the other hand, in the case of distant recognition, it is achieved that feature points of blobs as well as circular features of SURF are detected simultaneously. As a result, the recognition rate was improved because the false detection of the background and the true detection points could be differentiated for the obtained candidates. Thus, although the overall improvement rate is small, the introduction of the proposed method improved the recognition rate in specific situations and thus it proves to be a robust method.

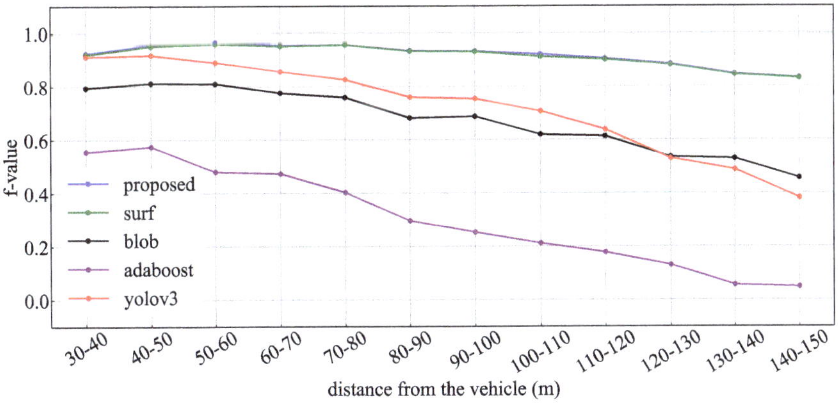

Figure 15. F values for whole data: different candidate extraction methods.

Table 6. Experimental results: different candidate extraction methods.

Scene	Method	Mean Precision	Mean Recall	Mean F-Value
Whole	proposed	0.938	0.899	**0.918**
	SURF	0.936	0.894	0.914
	blob	0.917	0.536	0.672
	AdaBoost	0.990	0.194	0.305
	YOLOv3	0.972	0.598	0.722
Daytime	proposed	0.930	0.902	**0.915**
	SURF	0.928	0.899	0.913
	blob	0.921	0.587	0.712
	AdaBoost	0.990	0.250	0.372
	YOLOv3	0.986	0.710	0.806
Dark	proposed	0.959	0.871	0.913
	SURF	0.958	0.875	**0.915**
	blob	0.902	0.372	0.523
	AdaBoost	0.583	0.007	0.013
	YOLOv3	0.840	0.226	0.345

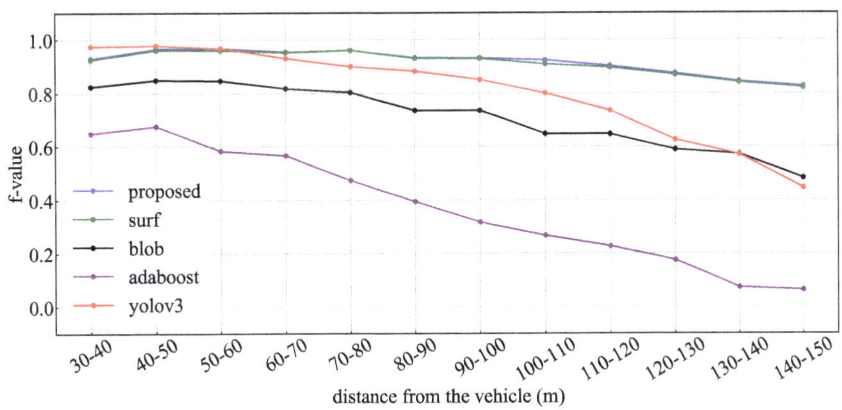

Figure 16. F values for daytime data: different candidate extraction methods.

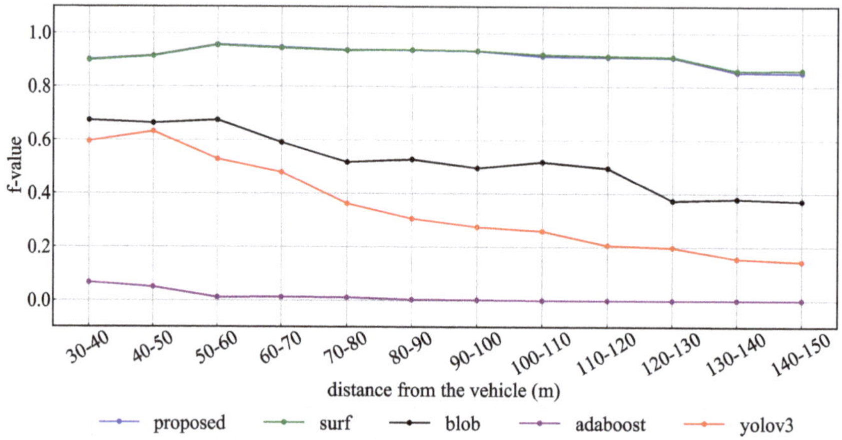

Figure 17. F values for dark data: different candidate extraction methods.

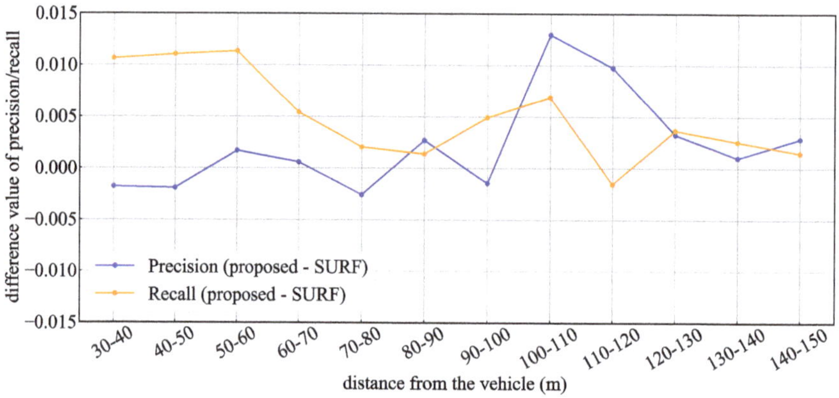

Figure 18. Difference value of precision/recall between the proposed method and sped up robust features (SURF).

Figure 19 shows typical scenes of true-positives for small arrow detection. By utilizing prior information, the proposed method can detect small arrow TLs of 10 pixels or less. As shown in Figure 19, it was confirmed that an arrow of 5 pixels could be recognized. However, according to the evaluated results, there were difficult scenes where false-negatives occurred. Figure 20 shows typical false-negative images. On suburban roads, old ramp signals are installed. Such traffic lights have very low brightness, and in some cases there are images where it is difficult for humans to see the lighting color. In such a situation, even if the overall shape of the TL could be detected, it was false-negative because the lighting color was not bright enough.

Finally, the processing time for each method has been evaluated. Table 7 shows the average processing time and standard deviation of each method. In this experiment, the images have been captured at 7.5 Hz Then, if that could be processed within 133 ms, it would be a method that can be said to operate in real-time. According to Table 7, the real-time operation is possible because the average processing time of all methods was within 100 ms. However, while YOLOv3 required processing on the GPU, other methods, including the proposed method, could be operated with only the CPU. Therefore the proposed method was practically easier. Automated vehicles are required to process many recognition and decision-making processes in a limited computer environment. In this respect, usefulness was shown by analyzing the recognition performance and processing performance of the proposed method. According to Table 7, there were cases where the proposed method took about 100 ms instantaneously. There was no critical problem on the decision of approaching the intersection, because the moving amount of the vehicle during that time was about 1 or 2 m. The validity of the TL recognition during automated driving was evaluated in the next section.

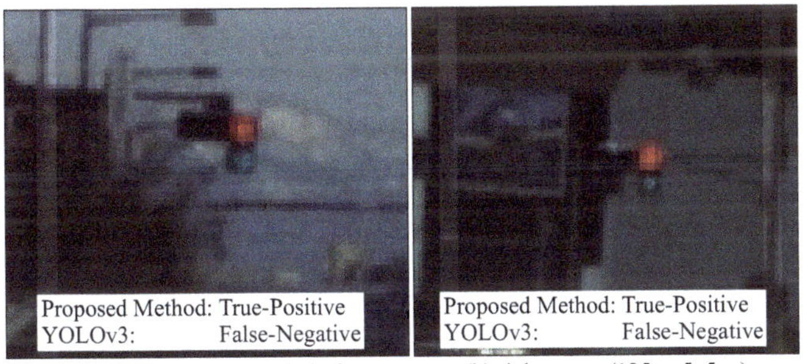

(a) right arrow (111m, 6x6px) (b) right arrow (125m, 5x5px)

Figure 19. Typical improvement for far arrow TLs.

(a) distant TL (Green, 150m, 5x12px) (b) near TL (Green, 12m, 68x200px)

Figure 20. Typical false-negative situations due to low brightness of lamp-type TLs.

Table 7. Experimental results: computational time.

Method	CPU/GPU	Average (ms)	Standard Deviation (ms)
proposed	CPU	64.278	19.374
SURF	CPU	61.036	16.890
blob	CPU	45.806	10.267
AdaBoost	CPU	71.663	22.322
YOLOv3	CPU/GPU	93.945	29.693

4.3. *Verification for Automated Driving*

We showed the superiority of the proposed method by evaluating a recognition distance for TLs and arrow TLs. Next, additional verification was carried out to determine if the proposed method has the required performance in actual automated driving. In the verification experiment, automated driving was performed in the Odaiba area of Tokyo as shown in Figure 21. In the automated driving, the role of the TL recognition is to determine the intersection approach following the traffic rules. If the signal state at the intersection is not properly recognized, the decision to enter the intersection will be incorrect and unnatural acceleration/deceleration will occur. In order to evaluate the effects of such a situation, it is necessary to investigate the stability of the TL recognition results and the vehicle acceleration when the automated vehicle is driving at the maximum velocity (the speed limit). Therefore, the transition of velocity and acceleration when passing through an intersection is verified while recognizing TLs.

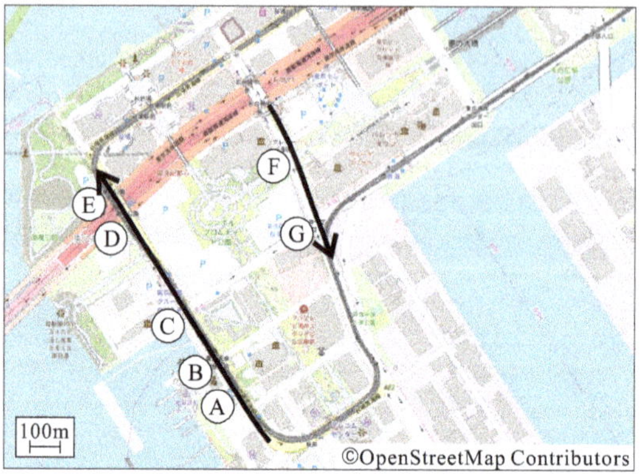

Figure 21. Driving routes for verification experiments.

Driving data were measured on the following two types of travel routes.

- Route passing through intersections A to E with only TLs in Figure 21
- Route passing through intersections F to G with TLs and arrow TLs in Figure 21.

In both routes, there was no preceding vehicle, because the automated vehicle should drive at the maximum velocity. Then it was verified whether the vehicle could drive smoothly while properly recognizing each TL. In this verification experiment, a camera different from Section 4.1 was used owing to a hardware issues. A camera with a resolution of 1920 × 1080 was installed in this experiment. It has a similar vertical field of view compared to the camera described in Section 4.1. Therefore, the recognition distance was extended by the ratio of the resolution ratio compared to the recognition distance in Section 4.2. Figures 22 and 23 show the velocity and acceleration of each driving data and the TL state recognized at each intersection as verification results. The bar graph shows the TL status recognized at each time, the color of the graph is the light color, and the numerical value below it is the distance to the corresponding TL. According to Figure 22, it can be confirmed that the TL state was recognized immediately within 150 m of the recognition range. At the intersections C and E, the vehicle stopped at a red light. In this case, the vehicle stopped smoothly with moderate deceleration. On the other hand, Figure 23 shows a driving scene passing through an intersection while recognizing an arrow light in the straight direction. As a result of the evaluation in Section 4.2, the recognition performance of the arrow light was deteriorated for distant objects. Therefore, the recognition result at a point 100 m away at the intersection G was unstable. From the experimental results, it can be confirmed that the recognition result became stable at a point 85 m away. Even in such a situation, it was shown that unnecessary deceleration did not occur according to the transition of speed and acceleration. Thus, it was shown that the recognition result obtained by the proposed method had the necessary performance to drive smoothly at intersections in the urban areas.

In addition, in our group's work, demonstration experiments of automated driving have been conducted in the area shown in Figure 21 since September 2019. In the half-year, there were no critical problems regarding the recognition distance and stability of the TL recognition. In such respect, the validity of the proposed method was qualitatively evaluated.

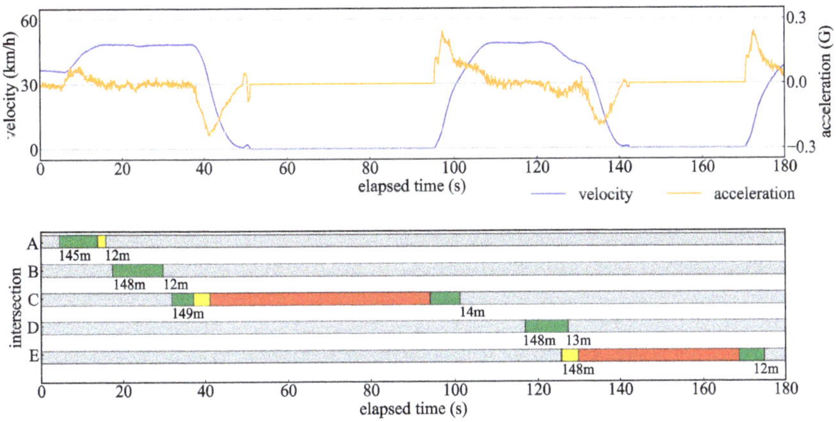

Figure 22. Verification results for automated driving from the intersection A to E.

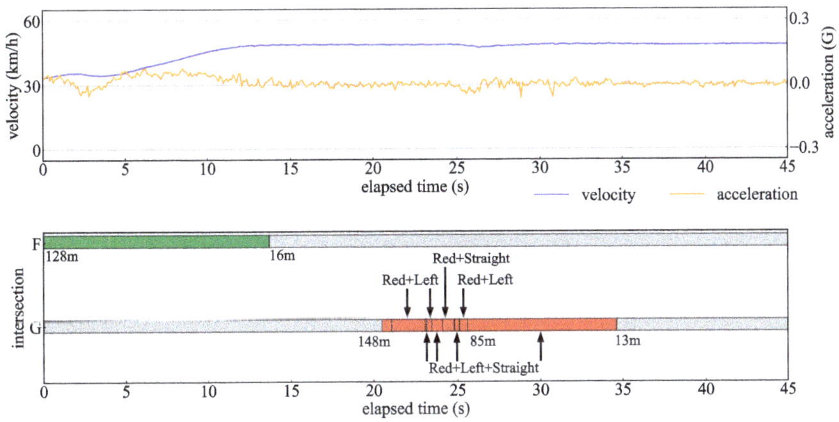

Figure 23. Verification results for automated driving from the intersection F to G.

5. Conclusions

This work has proposed a TL recognition algorithm for urban automated driving. We prepared the challenging dataset that includes objects with large and small pixel sizes of objects for traffic lights and arrow lights. The proposed method can be processed in real time by the CPU, and our work verified that the proposed method can recognize TLs within 150 m with an F-value of 91.8%. This f-value is the recognition rate in one frame. When approaching an intersection from a distance of 150 m, recognition process of about 100 frames is performed, then the state of the intersection can be estimated with high confidence. The evaluations verify the following as the contributions of the work.

- Robust recognition was achieved by integrating multiple types of detection methods to recognize TLs including the small size of objects with a few pixels.
- Arrow TL recognition using prior information obtained from the HD map was proposed, and it was shown that small arrow object can be detected even if their size is smaller than 10 pixel.
- It was verified that the proposed method satisfies the necessary requirements for smooth deceleration of approximately 0.2 G at intersection approaches in urban automated driving.

In the arrow recognition by the proposed method, the arrangement pattern of the signal light and the arrow light given in Figure 2 was used as prior information to improve the recognition rate for distant objects. The map information corresponding to such prior information can be used practically, because it is included as static information of the HD map that has already been studied [31]. On the other hand, the evaluation in this experiment showed a recognition rate of more than 90%, but there were cases where recognition became difficult. In addition to Figure 20, there are cases where it is relatively difficult to determine the lighting color of the lamp due to the influence of the surrounding brightness. For example, in the case of receiving the sunlight from behind the vehicle, all lamps may appear to be lit. Moreover, there are cases where yellow and red lighting colors can be visually recognized to the same color in the images. These cases will be a false-positive detection. In addition to the issues of software based recognition algorithms, but also the performance limit of hardware needs to be discussed from a practical point of view. As described in [32], there are situations in which it is difficult to view the TLs in the image in severe sunshine as a hardware performance limit. It is desirable to develop a practical recognition methods while discussing such software and hardware issues.

Author Contributions: Conceptualization, and methodology, K.Y., A.K. and N.S.; software and validation, K.Y., A.K. and T.A.; data collection, K.Y., A.K., N.S. and R.Y.; data analysis, K.Y., A.K., T.A. and M.A.; writing—original draft preparation, K.Y.; writing—review and editing, K.Y. and A.K.; supervision, N.S.; project administration, K.Y.; funding acquisition, N.S. All authors have read and agreed to the published version of the manuscript.

Funding: This research was funded by New Energy and Industrial Technology Development Organization (NEDO).

Acknowledgments: This work was supported by Council for Science, Technology and Innovation(CSTI), Cross-ministerial Strategic Innovation Promotion Program (SIP), "Research and development for recognition technologies and other technologies necessary for automated driving technologies(levels 3 and 4)" (funded by NEDO).

Conflicts of Interest: The authors declare no conflict of interest.

References

1. Franke, U.; Pfeiffer, D.; Rabe, C.; Knoeppel, C.; Enzweiler, M.; Stein, F.; Herrtwich, R.G. Making Bertha See. In Proceedings of the ICCV Workshop on Computer Vision for Autonomous Driving, Sydney, Australia, 8–12 December 2013.
2. Broggi, A.; Cerri, P.; Debattisti, S.; Laghi, M.C.; Porta, P.P.; Panciroli, M.; Prioletti, A. PROUD-Public ROad Urban Driverless test: Architecture and results. In Proceedings of the 2014 IEEE Intelligent Vehicles Symposium, Dearborn, MI, USA, 8–11 June 2014; pp. 648–654.
3. Kato, S.; Takeuchi, E.; Ishiguro, Y.; Ninomiya, Y.; Takeda, K.; Hamada, T. An Open Approach to Autonomous Vehicles. *IEEE Micro* **2015**, *35*, 60–69. [CrossRef]
4. Hoberock, L.L. A survey of longitudinal acceleration comfort studies in ground transportation vehicles. *J. Dyn. Syst. Meas. Control* **1977**, *99*, 76–84. [CrossRef]
5. Powell, J.P.; Palacin, R. Passenger Stability Within Moving Railway Vehicles: Limits on Maximum Longitudinal Acceleration. *Urban Rail Transit* **2015**, *1*, 95–103. [CrossRef]
6. Levinson, J.; Thrun, S. Robust Vehicle Localization in Urban Environments Using Probabilistic Maps. In Proceedings of the 2010 IEEE International Conference on Robotics and Automation, Anchorage, AK, USA, 3–7 May 2010; pp. 4372–4378.
7. Yoneda, K.; Hashimoto, N.; Yanase, R.; Aldibaja, M.; Suganuma, N. Vehicle Localization using 76 GHz Omnidirectional Millimeter-Wave Radar for Winter Automated Driving. In Proceedings of the 2018 IEEE Intelligent Vehicles Symposium, Changshu, China, 26–30 June 2018; pp. 971–977.
8. Wolcott, R.W.; Eustice, R.M. Visual Localization within LIDAR Maps for Automated Urban Driving. In Proceedings of the 2014 IEEE/RSJ International Conference on Intelligent Robots and Systems, Chicago, IL, USA, 14–18 September 2014; pp. 176–183.
9. Fairfield, N.; Urmson, C. Traffic Light Mapping and Detection. In Proceedings of the 2011 IEEE International Conference on Robotics and Automation, Shanghai, China, 9–13 May 2011; pp. 5421–5426.
10. Levinson, J.; Askeland, J.; Dolson, J.; Thrun, S. Traffic light mapping, localization and state detection for autonomous vehicles. In Proceedings of the 2011 IEEE International Conference on Robotics and Automation, Shanghai, China, 9–13 May 2011; pp. 5784–5791.

11. John, V.; Yoneda, K.; Qi, B.; Liu, Z.; Mita, S. Traffic Light Recognition in Varying Illumination Using Deep Learning and Saliency Map. In Proceedings of the 17th International IEEE Conference on Intelligent Transportation Systems, Qingdao, China, 8–11 October 2014; pp. 2286–2291.
12. John, V.; Yoneda, K.; Liu, Z.; Mita, S. Saliency Map Generation by the Convolutional Neural Network for Real-Time Traffic Light Detection using Template Matching. *IEEE Trans. Comput. Imaging* **2015**, *1*, 159–173. [CrossRef]
13. De Charette, R.; Nashashibi, F. Real Time Visual Traffic Lights Recognition Based on Spot Light Detection and Adaptive Traffic Lights Templates. In Proceedings of the 2009 IEEE Intelligent Vehicles Symposium, Xi'an, China, 3–5 June 2009; pp. 358–363.
14. Chen, Z.; Huang, X. Accurate and Reliable Detection of Traffic Lights Using Multiclass Learning and Multiobject Tracking. *IEEE Intell. Transp. Syst. Mag.* **2016**, *8*, 28–42. [CrossRef]
15. Chan, T.H.; Jia, K.; Gao, S.; Lu, J.; Zeng, Z.; Ma, Y. Pcanet: A simple deep learning baseline for image classification? *arXiv* **2014**, arXiv:1404.3606.
16. Hirabayashi, M.; Sujiwo, A.; Monrroy, A.; Kato, S.; Edahiro, M. Traffic light recognition using high-definition map features. *Robot. Auton. Syst.* **2019**, *111*, 62–72. [CrossRef]
17. Liu, W.; Anguelov, D.; Erhan, D.; Szegedy, C.; Reed, S.; Fu, C.Y.; Berg, A.C. SSD: Single Shot MultiBox Detector. In Proceedings of the European Conference on Computer Vision, Amsterdam, The Netherlands, 8–10 October 2016; pp. 21–37.
18. Behrendt, K.; Novak, L.; Botros, R. A Deep Learning Approach to Traffic Lights: Detection,Tracking, and Classification. In Proceedings of the 2017 IEEE International Conference on Robotics and Automation, Singapore, 29 May–3 June 2017; pp. 1370–1377.
19. Redmon, J.; Divvala, S.K.; Girshick, R.B.; Farhadi, A. You only look once: Unified, real-time object detection. In Proceedings of the 2016 IEEE Conference on Computer Vision and Pattern Recognition, Las Vegas, NV, USA, 27–30 June 2016.
20. Siogkas, G.; Skodras, E.; Dermates, E. Traffic Lights Detection in Adverse Conditions using Color, Symmetry and Spatiotemporal Information. In Proceedings of the VISAPP 2012, Rome, Italy, 24–26 February 2012; pp. 620–627.
21. Omachi, M.; Omachi, S. Traffic light detection with color and edge information. In Proceedings of the ICCSIT 2009, Beijing, China, 8–11 August 2009; pp. 284–287.
22. Trehard, G.; Pollard, E.; Bradai, B.; Nashashibi, F. Tracking both pose and status of a traffic light via an Interacting Multiple Model filter. In Proceedings of the 17th International Conference on Information Fusion, Salamanca, Spain, 7–10 July 2014.
23. Pon, A.D.; Andrienko, O.; Harakeh, A.; Waslander, S.L. A Hierarchical Deep Architecture and Mini-Batch Selection Method For Joint Traffic Sign and Light Detection. In Proceedings of the IEEE 15th Conference on Computer and Robot Vision, Toronto, ON, Canada, 8–10 May 2018; pp. 102–109.
24. Yoneda, K.; Suganuma, N.; Aldibaja, M. Simultaneous Sate Recognition for Multiple Traffic Signals on Urban Road. In Proceedings of the MECATRONICS-REM, Compiegne, France, 15–17 June 2016; pp. 135–140.
25. Bay, H.; Tuytelaars, T.; Gool, L.V. SURF: Speeded Up Robust Features. In Proceedings of the European Conference on Computer Vision 2006, Graz, Austria, 7–13 May 2006; pp. 404–417.
26. Kuramoto, A.; Kameyama, J.; Yanase, R.; Aldibaja, M.; Kim, T.H.; Yoneda, K.; Suganuma, N. Digital Map Based Signal State Recognition of Far Traffic Lights with Low Brightness. In Proceedings of the 44th Annual Conference of the IEEE Industrial Electronics Society, Washington, DC, USA, 21–23 October 2018; pp. 5445–5450.
27. Schapire, R.E.; Singer, Y. Improved Boosting Algorithms Using Confidence-rated Predictions. *Mach. Learn.* **1999**, *37*, 297–336. [CrossRef]
28. Redmon, J.; Farhadi, A. YOLOv3: An Incremental Improvement. *arXiv* **2018**, arXiv:1804.02767r.
29. Yamamoto, D.; Suganuma, N. Localization for Autonomous Driving on Urban Road. In Proceedings of the International Conference on Intelligent Informatics and BioMedical Sciences, Okinawa, Japan, 28–30 November 2015.
30. Yoneda, K.; Yanase, R.; Aldibaja, M.; Suganuma, N.; Sato, K. Mono-Camera based vehicle localization using Lidar Intensity Map for automated driving. *Artif. Life Robot.* **2018**, *24*, 147–154. [CrossRef]

31. Watanabe, Y.; Sato, K.; Takada, H. DynamicMap 2.0: A Traffic Data Management Platform Leveraging Clouds, Edges and Embedded Systems. *Int. J. Intell. Transp. Syst. Res.* **2018**, *18*, 77–89. [CrossRef]
32. Yoneda, K.; Suganuma, N.; Yanase, R.; Aldibaja, M. Automated driving recognition technologies for adverse weather conditions. *IATSS Res.* **2019**, in press. [CrossRef]

sensors

MDPI

Article

A Repeated Game Freeway Lane Changing Model

Kyungwon Kang and Hesham A. Rakha *

Center for Sustainable Mobility, Virginia Tech Transportation Institute, Blacksburg, VA 24061, USA; kwkang@vt.edu

* Correspondence: hrakha@vt.edu; Tel.: +1-540-231-1505

Received: 29 January 2020; Accepted: 5 March 2020; Published: 11 March 2020

Abstract: Lane changes are complex safety- and throughput-critical driver actions. Most lane-changing models deal with lane-changing maneuvers solely from the merging driver's standpoint and thus ignore driver interaction. To overcome this shortcoming, we develop a game-theoretical decision-making model and validate the model using empirical merging maneuver data at a freeway on-ramp. Specifically, this paper advances our repeated game model by using updated payoff functions. Validation results using the Next Generation SIMulation (NGSIM) empirical data show that the developed game-theoretical model provides better prediction accuracy compared to previous work, giving correct predictions approximately 86% of the time. In addition, a sensitivity analysis demonstrates the rationality of the model and its sensitivity to variations in various factors. To provide evidence of the benefits of the repeated game approach, which takes into account previous decision-making results, a case study is conducted using an agent-based simulation model. The proposed repeated game model produces superior performance to a one-shot game model when simulating actual freeway merging behaviors. Finally, this lane change model, which captures the collective decision-making between human drivers, can be used to develop automated vehicle driving strategies.

Keywords: lane-changing; merging maneuvers; game theory; decision-making; intelligent vehicles

1. Introduction

Driving behavior strongly affects the safety and throughput of the transportation system [1], Due to its interference with surrounding vehicles, lane-changing significantly affects traffic stream flow. Several studies have concluded that lane-changing produces a capacity drop, forming a bottleneck [2–4]. The impacts of lane-changing maneuvers have been modeled in several studies [5–8]. In particular, Liu et al. [9] argued that traffic conflicts between merging and through vehicles, which are common near freeway on-ramps, are notable for inducing shockwaves, resulting in congestion. In order to analyze traffic flow, therefore, the development of a state-of-the-art lane-changing model is important.

The applications of lane-changing models can be broadly classified into two groups: adaptive cruise control and microscopic traffic simulation [1]. Driving assistance models for adaptive cruise control consist of collision prevention models and automation models [10]. In addition, driving decision models focus on drivers' lane-changing decisions for different traffic conditions and for different situational and environmental characteristics [10]. Lane-changing models were proposed based on various methodologies, which are reviewed in the next section, and calibrated based on field data collected on freeways. These models are an important component of microscopic traffic simulation [11]. Most models, however, focus on only the lane-changing vehicle in decision-making and vehicle control, which could be detrimental in microscopic traffic simulation, as interaction with surrounding vehicles is also critical in lane-changing. Specifically, drivers of vehicles surrounding the lane-changing vehicle, especially the closest following vehicle in the target lane, react after recognizing a lane-changing vehicle's intention to change lanes. For example, a human driver will sometimes not allow a lane change. Even though this type of competitive lane-changing behavior is rarely observed, decision-making considering drivers' interaction when changing lanes should be studied in order to develop a precise lane-changing model.

In addition, modeling a driving strategy for automated vehicles (AVs) gives rise to a new application for lane-changing models. The introduction of AVs onto the roadway means that reasonable lane-changing decision-making can be conducted by an intelligent robot or a well-programmed machine. During the transition to fully autonomous transportation systems, harmonization with human drivers will be necessary for the operation of AVs. Therefore, the development of a realistic lane-changing model that can depict human drivers' decision-making is also required to enhance AVs' driving performance.

To model lane-changing behaviors considering realistic decision-making, we developed a game-theoretical decision-making model for merging maneuvers at a freeway on-ramp [12], and then proposed a repeated game model [13]. This paper enhances our repeated game lane-changing model proposed in [13] and evaluates the proposed model's performance. The paper begins by introducing the lane-changing models based on various methodologies, including a game theoretical approach. To enhance model efficiency and complement the multivariate function in the previous model, the payoff functions for a stage game are reformulated in Section 3. This study also applies the repeated game approach, which uses cumulative payoffs, in order to capture realistic human driver behavior at a freeway merging section. Both the repeated game model and the one-shot game model based on the reformed stage game are calibrated and validated using empirical data extracted from the Next Generation SIMulation (NGSIM) dataset [14,15] to demonstrate the prediction ability. In the rest of this paper, we present a sensitivity analysis to describe the stage game's efficiency. The simulation case study using an agent-based model (ABM) follows. Finally, we draw concluding remarks on this work, and point out areas of potential future research.

2. Literature Review

A comprehensive literature review is required to introduce previous research efforts and present the motivations for this study. This section begins with a review of lane-changing models, focusing on methodologies. Then, game theory-based models are introduced in detail. Based upon the literature review, the motivations for the study are presented.

2.1. Lane-Changing Decision-Making Models

In general, the lane-changing process can be categorized as a sequence of four steps: (1) checking for lane-change necessity, (2) lane selection to decide on a target lane, (3) gap choice in the target lane, and (4) lane-changing execution through gap acceptance. To model lane-changing behaviors, lane-changing models have been developed using various methodologies that can be grouped into four types: (1) rule-based models, (2) discrete-choice-based models, (3) artificial intelligence models, and (4) incentive-based models [1].

The first model type, the rule-based model, is one of the most popular driver-perspective-based methodologies [1]. Drivers' decisions in the lane-changing process are simply defined as the independent variable. Gipps [16] initially introduced a lane-changing model covering various urban driving situations, which was intended for microscopic traffic simulation tools [17]. Gipps' model represented the lane-changing process as a decision tree with a series of fixed conditions, where the final output of this rule-based triggered event is a binary choice (i.e., change or no change) [1]. The CORridor SIMulation (CORSIM) model classified lane changes into two types: (1) discretionary lane-changing (DLC), which occurs when a driver is unsatisfied with the driving situation in their current lane, while the target lane shows better driving conditions; (2) mandatory lane-changing (MLC), which is coercively required according to the route choice (i.e., lane change toward on-ramp or off-ramp) [18,19]. Rahman et al. [1] categorized the game theory-based model, which explains lane-changing when a traffic conflict arises between the merging vehicle and the closest following vehicle in the target lane, as a rule-based model. Game theory, which is used in this paper, is the study of mathematical models of conflict and cooperation between decision-makers [20]. It focuses on decision-making in consideration of the interaction between intelligent drivers. Using a game theoretical approach is advantageous in that it takes into account the behaviors of the following vehicle driver in the target lane, while the other approaches introduced above focus only on the lane-changing vehicle driver's decision.

The second model type, the discrete-choice model, relies on a logit or probit model to describe lane-changing maneuvers. Lane-changing is decided based on probabilistic results instead of binary answers. Ahmed [21] modeled lane-changing motivation (i.e., trigger to change a lane), target lane choice, and gap acceptance, presenting three categories of lane-changing: DLC, MLC, and forced merging (FM), in which a gap is not sufficient but a driver nonetheless executes a lane-changing maneuver in heavily congested traffic conditions. Ahmed [21] assumed that critical gaps follow a lognormal distribution to guarantee that they are nonnegative. Toledo et al. [22] developed a probabilistic lane-changing decision model by combining MLC and DLC through a single utility function. Both models developed by Ahmed [21] and Toledo et al. [22] considered drivers' heterogeneity, such as aggressiveness and driving skill level, using a random term as one of the explanatory variables.

The third model type, which includes fuzzy models and artificial neural network (ANN) models, is artificial intelligence models. The fuzzy model considers humans' imprecise perception and decision biases, and incorporates more variables than the common mathematical models [23]. However, the fuzzy model has disadvantages, such as unexpected difficulties and complexity in the fuzzy rules [23]. The ANN model processes information using functional architecture and mathematical models that are similar to the neuron structure of the human brain [1]. Hunt and Lyons [24] modeled the lane-changing decisions of drivers on dual carriageways. Since the neural network model is completely data-driven and requires field-collected traffic data, Hunt and Lyons used interactive driving simulation to train the model. As this example shows, one major disadvantage of the ANN model is that it requires a huge amount of data to be optimized and also requires a training period.

The last type of model, the incentive-based model, models lane-changing desire utilizing the defined incentive. In other words, this model assumes that a driver chooses to change lanes in order to maximize their benefits [1]. The minimizing overall braking induced by lane change (MOBIL) model, which was developed in Kesting et al. [11], is based on measuring both the attractiveness and the risk associated with lane changes in terms of acceleration. Therefore, both the incentive criterion and the safety constraint are formed using the acceleration function of the underlying car-following model. In addition, the model attempts to capture the degree of passive cooperation among drivers, using the politeness factor as a weight on the term for total advantage of the surrounding vehicles.

2.2. Game Theory-Based Lane-Changing Decision-Making Model

It is clear that lane-changing involves not only a driver of the subject vehicle (SV), who is motivated to change lanes, but also a driver of the lag vehicle (LV) in the target lane, who controls their own vehicle (i.e., the LV) after perceiving the lane-changing vehicle in the adjacent lane. Specifically, the driver of the SV controls their longitudinal and lateral movements to safely change lane in consideration of surrounding vehicles, and the driver of the LV responds by showing acceptance or non-acceptance of an SV's lane-changing intention. This decision-making process involving both drivers motivated previous studies to use a game theoretical approach. Game-theory-based models, therefore, were modeled as a two-player non-cooperative game.

Kita [25] modeled merging-giveway interaction between vehicles in a merging section based on a game theoretical approach. The action strategies of the driver of SV are merging or maintaining the current lane, while the strategies of the driver of LV in the target lane are giving way (i.e., yielding) or not. Kita [25] modeled interaction between drivers as a game under perfect information conditions. However, perfect information in game theory indicates that all players have perfect and instantaneous knowledge of their own utility and the events that have previously occurred. In a traditional transportation environment, in which a driver becomes aware of their surroundings through sight only, this assumption is irrational. Additionally, Kita's model assumed that vehicle speeds were constant during the merging process, which is likewise unrealistic [9].

Liu et al. [9] modeled merging and yielding behavior using modeled payoff functions about the drivers' objectives. In Liu et al. [9], the objective of the driver of SV is to minimize the time spent in an acceleration lane subject to safety constraints, while the objective of the driver of LV is to minimize speed variation. The payoffs of drivers of the SV and LV were formulated using acceleration level and time that the merging vehicle spends in the acceleration lane for each action strategy, respectively. However, the driver of SV occasionally showed different behaviors, which were assumed to be based on the objective of the driver of SV. Kondyli and Elefteriadou [26] found that all drivers want to reach a speed close to the freeway speed or the speed limit, if there is no lead vehicle. This speed synchronization process that causes drivers to accelerate when arriving at the beginning of an acceleration lane was observed at a merging section on a freeway [27]. To solve the game, Liu et al. [9] proposed a bi-level calibration framework, in which the upper level programming is an ordinary least square problem and the lower level programming is a linear complementarity problem, for finding the Nash equilibrium.

In [12], we modeled a decision-making game model for merging maneuvers using five decision factors and evaluated the proposed model using NGSIM data. In addition, we introduced a repeated game approach in order to avoid an instantaneous fluctuation in decisions in microscopic simulation [13]. Even though these models showed high prediction accuracy, there were limitations, namely that the number of data showing all action strategies sets was unbalanced due to data collection during the morning peak time, and the model validation results were unable to show the distinct performance of the repeated game approach in microscopic simulation.

The development of advanced vehicle technologies (e.g., vehicle-to-vehicle communication) and AVs, has led recent research efforts to focus on the cooperative interaction between vehicles [28,29]. Talebpour et al. [29], for instance, modeled both mandatory and discretionary lane-changing by applying the Harsanyi transformation [30] within a connected environment. Yu et al. [31] designed a human-like, game theory-based controller for AVs in consideration of mixed traffic.

2.3. Motivation and Contribution of the Paper

The following are the contributions of this paper. First, we enhance the payoff functions that were previously developed in [12,13] by taking into consideration multiple decision factors and normalizing the decision variables. Multivariate functions using variables, which have different units, may induce a trivial equilibrium solution when variables are correlated. To solve this issue, we reformulated the payoff function by considering dimensionless variables. Second, we validate and compare the previous and proposed models. Third, we conduct a sensitivity analysis of the proposed model performance. Fourth, we demonstrate the benefits of a repeated-game approach using a simulation tool. The repeated game model first introduced in [13], in which a stage game is repeatedly played taking into consideration previous game results, showed no evidence of benefits compared to a one-shot game model, played independently based on instantaneous data at every decision point. If there is competition between drivers due to an ambiguous merging situation—for example, not only small lag spacings but also similar vehicle speeds—the one-shot game model may be sensitive to instantaneous data, causing fluctuations in driver decisions during the decision-making process. On the other hand, the repeated game model's initial cooperative decision can be expected to remain the same when there is only a slight variation in payoffs. Furthermore, the game model can produce a change from a non-cooperative to a cooperative game. Even though this type of driver competition in merging seldom occurs, the robust game model can be integrated into a microscopic traffic simulation software in order to simulate stereotypical vehicle movement patterns. Consequently, in this study we adopt the previous repeated game approach with enhancements in the payoff function and then provide evidence of the repeated game model's benefits through a case study.

Lastly, a desired acceleration level, which is calculated to achieve the action set chosen by both players, should be an additional component of a vehicle acceleration model. A lane-changing model based on a game theoretical approach captures the decision-making process between two intelligent decision-makers. The model output is an action that will be conducted by two players at future time steps, rather than a decision to start lane-changing. To depict practical lane-changing behaviors in a microscopic traffic simulator, therefore, the game model should be integrated with other models, such as car-following, lane selection, and gap acceptance models. This study develops an A simulation model based on ABM, including a vehicle acceleration controller based on the game model and a car-following model, then conducts a simulation study to evaluate the performance of the repeated game model.

3. Merging Decision-Making Model Using a Repeated Game Concept

As previously noted, this study aims at developing a decision-making game for merging maneuvers on a freeway based on the repeated game concept. The following subsections describe, in detail, a stage game for merging decision-making and repeated game design and the development of the player payoff functions.

3.1. Stage Game Design

The game model defines the number of players, action strategies of each player, and corresponding payoff functions to describe the outcome for each player throughout the game [32]. This study adopts the decision-making game model structure for merging maneuvers proposed by the authors in 2017, which consists of two players: the drivers of the SV and the LV. The driver of SV, who wants to make a lane change, has three action strategies (see Figure 1a): (1) change lane (s_1), (2) wait for the LV's overtaking maneuver (s_2), or (3) overtake the LV and use a forward gap to merge (s_3). The opposite player, the driver of LV, has two action strategies (see Figure 1b): (1) yield to allow the lane change maneuver of the driver of SV (l_1) or (2) block the SV's merging maneuver by decreasing the spacing available for the SV (l_2) [12]. In real life situations, the driver of LV can choose lane-changing to the left lane to avoid potential collision or considerable deceleration [33], and this lane-changing behavior was considered as an action strategy of the driver of LV in [29]. Freeway vehicles on the rightmost lane generally change lanes from the rightmost lane upstream of the merging section after perceiving the approach of the merging vehicle in order to maintain their speed. Since this mainline vehicle's lane change is conducted earlier and thus does not involve interaction with the merging vehicle, this study does not include a lane-changing action as one of the actions of the driver of the LV in the proposed merging game.

Let $S = \{s_1, s_2, s_3\}$ and $L = \{l_1, l_2\}$ denote the set of pure strategies for the drivers of the SV and LV, respectively. In addition, $a = \left(s_i,\ l_j\right)$ denotes a set of actions ($a \in S \times L$) where i and j indicate the index of action strategies of the drivers of the SV and LV (i.e., i = 1, 2, 3 and j = 1, 2). As such, a total of six sets of action strategies were defined for the non-cooperative decision-making stage game. In these action strategies, (s_1, l_1), (s_2, l_2), and (s_3, l_1) are cooperative action strategies, whereas both (s_1, l_2) and (s_2, l_1) are non-cooperative strategies in which both players compete to achieve their objectives. The action strategy (s_3, l_2) is neither cooperative nor competitive. The proposed stage game with imperfect information, which captures the fact that players are simply unaware of the actions chosen by other players, is represented in Figure 2. In the figure, a dashed line uniting three nodes, which implies imperfect information, indicates that the players do not know which node they are in. This means that there is no sequence in making a decision, and thus the driver of LV does not know the SV's movement. Moreover, P_{ij} and Q_{ij} denote the payoff for the drivers of the SV and LV for each action strategy a_{ij}, respectively.

The drivers initially play the stage game to decide on an individual action at the moment when an SV, an LV, and a preceding (lead) vehicle (PV) are identified ([12]). It was assumed that the initial game is played when the driver of the SV reaches the start of an acceleration lane. Additional stage games are formed by overtaking the PV or waiting to be overtaken by the LV. In other words, the stage game is re-built when a change in surrounding vehicles occurs, i.e., PV or LV, in the target lane.

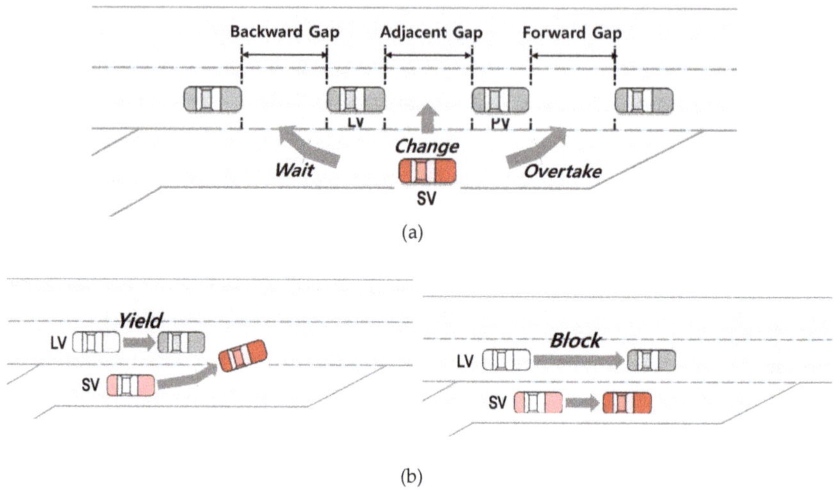

Figure 1. Players' strategies for merging maneuver: (**a**) the driver of subject vehicle (SV); (**b**) the driver of lag vehicle (LV).

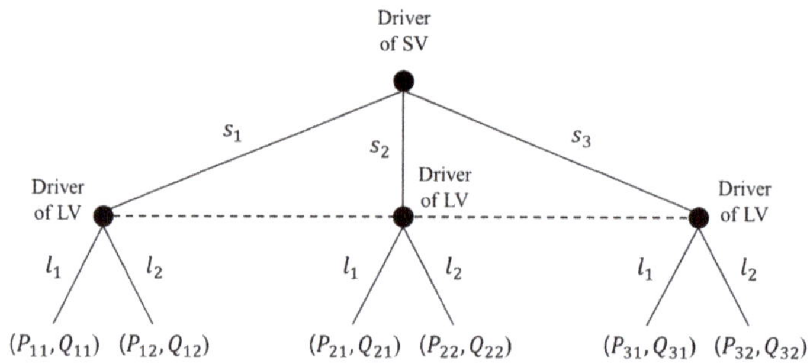

Figure 2. Merging decision-making game in the extensive form.

3.2. Repeated Game Design

In the game model, one of the characteristics to be specified is the number of games to be repeated [25]. In the authors' previous study, a repeated game approach was used in order to depict a practical decision-making process for merging maneuvers [13]. In real life, at a freeway merging section in a traditional transportation environment, a driver continuously makes a decision using the data taken in by sight and controls the vehicle to fulfill their decision. When the merging vehicle enters the acceleration lane, the driver of the SV selects a gap type to change a lane and then directs their vehicle accordingly. The driver controls the acceleration level to synchronize the vehicle speed with the freeway vehicles and ensure a safe gap distance [27,33]. During this lane-changing preparation process, the driver of SV repeatedly checks surroundings to judge if their decision can be fulfilled and tries to follow-up on their decision. In this study, therefore, this repetition in decision-making for merging maneuvers prior to lane-changing execution was regarded as playing the game repeatedly.

The repeated game concept implies that a stage game with identical structure is repeatedly played until termination of the game, which is divided into two classes, finite or infinite, depending on the players' beliefs about the number of repetitions. In this study, the decision-making game for merging was regarded as an infinitely repeated game because the players in the game do not know how many times the game will be repeated. Note that, for an infinitely repeated game, the stage game will not necessarily be repeated an infinite number of times.

Drivers (i.e., players) interact by playing a stage game multiple times. As a summary explanation about the game model type, the one-shot game model implies that previous game results do not affect the present game, while the decision-makers take previous game results into account in the repeated game model, as illustrated in Figure 3. This study adopts the repeated decision-making game approach using the cumulative payoffs to prevent repeated fluctuations in payoffs, as proposed in [13]. The stage decision-making game is conducted periodically and repeatedly over discrete time periods $T \in [t_1, t_n]$. Time preference is considered by assuming that future payoffs are weighted proportionately at a constant rate δ, called the rate factor. Cumulative payoffs of the driver d for action strategy a_{ij}, i.e., $U_{ij}^d = P_{ij}$ or Q_{ij}, are presented in Equation (1).

$$U_{ij}^d(T) = \sum_{t_1}^{t_n} \delta^{t-1} u_{ij}^d(t). \tag{1}$$

Here, $u_{ij}^d(t)$ is a utility of a driver d for an action strategy set (s_i, l_j) at time step t; T is the number of decision-making time steps; and d denotes a driver, i.e., a player in a game, the driver of SV or DL. If $\delta > 1$, it implies that the current payoffs are more important than the past payoffs. Otherwise, the previous game results could significantly affect the decision-making in a future game.

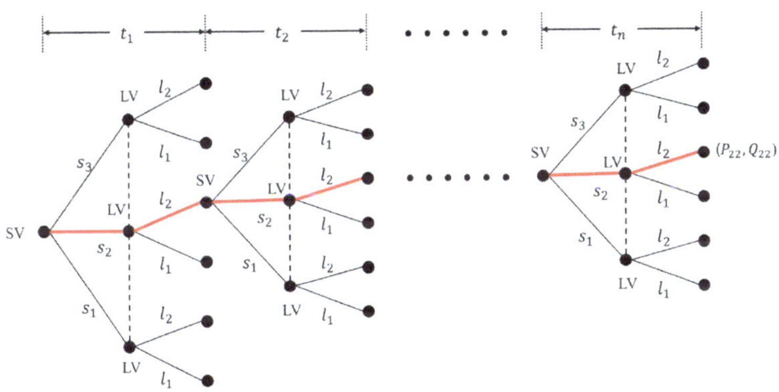

Figure 3. Decision-making game based on the repeated game approach in extensive form.

3.3. *Reformulated Payoff Functions*

In previous game theory-based-models, the payoff functions for two players were formulated using the significant decision factors, such as safety, spacing (or gap), relative speed, travel time, expected acceleration level, and remaining distance to reach the end of acceleration lane [11–13,25,29,31]. In [12], we initially proposed the payoffs using five decision factors: minimization of travel time, avoidance of collisions (i.e., safety), travel efficiency, the LV's expected acceleration, and the remaining distance to execute the maneuver. In a following study [13], the payoffs of the driver of SV were formulated as the expected gap and remaining distance, and the expected relative speed was considered as the other driver's main decision variable. Both previous studies used multiple dimensioned variables, meaning the payoffs are only interpreted as a qualitative outcome to represent the player's preference. In addition, an error term was considered to capture unobserved variables, assumed to be a constant, resulting in minimal consideration of a driver's randomness. As described previously, therefore, this study updates the payoff functions to use efficient decision variables including a random error term and proposes monotone (dimensionless) functions by the transformation of quantitative variables.

This section introduces the decision variables, and then presents the reformulated payoff functions for each driver.

3.3.1. Safety Payoff

Among various decision factors, safety is a key factor for human drivers' decision to avoid a potential collision or not induce a dangerous situation. Yu et al. [31] used the time headway as a safety payoff, as presented in Equation (2).

$$h_{PV,\,SV}(t) = \frac{x_{PV}(t) - x_{SV}(t)}{v_{SV}(t)}, \tag{2}$$

Here, $x_{PV}(t)$ and $x_{SV}(t)$ are the positions of the (potential) PV and SV at instant time t, respectively; and $v_{SV}(t)$ is speed of the SV at time t. However, they did not take the speed of a PV into account. In [13], the expected spacing between vehicles, indicating the possibility of ensuring a safe distance with consideration of vehicles' speed and acceleration levels, was used. Additionally, Wang et al. [34] used a penalty formulated using relative speed and the gap distance. Kita [25] used the Time-To-Collision (TTC) between vehicles as the main payoff, as defined in Equation (3).

$$TTC_{PV,SV}(t) = \frac{x_{PV}(t) - x_{SV}(t) - l_{PV}}{v_{SV}(t) - v_{PV}(t)} \qquad \text{if } v_{SV}(t) > v_{PV}(t), \tag{3}$$

Here, l_{PV} denotes the length of the PV; and $v_{PV}(t)$ is the speed of the PV at instant time t.

The interactive effects of relative speed and gap distance are contained in the single measure TTC [35]. Brackstone et al. [36] collected realistic data using an instrumented vehicle equipped with relative distance- and speed-measuring sensors. Observations of vehicle trajectories from five participants showed that TTC is a major factor in lane-changing decisions. Most collision avoidance systems (or pre-crash safety systems) applied in a vehicle use the instantaneous TTC to evaluate collision risk [37]. Moreover, Vogel [38] recommended the use of TTC for the evaluation of safety because it indicates the actual occurrence of dangerous situations. Vogel also noted that a situation with a small TTC is imminently dangerous and that a situation with a small headway and relatively large TTC is a potentially dangerous situation. Therefore, this study proposes the integrated safety payoff function A^S with consideration of not only TTC but also headway, which was formulated using the hyperbolic tangent function, as presented in Equations (4) and (5).

$$A^S_{PV,SV} = \begin{cases} \left(tanh\left(\frac{TTC_{PV,SV}(t)}{t^S} - 1\right) + tanh\left(\frac{h_{PV,SV}(t)}{t^S} - 1\right)\right) \times 0.5, & \text{if } v_{SV}(t) > v_{PV}(t) \\ \left(1 + tanh\left(\frac{h_{PV,SV}(t)}{t^S} - 1\right)\right) \times 0.5, & o.w. \end{cases} \tag{4}$$

$$A^S_{SV,LV} = \begin{cases} \left(tanh\left(\frac{TTC_{SV,LV}(t)}{t^S} - 1\right) + tanh\left(\frac{h_{SV,LV}(t)}{t^S} - 1\right)\right) \times 0.5, & \text{if } v_{LV}(t) > v_{SV}(t) \\ \left(1 + tanh\left(\frac{h_{SV,LV}(t)}{t^S} - 1\right)\right) \times 0.5, & o.w. \end{cases} \tag{5}$$

Here, $t^S = \min\left(\frac{RD_{SV}}{v_{SV}(t)}, 3\right)$ denotes the minimum safe time headway between the 3-second rule recommended by the National Safety Council [39] and the time headway to reach the end of the acceleration lane.

The safety payoffs of both drivers for the action strategies were formulated to satisfy $U^S \in [-1, 1]$, as shown in Equations (6) to (9).

$$U^S_{SV}(s_1) = 0.5\left(A^S_{PV,SV} + A^S_{SV,LV}\right), \tag{6}$$

$$U^S_{SV}(s_2) = -A^S_{SV,LV}, \tag{7}$$

$$U^S_{SV}(s_3) = -A^S_{PV,SV}, \tag{8}$$

$$U^S_{LV}(l_1) = A^S_{SV,LV} = -U^S_{LV}(l_2). \tag{9}$$

For the 'change (s_1)' action of the driver of SV, $U^S_{SV}(s_1)$ was formulated as the average of safety payoffs, taking both the PV and LV in the target lane into account. For the 'wait (s_2)' and 'overtake (s_3)' action of the driver of SV, on the other hand, the driver's safety payoffs were formulated to consider only the corresponding vehicle related to each action strategy. Likewise, it was assumed that the driver of LV also evaluates their safety in consideration of the SV only.

As shown in the safety payoff formulation, the safety payoffs vary by the spacing between vehicles and each vehicle's speed. Figure 4 shows the prospective safety payoffs of the driver of SV at the various speeds of the three vehicles (i.e., PV, SV, and LV), with the SV in different positions between the PV and LV. In this example, spacing between the PV and LV is constant at 77 m. Figure 4a presents a case in which the SV is located close to the PV. In other words, the lead gap $\Delta x_{PV,SV}$ is small and the lag gap $\Delta x_{SV,LV}$ is large. If $v_{PV} > v_{SV}$, $U^S_{SV}(s_1)$ is greater than $U^S_{SV}(s_3)$. Otherwise, the driver of SV is attracted to choosing the 'overtake (s_3)' action in consideration of safety. In the second case, described in Figure 4b, the SV is located at the middle position between the PV and LV. Therefore, the 'change (s_1)' action is relatively attractive, i.e., $U^S_{SV}(s_1) > U^S_{SV}(s_2)$ and, $U^S_{SV}(s_1) > U^S_{SV}(s_3)$ even if v_{SV} is slightly less than v_{PV} and v_{LV}. The 'overtake (s_3)' action is attractive when $v_{SV} \gg v_{PV}$, and $U^S_{SV}(s_2)$ are greater than $U^S_{SV}(s_1)$ when $v_{SV} \ll v_{LV}$. The last case, in which the SV is close to the LV, represents the case where the driver of SV is drawn to choosing the 'wait (s_2)' action if $v_{LV} > v_{SV}$. If $v_{SV} > v_{LV}$, the 'change (s_1)' action is more attractive. From these cases, transformed safety payoffs are reasonable to represent the general decision-making results of the driver of SV.

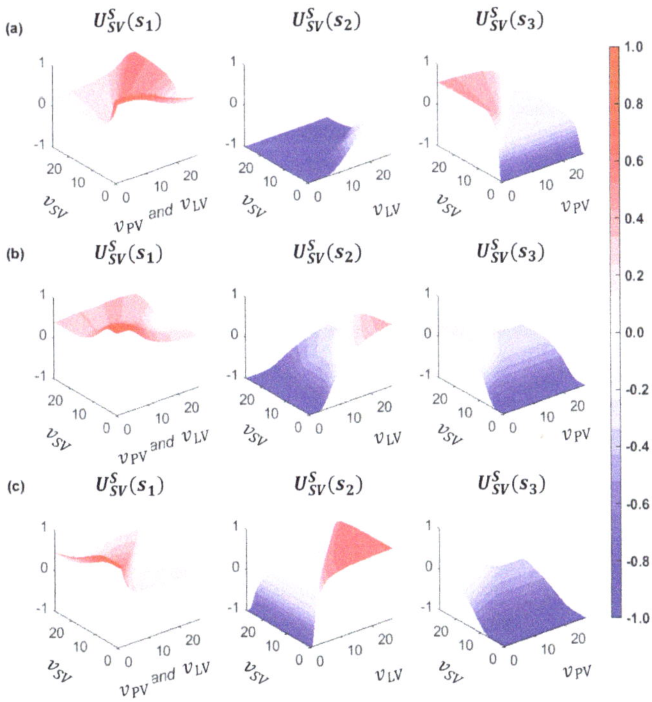

Figure 4. Safety payoffs of the driver of SV for the s_1, s_2, and s_3 action: (**a**) close to the preceding vehicle (PV) ($\Delta x_{SV,LV} = 67$ m, $\Delta x_{PV,SV} = 10$ m); (**b**) middle position between PV and LV ($\Delta x_{SV,LV} = 38$ m, $\Delta x_{PV,SV} = 39$ m); (**c**) close to the LV ($\Delta x_{SV,LV} = 10$ m, $\Delta x_{PV,SV} = 67$ m).

Figure 5 presents the safety payoffs for the driver of LV in the three cases described above. In Figure 5a, which shows that $\Delta x_{SV,LV}$ is considerably large, the driver of LV desires to choose the 'yield (l_1)' action, except in the case where $v_n \ll v_{n+1}$. These payoffs seem to be reasonable because the LV is far away from the SV. In the second case, the 'yield (l_1)' action is attractive as well. This case is similar to a real field situation, where the lane-changing action of the following vehicle in the target lane mostly shows cooperation in order to accept the merging vehicle's lane change. In the third case, the huge deceleration is expected to provide a gap to the SV because the LV is close to the SV. Therefore, the safety payoffs of the driver of LV for the 'block (l_2)' action are higher than for the l_1 action if $v_{SV} < v_{LV}$. Otherwise, the safety payoff of the driver of LV for the 'yield (l_1)' action is slightly higher, except in a freeway congested traffic condition (i.e., $v_{SV} \gg v_{LV}$).

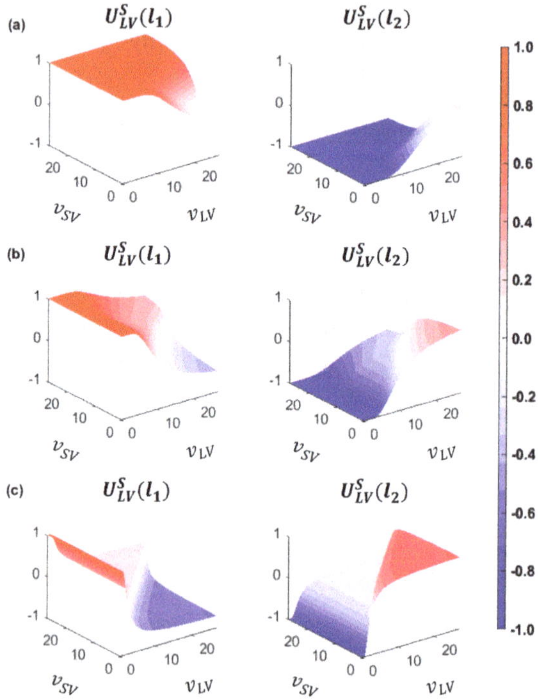

Figure 5. Safety payoffs of the driver of LV for the l_1 and l_2 action: (**a**) close to the PV ($\Delta x_{SV,LV} =$ 67 m, $\Delta x_{PV,SV} =$ 10 m); (**b**) middle position between PV and LV ($\Delta x_{SV,LV} =$ 38 m, $\Delta x_{PV,SV} =$ 39 m); (**c**) close to the LV ($\Delta x_{SV,LV} =$ 10 m, $\Delta x_{PV,SV} =$ 67 m).

3.3.2. Forced Merging Payoff for the Driver of SV

According to the empirical field data collected at a freeway merging section, the driver of a vehicle entering through an on-ramp usually accelerates for speed-harmonization with freeway vehicles. The driver of SV then selects a gap to merge onto the freeway. In congested traffic conditions, however, the merging vehicles travel at a higher speed than the surrounding vehicles on the freeway. Thus, the driver occasionally rejects the initial gap and then uses a farther forward gap, close to the end of the acceleration lane. Wan et al. found that merging vehicles pass freeway vehicles and try to find an acceptable gap to merge onto the freeway after traveling longer than the normal merging cases in congested traffic conditions [27]. Marczak et al. [40] analyzed data collected at two sites to find variables related to gap acceptance, concluding that the distance to the end of the acceleration lane is a significant variable. Hwang and Park [41] also concluded that the remaining distance is the most important factor for determining gap acceptance; the driver will most likely accept a smaller gap if the remaining distance to the end of the acceleration lane is smaller. In order to consider the case in which a vehicle merges close to the end of the acceleration lane, the payoff function of the driver of SV should include a term called the forced merging payoff, which relates the remaining distance to the end of the acceleration lane. This affects cases where the driver decides the 'change (s_1)' action at the decision point where the remaining distance is considerably short.

This study formulated the forced merging payoff as a function of the remaining distance and $v_{SV}(t)$. There is an assumption that the end of the acceleration lane is an imaginary preceding vehicle that is stopped. The presence of this imaginary vehicle, which is also considered as a hard wall, means the driver of SV cannot drive further, due to the restricted length of the acceleration lane. Thus, the expected safety distance to maintain the instant speed of the SV, $v_{SV}(t)$, was estimated by a car-following model. This study used the Rakha-Pasumarthy-Adjerid (RPA) car-following model, which was first developed by Rakha et al. [42]. The performance of the RPA car-following model has been validated against naturalistic driving data [43]. This study estimated the safety distance for the SV, $x_{SV}^{CF}(t)$ using the RPA model's two components: steady-state traffic stream behavior and collision avoidance. The steady-state modeling applies the Van Aerde's steady state car-following model [44,45], which is a non-linear single regime function of vehicle speed and spacing. The first safe spacing (i.e., safety distance) provided by the steady-state model is

$$x_{SV}^{CF_1}(t) = c_1 + c_3 \cdot v_{SV}(t) + \frac{c_2}{v_f - v_{SV}(t)}. \tag{10}$$

Here, v_f indicates the free-flow speed. The model coefficients can be computed as

$$c_1 = \frac{v_f}{k_j {v_c}^2}\left(2v_c - v_f\right), \tag{11}$$

$$c_2 = \frac{v_f}{k_j {v_c}^2}\left(v_f - v_c\right)^2, \tag{12}$$

$$c_3 = \frac{1}{q_c} - \frac{v_f}{k_j {v_c}^2}. \tag{13}$$

Here, k_j, v_c, and q_c indicate the jam density, speed-at-capacity, and saturation flow rate, respectively. The detailed definition of these coefficients is described in [44].

As the second component of the RPA model, collision avoidance was modeled to avoid incidents at non-steady-state conditions [43]. The second safe spacing estimated by collision avoidance is defined as

$$x_{SV}^{CF\_2}(t) = \frac{v_{SV}(t)^2}{2 \cdot a_{min}} + x_j. \tag{14}$$

Here, a_{min} and x_j denote the minimum acceleration (i.e., maximum deceleration) and the jam spacing, respectively.

The maximum value of two safe spacings, $x_{SV}^{CF\_1}(t)$ and $x_{SV}^{CF\_2}(t)$, is considered as the expected safe spacing to maintain current speed.

$$x_{SV}^{CF}(t) = \max\left(x_n^{CF\_1}(t),\, x_n^{CF\_2}(t),\, x_{max}^{RD}\right). \tag{15}$$

Here, x_{max}^{RD} is the maximum of the remaining distance, i.e., the longitudinal length of the acceleration lane.

To balance each payoff, this study re-formulated the forced merging payoff of the driver of SV, U_{SV}^{FM}.

$$U_{SV}^{FM} = \left[\frac{\max\left(x_{SV}^{CF}(t) - x_{SV}^{RD}(t), 0\right)}{x_{SV}^{CF}(t)}\right]^2. \quad (16)$$

Here, $x_{SV}^{RD}(t)$ indicates the remaining distance for the SV in the acceleration lane at time t. This formulation satisfies $U_{SV}^{FM} \in [0, 1]$ as shown in Figure 6. If the remaining distance is shorter than $x_{SV}^{CF}(t)$, U_{SV}^{FM} begins to have positive payoffs, inducing a preference for the 'change (s_1)' action. This term presents greater payoffs when $v_{SV}(t)$ is faster.

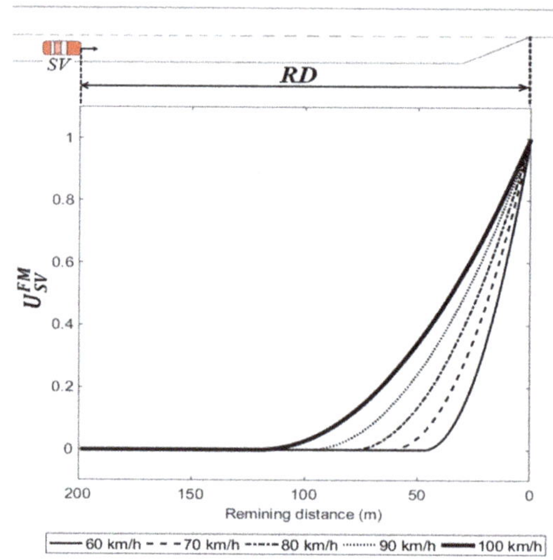

Figure 6. Forced merging payoff by the remaining distance at various speeds.

3.3.3. Payoff Functions for the Drivers of the SV and LV

Table 1 represents the updated merging decision-making model in normal form. The payoff functions of the driver of SV consist of both the safety and forced merging payoffs, and those of the driver of LV include the safety payoffs only. In order to capture unobserved utility, both players' payoff functions also have an error term, which was assumed to be normally distributed as $\varepsilon_{ij}^{SV\ or\ LV} \sim N(0, 1)$. The parameters in the payoff functions, i.e., set of α_{ij} and β_{ij} (i = 1,2,3 and j = 1, 2), are parameters to be estimated.

Table 1. Game Structure and Payoff Functions of the Merging Decision-Making Game in Normal Form.

Player & Actions		Driver of LV: Yield $[l_1(q_1)]$ [2]	Driver of LV: Block $[l_2(q_2)]$
Driver of SV	Change $[s_1(p_1)]$ [1]	$P_{11} = \alpha_{11}^1 + \alpha_{11}^2 U_{SV}^S(s_1) + \alpha_{11}^3 U_{SV}^{FM} + \varepsilon_{11}^{SV}$ $Q_{11} = \beta_{11}^1 + \beta_{11}^2 U_{LV}^S(l_1) + \varepsilon_{11}^{LV}$	$P_{12} = \alpha_{12}^1 + \alpha_{12}^2 U_{SV}^S(s_1) + \alpha_{12}^3 U_{SV}^{FM} + \varepsilon_{12}^{SV}$ $Q_{12} = \beta_{12}^1 + \beta_{12}^2 U_{LV}^S(l_2) + \varepsilon_{12}^{LV}$
	Wait $[s_2(p_2)]$	$P_{21} = \alpha_{21}^1 + \alpha_{21}^2 U_{SV}^S(s_2) + \varepsilon_{21}^{SV}$ $Q_{21} = \beta_{21}^1 + \beta_{21}^2 U_{LV}^S(l_1) + \varepsilon_{21}^{LV}$	$P_{22} = \alpha_{22}^1 + \alpha_{22}^2 U_{SV}^S(s_2) + \varepsilon_{22}^{SV}$ $Q_{22} = \beta_{22}^1 + \beta_{22}^2 U_{LV}^S(l_2) + \varepsilon_{22}^{LV}$
	Overtake $[s_3(p_3)]$	$P_{31} = \alpha_{31}^1 + \alpha_{31}^2 U_{SV}^S(s_3) + \varepsilon_{31}^{SV}$ $Q_{31} = \beta_{31}^1 + \beta_{31}^2 U_{LV}^S(l_1) + \varepsilon_{31}^{LV}$	$P_{32} = \alpha_{32}^1 + \alpha_{32}^2 U_{SV}^S(s_3) + \varepsilon_{32}^{SV}$ $Q_{32} = \beta_{32}^1 + \beta_{32}^2 U_{LV}^S(l_2) + \varepsilon_{32}^{LV}$

[1] p_i in parentheses denotes the probability assigned to the pure strategy of the driver of SV, s_i; $\sum_{i=1}^{3} p_i = 1$. [2] q_j in parentheses denotes the probability assigned to the pure strategy of the driver of LV, l_j; $\sum_{j=1}^{2} q_j = 1$.

4. Model Calibration and Validation

Model evaluation was conducted to prove the efficiency of the game models using the stage game based on the newly formulated payoff functions. This section introduces the observation dataset for model evaluation and calibration methodology. In addition, the calibration and validation results of our previous model and the updated repeated game models are presented.

4.1. Preparation of Observation Dataset

This study used NGSIM vehicle trajectory data from a segment of U.S. Highway 101 (Hollywood Freeway) in Los Angeles, California, collected between 7:50 and 8:35 a.m. on June 15, 2005 [14,15]. Reasonable classification of the action strategies chosen by the drivers of the SV and LV is a critical issue, as it is directly related to the results of the game model [13]. There is a limitation on the classification of drivers' decisions based on trajectories and speed profile data. This study used a total of 1504 observations extracted from NGSIM data in [13]. For classification of the SV's maneuvers observed in the field, this study used the types of gap that were selected at game-playing moments among the three following gap types (as illustrated in Figure 1a): (1) forward (lead) gap, (2) adjacent (current) gap, or (3) backward (lag) gap. In addition, the spacing between the SV and LV was used for the classification of the LV's maneuvers. A detailed classification methodology is described in [13]. Next, all data were reviewed to judge whether the classification results were reasonable to show drivers' intentions. If the specific data were regarded as improper classification, these data were modified. Decisions made by drivers in all observations were classified using this process.

4.2. Model Calibration

4.2.1. Calibration Approach

In the game model, each player chooses an action to achieve the goal of the game. In game theory, the Nash equilibrium is a solution to find the optimal set of strategies for both drivers where they have no incentive to deviate from their initial strategy. If the Nash equilibrium exists, it implies that each player will choose the strategy that maximizes their own payoff while considering an opponent who also wants to maximize their payoff. The Nash equilibrium defines pure strategy as

$$\begin{cases} P(s^*,\ l^*) \geq P(s_i,\ l^*), & \forall\, s_i \in S,\ i = 1,2,3 \\ Q(s^*,\ l^*) \geq Q\left(s^*,\ l_j\right), & \forall\, l_j \in L,\ j = 1,2 \end{cases}, \tag{17}$$

where s^* and l^* indicate the equilibrium action strategy of the drivers of the SV and LV, respectively. In this study, if a pure strategy Nash equilibrium does not exist, a mixed strategy Nash equilibrium involves at least one player playing a randomized strategy and no player being able to increase their expected payoff by playing an alternate strategy. A probability for each player's strategy is assigned with consideration of each player's expected payoff from the different strategies [28]. This paper used the MATLAB function N-Person Game (NPG), developed by Chatterjee [46], to solve a two-player, finite, non-cooperative game. Chatterjee's algorithm [46] solves the game by computing the Nash equilibrium in mixed strategies based on the estimated parameters and expected payoffs (i.e., P_{ij} and Q_{ij}). The algorithm provides the probabilities of the choice of pure action strategies for each driver (i.e., p_i and q_j) in each observation.

In order to calibrate the merging decision-making model, this study followed the calibration method developed by Liu et al. [9], who proposed a parameter estimation method by solving a bi-level programming problem. As illustrated in Figure 7, the lower-level programming is to find the Nash equilibrium using Chatterjee's function [46]. The upper level is a non-linear programming problem that minimizes the total deviation in probabilities in the system in order to choose actual observed actions using the following function

$$\min \sum_{k=1}^{n}\left(1 - p_{a^k} \cdot q_{a^k}\right), \tag{18}$$

where k denotes the index of observations; a^k is the observed action strategy set $(s_i^{\,k},\ l_j^{\,k})$ in observation k; and p_{a^k} and q_{a^k} are the probabilities that drivers of the SV and LV, respectively, choose the observed action in a_k. Here, A^k and B^k denote all parameters to be estimated for each driver's payoff functions.

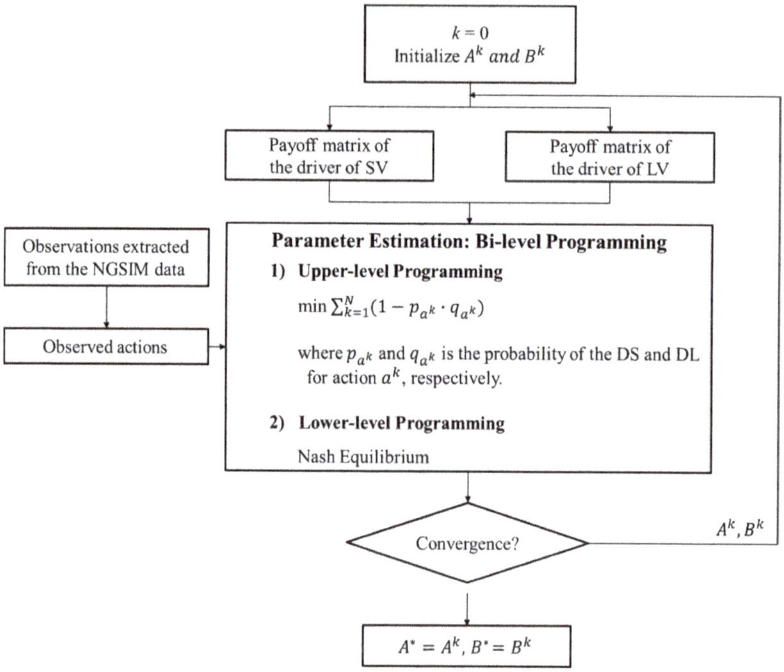

Figure 7. Schematic workflow for bi-level programming.

4.2.2. Calibration Results

As mentioned earlier, this study calibrated a total of two types of game model: (1) the one-shot game model, in which the developed stage game is played independently at every game point based on instantaneous status only; (2) the repeated game model using the cumulative payoffs with factor δ of various rates conducted every 0.5s. To verify the performance of the updated payoff functions in predicting human drivers' decisions in merging situations, the first type of model was subdivided into two models according to the payoff functions used in model calibration, as below.

- One-shot game model based on the stage game using the payoff functions developed in [13];
- One-shot game model based on the stage game using the reformulated payoff functions in Section 3.3

Here, the former and latter models were called the 'previous one-shot game model' and the 'one-shot game model', respectively. For model calibration, an NGSIM dataset observed between 7:50 and 8:20 a.m. was used. The number of observations used in model calibration was 685 (out of 1504). Table 2 shows the estimated parameters of the payoff functions of the drivers of the SV and LV.

Table 2. Estimated Parameters of the Payoff Functions for Game Models.

Payoff Function	Parameters	One-Shot Game Model	Repeated Game Models					
			Model 1 ($\delta = 0.6$)	Model 2 ($\delta = 0.8$)	Model 3 ($\delta = 1.0$)	Model 4 ($\delta = 1.2$)	Model 5 ($\delta = 1.4$)	Model 6 ($\delta = 1.6$)
P_{11}	α^1_{11}	9.64	5.10	2.88	6.69	−1.77	7.08	7.11
	α^2_{11}	23.51	74.83	48.38	96.45	9.20	27.34	8.38
	α^3_{11}	32.69	59.51	69.45	1.00	5.16	97.08	2.75
P_{12}	α^1_{12}	9.43	8.83	3.58	7.87	8.64	7.27	−6.26
	α^2_{12}	87.57	77.60	44.40	86.30	3.11	50.13	4.25
	α^3_{12}	10.98	43.84	1.80	71.19	5.73	84.75	7.34
P_{21}	α^1_{21}	0.63	−9.78	−7.49	−6.91	−8.88	−6.65	−8.13
	α^2_{21}	3.35	26.60	10.68	62.49	3.18	31.94	1.75
P_{22}	α^1_{22}	−7.88	−8.50	−3.42	−6.19	9.73	−8.98	5.56
	α^2_{22}	42.64	20.75	5.21	65.72	6.22	19.43	7.16
P_{31}	α^1_{31}	−0.66	6.07	−9.38	−6.21	−2.84	−5.18	6.41
	α^2_{31}	67.24	48.05	78.92	94.59	11.19	25.08	7.53
P_{32}	α^1_{32}	−0.53	-3.10	−5.39	−0.44	2.75	−3.69	8.35
	α^2_{32}	16.91	52.79	95.22	59.86	2.21	30.06	4.79
Q_{11}	β^1_{11}	9.93	3.78	6.96	9.80	−1.99	7.97	−3.75
	β^2_{11}	13.30	17.29	6.64	25.06	6.88	5.86	10.22
Q_{12}	β^1_{12}	−1.26	−8.39	-6.24	−5.83	−7.03	−8.90	−8.36
	β^2_{12}	3.70	0.29	19.40	23.84	10.20	18.49	1.89
Q_{21}	β^1_{21}	5.78	7.64	8.05	8.74	5.52	8.25	0.27
	β^2_{21}	89.18	57.76	58.65	78.06	2.76	82.45	4.12
Q_{22}	β^1_{22}	7.73	−4.36	−4.36	0.63	0.34	−8.66	−5.95
	β^2_{22}	57.97	6.64	55.26	14.12	7.43	38.74	7.61
Q_{31}	β^1_{31}	3.88	−4.02	-6.99	6.38	9.39	−0.82	3.68
	β^2_{31}	55.87	96.95	98.01	1.12	4.35	46.49	9.22
Q_{32}	β^1_{32}	4.26	−9.75	1.08	−8.01	6.78	1.53	−4.85
	β^2_{32}	27.87	26.74	22.93	74.89	2.20	86.19	7.83

Note that the previous one-shot game model using the payoff functions in [13] was calibrated using the same calibration methodology, but the estimated parameters are not shown in the table because of the different formulation for payoff functions.

In order to compare the models' prediction accuracy, the mean absolute error (MAE) was calculated using Equation (19)

$$MAE = \frac{1}{N}\sum_{k=1}^{N}\left|1 - 1(\hat{x}_k - x_k)\right|, \tag{19}$$

where N, $\hat{x}_k$, and x_k denote the number of observations, model prediction, and actual observations, respectively. Note that $1(\hat{x}_k - x_k)$ is equal to one if $\hat{x}_k = x_k$, and is zero otherwise. The model prediction $\hat{x}_k$ was estimated by probabilities calculated using Chatterjee's algorithm [46]. Table 3 shows the calibration results for the MAEs of the three types of models. In comparison with our previous model, the one-shot game model using the updated payoff functions shows a higher prediction capacity in merging decision-making. In the repeated game models, the models with $\delta > 1.0$ were calibrated with lower MAEs than those with $\delta \leq 1.0$.

Table 3. Calibration Results.

Models	Previous One-Shot Game Model (2018)	One-Shot Game Model	Repeated Game Models					
			Model 1	Model 2	Model 3	Model 4	Model 5	Model 6
Rate factor, δ	*na* [1]	*na*	0.6	0.8	1.0	1.2	1.4	1.6
MAE [2]	0.2555 (74.45 %)	0.1241 (87.59 %)	0.1708 (82.92 %)	0.1606 (83.94 %)	0.1606 (83.94 %)	0.1372 (86.28 %)	0.1358 (86.42 %)	0.1460 (85.40 %)

[1] Not applicable. [2] The number in parentheses indicates prediction accuracy.

4.3. Model Validation

The rest of the data, 819 observations out of 1504, collected between 8:20 and 8:35 a.m., were used for validating the model, and the validation results are shown in Table 4. Model validation results, which show the same trends as the calibration results, are summarized as follows. First, when comparing the results of the stage game developed in the previous study [13] and this study, the prediction accuracy increases by about 12% when the third stage game is used. Thus, this study enhances the decision-making game model's performance by using the reformulated payoff functions to represent merging maneuvers. Next, in the validation results, the repeated game models with $\delta \geq 1.0$ show a prediction accuracy of higher than 85%. In particular, the repeated game model shows the highest prediction accuracy when $\delta = 1.4$. Both the one-shot game and repeated game model with $\delta = 1.4$ show a considerably high prediction accuracy of more than 86%. Due to the limitations of unbalanced observation data [12], nevertheless, model validation using field data cannot provide evidence that is beneficial using the repeated game. It is also hard to show the apparent difference between the one-shot game and the repeated game model. In the following sections, therefore, the game models are additionally evaluated through sensitivity analysis and simulation study.

Table 4. Validation results.

Models	Previous One-Shot Game Model (2018)	One-Shot Game Model	Repeated Game Models					
			Model 1	Model 2	Model 3	Model 4	Model 5	Model 6
Rate factor, δ	*na*	*na*	0.6	0.8	1.0	1.2	1.4	1.6
MAE [1]	0.2418 (75.82%)	0.1197 (88.03%)	0.1954 (80.46%)	0.1758 (82.42%)	0.1465 (85.35%)	0.1368 (86.32%)	0.1307 (86.94%)	0.1355 (86.45%)

[1] The number in parentheses indicates prediction accuracy.

5. Sensitivity Analysis of the Calibrated Stage Game

In this section, this study describes the sensitivity analysis conducted to observe how factor changes related to the proposed payoffs impact the stage game results. In reality, drivers' merging behavior to select an acceptable gap size and speed difference between the freeway mainline vehicles and the merging vehicle is different depending on the merging point [27,40]. Hence, this sensitivity analysis is required to demonstrate whether the developed stage game model represents merging behaviors observed in the field in various conditions. To show the decision-making model's sensitivity, the stage game is independently played in diverse scenarios varied by three input factors: game location, relative speed, and spacing. Preparation for the sensitivity analysis is presented first in the following sections, then results and corresponding discussions are provided.

5.1. Sensitivity Analysis Setting

As shown in Figure 8, a freeway segment that included an on-ramp was used for the analysis, with locations to play a game classified into two areas: the beginning of the acceleration lane and the end of the acceleration lane. For the spacing factor test, the SV changed its position between the PV and LV. For the speed profile test, the freeway mainline vehicles' speed was basically categorized into five scenarios: 60 km/h, 70 km/h, 80 km/h, 90 km/h, and 100 km/h. In each speed scenario, the SV's speed varied from 60 km/h to 100 km/h. The freeway testbed and calibrated stage game were modeled on MATLAB, and other simulation settings are described below.

1. The length of the acceleration lane was 250 m;
2. Based on initial longitudinal coordination, $n-1$, n, and $n+1$ denote the PV, SV, and LV, respectively;
3. It was assumed that spacing between the PV and LV, $\Delta x_{n-1,n+1}$, was constant as 40 m: in the game played at the beginning of the acceleration lane, the PV and LV were located at 70 m and 30 m from the beginning of the acceleration lane, respectively. In the game played at the end of the acceleration lane, the longitudinal position of the PV and LV were 230 m and 190 m from the beginning point, respectively;
4. The length of all vehicles was assumed as constant at 4.8 m;
5. Link properties for the freeway are as follows. Saturation flow rate was 2400 veh/h/lane. Jam density was 160 veh/km/lane. Free-flow speed and speed-at-capacity were 100 km/h and 80 km/h, respectively;
6. Calibrated parameters of payoff functions for the repeated game model with $\delta = 1.4$ were used.

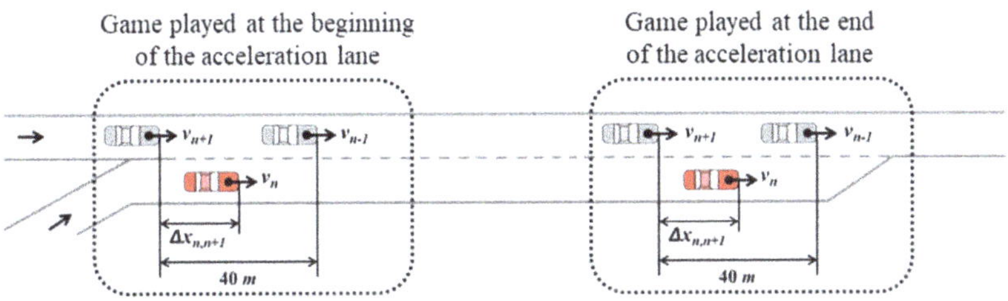

Figure 8. Topology of freeway merging section for sensitivity analysis.

5.2. Sensitivity Analysis Results

Based on the results of the stage game played at two locations in various lag spacing and relative speed scenarios, the impact of input factors and other findings revealed by the sensitivity analysis are provided. Figure 9a–e show the results after playing games near the beginning of the acceleration lane, and Figure 9f–j reveal the game results after playing the game near the end of the acceleration lane. The Chatterjee function for finding the Nash equilibrium was used to decide these game results [46]. If the game result in each case is a pure strategy Nash equilibrium, the corresponding action set is a dominant decision made by two drivers, i.e., the probability of one of six action strategies ($p_{ij} \times q_{ij}$) is one. Otherwise, when a mixed strategy Nash equilibrium exists, the game result is randomly chosen by probabilities.

Differences in drivers' behaviors based on the merging point are distinct in merging maneuver decisions. Near the beginning of the acceleration lane, a merging vehicle driver usually passes a lead vehicle when $v_n > v_{n-1}$ and when lead spacing ($\Delta x_{n-1,n}$) is quite small [27]. The higher psychological pressure related to merging makes drivers accept smaller gaps as they arrive nearer to the end of the auxiliary lane compared to cases where they can take an original gap near the beginning of the acceleration lane [27]. In other words, field data show that the driver of SV tried a forced merging maneuver at close to the end of the acceleration lane [27,33]. When $v_n < v_{n+1}$ and the lag spacing ($\Delta x_{n,n+1}$) is quite small, the driver of SV waits until the LV passes the SV and then may merge using a backward gap. In Figure 8, the calibrated stage game results show these behaviors in choosing an 'overtake (s_3)' and 'wait (s_2)' action according to the game location.

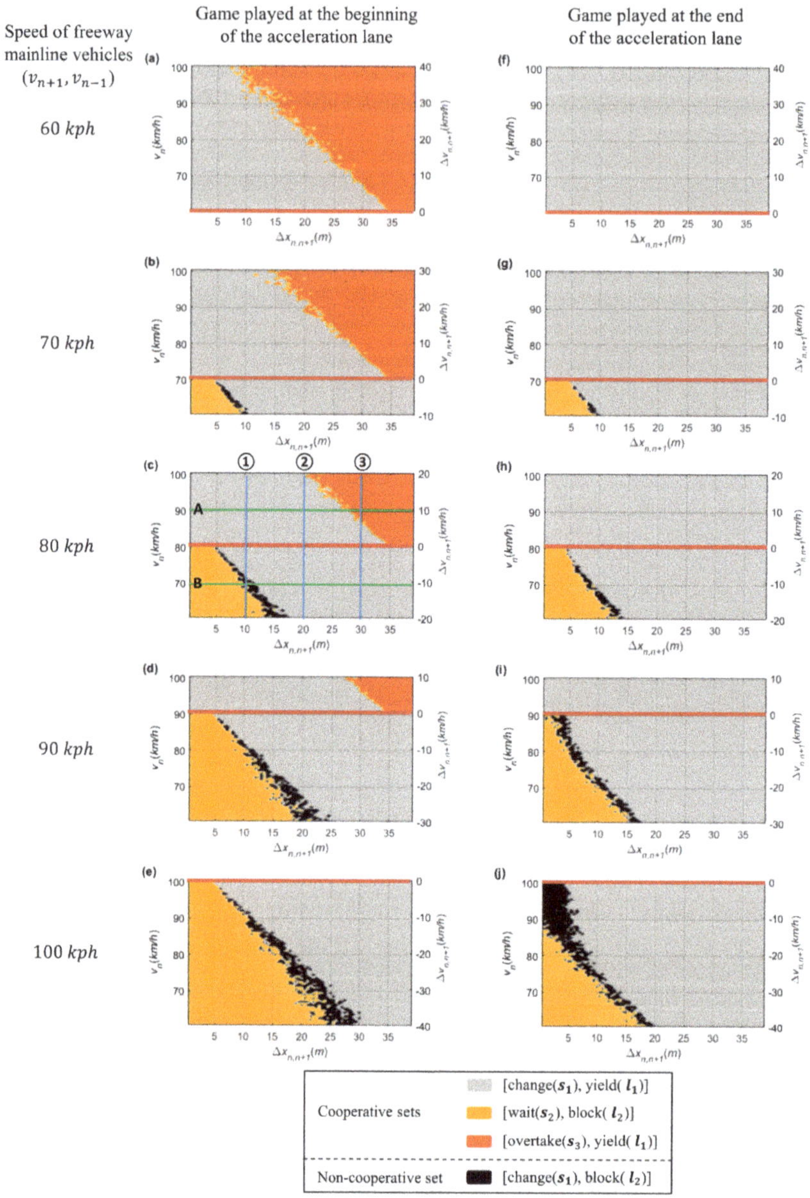

Figure 9. Graphical representation of the one-shot game results depending on game locations, spacing between vehicles ($\Delta x_{n,n+1}$), and speed of the SV (v_n): (**a–e**) game played at the beginning of the acceleration lane with mainline vehicles driving at 60 km/h to 100 km/h, respectively; (**f–j**) game played at the end of the acceleration lane with mainline vehicles driving at 60 km/h to 100 km/h, respectively. Note that a red line parallel to the x-axis on each graph indicates the speed of the freeway mainline vehicles (v_{n-1}, v_{n+1}).

Near the beginning of the lane, as illustrated in Figure 9a–d, the game results show that the driver of SV chooses the 'overtake (s_3)' action in conditions indicative of higher relative speed and short lead spacing. In contrast, the game results (as illustrated in Figure 9f–i) show that the driver of SV intentionally changes a lane due to a short remaining distance in the acceleration lane. For the 'wait (s_2)' action, differences in the results of the stage game for merging decision-making are revealed according to game location. These results prove that the forced merging utility works correctly when the SV is close to the end of the acceleration lane. Consequently, the stage game developed in this study accurately depicts realistic decisions made by human drivers according to game location.

As discussed in Section 3.3.3, TTC is critical in making lane-changing decisions. Since TTC is comprised of spacing (i.e., space headway) and relative speed, both are important in human drivers' decision-making for merging maneuvers at freeway merging sections. Hence, this study also analyzed the impacts of these factors. In Figure 9c, blue lines parallel to the y-axis (as marked with ① to ③) and green lines parallel to the x-axis (as marked with A and B) denote test cases for sensitivity analysis on relative speed and spacing, respectively.

In the sensitivity analysis on relative speed, the PV and LV are supposed to drive at 80 km/h, and the SV's speed varies from 60 km/h to 100 km/h. Scenarios were prepared with three lag spacings: 10 m, 20 m, and 30 m, and the game results of all scenarios are shown in Figure 10. Game results clearly show that the relative speed affects decision-making. When lag spacing ($\Delta x_{n,n+1}$) is 10 m (as shown in Figure 10a), the drivers of the SV and LV decide on a 'wait (s_2) and block (l_2)' action set if $\Delta v_{n,n+1} \leq -10$ km/h. In addition, both drivers are willing to choose a 'change (s_1) and yield (l_1)' action set through the stage game if $\Delta v_{n,n+1} \geq -7$ km/h. These cooperative action strategy sets are results of both drivers' common consent subject to safety. In a certain range, i.e., -10 km/h < $\Delta v_{n,n+1}$ < -7 km/h, drivers' desired actions are competitive; in these conditions, the non-cooperative behaviors, 'change (s_1) and a block (l_2)' action, will be carried out.

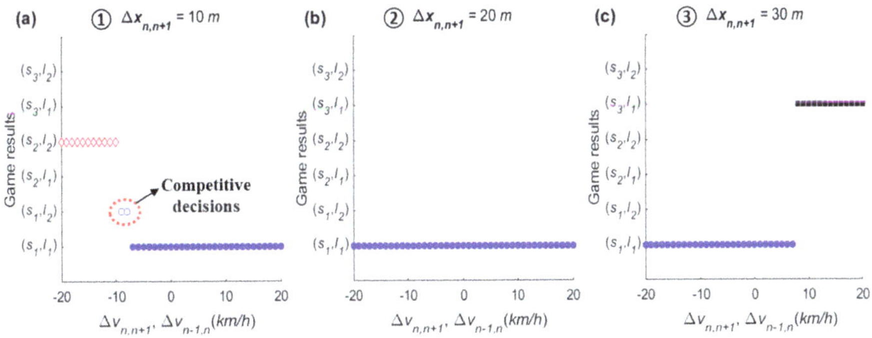

Figure 10. Game results on relative speed: (**a**) $\Delta x_{n,n+1} = 10$ m; (**b**) $\Delta x_{n,n+1} = 20$ m; (**c**) $\Delta x_{n,n+1} = 30$ m.

When $\Delta x_{n,n+1} = 20$ m, in Figure 10b, the driver of the SV and LV choose a cooperative action strategy (s_1, l_1) even if $\Delta v_{n,n+1} = -20$ km/h. This means that the relative speed is largely irrelevant in influencing the driver of SV to choose a lane-changing action if there is sufficient spacing between vehicles. If there is enough space headway, real-life experience generally shows that a driver of a merging vehicle will change lane upon reaching an acceleration lane even though a speed harmonization process is required. In response to the merging vehicle's lane change, the driver of LV decreases speed to adjust to the new preceding vehicle (i.e., the SV) or changes a lane to the left to maintain its speed. When $\Delta x_{n,n+1} = 30$ m (i.e., $\Delta x_{n-1,n} = 10$ m), moreover, the game results show a distinct feature depending on the relative speed. The cooperative action strategy (s_1, l_1) is chosen by the stage game until v_n is slightly higher than v_{n-1}. If $\Delta v_{n,n-1} \geq 8$ km/h, the driver of SV chooses an 'overtake (s_3)' action due to a relatively small TTC in order to avoid harsh braking. Of the overtaking vehicles, 97.7% were found to have a speed higher than the freeway mainline vehicles [27]. Thus, this game model can reasonably represent decision-making results according to relative speed.

For the sensitivity analysis of spacing, the stage game was played with various lag spacing from 0 m to 40 m. The PV and LV are supposed to drive at 80 km/h, and the SV's speed is 70 km/h and 90 km/h. Game results of all scenarios are shown in Figure 11. In the figure, the x-axis indicates the lag spacing ($\Delta x_{n,n+1}$), and hence an increase in $\Delta x_{n,n+1}$ means a decrease in lead spacing ($\Delta x_{n-1,n}$).

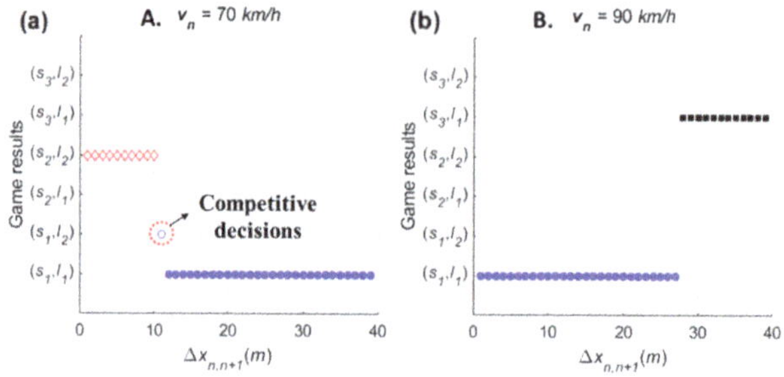

Figure 11. Game results on spacing: (**a**) $\Delta v_n = 70$ km/h; (**b**) $\Delta v_n = 90$ km/h.

When $v_n < v_{n-1}$, as shown in Figure 11a, the stage game results show that the driver of SV decides on a 'wait (s_2)' action in cases in which lag spacing is less than 10 m. In other words, results indicate that a slower SV requires spacing of more than 10 m to choose a 'change (s_1)' action. Depending on the spacing, competitive decision-making is also expected. This trend is also found in choosing an 'overtake (s_3)' action when $v_n > v_{n-1}$. In Figure 11b, the driver of SV decides to overtake at $\Delta x_{n-1,n} \leq 12$ m. Therefore, the sensitivity results indicate that the stage game reasonably explains the difference in drivers' choices according to spacing.

In the results, decisions included in a non-cooperative action strategy set, i.e., (s_1, l_2), are found in a specific decision-making region, as colored black in Figure 9. This region implies that this strategy set, which is decided simultaneously by drivers, puts them into competition. This result means that the driver of SV wants to change a lane after trying to ensure a safe lead and lag gap and the driver of LV does not allow the SV to merge. During the game period, one driver should change their initial decision to avoid a potential collision, and the final decision set would be a cooperative set. In addition, due to an unbalance in the number of observations indicating each action strategy, the (s_2 , l_1) action cannot be determined in this sensitivity analysis. From field data, including NGSIM data, it is clear that merging maneuvers are usually cooperative, as the driver of LV perceives the SV's lane-changing intention. Compared to cooperative merging, non-cooperative cases are only occasionally observed. The stage game results describe cooperative behaviors, and competition between drivers can be found at certain relative speed and spacing profiles. Consequently, the stage game model proposed in this study successfully explains rational human drivers' decision-making results.

6. Simulation Case Study

In this chapter, a simulation study is presented to demonstrate the performance of the game model based on the developed stage game for merging. For this case study, a microscopic simulation model based on an ABM method that included a vehicle acceleration controller was developed. To verify the performance of the ABM, a comparison between NGSIM data and simulation results is provided. The simulation setting is defined, and then various merging scenarios representing both cooperative and non-cooperative cases are explained. Next, simulation results for each scenario are presented.

6.1. Simulation Model Development

To investigate whether the repeated game model is efficient to use in microscopic traffic simulation, we used an ABM approach. ABM is a powerful method for making simulations that is widely applied across real-life problems [47–49]. This study developed a simulation model that was built on MATALB using the ABM method combined with the game model. ABM is a suitable approach for simulating the actions and interactions of intelligent entities, which includes individual people. Collaboration and competition, in particular, are major concerns in game theory; these are two typical types of human interactions addressed in several ABM methods [50]. One of the applicable situations for using ABM is when interactions among agents are heterogeneous and can lead to network effects [48,51]. Thus, this study develops a simulation model to explain merging interactions.

According to Zheng et al. [49], the ABMs explored for the existing transportation system in today's literature, in general, have the distinguishing feature of integration, combining three components: drivers' action decisions, drivers' route decisions, and microsimulation. As a microsimulation component, the simulation model developed in this study basically simulates vehicle movements based on position and by speed profile, as determined by an acceleration controller at each time step. As shown in Figure 12, the controller consists of a game module and a car-following module. According to the game model for the drivers' action decision component, a driver of SV plays a stage game with a driver of LV in the target lane. Depending on the action strategies at each game time, both drivers determine the acceleration level to accomplish their own strategy. In the car-following module, in addition, the desired acceleration level is decided by the RPA car-following model. In this acceleration controller, neither the individual demographic nor the travel characteristics of either agent are considered.

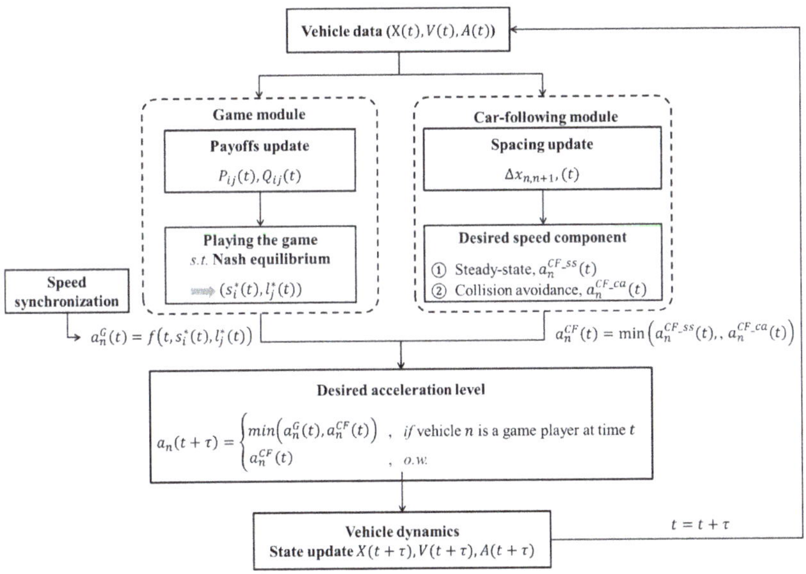

Figure 12. Vehicle acceleration controller structure in the developed simulation model.

As the game results show, when the driver of SV chooses a 'change (s_1)' action, they evaluate lead and lag spacing for gap acceptance to satisfy sufficient spacing and avoid collision. If the instantaneous gap is enough to change lane, the SV begins merging onto the freeway, and the driver of LV determines the acceleration level to follow the SV in the car-following model in response to recognition of the SV's lane-change. In addition, a route decision module is not required because merging scenarios are tested on the one-lane freeway network, which includes a merging ramp.

The car-following module estimates a desired acceleration level based on instantaneous spacing between vehicles and speed at each time step t. This study used two components, i.e., steady-state and collision avoidance, of the RPA car-following model for the module [43]. The detailed definition and formulas of the components in the RPA model are described in [43]. Figure 13 shows the performance of car-following module in a case in which five vehicles formed a platoon. Vehicles decide an acceleration level to follow the preceding vehicle by the RPA car-following model. Here, it was assumed that vehicles were located with shorter spacing than the steady-state spacing of Van Aerde's car-following model [44] at simulation time 0. As illustrated in Figure 13, therefore, following vehicles initially decreased speed to ensure proper spacing between vehicles. Then, they began to accelerate after ensuring the sufficient spacing by sequence in the platoon. In conclusion, acceleration level and speed oscillated for a while, and then they were stabilized.

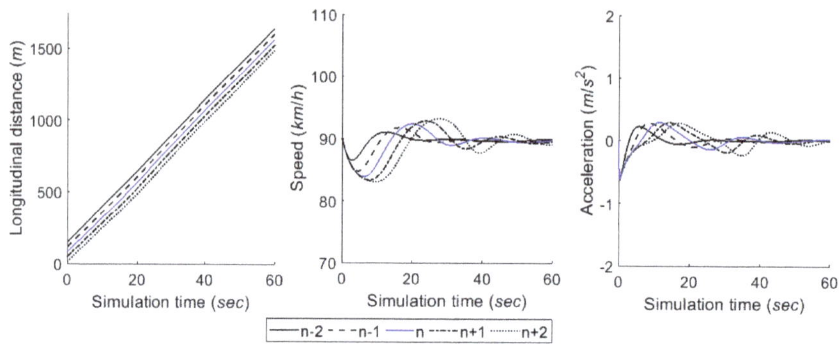

Figure 13. Performance of the car-following module.

The game module begins operating as soon as the SV enters the acceleration lane. The nearest following vehicle in the target lane becomes the opposite player. In this module, there are two types of merging game: (1) the one-shot game; (2) the repeated game. In detail, the one-shot game uses instantaneous payoffs, which are computed based on spacing and speed profile at time t, for each action strategy set, i.e., $P_{ij}(t)$, $Q_{ij}(t)$. In the repeated game, on the other hand, the cumulated payoffs are utilized. Regardless of the game type, two players decide an action strategy set subject to the Nash equilibrium. Based upon the action chosen at time t, the desired acceleration level for each vehicle is calculated to execute that vehicle's individual action strategy. For the SV, the desired acceleration level is determined as stated below:

- For 'change (s_1)' action, the driver of SV determines acceleration level in consideration of not only speed synchronization but also gap acceptance. If $v_n(t) \ll v_{n+1}(t)$, an acceleration level for speed harmonization is additionally calculated. By gap acceptance rule, another acceleration level is calculated to ensure a sufficient gap for lead and lag spacing;
- For 'wait (s_2)' action, a required acceleration level to wait in acceleration lane until the lag vehicle passes the SV is computed. Generally, waiting cases are observed when $v_n(t) \ll v_{n+1}(t)$ and $\Delta x_{n,n+1}$ is not sufficient. If $v_n(t) \ll v_{n+1}(t)$ and the remaining distance to the end of the acceleration lane at time t, $RD_n(t)$, is sufficient to not require deceleration, the SV slightly accelerates to harmonize the speed with freeway vehicles during waiting time;
- Lastly, it needs to calculate the required acceleration level to use the forward gap for 'overtake (s_3)' action. This case is observed when $v_n(t) \gg v_{n+1}(t)$ and $\Delta x_{n-1,n}$ is not sufficient. For this strategy, therefore, speed harmonization is excluded as an acceleration component.

In addition, the driver of LV decides the acceleration level for a 'yield (l_1)' action by accepting the SV's merging intention. To provide safe spacing for merging, the LV's acceleration level was calculated based on the car-following model with an assumption that the SV became a potential lead vehicle. For a 'block (l_2)' action, on the other hand, the driver of SV shows an acceleration to pass the SV by decreasing spacing. This decrease in spacing is regarded as blocking intention.

6.2. Simulation Model Validation

Prior to conducting a case study, validation of the simulation model developed in this study was required to determine whether the conceptual model is a reasonably accurate representation of the real world [52] and whether the output of simulations is consistent with real-world output [53]. To validate the simulation model, this study used the graphical comparison technique, in which the graphs of values derived from the simulation model over time are compared with the graphs of values collected in a real system. It is a subjective, yet practical approach, and is especially useful as a preliminary approach [54]. Since the objective of the case study was to verify the repeated game's efficiency, the simulation focuses on presenting microscopic vehicle movements based on rational drivers' decision-making without consideration of individual characteristics. Considering this objective, a mathematical approach, such as statistical testing of simulation results, was not selected for model validation. Therefore, this study provides a graphical comparison between NGSIM data and the results derived from the simulation model to investigate similarity of trend in vehicle position and corresponding spacing.

This study extracted game cases from NGSIM data in which there was no interference by other surrounding vehicles except for the three main vehicles (i.e., the SV, PV, and LV). Next, instantaneous vehicles' location and speed prior to 1.0 seconds in each case were prepared as input data for simulation. The graphical comparison results showing longitudinal vehicle position and spacing are shown in Figure 14. In an example, to show changing situation (see Figure 14a), vehicle position and corresponding lead and lag spacing are almost identical. In an example showing an overtaking situation (see Figure 14b), considerable similarity is observed. The results show that the simulation model based on the ABM represents values similar to those found in the NGSIM data in longitudinal vehicle position and spacing. Consequently, it was possible to conclude that the developed simulation model could be utilized in the case study.

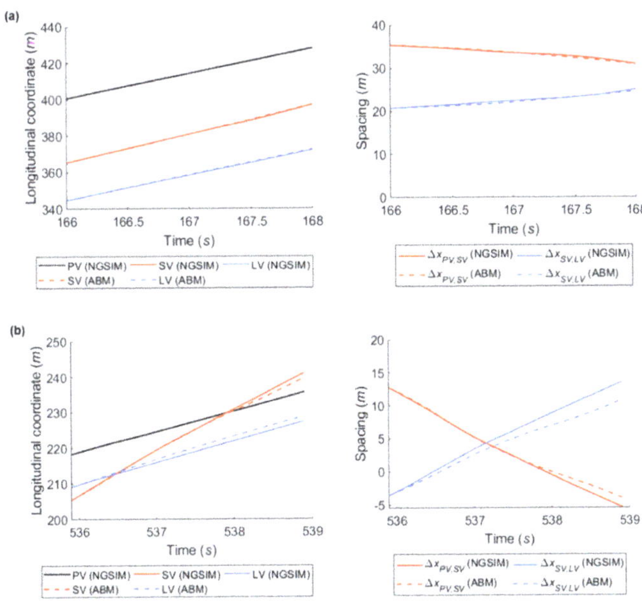

Figure 14. Simulation model validation results based on the graphical comparison method: (**a**) changing situation (SV ID: 268, PV ID: 258, and LV ID: 269 in the US101 data collected from 8:05 to 8:20 a.m.) and (**b**) overtaking situation (SV ID: 1108, PV ID: 1112, and LV ID: 1118 in the US101 data collected from 8:20 to 8:35 a.m.).

6.3. Simulation Setting and Cases

This study conducted case studies in various merging scenarios simulated for a total of five vehicles, including a merging vehicle. Simulation experiments were executed using both the one-shot game model and the repeated game model. As described above, the one-shot game herein is played independently without consideration of previous results at every decision-making point. The repeated game is played based on the cumulative payoffs proposed in Section 3.2. In addition, a freeway segment, including one merging section, was modeled on MATLAB, as illustrated in Figure 15. The length of the freeway mainline was 1.0 km and the 250 m acceleration lane was located 80 m downstream of the beginning of the network. The details of the simulation settings are defined as follows.

1. Link properties for the freeway are as follows. Saturation flow rate was 2400 veh/h/lane. Jam density was 160 veh/km/lane. Free-flow speed and speed-at-capacity were 100 km/h and 80 km/h, respectively;
2. Based on initial longitudinal coordination, vehicles on the network were designated as $n-2$, $n-1$, n, $n+1$, and $n+2$, respectively. Here, the vehicle n denotes the SV;
3. It was assumed that the average initial speed of freeway vehicles was v_{fwy}. The initial speeds of four vehicles on the freeway mainline (i.e., $n-2$, $n-1$, $n+1$, $n+2$) were randomly determined using the normal distribution with a mean of v_{fwy} and standard deviation of 0.2 at simulation start time;
4. The initial spacing between freeway vehicles, i.e., $\Delta x_{n-2,n-1}$, $\Delta x_{n-1,n+1}$, $\Delta x_{n+1,n+2}$, was determined using the Van Aerde's steady-state model according to instantaneous speed of corresponding following vehicle at time-step 0;
5. With regard to the game, the time interval for playing the game was 0.5 s. The stage game would be newly formed if the LV or PV changed;
6. The rate factor (δ) of 1.4 and corresponding calibrated parameters of payoff functions, as shown in Table 2, were used for the repeated game model;
7. Maximum and minimum accelerations are 3.4 m/s^2 and -3.4 m/s^2, respectively, as determined with reference to the NGSIM data. The length of all vehicles was assumed as constant as 4.8 m;
8. In this simulation model, the freeway mainline vehicles' behaviors to avoid a potential collision with the merging vehicle, i.e., lane change to left or deceleration before arriving at the merging section, were excluded. These behaviors could not be modeled for an individual vehicle's driving maneuvers in traffic simulator because they are a result of vehicles' independent decisions rather than any interaction with the merging vehicle after recognizing the merging vehicle.

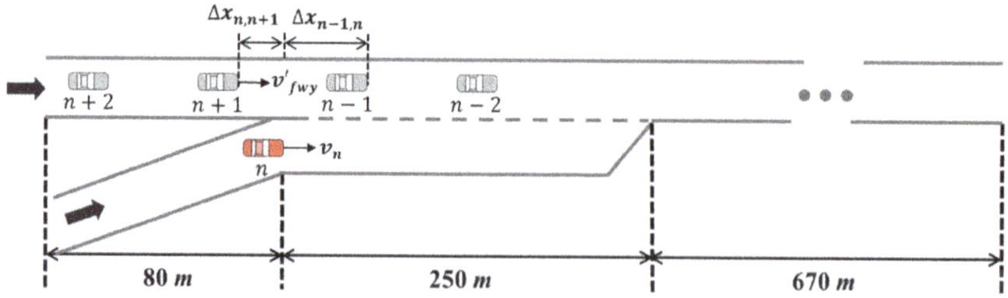

Figure 15. Simulation network configurations.

A total of five simulation cases were prepared, as summarized in Table 5, to represent plausible merging cases as defined by the diverse input values of three factors: freeway mainline vehicles' average speed (v_{fwy}), initial SV's speed (v_n), and initial lag spacing ($\Delta x_{n,n+1}$). There are two main categories in merging: cooperative and competitive merging. Cooperative merging cases, in which the drivers' decision set would be collaborative by the common consent of both drivers, indicate typical cases to select a gap type among three types: a forward gap, an adjacent gap, and a backward gap. In contrast, a competitive merging case represents an example showing a conflict in both drivers' behavior. For example, the driver of SV who wants to use an adjacent gap is willing to prepare to merge onto the freeway by turning a signal on, and then executing a lane change. In that time, the driver of LV decides not to allow the cut-in to avoid the expected considerable deceleration. One of the drivers should change their initial decision in order to avoid a potential collision. This competitive situation is not common, but many drivers may have had an experience of this type. Thus, we picked two cases in order to show not only the game model's performance in non-cooperative cases but also differences between the two game models in competitive scenarios.

Table 5. Initial Conditions of Merging Scenarios for Case Study.

Index	Scenarios	Gap Type Used for Merging	$\overline{v_{fwy}}$	$\overline{v_n}$	$\Delta x_{n,n+1}$
1	Cooperative	Adjacent gap	90 km/h	75 km/h	20.0 m
2		Backward (lag) gap	90 km/h	65 km/h	15.0 m
3		Forward (lead) gap	50 km/h	65 km/h	15.0 m
4	Competitive	Adjacent gap or backward gap (Initial decision: non-cooperative)	85 km/h	72 km/h	14.0 m
5		Adjacent gap or backward gap (Initial decision: cooperative)	90 km/h	75 km/h	7.5 m

6.4. Case Study Results

Cooperative and competitive cases were tested using the developed simulation model. In order to validate the repeated game model's performance, the simulation results using the repeated model are compared with results using the calibrated stage game model played independently, i.e., one-shot game model at every decision-making point.

In cooperative scenarios, a dominant action strategy is found in rational decision-making due to the apparent situation. The simulation model using the repeated game model shows a very close performance with the model using the one-shot game as the game results are same in each game point. Since there is a mixed strategy Nash equilibrium in the competitive cases, both drivers decide an action strategy depending on the probability of actions. For case study results, this study provides the typical outcome of each scenario if there is no distinct difference in decision-making using the two game models. Otherwise, especially in the competitive scenario, the decision-making output simulation results of each game model are individually presented.

6.4.1. Case 1: Cooperative Merging Scenario Using an Adjacent Gap

In simulation results for the first case, Figure 16 presents that the SV smoothly merged onto the freeway. As described in the sensitivity analysis, the developed game model has the ability to represent drivers' decisions in normal cooperative merging cases. According to the game results, as shown in Figure 17, drivers chose a 'change (s_1) and yield (l_1)' action set during the game period. The SV slightly accelerated by speed harmonization rules in preparation for merging while the LV decelerated in order to accept the SV's lane change. When a lead and lag gap was acceptable, the SV merged onto the freeway mainline. In simulation, the driver of SV controlled the vehicle's speed via the car-following rule as soon as it executed the lane change and its following vehicles also showed oscillation in their speed profiles to ensure a safe gap.

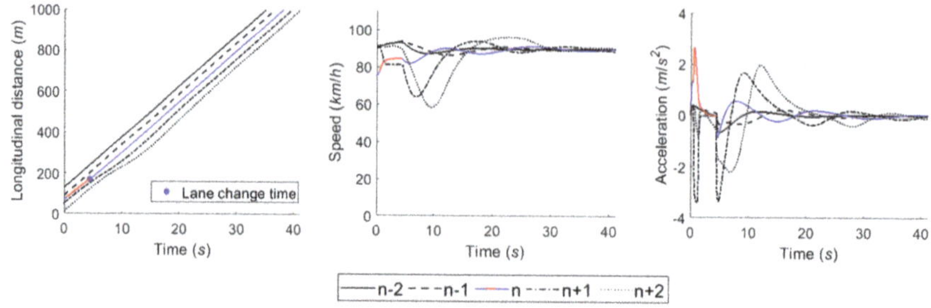

Figure 16. Graphical representation of simulation results in case 1. Note that a red solid line indicates simulation data of the SV (vehicle n) during game period, whereas a blue solid line shows the SV's data in simulation time except game period.

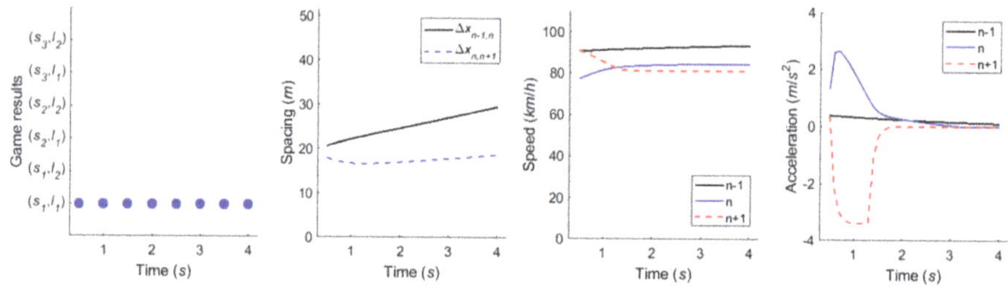

Figure 17. Decision-making game results in case 1.

6.4.2. Case 2: Cooperative Merging Scenario Using a Backward Gap

Simulation results for the second case, as shown in Figure 18, indicate that the driver of SV used the backward gap after the initial LV to overtake the SV. In Figure 19a, the drivers decided on a 'wait (s_2) and block (l_2)' action strategy, respectively. The LV accelerated to block merging, and the SV also accelerated for speed synchronization even though the driver of SV decided to take a 'wait (s_2)' action. As soon as the initial LV overtook the SV, a new merging decision-making game was identified in which the vehicle $n+2$ became the new LV. The results of the second game are shown in Figure 19b. The SV continuously chose a 'change (s_1)' action until the gap acceptance rule was satisfied, then moved to the freeway mainline in consideration of gap size and relative speed. The LV, i.e., the vehicle $n+2$, in the second game decelerated in a yielding action in response to the SV's intention to merge. In conclusion, the merging decision-making model was shown to depict a typical waiting scenario for both game models.

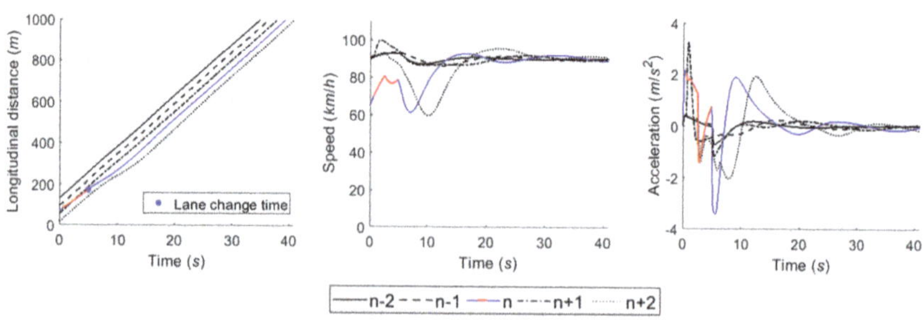

Figure 18. Graphical representation of simulation results in case 2. Note that a red solid line indicates simulation data of the SV (vehicle n) during game period, whereas a blue solid line shows the SV's data in simulation time except game period.

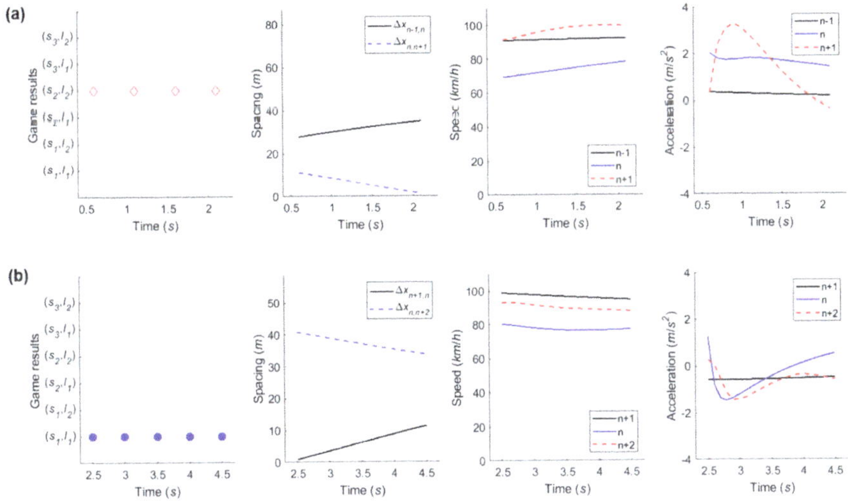

Figure 19. Decision-making game results in case 2: (**a**) Initial game with $n+1$; (**b**) additional game with $n+2$.

6.4.3. Case 3: Cooperative Merging Scenario Using a Forward Gap

In overtaking scenario, the time–space diagram in Figure 20 shows that the SV took the forward gap and then merged onto the freeway. When the SV entered the acceleration lane, as presented in Figure 21a, the SV and LV chose the 'overtake (s_3) and yield (l_1)' action set. Although the LV decided the yielding action, it was observed that the LV maintained its speed during the first game period due to observing the SV's passing. After overtaking the lead vehicle, the SV began to decrease speed to harmonize with that of freeway vehicles. As shown in Figure 21b, a new LV, i.e., one which had been the lead vehicle in the first game period, selected the yielding action in interaction with the SV. It therefore showed a deep deceleration during the second game period. The SV maintained on the acceleration lane, then changed lane as soon as the gap acceptance rule was satisfied. As described in the simulation setting, the overtaking scenario is usually observed in congested traffic conditions. Thus, this lane-changing by overtaking action caused a huge oscillation in speed profile because, generally, spacing between vehicles is small under congested traffic conditions. It is concluded that this simulation model based on the proposed game model well represents the induction of a backward-forming shockwave by merging traffic in congested conditions.

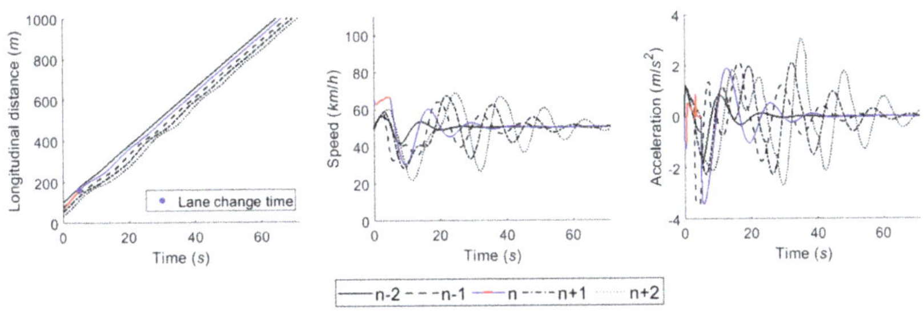

Figure 20. Graphical representation of simulation results in case 3. Note that a red solid line indicates simulation data of the SV (vehicle n) during game period, whereas a blue solid line shows the SV's data in simulation time except game period.

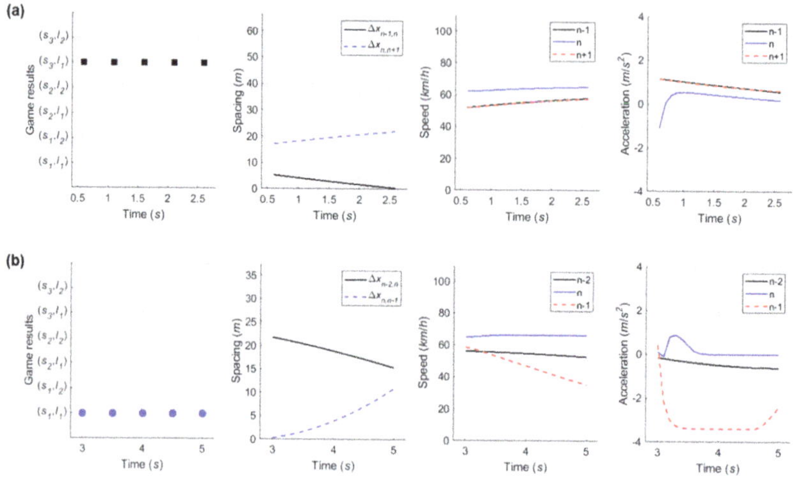

Figure 21. Decision-making game results in case 3: (**a**) Initial game with $n+1$; (**b**) additional game with $n-1$.

6.4.4. Case 4: Competitive Merging Scenario Choosing an Adjacent Gap or a Backward Gap (1)

In the fourth competitive merging case, as presented in Figure 22, the SV spent relatively longer time in playing decision-making game than previous three cases. The initial game result of (s_1, l_2) is observed in Figure 23a. As a non-cooperative action strategy set, both drivers are in competition to achieve their own objective. At the third decision-making point, a decision they make becomes (s_2, l_2) as a cooperative action strategy set. Although the driver of SV initially wanted to change a lane using an adjacent gap as soon as entering an acceleration lane, they change the initial decision in order to avoid collision after recognizing the opposite driver's aggressive behavior. Thus, the driver finally uses the backward gap for merging onto the freeway. From this case, this study concludes that the repeated game model can depict practical changes in drivers' decisions in competitive decision-making, even using the cumulative function.

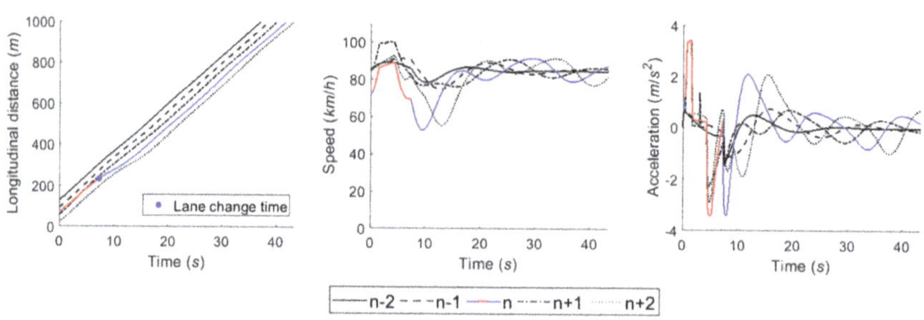

Figure 22. Graphical representation of simulation results in case 4. Note that a red solid line indicates simulation data of the SV (vehicle n) during game period, whereas a blue solid line shows the SV's data in simulation time except game period.

6.4.5. Case 5: Competitive Merging Scenario Choosing an Adjacent Gap or a Backward Gap (2)

In Case 5, the simulation results show the SV used the backward gap for merging onto the freeway whichever game model is used, as illustrated in Figures 24 and 25. This example shows a competition to choose an adjacent gap or a backward gap, as in Case 4. However, there is a difference in that the initial decision is a cooperative action strategy in Case 5.

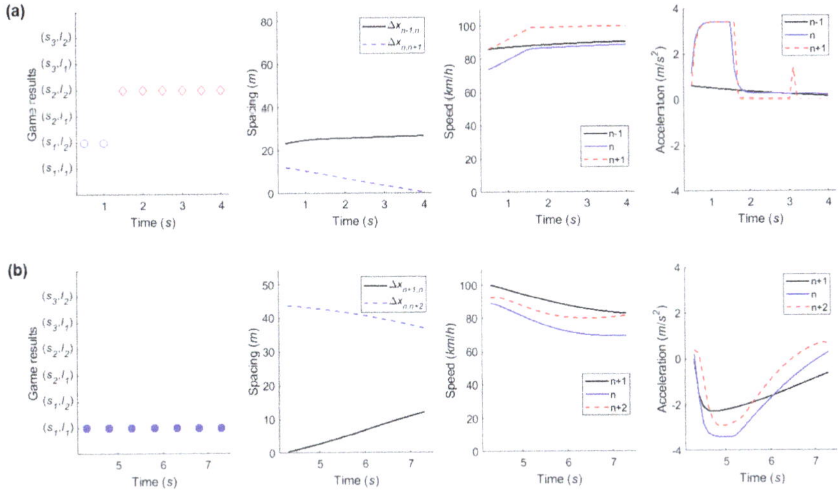

Figure 23. Decision-making game results in case 4: (**a**) Initial game with $n+1$; (**b**) additional game with $n+2$.

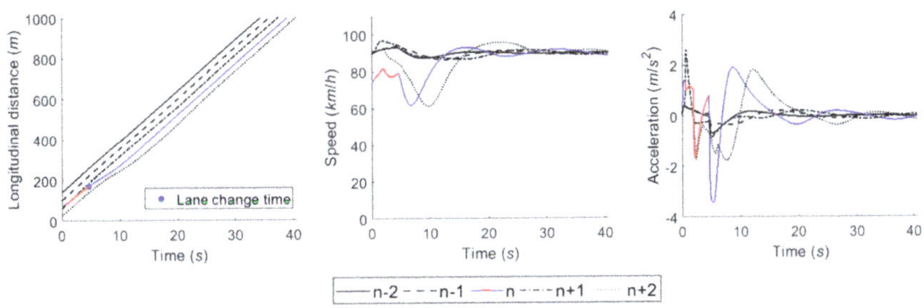

Figure 24. Graphical representation of simulation results in case 5 using the repeated game model. Note that a red solid line indicates simulation data of the SV (vehicle n) during game period, whereas a blue solid line shows the SV's data in simulation time except game period.

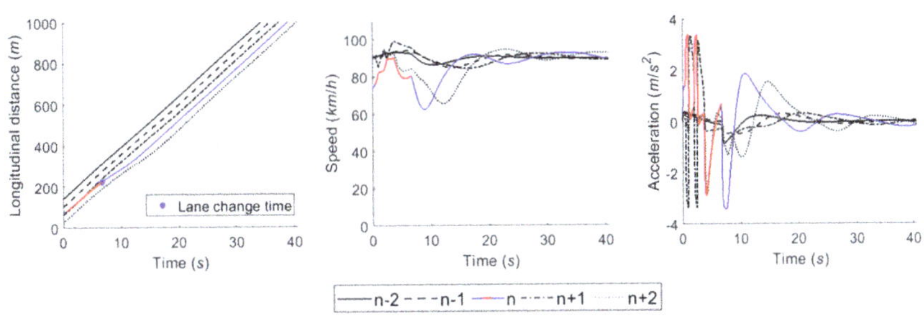

Figure 25. Graphical representation of simulation results in case 5 using the one-shot game model. Note that a red solid line indicates simulation data of the SV (vehicle n) during game period, whereas a blue solid line shows the SV's data in simulation time except game period.

In Figure 26a, when the repeated game model was used, the driver of SV chose a 'wait (s_2)' action during the first game period and then decided to change lane in the second game period. While decision-making results were maintained using the repeated game model, an oscillation in decision-making is revealed when the one-shot game is used, as shown in Figure 27a. One reason the one-shot game model causes unstable decision results is that the stage game decides a driver's action in a merging situation based on instantaneous vehicle location, speed, and acceleration data without consideration of previous game results (i.e., decisions made at previous game points). Considering the goal of each action, a change from a non-cooperative strategy set to a cooperative strategy is required in order to avoid a collision (if (s_1, l_2) is chosen) or unnecessary deceleration (if (s_2, l_1) is selected). However, changes between cooperative action strategy sets (i.e., (s_1, l_1) and (s_2, l_2)) are not realistic except when there is a surrounding vehicle intervention. This case shows a distinct difference observed in simulation results depending on which type of the two game models is used. Oscillation in decision-making may reduce the performance of microscopic traffic simulation models even though it is only observed in specific competitive merging situations.

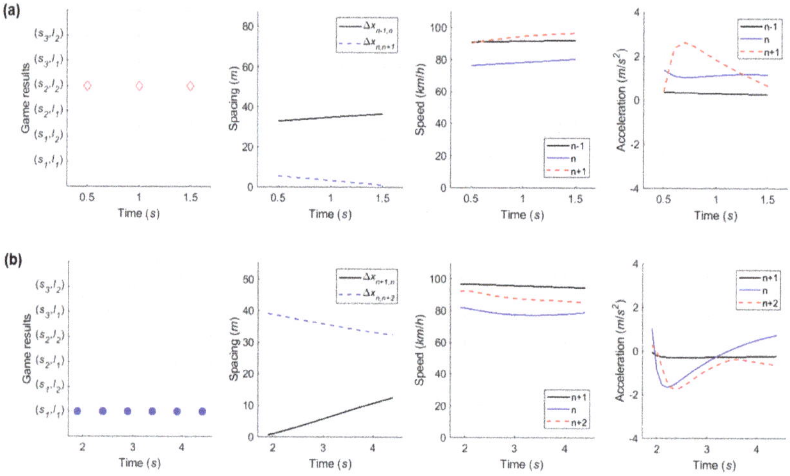

Figure 26. Decision-making game results in case 5 using the repeated game model: (**a**) Initial game with $n + 1$; (**b**) additional game with $n + 2$.

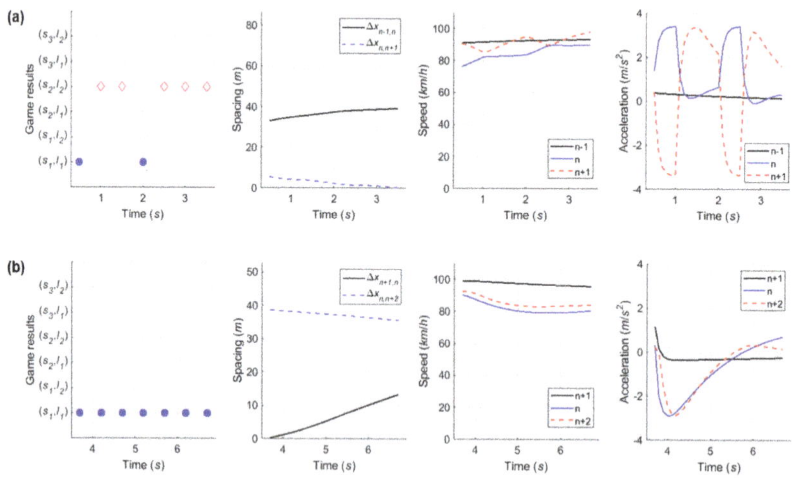

Figure 27. Decision-making game results in case 5 using the one-shot game model: (**a**) Initial game with $n + 1$; (**b**) additional game with $n + 2$.

7. Conclusions

Drivers' behavior has a big impact on the safety and throughput of the transportation system. This is especially true for traffic conflicts between merging and through vehicles, in that merging vehicles induce shockwaves, which result in a reduction in the roadway capacity resulting in traffic congestion. Consequently, modeling driving behavior thoroughly and accurately is critical when analyzing traffic flow in microscopic traffic simulation and in taking advantage of the advanced vehicle-driving technologies and strategies in AVs. The purpose of this study is to update the repeated game lane-changing model proposed in [13]. This game model has a feature that interprets interaction between drivers, as compared to most lane-changing models, which are focused on the lane-changing vehicle only. In this study, the payoff functions were newly formulated, focusing on not only improvements in prediction performance but also use in microscopic traffic simulators. In the model evaluation, the developed model captured drivers' merging behaviors with a prediction accuracy of about 86%, showing an improvement of about 12% compared to [13]. This study also presented a sensitivity analysis to indicate that the developed model can depict rational merging decision-making according to variations in the related factors: game location, relative speed, and gap size. In order to demonstrate why the repeated game is required in microscopic traffic simulation, moreover, a case study was conducted using the ABM developed to simulate merging situations. Using the repeated game model showed that it had a superior performance compared to a one-shot game model, in which the stage game is independently played, in terms of representing practical merging behaviors in cooperative and competitive merging scenarios.

In order to elaborate on this study as a state-of-the-art lane-changing model, the decision-making model based on the game theoretical approach needs to be expanded as a decision-making model for both mandatory and discretionary lane changing. Since lane-changing-related decision making can be affected by several factors (e.g., road design, traffic stream condition, driving skill, driver's aggressiveness), the model should be calibrated based on the field data collected in various conditions. Lastly, the game model can be applied to advanced vehicle systems, such as AVs, which coexist with human-operated vehicles on the roadway. The model based on the game theoretical approach is anticipated to become an appropriate model to decide lane-changing maneuvers and predict surrounding vehicle drivers' behaviors.

Author Contributions: Conceptualization, K.K.; methodology, K.K. and H.A.R; validation, K.K.; simulation, K.K.; formal analysis, K.K. and H.A.R; writing—original draft preparation, K.K.; writing—review and editing, H.A.R.; visualization, K.K.; supervision, H.A.R. All authors have read and agreed to the published version of the manuscript.

Funding: This research was funded partially by the University Mobility and Equity Center (UMEC) and a gift from the Toyota InfoTechnology Center.

Conflicts of Interest: The authors do not have any conflict of interest with other entities or researchers.

References

1. Rahman, M.; Chowdhury, M.; Xie, Y.; He, Y. Review of Microscopic Lane-Changing Models and Future Research Opportunities. *IEEE Trans. Intell. Transp. Syst.* **2013**, *14*, 1942–1956. [CrossRef]
2. Cassidy, M.J.; Bertini, R. Some traffic features at freeway bottlenecks. *Transp. Res. Part B: Methodol.* **1999**, *33*, 25–42. [CrossRef]
3. Bertini, R.; Leal, M.T. Empirical Study of Traffic Features at a Freeway Lane Drop. *J. Transp. Eng.* **2005**, *131*, 397–407. [CrossRef]
4. Laval, J.A.; Daganzo, C.F. Lane-changing in traffic streams. *Transp. Res. Part B Methodol.* **2006**, *40*, 251–264. [CrossRef]
5. Coifman, B.; Mishalani, R.; Wang, C.; Krishnamurthy, S. Impact of Lane-Change Maneuvers on Congested Freeway Segment Delays: Pilot Study. *Transp. Res. Rec. J. Transp. Res. Board* **2006**, *1965*, 152–159. [CrossRef]

6. Ahn, S.; Cassidy, M.J. Freeway Traffic Oscillations and Vehicle Lane-Change Maneuvers. In *Transportation and Traffic Theory 2007*; Allsop, R.E., Bell, M.G.H., Heydecker, B., Eds.; Elsevier: Amsterdam, The Netherlands, 2007; pp. 691–710.
7. Pan, T.; Lam, W.; Sumalee, A.; Zhong, R. Modeling the impacts of mandatory and discretionary lane-changing maneuvers. *Transp. Res. Part C Emerg. Technol.* **2016**, *68*, 403–424. [CrossRef]
8. Li, X.; Sun, J.-Q. Studies of Vehicle Lane-Changing Dynamics and Its Effect on Traffic Efficiency, Safety and Environmental Impact. *Phys. A Stat. Mech. its Appl.* **2017**, *467*, 41–58. [CrossRef]
9. Liu, H.X.; Xin, W.; Adam, Z.M.; Ban, J.X. A game theoretical approach for modeling merging and yielding behavior at freeway on-ramp section. In *Transportation and Traffic Theory 2007*; Allsop, R.E., Bell, M.G.H., Heydecker, B., Eds.; Elsevier: Amsterdam, The Netherlands, 2007; pp. 196–211.
10. Moridpour, S.; Sarvi, M.; Rose, G. Lane changing models: A critical review. *Transp. Lett.* **2010**, *2*, 157–173. [CrossRef]
11. Kesting, A.; Treiber, M.; Helbing, D. General Lane-Changing Model MOBIL for Car-Following Models. *Transp. Res. Rec. J. Transp. Res. Board* **2007**, *1999*, 86–94. [CrossRef]
12. Kang, K.; Rakha, H.A. Game Theoretical Approach to Model Decision Making for Merging Maneuvers at Freeway On-Ramps. *Transp. Res. Rec. J. Transp. Res. Board* **2017**, *2623*, 19–28. [CrossRef]
13. Kang, K.; Rakha, H.A. Modeling Driver Merging Behavior: A Repeated Game Theoretical Approach. *Transp. Res. Rec. J. Transp. Res. Board* **2018**, *2672*, 144–153. [CrossRef]
14. FHWA. Fact Sheet: Next Generation Simulation US101 Dataset, FHWA-HRT-07-030. Available online: http://www.fhwa.dot.gov/publications/research/operations/07030/ (accessed on 25 January 2016).
15. FHWA. Next Generation Simulation: US101 Freeway Dataset. Available online: http://ops.fhwa.dot.gov/trafficanalysistools/ngsim.htm (accessed on 25 January 2016).
16. Gipps, P. A model for the structure of lane-changing decisions. *Transp. Res. Part B Methodol.* **1986**, *20*, 403–414. [CrossRef]
17. Toledo, T.; Koutsopoulos, H.N.; Ben-Akiva, M.E. Modeling Integrated Lane-Changing Behavior. *Transp. Res. Rec. J. Transp. Res. Board* **2003**, *1857*, 30–38. [CrossRef]
18. Halati, A.; Lieu, H.; Walker, S. CORSIM—Corridor traffic simulation model. In Proceedings of the Traffic Congestion and Traffic Safety in the 21st Century: Challenges, Innovations, and Opportunities, Chicago, IL, USA, 8–10 June 1997; American Society of Civil Engineers: New York, NY, USA, 1997; pp. 570–576.
19. FHWA. *CORSIM User Manual, Version 1.04*; U.S. Department of Transportation: McLean, VA, USA, 1998.
20. Myerson, R.B. *Game Theory: Analysis of Conflict*; Harvard University Press: Cambridge, MA, USA, 1997.
21. Ahmed, K.I. Modeling Drivers' Acceleration and Lane-Changing Behavior. Ph.D. Thesis, Dept. Civil and Environmental Eng., Massachusetts Institute of Technology, Cambridge, MA, USA, 1999.
22. Toledo, T.; Koutsopoulos, H.N.; Ben-Akiva, M. Integrated driving behavior modeling. *Transp. Res. Part C Emerg. Technol.* **2007**, *15*, 96–112. [CrossRef]
23. Ma, X. Toward an integrated car-following and lane-changing model based on neural-fuzzy approach. In Proceedings of the Helsinki Summer Workshop, Espoo, Finland, 6–13 November 2004.
24. Hunt, J.; Lyons, G. Modelling dual carriageway lane changing using neural networks. *Transp. Res. Part C Emerg. Technol.* **1994**, *2*, 231–245. [CrossRef]
25. Kita, H. A merging–giveway interaction model of cars in a merging section: A game theoretic analysis. *Transp. Res. Part A Policy Pr.* **1999**, *33*, 305–312. [CrossRef]
26. Kondyli, A.; Elefteriadou, L. Driver Behavior at Freeway-Ramp Merging Areas. *Transp. Res. Rec. J. Transp. Res. Board* **2009**, *2124*, 157–166. [CrossRef]
27. Wan, X.; Jin, P.J.; Zheng, L.; Cheng, Y.; Ran, B. Speed Synchronization Process of Merging Vehicles from the Entrance Ramp. *Transp. Res. Rec. J. Transp. Res. Board* **2013**, *2391*, 11–21. [CrossRef]
28. Kim, C.; Langari, R. Game theory based autonomous vehicles operation. *Int. J. Veh. Des.* **2014**, *65*, 360. [CrossRef]
29. Talebpour, A.; Mahmassani, H.S.; Hamdar, S.H. Modeling Lane-Changing Behavior in a Connected Environment: A Game Theory Approach. *Transp. Res. Procedia* **2015**, *7*, 420–440. [CrossRef]
30. Harsanyi, J.C. Games with Incomplete Information Played by "Bayesian" Players, I–III Part I. The Basic Model. *Manag. Sci.* **1967**, *14*, 159–182. [CrossRef]
31. Yu, H.; Tseng, H.E.; Langari, R. A human-like game theory-based controller for automatic lane changing. *Transp. Res. Part C Emerg. Technol.* **2018**, *88*, 140–158. [CrossRef]

32. Nash, J. Non-Cooperative Games. *Ann. Math.* **1951**, *54*, 286. [CrossRef]
33. Kondyli, A.; Elefteriadou, L. Driver behavior at freeway-ramp merging areas based on instrumented vehicle observations. *Transp. Lett.* **2012**, *4*, 129–142. [CrossRef]
34. Wang, Z.; Wu, G.; Barth, M. Distributed Consensus-Based Cooperative Highway On-Ramp Merging Using V2X Communications. In Proceedings of the WCX: SAE World Congress Experience, Detroit, MI, USA, 3 April 2018.
35. Lee, S.E.; Olsen, E.C.; Wierwille, W.W. *A Comprehensive Examination of Naturalistic Lane-Changes*; American Psychological Association (APA): Washington, DC, USA, 2013.
36. Brackstone, M.; McDonald, M.; Sultan, B. Dynamic Behavioral Data Collection Using an Instrumented Vehicle. *Transp. Res. Rec. J. Transp. Res. Board* **1999**, *1689*, 9–16. [CrossRef]
37. Kusano, K.D.; Gabler, H. Method for Estimating Time to Collision at Braking in Real-World, Lead Vehicle Stopped Rear-End Crashes for Use in Pre-Crash System Design. *SAE Int. J. Passeng. Cars-Mech. Syst.* **2011**, *4*, 435–443. [CrossRef]
38. Vogel, K. A comparison of headway and time to collision as safety indicators. *Accid. Anal. Prev.* **2003**, *35*, 427–433. [CrossRef]
39. National Satety Council: Maintaining a Safe Following Distance While Driving. Available online: https://www.nsc.org/Portals/0/Documents/TeenDrivingDocuments/DriveItHome/Lesson48-English.pdf (accessed on 11 December 2018).
40. Marczak, F.; Daamen, W.; Buisson, C. Key Variables of Merging Behaviour: Empirical Comparison between Two Sites and Assessment of Gap Acceptance Theory. *Procedia-Soc. Behav. Sci.* **2013**, *80*, 678–697. [CrossRef]
41. Hwang, S.Y.; Park, C.H. Modeling of the Gap Acceptance Behavior at a Merging Section of Urban Freeway. In Proceedings of the 2005 Eastern Asia Society for Transportation Studies, Bangkok, Thailand, 21–24 September 2005; pp. 1641–1656.
42. Rakha, H.A.; Pasumarthy, P.; Adjerid, S. A simplified behavioral vehicle longitudinal motion model. *Transp. Lett.* **2009**, *1*, 95–110. [CrossRef]
43. Sangster, J.D.; Rakha, H.A. Enhancing and Calibrating the Rakha-Pasumarthy-Adjerid Car-Following Model using Naturalistic Driving Data. *Int. J. Transp. Sci. Technol.* **2014**, *3*, 229–247. [CrossRef]
44. Van Aerde, M. Single Regime Speed-Flow-Density Relationship for Congested and Uncongested Highways. In Proceedings of the 74th Annual Meeting of the Transportation Research Board, Washington, DC, USA, 27 January 1995.
45. Van Aerde, M.; Rakha, H. Multivariate calibration of single regime speed-flow-density relationships [road traffic management]. In Proceedings of the Pacific Rim TransTech Conference. 1995 Vehicle Navigation and Information Systems Conference Proceedings. 6th International VNIS. A Ride into the Future, Seattle, WA, USA, 30 July–2 August 1995.
46. Chatterjee, B. An optimization formulation to compute Nash equilibrium in finite games. In Proceedings of the 2009 International Conference on Methods and Models in Computer Science (ICM2CS), Delhi, India, 14–15 December 2009.
47. Bonebau, E. Agent-based modeling: Methods and techniques for simulating human systems. *Proc. Natl. Acad. Sci. USA* **2002**, *99*, 7280–7287. [CrossRef] [PubMed]
48. Macal, C.; North, M.J. Tutorial on Agent-Based Modeling and Simulation PART 2: How to Model with Agents. In Proceedings of the 2006 Winter Simulation Conference, Monterey, CA, USA, 3–6 December 2006.
49. Zheng, H.; Son, Y.; Chiu, Y.; Head, L.; Feng, Y.; Xi, H.; Kim, S.; Hickman, M. *A Primer for Agent-Based Simulation and Modeling in Transportation Applications (FHWA-HRT-13-054)*; Federal Highway Administration: McLean, VA, USA, 2013.
50. Elliott, E.; Kiel, D.P. Exploring cooperation and competition using agent-based modeling. *Proc. Natl. Acad. Sci. USA* **2002**, *99*, 7193–7194. [CrossRef] [PubMed]
51. Ljubovic, V. Traffic simulation using agent-based models. In Proceedings of the 2009 XXII International Symposium on Information, Communication and Automation Technologies, Bosnia, Serbia, 29–31 October 2009.
52. Law, A.M.; Kelton, W.D. *Simulation Modelling and Analysis*, 2nd ed.; McGraw-Hill: New York, NY, USA, 1991.

53. Xiang, X.; Kennedy, R.; Madey, G.; Cabaniss, S. Verification and Validation of Agent-Based Scientific Simulation Models. In Proceedings of the Agent-directed Simulation Conference, San Diego, CA, USA, April 2005; pp. 47–55.
54. Balci, O. Verification, Validation, and Testing. In *Handbook of Simulation*; Wiley: New York, NY, USA, 2007; pp. 335–393.

sensors

Article

Model Predictive Controller Based on Online Obtaining of Softness Factor and Fusion Velocity for Automatic Train Operation

Longda Wang [1], Xingcheng Wang [1,*], Zhao Sheng [2] and Senkui Lu [1]

1 School of Marine Electrical Engineering, Dalian Maritime University, Dalian 116026, China; wanglongda@dlmu.edu.cn (L.W.); dlm@dlmu.edu.cn (S.L.)
2 School of Electronic and Information Engineering, Beijing Jiaotong University, Beijing 100044, China; zhaosheng@bjtu.edu.cn
* Correspondence: dmuwxc@dlmu.edu.cn; Tel.: +86-411-8472-7801

Received: 8 February 2020; Accepted: 16 March 2020; Published: 19 March 2020

Abstract: This paper develops an improved model predictive controller based on the online obtaining of softness factor and fusion velocity for automatic train operation to enhance the tracking control performance. Specifically, the softness factor of the improved model predictive control algorithm is not a constant, conversely, an improved online adaptive adjusting method for softness factor based on fuzzy satisfaction of system output value and velocity distance trajectory characteristic is adopted, and an improved whale optimization algorithm has been proposed to solve the adjustable parameters; meanwhile, the system output value for automatic train operation is not sampled by a normal speed sensor, on the contrary, an improved online velocity sampled method for the system output value based on a fusion velocity model and an intelligent digital torque sensor is applied. In addition, the two improved strategies proposed take the real-time storage and calculation capacities of the core chip of the controller into account. Therefore, the proposed improved strategies (I) have good performance in tracking precision, (II) are simple and easily conducted, and (III) can ensure the accomplishing of computational tasks in real-time. Finally, to verify the effectiveness of the improved model predictive controller, the Matlab/simulink simulation and hardware-in-the-loop simulation (HILS) are adopted for automatic train operation tracking control, and the tracking control simulation results indicate that the improved model predictive controller has better tracking control effectiveness compared with the existing traditional improved model predictive controller.

Keywords: model predictive controller; automatic train operation; softness factor; fusion velocity; online obtaining; hardware-in-the-loop simulation

1. Introduction

The urban rail transit system with automatic train operation system has the advantages of safety, stability, economy, and comfort, and it has become one of the most popular and efficient means of the urban public transportation [1]. The tracking control functional module makes the velocity trajectory track at the optimal target speed obtained by the upper-layer optimal loop, and according to the appropriate and efficient tracking control algorithm, it is an indispensable crucial system and necessary to ensure optimal safety, comfort, energy-efficiency, punctuality, and parking accuracy for train operation process, which requires the corresponding algorithm to possess good control performance [2]. Therefore, aiming at improving the multi-objective performance index of the train operation process, an automatic train operation system has been developed rapidly and is widely applied in urban rail trains operation [3–5]. Meanwhile, various improved intelligent optimization control algorithms have been proposed and applied for the automatic train operation system [6–8].

In recent years, many improved algorithms have been applied in the automatic train operation tracking control field, such as robust adaptive automatic control, model predictive control, online learning, iterative learning control, matter-element model, etc. [9–13]. An online approximation-based robust adaptive automatic train control method is proposed for the automatic train operation (ATO) system [9]. A fuzzy model predictive control approach is proposed to provide locomotive operation instructions for mainline railways continuously, and extensive simulations show that the proposed approach can provide sufficient solution optimality in reasonable computational time and energy consumption in train operations is reduced [10]. A novel online learning control strategy is proposed to solve the train automatic stop control (TASC) problem [11]. An iterative learning control based on automatic train operation is proposed to deal with the trajectory tracking control problem under certain velocity constrains [12]. Matter-element theory is applied to the established models to optimize speed trajectory for achieving multi-objective optimization, and the relative performance indices weighting is determined in different stages so that the more satisfied decision speeds could be calculated with the goodness evaluation method [13]. The above research can improve the tracking control performance of the traditional control algorithms.

Among numerous algorithms, Model Predictive Control (MPC) is one of the most effective control algorithms, which is characterized by good robustness, fast tracking speed, accurate tracking target speed, etc. [14]. A linear time-varying MPC is used to obtain the power split between the combustion engine and electrical machines and the system operating points at each sample time [15]. A coordinated energy dispatch based on Distributed model predictive control (DMPC) is proposed, and the corresponding simulation results show the effectiveness of the proposed method [16]. A co-design of the self-triggered mechanism and distributed model predictive control (DMPC) is proposed to achieve the cooperative objectives while efficiently exploiting communication network [17]. A model predictive control-based droop current regulator to interface PV in smart dc distribution systems is proposed [18]. From the various model predictive control algorithms, the dynamic matrix control model predictive control (DMC MPC) is an effective algorithm among them due to its characteristics of strong robustness, fast tracking speed, high precision for tracking control, avoiding the parameter identification for the transfer function model, and solving the problem of delay process effectively. A new method that linearizes the RC equivalent circuit model and predicts available battery power according to original Dynamic Matrix Control algorithm is proposed [19]. An application of dynamic matrix control (DMC) to a drum-type boiler–turbine system is proposed [20]. Of particular note is the use of the improved DMC MPC for automatic train operation tracking control scenario [21].

It is necessary to conduct further study on the basis of previous research findings, and key parameters adjusting and improving the sampling accuracy should be taken seriously. A method for calculating the traction characteristics of a traction motor is proposed [22]. A new method to identify the train key design variables against the running performance indicators based on the sensitivity analysis is proposed, which in turn bases itself on simulation-oriented surrogate models [23]. A novel adaptive sampling algorithm for power management in the automated monitoring of the quality of water in an environment is devised and applied [24].

Traditional simulations based on a pure software environment cannot truly reflect the actual automatic train operation process, and the situation representing the actual automatic train operation experiment is difficult to implement because it is expensive, has restricted experimental conditions, high construction difficulties, and high security protection requirements. Hardware-in-the-loop simulation (HILS) is a new simulation technology for solving this difficult issue [25,26]. At present, numerous relative research findings have achieved improvements in the traction control system [27,28].

An improved model predictive controller based on online obtaining of softness factor and fusion velocity for automatic train operation is developed. The following summarizes the main contributions of this paper.

(I) An improved whale optimization algorithm (IWOA) based on the Tchebycheff decomposition method, convergence factor nonlinear decline, and genetic evolution measurement is proposed to solve the optimization of the softness factor adaptive adjusting parameters appropriately.
(II) Aiming at improving the tracking control performance for automatic train operation, an improved model predictive controller based on online obtaining of the softness factor and fusion velocity is developed for automatic train operation tracking control effectively.
(III) To further verify the effectiveness of the developed model predictive control controller, the scenario about rail transit line No.12 in Dalian, China is chosen for simulation test. The results of the Matlab simulation and hardware-in-the-loop simulation (HILS) show that the tracking controller proposed in this paper has good tracking control performance.

The paper is organized as follows. Section 2 introduces the model predictive controller for automatic train operation tracking control. Section 3 illustrates the improved DMC model predictive controller based on online obtaining of softness factor and fusion velocity developed in this paper. Section 4 provides the Matlab/simulation results and hardware-in-the-loop simulation (HILS) results to illustrate the proposed method. Section 5 concludes this article.

2. Model Predictive Controller for Automatic Train Operation Tracking Control

2.1. Evaluation Index for Automatic Train Operation Tracking Control

The integral of time multiplied by the absolute value of error (ITAE) is the frequently used evaluation index for tracking control performance [29]. The specific formula for the evaluation index ITAE is as follows,

$$ITAE = \int t\,|e(t)|dt \tag{1}$$

where t represents the sample time of control process, and $|e(t)|$ represents the absolute value of error between target speed and actual tracking control speed.

As automatic train operation has its own unique characteristics and requirements, the multi-objective performance index is more appropriative, and it used universally. The computation model of multi-objective performance index P_k is as follows,

$$\begin{cases} P_k = \sum\limits_{i=1}^{4} \omega_i \times \frac{f_i - \min(f_i)}{\max(f_i) - \min(f_i)} \times f_i \\ (f_1, f_2, f_3, f_4) = (\Delta s, \Delta t, K_{Jerk}, E) \\ Mv\frac{dv}{ds} = F(u,v) - R(v,s) - B(u,v) \\ \frac{dt}{ds} = \frac{1}{v} \\ v(s) \le v_{\lim}(s) \\ R(v,s) = r(v) + R_l(s) \\ \Delta s = |s_z - D| < \Delta s_{\max} \\ \Delta t = |\bar{T} - T_r| < \Delta t_{\max} \\ K_{Jerk} = \int |\Delta a|\, d_s / D \\ E = \int (Ma - R)\, d_s \end{cases} \tag{2}$$

where ω_i represents the index importance weight factor ($\sum\limits_{i=1}^{k} \omega'_i = 1$), which reflects the relative importance of the i th optimization index; t represents the actual running time of the train; s represents the actual position of the train; a represents the actual acceleration of the train; $|\Delta a|$ represents the actual impingement rate of the train; M is the mass of the train; $F_t(u,v)$ and $B_r(u,v)$ are the traction force and braking force of the current velocity, respectively; $R(v,s)$ is the resistance of the train determined by the current speed and line position; s_z is the terminal position; T_r is the actual running time; D is the actual running distance; $v(s)$ represents the instantaneous velocity in the position s; $\bar{T}$ represents the prospective running time; $v_{lim}(s)$ represents the upper limit velocity in the position s; $\Delta s_{\max}$ represents the allowable maximum parking error; $\Delta t_{\max}$ represents the allowable maximum time error; Δs and Δt represent the actual parking error and time error, respectively; u represents the train control quantity; K_{Jerk} represents the comfort performance index; and E represents the energy consumption during the train operation process [2,30].

In addition, security index should be taken into account as well. Traveling over the velocity limit is the main risk and non-negligible factors can cause an unsafe environment. The computation formula of security index K_{safe} is as follows,

$$\begin{gathered}
\min K_{safe} \\
K_{safe} = \sum_{is=1}^{sn} YS(is) \Big/ sn \\
YS(is) = \begin{cases} 0 & v(is) > v_{\lim}(is) \\ 1 & v(is) \leq v_{\lim}(is) \end{cases}
\end{gathered} \tag{3}$$

where is represents the index of sampling point, $YS(is)$ represents the security evaluation value of the is-th sampling point, and sn represents the number of sampling points [1].

2.2. Conventional Dynamic Matrix Control Model Predictive Control

DMC MPC uses three methods, including the DMC predictive model, rolling optimization, and feedback correction, to control the controlled object [31].

2.2.1. DMC Predictive Model

The DMC predictive model is one of significant models for DMC MPC. The unit step response model reflecting the dynamic performance is adopted as the DMC predictive model for controlled object, and the predictive value of system output is obtained by the step response characteristic for controlled object.

If the model length is N, then the N sampled values of the controlled object unit step response can be used to describe the dynamic response characteristics of the system. The specific calculation formula for the predictive value of system output is as follows,

$$Y_p(k) = Y_0(k) + A\Delta U(k) \tag{4}$$

where $Y_p(k) = [y_p(k+1|k), y_p(k+2|k), ..., y_p(k+N|k)]^T$ represents the predictive value of system output, $Y_0(k) = [y_0(k+1|k), y_0(k+2|k), ..., y_0(k+N|k)]^T$ represents the predictive value of predictive model, $\Delta U(k) = [\Delta u(k), \Delta u(k+1), ..., \Delta u(k+N-1)]^T$ represents the incremental sequence for control, and A represents the dynamic matrix. The specific dynamic matrix A and the specific calculation formula for the element of Y_p are as follows,

$$A = \begin{bmatrix} a_1 & 0 & 0 & \cdots & 0 \\ a_2 & a_1 & 0 & \cdots & 0 \\ \vdots & & \vdots & \ddots & \vdots \\ a_N & a_{N-1} & & \cdots & a_1 \end{bmatrix} \tag{5}$$

$$y_p(k+i|k) = y_0(k+i|k) + \sum_{j=1}^{i} a_{i-j+1}\Delta u(k+j-1) \tag{6}$$

where i represents the element index of Y_p, $i \in \{1,2,...,N\}$; k represents the initial point of DMC predictive model [31,32].

2.2.2. Rolling Optimization

With the aim of avoiding the violent fluctuations in the control process effectively, it is necessary to make the final output value y_f reach to the reference target value y_r along the predetermined smooth path by the DMC MPC system, so as to enhance the system robustness. Thus, a popular reference path used in DMC MPC is as follows,

$$y_f(k+i) = \alpha^i y(k) + \left(1-\alpha^i\right) y_r \tag{7}$$

where $y_f(k+i)$ represents the final output value expected, α^i represents the ith softness factor ($0 < \alpha^i < 1$), $y(k)$ represents the actual output value of the system, and y_r represents the reference target value of the system.

The quadratic rolling optimization object of the system is necessary for rolling optimization. If the predictive length is M and control length is L, in general, $L \le M \le N$. The specific quadratic rolling optimization object of system is as follows,

$$\begin{aligned} J &= \left\| Y_p(k) - Y_f(k) \right\|_Q^2 + \left\| \Delta U_L(k) \right\|_R^2 \\ &= \sum_{i=1}^{M} q_i \left[y_p(k+i|k) - y_f(k+i) \right] + \sum_{i=1}^{L} r_i \Delta u(k+i-1) \end{aligned} \tag{8}$$

where $Y_f(k) = \left[y_f(k+1), y_f(k+2), \ldots, y_f(k+M) \right]^T$ represents the control sequence of the system, $R = diag[r_1, r_2, \ldots, r_L]^T$ represents the weight coefficient matrix of constraint for error revise, $Q = diag[q_1, q_2, \ldots, q_M]^T$ represents the weight coefficient matrix of constraint for error increment revise, and $diag$ represents the diagonal matrix.

The necessary condition for obtaining the minimum value of objective function J is $\frac{\partial J}{\partial \Delta U_L(k)} = 0$ through extreme value theory under unconstrained conditions. Therefore, the control sequence optimal solution can be obtained by rolling optimization. The specific calculation formula of the control sequence optimal solution is as follows.

$$\Delta U_L(k-1) = \left(A^T Q A + R \right)^{-1} A^T Q \left[Y_f(k) - Y_0(k) \right] \tag{9}$$

Then, the actual control quantity $u(k)$ can be obtained. The specific calculation formula of the actual control quantity $u(k)$ is as follows,

$$u(k) = u(k-1) + \Delta u(k-1) \tag{10}$$

In the next control period, i.e., the $k+1$ th control period, the corresponding $\Delta u(k)$ and $u(k+1)$ can be obtained by the above way. Thus, it can realize the rolling optimization of the actual control quantity in the iterative control process [21,33].

2.2.3. Feedback Correction

Feedback correction is an important component of DMC MPC; it is used to reduce the influence of system disturbance for the control system, so as to achieve the ideal control effectiveness. The specific calculation formula of the error between actual system output value and the predicted output value in the present control period (the k th control period) is as follows.

$$e(k) = y(k) - y_p(k) \tag{11}$$

After feedback correction calculation, the predicted output value can be corrected to certain extent [21,33,34]. The specific corrected calculation formula is as follows,

$$Y_{p2}(k) = Y_0(k) + A\Delta U(k-1) + HC \tag{12}$$

where C represents the error corrected matrix, and its length is N; H represents the corresponding transformed matrix.

2.3. Fuzzy DMC Model Predictive Controller for Automatic Train Operation

Aiming at improving the precision of the automatic train operation tracking control for DMC MPC, using the fuzzy model prediction based on the train operation mechanism is a good choice. The slope and velocity error are the most important train operation information. For example, when the train runs in steep uphill and current velocity is far less than target velocity, the conversion degree for train operation is "PB", that is, the maximum extent to drew train is adopted to assist the climb by accelerating or keeping the velocity. In this time, if the maximum traction is used according to the intrinsic DMC MPC, the addition traction incremental quantity is not necessary; otherwise, the appropriate addition traction incremental quantity should be used to correct this error. The fuzzy sets are divided into ['N4',.......,'Z',.......,'P4'] [10,35]. The specific calculation model for fuzzy model prediction is as follows,

$$\begin{aligned} u_{f\_p}(k) &= C_{fuzzy1}(\omega(k), e(k)) \\ \Delta u_{f\_p}(k) &= C_{fuzzy2}\left(u_{f\_p}(k), u(k)\right) \\ u_c(k) &= u(k) + \Delta u_{f\_p}(k) \end{aligned} \tag{13}$$

where C_{fuzzy1} and C_{fuzzy2} represent the fuzzy inference functions by using two kinds of fuzzy rules, respectively; $u_{f\_p}(k)$ represents the calculated control quantity by using fuzzy rule about slope and velocity error; $\Delta u_{f\_p}(k)$ represents the calculated control quantity by using fuzzy rule about control quantity calculated by intrinsic DMC MPC and control quantity calculated by fuzzy logic and train operation information; and $u_c(k)$ represents the final calculated control quantity for the automatic train operation tracking control.

The fuzzy rules for fuzzy model prediction and partial membership function for control quantity are shown in Figure 1.

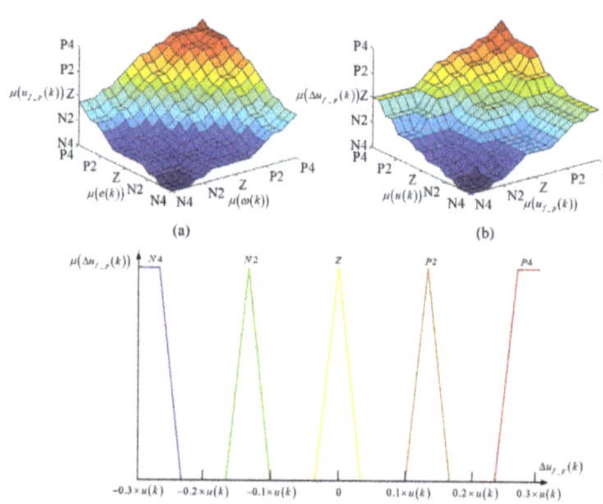

Figure 1. The fuzzy rules for fuzzy model prediction and partial membership function for control quantity. (**a**) Fuzzy rule for train operation information. (**b**) Fuzzy rule for control quantity prediction. (**c**) Partial membership functions for control quantity.

Fuzzy dynamic matrix control model predictive control (Fuzzy DMC MPC) is a control method that considers the step response characteristics and fuzzy logic for the train operation mechanism of the control object. The fuzzy DMC model predictive controller is widely used in automatic train operation due to its characteristics of simple design scheme and high tracking precision. The fuzzy DMC model prediction controller is mainly composed of four function chips (Fuzzy model prediction function chip, DMC model prediction function chip, rolling optimization function chip, and feedback correction function chip), and it is used to realize four function modules of Fuzzy DMC MPC (Fuzzy model prediction, DMC model prediction, rolling optimization, and feedback correction). The schematic diagram of the Fuzzy DMC model predictive controller for automatic train operation is shown in Figure 2.

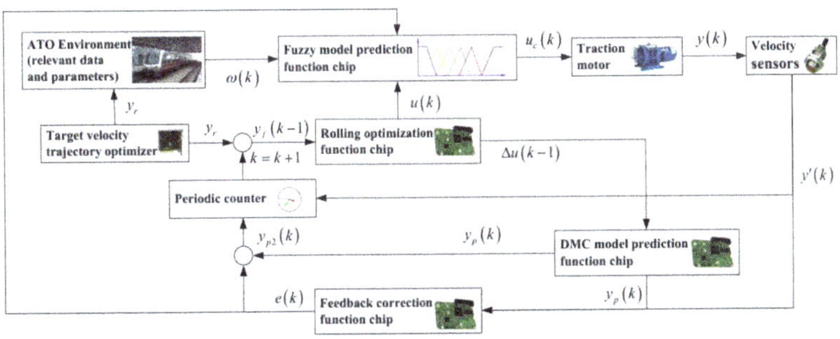

Figure 2. Schematic diagram of the Fuzzy DMC model predictive controller for automatic train operation.

As can be seen from Figure 3, it is impossible to obtain the actual output value (real-time velocity) for automatic train operation tracking control system. In addition, as can be seen from Formula (7), the real-time softness factor is also an important factor that impacted the multi-objective performance index for automatic train operation tracking control. Therefore, it is necessary to improve the real-time velocity sampling accuracy and softness factor accuracy for automatic train operation tracking control as much as possible.

3. Model Predictive Controller Based on Online Obtaining of Softness Factor and Fusion Velocity

3.1. Fusion Velocity Computation Model and Corrected Model Based on Online Obtaining

3.1.1. Fusion Velocity Computation Model Based on Online Obtaining

According to the multi-objective performance index for automatic train operation tracking control, the fusion velocity model based on online obtaining is necessary to take into account the energy consumption, running time, comfort, and parking accuracy. Taking into account the sampling effect, hardware technology (storage and computing ability), funds, space, and other factors, three kinds of velocity sampling sources are sufficient (motor speed, motor torque and train instantaneous displacement) and are selected and synthesized. The fusion velocity computation model based on online obtaining is established as follows,

$$\begin{cases} v_{is,v} = n_{is} \times tr_{ntv} \times \eta_g \times \eta_{is,T} \times \eta_{is,itc} \\ v_{is,F} = v_{is-1,a} + {F_{is} - w_{is}}/{M} \cdot \Delta t \\ v_{is,s} = {\Delta s}/{\Delta t} = {s_{is} - s_{is-1}}/{\Delta t} \end{cases} \quad (14)$$

$$v_{is,a} = \lambda_{is,v} \times v_{is,v} + \lambda_{is,F} \times v_{is,F} + \lambda_{is,s} \times v_{is,s} \quad (15)$$

where $v_{is,a}$ represents the final calculated velocity by speed analyzer ultimately of the i-th sampling point; $v_{is,v}$ represents the velocity calculated based on actual motor speed sampled of the i-th sampling point; $v_{is,F}$ represents the velocity calculated based on actual motor torque sampled of the i-th sampling point; $v_{is,s}$ represents the velocity calculated based on actual train instantaneous displacement sampled of the i-th sampling point; n_{is} represents the actual motor speed sampled by speed sensors of the i-th sampling point; tr_{ntv} represents the transmission ratio of the motor speed to train velocity; η_g represents the degree of tooth engagement between the gears; $\eta_{is,T}$ represents the speed transmission efficiency of the i-th sampling point; $\eta_{is,itc}$ represents the efficiency for the train to overcome idling, taxiing, and creep sliding of the i-th sampling point; F_{is} represents the force calculated based on actual motor torque sampled of the i-th sampling point; $F_{is} = \eta_{is,F} \times {}^{T_{is}}/_{R_{mr}}$; T_{is} represents the actual motor torque sampled by torquemeter of the i-th sampling point; $\eta_{is,F}$ represents the force comprehensive transmission efficiency of the i-th sampling point; R_{mr} represents the radius of motor rotor; $w_{is}(v,s)$ represents the actual resistance of the i-th sampling point (v,s); Δt represents the sampling time-interval, Δt is 500 μs in this paper; Δs represents the sampling displacement-interval; s_{is} represents the actual train instantaneous displacement sampled by displacement pickup of the i-th sampling point; and $\lambda_{is} = \{\lambda_{is,v}, \lambda_{is,F}, \lambda_{is,s}\}$ represents the synthetic weight of the velocity sampled by different ways.

Synthetic weight is vital for real-time sampling precision in the tracking control process, the importance of each velocity sampling sources needs to be considered, so as to give the appropriate synthetic weight. The synthetic weight indicates the importance of the real-time velocity obtained by different speed sampling sources. Yet, the selection of the synthetic weight by traditional methods lacks the specific theoretical basis, so there is certain subjective limitation in actual applied. As automatic train operation tracking control is an extremely complex issue, there is a trajectory characteristic for automatic train operation tracking control curve dominated by velocity target curve, train parameters, line conditions and running requirements, and the traditional methods for setting synthetic weight based on experience empower is subjective and blind, so it is necessary to be improved. In this paper, an synthetic weight empower using entropy method is applied for automatic train operation tracking control. First, the whole tracking control curve is divided by a position according to trajectory characteristic and line conditions; second, a large number of real-time data, including velocity, force, and position information for the whole tracking control process, should be sampled to prepare for calculation; finally, the entropy method is used to calculate the synthetic weight of each divided subinterval of tracking control curve.

Entropy is a measure of uncertainty for information calculation. The entropy weight method is utilizes the entropy characteristics and assigns a weight to each index in an event by calculating the entropy value. The entropy weight method is an objective weight empower method, because it simply depends on the discreteness of data itself. The specific steps of computational process for entropy weight method are as follows.

A certain number of samples (as many as possible) must be collected to prepare for the calculation, and their index values also needed to be recorded.

To eliminate the negative influences caused by the difference between units and magnitude orders, the index values must be normalized. The calculation formulas for the normalization can be expressed as follows,

$$x_{ij}{}' = \frac{x_{ij} - \min\left\{x_{ij}, x_{2j}, \ldots, x_{nj}\right\}}{\max\left\{x_{ij}, x_{2j}, \ldots, x_{nj}\right\} - \min\left\{x_{ij}, x_{2j}, \ldots, x_{nj}\right\}} \tag{16}$$

$$x_{ij}{}' = \frac{\max\left\{x_{ij}, x_{2j}, \ldots, x_{nj}\right\} - x_{ij}}{\max\left\{x_{ij}, x_{2j}, \ldots, x_{nj}\right\} - \min\left\{x_{ij}, x_{2j}, \ldots, x_{nj}\right\}} \tag{17}$$

where $j = (1, 2, \cdots, m), i = (1, 2, \cdots, n)$; m represents index number; n represents the number of samples; x_{ij}' represents the j-th processed index value of the i-th sample after normalization; x_{ij} represents the j-th original index value of the i-th sample before normalization; max and min, respectively, represent the maximum and minimum values of the array. If index value x_{ij} is a positive number, Formula (16) is used to normalize; otherwise, Formula (17) is used to normalize.

The normalized index values are necessary to filtered out the zero value further, so as to avoid illegal logarithmic function (ln(0)) in next subsequent calculation processes. The specific formula for filtered out zero value is as follows,

$$x_{ij}'' = \Delta z + (1 - 2 \times \Delta z) \times x_{ij}' \tag{18}$$

where x_{ij}'' represents the j-th processed index value of i-th sample after filtering out the zero value; Δz represents the tiny value reasonable; Δz is 0.01 in this paper.

Then, the entropy values of each index values are necessary to be calculated. The specific formulas for calculating the entropy values are as follows,

$$p_{ij} = \frac{x_{ij}''}{\sum\limits_{i=1}^{n} x_{ij}''} \tag{19}$$

$$e_j = -k \times \sum_{i=1}^{n} \left(p_{ij} \times \ln\left(p_{ij}\right)\right) \tag{20}$$

where p_{ij} represents the j-th index weight value of i-th sample in j-th index; e_j represents the j-th index entropy value; k represents the entropy coefficient, the value is the reciprocal of $\ln(n)$, $k = {}^{1}\!/\!_{\ln(n)}$.

Finally, the index weight value could be calculated. The specific formula for calculating the index weight values is as follows,

$$d_j = 1 - e_j \tag{21}$$

$$\lambda_j = \frac{d_j}{\sum\limits_{j=1}^{m} d_j} \tag{22}$$

where d_j represents the j-th entropy redundancy value of j-th index, it indicates the difference degree of this index; e_j represents the j-th index entropy value; λ_j represents the j-th weight value.

3.1.2. Fusion Velocity Corrected Model Based on Online Obtaining

If the factor of the velocity sampling sources sampled inaccurately is not considered, the improvement effect for automatic train operation tracking control will inevitably be restricted. To improve the real-time sampling velocity precision, a corrected method of real-time sampling velocity for automatic train operation tracking control using auxiliary corrective velocity sampling source is popularly applied in various types of urban rail vehicles. The specific evaluation and corrected formulas are as follows,

$$\Delta v_{is,c} = \left| v_{is,x} - v_{is,ref} \right| \leq \Delta v_{is,p} \tag{23}$$

$$v_{is,c} = \frac{\Delta v_{is,p}}{\Delta v_{is,c}} \times v_{is,x} + \frac{\Delta v_{is,c} - \Delta v_{is,p}}{\Delta v_{is,c}} \times v_{is,ref} \tag{24}$$

where $v_{is,x}$ represents a velocity sampling source x from $v_{is,v}, v_{is,F}, v_{is,s}$ using to synthetic final calculated velocity $v_{is,a}$; $v_{is,ref}$ represents the reference velocity (auxiliary corrective velocity sampling source) based on actual sampling data sampled by specific auxiliary sensor; $\Delta v_{is,c}$ represents the actual error value between $v_{is,x}$ and $v_{is,ref}$; $\Delta v_{is,p}$ represents the maximum permit satisfied error value between $v_{is,x}$ and $v_{is,ref}$; $v_{is,c}$ represents the final corrected value calculated by Formula (24) when correctness condition (Formula (23)) is not satisfied.

Reference velocity $v_{is,ref}$ will exert a measure of influence over the velocity-corrected effect. Thus, the choice of auxiliary corrective velocity sampling source is significant. The specific gear speed on the vehicle wheel side of the gear box is a good choice.

3.2. *Softness Factor Adaptive Adjusting Model Based on Online Obtaining*

The softness factor is a key parameter for DMC MPC; it plays plays an important role in balancing the degree of robustness and rapidity for the DMC MPC tracking control system. If softness factor α is chosen as a larger value, the system will have slower response speed and stronger robustness; by contrast, if softness factor α is chosen as a smaller value, the system will have faster response speed and worse robustness [36]. Thus, both response speed and robustness must be taken into account for softness factor α setting.

Considering the trajectory characteristic and tracking control condition for automatic train operation, the softness factor adaptive adjusting model based on online obtaining is established as follows,

$$\alpha = \lambda_{\alpha Ts} \times \alpha_{Ts}(s) + \lambda_{\alpha_{\mu y}} \times \alpha_{\mu y}(y(k), y_r) \tag{25}$$

where α represents the final calculated real-time softness factor; α_{Ts} represents the real-time softness factor calculated based on the trajectory characteristic of the present position; $\alpha_{\mu y}(y(k), y_r)$ represents the real-time softness factor calculated based on tracking control condition of the present system output $y(k)$; $\lambda_{\alpha Ts}$ and $\lambda_{\alpha_{\mu y}}$ represent the fusion weights of α_{Ts} and $\alpha_{\mu y}(y(k), y_r)$, respectively; and $\lambda_{\alpha Ts} + \lambda_{\alpha_{\mu y}} = 1$.

The whole tracking control curve must be divided into several different types of subintervals by position according to the trajectory characteristic and line conditions. The specific types of subintervals are described as follows.

Type 1: The vibrating area nearby inflection point of tracking control curve.

In this area, there is the strong velocity fluctuation in the velocity trajectory. Thus, aiming at improving the system robustness as much as possible, softness factor α_{Ts} should be an appropriate larger value at the cost of reduce acceptable system rapidity.

Type 2: The smooth area of tracking control curve.

In this area, there is no obvious velocity fluctuation in the velocity trajectory. Thus, aiming at improving the system rapidity as much as possible, softness factor α_{Ts} should be chosen as an appropriate smaller value at the cost of reduce acceptable system robustness.

Type 3: The connected area in the middle of smooth area and the vibrating area of the tracking control curve.

In this area, the system rapidity and rapidity are taken into account for softness factor α_{Ts} setting. Thus, softness factor α_{Ts} should be choose a appropriate intermediate value.

In addition, although in the same type of subintervals, the softening factor α_{Ts} almost varies because of the different intensity degrees of velocity fluctuation. The specific calculation formula for softness factor α_{Ts} based on trajectory characteristic of the present position is described as follows,

$$\alpha_{Ts}(s) = \begin{cases} \alpha_{Tr,si} + (\alpha_{Tr,si} - \alpha_{Tr,si-1}) \times \frac{((S_{si}+S_1)-s)}{S_1+S_2} & s < S_{si}+S_1 \\ \alpha_{Tr,si} & S_{si}+S_1 \le s \le S_{si+1}-S_2 \\ \alpha_{Tr,si} + (\alpha_{Tr,si+1} - \alpha_{Tr,si}) \times \frac{(s-(S_{si}-S_2))}{S_1+S_2} & s > S_{si+1}-S_2 \end{cases} \tag{26}$$

where si represents the subinterval index $si \in \{1,2,...,si_{max}\}$; si_max represents the number of subintervals; S_{si} represents the starting position of the si-th subinterval; S_{si_max+1} represents the terminal position of tracking control curve, it is a target parking position; $\alpha_{Tr,si}$ represents the reference value of soften factor in the si-th subinterval; $\alpha_{Ts,0} = \alpha_{Ts,1}$, $\alpha_{Ts,si\_max+1} = \alpha_{Ts,si\_max}$; S_1 and S_2 represent the connected length, in the connected area; the softness factor α_{Ts} is reduced or increased linearly and smoothly, so as to avoid the instability of tracking control system.

Aiming at solving this control problem with fuzzy characteristic, an fuzzy adaptive adjusting method for online obtaining softness factor $\alpha_{\mu y}$ is applied. First of all, the satisfaction degree of control is defined, so as to the automatic train operation tracking control problem can be transformed into an optimization decision-making problem by fuzzy reasoning; then, the corresponding real-time parameters of the controller are adjusted online to meet the requirements of the system control quality, so as to achieve the purpose of system optimization control. The specific calculation formula for fuzzy satisfaction degree $\mu_{y(k)}$ of system output $y(k)$ is as follows,

$$\mu_{y(k)} = \begin{cases} 0 & y(k) < y_{\min} - s_1 \\ 1 + \frac{y(k) - y_{\min}}{s_1} & y_{\min} - s_1 \leq y(k) < y_{\min} \\ 1 & y_{\min} \leq y(k) < y_{\max} \\ 1 + \frac{y(k) - y_{\max}}{s_2} & y_{\max} \leq y(k) < y_{\max} + s_2 \\ 0 & y_{\min} - s_1 \geq y(k) \end{cases} \tag{27}$$

where s_1 and s_2 represent the blur width, which can indicate the requirement of designer, if $s_1 = s_2 = 0$, the requirements for the control system are strict, and the automatic train operation tracking control is not so, and this represents a combination of the practical situation; $y_{\max}$ and $y_{\min}$ represent the maximum and minimum value of design expectation, respectively, if $y_{\max} = y_{\min}$, it will be shown as trigonometric membership function; otherwise, it will be shown as trapezoid membership function. The corresponding diagram for fuzzy satisfaction degree calculation $\mu_{y(k)}$ of system output $y(k)$ is shown in Figure 3.

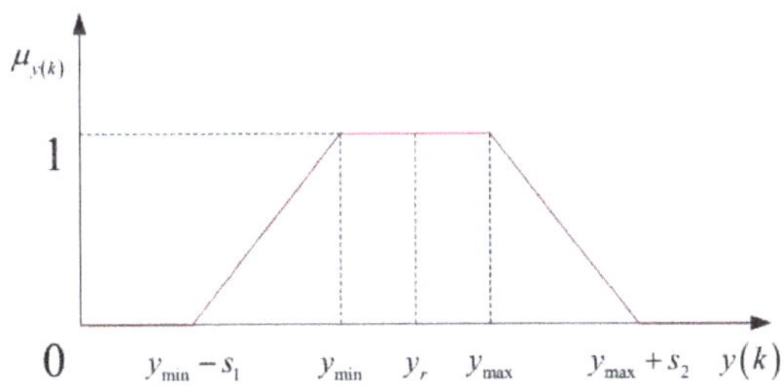

Figure 3. Diagram for fuzzy satisfaction degree calculation $\mu_{y(k)}$ of system output $y(k)$.

The error between the output value and the reference target value of system (i.e., fuzzy satisfaction degree $\mu_{y(k)}$ of system output $y(k)$) should also be considered. If the fuzzy satisfaction degree $\mu_{y(k)}$ is larger, it can indicate that the error between the output value and the reference target value of system is smaller, at this time, there is a small overshoot of the system and softness factor $\alpha_{\mu y}$ so that an appropriate lager value needs to chosen to increase the system rapidity; by contrast, there is an obvious overshoot of the system and softness factor $\alpha_{\mu y}$ so that an appropriate small value needs to be chosen to reduce the system rapidity to ensure system robustness [36]. According to the influence of softness factor $\alpha_{\mu y}$ for the system dynamic response, the specific exponential calculation formula for softness factor $\alpha_{\mu y}$ by fuzzy satisfaction degree $\mu_{y(k)}$ of system output $y(k)$ is as follows,

$$\alpha_{\mu y}(y(k), y_r) = \alpha_{\max} + \left(\alpha_{\max} \times e^{-(b \times \mu_{y(k)})} - \alpha_{\max}\right) \times \mu_{y(k)} \tag{28}$$

where $\alpha_{\max}$ represents the maximum value of softness factor $\mu_{y(k)}$; b represents the gain coefficient; it determines the shape of the softening factor $\alpha_{\mu y}(y(k), y_r)$ function curve.

The corresponding diagram for softness factor $\alpha_{\mu y}$ of the fuzzy satisfaction degree calculation $\mu_{y(k)}$, and softness factor $\alpha_{\mu y}$ of system output $y(k)$ are shown in Figures 4 and 5.

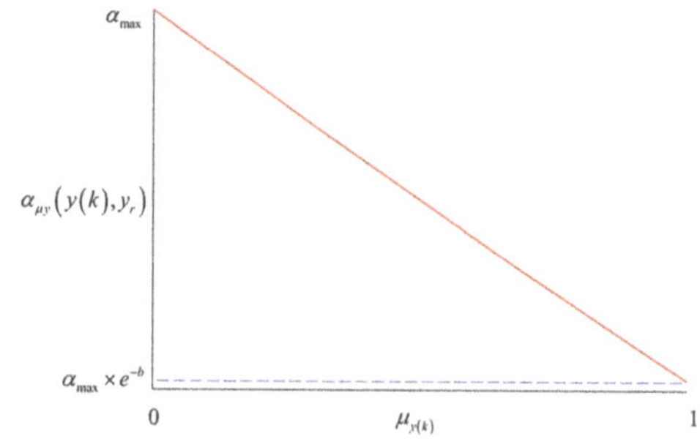

Figure 4. Diagram for fuzzy satisfaction degree calculation $\mu_{y(k)}$ of fuzzy satisfaction degree calculation $\mu_{y(k)}$.

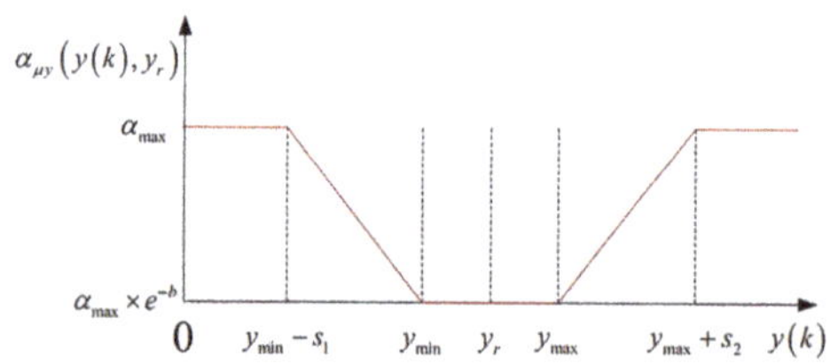

Figure 5. Diagram for fuzzy satisfaction degree calculation $\mu_{y(k)}$ of system output $y(k)$.

3.3. Improved Whale Optimization Algorithm for Softness Factor Adaptive Adjusting Parameters Optimization

Optimization algorithms are used to obtain a set of adjustable parameters for the satisfactory tracking control effect in actual automatic train operation scenarios. The specific softness factor adaptive adjusting parameters optimization model is as follows,

$$\begin{array}{ll} \min & F(x) = (P_k, \frac{ITAE}{\max(ITAE)}, \frac{K_{safe}}{\max(K_{safe})}) \\ s.t. & x = (\lambda_{\alpha Ts}, \lambda_{\alpha \mu y}, \alpha_{Tr}, \alpha_{\max}, b) \\ & g_{ig}(x) \leq 0,\ ig = 1, 2, \cdots, ng \\ & x \in \Omega'' \end{array} \tag{29}$$

where x represents the solution vector; $F(x)$ represents the target vector; Ω'' represents feasible solution space of x; $g_{ig}(x)$ represents the ig-th equality or inequality constraint for automatic train operation tracking control problem, ng represents the number of equalities and inequality constraints; the five adjustable parameters $(\lambda_{\alpha Ts}, \lambda_{\alpha \mu y}, \alpha_{Tr}, \alpha_{\max}, b)$ are decision variables.

Objective decomposition is an effective method to solve the multi-objective optimization problems. The Tchebycheff decomposition method is selected in this paper among many objective decomposition methods [37]. The specific calculation formula for the aggregate function value of the Tchebycheff decomposition method is as follows,

$$\begin{array}{l} \min g^{te}(x|\lambda, z^*) = \max\limits_{1 \leq i \leq m} \{\lambda_i |f_i(x) - z_i^*|\} \\ s.t. x \in \Omega'' \end{array} \tag{30}$$

where z^* represents the reference point, ($z_i^* = \min\{f_i(x)|x \in \Omega\}, i = 1, ..., m$), which is the optimal solution of each objective function at present; λ_i is the weight of the i th objective, $\sum_{i=1}^{m} \lambda_i = 1$.

Whale optimization algorithm with strong global optimization ability is chosen in this paper. Whale optimization algorithm (WOA) is a new metaheuristic optimization algorithm learned from whale predatory behavior. There are two operators (position update and prey searching) in the computation process of the whale optimization algorithm [38]. The specific calculation formula for the position update of the basic whale optimization algorithm is as follows,

$$X(t+1) = \begin{cases} X^*(t) - A \cdot D & p < P_s \\ X^*(t) + D_p \cdot e^{Bl} \cdot \cos(2\pi l) & p \geq P_s \end{cases} \tag{31}$$

where $X^*(t)$ represents the optimal position vector obtained by the current optimization; $D_p = |X^*(t) - X(t)|$ represents the distance between humpback whales and their prey; p represents the behavioral selection probability of humpback whales, $p \in [0,1]$; P_s represents the probability of surrounding prey of humpback whales, $P_s \in [0,1]$; the probability of spiral hunting is $1 - p_s$; B represents a constant, which is used to define the shape of spiral; l represents the random number in $(-1,1)$; t is the current iteration number; T_{max} is the maximum number of iterations; a represents convergence factor; A and C represent the correlation coefficients respectively; r_1 and r_2 are random numbers, $r_1 \in [0,1]$, $r_2 \in [0,1]$.

The specific calculation formulas for convergence factor a, correlation coefficients A, and C is as follows,

$$a = 2 - 2 \times t/T_{\max} \tag{32}$$

$$A = 2a \times r_1 - a \tag{33}$$

$$C = 2 \times r_2 \tag{34}$$

After the position updated, prey searching is implemented by means of random individual positions. The specific calculation formula for prey searching of the basic whale optimization algorithm is as follows,

$$D = |CX_{rand} - X(t)| \tag{35}$$

$$X(t+1) = X_{rand} - A \cdot D \tag{36}$$

where X_{rand} is the position vector of randomly selected whales. If $A \geq 1$, a search leader individual is randomly selected, and the position of other whales is updated based on the whale position of the leader individual, so as to guide the whales to leave the prey and find a more suitable prey to enhance the global search ability of the algorithm.

The relatively fixed method of linear decline of convergence factor a will reduce the population diversity maintenance ability, so that the algorithm can easy to fall into local convergence in the late iteration. Aiming at solving this problem, the strategy of cosine decline combined with chaotic random method for convergence factor nonlinear decline is proposed in this paper. The specific calculation formula is as follows,

$$\begin{array}{ll} a = 2 \cdot \cos\left(\frac{\pi}{2} \cdot \frac{t}{T_{\max}}\right) & p_a(t) < P_a \\ a = 2 \cdot rand^2 \cdot \sin(\pi \cdot rand) & p_a(t) > P_a \end{array} \tag{37}$$

where $p_a(t)$ represents the behavioral selection probability of the convergence factor a, $p_a(t) \in [0,1]$; P_a represents the probability of cosine decline of the convergence factor a, $P_a \in [0,1]$; and the probability of chaotic random is $1 - P_a$.

Compared with the linear decline strategy, the decline rate of the convergence factor is significantly different in the whole iteration cycle caused by the nonlinear decline strategy with a certain degree of chaos uncertainty for convergence factor a, and it is helpful for maintaining the population diversity, thus the algorithm global convergence performance will be improved [39].

According to the Tchebycheff decomposition method, the aggregate function value is the fitness index for the multi-objective optimization algorithm. After the computation process of each iteration, the newly generated non-dominated solutions of the current population are put into the elite archive. The archive must kept within a certain size by some elite individuals with small differences from other elite individuals, so as to avoid computational burden of the algorithm. According to the updating rules of the whale optimization algorithm, the reference point z^* plays an important role in guiding the direction of global convergence, and a certain degree of local convergence due to this fixed foraging behavior. Meanwhile, the selector, crossover, and mutation of the genetic algorithm can generate a large number of new solutions with great differences for the whale optimization algorithm based on evolutionary processes, so as to further improve the global convergence performance due to more powerful population diversity maintenance ability.

The specific steps of improved whale optimization algorithm proposed in this paper are as follows.

Step 1: Initialization.

Initialize the whale population (the size is Nw), and the Tchebycheff aggregation function values of each whale individual are calculated.

Step 2: Iterative computations.

the *Archive* is obtained;

$Archive = \varnothing$, the reference point $z^* = (z_1^*, z_2^*, \ldots, z_m^*)$, and $z_j^* = \min(f_j(x))$, $j = 1, 2, \ldots, m$, m represents the number of objectives, a uniformly distributed weight vector set λ_0 is generated, and $\lambda^1 = \lambda_0$.

If the current iteration number is greater than 1, the weight $\lambda^{t,i}$ for solution $x^{t,i}$ are need to be recalculated. According to the literature [40], in the t-th iteration, the specific calculation formula for weight $\lambda^{t,i,k}$ of the k-th optimization index of the i-th individual (solution) $x^{t,i}$ in the population is as follows,

$$\lambda^{t,i,k} = \frac{1}{f(x^{t,i})^k - Z^{ref,k}} \left(\sum_{ik=1}^{m} \frac{1}{f(x^{t,i})^{ik} - Z^{ref,ik}} \right)^{-1} \tag{38}$$

where $i \in \{1, 2, \ldots, Nw\}$, $k \in \{1, 2, \ldots, m\}$.

For any solution target $z^c = (z_1^c, z_2^c, \ldots, z_m^c)$ of Pareto front, its weight vector is $\frac{1}{f(x)-Z^c} \left(\sum_{ik=1}^{m} \frac{1}{f(x)-Z^{c,ik}} \right)^{-1}$. Because the Pareto front is not easily available, it is replaced by the nearest solution target Z^{ref} in *Archive*;

The strategy of cosine decline combined with chaotic random method is used to calculate convergence factor a;

The updating rules of the whale optimization algorithm are used to update each individual whale.

Step 3: The archive and genetic evolution mechanism.

The Pareto front of the current whale population is obtained, and it is used to expand the *Archive*.

Some elite individuals with small differences from other elite individuals are deleted, until the size of *Archive* is not exceed allowed limit archive size NA.

Three operators (selection, crossover, and mutation) of the genetic algorithm are applied in the whole whale population, so as to further improve the population diversity maintenance ability.

Step 4: Termination judging.

The hypervolume indicator is chosen as termination judging indicator. Using the hypervolume indicator of the dominated portion of the objective space as a measure for the quality of Pareto set approximations is appropriate and effective. The specific formula for hypervolume indicator for objective vector set A with reference point (0,0,...,0) is as follows.

$$I^*{}_H(A) = \int_{(0,0,\ldots,0)}^{(1,1,\ldots,1)} \alpha_A(z)dz \tag{39}$$

where A is any objective vector set in objective space Ω; if there is an objective vector a, Pareto superior z and a belongs to A, $\alpha_A(z) = 1$; otherwise, $\alpha_A(z) = 0$ [41].

Thus, the hypervolume indicator indicates the dominated portion in objective space Ω. Generally, Hypervolume as Klee's Measure Problem (HKMP) is an effective calculation method for hypervolume indicator [42]. As only 3 objects $\left(P_k, \frac{ITAE}{\max(ITAE)}, \frac{K_{safe}}{\max(K_{safe})}\right)$ must be taken into account, the calculation method by using equivalent volume model for the volume of irregular objects can be used. The specific formulas for the hypervolume indicator for 3 objects by using equivalent volume model is as follows,

$$I^*{}_H(A) = \frac{\sum\limits_{ia=1}^{na}\sum\limits_{ib=1}^{nb}\sum\limits_{ic=1}^{nc} \alpha_A(cp(ia,ib,ic))}{na \times nb \times nc} \tag{40}$$

where na, nb, and nc are the split numbers for each normalization objective domain (0,1); $cp(ia,ib,ic)$ represents the central point of the (ia,ib,ic)th cube of normalization objective space.

The schematic diagram of the hypervolume indicator for 3 objects by using equivalent volume model is shown in Figure 6.

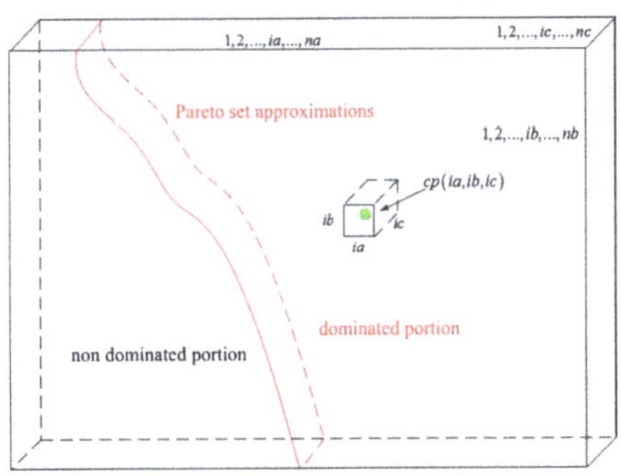

Figure 6. Schematic diagram of hypervolume indicator for 3 objects by using equivalent volume model.

If the hypervolume indicator $I^*{}_H(A)$ is reached beforehand, an unchanged number of hypervolume indicators nH will be resetted; otherwise, make $nH = nH + 1$.

If the maximum unchanged number of the hypervolume indicator $nHmax$ is reached, the calculation will be terminated; otherwise, return Step 2.

The flowchart of improved whale optimization algorithm proposed in this paper is shown in Figure 7.

Aiming at improving the global searching ability, the evolution law is must be considered in the computation process. The excellent individuals should have a small mutation probability, so that they can accumulate optimization results effectively, and the poor individuals should choose a large mutation probability, which can be fully eliminated, so as to enhance the capacity of exploration [43]. The mutation probability calculation formula based on sigmoid function $y = \frac{1}{1+e^{-x}}$ is as follows,

$$p_m = p_{m\_\max}\frac{1}{1+e^{-ap\left(N_s - fit(x)'\right)}} \tag{41}$$

where P_m is the mutation probability value; $p_{m\_\max}$ is the maximum mutation probability; ap is the shape factor for the sigmoid function of mutation probability; N_s is the demarcation point of the whale population; $fit(x)'$ is the normalized value of fitness function value $fit(x)$ of whale individual x in the whale population.

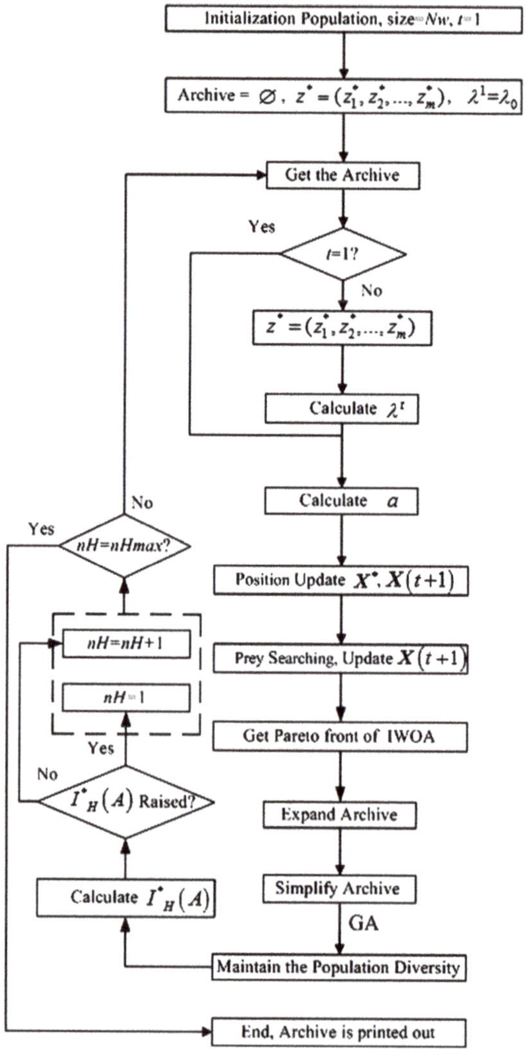

Figure 7. The flowchart of the improved whale optimization algorithm proposed in this paper.

This mutation probability calculation method has certain fairness, and the whale individuals have appropriate mutation probability according to fitness function value, so as to prevent the population controlled by advantage individuals and persist evolution opportunity for disadvantaged individuals.

A multimodal crossover method is conducive to finding a more satisfactory optimal solution for the complex optimization issue [44,45]. The multimodal crossover combining popular blended crossover and unimodal normal distribution crossover is applied in this paper. The specific calculation formula for multimodal crossover is as follows,

$$\begin{array}{ll} x_c = r_{p1} \times x_{p1} + (1 - r_{p1}) \times x_{p2} & pc < pc_b \\ x_c = x_p + \varepsilon d + \sum\limits_{ic=1, ic \neq p}^{\mu} v_{ic} e_{ic} & pc_b < pc < (pc_b + pc_u) \end{array} \quad (12)$$

where x_c is the solution after multimodal crossover operation; x_{p1} and x_{p2} are two parent solutions for blended crossover; pc is the behavioral selection probability about multimodal crossover operation; pc_b and pc_u are the crossover probabilities for blended crossover and unimodal normal distribution crossover, respectively; x_p is the midpoint for μ parent solutions for unimodal normal distribution crossover; d is the differential vector; e_{ic} is the ic th orthogonal basis; ε and v_{ic} are the random numbers obey normal distribution $N(0, \sigma_1^2)$ and $N(0, \sigma_2^2)$.

3.4. Performance Analysis of Optimization Algorithms Based on Standard Test Functions

Aiming at verifying the effectiveness of IWOA proposed in this paper, standard test functions (ZDT1, ZDT3, and DTLZ2) are selected as optimization objects, and multi-objective particle swarm optimization based on decomposition (dMOPSO) [46] and multi-objective evolutionary algorithm based on decomposition (MOEA/D) [47] are selected as contrasted optimization algorithms. The improved whale optimization algorithm parameters are set as follows; maximum number of iterations is 100; probability of surrounding prey of humpback whales P_s is 0.6; population size Nw is 50; allowed limit archive size NA is 30; probability of cosine decline of the convergence factor P_a is 0.9; shape constant of spiral b is 1; shape factor for sigmoid function of mutation probability ap is 4.9; demarcation point of whale population N_s is 0.8; probability of blended crossover pc_b and unimodal normal distribution crossover pc_u are 0.3 and 0.3, respectively; selection probability is 0.5; split number for each normalization objective na, nb, and nc are all 50; the maximum unchanged number of hypervolume indicator $nHmax$ is 25. The Matlab/simulink platform is used for verifying, and the Matlab/simulink revision and the computer processor type are 2016b, MathWorks and CPU Core i9-7920X @ 2.9GHZ. The specific optimization results (approximate Pareto solution set) for test functions of each optimization algorithms are shown in Tables 1 and 2 and Figure 8.

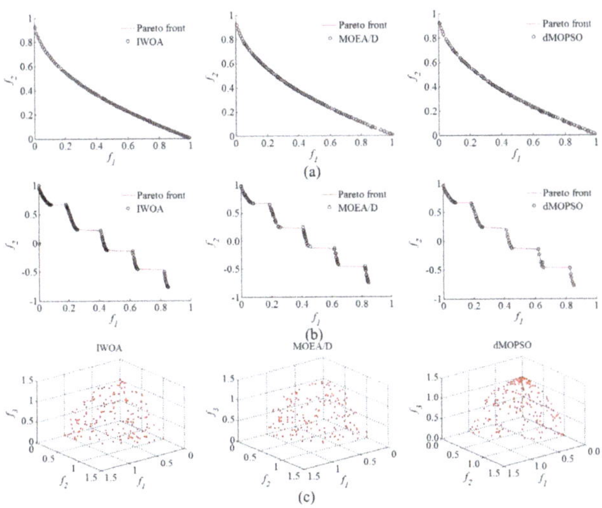

Figure 8. The optimization results for test functions of each optimization algorithms. (**a**) Optimization results for ZDT1. (**b**) Optimization results for ZDT3. (**c**) Optimization results for DTLZ2.

Table 1. The hypervolume indicator ratio $\frac{I^*_H(A(op))}{I^*_H(A(tp))}$ for test functions of each optimization algorithm.

Optimization Algorithm	ZDT1	ZDT1	DTLZ2
IWOA	99.81%	99.73%	99.04%
MOEA/D	99.72%	99.60%	98.73%
dMOPSO	99.54%	99.37%	98.57%

Table 2. The computation time for test functions of each optimization algorithm.

Optimization Algorithm	ZDT1	ZDT1	DTLZ2
IWOA	1279*s*	1541*s*	1895*s*
MOEA/D	1384*s*	1733*s*	2079*s*
dMOPSO	1423*s*	1694*s*	2104*s*

As can be seen from Tables 1 and 2, compared with other optimization algorithms (dMOPSO and MOEA/D), IWOA has been improved to a considerable extent, not only in the computation speed, but also in the optimization effect for approximate Pareto solution set reflected by the hypervolume indicator ratio $\frac{I^*_H(A(op))}{I^*_H(A(tp))}$, $A\,(op)$ and $A\,(tp)$ are the approximate Pareto solution sets obtained by optimization algorithms and real Pareto front set, respectively. According to Figure 9, compared with other optimization algorithms (dMOPSO and MOEA/D), only the better approximate Pareto solution sets closer to the Pareto fronts of ZDT1, ZDT3, and DTLZ2 were found using IWOA, but also the distribution for obtained approximate Pareto solution set was more evenly. This indicates that the IWOA proposed in this paper has better optimization effectiveness.

3.5. Improved DMC Model Predictive Controller and Hardware-In-The-Loop Simulation Platform

Based on the traditional Fuzzy DMC model predictive controller, a new functional module is necessary to be realized the function of online obtaining of softness factor and fusion velocity. The schematic diagram of improved DMC model predictive controller for automatic train operation designed in this paper is shown in Figure 9.

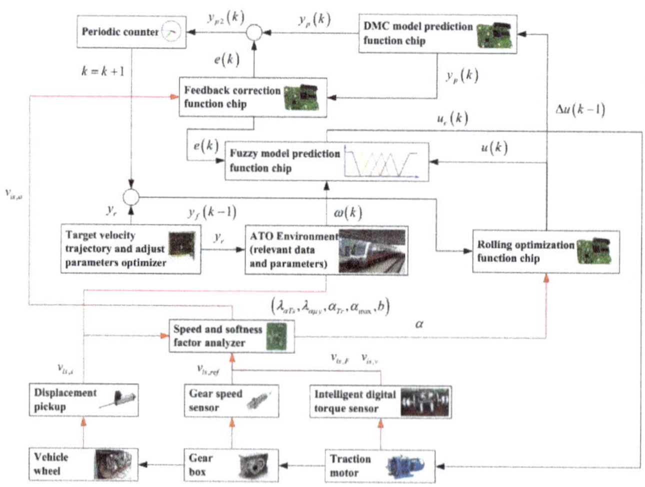

Figure 9. The schematic diagram of improved DMC model predictive controller for automatic train operation designed in this paper.

In Figure 9, the improved DMC model predictive controller could provide control commands for the corresponding equipments in real-time using fuzzy DMC MPC based on online obtaining of softness factor and fusion velocity, enabling the the urban rail vehicle to track the target velocity trajectory; the Speed and softness factor analyzer could provide the precision instantaneous fusion velocity $v_{is,a}$ and softness factor α for rolling optimization and feedback correction based on three kinds of velocity sampling sources ($v_{is,v}$, $v_{is,F}$, and $v_{is,s}$) and a set of softness factor α adjustable parameters. The intelligent digital torque sensor, gear speed sensor, and displacement pickup are data acquisition devices. The physical diagram of the intelligent digital torque sensor is shown in Figure 10.

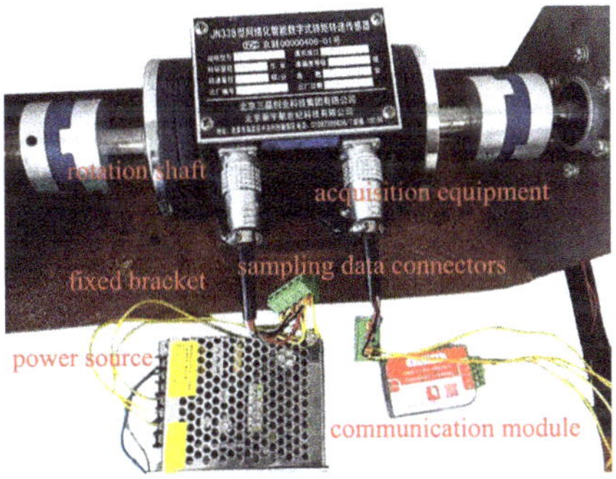

Figure 10. The physical diagram of the intelligent digital torque sensor.

In Figure 10, the acquisition equipment is an intelligent digital torque sensor of model No. JN338; the fixed bracket is made of iron and has two functions of fixing and preventing jitter; the power source supplies electricity to torque sensor of model No. JN338; the communication module transmits the real-time value of motor speed and torque using the 485 communication protocol; the sampling data connectors are connectors for sampling data (motor speed and torque); and can be connected to communication module, controller, or other equipment; rotation shaft of intelligent digital torque sensor must be connected to rotation shaft of traction motor and load (breaker or rheostat box).

To more effectively test the performance of the tracking control algorithm in actual automatic train operation tracking control scenarios, the dSPACE HILS technology is adopted. In this way, the optimization algorithm or control algorithm needed to be verified is written into the chip of the optimizer or controller. The structure diagram of HILS platform used in this paper for automatic train operation tracking control scenario, and the physical diagram of controller cabinet and simulation cabinet for HILS platform are shown in Figures 11 and 12.

In Figure 11, the Displacement generator is used to generate the train instantaneous displacement, so that the static (no actual displacement) HILS platform can truly reflect the actual automatic train operation tracking control scenario; various sensors are used to feed electrical waves of sampling sources back to the Controller in real-time; the Conditioning circuit can regulate electrical signals properly for the Tracking controller appropriately; the Motor controller could provide electrical control commands for Traction moto and other corresponding equipments of Electrical loop in real-time using a proper electrical control algorithm. DC power source, Converter system, Traction motor, Digital rheostat box, and Gear box are simulation electric hardware equipments.

In Figure 12, the 'train controller cabinet' and 'train emulator cabinet' are the vital equipments for automatic train operation HILS, except for the controller and emulator, the conditioning circuit, signal processing unit, and corresponding switch groups are included. The 'emulator' provides some correlative simulation environments for the automatic train operation HILS, the related models included such as accurate braking model, traction transformer model, running line model, velocity fluctuation model, etc. The 'conditioning circuit' can regulate electrical signals properly for 'Controller'. The 'signal processing unit' can regulate net signals properly for 'Optimizer' appropriately.

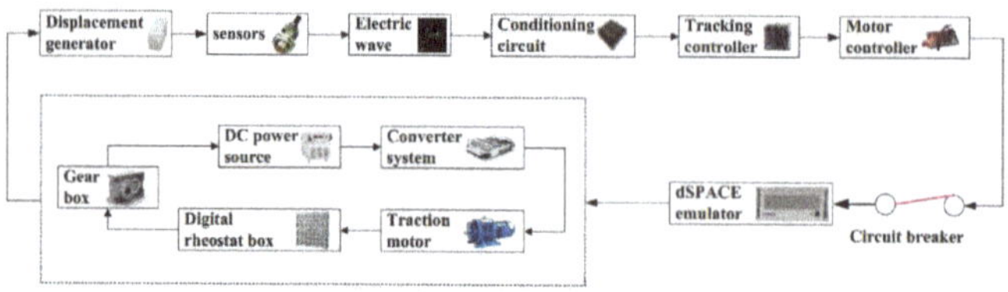

Figure 11. The structure diagram of the hardware-in-the-loop simulation (HILS) platform used in this paper for automatic train operation tracking control scenario.

Figure 12. The physical diagram of controller cabinet and simulation cabinet for HILS platform.

4. Simulation for Automatic Train Operation Tracking Control Scenario

4.1. Data and Parameters for Automatic Train Operation Tracking Control Scenario

The automatic train operation tracking control scenario for rail transit line No.12 in Dalian, China is chosen as the experimental simulation object. Rail transit line No.12 is a significant urban rail transit line with 40.38 kilometers from Hekou station to Lvshun New Port. The running simulation line of scenario about rail transit line No.12 is from Lvshun New Port to Tieshan Town, there are three long steep ramps and three velocity limit subintervals in running interval. The main parameters of the automatic train operation tracking control scenario are shown in Table 3, and the target velocity trajectory, slopes, and limited velocity curves for automatic train operation are shown in Figure 13.

Table 3. The main parameters of the automatic train operation tracking control scenario.

Parameter Name	Parameter Characteristics
Maximum limited velocity (km/h)	75
Running interval distance (m)	2940
Prospective running time (s)	180
Maximum allowed parking error (m)	0.4
Maximum allowed punctual time error (s)	0.5

The basic DMC MPC parameters are set as follows; sampling time is 500 μs; model length N is 60; control length is 15; predictive length is 15. The addition adjustable parameters for online obtaining of softness factor are set by practical experience as follows; blur width $s_1 = s_2 = 0.05$ km/h; maximum and minimum value of design expectation $y_{\max} = y_r + 0.05$ km/h and $y_{\min} = y_r - 0.05$ km/h; connected length $S_1 = S_2 = 0.4$ m. Considering the online real-time calculation efficiency and tracking control effect of the improved DMC MPC proposed in this paper, the following parameters are given based on the relevant scientific literature, field experience, and simulation results of multiple experiments. The Matlab/simulink simulation platform is used for softness factor adaptive adjusting parameters optimization and the experience parameters are as follows; maximum allowed multi-objective performance index, $ITAE$ index, and security index are 0.8, 750, and 6%, respectively; ideal multi-objective performance index, $ITAE$ index, and security index are 0.2, 250, and 3%, respectively. The chosen addition adjustable parameters for online obtaining of softness factor obtained by each optimization algorithms (improved whale optimization algorithm, MOEA/D, dMOPSO) are as follows; fusion weights of $\lambda_{\alpha Ts}$ and $\lambda_{\alpha_{\mu y}}$, $\lambda_{\alpha Ts} = 0.582, 0.592, 0.573$ and $\lambda_{\alpha_{\mu y}} = 0.418, 0.408, 0.427$; maximum value of softness factor α_{max} is 0.939, 0.941, and 0.930; gain coefficient b is 0.909, 0.911, and 0.913. The subinterval range obtained by practical experience, reference value of soften factor obtained by each optimization algorithms, synthetic weight of fusion velocity obtained by entropy weight method are shown in Table 4. The specific optimization results (approximate Pareto solution set) for softness factor adaptive adjusting parameters optimization of each optimization algorithms are shown in Table 5 and Figure 14.

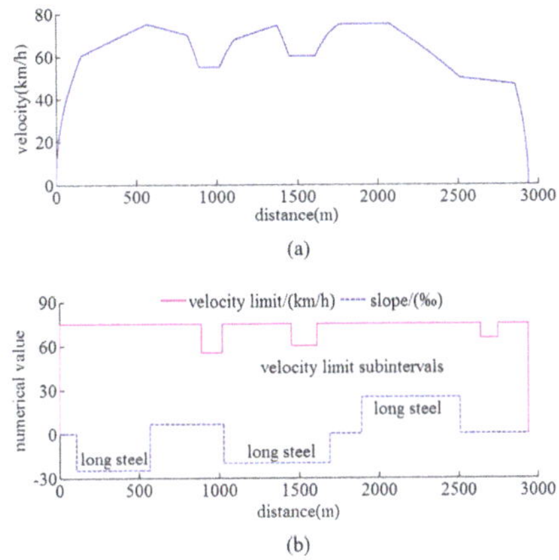

Figure 13. The target velocity trajectory, slopes, and limited velocity curves for automatic train operation tracking control scenario. (**a**) Target velocity trajectory. (**b**) Slopes and limited velocity curves.

Table 4. The optimization results for subinterval range, reference value of soften factor α_{Tr}, and synthetic weight of fusion velocity λ_{is}.

Subinterval Index	Subinterval Range s (m)	α_{Tr} Obtained by (IWOA, MOEA/D, dMOPSO)	Synthetic Weight λ_{is}
1	0–140	0.892, 0.896, 0.897	0.76, 0.15, 0.09
2	140–210	0.924, 0.926, 0.919	0.20, 0.71, 0.09
3	210–753	0.885, 0.883, 0.882	0.83, 0.11, 0.16
4	753–830	0.924, 0.922, 0.926	0.20, 0.61, 0.19
...	...	...	...
14	1430–1490	0.930, 0.928, 0.927	0.33, 0.51, 0.16
...	...	...	...
24	2910–2940	0.914, 0.911, 0.912	0.09, 0.08, 0.83

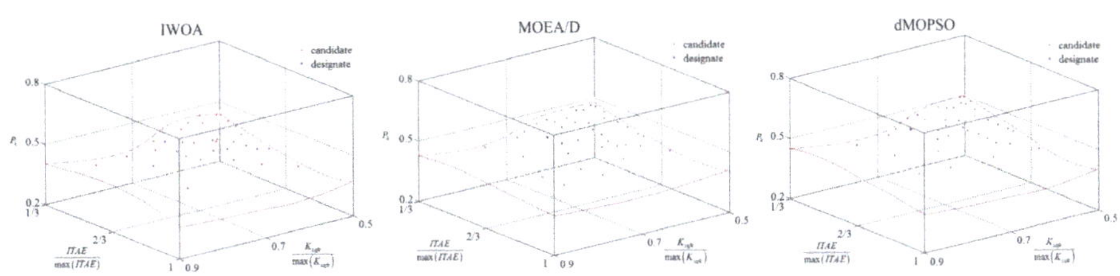

Figure 14. The optimization results for the softness factor adaptive adjusting parameters optimization of each optimization algorithm.

Table 5. The hypervolume indicator ratio $\frac{I^{*}_{H}(A(op))}{V(\Omega_S)}$ and computation time for softness factor adaptive adjusting parameters optimization of each optimization algorithms.

Optimization Algorithm	Computation Time	Hypervolume Indicator Ratio
IWOA	3726*s*	60.94%
MOEA/D	4582*s*	63.37%
dMOPSO	5047*s*	71.84%

In Figure 14, the approximate Pareto solution sets have been obtained by optimization algorithms (IWOA, dMOPSO, and MOEA/D), and one of the approximate Pareto solutions has been chosen for automatic train operation tracking control scenario. As can be seen from Figure 14, compared with other optimization algorithms (dMOPSO and MOEA/D), the wider dominated portion for approximate Pareto solution sets obtained by using IWOA. This indicates that the IWOA proposed in this paper has better optimization effectiveness. As can be seen from Table 5, compared with other optimization algorithms (dMOPSO and MOEA/D), IWOA has been improved to a considerable extent not only in the computation speed but also in the optimization effect for approximate Pareto solution set reflected by the hypervolume indicator ratio $\frac{I^{*}_{H}(A(op))}{V(\Omega_S)}$. $V(\Omega_S)$ is the volume of selected objective space Ω_S by experience, and it is 0.16 ($\frac{3}{5} \times \frac{2}{3} \times \frac{2}{5}$) in this paper.

4.2. Matlab/simulink Simulation Results for Automatic Train Operation Tracking Control Scenario

The sampling process is not necessary in the Matlab/simulink simulation platform. According to the automatic train operation tracking control scenario of rail transit line No.12 in Dalian, China, the Matlab/simulink simulation results are obtained by using the fuzzy DMC MPC based on online obtaining of softness factor α, and the softness factor adaptive adjusting parameters optimization using IWOA proposed in this paper (denoted as DMC MPC II); the fuzzy DMC MPC is based on online obtaining of softness factor α, which uses softness factor adaptive adjusting parameters optimization using MOEA/D (denoted as DMC MPC I) and traditional fuzzy DMC MPC. The specific configuration of the Matlab/simulink platform used in this paper is described as follows; the Matlab/simulink revision is 2016b, MathWorks; the type of computer processor is CPU Core i9-7920X @ 2.9GHZ. The specific Matlab/simulink results are shown in Figures 15–18 and Tables 6 and 7.

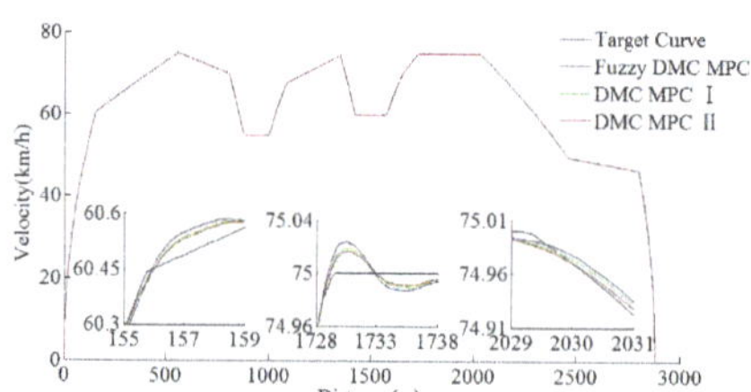

Figure 15. The Matlab/simulink velocity trajectory curves of different DMC MPC algorithms for automatic train operation tracking control scenario.

As can be seen from Tables 6 and 7, the tracking control results obtained by the DMC MPC II are superior to that of DMC MPC I and traditional fuzzy DMC MPC, and four indexes of multi-objective performance index (energy saving, punctuality, parking precision, and comfort), ITAE index, and security index for automatic train operation have been improved considerably. As can be seen from the Figures 15 and 16, the DMC MPC II can make the tracking control curves closer to target curves, so as to obtain the ideal tracking control results as smooth as possible. As can be seen from the six enlarged areas of the velocity trajectory curves in Figures 15 and 16, the velocity fluctuation degree is weaker and the velocity trajectory is closer to the target by using DMC MPC II. As can be seen from the one enlarged areas of time distance curves of Figure 16, compared with DMC MPC I and traditional fuzzy DMC MPC, the time traceability of DMC MPC II is more powerful, so as to obtained the time distance curve closer to target. As can be seen from Figure 17, the smaller velocity error effect can be obtained by using DMC MPC II. As can be seen from Figure 18, the more ideal parking results can be obtained by using DMC MPC II; both the distance and time errors of parking are reduced to certain extent, so as to improve the punctuality and fixed position effect.

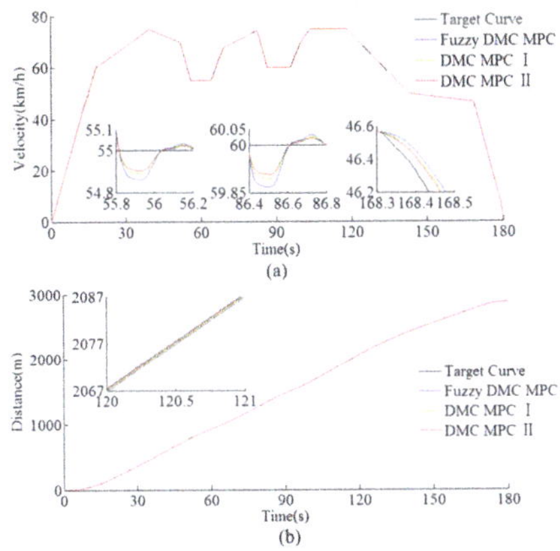

Figure 16. The Matlab/simulink time traceability curves of different DMC MPC algorithms for automatic train operation tracking control scenario. (**a**) Time–velocity curves. (**b**) Time–distance curves.

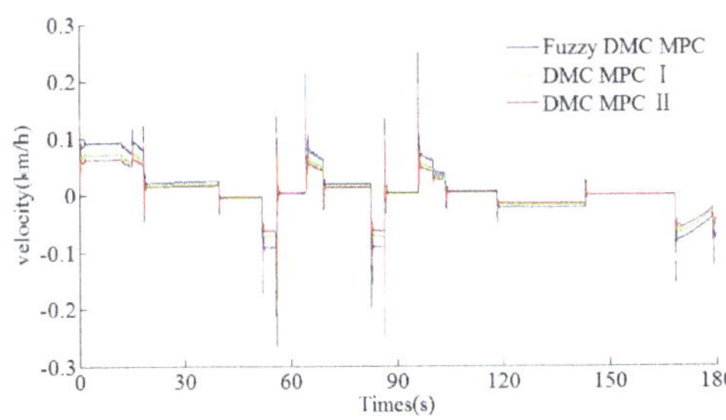

Figure 17. The Matlab/simulink time velocity error curves of different DMC MPC algorithms for automatic train operation tracking control scenario.

Table 6. The Matlab/simulink tracking control results of energy saving, punctuality, parking precision, and comfort for automatic train operation.

Algorithm	Energy Consumption	Actual Time	Parking Position	Comfort Level
Target curve	98615 KJ	179.914 s	2939.884	5.517 $m/s^2/km$
Fuzzy DMC MPC	112094 KJ	179.871 s	2939.774	30.725 $m/s^2/km$
DMC MPC I	110844 KJ	179.892 s	2939.781	28.754 $m/s^2/km$
DMC MPC II	108759 KJ	179.032 s	2939.812	24.339 $m/s^2/km$

Table 7. The Matlab/simulink tracking control results of multi-objective performance index, ITAE index, and security index for automatic train operation.

Algorithm	Multi-Objective Performance Index	*ITAE* Index	Security Index
Fuzzy DMC MPC	0.507	552.69	5.12%
DMC MPC I	0.467	454.87	4.89%
DMC MPC II	0.370	380.54	4.63%

Compared with traditional fuzzy DMC MPC and DMC MPC I, DMC MPC II has several obvious superiorities in the matlab/simulation environment. However, as there is no hardware equipment in the actual automatic train operation tracking control scenario in matlab/simulation environment, the effectiveness of DMC MPC proposed in this paper must be further tested and verified.

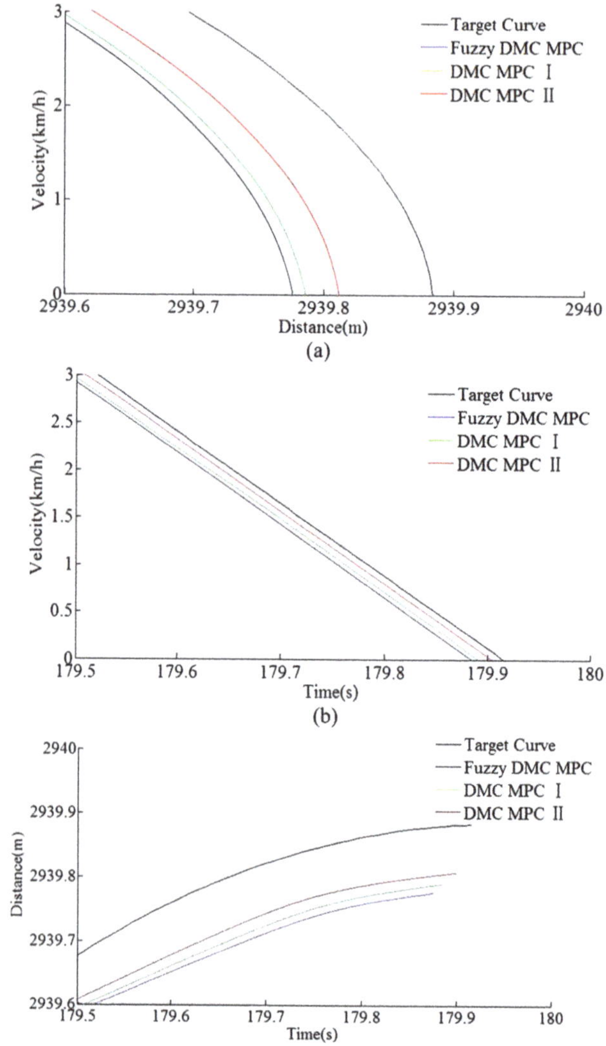

Figure 18. The Matlab/simulink parking error curves of different DMC MPC algorithms for automatic train operation tracking control scenario. (**a**) Distance–velocity curves. (**b**) Time–velocity curves. (**c**) Time–distance curves.

4.3. HILS Results for Automatic Train Operation Tracking Control Scenario

In this way, sampling accuracy must be taken into account. To further verify the effectiveness of the algorithm, according to the automatic train operation tracking control scenario of rail transit line No.12 in Dalian, China, the HILS results are obtained by using the fuzzy DMC MPC based on online obtaining of softness factor α and fusion velocity, which uses the softness factor adaptive adjusting parameters optimization using IWOA proposed in this paper (denoted as DMC MPC V), the fuzzy DMC MPC based on online obtaining of softness factor α and fusion velocity, which softness factor adaptive adjusting parameters optimization using dMOPSO (denoted as DMC MPC IV) and the fuzzy DMC MPC based on online obtaining of fusion velocity (denoted as DMC MPC III). The specific configuration of the automatic train operation HILS platform used in this paper is described as follows; the Matlab/simulink revision is "2016b, MathWorks"; the type of computer processor is "CPU Core i9-7920X @ 2.9GHZ"; the core chip of "Tracking controller" and "Motor optimizer" is "TMS320F28335"; the simulation software of "dSPACE emulator" is dSPACE control desk (revision is control desk 6.1); the communication protocol of the HILS platform is MVB (multifunction vehicle bus); the fuzzy PID (proportion integration differentiation) algorithm is adopted as motor control algorithm; vehicle velocity proportion is (0.83 × 400 rad/min)/(80 km/h). The specific HILS results are shown in Figures 19–22 and Tables 8 and 9.

Table 8. The HILS tracking control results of energy saving, punctuality, parking precision, and comfort for automatic train operation.

Algorithm	Energy Consumption	Actual Time	Parking Position	Comfort Level
Target curve	98703 KJ	179.874 s	2939.823	5.429 $m/s^2/km$
DMC MPC III	119,192 KJ	179.809 s	2939.704	35.403 $m/s^2/km$
DMC MPC IV	115,048 KJ	179.824 s	2939.754	32.935 $m/s^2/km$
DMC MPC V	113,784 KJ	179.835 s	2939.778	31.354 $m/s^2/km$

Table 9. The HILS tracking control results of multi-objective performance index, ITAE index, and security index for automatic train operation.

Algorithm	Multi-Objective Performance Index	*ITAE* Index	Security Index
DMC MPC III	0.712	821.57	5.68%
DMC MPC IV	0.585	744.23	5.14%
DMC MPC V	0.530	690.43	5.09%

According to the HILS results of different algorithms from Tables 8 and 9, compared with DMC MPC III and DMC MPC IV, DMC MPC V has an obvious performance improvement effectiveness, the multi-objective performance index (energy saving, punctuality, parking precision, and comfort) of the tracking control trajectory has been improved considerably; meanwhile, the ITAE index and security index have also been reduced considerably. In Figure 19, during the automatic train operation tracking control experiment simulation, all the pilot lights and buttons are in normal. As can be seen from Figures 19 and 20, the DMC MPC V can bring the tracking control curves closer to the target curves, so as to obtain the ideal tracking control results as smooth as possible. As can be seen from the six enlarged areas of velocity trajectory curves of Figures 19 and 20, the velocity trajectory curves obtained by DMC MPC V were smoother; compared with the traditional improved tracking control algorithm (DMC MPC), DMC MPC V enables the train to be in the optimal working state as much as possible, so as to reduce the velocity fluctuation degree and obtain more ideal tracking control results. As can be seen from the one enlarged areas of the time–distance curves in Figure 20, compared with DMC MPC III and DMC MPC IV, the time traceability of DMC MPC V is more powerful, so as to obtain a time–distance curve closer to target. As can be seen from Figure 21, the velocity error obtained by using DMC MPC V is smaller. As can be seen from Figure 22, the more ideal parking point (parking time and position) can be obtained by using DMC MPC V, its parking point is closer to prospective parking point (180 s and 2940 m).

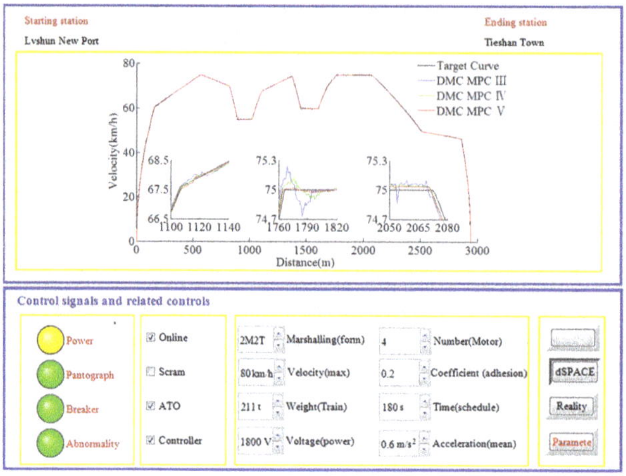

Figure 19. The HILS velocity trajectory curves of different DMC MPC algorithms for automatic train operation tracking control scenario.

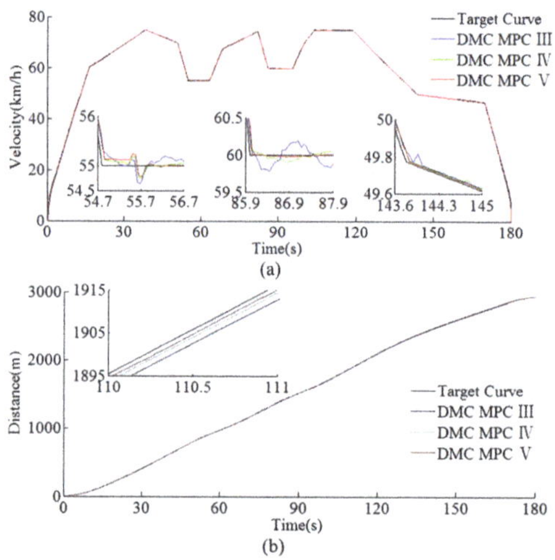

Figure 20. The HILS time traceability curves of different DMC MPC algorithms for automatic train operation tracking control scenario. (**a**) Time–velocity curves. (**b**) Time–distance curves.

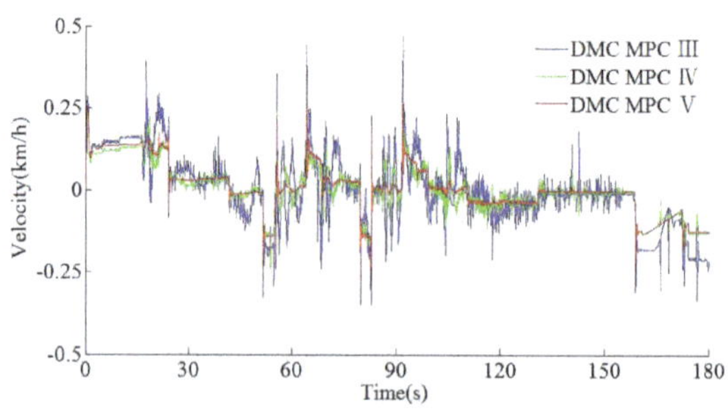

Figure 21. The HILS time–velocity error curves of different DMC MPC algorithms for automatic train operation tracking control scenario.

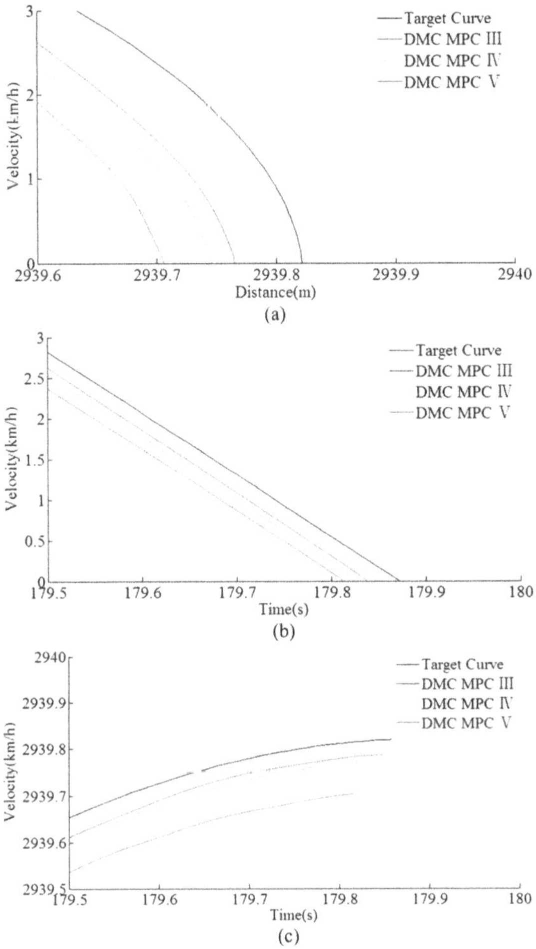

Figure 22. The HILS parking error curves of different DMC MPC algorithms for automatic train operation tracking control scenario. (**a**) Distance–velocity curves. (**b**) Time–velocity curves. (**c**) Time–distance curves.

The above HILS results show that DMC MPC V is a tracking control algorithm with good practical tracking control effect for automatic train operation tracking control scenario.

5. Conclusions

Tracking control optimization for automatic train operation is a sophisticated optimization problem, and the model predictive controller is widely used to solve this problem due to its advantages of strong robustness and good performance in tracking speed and tracking precision. Aiming at obtaining a more ideal tracking control performance for automatic train operation, an improved model predictive control algorithm and corresponding controller based on online obtaining of softness factor and fusion velocity for automatic train operation are proposed and developed, and an improved whale optimization algorithm based on Tchebycheff decomposition method was proposed for softness factor adaptive adjusting parameters optimization. This clearly shows that model predictive controller has been improved to a considerable extent in tracking control optimization for automatic train operation not only in the pure software scenarios but also in the hardware-in-the-loop simulation scenarios (the

$ITAE$ index obtained by DMC MPC II is 16.3% and 31.2% lower than that of DMC MPC I and Fuzzy DMC MPC, and that obtained by the DMC MPC V is 7.3% and 16.1% lower than that of DMC MPC IV and DMC MPC III). The specific advantages are described below.

(I) Aiming at improving the efficiency of the whale optimization algorithm based on the Tchebycheff decomposition method, the strategy of cosine decline combined with chaotic random method for convergence factor nonlinear decline is proposed, so as to obtain more satisfactory softness factor adaptive adjusting parameters for tracking control.
(II) Not only is an improved online adaptive adjusting method for softness factor based on fuzzy satisfaction of system output value and velocity distance trajectory characteristic adopted, but also a fusion velocity model and a corrected model of real-time sampling for automatic train operation tracking control are adopted. Thus, compared with traditional improved model predictive controller, the improved model predictive controller developed in this paper based on online obtaining of softness factor and fusion velocity could enable the train in the optimal working state as much as possible, so as to obtain a more ideal tracking control result with more satisfactory performance indexes, including energy saving, punctuality, parking precision and comfort, ITAE, and security index.
(III) For any tracking control system, the accomplishing capacity of computational tasks real-time is significant important. The only purpose of the improved strategies is the online obtaining of optimal softness factor and fusion velocity, so as to obtain the more reasonable real-time control quantity $u(k)$, and enable the automatic train operation tracking control system robustness and rapidity as much as possible. Thus, the quantity of additional computational tasks is not very large. In addition, some complex computational tasks for adjustable parameters optimization for softness factor adaptive adjusting and setting synthetic weight of the velocity sampled according to entropy weight method are achieved offline. Then, the advanced urban rail vehicle velocity monitoring device and the additional function chip of online obtaining of softness factor and fusion velocity are applied to improved the accomplishing capacity of computational tasks real-time is significant important. Finally, some complex functions such as logarithm or exponential function are avoided in online computation. Thus, it has advantage of simple and easily conducted.

According to the Matlab/simulink results and ATO HILS results (compare with the other DMC MPC algorithms for comparison), the improved DMC MPC based on online obtaining of softness factor and fusion velocity proposed in this paper has better tracking control performance, so it can obtain more ideal tracking control results. In actuality, it can also be designed by other ways for online obtaining of softness factor and fusion velocity. It is important to note that, as the computational tasks such as various matrix computation of basic DMC MPC are onerous, the designed scheme should be simple and appropriate caused by the limited computation margin in real-time.

Author Contributions: The work presented here was performed in collaboration among all authors. L.W. designed, analyzed, and wrote the paper, and completed the simulation experiment. X.W. guided the full text and provided simulation conditions. Z.S. conceived the idea and involved simulation experiment. S.L. involved simulation experiment and analyzed the data. All authors have contributed to and approved the manuscript.

Funding: This research was funded by the Nature Science Foundation of China (grant number 60574018).

Conflicts of Interest: The authors declare no conflicts of interest.

Abbreviations

The following abbreviations are used in this manuscript.

ATO	automatic train operation
HILS	hardware-in-the-loop simulation
DMC MPC	dynamic matrix control model predictive control

DMC MPC I	fuzzy DMC MPC based on online obtaining of softness factor which softness factor adaptive adjusting parameters optimization using MOEA/D
DMC MPC II	fuzzy DMC MPC based on online obtaining of softness factor which softness factor adaptive adjusting parameters optimization using IWOA
DMC MPC III	fuzzy DMC MPC based on online obtaining of fusion velocity
DMC MPC IV	fuzzy DMC MPC based on online obtaining of softness factor and fusion velocity which softness factor adaptive adjusting parameters optimization using dMOPSO
DMC MPC V	fuzzy DMC MPC based on online obtaining of softness factor and fusion velocity which softness factor adaptive adjusting parameters optimization using IWOA

References

1. Jiateng, Y.; Chen, D.; Li, L. Intelligent Train Operation Algorithms for Subway by Expert System and Reinforcement Learning. *IEEE Trans. Intell. Transp. Syst.* **2014**, *15*, 2561–2571.
2. Liang, Y.; Liu, H.; Qian, C. A Modified Genetic Algorithm for Multi-Objective Optimization on Running Curve of Automatic Train Operation System Using Penalty Function Method. *Int. J. Intell. Transp. Syst. Res.* **2019**, *17*, 74–87. [CrossRef]
3. Adrián, F.; Antonio, F.; Asunción, P.C.; Marı'a, D.; Tad, G. Design of Robust and Energy-Efficient ATO Speed Profiles of Metropolitan Lines Considering Train Load Variations and Delays. *IEEE Trans. Autom. Sci. Eng.* **2015**, *16*, 2061–2071.
4. Zhou, M.; Dong, H.; Zhao, Y.; Ioannou, P.A.; Wang, F. Optimization of Crowd Evacuation With Leaders in Urban Rail Transit Stations. *IEEE Trans. Intell. Transp. Syst.* **2019**, *20*, 4476–4487.. [CrossRef]
5. Watanabe, S.; Koseki, T.; Isobe, E. Evaluation of Automatic Train Operation Design for Energy Saving Based on the Measured Efficiency of a Linear-Motor Train. *IEEE Trans. Intell. Transp. Syst.* **2017**, *137*, 460–468. [CrossRef]
6. Marı'a, D.; Antonio, F.; Asunción, P.C.; Ramón, R.P. Energy Savings in Metropolitan Railway Substations Through Regenerative Energy Recovery and Optimal Design of ATO Speed Profiles. *IEEE Trans. Autom. Sci. Eng.* **2012**, *9*, 496–504.
7. Cheng, R.; Yu, W.; Song, Y.; Chen, D.; Ma, X.; Cheng, Y. Intelligent Safe Driving Methods Based on Hybrid Automata and Ensemble CART Algorithms for Multihigh-Speed Trains. *IEEE Trans. Cybern.* **2019**, *49*, 3816–3826. [CrossRef]
8. Shangguan, W.; Yan, X.; Cai, B.; Wang, J. Multiobjective Optimization for Train Speed Trajectory in CTCS High-Speed Railway With Hybrid Evolutionary Algorithm. *IEEE Trans. Intell. Transp. Syst.* **2015**, *16*, 2215–2225. [CrossRef]
9. Gao, S.; Dong, H.; Chen, Y.; Ning, B.; Chen, G. Approximation-Based Robust Adaptive Automatic Train Control: An Approach for Actuator Saturation. *IEEE Trans. Intell. Transp. Syst.* **2013**, *14*, 1733–1742. [CrossRef]
10. Bai, Y.; Tin, K.H.; Mao, B.; Ding, Y.; Chen, S. Energy-Efficient Locomotive Operation for Chinese Mainline Railways by Fuzzy Predictive Control. *IEEE Trans. Intell. Transp. Syst.* **2014**, *15*, 938–948. [CrossRef]
11. Chen, D.; Chen, R.; Li, Y.; Tang, T. Online Learning Algorithms for Train Automatic Stop Control Using Precise Location Data of Balises. *IEEE Trans. Intell. Transp. Syst.* **2013**, *14*, 1526–1535. [CrossRef]
12. Li, Z.; Hou, Z.; Yin, C. Iterative learning control for train trajectory tracking under speed constrains with iteration-varying parameter. *Trans. Inst. Meas. Control* **2015**, *34*, 485–493. [CrossRef]
13. Meng, J.; Xu, R.; Li, D.; Chen, X. Combining the Matter-Element Model With the Associated Function of Performance Indices for Automatic Train Operation Algorithm. *IEEE Trans. Intell. Transp. Syst.* **2019**, *20*, 253–263. [CrossRef]
14. Ma, Y.; Cai, Y. A Fuzzy Model Predictive Control Based Upon Adaptive Neural Network Disturbance Observer for a Constrained Hypersonic Vehicle. *IEEE Access* **2018**, *6*, 5927–5938. [CrossRef]
15. Borhan, H.; Vahidi, Y.; Phillips, A.M.; Kuang, M.; Kolmanovsky, I.V.; Cairano, S. MPC-Based Energy Management of a Power-Split Hybrid Electric Vehicle. *IEEE Trans. Control Syst. Technol.* **2012**, *20*, 593–603. [CrossRef]
16. Du, Y.; Wu, J.; Li, S.; Long, C.; Onori, S. Coordinated Energy Dispatch of Autonomous Microgrids with Distributed MPC Optimization. *IEEE Trans. Ind. Inform.* **2019**, *15*, 5289–5298. [CrossRef]

17. Mi, X.; Zou, Y.; Li, S.; Karimi, H. Self-triggered DMPC Design for Cooperative Multi-agent Systems. *IEEE Trans. Ind. Inform.* **2020**, *67*, 512–520. [CrossRef]
18. Shadmand, M.B.; Balog, R.S.; Abu-Rub, H. Model Predictive Control of PV Sources in a Smart DC Distribution System: Maximum Power Point Tracking and Droop Control. *IEEE Trans. Energy Convers.* **2014**, *29*, 913–921. [CrossRef]
19. Wang, L.; Cheng, Y.; Zou, J. Battery available power prediction of hybrid electric vehicle based on improved Dynamic Matrix Control algorithms. *J. Power Sources* **2014**, *261*, 337–347. [CrossRef]
20. Moon, U.C.; Lee, K.Y. Step-Response Model Development for Dynamic Matrix Control of a Drum-Type Boiler-Turbine System. *IEEE Trans. Energy Convers.* **2009**, *24*, 423–430. [CrossRef]
21. Liu, K.; Wang, X.; Qu, Z. Research on Multi-Objective Optimization and Control Algorithms for Automatic Train Operation. *Energies* **2018**, *12*, 3842. [CrossRef]
22. Fu, Q.; Zhu, J.; Mao, Z.; Zhang, G; Chen, T. Online Condition Monitoring of Onboard Traction Transformer Core Based on Core-Loss Calculation Model. *IEEE Trans. Ind. Electron.* **2018**, *65*, 3499–3508. [CrossRef]
23. Zhang, J.; Ding, G.; Zhou, Y.; Jiang, J.; Ying, X.; Qin, S. Identification of key design parameters of high-speed train for optimal design. *Int. J. Adv. Manuf. Technol.* **2014**, *73*, 251–265. [CrossRef]
24. Tongxin, S.; Min, X.; Chen, J.; Clarence, D. An Energy Efficient Adaptive Sampling Algorithm in a Sensor Network for Automated Water Quality Monitoring. *Sensors* **2017**, *17*, 2551–2565.
25. Mayet, C.; Delarue, P.; Bouscayrol, A.; Chattot, E. Hardware-In-the-Loop Simulation of Traction Power Supply for Power Flows Analysis of Multi-Train Subway Lines. *IEEE Trans. Veh. Technol.* **2017**, *66*, 5564–5571. [CrossRef]
26. Hasanzadeh, A.; Edrington, C.S.; Stroupe, N.; Bevis, T. Real-Time Emulation of a High-Speed Microturbine Permanent-Magnet Synchronous Generator Using Multiplatform Hardware-in-the-Loop Realization. *IEEE Trans. Ind. Electron.* **2014**, *61*, 3109–3118. [CrossRef]
27. Terwiesch, P.; Keller, T.; Scheiben, E. Rail vehicle control system integration testing using digital hardware-in-the-loop simulation. *IEEE Trans. Control Syst. Technol.* **1999**, *7*, 352–362. [CrossRef]
28. Yang, X.; Yang, C.; Peng, T.; Liu, B.; Gui, W. Hardware-in-the-Loop Fault Injection for Traction Control System. *IEEE J. Emerg. Sel. Top. Power Electron.* **2018**, *6*, 696–706. [CrossRef]
29. Qiu, B.; Wang, G.; Fan, Y.; Mu, D.; Sun, X. Robust Adaptive Trajectory Linearization Control for Tracking Control of Surface Vessels With Modeling Uncertainties Under Input Saturation. *IEEE Access* **2018**, *7*, 5057–5070. [CrossRef]
30. Howlett, P.; Cheng, J. Optimal driving strategies for a train on a track with continuously varying gradient. *J. Aust. Math. Soc.* **1997**, *38*, 388–410. [CrossRef]
31. Aufderheide, B.; Bequette, B.W. Extension of dynamic matrix control to multiple models. *Comput. Chem. Eng.* **2003**, *27*, 1079–1096. [CrossRef]
32. Ravi, V. R.; Thyagarajan, T.; Maheshwaran, G.U. Dynamic Matrix Control of a Two Conical Tank Interacting Level System. *Procedia Eng.* **2012**, *38*, 2601–2610. [CrossRef]
33. Wang, D.; Li, C. Self-Adaptive Dynamic Matrix Control for High-Speed Machining Servo Control. *Int. J. Adv. Manuf. Technol.* **2003**, *21*, 733–738. [CrossRef]
34. Kozlik, C.; Geringer, B.; Schirrer, A.; Jakubek, S. Dynamic matrix control applied to emission control of a diesel engine. *Int. J. Engine Res.* **2016**, *17*, 1–20. [CrossRef]
35. Wang, L.; Wang, X.; Sun, D.; Hao, H. Multi-objective optimization improved GA algorithm and fuzzy PID control of ATO system for train operation. *Commun. Comput. Inf. Sci.* **2017**, 13–22._2. [CrossRef]
36. Li, S.; Du, G. Online Parameter Tuning of Generalized Predictive Controller based on Fuzzy Satisfying Degree Function. *Control Decis.* **2002**, *17*, 852–863.
37. Lu, H.; Zhang, M.; Fei, Z.; Mao, K. Multi-Objective Energy Consumption Scheduling in Smart Grid Based on Tchebycheff Decomposition. *IEEE Trans. Smart Grid* **2015**, *6*, 2869–2883. [CrossRef]
38. Mirjalili, S.; Lewis, A. The Whale Optimization Algorithm. *Adv. Eng. Softw.* **2016**, *95*, 51–67. [CrossRef]
39. Gao, Y.; An, X.; Liu, J. A Particle Swarm Optimization Algorithm with Logarithm Decreasing Inertia Weight and Chaos Mutation. *Int. Conf. Comput. Intell. Secur.* **2008**, *1*, 61–65.
40. Miettinen, K. *Nonlinear Multiobjective Optimization*; Kluwer Academic Publishers: Norwell, MA, USA, 2017.
41. Zitzler, E.; Dimo, B.; Lothar, T. The hypervolume indicator revisited: On the design of Pareto-compliant indicators via weighted integration. In *International Conference on Evolutionary Multi-Criterion Optimization*; Springer: Berlin/Heidelberg, Germany, 2007.

42. Beume, N.; Rudolph, G. S-Metric Calculation by Considering Dominated Hypervolume as Klee's Measure Problem. *Evol. Comput.* **2009**, *17*, 477–492. [CrossRef]
43. Prasad, K.; Ranjan, R.; Sahoo, N.C.; Chaturvedi, A. Optimal Reconfiguration of Radial Distribution Systems Using a Fuzzy Mutated Genetic Algorithm. *IEEE Trans. Power Deliv.* **2005**, *20*, 1211–1213. [CrossRef]
44. Ariyarit, A.; Kanazaki, M. Multi-modal distribution crossover method based on two crossing segments bounded by selected parents applied to multi-objective design optimization. *J. Mech. Sci. Technol.* **2015**, *29*, 1443–1448. [CrossRef]
45. Herrera, F.; Lozano, M.; Pére, E.; Sánchez, A.M.; Villar, P. Multiple Crossover per Couple with Selection of the Two Best Offspring An Experimental Study with the BLX$-\alpha$ Crossover Operator for Real-Coded Genetic Algorithms. *Ibero-Am. Conf. Artif. Intell.* **2002**, 392–401._40. [CrossRef]
46. Peng, H.; Li, R.; Cao, L.; Li, L. Multiple Swarms Multi-Objective Particle Swarm Optimization Based on Decomposition. *Procedia Eng.* **2011**, *15*, 3371–3375.
47. Zhang, Q.; Li, H. MOEA/D: A Multiobjective Evolutionary Algorithm Based on Decomposition. *IEEE Trans. Evol. Comput.* **2008**, *11*, 712–731. [CrossRef]

sensors

MDPI

Article

Intelligent Driving Assistant Based on Road Accident Risk Map Analysis and Vehicle Telemetry

José Terán, Loraine Navarro, Christian G. Quintero M. * and Mauricio Pardo

Department of Electrical and Electronics Engineering, Universidad del Norte, Barranquilla 081007, Colombia; mjteran@uninorte.edu.co (J.T.); lorainen@uninorte.edu.co (L.N.); mpardo@uninorte.edu.co (M.P.)
* Correspondence: christianq@uninorte.edu.co

Received: 19 February 2020; Accepted: 18 March 2020; Published: 22 March 2020

Abstract: Through the application of intelligent systems in driver assistance systems, the experience of traveling by road has become much more comfortable and safe. In this sense, this paper then reports the development of an intelligent driving assistant, based on vehicle telemetry and road accident risk map analysis, whose responsibility is to alert the driver in order to avoid risky situations that may cause traffic accidents. In performance evaluations using real cars in a real environment, the on-board intelligent assistant reproduced real-time audio-visual alerts according to information obtained from both telemetry and road accident risk map analysis. As a result, an intelligent assistance agent based on fuzzy reasoning was obtained, which supported the driver correctly in real-time according to the telemetry data, the vehicle environment and the principles of secure driving practices and transportation regulation laws. Experimental results and conclusions emphasizing the advantages of the proposed intelligent driving assistant in the improvement of the driving task are presented.

Keywords: driving assistant; driving diagnosis; accident risk maps; driving safety

1. Introduction

Currently, road traffic accidents are one of the main causes of mortality according to the World Health Organization (WHO). Over 1.2 million people die every year on the world's roads and millions more live with serious injuries or long-term adverse consequences. Traffic accidents are the ninth most common cause of death worldwide, and the main cause of death for young-adults between 15 and 29 years old. In Colombia, about 8000 people are victims of those accidents, including drivers, passengers, and pedestrians [1]. As a result, governments have taken preventive measures to mitigate the number of accidents, implementing traffic regulation laws considering risk factors (driving under alcohol influence or psychoactive substances, lack of driving skills, speeding, reckless drives, among others). However, these laws are not enough to cover the risks caused by bad driving practices. Therefore, over the years, intelligent driving assistance systems (I-DAS) have been created to mitigate incidents because most of the accidents frequently rely on driver performance [2]. Thus, these systems assist the driver in different tasks, such as getting oriented, increasing fuel-consumption efficiency, and proving useful information about both vehicle tracking and motion. The goal is to maintain the driver attention in the road and with that improve safety on the roads [3–5].

Although the term driving assistant can be very broad (due to the diversity of active and passive applications), the present paper emphasizes the concept of an expert advisor assistant that accompanies the driver while is operating a vehicle. The proposed system comprises on-board telemetry data and a road accident risk map analysis integrated jointly through fuzzy logic in an abstraction of driving regulations and secure driving techniques for real-time intelligent driving assistance. As commented, a real I-DAS employs telemetry and in-vehicle data acquisition for proper driving assistance. Vehicular telemetry systems focus on those variables that can be taken directly from the vehicle, such as the variation of yaw angle, pedaling, steering, speed, acceleration, and engine conditions, among others. Depending on how they have been implemented, these telemetry systems can be acquisition systems designed specifically for a particular vehicle, on board diagnostics (OBD) systems, or standard systems (adaptable telemetry devices), which can be used for different vehicles [6–8]. Vehicle telemetry data acquisition is an indispensable step before driving analysis. This allows observing the behavior of each monitored signal and identifying the relationship between these variables and the maneuver performed by the driver.

At the same time, the road accident risk map analysis allows estimating the risk level of each road, so that it is possible to identify which roads are more dangerous than others. Consequently, it is most common to find road accident studies for a certain region or sector [9,10]. Here, the main goal is to identify the factors involved in road traffic accidents and analyze the relationship between them. The typical procedure in this type of work is to take a certain amount of accident data, analyze it through multiple statistical parameters; and then, organize and classify the accidents, roads, and sectors according to defined criteria (i.e., road accident risk maps). Thanks to the information obtained from the in-vehicle acquisition systems and the statistical studies of traffic accidents, different applications related to driving assistance can be developed. In this sense, the efficiency in the prevention of accidents depends significantly on the reliability of the collected data and the use of the appropriate methods of analysis. Today, the scientific community has taken the initiative in developing vehicular measurement devices, as well as tools that seek to assess driver performance. The aim is to establish all possible causes that lead to these accidents. Scientific research has mainly converged on the following topics: audiovisual records for the analysis of driving behavior (driver supervision, obstacle detection, proximity between vehicles, and pattern recognition for drowsiness, gaze tracking, road lanes or road signals) [11–16], driving modeling and driver behavior analysis (bad driving practices, careless or reckless driving) [17–22], driving style recognition (erratic driving assessment) [23,24].

In addition, intelligent systems applied in real vehicles have been motivated by the positive results obtained from simulators. The availability of resources and the multiple driving scenarios for experimentation (vehicle variables, types of road, types of vehicles, drivers, weather conditions, etc.) are the main advantages offered by them. In addition, they facilitate the study of new approaches in cases where implementing them directly in a real vehicle and a real environment can cause highly risky situations [19,25–29]. However, no matter how sophisticated the simulators are, they cannot provide all the physical aspects related to the actual driving process. Therefore, it is equally important to extend the studies to real scenarios [19].

In this context, the development of an intelligent driving assistant based on vehicular telemetry and road accident risk maps analysis in a real environment is proposed. The aim is to alert the driver by suggesting actions while the driving process is being carried out; therefore, careless situations can be avoided and with these, traffic accidents. While there have been studies that have developed intelligent driving assistants in real vehicles, as in the case of Yay et al. [5] and Kazuaki et al. [4], there are not many studies combining the road accident risk in the assistance process. This is particularly useful because, for example, some studies have shown that at higher speeds, the probability of road accidents becomes greater [1]. Nevertheless, it cannot be implied that speeding is a bad driving practice, because there are roads specially designed for high-speed driving and those not necessarily have high accident rates. At the same time, the acceleration and steering maneuvers are strongly related to the speed of the vehicle. However, what can be considered an aggressive steering maneuver at high speed may not be considered as such at low speed. This same concept, applied to the risk of accidents, implies that, for a road with a low accident rate, there would be no problem in performing maneuvers with a certain level of risk. This flexibility, when detecting and evaluating maneuvers, is one of the main goals in the development of this proposed intelligent driving assistant.

2. Method

The implementation of an intelligent driving assistant in a real environment looks for generating alerts to support the driver in real-time. Figure 1 shows the system components of the proposed assistant. This architecture considers the expert knowledge abstraction of driving laws and secure driving practices (IDA), along with the data related to the vehicle motion state (VTS) and its surroundings (RARM).

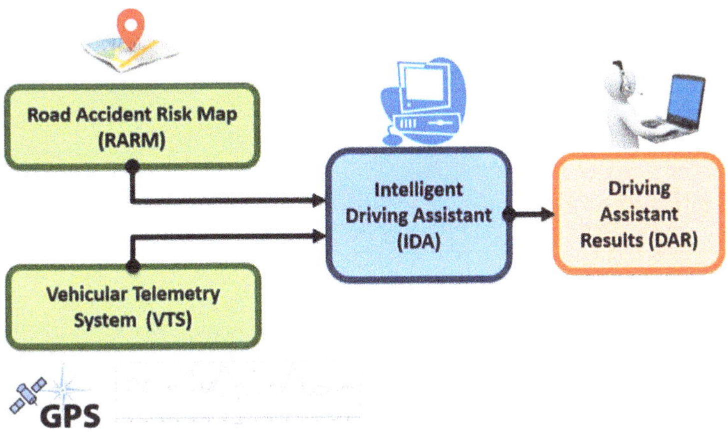

Figure 1. Intelligent Driving Assistant. System scheme.

2.1. *Road Accident Risk Map (RARM)*

The risk maps are usually constructed by considering the statistics related to accidents in a specific area or road. According to the manual "roadway safety information analysis" [30], the crash rate by roadway mileage is an assessment parameter used to identify which roads present a higher risk than others. Equation 1 allows the calculation of this rate as the road accident risk (RAR),

$$RAR = \frac{C}{N \times L}. \tag{1}$$

Building the RARM involves first calculating the RAR of each road (length L in kilometers), for a number of accidents (C) in a period of time (N). A database is produced with the resulting information, and is incorporated into the driving assistant. The concept of "tags" is used to classify the roads. Roads are categorized into three types according to their function as [31,32]: "highway", designed for high-speed traffic flow, "urban", designed to deliver traffic to highways (in-city roads), and "local" designed for low-speed traffic flow.

2.2. *Vehicular Telemetry System (VTS)*

The vehicular telemetry system consists of an electronic device easily adaptable to any car. It is responsible for monitoring variables related to the vehicle movement (driver maneuvers), and also for recording inside and outside the vehicle using video cameras to supervise the driving process. The data (signals) acquired from the VTS are geo-referenced via GPS standard protocol (NMEA standard). When the signals are acquired, they are continuously processed by the vehicle on-board computational system to provide real-time assistance. In this work, three risky maneuvers are categorized:

- Speeding: corresponds to those time instants in which the driver exceeded the speed limit allowed on the road, so it could be directly determined by observing the vehicle "speed" [km/h] (SPD). In Colombia, the maximum SPD is 40 km/h for local roads, 60 km/h for urban roads and 90 km/h for highways [33,34]. Figure 2a shows an example for local roads of how speed behaves in time.

- Bad pedaling: corresponds to the incorrect pedal handling (throttle and brake), when a driver performs abrupt or sudden acceleration/deceleration actions. It can be directly determined by observing the vehicle "longitudinal acceleration" [G] (LA). Best practitioners establish that the acceleration process should be progressive over time; which represents values between 0.1 G and 0.23 G in magnitude [35]. A maximum LA of 0.23 G for local roads, and 0.17 G for urban and highways is chosen. Figure 2b shows an example for local roads of how longitudinal acceleration behaves in time along with speed. It can be seen that, for abrupt deceleration maneuvers, the longitudinal acceleration perceives such variation, whereas for gradual decelerations it does not.
- Bad steering: corresponds to the incorrect handling of the steering shaft (also known as the steering column), when the driver performs abrupt or sudden turning maneuvers (changes in orientation). It can be determined by observing, the "heading" [°], the "yaw angle rate" [°/s] (YAR) or the "lateral acceleration" [G] of the vehicle, however, it is decided to use only the YAR as a measure of bad steering. A low YAR magnitude means normal use and not dangerous driving. YAR values close to 30 °/s are considered to be sharp fluctuations, and hence aggressive steering maneuvers [36,37]. In this study, a fixed maximum YAR value of 27.5°/s for local roads and 21°/s for urban and highways is used. Figure 2c shows a sample of these signals for local roads. The yaw angle rate presented the largest range of variation, being the most sensitive signal to detect steering maneuvers.

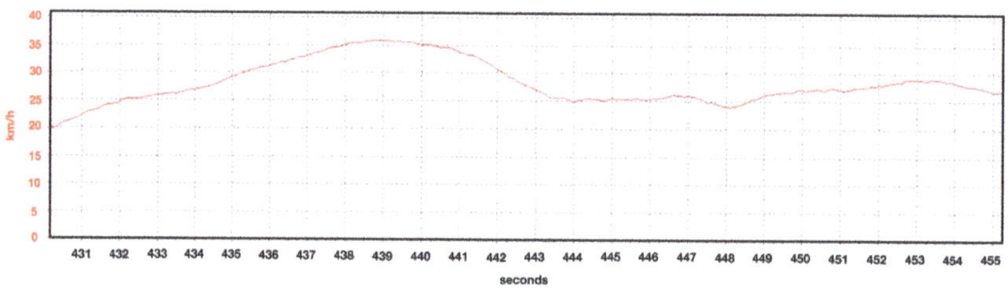

(a)

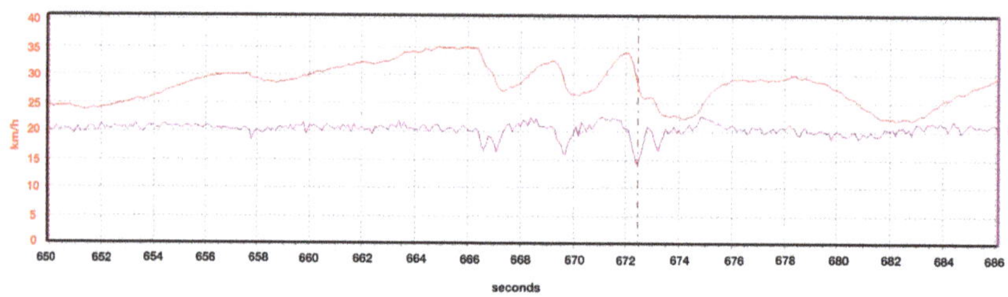

(b)

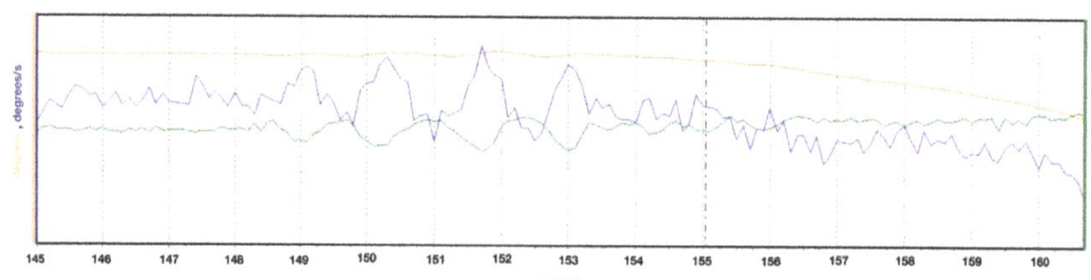

(c)

Figure 2. VTS signals: (**a**) Speed (red); (**b**) Speed (red) and Longitudinal Acceleration (purple); and (**c**) Heading (orange), Yaw Angle Rate (blue) and Lateral Acceleration (green).

The RACELOGIC VBOX, a device based on GPS, is a high precision performance meter used in real vehicles for automotive testing and provides considerable data related to the vehicle motion state [38]. Thus, this device is adopted as the VTS. Figure 3 shows the VBOX equipment.

Figure 3. Vehicular Telemetry System (VTS).

2.3. *Intelligent Driving Assistant (IDA)*

The proposed approach implements fuzzy logic in the design of the intelligent agent, due to the advantages provided related to adaptable evaluation criteria and similarity to human reasoning. Such an approach is particularly suitable since the goal is to develop an intelligent agent that approximates the assessment that an expert driver would make concerning driving maneuvers.

The abstraction of the expert knowledge, based on traffic regulation laws and secure driving practices, is mapped in a set of Mamdani fuzzy inference rules, along with the membership functions of the input and output variables. For signal analysis, it is decided to process the normalized value of the input variables, instead of the variable directly. This normalization is done based on the limits allowed for each signal. The main reason to employ this standardization is that the approach is to implement a single intelligent assistant adaptable to different roads by updating the maximum limit value for a specific road, instead of implementing an intelligent assistant for each road type. The agent processes four input signals. Figure 4 shows their membership functions. The defuzzifying process allows getting a magnitude (by centroid of the area) in order to determine which driving assistance is emitted. Figure 5 shows the three fuzzy variables (outputs) for risky maneuvers: speeding, bad pedaling, and bad steering. The same function set is used for each maneuver.

Then, the next step is the association of the previous fuzzy variables. The selection process of the fuzzy rule set is made through the schemes shown in Figure 6. The detection of risky maneuvers is based on the following associations: speeding depending on speed (SPD) and road accident risk (RAR); bad pedaling depending on longitudinal acceleration (LA), SPD, and RAR; and bad steering depending on yaw angle rate (YAR), SPD, and RAR. The structure of the diagrams in Figure 6 consists of locating the input variables on the X- and Y-axes, and the output variable on the Z-axis (out of the paper). The dynamics for the rule selection are presented below.

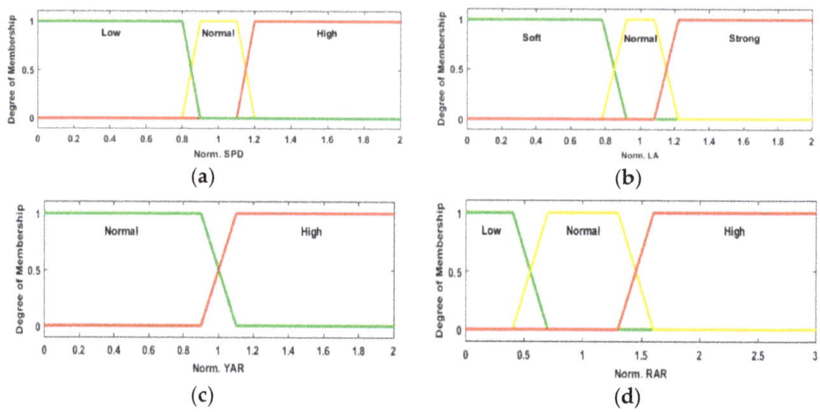

Figure 4. Proposed input variables membership functions: (**a**) Normalized Speed (SPD_{Norm}), (**b**) Normalized Long. Acceleration (LA_{Norm}), (**c**) Normalized Yaw Angle Rate (YAR_{Norm}) and (**d**) Normalized Road Accident Risk (RAR_{Norm}).

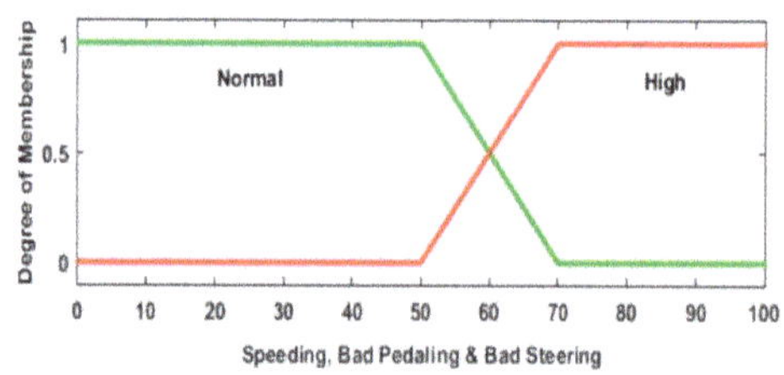

Figure 5. Proposed output variables membership functions: *Speeding*, *Bad Pedaling* and *Bad Steering*.

First, the rules to detect the speeding maneuver are selected (see Figure 6a). Speeding is defined in these circumstances: (1) "high" SPD and "low" RAR; (2) "high" SPD and "normal" RAR; (3) "normal" SPD and "high" RAR; and (4) "high" SPD and "high" RAR. The selected circumstances produce Rules 1, 2, 3 and 4, respectively.

Then, the rules to detect bad pedaling are selected (see Figure 6b–d for the different RARs). For cases with "low" RAR (Figure 6b), bad pedaling is defined in these circumstances: (5) "strong" LA and "low" SPD; and (6) "strong" LA and "normal" SPD. The selected circumstances produce Rules 5 and 6, respectively. Since cases with "high" SPD are already covered by Rule 1 (speeding detection), they are not considered. The same procedure is followed in the cases of "normal" RAR for Rules 7, 8 and 9 (Figure 6c), and "high" RAR for Rules 10 and 11 (Figure 6d).

Similarly, the rules to detect bad steering are shown in Figure 6e, f and g for the different RARs. For "Low" RAR (Figure 6e), bad steering is defined in these circumstances: (12) "high" YAR and "normal" SPD, giving rule 12. Since the case with "high" SPD is already covered by Rule 1 (speeding detection), it is not considered. The same procedure is followed for "normal" RAR (Figure 6f), giving Rules 13 and 14; and "high" RAR (Figure 6g), giving Rule 15.

According to these selection dynamics, the speeding maneuver is the most important to detect (hierarchically). The cases of bad pedaling and bad steering, which are not considered because of the speeding detection, are situations that are not contemplated due to the extreme risk involved, and it would most likely result in traffic accidents.

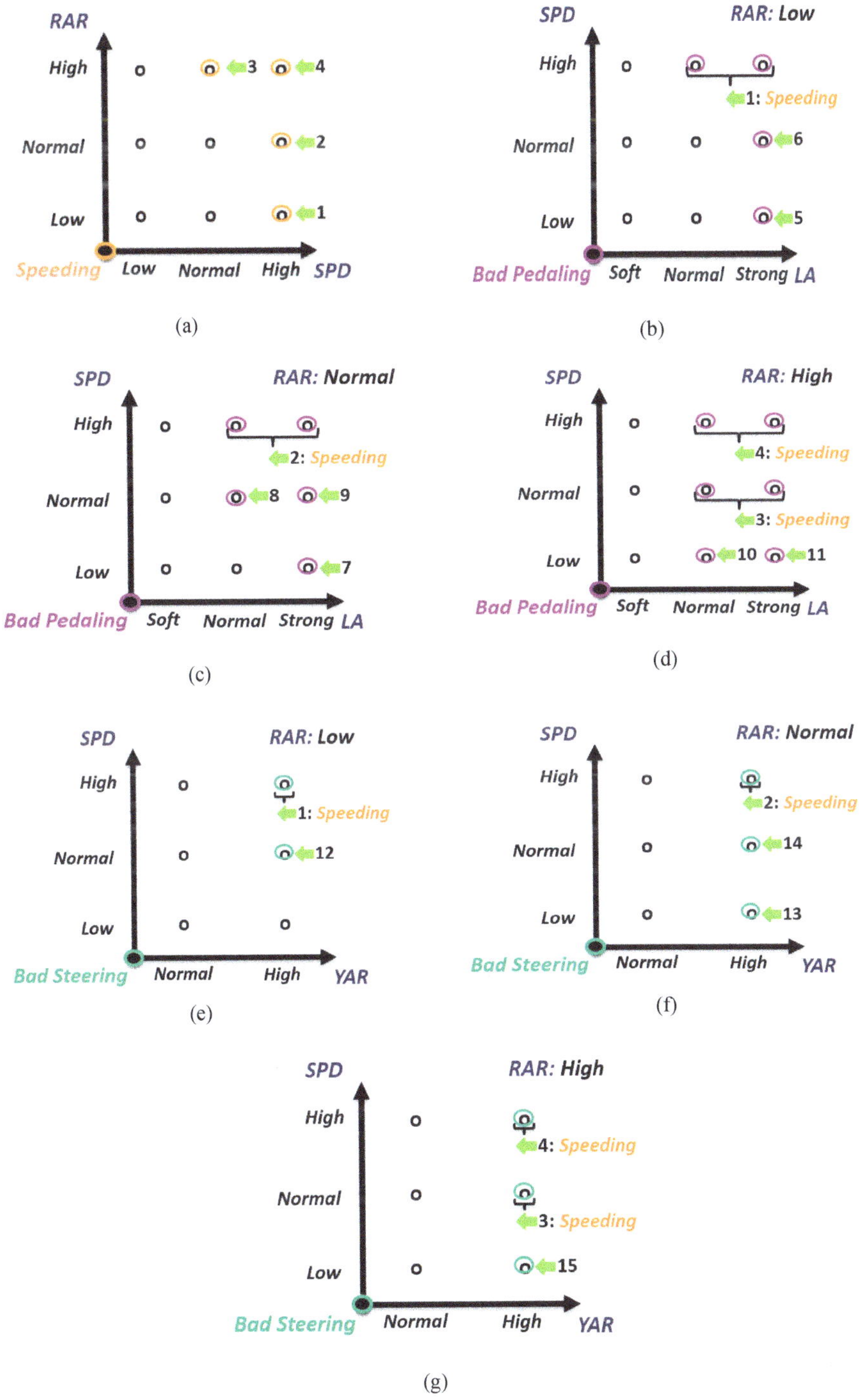

Figure 6. Fuzzy rule set selection process. (**a**) *Speeding* detection; (**b**), (**c**), (**d**) *Bad Pedaling* detection; and (**e**), (**f**), (**g**) *Bad Steering* detection.

Table 1 presents the proposed rules set for the fuzzy inference system.

Table 1. Proposed inference rules for the Intelligent Driving Assistant (IDA).

Rule	Input				Output		
	Norm SPD	Norm LA	Norm YAR	Norm RAR	Speeding	Bad Pedaling	Bad Steering
1	High			Low	High		
2	High			Normal	High		
3	Normal			High	High		
4	High			High	High		
5	Low	Strong		Low		High	
6	Normal	Strong		Low		High	
7	Low	Strong		Normal		High	
8	Normal	Normal		Normal		High	
9	Normal	Strong		Normal		High	
10	Low	Normal		High		High	
11	Low	Strong		High		High	
12	Normal		High	Low			High
13	Low		High	Normal			High
14	Normal		High	Normal			High
15	Low		High	High			High

2.4. Driving Assistant Results (DAR)

Figure 7 shows the variables dependency followed by the fuzzy rules (Table 1) and the driving assistant alerts. The driving advice is set for each detected maneuver, generating a total of three alerts. The established alerts are: "slow down", "brake slowly" and "soften steering". The system issued advice in the order shown in Figure 7, if more than one maneuver was detected.

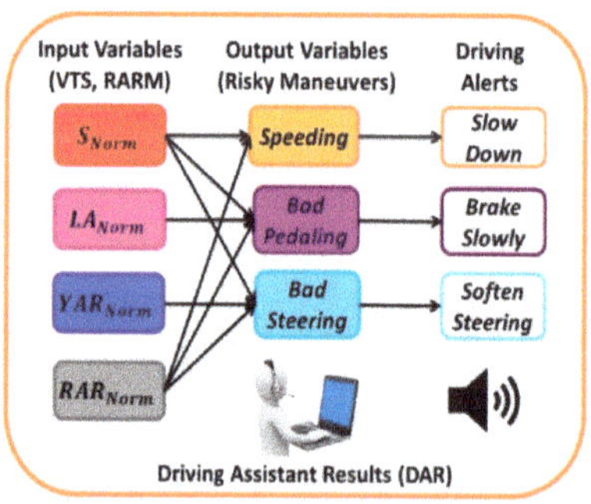

Figure 7. Driving assistant alerts.

2.5. Developed Computational System

The developed computational system allows two types of analysis to be performed: online analysis for evaluation in real-time, and offline analysis to evaluate driving registers in a post-driving analysis. Therefore, a software interface is developed to achieve friendly user interaction. Here, all the information handled by the assistant (vehicle motion state, environment, and intelligent assistance) is presented for both types of analysis. Figure 8 shows the interface designed for the computational system.

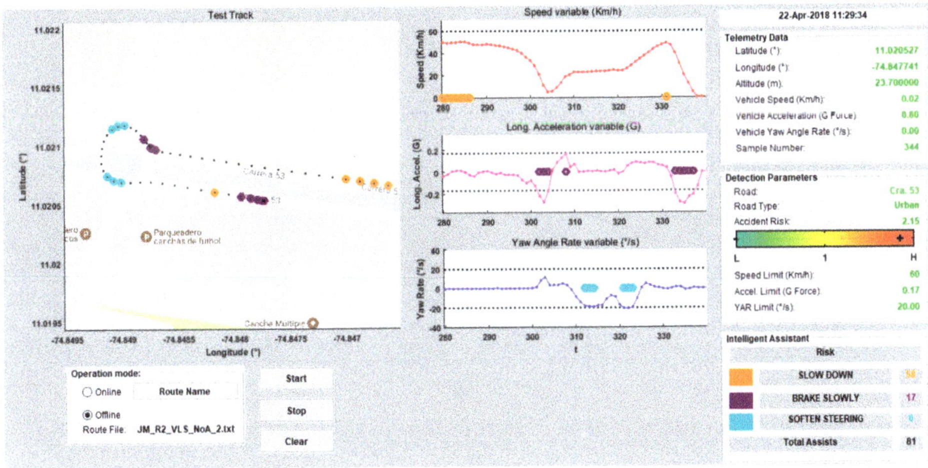

Figure 8. Software interface for Intelligent Driving Assistant (IDA).

3. Experimental Results and Discussion

The results presented in this section illustrate the proper functionality of the intelligent assistant in a real environment, as well as how the abstraction of expert knowledge is correctly applied in the driving maneuvers assessment. Three types of tests are made to quantitatively assess the performance of the IDA. The first one consists of a functionality test, the second one verified its effectiveness, and the third one seeks to evaluate the incidence of the use of the assistant in the driving process. It is important to note that if none of the considered risky maneuvers are detected; then, the IDA concludes an acceptable driving practice, and the system does not generate any alert (assistance).

Before presenting the test results, the parameters and road scenarios in which they are carried out are discussed.

3.1. Road Scenarios

The system developed in this study could adapt to driving environments and is also suitable for any vehicle. It is desirable to demonstrate the environmental adaptability with different road types, therefore a total of three routes are selected. Such selection is made based on the characteristics of the three types of road infrastructure: straight sections, curved sections, and intersections. Figure 9 shows the routes and Table 2, their characteristics. It is also important to demonstrate compatibility and system management with different types of vehicles. Hence, the vehicle weight and dimensions are used as classification criteria, and two types of vehicle are used:

- Light vehicles (V_1): Those with a weight of less than 1000 kg. This category includes short and small vehicles of the hatchback type.
- Heavy vehicles (V_2): Those with a weight greater than 1000 kg. This category includes long and large vehicles of sedan type.

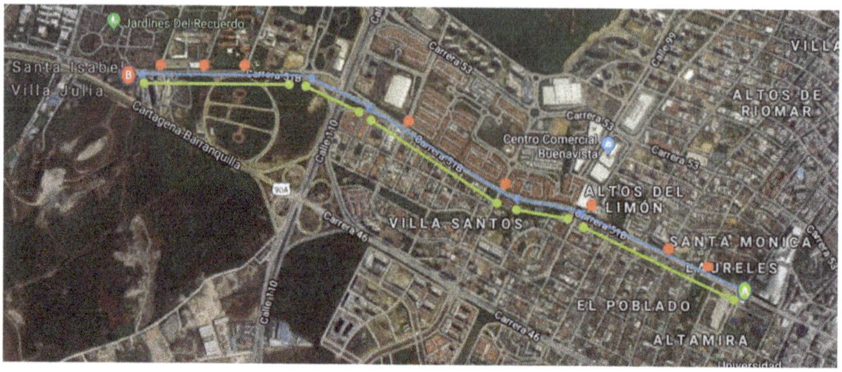

(a)

(b)

(c)

Figure 9. Types of Road (A: Start point, B: End point): (**a**) Route 1, (**b**) Route 2 and (**c**) Route 3.

Table 2. Road details: Route 1, Route 2 and Route 3.

Route 1	
Approx. Distance	3.5 km
Approx. Number of Straights	5 (long straights)
Approx. Number of Curves	0
Approx. Number of Intersections	8
Route 2	
Approx. Distance	3 km
Approx. Number of Straights	3
Approx. Number of Curves	5
Approx. Number of Intersections	2
Route 3	
Approx. Distance	1.76 km
Approx. Number of Straights	9
Approx. Number of Curves	6
Approx. Number of Intersections	9

3.2. Functionality Test

The purpose of the functionality test is assuring that the system detects risky maneuvers and issues alerts coherently, according to the behavior of the monitored variables (SPD, LA, YAR) and the environment (RAR). The following example illustrates the system operation on one of the routes (Figures 10 and 11).

The assessment parameters remain constant throughout the trip, as Route 2 is always on the same road section. Figure 10 shows the end of Route 2, and Figure 11a presents the assessment parameters for this route. The alerts given are three speeding alerts (orange marks), four bad pedaling alerts (purple marks), and one bad steering alert (cyan marks). The vehicle is on a road with a "high" RAR in each of these cases. speeding alerts are given when a speed of 49.47 km/h is reached (which corresponds to a "low-normal" SPD level) activating fuzzy Rule 3 (Table 1). For bad pedaling, alerts are given if accelerations up to −0.15 G (braking) are reached ("normal" LA) at speeds of approximately 28 km/h ("low" SPD) activating Rule 10. Bad steering alerts are given when a turning of −19°/s (left turn, "normal-high" YAR) is reached at a speed of 20 km/h ("low" SPD) activating Rule 15. Figure 11b shows the total assists/risky maneuvers obtained on Route 2.

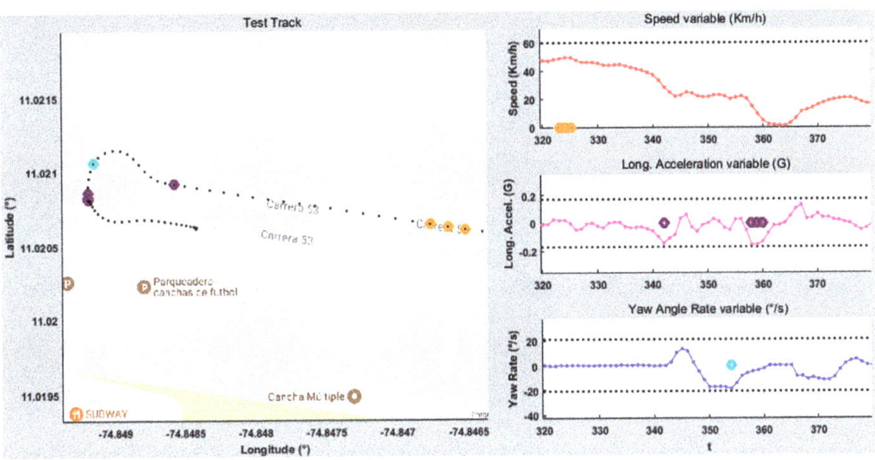

Figure 10. End of *Route* 2 for driver D_1 assistance results.

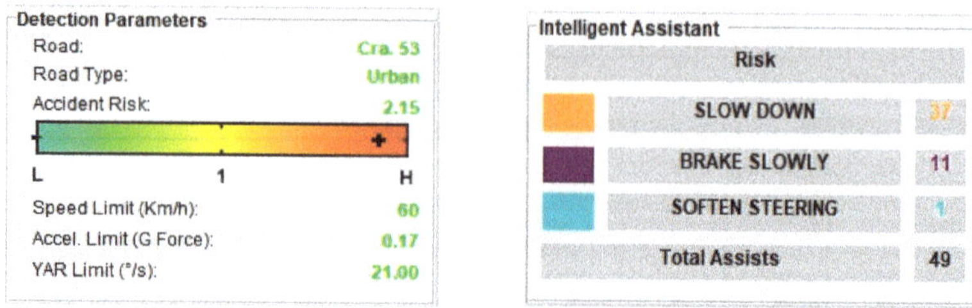

(a) (b)

Figure 11. Route 2: (**a**) Assessment parameters, (**b**) Assistance results.

This rational behavior is consistently exhibited on the other routes, suggesting that the intelligent assistant works properly and fulfills its task of assisting the driver real-time during the driving process.

3.3. *Efficiency Test*

An efficiency test is carried out to determine the overall effectiveness of the driving assistant. The methodology consists of taking a set of driving samples (three routes, two vehicles, eight drivers, and three repetitions per driver in each scenario, making a total of 144 experiments), and applying to each one the analysis made by the IDA. The objective is to compare the number of correct alerts versus the total number of alerts issued. The average correct alerts provided by the driving assistant is adopted as the assessment metric.

The results obtained in this test are organized according to the type of vehicle. Figure 12 shows the average correct alerts obtained for each driver on all routes for vehicle V_1. Table 3 presents the percentage of correct alerts obtained for each driver from Figure 12, as well as those obtained on each route separately. Similar results are shown for vehicle V_2 in Figure 13 and Table 4, and for the global case in Figure 14 and Table 5.

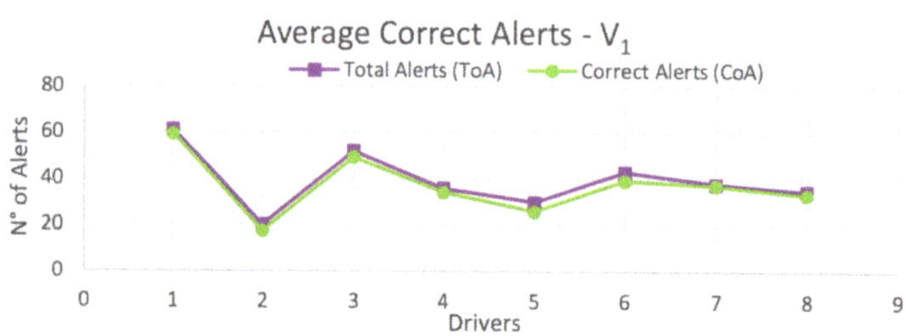

Figure 12. Average total correct alerts on all routes for V_1 (Light Vehicle).

Table 3. Percentage of correct alerts for V_1 (Light Vehicle): Route 1, Route 2, Route 3 and all routes.

Percentage of Correct Alerts [%]									
Drivers Routes	**D_1**	**D_2**	**D_3**	**D_4**	**D_5**	**D_6**	**D_7**	**D_8**	**Avg.**
Route 1	77.8	66.7	75.3	88.9	66.7	88.9	88.9	66.7	77.5
Route 2	97.8	84.6	91.7	94.4	79.2	92.6	100	92.6	91.6
Route 3	100	100	98.9	100	100	83.3	97.6	100	97.5
All Routes	97.3	86.7	94.9	95.4	86.7	91.5	98.2	95.2	93.2

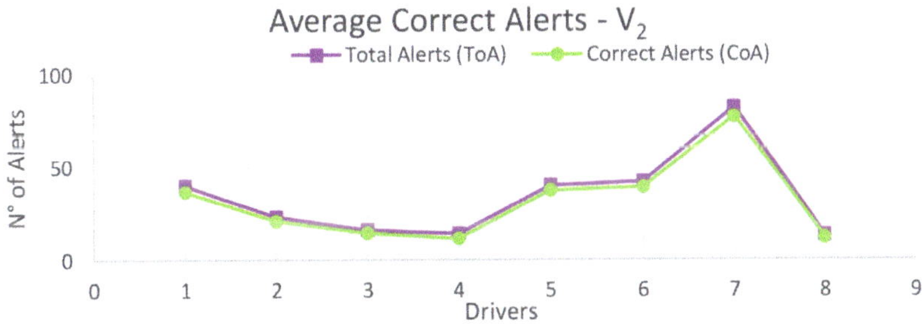

Figure 13. Average total correct alerts on all routes for V_2 (*Heavy Vehicle*).

Table 4. Percentage of correct alerts for V_2 (Heavy Vehicle): Route 1, Route 2, Route 3 and all routes.

Percentage of Correct Alerts [%]									
Drivers Routes	**D_1**	**D_2**	**D_3**	**D_4**	**D_5**	**D_6**	**D_7**	**D_8**	**Avg.**
Route 1	50.0	55.6	66.7	66.7	83.3	83.3	86.7	100	74.0
Route 2	86.3	83.3	87.5	83.3	92.3	87.3	90.2	83.3	86.7
Route 3	100	100	95.2	83.3	97.2	100	97.6	100	96.7
All Routes	91.7	89.9	89.6	81.0	93.3	92.9	93.9	89.7	90.2

Figure 14. Average total correct alerts on all routes for all vehicles (V_1 & V_2).

Table 5. Percentage of correct alerts on all routes for all vehicles (V_1 & V_2).

Percentage of Correct Alerts [%]										
Drivers	**D_1**	**D_2**	**D_3**	**D_4**	**D_5**	**D_6**	**D_7**	**D_8**	**Avg.**	**Std. Dev.**
All Routes	95.0	88.4	93.5	91.2	90.6	91.3	95.2	93.8	92.4	2.4

In the case of V_1, high percentages of correct alerts (77.5%, 91.6%, 97.5%) are obtained for Routes 1–3, respectively; and a total of 93.2% for all drivers on all routes (Table 3). In the case of V_2, the percentages of correct alerts are 74%, 86.7%, 96.7% for Routes 1–3, respectively; and a total of 90.2% for all drivers on all routes (Table 4). This results in a global average of 92.4% of correct alerts (Table 5). In most cases, and also in the global case, a high percentage of correct alerts being issued is observed. Thus, it is suggested a high efficiency in the detection of risky maneuvers, and consequently, in providing driving assistance.

3.4. Driving Performance Test

The purpose of this test is to evaluate the incidence of using the assistant in the driving process. The methodology consists of taking driving journeys (the same as those used in the efficiency test), but this time for two cases: with and without the assistant (making a total of 288 experiments). The evaluation metric is the average decrement in the number of risky maneuvers. First, the routes are driven with the assistance system in operation but disabling the audio alerts. Hence, the risky maneuvers detected are still recorded. Later, the same route is driven again but with the audio alerts on. It should be noted that, although the assistant is responsible for issuing driving advice, it is the driver responsibility to follow or ignore the assistance. The goal is to compare the number of risky maneuvers/alerts obtained with and without assistance.

The results obtained from this test are presented in Figures 15 and 16, and Tables 6 and 7, according to the type of vehicle. Figure 15 shows the average total number of risky maneuvers made by each driver on all routes for vehicle V_1. Table 6 presents the decrement in the percentage of risky maneuvers for each driver, as well as those obtained on each route separately. Analogous data for vehicle V_2 are shown in Figure 16 and Table 7, and for the global case in Figure 17 and Table 8.

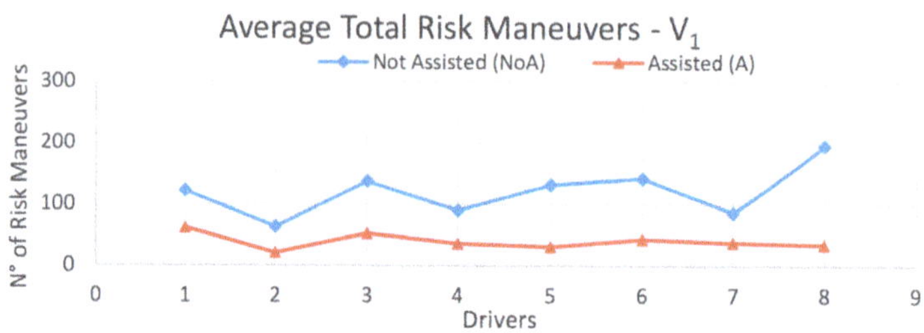

Figure 15. Average total risk maneuvers on all routes for V_1 (Light Vehicle).

Table 6. Decrement percentage of risky maneuvers for V_1 (Light Vehicle): Route 1, Route 2, Route 3 and all routes.

	Percentage of Correct Alerts [%]								
Drivers Routes	D_1	D_2	D_3	D_4	D_5	D_6	D_7	D_8	**Avg.**
Route 1	33.3	40.0	33.3	30.0	78.6	11.1	11.1	42.9	32.0
Route 2	38.2	15.2	77.6	55.8	77.3	52.8	43.1	87.4	55.9
Route 3	71.4	88.6	32.3	71.6	76.3	93.6	69.3	69.9	71.6
All Routes	49.6	68.1	62.3	60.4	77.0	70.0	56.2	82.2	65.7

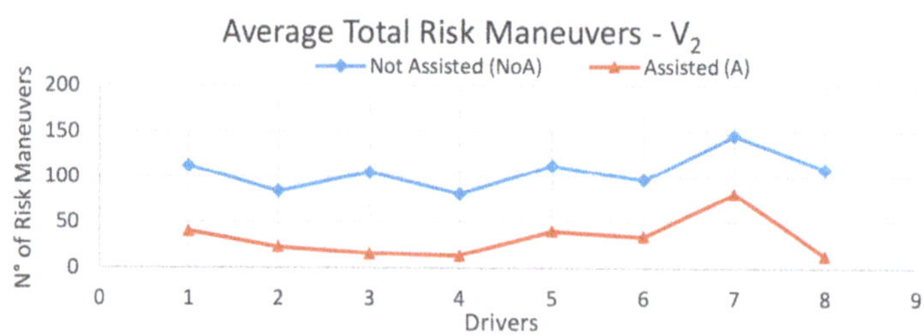

Figure 16. Average total risky maneuvers on all routes for V_2 (*Heavy Vehicle*).

Table 7. Decrement percentage of risky maneuvers for V_2 (Heavy Vehicle): Route 1, Route 2, Route 3 and all routes.

	Percentage of Correct Alerts [%]								
Drivers Routes	D_1	D_2	D_3	D_4	D_5	D_6	D_7	D_8	**Avg.**
Route 1	50.0	0.0	0.0	12.5	80.0	60.0	67.4	75.0	43.1
Route 2	70.0	85.7	86.1	84.7	50.3	62.0	59.1	87.9	73.2
Route 3	59.1	66.1	85.0	87.4	76.0	69.9	8.6	87.8	67.5
All Routes	64.3	73.0	85.1	83.3	64.2	64.9	43.9	87.7	70.8

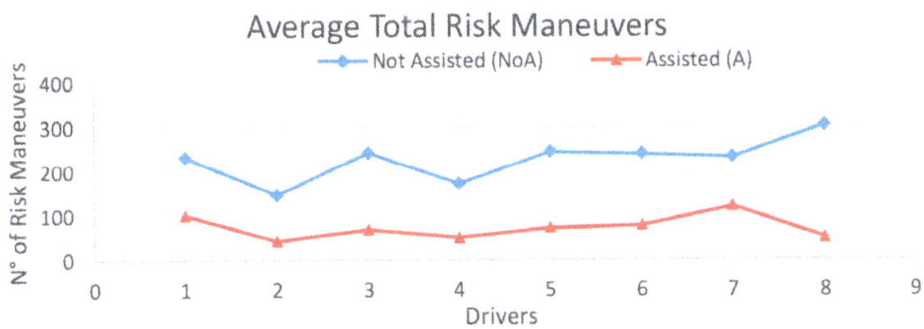

Figure 17. Average total risky maneuvers on all routes for all vehicles (V_1 & V_2).

Table 8. Decrease percentage of risky maneuvers on all routes for all vehicles (V_1 & V_2).

	Percentage of Correct Alerts [%]									
Drivers	D_1	D_2	D_3	D_4	D_5	D_6	D_7	D_8	**Avg.**	**Std. Dev.**
All Routes	56.7	70.9	72.2	71.3	71.1	67.9	48.5	84.2	67.8	10.8

In the case of vehicle V_1, the percentage decrement in risky maneuvers is 35%, 55.9%, and 71.6% for Routes 1–3, respectively; and a total of 65.7% for all routes (Table 6). The corresponding results for vehicle V_2 are 43.1%, 73.2%, and 67.5%, for Routes 1–3, respectively: and 70.8% for all routes (Table 7). The decrement in risky maneuvers for all vehicles over all routes is 67.8% (Table 8). In most cases, the number of risky maneuvers decreases after driving the routes with the IDA assistance enabled. This suggests that there is a positive influence of using the assistant.

4. Statistical Validation

Statistical analysis is carried out to verify that the results obtained from the selected samples are representative of other drivers in the population. In the current work, this statistical validation is carried out through hypothesis testing, in which a null hypothesis (H_0) and an alternative hypothesis (H_1) are proposed. According to the structure of this type of test, the term H_0 corresponds to the hypothesis tested with the goal to be rejected leading to the acceptance of hypothesis H_1. H_1 usually corresponds to the question or theory that is desired and opposite to H_0 [35].

In the case of the efficiency test, the following hypotheses are established:

- Null Hypothesis (H_0): The average percentage of correct alerts is less than or equal to 90% ($\mu \leq 0.90$).
- Alternative Hypothesis (H_1): The average percentage of correct alerts is greater than 90% ($\mu > 0.90$).

It is decided to use the right tail of the t-student distribution to determine the veracity of H_0 with a confidence level of 95% ($1 - \alpha = 0.95$), considering the hypothesis and the number of samples (repetitions per driver) in this test. Equation (2) describes the calculation of the statistical value t_α to evaluate H_0 [35]:

$$t_\alpha = \frac{\overline{X} - \mu}{S \ / \ \sqrt{N}} \tag{2}$$

where t_α is the distribution value for a certain level of significance (α), and a certain degree of freedom ($v = N - 1$), $\overline{X}$ is the sample mean, μ is the population mean, S is the sample standard deviation, and N is the number of samples. Rearranging Equation (2) gives Equation (3):

$$\mu = \overline{X} - \left(t_\alpha * S \ / \ \sqrt{N}\right) \tag{3}$$

$$\mu = 0.924 - \left(1.714 * 0.024 \ / \ \sqrt{24}\right) = 0.92 \tag{4}$$

Using the following values obtained from the experimental results ($N = 24$, $\alpha = 0.05$, $v = N - 1 = 23$, $t_{0.05} = 1.714$, $\overline{X} = 0.924$ and $S = 0.024$), Equation (4) produce a $\mu > 0.9$ (Table 5). Thus, the rejection of H_0 is implied, and therefore the veracity of H_1, allowing the validation of these results for other drivers.

For the driving performance test, the following hypotheses are established:

- Null Hypothesis (H_0): The average percentage decrease in risky maneuvers is less than or equal to 50% ($\mu \leq 0.5$).
- Alternative Hypothesis (H_1): The average percentage decrease in risky maneuvers is greater than 50% ($\mu > 0.5$).

Similarly, as in the previous analysis, Equation (3) is used to verify H_0 with a confidence level of 95% ($1 - \alpha = 0.95$). Using the following values obtained from the experimental results ($N = 24$, $\alpha = 0.05$, $v = N - 1 = 23$, $t_{0.05} = 1.714$, $\overline{X} = 0.678$ and $S = 0.108$) (Table 8), Equation (5) becomes:

$$\mu = 0.678 - \left(1.714 * 0.108 \ / \ \sqrt{24}\right) = 0.64 \tag{5}$$

As $\mu > 0.5$ implying the rejection of H_0, and therefore the veracity of H_1. This allows the validation of these results for other drivers.

5. Conclusions

In this work, an intelligent driving assistant based on road accident risk map analysis and vehicle telemetry is implemented in a real environment. The results demonstrate the relevance of the design and implementation of vehicle safety systems within the intelligent transport system (ITS) framework. An intelligent agent capable of assisting the driver in situations of driver carelessness is developed. The present fuzzy-logic-based IDA becomes a useful support element for the driver during the driving process. This IDA provides suggestions when risky maneuvers are detected for different road scenarios. The driving assistant uses an adaptive evaluation criterion that allows proper detection of risky maneuvers, based on information related to the VTS, the RARM, and supervision by video recording.

The performance of the system is evaluated in three tests. The first consists of a functionality test, which verifies that the assistant issued coherent driving alerts when the user is at risk by performing an established risky maneuver in real-time (system proper operation). The second consists of an efficiency test, which compares the number of correct alerts versus the number of total alerts issued. The system achieves an efficiency above 90% for correct alerts for the test routes. This suggests an optimal behavior within the ITS framework. Finally, the third test evaluates the influence of the assistant on driver performance comparing the number of risk maneuvers performed with the driving assistance enabled and disabled. A decrement in the number of risky maneuvers above 50% is observed in most cases, considering the variables, factors, road scenarios, and driver behavior. This demonstrates a positive influence of using the assistant and shows the importance of driver behavior in improving road safety [4,5]. In addition, statistical validation (confidence level of 95%) is carried out to verify that the results obtained for the driving study in a real environment, are equally valid for other drivers (population) [39].

The adaptability of the intelligent assistant for different types of vehicles and different types of roads is one of the most important aspects to consider for real environments. To address this, two vehicles are used and three routes are selected for the experiments. The vehicle dimensions and weights and the route characteristics (straights, curves, and intersections) are considered. The consistent behavior of the results shows that the adaptability requirement is met due to the telemetry system characteristics (an easily adaptable device for any car) and the constant updating of the assessment parameters according to the vehicle location (adaptive evaluation criteria for each type of road). In all experiments, it is assumed that the driver behaved rationally and followed the recommendations provided by the assistant.

As an application in a real environment, based on the analysis of the results, it is possible to establish future improvements and propose future studies to continue this line of research. One of the most important aspects to study is the elimination of false warnings. Although a high efficiency in issuing correct alerts is achieved, there is still the possibility of eliminating false alerts entirely. The false alerts occur because of the inaccuracy of the data received from the VTS (due to the interruption of the satellite signal by external factors). Possible countermeasures are to opt for an Inertial Navigation System, as a support for the GPS signal (INS GPS integration), which enables tracking of the vehicle to be maintained during those short periods of time in which the satellite signal is disabled (hardware alternative) [40], or opt for a statistical treatment of the GPS signal (software alternative), to mitigate the lack of precision at those moments of time [41].

Another aspect to improve is the development of a higher resolution accident risk map, that is, not only having the risk of an accident by road section, but by sub-section or intersection, therefore increasing the adaptability of the assistant for the different road scenarios [30]. Similarly, an alternative method to study the influence of the assistant on the driver, it is proposed to use other assessment criteria, from that used in this study (number of risk maneuvers performed), to evaluate driving performance.

It is expected that, in the future, as new technologies emerge, these auxiliary schemes will continue in process of improvement and that most commercial cars (even connected autonomous vehicles) will have such intelligent driving assistance systems implemented.

Finally, the exploration of other computational intelligence techniques to develop new assessment approaches for intelligent driving assistants in the framework of ITS is desirable.

Author Contributions: Formal analysis, J.T.; Investigation, J.T.; L.N. and C.G.Q.M.; Methodology, J.T.; C.G.Q.M. and M.P.; Software, J.T. and C.G.Q.M.; Supervision, M.P.; Validation, J.T.; L.N.; C.G.Q.M. and M.P.; Writing—original draft, J.T.; L.N.; C.G.Q.M. and M.P.; Writing—review & editing, L.N.; C.G.Q.M. and M.P. All authors have read and agreed to the published version of the manuscript.

Funding: This research received no external funding.

Acknowledgments: This work was supported by Universidad del Norte, Barranquilla-Colombia.

Conflicts of Interest: The authors declare that there are no conflicts of interest regarding the publication of this paper.

References

1. World Health Organization (WHO). *Global Status Report on Road Safety 2015*; World Health Organization: Geneva, Switzerland, 2015; p. 12.
2. Verband der Automobilindustrie (VDA). *From Driver Assistance Systems to Automated Driving*; VDA Magazine–Automation: Berlin, Germany, 2015; p. 16.
3. European Field Operational Test (euroFOT). *euroFOT Study Demonstrates How Driver Assistance Systems Can Increase Safety and Fuel Efficiency Across Europe*; European Field Operational Test (euroFOT): Brussels, Belgium, 2012.

4. Goshi, K.; Hayashi, M.; Matsunaga, K. Safe Driving Education System ASSIST—Education Anywhere, Anytime. In Proceedings of the 2012 9th International Conference on Ubiquitous Intelligence and Computing and 9th International Conference on Autonomic and Trusted Computing, Institute of Electrical and Electronics Engineers (IEEE), Fukuoka, Japan, 4–7 September 2012; pp. 402–404. [CrossRef]
5. Yay, E.; Martínez, N. SEEDrive—An Adaptive and Rule Based Driving System. In Proceedings of the 2013 9th International Conference on Intelligent Environments, Athens, Greece, 16–17 July 2013; pp. 262–265. [CrossRef]
6. Khan, A.; Mishra, R. GPS-GSM Based Tracking System. *Int. J. Eng. Trends Technol.* **2012**, *3*, 161–164.
7. Singh, J.; Rajendra, B.R.; Swetha, K.S. Cost Effective Real-Time Ambulance Tracking System Prototype. *Int. J. Comput. Sci. Inf. Technol. Res.* **2015**, *3*, 58–61.
8. Ya-Wen, H.; Jau-Woei, P.; Zong-Han, W. Design and Implementation of an Intelligent Road Detection System with Multisensor Integration. *Int. Conf. Mach. Learn. Cybern.* **2016**, *1*, 219–225. [CrossRef]
9. Al-Khateeb, G. Analysis of Accident Data and Evaluation of Leading Causes for Traffic Accidents in Jordan. *Jordan J. Civ. Eng.* **2010**, *4*, 76–94.
10. Minnesota Department of Transportation (MnDOT). *Accident Analysis of Significant Crash Rates for Low to Very Low Volume Roadways in 10 Minnesota Counties*; Report No: MN/RC–2004-22; Minnesota Department of Transportation (MnDOT): St. Paul, MI, USA, 2004.
11. Bagus, G.; Igi, A.; Teguh, B.A. A Review on Driver Drowsiness Based on Image, Bio-Signal, and Driver Behavior. In Proceedings of the 3rd International Conference on Science and Technology—Computer, Yogyakarta, Indonesia, 11–12 July 2017; pp. 70–75. [CrossRef]
12. Bounini, F.; Gingras, D.; Lapointe, V.; Pollart, H. Autonomous Vehicle and Real Time Road Lanes Detection and Tracking. In Proceedings of the IEEE Vehicle Power and Propulsion Conference (VPPC), Montreal, QC, Canada, 19–22 October 2015. [CrossRef]
13. Prajapati, N.; Bhatt, P. Driver Drowsiness Detection with Audio-Visual Warning. *Int. J. Innov. Res. Sci. Technol.* **2016**, *3*, 294–300.
14. Surendra, M.; Bhavana, A.; Pooja, S.; Ashish, M. Eye Tracking Based Driver Drowsiness Monitoring and Warning System. *Int. J. Tech. Res. Appl.* **2015**, *3*, 190–194.
15. Yang, Y.; Luo, H.; Xu, H.; Wu, F. Towards Real-Time Traffic Sign Detection and Classification. *IEEE Trans. Intell. Transp. Syst.* **2016**, *17*, 2022–2031. [CrossRef]
16. Khan, M.Q.; Lee, S. Gaze and Eye Tracking: Techniques and Applications in ADAS. *Sensors* **2019**, *19*, 5540. [CrossRef] [PubMed]
17. Jain, C.; Abhishek, R.; Pawar, R. Intelligent Driver Assistant System. *Int. J. Eng. Technol.* **2014**, *3*, 320–326.
18. Cuervo, A.; Quintero, C.; Premachandra, C. Intelligent Driving Diagnosis Based on a Fuzzy Logic Approach in a Real Environment Implementation. In Proceedings of the IEEE Intelligent Vehicles Symposium (IV), Dearborn, MI, USA, 8–11 June 2014; pp. 102–107. [CrossRef]
19. Oviedo, O.; Haque, M.; King, M.; Washington, S. Effects of road infrastructure and traffic complexity in speed adaptation behavior of distracted drivers. *Accid. Anal. Prev.* **2017**, *101*, 67–77. [CrossRef] [PubMed]
20. Rakhshan, A.; Ray, E.; Pishro-Nik, H. Real-Time Estimation of the Distribution of Brake Response Times for an Individual Driver Using Vehicular Ad Hoc Network. In Proceedings of the 2014 IEEE Intelligent Vehicles Symposium Proceedings, Dearborn, MI, USA, 8–11 June 2014.
21. Scott, P.B.; Oviedo, T.O. Young driver risky behavior and predictors of crash risk in Australia, New Zealand and Colombia: Same but different? *Accid. Anal. Prev.* **2017**, *99*, 30–38. [CrossRef] [PubMed]
22. Zhao, M.; Käthner, D.; Jipp, M.; Söffker, D.; Lemmer, K. Modeling Driver Behavior at Roundabouts: Results from a Field Study. In Proceedings of the IEEE Intelligent Vehicle Symposium (IV), Los Angeles, CA, USA, 11–14 June 2017; pp. 908–913. [CrossRef]
23. Khan, M.Q.; Lee, S. A comprehensive survey of driving monitoring and assistance systems. *Sensors* **2019**, *19*, 2574. [CrossRef] [PubMed]
24. Mata-Carballeira, Ó.; Gutiérrez-Zaballa, J.; del Campo, I.; Martínez, V. An FPGA-Based Neuro-Fuzzy Sensor for Personalized Driving Assistance. *Sensors* **2019**, *19*, 4011. [CrossRef] [PubMed]
25. Quintero, C.; Cuervo, A. Intelligent driving assistant based on accident risk maps analysis and intelligent driving diagnosis. In Proceedings of the IEEE Intelligent Vehicle Symposium (IV), Los Angeles, CA, USA, 11–14 June 2017; pp. 914–919. [CrossRef]

26. Chang, S.; Lin, C.; Fung, C.; Hwang, J.; Doong, J. Driving performance assessment: Effects of traffic accident location and alarm content. *Accid. Anal. Prev.* **2008**, *40*, 1637–1643. [CrossRef] [PubMed]
27. Etzioni, S.; Erev, I.; Ishaq, R.; Elias, W.; Shiftan, Y. Self-monitoring of driving speed. *Accid. Anal. Prev.* **2017**, *106*, 76–81. [CrossRef] [PubMed]
28. Hajiseyedjavadi, F.; Zhang, T.; Agrawal, R.; Knodler, M.; Fisher, D.; Samuel, S. Effectiveness of visual warnings on young drivers hazard anticipation and hazard mitigation abilities. *Accid. Anal. Prev.* **2018**, *116*, 41–52. [CrossRef] [PubMed]
29. Koglbauer, I.; Hulzinger, J.; Eichberger, A.; Lex, C. Driver's Interaction with Adaptive Cruise Control on Dry and Snowy Roads with Various Tire-Road Grip Potentials. *J. Adv. Transp.* **2017**, *2017*, 5496837. [CrossRef]
30. Federal Highway Administration (FHWA). *Roadway Safety Information Analysis: A Manual for Local Rural Road Owners*; U.S. Department of Transportation (USDOT): Washington, DC, USA, 2011; pp. 13–24.
31. Federal Highway Administration (FHWA). *Highway Functional Classification Concepts, Criteria and Procedures*; U.S. Department of Transportation (USDOT): Washington, DC, USA, 2013; p. 11.
32. Ministerio de Transporte (MINTRANSPORTE). *Decreto Número 000015 de 2011*; Ministerio de Transporte (MINTRANSPORTE): Bogotá, Colombia, 2011; p. 2.
33. Federal Highway Administration (FHWA). *Methods and Practices for Setting Speed Limits: An Information Report*; U.S. Department of Transportation (USDOT): Washington, DC, USA, 2012; p. 15.
34. Mehar, A.; Chandra, S.; Velmurugan, S. Speed and Acceleration Characteristics of Different Types of Vehicles on Multi-Lane Highways. *Eur. Transp.* **2013**, *55*, 1–12.
35. Liu, X.; Liao, S.; Rong, P. Vehicle Stability Control Based on RBF Adaptive Terminal Sliding Mode Controller. In Proceedings of the IEEE MIT Undergraduate Research Technology Conference (URTC), Cambridge, MA, USA, 3–5 November 2017; pp. 1–5. [CrossRef]
36. Pan, S.; Zhou, H. An Adaptive Fuzzy PID Control Strategy for Vehicle Yaw Stability. In Proceedings of the IEEE 2nd Information Technology, Networking, Electronic and Automation Control Conference (ITNEC), Chengdu, China, 15–17 December 2017; pp. 642–646. [CrossRef]
37. RACELOGIC. *VBOX Mini User Guide*; RACELOGIC: Buckingham, UK, 2014.
38. Wapole, R.; Myers, R.; Myers, S.; Ye, K. *Probabilidad Y Estadística Para Ingeniería Y Ciencias*, 9th ed.; PEARSON EDUCACIÓN: Naucalpan de Juárez, México, 2012; pp. 561–564.
39. Rahman, M.T.; Karamat, T.; Givigi, S.; Noureldin, A. Improving Multisensor Positioning of Land Vehicles with Integrated Visual Odometry for Next-Generation Self-Driving Cars. *J. Adv. Transp.* **2018**, *2018*, 6513970. [CrossRef]
40. Garcia, J.; Zhou, C. Improving GPS Precision and Processing Time using Parallel and Reduced-Length Wiener Filters. *J. Telecommun.* **2010**, *2*, 91–98.
41. Martins, V.; Hoiti, H.; Fonseca, R. Filtering GPS Navigation Solutions for Static Positioning. In Proceedings of the 17th International Congress of Mechanical Engineering, Sao Paulo, Brazil, 10–14 November 2003.

sensors

MDPI

Article

Research on a Simulation Method of the Millimeter Wave Radar Virtual Test Environment for Intelligent Driving

Xin Li [1,2], Xiaowen Tao [1], Bing Zhu [1,*] and Weiwen Deng [3]

1 State Key Laboratory of Automotive Simulation and Control, Jilin University, Changchun 130022, China
2 Aviation University of AF, Changchun 130022, China
3 School of Transportation Science & Engineering, Beihang University, Beijing 100191, China
* Correspondence: zhubing@jlu.edu.cn; Tel.:+86-135-0446-5260

Received: 3 February 2020; Accepted: 25 March 2020; Published: 30 March 2020

Abstract: This study addresses the virtual testing of intelligent driving, examines the key problems in modeling and simulating millimeter wave radar environmental clutter, and proposes a modeling and simulation method for the environmental clutter of millimeter wave radar in intelligent driving. First, based on the attributes of intelligent vehicle millimeter wave radar, the classification characteristics of the traffic environment of an intelligent vehicle and the generation mechanism of radar environmental clutter are analyzed. Next, the statistical distribution characteristics of the clutter amplitude, the distribution characteristics of the power spectrum, and the electromagnetic dielectric characteristics are analyzed. The simulation method of radar clutter under environmental conditions such as road surface, rainfall, snowfall, and fog are deduced and designed. Finally, experimental comparison results are utilized to validate the model and simulation method.

Keywords: intelligent driving; virtual test environment; millimeter wave radar

1. Introduction

Compared to traditional driving, intelligent driving can effectively solve issues such as human and vehicle safety and shared travel and has become a focus of research in the automotive industry and vehicle engineering. Intelligent driving has also become a core area of competition for high-tech enterprises working with artificial intelligence and the Internet, and the world's technological powers have incorporated intelligent driving in their science and technology development plans [1]. In the research and development of intelligent driving, virtual simulation tests can bypass bottlenecks such as the long cycle, high cost, and low safety of actual vehicle tests. Repeated testing in complex traffic scenarios is necessary for intelligent vehicles to be accepted by the public and ultimately to be safe on the road [2,3]. Millimeter wave radar has the advantages of technological maturity, wide application, low cost, high precision, and good stability in the traffic environment, and forms the basis of the indispensable sensors used in intelligent driving [1,4].

In the millimeter wave radar virtual testing of intelligent driving at both domestic and foreign sites, the simulation model of the millimeter wave radar test environment usually does not consider the mechanisms responsible for the generation of radar clutter, and hence the radar environmental clutter cannot change dynamically with the traffic scene. As a result, the simulation results of millimeter wave radar intelligent driving tend to be idealized, which does not objectively reflect the actual radar detection mechanism. This is an important issue that urgently needs to be solved in the virtual testing of intelligent driving.

To perform such virtual testing requires a complete set of high-fidelity millimeter wave radar system simulation model inputs. Among them, radar environmental clutter is the key factor affecting radar detection and measurement. Strong clutter background can lead to problems such as radar false alarms, missed detection, and measurement error [1,5,6]. Ignoring environmental clutter in the modeling and simulation research of millimeter wave radar will greatly reduce the fidelity of the model, which will seriously affect the credibility of virtual testing. This study addresses intelligent driving virtual tests in terms of the modeling and simulation of millimeter wave radar environmental clutter. This investigation has significant advantages, which can effectively solve the problem that the existing radar virtual test environment cannot change dynamically with the traffic scene, which results in a tendency for the radar simulation results to be idealized.

2. Analysis of Environmental Clutter Mechanism

Millimeter wave radar works in actual traffic scenes. In addition to radar-focused targets such as vehicles and pedestrians, there are non-radar target elements such as roads, buildings, transportation facilities, and weather conditions. These elements reflect radar electromagnetic waves, and these reflections are collectively referred to as radar clutter. The environmental clutter of the millimeter wave radar of intelligent vehicles consists mainly of ground clutter and weather clutter.

2.1. *Analysis of Ground Clutter*

The surface traffic depicted in Figure 1 generates ground clutter, which is the distributed scattering echo of the incident electromagnetic wave of the radar, a phenomenon that exerts a great influence on intelligent driving millimeter wave radar. In general, ground clutter is extremely unstable. For example, wind causes micro-motions of objects such as trees and grass, which can cause amplitude fluctuations and spectral broadening of ground clutter [7,8].

Figure 1. Surface traffic scene.

When examining the ground clutter mechanism, we should focus on the amplitude and frequency domain characteristics of ground clutter, which are affected by factors such as the wavelength of incident electromagnetic waves, surface area of radar radiation, incident angle of radar, polarization mode of incident electromagnetic waves, complex dielectric properties of the surface, and ground roughness [9–11].

In the study of the amplitude characteristics of ground clutter, it is usually necessary to solve the probability distribution of the amplitude based on the actual surface characteristics. Common amplitude distributions that can be used for ground clutter include the Lognormal, Weibull, and K distributions. The probability density distribution of the radar clutter amplitude describes the amplitude characteristics of the clutter signal in the time domain. To better describe the distribution characteristics of ground clutter generally requires an analysis of its spectral distribution characteristics. Common spectral distributions include the Gaussian, Cauchy, and Omnipolar distributions [10,12]. The distribution should be determined by fitting the scene surface data.

2.2. Analysis of Weather Clutter

Since the operating wavelength of millimeter wave radar for intelligent driving is on the order of millimeters, the wavelength is similar to the diameter of meteorological particles such as rain and snow. According to theory, the phase change of the incident field along the target length is more significant when the wavelength of the incident electromagnetic wave and the target size are of the same order of magnitude. Within the scattering region, each part of the weather scatterer affects the other parts [7,13,14]. The field strength at each point on the scatterer is the superposition of the scattered field strengths caused by the incident point and the remaining points in the scatterer, and the total effect of the interactions between the various parts of the scatterer determines the density distribution of the final current. For this scattering method, the exact Stratton–Zhulan integral equation must be solved in order to obtain the scattering field solution, usually with the moment solution [15–17].

The more common weather scenes of intelligent vehicles include rainy days, snowy days, and foggy days, as shown in Figure 2.

Figure 2. Traffic scene on rainy, snowy, and foggy days.

For weather scenes such as rain, snow, and fog, during radar detection there are many weather particles in each radar resolution unit. Assuming that the radar cross-section (RCS) of each weather particle is σ_i, the total RCS of the weather in the radar resolution unit is the sum of the RCS of all weather particles [10,12],

$$\sigma_c = V_c G_\eta = V_c \Sigma_i \sigma_i, \quad (1)$$

where G_η is the radar cross-sectional area of the weather particles per unit volume, and V_c is the total volume of the radar spatial resolution unit, which can be expressed as

$$V_c = \frac{\pi}{4}(R\theta_B)(R\phi_B)\left(\frac{cT}{2}\right)\frac{1}{2ln2}, \quad (2)$$

where R is the radial distance corresponding to the radar spatial resolution unit; T is the single-frequency modulation time of the radar; θ_B and ϕ_B are, respectively, the horizontal and vertical half-power beam widths of the radar antenna; and c is the propagation speed of the radar wave. If the diameter of the weather particle is D_i, then the cross-sectional area of the radar can be expressed as

$$\sigma_i = \frac{\pi^5 D^5{}_i}{\lambda^5}|K|^2, \quad (3)$$

where $|K|^2 = (\varepsilon - 1)/(\varepsilon + 2)$, and ε is the dielectric constant of the weather particles. Since ε is temperature dependent, $|K|^2$ changes with temperature.

The radar cross-sectional area η of weather particles per unit volume is

$$\eta = \Sigma_i \sigma_i = Tf^4 r^{1.6} \times 10^{-12}\ \mathrm{m}^2/\mathrm{m}^3, \quad (4)$$

The radar cross-sectional area of the weather particles is proportional to the 1.6th power of its diameter D_i. Let $\Sigma_i D_i$ be the radar reflectivity factor of the weather particles, denoted by Z.

Its mathematical description varies with rain, snow, fog, and other weather conditions. The radar reflectivity factor of rain is

$$Z_{rain} \approx 200r^{1.6}, \tag{5}$$

For dry snowfall, snow particles are mainly composed of ice crystals, single crystals, or synthetic crystals. The radar reflectivity factor of snow is

$$Z_{snow} \approx 1780r^{2.21}, \tag{6}$$

Mist is a floating combination of tiny water droplets or ice crystals close to the ground. Inland fog is usually radiation fog, whose average droplet diameter is generally < 20 μm [14]. The radar reflectivity factor of fog is

$$Z_{frog} = 4.62 \times 10^{-4.16}(V)^{-3.16}, \tag{7}$$

where r is the rainfall or snowfall rate in mm/h and V is the visibility of the fog in m.

3. Modeling of Environmental Clutter

3.1. Modeling of Ground Clutter

Ground clutter can be seen as a multi-point scattering set on the road surface, in which there is serious random mutual interference between the scattered signals. Ground clutter is a non-stationary random signal that changes with time.

The modeling of ground clutter must consider the strong scattering echoes of both ground clutter and ground stationary objects. The architecture of ground clutter modeling is shown in Figure 3.

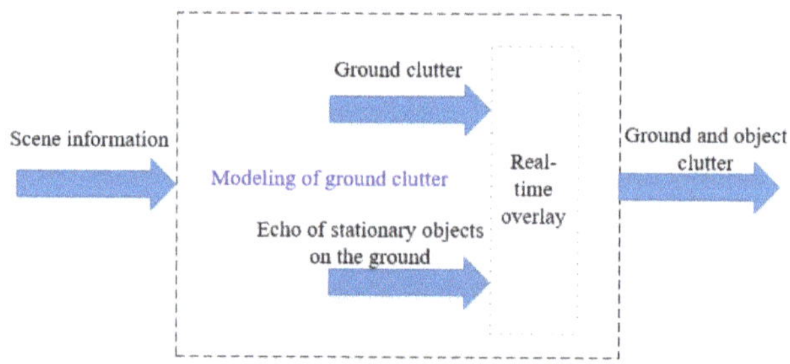

Figure 3. Architecture of ground clutter modeling.

In the study of ground clutter modeling, as shown in Figure 4, we distinguish three types of road environments: highways, urban roads, and rural roads.

A large number of ground clutter data measurements were carried out using self-developed millimeter-wave raw data radar.

(a) **(b)** **(c)**

Figure 4. Road environment for three types of traffic: (**a**) Highways; (**b**) Urban roads; (**c**) Rural roads.

According to the analysis of the ground clutter mechanism, the usual power spectrum distributions of ground clutter include the Gaussian, Cauchy, and Omnipolar distributions. For intelligent driving millimeter wave radar, we used the statistical analysis toolbox of MATLAB 2019 software. The ground clutter statistical fitting analysis of intelligent driving millimeter wave radar was performed using the measurement data from Delphi radar, TRW radar, and Continental radar in the multi-traffic road scene, as shown in Figure 5. The data used in the fitting analysis were from 266 actual road tests of the millimeter wave radar conducted from 2017–2019, and the test scenes consisted of highways, urban roads, and rural roads.

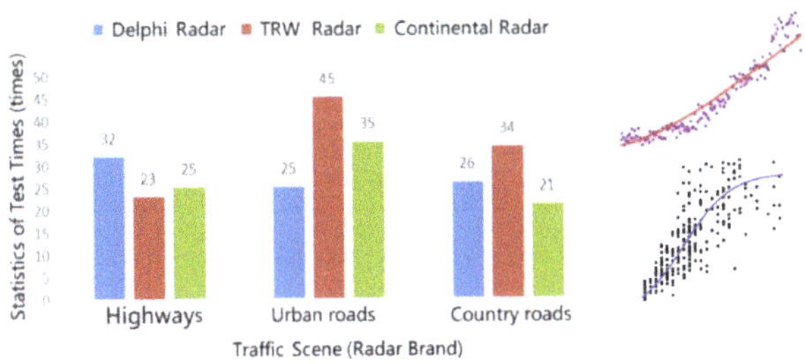

Figure 5. Analysis of ground clutter data fitting.

Using the abundant statistical analysis function library of MATLAB software, the Gaussian, Cauchy, and Omnipolar distributions were fitted in turn, and the data fitting results were analyzed synthetically. From the statistical results, we deduced that the ground clutter power spectrum of the intelligent driving millimeter wave radar in three types of traffic scenes most closely obeys a Gaussian distribution. Therefore, we utilized the Gaussian distribution power spectrum function to model the ground clutter of the intelligent driving millimeter wave radar.

The power spectrum of the Gaussian distribution is shown in Figure 6 and can be described mathematically as

$$S(f) = exp(-\frac{(f-f_d)^2}{2\sigma_f^2}), \tag{8}$$

where f_d is the average Doppler shift of the clutter power spectrum, and σ_f is its standard deviation. f_d is mainly affected by the speed of the vehicle on which the radar is mounted, i.e., $f_d = 2v_c/\lambda$, where v_c is the speed of the radar's vehicle, and λ is the working wavelength of the radar.

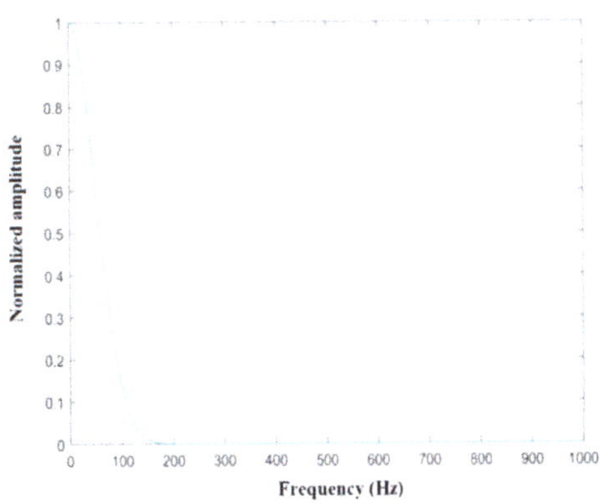

Figure 6. Spectral density curve of the Gaussian power spectrum.

The commonly used statistical distributions of the amplitude probability density of ground clutter include Lognormal, Weibull, and K distributions. Similarly, using the statistical analysis toolbox of MATLAB, the amplitude probability density distributions commonly used in ground clutter were fitted with the measurement data of the multi-radar in multi-class road traffic scenes. From a comprehensive analysis of the fitting results, we concluded that the probability density distribution of the ground clutter amplitude of intelligent driving millimeter wave radar in traffic scenes such as highways, urban roads, and rural roads is more similar to the Weibull distribution. Therefore, we used the amplitude probability density of the Weibull distribution to model the ground clutter of the intelligent driving millimeter wave radar.

The Weibull distribution of x, which is the amplitude of the clutter echo, is

$$f(x) = \frac{px^{p-1}}{q} exp\left[-(x/q)^{p}\right], \tag{9}$$

where p is related to factors such as the degree of undulation and continuity of the road surface, and q is related to the reflection intensity and echo power of the ground clutter.

For highways, urban roads, and rural roads, we carried out statistical analysis and fitting experiments for the ground clutter data of 30 different traffic scenes using MATLAB software. We substituted the measured ground clutter data into the Weibull distribution function in order to solve the parameters p and q and then averaged the parameter values obtained from 30 parameter solution processes in order to obtain the recommended values of p and q in the three traffic road scenes, as shown in Figures 7–9. In these figures, the blue dots represent the solution values of p and q in each experiment, and the red dotted lines represent the estimated average values of p and q, which were used as the recommended values for p and q.

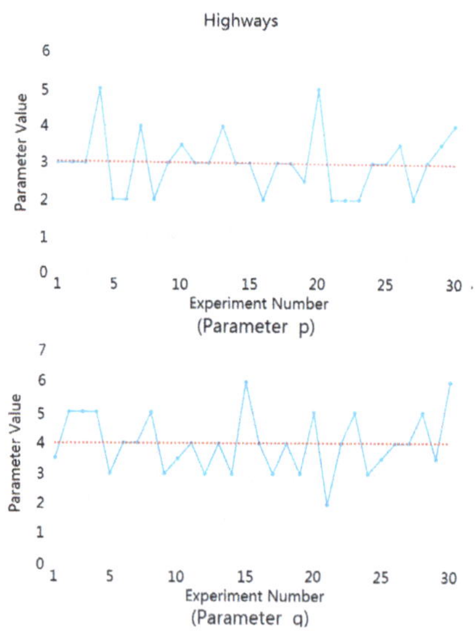

Figure 7. Solution process of parameters p and q for highways.

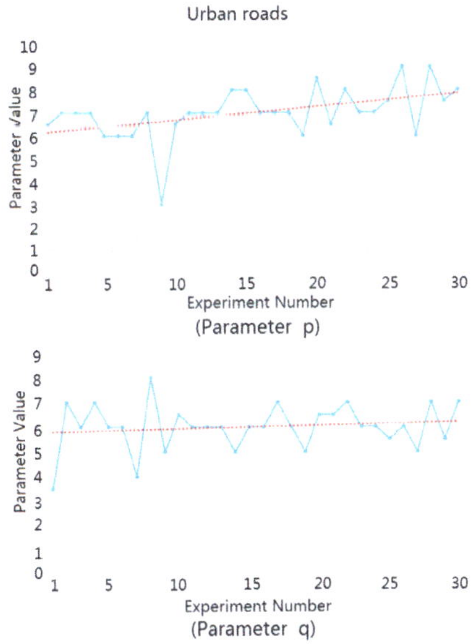

Figure 8. Solution process of parameters p and q for urban roads.

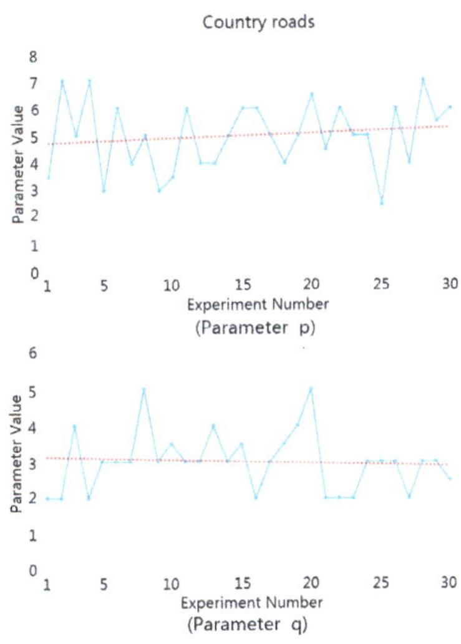

Figure 9. Solution process of parameters p and q for rural roads.

The estimated average values of parameters p and q for highways, urban roads, and rural roads are listed in Table 1.

Table 1. Recommended values of ground clutter parameters.

Parameter	Highway	Urban Road	Rural Road
p	3	7	5
q	4	6	3

In addition, we compared and analyzed the statistical error between the actual road clutter data and the ground clutter simulation data, which obeys the power spectrum of the Gaussian distribution and the amplitude probability density of the Weibull distribution. After randomly selecting 90 sets of actual road clutter data, each with a time period of 5 s, we calculated the average amplitude of the clutter time domain signal for each time period. At the same time, we utilized the above statistical distribution characteristics to generate simulated clutter data and to calculate the average amplitude of the simulation clutter data. On this basis, we normalized the amplitude data, and the error between the actual ground clutter and the simulated ground clutter was calculated. The error statistics comparison results are presented in Figure 10. From this comparison, we can see that the error between the actual clutter and the simulated clutter is smaller and the consistency is better.

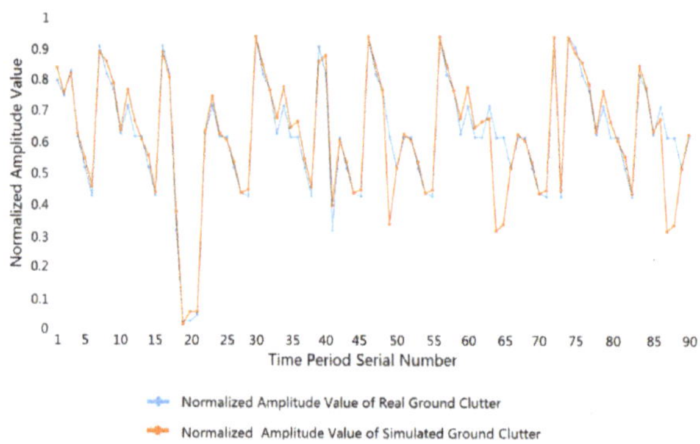

Figure 10. Comparison of amplitude values between actual clutter and simulated clutter.

Considering the echo of the ground stationary to be strongly scattered, and superimposing it on the ground clutter, the simulated flow of the obtained ground clutter is shown in Figure 11.

Let $w = u + jv$ be a complex variable of a Weibull distribution, which can be generated by the transformation of the complex Gaussian random variable $m = x + jy$:

$$\begin{cases} u = \frac{x(x^2+y^2)^{\frac{1}{p}}}{\sqrt{(x^2+y^2)}} \\ v = \frac{y(x^2+y^2)^{\frac{1}{p}}}{\sqrt{(x^2+y^2)}} \end{cases}, \tag{10}$$

where x and y are Gaussian variables distributed as $N\left(0, \sigma^2\right)$ $(\sigma^2 = q^p/2)$.

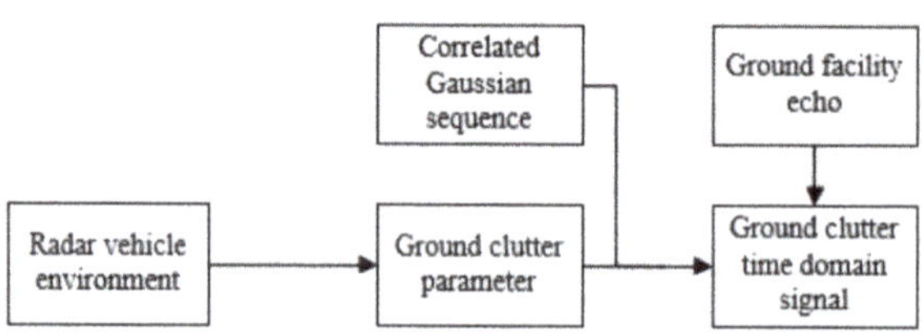

Figure 11. Simulated flow of ground clutter.

In Figure 11, the strong ground scattering facility is an arbitrary strong scattering target that is stationary relative to the road surface. The radar echo is

$$s_r(t) = e^{j2\pi[f_0(t-\frac{2R}{c})+\pi(t-\frac{2R}{c})^2/\lambda]}, \tag{11}$$

where f_0 is the radar carrier frequency, λ is the working wavelength of the radar, and R is the distance from the stationary target to the radar.

For example, in the traffic scene of a city road, let $p = 7, q = 6$. When two stationary parked vehicles are added to the simulation, the time domain signal of the complex Weibull clutter generated by the above modeling method is the result shown in Figure 12.

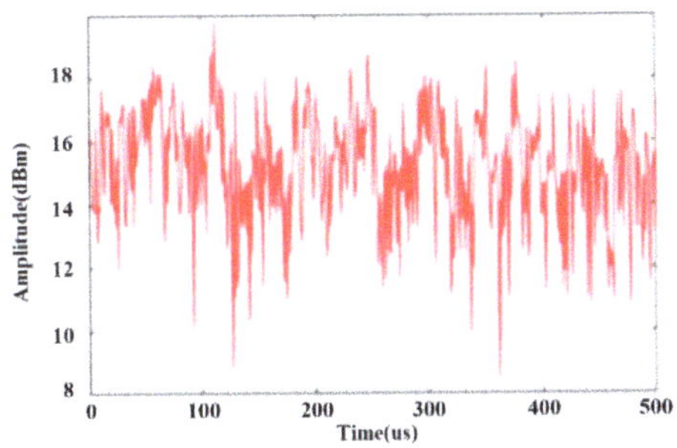

Figure 12. Ground clutter modeling method used to generate the Weibull distribution time domain signal.

The comparison between the complex Weibull clutter generated by the ground clutter modeling method and the ideal Weibull curve is shown in Figure 13. It can be seen from this figure that the curve fitting effect of the model is better, and the injection effect of the strong scattering target of the two vehicles is also significant.

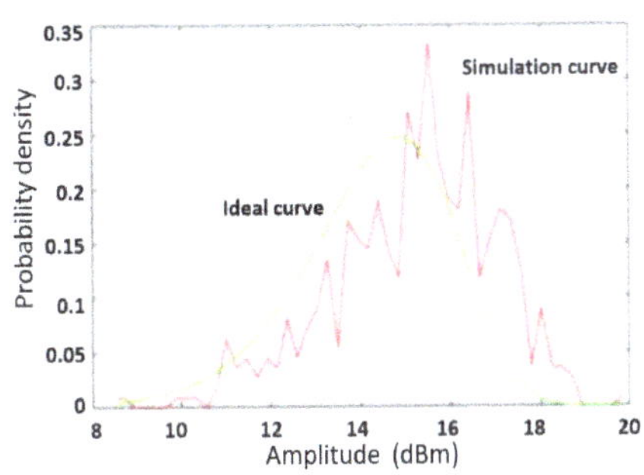

Figure 13. Comparison of complex Weibull clutter and ideal Weibull curve generated by the ground clutter modeling method.

3.2. Modeling of Weather Clutter

Considering the rainfall attenuation rate, reflectivity, amplitude, and phase distribution of factors such as rain clutter and radar transmission power, the rain clutter modeling method was designed, as shown in Figure 14.

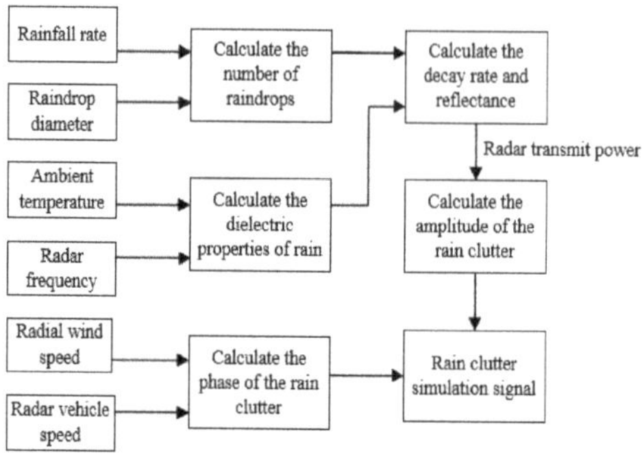

Figure 14. Rain clutter modeling method.

Based on the analysis of the rain clutter mechanism, combined with the physical characteristics of the rain in a traffic scene, the motion characteristics, and the millimeter wave radar characteristics, the simulation method for rain clutter in the time domain signal was constructed. The size distribution of raindrops varies with the type of rainfall, i.e., light, moderate, or heavy rain. The raindrop size distribution of each type of rainfall is

$$N(d_{min}) = \begin{cases} 60000e^{-5.7{R_{rain}}^{-0.21}d_{rain}} \text{ light rain} \\ 14000e^{-41{R_{rain}}^{-0.21}d_{rain}} \text{ moderate rain} \\ 2800e^{-3.0{R_{rain}}^{-0.21}d_{rain}} \text{ heavy rain,} \end{cases} \tag{12}$$

where d_{rain} is the raindrop diameter in mm, and R_{rain} is the rainfall rate in mm/h. The dielectric constant of rainwater can be described by a complex number whose real and imaginary parts are

$$\varepsilon_{rain1} = \varepsilon_\infty + \frac{(\varepsilon_x - \varepsilon_\infty)\left(1 + \frac{\lambda_s}{\lambda}\right)^{1-\alpha}\sin\left(\alpha\frac{\pi}{2}\right)}{1 + 2(\lambda_s/\lambda)^{1-\alpha}\sin\left(\alpha\frac{\pi}{2}\right) + (\lambda_s/\lambda)^{2(1-\alpha)}}, \tag{13}$$

$$\varepsilon_{rain2} = \frac{\sigma\lambda}{18.8496 \times 10^{10}} + \frac{(\varepsilon_x - \varepsilon_\infty)\left(\frac{\lambda_x}{\lambda}\right)^{1-\alpha}\cos\left(\alpha\frac{\pi}{2}\right)}{1 + 2(\lambda_s/\lambda)^{1-\alpha}\sin\left(\alpha\frac{\pi}{2}\right) + (\lambda_s/\lambda)^{2(1-\alpha)}}. \tag{14}$$

Let T (°C) be the ambient temperature. The parameters in the above equations are

$$\varepsilon_x = 78\left[1 - 4.6 \times 10^{-3}(T-25) + 1.2 \times 10^{-5}(T-25)^2 - 2.8 \times 10^{-8}(T-25)^3\right], \tag{15}$$

$$\varepsilon_\infty = 5.3 + 2.2 \times 10^{-2}T - 1.3 \times 10^{-3}T^2, \tag{16}$$

$$\lambda_s = 3.4 \times 10^{-4}e^{\left(\frac{2513}{T+273}\right)}, \tag{17}$$

$$\alpha = -\frac{16.8}{T+273} + 6.1 \times 10^{-2}, \tag{18}$$

$$\sigma = 12.6 \times 10^8. \tag{19}$$

Using Mie scattering theory, the analytical solution of rain particle scattering was obtained by solving Maxwell's equations of wavelength-sized particles. The radar cross-section of the total rainfall attenuation is

$$\sigma_t(d_{rain}) = \frac{2\pi c^2}{{f_c}^2}\sum_{n=1}^{\infty}(2n+1)Re(a_n + b_n), \tag{20}$$

and the backscattering segment of the rainfall is

$$\sigma_b(d_{rain}) = \frac{\pi c^2}{{f_c}^2}\left|\Sigma_{n=1}^{\infty}(2n+1)(-1)^n(a_n - b_n)\right|^2, \tag{21}$$

where

$$a_n = \frac{\psi_n(\alpha)\psi_n'(\beta) - m\psi_n(\beta)\psi_n'(\alpha)}{\xi_n(\alpha)\psi_n'(\beta) - m\psi_n(\beta)\xi_n'(\alpha)}, \tag{22}$$

$$b_n = \frac{m\psi_n(\alpha)\psi_n'(\beta) - \psi_n(\beta)\psi_n'(\alpha)}{m\xi_n(\alpha)\psi_n'(\beta) - \psi_n(\beta)\xi_n'(\alpha)}. \tag{23}$$

Here, α and β are related to the dielectric constant of rain, the size of the raindrop particles, the carrier frequency of the radar wave, and the propagation speed. The quantitative relationships are

$$\alpha = \frac{f_c}{c}d_{rain}, \tag{24}$$

$$\beta = \frac{f_c}{c}d_{rain}\sqrt[4]{{\varepsilon^2}_{rain1} + {\varepsilon^2}_{rain2}} \tag{25}$$

Let us set $\psi_n(x)$ as the first type of Bessel function and $\zeta_n(x)$ as the second type of Hankel function,

$$\psi_n(x) = xj_n(x) = \sqrt{\pi x/2}J_{n+\frac{1}{2}}(x), \tag{26}$$

$$\xi_n(x) = xh_n^{(1)}(x) = \sqrt{\pi x/2}H_{n+\frac{1}{2}}^{(1)}(x), \tag{27}$$

We find that the respective decay rate and reflectance of rainfall are

$$\gamma_{rain} = 10^{(4.343\times10^2\int\sigma_t(d_{rain})N(d_{rain})dd_{rain})-3}, \tag{28}$$

$$\eta_{rain} = \int\sigma_b(d_{rain})N(d_{rain})dd_{rain}. \tag{29}$$

The amplitude variation of rain clutter obeys the Rayleigh distribution, and the phase change is uniformly distributed. When there is wind in the ambient environment, the clutter spectrum becomes $f_d = f_w + f_0$, where f_w and f_0 are the Doppler shifts of the wind speed and radar, respectively.

Similarly, considering the snow attenuation rate, reflectivity, amplitude and phase distribution of snow clutter, and radar transmission power, the modeling of snow clutter was determined.

Our analysis revealed that snowfall is more sensitive to the influence of ambient temperature. At the same time, the correlation between light snow, moderate snow, and heavy snow is relatively high. Therefore, we no longer classify the snowfall based on snowfall intensity, but according to the temperature of the environment. The size distribution of the snowfall particles with diameter d_{snow} is

$$N_s(d_{snow}) = \begin{cases} 2.5\times10^3{R_{snow}}^{-0.94}d_s^{1/3}e^{-2.29{R_{snow}}^{-0.45}d_{snow}^{4/3}} & T \le -10\,^{\circ}\mathrm{C} \\ 9.25\times10^2{R_{snow}}^{-0.94}d_s^{1/3}e^{-2.29{R_{snow}}^{-0.45}d_{snow}^{4/3}} & T > -10\,^{\circ}\mathrm{C} \end{cases}, \tag{30}$$

where d_{snow} is the snow particle diameter in mm, and R_{snow} is the snowfall rate in mm/h. The calculation of the complex permittivity of snowflake particles is similar to that of rain particles, but the real and imaginary parts are different:

$$\varepsilon_{snow1} = \varepsilon_\infty + \frac{(\varepsilon_x - \varepsilon_\infty)\left(1+\frac{\lambda_s}{\lambda}\right)^{1-\alpha}sin\left(\alpha\frac{\pi}{2}\right)}{1+2(\lambda_s/\lambda)^{1-\alpha}sin\left(\alpha\frac{\pi}{2}\right)+(\lambda_s/\lambda)^{2(1-\alpha)}}, \tag{31}$$

$$\varepsilon_{snow2} = \frac{\sigma\lambda}{19\times10^{10}} + \frac{(\varepsilon_x - \varepsilon_\infty)\left(\frac{\lambda_x}{\lambda}\right)^{1-\alpha} cos\left(\alpha\frac{\pi}{2}\right)}{1+2(\lambda_s/\lambda)^{1-\alpha} sin\left(\alpha\frac{\pi}{2}\right) + (\lambda_s/\lambda)^{2(1-\alpha)}}, \tag{32}$$

where

$$\varepsilon_x = 203 + 2.5T + 0.15T^2, \tag{33}$$

$$\varepsilon_\infty = 3.168, \tag{34}$$

$$\alpha = 0.288 + 0.0052T + 0.00023T^2, \tag{35}$$

$$\lambda_s = 10^{-4} e^{\left(\frac{1320000}{2(T+273)}\right)}, \tag{36}$$

$$\sigma = 1.26e^{\{-12500(T+273)\}}. \tag{37}$$

Mist consists of tiny water droplets or ice crystals. Fog droplets near the surface of the Earth are usually meteorological particles with an average diameter < 20. Considering the transmission power of the intelligent vehicle millimeter wave radar, the attenuation rate of fog, reflectivity, amplitude of the fog clutter, and phase distribution, the modeling method of the fog clutter was determined.

The size distribution of the droplets is highly correlated with visibility and is described as

$$N\left(d_{fog}\right) = \frac{9.8}{\left(42000Y_{fog}\right)^{-1.7}} \cdot 10^9 \cdot \frac{d_{fog}}{2} e^{\left(-\frac{6.25}{(42000Y_{fog})^{-0.5}} d_{fog}\right)}, \tag{38}$$

where d_{fog} is the raindrop diameter in mm, and Y_{fog} is the fog visibility in m. The real and imaginary parts of the dielectric constant of the mist particle are calculated respectively as

$$\varepsilon_{fog1} = \frac{\varepsilon_x - \varepsilon_a}{1+\left(c/\lambda f_p\right)^2} + \frac{\varepsilon_a - \varepsilon_b}{1+(c/\lambda f_s)^2} + \varepsilon_b, \tag{39}$$

$$\varepsilon_{fog2} = \frac{c(\varepsilon_x - \varepsilon_a)}{\lambda f_p\left[1+\left(c/\lambda f_p\right)^2\right]} + \frac{c(\varepsilon_a - \varepsilon_b)}{\lambda f_s\left[1+(c/\lambda f_s)^2\right]}, \tag{40}$$

where

$$\varepsilon_x = 77 + 103(\theta - 1), \tag{41}$$

$$\varepsilon_a = 5.48, \varepsilon_b = 3.51, \tag{42}$$

$$f_p = 20 - 142(\theta - 1) + 294(\theta - 1)^2, \tag{43}$$

$$f_s = 590 - 1500(\theta - 1), \tag{44}$$

$$\theta = 300/T \tag{45}$$

4. Simulation Application of Environmental Clutter

4.1. Simulation Application of Ground Debris Wave

In order to verify the effectiveness of the ground clutter modeling method, we set up a traffic simulation scenario to simulate the application of ground clutter.

4.1.1. Parameters of Traffic Simulation Scenario

The center frequency f_c of the radar carrier was 77 GHz, the sampling rate F_s of the radar was 50 MHz, the frequency modulation of the radar was 16.7 μs wide, the bandwidth was 500 MHz, and the transmission power was 25 dBm. The antenna was set to single-shot and multi-receiver, the receiving antenna was a line array, the number of channels was 6, and there were 128 distance-dimension fast Fourier transform (FFT) processing points and 128 speed-dimension FFT processing points. The spatial positions of the two stationary strong scattering targets (cars) are illustrated in Figure 15. In this figure, the radar vehicles are blue, the stationary vehicles are yellow and orange, and the stationary vehicles are 37 and 44 m away from the radar vehicles. The parameters of ground clutter, p and q, are 3 and 4, respectively. The radar has speeds of 10, 20, and 30 m/s.

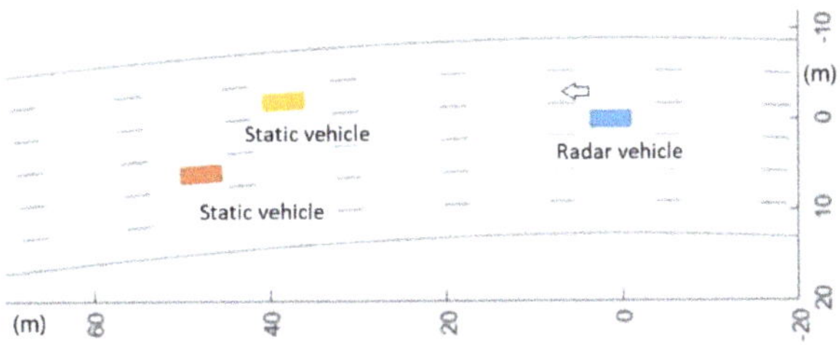

Figure 15. Schematic diagram of ground clutter scene.

4.1.2. Simulation Results

The simulation was carried out using the above ground clutter modeling method. The spectrum formed by the range and relative velocity (i.e., the Range Doppler (RD) spectrum) of the ground clutter at different speeds of the radar is shown in Figure 16. It can be seen from this figure that due to the motion speed of the radar carrier, the ground clutter spectrum migrates from the zero-intermediate frequency to the radial speed of the radar carrier. At the same time, since there are two stationary vehicles in the scene, the radar reflection intensity of a vehicle is higher than the road reflection intensity, so two strong scattering bright spots appear at 37 and 44 m on the spectrum.

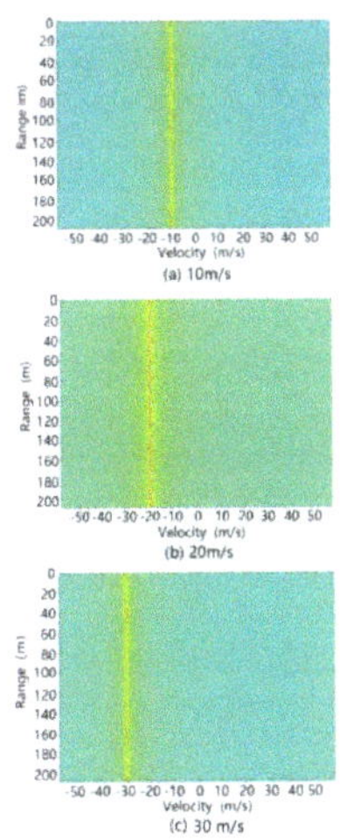

Figure 16. RD spectrum of simulated ground clutter at different speeds of radar-carrying vehicles.

From the ground clutter modeling simulation, the position of the stationary vehicle in the RD spectrum of the ground clutter was determined, as shown in Figure 17.

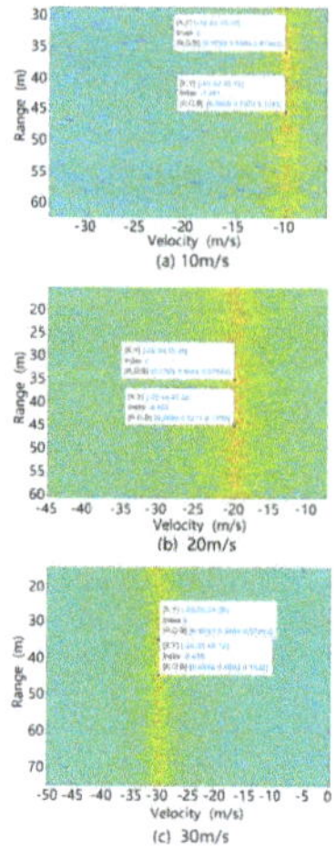

Figure 17. Position of stationary vehicle in RD spectrum of ground clutter during simulation.

Using the classical processing algorithm of the actual vehicle intelligent driving millimeter wave radar on the original simulation results of the ground clutter RD spectrum, we successively carried out moving target detection (MTD) processing, constant false-alarm rate (CFAR) detection and processing, and digital beam forming (DBF) processing.

The original simulation results for the ground clutter RD spectrum at different relative speeds after MTD processing are shown in Figure 18. It can be seen from this figure that the simulation results for the ground clutter RD spectrum after MTD processing not only reflected the real-time distribution characteristics of the ground clutter simulation scene but also introduced inherent frequency processing errors such as spectrum hollowing, spectrum broadening, spectrum shifting, and so on. Moreover, after MTD algorithm processing, the ground clutter simulation data did not cause additional unreasonable frequency interference components to the RD spectrum. Thus, for the MTD processing link, the theoretical verification effect of the ground clutter modeling method is good.

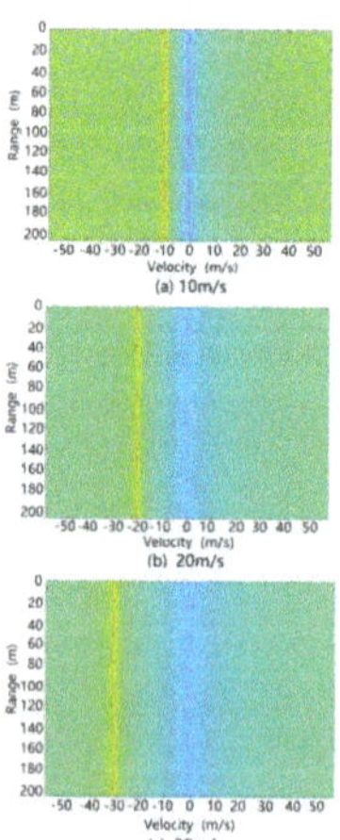

Figure 18. RD spectrum of ground clutter treated by moving target detection (MTD) processing during simulation.

The simulation results of the ground clutter RD spectrum after MTD processing also need to be processed using radar CFAR detection. The simulation results after radar CFAR detection processing at relative speeds of 10, 20, and 30 m/s are shown in Figure 19.

It can be seen from this figure that the simulation results after CFAR detection and processing not only reflected the real-time range characteristics and speed characteristics of each target in the ground clutter simulation scene but also introduced the false alarm target, range error, and speed error associated with the real CFAR detection and processing. After CFAR detection and processing, the ground clutter simulation data did not cause additional unreasonable redundant false target interference to the radar target detection. Therefore, the ground clutter modeling method exhibits a good theoretical verification effect for CFAR detection and processing.

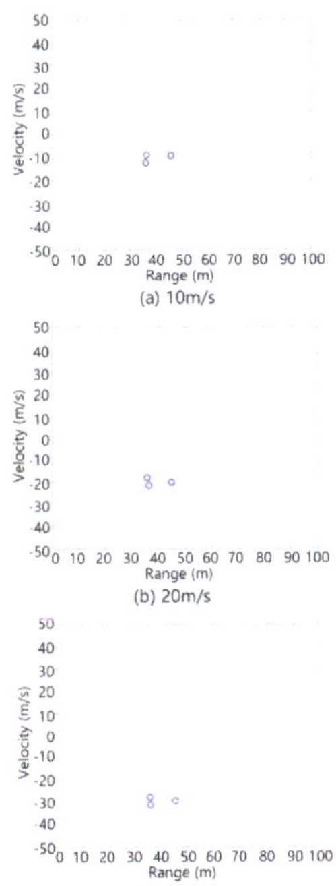

Figure 19. Results of ground clutter after constant false-alarm rate (CFAR) detection and processing during simulation.

The simulation results of radar detection after CFAR detection processing also need to be processed by radar angle DBF. The simulation results after radar angle DBF processing at relative speeds of 10, 20, and 30 m/s are shown in Figure 20.

It can be seen from this figure that the simulation results after DBF processing not only reflected the real-time relative angle characteristics of each target in the ground clutter simulation scene but also introduced the angle error of the false alarm target associated with the actual DBF processing. Moreover, after DBF processing, the ground clutter simulation data did not add unreasonable redundant false target interference to the radar target detection. Therefore, the theory of ground clutter modeling was proven to be effective for DBF processing.

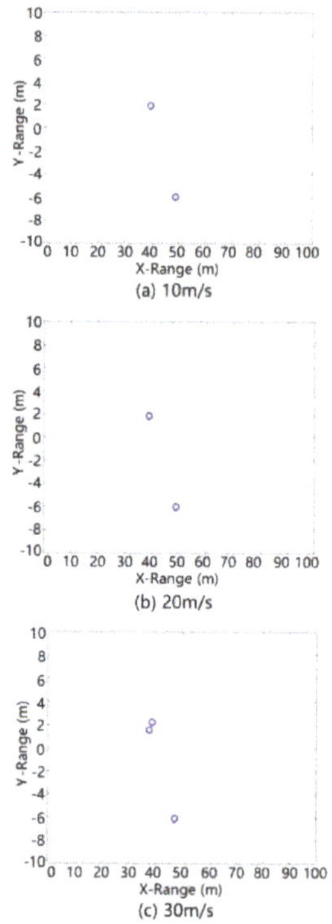

Figure 20. Results of ground clutter after digital beam forming (DBF) processing during simulation.

In addition, considering the direct influence of the relative amplitude of ground clutter on the radar detection results, we selected the rectangular window processing function, Hanning window processing function, and Hemingway window processing function, all of which are commonly used in actual radar, and applied the above ground clutter modeling method and the ground clutter simulation data generated by the simulation scene in order to calculate and plot the influence curve of the ground clutter amplitude on the number of radar detection targets, as shown in Figure 21.

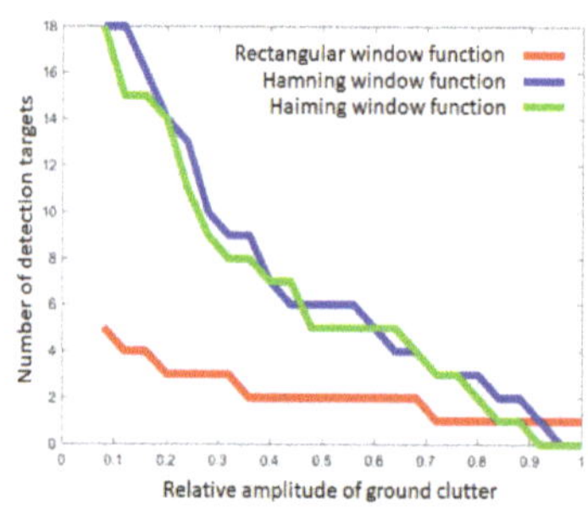

Figure 21. Effect of relative amplitude of surface clutter on test results during simulation.

As can be seen from this figure, the simulation results after rectangular window function processing indicate that the influence curve of the ground clutter amplitude on the number of radar detection targets was relatively weak, and the number of detection false alarm targets was small. On the other hand, the simulation results after Hanning window function processing and Hemingway window function processing reveal that the influence curve of the ground clutter amplitude on the number of radar detection targets was relatively strong, and there were more detected false alarm targets. The above characteristics are consistent with the processing results of the window function in actual radar detection. Therefore, the theoretical validation of ground clutter modeling method for the relationship between the amplitude of ground clutter and the number of targets detected by radar is good.

Overall, the ground clutter simulation data exhibited good consistency with actual radar in terms of radar RD spectrum generation, MTD processing, CFAR processing, DBF processing, and detection under different time domain window functions.

4.2. Simulation Application of Weather Clutter

In order to verify the effectiveness of the modeling method on weather clutter, we set up a traffic simulation scenario in which we could conduct simulation application experiments of weather clutter. We realized, of course, that weather clutter and ground clutter are usually present at the same time. Therefore, in the simulation scenario described below, the ground clutter simulation data were also generated synchronously. We distinguished the simulation scenes of highways with heavy rain, heavy snow, and dense fog. The simulation application test of weather clutter was performed separately.

The highway scene with heavy rain is shown in Figure 22.

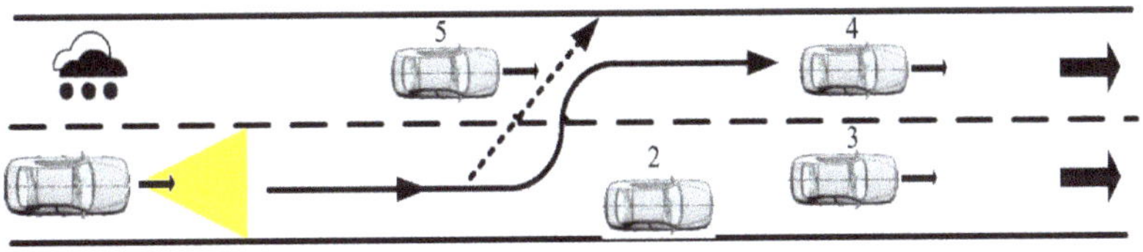

Figure 22. Highway scene with heavy rain.

The vehicle parameter settings in the highway scene with heavy rain are listed in Table 2.

Table 2. Vehicle information in highway scene with heavy rain.

Parameter	Radar Car	Car Number 2	Car Number 3	Car Number 4	Car Number 5
Driving speed	20 m/s	0 m/s	20 m/s	30 m/s	10 m/s
Relative distance	0 m	60 m	100 m	100 m	40 m

The radar parameter settings in the highway scene with heavy rain are listed in Table 3.

Table 3. Radar parameter settings in highway scene with heavy rain.

Parameter	Parameter Value
Radar center carrier frequency	77 GHz
Transmission power	25 dbm
Transmitting antenna gain	27 dB
Receiver antenna gain	27 dB

The road surface parameters in the highway scene with heavy rain are listed in Table 4.

Table 4. Pavement parameters in highway scene with heavy rain.

Parameter	Parameter Value
Pavement parameter p	3
Pavement parameter q	4

The weather parameters in the highway scene with heavy rain are listed in Table 5.

Table 5. Weather parameters in highway scene with heavy rain.

Parameter	Parameter Value
Rainfall rate	10 mm/h
Raindrop diameter	2 mm
Ambient temperature	27 °C
Wind speed	10 m/s

In the high-speed road scene with heavy rain, the synthetic RD spectrum of the target echoes and environmental clutter received by the radar are shown in Figure 23.

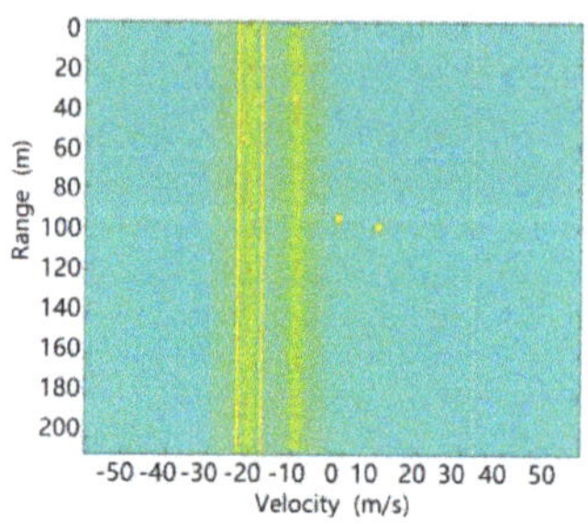

Figure 23. RD spectrum of radar echoes in highway scene with heavy rain.

The target output after radar processing is shown in Figure 24.

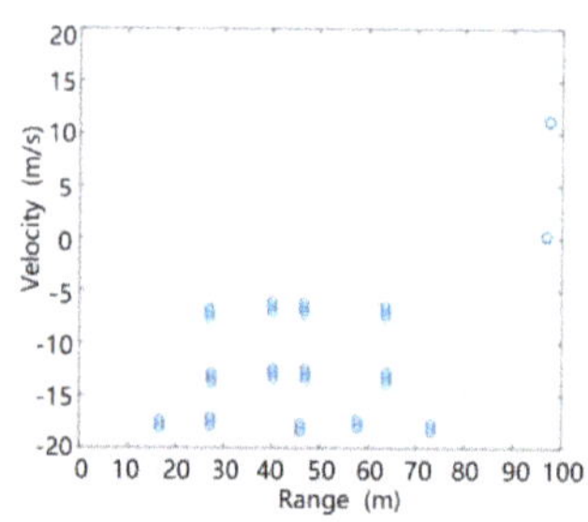

Figure 24. Target output of radar in highway scene with heavy rain.

It can be seen from this figure that in the highway scene with heavy rain, both the No. 3 car and the No. 4 car were normally detected by the radar. Due to the combined effects of roads and heavy rainfall, neither the No. 2 car nor the No. 5 car was detected normally, causing missing radar reports. At the same time, there were many false targets in the radar output, and there were certain distance measurement errors and speed measurement errors for the target, which were related to environmental clutter interference. The simulation results were in good agreement with the actual radar processing results.

We constructed the simulation test scenario for the autonomous emergency braking (AEB) system, as shown in Figure 25, and used MATLAB and PanoSim software to generate the radar environment simulation data using the aforementioned environmental clutter modeling method. We then tested the intelligent driving AEB decision algorithm, as shown in Figure 26.

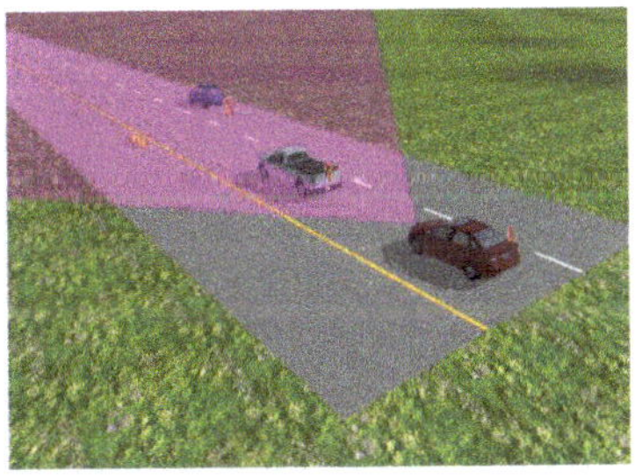

Figure 25. Test scenario for autonomous emergency braking (AEB) decision algorithm.

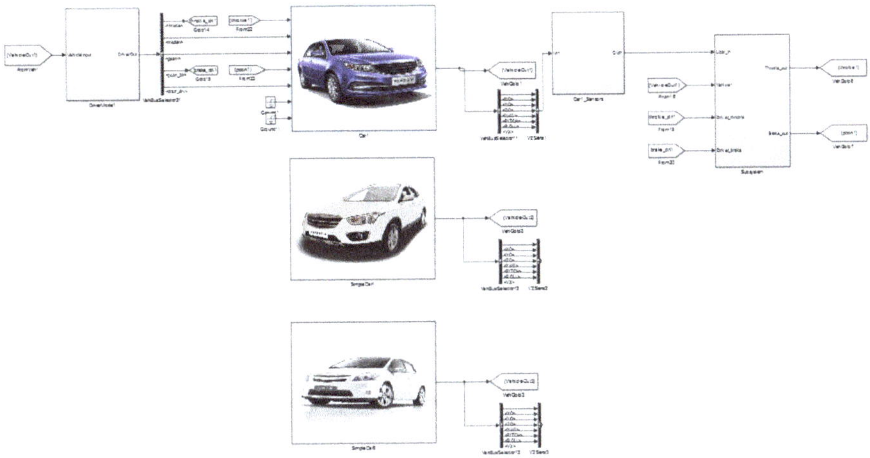

Figure 26. Simulink model file of AEB decision algorithm based on MATLAB and PanoSim software.

The test results are shown in Figures 27 and 28 below, in which 1 indicates that the AEB system has not started, 2 indicates that the AEB system is in the warning state, and 3 indicates that the AEB system is in the braking state. The test results of the AEB decision algorithm without the support of the radar environmental clutter simulation data were then compared and analyzed.

From the comparison results, we can see that the radar environmental clutter simulation method presented in this study exhibited obvious advantages. When there was no clutter in the simulation data, the test results of the AEB decision algorithm were idealized, the transition of the AEB system working state was stable, and it could not reflect the dynamic changes that occur during an actual road test. After the radar clutter simulation method was added, however, the potential defects and shortcomings of the AEB algorithm were clearly exposed, and the AEB state in the test results was closer to the actual road test results. Compared with other test methods, the simulation method of radar environmental clutter can support the repeatability test of the decision algorithm more effectively and improve the optimization of the decision algorithm.

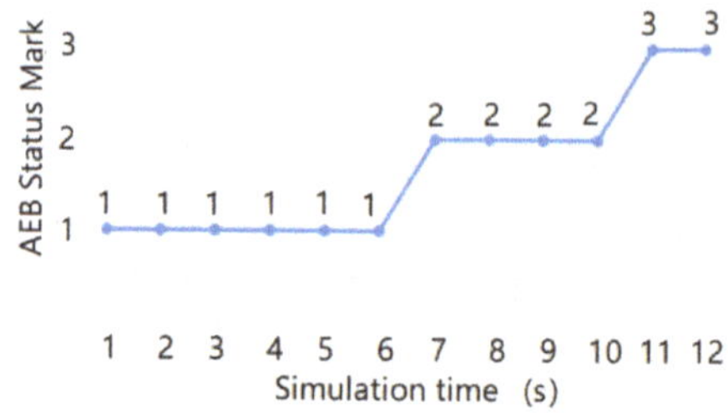

Figure 27. AEB state change without radar environmental clutter simulation data.

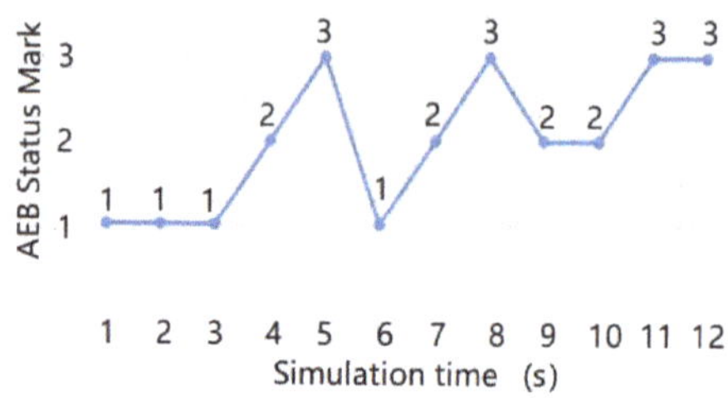

Figure 28. AEB state change with radar environmental clutter simulation data.

5. Discussion and Conclusions

This study proposed a simulation method of the millimeter wave radar virtual test environment for intelligent driving. The work and innovations of this project are summarized as follows:

1. According to the characteristics of intelligent driving millimeter wave radar, and based on the principles of statistics and electromagnetism, millimeter wave radar for intelligent driving was analyzed for the first time, and the mechanism of environmental clutter was examined in detail.
2. Based on the surface characteristics in intelligent vehicle traffic scenes, the surface differences between highway traffic roads, urban traffic roads, and rural traffic roads were investigated. A simulation method for the ground clutter in millimeter wave radar for intelligent driving was proposed.
3. In terms of the weather characteristics of various intelligent vehicle traffic scenes, an analysis of the statistical distribution characteristics of rain, snow, and fog provided us with the capability to distinguish weather particle size, inter-particle density, wind speed, and wind direction. A simulation method for rain, snow, and fog weather clutter of millimeter wave radar for intelligent driving was proposed.
4. The surface distribution characteristics under typical scenes were designed. Based on the signal processing and data processing algorithms of millimeter wave radar, the effectiveness of the ground clutter simulation method was verified.
5. The distribution characteristics of rain, snow, and fog in typical scenes were designed. Based on the radar signal processing and data processing algorithms, the effectiveness of the weather clutter simulation method was verified.
6. The research content of this study is an important part of the simulation model of intelligent vehicle millimeter wave radar, since it supplies the missing environmental clutter modeling and simulation method in the simulation model of intelligent vehicle millimeter wave radar, thereby solving a key problem and shortcoming.

The research results and conclusions of this study are significant to the field of intelligent driving simulation testing based on millimeter wave radar.

Author Contributions: X.L. was responsible for the concept and progress of the project, the establishment of the model, and the application of the model. X.T. was responsible for creating the research design, writing, and editing. W.D. assisted in the field. B.Z. was responsible for funding and editing. All authors have read and agreed to the published version of the manuscript.

Funding: This research was funded by the National Key Research and Development Project (2018YFB0105103).

Acknowledgments: The authors wish to thank the support team who assisted with fieldwork and offered external guidance and editorial comments.

Conflicts of Interest: The authors declare no conflict of interest.

References

1. Deng, W. Electrification and intelligent technology-the driving force of the future automobile. *J. Automob. Saf. Energy Conserv.* **2010**, *1*, 179–189.
2. Song, W.; Yang, Y.; Fu, M.; Qiu, F.; Wang, M. Real-Time Obstacles Detection and Status Classification for Collision Warning in a Vehicle Active Safety System. *IEEE Trans. Intell. Transp. Syst.* **2018**, *19*, 758–773. [CrossRef]
3. Seo, M.; Yoo, C.; Park, S.-S.; Nam, K. Development of Wheel Pressure Control Algorithm for Electronic Stability Control (ESC) System of Commercial Trucks. *Sensors* **2018**, *18*, 2317. [CrossRef] [PubMed]
4. Gashinova, M.; Hoare, E.; Stove, A. Predicted sensitivity of a 300GHz FMCW radar to pedestrians. In Proceedings of the 2016 46th European Microwave Conference (EuMC), London, UK, 4–6 October 2016.
5. Knudde, N.; Vandersmissen, B.; Parashar, K.; Couckuyt, I.; Jalalvand, A.; Bourdoux, A.; De Neve, W.; Dhaene, T. Indoor tracking of multiple persons with a 77 GHz MIMO FMCW radar. In Proceedings of the 2017 European Radar Conference (EURAD), Nuremberg, Germany, 11–13 October 2017.
6. Herzel, F.; Kissinger, D.; Ng, H.J. Analysis of Ranging Precision in an FMCW Radar Measurement Using a Phase-Locked Loop. *IEEE Trans. Circuits Syst. I Regul. Pap.* **2017**, *65*, 783–792. [CrossRef]
7. Monakov, A.; Nesterov, M. Statistical Properties of FMCW Radar Altimeter Signals Scattered from a Rough Cylindrical Surface. *IEEE Trans. Aerosp. Electron. Syst.* **2017**, *53*, 323–333. [CrossRef]
8. Maitra, A.; Dan, M. Propagation of Pulses at Optical Wavelengths Through Fog-filled Medium. *Radio Sci.* **1996**, *31*, 469–475. [CrossRef]
9. Sun, F.; Zhu, L.; Zhang, C.; Duan, H. The research of J-Quinn ranging algorithm for LFMCW Radar. In Proceedings of the 2016 IEEE Advanced Information Management, Communicates, Electronic & Automation Control Conference, Xi'an, China, 3–5 October 2016.
10. Zhang, Z.; Gong, T.; Zhu, D.; Liu, Y. High-precision ranging for radar micro-motion target based on envelope migration. In Proceedings of the 2016 CIE International Conference on Radar, Guangzhou, China, 10–13 October 2016.
11. Rambach, K.; Yang, B. Direction of arrival estimation of two moving targets using a time division multiplexed colocated MIMO radar. In Proceedings of the IEEE Radar Conference, Cincinnati, OH, USA, 19–23 May 2014; pp. 1118–1123.
12. Zhang, Y.; Li, Q.; Huang, L.; Song, J. Waveform design for joint radar-communication system with multi-user based on MIMO radar. In Proceedings of the 2017 IEEE Radar Conference, Seattle, WA, USA, 8–12 May 2017.
13. Wu, L.; Wang, L.; Min, L.; Hou, W.; Guo, Z.; Zhao, J.; Li, N. Discrimination of Algal-Bloom Using Spaceborne SAR Observations of Great Lakes in China. *Remote Sens.* **2018**, *10*, 767. [CrossRef]
14. Vu, P.; Haimovich, A.M.; Himed, B. Direct tracking of multiple targets in MIMO radar. In Proceedings of the 2016 50th Asilomar Conference on Signals, Systems & Computers, Pacific Grove, CA, USA, 6–9 November 2016.
15. Sediono, W.; Sediono, W. Method of measuring Doppler shift of moving targets using FMCW maritime radar. In Proceedings of the 2013 IEEE International Conference on Teaching, Bali, Indonesia, 26–29 August 2013.

16. Pfeffer, C.; Feger, R.; Wagner, C.; Stelzer, A. FMCW MIMO Radar System for Frequency-Division Multiple TX-Beamforming. *IEEE Trans. Microw. Theory Tech.* **2013**, *61*, 4262–4274. [CrossRef]
17. Yin, J.; Unal, C.; Russchenberg, H. Narrow-Band Clutter Mitigation in Spectral Polarimetric Weather Radar. *IEEE Trans. Geosci. Remote Sens.* **2017**, *55*, 4655–4667. [CrossRef]

sensors

MDPI

Article

Human-Like Lane Change Decision Model for Autonomous Vehicles that Considers the Risk Perception of Drivers in Mixed Traffic

Chang Wang, Qinyu Sun *, Zhen Li and Hongjia Zhang

School of Automobile, Chang'an University, Xi'an 710064, Shaanxi, China; wangchang@chd.edu.cn (C.W.); lizhen@chd.edu.cn (Z.L.); zhanghongjia@chd.edu.cn (H.Z.)

* Correspondence: sunqinyu@chd.edu.cn

Received: 22 March 2020; Accepted: 15 April 2020; Published: 16 April 2020

Abstract: Determining an appropriate time to execute a lane change is a critical issue for the development of Autonomous Vehicles (AVs).However, few studies have considered the rear and the front vehicle-driver's risk perception while developing a human-like lane-change decision model. This paper aims to develop a lane-change decision model for AVs and to identify a two level threshold that conforms to a driver's perception of the ability to safely change lanes with a rear vehicle approaching fast. Based on the signal detection theory and extreme moment trials on a real highway, two thresholds of safe lane change were determined with consideration of risk perception of the rear and the subject vehicle drivers, respectively. The rear vehicle's Minimum Safe Deceleration (MSD) during the lane change maneuver of the subject vehicle was selected as the lane change safety indicator, and was calculated using the proposed human-like lane-change decision model. The results showed that, compared with the driver in the front extreme moment trial, the driver in the rear extreme moment trial is more conservative during the lane change process. To meet the safety expectations of the subject and rear vehicle drivers, the primary and secondary safe thresholds were determined to be 0.85 m/s^2 and 1.76 m/s^2, respectively. The decision model can help make AVs safer and more polite during lane changes, as it not only improves acceptance of the intelligent driving system, but also further ensures the rear vehicle's driver's safety.

Keywords: autonomous vehicles; lane-change decision; risk perception; mixed traffic; minimum safe deceleration

1. Introduction

Intelligent driving technologies are designed for the purpose of facilitating driving strategies that improve driving safety and reduce driver work load. Examples include, but are not limited to the: Lane-Change Decision Aid System (LCDAS), Adaptive Cruise Control (ACC), and Lane Departure Warning (LDW) [1–4]. Changing lanes is one of the most dangerous driving maneuvers, and accounts for about 5% of traffic collisions in China and Europe [5,6]. Dangerous lane change behavior is the leading cause of two-vehicle collisions, and seriously affects the safety of both the subject vehicle and rear vehicle in the target lane. Therefore, the safety of both the subject vehicle and the rear vehicle should be considered when developing Autonomous Vehicles (AVs).

Before establishing a lane changing decision strategy for AVs, it is necessary to understand the decision making process and subsequent behavior associated with lane changes. Considering how the driving environment effects driving behavior, Gipps [7] first proposed a framework governing lane-change decisions, in which the possibility, necessity, and desirability of the lane change were the main factors in determining if, when, and how the lane change was performed. Based on this decision-making framework, a lane change model was suggested by Halati et al. [8], which states that most lane change behaviors can be classified as either mandatory lane changes or discretionary lane changes; and that in general, lane-change behavior is in response to motivation, advantage, and urgency. For example, if the subject vehicle (S) cannot maintain an acceptable distance from the vehicle in front of it, S executes a lane change. Hidas [9] identified the gap between the rear vehicle (R) and S as a pivotal factor in the lane change process, and classified lane change behavior into three groups: free (discretionary lane change), forced (mandatory lane change), and cooperative. The cooperative lane change accounts for cooperation between R and S, in which R willingly decelerates, thereby positively impacting lane change feasibility. Similarly, the lane change process was described by game theory [10,11], which stated that R and S can influence each other's driving behavior.

Nilsson et al. [12] reported that determining the appropriate time to execute a lane change is a critical issue for the development of AVs. A classic safety lane change model was proposed by Jula et al. [13], in which the Minimum Safe Spacing (MSS) (the minimum relative distance required to avoid a collision) between S and R was selected as the indicator for evaluating the feasibility of S performing a lane change. Kamal et al. [14] built on the MSS concept, and designed a lane change control algorithm for the connected vehicle. Balal et al. [15] proposed a fuzzy inference system to model lane change behavior that considers the gap between the F and surrounding vehicles. Over time, an increasing number of naturalistic driving experiments were conducted to examine the gap acceptance and obtain empirical evidence that supports that MSS theoretical research. [16–18]. The results showed that drivers' gap acceptances of lane changes are basically the same.

In addition, Time to Collision (TTC) (the ratio of relative distance to relative speed between two vehicles) was extensively used in developing lane change maneuver algorithms. Lee et al. [19] used TTC to classify the lane change process into four groups, based on motivation, Lane Change Duration (LCD) (the time from the beginning of the lane change to the end of the lane change), and relative distance. Wakasugi [20] reported that the driver is able to perform a lane change when the TTC is more than six seconds. The International Organization for Standardization [21] proposed a multi-level warning model for different relative speeds of S and R, in which the lane change warning system uses TTC as the warning indicator. Bordes [22] established a similar multi-level warning model that takes into account the relative distance between S and R. While numerous studies have employed TTC as an indicator to assess lane change associated risks, there are different interpretations of the data, which have resulted in different recommendations for lane-change safety thresholds. Dijck [23] suggested that the driver may consider the lane change safe when the TTC is higher than three seconds or the gap is longer than five meters. This value is lower than the TTC finding in Hirst [24]; while Saunier and Sayed [25] suggested that the TTC can be lower than three seconds, which is lower than Dijck's recommendation.

However, while determining a safe threshold for changing lanes, the rear and front vehicle drivers' risk perceptions are underestimated. According to the game theory, R and S influence the driving behavior of each other. While an inappropriate lane change may directly trigger a rear-end crash between R and S, it can also stimulate a negative emotional response, like anger or anxiety, in the rear vehicle driver [26,27], who then deliberately chooses not to cooperate with F's attempt to change lanes [28,29]. These scenarios are more dangerous for R, as they may result in R colliding with other surrounding vehicles.

There are scientists and engineers who envision that in the future, roads will be populated by mixed traffic consisting of AVs and conventional vehicles [30,31]. Many authorities are of the opinion that in order to improve trust and acceptance, AVs should mimic human-like driving behavior, which satisfies the driver's subjective expectation [32–34]. Contrarily, AV driving behavior significantly benefits from being as polite as possible, since the negative impact on surrounding conventional vehicles is reduced.

To address the deficiencies in the lane-change decision model for AVs that are inconsistent with the driver's cognition, the present work proposed a theoretical lane-change decision model with a two-level threshold by considering the different drivers' risk perceptions and data from previous studies. Combining the MSS model with Gipps's lane-change safety theory, the proposed lane-change decision model confirmed the Minimum SafeDeceleration (MSD) of a fast approaching rear vehicle as the safety indicator. According to the game theory, the deceleration of R, as an intuitive indicator for our model to evaluate the lane change safety, could directly relates to R's driving behavior and willingness to cooperate. Then we implemented a naturalistic driving experiment to explore the lane change behavior and calibrate the proposed lane-change decision model. In order to acquire the discrepant safety levels of a lane change, two extreme experiments, including the front extreme moment trial and the rear extreme moment trial, were conducted, and the two-level safety threshold was determined by evaluating the risk perception of different drivers, using Signal Detection Theory (SDT). Finally, we determined the primary and secondary safe thresholds for our proposed lane-change decision model.

2. Related Works

The formulation of a lane-change safety indicator is the kernel of establishing a lane-change decision model for AVs, and numerous empirical studies have investigated different lane-change safety indicators. Gipps [7] first structured a lane-change decision model for urban roads that employed the requisite deceleration value of the rear vehicle in the target lane to evaluate lane-change safety. Kestinget al. [35] proposed the Minimizing Overall Braking Induced by Lane-change (MOBIL) model, and the lane-change safety criteria for discretionary lane-changes and mandatory lane-changes based on different incentives were derived. These safety criteria could reflect lane-change safety more factually on account of considering the advantages and disadvantages of other drivers. Schakelet al. [36] established an integrated Lane-change Model with Relaxation and Synchronization (LMRS) according to the naturalistic driving data. The model integrated lane-change desire and incentives with a car-following model to assess lane-change safety. The MSS model was first established by Jula et al. [13] and this model used the calculated value of the minimum longitudinal safe distance between a subject vehicle and a rear vehicle in the target lane to evaluate the safety boundary during the lane-change process. Wang et al. [37] proposed a lane-change decision model based on the Minimum Safe Deceleration (MSD) (the minimum deceleration of rear vehicle required to avoid collision) model and the deceleration of a rear vehicle in the target lane. The Lane-Change Risk Index (LCRI) model was put forward by Hyunjin et al. [38] based on the actual traffic accident data, and the model could quantize and estimate the collision risk during the lane-change process.

According to the different assessment indicators of lane-change safety, various approaches have been pursued to establish the most appropriate threshold for a lane-change decision model that can conform to the driver's lane-change safety cognition. Gipps [7] determined the safe deceleration of rear vehicle as –4 m/s^2, and if the calculated value was less than that threshold value, the lane-change operation was considered to be terminated. Wang et al. [37] established a lane-change decision model with a two-level threshold by calculating the deceleration of the rear-approaching vehicle, the primary and secondary thresholds were determined as 1.5 m/s^2 and 2.7 m/s^2, respectively. Considering that drivers with different driving styles may possess discrepancies in their cognitions of lane-change safety, Wang et al. [39] divided drivers into four different driving styles denoted as prudent drivers, lees prudent drivers, less aggressive drivers, and aggressive drivers, then different thresholds were determined for each driving style.

Currently, the TTC indicator, deduced from MSS model, has been generally applied in lane-change decision models. According to the relative speed between the rear vehicle and the subject vehicle, International Standards Organization (ISO) 17387:2008 [21] divided the TTC threshold into three different levels, and the TTC thresholds were confirmed as 2.5 s, 3.0 s, and 3.5 s, respectively, corresponding to the relative speed interval less than 10 m/s, from 10 m/s to 15 m/s, and from 15 m/s to 20 m/s. Similarly, a patent applied by BOSCH Company (Stuttgart, Germany) [22] divided TTC threshold based on different relative distance between the rear vehicle and subject vehicle, and the TTC thresholds were determined as 2.5 s, 3.0 s, and 3.5 s, respectively, corresponding to the relative distance interval from 3 m to 25 m, from 25 m to 45 m, and from 45 m to 70 m. Wakasugi [20] recommended a two-level TTC threshold of 3 s and 5 s, respectively. However, the computed TTC threshold would be easily influenced by the relative speed and distance between the rear vehicle and the subject vehicle.

The remainder of the paper is organized as follows. Related works on lane-change decision models and safety indicator thresholds are introduced in Section 2. Section 3 introduces the naturalistic lane-change trial and extreme moment trials. Section 4 presents the lane-change decision model and the calibration parameters of the model. The two-level threshold based on the proposed model is determined in Section 5. Finally, a discussion and conclusion are presented in Section 6.

3. Method

On-road experiment is the main research method used in this paper, and the experiments include the naturalistic lane-change trial and extreme moment trials. The purpose of the naturalistic trial is to calibrate the parameters of the proposed lane-change decision model. The extreme moment trials are used to accurately capture the variation of driver cognition characteristics of lane-change safety. In this section, we willintroduce the required equipment, participants, test route and procedures for the experiments in detail.

3.1. Apparatus

The test vehicle is depicted in Figure 1. The test vehicle used in our experiments was a 2008 Volkswagen Touran, equipped with a Lane Mark Recognition system (Mobileye C2-170, made by Mobileye Company, Jerusalem, Israel), two millimeter-wave radars for measuring the relative speed and distance between the subject vehicle (S) and the surrounding vehicles, a video monitoring system for collecting the head motion and eye movement of drivers and the driving environment, and a VBOX (a piece of equipment that can obtain vehicle's GPS coordinate, made by Racelogic Company, London, England) to collect the driving speed and acceleration. A wireless button was fixed on the left side of the steering wheel, the button press time can be recorded.

Figure 1. The test vehicle.

3.2. Participants and Driving Route

Thirty experienced drivers participated in the two experiments. Drivers' ages ranged from 27 to 50 years old, with an average age of 39.8 years (Standard Deviation = 7.17). Their driving experience ranged from 2 to 28 years (mean = 14.2, Standard Deviation = 8.3). All the participants were non-professional drivers with a valid driver's license, normal or corrected vision, and experienced no traffic accidents over the past two years.

The drivers were required to drive the test vehicle on a section of highway, from Sanqiao to Xinzhu, Xi'an, China, as shown in Figure 2. The route was a 38 km, two-way six-lane road, with a 3.75 m lane width, and speed limit of 100 km/h. To keep the drivers safe while driving at a high speed, the test was carried out during non-peak hours and in clear weather conditions. To reduce driving workload and to ensure driving safety, the driving route had a zero gradient and most of highway section was straight road. Participants were paid ¥300 for their participation after they had finished all the experiments.

Figure 2. The experiment route map.

3.3. *Naturalistic Lane-Change Trial*

To investigate the naturalistic lane-change behavior, participants were required to drive on the test road using their own driving style, without any instructions or requirements.

To maintain driving safety, a staff member, who is an experienced driver, accompanied the driver in the front passenger seat and alerted the driver if he detected any potential risk. A second staff member was seated in the back to ensure the monitoring equipment in the test vehicle was working properly. The two staff members were instructed not to converse during the experiment except to make the driver aware of a hazardous condition or to address an equipment problem.

Data on successful lane-change maneuvers and failed lane-change maneuvers were collected. A failed lane change is when the driver intended to change lanes, but failed to perform the lane-change maneuver. A video monitoring system recorded the driver during the duration of the experiment. The intention to change lanes was detected by observing the driver's eye and head movement, the use of turn signal lamp, and the driving environment [40–43], which was recorded by the monitoring system.

3.4. *Extreme Moment Trials*

The extreme moment trials were divided into two parts: the front extreme moment trial and the rear extreme moment trial.

The front moment is the last possible moment that the front vehicle (F), on the host lane, can safely change to the target lane without colliding with the rear vehicle (R). In the real lane-change processes, when R is driving on the target lane and is quickly approaching F, the relative distance between R and F shortens; thus, the lane-change process will gradually change from a safe to a dangerous state. In this work, the moment between the safe state and the dangerous state, i.e., the last chance for the driver to perform a safe lane change and have no negative effect on the R, is called the front extreme moment. A lane change anytime up to the front extreme moment produces no danger and the rear vehicle driver can drive normally without feeling the need to take evasive action.

Similar to the front moment, when R is approaching F quickly, there is a moment between when the rear driver feels safe and when he feels endangered. In this work, this moment is referred to as the rear extreme moment.

Figure 3 is a schematic diagram of the front extreme moment trial. In this part of the experiment, the participants were required to drive our test vehicle (T) on the target lane, and estimate the front extreme moment. Participants were instructed to quickly approach the front vehicle, and indicate the extreme moment during their approach by pressing the wireless button.

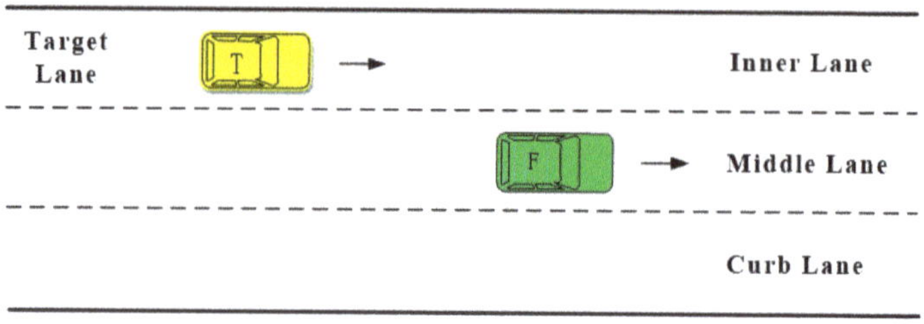

Figure 3. Front extreme moment trial.

Figure 4 is a schematic diagram of the rear extreme trial. In this part of the experiment, the participants were required to drive our test vehicle (T), and estimate the rear extreme moment based on their observation of R. Participants were instructed to drive on the middle lane and to indicate the extreme moment by pressing the wireless button when R was fast approaching the test vehicle (T) from the target lane.

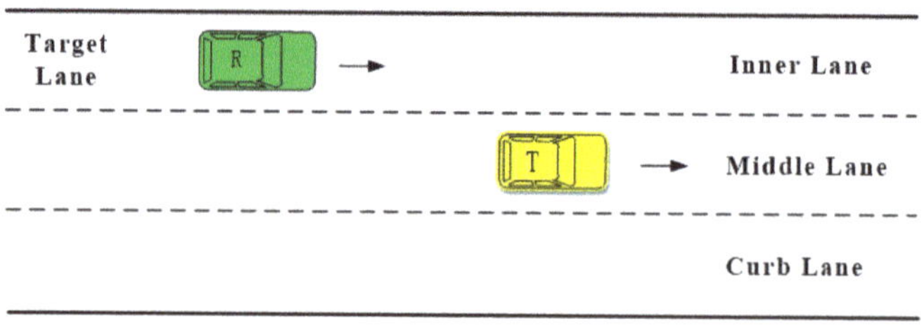

Figure 4. Rear extreme moment trial.

As with the naturalistic lane-change experiment, two staff members were seated in the front and back, respectively. During the experiment, the staff member in the back seat focused on R. When R, driving on the adjacent lane, was quickly approaching T, the staff member would ask the driver to observe R and indicate the extreme moment by pressing the button. The front staff member's responsibility was to observe the driving environment and ensure the experiment safety.

Due to the drivers needed to frequently observe R, the workload in this test was heavier than normal driving. To reduce the workload, the experiment was carried out in cruise control mode at speeds of 60 km/h, 70 km/h, 80 km/h, and 90 km/h.

3.5. Procedures

Before the experiment, the drivers were asked to participate in a practice round for approximately 10 min to familiarize them with the test vehicle and the road. Next, the participants began the naturalistic lane-change experiment. To ensure the drivers drove with their personal style, the staff member asked them to first drive along the test route and did not give them any instructions during the experiment. Following completion of the naturalistic lane-change experiment and a 10 min break, the two extreme moment experiments were carried out in a random order. Participants were given a second 10 min break between front and rear extreme moment trials.

4. Lane-Change Behavior Analysis

4.1. Lane-Change Process

To investigate the lane-change behavior, the speed of S, relative speed, distance between S and surrounding vehicles, Lane-Change Duration (LCD), Time to Line Crossing (TLC), and deceleration behavior of S were assessed.

Figure 5 exhibits a complete lane-change process. t_0 marks the time the driver decides to change lanes. At t_1, the lane change begins (start moment); at t_2, the front wheel touches the lane marking; at t_3, the whole vehicle is in the adjacent lane; and at t_4, the lane-change process is complete.

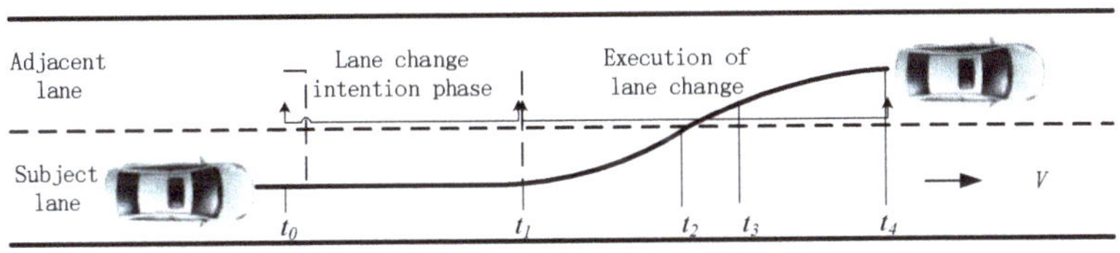

Figure 5. Lane-change process.

According to the LCD definition [44], the LCD started and ended when the vehicle began and stopped moving in a lateral direction, namely t_1 to t_4. Based on the TLC definition [45], the TLC started when the vehicle began moving in a lateral direction and ended when the front wheel touched the lane marker, namely t_1 to t_2.

4.2. Lane-Change Decision Model

Two safe level thresholds were proposed based on the driver's subjective extreme moment. In this model, once it is observed that S started changing lanes, the lane-change model predicts the safety deceleration of R, which is the minimum deceleration to ensure that R does not collide with S.

As shown in Figure 6, in real lane-change processes, most collisions occur after S crosses the lane marking, namely after t_2. Thus, t_2 to t_4 is the high-risk traffic conflict period during the lane-change process. Therefore, the lane-change safety evaluation should be completed before t_2 at the latest to effectively reduce the accident rate.

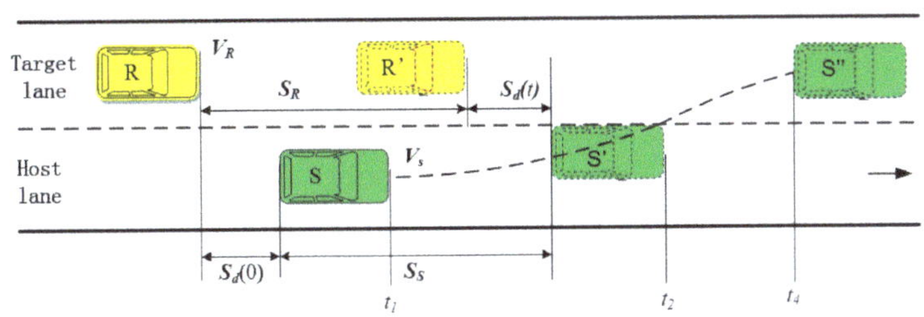

Figure 6. Lane change behavior.

To establish an intuitive lane-change decision model, the MSD of R during the lane change of S was selected as the indictor, which ensures R does not collide with S. This proposed model asserts that lane changes can be safely performed when the MSD is lower than a specific safe threshold. In some safety distance models, the safety distance was calculated according to the vehicle's maximum deceleration. However, on a real high-speed road, if R brakes with the maximum deceleration, the collision occurrence between R and the car behind it will increase. In reality, the driver usually does not break with maximum deceleration on a high-speed road. Therefore, in this study, two levels of safe thresholds were selected based on naturalistic driving behavior on a real highway.

The key parameters of lane-change behavior derived from the naturalistic lane-change trial include the: velocity of S (V_S), velocity of R (V_R), relative velocity (V_r), deceleration of R, TLC, and relative longitudinal distance between S and R (S_d).

Considering t_1 as the initial time of the lane change, and t_2 as the TLC of F, for any time before the line crossing, the longitudinal displacement of S ($S_S(t)$) can be calculated as:

$$S_S(t) = V_S(t_1)t - \int_{t_1}^{t}\int_{t_1}^{t} a_S(\tau)d\tau dt \quad t \in [t_1, t_2] \tag{1}$$

where $a_S(\tau)$ is the longitudinal deceleration of S at time t after the lane-change start moment and $V_S(t_1)$ is the speed of S at t_1.

Moridpour et al. [43] reported that drivers (someone who wants to change lane) keep a constant speed during the lane-change process. Therefore, Equation (1) is simplified as:

$$S_S(t) = V_S(t_1)t \quad t \in [t_1, t_2] \tag{2}$$

The longitudinal displacement of R ($S_R(t)$) can be calculated as:

$$\begin{cases} S_R(t) = V_R(t_1)t \quad t \le T \\ S_R(t) = V_R(t_1)t - \int_{t_1}^{t-T}\int_{t_1}^{t-T} a_R(\tau)d\tau dt \quad t > T \end{cases} \quad t \in [t_1, t_2] \tag{3}$$

where $a_R(\tau)$ is the longitudinal deceleration of R during the lane change and $V_R(t_1)$ is the speed of S at t_1. Rear vehicle driver's reaction time (T) is an important factor in the lane-change process. Many researchers [46,47] have investigated the reaction time during brake behavior, and found the reaction time for deceleration is about 1 s.

In Equation (3) describes the total longitudinal displacement of R from t_1 to $t.t \le T$ means the driver of R has not yet responded, and R continues to drive at a constant speed of $V_R(t_1)$. During this period, the total longitudinal displacement of R is $V_R(t_1) * t.t > T$ means that the driver of R has responded to the braking of S, R maintains a deceleration of $a_R(\tau)$. During this period, the total longitudinal displacement of R from t_1 to t is $S_R(t) = V_R(t_1)t - \int_{t_1}^{t-T}\int_{t_1}^{t-T} a_R(\tau)d\tau dt$.

At the any moment of t during a lane change, $S_d(t)$ can be calculated as:

$$S_d(t) = S_d(t_1) + [S_S(t) - S_R(t)] \quad t \in [t_1, t_2] \tag{4}$$

where $S_d(t_1)$ is the relative longitudinal distance between S and R at t_1.

At t_2, if the relative velocity $V_r(t_2)$ is close to or lower than 0 m/s, even if the distance between S and R is small, the TTC will be small, and the moment is considered a safe stage [21]. However, drivers tend to keep a minimum safe distance (D_{t2}) at t_2. Therefore, $S_d(t)$ at t_2 can be calculated as:

$$S_d(t_2) = S_d(t_1) - V_r(t_1)t_2 + \frac{1}{2}a_R(t_2 - T)^2 - D_{t2} \ge 0 \quad V_r(t_2) \le 0 \tag{5}$$

$$V_r(t_2) = V_R - a_R(t_2 - T) - V_S \tag{6}$$

Furthermore, if $V_r(t_2)$ is higher than 0 m/s, to avoid a collision with S, the rear vehicle driver will maintain the previous or a greater deceleration until V_r is equal to 0 m/s. If $V_r(t_2)$ is higher than 0 m/s,

$S_d(t_2)$ should be higher than a safe distance (D_S), the D_S can help ensures safety before the relative speed is reduced to 0 m/s, the D_S can be calculated as:

$$D_S = \frac{V_r^2(t_2)}{2a_R} \leq S_d(t_2)V_r(t_2) \leq 0 \tag{7}$$

Therefore, $S_d(t_2)$ can be calculated as:

$$S_d(t_2) = S_d(t_1) - \frac{V_r^2(t_1)}{2a_R} - V_r(t_1)T - D_{t2} \geq 0 \quad V_r(t_2) \leq 0 \tag{8}$$

In addition, if the relative speed is small, a lane change under these conditions is theoretically safe. However, the driver will consider the lane change unsafe when the relative distance at t_1 is small. That is, the lane-change decision model will only perform the lane change if the $S_d(t_1)$ is greater than a minimum acceptance distance (D_{t1}). Therefore, to perform a safe lanechange, the following two conditions must be met:

$$\begin{cases} S_d(t_1) \geq D_{t1} \\ S_d(t_2) \geq 0 \end{cases} \tag{9}$$

$S_d(t_1)$ and $V_r(t_1)$ are collected using the millimeter wave radar; t_2, D_{t1}, and D_{t2} are analyzed based on the naturalistic driving data, and a_R is calculated using Equations (5) and (8).

4.3. Parameter Calibration

During the naturalistic lane-change trial, 895 lane-change processes were recorded; 317 of which had R approaching quickly. A statistical analysis on all the lane-change processes was conducted and the LCD distribution is shown in Figure 7. The LCD ranged from 1.6 to 20.0 s, with a median of 6.6 s, a mean of 7.0 s, and a standard deviation of 2.1 s. After the lane-change start moment, some drivers slowly changed lanes, waiting to be overtaken by R, which resulted in the LCDs of this experiment being higher than LCDs recorded in the previous studies [48,49].

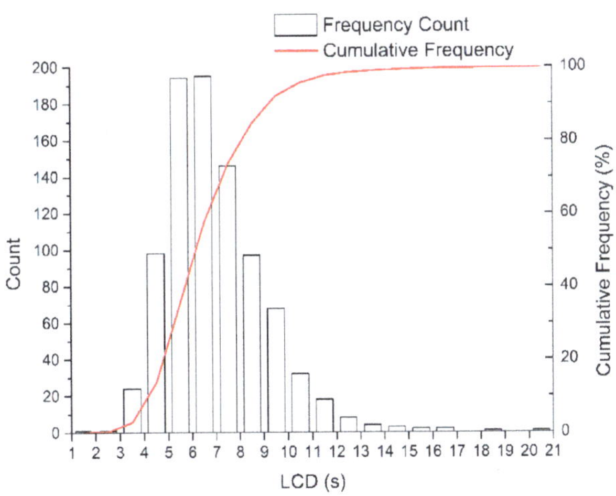

Figure 7. Lane-Change Duration (LCD) distribution.

Based on the definition of TLC, this paper calculated TLC using the distance between the vehicle's front wheel and the lane mark collected by the lane mark recognition system (Mobileye C2-170).

The TLC distribution is shown in Figure 8. The TLC ranged from 0.2 to 8.3 s, with a median of 1.6 s, a mean of 1.7 s, and a standard deviation of 1.0 s. A total of 2.8% of the TLC during the lane-change processes were > 4 s, the result of some drivers slowly changing lanes to allow R to overtake, which helps avoid a collision with R. Furthermore, 85.8% of the TLC ranged between 0.2 and 2.5 s. In this model, we calibrated the TLC as 1.6 s.

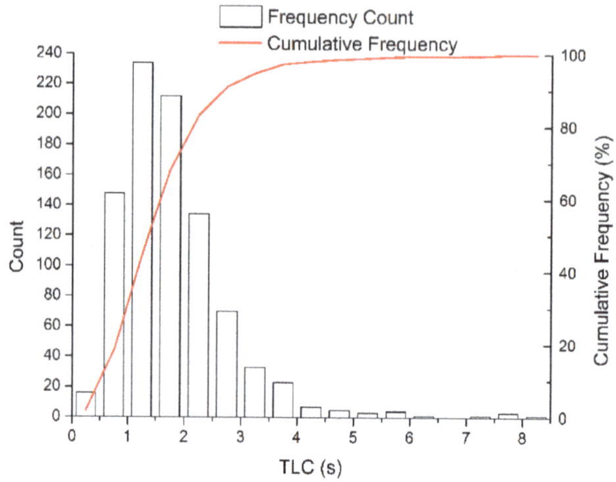

Figure 8. Time to Line Crossing (TLC) distribution.

The distribution of the test vehicle acceleration during the lane changes is shown in Figure 9. Acceleration ranged from −1.02 to 1.16 m/s^2, with a median of 0.06 m/s^2, a mean of 0.07 m/s^2, and a standard deviation of 0.27 m/s^2. The acceleration was close to 0 m/s^2, which implies that the driving speed is maintained during the lane change, which confirms the previous study [43].

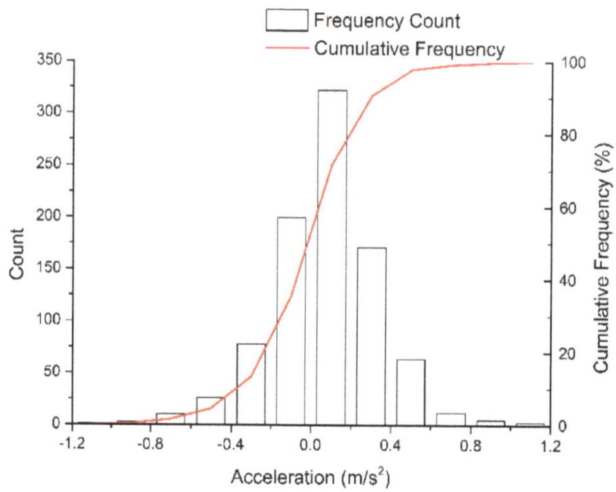

Figure 9. Acceleration distribution.

4.4. Minimun Acceptance Distance (D_{t1}) and Minimum Safe Distance (D_{t2})

On the highway, a driver may not perform a lane change when the relative distance at t_1 is small. To investigate $\boldsymbol{D_{t1}}$, we collected the relative distance before t_2 from the data set of successful lane changes. The results showed a maximum is 204.60 m, a minimum is 4.59 m, a mean of 42.33 m, and a standard deviation of 32.51 m. Therefore, $\boldsymbol{D_{t1}}$ was determined as 4.59 m.

After t_2, even when the relative speed is close to 0 m/s, the front and rear drivers try to maintain a suitable safe distance. Sultan et al. [50] suggested that the relative speed is considered low when in the range of [−1.5 m/s, 1.5m/s]. To determine D_{t2}, we collected the relative distance when relative speed after t_2 was low. The results showed a maximum of 175.20 m, a minimum of 3.25, a mean of 32.12 m, and standard deviation of 27.60 m. Therefore, D_{t2} was determined as 3.25 m.

4.5. Extreme Moment Data

In the extreme moment trials, to ensure the safety of the driver, lane changes were not performed during the experiment. Participants were instructed only to push the wireless button to indicate the extreme moment. Based on the relative longitudinal distance and relative speed between R and T (F and T) at the subjective extreme moment, the MSD of R was calculated using the proposed lane-change decision model in Section 4.2. The MSDs at the front extreme moment and the rear extreme moment from different participants are shown in Figure 10.

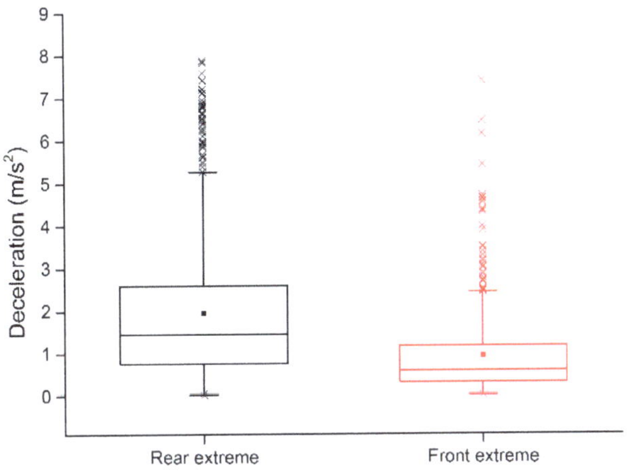

Figure 10. Minimum Safe Deceleration (MSD) for the extreme moment.

In the rear extreme moment trial, 1300 rear extreme moments were collected, and the corresponding MSDs of R were calculated using the lane-change decision model. As shown in Figure 10, the decelerations ranged from 0.02 to 7.88 m/s^2; the 25th, median, and 75th percentiles were 0.75 m/s^2, 1.42 m/s^2, and 2.58 m/s^2, respectively; the mean was 1.95 m/s^2, and the standard deviation was 1.66 m/s^2. The result showed that from the perspective of the participants in the rear extreme trial, half of the participants in the rear extreme trial felt they could safely change lanes while the MSD is higher than 1.42 m/s^2, and 75% of participants in the rear extreme trial cannot accept a MSD that is more than 2.58 m/s^2.

Based on the statistical analysis of the rear extreme moment trial results, the second safe threshold (ST_2) was initially set between the 75th percentile and the median, namely 1.42 to 2.58 m/s^2.

In the front extreme moment trial, 912 front extreme moments were collected, and the corresponding MSDs of R were calculated using the lane-change decision model. As shown in Figure 10, the decelerations ranged from 0.02 to 7.40 m/s^2; the 25th, median, and 75th percentiles were 0.29 m/s^2, 0.50 m/s^2, and 1.12 m/s^2, respectively; the mean was 0.93 m/s^2, and the standard deviation was 1.01 m/s^2. The results showed that the subjective MSD of half the participants in the front extreme trial was higher than 0.50 m/s^2, and 75% of the participants in the front extreme trial cannot accept a MSD of more than 1.12 m/s^2, as it likely triggers a high level of anxiety in the drivers [51].

Based on the statistical analysis of the front extreme moment trial results, the primary safe threshold (ST_1) was initially set between the 75th percentile and the median, namely 0.50 to 1.12 m/s^2.

The t-test was used to compare the subjective perception of the drivers in the front and rear extreme trial on lane-change safety. The Levene's Test for equality of variances results showed that F(912, 1300) = 216.515, $p < 0.001$, which means the variances among the two driver types is not equal. The Welch's t-test for equality of means results shown that $t(2117.681) = 26.619$, $p < 0.001$. The subjective perception of lane-change safety between participants in the rear extreme trial and participants in the front extreme trial have significant differences; in particular, the required deceleration in the rear extreme trial was higher than that of the front extreme trial. These results indicate that during the lane change, the participants in the front extreme trial were more aggressive, and the required deceleration safety level was higher in the front extreme trial than that in the rear extreme.

5. Threshold Determination

5.1. Signal Detection Theory

In order to establish the two-level safe lane-change thresholds that consider the driver's risk perception in different trials, we used the SDT, which is widely applied in the determination of the optimal threshold for human perception [52,53].

Signals and noise have different definitions among various psychology fields, and SDT was used to discriminate between them. Before performing the lane-change maneuver, the signal and noise was defined as a safe and unsafe signal. When the MSD is lower than the safe threshold, the lane-change decision system permits the lane change to be executed; if not, then the system decides to wait for the proper time to perform the lane change. In the natural driving experiment, the lane-change data was used to verify the safe threshold. The safety lane-change process in the natural driving experiment was defined as the safe lane change, and the data of the failed lane change was defined as the unsafe lane change. The lane-change decision matrix is shown in Table 1.

Table 1. The lane-change decision matrix.

	Safe Lane Change	**Unsafe Lane Change**
Safe signal	Hit	False alarm
Unsafe signal	False negative	Correct rejection

Performing a lane change under the safe condition is correct and is termed Hit; performing a lane change in the unsafe condition is incorrect and is defined as a False alarm. When the lane-change decision system neglects the safe lane-change situation, it is incorrect and is termed as a False negative; waiting in the unsafe lane-change situation is correct and is defined as a Correct rejection. Both Hit and Correct rejection are correct signals, and the correct rate is termed the accuracy (P_A). In this paper, the P_A, False negative rate (P_{FA}), and False alarm rate (P_{FN}) were used to evaluate the decision model's performance. The same P_{FN}, a higher P_A, and lower P_{FA} indicate the better performance by the decision model.

The accuracy is calculated as:

$$P_A = 1 - \frac{N_{FA} + N_{FN}}{N_S + N_U} \tag{10}$$

The False alarm rate is calculated as:

$$P_{FA} = \frac{N_{FA}}{N_S} \tag{11}$$

The False negative rate is calculated as:

$$P_{FN} = \frac{N_{FN}}{N_U} \tag{12}$$

where the N_S is the total number of safe lane changes, N_U is the total number of unsafe lane changes, N_{FN} is the number of False negatives, and N_{FA} is the number of False alarms.

We selected the optimal safe thresholds within the range of ST_1 and ST_2 obtained in the previous section. The ST_1 ranged from 0.50 to 1.12 m/s^2, and the ST_2 ranged from 1.42 to 2.58 m/s^2. To calculate P_A, P_{FA}, and P_{FN} at different MSD, the 0.01 m/s^2 was selected as the step length. Two level safe thresholds were determined by considering the P_A, P_{FA}, and P_{FN}.

5.2. Primary Safe Threshold (ST_1) Selection

In our two-level lane-change decision model, the emphasis on the two safe thresholds are different. For the ST_1, the major aim is to ensure that the S lane-change maneuver will not have a serious negative impact on the rear vehicle driver's driving behavior, e.g., accelerating to avoid being cut-in, emergency braking, or anxiety and/or road rage. Therefore the primary deceleration threshold was selected within 0.50 to 1.12 m/s^2, which meets most rear vehicle drivers' expectations for safe car-following after cut-in events.

P_A, P_{FA}, and P_{FN}, at different ST_1 are shown in Figure 11. The results showed that P_A and P_{FN} increase in parallel with MSD, while P_{FA} decreases as a function of increasing MSD. The purpose of ST_1 in the proposed model is to minimize the effect of S lane-change behavior on the R driver, and to avoid all potential dangerous situations to the highest extent possible. The higher P_{FN} may trigger to the more potential risk. Therefore, the major aim of ST_1 is to reduce the P_{FN}. Within the range of ST_1, all the P_{FN} ranged from 3.2% to 8.9%; the P_{FN} was lower than 5% when the safe threshold was less than 0.85 m/s^2.

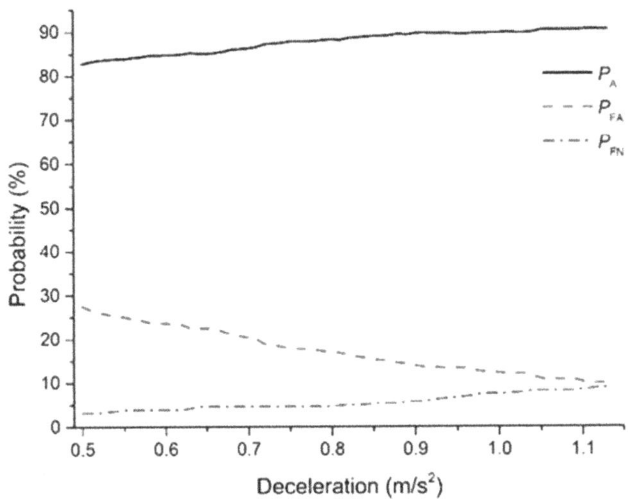

Figure 11. Signal Detection Theory (SDT) for the primary safe threshold (ST_1).

In addition, the higher the threshold, the higher the P_A and the lower the P_{FA}. This trend indicates better performance of the lane-change decision system. When the threshold was 0.85 m/s^2, P_A and P_{FA} reached 88.8% and 15.7%, respectively.

Therefore, by considering P_A, P_{FA}, and P_{FN}, ST_1 was determined as 0.85 m/s^2; while P_A, P_{FA}, and P_{FN} were 88.8%, 15.7%, and 5.0%, respectively. The primary safety threshold minimizes the occurrence of potential risks while ensuring accuracy, and satisfies the expectations of more than half of the rear vehicle drivers for safe car-following.

5.3. *Secondary Threshold (ST_2) Selection*

Unlike ST_1, which focuses on rear vehicle drivers and all the associated potential risks, the main target of ST_2 is the subject vehicle-driver's risk perception and ensuring the decision system reliability. Whether the safe threshold is higher or lower than the driver's perceived safe acceleration, it still could possibly reduce the driver's acceptance and trust of the intelligent assistance system. Therefore, ST_2 was selected by considering the risk assessment of the subject vehicle driver in the rear extreme trial, which ranged from 1.42 m/s^2 to 2.58 m/s^2.

P_A, P_{FA}, and P_{FN} at different ST_2 are shown in Figure 12. Within the range of ST_2, changes of ST_2 had little effect on P_A, which was generally stable at around 90%. P_{FN} rapidly increased with the increase of ST_2, and P_{FA} was lower than 7%. According to the purpose of the ST_2 in the proposed model, the major aim is to increase the decision system accuracy, which could improve the acceptance and trust in the lane-change decision system. Thus, the P_A needs to be considered first while selecting the secondary safe threshold.

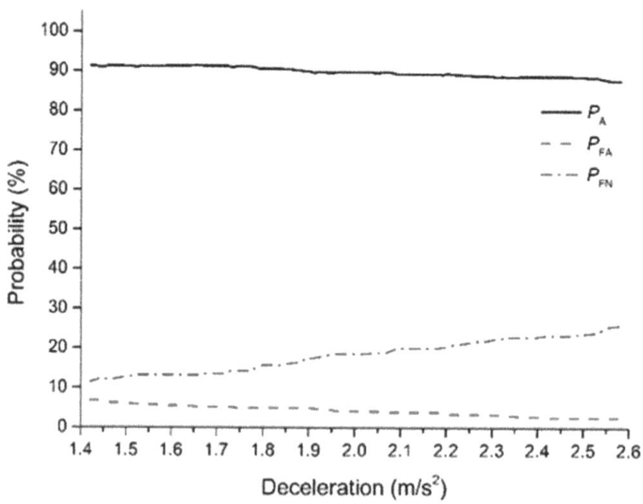

Figure 12. SDT for the second safe threshold (ST_2).

Within the range of the secondary safe threshold, P_A fluctuated between 87.6% and 91.4%. When ST_2 was less than 1.87 m/s^2, the P_A exceeded 90%.

In addition, P_{FA} and P_{FN} is an important factor for the improvement of the driver trust and acceptance. False alarm denotes performing a lane change in unsafe condition, which may reduce the trust, and false negative denotes waiting in a safe condition, which may reduce the acceptance. Thus, the P_{FA} and P_{FN} were taken into consideration while determining ST_2. When the safe threshold was higher than 1.76 m/s^2, the growth rate of P_{FN} increased. In addition, the P_{FA} was less than 5.0% when the threshold was 1.76 m/s^2.

Therefore, by considering P_A, P_{FA}, and P_{FN}, ST_2 was determined as 1.76 m/s^2; P_A, P_{FA}, and P_{FN} were 91.1%, 4.9%, and 14.3%, respectively. ST_2 can improve the trust and acceptance of the subject vehicle driver in the lane-change decision system.

5.4. Summary of the Lane-Change Decision Model

According to the two-level safe thresholds, the decision rule is:

$$Decision = \begin{cases} \text{Safe and polite} & \text{if MSD} \leq \text{ST}_1 \\ \text{Safe but impolite} & \text{if ST}_1 < \text{MSD} \leq \text{ST}_2 \\ \text{Waiting} & \text{if MSD} > \text{ST}_2 \end{cases} \quad (13)$$

In the decision model, "Safe and good" denotes an opportunity in which AVs can perform a safe and polite lane change, which will not negatively affect the rear vehicle. "Safe but impolite" means that the AVs can safely execute a lane change, but this maneuver may disturb the driving behavior of the rear vehicle. "Waiting" means a rear-end collision may occur if AVs change lane at this moment.

To verify the lane-change decision model's safety, the ISO model was compared with the decision model. ISO [21] considered TTC as the indictor and proposed a multi-level lane-change safe threshold for different relative speeds while R was quickly approaching S. It suggested that the TTC threshold was 2.5 s when the relative speed less than 10 m/s; the TTC threshold was 3.0 s when the relative speed ranged from 10 to 15 m/s; and the TTC threshold was 3.5 s when the relative speed ranged from 15 to 20 m/s.

The dataset of safe lane changes and unsafe lane changes was used to evaluate the lane-change decision model's performance and the ISO model. The result is shown in Figure 13.

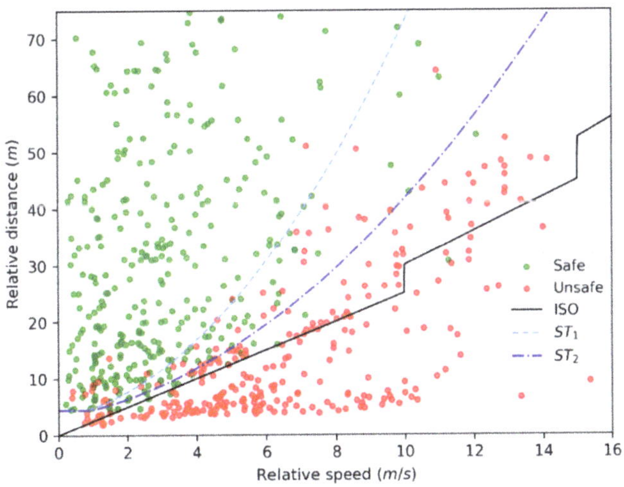

Figure 13. The comparison between the International Standards Organization (ISO) model and lane-change decision model.

A summary of P_A, P_{FA}, and P_{FN} of the ISO model and the proposed model are shown in Table 2. Compared with the ISO model, P_A for both ST_1 and ST_2 was higher. Although P_{FA} of the ISO model is small, P_{FN} of the ISO model is far higher than that of lane-change decision model. The result indicates that the lane-change decision model has high safety reliability.

Table 2. The lane-change decision matrix.

	P_A (%)	P_{FA} (%)	P_{FN} (%)
ST_1	88.8	15.7	5.0
ST_2	91.1	4.9	14.3
ISO model	82.0	1.3	40.1

6. Discussion and Conclusions

In this work, we established a lane-change decision model with a two-level safe threshold in mixed traffic, which considers the rear vehicle's deceleration behavior. The parameters in this model were calibrated based on the naturalistic lane-change behaviors. The two-level safe thresholds were determined according to the driver's various risk perceptions—as the driver of R and driver of F—which were investigated in the front and rear extreme trial. The decision model was evaluated using SDT, which takes into account various risk perception of different drivers and real lane-change data. ST_1 and ST_2 were determined as 0.85 m/s^2 and 1.76 m/s^2, respectively.

The rear vehicle's MSD was selected as an indicator to evaluate the lane-change safety, which was different from other widely used indicators, such as TTC, Time Headway (THW) (the ratio of relative distance between subject vehicle and front vehicle to the speed of subject vehicle), and TTCi (the reciprocal of TTC value) [54–56]. The MSD is an intuitive indicator, which is directly related to the maneuverability and willingness of the rear vehicle driver. A progressively higher MSD means the rear vehicle driver has to make a faster and greater response, which may increase the occurrence of the rear-end collision.

Two different extreme moment experiments were conducted on a highway to examine the lane-change behavior and risk perception of the driver. In the experiments, a vehicle outfitted with instruments was used to collect the information of surrounding vehicles and the subject vehicle's driving data.

The lane change behavior was investigated based on the naturalistic lane-change experiment. The lane-change behavior measures assessed for the decision model were: LCD, TLC, and deceleration. Compared with the previous studies [47,48,57], both LCD and TLC were slightly higher, for two primary reasons. This experiment was carried out on the highway, a more dangerous environment for a vehicle to be fast approaching the test vehicle. Some drivers tended to slowly change lanes and let the rear vehicle overtake them to avoid collisions. Moreover, some researchers [58–60], found that the vehicle and driving environment model effect the driving behavior. Thus, the test vehicle and driving route model in this study may impact the LCD and TLC. Furthermore, the acceleration behavior during the lane change was investigated. The results validated those of a previous study [40], which reported that the driving speed is kept constant during the lane change. Therefore, we simplified the lane-change decision model by ignoring the car's speed change during the lane-change process.

The various risk perceptions of the rear vehicle driver and the front vehicle driver were explored using front and rear extreme trials. The MSD was used to characterize the driver's risk perception, which was calculated using the proposed model. Compared to the driver of the preceding car, when the participant is a driver of the rear car, the perceived MSD is significantly smaller, which indicates that the rear car driver is more cautious during the lane-change process. Using SDT, the two-level thresholds were determined by considering the different perceived MSD in the two extreme trials.

The major benefit of this study is the determination of two safe thresholds, each of which has a different role. The rear vehicle driver's safety and risk perception is primarily accounted for by ST_1, which was examined in the front extreme trial. The lane-change process can be viewed as the game process between the front vehicle and the rear vehicle [61], in which the preceding vehicle's lane-change behavior greatly affects the rear vehicle driver's driving behavior and emotion [62,63]. This implies that dangerous lane-change behavior not only leads to a potential rear-end collision between the front and rear vehicles, but also triggers the rear vehicle driver to engage in dangerous driving behavior, which increases the likelihood of the rear car colliding with other surrounding vehicles. Therefore, determination of ST_1 based on the rear vehicle driver's risk perception alleviates the driver's anxiety and improves the rear vehicle's safety.

The proposed ST_1 was determined based on the rear extreme trial results analysis. Compared with the existing lane-change decision system, the secondary safe threshold not only ensures lane-change safety, but also accounts for the driver's expectation, which minimizes the decision model interference on the driving behavior. ST_1, which is established by the driver's subjective perception, enhances the driver's acceptance and trust of the intelligent driving system and contributes to the intelligent assistant system popularization [33,64].

Although this model's thresholds were based on the subjective feelings of different drivers, the accuracy was guaranteed. Compared with the ISO model, this model's accuracy was higher at both two-level thresholds, which indicates that this model improves lane-change safety, while ensuring the driver's comfort.

The decision model can be used in different driving situations. In the free condition [9], AVs can perform a safe and polite lane-change maneuver. In the force condition or in a hurry, AVs can perform the safe but not polite lane-change maneuver, which can save passengers time on the basis of safety.

The novel human-like lane-change decision model can find a more suitable time to change lanes in mixed traffic, which ensures the subject vehicle's safety and reduces its interferences with the rear vehicle, thus further ensuring safety all around.

A few aspects of our work need to be improved in future research. Different driving styles may reflect different subjective perceptions of the ability to safely change lanes. Assuming that the numbers of samples are adequate, future study will focus on establishing diverse thresholds on the basis of different driving styles to enhance the acceptability of the proposed lane-change decision model. In addition, model parameters need to be calibrated based on a more sufficient number of samples.

Author Contributions: C.W., Q.S., and Z.L. conceived of and designed the research. H.Z., Q.S. and Z.L. conducted the experiments. C.W., H.Z., Q.S. and Z.L. wrote the manuscript. All authors discussed and commented on the manuscript. All authors have read and agreed to the published version of the manuscript.

Funding: This work was supported in part by the National Key Research and Development Program of China under Grant 2019YFB1600500, in part by the National Natural Science Foundation of China under Grant 51908054, in part by the Key Research and Development Program of Shannxi under Grant 2020GY-163, and in part by the Fundamental Research Funds for the Central Universities, CHD 300102220202.

Conflicts of Interest: The authors declare that they have no competing interest.

References

1. Cicchino, J.B. Effects of lane departure warning on police-reported crash rates. *J. Saf. Res.* **2018**, *66*, 61–70. [CrossRef] [PubMed]
2. Dey, K.C.; Yan, L.; Wang, X.; Wang, Y.; Shen, H.; Chowdhury, M.; Soundararaj, V. A review of communication, driver characteristics, and controls aspects of cooperative adaptive cruise control (CACC). *IEEE Trans. Intell. Transp. Syst.* **2015**, *17*, 491–509. [CrossRef]
3. Gong, S.; Du, L. Optimal location of advance warning for mandatory lane change near a two-lane highway off-ramp. *Transp. Res. Part B Methodol.* **2016**, *84*, 1–30. [CrossRef]
4. Hou, Y.; Edara, P.; Sun, C. Situation assessment and decision making for lane change assistance using ensemble learning methods. *Expert Syst. Appl.* **2015**, *42*, 3875–3882. [CrossRef]
5. Traffic Management Bureau of the Public Security Ministry. *Annual Statistic Yearbook of Road Traffic Accidents in China (2015)*; Traffic Management Bureau of the Public Security Ministry: Beijing, China, 2016.
6. Carl, J.A.; Karsten, H. European Accident Research and Safety Report 2013. Report, Volvo Trucks. Driving Progress, Gothenburg. Available online: https://www.volvogroup.com/content/dam/volvo/volvo-group/markets/global/en-en/about-us/traffic-safety/ART-report-2013.pdf (accessed on 9 January 2013).
7. Gipps, P.G. A model for the structure of lane-changing decisions. *Transp. Res. Part B Methodol.* **1986**, *20*, 403–414. [CrossRef]

8. Halati, A.; Lieu, H.; Walker, S. CORSIM-corridor traffic simulation model. In *Proceedings of the Traffic Congestion and Traffic Safety in the 21st Century: Challenges, Innovations, and Opportunities, Chicago, IL, USA, 8–11 June 1997*; ASCE: New York, NY, USA, 1997.
9. Hidas, P. Modelling vehicle interactions in microscopic simulation of merging and weaving. *Transp. Res. Part C Emerg. Technol.* **2005**, *13*, 37–62. [CrossRef]
10. Kita, H. A merging–giveway interaction model of cars in a merging section: A game theoretic analysis. *Transp. Res. Part A Policy Pract.* **1999**, *33*, 305–312. [CrossRef]
11. Arbis, D.; Dixit, V.V. Game theoretic model for lane changing: Incorporating conflict risks. *Accid. Anal. Prev.* **2019**, *125*, 158–164. [CrossRef]
12. Nilsson, J.; Silvlin, J.; Brannstrom, M.; Coelingh, E.; Fredriksson, J. If, when, and how to perform lane change maneuvers on highways. *IEEE Intell. Transp. Syst. Mag.* **2016**, *8*, 68–78. [CrossRef]
13. Jula, H.; Kosmatopoulos, E.B.; Ioannou, P.A. Collision avoidance analysis for lane changing and merging. *IEEE Trans. Veh. Technol.* **2000**, *49*, 2295–2308. [CrossRef]
14. Kamal, M.A.S.; Taguchi, S.; Yoshimura, T. Efficient vehicle driving on multi-lane roads using model predictive control under a connected vehicle environment. In Proceedings of the 2015 IEEE Intelligent Vehicles Symposium (IV), Seoul, Korea, 28 June–1 July 2015; pp. 736–741.
15. Balal, E.; Cheu, R.L.; Sarkodie-Gyan, T. A binary decision model for discretionary lane changing move based on fuzzy inference system. *Transp. Res. Part C Emerg. Technol.* **2016**, *67*, 7–61. [CrossRef]
16. Hill, C.; Elefteriadou, L. Exploration of lane changing behavior on freeways. No. 13-2199. In Proceedings of the Transportation Research Board 92nd Annual Meeting, Washington, DC, USA, 13–17 January 2013.
17. Nobukawa, K.; Bao, S.; LeBlanc, D.J.; Zhao, D.; Peng, H.; Pan, C.S. Gap acceptance during lane changes by large-truck drivers—An image-based analysis. *IEEE Trans. Intell. Transp. Syst.* **2015**, *17*, 772–781. [CrossRef] [PubMed]
18. Yang, M.; Wang, X.; Quddus, M. Examining lane change gap acceptance, duration and impact using naturalistic driving data. *Transp. Res. Part C Emerg. Technol.* **2019**, *104*, 317–331. [CrossRef]
19. Lee, S.E.; Olsen, E.C.; Wierwille, W.W. *A Comprehensive Examination of Naturalistic Lane-Changes (No. FHWA-JPO-04-092)*; National Highway Traffic Safety Administration: Washington, DC, USA, 2004.
20. Wakasugi, T. A study on warning timing for lane change decision aid systems based on driver's lane change maneuver. In *Proceedings of the International Technical Conference on the Enhanced Safety of Vehicles*; National Highway Traffic Safety Administration: Washington, DC, USA, 2005; Volume 2005, p. 7.
21. International Standards Organization (ISO). *Intelligent transport systems—Lane Change Decision Aid Systems (LCDAS) 17387:2008*; International Standards Organization: Geneva, Switzerland, 2008.
22. Bordes, M.E.G. Combined lane change assist and rear, cross-traffic alert functionality. U.S. Patent Application No.12/855,238, 12 February 2012.
23. Van Dijck, T.; van der Heijden, G.A. VisionSense: An advanced lateral collision warning system. In Proceedings of the IEEE Intelligent Vehicles Symposium, Las Vegas, NV, USA, 6–8 June 2005; pp. 296–301.
24. Hirst, S. Of Collision Warnings. In *Ergonomics and Safety of Intelligent Driver Interfaces*; Loughborough University: Loughborough, UK, 1997; p. 203.
25. Saunier, N.; Sayed, T. Automated analysis of road safety with video data. *Transp. Res. Rec.* **2007**, *2019*, 57–64. [CrossRef]
26. Berdoulat, E.; Vavassori, D.; Sastre, M.T.M. Driving anger, emotional and instrumental aggressiveness, and impulsiveness in the prediction of aggressive and transgressive driving. *Accid. Anal. Prev.* **2013**, *50*, 758–767. [CrossRef]
27. Roidl, E.; Frehse, B.; Höger, R. Emotional states of drivers and the impact on speed, acceleration and traffic violations—A simulator study. *Accid. Anal. Prev.* **2014**, *70*, 282–292. [CrossRef]
28. Duan, J.; Li, R.; Hou, L.; Wang, W.; Li, G.; Li, S.E.; Gao, H. Driver braking behavior analysis to improve autonomous emergency braking systems in typical Chinese vehicle-bicycle conflicts. *Accid. Anal. Prev.* **2017**, *108*, 74–82. [CrossRef]
29. Feng, Z.; Ma, X.; Zhu, X.; Ma, Z. Analysis of Driver Brake Behavior under Critical Cut-in Scenarios. In Proceedings of the 2018 IEEE Intelligent Vehicles Symposium (IV), Changshu, China, 26–30 June 2018; pp. 2054–2059.
30. Bhavsar, P.; Das, P.; Paugh, M.; Dey, K.; Chowdhury, M. Risk analysis of autonomous vehicles in mixed traffic streams. *Transp. Res. Rec. J. Transp. Res. Board* **2017**, *2625*, 51–61. [CrossRef]

31. Zhu, W.X.; Zhang, H.M. Analysis of mixed traffic flow with human-driving and autonomous cars based on car-following model. *Phys. A Stat. Mech. Appl.* **2018**, *496*, 274–285. [CrossRef]
32. Levinson, J.; Askeland, J.; Becker, J.; Dolson, J.; Held, D.; Kammel, S.; Kolter, J.Z.; Langer, D.; Pink, O.; Pratt, V.; et al. Towards fully autonomous driving: Systems and algorithms. In Proceedings of the 2011 IEEE Intelligent Vehicles Symposium (IV), Baden, Germany, 5–9 June 2011; pp. 163–168.
33. Guo, C.; Kidono, K.; Machida, T.; Terashima, R.; Kojima, Y. Human-like behavior generation for intelligent vehicles in urban environment based on a hybrid potential map. In Proceedings of the 2017 IEEE Intelligent Vehicles Symposium (IV), Los Angeles, CA, USA, 11–14 June 2017; pp. 197–203.
34. Li, L.; Ota, K.; Dong, M. Humanlike driving: Empirical decision-making system for autonomous vehicles. *IEEE Trans. Veh. Technol.* **2018**, *67*, 6814–6823. [CrossRef]
35. Kesting, A.; Treiber, M.; Helbing, D. General lane-changing model MOBIL for car-following models. *Transp. Res. Rec.* **2007**, *1999*, 86–94. [CrossRef]
36. Schakel, W.J.; Knoop, V.L.; Arem, B.V. Integrated lane change model with relaxation and synchronization. *Transp. Res. Rec.* **2012**, *2316*, 47–57. [CrossRef]
37. Chang, W.; Qinyu, S.; Rui, F.; Zhen, L.; Qiong, Z. Lane change warning threshold based on driver perception characteristics. *Accid. Anal. Prev.* **2018**, *117*, 164–174.
38. Park, H.; Oh, C.; Moon, J.; Kim, S. Development of a lane change risk index using vehicle trajectory data. *Accid. Anal. Prev.* **2018**, *110*, 1–8. [CrossRef] [PubMed]
39. Wang, C.; Sun, Q.; Guo, Y.; Fu, R.; Yuan, W. Improving the User Acceptability of Advanced Driver Assistance Systems Based on Different Driving Styles: A Case Study of Lane Change Warning Systems. *IEEE Trans. Intell. Transp. Syst.* **2019**. [CrossRef]
40. Peng, J.; Guo, Y.; Fu, R.; Yuan, W.; Wang, C. Multi-parameter prediction of drivers' lane-changing behaviour with neural network model. *Appl. Ergon.* **2015**, *50*, 207–217. [CrossRef]
41. Doshi, A.; Trivedi, M.M. On the roles of eye gaze and head dynamics in predicting driver's intent to change lanes. *IEEE Trans. Intell. Transp. Syst.* **2009**, *10*, 453–462. [CrossRef]
42. Henning, M.J.; Georgeon, O.; Krems, J.F. The quality of behavioral and environmental indicators used to infer the intention to change lanes. In Proceedings of the Fourth International Driving Symposium on Human Factors in Driver Assessment, Training and Vehicle Design, Washington, DC, USA, 9–12 July 2007.
43. Green, M. "How long does it take to stop?" Methodological analysis of driver perception-brake times. *Transp. Hum. Factors* **2000**, *2*, 195–216. [CrossRef]
44. Jin, W.L. A kinematic wave theory of lane-changing traffic flow. *Transp. Res. Part B Methodol.* **2010**, *44*, 1001–1021. [CrossRef]
45. Li, P.; Merat, N.; Zheng, Z.; Markkula, G.; Li, Y.; Wang, Y. Does cognitive distraction improve or degrade lane keeping performance? Analysis of time-to-line crossing safety margins. *Transp. Res. Part F Traffic Psychol. Behav.* **2018**, *57*, 48–58. [CrossRef]
46. Young, M.S.; Stanton, N.A. Back to the future: Brake reaction times for manual and automated vehicles. *Ergonomics* **2007**, *50*, 46–58. [CrossRef] [PubMed]
47. Salvucci, D.D.; Liu, A. The time course of a lane change: Driver control and eye-movement behavior. *Transp. Res. Part F Traffic Psychol. Behav.* **2002**, *5*, 123–132. [CrossRef]
48. Yuan, J.; Abdel-Aty, M.; Cai, Q.; Lee, J. Investigating drivers' mandatory lane change behavior on the weaving section of freeway with managed lanes: A driving simulator study. *Transp. Res. Part F Traffic Psychol. Behav.* **2019**, *62*, 11–32. [CrossRef]
49. Moridpour, S.; Rose, G.; Sarvi, M. Effect of surrounding traffic characteristics on lane changing behavior. *J. Transp. Eng.* **2010**, *136*, 973–985. [CrossRef]
50. Sultan, B.; Brackstone, M.; McDonald, M. Drivers' use of deceleration and acceleration information in car-following process. *Transp. Res. Rec. J. Transp. Res. Board* **2004**, *1883*, 31–39. [CrossRef]
51. Lucidi, F.; Giannini, A.M.; Sgalla, R.; Mallia, L.; Devoto, A.; Reichmann, S. Young novice driver subtypes: Relationship to driving violations, errors and lapses. *Accid. Anal. Prev.* **2010**, *42*, 1689–1696. [CrossRef]
52. Lees, M.N.; Lee, J.D. The influence of distraction and driving context on driver response to imperfect collision warning systems. *Ergonomics* **2007**, *50*, 1264–1286. [CrossRef]
53. Phan, M.T.; Fremont, V.; Thouvenin, I.; Sallak, M.; Cherfaoui, V. Estimation of driver awareness of pedestrian based on Hidden Markov Model. In Proceedings of the 2015 IEEE Intelligent Vehicles Symposium (IV), Seoul, Korea, 28 June–1 July 2015; pp. 970–975.

54. Itoh, M.; Horikome, T.; Inagaki, T. Effectiveness and driver acceptance of a semi-autonomous forward obstacle collision avoidance system. *Appl. Ergon.* **2013**, *44*, 756–763. [CrossRef]
55. Zhao, D.; Lam, H.; Peng, H.; Bao, S.; LeBlanc, D.J.; Nobukawa, K.; Pan, C.S. Accelerated evaluation of automated vehicles safety in lane-change scenarios based on importance sampling techniques. *IEEE Trans. Intell. Transp. Syst.* **2016**, *18*, 595–607. [CrossRef]
56. Yurtsever, E.; Yamazaki, S.; Miyajima, C.; Takeda, K.; Mori, M.; Hitomi, K.; Egawa, M. Integrating driving behavior and traffic context through signal symbolization for data reduction and risky lane change detection. *IEEE Trans. Intell. Veh.* **2018**, *3*, 242–253. [CrossRef]
57. Amditis, A.; Bimpas, M.; Thomaidis, G.; Tsogas, M.; Netto, M.; Mammar, S.; Etemad, A. A situation-adaptive lane-keeping support system: Overview of the safelane approach. *IEEE Trans. Intell. Transp. Syst.* **2010**, *11*, 617–629. [CrossRef]
58. Wang, C.; Quddus, M.A.; Ison, S.G. The effect of traffic and road characteristics on road safety: A review and future research direction. *Saf. Sci.* **2013**, *57*, 264–275. [CrossRef]
59. Toledo, T.; Zohar, D. Modeling duration of lane changes. *Transp. Res. Rec.* **2007**, *1999*, 71–78. [CrossRef]
60. Tijerina, L.; Garrott, W.R.; Glecker, M.; Stoltzfus, D.; Parmer, E. *Van and Passenger Car Driver Eye Glance Behavior during Lane Change Decision Phase, Interim Report*; Transportation Research Center Report; National Highway Transportation Safety Administration: Washington, DC, USA, 1997.
61. Talebpour, A.; Mahmassani, H.S.; Hamdar, S.H. Modeling lane-changing behavior in a connected environment: A game theory approach. *Transp. Res. Procedia* **2015**, *7*, 420–440. [CrossRef]
62. Wiesenthal, D.L.; Hennessy, D.; Gibson, P.M. The Driving Vengeance Questionnaire (DVQ): The development of a scale to measure deviant drivers' attitudes. *Violence Vict.* **2000**, *15*, 115. [CrossRef]
63. Jurecki, R.; Poliak, M.; Jaśkiewicz, M. Young adult drivers: Simulated behaviour in a car-following situation. *Promet-Traffic Transp.* **2017**, *29*, 381–390. [CrossRef]
64. Shin, D.; Kim, B.; Yi, K.; Carvalho, A.; Borrelli, F. Human-Centered Risk Assessment of an Automated Vehicle Using Vehicular Wireless Communication. *IEEE Trans. Intell. Transp. Syst.* **2018**, *20*, 667–681. [CrossRef]

sensors

MDPI

Article

Reliable Road Scene Interpretation Based on ITOM with the Integrated Fusion of Vehicle and Lane Tracker in Dense Traffic Situation

Jinhan Jeong, Yook Hyun Yoon and Jahng Hyon Park *

Department of Automotive Engineering, Hanyang University, Seoul 04763, Korea; pegasus815@hanyang.ac.kr (J.J.); yyh6920@hanyang.ac.kr (Y.H.Y.)

* Correspondence: jpark@hanyang.ac.kr

Received: 26 March 2020; Accepted: 24 April 2020; Published: 26 April 2020

Abstract: Lane detection and tracking in a complex road environment is one of the most important research areas in highly automated driving systems. Studies on lane detection cover a variety of difficulties, such as shadowy situations, dimmed lane painting, and obstacles that prohibit lane feature detection. There are several hard cases in which lane candidate features are not easily extracted from image frames captured by a driving vehicle. We have carefully selected typical scenarios in which the extraction of lane candidate features can be easily corrupted by road vehicles and road markers that lead to degradations in the understanding of road scenes, resulting in difficult decision making. We have introduced two main contributions to the interpretation of road scenes in dense traffic environments. First, to obtain robust road scene understanding, we have designed a novel framework combining a lane tracker method integrated with a camera and a radar forward vehicle tracker system, which is especially useful in dense traffic situations. We have introduced an image template occupancy matching method with the integrated vehicle tracker that makes it possible to avoid extracting irrelevant lane features caused by forward target vehicles and road markers. Second, we present a robust multi-lane detection by a tracking algorithm that incudes adjacent lanes as well as ego lanes. We verify a comprehensive experimental evaluation with a real dataset comprised of problematic road scenarios. Experimental result shows that the proposed method is very reliable for multi-lane detection at the presented difficult situations.

Keywords: automated driving system (ADS); sensor fusion; multi-lane detection

1. Introduction

The Euro NCAP, NHTSA and ISO have published assessment protocols to meet the proper criteria of the well-known Advance Driver Assistance System (ADAS). The assessment protocols have been renewed almost every year and have become more sophisticated [1]. These documents contain the main fundamental functions of road environment perception to perform ADAS functionalities, like Adaptive Cruise Control, Automatic Emergency Braking and Lane Keeping System. However, each level of the automated driving system (ADS) has an Operational Design Domain (ODD) which means the system is not capable of constant full driving automation since these protocols are subject to the current production line for automakers [2]. To move towards advanced ADS, enormous surges of autonomous driving technologies have focused on the perception of the road environment as well as vehicle control.

Among the autonomous driving technologies, road environment perception requires the ability to use onboard sensors to extract lane markers, road users (like vehicles and pedestrians), and infrastructure (like traffic signs and traffic structures). Especially, the most important but basic research fields are lane and vehicle detection and tracking. We can currently utilize these two states of ego lane and vehicles to adjust the spacing/speed control and the lane keeping system which corresponds to level-2 driving automation out of the 6 levels of driving automation [3]. To advance to level-3 or ultimately full AD, the system requires the ability to overcome issues with reliable and robust multi-lane detection and tracking in complex environments. Especially, the adjacent lanes deliver more specific information to the ADS module to reliably be able to perform accurate and comfortable driving. For example, if the system can predict paths of the surrounding vehicles through the road geometry from lanes, including adjacent lanes, the system can take more safety actions in advance depending on the road scene interpretation. Therefore, to realize highly-advanced ADS technologies, such as planning driving strategies in a complex environment, the perception module must provide key driving information about the accurate lane information, including adjacent lanes, to situational awareness processes.

Most existing lane detection research has concentrated on securing robustness to environmental factors, such as various challenging road types, and shadow and light conditions. Several important studies have explored to multi-lane detection [4–11]. However, there is still a lack of research on adjacent lane detection in the cases that lane features are disturbed by heavy traffic situations or poor visibility. These factors are a critical problem for the previous lane tracker frameworks since most tracking processes rely on the given sequence of image frames. In this paper, we focus on resolving the problems in the extraction of lane candidate features due to occlusion and disturbing features caused by the presence of traffic, road markers, and shadows. We describe the poor extraction of lane candidate features originating from traffic, especially, when the adjacent lane of the road and road markers pose bad influential factors on the process of lane model fitting.

The proposed framework for multi-lane tracking is divided into two main parts: one is an image template occupancy matching (ITOM)-based lane tracker, the other is a frame-level detection approach [6]. Each part was applied with our practical data fusion techniques that combine the vehicle tracker and lane tracker to obtain improvement in managing the adjacent lane as well as the ego lane with an experimental dataset acquired from onboard sensors. Our approach has main contributions of a novel framework of multi-sensor data fusion for a maximum four-lane tracker in the following aspects:

- The state-of-the-art research regarding multi-lane detection focuses on a variety of challenging multi-lane detection scenarios. However, there is a lack of intense research on multi-lane detection and tracking regarding complex road environments for the situational awareness of AD applications. We define three bad influential factors, which are the presence of traffic, road markers, and shadowy road conditions for the four-lane tracker. To overcome this problem, we propose a novel framework for the lane-tracker method integrated with a vehicle tracker consisting of a camera and a radar.
- To assure the robustness of the proposed method to extract lane candidate features, we present two main contributions to the framework for multi-lane tracking. The first is to remove those two key influential factors with the integration of the vehicle tracker and the multi-lane tracker through the ITOM we have introduced. The other is to enhance the performance of the algorithm for extracting lane candidate features in the case of shadowy conditions through a frame level detection approach.
- To assure consistency for the states with multi-lanes, a method for frame-level management of multi-lane detection and tracking is introduced. For multi-lane detection, we design a specified feature extraction function and present a method to select the right feature among the lane candidate features in the current frame. Finally, the states of the multi-lane scenario are framed in a managed condition that it can adjust parameter changes and manage the lane track histories for the multi-lane tracker.

This specialized road scene interpretation method will be discussed throughout the paper. The rest of this paper is organized as follows. Section 2 will explore and summarize related works. Section 3 describes the system overview, including the problem definition, algorithm overview, and problem formulation. The main algorithm is divided into two parts, with the method described in detail in Sections 3 and 4. The experimental results are presented in Section 5. Finally, the conclusions can be found in Section 6.

2. Related Work

Recently, as the ADS techniques have advanced, the need for reliable road scene interpretation, including multi-object detection and tracking methods, has steadily increased. Among these methods, vision-based lane detection and tracking have been stated as one fundamental component of ADS functionalities [2]. In brief, previous research has mainly focused on robust and accurate lane detection. Regarding the specific details in the related works, the lane detection scope can be divided into two parts; single-lane detection and multi-lane detection, as shown in Table 1. According to the detection scope, various approaches for lane detection have been conducted considering road types and conditions with different strategies and characteristics. In single-lane detection, there are various driving situations including from simple road condition [12] to urban areas [13,14], and road environmental difficulties with more complex scenarios [15–18]. For robustness in lane detection and tracking, parameter estimation based on probabilistic approaches in the form of tracking, using the well-known Kalman Filter (KF) and particle filter, should be conducted. The use of the KF, which is a well-known clothoid lane model-based approach that uses lane features to fit lane models, offers prediction and the smoothing of outliers. However, it is clear that the single model-based approach is not the best choice due to the frequently poor quality of lane features. Therefore, a combination with a multi-model tracking algorithm [12,19] or a feature-based approach, like the ones presented using particle filters [19,20], integrated detection and tracking in the usage of particle filters in [13,16], provides a way to handle the discontinuous road conditions. In [16], Kim introduced a method to find robust lane-boundary hypotheses combining the Random sample consensus (RANSAC) algorithm for detection and a particle filter for tracking. Some other approaches have also proposed a robust lane detection based on the M-estimator Sample Consensus (MSAC) [17], which has a better performance than the RANSAC method for lane model fitting [15]. Recently, several methods based on deep learning for single lane detection have been proposed in [18,21]. Multi-lane detection needs more sophisticated algorithm due to the unknown number of lanes and increase complexities. For a multi-lane feature clustering and association, a method based on conditional random field (CRF) [4,5], lane stability optimization [6], and homgraphy matrix estimation [7,10] are presented. For a multi-lane tracking, particle filter is used considering the dependencies of multiple lane geometric constraints in [8,11], and spline model based KF is also presented [9]. On the other hand, the previous study deal with a robustness on the occlusion and shadow for a short time. For a consideration of complex road environment in practical dense traffic, multi-sensor data fusion to combine data that is not accessible from an individual sensor is inevitable [22–25]. Over the past few decades, researchers have made substantial progress in multiple research fields studying object tracking. In this paper, we propose a novel framework of multi-sensor data fusion for a multi-lane tracker that works well in a dense road scene interpretation.

Table 1. Various approaches for lane detection.

Research Target Detection Scope	Road Types and Conditions	Strategies and Characteristics
Single-Lane [12–21]	Challenge scenarios [13–16,20]	Particle filter [14,19,20], Kalman [12,16], Robust strategies [15–17], Deep learning [18,21]
Multi-Lane [4–11]	Challenge scenarios [4–6] Adjacent lane [7–10]	CRF association [4,5], Robust strategies [6,7,10], Particle filter [8,11], Spline EKF [9]

3. System Overview

3.1. Multi-Sensor Data Fusion Framework

The main goal of this research is to detect and track a maximum of four-lanes through an integration of road scene information including forward vehicle tracking. As related works have shown in Section 2, it is difficult to handle an algorithm for multi-lane detection that covers all of the various scenarios. Most studies solve the difficulties to combine detection and tracking or apply methods with known robust estimation methods under the tracking process [12–16]. In this paper, the proposed multi-lane tracker framework has key processes for the robust extraction of lane candidate features, including the preprocessing of multi-sensor data, with the integration and fusion of the vehicle tracker and lane tracker. The integrated fusion system with the vehicle tracker compensate for the regions of unwanted feature points that make multi-lane detection difficult in the top-view image frame of the flowchart in Figure 1, which will be explained in detail in Section 4.

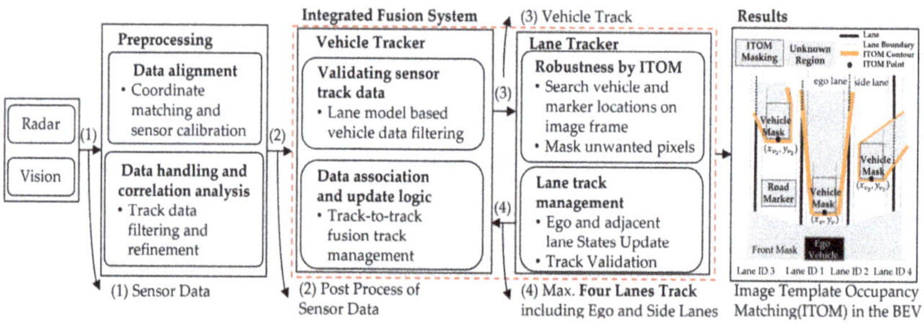

Figure 1. System overview of the multi-sensor data fusion framework.

- Preprocessing

The coordinate systems are aligned with the point of origin of the front-centered-ground of the ego vehicle which follows a standard SAE coordinate system, but differentiates the center of the rear axis from the front-centered-ground to handle all of the sensor data intuitively in Figure 2a. The *X*-axis is positive toward the front and the *Y*-axis is positive toward the left side on the vehicle. The coordinate matching for each sensor should be conducted since they are installed at different mounting positions. For a radar sensor in Figure 2b, it is easy to employ the coordinate by a simple transition from polar coordinates to vehicle coordinates. However, for the camera in Figure 2c, the image frame should be rectified into the top-view image, which can be aligned with the same vehicle coordinates by Inverse Perspective Mapping (IPM) [15–17]. Through the image transformation, we intuitively interpret a road scene situation with the integrated vehicles and lanes information at the same time. To obtain a top-view image or bird-eye view (BEV), we need to find a homography matrix by the camera calibration process in Figure 2d. For clarity, each sensor transformation is constructed as follows:

- The transformation from each sensor frame to the vehicle local frame:

$$vehicle\ local\ frame:\ \{X_{local}, Y_{local}\}, \tag{1}$$

- Typical installation positions of the camera and radar sensor:

$$\text{Radar}: \begin{bmatrix} X_{r\_local} \\ Y_{r\_local} \end{bmatrix} = \begin{bmatrix} R_r cos(\theta_r) \\ R_r sin(\theta_r) \end{bmatrix}, \tag{2}$$

$$\text{Camera}: \begin{bmatrix} X_{v\_local} \\ Y_{v\_local} \end{bmatrix} = \begin{bmatrix} x_{BEV} + l_{offset} \\ y_{BEV} \end{bmatrix}, \tag{3}$$

- Image transformation

$$\text{A camera matrix}: H_{3\times4} = \begin{bmatrix} fx & 0 & cx \\ 0 & fy & cy \\ 0 & 0 & 1 \end{bmatrix} [R|T]_{3x4}, \tag{4}$$

$$\text{A homography transformation matrix}: Trans = H_{3\times4}(:, [1\ 2\ 4]), \tag{5}$$

The camera intrinsic parameters $(fx, fy,\ cx,\ and\ cy)$ and extrinsic parameters (R: rotation, T: Translation matrix) are obtained from the process of the camera calibration in Figure 2d.

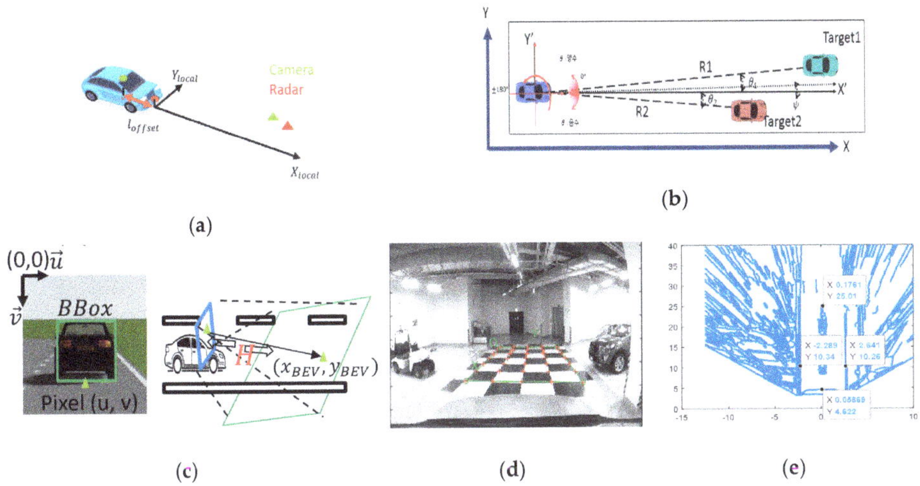

Figure 2. Multi-sensor coordination alignment process: (**a**) multi-sensor coordinate matching, (**b**) radar coordinates, (**c**) transformation of image frame to top-view frame, (**d**) camera calibration, (**e**) transformation into vehicle coordinates.

Once we obtain the homography matrix through the Equations (4) and (5), we can find a point (u, v) in the image frame corresponding to the point (x_{BEV}, y_{BEV}) which indicates Figure 2c,e as follows:
Vehicle to image transformation:

$$(u', v', \alpha)^T = Trans * (x_{BEV}, y_{BEV}, 1)^T (u, v)^T = (u'/\alpha, v'/\alpha)^T, \tag{6}$$

Image to vehicle transformation:

$$(x', y', \alpha)^T = Trans^{-1} * (u, v, 1)^T (x_{BEV}, y_{BEV})^T = (x'/\alpha, y'/\alpha, 1)^T, \tag{7}$$

3.2. Experimental Setup

The test vehicle is equipped with a radar, vision module, and a camera. The self-collected datasets are acquired for evaluating the proposed multi-lane tracker in a variety of complex road conditions and the validation process was conducted in post-process. The onboard perception module obtains the radar object tracks, vision lane/object tracks, and ego vehicle data from the CAN bus and the current image frame at different cycle times. As shown in Figure 3, the multi-rate and multi-sensor fusion flow chart depicts the test sensor setup and how the system works with a different sensor set.

- A 76-77GHz (174 m, +/−10°) and mid-range (60 m, +/−45°) delphi-ESR1.1 radar provides simultaneous multimode electronic scanning simultaneously. However, we are only concerned with the object's range (R_r), azimuth (θ_r), and speed (v_r) data up to 64 targets within the region of interest of the extent of the image frame (35 m).

- A Mobileye 630 vision module detects and tracks four-lane and vehicle positions up to 64 target tracks with a horizontal FOV of 47°. We take object position track (x_V, y_v) for the sake of facile fusion with the radar track instead of vehicle detection from the image frame. Furthermore, we can easily compare four-lane tracks with our integrated lane tracker method.
- A RealSense SR-300 camera offers 1920 × 1080 pixels with a horizontal FOV of 73°, with which we interpret and understand the road scene. We have downsized the image by 640 × 480 pixels for improving computational efficiency with performing appropriate performance for the situational awareness. The image frames are mainly used for multi-lane detection and integration with the vehicle tracker. The feature points in the vehicle coordinates are derived from the rectified top-view image frame as described in Section 3.1.

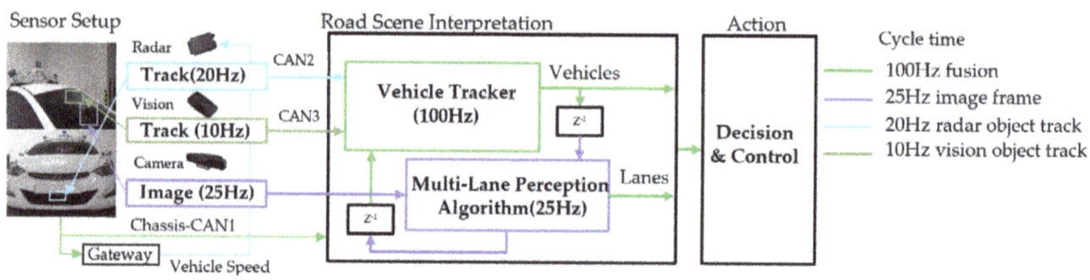

Figure 3. The multi-rate and multi-sensor fusion flow chart.

4. Integrated Fusion of the Vehicle and Lane Tracker

In the case of a presence of vehicles to the front or side, or road markers and shadows, it is impossible to consistently track multiple lanes consistently without any complementary solutions. We present an integrated fusion method for the multi-lane tracker and vehicle tracker that are complementary to the process of updating each track state. The robust multi-lane detection and tracking algorithm consists of four main steps: a vehicle tracker with vision and radar, robustness under complex road conditions with ITOM, frame level multi-lane detection, and multi-lane track management as shown in Figure 4.

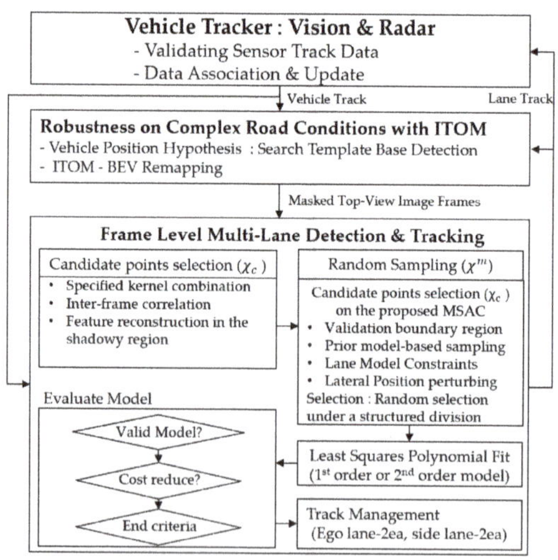

Figure 4. The robust multi-lane detection and tracking algorithm.

4.1. Vehicle Tracker

In this research, the vehicle tracker is responsible for combining the multi-rate multi-sensor track data fusion. The issues in the vehicle tracker process can be categorized into three steps: validating the track data, data association, and lowered delay of track fusion, which is part of the track-to-track fusion with the radar and vision system [24]. In other words, these datasets have already tracked the object states. One major characteristic is that the radar track includes clutter and noisy radar data, from guardrails, bridges, tunnels, etc., besides road users within the radar sensor range while the vision track only indicates certain vehicles in front of the ego vehicle. Therefore, we should consider the two issues of validating the track data and data association. First, a region of interest (ROI) is set up for validating the track data. The ROI is based on a lane boundary, which is determined by a lane tracker, as shown in Figure 5. In the case of Figure 5, where the lane tracker has four active lanes, the region is restricted to along the adjacent lane boundaries for the lateral direction and a distance limitation for longitudinal direction. The plot describes that the irrelevant radar data, such as guardrails and clutter, can be filtered out with the ROI. Secondly, the track data from each sensor decides whether fused track state and the new incoming track data is admissible inside gating using the Mahalanobis distance which is derived from χ^2-distribution [25]. Finally, the track update process follows the work well accomplished by the use of the Kalman Filter [26]. In the track update process, there are three states which are track creation, standby, and confirmed. We take the confirmed track when the vision track comes at the first time of track update states, while we take the standby track when the radar track arrives first due to the uncertainty. That is, the track update rules give priority to the vision track for lowered delay time while updating vehicle track states.

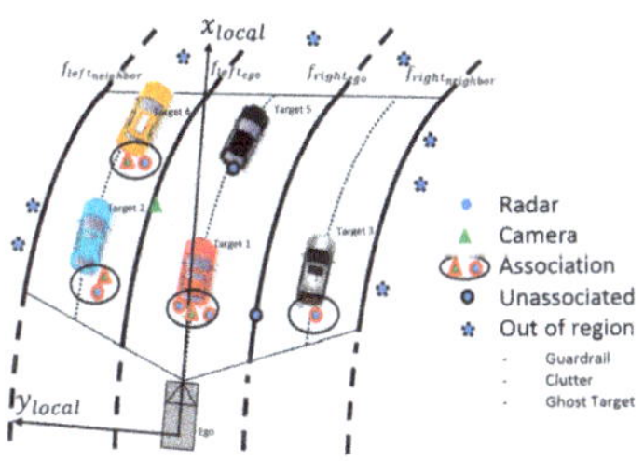

Figure 5. Multi-object data association within region of interest.

4.2. Lane Tracker

As mentioned in the related works, many unfortunate influential factors make lane detection surprisingly difficult, even if the extraction of lane candidate features from a road may seem straightforward. Since most tracking processes depend on the lane candidate features, the previous lane tracker frameworks are easily destroyed by consistently wrong or absent lane candidate features that occur from occlusion, poor visibility, or shadows in dense traffic conditions as shown in Figure 6. While these road conditions can be handled by previous research methods for a short time. ADS must catch adjacent lanes to be able to aware of the current traffic situation in the case of complex road conditions. To accomplish this special case, we present a robust lane tracker that secures adjacent lanes as well as ego lanes while integrating the vehicle tracker. In brief, the proposed lane tracker consists of two main parts: frame-level multi-lane detection and tracking, and securing robustness in complex road condition.

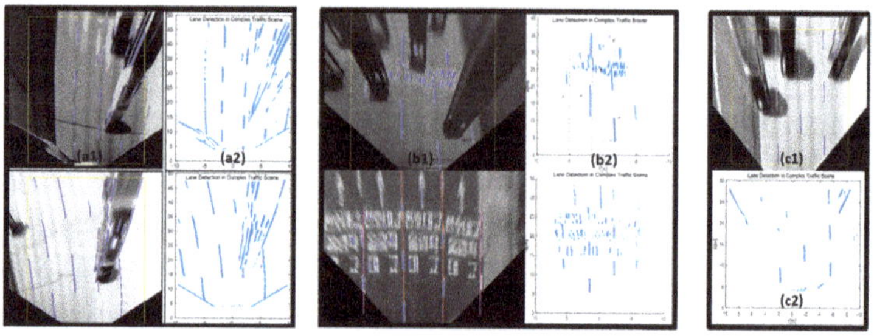

(a) (b) (c)

Figure 6. Three bad influential factors for lane detection: (**a**) a challenging scenario for road adjacent lane detection due to a vehicle in the side lane, (**b**) a challenging scenario for road lane detection due to road markers, (**c**) a shadowy road condition, with (a1,b1,c1) lane candidate points from a bird's-eye view and (a2, b2, c2) lane candidate points in the vehicle coordinates.

4.2.1. Frame Level Multi-Lane Detection and Tracking

- Lane Candidate Feature Extraction

This method requires initial lane candidate feature extraction. The features are extracted using a designed kernel edge filter based on the dark-light-dark (DLD) intensity transition characteristic [14]. For the kernel design, we first transform the image frame to a bird's-eye view image with IPM. From the rectified image frame, we analyze the lane width in the bird's-eye view image and design the specified kernel function, which consists of emphasizing the lane feature function (f_E) and shrinking the road region feature function f_S. Finally, we can extract the lane candidate features from the specified kernel combination function f_c, as shown in Figure 7a–c and Equations (8)–(10).

$$f_E = conv(I, [-1\ 0_{1\times n}\ 2\ \ 0_{1\times n} - 1]), \tag{8}$$

$$f_S = conv(I, [1\ 0_{1\times n}\ 0\ \ 0_{1\times n} - 1]) - conv(I, [1\ 0_{1\times n}\ 0\ 0_{1\times n}\ 1]), \tag{9}$$

$$f_c = \alpha f_E + f_S \tag{10}$$

where $conv(I, kernel)$ means a convolution with image frame I (a 640 × 480 pixel image) and *kernel*. The α scale factor indicates the amplification of the brightness. However, this extraction method is very unstable under shadows because of the gradient changes over different road environments. To handle the shadow problem, an adaptive threshold has been employed in this work. The process of reconstruction in the shadow region is conducted as shown in Figure 7d–g. Figure 7d represents the multi-lane ROI derived with the information from the front vehicle states and previous lane track states. Within the multi-lane ROI, we take an adaptive threshold for each lane to reconstruct lane candidate features as shown in Figure 7e. Finally, we can obtain well-resolved features of the current frame and we can compare the results in Figure 7f,g.

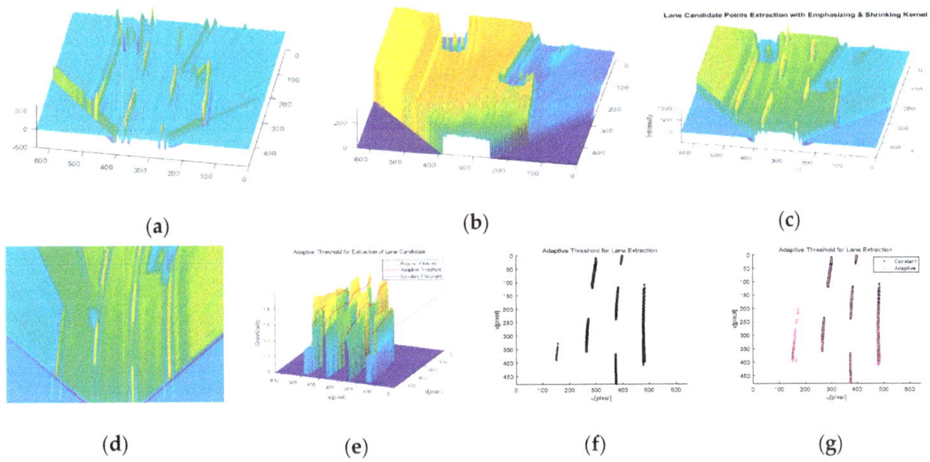

Figure 7. A robust lane feature extraction method, (**a**) emphasizing lane features (indicated by brightness), (**b**) shrinking the road region (darkness), (**c**) combining the two kernels, (**d**) frame level multi-lane region of interest (ROI), (**e**) reconstruction of the shadowy region features with a lane-level adaptive threshold, (**f**) lane feature extraction results without an adaptive threshold, (**g**) lane feature extraction results with an adaptive threshold.

- Robust Lane Detection and Tracking

The lane model described in [27] is a third order polynomial, clothoid model. In this work, we reduce the order of model to a parabolic or linear model according to feature quality because these models are more stable in complex traffics, and the model can be formularized as follows;

$$\text{Clothoidal model}: \; f_{index}^{3nd}(l) = y_0 + \psi l + \frac{C_0}{2}l^2 + \frac{C_1}{6}l^3 \tag{11}$$

$$Parabolic\ modle: \; f_{index}^{2nd}(l) = y_0 + \psi l + \frac{C_0}{2}l^2 \tag{12}$$

$$\text{Linear model}: \; f_{index}^{1st}(l) = \psi l + y_0 \tag{13}$$

where $index$ indicates the $left_{ego}$, $right_{ego}$, $left_{adjacent}$, $right_{adjacent}$ lanes, respectively, and the lane parameters consist of y_0, ψ, C_0, C_1 which are the lateral offset, lane heading, lane curvature, curvature rate, and l is a longitudinal distance.

For multi-lane detection, the proposed approach has been designed to handle more sophisticated individual lane detection based on the MSAC [28,29] under strictly controlled feature selection methods using the frame-level track coherence. Taking advantage of this characteristic, we applied an individual lane feature selection strategy to find the feature cluster more quickly and efficiently as shown in Figure 4. The individual four lanes have their own constraints based on a validation boundary region of the lane model parameters in Equation (12). Since the lane features cannot be changed abruptly, we take advantage of the validation boundary regions defined by the estimated lane parameters in the previous frame where the candidate points are located in the current frame. Through multi-lane track management, we group the subject multi-lane parameters ($left_{ego}$, $right_{ego}$, $left_{adjacent}$, $right_{adjacent}$) and manage them separately, as shown in Figure 8a. Every lane tracker cycle time increment, we check for adjacent lane validations, as shown in Figure 8b,c. We make sure of two check points (lane types and *Lifetime*) for clarity to validate the adjacent lane: a check on whether each ego lane is a dashed or solid lane, and the life time, which represents the number of track update after track creation, of the adjacent lane (*Lifetime* > 4). The reason for finding lane types is to determine whether the ego vehicle is in a driving edge lane, which means the lane no longer exists over the lane boundary. We also double-check any vehicle that has the same direction and velocity with the ego vehicle for third check.

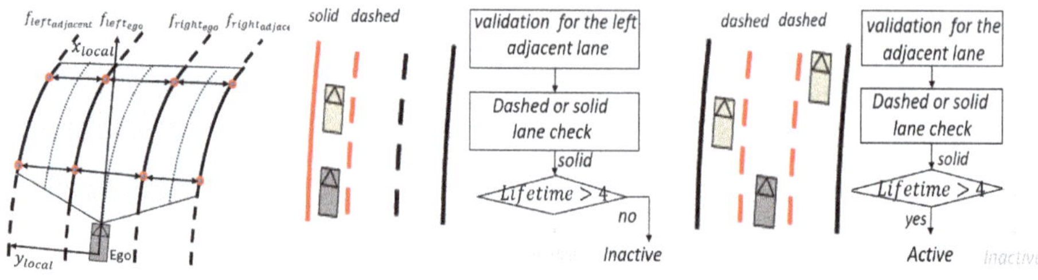

(a) (b) (c)

Figure 8. Lane track management, (**a**) subject multi-lane ($left_{ego}$, $right_{ego}$, $left_{adjacent}$, $right_{adjacent}$), (**b**,**c**) two cases of adjacent left lane with lane validation logic.

4.2.2. Robustness under Complex Road Conditions

In feature-based methods, Otsu thresholding segmentation has often been used and proven in preprocessing [23]. This thresholding method is also useless when the whole image frame is full of bright objects, when most of the image frame region is occupied by road markers, or in the presence of bright cars on the road. Therefore, we intend to mask all these factors to extract brightness intensities only corresponding to the lane features. In order to eliminate these factors, we introduce a novel concept of image template occupancy matching (ITOM). This concept finds a literally occupied region that matches with vehicles or road markers using predefined templates, as shown in Figure 9a. These templates consist of base templates and vehicle templates based on each lane geometry. The base template is responsible for searching hypotheses for road markers or vehicles, as shown in Figure 9b. After finding a hypothesis location, the occupancy-matching procedure is conducted. Each base template determines the occupancy of the vehicle position on the image from the front vehicle states from the vehicle tracker. If there are no matching results with a vehicle, it regards the candidate as a road marker, as shown in Figure 1 (results). The ITOM-based multi-lane tracker results are shown in Figures 10 and 11. In Figure 10a depicts the results of multi-lane detection on an image frame, Figure 10b displays the lane candidate features, Figure 10c depicts noisy lane candidate features in the vehicle coordinates, Figure 10d shows base templates occupied with road markers and the lane geometry, Figure 10e shows the elimination of road makers for each lanes, Figure 10f shows the lane candidate feature with removed road markers, and Figure 10g shows the multi-lane detection results for current frame. In Figure 11, when vehicles and road markers appear at the same time, the ITOM-based multi-lane tracker has successfully eliminated the disturbance factors and parameterized the driving environment.

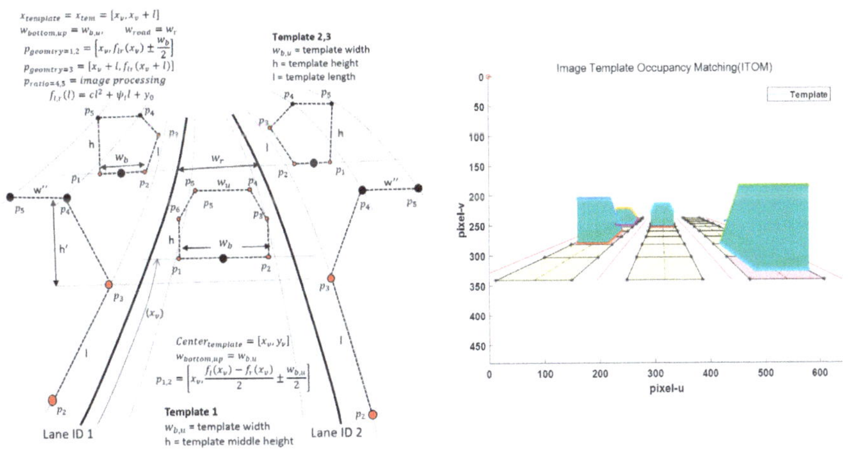

(a) (b)

Figure 9. Image template occupancy matching (ITOM) concept plot, (**a**) template design, (**b**) search base template and match on image frame.

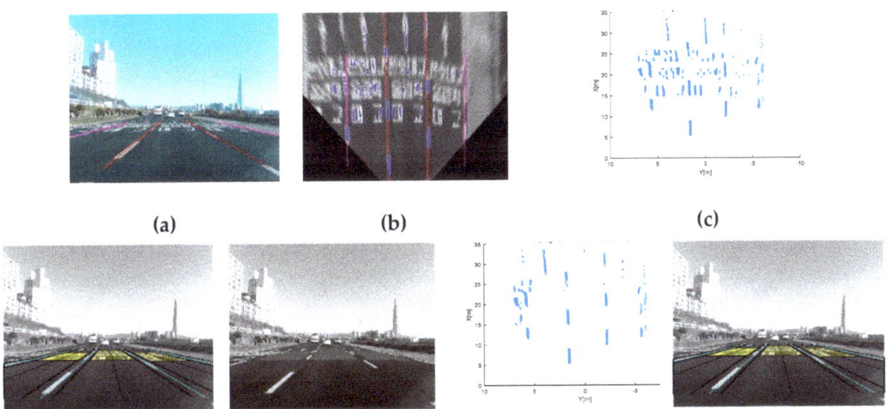

(a) (b) (c)

(d) (e) (f) (g)

Figure 10. Complex road conditions: ITOM-based road markers elimination. (**a**) Previous multi-lane states of the image frame, (**b**) top-view frame, (**c**) origin features, (**d**) ITOM process, (**e**) remove road marker, (**f**) features in the vehicle coordinates, (**g**) multi-lane detection.

(a) (b) (c)

Figure 11. Complex road conditions: (**a**) template-based detection, (**b**) ITOM remapping on the origin image frame, (**c**) robustness in complex road conditions.

5. Experimental Results

5.1. Experimental Methodology

The performance of the reliable road scene interpretation based on ITOM with the integrated fusion of the vehicle and lane tracker in a dense traffic situation is evaluated through mainly self-collected dataset from real driving data in dense traffic conditions. The purpose of this experiment mainly focuses on the specific driving conditions which fall under the consistent traffic occlusion from forward vehicles and the existence of road markers and shadows that yield incorrect lane parameter estimations. The video with a resolution of 640×480 from the evaluation dataset are manually labeled using the Matlab ground truth labeler. If the differences between the tracked multi-lanes and the ground truth are within a pre-defined threshold, a true positive detection is counted, otherwise a false negative [9]. The performance metric is as follows: A true positive (TPR) = (the number of detected lanes)/(the number of target lanes), a false positive rate(FPR) = (the number of false positives)/(the number of target lanes) [5,7].

5.2. Experimental Results

We test and verify the multi-lane tracker performance on the normal PC, Intel(R) Core i9-9900KF CPU @ 3.60GHz and 32.00GB RAM. The processing speed is around 25 frame per second (fps). It is a little slow, but it has enough possibility to optimize the proposed algorithm which is reasonable speed for ADs in a real time processor. A total of 3350 frames were tested and the target four-lanes are considered within 35m. The performance metric shows that the overall TPR of the algorithm for identifying lanes, including adjacent lanes, is 99.5 percent and the adjacent lane TPR is 96.8 percent, as shown in Table 2. This is a significant improvement compared to the Mobileye 630 lane tracker results, which has a TPR of only 90 percent and 88.6 percent respectively. We show a variety of the experimental results and validation in Figure 12. The figure describes presence of road marker and multiple vehicle ahead of ego vehicle. We picked specific difficult scenarios in Figure 12 which depict a road marking scenario in first row of (a), vehicle appearance on right or left lane in the rest rows of (a), mixed scenarios of complex road marking and vehicle ahead of ego vehicle in (b), and finally validation with the tracked multi-lanes and ground truth in (c). It is shown that the proposed multi-lane tracker outperformed in the ego lanes and adjacent lanes through disturbing features elimination.

Table 2. Various approaches for lane detection.

Measure		TPR		FPR
Type	**Ego Lane**	**Adjacent**	**Total**	**Total**
ITOM Lane Tracker	0.995	0.968	0.98	0.019
Mobileye 630	0.914	0.886	0.9	0.1

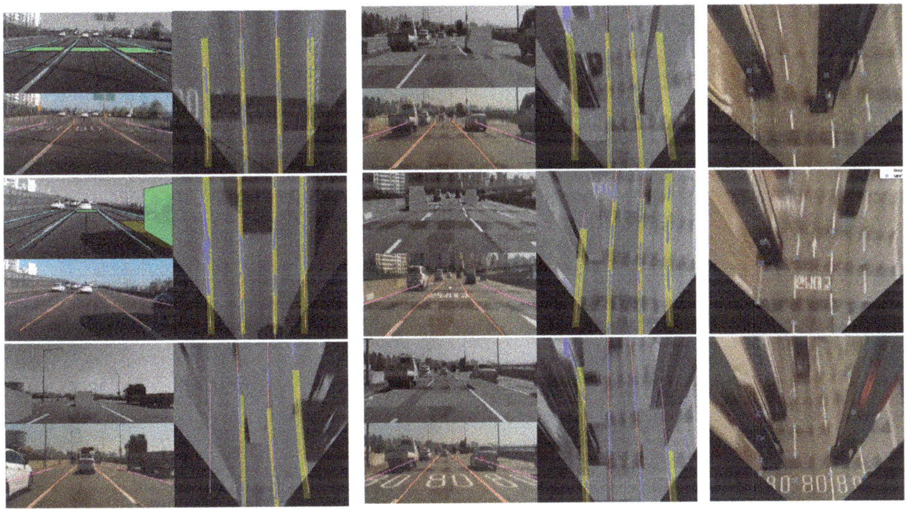

(a) (b) (c)

Figure 12. Experimental results and validation (**a**) scenario I: successful masking of the road marker, side vehicle emergence, front, and side vehicles. (**b**) Scenario II: successful masking of a complex scenario, (**c**) ground truth and results from the top view for validation.

6. Conclusions

This work proposed and evaluated a road scene interpretation algorithm based on ITOM with the integrated fusion of the vehicle and lane tracker in a dense traffic situation. The system mainly focused on resolving the problems in the extraction of lane candidate features due to three bad influential factors in complex road conditions under the process of a multi-lane tracker. To overcome this problem, the framework for a multi-lane tracker presents two key processes for the robust extraction of lane candidate features. The first process introduces an elimination process of bad influential factors including the presence of vehicles and road markers in the current top-view image frame through the proposed ITOM method. In this process, we consider validating track data, data association, and the radar track-to-vision track fusion method which gives priority to vision track for a lower delay time while updating vehicle track states. In the second process, the proposed robust methods for frame-level multi-lane detection and tracking are presented, including the extraction of lane candidate features under shadows, a frame-level lane hypothesis region, and managing the multi-lane tracker using the MSAC algorithm. In the experiment, we have carefully selected dataset to verify specific bothersome scenarios in which the study is interested. The test results showed that the proposed method can significantly improve the detection rates when it comes to the adjacent lanes under complex road conditions. Through this framework, we ensure multi-lane detection and tracking that make further driving strategies with situational awareness in ADs. However, this work needs to check the reliability in much more road scenarios and a variety of road types as well since this work mainly focus on the methodology of robust lane feature extraction and managing in the presence of road marking, vehicles on the road and shadows.

Author Contributions: Conceptualization, J.J.; Methodology, J.J.; Software, J.J.; Supervision, J.H.P.; Validation, J.J. and Y.H.Y.; Visualization, J.J.; Writing – original draft, J.J.; Writing – review & editing, J.J. and J.H.P. All authors have read and agreed to the published version of the manuscript.

Funding: This work was supported by the Technology Innovation Program (Fault Detection and Diagnosis for ADAS-Sensors and Hazardous Analysis and Development of Fault Management Strategy, 10076338) funded by the Ministry of Trade, Industry & Energy (MOTIE, Korea).

Acknowledgments: The experiment sensors and equipment were provided and supported by the Hyundai motor company.

Conflicts of Interest: The authors declare no conflict of interest.

References

1. Euro NCAP. Available online: https://www.euroncap.com/en/for-engineers/protocols/safety-assist/ (accessed on 16 October 2019).
2. SAE International. *Taxonomy and Definitions for Terms Related to Driving Automation Systems for on-Road Motor Vehicles*; SAE International: Warrendale, PA, USA, 2018.
3. Louw, T.; Merat, N.; Jamson, H. Engaging with highly automated driving: To be or not to be in the loop? In Proceedings of the Eighth International Driving Symposium on Human Factors in Driver Assessment, Training, and Vehicle Design, Salt Lake City, UT, USA, 22–25 June 2015.
4. Zhou, H.; Zhang, H.D.; Hasith, K.; Wang, H. Real-time Robust Multi-lane Detection and Tracking in Challenging Urban Scenarios. In Proceedings of the 2019 IEEE 4th International Conference on Advanced Robotics and Mechatronics (ICARM), Osaka, Japan, 3–5 July 2019; pp. 936–941.
5. Hur, J.; Kang, S.N.; Seo, S.W. Multi-Lane Detection in Urban Driving Environments Using Conditional Random Fields. In Proceedings of the 2013 IEEE Intelligent Vehicles Symposium (IV), Gold Coast, Australia, 23–26 June 2013; pp. 1297–1302.
6. Xuan, H.; Liu, H.Z.; Yuan, J.Z.; Li, Q. Robust Lane-Mark Extraction for Autonomous Driving under Complex Real Conditions. *IEEE Access* **2017**, *6*, 5749–5765. [CrossRef]
7. Kang, S.N.; Lee, S.; Hur, J.; Seo, S.W. Multi-Lane Detection Based on Accurate Geometric Lane Estimation in Highway Scenarios. In Proceedings of the IEEE Intelligent Vehicles Symposium Proceedings, Dearborn, MI, USA, 8–11 June 2014; pp. 221–226.
8. Deusch, H.; Jürgen, W.; Stephan, R.; Magdalena, S.; Marcus, K.; Klaus, D. A Random Finite Set Approach to Multiple Lane Detection. In Proceedings of the 2012 15th International IEEE Conference on Intelligent Transportation Systems, Anchorage, AK, USA, 16–19 September 2012; pp. 270–275.
9. Zhao, K.; Mirko, M.; Christian, N.; Dennis, M.; Stefan, M.S.; Josef, P. A Novel Multi-Lane Detection and Tracking System. In Proceedings of the IEEE Intelligent Vehicles Symposium, Alcala de Henares, Spain, 3–7 June 2012; pp. 1084–1089.
10. Nieto, M.; Salgado, L.; Jaureguizar, F.; Arróspide, J. Robust multiple lane road modeling based on perspective analysis. In Proceedings of the 15th IEEE International Conference on Image Processing, San Diego, CA, USA, 12–15 October 2008; pp. 2396–2399.
11. Vacek, S.; Bergmann, S.; Mohr, U.; Dillmann, R. Rule-based tracking of multiple lanes using particle filters. In Proceedings of the 2006 IEEE International Conference on Multisensor Fusion and Integration for Intelligent Systems, Heidelberg, Germany, 3–6 September 2006; pp. 203–208.
12. Meuter, M.; Stefan, M.S.; Adalbert, M.; Stephanie, H.; Christian, N.; Anton, K. A Novel Approach to Lane Detection and Tracking. In Proceedings of the 12th International IEEE Conference on Intelligent Transportation Systems, St. Louis, MO, USA, 3–7 October2009; pp. 582–587.
13. Minchae, L.; Chulhoon, J.; Myoungho, S. Probabilistic Lane Detection and Lane Tracking for Autonomous Vehicles Using a Cascade Particle Filter. *Proc. Inst. Mech. Eng. Part D J. Automob. Eng.* **2015**, *229*, 1656–1671.
14. Sehestedt, S.A.; Sarath, K.; Alen, A.; Gamini, D. Efficient Lane Detection and Tracking in Urban Environments. In Proceedings of the European Conference on Mobile Robots: ECMR, Freiburg, Germany, 19–21 September 2007; pp. 19–21.
15. Haloi, M.; Dinesh, B.J. A robust lane detection and departure warning system. In Proceedings of the IEEE Intelligent Vehicles Symposium (IV), Seoul, Korea, 28 June–1 July 2015; pp. 126–131.
16. Kim, Z.W. Robust lane detection and tracking in challenging scenarios. *IEEE Trans. Intell. Transp. Syst.* **2008**, *9*, 16–26. [CrossRef]
17. Tapia-Espinoza, R.; Torres-Torriti, M. Robust lane sensing and departure warning under shadows and occlusions. *Sensors* **2013**, *13*, 3270–3298. [CrossRef] [PubMed]

18. John, V.; Liu, Z.; Mita, S.; Guo, C.; Kidono, K. Real-time road surface and semantic lane estimation using deep features. Signal. *Image Video Process.* **2018**, *12*, 1133–1140. [CrossRef]
19. Southall, B.; Camillo, J.T. Stochastic Road Shape Estimation. In Proceedings of the Eighth IEEE International Conference on Computer Vision (ICCV 2001), Vancouver, BC, Canada, 7–14 July 2001; pp. 205–212.
20. Danescu, R.; Sergiu, N. Probabilistic Lane Tracking in Difficult Road Scenarios Using Stereovision. *IEEE Trans. Intell. Transp. Syst.* **2009**, *10*, 272–282. [CrossRef]
21. Zhang, W.; Liu, H.; Wu, X.; Xiao, L.; Qian, Y.; Fang, Z. Lane marking detection and classification with combined deep neural network for driver assistance. *Proc. Inst. Mech. Eng. Part D J. Automob. Eng.* **2019**, *233*, 1259–1268. [CrossRef]
22. Bakr, M.A.; Lee, S. Distributed multisensor data fusion under unknown correlation and data inconsistency. *Sensors* **2017**, *17*, 2472. [CrossRef] [PubMed]
23. Chavez-Garcia, R.O.; Aycard, O. Multiple sensor fusion and classification for moving object detection and tracking. *IEEE Trans. Intell. Transp. Syst.* **2015**, *17*, 525–534. [CrossRef]
24. Matzka, S.; Richard, A. A comparison of track-to-track fusion algorithms for automotive sensor fusion. In *Multisensor Fusion and Integration for Intelligent Systems*; Springer: Berlin/Heidelberg, Germany, 2009; pp. 69–81.
25. Stiller, C.; Fernando, P.L.; Marco, K. Information fusion for automotive applications–An overview. *Inf. Fusion* **2011**, *12*, 244–252. [CrossRef]
26. Yu, Z.; Bai, J.; Chen, S.; Huang, L.; Bi, X. Camera-Radar Data Fusion for Target Detection via Kalman Filter and Bayesian Estimation (No. 2018-01-1608). In *SAE Technical Paper, Proceedings of the Intelligent and Connected Vehicles Symposium, Kunshan City, China, 14–15 August 2018*; SAE International: Warrendale, PA, USA, 2018; pp. 1–8.
27. Dickmanns, E.D.; Mysliwetz, B.D. Recursive 3-d road and relative ego-state recognition. *IEEE Trans. Pattern Anal. Mach. Intell.* **1992**, *2*, 199–213. [CrossRef]
28. Küçükmanisa, A.; Tarım, G.; Urhan, O. Real-time illumination and shadow invariant lane detection on mobile platform. *J. Real Time Image Process.* **2017**, *16*, 1–14. [CrossRef]
29. Torr, P.H.; Zisserman, A. MLESAC: A new robust estimator with application to estimating image geometry. *Comput. Vis. Image Underst.* **2000**, *78*, 138–156. [CrossRef]

sensors

MDPI

Article

Self-Driving Car Location Estimation Based on a Particle-Aided Unscented Kalman Filter

Ming Lin, Jaewoo Yoon and Byeongwoo Kim *

Department of Electrical Engineering, University of Ulsan, 93 Daehak-ro, Nam-gu, Ulsan 44610, Korea; flaaud159@naver.com (M.L.); jaewoo127@naver.com (J.Y.)

* Correspondence: bywokim@ulsan.ac.kr

Received: 27 March 2020; Accepted: 28 April 2020; Published: 29 April 2020

Abstract: Localization is one of the key components in the operation of self-driving cars. Owing to the noisy global positioning system (GPS) signal and multipath routing in urban environments, a novel, practical approach is needed. In this study, a sensor fusion approach for self-driving cars was developed. To localize the vehicle position, we propose a particle-aided unscented Kalman filter (PAUKF) algorithm. The unscented Kalman filter updates the vehicle state, which includes the vehicle motion model and non-Gaussian noise affection. The particle filter provides additional updated position measurement information based on an onboard sensor and a high definition (HD) map. The simulations showed that our method achieves better precision and comparable stability in localization performance compared to previous approaches.

Keywords: particle filter; sensor fusion; self-driving car; unscented Kalman filter; vehicle model; Monte Carlo localization

1. Introduction

In recent years, research on self-driving cars has gained much prominence. The ultimate goal of self-driving cars is to transport people from one place to another without any help from a driver. A self-driving system must control numerous parameters, including speed, orientation, acceleration, and maneuvering, in order to drive without any human assistance. All of these control parameters are controlled by the decision-making module, which handles all perception data from the vehicle and sensors. The perception module determines the relationship between the ego vehicle and the surrounding environment. One of the most important algorithm modules is vehicle localization because all the sensors sense the environment based on local vehicle coordinates [1]. The typical perception sensors of a self-driving car are the camera, radar, light detection and ranging (LIDAR), 2D laser scanner, global positioning system (GPS), and inertial measurement unit (IMU) [2]. GPS is the most commonly used navigation system in self-driving cars. However, because of issues with multipath routing and poor signal availability in cities, relying entirely on GPS is not suitable for localizing vehicles in urban environments. Although differential GPS systems can be used, the high cost and size of these systems limit their implementation [3]. Furthermore, GPS systems cannot be used in tunnels or indoor environments. Vision-based localization has been proposed as a method for localizing vehicles using a low-cost camera. However, the vision-based localization algorithm is easily affected by weather and light conditions, leading to insufficient accuracy and stability [4–8]. LIDAR-based map matching can yield highly precise results with the help of a high definition (HD) map. By contrast, matching requires the environment to be accurately mapped such that the point cloud of the environment does not change. Point cloud matching is expensive and requires considerable power and computation resources [9–16]. Thus, developing a low-cost localization method that can utilize the vehicle's sensors and road infrastructure to achieve precise and stable location performance is necessary for furthering research on self-driving cars. One of the localization solutions that can be used in complex urban environments is vehicle localization based on local sensor systems and information from the HD map. Matching an entire point cloud with an HD map is inefficient; therefore, only the ground truth location map of infrastructures is used in this study, for computational efficiency. In previous research, vehicle localization based on vehicle-to-vehicle (V2V) and vehicular ad-hoc network (VANET) communication was proposed. The basic condition is that these algorithms require surrounding vehicles to be equipped with V2V communication equipment, which are then referred to for the infrastructures [17–23]. In this study, we only consider the case of a single vehicle. Moreover, because the vehicle data contain a large amount of noise, an efficient filtering algorithm is needed to obtain precise localization results.

The methods for vehicle localization have improved considerably over the years. The primary methodology that was used is the probabilistic approach. The Kalman filter (KF) is an optimal estimator that is designed for processing Gaussian noise with mean and variance, and it is an important component in several such approaches [24]. One of the assumptions of the KF is that the noise should be Gaussian. However, in practice, a function like the trigonometric filter renders the Gaussian noise non-Gaussian. Therefore, an extended Kalman filter (EKF), which uses a low order Taylor expansion to linearize the nonlinear (e.g., trigonometric) function, has been proposed. It uses a partial derivative to represent the rate of change of the nonlinear functions, which aims to keep the noise Gaussian. If the state is a vector, then the partial derivative parameters can be assembled into a new matrix, which is called a Jacobian matrix. Generally, in order to localize the vehicle's position, researchers derive the Jacobian matrix based on the transition and measurement models for handling the vehicle's noisy sensor data [25–30]. If an EKF based on a Jacobian matrix approximates a nonlinear function using a high order of Taylor series, it also works well in transforming nonlinear functions into linear ones. The critical problem, however, is that the Jacobian matrix is difficult to derive for complex dynamics. Therefore, a new, sample region-based Kalman filter, which is called the unscented Kalman filter (UKF), was proposed. The UKF performs better than the EKF and KF when the system model is highly nonlinear [31–33]. The UKF uses some key points, which are called sigma points, to approximate the non-Gaussian noise into Gaussian based on the unscented transform. In this way, it can properly capture the nonlinearity. Furthermore, because the UKF approximates the non-Gaussian noise with sigma points, it is easy to combine other information when selecting the sigma points and there is no need to calculate the Jacobian matrix.

The basic assumption of the Kalman filter family is that noise is Gaussian. In the real world, most noise does not have a Gaussian property. For processing non-Gaussian noise, a Monte Carlo-based localization approach, called particle filter (PF), has been proposed [34,35]. The particle filter uses several samples, referred to as particles, to approximate the non-Gaussian property. Because the particles are generated randomly, they can represent the properties of non-Gaussian noise precisely if there are sufficient numbers. However, a vehicle has limited computational resources; therefore, it cannot allow the particle filter to approximate the number of particles. Therefore, there is a trade-off between precision and computational resources when generating an effective particle-based system model. Thus, an extended Kalman filter-aided particle filter, called an extended particle filter (EPF), and an unscented particle filter (UPF), called the Kalman filter-aided particle filter, have been proposed [36–38]. Both the EPF and the UPF use system models to generate and update the particles. It should be noted that each particle should compute the sigma points or Jacobian matrix; therefore, both the EPF and UPF are computationally inefficient and difficult to implement [39–41].

In this study, we propose a new method, the particle-aided unscented Kalman filter (PAUKF), for vehicle localization. With the help of the particle filter, the unscented Kalman filter can estimate a system with high nonlinearity and various sources of nonlinear noise more precisely. Because each particle does not have to update the sigma points or share the prediction model, this method requires fewer computational resources. The computational burden and precision of PAUKF can be easily tuned by tuning the quantity of the sigma points and particles. The results of the simulation show that the PAUKF estimates the vehicle's position and state more accurately than other methods that use a limited number of particles. Section 2 illustrates the methodology of the PAUKF. Section 3 details the simulation conditions, and Section 4 presents the analysis of the simulation results. Finally, Section 5 presents the conclusion of this paper.

2. Particle-Aided Unscented Kalman Filter

This section describes the implementation of the PAUKF, including particle implementation and PAUKF implementation. Both the PF and UKF are Bayesian-based filters, and the environment is assumed to be Markov, which means that the PAUKF also has a Markov assumption.

2.1. Particle Filter Algorithm

The particle filter is a Monte Carlo-based method that can handle both Gaussian noise and non-Gaussian noise [42]. Because the vertical movement of the vehicle is small, we only consider the vehicle in a two-dimensional Cartesian space with the vehicle heading θ. A bicycle model is used in this study to represent the motion of the vehicle because a complex vehicle model aggravates the computational burden, and many parameters cause additional noise [43]. The state of the vehicle is represented by <x, y, and θ>, as shown in Figure 1.

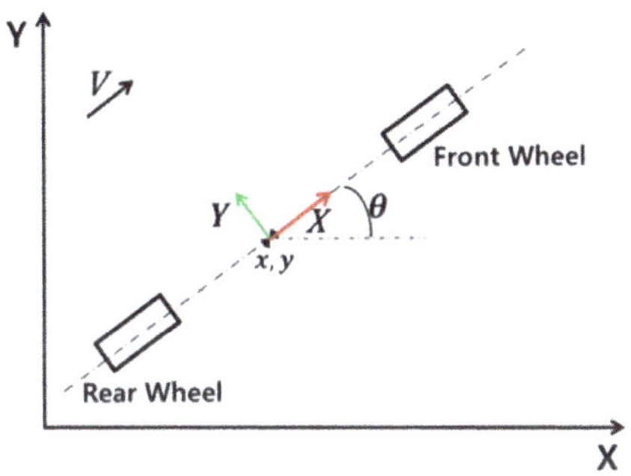

Figure 1. Vehicle state in two-dimensional Cartesian space.

The inputs that we use are the range sensor and the ground truth of the infrastructures in the HD map. The final position of the vehicle should be the best posterior belief based on past data and the current state. The particle filter is a nonparametric implementation of the Bayes filter, which uses a finite number of samples to approximate the posterior. Thus, the final belief bel(x) should be generated for each particle by using each important factor (weight), as shown in Equation (1). The $x_{[1,2\ldots N]}$ means the state vector of each particle and $w_{[1,2\ldots N]}$ is the weight of each particle. The size of each particle X was 3 × 1. W is a non-negative factor termed as the importance factor. In this study, we used 100 particles for simulation, which means that N is 100. The larger the importance factor, the more it affects the final estimation result.

$$\mathrm{bel}(x) = \sum\nolimits_{i=1}^{N} x_N w_N. \tag{1}$$

The particle filter for localization can be divided into four parts, which are introduced in the following subsections.

2.1.1. Initialization

To localize the vehicle in global coordinates, it is essential to provide an initial location to the vehicle. Otherwise, the vehicle will search for its position over the entire world. Therefore, we use the GPS sensor for initialization. Even though the GPS signal is poor due to multipath and blocking issues, it still provides a limited area for the vehicle to localize. Because the particle filter is recursive, after initialization, the noisy GPS signal data are filtered recursively. When a particle filter receives GPS data, it generates N random particles for initialization.

2.1.2. Prediction

To obtain the prior belief, each particle at timestamp $k-1$ should predict the current state based on the system prediction model. The prediction model was constructed based on the vehicle model. Complex models, such as a dynamic model with tires, can also be included. However, complex vehicle models reduce computation efficiency. Such models also require detailed vehicle parameters, which are difficult to set. Incorrect parameters can cause noisy estimations. Considering the computational burden and precision, a kinematic model was used in this study [43]. We ignore the slip angle because vehicle travel in cities is typically not fast.

$$\overline{X}_{k+1} = \begin{bmatrix} \overline{x} \\ \overline{y} \\ \overline{\theta} \end{bmatrix}_{k+1} = \begin{bmatrix} \frac{v_k}{\theta_k(\dot{\Delta t})}[\sin(\theta_k + \theta_k(\dot{\Delta t})) - \sin(\theta_k)] \\ \frac{v_k}{\theta_k(\dot{\Delta t})}[\cos(\theta_k) - \cos(\theta_k + \theta_k(\dot{\Delta t}))] \\ \theta_k(\dot{\Delta t}) \end{bmatrix} + \begin{bmatrix} x \\ x \\ \theta \end{bmatrix}_k . \tag{2}$$

As Equation (2) shows, the prediction contains several trigonometric functions, which correspond to a highly nonlinear prediction model. The theta angle of each particle is critical because it changes according to the local vehicle coordinates. The kinematic model incorporates several assumptions such as the value of $\dot{\theta}$ being equal to zero.

2.1.3. Weight Calculation

Weight is also an important factor that can heavily influence particle motion. Measurements from the range sensor were used to calculate the weight. We assume that the vehicle can receive all the data from the vehicle-to-everything (V2X), HD map, and range sensor. Thus, the vehicle can receive the distance data and orientation between the vehicle and every exterior infrastructure. Here, d_i is the measured distance between the vehicle and infrastructure i, ϵ_d is the distance measurement noise, and $\Delta\theta$ is the relative orientation angle of the vehicle and infrastructure i. As Equations (3) and (4) show, the measurement model is nonlinear in nature.

$$Z_{k+1} = f(\overline{X}_{k+1}) + \text{Noise}_{k+1}, \tag{3}$$

$$Z_{k+1} = \begin{bmatrix} d_{[i]} \\ \Delta\theta_{[i]} \end{bmatrix}_{k+1} = \begin{bmatrix} \sqrt{(\overline{x}_k - x_{b,i})^2 + (\overline{y}_k - y_{b,i})^2} \\ \arctan\left(\frac{\overline{x}_k - x_{b,i}}{\overline{y}_k - y_{b,i}}\right) \end{bmatrix}_{k+1} + \begin{bmatrix} \epsilon_d \\ -\theta_v + \epsilon_{\Delta\theta} \end{bmatrix}_{k+1} . \tag{4}$$

To evaluate the weight, a multivariable normal distribution function was used to assess the importance of each particle. Thus, a multivariable normal distribution function returns the weight of a particle based on the newest sensor measurement values and the predicted values from the model as

$$w_i = p(x_i, y_i) = \frac{1}{2\pi\sigma_x\sigma_y} e^{-\left(\frac{(x-\mu_x)^2}{2\sigma_x^2} + \frac{(y-\mu_y)^2}{2\sigma_y^2}\right)} \quad i = 1, 2 \ldots N. \tag{5}$$

2.1.4. Resampling

After calculating the weight and prediction values, the particle filter should select the particle along with its corresponding weight, which is the resampling step.

The entire PF part of the algorithm's pseudo code is shown in Table 1.

Table 1. Pseudo code of the particle filter.

Order	Process
1	Start one sample time iteration
2	Initialization $X_{1,2...N}$ particles
3	For 1 to N do
4	$\overline{X}_{k+1}$ = prediction model(X_k, u_t)
5	End for
6	Z_{k+1} = measurent input
7	$w_{[1,2...N]}$ = multivariable normal distribution$\left(\overline{X}_{k+1}, Z_{k+1}, \sigma_{distance}, \sigma_{orientation}\right)\overline{X}_{k+1}$
8	Return $\hat{X}_{PF} = f\left(X_{k+1}^{[1,2...N]}, w_{[1,2...N]}\right)$
9	End one sample time iteration

2.2. Particle-Aided Unscented Kalman Filter Algorithm

The particle filter algorithm is introduced in Section 2.1. The particle estimates the position of the vehicle by using the range sensor. The final results for each particle contain information about the surrounding infrastructures and the position of the ego vehicle. Therefore, it can be concluded that this sensor provides more accurate results. When the UKF estimates the state, the results from the particle filter will be the measurement value of the vehicle. Subsequently, the PAUKF can extract a more precise result based on the particle filter estimation results. A flowchart of the PAUKF is shown in Figure 2.

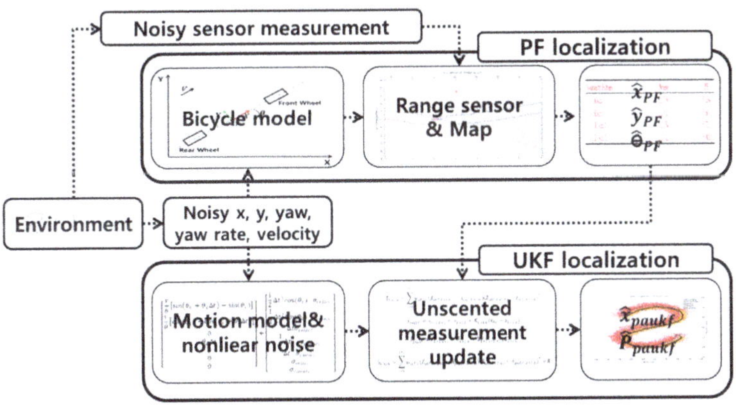

Figure 2. Particle-aided unscented Kalman filter algorithm flowchart.

The UKF is a Bayesian filter that has better performance than the EKF when estimating the state of a discrete-time nonlinear dynamic system. Because it is based on a Kalman filter, the framework of a UKF is almost the same as that of a KF. The difference is that a UKF performs stochastic linearization by using the weighted statistical linear regression process, known as an unscented transform. Instead of using a Taylor expansion, a UKF deterministically extracts the mean and covariance using the sigma points.

The sigma points are predefined by an empirical parameter λ that is calculated using Equation (6). The sigma point is a symmetrical region around the mean value. $P_{k|k}$ is the covariance matrix of the state, which updates at every iteration. The state vector of the vehicle is x_k, which is a 5×1 vector, as shown in Equation (7). The state vector value is the mean of the sigma matrix. n_x is the quantity of the state vector with a size of 5. Then, the sigma points are generated using Equation (8).

$$\lambda = 3 - n_x, \tag{6}$$

$$x_{paukf,\ k} = \begin{bmatrix} x & y & v & \theta & \dot{\theta} \end{bmatrix}^T, \tag{7}$$

$$X_{paukf,\ k} = (\mu_k,\ \mu_k + \sqrt{(\lambda + n_x)P_k},\ \mu_k - \sqrt{(\lambda + n_x)P_k}). \tag{8}$$

After it generates the sigma points, a UKF needs a prediction model to determine the prior probability of the state. To obtain a more precise position regarding position, a UKF considers more states of the vehicle and the effect of nonlinear noise on the states.

The constant turn rate and velocity magnitude (CTRV) vehicle motion model was used in this study [44]. Based on the CTRV model, the discrete state transition model is derived by integrating the differential equation of the state, which considers the nonlinear process noise vector. Considering the underlying structure of the vehicle for a real-world test, the acceleration and yaw acceleration are considered to be noise. In particular, the acceleration and yaw acceleration noise effects are nonlinear. Therefore, the process noise cannot be handled by addition alone. In order to handle nonlinear noise, it is considered to be a state, as shown in Equation (9). This means that the size of x_{ukf} becomes $x_{k,\ aug}$, which is a 7×1 vector.

$$x_{paukf,\ k,aug} = \begin{bmatrix} x & y & v & \theta & \dot{\theta} & w_{velacc} & w_{yawacc} \end{bmatrix}^T. \tag{9}$$

The process noises w_{velacc} and w_{yawacc} are set as normal Gaussian distributions with variances of ${\sigma_{velacc}}^2$ and ${\sigma_{yawacc}}^2$, respectively, as shown in Equations (10) and (11).

$$w_{velacc} \sim N(0,\ {\sigma_{velacc}}^2), \tag{10}$$

$$w_{yawacc} \sim N(0,\ {\sigma_{yawacc}}^2). \tag{11}$$

The covariance matrix P_k is also augmented into $P_{k,aug}$, which has a size of 7×7, as shown in Equation (12).

$$P_{k,aug} = \begin{bmatrix} P_k & 0 & 0 \\ 0 & {\sigma_{velacc}}^2 & 0 \\ 0 & 0 & {\sigma_{yawacc}}^2 \end{bmatrix}. \tag{12}$$

The process model is derived based on the CTRV assumption and noise, as shown in Equation (13). The model that we derive has highly nonlinear properties, as indicated in Equation (14).

$$x_{paukf,k+1} = x_{paukf,k} + \int_k^{k+1} \dot{x}_{paukf,k} dt, \tag{13}$$

$$\begin{aligned} x_{paukf,k+1,aug} = f(x_{paukf,k,aug}, \sigma_{velacc}, \sigma_{yawacc}) = x_{paukf,k,aug} + \begin{bmatrix} \frac{v}{\dot{\theta}}[\sin(\theta_k + \dot{\theta}_k\Delta t) - \sin(\theta_k)] \\ \frac{v}{\dot{\theta}}[\cos(\theta_k) - \cos(\theta_k + \dot{\theta}_k\Delta t)] \\ 0 \\ \dot{\theta}_k\Delta t \\ 0 \\ 0 \\ 0 \end{bmatrix} \\ + \begin{bmatrix} \frac{v}{\dot{\theta}}[\sin(\theta_k + \dot{\theta}_k\Delta t) - \sin(\theta_k)] \\ \frac{v}{\dot{\theta}}[\cos(\theta_k) - \cos(\theta_k + \dot{\theta}_k\Delta t)] \\ 0 \\ \dot{\theta}_k\Delta t \\ 0 \\ 0 \\ 0 \end{bmatrix} + \begin{bmatrix} \frac{1}{2}\Delta t^2 \cos(\theta_k)\cdot\sigma_{velacc} \\ \frac{1}{2}\Delta t^2 \sin(\theta_k)\cdot\sigma_{velacc} \\ \Delta t\sigma_{velacc} \\ \frac{1}{2}\Delta t^2\cdot\sigma_{yawacc} \\ \Delta t\cdot\sigma_{yawacc} \\ \sigma_{velacc} \\ \sigma_{yawacc} \end{bmatrix}. \end{aligned} \tag{14}$$

In the augmented prediction step, n_x, should be the quantity of the augmented state $n_{x,aug}$ with a size of 7, and λ also needs to be calculated as in Equation (15). Subsequently, the sigma points are predicted using Equation (16).

$$\lambda = 3 - n_{x,aug}, \tag{15}$$

$$X_{paukf,k,aug} = \left(\mu_{paukf,k,aug}, \mu_{paukf,k,aug} + \sqrt{(\lambda + n_{x,aug})P_{paukf,\ k,aug}}, \mu_{paukf,k,aug} - \sqrt{(\lambda + n_{x,aug})P_{paukf,k,aug}}\right). \tag{16}$$

The weight of each sigma point is calculated based on Equations (17) and (18). The predicted mean and covariance were calculated using Equations (19) and (20). The predicted value is the prior of the Bayesian distribution model. These predicted values should be updated when the measurement data are incoming.

$$w_{paukf,i} = \frac{\lambda}{\lambda + n_{x,aug}}, \text{ when } i = 0, \tag{17}$$

$$w_{paukf,i} = \frac{1}{2(\lambda + n_{x,aug})}, \text{ when } i = 1 \ldots n_{x,aug}, \tag{18}$$

$$\overline{x}_{paukf,k+1|k} = \sum_{i=0}^{n_a} w_{paukf,i} X_{paukf,k+1|k,i}, \tag{19}$$

$$\overline{P}_{k+1|k} = \sum_{i=0}^{2n_a} w_{paukf,i}\left(X_{paukf,k+1|k,i} - \overline{x}_{paukf,k+1|k}\right)\left(X_{paukf,k+1|k,i} - \overline{x}_{paukf,k+1|k}\right)^T. \tag{20}$$

After predicting the new mean and covariance matrix based on the augmented sigma point, the algorithm no longer needs to consider noise as the acceleration noise information is already included in the state. Therefore, the augmented state changes back to a normal state with a size of 5. Until now, the prior probability was calculated based on sigma points. When using a Bayesian filter, measurement prediction can be implemented. Instead of using the original range sensor with noise from the vehicle, $\hat{x}_{PF}$ is used in this step. This means that $\hat{x}_{PF}$ becomes a virtual sensor, which is more precise than the original sensor. Since $\hat{x}_{PF}$ already includes sensor information based on the particle filter, it optimally provides a more precise belief of the state. The measurement vector of the sensor is shown in Equation (21).

$$z = \begin{bmatrix} x \\ y \\ \theta \end{bmatrix}. \tag{21}$$

The measurement model is shown in Equations (22) and (23). The particle filter measurement provides x, y, and yaw data. This means that the number of rows in matrix A is 3. Because the augmented state information is included in the state, the state of the UKF recovers to 5. This means that the number of columns in matrix A is 5.

$$Z_{paukf,k+1|k,i} = AX_{paukf,k+1|k,i} + \omega_{paukf,k+1}, \tag{22}$$

$$A = \begin{bmatrix} 1 & 0 & 0 & 0 & 0 \\ 0 & 1 & 0 & 0 & 0 \\ 0 & 0 & 0 & 1 & 0 \end{bmatrix}. \tag{23}$$

The predicted measurement mean is calculated based on the weight of each measurement's sigma points, as shown in Equation (24).

$$\overline{z}_{paukf,k+1|k} = \sum_{i=1}^{n_x} w_{paukf,i} Z_{paukf,k+1|k,i}. \tag{24}$$

The predicted measurement covariance is calculated using Equation (25). R is the measurement noise covariance, as shown in Equation (26). The covariance is tuned according to the particle filter estimation results.

$$S_{k+1|k} = \sum_{i=0}^{2n_x} w_{paukf,i}\left(Z_{paukf,k+1|k,i} - \bar{z}_{paukf,k+1|k}\right)\left(Z_{paukf,k+1|k,i} - z_{paukf,k+1|k}\right)^T + R, \quad (25)$$

$$R = \begin{bmatrix} {\sigma_{x_{PF}}}^2 & 0 & 0 \\ 0 & {\sigma_{y_{PF}}}^2 & 0 \\ 0 & 0 & {\sigma_{\theta_{PF}}}^2 \end{bmatrix}. \quad (26)$$

At this time, a measurement value is needed to calculate the posterior probability. The update step is similar to that of the Kalman filter. The only difference is that the UKF needs to calculate the cross-correlation value, according to Equation (27), between the sigma points in the state space and the measurement space.

$$T_{k+1|k} = \sum_{i=0}^{2n_x} w_{paukf,i}\left(X_{paukf,k+1|k,i} - x_{paukf,k+1|k}\right)\left(Z_{paukf,k+1|k,i} - z_{paukf,k+1|k}\right)^T. \quad (27)$$

Based on the cross-correlation matrix and the measurement covariance, the Kalman gain is then calculated as

$$K_{k+1|k} = T_{k+1|k}S_{k+1|k}^{-1}. \quad (28)$$

The state is updated using the measurement value $\hat{x}_{PF}$, which is obtained from the particle filter estimation as

$$\hat{x}_{paukf} = x_{k+1|k+1} = x_{k+1|k} + K_{k+1|k}\left(\hat{x}_{PF} - z_{k+1|k}\right). \quad (29)$$

The covariance matrix is then updated based on the updated Kalman gain and the measurement covariance matrix as

$$\hat{P}_{paukf} = P_{k+1|k+1} = \bar{P}_{k+1|k} - K_{k+1|k}S_{k+1|k}K_{k+1|k}^T. \quad (30)$$

The terms $x_{k+1|k+1}$ and $P_{k+1|k+1}$ are the final estimation results of the PAUKF, which combines the bicycle model, CTRV motion model, Monte Carlo-based estimation, and unscented Kalman filter-based estimation. The complete pseudo code of the PAUKF algorithm is shown in Table 2.

Table 2. Pseudocode of the particle-aided unscented Kalman filter (PAUKF).

Order	Process
1	Start one sample time iteration
2	Initialization $X_{1,2...N}$ particles
3	For 1 to N do
4	$\bar{X}_{k+1} =$ prediction model(X_k, u_t)
5	End for
6	$\hat{x}_{PF} = f\left(\bar{X}_{k+1}, w_{[1,2...N]}\right)$
7	For 1 to n_{aug} do
8	X_{k+1} = unscented transform (λ, x_k, σ)
9	$\bar{x}_{paukf,k+1\|k}$ = CTRV model(based state prediction
10	$\bar{z}_{paukf,k+1\|k} = A(\bar{x}_{paukf,k+1\|k})$ for measurement prediction
11	$\hat{x}_{paukf}, \hat{P}_{paukf} =$ state update$\left(T_{k+1\|k}, S_{k+1\|k}, \bar{z}_{paukf,k+1\|k}, x_{paukf,k+1\|k}, \hat{x}_{PF}, R\right)$
12	End one sample time iteration

3. Simulation Environment

The simulation of the PAUKF algorithm was performed using MATLAB. The autonomous driving toolbox is used for constructing the infrastructure, road structure, and vehicle kinematic model. The update frequency of the GPS is 10 Hz, and the frequency of the range sensor and gyroscope is 100 Hz. The noise of the GPS and the range sensor is simulated using the ground truth data appended with

Gaussian noise and non-Gaussian noise. Gaussian noise is generated by using the normrnd function in MATLAB, and non-Gaussian noise is generated using a sinusoidal function (we assume the noise affected by the sinusoidal function) and a random number generator, as shown in Table 3 [45,46]. The seed of the random number is set as 50, and the sample time is set as 0.01 s. The variances fed into the PAUKF should be carefully tuned when applied to specific cases. It should be noted that even if the random seed is the same, the random number is generated depending on the number of times that it has been called. This means that any change in parameter changes the result because the input value changes.

Table 3. Simulated noisy environment setting.

Noise Name	Generate Method
GPS x error (Gaussian)	$\sim N(9.65,\ 12.2)$ [46]
GPS x error (Non_Gaussian)	$\sim 15\sin(N(0,\ 1)) + N(9.65,\ 12.2) + 5)$
GPS y error (Gaussian)	$\sim N(8.34,\ 12.33)$ [46]
GPS y error (Non_Gaussian)	$\sim 15\sin(N(0,\ 1)) + N(8.34,\ 12.33) + 5)$
Velocity error	$\sim \sin(N(0,\ 1))$
Yaw error	$\sim \sin(N(0,\ 10))$
Yaw rate error	$\sim \sin(N(0,\ 10))$
Random seed	50

The road simulated with three geometries is shown in Figure 3. There are S-shaped roads and a straight road in the X direction. An S-shaped road is used to verify the performance of each filter on a curved road. The straight-line road in the X direction is used in order to verify the performance of each filter on a straight road. There are 12 infrastructures around the road, and their positions are fixed, even when the map changes. In order to prevent the position of the infrastructures from affecting the performance of the filters, all the infrastructures are symmetrical. The velocity is set to a constant value, and the values are 60, 80, 100, and 120 km/h.

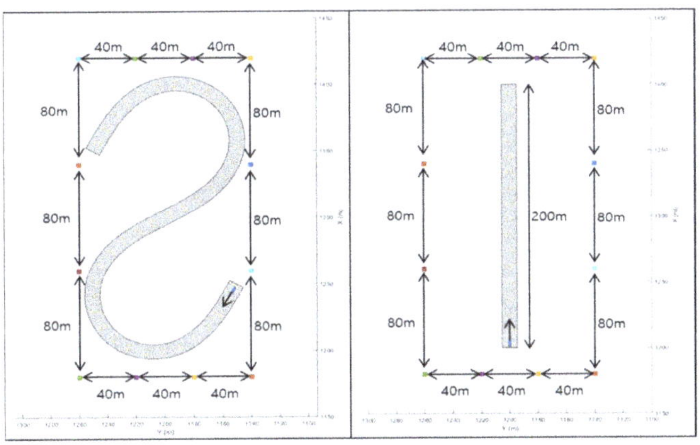

Figure 3. Simulation model.

4. Analysis of Simulation Results

The simulation results are compared to evaluate the performance of the PAUKF. The evaluation parameter is based on the Root Mean Square Error (RMSE) as Equation (31) shows. We choose RMSE as an assessment parameter because the estimation performance of the filter can be compared intuitively by the numerical value of RMSE alone. In Equation (31), N indicates the number of data points. The trajectory of the estimated results and the ground truth of the vehicle's trajectory are compared to verify the algorithm. The effect of the yaw angle is considered for both the x and y directions; therefore, there is no additional comparison of the yaw angle. The unit for all position parameters is meters.

$$\begin{bmatrix} \text{RMSE}_{\text{est}} \\ \text{RMSE}_{\text{noise}} \end{bmatrix} = \begin{bmatrix} \sqrt{\left[\sum_{i=1}^{N}\left(\text{Position}_{\text{est}_i} - \text{Position}_{\text{mean\_est}_i}\right)^2\right]/N} \\ \sqrt{\left[\sum_{i=1}^{N}\left(\text{Position}_{\text{noise}_i} - \text{Position}_{\text{mean\_noise}_i}\right)^2\right]/N} \end{bmatrix}. \quad (31)$$

4.1. *Filter Performance on the S-Shaped Road*

Figure 4 shows the trajectory results of the PF, UKF, and PAUKF, and noise in the S-shaped road. As the legend shows, the green line with a green circle is the ground truth trajectory, the dashed line with a red upward-pointing triangle is the noisy vehicle trajectory, the black dashed line with a black square is the PF estimated trajectory, the blue dashed line with a blue square is the UKF estimated trajectory, and the yellow dashed line with the yellow star marker is the PAUKF estimated trajectory. The data in Figure 4 are generated when the vehicle velocity is 60 km/h, and the noise is Gaussian, as shown in Table 3. The PF estimated trajectory is near the ground truth trajectory. However, the PF-estimated trajectory is not smooth, and the error is still large. This is because the PF localizes the vehicle position with noisy relative distance to each infrastructure and noisy vehicle data. Since there is no other measurement, it must be considered that the measurement is correct. Compared to that with the PF, the UKF-estimated trajectory is relatively smooth; however, it cannot filter the noise of the GPS data. Because the GPS measurement of the UKF has high variance and the UKF does not use range sensor data, the UKF believes the vehicle model more than the measurement. The noisy measurement also makes the UKF less sensitive to the changes in the position and yaw. Compared to that with the PF and UKF, the trajectory estimated by the PAUKF is more accurate and smoother. As it combines the smoothness of the UKF and the accuracy of the PF, the PAUKF reacts more quickly and precisely when the position and yaw change. Moreover, the PAUKF does not depend completely on either of the filters, trades off the filters, and generates even better results.

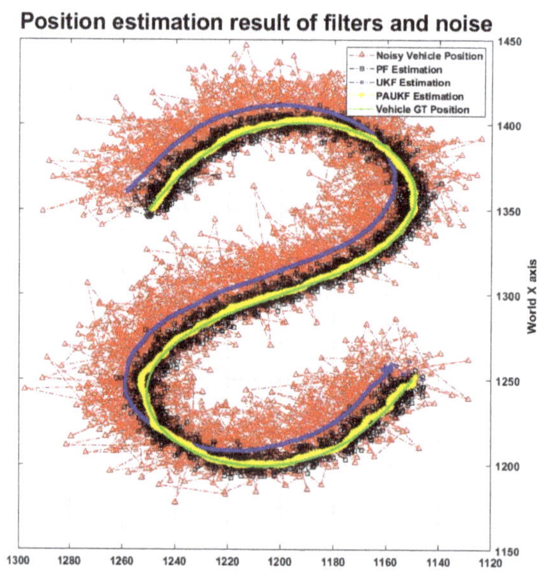

Figure 4. Position estimation result of the filters in the S curve road.

The filter performance results are shown in Table 4. Since the UKF does not use range sensor information, it is not appropriate to compare it with the PF and PAUF. Thus, there are no RMSEs for the UKF in Table 4. To compare with other literature, we calculate the mean value of estimation. The mean of estimation error for the PAUKF is 1.08 m and the variance is 0.7147 m, which is more precise than the mean of 1.69 m and variance of 1.63 m obtained by GANGNAM for similar noise [47]. In order to determine the performance of the filters in an extreme environment, the algorithm is tested under different velocity and noise environments. As mentioned in Section 3, even if the random seed is the same, the random number still changes depending on the number of times it has been called. Therefore, we analyzed the trend of every filter. It can be observed that the PF and PAUKF estimation errors increase slightly when the velocity increases. However, if we consider the magnitude of the RMSE of the changes in noise from 21.336 to 21.712 m, it can be found that the RMSE of the estimation error does not change even when the velocity increases from 60 to 120 km/h. Compared to the Gaussian noise, the non-Gaussian noise generated a larger mean value. Even so, the precision of the PF does not change even when the noise increases, and the precision is almost the same as the RMSE range of 5.489–5.959 m. The PAUKF has an RMSE range of 1.440–1.772 m, even when the noise increases and velocity increases. This is because PAUKF takes the PF estimation results as input and trades off the measurement and predicted value from the UKF. The trade-off is done using the cross-correlation function in Equation (27). Therefore, the PAUKF combines the recursiveness of the UKF and the location information of the infrastructures based on the PF. The PAUKF improves the accuracy by 4.028–4.049 m compared to the PF.

Table 4. Total Root Mean Square Error (RMSE) of filters in different conditions (unit: m).

	With Gaussian Noise			With Non-Gaussian Noise		
Velocity	**Noise**	**PF**	**PAUKF**	**Noise**	**PF**	**PAUKF**
60 km/h	21.336	5.634	1.451	29.796	5.959	1.655
80 km/h	21.310	5.600	1.651	29.730	5.579	1.440
100 km/h	21.154	5.720	1.501	29.430	5.631	1.616
120 km/h	21.712	5.800	1.772	29.934	5.489	1.454

4.2. Filter Performance on a Straight Road in the X Direction

Figure 5 shows the trajectory results of the PF, UKF, and PAUKF, and the noise on a straight road in the X direction. The data in Figure 5 are generated when the vehicle velocity is 60 km/h, and the noise is Gaussian. The algorithm for a straight line is used to determine the performance when the vehicle only moves in the X direction. From Figure 5, it can be seen that the PAUKF converges to the ground truth better than the PF and UKF. Because the vehicle only moves in the X direction, there is no information about the movement in the Y direction. Therefore, even though the PAUKF estimation is better than that of the UKF, in order to improve the response time, the PAUKF tends to believe more about noise in the Y direction. As a result, the PAUKF is not sufficiently precise in the Y direction, as Figure 5 shows.

The results for filter performance are shown in Table 5 when the vehicle moves in the X direction. It can be found that the PF and PAUKF estimation properties are the same as those of the vehicle when it runs on an S-shaped road. The estimation results show that the performance of the algorithm does not change even when the map changes. The RMSE of the PF is 5.384–5.692 m and the PAUKF has an RMSE of 1.312–1.800 m even when the noise increases and the velocity increases. The PAUKF improves the accuracy by 3.892–4.072 m compared to the PF.

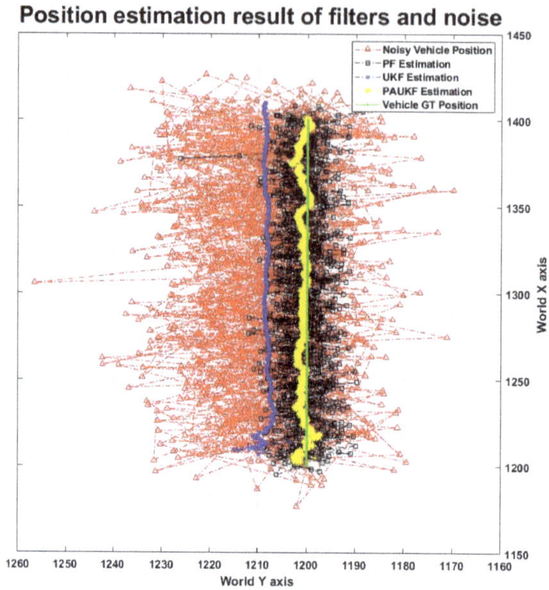

Figure 5. Position estimation result of the filters.

Table 5. RMSE of filters in different conditions (unit: m).

	With Gaussian Noise			With Non-Gaussian Noise		
Velocity	Noise	PF	PAUKF	Noise	PF	PAUKF
60 km/h	21.411	5.655	1.486	29.376	5.658	1.659
80 km/h	21.518	5.546	1.312	28.987	5.491	1.526
100 km/h	21.848	5.692	1.800	29.274	5.615	1.482
120 km/h	21.363	5.383	1.710	30.078	5.617	1.719

5. Conclusions

In this work, we propose a novel approach for a vehicle estimation algorithm, called the PAUKF, which combines the advantages of the PF and the UKF. The PAUKF combines the unscented transform property of a UKF with a sample-based PF to handle the localization problem in a bad GPS environment by using the range sensor and ground truth data of the infrastructure in an HD map. Owing to properties of the UKF, the PAUKF becomes more robust and precise compared to the original PF, given the same quantity of particles. The performance of the algorithm is stable and accurate (minimum RMSE: 1.44 m) when vehicles move along an S curve or any straight road at speeds of 60 to 120 km/h. The results of the simulation showed that the PAUKF has a significantly higher precision and stability than the PF and in previous research. In future work, we will try to implement the PAUKF in a real vehicle and incorporate the 3D range sensor data to upgrade the algorithm in the real world.

Author Contributions: Conceptualization, M.L. and J.Y.; Software, M.L.; Methodology, M.L. and J.Y.; Validation, M.L.; Writing—original draft preparation M.L.; B.K. Writing—review & editing. All authors have read and agreed to the published version of the manuscript.

Funding: This work was supported by KoreaHydro & Nuclear Power company through the project "Nuclear Innovation Center for Haeoleum Alliance" and Technology Innovation Program 2020 funded By the Ministry of Trade, industry & Energy (MI, Korea).

Conflicts of Interest: The authors declare no conflicts of interest.

Nomenclature

$bel(x)$	Belief of state
$x_{[1,2,3,\ldots i]}$	States of particle1, Particle 2, … Particle i
$w_{[1,2,3,\ldots i]}$	Weights of particle1, particle 2, … particle i
$\overline{X}_{k+1}$	State at sample time k + 1
K	Timestamp
x_k, y_k	Vehicle position in the x, y dimension at time k
v_k	Vehicles position in the x dimension at time k
θ_k	Yaw angle at time k
$\dot{\theta_k}(dt), \dot{\theta}$	Yaw rate of vehicle at time k
Δt, dt	Sample time
Z_{k+1}	Measurement vector at time k + 1
$d_{[i]}$	Distance of ego vehicle to ith beacon
$\Delta\theta_{[i]}$	Relative angle of vehicle orientation and ith beacon
$x_{b,i}, y_{b,i}$	Relative distance of vehicle and ith beacon
ϵ_d	Noise distance measurement
$\epsilon_{\Delta\theta}$	Noise of angle measurement
$p(x_i, y_i)$	Multivariable normal distribution
σ_x, σ_y	Covariance of sensor range noise in the x- and y-directions
$x_{paukf,\ k,aug}$	State of PAUKF
w_{velacc}	Noise of vehicle acceleration
w_{yawacc}	Noise of vehicle yaw acceleration
σ_{velacc}	Variance of noise of vehicle acceleration
σ_{velacc}	Variance of noise in vehicle yaw acceleration
$P_{k,aug}$	Variance matrix of PAUKF.
$X_{paukf,\ k+1,aug}$	Augmented state with sigma points of PAUKF at time k + 1
$\mu_{paukf,k,aug}$	Mean value of augmented state of PAUKF at time k
$n_{x,aug}$	Number of augmented states
$w_{paukf,i}$	Weight of ith sigma point
λ	Sigma point design parameter
$\overline{x}_{paukf,k+1\mid k}$	Predicted state based on the weight of sigma points and states
$P_{k+1\mid k}$	Predicted variance based on sigma points and predicted state mean
$\omega_{paukf,k+1}$	Measurement noise of PAUKF.
$Z_{paukf,k+1\mid k,i}$	Measurement prediction based on sigma points.
$X_{paukf,k+1\mid k,i}$	Sigma points of state
A	Measurement transition model.
$z_{paukf,k+1\mid k}$	Predicted measurement based on sigma points and weights
$S_{k+1\mid k}$	Predicted measurement covariance matrix.
R	Variance matrix of the measurement noise.
$\sigma_{x_{pf}}$	Covariance of PF estimation in the x dimension
$\sigma_{y_{pf}}$	Covariance of PF estimation in the y-dimension
$T_{k+1\mid k}$	Cross-correlation matrix of PAUKF
$K_{k+1\mid k}$	Kalman gain of PAUKF
$\hat{x}_{PAUFK}$	Final state estimation of PAUKF.
$\hat{P}_{PAUFK}$	Final state variance matrix of PAUKF

References

1. Levinson, J.; Montemerlo, M.; Thrun, S. Map-based precision vehicle localization in urban environments. In Proceedings of the Robotics: Science and Systems, Cambridge, MA, USA, 27–30 June 2007; Volume 4, p. 1.
2. Samyeul, N.; An, K.; Wooyong, H. High-Level Data Fusion based Probabilistic Situation Assessment for Highly Automated Driving. In Proceedings of the 2015 IEEE 18th International Conference on Intelligent Transportation Systems, Las Palmas, Spain, 15–18 September 2015; pp. 1587–1594.

3. Motilal, A.; Kurt, K. Real-time Localization in Outdoor Environments using Stereo Vision and Inexpensive GPS. In Proceedings of the 18th International Conference on Pattern Recognition (ICPR'06), Hong Kong, China, 20 August 2006; pp. 1063–1068.
4. Mattern, N.; Wanielik, G. Vehicle localization in urban environments using feature maps and aerial images. In Proceedings of the 14th International IEEE Conference on Intelligent Transportation Systems (ITSC), Washington, DC, USA, 5–7 October 2011; pp. 1027–1032.
5. Li, C.; Dai, B.; Wu, T. Vision-based precision vehicle localization in urban environments. In Proceedings of the 2013 Chinese Automation Congress, Changsha, China, 7–8 November 2013; pp. 599–604.
6. Parra, I.; Sotelo, M.; Llorca, D.; Ocaña, M. Robust visual odometry for vehicle localization in urban environments. *Robotica* **2010**, *28*, 441–452. [CrossRef]
7. KAMIJO, S.; Gu, Y.; Hsu, L. Autonomous vehicle technologies: Localization and mapping. *IEICE Fundam.* **2015**, *9*, 131–141. [CrossRef]
8. Vivacqua, R.; Vassallo, R.; Martins, F. A Low Cost Sensors Approach for Accurate Vehicle Localization and Autonomous Driving Application. *Sensors* **2017**, *17*, 2359. [CrossRef] [PubMed]
9. Yu, Y.; Li, J.; Guan, H.; Jia, F.; Wang, C. Three-dimensional object matching in mobile laser scanning point clouds. *IEEE Geosci. Remote Sens. Lett.* **2014**, *12*, 492–496.
10. Hata, A.Y.; Wolf, D.F. Feature detection for vehicle localization in urban environments using a multilayer LIDAR. *IEEE Trans. Intell. Transp. Syst.* **2016**, *17*, 420–429. [CrossRef]
11. Levinson, J.; Thrun, S. Robust vehicle localization in urban environments using probabilistic maps. In Proceedings of the IEEE International Conference on Robotics and Automation, Anchorage, AK, USA, 3–7 May 2010.
12. Castorena, J.; Agarwal, S. Ground-edge-based LIDAR localization without a reflectivity calibration for autonomous driving. *IEEE Robot. Autom. Lett.* **2017**, *3*, 344–351. [CrossRef]
13. Kim, H.; Liu, B.; Goh, C.Y.; Lee, S.; Myung, H. Robust vehicle localization using entropy-weighted particle filter-based data fusion of vertical and road intensity information for a large scale urban area. *IEEE Robot. Autom. Lett.* **2017**, *2*, 1518–1524. [CrossRef]
14. Wolcott, R.W.; Eustice, R.M. Robust LIDAR localization using multiresolution Gaussian mixture maps for autonomous driving. *Int. J. Robot. Res.* **2017**, *36*, 292–319. [CrossRef]
15. Wolcott, R.W.; Eustice, R.M. Visual localization within lidar maps for automated urban driving. In Proceedings of the IEEE/RSJ International Conference on Intelligent Robots and Systems, Chicago, IL, USA, 14–18 September 2014; pp. 176–183.
16. Sampo, K.; Saber, F.; Konstantinos, K.; Mehrdad, D.; Francis, M.; Alexandros, M. A Survey of the State-of-the-Art Localization Techniques and Their Potentials for Autonomous Vehicle Applications. *IEEE Internet Things J.* **2018**, *5*, 829–846.
17. Fujii, S.; Fujita, A.; Umedu, T.; Kaneda, S.; Yamaguchi, H.; Higashino, T.; Takai, M. Cooperative vehicle positioning via V2V communications and onboard sensors. In Proceedings of the 2011 IEEE Vehicular Technology Conference (VTC Fall), San Francisco, CA, USA, 5–8 September 2011; pp. 1–5.
18. Rohani, M.; Gingras, D.; Vigneron, V.; Gruyer, D. A new decentralized Bayesian approach for cooperative vehicle localization based on fusion of GPS and VANET based inter-vehicle distance measurement. *IEEE Intell. Transp. Syst. Mag.* **2015**, *7*, 85–95. [CrossRef]
19. Mattern, N.; Obst, M.; Schubert, R.; Wanielik, G. Co-operative vehicle localization algorithm—Evaluation of the CoVeL approach. In Proceedings of the International Multi-Conference on Systems, Signals & Devices, Chemnitz, Germany, 20–23 March 2012; pp. 1–5.
20. Ordóñez-Hurtado, R.H.; Griggs, W.M.; Crisostomi, E.; Shorten, R.N. Cooperative Positioning in Vehicular Ad-hoc Networks Supported by Stationary Vehicles. Available online: https://www.hamilton.ie/smarttransport/publications/arXivCooperativePositioningFeb2015.pdf (accessed on 29 April 2020).
21. Golestan, K.; Seifzadeh, S.; Kamel, M.; Karray, F.; Sattar, F. Vehicle localization in VANETs using data fusion and V2V communication. In Proceedings of the Second ACM International Symposium on Design and Analysis of Intelligent Vehicular Networks and Applications, Paphos, Cyprus, 21–22 October 2012; pp. 123–130.
22. Lina, A.; Imad, M.; Monika, R. Weighted localization in Vehicular Ad Hoc Networks using vehicle-to-vehicle communication. In Proceedings of the 2014 Global Information Infrastructure and Networking Symposium (GIIS), Montreal, QC, Canada, 15–19 September 2014; pp. 1–5.

23. Altoaimy, L.; Mahgoub, I. Fuzzy logic based localization for vehicular ad hoc networks. In Proceedings of the 2014 IEEE Symposium on Computational Intelligence in Vehicles and Transportation Systems (CIVTS), Orlando, FL, USA, 9–12 December 2014; pp. 121–128.
24. Kalman, R.E. A new approach to linear filtering and prediction problems. *J. Basic Eng.* **1960**, *82*, 35–45.
25. Andrew, H.J. *Stochastic Processes and Filtering Theory*; Academic Press: San Diego, CA, USA, 1970.
26. Hamzah, A.; Toru, N. Extended Kalman filter-based mobile robot localization with intermittent measurements. *Syst. Sci. Control Eng.* **2013**, *1*, 113–126.
27. Xu, Y.; Shmaliy, Y.S.; Ahn, C.K.; Tian, G.; Chen, X. Robust and accurate UWB-based indoor robot localisation using integrated EKF/EFIR filtering. *IET Radar Sonar Navig.* **2018**, *12*, 750–756.
28. Luo, Y.H.; Jiang, P.; Hu, W.W.; Xie, Y.Q. Application of EKF in Laser/Inertial Sensors Localization System. In Proceedings of the 2010 International Conference on Electrical and Control Engineering, Wuhan, China, 25–27 June 2010; pp. 3369–3372.
29. Jaewoo, Y.; Byeongwoo, K. Vehicle position estimation using nonlinear tire model for autonomous vehicle. *J. Mech. Sci. Technol.* **2016**, *30*, 3461–3468.
30. Zhao, S.; Gu, J.; Ou, Y.; Zhang, W.; Pu, J.; Peng, H. IRobot self-localization using EKF. In Proceedings of the 2016 IEEE International Conference on Information and Automation (ICIA), Ningbo, China, 1–3 August 2016; pp. 801–806.
31. Xiao, Y.; Ou, Y.; Feng, W. Localization of indoor robot based on particle filter with EKF proposal distribution. In Proceedings of the 2017 IEEE International Conference on Cybernetics and Intelligent Systems (CIS) and IEEE Conference on Robotics, Automation and Mechatronics (RAM), Ningbo, China, 19–21 November 2017; pp. 568–571.
32. Julier, S.J.; Uhlmann, J.K. Unscented filtering and nonlinear estimation. *Proc. IEEE* **2004**, *92*, 401–422. [CrossRef]
33. Wan, E.A.; Van Der Merwe, R. The unscented Kalman filter for nonlinear estimation. In Proceedings of the IEEE 2000 Adaptive Systems for Signal Processing, Communications, and Control Symposium, Lake Louise, AB, Canada, 4 October 2000; pp. 153–158.
34. Julier, S.J.; Uhlmann, J.K. New extension of the Kalman filter to nonlinear systems. In Proceedings of the Signal Processing, Sensor Fusion, and Target Recognition VI, Orlando, FL, USA, 28 July 1997; Volume 3068, pp. 182–193.
35. Doucet, A.; Johansen, A.M. A tutorial on particle filtering and smoothing: Fifteen years later. *Handb. Nonlinear Filter.* **2009**, *12*, 3.
36. Thrun, S. Particle filters in robotics. In Proceedings of the Eighteenth Conference on Uncertainty in Artificial Intelligence, Edmonton, AB, Canada, 1–4 August 2002; pp. 511–518.
37. Dan, C. *Particle Filters—A Theoretical Perspective. Sequential Monte Carlo Methods in Practice*, 1st ed.; Doucet, A., De Freitas, N., Gordon, N., Eds.; Springer: New York, NY, USA, 2001; pp. 17–38.
38. Aggarwal, P.; Syed, Z.; El-Sheimy, N. Hybrid extended particle filter (HEPF) for integrated inertial navigation and global positioning systems. *Meas. Sci. Technol.* **2009**, *20*, 055203–055212. [CrossRef]
39. Aggarwal, P.; Gu, D.; Nassar, S.; Syed, Z.; El-Sheimy, N. Extended particle filter (EPF) for INS/GPS land vehicle navigation applications. In Proceedings of the Institute of Navigation Satellite Division Technical Meeting (ION GNSS 2007), Fort Worth, TX, USA, 25–28 September 2007; pp. 2619–2626.
40. Wan, Y.; Wang, S.; Qin, X. IMM Iterated ExtendedH∞ Particle Filter Algorithm. *Hindawi Publ. Corp. Math. Probl. Eng.* **2013**, *2013*, 8.
41. Van, D.M.R.; Doucet, A.; De, F.N.; Wan, E.A. The unscented particle filter. In Proceedings of the Advances in Neural Information Processing Systems, Denver, CO, USA, 27–30 November 2000; pp. 584–590.
42. Yu, W.; Peng, J.; Zhang, X.; Li, S.; Liu, W. An adaptive unscented particle filter algorithm through relative entropy for mobile robot self-localization. *Math. Probl. Eng.* **2013**. [CrossRef]
43. Pak, J.M.; Ahn, C.K.; Shmaliy, Y.S.; Shi, P.; Lim, M.T. Accurate and reliable human localization using composite particle/FIR filtering. *IEEE Trans. Hum. Mach. Syst.* **2016**, *47*, 332–342. [CrossRef]
44. Rajamani, R. *Vehicle Dynamics and Control*; Springer Science & Business Media: Boston, MA, USA, 2011; pp. 20–26.
45. Sebastian, T.; Wolfram, B.; Dieter, F. *Probabilistic Robotics*; MIT Press: Cambridge, MA, USA, 2006; pp. 136–164.

46. Sun, L.; Shen, M.; Xu, W.; Li, Z.; Beadle, P. Model for Generating Non-gaussian Noise Sequences Having Specified Probability Distribution and Spectrum. In *Parallel and Distributed Computing: Applications and Technologies*; Liew, K.M., Shen, H., See, S., Cai, W., Fan, P., Horiguchi, S., Eds.; Springer: Berlin/Heidelberg, Germany, 2004; pp. 795–798.
47. Suhr, J.K.; Jang, J.; Min, D.; Jung, H.G. Sensor Fusion-Based Low-Cost Vehicle Localization System for Complex Urban Environments. *IEEE Trans. Intell. Transp. Syst.* **2016**, *18*, 1078–1086.

sensors

MDPI

Article

Motion State Estimation of Target Vehicle under Unknown Time-Varying Noises Based on Improved Square-Root Cubature Kalman Filter

Shiping Song and Jian Wu *

State Key Laboratory of Automotive Simulation and Control, Jilin University, Changchun 130022, China; songsp17@mails.jlu.edu.cn

* Correspondence: wujian@jlu.edu.cn

Received: 17 March 2020; Accepted: 2 May 2020; Published: 4 May 2020

Abstract: In the advanced driver assistance system (ADAS), millimeter-wave radar is an important sensor to estimate the motion state of the target-vehicle. In this paper, the estimation of target-vehicle motion state includes two parts: the tracking of the target-vehicle and the identification of the target-vehicle motion state. In the unknown time-varying noise, non-linear target-vehicle tracking faces the problem of low precision. Based on the square-root cubature Kalman filter (SRCKF), the Sage–Husa noise statistic estimator and the fading memory exponential weighting method are combined to derive a time-varying noise statistic estimator for non-linear systems. A method of classifying the motion state of the target vehicle based on the time window is proposed by analyzing the transfer mechanism of the motion state of the target vehicle. The results of the vehicle test show that: (1) Compared with the Sage–Husa extended Kalman filtering (SH-EKF) and SRCKF algorithms, the maximum increase in filtering accuracy of longitudinal distance using the improved square-root cubature Kalman filter (ISRCKF) algorithm is 45.53% and 59.15%, respectively, and the maximum increase in filtering the accuracy of longitudinal speed using the ISRCKF algorithm is 23.53% and 29.09%, respectively. (2) The classification and recognition results of the target-vehicle motion state are consistent with the target-vehicle motion state.

Keywords: millimeter-wave radar; square-root cubature Kalman filter; Sage-Husa algorithm; target tracking; stationary and moving object classification

---

1. Introduction

Millimeter-wave radar is an important sensor that constitutes an advanced driver assistance system (ADAS). The estimation of the moving state of the target vehicle based on the on-board millimeter-wave radar is essential for predicting the future trajectory of the target vehicle and determining the degree of danger of the target vehicle to the ego vehicle. In this paper, the motion state estimation of the target vehicle includes target-vehicle tracking and target-vehicle motion state classification.

The motion state information of the target vehicle (relative radial distance, azimuth, and relative radial rate) measured by millimeter-wave radar is obtained from the polar coordinate system. However, in the process of target tracking, the target motion model is usually established in the Cartesian coordinate system. As can be seen from the radar target-tracking process, the state equation is linear and the measurement equation is non-linear. Since the measurement equation is non-linear, the target-tracking system based on the millimeter-wave radar is a non-linear system.

Extended Kalman filter (EKF) [1,2], unscented Kalman filter (UKF) [3,4], particle filter (PF) [5] and cubature Kalman filter (CKF) [6] are common non-linear filtering state estimation algorithms.

The basic idea of EKF is: the non-linear system is linearized by Taylor series expansion, and then Kalman filtering is performed. Inaccurate modeling of system noise and changes in model parameters due to environmental factors will cause the decrease of the EKF estimation accuracy, and even the filter divergence will be caused. To ensure the accuracy and stability of EKF under unknown and time-varying conditions, many scholars carry out research on the adaptive extended Kalman filter algorithm, adaptive extended Kalman filter (AEKF) algorithm, such as Sage–Husa's maximum a posteriori estimation [7], fictitious noise compensating [8], dynamic bias decoupling estimation [9], etc. To solve the filtering divergence problem caused by system modeling errors, some scholars proposed an EKF with a suboptimal fading factor [10]. Meanwhile, to improve the estimation accuracy of the EKF algorithm, some scholars proposed an AEKF algorithm based on a neural network [11]. The adaptive learning characteristics of neural networks are used to identify the non-linear system model online, to overcome the influence of unmodeled dynamic characteristics of the filter. However, due to the following shortcomings of the EKF algorithm, its development in engineering practice is limited:

(1) The higher-order terms are truncated (with truncation error) and only the first-order terms retained during the Taylor series are expanded, and the accuracy of EKF is the first-order Taylor series;

(2) In many engineering applications, the Jacobian matrix of the measurement equation is difficult to solve.

The basic idea of UKF is: "It is easier to approximate the probability density distribution of non-linear functions than the approximation of non-linear functions" [12]. UKF uses the unscented transformation to approximate the posterior distribution of the state of the non-linear system. The most important part of the UKF algorithm is the sampling strategy. Different sampling strategies differ in the number, location, and corresponding weights of the extracted Sigma points [13]. Compared with EKF, UKF has the following two advantages:

(1) The accuracy of UKF can reach at least two orders under the condition of EKF, and UKF algorithm takes the same order of magnitude;

(2) In UKF, it is not necessary to calculate the Jacobian matrix of the measurement equation.

The above two advantages of UKF expand the application range of the EKF algorithm. However, certain sigma point weights $\omega < 0$ will cause the covariance matrix to non-positive definite condition when the dimension is too high ($N \geq 4$). This situation will lead to the following two effects: firstly, the filter value is not stable or even divergent; secondly, the dimension disaster problem will occur [14]. Therefore, some scholars through theoretical analysis and experiments have proved that UKF has high accuracy for low-dimensional ($N \leq 2$) non-linear systems [15].

The basic idea of PF is to approximate the posterior probability density function of the system state by random particles. PF uses the particle mean value instead of the integral operation to obtain the minimum variance of state. With the increase in the number of particles, the probability density function of particles gradually approximates the probability density function of the real state. However, PF has the following two shortcomings, which restrict the development of PF [16]:

(1) the particle degradation problem;

(2) It is difficult to realize online estimation due to the large amount of computation.

To better solve the problem of poor filtering performance and even divergence in high-dimensional non-linear filtering estimation, Arasaratnam and Haykin proposed a third-degree spherical-radial rule CKF [17]. After CKF was proposed, it was widely used in target tracking [18] and navigation [19]. Compared with UKF, CKF has the following advantages:

(1) The UKF algorithm selects 2n+1 sampling points with different weights, while the CKF algorithm selects 2n sampling points with the same weight. The CKF algorithm has fewer sampling points than UKF, so the CKF algorithm takes less time than UKF.

(2) Since the weight coefficients of the sampling points of the CKF algorithm are all positive, the robustness of the CKF algorithm is high when the dimension of the observed variable is too high ($N \geq 4$).

In conclusion, compared with EKF and UKF, CKF has higher estimation accuracy. During the operation of the standard CKF algorithm, the following two conditions should be met: (1) symmetry, (2) positive qualitative. However, in practical engineering, these two conditions are sometimes difficult to meet. Therefore, scholars proposed the SRCKF algorithm based on the CKF algorithm [20]. SRCKF has the following two advantages. On the one hand, SRCKF avoids computing the square root of a matrix by directly calculating the square root of the predicted value of error covariance and the estimated value of error covariance. On the other hand, in the SRCKF algorithm, the symmetry and positive qualitative value of the error covariance matrix can always be guaranteed.

During the derivation of the CKF algorithm, it is generally assumed that the statistical characteristics of system noise and measurement noise are known [21]. However, in practice, the statistical characteristics of noise are often unknown and time-varying. The Sage–Husa estimator is often used to estimate the statistical characteristics of noise because of its simplicity and good real-time performance [22]. However, the conventional Sage–Husa estimator is suitable for estimating the statistical properties of constant coefficient noise in linear systems [23]. Based on the conventional Sage–Husa algorithm, an adaptive noise statistical estimator for non-linear systems is derived by using the cubature rule.

For time-varying noise statistics, the real-time updated data play a leading role, while the old data play a small role compared with the new data. Therefore, we should gradually reduce the weight of old data and increase the weight of new data. The exponential weighted attenuation method for fading memory is introduced to estimate time-varying noise. The exponential weighted attenuation method has the characteristic of remembering the past historical data, but the weighted coefficient of the old data is small [24].

The current international standard "ISO/DIS15622 Intelligent transportation systems-adaptive cruise control systems-performance requirements and test procedures" clearly states that adaptive cruise control (ACC) may ignore stationary targets or do not respond to stationary targets. At the same time, for full-speed ACC and autonomous emergency braking (AEB) systems, it is necessary to accurately identify the target-vehicle as a stationary target-vehicle or a moving target-vehicle.

The recognition of the target motion state has the following two functions for the ADAS. On the one hand, it can predict the future trajectory of the target-vehicle; On the other hand, it can determine the degree of danger of the target-vehicle to the ego-vehicle. Therefore, it is essential to study the classification of the target-vehicle motion state. In the literature [25], targets detected by radar are divided into high-altitude targets, stationary targets, moving targets, and road targets, but the basis for target classification is not discussed in detail. Therefore, a method of classifying the motion state of the target-vehicle based on a time window is proposed by analyzing the transfer mechanism of the motion state of the target-vehicle.

The motivation of writing the paper is as follows: (1) For the on-board millimeter-wave radar in the unknown and time-varying noise environment, the accuracy of a high-dimensional non-linear target tracking process is low. The ISRCKF algorithm based on SRCKF is proposed to accurately estimate the unknown and time-varying noise statistics. (2) To accurately predict the future trajectory of the target vehicle and determine the danger degree of the target vehicle to the ego vehicle. We present a classification method for moving objects and stationary objects based on the mechanism of moving state transfer in a time window. The vehicle test results show: (1) The filter accuracy of the ISRCKF algorithm is higher than that of SRCKF and SH-EKF. (2) The classification and recognition results of the target-vehicle's motion state are consistent with the target-vehicle's motion state.

The rest of the paper is organized as follows. Section 2, based on the Cartesian coordinate system of millimeter-wave radar, the target-vehicle motion state model is established; In Section 3, based on the SRCKF, an adaptive square-root cubature Kalman filter (ASRCKF) is derived. In Section 4, based on the analysis of the motion state and transfer principle of the target, a classification method of moving target and stationary target based on the motion state transfer mechanism in a time window is proposed. In Section 5, the algorithm is validated and its results are analyzed by establishing a vehicle test platform. Section 6 presents the conclusions.

2. Motion Model of Target-Vehicle

2.1. Coordinate System

It can be learned from the dynamics that the description of the same target motion state varies in different reference coordinate systems. Therefore, it is meaningful to clarify the coordinate system that describes the target motion.

The target measurement information (relative radial distance, azimuth, and relative radial speed) measured by the millimeter-wave radar is obtained from the millimeter-wave radar polar coordinate system. This paper studies the target tracking algorithm based on the millimeter-wave radar Cartesian coordinate system, to verify the performance and accuracy of the proposed algorithm in the target tracking process.

The three coordinate systems herein are respectively, the geodetic coordinate system $x_o y_o z_o$, on the horizontal ground, the vehicle motion coordinate system $x_v y_v z_v$ with its origin coinciding with the center of gravity vehicle, the millimeter-wave radar coordinate system $x_r o y_r$, as shown in the Figure 1. The millimeter-wave radar is fixedly mounted on the front of the vehicle and the radar beam is aligned with the longitudinal axis of the vehicle. Therefore, the millimeter-wave radar coordinate system $x_r o y_r$ is parallel to the vehicle motion coordinate system $x_v y_v z_v$. The radar x_r axis direction is identical with the x_v direction in the vehicle motion coordinate system. The radar y_r axis direction is identical with the y_v direction in the vehicle motion coordinate system.

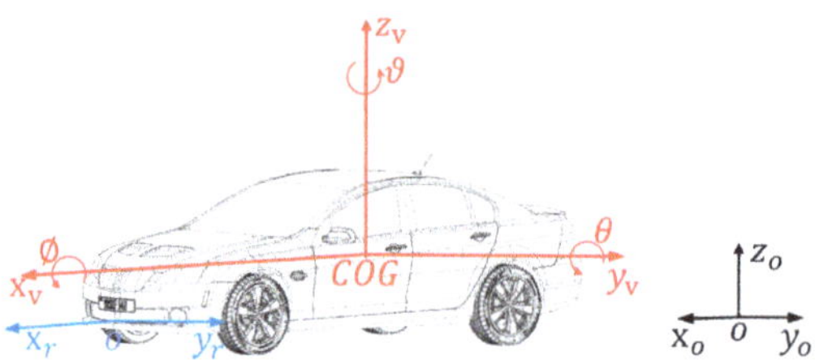

Figure 1. Coordinate system.

2.2. Equation of Motion

The forward millimeter-wave radar is installed in the middle of the front bumper and fixed to the vehicle body. As the vehicle travels, the millimeter-wave radar detects the target in front of the vehicle, mainly referring to the target-vehicles. These target-vehicles have no vertical motion or small moving speed in the vertical direction, so only the movement of the target-vehicles in the XY plane needs to be considered. Since the target-vehicles motion state has the characteristics of small mobility and low speed, a constant velocity model is established based on the millimeter-wave radar Cartesian coordinate system to describe the motion state of the front target-vehicles.

The model of the constant velocity of the target-vehicle can be obtained from the millimeter-wave radar Cartesian coordinate system:

$$\begin{cases} x(k+1) = x(k) + \dot{x}(k) * T + \frac{1}{2} w_x(k) * T^2 \\ \dot{x}(k+1) = \dot{x}(k) + w_x(k) * T \\ y(k+1) = y_{(k)} + \dot{y}(k) * T + \frac{1}{2} w_y(k) * T^2 \\ \dot{y}(k+1) = \dot{y}(k) + w_y(k) * T \end{cases} \tag{1}$$

where $(x(k+1), y(k+1))$ represents the longitudinal distance and the lateral distance from the forward target-vehicles in the millimeter-wave radar Cartesian coordinate system at $k+1$, respectively. $\left(\dot{x}(k+1), \dot{y}(k+1)\right)$ is the longitudinal velocity and the lateral velocity of the target-vehicles relative to the millimeter-wave radar Cartesian coordinate system movement at $k+1$. T is the sampling time.

The model of the target motion state is as follows:

$$X(k+1) = A * X(k) + B * w(k) \tag{2}$$

The target motion state vector $X = [x, \dot{x}, y, \dot{y}]^T$ here is designed to describe the motion state of the target-vehicle in the millimeter-wave radar Cartesian coordinate system; where, A is the state transition matrix, B noise-driven matrix, and $w(k)$ the measurement noise at k;

$$A = \begin{bmatrix} 1 & T & 0 & 0 \\ 0 & 1 & 0 & 0 \\ 0 & 0 & 1 & T \\ 0 & 0 & 0 & 1 \end{bmatrix} \quad B = \begin{bmatrix} T^2/2 & 0 \\ T & 0 \\ 0 & T^2/2 \\ 0 & T \end{bmatrix}$$

The target motion state information of millimeter-wave radar in a polar coordinate system is transformed to the front target vehicle in the millimeter-wave radar Cartesian coordinate system by coordinate transformation, the conversion formula is:

$$\begin{cases} x = R * \cos\theta \\ y = R * \sin\theta \end{cases} \tag{3}$$

where (x, y) is the position of the target-vehicle in the millimeter-wave radar Cartesian coordinate system θ is the azimuth.

The measurement equation of the target-vehicle in a millimeter-wave radar Cartesian coordinate system:

$$Z(k+1) = \begin{cases} R(k+1) = \sqrt{x^2(k+1) + y^2(k+1)} + v_1(k+1) \\ V(k+1) = \sqrt{\dot{x}^2(k+1) + \dot{y}^2(k+1)} + v_2(k+1) \end{cases} \tag{4}$$

where $v_i(k), i = 1, 2$ are the observation noises at k.

Equations (2) and (4) are the state equation and the measurement equation, respectively. As can be seen from the system model, the state equation is linear and the measurement equation is non-linear. Since the measurement equation is non-linear, the target tracking based on the millimeter-wave radar Cartesian coordinate system in a non-linear system.

2.3. *Parameters of the Millimeter-Wave Radar*

There are two main frequency bands for automobile millimeter-wave radar: 77 GHz and 24 GHz. The 24 GHz millimeter-wave radar is usually installed on the side of the vehicle, the detection range is small and mainly used for blind spot detection (BSD), lane change assistance (LCA). The 77 GHz millimeter-wave radar has a large detection range and is usually installed in front of the vehicle. It is mainly used for forward collision warning (FCW) and AEB. In order to verify the accuracy of the algorithm's estimation of the motion state of the target vehicle ahead, a 77 GHz millimeter-wave radar is used.

Table 1 shows the specific technical specifications of the on-board millimeter-wave radar provided by the supplier. Its main technical parameters such as detection range, measurement accuracy, and resolution are usually provided in its specification. A 77 GHz millimeter-wave radar is used herein provided by a supplier. The radar data rate is 40 ms. There are long-distance radar and medium-distance radar working states, which are subject to the driving speed of the ego-vehicle. When the travel speed of ego-vehicle is greater than 80 km/h and the millimeter-wave radar is in the long-distance radar working mode, it achieves the farthest detection distance of 120 m, and the angle detection range of around 30°. When the travel speed of the ego-vehicle is less than 80 km/h and the millimeter-wave radar is in the medium-distance radar working mode, it reaches the farthest detection distance of 100 m, and the angle detection range of around 50°.

Table 1. Main technical parameters of the millimeter-wave radar.

Parameter	Long-Distance Mode	Medium-Distance Mode
Ranging	1–120/(m)	1–100/(m)
Ranging resolution	0.6(m)	0.2(m)
Angle	±30/(°)	±50/(°)
Speed resolution	0.75/(km/h)	0.6/(km/h)
Speed	−50–55/(m/s)	−50–55/(m/s)

Remark: ① The horizontal angle range is negative when the target is on the left side of the radar and positive when on the right side. ② The relative speed is positive when the target is far from the radar, and negative when close to the radar.

3. Adaptive Square-Root Cubature Kalman Filter

This section is based on SRCKF. The Sage–Husa noise statistic estimator is extended from the linear constant noise statistic estimator to the non-linear time-varying noise statistic estimator.

3.1. Cubature Rule

CKF is derived from Bayesian filter theory in the Gaussian field. Under the framework of Bayesian filter theory in Gaussian domain, the non-linear filter problem can be summarized as a weighted integral in the form of "*non-linear* × *Gaussian density*". that is:

$$I(f) = \int_{\Omega} f(x)\omega(x)dx \tag{5}$$

where $f(x)$ is a non- linear function, $\Omega \subseteq \boldsymbol{R}^n$ is the region of integration, $\omega(x) = \exp\left(-x^T x\right)$ is a Gaussian density and satisfies the non-negativity condition in the entire region.

CKF uses the third-degree spherical-radial rule to calculate the non-linear filter problem [17]. Based on the third-degree spherical-radial rule, a set of 2n equal weight sampling points is used to approximate the integral value. that is:

$$I(f) \approx \sum_{i=1}^{2n} \omega_i f(\xi_i) \tag{6}$$

where $\xi_i = \sqrt{n}[\mathbf{1}]_i$ is the cubature point, $\omega_i = \frac{1}{2n}$ is the corresponding weight, $i = 1, \ldots, 2n$, $[\mathbf{1}]_i$ is the no.i element of the cubature points set:

$$[\mathbf{1}]_i = \left\{ \begin{bmatrix} 1 \\ 0 \\ \vdots \\ 0 \end{bmatrix}, \begin{bmatrix} 0 \\ 1 \\ \vdots \\ 0 \end{bmatrix} \cdots \begin{bmatrix} 0 \\ 0 \\ \vdots \\ 1 \end{bmatrix}, \begin{bmatrix} -1 \\ 0 \\ \vdots \\ 0 \end{bmatrix}, \begin{bmatrix} 0 \\ -1 \\ \vdots \\ 0 \end{bmatrix} \cdots \begin{bmatrix} 0 \\ 0 \\ \vdots \\ -1 \end{bmatrix} \right\} \tag{7}$$

3.2. Square-Root Cubarure Kalman Filter

In CKF, the error covariance matrix needs to satisfy two conditions: (1) symmetry (2) positive qualitative. It is important to ensure these two points in the algorithm iteration process to improve the robustness of the filter. The robustness of algorithm is defined as free of failure that the parameters in the algorithm violate the constraints, which makes the algorithm unable to continue running.

In order to ensure the symmetry and positive qualitative of the error covariance matrix, Arasaratnam proposed the SRCKF algorithm based on the CKF algorithm [17]. The SRCKF algorithm process is as follows:

(1) Initialization

$\hat{x}_k$ is the state variable, $\boldsymbol{P}_k$ is the error covariance matrix, $\boldsymbol{Q}_k$ is the process noise, $\boldsymbol{R}_k$ is the measurement noise.

(2) Evaluate the cubature points:

$$\boldsymbol{X}_{j,k} = S_k \xi_j + \hat{x}_k;\ j = 1, 2, \ldots, 2n \tag{8}$$

where ξ_j is the cubature points set, as shown below:

$$\xi_j = \begin{cases} \sqrt{n}[\mathbf{1}]_i & i = 1, 2, \ldots, n \\ -\sqrt{n}[\mathbf{1}]_i & i = n+1, n+2, \ldots, 2n \end{cases}$$

where $[\mathbf{1}]$ is the unit matrix.

(3) Spread cubature points and calculate the state prediction:

$$\boldsymbol{X}^*_{j,k+1} = f\left(\boldsymbol{X}_{j,k}, u_k\right) \tag{9}$$

$$\overline{x}_{k+1} = \frac{1}{2n} * \sum_{j=1}^{2n} \boldsymbol{X}^*_{j,k+1} \tag{10}$$

(4) Estimate the square-root factor of the predicted error covariance

$$\overline{S}_{k+1} = Tria\left(\left[\chi^*_{k+1} S_{Q,k+1}\right]\right) \tag{11}$$

where $Tria()$ represents a triangular operation. $\boldsymbol{B} = Tria(\boldsymbol{A})$ means: $\boldsymbol{B}$ is a matrix of $\boldsymbol{A}^T$ upper triangular matrix obtained $\boldsymbol{QR}$ decomposition.

χ^*_{k+1} and $S_{Q,k+1}$ are as follows:

$$\chi^*_{k+1} = 1/\left(\sqrt{2n}\right)\left[\boldsymbol{X}^*_{1,k+1} - \overline{x}_{k+1}, \boldsymbol{X}^*_{2,k+1} - \overline{x}_{k+1}, \cdots, \boldsymbol{X}^*_{2n,k+1} - \overline{x}_{k+1}\right]$$
$$S_{Q,k+1} = \mathrm{chol}\left(Q_{k+1}\right)$$

where chol() is Cholesky decomposition.

(5) Recalculate the cubature points

$$\boldsymbol{X}_{j,k+1} = \overline{S}_{k+1}\xi_j + \overline{x}_{k+1} \tag{12}$$

(6) Spread cubature points

$$\boldsymbol{Z}_{j,k+1} = h\left(\boldsymbol{X}_{j,k+1}, \boldsymbol{u}_{k+1}\right) \tag{13}$$

(7) Observation Prediction:

$$\overline{z}_{k+1} = \frac{1}{2n}\sum_{j=1}^{2n} \boldsymbol{Z}_{j,k+1} \tag{14}$$

(8) Estimate the square-root of the innovation covariance matrix:

$$S_{ZZ,k+1} = Tria\left(\left[\gamma_{k+1}\ S_{R,k+1}\right]\right) \tag{15}$$

where γ_{k+1} and S_R are as follows:

$$\gamma_{k+1} = 1/\left(\sqrt{2n}\right)\left[Z_{1,k+1} - \bar{z}_{k+1},\ Z_{2,k+1} - \bar{z}_{k+1}, \cdots, Z_{2n,k+1} - \bar{z}_{k+1}\right]$$
$$S_{R,k+1} = chol(R_{k+1})$$

(9) Estimate the cross-covariance matrix

$$P_{XZ,k+1} = \chi_{k+1}\gamma_{k+1}^T \tag{16}$$

where χ_{k+1} is as follows:

$$\chi_{k+1} = 1/\left(\sqrt{2n}\right)\left[X_{1,k+1} - \bar{x}_{k+1}, X_{2,k+1} - \bar{x}_{k+1}, \cdots, X_{2n,k+1} - \bar{x}_{k+1}\right]$$

(10) Estimate the Kalman gain

$$K_{k+1} = (P_{XZ,k+1}/S_{zz,k+1}^T)/S_{zz,k+1} \tag{17}$$

(11) Estimate the updated state

$$\hat{x}_{k+1} = \bar{x}_{k+1} + K_{k+1}(z_{k+1} - \bar{z}_{k+1}) \tag{18}$$

(12) Estimate the square-root factor of the corresponding error covariance

$$S_{k+1} = Tria([\chi_{k+1} - K_{k+1}\gamma_{k+1}K_{k+1}S_R]) \tag{19}$$

3.3. Improved Square-Root Cubature Kalman Filter

The on-board millimeter-wave radar is driving in the roads, and the statistical parameters of measurement noise are often unknown and time-varying. If the values of measurement noise and process noise are not consistent with the actual noise statistics, the filter will diverge. Therefore, constructing the ASRCKF for online estimation of unknown noise statistics is of great significance for improving the accuracy of filters. Sage-Husa is often used to estimate the statistical characteristics of noise online because of its simplicity and good real-time performance. However, the conventional Sage–Husa noise statistical estimator is suitable for the constant coefficient noise statistical characteristic estimation of linear systems. In the literature [26], the Sage–Husa noise statistic estimator in linear system is extended to the non-linear Sage–Husa noise statistic estimator. Therefore, based on the literature [26], we combine the Sage–Husa noise statistical estimator and cubature rules to derive a time-varying noise statistical estimator suitable for nonlinear systems.

The noise statistical estimator based on Sage–Husa in a non-linear system is:

Recursive formula for the process noise means:

$$\hat{q}_{k+1} = \frac{1}{k+1}[k * q_k + \hat{x}_{k+1} - \frac{1}{2n}\sum_{i=1}^{2n} f_k\left(X_{i,k|k}\right)] \tag{20}$$

Recursive formula for the measurement noise means:

$$\hat{r}_{k+1} = \frac{1}{k+1}[k * \hat{r}_k + z_{k+1} - \frac{1}{2n}\sum_{i=1}^{2n} h\left(X_{i,k+1|k}\right)] \tag{21}$$

The process noise variance is written in the form of a recursive estimation formula as:

$$\hat{Q}_{k+1} = \frac{1}{k+1}[k * \hat{Q}_k + K_{k+1} v_{k+1} v_{k+1}^T K_{k+1}^T + P_{k+1} \\ (\frac{1}{2n} \sum_{i=1}^{2n} X_{i,k+1|k}^{*} X_{i,k+1|k}^{*T} \quad \hat{x}_{k+1|k} \hat{x}_{k+1|k}^T)] \tag{22}$$

The measurement noise variance is written in the form of a recursive estimation formula as:

$$\hat{R}_{k+1} = \frac{1}{k+1}[k * \hat{R}_k + v_{k+1} v_{k+1}^T - (\frac{1}{2n} \sum_{i=1}^{2n} Z_{i,k+1|k} Z_{i,k+1|k}^T - \hat{z}_{k+1|k} \hat{z}_{k+1|k}^T)] \tag{23}$$

Including: $\hat{q}, \hat{r}, \hat{Q}, \hat{R}$ are the maximum a posterior of q, r, Q, R respectively.

As can be seen from (20), (21), (22), (23), the weight coefficient of each factor in $k+1$ moments, every factor of the weighted coefficient of $\hat{q}, \hat{r}, \hat{Q}, \hat{R}$ is $1/(k+1)$. For time-varying noise, the role of new data should be increased, while the role of old data should be gradually forgotten. Therefore, a different weighting factor should be multiplied for each factor in the system noise. That is: the weighting coefficient of new data should be greater than that of the old data. On the basis of the constant noise statistic estimator, the time-varying noise statistic estimator suitable for SRCKF is deduced by using the method of fading memory index weighting.

Weighted coefficient: $\lambda_i = \lambda_{i-1} * b, 0 < b < 1, \sum_{i=1}^{k+1} \lambda_i = 1$.

Therefore, the weighted index of fading memory is:

$$\begin{cases} \lambda_i = d_k b^{i-1} \\ d_k = \frac{1-b}{1-b^k} \end{cases} \tag{24}$$

where b is the forgetting factor, and its value range is usually between 0.95 and 0.99. A time-varying noise statistical estimator is obtained by replacing λ_{k+1-i} with the weight factor of $1/(k+1)$ in the constant noise statistical estimator:

Recursive formula for the process noise means:

$$\hat{q}_{k+1} = (1 - d_{k+1})\hat{q}_k + d_{k+1}[\hat{x}_{k+1|k+1} - \frac{1}{2n} \sum_{i=1}^{2n} f_k(X_{i,k|k})] \tag{25}$$

Recursive formula for the measurement noise means:

$$\hat{r}_{k+1} = (1 - d_{k+1})\hat{r}_k + d_{k+1}[z_{k+1} - \frac{1}{2n} \sum_{i=1}^{2n} h_{k+1}(X_{i,k+1|k})] \tag{26}$$

The process noise variance is written in the form of a recursive estimation formula as:

$$\hat{Q}_{k+1} = (1 - d_{k+1})\hat{Q}_k + d_{k+1}[K_{k+1} \varepsilon_{k+1} \varepsilon_{k+1}^T K_{k+1}^T + P_{k+1} \\ -(\frac{1}{2n} \sum_{i=1}^{2n} X_{i,k+1|k}^{*} X_{i,k+1|k}^{*T} - \hat{x}_{k+1|k} \hat{x}_{k+1|k}^T)] \tag{27}$$

where

$$\varepsilon_k = z_k - \hat{z}_{k|k-1}$$

The measurement noise variance is written in the form of a recursive estimation formula as:

$$\hat{R}_{k+1} = (1 - d_{k+1})\hat{R}_k + d_{k+1}[\varepsilon_{k+1} \varepsilon_{k+1}^T - (\frac{1}{2n} \sum_{i=1}^{2n} Z_{i,k+1|k} Z_{i,k+1|k}^T - \hat{z}_{k+1|k} \hat{z}_{k+1|k}^T)] \tag{28}$$

4. Classification of Target-Vehicle Motion State

Due to the influence of the ego-vehicle speed sensor and millimeter-wave radar measurement error, the direct use of the current moment of the ego vehicle and target vehicle motion relationship to identify the target-vehicle motion state, will lead to the vibration and even inaccurate motion state classification results. A method of classifying the motion state of the target vehicle based on time window is proposed by analyzing the transfer mechanism of the motion state of the target-vehicle. According to the absolute velocity of the target vehicle, the motion state of the target vehicle is divided into stationary target vehicle, moving target vehicle, oncoming target vehicle, start-stop target vehicle, and unclassified target vehicle.

(1) Unclassified target vehicle: the motion state of the target vehicle obtained at the initial moment of radar is the unclassified target vehicle;

(2) Stationary target vehicle: the target vehicle whose absolute speed stays near zero for a long time;

(3) Moving the target vehicle in the same direction: the movement direction of the target vehicle is the same as that of the ego vehicle;

(4) Oncoming target vehicle: the movement direction of the target vehicle is opposite to that of the ego vehicle;

(5) Start-stop target vehicle: the speed of the moving target vehicle (or the oncoming target vehicle) is reduced to near zero.

Since the velocity measured by the on-board millimeter-wave radar is the relative motion velocity of the target vehicle relative to the ego vehicle. Therefore, the absolute velocity of the target vehicle relative to the geodetic coordinate system can be deduced:

$$V_1 = V_r + V_v \tag{29}$$

where:

V_1: The absolute velocity of the target vehicle;
V_r: The relative velocity of the target vehicle;
V_v: The speed of the ego vehicle.

Figure 2 shows the flow chart of movement state transfer of the target vehicle.

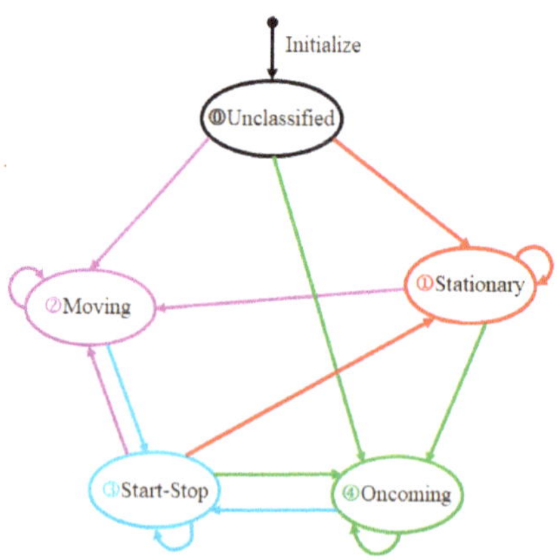

Figure 2. The flow chart of movement state transfer of the target vehicle.

According to the absolute speed of the target vehicle, the determination of the target motion state is mainly influenced by the following two factors: first, the measurement error of the ego vehicle speed sensor; Second, millimeter-wave radar speed error. Due to the above two measurement errors, the stationary target may also return a non-zero velocity value. Therefore, it is essential to determine the appropriate threshold value to judge the target motion state. The influence of velocity sensor measurement error and millimeter-wave radar measurement error is considered. In this paper, the reference ranges of the velocity threshold are $[0.5 \sim 1\ \text{m/s}]$ and $[-1 \sim -0.5\ \text{m/s}]$. Because the fluctuation range of the ego-vehicle velocity is 0.3 m/s. Therefore, the threshold value of the velocity of the stationary target-vehicle is set to $\pm 0.5 m/s$. The moving state transition rules of the target vehicle are as follows:

(1) The motion state of the target vehicle obtained during the initial operation of the radar is unclassified;

(2) The absolute speed of the target vehicle is between $[-0.5 \sim 0.5 m/s]$ for N consecutive cycles. The target motion state transition can be divided into the following four conditions:

① Switch from unclassified to stationary;

② Keep stationary;

③ Switch from oncoming to start-stop;

④ Switch from moving to starting-stopping;

(3) When the absolute speed of the target-vehicle is greater than 0.5m/s for N consecutive cycles, the target motion state transition has the following four conditions:

① Switches from unclassified to moving;

② Keep moving;

③ Switch from stationary to moving;

④ Switch from start-stop to moving;

(4) When the absolute speed of the target-vehicle is less than –0.5 m/s for N consecutive cycles, the target motion state transition has the following four conditions:

① Switch from unclassified to oncoming;

② Keep oncoming;

③ Switch from stationary to oncoming;

④ Switch from start-stop to oncoming;

(5) By recording the time T when the target vehicle is recognized as the start-stop motion state, the transition relationship between the start-stop motion state and the stationary state is identified.

① If T is greater than or equal to M, the target vehicle is switched from start-stop motion state to the stationary state.

② If T is less than M, the target vehicle keeps start-stop motion state.

As the length of time window N is longer, the delay of target-vehicle motion state recognition is more serious. The value of M has a significant influence on decision-making and control of the vehicle. In this paper, $N = 3$, $M = 2s$.

Because the target vehicle has inertia, there is no sudden change in the speed of the target-vehicle. In the process of state transfer between moving target vehicles in the same direction and moving target vehicles in the opposite direction, it is necessary to go through the start-stop motion state.

5. Experiment and Discussion

5.1. Construction of the Experimental Platform

As shown in the Figure 3, to check the accuracy and reliability of the proposed algorithm, the Sagitar vehicle is applied herein in the test. Figure 3 (a) is the Sagitar test vehicle platform; (b) is the experimental platform communication.

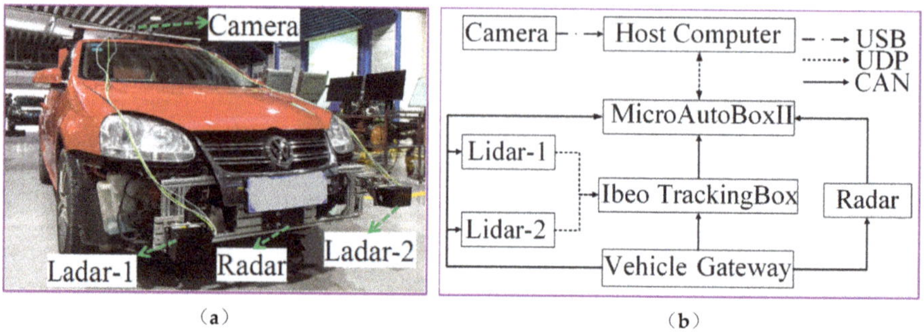

Figure 3. Experimental vehicle. (**a**) Test platform equipment, (**b**) experimental platform communication.

The ego-vehicle is equipped with a 77 GHz millimeter-wave radar as shown in (a) and (b) of Figure 3. The 77 GHz millimeter-wave radar is installed directly above the license plate on the front of the ego-vehicle. The 77 GHz millimeter-wave radar is provided with two-way controller area network (CAN). One CAN is connected to the vehicle gateway. The other is connected to MicroAutoBoxII through CAN. The AR023Z-1080p camera is installed in the bracket above the ego-vehicle to synchronously record the scene of the vehicle.

We use the light detection and ranging (lidar) produced by Ibeo to detect the motion state information of the target vehicle and take it as the ground truth, which is used to verify the performance of the target tracking algorithm of the millimeter-wave radar. In order to make the sensors smarter, Ibeo will provide point cloud processing algorithm software for the lidar. At present, the algorithm provided by Ibeo supports target recognition and tracking. The motion state information of the target vehicle includes longitudinal distance and longitudinal speed. As shown in Figure 4a, the ego-vehicle is equipped with two lidars, which are lidar-1 and lidar-2. The Ibeo TrackingBox produced by Ibeo is responsible for data fusion of the two lidars. Table 2 shows the specific technical specifications of the four scan levels lidar provided by Ibeo.

Table 2. Main technical parameters of the lidar.

Parameter	Value/(Units)
Ranging	200(m)
Ranging resolution	0.04(m)
Fov(H*V)	110*3.2/(°)
Vertical angle resolution	0.8/(°)
update rate	25/(Hz)

The ego-vehicle is equipped with MicroAutoBoxII and connected to the vehicle gateway port via the CAN to obtain the longitudinal velocity and steering wheel angle information of the vehicle measured by the electronic stability controller (ESC). MicroAutoBoxII is connected to the Ibeo TrackingBox through CAN and obtains the target vehicle movement status after data fusion.

The motion state estimation algorithm of the target vehicle is written in the environment of MatlabR2018a/Simulink in the host computer. The automatic code generation software provided by the dSPACE company is used to download the code to the MicroAutoBoxII1401/ 1505/1507 rapid prototyping controller through user datagram protocol (UDP) to run.

In order to exclude the randomness of the experiments, we conducted multiple sets of experiments. As shown in Figure 4, four test environments are selected, and three groups of experiments are carried out for each test environment.

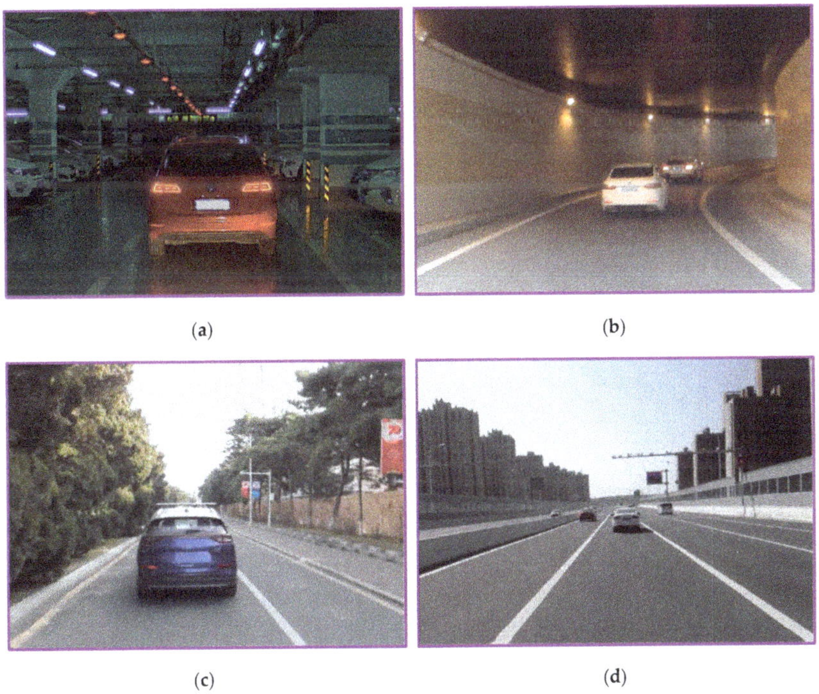

Figure 4. Experimental environment. (**a**) underground parking, (**b**) tunnel, (**c**) campus, (**d**) expressways.

5.2. Analysis of Experiment Results

Since the test results for the 12 groups of experiments are similar, one set of test data is selected for discussion and analysis. Figure 5 to Figure 6 show the experimental results of the effect of verifying the ISRCKF target tracking algorithm.

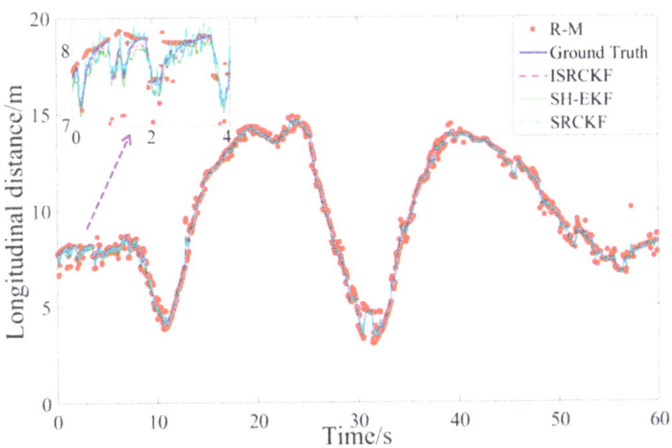

Figure 5. Target-vehicle longitudinal distance time history curve.

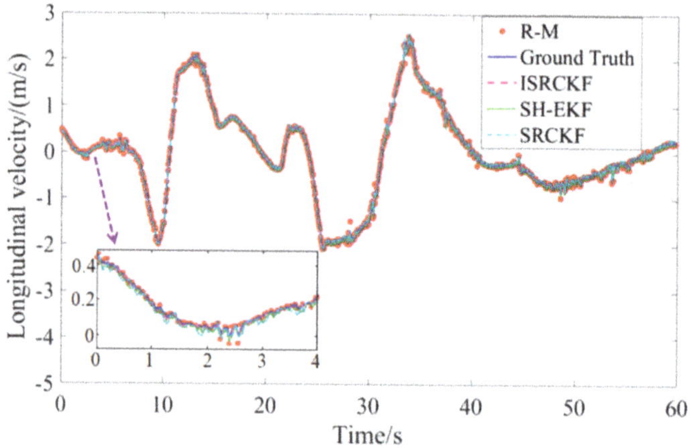

Figure 6. Target-vehicle longitudinal velocity time history curve.

The measurement data of the target vehicle are indicated by the red circle, without the ISRCKF algorithm optimization and containing noise (R-M). The solid blue line represents the ground truth of the motion state of the target-vehicle measured by lidar. The measurement data of the target vehicle are indicated by the magenta dashed line, with the ISRCKF algorithm optimization. The measurement data of the target vehicle are indicated by the green solid line, with the SH-EKF algorithm optimization. The measurement data of the target vehicle are indicated by the cyan dot-dash line, with the SRCKF algorithm optimization.

Figures 5 and 6 show the time history curves of the longitudinal distance and longitudinal velocity of the target-vehicle relative to the ego-vehicle. It can be concluded from the figures that the data fluctuation of the ISRCKF algorithm during tracking the target is small, and the target tracking performance of the ISRCKF algorithm is significantly better than the SH-EKF and SRCKF algorithms. Through references [27–29], it is found that the root mean square error (RMSE) can be used to quantitatively analyze the filtering accuracy. In order to quantitatively analyze the filtering accuracy of the ISRCKF, SH-EKF, and SRCKF algorithms, we counted the RMSE of the 12 groups of experiments. and the results are shown in Figures 7 and 8. The test environment of the statistical value for RMSE of groups 1, 2, and 3 is the underground parking. The test environment of the statistical value for RMSE of groups 4, 5, and 6 is the tunnel. The test environment of the statistical value for RMSE of groups 7, 8, and 9 is the campus. The test environment for the statistical value for RMSE of groups 10, 11, and 12 is the expressway.

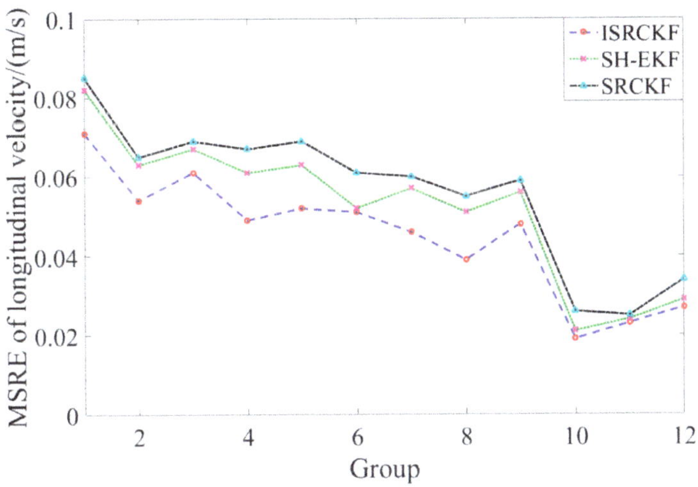

Figure 7. MSRE curves of longitudinal velocity for various filters.

It can be concluded from Figures 7 and 8 that in the driving environment where the system noise is unknown and time-variant, the RMSE of the longitudinal speed and longitudinal distance of target tracking using SRCKF and SH-EKF algorithm is larger than that of ISRCKF algorithm. Compared with SH-EKF and SRCKF algorithms, the maximum increase in filtering accuracy of longitudinal distance using the ISRCKF algorithm is 45.53% and 59.15%, respectively, and the maximum increase in filtering accuracy of longitudinal speed using the ISRCKF algorithm is 23.53% and 29.09%, respectively. Through the above analysis, it can be concluded that using ISRCKF algorithm to track can effectively suppress the divergence of target tracking, thereby reducing the tracking error and improving the tracking accuracy.

As shown in Figure 9, the time consumption of different algorithms is measured by the mean time of the algorithm running once in the MicroAutoBoxII.

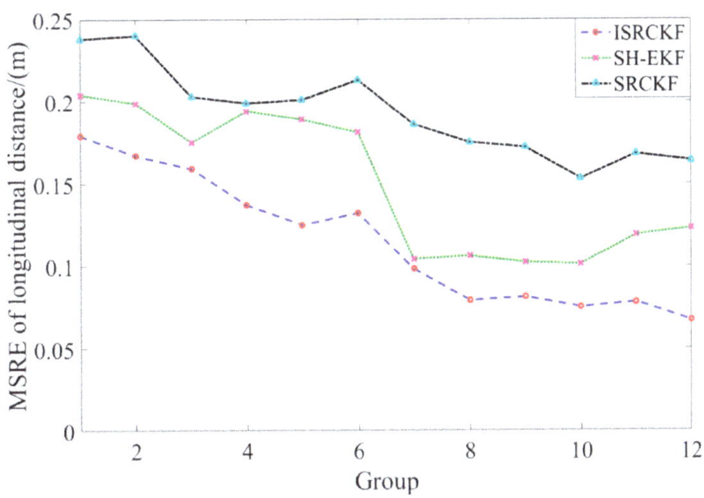

Figure 8. MSRE curves of longitudinal distance for various filters.

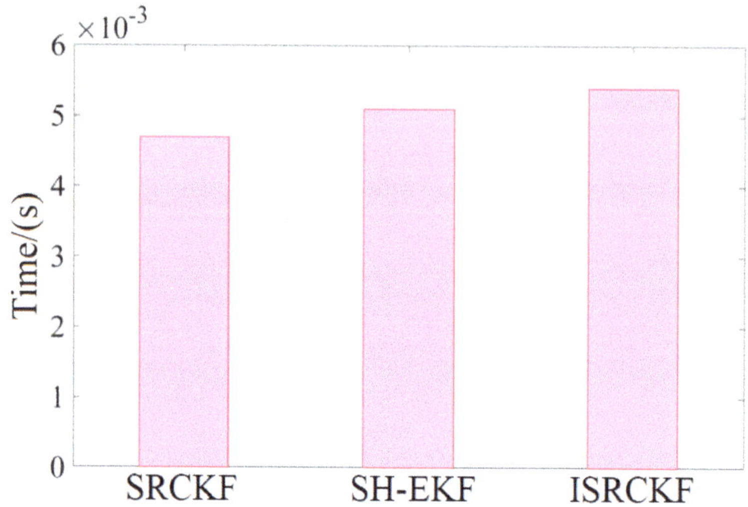

Figure 9. Time mean of the algorithm running once.

From Figure 9, it can be concluded that the target tracking algorithm proposed herein takes 0.0053 s on average, slightly up that of SH-EKF and SRCKF. SH-EKF and SRCKF take 0.0051 s and 0.0047 s on average, respectively.

The vehicle test results show that the ISRCKF algorithm has the highest accuracy when the time consumption is not increased much compared with SH-EKF and SRCKF.

In order to ensure the safety and rigorous of the experiment, we conducted 10 groups of experiments on campus. During the test, the driving process of the target vehicle was divided into the following stages: parking, accelerated reverse, braking to stop, starting acceleration, and decelerating to stop. Since the test results of the 10 groups are similar, we choose one of the data for analysis.

Figure 10 is the time history curve of the target-vehicle's absolute velocity. The meanings of ①, ②, ③, and ④ in Figure 10 represent the stationary, moving, start-stop, oncoming movement states in Figure 2, respectively. We have used the same colors to indicate the same motion state in Figures 2 and 10.

Figure 11 is the transition process of the target vehicle's motion state. The initial motion state of the target vehicle is the unclassified motion state. As the reversing speed of the target vehicle increases at 2.3 s, the motion state of the target-vehicle changes from the stationary motion state to the oncoming motion state. As the reversing speed of the target vehicle decreases, the motion state of the target vehicle changes from the oncoming motion state to the start-stop motion state (at 14.5 s). As the time T of the start-stop motion state increases, the status of the target vehicle changes from the start-stop motion state to the stationary state at the time of 16.5 s. As the forward speed of the target vehicle increases, the motion state of the target vehicle changes from the stationary state to the same motion state (at 19.4 s). As the forward speed of the target vehicle decreases, the motion state of the target vehicle changes from the same motion state to the start-stop motion state at 28.4 s.

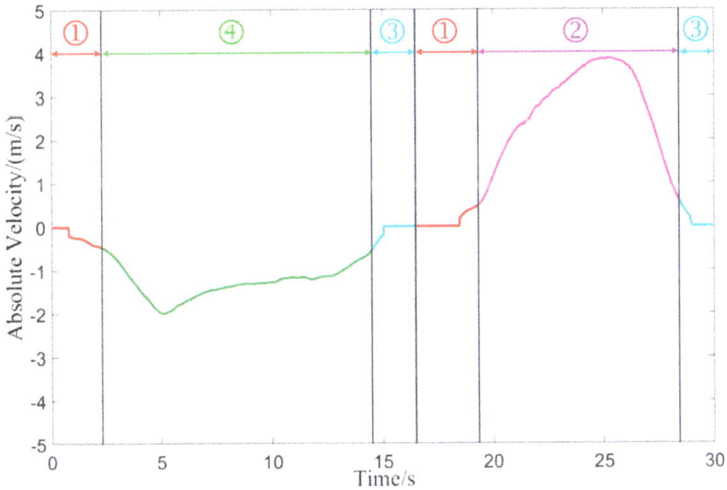

Figure 10. Target-vehicle absolute velocity time history curve.

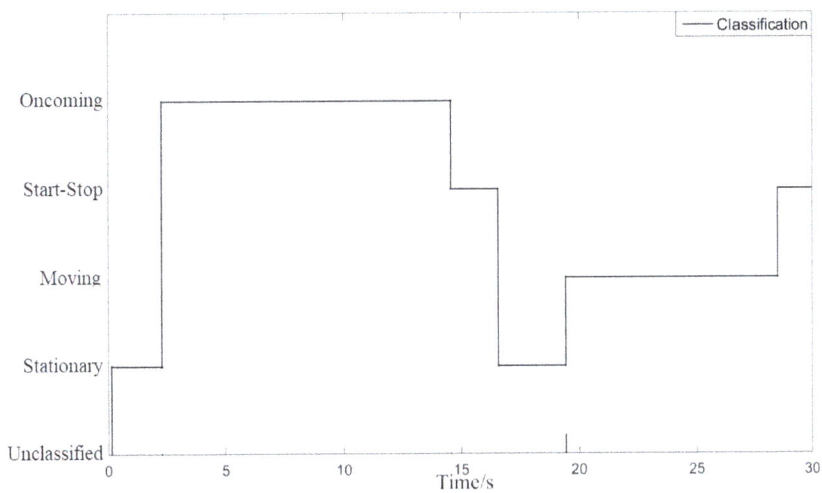

Figure 11. The movement state transfer of the target-vehicle.

In the two sports periods of 14.5 s ~ 16.5 s and 28.4 s ~ 30 s, the classification method proposed classifies target vehicles as start-stop targets. The classification method shows that it has a certain memory effect on the motion state of the target vehicle. The test results show that the classification and recognition results are consistent with the motion state of the target vehicle.

6. Conclusions

In the process of on-board millimeter-wave radar target tracking, the characteristics of unknown time-varying noise cannot be accurately counted, which will cause the filter accuracy to decline or even diverge. In response to this question, based on the SRCKF, the Sage–Husa noise statistic estimator and the fading memory exponential weighting method are combined to derive a time-varying noise statistic estimator for non-linear systems. ISRCKF can effectively overcome the problem of low accuracy and even divergence of SRCKF filtering-varying noise. When the motion relation between the ego vehicle and the target vehicle in the current is used to identify the motion state of the target vehicle, there exists the problem that the classification result of motion state is vibration or even inaccurate. A method of classifying the motion state of the target vehicle based on the time window is proposed by analyzing the transfer mechanism of the motion state of the target vehicle. According to the absolute velocity of the target vehicle, the motion state of the target vehicle is divided into stationary target vehicle, moving target vehicle, oncoming target vehicle, start-stop target vehicle, and unclassified target vehicle. Because the target vehicle has inertia, there is no sudden change in the speed of the target vehicle. In this classification method, the target vehicle needs to go through the start-stop motion state during the state transfer between the same motion and the reverse motion. Since the starting and stopping motion state of the target vehicle is different from the stationary motion state of the target vehicle, this method reflects that the target vehicle has a certain memory effect on its motion state. The results of the vehicle test show that: (1) the accuracy of the ISRCKF algorithm is significantly improved; (2) the classification and recognition results of the target vehicle motion state are consistent with the target vehicle motion state.

Author Contributions: Funding acquisition, J.W.; Methodology, S.S.; Writing—original draft, S.S.; Writing—review and editing, S.S. and J.W. All authors have read and agreed to the published version of the manuscript.

Funding: This research was funded by the Research on Construction and Simulation Technology of Hardware in Loop Testing Scenario for Self-driving Electric Vehicle of China, grant number 2018YFC0105103 and the National Natural Science Foundation of China, grant number U1564211.

Conflicts of Interest: The authors declare no conflict of interest.

References

1. Zhang, R.; Cao, S. Extending reliability of mmwave radar tracking and detection via fusion with camera. *IEEE Access* **2019**, *7*, 137065–137079. [CrossRef]
2. Chen, B.; Pei, X.; Chen, Z. Research on target detection based on distributed track fusion for intelligent vehicles. *Sensors* **2020**, *20*, 56. [CrossRef] [PubMed]
3. Wan, L.; Liu, Y.; Pi, Y. Comparing of target-tracking performances of EKF, UKF and PF. *Radar Sci. Technol.* **2007**, *5*, 13–16.
4. Liu, J.; Wang, Z.; Xu, M. A Kalman estimation based rao-blackwellized particle filtering for radar tracking. *IEEE Access* **2017**, *5*, 8162–8174. [CrossRef]
5. Morelande, M.R.; Challa, S. Manoeuvring target tracking in clutter using particle filters. *IEEE Trans. Aerosp. Electron. Syst.* **2005**, *41*, 252–270. [CrossRef]
6. Roy, A.; Mitra, D. Multi-target trackers using cubature Kalman filter for Doppler radar tracking in clutter. *IET Signal Process.* **2016**, *10*, 888–901. [CrossRef]
7. Sage, A.P.; Husa, G.W. In Adaptive filtering with unknown prior statistics. *IEEE Trans. Autom. Control* **1969**, 760–769. [CrossRef]
8. Deng, Z.L.; Wang, J.G. Adaptive extended Kalman filtering for nonlinear systems. *Acta Autom. Sin.* **1987**, *5*, 375–379.
9. Ignagni, M. An alternate derivation and extension of Friendland's two-stage Kalman estimator. *IEEE Trans. Autom. Control* **1981**, *26*, 746–750. [CrossRef]
10. Zhou, D.H.; Xi, Y.G.; Zhang, Z.J. Suboptimal fading extended Kalman filtering for nonlinear systems. *Control Decis.* **1990**, *5*, 1–6.
11. Chao, C.T.; Teng, C.C. A fuzzy neural network based extended Kalman filter. *Int. J. Syst. Sci.* **1996**, *27*, 333–339. [CrossRef]
12. Julier, S.J.; Uhlmann, J.K. New extension of the Kalman filter to nonlinear systems. In *Signal Processing, Sensor Fusion, and Target Recognition VI*; Society of Photo Optical: Bellingham, WA, USA, 1997; pp. 182–193.
13. Ge, B.; Zhang, H.; Jiang, L.; Li, Z.; Butt, M.M. Adaptive unscented Kalman filter for target tracking with unknown time-varying noise covariance. *Sensors* **2019**, *19*, 1371. [CrossRef]
14. Wu, Y.; Hu, D.; Wu, M.; Hu, X. A numerical-integration perspective on Gaussian filters. *IEEE Trans. Signal Process.* **2006**, *54*, 2910–2921. [CrossRef]

15. Sun, F.; Tang, L.J. Estimation precision comparison of cubature Kalman filter and unscented Kalman filter. *Control Decis.* **2013**, *28*, 303–308.
16. Doucet, A.; Godsill, S.; Andrieu, C. On sequential Monte Carlo sampling methods for Bayesian filtering. *Stat. Comput.* **2000**, *10*, 197–208. [CrossRef]
17. Arasaratnam, I.; Haykin, S. Cubature Kalman filters. *IEEE Trans. Autom. Control* **2009**, *54*, 1254–1269. [CrossRef]
18. Zhu, W.; Wang, W.; Yuan, G. An improved interacting multiple model filtering algorithm based on the cubature Kalman filter for maneuvering target tracking. *Sensors* **2016**, *16*, 805. [CrossRef]
19. Bao, W.; Lv, Y.; Zhu, R.; Pan, X.; Wang, J.; Zhou, H. Autonomous navigation algorithms based on improved CKF filters. In Proceedings of the Selected Papers of the Photoelectronic Technology Committee Conferences, Hefei, Suzhou, and Harbin, China, 14–19 June 2015; p. 9796.
20. Arasaratnam, I.; Haykin, S. Square-root quadrature Kalman filtering. *IEEE Trans. Signal Process.* **2008**, *56*, 2589–2593. [CrossRef]
21. Bhaumik, S. Cubature Kalman filter with risk sensitive cost function. In Proceedings of the 2011 IEEE International Conference on Signal and Image Processing Applications (ICSIPA), Kuala Lumpur, Malaysia, 16–18 November 2011; pp. 144–149.
22. Wei, W.; Qin, Y.Y.; Zhang, X.D.; Zhang, Y.C. Amelioration of the sage-husa algorithm. *J. Chin. Inert. Technol.* **2012**, *20*, 6.
23. Zhou, Y.; Zhang, Y.F.; Zhang, C.; Zhang, J.Z. A novel algorithm of linear adaptive square-root Kalman filtering based on sage-husa. *J. Northwestern Polytech. Univ.* **2013**, *31*, 89–93.
24. Deng, Z.L. *Self-tuning Filtering Theory with Applications-modern Time Series Analysis Method*; Press of Harbin Institute of Technology: Harbin, China, 2003; pp. 161–165.
25. Polychronopoulos, A.; Amditis, A.; Floudas, N.; Lind, H. Integrated object and road border tracking using 77 GHz automotive radars. *IEEE Proc. Radar Sonar Navig.* **2004**, *151*, 375–381. [CrossRef]
26. Li, Q.R. Research on Improved Cubature Kalman Filter and its Application in Navigation. Ph.D. Thesis, School of Automation, Harbin Engineering University, Harbin, China, 2015.
27. Wang, C.L.; Xiong, X.; Liu, H.J. Target tracking algorithm of automotive radar based on Iterated Square-root CKF. *J. Phys. Conf. Ser.* **2018**, *976*, 012010. [CrossRef]
28. Xiong, X.; Wang, C.L.; Liu, H.J. Target tracking algorithm for automotive radar based on Iterated Square-root CKF by noise compensation. *J. Nanjing Univ. Posts Telecommun. (Nat. Sci. Ed.)* **2018**, *38*, 113–118.
29. Liu, H.J.; Lai, S.F. Fast square root CKF for automotive millimeter-wave radar target tracking. *J. Nanjing Univ. Sci. Technol.* **2016**, *40*, 56–60.

sensors

MDPI

Article

Improved LiDAR Probabilistic Localization for Autonomous Vehicles Using GNSS

Miguel Ángel de Miguel *, Fernando García and José María Armingol

Intelligent Systems Laboratory, Universidad Carlos III de Madrid, Av. de la Universidad, 30, 28911 Leganés, Madrid, Spain; fegarcia@ing.uc3m.es (F.G.); armingol@ing.uc3m.es (J.M.A.)

* Correspondence: mimiguel@ing.uc3m.es

Received: 27 April 2020; Accepted: 30 May 2020; Published: 2 June 2020

Abstract: This paper proposes a method that improves autonomous vehicles localization using a modification of probabilistic laser localization like Monte Carlo Localization (MCL) algorithm, enhancing the weights of the particles by adding Kalman filtered Global Navigation Satellite System (GNSS) information. GNSS data are used to improve localization accuracy in places with fewer map features and to prevent the kidnapped robot problems. Besides, laser information improves accuracy in places where the map has more features and GNSS higher covariance, allowing the approach to be used in specifically difficult scenarios for GNSS such as urban canyons. The algorithm is tested using KITTI odometry dataset proving that it improves localization compared with classic GNSS + Inertial Navigation System (INS) fusion and Adaptive Monte Carlo Localization (AMCL), it is also tested in the autonomous vehicle platform of the Intelligent Systems Lab (LSI), of the University Carlos III de of Madrid, providing qualitative results.

Keywords: localization; LiDAR; GNSS; Global Positioning System (GPS); monte carlo; particle filter; autonomous driving

---

1. Introduction

Autonomous vehicle localization is the problem of estimating its position, determined by the x and y coordinates in a map and its orientation. This localization must be as accurate as possible since many vehicle's modules, such as control or path planning strongly depend on how good it is. Errors in localization can cause the vehicle to have an undesirable behavior or to even not being able to follow the desired path. Localization techniques can be divided mainly in mapping or sensor based [1]. Global Navigation Satellite System (GNSS) information is commonly used to solve localization problems using a sensor. It provides a good global localization with no drift but it has to deal with some errors from different sources. Those errors can be generated due to the satellites themselves (e.g., clock inaccuracies or dilution of precision), interference in the satellite signal (e.g., signal jamming, satellite occlusion) or signal propagation errors. That last error source includes inaccuracies produced by different weather conditions in the ionosphere and troposphere earth layers, and by multipath interference, caused by the reflection of the satellite waves when the vehicle is surrounded by large obstacles (high buildings or trees) [2,3].

High precision GNSS receivers, like Real Time Kinematic (RTK) or differential GPS, improves accuracy considerably but they have commonly a very high cost and they don't solve some accuracy errors. On the other hand, laser probabilistic localization methods, like Monte Carlo Localization (MCL) [4] can compare a precomputed map with the laser readings to acquire a precise position and orientation of the vehicle. The problem presented by this kind of algorithms is the opposite that with GNSS; in open environments with less map features, their accuracy decreases significantly. Here, particles of the filter would disperse generating different clusters of particles far away from the actual localization. This is known as the kidnapped robot problem and is a common localization error for probabilistic methods like MCL [5].

To solve those autonomous vehicle's localization problems, many different solutions have been proposed. One of the most known solution includes the fusion of different localization sources using Kalman filter based methods, i.e., Extended Kalman Filter or Unscented Kalman Filter [6]. This method commonly fuses GNSS, IMU and other different odometry sources like the one generated by the wheels encoders or LiDAR/camera odometry.

Fusion filters based on Kalman fuses all the localization sources and generates a more accurate estimation of the position, but they depend strongly on the accuracy of the GNSS measurements, as it is the only absolute localization source i.e. the only one that does not have drift.

Furthermore, GNSS can suffer different problems, being one of the most common the multipath problem, which is a common reaserch topic where several works try to reduce it. In [7] , a digital map of the environment is generated with OpenStreetMaps to prevent these errors and in [8], multipath is estimated and mitigated using a particle filter. However, the results don't give enough accuracy and they depend on the information in OpenStreetMaps which is not always available.

Another solution adopted by some researchers consists in probabilistic localization, which can be performed using different sensors such as LiDAR [9], cameras [10,11] or magnetometers [12]. One of the most common method for this kind of solution is Monte Carlo Localization (MCL) [4]. MCL works as a probabilistic particle filter that uses the match between the LiDAR sensor and the map as a feature for each particle that determines the probability of existing in the next iteration. As an improvement for MCL, Adaptive Monte Carlo Localization (AMCL) outperforms classic (MCL) [13] as it uses Kullback-Leibler Distance (KLD) sampling to make the filter converge faster. Particle filter localization methods can be applied to autonomous vehicles, like in [9], that uses a 2D LiDAR and a map with the features of the road.

Both previously commented algorithms, GNSS and MCL have a good performance, but also have some drawbacks. To get the best from every algorithm, several works explore the idea of combining the two sources of information to improve the localization. Most of the works available in the literature, improve MCL by resetting the calculated position when it differs too much from the GNSS position, in other words, the kidnapped robot problem is corrected once it is detected. Those improvements consists of resetting methods, which makes MCL more robust. An example of this is [14], where the kidnapped robot is detected and then solved using the Expansion resetting method.

Different localization sensors are used to solve this problem like in [15], where GNSS is used to detect and solve it, or in [16], where WiFi signal detection provides information about the localization error. These methods give good results as they are able to detect and solve the kidnapped robot error in most of the cases. However, the time necessary to detect and solve this problem adds further error to the system, as during this time, the vehicle is driving with a wrong localization. That localization error is unacceptable for autonomous driving vehicles as it would result in wrong control or path planning commands and thus, incorrect behavior. Consequently, it seems a reasonable assumption that the prevention of the kidnapped problem, as it is intended in this work, would lead to better results.

The work of [17] gives a solution based on replacing particles with a low degree of laser fit on the map with new particles according to the probability density given by the sensor, but it can have localization problems in empty maps as no GNSS is used. In [18], GNSS data are used to generate new particles on the filter and to eliminate distant ones. However, this method doesn't handle orientation, it only considers the position. Furthermore introducing particles based on GNSS data in all the cycles of the filter, would increase the noise of the localization and unfortunately no quantitative results of its performance are provided. As shown in [19], fusion of both sensors based on particle weight gives better results than adding new particles based on GNSS data. However, that work does not handle the kidnapped robot problem, as no strategies to recover are performed when the GNSS and particle filter positions differ considerably.

In contrast to all the approaches mentioned above, our method continuously uses GNSS data in the filter in both ways: modifying the weight of the particles, and injecting new ones if needed, avoiding kidnapped robot problem and making unnecessary to detect and reset states where the robot is badly localized. Furthermore, the method is designed to not replace directly the particles with new ones based on GNSS data, but to calculate its probability considering multiple parameters of both localization sources, making the particle filter more stable, an reducing noise generated by adding directly new particles on every cycle of the filter. Besides, we also consider orientation error when calculating the new probability, which is not done in most of the reviewed works.

The rest of the paper is structured as follows: Sections 2 and 3 describe the software architecture and the method proposed respectively, Section 4 evaluates the localization with real data and in Section 5 the conclusions are exposed.

The proposed method is coded and tested with the KITTI Dataset [20]. The source code of this method is publicly available at https://github.com/midemig/gps_amcl so anyone can replicate the experiments described in this work.

2. Sensors and Software Architecture

This section describes the software architecture of the tests, including the configuration of the different modules used. This architecture includes the AMCL implementation for Robot Operating System (ROS) [21] from [4], the GNSS/INS based localization module [22] and the map generation module, in charge of the generation of the map, used by AMCL module.

2.1. Software Modules

In this section, the specific version of the software modules used in the comparison are described.

2.1.1. AMCL

This algorithm outperforms original MCL [13] and is chosen to be the probabilistic LiDAR localization used. Specifically, it is used the AMCL ROS node implementation as it is a stable and maintained version of the algorithm. Most of the parameters' configuration is set to the default values values, however, the most relevant ones are shown in Table 1, were the odometry model defines the equations that better describe the movement of the vehicle, the laser model describes the method used to calculate the probability of being at a certain position given a laser measurement and max particles and beams represents the maximum number of possible positions and laser beams around the vehicle respectively (more particles and beams can increase localization accuracy, but also computation time).

Table 1. Adaptive Monte Carlo Localization (AMCL)'s most relevant parameters.

Parameter	Value
Odometry model	differential
Laser model	likelihood filed
Max particles	2000
Max beams	360

When the AMCL module is used in a map with a small number of reference obstacles surrounding the vehicle, the kidnapped robot localization error can easily appear, as shown in Figure 1, where multiple particles clusters appear due to the lack of features in the map, and the localization obtained by AMCL is inaccurate.

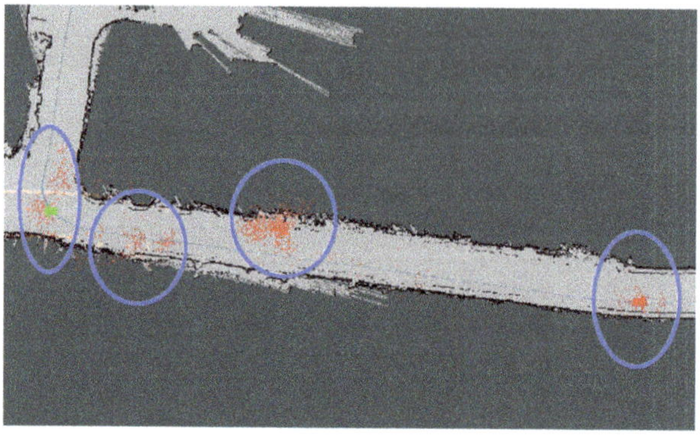

Figure 1. AMCL kidnapped robot localization error. Different clusters rounded in blue, red dots are the filter's particles and red and green arrows are ground truth.

2.1.2. Fusing Method

Kalman filter is commonly used for fusing localization data from different sources giving, as a result, a more accurate one. The software version used in this work is the robot localization ROS implementation [22]. Unscent Kalman filter is used as it is known to deal better with non-linearities in the filtering process [23]. In order to improve GNSS localization, it is fused with IMU data.

Robot localization package can fuse n different localization sources enhancing single-sensor based localization, and providing useful information such as the covariance of the position, which is of great relevance in this work.

2.1.3. Map Generation

To use AMCL localization, a pre-generated map of the environment is generated using different Simultaneous Localization and Mapping algorithms [24]. As later on, GNSS information will be added to the localization step, this map must be generated with GNSS localization data. That means that a high precision GNSS receiver is needed, but only for the map generation. The Laser information is transformed accordingly to the GNSS position at each time step and then accumulated into the map using gmapping Simultaneous Localization and Mapping (SLAM) method [25]. The generated map (Figure 2) has all the features needed later by the AMCL algorithm to match a new laser scan with it.

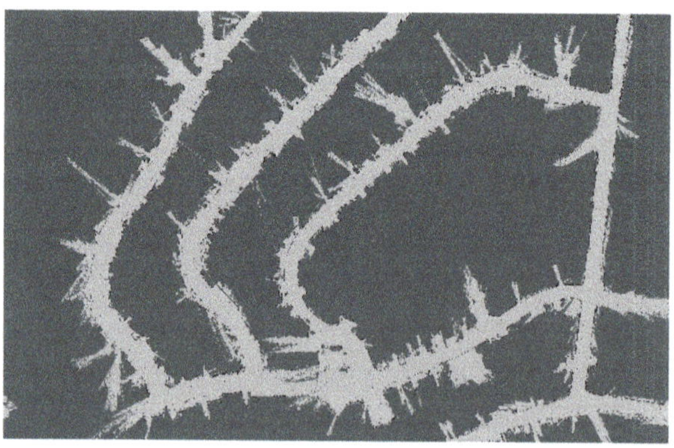

Figure 2. Map generated of one of the KITTI sequences.

2.2. Vehicle Sensor Setup

Although the results presented in the test section were performed using KITTI dataset, this algorithm can be implemented in any vehicle that fulfills some specified sensor requirements that are explained in this subsection:

2.2.1. LiDAR

LiDAR Information is necessary to provide map information, although other 3D output may be used, the recommendation for this application, where accuracy is a key point, it is to use 3D Laser scanner technology. KITTI data provides a 64 layers 3D laser scanner. As 360 degrees 2D LiDAR scanner is needed to get environment information and match it with the pre-generated map using AMCL, the 3D laser scanner information is converted to 2D. This is done by choosing a minimum and a maximum height from the LiDAR, the multiple layers can be converted into a single 2D one. These parameters are selected to incorporate the maximum amount of features from the environment, but removing the ground plane, as it would only increase the noise in the particle filter. Other LiDAR configuration can be used, the test vehicle of LSI where this algorithm was implemented and tested is based on a 32 layers LiDAR, which provide similar results.

2.2.2. GNSS Receiver

For this application, a Low-cost GNSS receiver can be used. Instead of high-cost RTK or differential GPS receivers, our method can work with lower accuracy in GNSS localization, making possible the generalization of these applications [26]. As the only localization information provided by KITTI dataset is the ground truth, it is considered to not have any zero error. Here, a low-accuracy GNSS receiver can be simulated by adding Gaussian noise to every ground truth measure, providing output similar to low-cost sensors [27]. The qualitative results provided on this test were performed using a low-accuracy GPS receiver, based on PixHawk technology.

2.2.3. Inertial Measurement Unit (IMU) Sensor

IMU is the most common sensor fused with GNSS data, as it can provide position and orientation data. KITTI database provides IMU data with extrinsic calibration information, needed for the fusion. The PixHawk sensor unit used in the qualitative tests were performed using the PixHawk unit.

3. Method Description

Based on the original AMCL algorithm, several modifications are made to integrate GNSS data into the loop.

In this section, all those modifications are detailed and justified. Furthermore, the resulting algorithm is presented in Algorithm 1.

3.1. LiDAR Likelihood

The proposed solution computes a weight for each particle in AMCL by comparing the laser data transformed to each particle position with the map. In addition to that weight, our method computes a score of how accurate this particle is matched with the map. This score s_i is computed using a Gaussian model [28,29] for the LiDAR data, following the expression:

$$s_i = \frac{1}{N} \cdot \sum_{n=1}^{N} \frac{1}{\sigma_{hit} \cdot \sqrt{2 \cdot \pi}} \cdot e^{\frac{-z_n^2}{2 * \sigma_{hit}^2}} \tag{1}$$

where N is the number of lasers of a laser scan, z is the distance from the laser hit point to the closest map occupied cell and σ_{hit} is the standard deviation of the laser. With this expression, we can evaluate how good the LiDAR measurements match with the map in every particle position, and is later used to modify the probability of that particle to exist.

3.2. GNSS likelihood Estimation

In addition to weight calculated based on the matching of the sensor with the map, we add a second weight based on the GNSS Kalman filtered position estimation. As for the LiDAR likelihood, a Gaussian model is used to estimate the GNSS based weight of each particle d_i, but in this case, as the position received and each particle of the filter are three dimensional (x, y and ψ), the n dimensional Gaussian model is used,

$$d_i = \frac{1}{(2 \cdot \pi)^{\frac{3}{2}} \cdot |\Sigma|^{\frac{1}{2}}} \exp\left(-\frac{1}{2}(x_i - \mu_k)^T \Sigma_k^{-1} (x_i - \mu_k)\right) \tag{2}$$

where the received position is $\mu_k = (x_k, y_k, \psi_k)$ with covariance matrix Σ_k and the position and orientation of each particle is defined with $x_i = (x_i, y_i, \psi_i)$. Using this model, orientation error is also considered when computing the GNSS based weight. Furthermore multiplying all the errors by the inverse of the covariance matrix in the exponential part, makes position and orientation errors scale-invariant. Here it is important to remark that the use of the covariance of the Kalman output allows reducing the importance of this weight when the quality of the information is low.

3.3. New Particle Weight Computed

To modify the weight of every particle so that Kalman filtered GNSS data are incorporated into the filter, the new weight of every particle is calculated by making use of the weights computed in (1), (2), and the following equation:

$$w_{i-new} = w_i \cdot s_i \cdot k_l + d_i \tag{3}$$

where w_i, the weight calculated by original AMCL, is modified in order to incorporate GNSS probabilistic data. The weight k_l is a constant and it is added to balance the importance of each source of information. It is empirically set to 200, where it gives the best results in all the environments tested. After this weight modification, they are normalized to make the sum of all the weights equal 1.

3.4. New Particles Generation

As the particle filter eliminates particles accordingly to its probability, the original AMCL method adds new particles randomly distributed in the map, based on two parameters that define how often it is needed to add those particles. A different function was defined to generate new particles X_{new} near GNSS Kalman position and to determine the probability of generating those new particles. The following expression defines how the new particles are generated to follow a normal distribution centered in μ_k with covariance Σ_k.

$$\begin{pmatrix} x_{new} & y_{new} & \psi_{new} \end{pmatrix}^T = \lambda_k^{\frac{1}{2}} \phi_k \cdot R + \mu_k \tag{4}$$

where R is a random vector with distribution $N[0,1]$ and λ, ϕ are the diagonal matrix of eigenvalues and the eigenvectors matrix of the covariance Σ_k respectively. Besides, the probability p of generating a new GNSS based particle at each cycle is determine as:

$$p(X_{new}|x, \mu) = \begin{cases} d_{mean}, & \text{if } d_{mean} > 0 \\ 0, & \text{otherwise} \end{cases} \tag{5}$$

with

$$d_{mean} = p_{max} - \frac{1}{N} \cdot \sum_{n=1}^{N} d_i \tag{6}$$

where p_{max} is the maximum probability allowed to generate new particles at each filter cycle and is experimentally set to 0.01.

The addition of particles does not increase the noise in the final localization given by the particle filter, as they are only added when the filter particles begin to considerably differ from the GNSS localization, and the newly generated number never goes beyond the limit of 1%.This situation avoids the creation of GNSS based particles when the filter provides accurate detection, these are generally particularly difficult situations for the GNSS such as urban canyons.

Algorithm 1: New weight and resample of filter particles

```
Input: x, μ_k, w, z
Output: x_new
for i = 0 to N do
    s_i = lidarLikelihood(z_i);
    d_i = gnssLikelihood(x_i, μ_ki);
    w_i-new = w_i · s_i · k_l + d_i;
end
w_new = normalizeWheigt(w_new);
for i = 0 to N do
    p(X_new|x, μ) = probOfNewParticle(d_i);
    if rand() < p(X_new|x, μ) then
        x_i-new = generateNewParticlePosition(μ_k);
    else
        x_i-new = sampleParticle(x, w_new) ;          // Original function from AMCL
    end
end
```

4. Experimental Results and Discussion

Two different group of experiments were performed. On the one hand, the proposed method is tested using the KITTI odometry dataset for quantitative results. On the other hand, the LSI platform for autonomous driving is used for qualitative results [30].

4.1. Dataset

The KITTI dataset is commonly used to test different odometry algorithms, and compare them as it includes calibrated data from different sensors (cameras, LiDAR and IMU) and precise ground truth for localization. The tests performed using this dataset compare the ground truth localization accuracy with the three following methods:

- Kalman filtered GNSS and IMU. As covariance of GNSS localization is set fixed, the mean error value is displayed for visualization purpose.
- AMCL original implementation.
- Proposed method, having the same odometry source as original AMCL and the same parameters configuration.

For every position and orientation, euclidean distance and orientation error are compared with the closest one in time of the ground truth (interpolating if necessary). KITTI dataset gives multiple sequences in different scenarios. For this evaluation residential sequences were selected, since they combine narrow and empty streets, including both types of scenarios. For every sequence, first, a map is generated using the localization ground truth and the LiDAR data described in Section 2.1.3. The experiments are designed to compare the proposed method with the other two considering the worst-case scenario for our method where, at least, it needs to give similar results to the comparing methods. Then a normal situation is tested to quantify the improvements of the proposed method.

Considering this, the following three scenarios are tested.

4.2. Empty Map

This scenario refers to situations with a low number of obstacles, i.e. lack of references for the LiDAR points to match. This is considered to be the worst scenario for AMCL. However, GNSS has better accuracy when it is not surrounded by obstacles that interfere with observations. Localization error is compared for Kalman filtered GNSS and the proposed method. As it is shown in Figure 3, localization error provided by our approach is very similar to the GNSS based, giving almost the same values for position and orientation. This is according to the expected as the proposed method can not improve localization accuracy with no laser information, but it proves that in empty environments it would perform as good as GNSS based localization.

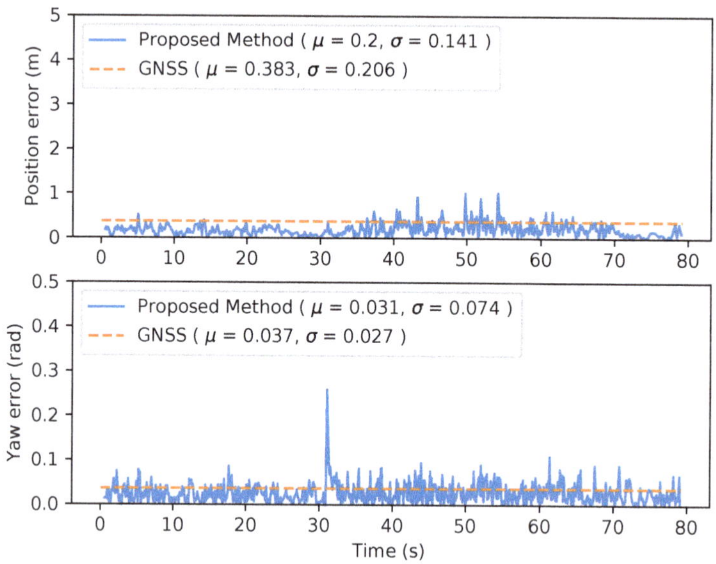

Figure 3. Comparison between proposed method with no LiDAR information but Global Navigation Satellite System (GNSS) localization. (GNSS is plotted as constant of mean value for visualitation purpose.)

4.3. GNSS Challenging Scenarios

This scenario describes the opposite problem. In an environment full of objects or an urban environment with urban canyons, GNSS localization would fail or would give noisy measurements with high covariance. In these cases, AMCL algorithm gives better results thanks to a map with numerous features and objects to match with the laser points. Localization is compared in this scenario for AMCL and the proposed method. As shown in Table 2, localization errors are similar. When no GNSS localization is received, or it has a high covariance, the proposed method, as it is designed, has a very similar behavior as original AMCL.

Table 2. Comparison between proposed method and AMCL with no GNSS information.

Method	Position Mean (m)	Position Std (m)	Yaw Mean	Yaw Std
Proposed	0.513	0.856	0.033	0.025
AMCL	0.566	0.742	0.023	0.023

4.4. Mixed Environments

The last scenario tested is a more common one where GNSS data are received with an acceptable covariance, and the map has features unequally distributed on it, generating places with more objects and empty places. The three algorithms are tested and compared in this scenario using the KITTI dataset residential sequences, that include more than 45 min of recorded data in a residential environment, but only the most relevant ones are discussed in this section.

4.4.1. AMCL

When evaluating AMCL performance two different localization issues can be identified. The first one, as shown in Figure 4, presents an increase in the error due to multiple particles' clusters that make the filter jump from one to another, increasing the error but finally converging again to the true localization. The second problem can be seen in Figure 5, where the kidnapped robot problem appears around the second 180, where the localization jumps to a similar place in the map.

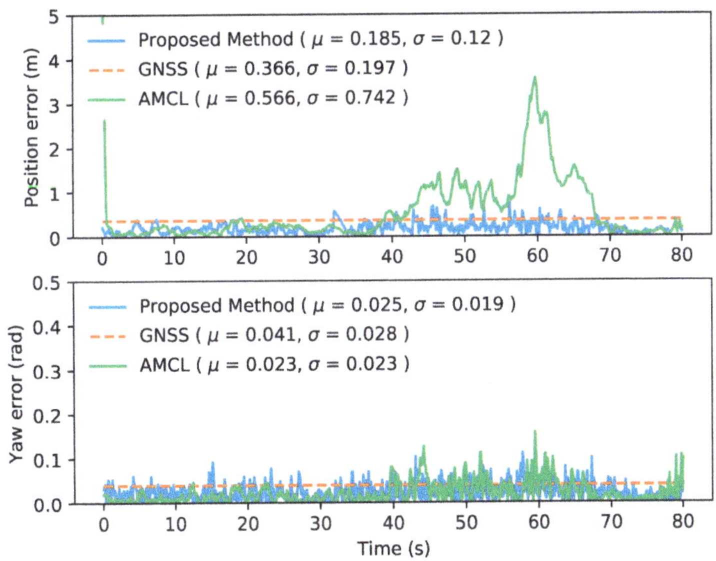

Figure 4. Localization error of the three methods described in KITTI sequence 36. (GNSS is plotted as constant of mean value for visualitation purpose).

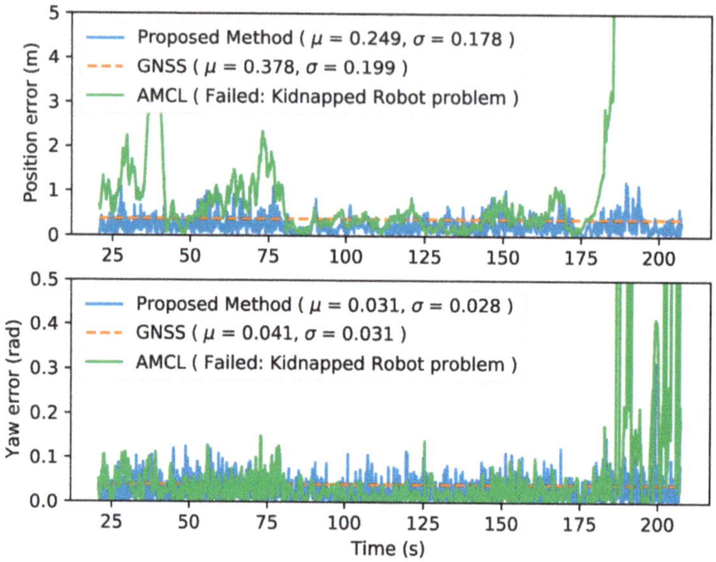

Figure 5. Localization error of the three methods described in KITTI sequence 34. (GNSS is plotted as constant of mean value for visualitation purpose).

4.4.2. GNSS

It is set to have a constant covariance error, so this localization maintains the same localization error along all the sequence. In real environments, Kalman results could be even worse as GNSS position error could be even higher in narrow streets.

4.4.3. Proposed Method

The proposed method, gives a better performance along all the path, improving localization mean error compared to AMCL and GNSS. It outperforms the other algorithms in terms of accuracy and stability as it has a localization error better or equal to Kalman localization method, and does not present kidnapped robot localization problem.

Figures 4 and 5 present the localization error along the sequence and the mean error values for sequences 36 and 34 respectively, showing how the proposed method outperforms original AMCL avoiding the kidnapped robot problem, while maintaining the localization error below the GNSS one.

Moreover, Table 3 compares the error of the proposed method as the accuracy of the GNSS localization decreases. As it is shown, for high GNSS accuracy, the error is close, but as the GNSS error increases, our method is able to reduce it using the laser information, giving always better results than the original AMCL algorithm.

Table 3. Evaluation of the proposed method errors for different GNSS mean errors in sequence 34.

GNSS Mean (m)	Position Mean (m)	Position Std (m)	Yaw Mean	Yaw Std
0.127	0.141	0.137	0.015	0.016
0.374	0.186	0.11	0.028	0.024
1.256	0.367	1.767	0.029	0.129
6.256	0.496	1.963	0.023	0.116
12.600	0.554	2.115	0.026	0.151
37.581	0.593	2.495	0.032	0.196
AMCL		Robot kidnapped error		

4.5. Real Platform Qualitative Results

In addition to the evaluation using KITTI database, the proposed method is also tested in our research platform as shown in Figure 6. As this platform does not include ground truth localization information, only a qualitative analysis of the performance of the proposed method is possible giving, as a result, a stable localization accurate enough to control the vehicle, solving the kidnapped robot that this platform showed before the implementation of this method when using AMCL, and improving the localization accuracy of the GNSS receiver.

Figure 6. Our autonomous vehicle platform testing the proposed method (left), and a visualization of the LiDAR pointcloud and the map (right).

5. Conclusions

In this work, a novel localization method for autonomous vehicles is proposed. It consists of a modified version of AMCL where GNSS data are integrated. As it is shown in Section 4, this method combines the strengths of the two combined localization methods, without compromising accuracy at any scenario, urban and non-urban. The results prove a good performance and a smooth transition from using GNSS data when the map is featureless to using LiDAR and map data when GNSS localization is not accurate, improving localization when both sources are available. The improvements depend on the environment but the proposed method always gives better results, or, in the worst case, the same results. It also gives a low-cost solution for different enviroments using low-cost sensors such as one layer LiDAR or low accuracy GNSS receiver compared to systems that use High-cost RTK or differential GNSS receivers and multiple layer LiDARs, making possible the generalization of these applications. The method is tested and evaluated with a well-known database (KITTI) which is usually used to evaluate autonomous vehicles perception and localization algorithms and with a real platform, improving its performance thanks to a better localization. Besides, the source code of the method is published so it can be tested and improved by anyone.

Author Contributions: M.Á.d.M. wrote the manuscript and carried out the investigation and methodology, F.G. reviewed the manuscript and the methodology and J.M.A. reviewed the manuscript. All authors have read and agreed to the published version of the manuscript.

Funding: Research supported by the Spanish Government through the CICYT projects (TRA2016-78886-C3-1-R and RTI2018-096036-B-C21), Universidad Carlos III of Madrid through (PEAVAUTO-CM-UC3M) and the Comunidad de Madrid through SEGVAUTO-4.0-CM (P2018/EMT-4362).

Conflicts of Interest: The authors declare no conflicts of interest.

References

1. Kuutti, S.; Fallah, S.; Katsaros, K.; Dianati, M.; Mccullough, F.; Mouzakitis, A. A survey of the state-of-the-art localization techniques and their potentials for autonomous vehicle applications. *IEEE Internet Things J.* **2018**, *5*, 829–846. [CrossRef]
2. Karaim, M.; Elsheikh, M.; Noureldin, A. GNSS Error Sources. In *Multifunctional Operation and Application of GPS;* IntechOpen: London, UK, 2018.
3. Huang, D.; Xiong, Y.; Yuan, L. *Global Positioning System (GPS)-Theory and Practice;* Xi'an Jiaotong University Press: Chengdu, Chain, 2006; pp. 71–73.
4. Fox, D.; Burgard, W.; Dellaert, F.; Thrun, S. Monte carlo localization: Efficient position estimation for mobile robots. *AAAI/IAAI* **1999**, *1999*, 1–7.
5. Gutmann, J.S.; Fox, D. An experimental comparison of localization methods continued. In Proceedings of the IEEE/RSJ International Conference on Intelligent Robots and Systems, Lausanne, Switzerland, 30 September–4 October 2002; Volume 1, pp. 454–459.
6. Martí, E.D.; Martín, D.; García, J.; De la Escalera, A.; Molina, J.M.; Armingol, J.M. Context-aided sensor fusion for enhanced urban navigation. *Sensors* **2012**, *12*, 16802–16837. [CrossRef] [PubMed]
7. Obst, M.; Bauer, S.; Reisdorf, P.; Wanielik, G. Multipath detection with 3D digital maps for robust multi-constellation GNSS/INS vehicle localization in urban areas. In Proceedings of the 2012 IEEE Intelligent Vehicles Symposium, Alcala de Henares, Spain, 3–7 June 2012; pp. 184–190.
8. Qin, H.; Xue, X.; Yang, Q. GNSS multipath estimation and mitigation based on particle filter. *IET Radar Sonar Nav.* **2019**, *13*, 1588–1596. [CrossRef]
9. Ahn, K.; Kang, Y. A Particle Filter Localization Method Using 2D Laser Sensor Measurements and Road Features for Autonomous Vehicle. *Int. J. Rail Transp.* **2019**, *2019*, 1–11. [CrossRef]
10. Sefati, M.; Daum, M.; Sondermann, B.; Kreisköther, K.D.; Kampker, A. Improving vehicle localization using semantic and pole-like landmarks. In Proceedings of the 2017 IEEE Intelligent Vehicles Symposium (IV), Los Angeles, CA, USA, 11–14 June 2017; pp. 13–19.
11. Zhou, X.; Su, Z.; Huang, D.; Zhang, H.; Cheng, T.; Wu, J. Robust global localization by using global visual features and range finders data. In Proceedings of the 2018 IEEE International Conference on Robotics and Biomimetics (ROBIO), Kuala Lumpur, Malaysia, 12–15 December 2018, pp. 218–223.
12. Fentaw, H.W.; Kim, T.H. Indoor localization using magnetic field anomalies and inertial measurement units based on Monte Carlo localization. In Proceedings of the 2017 Ninth International Conference on Ubiquitous and Future Networks (ICUFN), Milan, Italy, 4–7 July 2017; pp. 33–37.
13. Pfaff, P.; Burgard, W.; Fox, D. Robust monte-carlo localization using adaptive likelihood models. In Proceedings of the 1st European Robotics Symposium (EUROS-06), Palermo, Italy, 16–18 March 2006; pp. 181–194.
14. Ueda, R.; Arai, T.; Sakamoto, K.; Kikuchi, T.; Kamiya, S. Expansion resetting for recovery from fatal error in monte carlo localization-comparison with sensor resetting methods. In Proceedings of the 2004 IEEE/RSJ International Conference on Intelligent Robots and Systems (IROS) (IEEE Cat. No. 04CH37566), Sendai, Japan, 28 September–2 October 2004; pp. 2481–2486.
15. Goto, H.; Ueda, R.; Hayashibara, Y. Resetting Method using GNSS in LIDAR-based Probabilistic Self-localization. In Proceedings of the 2018 IEEE International Conference on Robotics and Biomimetics (ROBIO), Kuala Lumpur, Malaysia, 12–15 December 2018; pp. 1113–1118.
16. Seow, Y.; Miyagusuku, R.; Yamashita, A.; Asama, H. Detecting and solving the kidnapped robot problem using laser range finder and wifi signal. In Proceedings of the 2017 IEEE international conference on real-time computing and robotics (RCAR), Okinawa, Japan, 14–18 July 2017; pp. 303–308.
17. Lenser, S.; Veloso, M. Sensor resetting localization for poorly modelled mobile robots. In Proceedings of the 2000 ICRA. Millennium Conference. IEEE International Conference on Robotics and Automation. Symposia Proceedings (Cat. No. 00CH37065), San Francisco, CA, USA, 24–28 April 2000; pp. 1225–1232.
18. Hentschel, M.; Wulf, O.; Wagner, B. A GPS and laser-based localization for urban and non-urban outdoor environments. In Proceedings of the 2008 IEEE/RSJ International Conference on Intelligent Robots and Systems, Nice, France, 22–26 September 2008; pp. 149–154.
19. Perea, D.; Morell, A.; Toledo, J.; Acosta, L. GNSS integration in the localization system of an autonomous vehicle based on Particle Weighting. *IEEE Sens. J.* **2019**, *20*, 3314–3323. [CrossRef]

20. Geiger, A.; Lenz, P.; Stiller, C.; Urtasun, R. Vision meets Robotics: The KITTI Dataset. *Int. J. Rob. Res.* **2013**, *32*, 1231–1237. [CrossRef]
21. Quigley, M.; Conley, K.; Gerkey, B.; Faust, J.; Foote, T.; Leibs, J.; Wheeler, R.; Ng, A.Y. ROS: an open-source Robot Operating System. In Proceedings of the ICRA Workshop on Open Source Software, Kobe, Japan, 12–17 May 2009.
22. Moore, T.; Stouch, D. A Generalized Extended Kalman Filter Implementation for the Robot Operating System. In Proceedings of the 13th International Conference on Intelligent Autonomous Systems (IAS-13), Padua, Italy, 15–18 July 2014; pp. 335–348.
23. Wan, E.A.; Van Der Merwe, R. The unscented Kalman filter for nonlinear estimation. In Proceedings of the IEEE 2000 Adaptive Systems for Signal Processing, Communications, and Control Symposium (Cat. No. 00EX373), Lake Louise, Alberta, Canada, 4 October 2000; pp. 153–158.
24. Bresson, G.; Alsayed, Z.; Yu, L.; Glaser, S. Simultaneous localization and mapping: A survey of current trends in autonomous driving. *IEEE Trans. Intell. Veh.* **2017**, *2*, 194–220. [CrossRef]
25. Balasuriya, B.; Chathuranga, B.; Jayasundara, B.; Napagoda, N.; Kumarawadu, S.; Chandima, D.; Jayasekara, A. Outdoor robot navigation using Gmapping based SLAM algorithm. In Proceedings of the 2016 Moratuwa Engineering Research Conference (MERCon), Moratuwa, Sri Lanka, 5–6 April 2016; pp. 403–408.
26. Marti, E.; de Miguel, M.A.; Garcia, F.; Perez, J. A Review of Sensor Technologies for Perception in Automated Driving. *IEEE Intell. Transp. Syst. Mag.* **2019**, *11*, 94–108. [CrossRef]
27. Milošević, M.; Stefanović, M. Performance Loss Due to Atmospheric Noise and Noisy Carrier Reference Signal in Qpsk Communication Systems. *Elektron. ir Elektrotechnika* **2005**, *58*, 5–9.
28. Thrun, S.; Burgard, W.; Fox, D. *Probabilistic Robotics*; MIT Press: Cambridge, UK, 2000.
29. Burgard, W.; Stachniss, C.; Bennewitz, M.; Grisetti, G.; Arras, K. Introduction to Mobile Robotics. Available online: http://domino.informatik.uni-freiburg.de/teaching/ss13/robotics/slides/12-pf-mcl.pdf (accessed on 1 June 2020).
30. de Miguel, M.A.; Moreno, F.M.; Garcia, F.; Armingol, J.M. Autonomous vehicle architecture for high automation. In Proceedings of the International Conference on Computer Aided Systems Theory, Las Palmas de Gran Canaria, Spain, 17–22 February 2019; pp. 145–152.

sensors

MDPI

Article

Multiple-Target Homotopic Quasi-Complete Path Planning Method for Mobile Robot Using a Piecewise Linear Approach

Gerardo Diaz-Arango [1], Hector Vazquez-Leal [2,3,*], Luis Hernandez-Martinez [4], Victor Manuel Jimenez-Fernandez [2], Aurelio Heredia-Jimenez [5], Roberto C. Ambrosio [6], Jesus Huerta-Chua [7], Hector De Cos-Cholula [4] and Sergio Hernandez-Mendez [1]

1 Engineering School, University of Xalapa, Km. 2 Carretera Xalapa-Veracruz, Xalapa, Veracruz 91190, Mexico; guda.diaz.gd@gmail.com (G.D.-A.); sergio.h@ux.edu.mx (S.H.-M.)
2 Facultad de Instrumentacion Electronica, Universidad Veracruzana, Cto. Gonzalo Aguirre Beltran S/N, Xalapa, Veracruz 91000, Mexico; vicjimenez@uv.mx
3 Consejo Veracruzano de Investigacion Cientifica y Desarrollo Tecnologico (COVEICYDET), Av. Rafael Murillo Vidal No. 1735, Cuauhtemoc, Xalapa, Veracruz 91069, Mexico
4 Electronics Department, National Institute for Astrophysics, Optics and Electronics, Sta. María Tonantzintla, Puebla 72840, Mexico; luish@inaoep.mx (L.H.-M.); hdecos@inaoep.mx (H.D.C.-C.)
5 Electronics Department, UPAEP, 21 Sur 1103, Puebla 72410, Mexico; aureliohoracio.heredia@upaep.mx
6 Faculty of Electronics Science Meritorious University Autonomous of Puebla, 4 Sur 104 Centro, Puebla 72000, Mexico; roberto.ambrosio@correo.buap.mx
7 Instituto Tecnologico Superior de Poza Rica, Tecnologico Nacional de Mexico, Luis Donaldo Colosio Murrieta S/N, Arroyo del Maiz, Poza Rica, Veracruz 93230, Mexico; chua@itspozarica.edu.mx
* Correspondence: hvazquez@uv.mx

Received: 4 May 2020; Accepted: 27 May 2020; Published: 8 June 2020

Abstract: The ability to plan a multiple-target path that goes through places considered important is desirable for autonomous mobile robots that perform tasks in industrial environments. This characteristic is necessary for inspection robots that monitor the critical conditions of sectors in thermal, nuclear, and hydropower plants. This ability is also useful for applications such as service at home, victim rescue, museum guidance, land mine detection, and so forth. Multiple-target collision-free path planning is a topic that has not been very studied because of the complexity that it implies. Usually, this issue is left in second place because, commonly, it is solved by segmentation using the point-to-point strategy. Nevertheless, this approach exhibits a poor performance, in terms of path length, due to unnecessary turnings and redundant segments present in the found path. In this paper, a multiple-target method based on homotopy continuation capable to calculate a collision-free path in a single execution for complex environments is presented. This method exhibits a better performance, both in speed and efficiency, and robustness compared to the original Homotopic Path Planning Method (HPPM). Among the new schemes that improve their performance are the Double Spherical Tracking (DST), the dummy obstacle scheme, and a systematic criterion to a selection of repulsion parameter. The case studies show its effectiveness to find a solution path for office-like environments in just a few milliseconds, even if they have narrow corridors and hundreds of obstacles. Additionally, a comparison between the proposed method and sampling-based planning algorithms (SBP) with the best performance is presented. Furthermore, the results of case studies show that the proposed method exhibits a better performance than SBP algorithms for execution time, memory, and in some cases path length metrics. Finally, to validate the feasibility of the paths calculated by the proposed planner; two simulations using the pure-pursuit controlled and differential drive robot model contained in the Robotics System Toolbox of MATLAB are presented.

Keywords: robot motion; path planning; piecewise linear approximation; multiple-target path planning; autonomous mobile robot; homotopy based path planning

1. Introduction

Autonomous mobile robot is an entity capable of performing a wide variety of tasks that involve displacement in its workspace, such as home service, pickup and delivery assistance in offices, monitoring factories, and so forth. An autonomous robot develops its assignment without human intervention because it commonly implies a certain degree of risk for a human being or environmental conditions make the use of teleoperation impossible. To execute its task, the robot must be capable to plan a path from an initial position to a target-position. In principle, path planning is a geometric process in which an autonomous agent must find a collision-free path in its workspace, without considering its kinematics and dynamics restrictions [1–4]. Then, once a path is specified, another process is executed to calculate the motion plan using the kinodynamic properties of the robot. Usually, for a path planning process, the robot is considered as a point in the configuration space (C_{space}), which is the space generated by all feasible positions that it can reach [3,5,6]. Then, C_{space} is divided into free configuration space (C_{free}) for valid positions and obstacles space (C_{obs}) for all forbidden configurations.

1.1. Planning Algorithms

Finding a collision-free path in C_{free} may seem like an easy task for a human agent, nevertheless, it is a very complex issue for artificial intelligence. This has been the main motivation for researchers and developers community which have worked in this area during the last five decades [2–4,6–11]. These efforts have generated diverse planning algorithms with particular characteristics that define its degree of completeness, approach, and configuration of problems that each planner is capable of solve. In general, the planners can be classified into three categories.

1. Planners classified according to its completeness are: (I) Complete. These algorithms can find one solution, if it exists; otherwise, reports a failure. The most common algorithms in this category are visibility graph, Voronoi diagrams, Delaunay triangulations, among others graph based planners [12–15]. (II) Semi-complete. These algorithms can find a solution if one exists; otherwise, it may run forever while a stop criterion is not reached. The most relevant planners in this category are the method of artificial potential fields (APF) [6,16–18] and homotopic path planning method (HPPM) [10,19–21]. (III) Resolution complete. For any algorithm in this category; the completeness is strongly related to the resolution, size, and shape of the cells in the grid. Here, if a solution exists, any of these algorithms can obtain one; otherwise, terminates and reports that no solution exists for the specified resolution. All planners in this category are based on a cell decomposition which uses a search algorithm to find the collision-free path. The most used search algorithms are Dijkstra's, A*, the local current comparison, and any of its variants [22–28]. (IV) Probabilistically complete. The degree of completeness for any algorithm in this category is considered probabilistic, because, if a solution exists, the probability tends to one hundred percent as long as the number of iterations of this process tends to infinity. The most effective algorithms, in this category, are the sampling-based planners (SBP).
2. Planners classified according to its formulation or approach are: graph-based, cells-decomposition, artificial potential fields, sampling-based, and homotopy continuation methods. (a) Graph-based contains algorithms for whose their roadmap is modeled by a graph and a search algorithm is employed to obtain the valid path. (b) Cells decomposition contains all discrete-space based algorithms, generally, these are the best choice to obtain a valid path for maze-type environments. Nevertheless, for non-structured environments, its performance depends on the resolution of the grid. (c) Artificial potential fields contains little variants of the original method, in this classification a random optimization or another technique is implemented to deal with the local minimum problem. (d) Sampling-based

contains all probabilistic-complete algorithms such as: probabilistic road maps (PRM) [7,29], expansive spaces trees (EST) [30], rapidly-exploring random trees (RRT) [3,7,31], bidirectional RRT (Bi-RRT, also named RRT-Connect) [32], RRT combined with a shortest path approach A (RRT*) [7], kynodinamic motion planning by interior-exterior cells of exploration (KPIECE) [1], and collision checking efficient algorithms (LazyPRM and LazyRRT) [33,34] which have the common modules; uniform distribution samples generator, collision checker, local planner, and smoothing post-processing algorithm [2,3,35]. (e) Homotopy continuation based planners are a new category that contains some variants of the original method introduced in Reference [19]. These variants have been proposed to improve performance, minimize computation time, and obtain the shortest path (reported in References [10,20,36,37]).

3. Planners are classified according to their ability to reuse pre-processed data for solving problems into: (a) single-query and (b) multiple-query. In this way, the algorithms and methods mentioned previously, as well as their variants, fall in one of these categories. On the one hand, multiple-query algorithms are commonly applied to solve static environments and the roadmap generated can be reused as many times as needed. Therefore, queries are very fast, nevertheless, the computational cost and time to generate the roadmap are impractical for dynamic environments; algorithms like PRM and graph-based (Visibility graphs and A*) have this property. On the other hand, for single-query planners, roadmap generation and extend function are developed in parallel to reduce the high computational cost of analyzing the entire environment. This characteristic makes these algorithms faster, nevertheless, the resulting roadmap is only useful for the current query. Some algorithms contained in this category are RRT, EST, KPIECE, artificial potential fields method, HPPM, among others (see Table 1).

Table 1 shows the main characteristics of the most used collision-free path planners. The last row presents the advantages of HPPM (the method of interest in this work) which makes it one of the most versatile and reliable planners, based on its single-query execution, deterministic approach, and quasi-complete success. Furthermore, this table remarks the main issue, or bottleneck, found in each sampling-based algorithm, APF method, visibility graph approach, A* algorithm, and HPPM. On the one hand, for SBP algorithms, the bottleneck is the computing time spent during the collision query procedure which consumes about sixty percent of the total time [9]. On the other hand, the main feature of LazzyRRT and LazzyPRM over RRT and PRM is the lower time for the collision query. Nevertheless, for these planners, the number of samples (density of nodes) has a high impact on its success to find a feasible path. For A* algorithm, its success lies in the resolution of the discretized workspace (better resolution implies higher memory consumption and execution time). For AFP, the main issue is the local minimum, which confine the robot and prevents the method to continue. For the visibility graph approach, the bottleneck is the calculation time, which increases according to the number of obstacles. As for HPPM, the main issue lies in the selection of the repulsion parameter for obstacles. This parameter plays an important role in this method because it determines the length of the path and, in consequence, the execution time. Although, the repulsion parameter selection for HPPM is considered solved for particular applications in References [38,39] and References [20,21,36,37]. A generalized strategy, at this moment, it is currently considered an open problem. One of the main contributions in the present work is the implementation of a systematic criterion to select the repulsion parameter to generate an optimal path, in terms of length and execution time, for HPPM.

Table 1. Main characteristics of collision-free path planners.

Planner	Query	Completeness	Approach	Main Issue/Bottleneck
RRT [3,7,31]	Single	Probabilistically	Sampling-based	Collision query
Bi-RRT [32]	Single	Probabilistically	Sampling-based	Collision query
RRT* [7]	Single	Probabilistically	Sampling-based	Collision query
PRM [7,29]	Multiple	Probabilistically	Sampling-based	Collision query
EST [30]	Single	Probabilistically	Sampling-based	Collision query
KPIECE [1]	Single	Probabilistically	Sampling-based	Collision query
LazyRRT [33]	Single	Probabilistically	Sampling-based	Density of nodes
LazyPRM [34]	Multiple	Probabilistically	Sampling-based	Density of nodes
APF [16–18]	Single	Semi-complete	Potential fields	Local minima
A* [24–26]	Single	Resolution	Cells decomposition	Grid resolution
Visibility graphs [12–15]	Multiple	Complete	Graph-based	Obstacles density
HPPM [10,19,20]	Single	Semi-complete	HCM (Mathematical model)	Repulsion parameter selection

1.2. Multiple-Target Path Planning

Commonly, the single-query and multiple-query collision-free planners are developed to find a valid path from a starting point to a single target point. Nevertheless, for applications like: robot vacuum cleaner, service at home, pickup and delivery services in office, industrial process and inspection, victim rescue, museum guidance, and land mine detectors, robots need to move through a sequence of target points [40–45]. The multiple-target collision-free path planning problem lacks deeper study, since it is considered a trivial task by using a point-to-point strategy. It operates through a set of sub-paths which creates a full multiple-target path, which connects each target point to another. While this strategy operates properly, its main disadvantage is the unnecessary and redundant turnings generated by connecting two or more segments; causing the sub-paths to cross each other and, in consequence, the robot will cross through the same place more than once [46] (see Figure 1b). In this sense, a complete path found through the point-to-point strategy is not optimal in terms of length.

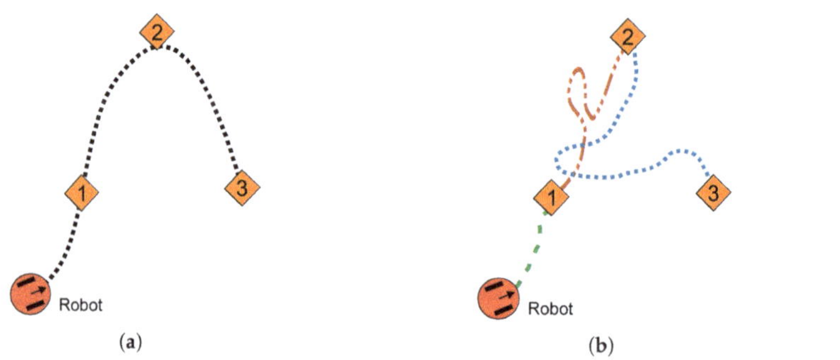

Figure 1. Efficient route and unnecessary turnings for multiple-target problem. (**a**) Efficient path for multiple-target problem. (**b**) Unnecessary turnings in multiple-target path (point-to-point strategy).

Figure 1a shows an example of desirable path (the shortest path) for a multiple-target problem. It can be noticed that the path is smooth without redundant configurations and pass very close to every target. On the other hand, Figure 1 depicts an example of point-to-point strategy for multiple-target problem. The unnecessary turnings and crossings between sub-paths are noticeable, making the trajectory inefficient in terms of distance. Figure 2 shows an example of a common office-like environment, this floor plan corresponds to a museum hall and the robot performs guidance tasks. In this case, it is considered that the robot calculates a path in some milliseconds, then, people and other dynamic agents could be represented as quasi-static obstacles (circular shapes in the environment map). The tasks of autonomous robots are to guide a group of people, provide relevant information about the displayed art, and visit all rooms while avoiding any collision with objects.

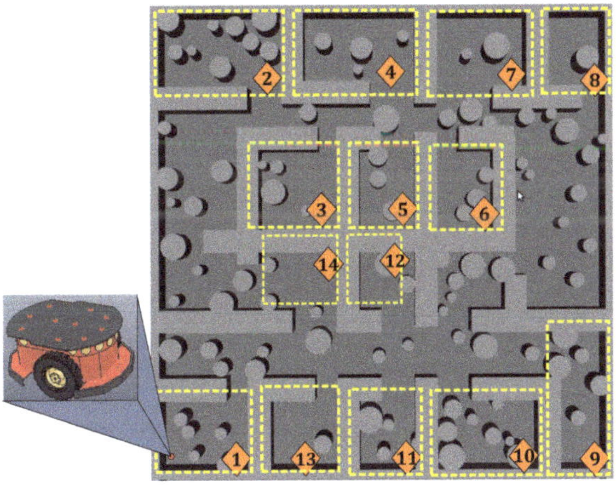

Figure 2. Multiple-target path planning issue in a complex environment (museum floor plan).

The museum environment in Figure 2 contains three common challenges for path planning algorithms. First, the map contains many obstacles that represent an issue for planners like BUG algorithm, artificial potential field method, and visibility graph methods. For these, the complexity and computing time increases accordingly to obstacles density. Second, the map displays several narrow corridors between walls and obstacles that are difficult to solve by SBP algorithms; while for artificial potential fields method this configuration exhibits local minimum problems. Third, for the layout in this map, the path should visit fourteen places, that is, the robot system must be capable to plan a path for multiple-target points. Although the first two issues have been thoroughly studied, multiple-targets is still considered as an open problem [40,44]. In this paper, a capable planner based on homotopy continuation to deal with narrow corridors, environment maps with hundreds of obstacles, and multiple-target issues is proposed. This paper is organized as follows. In Section 2, the bases of the homotopic path planning method are explained. The spherical algorithm applied to trace homotopy curves is presented in Section 3. Chua's canonical piecewise linear model is shown in Section 4. Section 5 provides the main contributions and formulation of our proposed multiple-target HPPM. Furthermore, in Section 6, a HPPM with visibility approach is explained. Some case studies and performance comparison between the proposed methodology and SBP algorithms are presented in Section 7. Path tracking examples applying the pure-pursuit algorithm are provided in Section 8. Finally, the concluding remarks are given in Section 9.

2. Homotopic Path Planning Method

Homotopic path planning method (HPPM) is a single-query planner capable to obtain a collision-free path by using a mathematical model emanated from the configuration space, initial and final positions [10,19]. This planner is based on homotopy continuation methods (HCM) which, frequently, are employed to find multiple solutions of non-linear algebraic equation system (NAES) of the form

$$f(X) = 0: \quad \mathbb{R}^n \longrightarrow \mathbb{R}^n. \tag{1}$$

The homotopy operates through a deformation of $f(X)$ by adding a function $G(X)$ and a homotopy parameter λ, such that

$$H(X, \lambda) = \lambda f(X) + (1 - \lambda) G(X) = 0, \tag{2}$$

where $H(f(X)) = H(X, \lambda) : \mathbb{R}^{n+1} \longrightarrow \mathbb{R}^n, \quad X \in \mathbb{R}^n \quad , \quad \lambda \in [0, 1]$; $G(X)$ is a function with a trivial or known solution. The homotopy system has the following properties

- For $H(X,0) = 0$, the trivial or known solution is obtained.
- For $H(X,1) = f(X)$, one solution of the original system is found.
- The homotopy curve is formed by the set of intersection points between the equations in the homotopy system (2), that is, $H(X,\lambda) = 0 : (X,\lambda) \in H^{-1}(0)$. Where, $H^{-1}(0)$ represents the set of intersections and is denoted, commonly, by γ [19,47–49]. Furthermore, all solutions of the original system $f(X) = 0$ are included in $H^{-1}(0)$; these are found during the continuous deformation at $\lambda = 1$.

HPPM process takes the system of equations that models the configuration space and, through the application of Newton's homotopy, a free collision path is obtained. For Newton's homotopy, the auxiliary function is $G(X) = f(X) - f(X_0)$; where X_0 is the known starting point. Then, HPPM employs the spherical algorithm (SA) to properly track the homotopy curve. For this system of equations, the curve represents a sequence of points that describes a continuous motion from a starting point to a target-point avoiding collisions with obstacles [10,19]. The configuration for a 2-D environment is presented as NAES $f_1(x,y) = 0$ and $f_2(x,y) = 0$. For both equations the unique solution lies on the target-point (x_T, y_T) [10,19]. The configuration space is modeled by the following equations

$$f_1(x,y) = L_1(x,y) = 0, \tag{3}$$

$$f_2(x,y) = L_2(x,y) + g(x,y) = 0, \tag{4}$$

where $g(x,y)$ model the obstacles on the map, which creates singular regions around them. $L_1(x,y)$ and $L_2(x,y)$ are two auxiliary straight lines that intersect at the target-point, these are represented by

$$L_i(x,y) = -y + m_i x + b_i = 0;\ i = 1,2, \tag{5}$$

then, applying Newton's homotopy to Equation (3) and Equation (4), the NAES is transformed into

$$H = \begin{cases} H_1(f_1(x,y),\lambda) = f_1(x,y) - (1-\lambda)f_1(x_0,y_0) = 0, \\ H_2(f_2(x,y),\lambda) = f_2(x,y) - (1-\lambda)f_2(x_0,y_0) = 0, \end{cases} \tag{6}$$

where (x_0, y_0) is the initial point. The obstacles in the configuration space can be modeled by circumferences, ellipsoids, and other closed curves. For this work, only circular and quasi-rectangular shapes modeled using circle and ellipsoid equations [19], respectively are employed. Furthermore, the solid obstacles representation proposed in Reference [36] is used, such that each circular obstacle is defined by

$$|C_i(x,y)| + C_i(x,y) = 0, \tag{7}$$

where $C(x,y)$ is the general equation of the circumference, modeled by

$$C_i(x,y) = (x - x_{C,i})^2 + (y - y_{C,i})^2 - r_{C,i}^2 = 0, \tag{8}$$

$i = 1,2,3,\ldots,k$, k is the number of circular obstacles in the map; $(x_{C,i}, y_{C,i})$ represents the center of the i-th circular obstacle, and $r_{C,i}$ its radius. The solid circle obstacle representation has solution for all points inside and at the contour of the circumference, that is, $\forall\ (x,y) \in \mathbb{R}^2$ such that $\sqrt{(x - x_{C,i})^2 + (y - y_{C,i})^2} \leq r_{C,i}$. Then, the expression representing all circular obstacles in the map is

$$W_C(x,y) = \sum_{i=1}^{i=k} \frac{p_{C,i}}{C_i(x,y) + |C_i(x,y)|}, \tag{9}$$

here, $p_{C,i}$ is the repulsion parameter of the i-th circular obstacle [10,19]. Likewise, this principle is applied to ellipsoidal obstacles using

$$R_j(x,y) = \left(\frac{x - x_{R,j}}{\alpha}\right)^{2\eta} + \left(\frac{y - y_{R,j}}{\beta}\right)^{2\nu} - 1 = 0, \tag{10}$$

where, the shape of each rectangular obstacle is defined by $R(x,y)$ in Equation (10); $j = 1,2,3,\ldots,l$, l is the number of rectangular obstacles in the map; $(x_{R,j}, y_{R,j})$ is the center for the j-th rectangular obstacle; α and β are width and length, respectively; η and ν are two integers that define the sharp of the vertex [19,36]. The solid representation for ellipsoidal obstacles is modeled as

$$|R_j(x,y)| + R_j(x,y) = 0, \tag{11}$$

in a similar way to circular obstacles, this representation has a solution for all points inside and at the contour of the ellipsoid. Then, the expression that represents all ellipsoidal obstacles in the map is

$$W_R(x,y) = \sum_{j=1}^{j=l} \frac{p_{R,j}}{R_j(x,y) + |R_j(x,y)|}, \tag{12}$$

$p_{R,j}$ is the repulsion parameter for the j-th ellipsoidal obstacle. It is important to notice that the points inside and at the contour of the obstacles result in divisions by zero for Equations (9) and (12). These divisions by zero creates singularities in the map, thus, HPPM uses them to automatically avoid collisions with the obstacles. The effect of all obstacles, whether they are ellipsoid or circular, are contained in the following expression

$$W(x,y) = W_C(x,y) + W_R(x,y), \tag{13}$$

where $W_C(x,y)$ represents the solid circular obstacles and $W_R(x,y)$ represents the solid ellipsoidal obstacles. The term $g(x,y)$ in Equation (4) is redefined as

$$g(x,y) = W(x,y) - W(x_T, y_T), \tag{14}$$

here, $W(x_T, y_T)$ is added to the NAES to cancel the effect of the obstacles at the target-point.

3. Homotopy Path Tracking Scheme

Spherical algorithm (SA) is a tool applied to track continuous curves resulting from a non-homogeneous NAES. Previous works have demonstrated the effectiveness of SA to trace homotopy curves γ [10,36,47–50]. The homotopy is an operator such that $H(f(X)) : \mathbb{R}^{n+1} \longrightarrow \mathbb{R}^n$, that is, homotopy system (6) is non-homogeneous. SA operates by adding an n-dimensional sphere equation $S_i(X,\lambda)$ (the dimension of sphere or hypersphere depends on the number of variables in the system of equations) that guarantees, at least, two cross points with the homotopy curve (6), as long as its center is at $H^{-1}(0)$ [47]. The homotopy process is described as follows: first step, the center of first sphere (S_1) is placed at $O_1 = (x_0, y_0, \lambda = 0)$ (see Figure 3). Then, the first root on the right is calculated by a predictor-corrector scheme; the root represents the intersection between the homotopy curve and the first hypersphere for values of $\lambda \geq 0$. The next step consists of the assignation of the calculated root on the right as the center (O_2) of a new hypersphere (S_2), as shown in Figure 3. Numerical path tracking scheme is based on

$$H_S = \begin{cases} H_1(x,y,\lambda) = 0, \\ H_2(x,y,\lambda) = 0, \\ S_i(x,y,\lambda) = 0, \end{cases} \tag{15}$$

for three dimensions (Bi-dimensional space and homotopy parameter λ), the sphere is represented as

$$S_i(x, y, \lambda) = (x - c_x)^2 + (y - c_y)^2 + (\lambda - c_\lambda)^2 - r_s^2 = 0, \tag{16}$$

where r_s is the radius and $O_i = (c_x, c_y, c_\lambda)$ is the center of the sphere for each step in the spherical tracking.

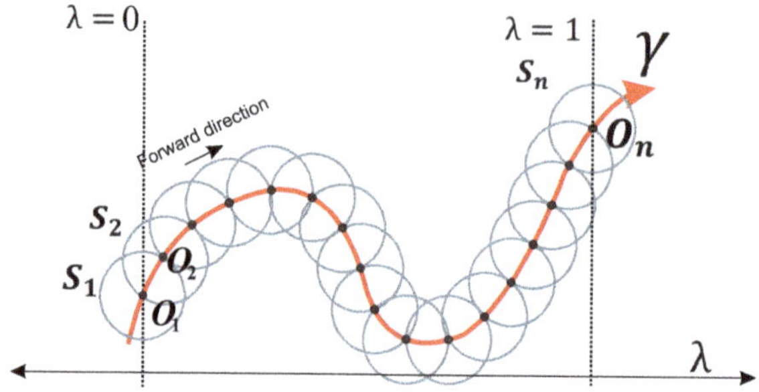

Figure 3. 2-D representation of spherical tracking algorithm process.

Homotopy curves tracking is a complex task, because the sharp turning points makes the convergence of the corrector step difficult [10,19,48,49]. In this regard, a proper predictor scheme is employed to improve the performance of the SA; in this work the Euler's predictor is applied. Moreover, this scheme has shown adequate performance for tracking homotopy curves making the corrector scheme to converge faster [10,49]. This predictor scheme operates using the tangent vector to calculate an approximation of the next point in the tracking curve, as explained in the following section.

3.1. Euler's Predictor Scheme

Euler's scheme is a complement of SA that provides a predictor point close to the intersection between homotopy curve and the respective sphere at each step of the tracking. Work in References [10,49] explain its operation and show its effectiveness as a predictor scheme in SA. For system equations with three variables, the Euler predictor point is calculated using

$$(x_p, y_p, \lambda_p) = (x_i, y_i, \lambda_i) + r_s \left(\frac{\vec{v}_p}{\|\vec{v}_p\|} \right), \tag{17}$$

where (x_p, y_p, λ_p) is the predictor point; (x_i, y_i, λ_i) is the center of the sphere; r_s is the radius of the sphere, and $\vec{v}_p$ is the tangent vector. Figure 4 shows a 2-D view of the predictor scheme operation in the SA. The predictor point (x_p, y_p, λ_p) is located at the intersection between the vector point and the sphere S_i.

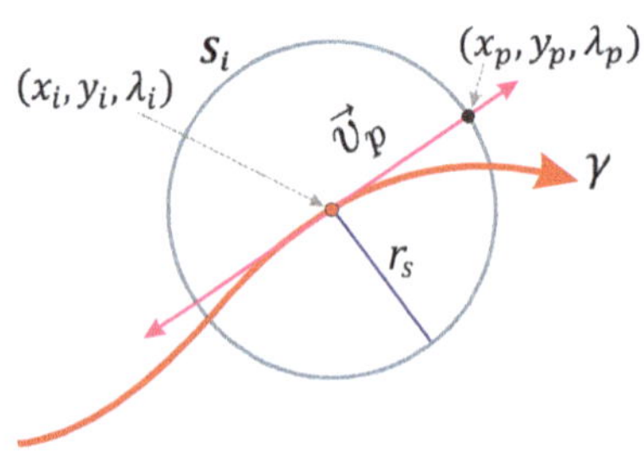

Figure 4. Euler's predictor scheme.

The tangent vector is obtained from the partial derivatives of the homotopy system (6), as explained in References [10,49]. This method uses partial derivatives with respect to an arbitrary parameter ρ, such that $(x(\rho), y(\rho), \lambda(\rho))$; then, the chain rule is employed to calculate the tangent vector for the homotopy system at the point (x_i, y_i, λ_i). Placing the tangent vector in Equation (17) the predictor point $((x_p, y_p, \lambda_p))$ is obtained

$$(x_p, y_p, \lambda_p) = (x_i, y_i, \lambda_i) + r_s \frac{(x_i'(\rho), y_i'(\rho), \lambda_i'(\rho))}{\|(x_i'(\rho), y_i'(\rho), \lambda_i'(\rho))\|}, \tag{18}$$

where,

$$(x'(\rho), y'(\rho), \lambda'(\rho)) = \left(\frac{\partial x_i(\rho)}{\partial \rho}, \frac{\partial y_i(\rho)}{\partial \rho}, \frac{\partial \lambda_i(\rho)}{\partial \rho} \right), \tag{19}$$

here, the partial derivative vector of each variable in the system with respect to ρ, evaluated at (x_i, y_i, λ_i), represents the tangent vector [10,49].

3.2. Broyden's Method as Corrector Scheme

The corrector scheme is the core of the HCM process, capable to obtain one solution for homogeneous NAES from an initial condition. Commonly, Newton-like correctors are described by the expression

$$X_{i+1} = X_i - [J(X_i)]^{-1} f(X_i), \tag{20}$$

here, $X \in \mathbb{R}^n$, $i = 0, 1, 2, \ldots, n$; X_{i+1} is the next point in the iterative process; X_i is the current point and represents the initial condition for $i = 0$, and $[J(X_i)]^{-1}$ is the Jacobian inverse matrix of the NAES $f(X)$. The iterative process ends when the stop criterion is achieved or the maximum number of iterations is reached; both parameters set by the user. Broyden's method is a quasi-Newton method employed to calculate the roots of non-linear algebraic equation systems. For this, the Jacobian matrix $J(X_i)$ is replaced by an approximation A_i to reduce its calculation complexity. A_i is calculated using

$$A_j = A_{j-1} + \frac{f(X_j) - f(X_{j-1}) - A_{j-1}[X_j - X_{j-1}]}{\|X_j - X_{j-1}\|^2} [X_j - X_{j-1}]^T, \tag{21}$$

for $j = 1, 2, 3, \ldots, n$; Broyden's method is represented by

$$X_{j+1} = X_j - \left[A_j(X_j)\right]^{-1} f(X_j), \tag{22}$$

where, in the first iteration ($j = 1$), X_1 is calculated using Equation (20) and the matrix $A_0 = J(X_0)$, that is, the Jacobian matrix only needs to be calculated once. Furthermore, the Jacobian matrix calculation is also employed to obtain the predictor point, hence, this matrix is strongly used in two complementary SA processes. In this work, Broyden's Method stop criterion is set at $i = 20$ or when the next criterion is fulfilled

$$\|f(X_i)\| < 1 \times 10^{-9}. \tag{23}$$

For the spherical algorithm, the initial point X_0 of Broyden's method procedure is the predictor point (x_p, y_p, λ_p) obtained using the Euler's scheme. Then, the Broyden's process is executed to solve the NAES system (15) for the current sphere (S_i) until the stop criterion is fulfilled. Which implies that the next solution point $(x_{i+1}, y_{i+1}, \lambda_{i+1})$ has been found. Figure 5 depicts, in 2-D, the entire operation of the predictor-corrector scheme for SA.

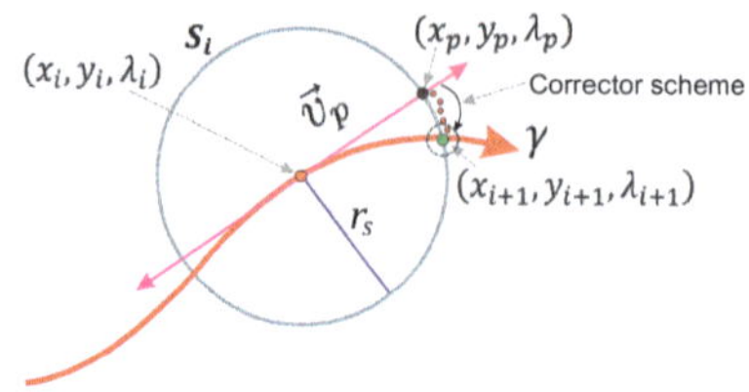

Figure 5. Corrector scheme for spherical tracking.

4. Canonical Piecewise Linear Representation

A piecewise linear model can be defined as a mathematical representation that collects linear, related, descriptions to approximate a nonlinear function. The main reason that motivates the use of this type of model is the simplicity of their structure which allows an efficient implementation in algorithms. Although there are many piecewise linear models reported in the literature [51–55], due to its compact formulation, reduced number of parameters, and low computational requirements, the most popular is the so-called Chua's model; it is described by a compact global representation named canonical piecewise linear function, given by

$$y(x) = a + bx + \sum_{i=1}^{\sigma} c_i |x - x_{B,i}| = 0, \tag{24}$$

where σ is the number of breakpoints. Model parameters a, b, and c_i, for $i = 1, 2, ..., \sigma$, can be determined as follows

$$a = y(0) - \sum_{i=1}^{\sigma} c_i |x_{B,i}|,$$

$$b = \frac{\mathcal{J}_1 + \mathcal{J}_{\sigma+1}}{2},$$

$$c_i = \frac{\mathcal{J}_{i+1} - \mathcal{J}_i}{2}.$$

These parameters are strongly related to the graph of the piecewise linear function $y(x)$. For instance, b and c_i are both described in terms of $\mathcal{J}_i$ which represents the slope of the i-th constitutive linear segment in the graph of $y(x)$. Moreover, the parameter a is computed considering the σ-breakpoints $(x_{B,i}, y_{B,i})$, for $i = 1, 2, ..., \sigma$ included in the graph of $y(x)$. To graphically illustrate this relation, Figure 6 shows a single-valued piecewise linear function constituted by σ breakpoints and $\sigma + 1$ segments.

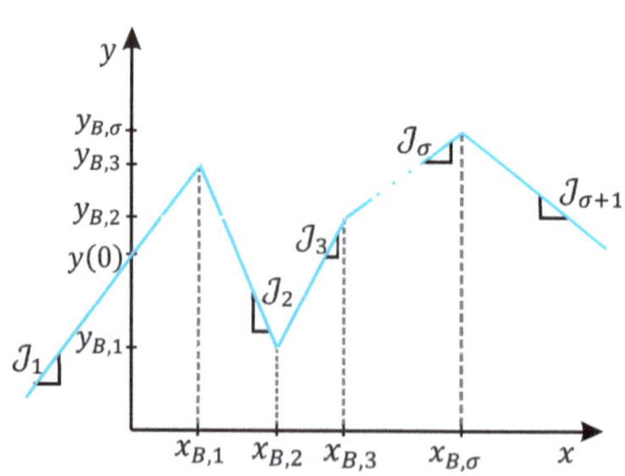

Figure 6. Single-valued piecewise linear function.

In this work, Chua's model serves as the base for the proposed homotopy scheme and the two-dimensional PWL model is described by

$$\mathrm{PWL}(x,y) = y - \left(a + bx + \sum_{i=1}^{\sigma} c_i|x - x_{B,i}|\right), \tag{25}$$

where $\mathrm{PWL}(x,y)$ is an implicit representation of Equation (24). It can notice, that depends on x values. The breakpoints $(x_{B,i}, y_{B,i})$ can take any value in the y-axis but, for x-axis, the values must be incremental such that $x_{B,i} < x_{B,i+1} < \ldots < x_{B,i+n}$.

5. Multiple-Target Homotopic Path Planning Method

This section is devoted to explain our proposed multiple-target homotopic path planning method (MTHPPM) and highlight the most relevant contributions of this work. MTHPPM is a planner based on the HPPM principle; for both, a set of auxiliary functions is necessary to generate a NAES in which the target-points are solutions. The multiple-target points in MTHPPM are provided by a set of two auxiliary functions that intersect in more of one point. In this work, target-points are generated by two single-valued PWL of the form Equation (24) with a similar formulation to the auxiliary functions in HPPM. These are represented by

$$f_{\mathrm{PWL}_1}(x,y) = \mathrm{PWL}_1(x,y), \tag{26}$$

$$\begin{aligned} f_{\mathrm{PWL}_2}(x,y) &= \mathrm{PWL}_2(x,y) \\ &- g(x,y)\left(|\mathrm{PWL}_2(x,y)| + |\mathrm{PWL}_1(x,y)|\right), \end{aligned} \tag{27}$$

where $g(x,y)$ is the expression that models all obstacles in the proposed MTHPPM, represented by

$$g(x,y) = W_C(x,y) + W_R(x,y), \tag{28}$$

$W_C(x,y)$ and $W_R(x,y)$ are the mathematical representation of circular and ellipsoidal obstacles, modeled by Equation (12) and Equation (9), respectively. Then, the homotopy system (6) is redefined by Equation (32), Equation (33), and Newton's homotopy formulation, obtaining

$$H = \begin{cases} H_1(f_{\mathrm{PWL}_1}(x,y),\lambda) &= f_{\mathrm{PWL}_1}(x,y) \\ &\quad -(1-\lambda)f_{\mathrm{PWL}_1}(x_0,y_0) = 0, \\ H_2(f_{\mathrm{PWL}_2}(x,y),\lambda) &= f_{\mathrm{PWL}_2}(x,y) \\ &\quad -(1-\lambda)f_{\mathrm{PWL}_2}(x_0,y_0) = 0, \end{cases} \tag{29}$$

where (x_0, y_0) is the initial point of the trajectory.

In addition to the PWL auxiliary lines, the second expression (27) is reformulated to keep the properties of obstacles for the multiple-target points approach. This reformulation guarantees that, when $\mathrm{PWL}_1(x,y) = 0$ and $\mathrm{PWL}_2(x,y) = 0$, the effect of the obstacles function $g(x,y)$ over the homotopy curve is cancelled. In other words, the effect of the obstacles over the homotopy system progressively vanishes as the robot approaches to any target point, that is, this effect decreases in proportion to the difference between $\mathrm{PWL}_1(x,y)$ and $\mathrm{PWL}_2(x,y)$. The absolute values are employed in this formulation because any change in the sign of $g(x,y)$ modifies the behavior of the homotopy curve, as it is presented in References [10,20]. It can be noticed that absolute value terms produce a singular Jacobian matrix during the iterative process. Then, in order to solve these issues and improve the performance of MGHPPM, a differentiable approximation of absolute value is used.

5.1. Approximation of Absolute Value Function to Improve the Piecewise Linear Approach

From definition of $f(x) = |x|$, $f(x)$ is continuous for all values of x. Nevertheless, it is not differentiable when $x = 0$. Then, the absolute value function is not differentiable at the breakpoints, that is, for all points (x, y) such that $x = x_{B,i}$ (see expression (24)). In order to remove this mathematical issue in the procedure, an approximation of the absolute value function is employed. For this work, the approximation of the absolute value function presented in Reference [56] is used to warrant the continuity of the PWL derivative function. This formulation is defined by

$$|x| \approx \xi(x, \alpha) = \frac{1}{\alpha}\left(\ln\left(1 + e^{-\alpha x}\right) + \ln\left(1 + e^{\alpha x}\right)\right), \tag{30}$$

where α is a parameter which reduces the error between $|x|$ and $\xi(x)$. Figure 7a shows the approximation of $\xi(x, \alpha)$ to $|x|$ when $\alpha \to \infty$, this can also can be observed in Figure 7c. The first derivation graphs for $\xi(x, \alpha)$ and $|x|$ are presented in Figure 7b. It shows that, like in the previous figure, the approximation $\xi(x, \alpha)$ is more similar to $|x|$ when $\alpha \to \infty$.

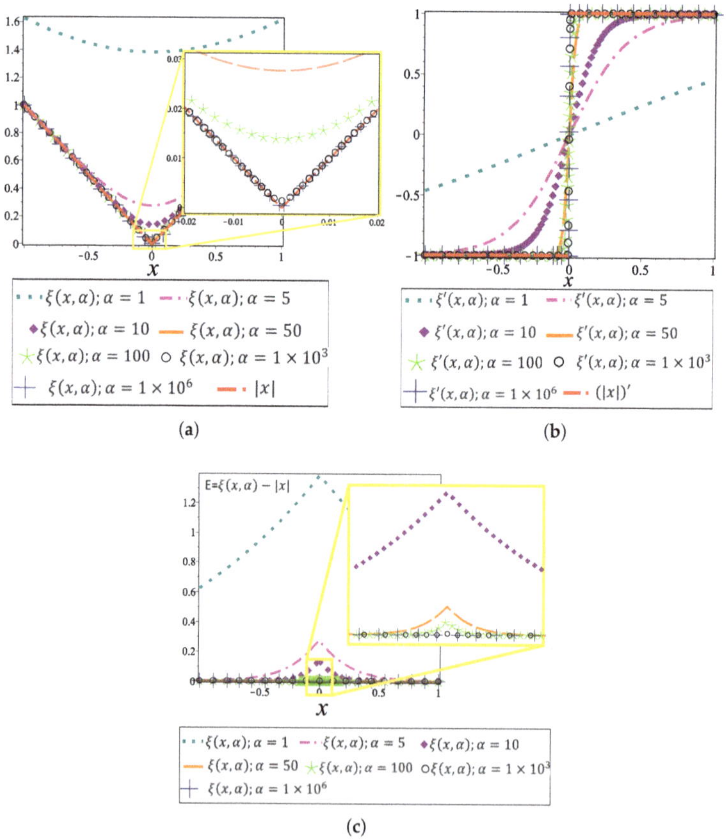

Figure 7. Approximation of absolute value function. (**a**) Approximation of absolute value function. (**b**) First derivation of the approximation of the absolute value function. (**c**) Error (E) between the approximation of absolute value function and $|x|$.

By using this approximation of absolute value function, the canonical $\mathrm{PWL}(x,y)$ representation is modified to generate smooth and differentiable auxiliary PWL functions ($\widetilde{\mathrm{PWL}}(x,y)$) for MGHPPM. The $\widetilde{\mathrm{PWL}}(x,y)$ representation is expressed as

$$\widetilde{\mathrm{PWL}}(x,y) = y - \left(a + bx + \sum_{i=1}^{i=\sigma} c_i \left(\xi(x - x_{B,i}, \alpha) \right) \right), \tag{31}$$

Figure 8a presents a $\widetilde{\mathrm{PWL}}$ function generated with the canonical representation for $\alpha = 1, 5, 10, 50, 1\times10^3, 1\times10^9$ in $\xi(x,\alpha)$. The first derivation of $\widetilde{\mathrm{PWL}}(x,y)$ is depicted in Figure 8b which shows the discontinuities present in $\mathrm{PWL}'(x,y)$. It is important to note that for values of $\alpha \leq 50$, the behaviour of $\widetilde{\mathrm{PWL}}(x,y)$ is unsuitable for the MTHPPM formulation due to the inaccuracy between $\mathrm{PWL}(x,y)$ and $\widetilde{\mathrm{PWL}}(x,y)$ for these values of α, as it is depicted in Figure 8a,b. Nevertheless, a good approximation of $\mathrm{PWL}(x,y)$ is reached with $\alpha = 1\times10^3$ and $\alpha = 1\times10^9$ in $\widetilde{\mathrm{PWL}}(x,y)$ with the advantage that these have continuous differentiation, as it is shown in Figure 8b.

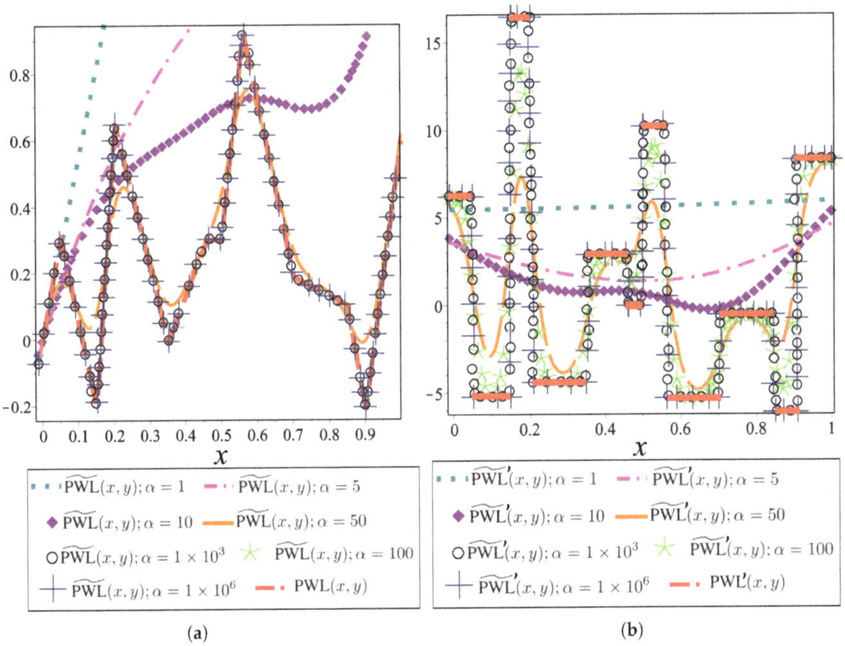

Figure 8. Smooth and differentiable piecewise linear function ($\widetilde{\mathrm{PWL}}$). (**a**) $\widetilde{\mathrm{PWL}}$ function generated with approximation of absolute value function. (**b**) First derivation of smooth $\widetilde{\mathrm{PWL}}$ representation.

For the MGHPPM, breakpoints are only used to generate intersections between auxiliary functions, in this sense, a good approximation that achieves this statement is when $\alpha = 1\times10^3$ (see Figure 8a). Then, this value is used for all case studies presented in the following section where, in order to simplify the notation, $\xi(x, 1\times10^3)$ is denoted by $\xi(x)$ and expresions (26), (27) and (29) are redefined as

$$f_{\widetilde{\mathrm{PWL}}_1}(x,y) = \widetilde{\mathrm{PWL}}_1(x,y), \tag{32}$$

$$\begin{aligned} f_{\widetilde{\mathrm{PWL}}_2}(x,y) &= \widetilde{\mathrm{PWL}}_2(x,y) \\ &- g(x,y)\left(\xi\left(\widetilde{\mathrm{PWL}}_2(x,y)\right) + \xi\left(\widetilde{\mathrm{PWL}}_1(x,y)\right)\right), \end{aligned} \tag{33}$$

$$H = \begin{cases} H_1(f_{\widetilde{\mathrm{PWL}}_1}(x,y),\lambda) & = f_{\widetilde{\mathrm{PWL}}_1}(x,y) \\ & \quad -(1-\lambda) f_{\widetilde{\mathrm{PWL}}_1}(x_0,y_0) = 0, \\ H_2(f_{\widetilde{\mathrm{PWL}}_2}(x,y),\lambda) & = f_{\widetilde{\mathrm{PWL}}_2}(x,y) \\ & \quad -(1-\lambda) f_{\widetilde{\mathrm{PWL}}_2}(x_0,y_0) = 0, \end{cases} \tag{34}$$

5.2. Breakpoints Selection for the Piecewise-Linear Functions

A set of proper auxiliary $\widetilde{\mathrm{PWL}}$ functions generate the least number of unnecessary turns over the homotopy path. For this work, the way to obtain these are from two different initial points $y(0)$ and the same breakpoints on the x-axis for both. This strategy guarantees that functions are different and intersect only at the desired points. Breakpoints are calculated from initial-points and target-points using the expressions

$$x_{B,i} = x_{T,i} + \frac{x_{T,i+1} - x_{T,i}}{\rho}, \; \rho = 2,3,4,\ldots,<\infty \tag{35}$$

$$y_{B,i} = \mathcal{J}_i \, x_{B,i} + y_{B,i-1} - \mathcal{J}_i(x_{B,i-1} - x_{B,i}), \tag{36}$$

$$\mathcal{J}_i = \frac{y_{T,i} - y_{B,i-1}}{x_{T,i} - x_{B,i-1}}, \tag{37}$$

where $\rho \in \mathbb{R}^+$ is a proportionality parameter and determines the proximity between the i-th solution point and the i-th breakpoint ($i = 1,2,3,\ldots,\sigma$); for σ number of breakpoints there are $\sigma + 1$ target-points; $(x_{B,i}, y_{B,i})$ is the position of the i-th breakpoint; $(x_{T,i}, y_{T,i})$ is the position of the i-th target-point; $\mathcal{J}_i$ is the slope of the i-th segment of the $\widetilde{\mathrm{PWL}}$ function, and $(x_{B,0} = 0, y_{B,0} = y(0))$ is the initial-point of the $\widetilde{\mathrm{PWL}}$ function. A simple environment map with two circular obstacles and two variants of $\widetilde{\mathrm{PWL}}_1$ and $\widetilde{\mathrm{PWL}}_2$ configurations is shown in Figure 9. For the $\widetilde{\mathrm{PWL}}$ functions in Figure 9a, the parameter is $\rho = 2$; for Figure 9b, the parameter $\rho = 20$ means that in this figure breakpoints are closer to the target-points.

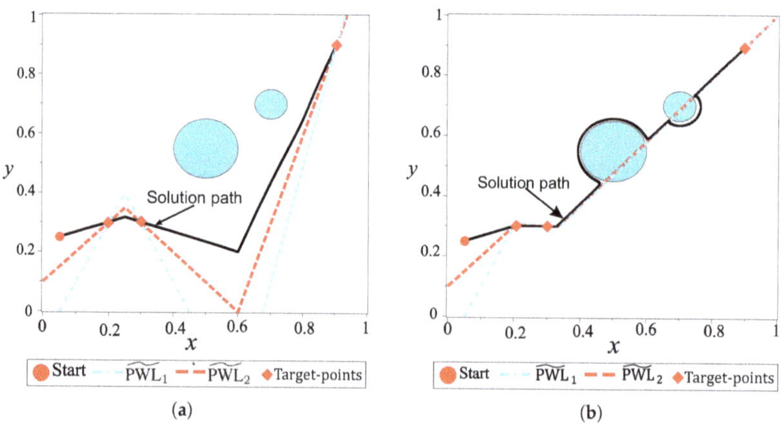

Figure 9. Selection of breakpoints for the $\widetilde{\mathrm{PWL}}$ functions. (**a**) Breakpoints placed far from target-points. (**b**) Breakpoints placed close to target-points.

Figure 9b shows the similarity between $\widetilde{\mathrm{PWL}}_1$ and $\widetilde{\mathrm{PWL}}_2$, nevertheless, these only intersect at target-points. Figure 9a,b show the effect of $\widetilde{\mathrm{PWL}}_1$ and $\widetilde{\mathrm{PWL}}_1$ in the homotopy formulation through Equation (33), where, the similarity of these in the homotopy formulation produces path with a similar tendency than $\widetilde{\mathrm{PWL}}$'s. Figure 9b shows two $\widetilde{\mathrm{PWL}}$'s which are overlapped to obstacles, this configuration produces that homotopy path pass very close to obstacles. This characteristic could be desirable in some cases, but when the map contains several grouped obstacles, the solution path could be inefficient in terms of length. In order to obtain paths very close to direct path between two points, for this work, the breakpoints are placed close to the target-points (similar form than Figure 9b).

5.3. Technique for Successful Convergence and Avoid Reversal Effect

Reversal effect is a phenomenon inherent to spherical path tracking and it is one of the most complex problems for the tracking techniques. Some works have proposed strategies to deal with this issue [10,47,48], nevertheless, these are inefficient when a $\widetilde{\text{PWL}}$ approach is attempted. This section introduces a new strategy to avoid reversal effect during the tracking and improve the convergence of the corrector scheme. The proposed strategy is named double spherical tracking (DST); it is a spherical tracking embedded into another spherical tracking. DST is executed when a reversal effect or a non-convergence is detected. On the one hand, the reversal effect is recognized using the directional cosine strategy as explained in References [10,48]. On the other hand, a non-convergence is detected when the maximum number of iterations is reached without meeting the stop criterion. DST procedure is explained using the non-convergence case shown in Figure 10. Figure 10a shows a simple map with one obstacle and two target-points. Figure 10b shows the position of the predictor point (x_p, y_p, λ_p) which is far from the intersection between $H_1(x,y,\lambda)$, $H_2(x,y,\lambda)$, and $S_i(x,y,\lambda)$. DST formulation is based on the principle that $\forall (x,y,\lambda)_{sol} \in H_S^{-1}(0) \Rightarrow (x,y,\lambda)_{sol} \in H_{1,S}^{-1}(0)$. Where, $H_s^{-1}(0)$ is the solution set for the system of Equations (15) and $H_{1,S}^{-1}(0)$ is the solution set for the system of equations

$$H_{1,S} = \begin{cases} H_1(x,y,\lambda) = 0, \\ S_i(x,y,\lambda) = 0, \end{cases} \tag{38}$$

The procedure steps of DST are described as follows:

1. The system of equations H_{DST} is established from $H_1(x,y,\lambda)$, $S_i(x,y,\lambda)$, and the sphere $\text{DST}(x,y,\lambda)$. DST has been formulated to track the curve of intersection between $H_1(x,y,\lambda)$ and $S_i(x,y,\lambda)$ using the SA algorithm.

$$H_{\text{DST}} = \begin{cases} H_1(x,y,\lambda) = 0, \\ S_i(x,y,\lambda) = 0, \\ \text{DST}_k(x,y,\lambda) = 0, \end{cases} \tag{39}$$

$$\begin{aligned} \text{DST}_k(x,y,\lambda) &= (x - \text{dc}_x)^2 + (y - \text{dc}_y)^2 \\ &\quad + (\lambda - \text{dc}_\lambda)^2 - r_{\text{dst}}^2 = 0, \end{aligned} \tag{40}$$

   where $k = 1,2,3,\ldots,n$, n is the number of DST steps; r_{dst} is the radius, and $(\text{dc}_x, \text{dc}_y, \text{dc}_\lambda)$ is the center of the sphere for every k-th step of the DST.
2. The first sphere $\text{DST}_i(x,y,\lambda)$ is placed at O_{i-1} point, that is, $(\text{dc}_x, \text{dc}_y, \text{dc}_\lambda)_1 = (x_{i-1}, y_{i-1}, \lambda_{i-1})$. Figure 10b indicates the position of point $(x_{i-1}, y_{i-1}, \lambda_{i-1})$, notice that it is at the intersection between all members of the system of Equation (15).
3. The SA algorithm is executed using the predictor and corrector schemes, explained above, for the non-linear system of Equation (39). Figure 10c depicts the schematic operation of DST; it can be noticed that the procedure starts at $(x_{i-1}, y_{i-1}, \lambda_{i-1})$ point and continues until its initial point is reached, it means that DST tracks a closed curve. It is important to note that the system of Equation (39) are easier to track with less computation cost than the system of Equation (15).

For his paper, the DST stop criterion is based on the distance between initial point of DST procedure $(x_{i-1}, y_{i-1}, \lambda_{i-1})$ and the current DST solution $(\mathrm{dc}_x, \mathrm{dc}_y, \mathrm{dc}_\lambda)_k$ as

$$\|(\mathrm{dc}_x, \mathrm{dc}_y, \mathrm{dc}_\lambda)_k - (x_{i-1}, y_{i-1}, \lambda_{i-1})\| < r_{\mathrm{dst}} \quad (41)$$

here, the radius of DST sphere is proposed as $r_{\mathrm{dst}} = (2\pi r_s)/n_{\mathrm{dst}}$, where r_s is the radius of the sphere S_i of the spherical tracking. n_{dst} is the minimum number of steps of DST which is set to 24.

4. Finally, all points $(\mathrm{dc}_x, \mathrm{dc}_y, \mathrm{dc}_\lambda)_k$, except $(\mathrm{dc}_x, \mathrm{dc}_y, \mathrm{dc}_\lambda)_1$, are evaluated in Equation (15). Then, the point for which the evaluation is closer to zero, $H_S((\mathrm{dc}_x, \mathrm{dc}_y, \mathrm{dc}_\lambda)_k) \approx 0$, is taken as a new predictor in the SA tracking of H_S. Figure 10d shows the new predictor point $(x_p, y_p, \lambda_p)_{\mathrm{new}}$ which is a point from the solution set $H_{\mathrm{DST}}^{-1}(0)$. Figure 10e provides a closer view where $(x_p, y_p, \lambda_p)_{\mathrm{new}}$ is close to the intersection between $H_1(x, y, \lambda)$, $H_2(x, y, \lambda)$ and $S_i(x, y, \lambda)$. Then, this point is taken as the predictor point in the Broyden's corrector scheme. The position of the new point $(x_{i+1}, y_{i+1}, \lambda_{i+1})$ can be seen in Figure 10f.

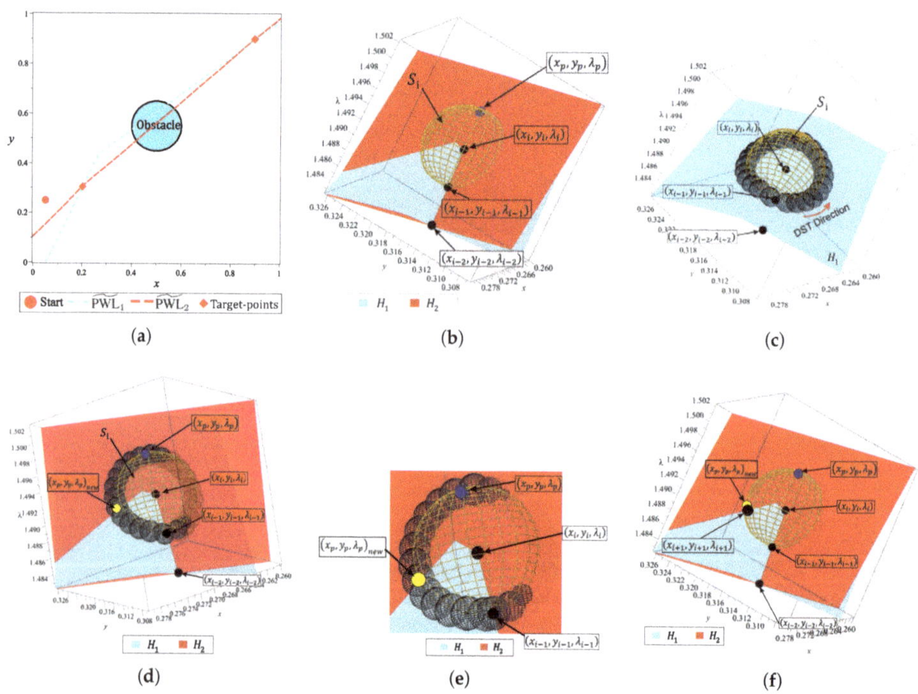

Figure 10. Double spherical tracking strategy. **(a)** Environment map with two target-points. **(b)** Non-convergence issue. **(c)** Double spherical tracking H_1 and S_i. **(d)** Double spherical tracking. **(e)** Double spherical tracking (zoom view). **(f)** New predictor-point.

It is important to remark that, DST is a backup technique for SA which is only applied when a non-convergence or reversal phenomenon is detected. Furthermore, the execution time of this technique is comparable with the execution time spent by one step of the SA. This because the H_{DST} system of equations does not contain the obstacles representation in its formulation. For the map shown in Figure 10a, the DST strategy is employed only once for the non-convergence point found during the tracking. Figure 11 shows the 2-D and 3-D representation of the NAES and homotopy surfaces, the region where is located the non-convergence point is indicated.

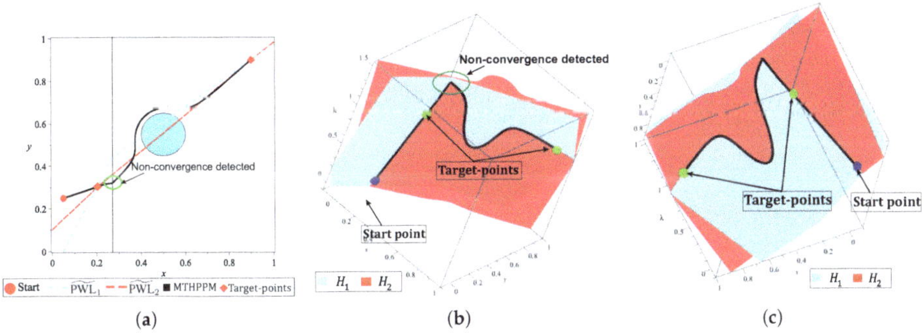

Figure 11. 2-D and 3-D representation of the solution path. (**a**) Solution path 2-D representation. (**b**) Solution path 3-D representation (View 1). (**c**) Solution path 3-D representation (View 2).

Figure 11a shows the successful solution path obtained once that non-convergence is solved. Figure 11b,c show two different 3-D views of the surfaces which correspond at each equation of H_S. In these figures, it can be noticed that the solution path represents the intersection between $H_1(x, y, \lambda)$ and $H_2(x, y, \lambda)$, crossing the two target-points. In this section, the DST technique for solving a non-convergence is explained and its ability to solve the reversion phenomenon is demonstrated.

5.4. A Dummy Obstacle to Improve the Spherical Algorithm Performance

Spherical tracking is an iterative process in which execution time and success depends on three main factors: predictor-corrector scheme, the radius of the sphere, and length of the path. First, the predictor-corrector scheme defines the total execution time, while the radius of the sphere and length of the path determines the number of predictor-corrector executions. For each SA step, the predictor is only employed one time, meanwhile, the number of iterations in Broyden's corrector depends on the predictor point. In other words, the success and fast convergence of the Broyden's corrector scheme depends on the quality of the predictor point. For quality, it means that proximity of the predictor point to the intersection between homotopy curve and the i-th sphere intersection (SA solution point). Then, the time for tracking an entire homotopy curve, from initial-point to target-point, is mostly determined by the total number of corrector scheme executions. In addition, the execution time for each corrector scheme iteration is the same for all SA steps; this implies a linear relationship between the total number of corrector iterations and the total computing time, as explained in Reference [10]. Second, if the radius of the sphere is reduced, the SA needs more steps to cover the same distance that could be traced by bigger spheres. Nonetheless, the sphere radius is restricted by the size of the obstacles, that is, if the sphere is bigger than the obstacles the SA could lose the path [10]. Third, the number of corrector executions increases as the homotopy curve length increases. The increase in length is due to the homotopy system formulation, as explained in Reference [10]. For homotopy with $\widetilde{\mathrm{PWL}}$ formulation, the length of the curve drastically increases caused by turnings when slopes changes. For the implementation in this work, the homotopy curves generated by $\widetilde{\mathrm{PWL}}$ formulation (34) are mapped in the space $x \in (0,1)$, $y \in (0,1)$ and $\lambda \in (-\infty, +\infty)$. The plane $x - y$ is delimited by the normalization of the workspace but λ is unlimited, nevertheless, for homotopy methods, the region of interest is $\lambda \in (0,1)$. Figure 11b,c show that displacement of the homotopy path ever the λ-axis is greater than over the $x - y$ plane. Then, in order to minimize the λ-axis variations, a small perturbation in $H_1(x, y, \lambda)$ is proposed to generate a significant modification in the surface H_1. This perturbation is performed by C_d which is a dummy obstacle placed very close to the initial point (x_0, y_0) with a radius smaller than the spheres employed in the SA. This obstacle generates a change of orientation of surface $H_1(x, y, \lambda)$, and their effect on the homotopy system is negligible for points far away from (x_0, y_0). By using the dummy obstacle, the surface $H_1(x, y, \lambda)$

is forced to change its orientation. This change relates to the nature of Newton's homotopy which implies two properties in the new formulation: (a) its representation of the obstacles in the 3-D space are cylinders and rectangular prisms projections which are parallel to λ-axis. This can be concluded from the projections of Equations (7) and (10) expressions from 2-D to 3-D. Then, the new function that includes the dummy obstacle is represented by

$$f_{\widetilde{\mathrm{PWL}}_1}(x,y) = \widetilde{\mathrm{PWL}}_1(x,y) - \left(\frac{p_{C,d}}{C_d}\right)\left(\xi\left(\widetilde{\mathrm{PWL}}_2(x,y)\right) + \xi\left(\widetilde{\mathrm{PWL}}_1(x,y)\right)\right), \tag{42}$$

where, $p_{C,d}$ is the repulsion parameter of C_d. For this work, $p_{C,d} = 0.1$, this value guarantees a minimal effect of the dummy obstacle in regions close to the starting point. b) The dummy obstacle integration in the homotopy system implies that $\widetilde{\mathrm{PWL}}_2(x,y)$ must be modified to guarantee a continuous homotopy curve from initial-point to target-points. Figure 12 shows the effect of the C_d on the surface $H_1(x,y,\lambda)$. This surface is flattened and their variations over λ-axis is small for points away from the dummy obstacle. The use of a dummy obstacle allowed to reduce, substantially, the sweep on λ-axis for the homotopy system in the region away from this obstacle.

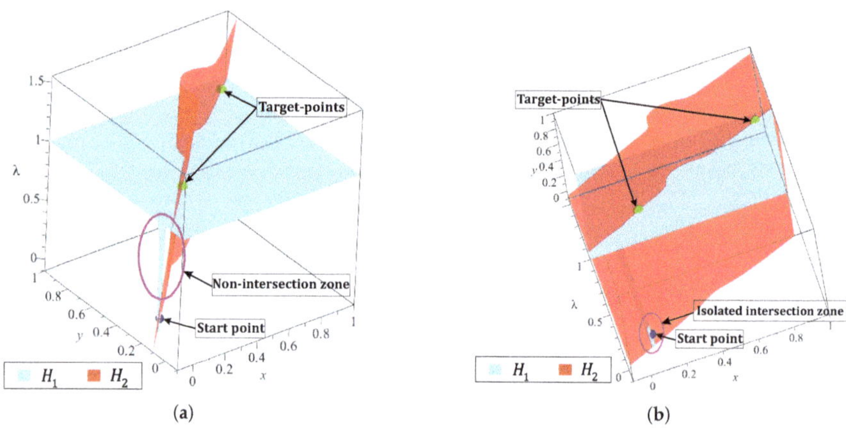

Figure 12. Map containing a dummy obstacle with isolated intersection zone. (**a**) Homotopy surfaces representation (View 1). (**b**)Homotopy surfaces representation (View 2).

The dummy obstacle represents a singularity in the flat surface, then for points closer to it, the variations over λ-axis are noticeable as shown in Figure 12a. It can observe that a non-intersection zone created an inclination on surface $H_2(x,y,\lambda)$, caused by its $\widetilde{\mathrm{PWL}}$ formulation. Figure 12b shows the isolated zone in which the spherical tracking should be confined. From the previous analysis, a solution for the isolated region is proposed and performed by the formulation of C_d

$$C_d = (x - x_{C,d})^2 + (y - y_{C,d})^2 - (r_{C,d})^2, \tag{43}$$

$$x_{C,d} = x_0 - r_s,$$
$$y_{C,d} = y_0 - \mathcal{J}_1\left(x_0 - x_{C,d}\right),$$
$$\mathcal{J}_1 = \frac{y_{G,1} - y_0}{x_{G,1} - x_0},$$

here, the radius of dummy obstacle $r_{C,d} = \frac{r_s}{100}$ is small enough to be disregarded by the spherical tracking; (x_0, y_0) is the start point, and $(x_{G,1}, y_{G,1})$ is the first target-point. From the previous analysis, the initial point $(x_{B,0}, y_{B,0})$ of $\widetilde{\mathrm{PWL}}_2(x,y)$ which guarantees a continuous solution path from start to every target-points is calculated using the next expression

$$(x_{B,0}, y_{B,0}) = (0, y_0 - \mathcal{J}_1\, x_0)\,, \tag{44}$$

Figure 13a depicts the dummy obstacle with center in $(x_{C,d}, y_{C,d})$ and using Equation (44) the point $(x_{B,0}, y_{B,0})$ for the $\widetilde{\mathrm{PWL}}_2(x,y)$ is calculated. The successful solution path for multiple-target problem is shown in Figure 13b. Figure 13c,d show two views of the homotopy surfaces, $H_1(x,y,\lambda)$ and $H_2(x,y,\lambda)$. In these, it can notice the successful path begins at the start point and visits the two projected target-points. Figure 13d shows the effect of the dummy obstacle and the point $(x_{B,0}, y_{B,0})$ of $\widetilde{\mathrm{PWL}}_2(x,y)$ to improve the homotopy path (in terms of length and number of SA stems). Furthermore, in this figure, the continuous curve denoted by the intersection between the homotopy surfaces H_1 and H_2 from the start point through all target-points is showed. The improvement of path (in terms of length) by using the dummy obstacle strategy can be validated from the visual comparison between the paths in the Figures 11 and 13.

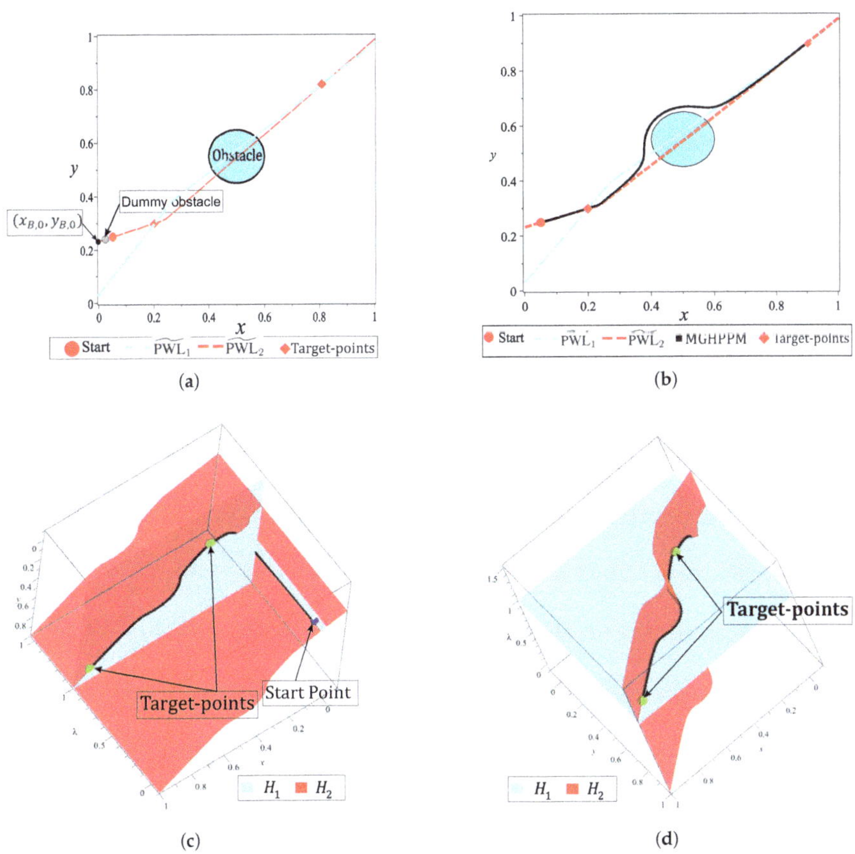

Figure 13. Map with added dummy obstacle. (**a**) Environment map with dummy obstacle and two target-points. (**b**) 2-D representation solution path. (**c**) [3-D representation (View 1) solution path. (**d**) 3-D representation (View 2) solution path.

5.5. Strategy to Simplified the Jacobian Matrix Based on Symbolic Manipulation

To reduce the computational complexity and provide a simple implementation in C++ programming language, without the use of scientific specialized libraries, a symbolic manipulation of the Jacobian matrix (45) is developed. Jacobian matrix is the highest cost processes embedded in the SA; it is employed by predictor ánd corrector schemes for each step of the tracking. Furthermore, some terms also are employed on every step of DST.

$$J(x,y,\lambda)=\begin{pmatrix}\frac{\partial H_1(x,y,\lambda)}{\partial x} & \frac{\partial H_1(x,y,\lambda)}{\partial y} & \frac{\partial H_1(x,y,\lambda)}{\partial \lambda}\\ \frac{\partial H_2(x,y,\lambda)}{\partial x} & \frac{\partial H_2(x,y,\lambda)}{\partial y} & \frac{\partial H_2(x,y,\lambda)}{\partial \lambda}\\ \frac{\partial S_i(x,y,\lambda)}{\partial x} & \frac{\partial S_i(x,y,\lambda)}{\partial y} & \frac{\partial S_i(x,y,\lambda)}{\partial \lambda}\end{pmatrix} \tag{45}$$

The terms corresponding to the sphere S_i and the partial derivatives to λ for H_1 and H_2 in the Jacobian matrix can be calculated by

1. From Equations (34) and (42)

$$\frac{\partial H_1(x,y,\lambda)}{\partial \lambda}=\widetilde{\mathrm{PWL}}_1(x_0,y_0)-\left(\frac{p_{C,d}}{C_d}\right)\left(\xi\left(\widetilde{\mathrm{PWL}}_2(x_0,y_0)\right)+\xi\left(\widetilde{\mathrm{PWL}}_1(x_0,y_0)\right)\right), \tag{46}$$

2. From Equations (34) and (33)

$$\frac{\partial H_2(x,y,\lambda)}{\partial \lambda}=\widetilde{\mathrm{PWL}}_2(x_0,y_0)-g(x_0,y_0)\left(\xi\left(\widetilde{\mathrm{PWL}}_2(x_0,y_0)\right)+\xi\left(\widetilde{\mathrm{PWL}}_1(x_0,y_0)\right)\right), \tag{47}$$

3. From Equation (16)

$$\frac{\partial S_i(x,y,\lambda)}{\partial x}=2(x-c_x), \tag{48}$$

$$\frac{\partial S_i(x,y,\lambda)}{\partial y}=2(y-c_y), \tag{49}$$

$$\frac{\partial S_i(x,y,\lambda)}{\partial \lambda}=2(\lambda-c_\lambda), \tag{50}$$

Using symbolic manipulation and the chain rule to obtain the derivative of composite functions, and knowing that evaluation of $f_{\widetilde{\mathrm{PWL}}_1}(x_0,y_0)$ and $f_{\widetilde{\mathrm{PWL}}_2}(x_0,y_0)$ are constants, the rest of terms can be reduced as

1. From Equations (34) and (42)

$$\begin{aligned}\frac{\partial\left(H_1(x,y,\lambda)\right)}{\partial x} &= \frac{\partial\left(\widetilde{\mathrm{PWL}}_1(x,y)\right)}{\partial x}-\frac{\partial\left(\frac{p_{C,d}}{C_d(x,y)}\right)}{\partial x}\left(\xi\left(\widetilde{\mathrm{PWL}}_2(x,y)\right)+\xi\left(\widetilde{\mathrm{PWL}}_1(x,y)\right)\right)\\ &\quad-\frac{\partial\left(\xi\left(\widetilde{\mathrm{PWL}}_2(x,y)\right)+\xi\left(\widetilde{\mathrm{PWL}}_1(x,y)\right)\right)}{\partial x}\left(\frac{p_{C,d}}{C_d(x,y)}\right),\end{aligned} \tag{51}$$

$$\begin{aligned}\frac{\partial\left(H_1(x,y,\lambda)\right)}{\partial y} &= \frac{\partial\left(\widetilde{\mathrm{PWL}}_1(x,y)\right)}{\partial y}\\ &-\frac{\partial\left(\frac{p_{C,d}}{C_d(x,y)}\right)}{\partial y}\left(\xi\left(\widetilde{\mathrm{PWL}}_2(x,y)\right)+\xi\left(\widetilde{\mathrm{PWL}}_1(x,y)\right)\right)\\ &-\frac{\partial\left(\xi\left(\widetilde{\mathrm{PWL}}_2(x,y)\right)+\xi\left(\widetilde{\mathrm{PWL}}_1(x,y)\right)\right)}{\partial y}\left(\frac{p_{C,d}}{C_d(x,y)}\right), \end{aligned} \tag{52}$$

2. From Equations (34) and (33)

$$\begin{aligned}\frac{\partial\left(H_2(x,y,\lambda)\right)}{\partial x} &= \frac{\partial\left(\widetilde{\mathrm{PWL}}_2(x,y)\right)}{\partial x}-\frac{\partial\left(g(x,y)\right)}{\partial x}\left(\xi\left(\widetilde{\mathrm{PWL}}_2(x,y)\right)+\xi\left(\widetilde{\mathrm{PWL}}_1(x,y)\right)\right)\\ &\frac{\partial\left(\xi\left(\widetilde{\mathrm{PWL}}_2(x,y)\right)+\xi\left(\widetilde{\mathrm{PWL}}_1(x,y)\right)\right)}{\partial x}\left(g(x,y)\right), \end{aligned} \tag{53}$$

$$\begin{aligned}\frac{\partial\left(H_2(x,y,\lambda)\right)}{\partial y} &= \frac{\partial\left(\widetilde{\mathrm{PWL}}_2(x,y)\right)}{\partial y}-\frac{\partial\left(g(x,y)\right)}{\partial y}\left(\xi\left(\widetilde{\mathrm{PWL}}_2(x,y)\right)+\xi\left(\widetilde{\mathrm{PWL}}_1(x,y)\right)\right)\\ &-\frac{\partial\left(\xi\left(\widetilde{\mathrm{PWL}}_2(x,y)\right)+\xi\left(\widetilde{\mathrm{PWL}}_1(x,y)\right)\right)}{\partial y}\left(g(x,y)\right), \end{aligned} \tag{54}$$

where $\frac{\partial(g(x,y))}{\partial x}$ and $\frac{\partial(g(x,y))}{\partial y}$ are

$$\frac{\partial\left(g(x,y)\right)}{\partial x}=\sum_{i=1}^{i=k}\frac{-p_{O,i}\left(\frac{\partial(O_i(x,y))}{\partial x}\right)}{\left(|O_i(x,y)|+O_i(x,y)\right)^2}\left(1+\frac{|O_i(x,y)|}{O_i(x,y)}\right), \tag{55}$$

$$\frac{\partial\left(g(x,y)\right)}{\partial y}=\sum_{i=1}^{i=k}\frac{-p_{O,i}\left(\frac{\partial(O_i(x,y))}{\partial y}\right)}{\left(|O_i(x,y)|+O_i(x,y)\right)^2}\left(1+\frac{|O_i(x,y)|}{O_i(x,y)}\right), \tag{56}$$

where $O_i(x,y)$ is the expression that describes the shape of the i-th obstacle; C_i for a circular obstacle, and R_i for an ellipsoidal obstacle. The parameter $p_{O,i}$ is the repulsion parameter of each obstacle; $p_{C,i}$ for a circular obstacle, and $p_{R,i}$ for an ellipsoidal obstacle.

$$\begin{aligned}\frac{\partial\left(\xi\left(\widetilde{\mathrm{PWL}}_2(x,y)\right)+\xi\left(\widetilde{\mathrm{PWL}}_1(x,y)\right)\right)}{\partial x} &= \frac{1}{\alpha}\frac{\partial\left(\widetilde{\mathrm{PWL}}_2(x,y)\right)}{\partial x}\left(-\frac{\alpha_{\mathrm{e}}{}^{-\alpha\widetilde{\mathrm{PWL}}_2(x,y)}}{1+{}_{\mathrm{e}}{}^{-\alpha\widetilde{\mathrm{PWL}}_2(x,y)}}+\frac{\alpha_{\mathrm{e}}{}^{\alpha\widetilde{\mathrm{PWL}}_2(x,y)}}{1+{}_{\mathrm{e}}{}^{\alpha\widetilde{\mathrm{PWL}}_2(x,y)}}\right)\\ &+\frac{1}{\alpha}\frac{\partial\left(\widetilde{\mathrm{PWL}}_1(x,y)\right)}{\partial x}\left(-\frac{\alpha_{\mathrm{e}}{}^{-\alpha\widetilde{\mathrm{PWL}}_1(x,y)}}{1+{}_{\mathrm{e}}{}^{-\alpha\widetilde{\mathrm{PWL}}_1(x,y)}}+\frac{\alpha_{\mathrm{e}}{}^{\alpha\widetilde{\mathrm{PWL}}_1(x,y)}}{1+{}_{\mathrm{e}}{}^{\alpha\widetilde{\mathrm{PWL}}_1(x,y)}}\right), \end{aligned} \tag{57}$$

$$\begin{aligned}\frac{\partial\left(\xi\left(\widetilde{\mathrm{PWL}}_2(x,y)\right)+\xi\left(\widetilde{\mathrm{PWL}}_1(x,y)\right)\right)}{\partial y} &= \frac{1}{\alpha}\frac{\partial\left(\widetilde{\mathrm{PWL}}_2(x,y)\right)}{\partial y}\left(-\frac{\alpha_{\mathrm{e}}{}^{-\alpha\widetilde{\mathrm{PWL}}_2(x,y)}}{1+{}_{\mathrm{e}}{}^{-\alpha\widetilde{\mathrm{PWL}}_2(x,y)}}+\frac{\alpha_{\mathrm{e}}{}^{\alpha\widetilde{\mathrm{PWL}}_2(x,y)}}{1+{}_{\mathrm{e}}{}^{\alpha\widetilde{\mathrm{PWL}}_2(x,y)}}\right)\\ &+\frac{1}{\alpha}\frac{\partial\left(\widetilde{\mathrm{PWL}}_1(x,y)\right)}{\partial y}\left(-\frac{\alpha_{\mathrm{e}}{}^{-\alpha\widetilde{\mathrm{PWL}}_1(x,y)}}{1+{}_{\mathrm{e}}{}^{-\alpha\widetilde{\mathrm{PWL}}_1(x,y)}}+\frac{\alpha_{\mathrm{e}}{}^{\alpha\widetilde{\mathrm{PWL}}_1(x,y)}}{1+{}_{\mathrm{e}}{}^{\alpha\widetilde{\mathrm{PWL}}_1(x,y)}}\right), \end{aligned} \tag{58}$$

where, $\alpha = 1 \times 10^3$. Finally,

$$\frac{\partial\left(\widetilde{\text{PWL}}_1(x,y)\right)}{\partial x} = -\left(b_1 x + \sum_{i_1=1}^{i_1=\sigma} c_{i_1}\left(\frac{\xi\left(x - x_{B,i_1}\right)}{x - x_{B,i_1}}\right)\right), \tag{59}$$

$$\frac{\partial\left(\widetilde{\text{PWL}}_2(x,y)\right)}{\partial x} = -\left(b_2 x + \sum_{i_2=1}^{i_2=\sigma} c_{i_2}\left(\frac{\xi\left(x - x_{B,i_2}\right)}{x - x_{B,i_2}}\right)\right), \tag{60}$$

$$\frac{\partial\left(\widetilde{\text{PWL}}_1(x,y)\right)}{\partial y} = 1, \tag{61}$$

$$\frac{\partial\left(\widetilde{\text{PWL}}_2(x,y)\right)}{\partial y} = 1, \tag{62}$$

where (x_{B,i_j}, y_{B,i_j}) is the i-th breakpoint of the j-th $\widetilde{\text{PWL}}$ functions $\widetilde{\text{PWL}}_1$ and $\widetilde{\text{PWL}}_2$, respectively. Every expression presented in this section has been implemented in our C++ program. The symbolic manipulation reduces the computing time because it only requires substitutions and evaluations.

5.6. A Systematic Criterion to Select the Repulsion Parameter

The selection of repulsion parameters for obstacles has been an open problem since the HPPM was proposed in Reference [19]. This is a complex task because the effect of each parameter has a great impact on the homotopy surfaces and, hence, in the solution curve. Advances in the characterization and creation of a criterion have been reported in previous works, where: A study of the effect produced by the sign and magnitude of each parameter over the homotopy curve was introduced in Reference [10], nevertheless, the criterion employed to set them was only based on empirical knowledge. Then, Reference [20] presents a criterion based on random selection and segmentation of the map to enhance the performance of the planner and find the shortest path. Also, a grouping obstacle strategy to reduce the number of repulsion parameters in the formulation was presented in Reference [20]. This strategy is applied to reduce the complexity of parameter selection tasks for maps with a large number of obstacles. Next, a first approach of systematic criterion based on the distance between obstacles and the diagonal path to properly select the repulsion parameter was presented in Reference [37]. Showed the effectiveness of a systematic criterion using examples, nonetheless, its performance takes a higher portion of the total execution time. This section is devoted to explaining a new systematic criterion based on results reported in Reference [10,20,37] and the properties of auxiliary $\widetilde{\text{PWL}}$ functions proposed in this work. The main property of the formulation that has an effect over the repulsion parameter is denoted in the auxiliary function (33) which contains all mathematical representations of obstacles in the map. This function produces a scaling effect in the value of every repulsion parameter due to its absolute value terms. These reduce the effect of repulsion parameters in the vicinity of target-points and proportionally changing its value with the proximity between obstacles and any $\widetilde{\text{PWL}}$ function. From the manipulation of Equation (33)

$$f_{\widetilde{\text{PWL}}_2}(x,y) = \widetilde{\text{PWL}}_2(x,y) - \left(\sum_{i=1}^{i=n} \frac{p_{O,i}\left(\xi\left(\widetilde{\text{PWL}}_2(x,y)\right) + \xi\left(\widetilde{\text{PWL}}_1(x,y)\right)\right)}{O_i(x,y) + |O_i(x,y)|}\right), \tag{63}$$

here, $i = 1, 2, 3, \ldots n$, n is the number of obstacles in the map; $O_i(x, y)$ is the mathematical representation of any i-th circular obstacle, and $p_{O,i}$ represents its respective repulsion parameter. Then, the effective value of the parameter is denoted as

$$p_{\text{eff},i} = p_{O,i}\left(\xi\left(\widetilde{\text{PWL}}_2(x,y)\right) + \xi\left(\widetilde{\text{PWL}}_1(x,y)\right)\right), \tag{64}$$

where $p_{\text{eff},i}$ is the effective value of repulsion parameter in Equation (33) for each i-th obstacle. It can be noticed that the value of $p_{\text{eff},i}$ dynamically varies according to SA tracking the solution path, then, only one value of p_{base} is selected for all circular obstacles. From the characterization presented in Reference [10], the next rank of magnitude for this parameter is employed in the case studies; $1 \geq |p_{\text{base}}| > 0$ for circular obstacles and $100 \geq |p_{\text{base}}| > 1$ for ellipsoidal obstacles. The expression (64) denotes that parameter magnitude of any obstacle decreases according to its proximity to any $\widetilde{\text{PWL}}$ function.

In this subsection, an environment map with three circular obstacles is used to explain the proposed systematic criterion for selecting the sign of repulsion parameters. As already explained above, solution path tends to follow the function $\widetilde{\text{PWL}}_2(x,y)$ as a consequence of using a dummy obstacle. In this sense, the criterion to set the sign of the repulsion parameter for any circular obstacle can be established from its center position with respect to $\widetilde{\text{PWL}}_2(x,y)$ (see Figure 14b). This criterion is based on that, for all points (a, b) located above $\widetilde{\text{PWL}}_2(x,y)$, the value of $\widetilde{\text{PWL}}_2(a,b)$ is greater than zero ; for points located below $\widetilde{\text{PWL}}_2(x,y)$, the value is less than zero (see Figure 14c). By using the dynamic value of repulsion parameter property provided by the formulation and relative position of the center $((x_{C,i}, y_{C,i}))$ for every i-th circular obstacle with respect to $\widetilde{\text{PWL}}_2(x,y)$, a new definition of the repulsion parameter is described by

$$p_{O,i} = \begin{cases} -p_{\text{base}}\text{sgn}\left(\widetilde{\text{PWL}}_2(x_{C,i}, y_{C,i})\right) & \text{if} \quad \widetilde{\text{PWL}}_2(x_{C,i}, y_{C,i}) \neq 0, \\ -p_{\text{base}} & \text{if} \quad \widetilde{\text{PWL}}_2(x_{C,i}, y_{C,i}) = 0, \end{cases} \tag{65}$$

where $p_{O,i}$ is the base parameter value with a sign that determines a lower or upper side path according to the i-th circular obstacle. Figure 14d shows the resulting solution path once the systematic criterion to set the repulsion parameter is executed. In this case study, the strategy is capable to properly set the signs of repulsion parameters which generate an efficient path with minimum length. This example validates the criterion effectiveness, which is addressed in the following sections.

The criterion introduced in this subsection has been formulated and characterized only for circular obstacles because of its symmetry. Nonetheless, it is assumed that ellipsoidal obstacles are only employed to represent walls or any other obstacle of big dimensions on the map. Then, the criterion to automatically assign the sign of repulsion parameter needs an additional step which will be explained in the next section.

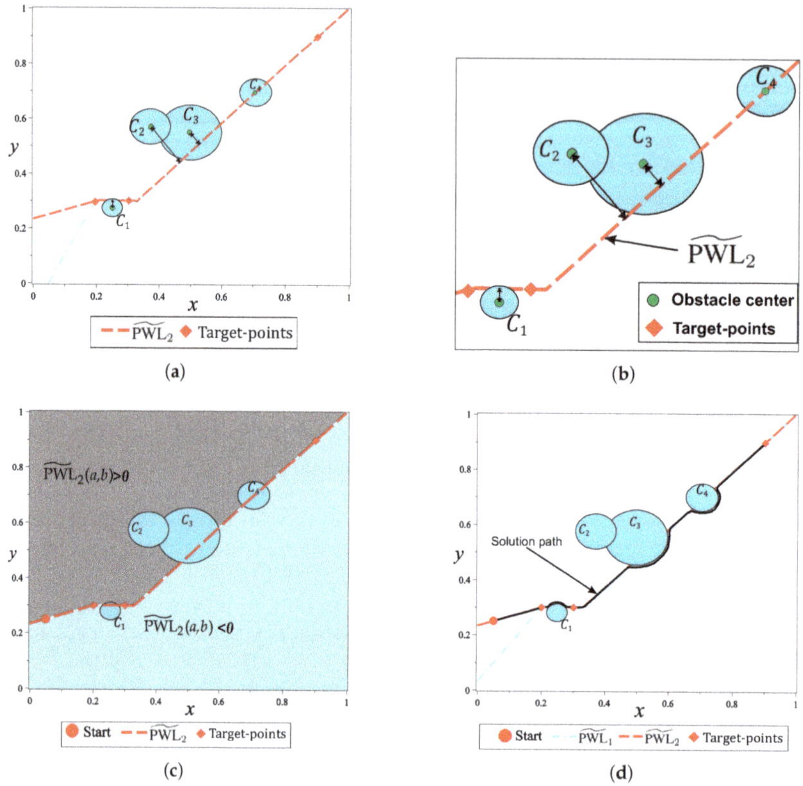

Figure 14. Criterion to select the repulsion parameter. (**a**) Environment map. (**b**) Position of obstacles with respect to $\widetilde{\mathrm{PWL}}_2(x, y, \lambda)$. (**c**) Positive and negative regions with respect to $\widetilde{\mathrm{PWL}}_2(x, y, \lambda)$. (**d**) Solution path using systematic criterion to select repulsion parameter.

6. Multiple-Target Homotopic Path Planning Method with Visibility Graph Approach

Visibility graph (VG) is one of the most widely used roadmap methods to find the shortest path. This method employs a geometrical map representation to generate a roadmap that contains all links between vertices of the polygonal obstacles, start-point, and target-point [12–14]. Although this algorithm can find the shortest path for maps of structured environments, it is only employed to solve configurations with few polygonal obstacles, because its execution time and memory consumption depends on the number of obstacles [12]. In this work, the visibility graph algorithm is applied to pre-process the map and obtain a first approximation of the solution path for closed and office-like environments. In this way, only big polygonal objects such as walls and office furniture (static objects) are considered. Then, MTHPPM uses the path provided by VG (commonly the shortest path) to compute a collision-free path for the full environment containing non-polygonal obstacles. Figure 15 shows the operation process of MTHPPM using the visibility approach applying (MTHPPM_VG) for an office-like environment with five hundred obstacles.

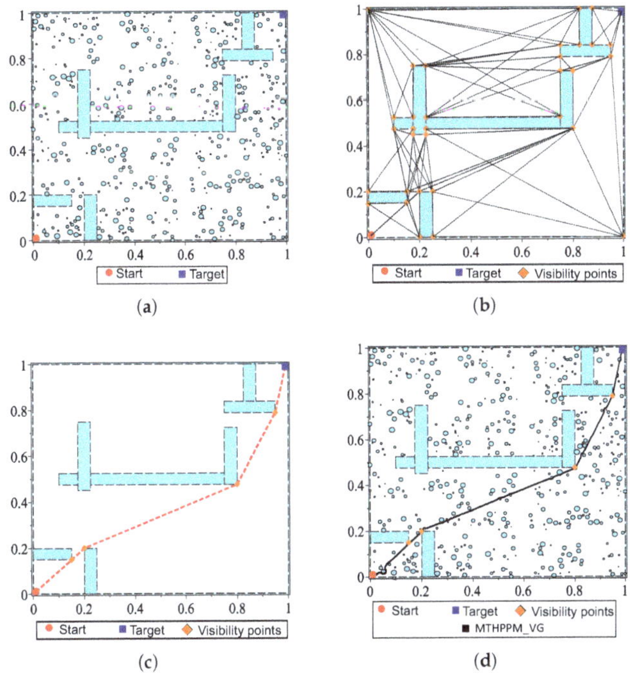

Figure 15. Visibility approach applied to Homotopic Path Planning Method. (**a**) Environment map with walls. (**b**) Visibility points and pre-processed path. (**c**) Visibility graph path without circular obstacles. (**d**) Solution path obtained by MTHPPM using the visibility approach.

For the example in Figure 15a, the main task of the mobile robot is to reach the target-point located on the opposite corner of the room, concerning to the initial position. Solving this example is done through the following steps; first, an approximation of the solution path is obtained by the visibility graph algorithm using a map, where only walls are considered. The configuration map without obstacles is reduced to a simple problem (see Figure 15b). The graph complexity is further reduced using the premise that $\widetilde{\mathrm{PWL}}$ formulation is a single-valued function. In this regard, every path with links that implies returns in x-axis is suppressed. Second, the first approximation of the path will be integrated by only links (corners of static obstacles and walls) in x-axis forward direction, as it is shown in Figure 15c. It can be noticed that, for environment maps with bug traps and maze layout, a path with visible vertices in forward direction might not exist. Two case studies with this problem are treated and solved in the next section. Third, the solution path provided by VG is employed to generate $\widetilde{\mathrm{PWL}}$ formulation which contains the information about the position of the ellipsoidal obstacles. Then, automatic sign assignation for this type of obstacle is executed, operating in the same way as circular obstacles. The function to obtain the sign of an ellipsoidal obstacle is

$$p_{O,i} = \begin{cases} -p_{\text{base}}\text{sgn}\left(\widetilde{\mathrm{PWL}}_2(x_{R,i}, y_{R,i})\right) & \text{if} \quad \widetilde{\mathrm{PWL}}_2(x_{R,i}, y_{R,i}) \neq 0, \\ -p_{\text{base}} & \text{if} \quad \widetilde{\mathrm{PWL}}_2(x_{R,i}, y_{R,i}) = 0, \end{cases} \tag{66}$$

where $p_{O,i}$ is the base parameter value with a sign that determine if the solution path pass below or above the i-th ellipsoidal obstacle with center at point $(x_{R,i}, y_{R,i})$ and p_{base} is the base value selected, arbitrarily, for all ellipsoidal obstacles; from empirical data presented in Reference [10], the value is in range $100 \geq |p_{\text{base}}| > 1$. Finally, the MTHPPM is executed for the configuration map containing obstacles, walls, and vertices of the visibility path as target-points; solution is the shortest collision-free path (see Figure 15d). The approach explained in this section can be used for problems with only one target-point or multiple-targets. For multiple-targets cases, the visibility points should be integrated in the set of targets as long as these do not represent a backward advance in the $\widetilde{\text{PWL}}$ formulation, in which case these points must be removed.

The procedure of the proposed multiple-target planner can be summarised using the flow chart presented in Figure 16. If the planning problem has only a single target the method is named HPPM and for multiple targets MGHPPM. Furthermore, when the planning problem is performed on a structured workspace with a single target point, the procedure is the same and the visibility graph approach is used to calculate the first approximation of the solution, and the homotopy based method is named MGHPPM_VG.

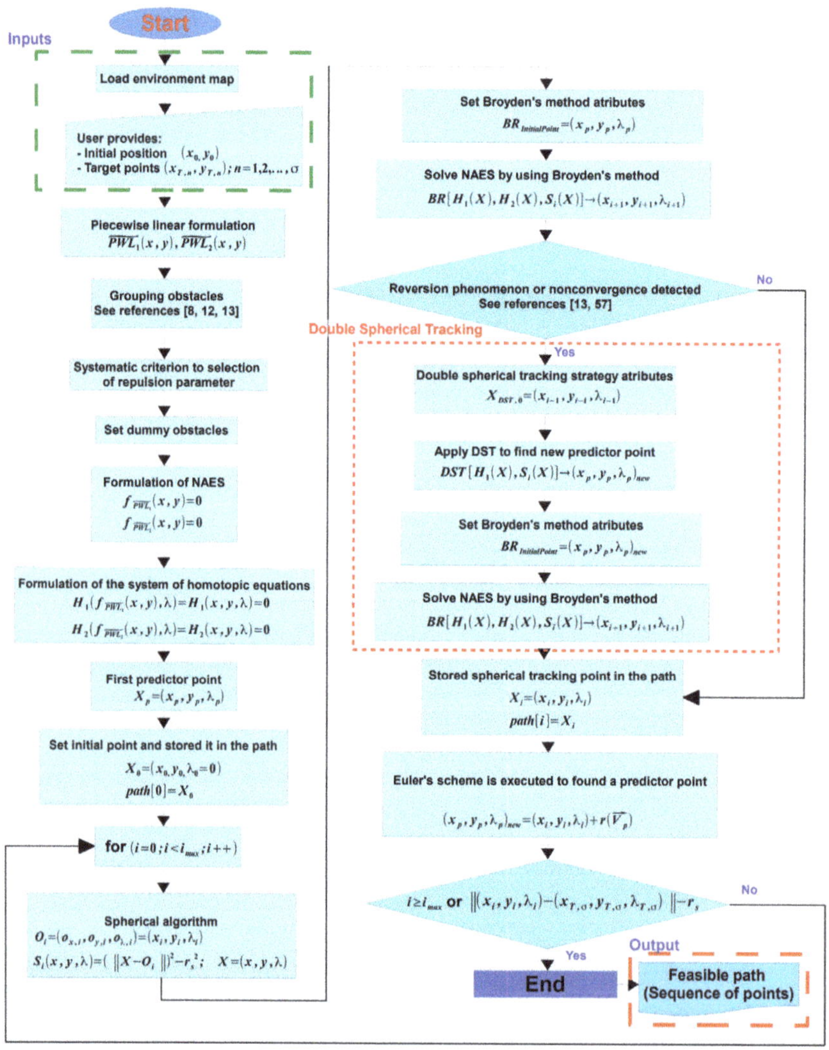

Figure 16. Multiple-target homotopic path planning method flow chart.

The flow chart (see Figure 16) contains all strategies explained previously and the addition of points generated by the visibility graph approach into the set of target-points. The proposed methodology considers that the environment map is known in advance or it is provided by sensors onboard (for real-time implementations) in the load environment map stage. In this sense, inputs in the proposed planner are a semi-algebraic representation of the environment map ($\mathcal{C}_{space}$), the initial position of the robot, and one or more target-points designed by the user or generated by a visibility-graph algorithm (because visibility graph planners have been treated strongly in others works such as References [12–14], in this work it is not integrated in the homotopic planner process, is only used as sub-process). Here, it is considered that the robot has a navigation module and control rules to guarantee the correct path following. Then, the iterative procedure of MTHPPM (for multiple-target points) or HPPM (for single target-point) starts. First, the mathematical model of the environment and data provided by the user is generated in the $\widetilde{\text{PWL}}$ formulation stage. Second, the techniques to enhance convergence, length of path, and automatic assignation of repulsion parameters are performed (grouping obstacles, systematic criterion to select repulsion parameter, and dummy obstacle addition). Third, homotopy system of non-linear equations to represent the configuration space is formulated in NAES formulation stage. Then, the first predictor point for the HPPM (or MGHPPM) is calculated from the initial position of the robot (first point of the path). Fourth, the iterative procedure of spherical algorithm is executed; within it, the predictor-corrector scheme and reversion phenomenon (or non-convergence) check are performed for each step. This procedure is executed as explained in Section 3, where the Broyden's method is employed to calculate a new point of path which represents a solution of the system of homotopic equations and Euler's scheme is executed to obtain a new predictor point. During the Spherical algorithm procedure, if a reversion or non-convergence is detected, the DST technique is enabled and employed to solve this issue as already explained in the Section 5.3. Finally, the procedure finishes when the SA reaches a target-point or the maximum number of steps is achieved; then, returns a set of points that represent the solution path. In the next section, some case studies are presented to validate the utility of this methodology.

7. Case Studies

This section provides five case studies to provide certainty about the usefulness of the proposed methodology and how it solves path planning problems like bug traps and narrow corridors. The first three simulations are focused on how to solve planning problems in sceneries with narrow corridors. Simulation four demonstrates its effectiveness to solve maps with bug traps. On the one hand, for these four case studies, a comparison between the proposed methodology in its two variants (HPPM formulation and MTHPPM with visibility graph approach) and six sampling-based planning algorithms is provided. This comparison is based on important aspects like: memory consumption, execution time, percentage of feasible paths found, and path length. On the other hand, an example of multiple-target path planning for a closed environment with narrow corridors is shown in simulation five. All simulations were performed using the following set-up: all planners have the same step size (spherical tracking radius for MTHPPM and HPPM; collision checking resolution parameter for BSP algorithms), MTHPPM code in C++, and executed on PC (Intel i7 2.6 GHz processor, RAM 16 GB, and 64-bit Ubuntu 16.04 operating system). SBP algorithms ran on the same PC using the well known Open Motion Planning Library v.1.3.1 [57] and OMPL Planer Arena [58] to obtain performance data. To highlight advantages and weaknesses of the proposed homotopy based planner, it is compared to eight SBP algorithms: Expansive Space Trees (EST), Kinematic Planning by Interior-Exterior Cell Exploration (KPIECE1), Probabilistic Roadmap Method (PRM), Rapidly-exploring Random Trees (RRT), Bidirectional Rapidly-exploring Random Trees (Bi-RRT, also named RRT-Connect), Rapidly-exploring Random Trees with A* approach (RRT*), RRT and PRM with Lazy collision method (LazyRRT), and (Lazy PRM). To obtain a significant performance data, each SBP algorithm was run one hundred times.

7.1. Case 1

This case study is devoted to show the ability of MTHPPM to deal with the narrow corridor problem. Figure 17a depicts the visual comparison between HPPM and MTHPPM_VG (MTHPPM with visibility graph approach) paths on a normalized office-like map in 2-D with five hundred circular obstacles. Here, the initial position of the robot is at point $(0, 0)$ and target-point at $(1, 1)$. The visual comparison between paths obtained by SBP algorithms is presented in Figure 17b.

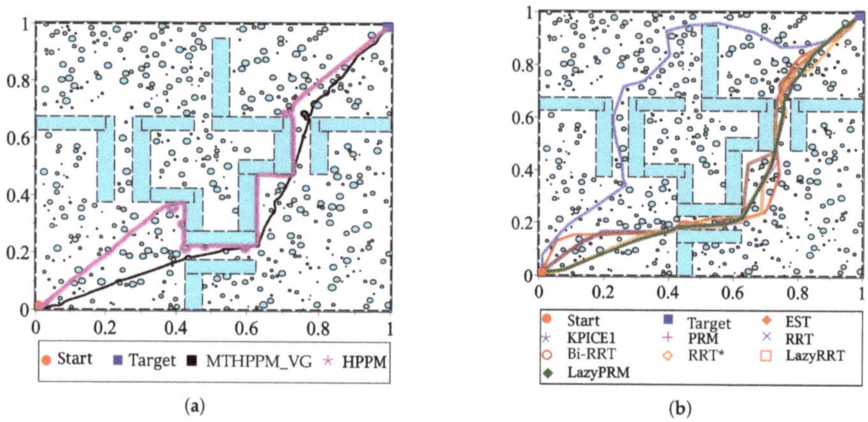

Figure 17. Visual comparative between MTHPPM and HPPM against best run of each SBP algorithm for case study 1. (**a**) Paths obtained with HPPM and MTHPPM_VG. (**b**) Paths obtained with SBP algorithms.

Figure 17a,b, the calculated paths by HPPM against MTHPPM_VG and SBP algorithms (shortest path after 100 runs) are contrasted. It is important to note that the path obtained using HPPM is one of the longest because it tends is to round all obstacles in direction to the diagonal line between initial and target-point. On the other hand, the path obtained by MTHPPM with visibility approach is one of the shortest. Figure 17a shows the effectiveness of HPPM and MTHPPM (MTHPPM_VG) to solve maps with narrow corridors and several obstacles. Furthermore, the ability of MTHPPM to calculate the shortest collision-free path using the visibility graph approach explained in Section 5 is validated in this case study. The box plots (box-and-whisker diagrams) in Figure 18 depicts a quantitative comparison between HPPM, MTHPPM_VG, and SBP algorithms for percentage of feasible paths found, execution time, path length, and memory consumption.

Figure 18 shows the summarized results which denote the following characteristics: First, Figure 18a shows that percentage of fails for PRM is higher than twenty percent, while for LazyRRT is closer to that percentage. For HPPM and MTHPPM_VG, the success rate is one hundred percent due to its deterministic formulation. Second, the execution time spent to solve the configuration map is presented in Figure 18c. In this box-and-whisker diagram, it can notice, that SBP algorithms are in the order of seconds, while MTHPPM_VG and HPPM their time is in the order of milliseconds. Third, the results in Figure 18b show that the path obtained by MTHPPM_VG is one of the shortest, RRT* provided the best. Finally, memory consumption comparison exhibits a big gap between HPPM and MTHPPM against SBP algorithms. The difference is about three orders of magnitude, from KB to MB. This is because SBP algorithms store a roadmap of collision-free points, while homotopy based methods only store the path and obstacle positions.

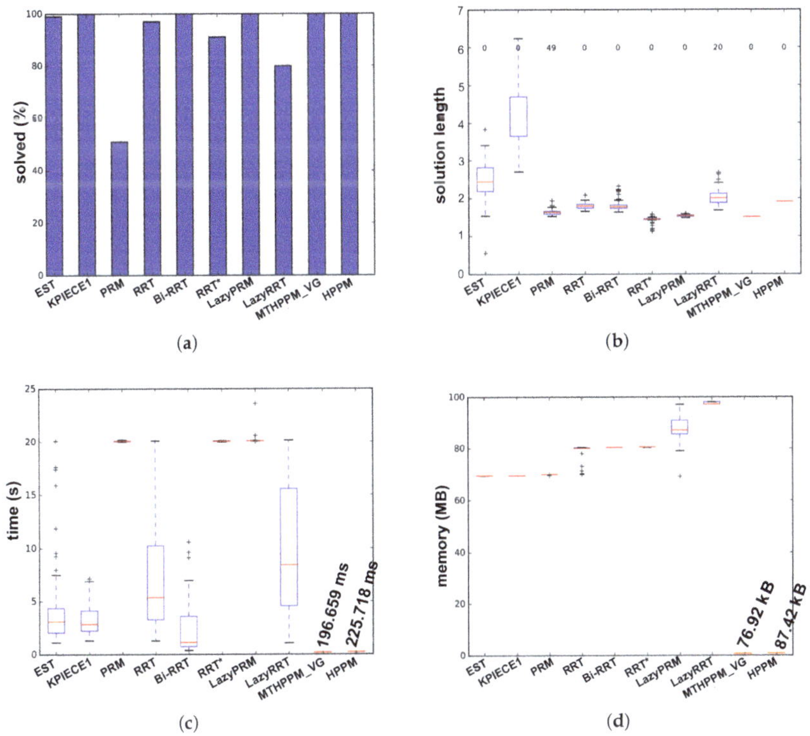

Figure 18. Comparative results for case study 1. (**a**) Successful paths. (**b**) Path length. (**c**) Execution time. (**d**) Memory.

7.2. Case 2

This case study presents a path planning problem of an office-like environment with narrow corridors and five hundred obstacles. This example shows one of the future applications of HPPM and MTHPPM, to integrate it in the navigation system of a parcel service robot. The main task of the robot, in this example, is to collect and deliver packages or documents from one cubicle to another. Figure 19a provides the floor plan of an office represented on a normalized 2-D space, where, the initial and target points are located in opposites corners.

Figure 19a,b show a visual comparison of paths generated by HPPM against MTHPPM_VG and between SBP algorithms, respectively. In these patterns, the difference between path lengths calculated by SBP algorithms, HPPM, and MTHPPM_VG are depicted. It is important to note that, for SBP algorithms, the most optimistic simulation is taken, that is, the shortest path after one hundred runs. Figure 20 shows the box-and-whisker diagrams of the performance results for SBP algorithms and homotopy based planners (HPPM and MTHPPM).

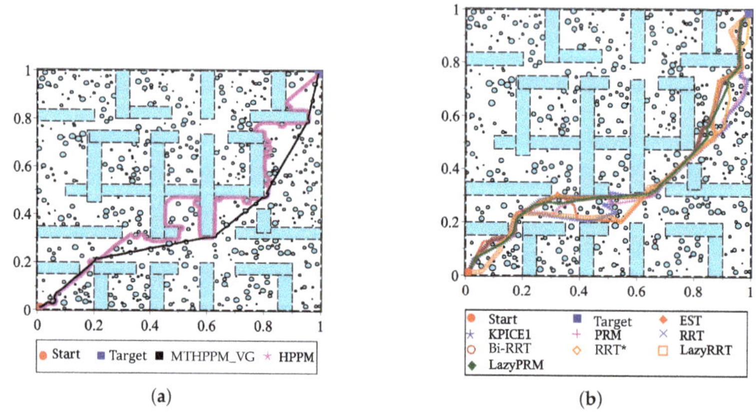

Figure 19. Visual comparative for case study 2 with five hundred obstacles; best run of each SBP algorithm. (**a**) Paths obtained with HPPM and MTHPPM_VG. (**b**) Paths obtained using SBP algorithms.

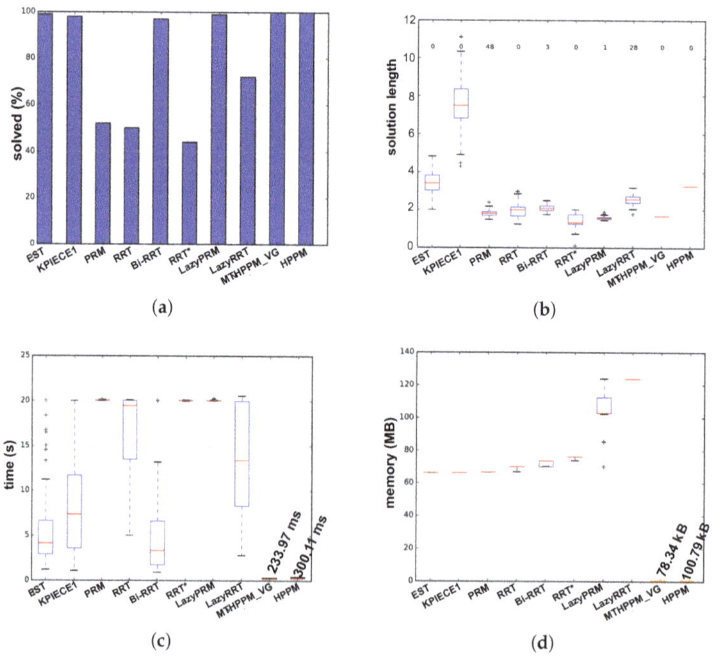

Figure 20. Comparative results for case study 2. (**a**) Successful paths. (**b**) Path length. (**c**) Execution time. (**d**) Memory.

From simulations, it can be concluded the following: First, RRT, PRM, RRT*, and LazyRRT have a poor (less than eighty percent) performance of success percentage metric against the homotopy based planners. Second, like the results of the previous case study, HPPMs spent less time, significantly, than sampling-based planners (see Figure 20c). Third, meanwhile HPPM found one of the longest paths, MTHPPM_VG obtained one of the three shortest; and the best compared to LazyPRM and

RRT*. Finally, memory used by homotopy based planners is about a thousand times smaller than the sampling algorithms, this is due to the formulation of each approach.

7.3. Case 3

This case study presents an example in which some visibility points are selected to create a forward direction sequence of target-points. This is adequate to be integrated in a single-value piecewise linear formulation employed in MTHPPM, as already explained in Section 5. Figure 21a depicts the environment map for this case, contains walls, and two hundred circular obstacles. The operation of the proposed MTHPPM_VG methodology for this configuration is similar to previous examples, furthermore, the visibility-points selection process is added. First, the approximation of the solution path is computed using the VG algorithm, then, an automatic criterion to discard nodes of the visibility path which implies backward advance in x-axis. Figure 21b shows nodes and paths after the visibility-point selection process is executed; the remaining nodes are employed as target-points for MTHPPM (dashed path). Figure 21c,d displays the resulting collision-free paths for HPPM against MTHPPM and between SBP algorithms (shortest path after one hundred runs for each SBP), respectively.

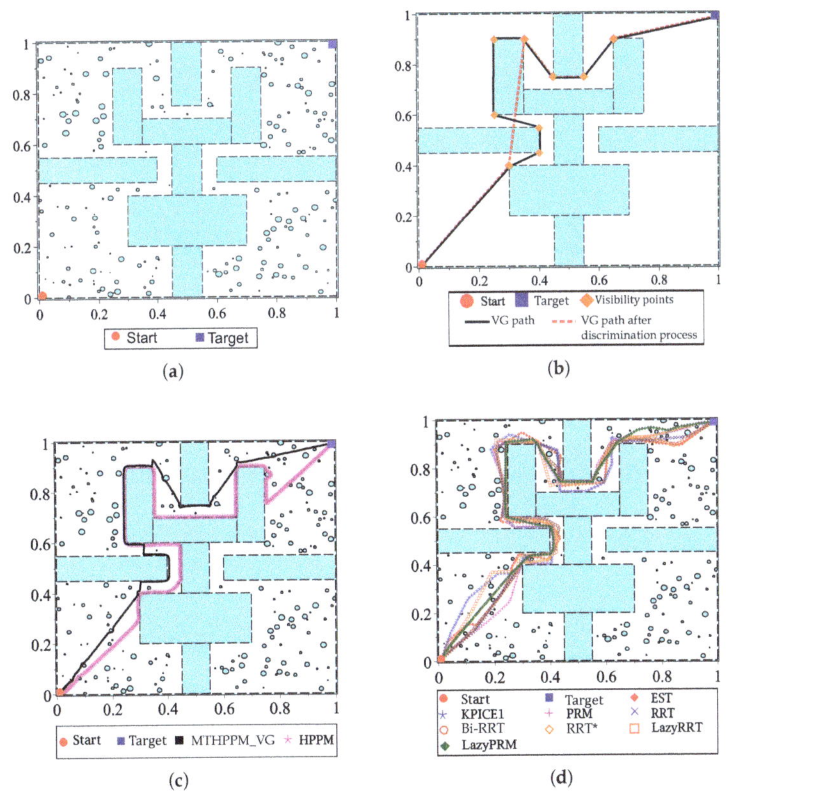

Figure 21. Comparative results for case study 3; best run of each SBP algorithm. (**a**) Path planning problem case study 3. (**b**) Visibility graph path without circular obstacles. (**c**) Paths obtained with HPPM and MTHPPM_VG. (**d**) Paths obtained with SBP algorithms.

Figure 21 shows that the path obtained by MTHPPM_VG is drastically shorter than the HPPM path. Figure 22 show the summarized results of HPPM, MTHPPM, and SBP algorithms for the most significant metrics.

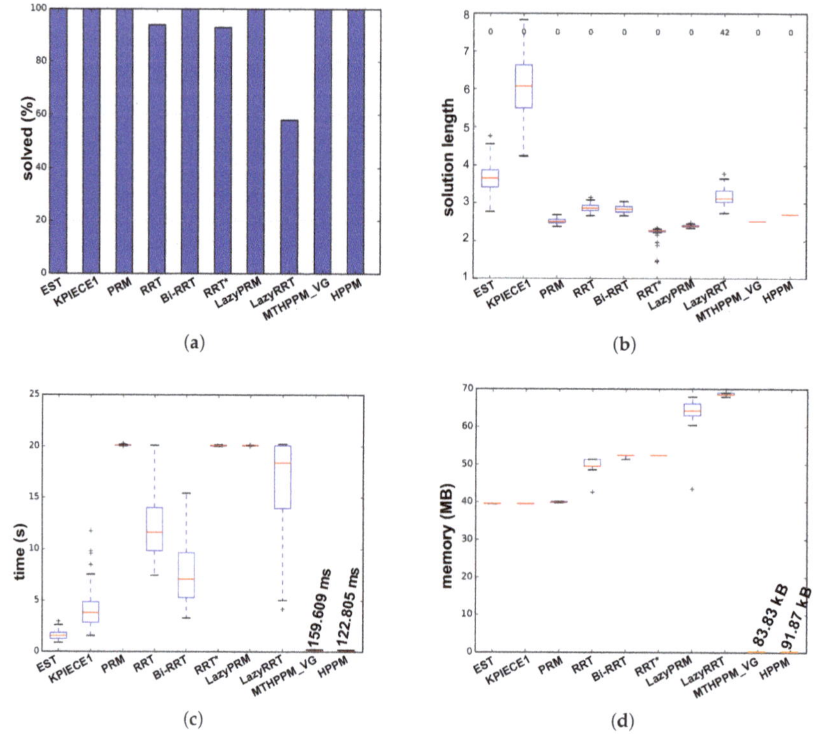

Figure 22. Comparative results for case study 3. (**a**) Successful paths. (**b**) Path length. (**c**) Execution time. (**d**) Memory.

By analyzing the simulations, it is important to note: First, lazyRRT has the worst performance of all SBP algorithms concerning to percentage of feasible paths found; EST, KPIECE1, PRM, Bi-RRT, LazyPRM have one hundred percent of success. Second, using results from previous case studies, homotopy based planners spend, notoriously, the shortest time compared to sampling-based planners (see Figure 22c). Third, MTHPPM_VG obtains a path very close to the shortest (RRT* calculated the best). Finally, consumed memory by homotopy based planners are around one thousand times lower than SBP algorithms.

7.4. Case 4

This case study focuses on the bug trap problem, additionally, it deals with visibility nodes issue introduced in the previous case study. Figure 23a shows the configuration map with three hundred circular obstacles, walls, and narrow corridors. The initial position of the robot is inside of the bug trap, then, the planners need to plan a path that allows them to surround it, avoid the trap, and pass through two narrow corridors to reach target-point placed at the corner on the map. Similar to the previous example, for MGGHPPM, the VG algorithm found a first approximation of the path considering only the walls. Afterward, the discrimination process is executed to delete nodes of the VG path, which implies tracking in the backward direction as drawn in Figure 23b (dashed path). Finally, the full map is solved by MTHPPM using the remaining nodes of VG path as target-points. The resulting collision-free paths for HPPM against MTHPPM and between SBP algorithms (shortest path after one hundred runs for each SBP) are presented in Figure 23c,d, respectively.

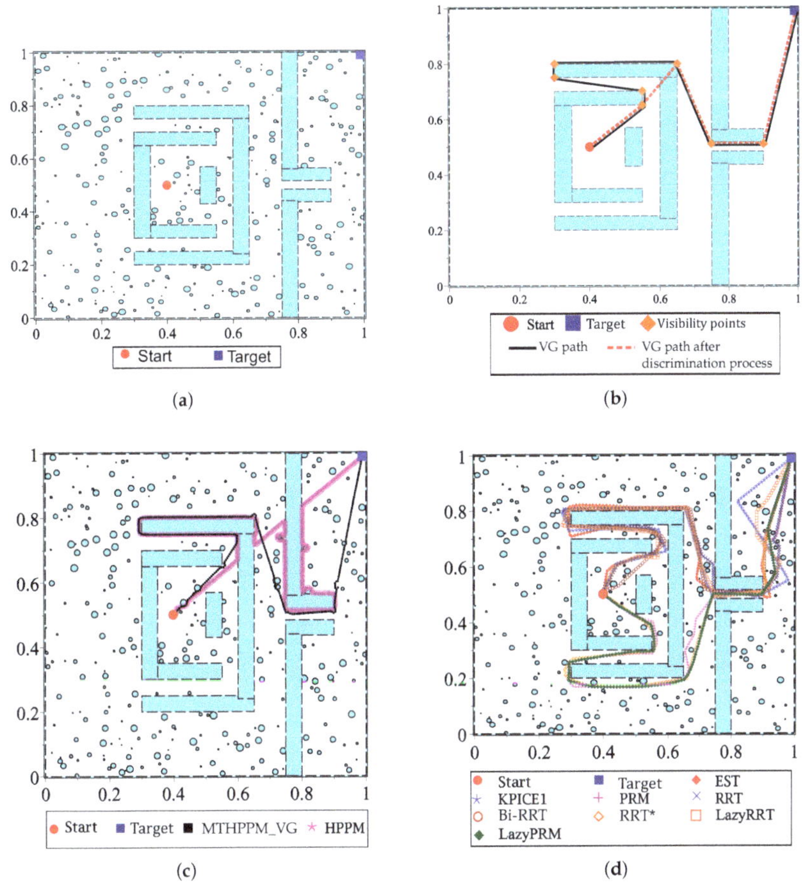

Figure 23. Comparative results for case study 4; best run of each SBP algorithm. (**a**) Path planning problem case study 4. (**b**) Visibility graph path without circular obstacles. (**c**) Paths obtained with HPPM and MTHPPM_VG. (**d**) Paths obtained with SBP algorithms.

Figure 23c shows the ability of HPPM and MGHPPM_VG to deal with a bug trap and narrow corridors. The tendency of HPPM to follow a direct path can be observed in Figure 23c. On the other hand, it also depicts an enhanced homotopy path since MTHPPM uses the points provided by the VG algorithm. Figure 24 shows a condensed performance information of the HPPM, MTHPPM, and SBP algorithms for time consumption, memory, percentage of feasible paths found, and length of the path.

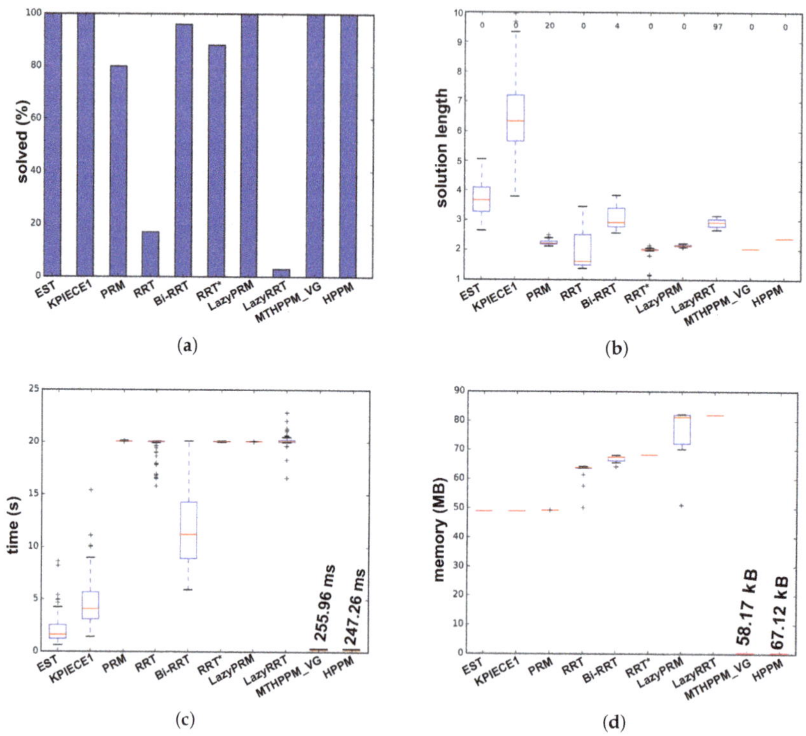

Figure 24. Comparative results for case study 4. (**a**) Successful paths. (**b**) Path length. (**c**) Execution time. (**d**) Memory.

Analyzing the obtained results, can be summarized as follows: First, RRT and LazyRRT exhibit poor performance to deal with bug traps; unlike EST, KPIECE1, PRM, HPPM, and MTHPPM_VG which have a one hundred percentage of success (see Figure 24a). Second, Figure 24c shows that homotopy based planners spent, noticeably, less execution time than sampling-based planners. Third, the HPPM creates a path with competitive length, while MTHPPM_VG obtained a path close to the shortest (RRT* calculated the best). For this metric, the results of RRT because the length of its best path corresponded to an unsuccessful run are discarded. Finally, homotopy planners exhibited execution times around three orders of magnitude lower than the best SBP algorithms.

7.5. Case 5

The utility of MTHPPM to solve the problem of multiple-target path planning in applications like pick-and-delivery activities, museum guidance robot, rescue task robot, and so forth, is presented. For this case study, target-points are previously selected by the user according to a sequence of rooms that the robot must visit. Figure 25a shows the environment with a sequence of fifteen milestones, including the start-position (point 1) and target-position (point 15). This map contains five hundred circular obstacles representing the configuration of people (considered as static obstacles for a given instant of time). The main objective of this case study is to show the ability of MTHPPM to calculate a collision-free path for multiple-targets in a map filled with obstacles. For this case, each SBP algorithm was executed one hundred times and applied the point-to-point strategy to generate a path to visit all fifteen target-points. Figure 25 depicts the solution paths of six SBP algorithms generated using the point-to-point technique. These figures show the best run (length of path) for each SBP algorithm between two points after one hundred runs.

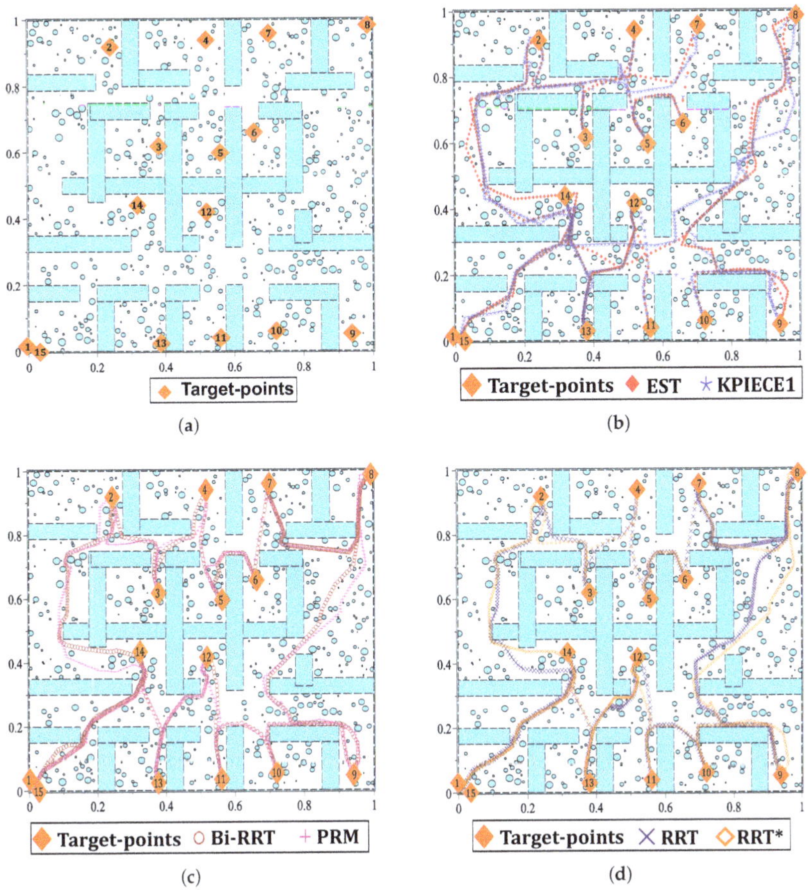

Figure 25. Multiple-target path problem solution using SBP algorithms with point-to-point strategy. (**a**) Multiple-target point problem (case study 5). (**b**) Multiple-target path using EST and KPIECE1. (**c**) Multiple-target path using RRTConnect and PRM. (**d**) Multiple-target path using RRT and RRT*.

Figure 25b shows that EST and KPIECE1 planners are not capable to solve the problem of the narrow corridor between target-points seven and eight. Here, it presents a visual comparison of full path length because benchmark results provided a similar outcome regarding memory consumption, time, and success percentage compared to previous experiments. It is important to remark that, the full path of each SBP algorithm is integrated by the shortest segments between two points obtained after one hundred executions. Figure 26a presents the single value $\widetilde{\text{PWL}}$ function employed to calculate two multiple-target paths through MTHPPM. The fist path is calculated for the motion in the forward direction (path from point 1 to 8) and the second in the backward direction (path from point 8 to 15), as can be seen in Figure 26b.

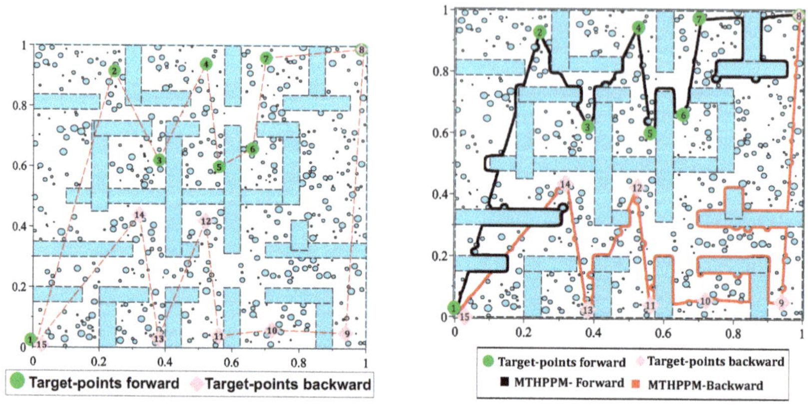

(a) (b)

Figure 26. Performance of MTHPPM for office-like environments. (**a**) Piecewise-linear representation of direct path for multiple-target approach. (**b**) Successful solution path.

Figure 26b shows the ability of MTHPPM to obtain a free collision path for a sequence of target-points. This planner calculates the full path in only two runs, one path in the forward and the other one in the backward direction. Although the path obtained by MTHPPM is not optimal, this case study shows the application of the proposed MGHPPM to solve multiple-target problems with hundreds of obstacles and narrow corridors. This characteristic is desirable for monitoring robots in oil platforms and forests or natural reserves.

8. Pure-Pursuit Controller Using Matlab Robotics System Toolbox

This section provides two simulations to validate the feasibility of paths obtained by HPPM, MTHPPM, and the pure-pursuit algorithm. The pure-pursuit algorithm procedure computes the angular velocity of the robot from the current position to reach a look-ahead point (see in Figure 27). This is not a common controller, although it is suitable to be employed for path tracking purposes [59,60]. Here, two tracking simulations of a waypoints set obtained by homotopy planners (HPPM and MTHPPM) are used. The tracking is simulated by the pure-pursuit class contained in MATLAB Robotics System Toolbox and the model of a differential drive robot. It is assumed that linear velocity is constant and angular velocity changes according to the instantaneous center of curvature (ICC). The ICC is calculated from the look-ahead parameter as explained in References [59,60]. Pure-pursuit operates as follows: First, the desired path is obtained by any path planning algorithm, a homotopy planner for this case. The path should be represented as a finite array of n-waypoints in the form (x_i, y_i), $i = 1, 2, 3, \ldots, n$. Second, using the look-ahead distance (value previously set), the algorithm calculates ICC and then the angular velocity to move the robot from one waypoint to another until the last point of the path is reached.

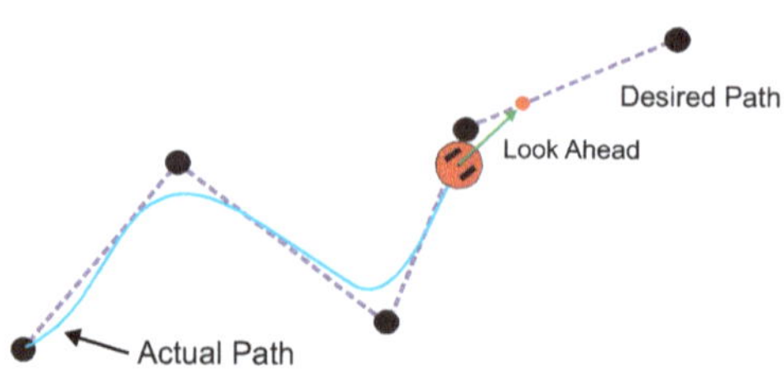

Figure 27. Pure-pursuit scheme.

Figure 27 presents the basic operation of the pure-pursuit controller. It can be noticed that the look-ahead distance shown in this figure is shorter than the distance between two waypoints. For these cases, the algorithm completes the desired path using segments of straight lines. For the next simulations, parameters of pure-pursuit controller are set as follows

- The dimensions of the environment map are 15 m × 15 m.
- Robot radius is 0.25 m.
- Look-ahead distance is 0.5 m.
- Linear velocity is 0.5 m/s.
- Maximum angular velocity is 2 rad/s.

In this example a configuration similar to case studies 3 and 4 in Section 7 is applied. Before the homotopy based planner is executed, the dimensions of the robot must be added to all obstacles, the robot dimensions are considered, as shown in Figures 28a and 29a. In these, the differential drive robot is modeled by a disc, then, its radius is added to dimensions of all obstacles as guard distance (dash line in Figures 28a and 29a), as explained in Reference [10]. The complete environment map, considering dimensions of the robot, is normalized to be introduced in HPPM. Figures 28b and 29b show the path followed by differential robot employing the pure-pursuit algorithm. These figures show that the difference between waypoints (homotopy path) and pure-pursuit path is minimal.

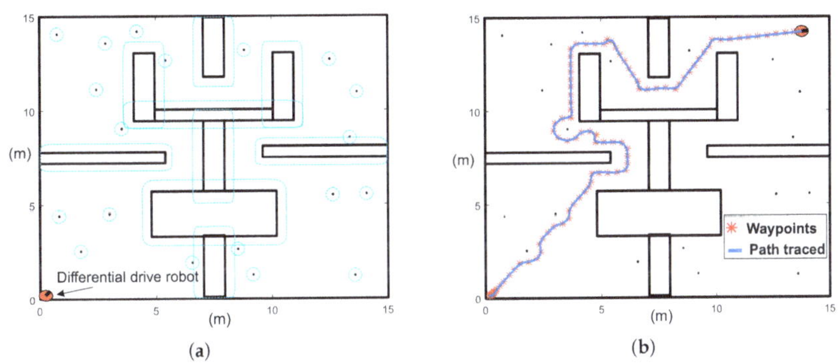

Figure 28. Pure-pursuit path tracking for the solution path of case study 3. (**a**) Floor plan considering dimensions of robot. (**b**) Path traced by differential drive robot.

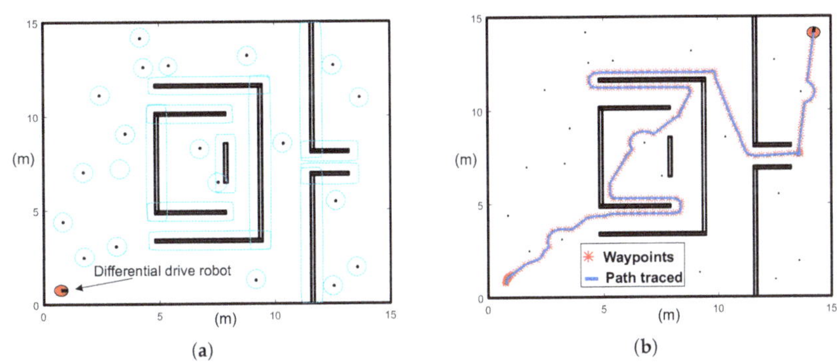

Figure 29. Pure-pursuit path tracking for the solution path of case study 4. (**a**) Floor plan considering dimensions of the robot. (**b**) Path traced by differential drive robot.

For the example of Figure 28, the initial point is placed at $(0, 0)$ and final point at $(14.5\text{ m}, 14.5\text{ m})$. Meanwhile, for the example of Figure 29, the initial point is placed at $(0.5\text{ m}, 0.5\text{ m})$ and final point at $(14.5\text{ m}, 14.5\text{ m})$.The apparent radius (dash line) of all circular obstacles is 0.26 m, which represents the sum of the obstacle radius (0.01 m) and the robot radius. The capability of HPPM to find solution of maps with narrow corridors can be observed in the example of Figure 29b.

Figures 28 and 29 validate that paths obtained by the proposed homotopy based planner (MGHPPM) can be applied to differential drive robots through the pure-pursuit controller. These simulations also denote that homotopy paths do not require post-processing or smoothed unlike SBP algorithms.

9. Discussion

This work presents a novel planner to obtain a collision-free path from a sequence of target-points which contains several improvements in the numerical implementation of the original homotopic path planner [19]. The main contributions of this work are outlined in four aspects.

1. A novel NAES formulation based on smooth PWL auxiliary functions ($\widetilde{\text{PWL}}$) generated from and approximation of absolute values function is presented. This formulation allows generating a multiple-target planner scheme, the MTHPPM, which uses the ability of the continuation homotopy to find more than one solution. Furthermore, the $\widetilde{\text{PWL}}$'s provides a scheme without mathematical discontinuities due to the integration of approximation of absolute value on the PWL formulation. For this scheme, the targets can be set by the user or using an automatic process, as explained in the case studies. Furthermore, this formulation does not imply a significant increase in execution time or memory consumption respect to original HPPM.
2. A dummy obstacle scheme to reduce the number of steps needed to generate a successful path was proposed. This scheme generates a modification in one of the homotopy surfaces, which reduces the distance of the solution path (points of intersection between homotopy surfaces). The effect of a dummy obstacle has a great impact on the number of steps in the procedure of homotopy based planner (HPPM and MTHPPM), thus, the execution time is significantly reduced.
3. The operation of the technique to solve the reversal phenomenon, found in spherical tracking, named double spherical tracking (DST) is presented and explained in Section 5.3. This technique also can be applied to improve the convergence of SA when the corrector scheme (Broyden's method) fails. The effectiveness of this technique is validated by numerical simulations presented in Section 7, in which, DST technique allowed the continuation of path planning. Table 2 presents the issues (reversal phenomena and non-convergences of corrector scheme), steps in which were detected, and a total of steps (number of points in the path) for case studies 1–4. These data validate the effectiveness of DST technique to solve the reversal phenomenon and enhance the convergence of HPPM and MTHPPM (MTHPPM_VG). The non-convergence and reversal effect issues during the iterative HPPM and MTHPPM processes for case studies 1–4 were between 0.082–0.633% of the total steps, that is, the impact over the execution time and memory consumption is hardly noticeable. It is important to remark that, although MTHPPM (MTHPPM_VG) is more susceptible to non-convergence and reversal effect issues than HPPM because of its formulation, the implementation of multiple-target strategy reduces the number of SA steps and enhance the homotopy path (in terms of length).

Table 2. Number of reversal effects and non-convergences solved in SA for HPPM and MTHPPM, case studies 1–4, and using DST technique.

Methodology		Non-Convergence Solved in SA (# Step)	Reversal Effects Solved in SA (# Step)	Total Steps
Case 1	HPPM	0	1940,1986,2974	3033
	MTHPPM_VG	1257,1927,1989,2373,2471,2492,2549	1825,1984,1994,2036	2602
Case 2	HPPM	1303	0	3594
	MTHPPM_VG	1248,1984,2012,2061, 2417,2441,2483,2558,2636	1318,1579,1908,1961 2018,2103,2457,2636	2689
Case 3	HPPM	0	0	3633
	MTHPPM_VG	1427,1491,2288,2316,2610	1777,2611	3192
Case 4	HPPM	17	36	2437
	MTHPPM_VG	17,210,245,611	565	2046

4. Automatic assignation of sign and magnitude of repulsion parameter for circular obstacles is introduced in Section 5.6. This formulation optimizes paths length because it forces the homotopy curves to stay close at the direct trajectory.
5. The multiple-target HPPM with visibility graph approach is provided in Section 6. For this method, a first approximation of the path is obtained by visibility graphs algorithm considering only walls in the environment, then, the visibility points (nodes of visibility graph path) are taken as targets by MTHPPM and solve the entire map for all obstacles and walls. This approach offers the best of both methods (Visibility graph and MTHPPM): (a) Shortest path (from visibility graph), (b) low time and memory consumption (from MTHPPM), and (c) ability to find the solution path if it exists (from the combination of MTHPPM and VG).
6. The symbolic manipulation of the Jacobian matrix proposed in this work is employed to simplify the complexity of the homotopy based planner implementations. This strategy allows a fast evaluation of the Jacobian matrix in the predictor and corrector schemes (process implicit at each SA step). Furthermore, this strategy avoids the use of specialized mathematical libraries and packages and provides an easy and cheap (in terms of memory) implementation in any programmable platform.
7. The feasibly of paths obtained with the MTHPPM and HPPM to be executed by a differential drive robot model is shown in Section 8. The simulations presented here proved that homotopy paths have a great compatibility with the pure-pursuit scheme due to the smoothness of the paths. Furthermore, Figures 28b and 29b showed, visually, that difference between homotopy path (waypoints) and traced path (using pure-pursuit controller) is very small. It implies that the homotopy planner does not need an additional stage of post-processing due to the smoothness of the paths, unlike the SBP algorithms which need an additional process to simplify the path found [2,3,7–10].

Table 3 shows the summarized results of case studies 1–4 for the maximum, minimum, and median values of performance metrics presented in its respective box diagrams. It is important to note that minimum length (dimensionless because map is normalized) for some SBP planners denotes the result of an unsuccessful run. Furthermore, this table shows the global execution time for all SBP algorithms, which considers the execution time of the algorithm and time spent to smooth the path (path simplification process). If the median value of each metric for every SBP planner is taken, then, it is possible to remark the following conclusions for the case studies: (I) The HPPM and MTHPPM have the best performance for memory consumption, about one-thousandth, compared to the amount spent by SBP algorithms; for instance, lazzyRRT and LazzyPRM for the second case study. (II) The execution of homotopy based planners is between ten and one hundred times faster than SBP algorithms. (III) The homotopy based planners presented in this paper have an adequate balance between the measured performance metrics; while for SBP algorithms, execution time and length of the path are inversely proportional parameters. In other words, faster planners (EST, KPIECE1, and RRT) find longer paths and shortest path calculations consume the maximum amount of time (RRT*). Results also show that the proposed methodology does not have any dependence on libraries, packages, or external functions because complex procedures are treated using symbolical analysis as it is presented in Section 5.5. Therefore, its implementation is simple and cheap in terms of computational resources. This allows an easy implementation in multiple platforms like embedded systems with microcontrollers, microprocessors up to implementations in PC's and FPGA's. In addition to these advantages, simulations presented in Section 8 show that the path obtained through homotopy based planners can be easily followed by a differential drive robot using the pure-pursuit algorithm.

Table 3. Summarized performance results for case studies 1–4. Metrics: memory consumption (M), execution time (T), and length of path (L).

Planner	Metric	Case 1			Case 2			Case 3			Case 4		
		Min	Med	Max	Min	Med	Max	Min	Med	Max	Min	Med	Max
EST	M(MB)	69.3	69.5	69.5	66.2	66.4	66.4	39.3	39.4	39.4	49	49	49
	T(s)	1.13	3.10	20.0	1.21	4.16	20.0	0.86	1.53	2.93	0.619	1.629	8.613
	L	0.613	2.679	4.219	2.002	2.711	3.418	2.769	3.650	4.764	2.658	3.677	5.057
KPIECE1	M(MB)	69.5	69.5	69.5	66.4	66.4	66.4	39.4	39.4	39.4	49	49	49
	T(s)	1.30	2.84	7.16	1.0	7.3	20.0	1.53	3.78	11.7	1.41	4.09	15.3
	L	2.969	4.545	6.847	3.147	4.749	6.563	4.235	6.074	7.834	3.795	6.340	9.955
PRM	M(MB)	69.5	69.9	70	66.7	66.8	66.8	39.7	39.9	40.0	49.2	49.3	49.3
	T(s)	20.0	20.0	20.1	20.00	20.10	20.18	20.0	20.1	20.2	20.0	20.0	20.1
	L	1.679	1.792	2.128	1.746	1.904	2.195	2.380	2.508	2.690	2.123	2.228	2.502
RRT	M(MB)	70.0	80.0	80.3	66.8	70.1	70.1	42.5	49.5	51.3	50.07	64	64.25
	T(s)	1.28	5.34	20.0	5.010	19.49	20.11	7.44	11.6	20.0	15.7	20.0	20.1
	L	1.824	1.971	2.293	1.626	2.004	2.488	2.666	2.861	3.144	1.382	1.620	3.463
Bi-RRT	M(MB)	80.3	80.3	80.3	70.1	74.0	74.0	51.3	52.4	52.4	64.2	67.5	68.1
	T(s)	0.37	1.14	10.59	0.925	3.368	20.07	3.27	7.09	15.4	5.91	11.2	20.1
	L	1.802	1.949	2.554	1.879	2.049	2.246	2.668	2.836	3.037	2.577	2.930	3.837
RRT*	M(MB)	80.3	80.5	80.5	74.0	76.4	76.4	52.4	52.4	52.4	68.3	68.3	68.3
	T(s)	20.0	20.0	20.1	20.00	20.07	20.14	20.0	20.0	20.1	20.0	20.0	20.1
	L	1.237	1.587	1.735	1.058	1.671	2.006	1.458	2.266	2.337	1.140	2.028	2.156
LazyPRM	M(MB)	69.2	87.2	97.1	69.9	103.1	123.9	43.4	64.2	67.9	51	81.4	82.2
	T(s)	20.0	20.0	23.5	20.02	20.07	20.33	20.0	20.0	20.1	20.0	20.0	20.1
	L	1.633	1.687	1.758	1.728	1.789	1.946	2.337	2.393	2.477	2.081	2.159	2.225
LazyRRT	M(MB)	97.1	97.1	98.0	123.9	123.9	123.9	67.9	68.4	68.9	50.0	64	64.25
	T(s)	1.06	8.43	20.14	2.825	13.41	20.57	4.16	18.4	20.2	15.79	20.07	20.11
	L	1.851	2.200	2.940	1.894	2.286	2.589	2.734	3.116	3.783	1.382	1.620	3.463
MTHPPM_VG	M(MB)		0.075			0.0765			0.08186			0.056807	
	T(s)		0.19			0.23397			0.159609			0.255962	
	L		1.654			1.851			2.708			2.058	
HPPM	M(MB)		0.085			0.0984			0.08972			0.06555	
	T(s)		0.22			0.30011			0.122805			0.247262	
	L		2.098			2.637			2.699			2.407	

Figures 30–33 show the minimum, median and maximum results for execution time, length of path, and memory of cases 1–4 (in number of times), according to the results of MGHPPM (MTHPPM_VG). These figures represent a visual interpretation of Table 3 which helps to denote the advantages of the MGHPPM where, a) the memory used in all case studies by the SBP algorithms is between hundreds and thousands of times greater than MGHPPM; b) the path obtained through MGHPPM is very close

to the shortest path found by SBP algorithms, and c) the execution time of MGHPPM for all case studies is between five and one hundred of times smaller than every SBP algorithms.

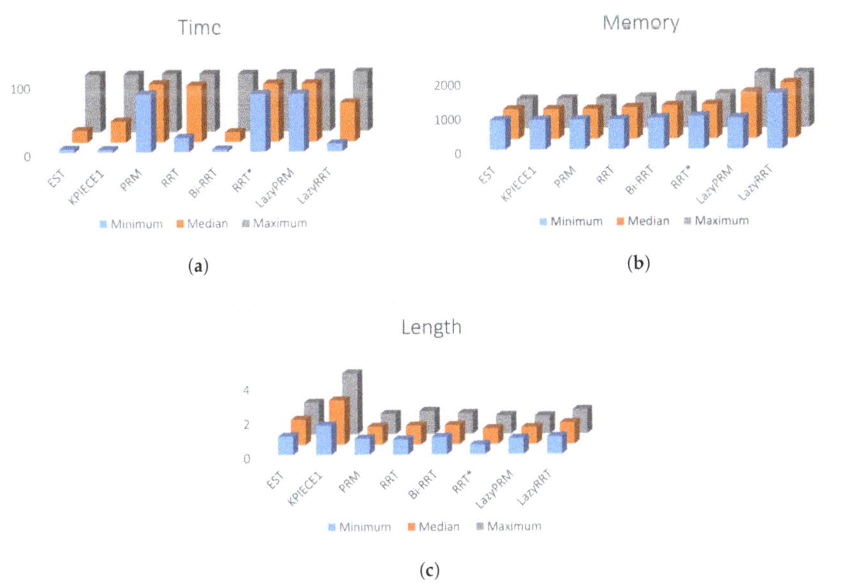

Figure 30. Bar graphs for time, length of path, and memory in number of times respect to the MGHPPM results of case 1. (**a**) Execution time. (**b**) Memory. (**c**) Length of path.

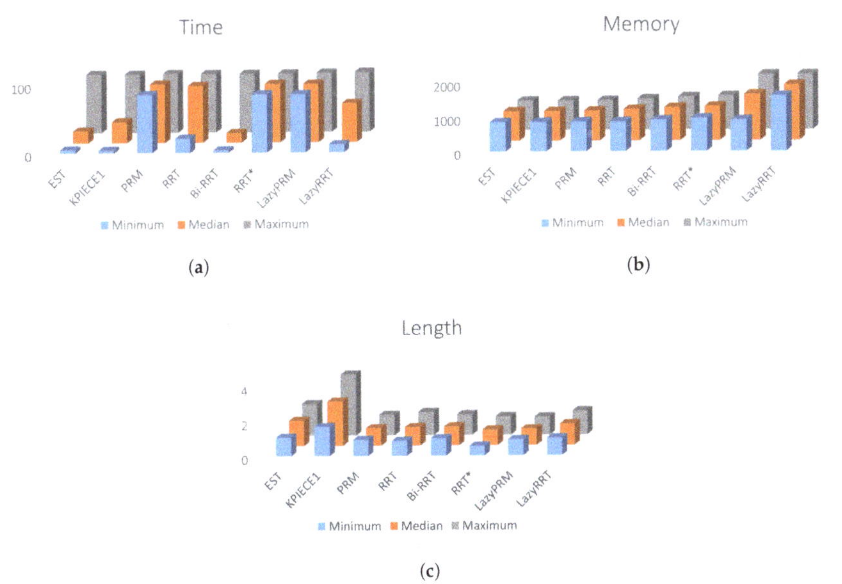

Figure 31. Bar graphs for time, length of path, and memory in number of times respect to the MGHPPM results of case 2. (**a**) Execution time. (**b**) Memory. (**c**) Length of path.

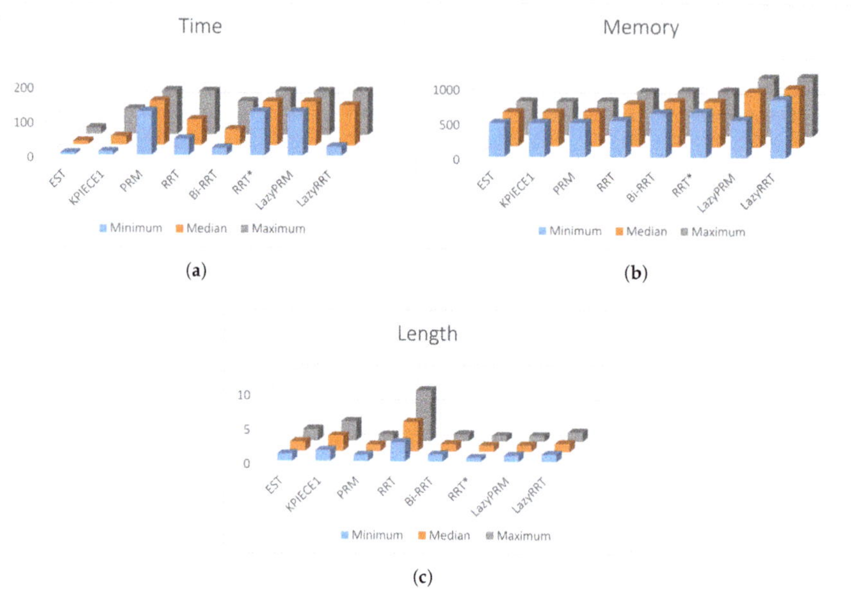

Figure 32. Bar graphs for time, length of path, and memory in number of times respect to the MGHPPM results of case 3. (**a**) Execution time. (**b**) Memory. (**c**) Length of path.

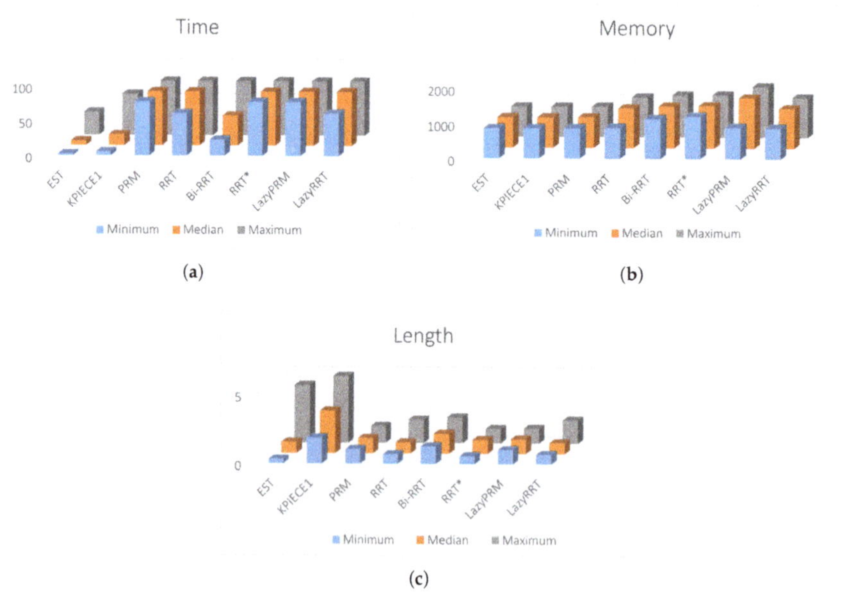

Figure 33. Bar graphs for time, length of path, and memory in number of times respect to the MGHPPM results of case 4. (**a**) Execution time. (**b**) Memory. (**c**) Length of path.

10. Conclusions and Future Work

In this work, a path planning method with multiple-target applications has been presented, it is capable of solving complex maps of hundreds of obstacles. This method contemplates in its computational core a series of novel and effective tools such as Double Spherical Tracking, the integration of a dummy obstacle to improve performance and reduce computing time and the number of iterations. As well as, a scheme of multiple solutions of a NAES formulated by approximations of PWL functions. In the same way, two new schemes are proposed to avoid discontinuities in the spherical tracking using patches and bridges in the plane. Furthermore, a simple and efficient solution is proposed as a criterion for the automatic selection of the repulsion parameter of the obstacles representation for the homotopy based planner methods.

From the case studies it can be denoted first, the proposed planner (MTHPPM) is faster; between five and one hundred times than the average of one hundred runs for each SBP algorithms. Second, the calculated path is very close to the shortest path; the difference is between ten and twenty percent. Third, the success rate is one hundred percent for all case studies, while some SBP algorithms achieved around thirty-five percent of the failure rate. Fourth, the proposed planner uses just one-thousandth of the memory than the best SBP algorithms employs for every case study. Fifth, the proposed methodology does not have any dependence on libraries, packages, or external functions for which its implementation is simple as well as cheap in terms of memory and computational resources. Therefore, these characteristics allow its implementation for real-time applications in multiple platforms from embedded systems with microcontrollers of low resources, such as it is presented in Reference [10], and microprocessors to PCs and FPGAs, as it is presented in Reference [21].

As future work, it is left for further research on use of multivalued piecewise linear function formulation, presented in Reference [51], which could generate two research lines. One for multiple-target planning of robots with displacement in 3-D or more dimensions like: drones, underwater robots, computer animation, robot manipulators, and molecular simulations [3,8,46]. The other one, employing multivalued or parametrized piece-wise linear functions like the ones presented in Reference [61,62], could enhance the performance of MGHHPP_VG in terms of path length, as it can be deducted from the cases studies, which would allow obtaining paths using visibility points in forward and backward movement on both axis. Besides, further work is needed to extend HPPM and MTHPPM to handle mechanical restrictions of non-holonomic robots, multi-agent systems, and path planning for dynamic environments with uncertainties. Finally, considering the advantages of the proposed homotopy based planner over SBP algorithms in terms of path length (close to the shortest path found by the best run of RRT*), execution time (between five and one hundred times faster than the median of SBP algorithms), and memory consumption (about a thousandth of that used by the SBP algorithms) makes of this a good choice to be implemented in practical applications. Besides, the linear relation between the number of obstacles and the complexity (in terms of memory and execution time) for homotopic planners, as it was presented in References [10,21] against the exponential complexity increase of SBP algorithms [2,3,8,9,63–65] allows to conclude that MTHPPM is the best option to planning collision-free robot motion paths. Especially, when it is needed to be implemented in a system with limited resources (like embedded systems) or required to solve complex environments where time constraints are tight.

Author Contributions: Investigation: G.D.-A. and H.V.-L.; Supervision: L.H.-M. and J.H.-C.; Methodology: V.M.J.-F.; Validation; H.D.C.-C.; Resources: A.H.-J., R.C.A. and S.H.-M.; writing—original draft preparation: G.U.D.-A.; writing—review and editing: L.H.-M. and H.V.-L.; All authors have read and agreed to the published version of the manuscript.

Funding: This research received no external funding

Conflicts of Interest: The authors declare no conflict of interest.

References

1. Şucan, I.A.; Kavraki, L.E. A Sampling-Based Tree Planner for Systems With Complex Dynamics. *IEEE Trans. Robot.* **2012**, *28*, 116–131. [CrossRef]
2. Elbanhawi, M.; Simic, M. Sampling-Based Robot Motion Planning: A Review. *IEEE Access* **2014**, *2*, 56–77. [CrossRef]
3. LaValle, S.M. *Planning Algorithms*; Cambridge University Press: Cambridge, UK, 2006. Available online: http://planning.cs.uiuc.edu/ (accessed on 29 March 2019).
4. Kalamian, N.; Niri, M.F.; Mehrabizadeh, H. Design of a Suboptimal Controller based on Riccati Equation and State-dependent Impulsive Observer for a Robotic Manipulator. In Proceedings of the 2019 6th International Conference on Control, Instrumentation and Automation (ICCIA), Sanandaj, Iran, 30–31 October 2019; pp. 1–6.
5. Xunyu, Z.; Jun, T.; Huosheng, H.; Xiafu, P. Hybrid Path Planning Based on Safe A* Algorithm and Adaptive Window Approach for Mobile Robot in Large-Scale Dynamic Environment. *J. Intell. Robot. Syst.* **2020**, 1–13. [CrossRef]
6. Patle, B.K.; Pandey, A.; Parhi, D.; Jagadeesh, A. A review: On path planning strategies for navigation of mobile robot. *Def. Technol.* **2019**, *15*, 582–606. [CrossRef]
7. Karaman, S.; Frazzoli, E. Sampling-based Algorithms for Optimal Motion Planning. *Int. J. Rob. Res.* **2011**, *30*, 846–894. [CrossRef]
8. Al-Bluwi, I.; Siméon, T.; Cortés, J. Motion planning algorithms for molecular simulations: A survey. *Comput. Sci. Rev.* **2012**, *6*, 125–143. [CrossRef]
9. Kleinbort, M.; Salzman, O.; Halperin, D. Collision Detection or nearest-neighbor search? On the computational bottleneck in sampling-based motion planning. *CoRR* **2016**. Available online: http://xxx.lanl.gov/abs/1607.04800 (accessed on 16 March 2019).
10. Diaz-Arango, G.; Vázquez-Leal, H.; Hernandez-Martinez, L.; Pascual, M.T.S.; Sandoval-Hernandez, M. Homotopy Path Planning for Terrestrial Robots Using Spherical Algorithm. *IEEE Trans. Autom. Sci. Eng.* **2017**, *15*, 567–585. [CrossRef]
11. Sharma, K.; Doriya, R. Path planning for robots: An elucidating draft. *Int. J. Intell. Robot. Appl.* **2020**. [CrossRef]
12. Nguyet, T.T.N.; Hoai, T.V.; Thi, N.A. Some Advanced Techniques in Reducing Time for Path Planning Based on Visibility Graph. In Proceedings of the 2011 Third International Conference on Knowledge and Systems Engineering, Hanoi, Vietnam, 14–17 October 2011; pp. 190–194.
13. Tran, N.; Nguyen, D.T.; Vu, D.L.; Truong, N.V. Global path planning for autonomous robots using modified visibility-graph. In Proceedings of the 2013 International Conference on Control, Automation and Information Sciences (ICCAIS), Ho Chi Minh City, Vietnam, 25–28 November 2013; pp. 317–321.
14. Jan, G.E.; Sun, C.C.; Tsai, W.C.; Lin, T.H. An $O(nlogn)$ Shortest Path Algorithm Based on Delaunay Triangulation. *IEEE/ASME Trans. Mechatron.* **2014**, *19*, 660–666. [CrossRef]
15. Foskey, M.; Garber, M.; Lin, M.C.; Manocha, D. A Voronoi-based hybrid motion planner. In Proceedings of the 2001 IEEE/RSJ International Conference on Intelligent Robots and Systems, Expanding the Societal Role of Robotics in the the Next Millennium (Cat. No.01CH37180), Maui, HI, USA, 29 October–3 November 2001; Volume 1, pp. 55–60.
16. Lee, M.C.; Park, M.G. Artificial potential field based path planning for mobile robots using a virtual obstacle concept. *Adv. Intell. Mechatron. IEEE/ASME Int. Conf.* **2003**, *2*, 735–740.
17. Laue, T.; Rofer, T. A behavior architecture for autonomous mobile robots based on potential fields. In *RoboCup 2004: Robot Soccer World Cup VIII*; Springer: Berlin/Heidelberg, Germany, 2005; pp. 122–133.
18. Rimon, E.; Koditschek, D.E. Exact robot navigation using artificial potential functions. *IEEE Trans. Robot. Autom.* **1992**, *8*, 501–518. [CrossRef]
19. Vazquez-Leal, H.; Marin-Hernandez, A.; Khan, Y.; Yildirim, A.; Filobello-Nino, U.; Castaneda-Sheissa, R.; Jimenez-Fernandez, V. Exploring collision-free path planning by using homotopy continuation methods. *Appl. Math. Comput.* **2013**, *219*, 7514–7532. [CrossRef]

20. Diaz-Arango, G.; Sarmiento-Reyes, A.; Hernandez-Martinez, L.; Vazquez-Leal, H.; Lopez-Hernandez, D.D.; Marin-Hernandez, A. Path optimization for terrestrial robots using Homotopy Path Planning Method. In Proceedings of the 2015 IEEE International Symposium on Circuits and Systems (ISCAS), Lisbon, Portugal, 24–27 May 2015; pp. 2824–2827.
21. De Cos-Cholula, H.E.; Díaz-Arango, G.U.; Hernandez Martinez, L.; Vazquez-Leal, H.; Sarmiento-Reyes, A.; Sanz-Pascual, M.T.; Herrera-May, A.L.; Castaneda-Sheissa, R. FPGA Implementation of Homotopic Path Planning Method with Automatic Assignment of Repulsion Parameter. *Energies* **2020**, *13*, 2623. [CrossRef]
22. Wang, H.; Yu, Y.; Yuan, Q. Application of Dijkstra algorithm in robot path-planning. In Proceedings of the 2011 Second International Conference on Mechanic Automation and Control Engineering, Hohhot, China, 15–17 July 2011; pp. 1067–1069.
23. Koziol, S.; Hasler, P.; Stilman, M. Robot path planning using Field Programmable Analog Arrays. In Proceedings of the 2012 IEEE International Conference on Robotics and Automation, St Paul, MN, USA, 14–19 May 2012; pp. 1747–1752.
24. Kopřiva, S.; Šišlák, D.; Pavlíček, D.; Pěchouček, M. Iterative accelerated A* path planning. In Proceedings of the 49th IEEE Conference on Decision and Control (CDC), Atlanta, GA, USA, 15–17 December 2010; pp. 1201–1206.
25. Soltani, A.; Tawfik, H.; Goulermas, J.; Fernando, T. Path planning in construction sites: Performance evaluation of the Dijkstra, A*, and GA search algorithms. *Adv. Eng. Inform.* **2002**, *16*, 291–303. [CrossRef]
26. Saian, P.O.N.; Suyoto; Pranowo. Optimized A-Star algorithm in hexagon-based environment using parallel bidirectional search. In Proceedings of the 2016 8th International Conference on Information Technology and Electrical Engineering (ICITEE), Yogyakarta, Indonesia, 5–6 October 2016; pp. 1–5.
27. Qiang, L.; Haibao, W.; Yan, Z.; Jingchang, H. Research on path planning of mobile robot based on improved ant colony algorithm. *Neural Comput. Appl.* **2019**, *32*, 1555–1566.
28. Reinoso, O.; Wu, S.; Du, Y.; Zhang, Y. Mobile Robot Path Planning Based on a Generalized Wavefront Algorithm. *Math. Probl. Eng. Hindawi* **2020**, *2020*, 6798798.
29. Kavraki, L.E.; Svestka, P.; Latombe, J.C.; Overmars, M.H. Probabilistic roadmaps for path planning in high-dimensional configuration spaces. *IEEE Trans. Robot. Autom.* **1996**, *12*, 566–580. [CrossRef]
30. Hsu, D.; Latombe, J.C.; Motwani, R. Path planning in expansive configuration spaces. In Proceedings of the International Conference on Robotics and Automation, Albuquerque, NM, USA, 25 April 1997; Volume 3, pp. 2719–2726.
31. LaValle, S.M. Rapidly-Exploring Random Trees: A New Tool for Path Planning. 1998. Available online: http://citeseerx.ist.psu.edu/viewdoc/summary?doi=10.1.1.35.1853 (accessed on 17 May 2019).
32. Kuffner, J.J.; LaValle, S.M. RRT-Connect: An efficient approach to single-query path planning. In Proceedings of the 2000 ICRA. Millennium Conference. IEEE International Conference on Robotics and Automation. Symposia Proceedings (Cat. No.00CH37065), San Francisco, CA, USA, 24–28 April 2000; Volume 2, pp. 995–1001.
33. Hauser, K. Lazy collision checking in asymptotically-optimal motion planning. In Proceedings of the 2015 IEEE International Conference on Robotics and Automation (ICRA), Washington, DC, USA, 26–30 May 2015; pp. 2951–2957.
34. Bohlin, R.; Kavraki, L.E. Path planning using lazy PRM. In Proceedings of the 2000 ICRA. Millennium Conference, IEEE International Conference on Robotics and Automation, Symposia Proceedings (Cat. No.00CH37065), San Francisco, CA, USA, 24–28 April 2000; Volume 1, pp. 521–528.
35. Wang, W.; Li, Y.; Xu, X.; Yang, S.X. An adaptive roadmap guided Multi-RRTs strategy for single query path planning. In Proceedings of the 2010 IEEE International Conference on Robotics and Automation, Anchorage, AK, USA, 3–8 May 2010; pp. 2871–2876.
36. Diaz-Arango, G.; Hernandez-Martinez, L.; Sarmiento-Reyes, A.; Vazquez-Leal, H. Fast and robust homotopy path planning method for mobile robotics. In Proceedings of the 2016 IEEE International Symposium on Circuits and Systems (ISCAS), Montreal, QC, Canada, 22–25 May 2016; pp. 2579–2582.
37. Cos-Cholula, H.E.D.; Díaz-Arango, G.U.; Hernández-Martínez, L.; Sarmiento-Reyes, A. An Homotopy Path Planning Method with automatic fixed value assignation of repulsion parameter for mobile robotics. In Proceedings of the 2016 13th International Conference on Electrical Engineering, Computing Science and Automatic Control (CCE), Mexico City, Mexico, 28–30 September 2016; pp. 1–6.

38. Park, M.G.; Lee, M.C. A new technique to escape local minimum in artificial potential field based path planning. *KSME Int. J.* **2003**, *17*, 1876–1885. [CrossRef]
39. Matoui, F.; Boussaid, B.; Abdelkrim, M.N. Local minimum solution for the potential field method in multiple robot motion planning task. In Proceedings of the 2015 16th International Conference on Sciences and Techniques of Automatic Control and Computer Engineering (STA), Monastir, Tunisia, 21–23 December 2015; pp. 452–457.
40. Luo, C.; Mo, H.; Shen, F.; Zhao, W. *Multi-Goal Motion Planning of an Autonomous Robot in Unknown Environments by an Ant Colony Optimization Approach*; Springer International Publishing: Cham, Switzerland, 2016; pp. 519–527.
41. Hernandez, K.; Bacca, B.; Posso, B. Multi-goal Path Planning Autonomous System for Picking up and Delivery Tasks in Mobile Robotics. *IEEE Lat. Am. Trans.* **2017**, *15*, 232–238. [CrossRef]
42. Bueckert, J.; Yang, S.X.; Yuan, X.; Meng, M.Q.H. Neural dynamics based multiple target path planning for a mobile robot. In Proceedings of the 2007 IEEE International Conference on Robotics and Biomimetics (ROBIO), Sanya, China, 15–28 December 2007; pp. 1047–1052.
43. Devaurs, D.; Siméon, T.; Cortés, J. A multi-tree extension of the transition-based RRT: Application to ordering-and-pathfinding problems in continuous cost spaces. In Proceedings of the 2014 IEEE/RSJ International Conference on Intelligent Robots and Systems, Chicago, IL, USA, 14–18 September 2014; pp. 2991–2996.
44. Faigl, J.; Váňa, P.; Deckerová, J. Fast Heuristics for the 3-D Multi-Goal Path Planning Based on the Generalized Traveling Salesman Problem With Neighborhoods. *IEEE Robot. Autom. Lett.* **2019**, *4*, 2439–2446. [CrossRef]
45. Ishida, S.; Rigter, M.; Hawes, N. Robot Path Planning for Multiple Target Regions. In Proceedings of the 2019 European Conference on Mobile Robots (ECMR), Prague, Czech Republic, 4–6 September 2019; pp. 1–6.
46. Petereit, J.; Emter, T.; Frey, C.W. Safe mobile robot motion planning for waypoint sequences in a dynamic environment. In Proceedings of the 2013 IEEE International Conference on Industrial Technology (ICIT), Cape Town, South Africa, 25–28 February 2013; pp. 181–186.
47. Yamamura, K. Simple algorithms for tracing solution curves. *IEEE Int. Symp. Circuits Syst.* **1992**, *6*, 2801–2804.
48. Torres-Muñoz, D.; Vazquez-Leal, H.; Hernandez-Martinez, L.; Sarmiento-Reyes, A. Improved spherical continuation algorithm with application to the double-bounded homotopy (DBH). *Comput. Appl. Math.* **2014**, *33*, 147–161. [CrossRef]
49. Oliveros-Munoz, J.M.; Jiménez-Islas, H. Hyperspherical path tracking methodology as correction step in homotopic continuation methods. *Chem. Eng. Sci.* **2013**, *97*, 413–429. [CrossRef]
50. Ramirez-Pinero, A.; Vazquez-Leal, H.; Jimenez-Fernandez, V.M.; Sedighi, H.M.; Rashidi, M.M.; Filobello-Nino, U.; Castaneda-Sheissa, R.; Huerta-Chua, J.; Sarmiento-Reyes, L.A.; Laguna-Camacho, J.R.; et al. Speed-up hyperspheres homotopic path tracking algorithm for PWL circuits simulations. *SpringerPlus* **2016**, *5*, 890. [CrossRef] [PubMed]
51. Chua, L.O.; Kang, S.M. Section-wise piecewise-linear functions: Canonical representation, properties, and applications. *Proc. IEEE* **1977**, *65*, 915–929. [CrossRef]
52. Chua, L.; Deng, A.C. Canonical piecewise-linear modeling. *IEEE Trans. Circuits Syst.* **1986**, *33*, 511–525. [CrossRef]
53. Julian, P.; Desages, A.; Agamennoni, O. High-level canonical piecewise linear representation using a simplicial partition. *IEEE Trans. Circuits Syst. Fundam. Theory Appl.* **1999**, *46*, 463–480. [CrossRef]
54. Julian, P.; Desages, A.; D'Amico, B. Orthonormal high-level canonical PWL functions with applications to model reduction. *IEEE Trans. Circuits Syst. Fundam. Theory Appl.* **2000**, *47*, 702–712. [CrossRef]
55. Guzelis, C.; Goknar, I.C. A canonical representation for piecewise-affine maps and its applications to circuit analysis. *IEEE Trans. Circuits Syst.* **1991**, *38*, 1342–1354. [CrossRef]
56. Schmidt, M.; Fung, G.; Rosales, R. Fast Optimization Methods for L1 Regularization: A Comparative Study and Two New Approaches. In *Machine Learning: ECML 2007*; Kok, J.N., Koronacki, J., Mantaras, R.L.d., Matwin, S., Mladenič, D., Skowron, A., Eds.; Springer: Berlin/Heidelberg, Germany, 2007; pp. 286–297.
57. Sucan, I.A.; Moll, M.; Kavraki, L.E. The Open Motion Planning Library. *IEEE Robot. Autom. Mag.* **2012**, *19*, 72–82. [CrossRef]
58. Moll, M.; Şucan, I.A.; Kavraki, L.E. Benchmarking Motion Planning Algorithms: An Extensible Infrastructure for Analysis and Visualization. *IEEE Robot. Autom. Mag.* **2015**, *22*, 96–102. [CrossRef]

59. Coulter, R.C. *Implementation of the Pure Pursuit Path Tracking Algorithm*; Technical Report; DTIC Document: Springfield, MA, USA, 1992.
60. Morales, J.; Martínez, J.L.; Martínez, M.A.; Mandow, A. Pure-Pursuit Reactive Path Tracking for Nonholonomic Mobile Robots with a 2D Laser Scanner. *EURASIP J. Adv. Signal Process.* **2009**, *2009*, 935237. [CrossRef]
61. Jimenez-Fernandez, V.M.; Jimenez-Fernandez, M.; Vazquez-Leal, H.; Filobello-Nino, U.A.; Castro-Gonzalez, F.J. Smoothing the High Level Canonical Piecewise-Linear Model by an Exponential Approximation of its Basis-Function. *Comput. Sist.* **2016**, *20*, 227–237. [CrossRef]
62. Jimenez-Fernandez, V.M.; Jimenez-Fernandez, M.; Vazquez-Leal, H.; Muñoz-Aguirre, E.; Cerecedo-Nunez, H.H.; Filobello-Nino, U.A.; Castro-Gonzalez, F.J. Transforming the canonical piecewise-linear model into a smooth-piecewise representation. *SpringerPlus* **2016**, *5*, 1612. [CrossRef] [PubMed]
63. Saha, M.; Latombe, J.C.; Chang, Y.C.; Prinz, F. Finding Narrow Passages with Probabilistic Roadmaps: The Small-Step Retraction Method. *Auton. Robot.* **2005**, *19*, 301–319. [CrossRef]
64. Sun, Z.; Hsu, D.; Jiang, T.; Kurniawati, H.; Reif, J.H. Narrow passage sampling for probabilistic roadmap planning. *IEEE Trans. Robot.* **2005**, *21*, 1105–1115.
65. Zhong, J.; Su, J. Narrow passages identification for Probabilistic Roadmap Method. In Proceedings of the 30th Chinese Control Conference, Yantai, China, 12 July 2011; pp. 3908–3912.

sensors

MDPI

Article

LiDAR-Based GNSS Denied Localization for Autonomous Racing Cars

Federico Massa [1,*], Luca Bonamini [1], Alessandro Settimi [1], Lucia Pallottino [1,2] and Danilo Caporale [1]

[1] Research Centre E. Piaggio, Università di Pisa, 56122 Pisa, Italy; lucabonamini28@gmail.com (L.B.); ale.settimi@gmail.com (A.S.); lucia.pallottino@unipi.it (L.P.); d.caporale@centropiaggio.unipi.it (D.C.)

[2] Dipartimento di Ingegneria dell'Informazione, Università di Pisa, 56122 Pisa, Italy

* Correspondence: f.massa@studenti.unipi.it

Received: 23 May 2020; Accepted: 14 July 2020; Published: 17 July 2020

Abstract: Self driving vehicles promise to bring one of the greatest technological and social revolutions of the next decade for their potential to drastically change human mobility and goods transportation, in particular regarding efficiency and safety. Autonomous racing provides very similar technological issues while allowing for more extreme conditions in a safe human environment. While the software stack driving the racing car consists of several modules, in this paper we focus on the localization problem, which provides as output the estimated pose of the vehicle needed by the planning and control modules. When driving near the friction limits, localization accuracy is critical as small errors can induce large errors in control due to the nonlinearities of the vehicle's dynamic model. In this paper, we present a localization architecture for a racing car that does not rely on Global Navigation Satellite Systems (GNSS). It consists of two multi-rate Extended Kalman Filters and an extension of a state-of-the-art laser-based Monte Carlo localization approach that exploits some a priori knowledge of the environment and context. We first compare the proposed method with a solution based on a widely employed state-of-the-art implementation, outlining its strengths and limitations within our experimental scenario. The architecture is then tested both in simulation and experimentally on a full-scale autonomous electric racing car during an event of Roborace Season Alpha. The results show its robustness in avoiding the robot kidnapping problem typical of particle filters localization methods, while providing a smooth and high rate pose estimate. The pose error distribution depends on the car velocity, and spans on average from 0.1 m (at 60 km/h) to 1.48 m (at 200 km/h) laterally and from 1.9 m (at 100 km/h) to 4.92 m (at 200 km/h) longitudinally.

Keywords: LiDAR signal processing; sensor and information fusion; advanced driver assistance systems; autonomous racing

1. Introduction

Aside from the classical uses of automated human transportation in urban scenarios, another exciting application of self-driving technologies comes with autonomous racing, intended as the competition between a self-driving race car and other vehicles, either human driven or autonomous. This scenario is particularly interesting as it represents an extreme yet safe condition in which to test autonomous driving algorithms: extreme because of high speeds and accelerations at the limit of friction, safe as the vehicle runs the artificial driver software in a racing track and does not transport humans. It is worth noticing that many of the problems faced in autonomous racing are the same as those found in more common scenarios, such as urban or highway autonomous driving, in particular regarding self state estimation tasks such as localization, which is the focus of this paper.

Any self-driving vehicle relies on localization systems to provide an accurate pose estimation to efficiently and safely navigate in the environment. Such estimates are obtained by means of sensor fusion algorithms that combine information coming from different sources, possibly at different rates. The key factors in evaluating these methods are accuracy, responsiveness, robustness and reliability in the presence of signal degradation [1,2].

In autonomous racing, these factors are inherently critical, as to achieve optimal behavior, two main problems arise:

- While facing a tight turn or passing through narrow corridors, the optimal trajectory is close to the internal edge of the turn or, more in general, the track border;
- The optimal speed profile is at the limit of friction, thus, small localization errors can lead to divergent behaviors.

An extensive literature exists on the more general problem of simultaneous localization and mapping (SLAM), where there are basically two approaches based on filtering or optimization methods, see the following surveys [2–6]. In [7,8], the authors proposed an efficient solution to implement Rao–Blackwellized particle filters to obtain occupancy grid maps of the environments. In [9], the use of probabilistic maps is extended for dynamic urban environments. In [10], the authors developed a complete navigation stack for the Defense Advanced Research Projects Agency (DARPA) Challenge using 3D Light Detection and Ranging (LiDAR), Global Positioning System (GPS) and inertial sensors for the localization task. In [11], an efficient (in terms of data storage) localization method based on 3D LiDARs without the use of GPS is described. In [12,13], LiDAR-based systems with combined GNSS data are described. In [14], the authors show that it is preferrable to have a LiDAR mounted on top of the vehicle with a 360 degrees field of view than multiple planar LiDARs mounted around the vehicle, as is the vehicle considered in this work, both for ease of operation and localization performance. Similar conclusions are drawn in [15], where the authors compare the performance of a localization system with an INS (Inertial Navigation System) and camera, with or without 2D/3D LiDAR, highlighting the important contribution of LiDAR sensors for this task. A survey on different localization methods for autonomous vehicles can be found in [16]. As for SLAM, two main approaches are used in autonomous driving: optimization-based and filter-based ones. In this work we leverage on a state-of-the-art implementation of Adaptive Monte Carlo Localization (AMCL) [17]. A common pitfall of approaches based on particle filters is the so-called kidnapping problem, which becomes particularly dangerous during a race. Deep-learning methods are generally used in camera-based systems, for instance in [18] for visual SLAM based on monocular camera input. Recently a deep-learning classifier has been proposed in [19] to detect kidnapping in particle filter based localization, and a Long-Short Term Memory network has been used in [20] to quickly recover from kidnapping of a mobile robot. It has been recently shown in [21] that a deep-learning and Monte Carlo Localization method can reduce localization time even in large maps. In that work, a training set of 35 h and over 150 km, which is still too far from our use case where generally less than 10 h are available and collected in the same week of the race. More generally, online diagnosis of failure of perception systems for autonomous driving is an open problem, and a recent mathematical framework has been proposed in [22]. Optimization-based approaches (see for example [23]) tend to suffer less from the kidnapping problem, at least in the authors' experience, but, on the other hand, are typically computationally intensive. A solution to achieve real time SLAM on Intel CPU using 3D LiDAR point clouds from a Velodyne sensor has been proposed in [24]. In our experience, both approaches are challenged in race tracks by the presence of straights. This is a problem especially when the LiDAR scans are matched to the map with the Iterative Closest Point (ICP) method [25]. An interesting solution was proposed in [26] in the context of train localization in tunnels by means of Rényi quadratic entropy. Normal Distribution Transform methods have been proposed in [27] for 3D mapping of the environment, and are promising for applications with 3D LiDARs even for self-driving applications.

The sensor setup is the first aspect to consider when designing a state estimation system. In a typical autonomous car, available sensors are optical speed sensors, wheel encoders, Inertial Measurement Unit (IMU) (propioceptive, working at frequencies of hundreds of Hz), LiDARs and cameras (exteroceptive, working at frequencies of tens of Hz). In order to achieve the required steering and traction control accuracy, it is necessary that the state estimation module outputs a high frequency signal (hundreds of Hz) with sufficient accuracy [28,29]. This multi-rate environment calls for specific solutions for state estimation that we address in this paper. Furthermore, as the racing context can be arbitrarily well structured, depending on the competition rules (pre-known track map, controlled weather, known type of track edges, known features of interacting agents), there is a lot of a priori information that can be exploited to enhance the quality of the state estimation [30–32].

Referring to GNSS, which can provide high frequency global pose estimates, is not always a viable solution. Despite being very reliable in open-sky environments, it can quickly stop being so in the presence of multiple sources of signal degradation [3,33]. Even in urban racing tracks such as those of Formula E (https://www.fiaformulae.com/) events, GNSS is affected by the presence of nearby buildings, trees, or tunnels. Referring to Real-Time Kinematic (RTK) antennas can mitigate this problem; however, the degradation can still be unacceptable in obstacle-dense scenarios and requires a widespread infrastructure. This is not a problem limited to autonomous cars, but it is very common for indoor navigation tasks. In [34,35], for example, the authors tackle the problem of navigation of a micro air vehicle in a GNSS-denied environment.

This motivated our choice of developing a system capable of not relying at all on any signal coming from GNSS for localization.

With respect to our previous works, where an optimization based approach was used [36,37], we adopted a method based on particle filtering due to the lower computational burden required on the specific hardware and, in perspective, because it is more amenable to parallelization than optimization based approaches. The aforementioned drawback of this method (the kidnapping problem) was solved by injecting a priori knowledge of the particular scenario into the filter, as will be explained in detail in Section 5, thus representing a valid alternative to optimization-based algorithms.

The testbed for the proposed architecture was the first Roborace (https://roborace.com/) Localization Challenge, which was created with the same goal in mind, which is to prove the capability of racing with degraded GNSS reception. To the best of our knowledge, this is the first autonomous racing challenge of this kind, an important step towards the realization of a more realistic racing challenge.

The race rules are simplified, consisting of a race with a single vehicle and no other moving obstacles on the car path, but still pose some interesting challenges as the car (2 m wide) was required to pass through several narrow gates (2.5 m wide) while driving at the maximum speed allowed by the track. A maximum speed of 100 km/h was imposed given the characteristics of the track, which is shown in Figure 1.

In this paper we report on the whole state estimation system, consisting of two multi-rate Kalman Filters for odometry estimation and smoothing, a Madgwick Filter for orientation estimation and a LiDAR processing algorithm that we named Informed Adaptive Monte Carlo localization (IAMCL), which was tested in simulation and on a real racing vehicle in the context of the localization challenge.

As the name suggests, this algorithm is an extension of the classical Monte Carlo localization algorithm and it is based on the famous AMCL implementation within the Robot Operating System (ROS) [17], where some a priori knowledge on the scenario is injected to enhance performances and prevent the kidnapping problem.

In the following, before presenting the proposed localization method in detail, we outline the problem formulation, the underlying sensors setup, and the mapping procedure. Finally, we report on simulation and experimental results.

Figure 1. Roborace DevBot 2.0 in the Zala Zone Circuit. Localization challenges were performed in this circuit, and the presented localization framework was used.

2. Problem Formulation

We focus on the development of a localization system relying on 2D LiDAR input in the absence of GNSS signal. We assume that an occupancy grid map is built before localization. We also assume that the LiDAR point cloud model is known and that, during localization, the occupancy grid map is noiseless.

This assumption is removed while generating the map, a process detailed in Section 4. More specifically, our goal is to provide a high frequency and smooth pose estimate for a car moving in a race track. To achieve this, we leverage upon a widely used implementation in the mobile robotics community (AMCL package in ROS) extending it to avoid known issues that hinder its usability in a racing setting: the need for manual initialization, which is prone to human error and is in general not efficient, and the kidnapped robot problem, which constitutes a safety concern for a racing car due to the possibility of sudden discontinuities in the pose estimate and the resulting feedback control actions.

Let the kinematic state of the vehicle be $q = (x, y, \varphi, u)$, for which we need to fuse a set of asynchronous proprioceptive sensor measurements with a LiDAR point cloud. We do not include, during localization, direct measurements of the vehicle pose (x, y, φ), as no GNSS/magnetometer data is available. The frame of reference for q is taken within a pre-built occupancy grid map that is also used for LiDAR scan matching. The vehicle control system requires a 250 Hz pose estimate [36]: while the on board velocity and acceleration sensors are able to provide signals at this (or higher) frequency, the LiDAR scans are provided at 25 Hz. Hence, a smooth pose estimate at high frequency has to be carefully computed from sources with multiple rates.

3. Sensors Setup

Roborace's DevBot 2.0 (London, UK), the car used during the race, has several sensors available, a subset of which is relevant to this work shown in Figure 2. For an overview of the vehicle's hardware architecture, see our previous work [36]. For the sake of this paper, the sensor data comes from the following three sources:

- **OxTS Inertial Navigation System (INS):** this commercial module consists of a dual-antenna GNSS and an IMU , which are pre-fused to obtain a high frequency (250 Hz) pose, velocity, and accelerations estimates;

- **Ibeo LiDAR range finders:** four LiDAR sensors are placed on the corners of the vehicle, each with 4 vertically stacked layers at 0.8 degrees spacing. The aggregate point cloud resulting from all the sensors is provided with a rate of 25 Hz.
- **Optical Speed Sensor (OSS):** this sensor provides direct longitudinal and lateral speed measurements through a Controller Area Network (CAN) interface at 500 Hz, and it is not affected by wheel drift.

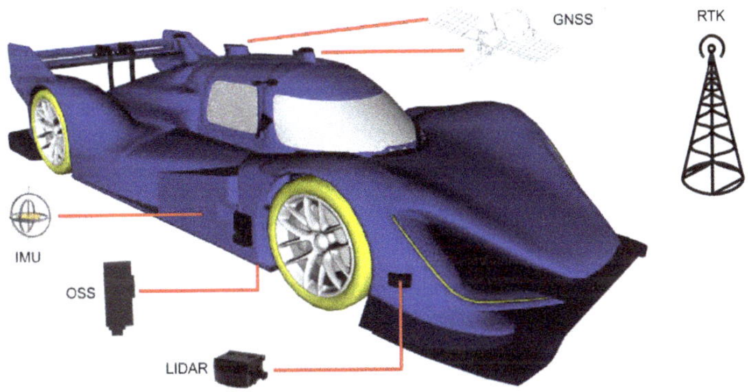

Figure 2. Overview of the DevBot sensors. Light Detection and Ranging (LiDARs) are mounted in the front, side and back of the car. Global Navigation Satellite Systems (GNSS) and Inertial Measurement Unit (IMU) sensors come from the OxTS system and the Optical Speed Sensor (OSS) measures longitudinal and lateral car velocities. The Real-Time Kinematic (RTK) base station is an optional system that allows extremely high positioning precision to the OxTS system.

Finally, an RTK base station is placed at a fixed position near the track, which can significantly improve the position accuracy provided by GNSS. Note that, when this system is on, the GNSS accuracy is so high (in our experimental context) that it solves the localization problem, as the measurement has an error of the order of a few millimeters. Thus, we consider the data coming from this system as a truth reference to compute the localization error metrics, while during the actual race this system is deactivated and the number of satellites is limited to simulate a GNSS-denied scenario.

We consider the reference frame depicted in Figure 3 with the x-axis of the velocity laying along the forward direction, and the y-axis on the left. The INS system provides also absolute position estimates with respect to the map origin and orientation φ relative to the cardinal east, longitudinal and lateral speeds u and v, longitudinal and lateral accelerations a_x and a_y, and angular velocity along the yaw axis ω.

Note that there are in theory two sensors that can provide vehicle speed measurements: INS and OSS. We decided to rely on OSS due to the fact that, while INS uses GNSS inputs to provide these measurements, thus producing unreliable estimates when that information is denied, OSS is totally independent from it. Thus, we rely on INS only for accelerations and angular rates. This way we obtained a more realistic simulation of a scenario with complete GNSS absence.

An estimate of the orientation φ is instead obtained from a Madgwick filter that fuse angular velocities and accelerations from the INS system, as will be described in Section 5.

4. Mapping Procedure

Mapping of the track is performed in dedicated sessions where the car is manually driven within the track borders for data acquisition and offline map generation. During these sessions it is permitted to use RTK-GNSS data that, being extremely accurate, make localization not an issue. Finally, the map is generated with OpenSlam's Gmapping [7], a particle filter based algorithm that builds grid maps through the use of LiDAR and odometry data. Note that, although localization is assumed accurate, the resulting map can still feature some distortions due to the point cloud noise, the LiDAR calibration error and the algorithm parameters tuning. In Figure 4 the resulting map of the racing track (Zala Zone circuit) is shown. In the following, we outline a procedure to qualitatively assess the accuracy of the map used while tuning the mapping algorithm.

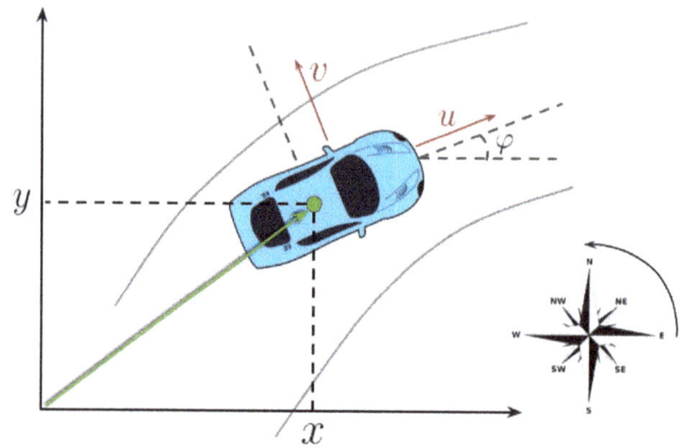

Figure 3. The adopted sensor measurement reference frame. u and v are the longitudinal and lateral velocities; x and y are the vehicle coordinates with respect to the map origin.

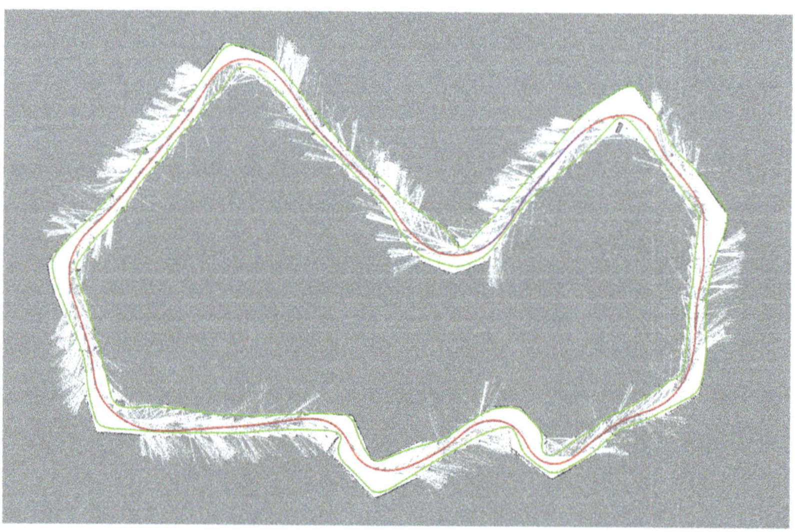

Figure 4. The map built for the race. The circuit track borders are represented by the two green lines, white areas are the obstacle free spaces, while the red line represents the racing line, to be followed during the run.

Map Quality Assessment

As described in Section 3, for localization we only rely on velocities, accelerations and on the LiDAR data. Of these, only LiDAR data provide an absolute pose estimate for the vehicle, which in our approach is obtained by matching the data with the offline built map, as will be described in Section 5. Thus, the localization performance is inherently limited by the map quality.

Issues such as distortions in the map result in a bias b of the final pose estimate that is not easily removable, thus we have:

$$\mathbb{E}\{\hat{\xi}(s)\} = \mathbb{E}\{\xi^*(s)\} + b(s)$$

where $\mathbb{E}$ is the statistical expectation operator, $\hat{\xi}$ is the car estimated pose, ξ^* is the car true pose, and s is the arc length distance along the racing track. This means that the empirical localization error with respect to GNSS consists of two contributions, the first being the actual localization error, and the second being influenced by the mapping error, which is not only unknown but also variable along the track.

The bias is qualitatively evaluated by measuring how well the LiDAR data matches the map when fixing the vehicle's pose at the GNSS/RTK position. To do this, we consider the average mapping error:

$$e_{\text{map}}\left(\xi^*(k)\right) = \frac{\sum_{i=1}^{\Lambda} \left| T_{\xi^*(k)}[p_i(k)] - m\left(T_{\xi^*(k)}[p_i(k)]\right)\right|}{\Lambda},$$

where k is the time instant, Λ is the number of laser scan points, ξ^* is the value of the GNSS/RTK pose estimate, p_i is the coordinate of a particular laser scan point, $T_q[\cdot]$ is a roto-translation defined by the pose q, and $m(\cdot)$ is a function that takes a laser scan point as input and returns the coordinate of the closest occupied point in the occupancy grid map. The procedure for a single given GNSS pose is illustrated in Figure 5a.

This procedure is iterated along the map, and its results are shown in Figure 5b, where colors ranging from green to red indicate a progressively worse value of the average mapping error. The result is compatible with the idea that mapping is more accurate near distinctive features of the track, i.e., elbows, whereas it shows worse results on straights, likely because of the LiDAR sensor giving mostly planar information. The proposed procedure gives an intuitive representation of the mapping quality, and it is used to evaluate different mappings and to tune mapping algorithms.

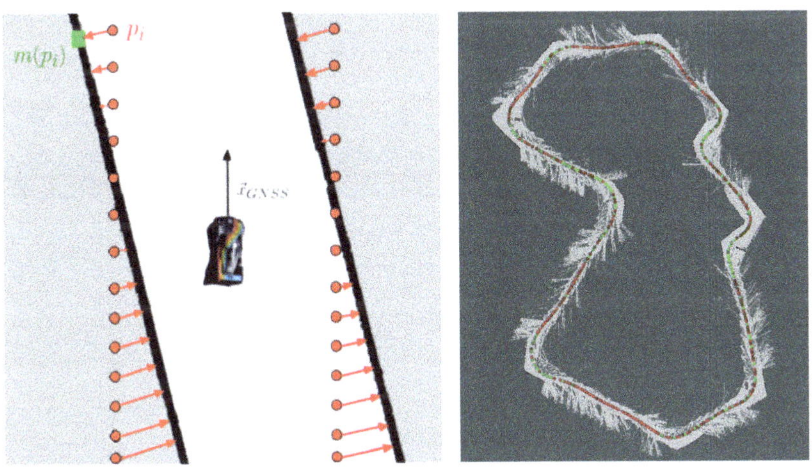

(a) (b)

Figure 5. (**a**) Computation of the mapping error $e_{map}(\xi^*(k))$. Given a LiDAR reading place on a GNSS pose, each laser scan point (p_i in red) is compared with its closest map occupied point (m_{p_i} in green); (**b**) Result of the map quality assessment procedure on the Zala Zone track map. The circular markers are placed on GNSS positions; the colors, from green to red, represent the values of e_{map} (green markers correspond to lower values of e_{map}).

5. State Estimation Algorithm

An overview of the proposed localization system is reported in Figure 6. Its main components are an Extended Kalman Filter (EKF1) used to compute a high frequency velocity estimate, a LiDAR-based particle filter called IAMCL used to compute a low frequency vehicle pose estimate by comparing the LiDAR data with the offline built map, and a final Extended Kalman Filter (EKF2) used to provide a smooth, high frequency and accurate pose estimate to the vehicle control system.

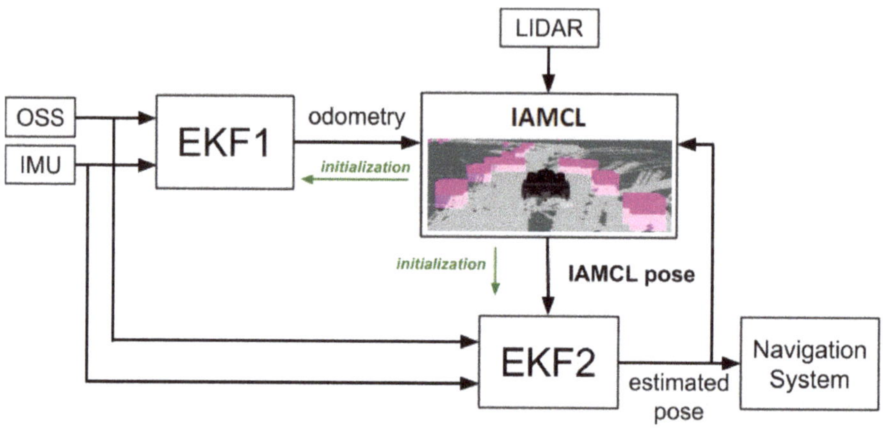

Figure 6. Localization pipeline overview. OSS and IMU data feed the Extended Kalman Filters (EKFs); EKF1 produces an odometry estimate which is sent to the Informed Adaptive Monte Carlo Localization (IAMCL) module. By comparing LiDAR data with the pre-built map, IAMCL outputs a pose estimate that is then sent to EKF2, a smoothing filter with a high-frequency output. The green arrows indicate the initialization procedure described in Section 5.2.1.

Some design choices were constrained by practical considerations related to the hardware available on the Roborace DevBot, specifically an NVIDIA DRIVE PX2 board with a non real-time Linux Kernel [36]. Thus, important constraints were in place in terms of:

- Computational burden: the NVIDIA Drive PX2 has a high number of CUDA cores, while CPU is rather limited; in this paper we propose a CPU implementation for simplicity and because the available computing power was enough for the particular experimental task;
- Flexibility: the particular race format affects not only strategy but also which sensors are available and what other modules must concurrently run on the board (e.g., interfaces with V2X race control infrastructure, planning software);
- Real-time requirements: the DevBot motion control module runs in real-time on a dedicated SpeedGoat board at 250 Hz. No patch was allowed to the standard Ubuntu kernel to make it real-time compliant. Thus, it needs to receive a pose estimate signal with high frequency.

In the following we describe the architecture of the three filters.

5.1. Odometry (EKF1)

This filter provides an odometry pose estimation. The goal of this filter is to provide the next filter (see Section 5.2) with a good velocity and orientation estimate. This filter consists of a Madgwick filter [38] for the orientation estimation and of an Extended Kalman Filter based on 4D state dynamics, in which the evolution follows a simple unicycle model (as in [39]):

$$\begin{aligned} x_{k+1} &= x_k + u_k \cos(\varphi_k)\Delta t \\ y_{k+1} &= y_k + u_k \sin(\varphi_k)\Delta t \\ \varphi_{k+1} &= \varphi_k + \hat{\omega}_k \Delta t \\ u_{k+1} &= u_k + \hat{a}_{x,k}\Delta t, \end{aligned} \tag{1}$$

where k is the time instant, x, y are the coordinates of the vehicle in a fixed world frame, φ is the vehicle yaw, and u is the longitudinal velocity, Δt is the time interval spent in the current filter iteration. The state initialization is provided by the procedure described in Section 5.2.1. The controls fed to the model are taken from the INS measurements of the angular velocity $\hat{\omega}$ and the longitudinal acceleration $\hat{a}_x$. We chose this simple model for approximating the vehicle dynamics as the longitudinal acceleration is also a control input for the vehicle, and the yaw rate is directly measured by the INS. Other models such as point mass or single track can also be employed.

During the update phase, the measurements used are the velocity estimate coming from the OSS sensor and the orientation estimate coming from the Madgwick filter. Note that, as explained in Section 3, in absence of GNSS data, the INS velocity estimate is unreliable, which is why we use OSS instead. Moreover, this filter does not receive any absolute pose estimate, thus that part of the output will tend to diverge, while still providing a smooth velocity, which is what the subsequent filter needs to work correctly.

5.2. *Lidar Scan Matching (IAMCL)*

The goal of this filter is to give an absolute pose estimate of the vehicle relative to the offline built map using LiDAR data processing. The output of this filter is rather slow, as the input data has an update frequency of 25 Hz.

While the Adaptive Monte Carlo localization (AMCL) algorithm [40] has been widely employed for many practical applications, we encountered some limitations when dealing with our specific use case. In particular we experienced that:

- The filter automatic initialization provided in the AMCL ROS package [17] takes too much time to converge; in the racing context, however, the initialization must be accurate and should be performed before the car starts driving;
- During the race, many particles are generated outside of the racing track boundaries or with opposite orientation (with respect to the race fixed direction), which is inefficient;
- Due to unavoidable, even small, map imperfections (mainly false positives in the occupancy grid), the algorithm exhibits a kidnapping problem pretty often.

To tackle these issues we introduced two improvements: an automatic initialization procedure and a localization algorithm that injects a priori knowledge into the particles' distribution.

5.2.1. Automatic Initialization Procedure

We designed an automatic procedure to initialize all filters based on the initial LiDAR measurements and the offline built map. The idea is to generate a cloud of particles along a discretized racing track the center line, aligning the LiDAR scan with each particle and looking for the best match with the occupancy grid map. Particles are only drawn among the ones within the track borders and with orientation compatible with the race direction, which is known a priori. The weights are still computed in the same way as the classic Monte Carlo localization approach [41], using the likelihood field model. This procedure, described in Algorithm 1, returns an estimate of the initial pose of the vehicle, which is then used to initialize all the filters, as shown in Figure 6. The procedure is depicted in Figure 7.

Algorithm 1: Automatic initialization algorithm (*init*).

Data: LiDAR scan $\vec{p}_i, i = 1, ..., N$ (with N number of scan points) at initial time t_0;
discretized racing line center $\vec{r} = \{(x_i, y_i, \varphi_i)\}, i = 1, ..., M$, M number of points in the center line, φ_i tangent to the line direction;
track occupancy grid map;
particle weight function $W(\cdot)$;
Result: An estimation of the vehicle's pose $\vec{q}$ in the map frame of reference.

```
Initialize a (best particle, best weight) pair;
for r_i in r⃗ do
    generate sample : generate P particles c⃗ = {c_j}, j = 1, ..., P with gaussian distribution
      N(r_i, σ), σ chosen as to cover a significant portion of the local racing track width and with
      a sufficient orientation spread;
    for c_j in c⃗ do
        while c_j lays outside of the track borders do
            c_j = generate sample()
        end
        compute particle weight w = W(c_j | p⃗, map, noise model);
        if w > best weight then
            (best particle, best weight) = (c_j, w)
        end
    end
end
return best particle;
```

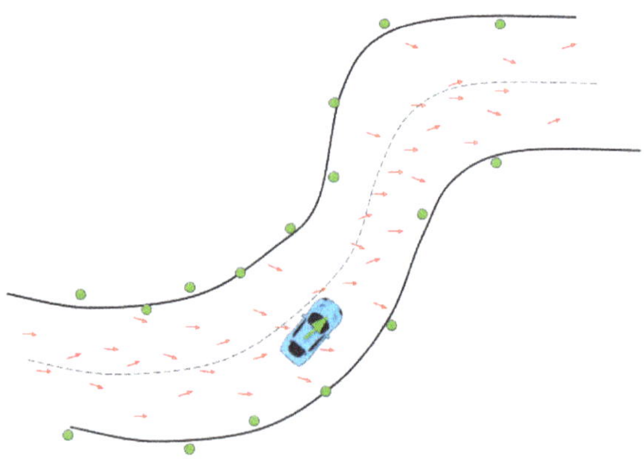

Figure 7. Representation of the automatic initialization procedure: a number of particles is generated with a Gaussian distribution around the track center line within the admissible state space (the inner area of the racing track), while limiting the particles orientation along the race direction. Each particle is represented by a red arrow; we rotate and translate the LiDAR point cloud around each red arrow, and compute the best matching particle, whose resulting match is shown in green.

5.2.2. Informed Prediction

The proposed particle filter is informed by sampling only in the positions allowed by the track, instead of sampling in the overall map space. A schematic view of the process is depicted in Figure 8.

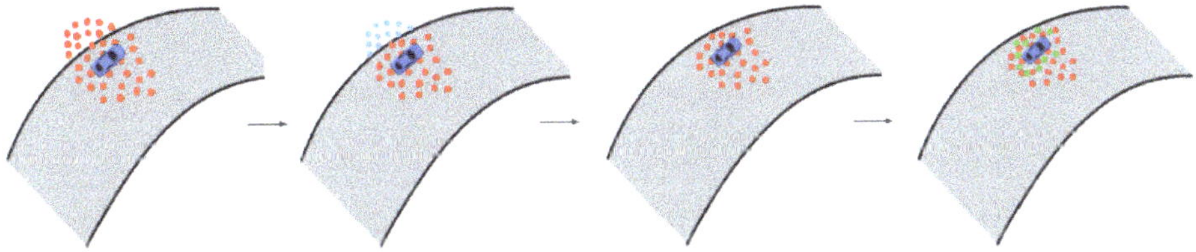

Figure 8. Illustrative schema of the IAMCL algorithm. The particles are first drawn with a distribution around an initial pose, then the ones with states that violate the track boundaries are eliminated and redrawn in the admissible state space.

A limit of a direct application of this idea is that, by constraining the particles to lie within the track borders, false positives are produced when the car happens to cross the track boundaries. This is not a concern for this application because, if this happened, a Race Control manager would call for an emergency stop. Nevertheless, we constantly monitor the weight of the particles and, in this case, if the value drops below a certain safety threshold, an emergency flag is raised. In this case, the control of the vehicle is handled by a separate emergency module in the real-time control board, which generates an emergency trajectory that allows the car to stop safely and avoiding wheel lock, even in the case of LiDAR failure, using as localization soures either GNSS (if available) or an Extended Kalman Filter based on velocity and acceleration measurements.

The theoretical framework of the particle filter for robot localization is described in [42]. A priori information can be injected into this framework during the *predict phase* by stating that

$$p(x_k|s^i_{k-1}, \mathbf{u}_{k-1}) = \begin{cases} \tilde{p}(x_k|s^i_{k-1}, \mathbf{u}_{k-1}) & \text{if } x_k \in I \\ 0 & \text{if } x_k \notin I \end{cases} \tag{2}$$

where I is the set of admissible robot states and $\tilde{p}(x_k|s^i_{k-1}, \mathbf{u}_{k-1})$ is the probability of the robot state being x_k at time k given the set of particles S at time $k-1$, and the controls $\mathbf{u}$ at time $k-1$. An explicit form of $\tilde{p}$ is not needed; instead, particles evolve according to the dynamic model when in an admissible state (see [41]), while they are redrawn around an external pose estimate when the state is not admissible, i.e., when the particle lies outside of the track boundaries. In the latter case, this pose estimate $\hat{x}_k$ is fed back from the smoothing filter described in Section 5.3, propagated forward in time using the model in Equation (1) to account for the asynchronicity among the filters:

$$s'^i_k = \hat{x}_k + \epsilon_k, \tag{3}$$

where s'^i_k is the newly generated i-th particle at time k, $\hat{x}_k$ is the latest smoothing filter pose estimate propagated at time k and ϵ_k is a gaussian noise. We choose $\hat{x}$ as output of the smoothing filter as it provides a robust localization estimate that is less affected by sudden pose jumps.

5.3. Smoothing Filter (EKF2)

The last element of the localization stack is an EKF that takes as input the OSS and the pose estimate from IAMCL. The goal of this filter is to provide a high frequency and smooth pose estimate to the vehicle control module that, in our experimental setup, runs on a real-time board. The filter is asynchronous: it iterates with a fixed frequency regardless of the sensor data received. New measurements from a sensor are stocked in a unit-length buffer, that is emptied as soon as they are used. This ensures a responsive output filter, the result of which is sent to the real-time board of the vehicle that computes the controls to the car. The system transition is based on the unicycle model (Equation (1)), where the initialization is provided by the procedure described in Section 5.2.1.

Due to the asynchronicity, with N_s independent sources of measurement (N_s = 2 in our case for OSS and pose estimate from IAMCL), there can be any of the 2^{N_s} combinations of measurements at each filter iteration, from no measurements at all, to all measurements concurrently available. The case with no measurements available involves the prediction step only. Each of these combinations requires a different observation model, defined as:

$$z(k) = h(q(k)) + v(k) \tag{4}$$

where $h : \mathbb{R}^m \rightarrow \mathbb{R}^n$ (m being the state dimensionality and n being the measurement dimensionality) is the observation model function, z the measurement vector, v the measurement noise, and

$$q(k) = \begin{bmatrix} x(k) \\ y(k) \\ \varphi(k) \\ u(k) \end{bmatrix} \tag{5}$$

is the overall filter state vector. In Table 1 we report the observation model jacobians (H) for some relevant cases.

Table 1. Some possible observation model Jacobians.

Case	H	$h(q(k))$
velocity not available	$\begin{bmatrix} 1 & 0 & 0 & 0 \\ 0 & 1 & 0 & 0 \\ 0 & 0 & 1 & 0 \end{bmatrix}$	$\begin{bmatrix} x \\ y \\ \varphi \end{bmatrix}$
pose estimate not available	$\begin{bmatrix} 0 & 0 & 0 & 1 \end{bmatrix}$	$\begin{bmatrix} u \end{bmatrix}$
all measurements available	$\mathbb{I}_{4x4}$	$\begin{bmatrix} x \\ y \\ \varphi \\ u \end{bmatrix}$

This strategy is only applied to the output filter so that the localization module is always guaranteed to produce an updated pose estimate, regardless of communication delays and sensor faults, thus ensuring that the control module always receives a high frequency and smooth signal. Moreover, to improve driving safety, the EKF2 can also raise a flag where the pose estimate is insufficiently accurate. This is performed by evaluating the covariance returned from the Kalman Filter. The output of this filter is fed back to IAMCL, as shown in Figure 6, the use of which is described in Section 5.2.2.

6. Results

Experimental results were conducted in the Zala Zone (https://zalazone.hu/) proving ground, in the course of testing for the first Roborace localization challenge that took place in August 2019. For this event, regulations allowed mapping to be performed before the race with accurate GNSS, with the availability of the RTK system. On the other hand, during the race the GNSS system was manually degraded to make it unusable for localization. This allowed us to simulate a realistic scenario and focus on the issues arising in localization, due to sensor filtering and map distortion.

In the following, we perform several comparisons between different methods and in different testing conditions, both experimental and simulated. We analyze these tests in light of the error definitions reported in Figure 9, where position error with respect to a reference trajectory is computed. Let $(\hat{x}, \hat{y})$ be the estimated position of car, and (x, y) the reference position, i.e., the GNSS signal when available. Since all signals are synchronized, it is possible to calculate the Euclidean distance between estimated and reference positions. Given the resulting vector, it can be broken down into a lateral and a longitudinal component; the first one is the so called lateral error (e_{LAT}), and the second one is the longitudinal error (e_{LON}). Also, the heading error e_{HEAD} is taken into account. These errors are computed as follows:

$$\begin{aligned} e_{\mathrm{LAT}}(k) &= |(\hat{x} - x)\sin(\psi) - (\hat{y} - y)\cos(\psi)| \\ e_{\mathrm{LON}}(k) &= (\hat{x} - x)\cos(\psi) + (\hat{y} - y)\sin(\psi) \\ e_{\mathrm{HEAD}}(k) &= \hat{\psi} - \psi \end{aligned} \tag{6}$$

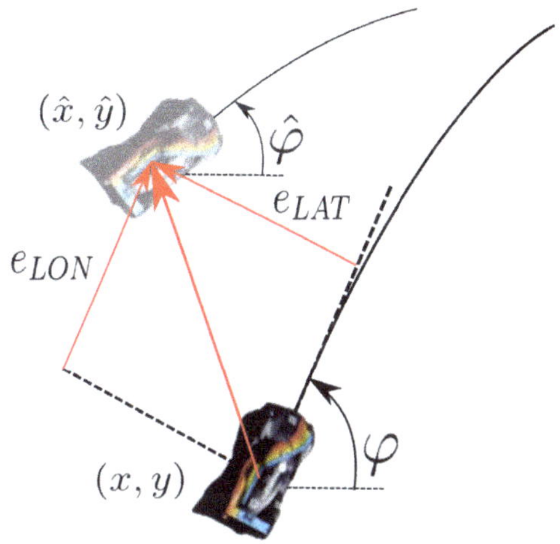

Figure 9. Longitudinal and lateral errors. These are computed as the distance error components along the GNSS defined car orientation direction.

The system was tested both in simulation and on a real racing vehicle, namely Roborace's DevBot 2.0. Simulations were performed on the *rFpro* [43] simulator, with the identified vehicle model of the DevBot car provided by Roborace running on a 3D representation of the Monteblanco circuit (La Palma Del Condado, Spain) (https://www.circuitomonteblanco.com/). Experimental results were conducted in the Zala Zone proving ground (Zalaegerszeg, Hungary) in August 2019 during the Roborace localization challenge mentioned in Section 1. The track is shown in Figure 10. The track features very narrow corridors called *gates* (see Figure 11), only 0.5 m wider than the car, used to demonstrate the localization system effectiveness. The track consists of several cones that delimits the borders, a number of water barriers and a couple of inflatable obstacles mimicking pedestrians or other vehicles.

In the following, we first present an experimental comparison between our method and the original AMCL algorithm, followed by the analysis of the results on both a simulated and a real scenario.

6.1. Comparison With State-of-the-Art

To evaluate the effect of the proposed improvements, we compare the proposed localization method using the original AMCL implementation as a baseline. First, we remark that the initialization procedure described in Algorithm 1 is a novelty with respect to the baseline, hence no quantitative comparison can be made. We focus on comparing the whole localization stack described in Section 5 with the only difference in the LiDAR scan matching filter, while keeping every tuning parameter identical, including the ones of the EKF1 and EKF2 filters.

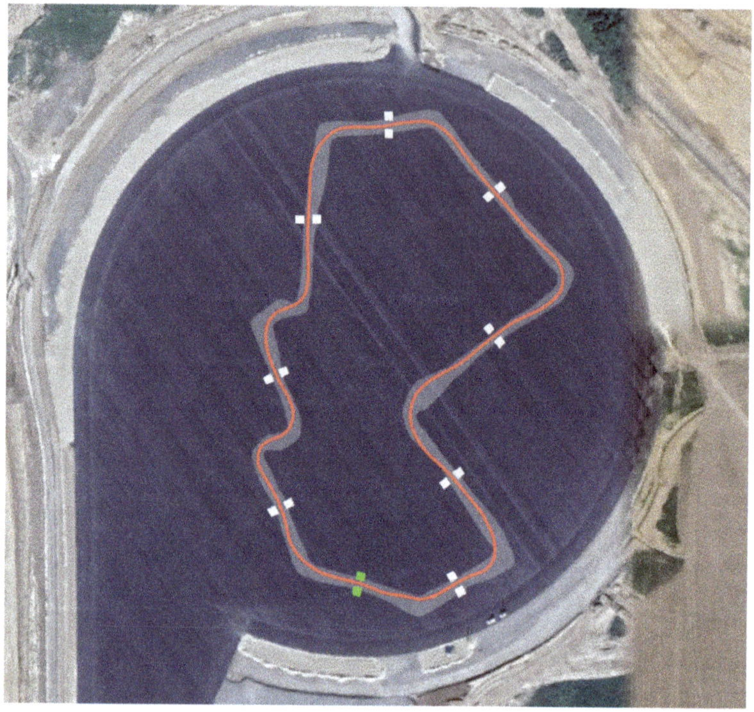

Figure 10. Satellite view of Zala Zone proving ground. Track boundaries and the racing line are shown in red. Gates are shown in white and the starting point in green.

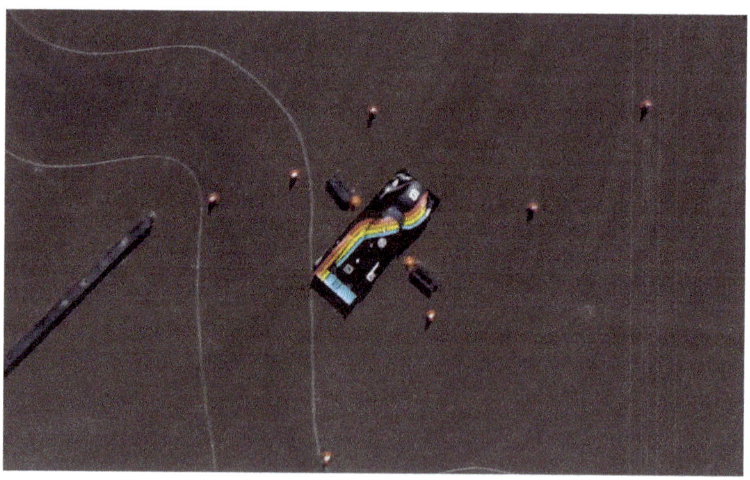

Figure 11. Top view of the Roborace DevBot 2.0 in Zala Zone Circuit. Narrow corridors (gates) were distributed along the track, the car is 2 meters wide, while the gates were 2.5 meters wide.

The comparison is performed on a dataset gathered during the trial week in the Zala Zone circuit, at a maximum speed of $v_{max} = 60$ km/h. It is worth noting that, as the stack is run offline, the localization performance has no effect on the vehicle control. We compare two variants of IAMCL with AMCL: the first IAMCL variant corresponds to the algorithm described in Section 5.2 as it is, the second one is an improvement of that same algorithm we developed after the Zala Zone race.

Indeed, during the race we observed a delay in the LiDAR scan matching module that mainly affected the longitudinal error, as it will be shown in Section 6.2. To tackle this problem, we measured the delay Δt_d between the laser scan timestamp t_l and the output pose generation and we extrapolated the actual pose at time $t_l + \Delta t_d$ with a forward Euler integration:

$$\begin{aligned} x(t_l + \Delta t_d) &= x(t_l) + v_x(t_l)\Delta t_d \\ y(t_l + \Delta t_d) &= y(t_l) + v_y(t_l)\Delta t_d \\ \theta(t_l + \Delta t_d) &= \theta(t_l) + \omega(t_l)\Delta t_d, \end{aligned} \tag{7}$$

where $(x(t_l), y(t_l), \theta(t_l))$ is IAMCL output pose, and $(v_x(t_l), v_y(t_l), \omega(t_l))$ are the Cartesian linear and angular velocities, as estimated from the odometry filter. In the following, we will refer to this improved version as IAMCL *with extrapolation*, whereas the original version (used during the tests) will be called IAMCL *without extrapolation*.

The dataset consists of several laps: here we report relevant results from the second and the beginning of the third lap, at the beginning of which we show an example of a situation where AMCL suffers from robot kidnapping and the proposed method does not. Results on the second lap are shown in Figure 12 where our method and AMCL achieve errors in the same order of magnitude. Relative to that figure, the average errors are reported in Table 2.

The localization errors obtained with the various methods indicate that: (i) extrapolation in IAMCL mitigates computational delays experienced during the tests, effectively bridging the gap with the baseline method in terms of Cartesian error.

Table 2. Comparison between the average longitudinal, lateral and heading average errors of the localization stack using IAMCL (with and without extrapolation) or AMCL.

	IAMCL (w/ extr.)	IAMCL (w/o extr.)	AMCL
Long. error (avg/max)	0.47/1.78 m	1.10/2.69 m	0.68/1.63 m
Lat. error (avg/max)	0.21/0.81 m	0.20/0.72 m	0.23/0.81 m
Heading error (avg/max)	0.51/1.39 deg	0.57/1.81 deg	0.29/1.29 deg

This is mainly due to the particle distribution being denser in more meaningful regions of the state space, and the correction of the estimate prediction with (7). On the contrary, the heading error is always better for AMCL, and this might be due to over-constraining the resampled particles that violate track boundaries with a distribution chosen by the user around the latest EKF2 pose.

We can observe a sharp difference in the performance of the various methods after the beginning of the third lap. In this lap there is an increase of speed and acceleration that triggers a kidnapping failure for AMCL, with the consequent drastic increase of the error. The kidnapping occurrence is visible in Figure 13 where it is evident that while the baseline localization fails, the proposed method does not.

We noted that when the kidnapping problem arose with AMCL, the estimation of the covariance produced was unrealistic, which made it virtually impossible for the subsequent Extended Kalman Filter to handle this emergency situation. The most likely reason for the kidnapping to happen is because of a local poor map quality, which violates the LiDAR model assumed by the AMCL algorithm. Although this effect could likely be minimized with a better mapping algorithm tuning, it nevertheless represents a huge risk in the context of racing, as the map is often unknown until shortly before the race, which makes a more robust method preferable.

In conclusion, we choose to rely on our method, both because it shows overall better positional errors, but mostly because it is more robust to local poor scan matching.

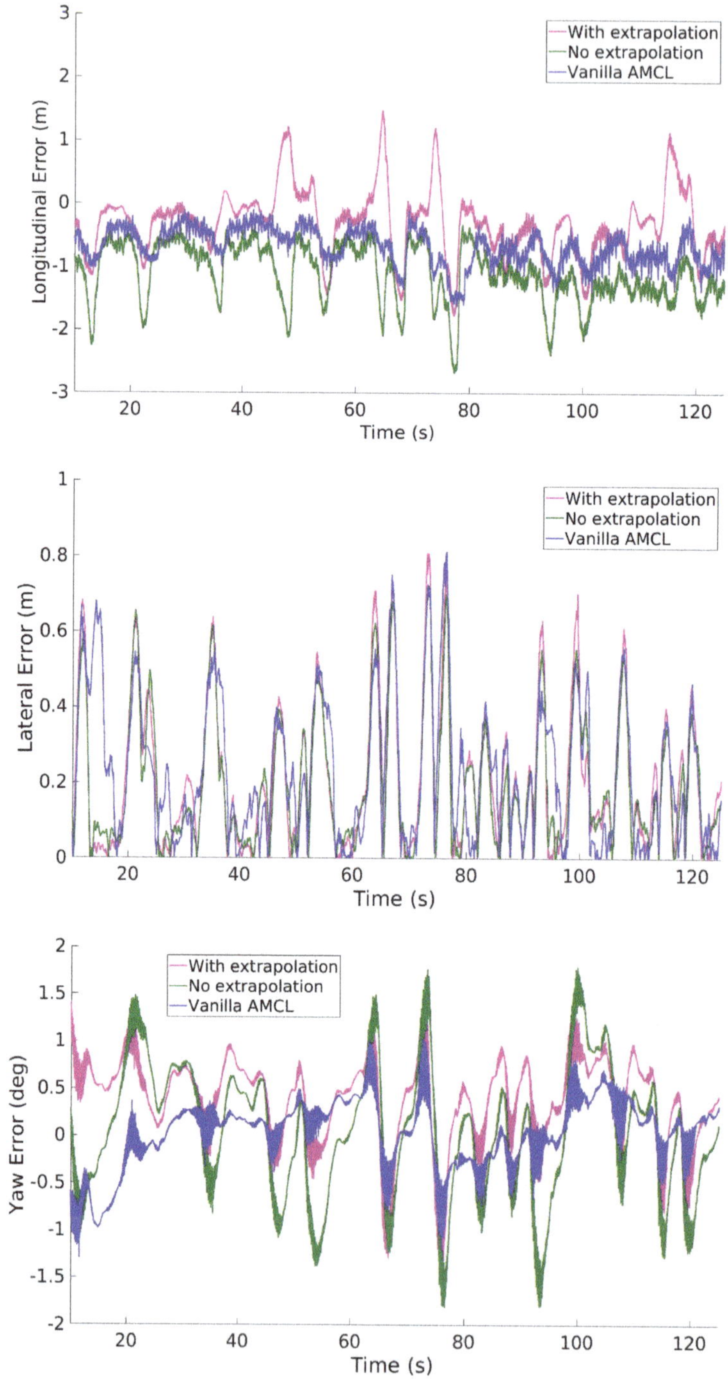

Figure 12. Comparison of the longitudinal, lateral and heading errors of two variants of IAMCL (with and without the extrapolation defined in Equation (7)) and the Adaptive Monte Carlo Localization (AMCL) algorithm during the second lap of the dataset.

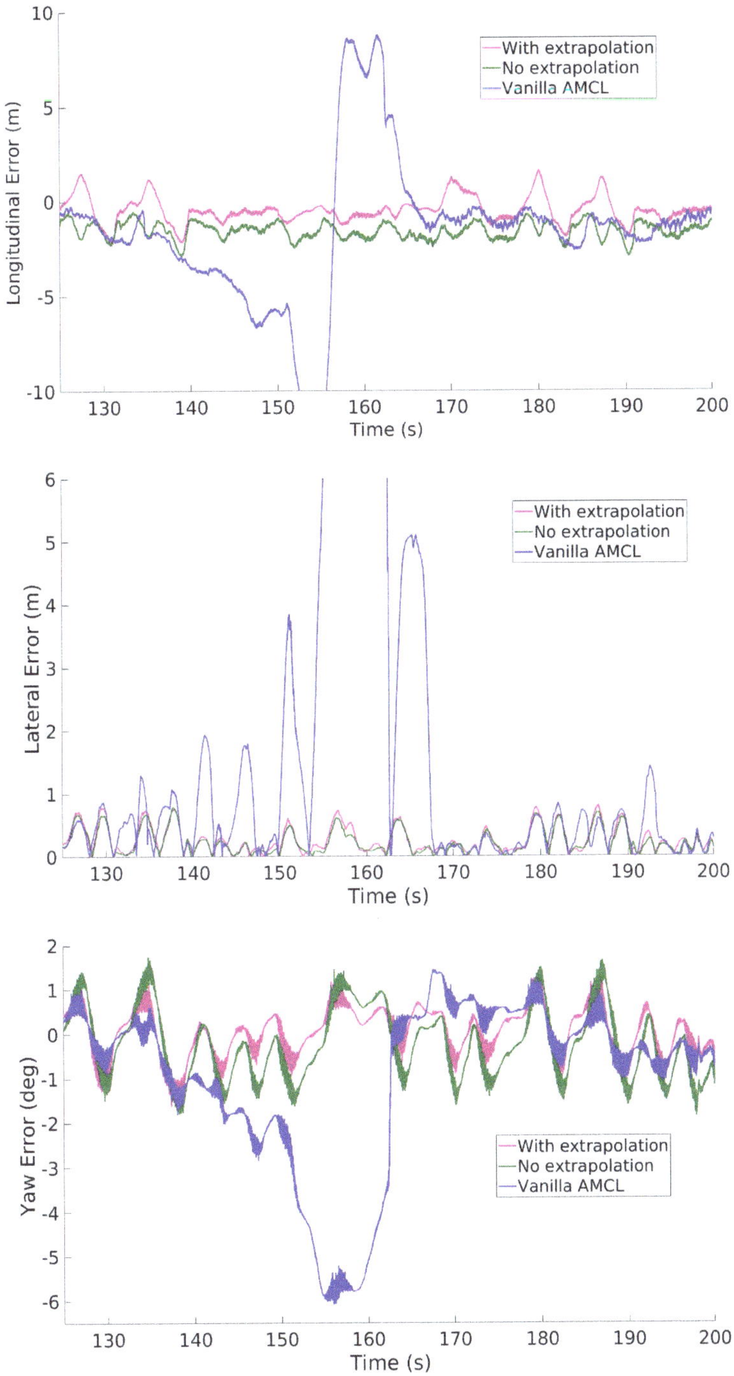

Figure 13. Comparison of the longitudinal, lateral and heading errors of two variants of IAMCL (with and without the extrapolation defined in Equation (7)) and AMCL during the third lap of the dataset, with focus on an AMCL failure (due to robot kidnapping). Note that IAMCL does not suffer from such situation even if they share the same underlying scan matching algorithm.

6.2. Experimental Tests

We present the results of our localization system in three different cases:

1. A run in autonomous mode at $v_{max} = 60$ km/h;
2. A run driven by a professional human driver at $v_{max} = 100$ km/h, the maximum speed allowed by the track;
3. A run in autonomous mode at $v_{max} = 200$ km/h on the simulator.

As the results presented in this paper are based on data collected during preparation of an official event, not every dataset has GNSS data available to be used as a truth reference. In those datasets where GNSS is available (datasets 2 and 3), we compare the localization module outputs with GNSS data, and when that is not available (dataset 1), we compare it against the racing line, i.e., the desired reference pose.

In the following, we will show the results relative to the three datasets, using the variant of IAMCL with no extrapolation (see Section 6.1), as that improvement was developed only after the race. Future tests will include that improvement in experimental tests.

6.3. Dataset 1: Autonomous Mode (Experimental)

In Figure 14, the localization performances of the autonomous mode are shown. The control law is based on [44], and the maximum achieved speed is about 60 km/h for safety reasons. In this case the GNSS signal was cut off as a race rule, so no signal to use as a truth reference was available.

Because of this, the errors in Equation (6) were computed with respect to the desired trajectory (i.e., the racing line, reference for the controller module), instead of the GNSS signal. Thus, these errors include contributions not only from the localization system, but also from the control module (although indirectly). Reported metrics are the offtrack error (analogous to the lateral error in (Equation (6)) but with the racing line as reference) (max 0.17 m, mean 0.1 m) and the heading error (max 6.8 deg, mean 1.2 deg), both computed with respect to the racing line. Despite GNSS data being unavailable during this run, the car completed the lap successfully. As the reference trajectory is not temporized, it makes no sense to compute the longitudinal error in this case.

6.4. Dataset 2: Manual Drive Mode (Experimental)

As a second result, we report on experiments performed during a run driven by a professional pilot at the maximum velocity allowed by the track. Figure 15 shows the results relative to the fourth and fastest lap of this run ($v_{max} = 100$ km/h). All errors are computed with respect to GNSS. The maximum lateral error is 0.64 m, with a mean of 0.18 m. The magnitude of the longitudinal error, about 10 times larger than the lateral (max 3.2 m, mean 1.9 m), is explained by the fact that the only absolute pose estimate provided to the output filter (EKF2) comes from IAMCL, which we measured having a response delay of 0.07 s. A car running at 100 km/h travels 2 meters in 0.07 s (a distance close to the observed longitudinal error), thus compatible with the time needed to complete an iteration of IAMCL. To compensate for this error, in future works we are going to use the IAMCL with the extrapolation improvement introduced in Section 6.1.

6.5. Dataset 3: Autonomous Mode (Simulation)

Finally, in Figure 16, localization performances of a simulation run in the Monteblanco circuit are reported. Maximum longitudinal velocity was 200 km/h, performed on the Roborace simulation system. In this case, the longitudinal error (max 8.4 m, mean 4.92 m) is even larger than in the previous case, nevertheless its magnitude is compatible with the measured IAMCL delay, as described in Section 6.4. The maximum lateral error is 1.48 m, with a mean of 0.25 m, the maximum heading error is 3.52 degrees, with a mean of 1.6 degrees. These results show the effectiveness of the proposed system also at high speeds, demonstrated by the magnitude of the lateral error that has maintained a reasonable level. The longitudinal error shows instead a significative increase, which confirms its dependency from the vehicle speed but that can be limited by using the IAMCL with extrapolation improvement described in Section 6.1, as for Dataset 2.

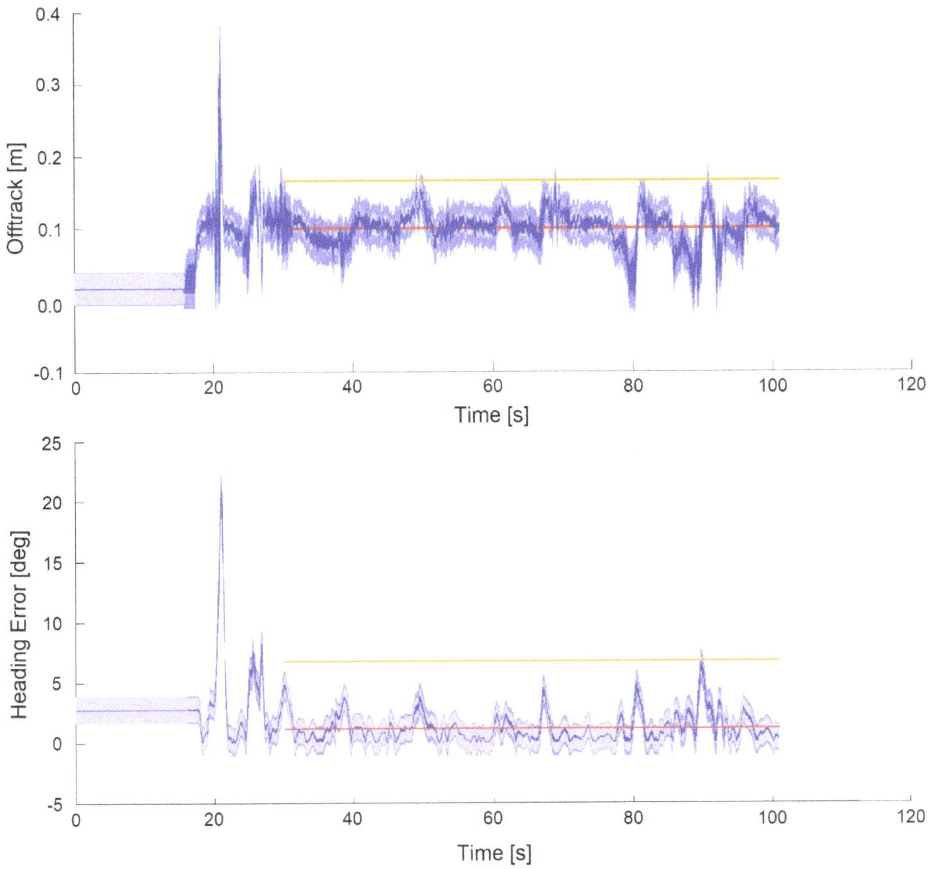

Figure 14. Autonomous lap at $v_{max} = 60$ km/h. Signals are plotted together with standard deviation (light blue), mean (red) and maximum error (yellow). The measurement of such metrics starts when the car actually starts driving after initialization.

7. Conclusions

This paper tackles the problem of localization in a GNSS-denied environment with a LiDAR-based system. The proposed method relies on two EKFs and a particle filter for LiDAR scan matching, the latter exploiting a priori information about the environment to build the particle set in a more efficient way. The envisioned application is that of autonomous racing vehicles. Reported results show good performance both in open loop (localization performed online during manual driving) and closed loop (with the pose estimate sent in feedback to the controller). Notably, our method shows robustness against the kidnapped robot failure with respect to a widely used state-of-the-art AMCL implementation. Future works will be devoted to further optimization by means of implementation in Compute Unified Device Architecture (CUDA) of the particle filter, to better exploit the GPU-based hardware available. Moreover, we aim at testing the system in a multi-vehicle racing scenario and at higher speeds.

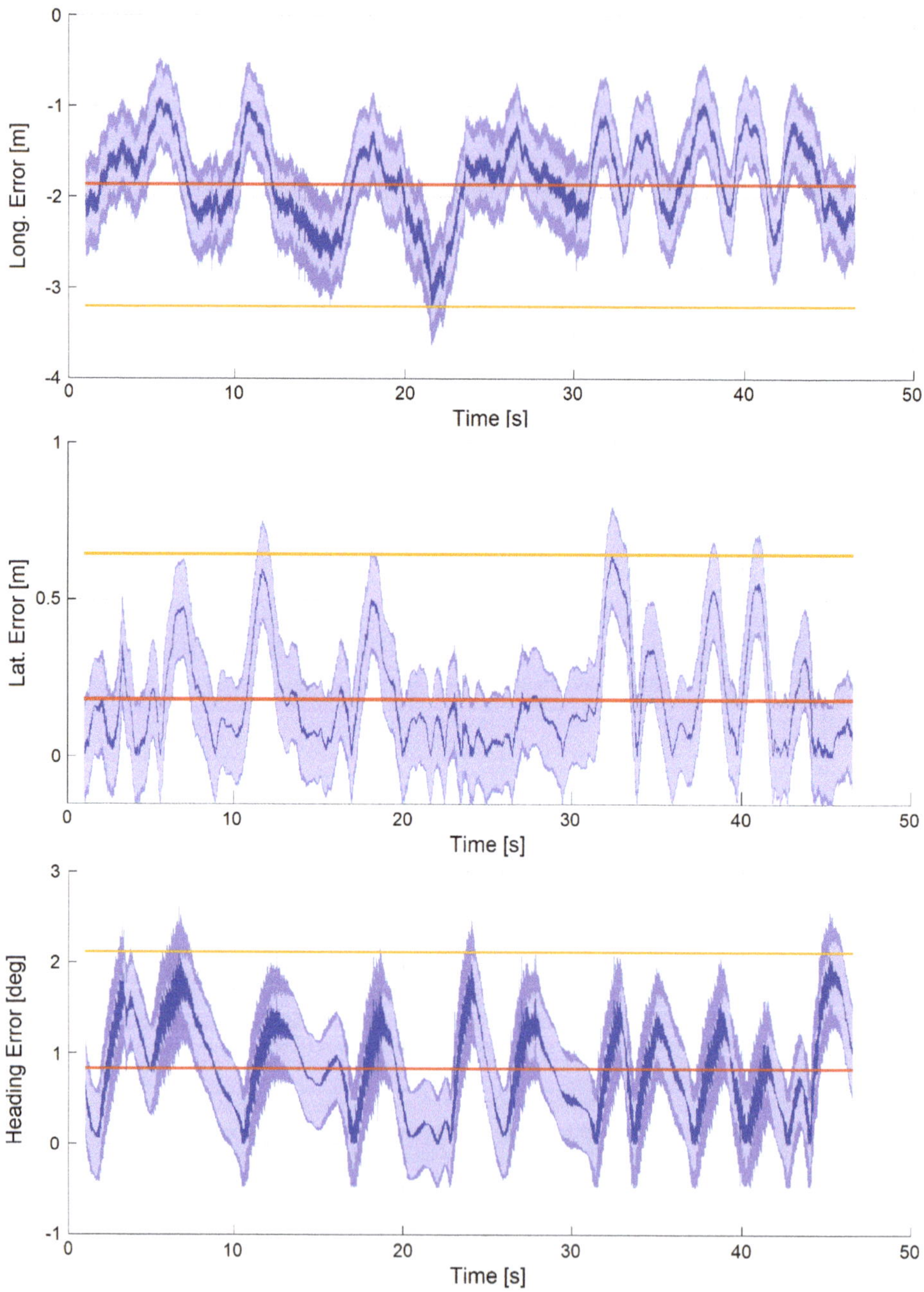

Figure 15. Manually driven lap at v_{max} = 100 km/h. Signals are plotted together with standard deviation (light blue), mean (red) and maximum error (yellow).

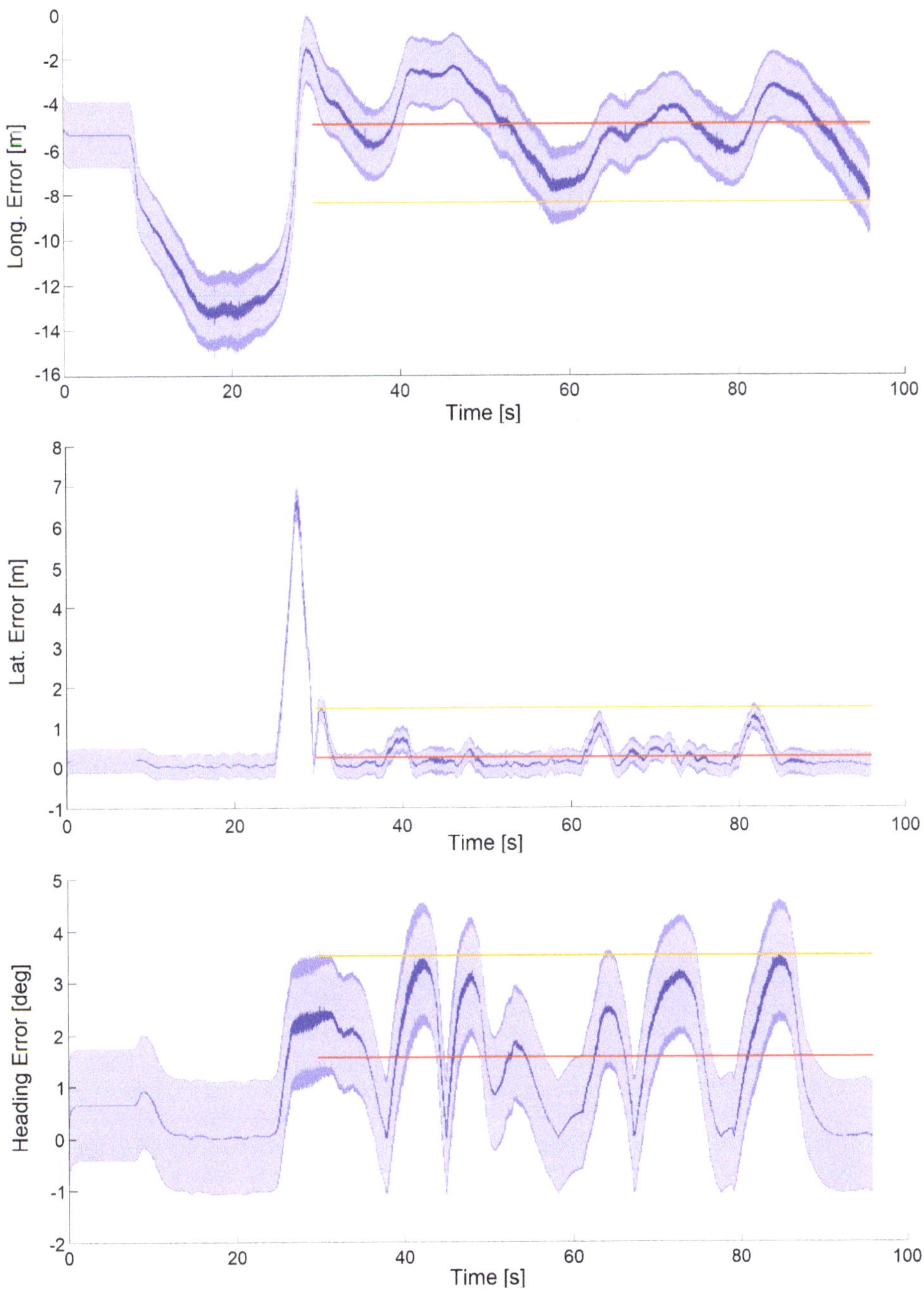

Figure 16. Autonomous lap in simulation at $v_{max} = 200$ km/h. Signals are plotted together with standard deviation (light blue), mean (red) and maximum error (yellow). The measurement of such metrics starts when the car actually starts driving after initialization.

Author Contributions: Conceptualization, F.M. and D.C.; methodology, F.M. and D.C.; software, F.M., L.B. and D.C.; validation, F.M., L.B. and D.C.; formal analysis, F.M. and D.C.; investigation, F.M. and D.C.; resources, A.S., L.P. and D.C.; data curation, F.M. and L.B.; writing—original draft preparation, F.M., L.B., A.S., and D.C.; writing—review and editing, A.S., L.P. and D.C.; visualization, F.M.; supervision, L.P. and D.C.; project administration, L.P.; funding acquisition, A.S., L.P. and D.C. All authors have read and agreed to the published version of the manuscript.

Funding: This research was partially funded by H2020-EU.2.1.1. ILIAD project, grant number 732737 and University of Pisa.

Acknowledgments: The authors would like to thank all the members of Roboteam Italia (https://www.youtube.com/watch?v=0jk1RVqIYYU) (https://www.roboteamitalia.it/) for the hard work, dedication and support provided to this project, in particular to Gioele Carignani and Andrea Giove. A special thanks goes to the Roborace team for the valuable support received during the preparation of this work and to Nvidia for the hardware support.

Conflicts of Interest: The authors declare no conflict of interest.

References

1. Karlsson, R.; Gustafsson, F. The Future of Automotive Localization Algorithms: Available, reliable, and scalable localization: Anywhere and anytime. *IEEE Signal Process. Mag.* **2017**, *34*, 60–69. [CrossRef]
2. Cadena, C.; Carlone, L.; Carrillo, H.; Latif, Y.; Scaramuzza, D.; Neira, J.; Reid, I.; Leonard, J.J. Past, present, and future of simultaneous localization and mapping: Toward the robust-perception age. *IEEE Trans. Robot.* **2016**, *32*, 1309–1332. [CrossRef]
3. Bresson, G.; Alsayed, Z.; Yu, L.; Glaser, S. Simultaneous Localization and Mapping: A Survey of Current Trends in Autonomous Driving. *IEEE Trans. Intell. Veh.* **2017**, *2*, 194–220. [CrossRef]
4. Durrant-Whyte, H.; Bailey, T. Simultaneous localization and mapping: Part I. *IEEE Robot. Autom. Mag.* **2006**, *13*, 99–110. [CrossRef]
5. Bailey, T.; Durrant-Whyte, H. Simultaneous localization and mapping (SLAM): Part II. *IEEE Robot. Autom. Mag.* **2006**, *13*, 108–117. [CrossRef]
6. Filatov, A.; Filatov, A.; Krinkin, K.; Chen, B.; Molodan, D. 2d slam quality evaluation methods. In Proceedings of the 2017 21st Conference of Open Innovations Association (FRUCT), Helsinki, Finland, 6–10 November 2017; pp. 120–126.
7. Grisetti, G.; Stachniss, C.; Burgard, W. Improving grid-based slam with rao-blackwellized particle filters by adaptive proposals and selective resampling. In Proceedings of the 2005 IEEE International Conference on Robotics and Automation, Barcelona, Spain, 18–22 April 2005; pp. 2432–2437.
8. Grisetti, G.; Stachniss, C.; Burgard, W. Improved techniques for grid mapping with rao-blackwellized particle filters. *IEEE Trans. Robot.* **2007**, *23*, 34. [CrossRef]
9. Levinson, J.; Thrun, S. Robust vehicle localization in urban environments using probabilistic maps. In Proceedings of the 2010 IEEE International Conference on Robotics and Automation, Anchorage, AK, USA, 3–7 May 2010; pp. 4372–4378. [CrossRef]
10. Levinson, J.; Askeland, J.; Becker, J.; Dolson, J.; Held, D.; Kammel, S.; Kolter, J.Z.; Langer, D.; Pink, O.; Pratt, V.; et al. Towards fully autonomous driving: Systems and algorithms. In Proceedings of the 2011 IEEE Intelligent Vehicles Symposium (IV), Baden, Germany, 5–9 June 2011; pp. 163–168. [CrossRef]
11. Javanmardi, E.; Javanmardi, M.; Gu, Y.; Kamijo, S. Autonomous vehicle self-localization based on multilayer 2D vector map and multi-channel LiDAR. In Proceedings of the 2017 IEEE Intelligent Vehicles Symposium (IV), Los Angeles, CA, USA, 11–14 June 2017; pp. 437–442. [CrossRef]
12. Meng, X.; Wang, H.; Liu, B. A robust vehicle localization approach based on GNSS/IMU/DMI/LiDAR sensor fusion for autonomous vehicles. *Sensors* **2017**, *17*, 2140. [CrossRef] [PubMed]
13. Wan, G.; Yang, X.; Cai, R.; Li, H.; Zhou, Y.; Wang, H.; Song, S. Robust and Precise Vehicle Localization Based on Multi-Sensor Fusion in Diverse City Scenes. In Proceedings of the 2018 IEEE International Conference on Robotics and Automation (ICRA), Brisbane, QLD, Australia, 21–25 May 2018; pp. 4670–4677.
14. Nobili, S.; Dominguez-Quijada, S.; Garcia, G.; Martinet, P. 6 channels Velodyne versus planar LiDARs based perception system for Large Scale 2D-SLAM. In Proceedings of the 7th Workshop on Planning, Perception and Navigation for Intelligent Vehicles, Hamburg, Germany, 18 September 2015.
15. Deilamsalehy, H.; Havens, T.C. Sensor fused three-dimensional localization using IMU, camera and LiDAR. In Proceedings of the 2016 IEEE SENSORS, Orlando, FL, USA, 30 October–3 November 2016; pp. 1–3. [CrossRef]

16. Kuutti, S.; Fallah, S.; Katsaros, K.; Dianati, M.; Mccullough, F.; Mouzakitis, A. A Survey of the State-of-the-Art Localization Techniques and Their Potentials for Autonomous Vehicle Applications. *IEEE Internet Things J.* **2018**, *5*, 829–846. [CrossRef]
17. AMCL ROS Package. Available online: https://github.com/ros-planning/navigation (accessed on 17 July 2020).
18. Xiao, L.; Wang, J.; Qiu, X.; Rong, Z.; Zou, X. Dynamic-SLAM: Semantic monocular visual localization and mapping based on deep learning in dynamic environment. *Robot. Auton. Syst.* **2019**, *117*, 1–16. [CrossRef]
19. Akail, N.; Moralesl, L.Y.; Murase, H. Reliability estimation of vehicle localization result. In Proceedings of the 2018 IEEE Intelligent Vehicles Symposium (IV), Changshu, China, 26–30 June 2018; pp. 740–747.
20. Xu, S.; Chou, W.; Dong, H. A Robust Indoor Localization System Integrating Visual Localization Aided by CNN-Based Image Retrieval with Monte Carlo Localization. *Sensors* **2019**, *19*, 249. [CrossRef] [PubMed]
21. Sun, L.; Adolfsson, D.; Magnusson, M.; Andreasson, H.; Posner, I.; Duckett, T. Localising Faster: Efficient and precise lidar-based robot localisation in large-scale environments. *arXiv* **2020**, arXiv:2003.01875.
22. Antonante, P.; Spivak, D.I.; Carlone, L. Monitoring and Diagnosability of Perception Systems. *arXiv* **2020**, arXiv:2005.11816.
23. Hess, W.; Kohler, D.; Rapp, H.; Andor, D. Real-time loop closure in 2D LIDAR SLAM. In Proceedings of the 2016 IEEE International Conference on Robotics and Automation (ICRA), Stockholm, Sweden, 16–21 May 2016; pp. 1271–1278.
24. Agrawal, P.; Iqbal, A.; Russell, B.; Hazrati, M.K.; Kashyap, V.; Akhbari, F. PCE-SLAM: A real-time simultaneous localization and mapping using LiDAR data. In Proceedings of the 2017 IEEE Intelligent Vehicles Symposium (IV), Los Angeles, CA, USA, 11–14 June 2017; pp. 1752–1757.
25. Besl, P.J.; McKay, N.D. Method for registration of 3-D shapes. In *Sensor Fusion IV: Control Paradigms and Data Structures*; International Society for Optics and Photonics: Bellingham, WA, USA, 1992; Volume 1611, pp. 586–606.
26. Daoust, T.; Pomerleau, F.; Barfoot, T.D. Light at the end of the tunnel: High-speed lidar-based train localization in challenging underground environments. In Proceedings of the 2016 13th Conference on Computer and Robot Vision (CRV), Victoria, BC, Canada, 1–3 June 2016; pp. 93–100.
27. Magnusson, M. The Three-Dimensional Normal-Distributions Transform: An Efficient Representation for Registration, Surface Analysis, and Loop Detection. Ph.D. Thesis, Örebro Universitet, Örebro, Sweden, 2009.
28. Woo, A.; Fidan, B.; Melek, W.; Zekavat, S.; Buehrer, R. Localization for Autonomous Driving. In *Handbook of Position Location: Theory, Practice, and Advances*, 2nd ed.; Wiley: Hoboken, NJ, USA, 2019; pp. 1051–1087. [CrossRef]
29. Reid, T.G.R.; Houts, S.E.; Cammarata, R.; Mills, G.; Agarwal, S.; Vora, A.; Pandey, G. Localization Requirements for Autonomous Vehicles. *arXiv* **2019**, arXiv:1906.01061.
30. Yomchinda, T. A method of multirate sensor fusion for target tracking and localization using extended Kalman Filter. In Proceedings of the 2017 Fourth Asian Conference on Defence Technology-Japan (ACDT), Tokyo, Japan, 29 November–1 December 2017; pp. 1–7. [CrossRef]
31. Hu, Y.; Duan, Z.; Zhou, D. Estimation Fusion with General Asynchronous Multi-Rate Sensors. *IEEE Trans. Aerosp. Electron. Syst.* **2010**, *46*, 2090–2102. [CrossRef]
32. Armesto, L.; Tornero, J.; Domenech, L. Improving Self-localization of Mobile Robots Based on Asynchronous Monte-Carlo Localization Method. In Proceedings of the 2006 IEEE Conference on Emerging Technologies and Factory Automation, Prague, Czech Republic, 20–22 September 2006; pp. 1028–1035. [CrossRef]
33. Lategahn, H.; Schreiber, M.; Ziegler, J.; Stiller, C. Urban localization with camera and inertial measurement unit. In Proceedings of the 2013 IEEE Intelligent Vehicles Symposium (IV), Gold Coast, QLD, Australia, 23–26 June 2013; pp. 719–724. [CrossRef]
34. Bachrach, A.; Prentice, S.; He, R.; Roy, N. RANGE—Robust autonomous navigation in GPS-denied environments. *J. Field Robot.* **2011**, *28*, 644–666, doi:10.1002/rob.20400. [CrossRef]
35. Achtelik, M.; Bachrach, A.; He, R.; Prentice, S.; Roy, N. Stereo vision and laser odometry for autonomous helicopters in GPS-denied indoor environments. In *Unmanned Systems Technology XI*; Gerhart, G.R., Gage, D.W., Shoemaker, C.M., Eds.; International Society for Optics and Photonics, SPIE: Bellingham, WA, USA, 2009; Volume 7332, pp. 336–345. [CrossRef]
36. Caporale, D.; Settimi, A.; Massa, F.; Amerotti, F.; Corti, A.; Fagiolini, A.; Guiggiani, M.; Bicchi, A.; Pallottino, L. Towards the design of robotic drivers for full-scale self-driving racing cars. In Proceedings of the 2019 International Conference on Robotics and Automation (ICRA), Montreal, QC, Canada, 20–24 May 2019; pp. 5643–5649.

37. Caporale, D.; Fagiolini, A.; Pallottino, L.; Settimi, A.; Biondo, A.; Amerotti, F.; Massa, F.; De Caro, S.; Corti, A.; Venturini, L. A Planning and Control System for Self-Driving Racing Vehicles. In Proceedings of the 2018 IEEE 4th International Forum on Research and Technology for Society and Industry (RTSI), Palermo, Italy, 10–13 September 2018; pp. 1–6. [CrossRef]
38. Madgwick, S. An efficient orientation filter for inertial and inertial/magnetic sensor arrays. *Rep. X-Io Univ. Bristol (UK)* **2010**, *25*, 113–118.
39. Paden, B.; Čáp, M.; Yong, S.Z.; Yershov, D.; Frazzoli, E. A survey of motion planning and control techniques for self-driving urban vehicles. *IEEE Trans. Intell. Veh.* **2016**, *1*, 33–55. [CrossRef]
40. Fox, D.; Burgard, W.; Dellaert, F.; Thrun, S. Monte carlo localization: Efficient position estimation for mobile robots. *AAAI/IAAI* **1999**, *1999*, 2.
41. Thrun, S.; Burgard, W.; Fox, D. *Probabilistic Robotics (Intelligent Robotics and Autonomous Agents)*; The MIT Press: Cambridge, MA, USA, 2005.
42. Dellaert, F.; Fox, D.; Burgard, W.; Thrun, S. Monte carlo localization for mobile robots. In Proceedings of the 1999 IEEE International Conference on Robotics and Automation (Cat. No.99CH36288C), Detroit, MI, USA, 10–15 May 1999; Volume 2, pp. 1322–1328.
43. rFpro Website. Available online: http://www.rfpro.com/ (accessed on 17 July 2020).
44. Kapania, N.R.; Gerdes, J.C. Design of a feedback-feedforward steering controller for accurate path tracking and stability at the limits of handling. *Veh. Syst. Dyn.* **2015**, *53*, 1687–1704. [CrossRef]

sensors

MDPI

Article

Real-Time Traffic Light Detection with Frequency Patterns Using a High-Speed Camera

Kento Yabuuchi [1,*], Masahiro Hirano [2], Taku Senoo [3], Norimasa Kishi [2] and Masatoshi Ishikawa [2]

[1] Department of Creative Informatics, Graduate School of Information Science and Technology, The University of Tokyo, Hongo 7-3-1, Bunkyo-ku, Tokyo 113-8656, Japan

[2] Information Technology Center, The University of Tokyo, Hongo 7-3-1, Bunkyo-ku, Tokyo 113-8656, Japan; hirano@ishikawa-vision.org (M.H.); kishi@ishikawa-vision.org (N.K.); ishikawa@ishikawa-vision.org (M.I.)

[3] Graduate School of Advanced Science and Engineering, Hiroshima University, 1-4-1 Kagamiyama, Higashi-Hiroshima City, Hiroshima 739-8527, Japan; taku-senoo@hiroshima-u.ac.jp

* Correspondence: kento.yabuuchi@pf.is.s.u-tokyo.ac.jp

Received: 31 May 2020; Accepted: 17 July 2020; Published: 20 July 2020

Abstract: LEDs are widely employed as traffic lights. Because most LED traffic lights are driven by alternative power, they blink at high frequencies, even at twice their frequencies. We propose a method to detect a traffic light from images captured by a high-speed camera that can recognize a blinking traffic light. This technique is robust under various illuminations because it can detect traffic lights by extracting information from the blinking pixels at a specific frequency. The method is composed of six modules, which includes a band-pass filter and a Kalman filter. All the modules run simultaneously to achieve real-time processing and can run at 500 fps for images with a resolution of 800 × 600. This technique was verified on an original dataset captured by a high-speed camera under different illumination conditions such as a sunset or night scene. The recall and accuracy justify the generalization of the proposed detection system. In particular, it can detect traffic lights with a different appearance without tuning parameters and without datasets having to be learned.

Keywords: traffic light detection; intelligent vehicles; high-speed camera; image processing; real-time systems

---

1. Introduction

Automobiles play an important role in modern society. Modern cars are cheaper, faster, and conbenient to use in many cases; however, many accidents occur every year. Statistics show that 94% of all traffic accidents are due to human error, and approximately, 38,000 vehicle accident deaths are reported each year in the United States [1]. There has been considerable research and development in autonomous vehicle and advanced driver-assistance systems (ADAS), which are expected to predict dangerous events and reduce traffic accidents. However, building automatic driving systems is challenging. An intersection that controls the flow of vehicles and pedestrians is an important aspect of for autonomous driving. In 2017, 890 people died in traffic collisions that involved running a red light in the United States [2]. An automatic driving system can make critical safety decisions in accordance with the state of the traffic lights. Therefore, it should be able to reliably recognize the state of a traffic light from a long distance and in real time. Such an automatic detection system has not been developed yet. As discussed in [3], traffic light detection for complex scenes is a significant challenge. Some of the factors contributing to these complex scenes include various illumination conditions; incomplete shapes due to occlusion; very few pixels for detecting distant traffic lights; and motion blurring due to high-speed driving.

The detection system requires a camera to recognize the state of the traffic light's lamp pattern. Therefore, many traffic light detection systems are based on image-processing techniques.

Numerous methods are based on vision techniques. These can be categorized into three approaches; heuristic model-based approaches, learning-based approaches and auxiliary sensor-based approaches. The heuristic model-based approach uses visual characteristics of a traffic light such as the color and shape [4–7]. This approach is intuitive, and parameter tuning is easy. The learning-based approach requires many traffic light images to construct a neural net that detects the traffic light. Because of the rapid development of machine learning techniques, this is currently one of the most popular approaches [8,9]. Some leaning based methods include not only traffic light detection but also car detection [8] and approaches that recognize which lane a traffic light belongs to have been developed [10,11]. The auxiliary sensor-based approach uses sensors other than the camera to perform accurate detection by integrating information [12]. Usually, this approach requires a prior map, which has the 3D location of the traffic light and the intersection [13,14]. Because creating a map requires a high cost, it is inconvenient to use it in a large area. Therefore, the scope of this approach is limited.

The appearance of the traffic light varies by country, region, and manufacturer. The different appearances make it difficult for the heuristic model-based and the learning-based approaches to detect it. In contrast, LED traffic lights are widely used because they can achieve better energy efficiency. Because the LED traffic light is driven by an alternate current (AC), it blinks in proportion to the input AC power. LED traffic lights blink at a high frequency, and neither the naked eye nor standard cameras can recognize it. A high-speed camera can capture images at several hundred fps, and it can recognize LED traffic lights. In [15], a hybrid traffic light detection system which combines frequency analysis and visual information with a high-speed camera was proposed. This approach encodes the variation of the brightness for the pixels. Afterwards, it detects the traffic light by extracting the area that shows the specific blinking pattern from the time-series image. By recognizing the blinking traffic light, this approach can achieve more robust detection in comparison to conventional methods that only use visual information.

Although previous work was performed with a high accuracy, it is far from practical use. False detection was observed in scenes that have irregularly reflecting objects or contain blinking self-luminous objects such as electronic bulletin boards. In addition, the conventional method cannot process several hundred images in real-time. This study proposes a real-time traffic light detection method based on blinking LED lights. This system can perform detections more robustly during severe illumination conditions. The contributions of this investigation include the achievement of a robust system with real-time performance that cannot be achieved with the conventional hybrid traffic light detector. The key elements to achieve improved robustness are extracting the blinking with a band-pass filter and state estimation using a Kalman filter. In the proposed method, all processing modules run in a parallel pipeline to enhance the throughput. Therefore, this investigation implemented a detection method with a concise algorithm that does not require a large amount of calculation to achieve real-time processing.

The next section will discuss the related works. The third section describes the proposed traffic light detection method, and the forth section presents the results to verify the performance of this method. Finally, the last section provides the conclusions of this study and our future work.

2. Related Works

Many previous studies have focused on visual information. We categorize the traffic light detection methods by the approaches used in previous studies.

2.1. Heuristic Model-Based Detection

The heuristic model-based approach uses the color, shape, location and edge information of the traffic light [4,5,7,16]. In some studies, a combination of multiple types of data increased the accuracy of detection [6,17]. The parameters and algorithms are very intuitive and can have a wide range of applications. Some methods that concentrate on the color of traffic lights perform detection by extracting a specific color in the RGB color space. Methods that utilize the HSV color space [6,16] or the LAB color space [4,7] have also been proposed. Meanwhile, some methods that focus on the edge and shape of traffic lights perform detection using Hough transform, which extracts circular regions [17]. In other similar methods, the spot-light detection or radial symmetry transform are used for extracting circular regions [4,17]. In [18], a method was proposed in which multiple cameras with different viewing angles are installed for traffic light detection over a wide range.

2.2. Learning-Based Detection

Learning-based approaches have become popular in the recent years owing to the rapid developments in machine learning and object detection techniques [19,20]. In comparison with heuristic model-based approaches, learning-based approaches requires much more training data and computation capacity. However, they are superior to the others owing to their higher robustness to variations and lower tendency to over-fitting. In [21], a method to reduce traffic light candidates based on position and size was proposed. In some detection methods, regions that are not traffic lights are often adopted as candidates. Moreover, these techniques are designed to reduce the computation on the recognition process. It is important to not only detect the traffic lights but also select them corresponding to the vehicle's current lane for practical application. Some methods detect traffic lights and classify which lane the traffic light corresponds to using CNN [10,11]. In [22], traffic light detection was based on a deep neural net and a prior map. They used a prior map to select traffic lights corresponding to the vehicle's current lane among the lights detected by the network. By estimating the location of the corresponding traffic lights in the image, the method achieved an efficient reduction of false positives. There are detectors that use a heuristic approach to select the region of interest to enable light weight and real-time detectors while still using CNN recognition [23]. Generally, learning-based approaches require costly network training. However, transfer learning can be used to reduce the computational and time resources during training [24,25]. In some cases, high recognition accuracy is achieved by fusing HOG features and features extracted by CNN [9]. Network models such as RetinaNet [26], and YOLO [8,22] have also been studied. Meanwhile, a method for designing an original background suppression filter and learning filter coefficients using numerous traffic light images without a neural network was proposed [27].

2.3. Auxiliary Sensor-Based Detection

The auxiliary sensor-based approach uses additional sensors such as GNSS, accelerometer, gyroscope, stereo-vision and LiDAR. Moreover, in some cases, a smartphone is used as a sensor to detect the traffic lights [12]. Efficient detection is achieved using the IMU and self-location, which help limit the search range in the image. In [13,14], methods that can predict the position of appearance of traffic lights in the camera view based on self-location estimated by LiDAR and GNSS and a prior map are proposed. These studies also proposed an efficient way to create the prior map using a stereo camera. The methods that used auxiliary sensors can detect the traffic light very accurately. However these methods generally require expensive sensors in addition to the cost of creating a prior map.

2.4. Hybrid Traffic Light Detection

Z. Wu demonstrated the effectiveness of the hybrid approach [15]. The greatest advantage of the hybrid approach is that it detects traffic lights in nighttime when housing can barely be recognized and lights are more difficult to distinguish from other lights. This is because the hybrid approach not only uses visual information but also uses the frequency information. Figure 1 shows the traffic lights driven by a 50-Hz AC taken using a 500 fps camera. This phenomenon is unnoticeable to the naked eye but can be detected by a high-speed camera. Because there are almost no objects blinking as fast as traffic lights, they can be detected by extracting the blinking area. Although previous studies had high accuracy in many scenes, it is still far from practical use for automatic driving systems or driver assistance systems.

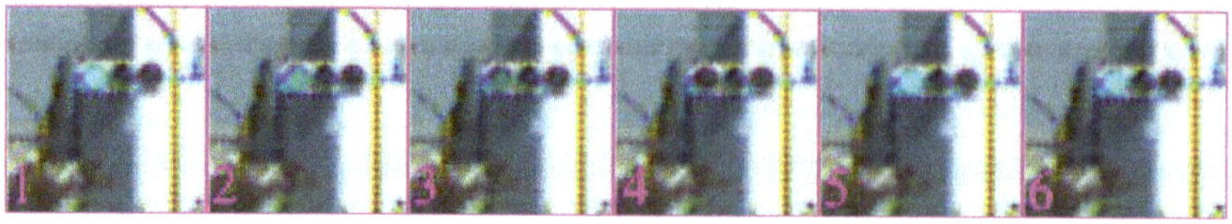

Figure 1. Example of the blinking of LED traffic light observed by a 500 fps camera. The time intervals are 2 ms.

This approach is not available in areas where such traffic lights are not installed or on cars that are not equipped with high-speed cameras. However, it is very useful to detect traffic lights at night or in the evening, which is difficult to do by other approaches. The use of a high-speed camera has other advantages. First, flicker problems caused by blinking traffic lights do not need to be addressed. In addition, the images are not blurred even if the vehicle moves at high speed.

3. Real-Time Traffic Light Detection System

3.1. Overview

This study considered a case where the traffic light was driven by a 50-Hz AC; thus, the lamp blinks at 100 fps. For this investigation, a high-speed camera that operates at 500 fps was mounted on a car because it could capture five images during a single period of lamp blinking. The proposed detection system consists of six modules that include loading, band-pass filter, binarization, buffering, detection, and classification. The overview of the proposed detection method is provided in Figure 2.

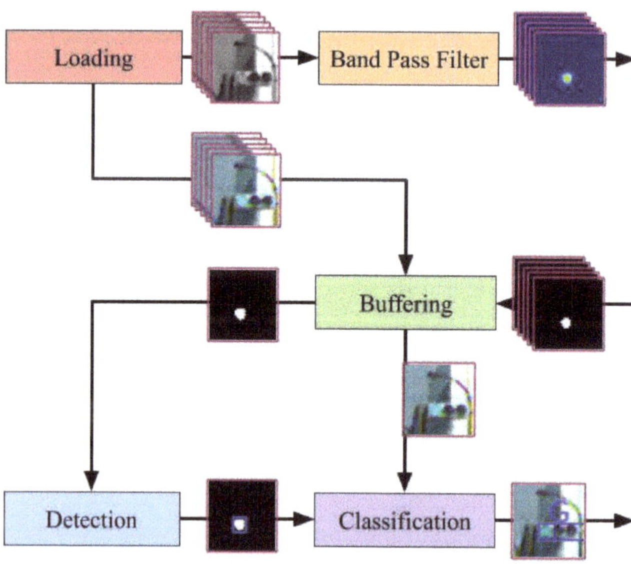

Figure 2. Procedure of the proposed traffic light detection method.

The loading module retrieves the images from the camera devices. In the experiment described in the fourth section, the image data were read from a stored video file. To emphasize, this data were not acquired directly from the on-vehicle camera. Moreover, the band-pass filter module applies the band-pass filter to the gray-scaled input image in the frequency domain and not in the spatial domain. This filter enhances the blinking area at a rate of 100 fps. The binarization module first estimates the state of the traffic light dynamics, which includes the blinking amplitude, offset, and phase. It uses the

Kalman filter for state estimation. Subsequently, it determines an appropriate threshold for binarizing the filtered image based on the estimated state and finally binarizes the filtered image. The buffering module relays the image, which has stronger signals compared to the previous images. This is because recognizing colors and areas from an image with non-maximum brightness is difficult. The detection module extracts the contours from the peak binarized image. Further, the size and shape are used to exclude candidates to prevent false detections. The classification module classifies the lamp color using the contours and RGB images. A support vector machine (SVM) was used for classification into three classes labeled red, yellow, and green. Figure 3 illustrates a summary of the processed images for each module. As depicted in the figure, the green traffic light was a successfully detected.

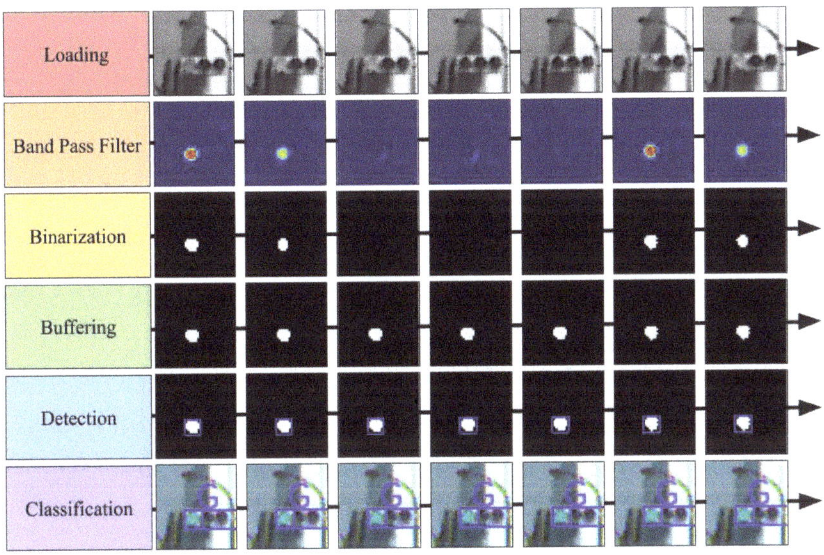

Figure 3. Sequence of the processed images by each module in the system.

3.2. Band-Pass Filter Module

This module enhances the area blinking at a specific frequency by applying the band-pass filter to the gray-scaled image over time. We used the IIR filter, which had steep frequency characteristics even in small dimensions, to reduce the amount of computation required for real-time processing. Table 1 shows the conditions for designing the band-pass filter.

Table 1. Condition of the band pass filter.

Parameter	Value
Sampling Rate	500 Hz
Pass Band	95–105 Hz
Filter Type	IIR (Butterworth)
Filter Order	4

An example of the images that were subjected to the filter is provided in the second row of Figure 3.

The band-pass filter module requires more computation than any other modules. Each filtering operation requires addition and multiplication of floating-point numbers in the order of several times the number of pixels. In addition, the operation must be performed several hundred times per second owing to the high-speed camera. If these calculations are performed in a straightforward manner, the real-time performance easily fails. Therefore, we devised several methods to increase the computational speed. In the field of view of an on-vehicle camera, the area in which the traffic light appears on the image is limited. Therefore, we decided not to process the area where no traffic lights would appear. The band-pass filters performed multiplication and summation of floating-point numbers. Consequently, there was negligible data dependency and no conditional branching. Based on this property, we implemented the filtering process by employing parallel processing with single instruction multiple data (SIMD) and OpenMP. In particular, using Intel's AVX512 instruction set, it was possible to multiply 16 pairs of single-precision floating-point objects simultaneously. By constructing an IIR filter with such instructions, the computation time was significantly reduced.

3.3. Binarization Module

For the investigation, binarizing the image that was applied to the band-pass filter was necessary to efficiently extract the blinking area. There are two methods of binarization: one is to use a common threshold for the entire image, and the other is to adaptively use the variable threshold for each pixel according to the surrounding pixels. We opted for the first method that used a common threshold as it reduced the computational cost. The appropriate threshold for binarization varies based on the conditions. In particular, the appearance of the traffic light on the image captured during daytime and night differs significantly. The state of traffic light was estimated. It included the amplitude, offset, and phase of the blinking traffic light. Calculating the appropriate threshold removed the disturbance while retaining the traffic lights in the image. For the state estimation, this study assumed that all the traffic lights in an image exist in the same state. This indicates that when there are traffic lights in the image, the blinking amplitude and phase for all traffic lights are the same. Figure 4 shows an example in which the phases of all traffic lights are aligned. In the figure, all the lamps seem to be turned off because all the amplitudes of all lights are the minimum simultaneously.

Figure 4. Example of the traffic lights with the same blinking phase.

After the image was passed through the band-pass filter, the pixel value corresponding to the blinking traffic light was obtained based on the absolute value of a sine wave. However, the pixel value in another area was suppressed. The brightness of each pixel in the image was modified, as shown in Figure 5.

In the figure, the red curve represents the pixel corresponding to the traffic light, and the orange curve represents the non-traffic light areas. Here, the envelope of the brightness can be approximated using the following equation:

$$\text{brightness} = A|\sin 2\pi f t| + b \quad (1)$$

where, A and b vary depending on the illumination conditions, and f is the frequency of the AC.

When the band-pass filter designed above was applied to videos with 8-bit color depth, A ranged from -40 to -100 and b ranged from 50 to 130. $|A|$ depends on how sharply the camera captures the blinking traffic lights in contrast to the background. For instance, owing to the dynamic range of the camera, blinking traffic lights in a bright scene do not appear as sharp as they do in a dark scene; therefore, $|A|$ is small. Next, $|b|$ is close to $|A|$, but $|b| - |A|$ varies depending on blinking disturbances. $|b| - |A|$ represents how strongly blinking noise is retained and this is equal to the length from 0 to the bottom of the envelope in Figure 5. If non-traffic lights are completely removed by a band-pass filter, then $|b|$ and $|A|$ are equal. In contrast, in scenes with many non-traffic lights, such night scenes, there is a large gap between $|b|$ and $|A|$. Within the same scene, A and b remain approximately constant over a short time period of 1 s. When binarization was performed using the blue dotted line as the threshold, the noise was removed without affecting the pixels of traffic light signals. An example of the binarized images is shown in the third row of Figure 3. The state was estimated as follows:

$$(A, b, \theta = 2\pi f t)^\top. \tag{2}$$

Subsequently, the threshold was set using $|b| - |A| \cdot k$, where k is a parameter that adjusts the severity of the threshold. The value of the envelope was obtained each time by searching for the brightest pixel in the image. We approximated the dynamics using a sufficiently simple function; therefore, this study used an extended Kalman filter (EKF) for the state estimation. The process formula of the EKF is as follows:

$$\begin{pmatrix} A_{t+1} \\ b_{t+1} \\ \theta_{t+1} \end{pmatrix} = \begin{pmatrix} A_t \\ b_t \\ \theta_t \end{pmatrix} + \begin{pmatrix} 0 \\ 0 \\ 2\pi f \Delta t \end{pmatrix}. \tag{3}$$

The observation formula is given by Equation (1).

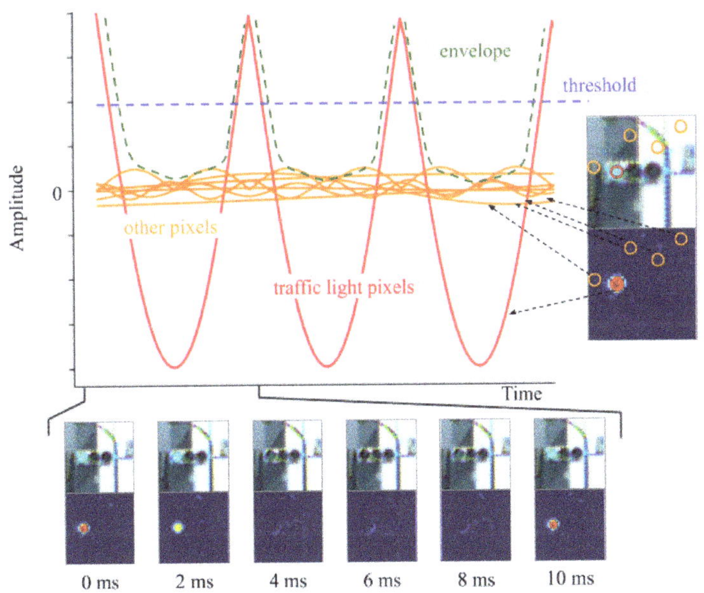

Figure 5. Brightness variation for each pixel.

We assumed that the process and measurement noise covariances are constant and there is no covariance between any of the variables. Specifically, the covariances of the process noise and the observed noise were expressed as Q and R, respectively, and we defined them as follows:

$$Q = \begin{pmatrix} \sigma_A^2 & 0 & 0 \\ 0 & \sigma_b^2 & 0 \\ 0 & 0 & \sigma_\theta^2 \end{pmatrix} \tag{4}$$

$$R = \left(\sigma_{\text{brightness}}^2\right). \tag{5}$$

Each self-covariance was adjusted manually. The EKF worked satisfactorily under these loose conditions described above, although more precise values could be set if the characteristics of the band-pass filter and the characteristics of the signal lights were considered. Using the EKF defined above, we estimated the current state of the blinking dynamics of the traffic lights. The threshold was adaptively determined using the estimated state; consequently, the binary image was obtained.

3.4. Buffering Module

This module performs two tasks: (1) it seeks the local maximum image from the last images, thus simplifying processing for subsequent modules; and (2) it compensates for the phase delay of the binarized image to aid its synchronization with the RGB image.

3.4.1. Searching for the Local Maximum Image

The blinking of the binary image and the RGB image makes it difficult to classify the color and to detect contours. This module selects an image that is easy to classify from the last five images and passes it to the subsequent modules. Specifically, the last five images are stored using a ring buffer, and the best image is selected using the phase estimated by the EKF in the binarization module.

3.4.2. Compensation for the Phase Delay Caused by the Band-Pass Filter

The binarized image passes through the band-pass filter. Therefore, the signal is delayed with respect to the RGB image. When the camera is moving, the pixel positions of the traffic light in the binary and the RGB images are different. In this case, when classifying the lamp color, the colored traffic light image could be incomplete, because of which its classification may fail. For this investigation, the data transfer of the RGB image was delayed to synchronize it with that of the binarized image. The phase delay owing to the band-pass filter is constant because the blinking frequency of the traffic light and the sampling rate of the camera are invariable. The appropriate delay was calculated in advance and the time consistency was adjusted in the images.

3.4.3. Handling Redundant Calculations Caused by Buffering

The traffic light achieves a maximum brightness once for every five inputs; consequently, the output image of this module is often duplicated. Therefore, the succeeding module uses the same image five times and produces the same output. As a measure to eliminate this redundant processing, we had an option to execute the subsequent modules only once for every five inputs. However, the duplicate output was not omitted to make the system straightforward.

3.5. Detection Module

An API called findContours was provided by the open source computer vision library OpenCV to detect the candidate. The contour extracted from the binary image consists of points that cover the periphery of the foreground area on an image. The extracted contours may contain noise that cannot be removed by a band-pass filter or binarization. Consequently, this was filtered using a circularity of the contour, the area, and the position on the image so that they could be removed. Circularity is defined as:

$$\text{circularity} = \frac{4\pi \cdot \text{Area}}{(\text{perimeter})^2}. \tag{6}$$

This indicates that a circle has a circularity of 1. That is, circularity indicates how close a contour is to a circle. Considering that the lamp of a traffic light is circular, contours that were not circular were removed.

3.6. Classification Module

An SVM was used to recognize the color of the traffic lights. There are three classification labels for SVM: "red", "yellow" and "green". The input image for the SVM had a resolution of 10 × 10 pixels, an 8-bit color-depth, and three channels. When an image was provided as input to the SVM, the bounding box of the contour was transformed into 10 × 10 pixels. The data for SVM training were generated using random numbers and were not extracted from the actual scene. The reason behind randomly generating images was to make the SVM less dependent on the appearance of the traffic lights. Figure 6 depicts an example of the generated training data. This module is only required to be able to classify three colors. Therefore, the classification in this method was implemented in a very concise manner. Specifically, 100 images were used for each label, and the training was performed using the default parameters of the SVM module in OpenCV.

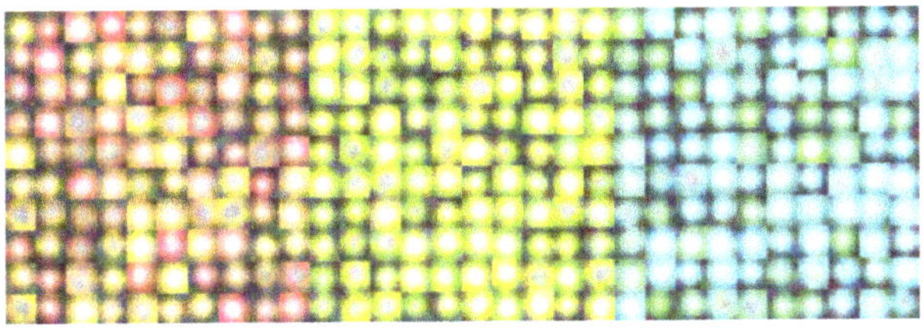

Figure 6. Examples of the training data for the support vector machine.

Traffic lights are always lined up in the order of red, yellow, and green. Based on this, the area of the traffic light housing was also estimated from the color, size, and position of the blinking area.

In Figure 2, there are two paths from buffering and detection to classification. The two paths refer to the data flow of the RGB image and the binarized image, respectively. This system is implemented to reduce the computational complexity and to avoid unnecessary copying of data as much as possible. RGB images are essentially only needed in the classification module, not in the other modules such as the band-pass filter, binarization, and detection. Therefore, the RGB images are passed through the modules in the order of loading, buffering, and classification.

3.7. Multi-Thread Processing

In this system, all modules depend on their previous modules, some of which require considerable computation to process an image. The proposed system must be calculated at 500 fps or more for its practical application. Therefore, this investigation adopted a parallel pipeline process to increase the throughput. Figure 7 provides an overview of the image processing. The modules start the process as soon as the required data are available in the queue. This does not reduce the amount of computation; therefore, the latency either remained constant or it may have been slightly increased owing to the effect of the data transfer between the threads. Moreover, the latency caused by pipeline processing of the six modules can be ignored during the operation because the system processes images at several hundred fps.

4. Experimental Results

The video sequences were taken in an urban street in East Japan using a Basler high-speed camera (model: acA800-510uc). It was mounted on a car and its output was an image with 800 × 600 pixel resolution, and the frame rate was 500 fps. The dataset collection was supported by Kotei Informatics Corporation. In all experiments, considering that the color depth of the camera image is 8 bits, we set the EKF parameters σ_A^2 and σ_b^2 to 65 , σ_θ^2 to 0.03 and $\sigma_{\text{brightness}}^2$ to 1600. Data in most public datasets are not captured by high-speed cameras; hence they could not be used to evaluate this system.

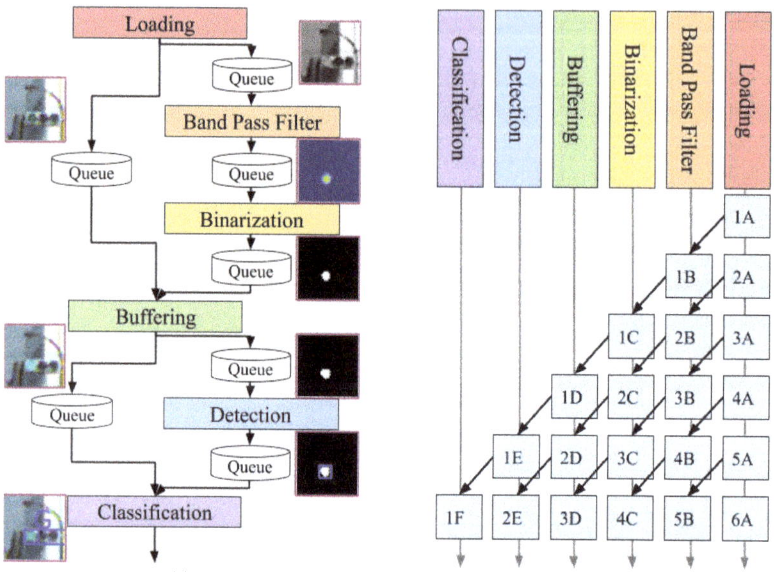

Figure 7. Overview of parallel processing in the system. (**a**) Illustration of the system flowchart with queues. (**b**) Demonstration of the transition of the processed image. The images in the same row were processed simultaneously.

4.1. *Accuracy Evaluation*

The accuracy evaluation experiment was performed in four scenes under different times and weather conditions. In each case, the brightness and the contrast between the background and the traffic light were different. The proposed method performed detection with fixed parameters in all scenes. We created a region-wise manually labeled ground truth for the dataset, where each LED traffic light region was represented by a bounding box. This bounding box was assigned to the entire housing of the traffic lights rather than to each lamp. This was because some traffic lights flashed multiple color lamps simultaneously or had arrow-shaped lamps that provide secondary instructions. Therefore, in practical application, it was necessary to recognize not only the lamp of a traffic light but also the entire signal.

Precision and recall were used to evaluate the accuracy, which is defined as

$$\text{Precision} = \frac{\text{TP}}{\text{TP} + \text{FP}} \tag{7}$$

$$\text{Recall} = \frac{\text{TP}}{\text{TP} + \text{FN}} \tag{8}$$

where TP, FP and FN indicate true positive, false positive, and false negative, respectively. The detection results were assigned to the ground truth objects and they were determined to be true or false positive by measuring the intersection over union (IoU). The threshold for the IoU was set to 0.4, which is a size that can be popped when the SVM classifies a candidate into the wrong color. This was done considering that in this experiment, we focus on whether the proposed method can detect the lights, and we were not interested in any amount of misalignment.

Table 2 summarizes the detection performance for each scene. The precision and recall values obtained in the daytime exceeded 90%, whereas the recall value was approximately 80% during the sunset and at night, when the detection was difficult. These results show that the proposed method is robust against changes in the illumination environment. We did not measure the accuracy of the color classification of SVM in this dataset because we were interested only in whether the detectors could detect traffic lights. However, if the color is misclassified, the recall values would be lower due to the failure to recognize the housing.

Table 2. Precision and Recall.

Scene	# of Frame	Precision	Recall
Morning	2000	0.98	0.98
Sunny	2000	0.96	0.91
Sunset	5000	0.89	0.79
Night	3600	0.91	0.84

Figure 8 provides the detection results for each scene. A traffic light that appears to be fairly small on the image was successfully detected. Moreover, a traffic light for pedestrians was also detected. In the sunset scene, even a traffic light covered by a smear of sunlight could be detected. In the night scene, the four traffic lights could be correctly recognized despite the confusion caused by streetlights and headlamps. In addition, the traffic light further away could not be recognized as a green lamp; however, they too were detected.

The intermediate results for all scenes are displayed in Figure 9. Some images were passed through the band-pass filter, as shown in the middle row of the figure. It was observed that the area other than the traffic lights is practically suppressed by the filter. In the rightmost center image of the figure, the streetlights and the electronic bulletin board were suppressed. However, in the binary image, they were deducted and only the traffic lights remained. This is because the threshold for binarization was appropriately set, as determined by the state estimation. Thus, this system could detect traffic lights in the same way without the need of parameter adjustment in daytime as well as during the night.

The dataset included many blinking lights other than traffic lights. Figure 10 shows images of the two scenes where the effects of disturbances are the most apparent. The sunny scene has an electronic bulletin board blinking at high speed. In the night scene, streetlights and, shop signs are blinking. Most of the disturbances caused by blinking lights are removed by the band-pass filter and binarization. The detection module also contributes to the removal of disturbances. There were not many round displays as bright as traffic lights, blinking at the same height and in the same position as traffic lights, that the proposed system could not distinguish.

However, in the proposed method, detection failed in some cases. First, because this method uses the blinking of a traffic light, the light that is reflected by the cars or buildings may be erroneously detected. This can be resolved by checking the detection position or the size; moreover, it is also possible to reduce the reflected light with a polarizing filter on the lens. Second, the shaking of the camera resulted in occasional failure to extract the blinking area. Because the band-pass filter treats the value of each pixel as an independent signal, it is not possible to obtain blinking on the pixel as the traffic light moved across the pixels frame by frame. Unless the blinking is detected, this system cannot identify traffic lights. In addition, detection may fail if a traffic light whose phase of blinking is shifted, enters the field of view. The phases are generally the same. However, they may be different owing to the differences in the drive circuits.

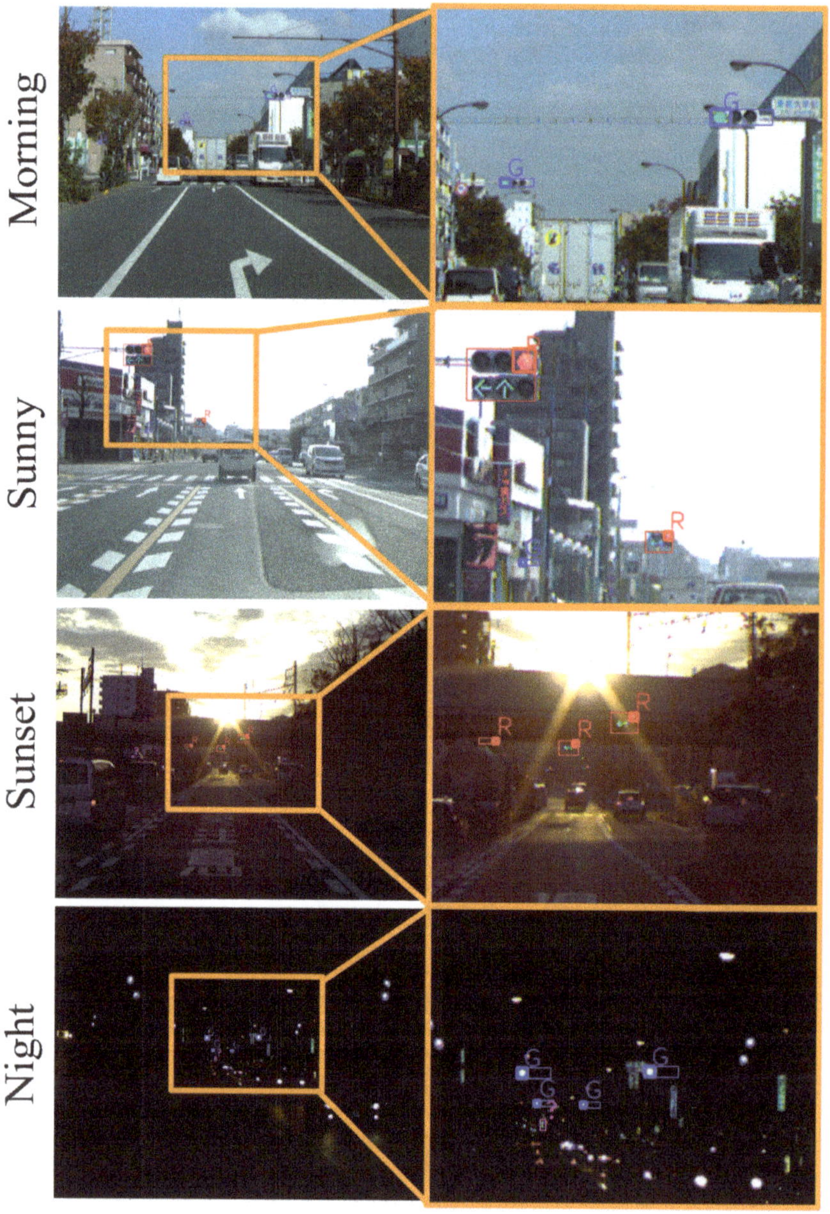

Figure 8. Detection results for all scenes. The left column is the full resolution image. The right column is an enlarged image of the yellow box.

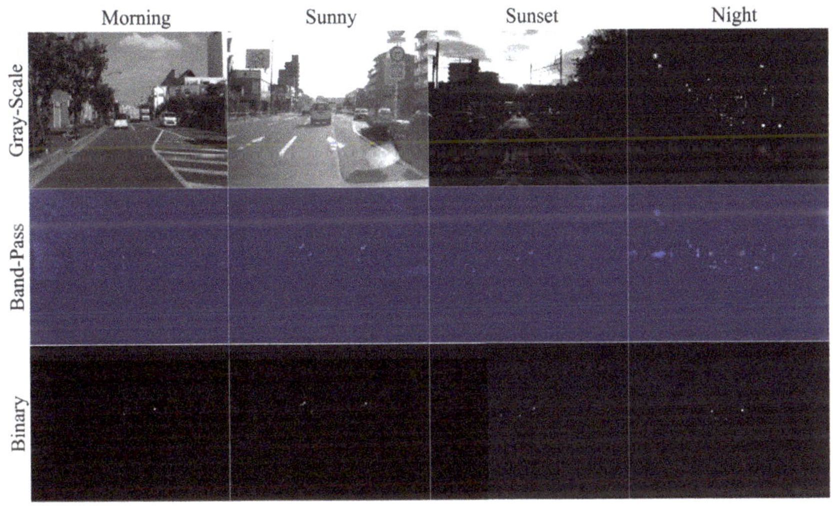

Figure 9. Intermediate results in all scenes. Each column corresponds to a scene. The upper row shows the gray-scaled images, the middle row displays the band-pass-filtered images, and the lower row illustrates the binarized images.

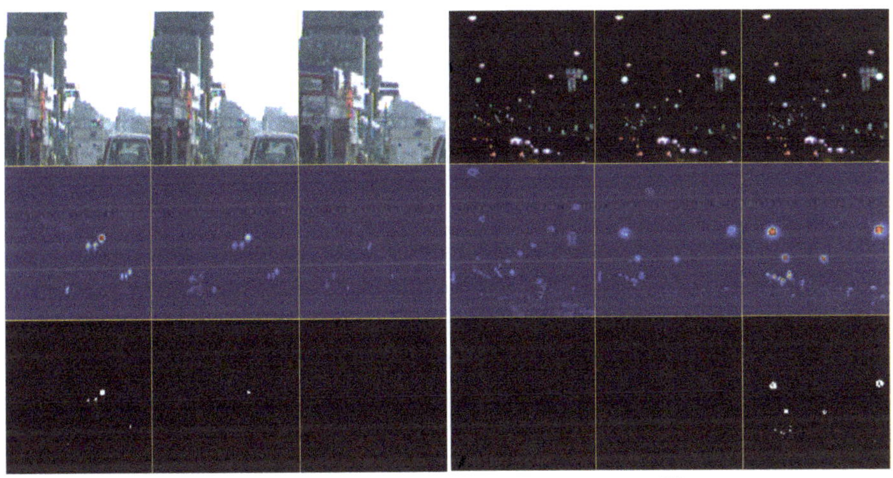

(**a**) (**b**)

Figure 10. The effects of a blinking disturbance. (**a**) Sunny scene with an electric bulletin board. (**b**) Night scene with many flashing lights, such as streetlights.

4.2. Comparison with the Conventional Hybrid Detection System

Our dataset taken by a high-speed camera includes many frames where the traffic light is completely off. Therefore, it is not possible to make valid comparisons with detection methods that use other approaches on the same dataset. Other methods that do not consider the blinking of traffic light cannot provide adequate performance. Therefore, we did not compare the accuracy with non-hybrid-based detection methods.

In this study, the results were compared with the conventional hybrid detection metho [15] by using a high-speed camera. However, unfortunately, no meaningful comparison can be made between the proposed method and the conventional hybrid-based method. This is because the conventional method only detects circular color lamps and does not perform color classification. Moreover, the method cannot detect the arrow-shaped lamps and entire light housing; thus, it cannot be compared under equal conditions with our proposed method that can detect all of them. Apart from that, the conventional method has limited functionality and is not as practical as the proposed method. From the above, this study analyzed the performance by comparing the output results. The same videos that were used in the accuracy evaluation experiment were used as the input. Figure 11 demonstrates the results of the traffic light detection using the comparison method, which uses the same parameters for detection in all the scenes.

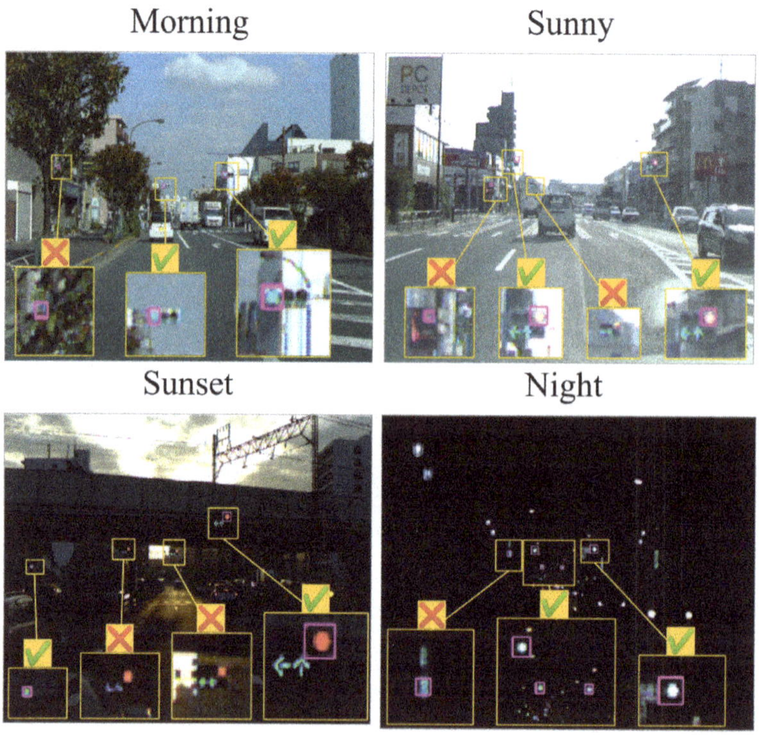

Figure 11. Example of the detection results by the conventional hybrid method. In the morning, the scattering of light by the tree was erroneously detected. When it was sunny, it failed to detect the electronic bulletin board. During the sunset, the traffic light under the viaduct could not be detected. During the night, all the traffic lights were detected; however, the light in a store was incorrectly detected.

This conventional method uses the blinking of the traffic light for detection; hence, it successfully identified some traffic lights during the day and nighttime. However, there were several false positives and false negatives. From the morning data, the light scattered by the leaves on the street was mistakenly recognized as a blinking traffic light. Similarly, in the case of sunny data, the electric bulletin board, and for the night data, parts of the illumination were mistakenly recognized as traffic light. Moreover, from the sunset data, some traffic lights were not detected even though the traffic lights were close enough.

In the method used for comparison, blinking is encoded based on whether the brightness changes for each pixel are greater than a threshold value. In addition, it employs regions where the time-series pattern of the code matches the specified pattern of traffic light candidates. Therefore, an area where the brightness change happens to match the specified pattern is erroneously recognized as a traffic light, or a traffic light that does not match the pattern due to the disturbance light will not be detected.

In particular, there were many false positives because of the scattering of light and the electronic bulletin board. In addition, false negatives occurred when the influence of sunlight was significant.

All these problems with the conventional method were noticeably improved by applying the proposed method. In the proposed method, the band-pass filter was used to extract data corresponding to the blinking area, and the region that was truly blinking represented the signal candidates. Thus, the proposed method can reject instantaneous brightness changes caused by scattering of light. The compared method requires a synchronization process of the blinking cycle every time it searches for an area encoded with a specific pattern. If the synchronization fails, the traffic light detection fails as well; however, in the proposed method, the band-pass filter extract the blinking area without synchronization. Furthermore, the adaptive binarization also contributes to an improvement in the performance. Although the compared method uses pre-calculated specific patterns, its robustness against the illumination change was insufficient because of the changes in brightness patterns that were based on the environment. Meanwhile, the proposed method recognized the blinking dynamics that rarely changed, and the thresholds of the blinking signals were estimated by the Kalman filter. This made it capable of detecting traffic lights in various environments without using specific patterns that may lack robustness.

4.3. Efficiency Evaluation

To verify whether this method can be processed in real-time, the calculation time of each module was measured. Because each module is executed in a parallel pipeline, the calculation time of any module must be less than 2 ms to be processed at 500 fps in real-time. The module that displays an image to confirm the detection result was not included in the calculation time because it is not related to traffic light detection. This study measured the computation time for traffic light detection in a 2000-frame video with a resolution of 800×600 pixels. The details of the computer that performed the detection are shown in Table 3. The detection parameters were set to the same values as those used in the detection accuracy evaluation experiment.

Table 3. PC specifications.

Parameter	Value
CPU	Intel Core i9-7900ZX
Clock	3.30 GHz
# of Core (Thread)	10 (20)
Memory	64 GB
OS (kernel)	Ubuntu 18.04 (4.15.0-72-generic)
GPU	NO-USED

The calculation times are summarized in Table 4. Even for the module that takes the longest calculation time, processing was completed in less than 1 ms. In all the modules, the average of the calculation time and the value obtained by adding the standard deviation (SD) to it was less than 2 ms. This indicates that the module has sufficient speed to process a 500-fps video in real-time.

Table 4. Time efficiency for each module.

Module	Average (ms)	SD (ms)	Best (ms)	Worst (ms)
Loading	0.78	0.27	0.39	7.64
Band Pass Filter	0.69	0.37	0.38	3.48
Binarization	0.37	0.08	0.20	0.74
Buffering	0.38	0.18	0.20	2.44
Detection	0.16	0.09	0.06	1.03
Classification	0.02	0.01	0.00	0.25

We confirmed that it is possible to run at 500 fps on a laptop PC (Core i7 9750). However, we did not perform detailed experiments on a laptop PC because it would be scarcely capable of performing the required calculations, which should be performed with a margin of computational resources in practical use. The PC used in the evaluation experiments can process the images at enough speed with extra computational power, and the extra resources can be used to conduct other processes. The power consumption of our high-speed camera is 3 W in catalog value. The PC used in the experiments did not include a GPU, and the CPU's thermal design power (TDP) was 140 W. We think they are small enough to be installed in an automated vehicle and can be practically applied.

5. Conclusions

This study proposes a real-time traffic light detection method through the recognition of LED blinking caused by AC. A band-pass filter was applied to the input images with the frequency of the AC to efficiently detect the blinking areas. The threshold of the binarization was set adaptively using the state estimated by the Kalman filter. Therefore, the difference in the environments can be detected in a similar manner. Although several previous studies have used visual information for detection such as the appearance of traffic lights, their results vary greatly depending on the illumination. Meanwhile, the proposed method barely depended on this, and it employed more robust frequency information for performing detections during severe conditions. No adjustment of the parameters, such as the heuristic thresholds or neural network training, was required. According to the experimental results, the proposed method performed detection with a high accuracy under various experimental scenes. It was confirmed that all the modules could be processed within 1 ms. Therefore, the proposed method could process more than 1000 fps. A future task is to improve the performance of the classification. On a real road, traffic lights that are not restricted to a particular lane may appear in the angle of view. This requires the ability to properly identify the traffic lights. In addition, the system is required to compensate for the image shaking caused by the vibration of car. If shaking correction can be achieved, the performance of the pixel-by-pixel band-pass filter would improve, and consequently, traffic light detection would become more accurate.

Author Contributions: Formal analysis, M.H.,T.S and N.K.; funding acquisition M.I.; investigation M.H., T.S. and N.K.; methodology K.Y., M.H. and T.S. All authors have read and agreed to the published version of the manuscript.

Funding: This research received no external funding.

Acknowledgments: Thanks for the help of reviewers and editors.

Conflicts of Interest: The authors declare no conflicts of interest.

References

1. Traffic Safety FACTS. Available online: https://crashstats.nhtsa.dot.gov/Api/Public/ViewPublication/812115 (accessed on 3 March 2020).
2. The Insurance Institute for Highway Safety (IIHS). Red light running. Available online: https://www.iihs.org/topics/red-light-running (accessed on 3 March 2020).
3. Jensen, M.B.; Philipsen, M.P.; Møgelmose, A.; Moeslund, T.B.; Trivedi, M.M. Vision for Looking at Traffic Lights: Issues, Survey, and Perspectives. *IEEE Trans. Intell. Transp. Syst.* **2016**, *17*, 1800–1815. [CrossRef]
4. Sooksatra, S.; Kondo, T. Red traffic light detection using fast radial symmetry transform. In Proceedings of the 2014 11th International Conference on Electrical Engineering/Electronics, Computer, Telecommunications and Information Technology (ECTI-CON), Nakhon Ratchasima, Thailand, 14–17 May 2014; pp. 1–6. [CrossRef]
5. Trehard, G.; Pollard, E.; Bradai, B.; Nashashibi, F. Tracking both pose and status of a traffic light via an Interacting Multiple Model filter. In Proceedings of the 17th International Conference on Information Fusion (FUSION), Salamanca, Spain, 7–10 July 2014; pp. 1–7.
6. Wonghabut, P.; Kumphong, J.; Ung-arunyawee, R.; Leelapatra, W.; Satiennam, T. Traffic Light Color Identification for Automatic Traffic Light Violation Detection System. In Proceedings of the 2018 International Conference on Engineering, Applied Sciences, and Technology (ICEAST), Phuket, Thailand, 4–7 July 2018; pp. 1–4. [CrossRef]

7. Ji, Y.; Yang, M.; Lu, Z.; Wang, C. Integrating visual selective attention model with HOG features for traffic light detection and recognition. In Proceedings of the 2015 IEEE Intelligent Vehicles Symposium (IV), Seoul, Korea, 28 June–1 July 2015; pp. 280–285.[CrossRef]
8. Du, L.; Chen, W.; Fu, S.; Kong, H.; Li, C.; Pei, Z. Real-time Detection of Vehicle and Traffic Light for Intelligent and Connected Vehicles Based on YOLOv3 Network. In Proceedings of the 2019 5th International Conference on Transportation Information and Safety (ICTIS), Liverpool, UK, 14–17 July 2019; pp. 388–392.
9. Wu, Y.; Geng, K.; Xue, P.; Yin, G.; Zhang, N.; Lin, Y. Traffic Lights Detection and Recognition Algorithm Based on Multi-feature Fusion. In Proceedings of the 2019 IEEE 4th International Conference on Image, Vision and Computing (ICIVC), Xiamen, China, 5–7 July 2019; pp. 427–432.
10. Weber, M.; Huber, M.; Zöllner, J.M. HDTLR: A CNN based Hierarchical Detector for Traffic Lights. In Proceedings of the 2018 21st International Conference on Intelligent Transportation Systems (ITSC), Maui, HI, USA, 4–7 November 2018; pp. 255–260. [CrossRef]
11. Bach, M.; Stumper, D.; Dietmayer, K. Deep Convolutional Traffic Light Recognition for Automated Driving. In Proceedings of the 2018 21st International Conference on Intelligent Transportation Systems (ITSC), Maui, HI, USA, 4–7 November 2018; pp. 851–858. [CrossRef]
12. Fregin, A.; Müller, J.; Dietmayer, K. Feature detectors for traffic light recognition. In Proceedings of the 2017 IEEE 20th International Conference on Intelligent Transportation Systems (ITSC), Yokohama, Japan, 16–19 October 2017; pp. 339–346. [CrossRef]
13. Fairfield, N.; Urmson, C. Traffic light mapping and detection. In Proceedings of the 2011 IEEE International Conference on Robotics and Automation, Shanghai, China, 9–13 May 2011; pp. 5421–5426. [CrossRef]
14. Levinson, J.; Askeland, J.; Dolson, J.; Thrun, S. Traffic light mapping, localization, and state detection for autonomous vehicles. In Proceedings of the 2011 IEEE International Conference on Robotics and Automation, Shanghai, China, 9–13 May 2011; pp. 5784–5791. [CrossRef]
15. Wu, Z.; Watanabe, Y.; Ishikawa, M. Hybrid LED traffic light detection using high-speed camera. In Proceedings of the 2016 IEEE 19th International Conference on Intelligent Transportation Systems (ITSC), Rio de Janeiro, Brazil, 1–4 November 2016; pp. 1235–1241. [CrossRef]
16. Chen, Z.; Shi, Q.; Huang, X. Automatic detection of traffic lights using support vector machine. In Proceedings of the 2015 IEEE Intelligent Vehicles Symposium (IV), Seoul, Korea, 28 June–1 July 2015; pp. 37–40. [CrossRef]
17. Shen, X.; Andersen, H.; Ang, M.H.; Rus, D. A hybrid approach of candidate region extraction for robust traffic light recognition. In Proceedings of the 2017 IEEE 20th International Conference on Intelligent Transportation Systems (ITSC), Yokohama, Japan, 16–19 October 2017; pp. 1–8. [CrossRef]
18. Müller, J.; Fregin, A.; Dietmayer, K. Multi-camera system for traffic light detection: About camera setup and mapping of detections. In Proceedings of the 2017 IEEE 20th International Conference on Intelligent Transportation Systems (ITSC), Yokohama, Japan, 16–19 October 2017; pp. 165–172. [CrossRef]
19. Franke, U.; Pfeiffer, D.; Rabe, C.; Knoeppel, C.; Enzweiler, M.; Stein, F.; Herrtwich, R.G. Making Bertha See. In Proceedings of the 2013 IEEE International Conference on Computer Vision Workshops, Sydney, NSW, Australia, 3–6 December 2013; pp. 214–221. [CrossRef]
20. Kim, J.; Cho, H.; Hwangbo, M.; Choi, J.; Canny, J.; Kwon, Y.P. Deep Traffic Light Detection for Self-driving Cars from a Large-scale Dataset. In Proceedings of the 2018 21st International Conference on Intelligent Transportation Systems (ITSC), Maui, HI, USA, 4–7 November 2018; pp. 280–285. [CrossRef]
21. Waisakurnia, W.; Widyantoro, D.H. Traffic light candidate elimination based on position. In Proceedings of the 2016 10th International Conference on Telecommunication Systems Services and Applications (TSSA), Denpasar, Indonesia, 6–7 October 2016; pp. 1–5.[CrossRef]
22. Possatti, L.C.; Guidolini, R.; Cardoso, V.B.; Berriel, R.F.; Paixão, T.M.; Badue, C.; De Souza, A.F.; Oliveira-Santos, T. Traffic Light Recognition Using Deep Learning and Prior Maps for Autonomous Cars. In Proceedings of the 2019 International Joint Conference on Neural Networks (IJCNN), Budapest, Hungary, 14–19 July 2019; pp. 1–8.
23. Ouyang, Z.; Niu, J.; Liu, Y.; Guizani, M. Deep CNN-Based Real-Time Traffic Light Detector for Self-Driving Vehicles. *IEEE Trans. Mob. Comput.* **2020**, *19*, 300–313. [CrossRef]
24. Kulkarni, R.; Dhavalikar, S.; Bangar, S. Traffic Light Detection and Recognition for Self Driving Cars Using Deep Learning. In Proceedings of the 2018 Fourth International Conference on Computing Communication Control and Automation (ICCUBEA), Pune, India, 16–18 August 2018; pp. 1–4.

25. Gupta, A.; Choudhary, A. A Framework for Traffic Light Detection and Recognition using Deep Learning and Grassmann Manifolds. In Proceedings of the 2019 IEEE Intelligent Vehicles Symposium (IV), Paris, France, 9–12 June 2019; pp. 600–605.
26. Aneesh, A.N.; Shine, L.; Pradeep, R.; Sajith, V. Real-time Traffic Light Detection and Recognition based on Deep RetinaNet for Self Driving Cars. In Proceedings of the 2019 2nd International Conference on Intelligent Computing, Instrumentation and Control Technologies (ICICICT), Kannur, Kerala, India, 5–6 July 2019; pp. 1554–1557.
27. Shi, Z.; Zou, Z.; Zhang, C. Real-Time Traffic Light Detection With Adaptive Background Suppression Filter. *IEEE Trans. Intell. Transp. Syst.* **2016**, *17*, 690–700. [CrossRef]

sensors

MDPI

Article

Fail-Aware LIDAR-Based Odometry for Autonomous Vehicles

Iván García Daza *, Mónica Rentero, Carlota Salinas Maldonado, Ruben Izquierdo Gonzalo, Noelia Hernández Parra, Augusto Ballardini and David Fernandez Llorca

Computer Engineering Department, Universidad de Alcalá, 28805 Alcalá de Henares, Spain; monica.rentero@edu.uah.es (M.R.); carlota.salinasmaldo@uah.es (C.S.M.); ruben.izquierdo@uah.es (R.I.G.); noelia.hernandez@uah.es (N.H.P.); augusto.ballardini@uah.es (A.B.); david.fernandezl@uah.es (D.F.L.)

* Correspondence: ivan.garciad@uah.es; Tel.: +34-918-856-622

Received: 8 May 2020; Accepted: 21 July 2020; Published: 23 July 2020

Abstract: Autonomous driving systems are set to become a reality in transport systems and, so, maximum acceptance is being sought among users. Currently, the most advanced architectures require driver intervention when functional system failures or critical sensor operations take place, presenting problems related to driver state, distractions, fatigue, and other factors that prevent safe control. Therefore, this work presents a redundant, accurate, robust, and scalable LiDAR odometry system with fail-aware system features that can allow other systems to perform a safe stop manoeuvre without driver mediation. All odometry systems have drift error, making it difficult to use them for localisation tasks over extended periods. For this reason, the paper presents an accurate LiDAR odometry system with a fail-aware indicator. This indicator estimates a time window in which the system manages the localisation tasks appropriately. The odometry error is minimised by applying a dynamic 6-DoF model and fusing measures based on the Iterative Closest Points (ICP), environment feature extraction, and Singular Value Decomposition (SVD) methods. The obtained results are promising for two reasons: First, in the KITTI odometry data set, the ranking achieved by the proposed method is twelfth, considering only LiDAR-based methods, where its translation and rotation errors are 1.00% and 0.0041 deg/m, respectively. Second, the encouraging results of the fail-aware indicator demonstrate the safety of the proposed LiDAR odometry system. The results depict that, in order to achieve an accurate odometry system, complex models and measurement fusion techniques must be used to improve its behaviour. Furthermore, if an odometry system is to be used for redundant localisation features, it must integrate a fail-aware indicator for use in a safe manner.

Keywords: LiDAR odometry; fail-operational systems; fail-aware; automated driving

1. Introduction

1.1. Motivation

At present, the concept of autonomous driving is becoming more and more popular. Therefore, new techniques are being developed and researched to help consolidate the reality of implementing the concept. As systems become autonomous, their safety must be improved to increase user acceptance. Consequently, it is necessary to integrate intelligent fault detection systems that guarantee the security of passengers and people in the environment. Sensors, perception, localisation, or control systems are essential elements for their development. However, they are also susceptible to failures and it is necessary to have fail-x systems, which prevent undesired or fatal actions. A fail-x system combines the following features: redundancy in design (fail-operational), ability to plan emergency manuevers and undertake safe stops (fail-safe), and monitoring the status of their sensors to detect failures or malfunctions in them (fail-aware). At present, in an urban environment where there are increasingly complex traffic elements such as multiple intersections, complex lane roundabouts, or tunnels, a localisation system based only on GPS may pose problems. Thus, autonomous driving will be a closer reality when LiDAR or Visual odometry systems are integrated to cover fail-operational functions. However, fail-aware behaviour has to be integrated into the global system also.

At present, the Global Positioning System (GPS) performs the main tasks of localisation due to its robustness and accuracy. However, GPS coverage problems derived from structural elements of the road (tunnels), GPS multi-path in urban areas, or failure in its operation, mean that this technology does not meet the necessary localisation requirements in 100% of use-cases, which makes it essential to design redundant systems based on LiDAR odometry [1], Visual odometry [2], Inertial Navigation Systems (INS) [3], Wifi [4], or a combination of the above, including digital maps [5]. However, LiDAR and Visual odometry systems suffer from a non-constant temporal drift, where the characteristics of the environment and the algorithm behaviour are determinants that improve or worsen this drift. Therefore, it is necessary to introduce, for those systems that have a non-constant temporal drift, a fail-aware indicator to discern when these can be used.

1.2. Problem Statement

Safe behaviour in a vehicle's control and navigation systems depends mostly on the redundancy and failure detections that these present. At the moment, when GPS-based localisation fails, either temporarily or permanently, the LiDAR and Visual odometry systems can start as redundant localisation systems, mitigating the erroneous behaviour of the GPS localisation. Redundant localisation based on 3D mapping techniques can be applied, as well. However, this is currently more widespread in robotic applications, as the 3D map accuracy in open environments is decisive for localisation tasks. However, companies such as Mapillary and Here have presented promising results for 3D map accuracy. Why is it challenging to build an accurate 3D map when relying only on GPS localisation? It is because the GPS angular error feature of market devices is close to 10^{-3} rad. This feature can place a 3D object with an error of 0.01 m when the object distance from the sensor is 100 m.

So, in the case of integrating redundancy into the localisation system with an odometry alternative, a fail-aware indicator has to be integrated into the odometry system, as a consequence of the non-constant drift error, in order for it to be used as a redundant system. The fail-aware indicator could be based on an estimated time window that satisfies a localisation error below the minimum requirements to planned emergency manoeuvring and placing the vehicle in a safe spot. Several alternatives can be presented to implement the fail-aware indicator. The first is to set a fixed time window in which the system is used. The second alternative is an adaptive time window, which is evaluated dynamically in the continuous localisation process to find the maximum time in which the redundant system can be used. At present, there have been no recent works focused on fail-aware LIDAR-based odometry for autonomous vehicles.

Therefore, it is necessary to look for an odometry process that maximises the time in exceeding the threshold that leads the system to a failure state and, for that purpose, we propose a robust, scalable, and precise localisation design that minimises the error in each iteration. Multiple measurement fusion techniques from both global positioning systems and odometry systems are used to make the system robust. Bayesian filtering guarantees an optimal fusion between the observation techniques applied in the odometry systems and improves the localisation accuracy by integrating (mostly kinematic) models of the vehicle's displacement, having either three or six degrees of freedom (DoF). The LiDAR odometry is based exclusively on the observations of the LiDAR sensor, where the emission of near-infrared pulses and the measurement of the reflection time allows us to represent the scene with a set of 3D points, called a Point Cloud. Thus, given a temporal sequence of measurements, we obtain the homogeneous transformation, rotation, and translation corresponding to two consecutive time instants, by applying iterative registering and optimisation methods. However, this process alone provides incorrect homogeneous transformations if the scene presents moving objects and, so, solutions based on feature detection must be explored in order to mitigate possible errors.

1.3. Contributions

The factors described previously motivated us to develop an accurate LiDAR odometry system with a fail-aware indicator ensuring its proper use as a redundant localisation system for autonomous vehicles, as shown in Figure 1. The accurate LiDAR odometry architecture is based on robust and scalable features. The architecture has a robust measurement topology as it integrates three measurement algorithms, two of which are based on Iterative Closest Point (ICP) variants, and the third one is based on non-mobile scene feature extraction and Singular Value Decomposition (SVD). Furthermore, our work proposes a scalable architecture to integrate a fusing block, which relies on the UKF scheme. Another factor taken into account to enhance the odometry accuracy was to incorporate a 6-DoF motion model based on vehicle dynamics and kinematics within the filter, where the variables of pitch and roll play a crucial impact on the precision. The proposed scalable architecture allows us to fuse any position measurement system or integrate into the LiDAR odometry system new measurement algorithms in a natural way. A fail-aware indicator based on the vehicle heading error is another contribution to the state-of-the-art. The fail-aware indicator introduces, in the system output, an estimated time to reach system malfunction, which enables other systems to take it into consideration.

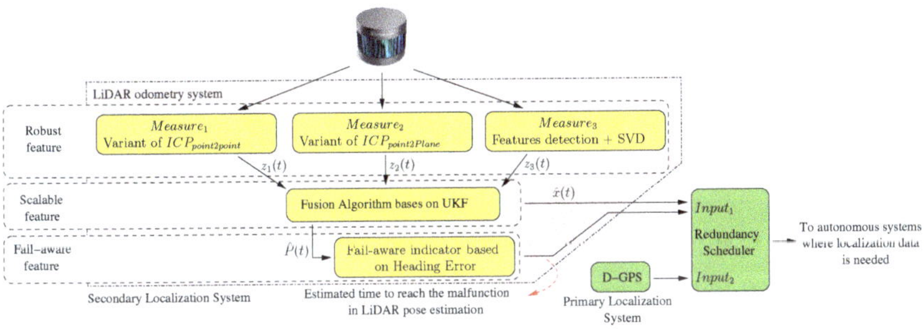

Figure 1. General diagram. The developed blocks are represented in yellow. The horizontal blue strips represent the main features of the odometry system. A framework where the LiDAR odometry system can be integrated within the autonomous driving cars topic is depicted with green blocks, such as a secondary localisation system.

The global system is validated by processing KITTI odometry data set sequences and evaluating the error committed in each of the available sequences, allowing for comparison with other state-of-the-art techniques. The variability in the available scenes allows us to validate the fail-aware functionality, by comparing sequences with low operating error with those with higher error, and observing how the temporal estimation factor increases for those sequences with worse results.

The rest of the paper is comprised of the following sections: Section 2 presents the state-of-the-art in the areas of LiDAR odometry and vehicle motion modelling. Section 3 details the integrated 6-DoF model, while Section 4 explains the global software architecture of the work, as well as the methodology applied to fuse the evaluated measures. Then, Section 5 describes the details of the proposed measurement systems. Section 6 describes the methodology to make the systems fail-aware. Section 7 describes, lists, and compares the results obtained by the developed system. Finally, Section 8 presents our conclusions and proposes a description of future work in the fields of odometry techniques, 3D mapping, and fail-aware systems.

2. Related Works

Many contemporary conceptual systems for autonomous driving require precise localisation systems. The geo-referenced localisation system does not usually satisfy such precision, as there are scenarios (e.g., tunnel or forest scenarios) where the localisation provided by GPS is not correct or has low accuracy, leading to safety issues and non-robust behaviours. For these reasons, GPS-based localisation techniques do not satisfy all the use-cases of driverless vehicles, and it is therefore mandatory to integrate other technologies into the final localisation system. Visual odometry systems could be candidates for such a technology, but scenarios with few characteristics or extreme environmental conditions lead to their non-robust performance, although considerable progress has been made in this area, as described in [6,7]. Nevertheless, LiDAR odometry systems mitigate part of the visual-related problems, but real-time features or accuracy in the algorithm remain issues. In the same way as Visual Odometry, significant advances and results have been obtained, in the last few years, in this topic.

The general challenge in odometry is to evaluate a vehicle's movement at all times without error, in order to obtain zero global localisation error; however, this issue is not reachable as the odometry measurement commits a small error in each iteration. These systems therefore have the weakness of drift error over time due to the accumulation of iterative errors; which is a typical integral problem. Drift error is a function of path length or elapsed time. However, these techniques have advanced in the last few decades, due to the improvement of sensor accuracies, achieving small errors (as presented in [8]).

LIDAR odometry is based on the procedures of point registration and optimisation between two consecutive frames. Many works have been inspired by these techniques, but they have the disadvantage of not ensuring a global solution, introducing errors in their performances. These techniques are called Iterative Closest Points (ICP), many of which have been described in [9], where modifications affecting all phases of the ICP algorithm were analysed, from the selection and registration of points to the minimisation approach. The most widely used are ICP point-to-point ICP_{p2p} [10,11] and ICP point-to-plane ICP_{p2p} [12,13]. For example, presented a point-to-point ICP algorithm based on two stages [10], in order to improve its exactness. Initially, a coarse registration stage based on KD-tree is applied to solve the alignment problem. Once the first transformation is applied, a second fine-recording stage based on KD-tree ICP is carried out, in order to solve the convergence problem more accurately.

Several optimisation techniques have been proposed for use when the cost function is established. Obtaining a rigid transformation is one of the most commonly used schemes, as has been detailed in the simplest ICP case [14], as well as in more advanced variants such as CPD [15]. This is easily achieved by decoupling the rotation and translation, obtaining the first using algebraic tools such as SVD (Singular Value Decomposition), while the second term is simply the mean/average translation. Other proposals, such as LM-ICP [16] or [11], perform a Levenberg–Marquardt approach to add robustness to the process. Finally, optimisation techniques such as gradient descent have been used in distribution-to-distribution methods like D2D NDT [17].

In order to increase robustness and computational performance, interest point descriptors for point clouds have recently been proposed. General point cloud or 2D depth maps are two general approaches to achieve this. The latter may include curvelet features, as analysed in [18], assuming the range data is dense and a single viewpoint is used in order to capture the point cloud. However, it may not perform accurately for a moving LiDAR—the objective of this paper. In a general point cloud approach, Fast Point Feature Histograms (FPFH) [19] and Integral Volume Descriptors (IVD) [20] are two feature-based global registration proposals of interest. The first one generates feature histograms, which have demonstrated great results in vision object detection problems, using such techniques as Histogram of Oriented Gradients (HOG), by means of computing some statistics about a point's neighbours relative positions and estimated surface normals. Feature histograms have shown the best

IVD performances and surface curvature estimates. However, neither of these methods offer reliable results in sparse point clouds and are slow to compute.

Once one correspondence has been established, using features instead of proximity, it can be used to initialize ICP techniques in order to improve their results. As described in previous paragraphs, this transformation can also be found by other techniques, such as PCA or SVD, which are both deterministic procedures. In order to obtain a transformation, three point correspondences are enough, as is shown in the proposal we introduce in this document. However, as many outlier points are typically present in a point cloud (such as those of vegetation), a random sample consensus (RANSAC) approach is usually used [19]. Other approaches include techniques tailored to the specific problem, such as the detection of structural elements of the scene [21].

In the field of observations or measurements, there are a large number of methods for measuring the homogeneous transformation between two moments or two point clouds. For this reason, many filtering and fusion systems have been applied to improve the robustness of systems. The two most widespread techniques to filter measurements are recursive filtering and batch optimisation [22]. Recursive filtering updates the status probabilistically, using only the latest sensor observations for status prediction and process updates. The Kalman filter and its variants mostly represent recursive filtering techniques. However, filtering based on batch optimisation maintains a history of observations to evaluate, on the basis of previous states, the most probable estimate of the current instant. Both techniques may integrate kinematic and dynamic models of the system under analysis to improve the process of estimating observations. In the field of autonomous driving, the authors of [23] justified the importance of applying models in the solution of driving problems, raising the need to work with complex models that correctly filter and fuse observations.

The best odometry system described in the state-of-the-art is VLOAM [8], which is based on Visual and LiDAR odometry. It is characterised by being a particularly robust method in the face of an aggressive movement and the occasional lack of visual features. The method starts with a stage of visual odometry, in order to obtain a first approximation of the movement. The final stage is executed with LiDAR odometry. The results shown applied to a set of ad-hoc tests and the KITTI odometry data set. The work presented as LIMO [24] also aimed to evaluate the movement of a vehicle accurately. Stereo images with LiDAR were used to provide depth information to the features detected by the cameras. The process includes a semantic segmentation, which is used to reject and weight characteristic points used for odometry. The results given were related to the KITTI data set. On the other hand, the authors of [25] presented a LiDAR odometry technique that models the projection of the point cloud in a 2D ring structure. Over the 2D structure, an unsupervised Convolutional Auto-Encoder (CAE-LO) system detects points of interest in the spherical ring (CAE-2D). It later extracts characteristics from the multi-resolution voxel model using 3D CAE. It was characterised as finding 50% more points of interest in the point cloud, improving the success rate in the cloud comparison process. To conclude, the system described in [26] proposed a real-time laser odometry approach, which presented small drift. The LiDAR system uses inertial data from the vehicle. The point clouds are captured in motion, but are compensated with a novel sweep correction algorithm based on two consecutive laser scans and a local map.

To the best of our knowledge, there have been no recent works focused on fail-aware LiDAR-based odometry for autonomous vehicles.

3. Kinematic and Dynamic Vehicle Model

Filters usually leverage mathematical models to better approximate state transitions. In the field of vehicle modelling, there are two ways to study the movement of a vehicle: with kinematic or dynamic models. In the field of kinematic vehicle modelling, one of the most-used models is the bicycle model, due to its ease of understanding and simplicity. This model requires knowledge of the slide angle (β) as well as the front wheel angle (δ) parameters. These variables are usually measured by dedicated car systems.

In this work, the variables β and δ are not registered in the data set, so the paper proposes an approach based on a dynamic model to evaluate them. The method proposed can be used as a redundancy system, replacing dedicated car systems. The technique relies on the application of LiDAR odometry and the application of vehicle dynamics models where linear and angular forces are taken into account and the variables β and δ are assessed during the car's movement. Figure 2 depicts the actuated forces in the x and y car axes, as well as the slip angle and the front-wheel angle. Given these variables, the bicycle model is applied to predict the car's movement.

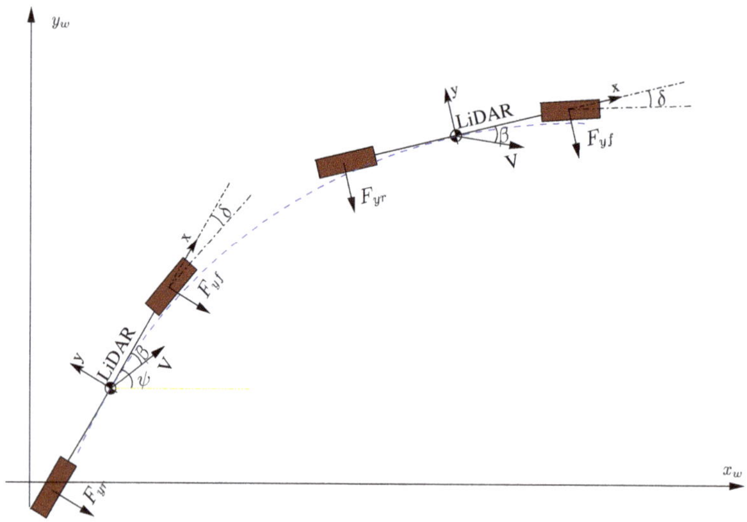

Figure 2. Vehicle representation by bicycle model. Using the vehicle reference system, our LiDAR-based odometry process assesses the vehicle forces between two instants of time, allowing for estimation of the β and δ variables.

From a technical perspective, the variables β and δ are evaluated using Equations (1) and (2). Equation (1) represents Newton's second law, applied on the car's transverse axis in a linear form and on the car's z-axis in an angular form.

$$\begin{array}{cc} Translational: & m(\ddot{y} + \dot{\psi} v_x) = F_{yf} + F_{yr} \\ Angular: & I_z \ddot{\psi} = l_f F_{yf} - l_r F_{yr} \end{array}, \tag{1}$$

where m is the mass of the vehicle, v_x is the projection of the speed car vector V on its longitudinal axis x, F_{yf} and F_{yr} are the lateral forces produced on the front and rear wheel, I_z is the inertia moment of the vehicle concerning to the z-axis, and l_f and l_r are the distances from the centre of masses of the front and rear wheels, respectively.

The lateral forces F_{yf} and F_{yr} are, in turn, functions of characteristic tyre parameters, cornering stiffnesses $C_{\alpha f}$ and $C_{\alpha r}$, the vehicle chassis configuration l_f and l_r, the linear and angular travel speed to which the vehicle is subjected to v_x, $\dot{\psi}$, the slip angle β, and the turning angle of the front wheel δ, as shown in Equation (2):

$$\begin{array}{c} F_{yf} = C_{\alpha f}(\delta - \beta - \frac{l_f \dot{\psi}}{v_x}) \\ F_{yr} = C_{\alpha r}(-\beta + \frac{l_r \dot{\psi}}{v_x}) \end{array} \tag{2}$$

Therefore, knowing the above vehicle parameters and assessing the variables $\ddot{y}$, v_x, $\ddot{\psi}$, and $\dot{\psi}$ from the LiDAR odometry displacement, with the method proposed in this work (see Figure 3), the variables β and δ can be derived by solving the two-equation system shown in Equation (1).

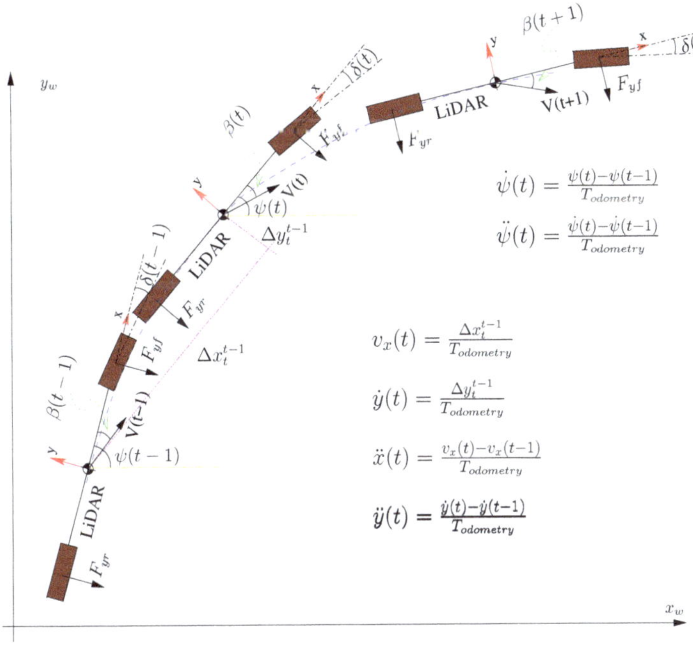

Figure 3. Evaluation of variables $\ddot{y}, v_x, \ddot{\psi}, \dot{\psi}$ with LiDAR odometry.

Finally, the variables β and δ are used in the kinematic bicycle model defined by Equation (3) to obtain the speeds $[\dot{X}, \dot{Y}, \dot{\psi}]$ which, in turn, are used to output the predicted vehicle pose at time $(t+1)$.

$$\begin{aligned} \dot{X} &= V\cos(\psi+\beta) \\ \dot{Y} &= V\sin(\psi+\beta) \\ \dot{\psi} &= \frac{V\cos(\beta)}{l_f+l_r}(\tan(\delta_f)-\tan(\delta_r)) \end{aligned} \quad . \tag{3}$$

However, the model mentioned above evaluates the vehicle's motion only in 3-DoF, while the LiDAR odometry gives us full 6-DoF displacement. Therefore, to assess the remaining 3-DoF, we propose to use another dynamic model based on the behaviour of the shock absorbers and the position of the vehicle's mechanical pitch (θ) and roll (α) axes; see Figure 4. Appling this second dynamical model, we can predict the car's movement in terms of its 6-DoF.

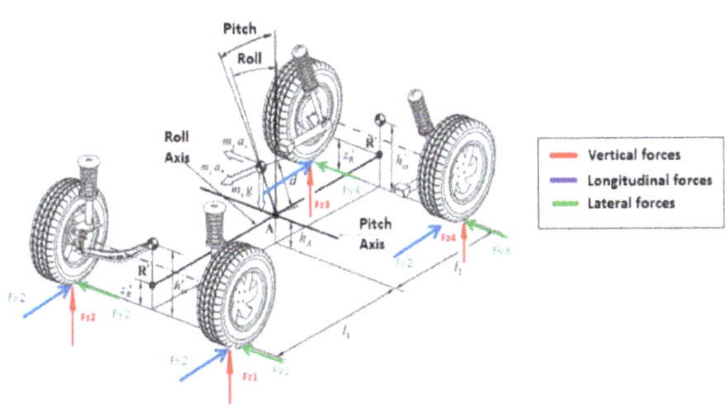

Figure 4. Detail of forces and moments applied to the vehicle. The distance d represented is broken down into d_{picth} and d_{roll}, concerning the pitch and roll axes of rotation, respectively. The figure references [27].

From a technical perspective, in order to evaluate these variables, we need to take into consideration the angular movement caused in the pitch and the roll axes.

First, regarding the pitch axis, the movement is due to the longitudinal acceleration suffered in the chassis, producing front and rear torsion on the springs and shock absorbers of the vehicle. Given the parameters D_{pitch} and K_{pitch}, which represent the distance between the centre of the pitch axis with respect to the centre of mass of the vehicle and the characteristics of the spring together with the shock absorber, respectively, Equation (4) defines the dynamics of the pitch angle, which represents the sum of the moments of forces applied to the pitch axis. The angular acceleration suffered by the vehicle chassis for the pitch axis is obtained by Equation (5), while the variables $\ddot{x}$, θ, and $\dot{\theta}$ are found in the LiDAR odometry process.

$$(I_y + m\, d_{pitch}^2)\ddot{\theta} - m\, d_{pitch}\ddot{x} + (K_{pitch} + m\, g\, d_{pitch})\theta + D_{pitch}\dot{\theta} = 0 \tag{4}$$

$$\ddot{\theta} = -\frac{-m\, d_{pitch}\ddot{x} + (K_{pitch} + m\, g\, d_{pitch})\theta + D_{pitch}\dot{\theta}}{I_y + m\, d_{pitch}^2} \tag{5}$$

where

$$\begin{aligned} D_{pitch} &= \frac{d_{shock\,f}\, l_f^2 + d_{shock\,r}\, l_r^2}{2} \\ K_{pitch} &= \frac{K_{spring\,f}\, l_f^2 + K_{spring\,r}\, l_r^2}{2} \end{aligned} \tag{6}$$

With the pitch acceleration and applying Equation (7), representing uniformly accelerated motion, the pitch of the vehicle can be predicted at time $(t+1)$.

$$\widetilde{\theta}(t+1) = \frac{1}{2}\ddot{\theta}(t)dt^2 + \dot{\theta}(t)dt + \theta(t) \tag{7}$$

On the other hand, the angular movement caused on the roll axis is due to the lateral acceleration or lateral dynamics suffered in the chassis. The parameter d_{roll} is the distance between the roll axis centre and the centre of mass of the vehicle, and mainly depends on the geometry of the suspension. The lateral forces multiplied by the distance d_{roll} generate an angular momentum, which is compensated for by the springs (K_{rollf}, K_{rollr}) and lateral shock absorbers of the vehicle (D_{rollf}, D_{rollr}), minimising the roll displacement suffered in the chassis. Equation (8) defines the movement dynamics of the roll angle, which represents the movement compensation effect with the sum of moments of forces applied on the axle.

$$(I_x + m\, d_{roll}^2)\ddot{\alpha} - m\, d_{roll}\ddot{y} + (K_{rollf} + K_{rollr} + m\, g\, d_{roll})\alpha + (D_{rollf} + D_{rollr})\dot{\alpha} = 0, \tag{8}$$

$$\ddot{\alpha} = -\frac{-m\, d_{roll}\ddot{y} + (K_{rollf} + K_{rollr} + m\, g\, d_{roll})\alpha + (D_{rollf} + D_{rollr})\dot{\alpha}}{I_x + m\, d_{roll}^2}, \tag{9}$$

where

$$\begin{aligned} D_{rollf} &= d_{shock\,f}\, t_f^2 & D_{rollr} &= d_{shock\,r}\, t_r^2 \\ K_{rollf} &= \frac{K_{spring\,f}\, t_f^2}{2}, & K_{rollr} &= \frac{K_{spring\,r}\, t_r^2}{2}. \end{aligned} \tag{10}$$

Given the roll acceleration and applying the uniformly accelerated motion Equation (11), the roll of the vehicle can be predicted at time $(t+1)$:

$$\widetilde{\alpha}(t+1) = \frac{1}{2}\ddot{\alpha}(t)dt^2 + \dot{\alpha}(t)dt + \alpha(t). \tag{11}$$

Finally, to complete the 6-DoF model parameterisation, we need to consider the vertical displacement of the vehicle, which is related to the angular movements of pitch and roll. Equation (12) represents the movement of the centre of masses concerning the vehicle z-axis, where COG_z is the height of the vehicle's centre of gravity at resting state:

$$\widetilde{z}(t+1) = COG_z + d_{pitch}(\cos(\widetilde{\theta}(t+1)) - 1) + d_{roll}(\cos(\widetilde{\alpha}(t+1)) - 1) \tag{12}$$

Table 1 lists the parameters and values used in the 6-DoF model. The values correspond to a Volkswagen Passat B6, and were found in the associated technical specs.

Table 1. Model parameters (chassis, tires, and suspension).

Name	Value
m = 1750 kg	Vehicle mass
$K_{spring\,f}$ = 30,800 $\frac{\text{N}}{\text{m}}$	Front suspension spring stiffness
$K_{spring\,r}$ = 28,900 $\frac{\text{N}}{\text{m}}$	Rear suspension spring stiffness
$D_{shock\,f}$ = 4500 $\frac{\text{Ns}}{\text{m}}$	Front suspension shock absorber damping coefficient
$D_{shock\,r}$ = 3500 $\frac{\text{Ns}}{\text{m}}$	Rear suspension shock absorber damping coefficient
d_{roll} = 0.1 m	Vertical distance between COG and roll axis
d_{pitch} = 0.25 m	Vertical distance between COG and pitch axis
I_x = 540 kg m^2	Vehicle's moment of inertia, with respect to the x axis
I_y = 2398 kg m^2	Vehicle's moment of inertia, with respect to the y axis
I_z = 2875 kg m^2	Vehicle's moment of inertia, with respect to the z axis
COG_z = 0.543 m	COG height from the ground
l_f = 1.07 m	Distance between COG and front axle
l_r = 1.6 m	Distance between COG and rear axle
t_f = 1.5 m	Front axle track width
t_r = 1.5 m	Rear axle track width

To deal with the imperfections of the kinematic model, we compared the output of the proposed 6-DoF model with the ground truth available in the KITTI odometry data set. The analysis was applied to all available sequences, in order to measure the uncertainty model in the best way.

By evaluating the pose differences (see Figure 5), the probability density function of the 6-DoF model was calculated, as well as the covariance matrix expressed in Equation (13).

$$Q = \begin{bmatrix} \sigma^2_{xx} & \sigma^2_{yx} & \sigma^2_{zx} & \sigma^2_{\phi x} & \sigma^2_{\theta x} & \sigma^2_{\psi x} \\ \sigma^2_{xy} & \sigma^2_{yy} & \sigma^2_{zy} & \sigma^2_{\phi y} & \sigma^2_{\theta y} & \sigma^2_{\psi y} \\ \sigma^2_{xz} & \sigma^2_{yz} & \sigma^2_{zz} & \sigma^2_{\phi z} & \sigma^2_{\theta z} & \sigma^2_{\psi z} \\ \sigma^2_{x\phi} & \sigma^2_{y\phi} & \sigma^2_{z\phi} & \sigma^2_{\phi\phi} & \sigma^2_{\theta\phi} & \sigma^2_{\psi\phi} \\ \sigma^2_{x\theta} & \sigma^2_{y\theta} & \sigma^2_{z\theta} & \sigma^2_{\phi\theta} & \sigma^2_{\theta\theta} & \sigma^2_{\psi\theta} \\ \sigma^2_{x\psi} & \sigma^2_{y\psi} & \sigma^2_{z\psi} & \sigma^2_{\phi\psi} & \sigma^2_{\theta\psi} & \sigma^2_{\psi\psi} \end{bmatrix}, \tag{13}$$

where σ_{xx} = 0.0485 m, σ_{yy} = 0.0435 m, σ_{zz} = 0.121 m, $\sigma_{\phi\phi}$ = 0.1456 rad, $\sigma_{\theta\theta}$ = 0.1456 rad, $\sigma_{\psi\psi}$ = 0.0044 rad, and the error covariance between variables has a zero value.

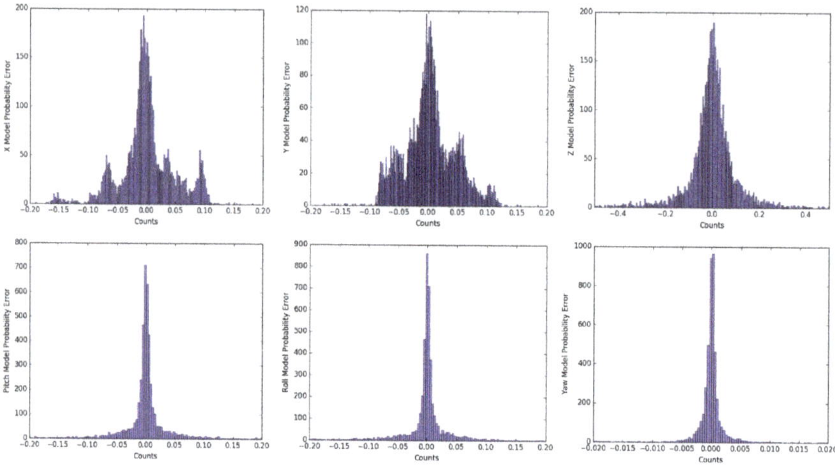

Figure 5. Probabilistic error distribution representation for each vehicle output variable.

4. Vehicle Pose Estimation System

This section details the architecture implemented to estimate the vehicle's attitude, by integrating the dynamic and kinematic model described in Section 3 and fusing the LiDAR-based measurement system described in Section 5. Several works have analysed the response of two of the most well-known filters for non-linear models, the Extended Kalman Filter (EKF) and the Unscented Kalman Filter (UKF), where the results were generally in favour of the UKF. For instance, in [28], the behaviour of both filters was compared to estimate the position and orientation of a mobile robot. Real-life experiments showed that UKF has better results in terms of localisation accuracy and consistency of estimation. The proposed architecture therefore integrates an Unscented Kalman Filter [29], which is divided into two stages: prediction and update (as shown in Figure 6).

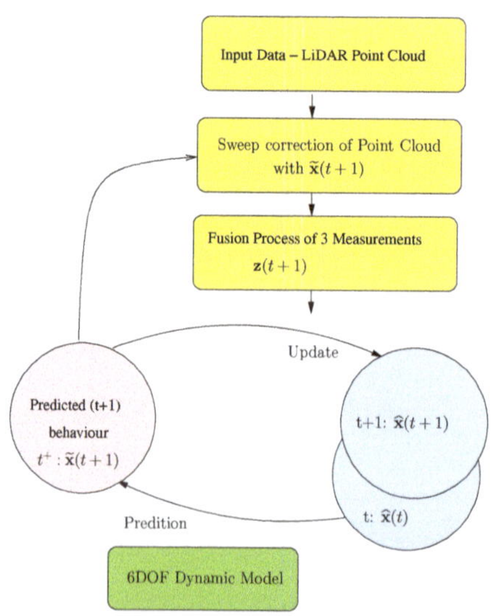

Figure 6. Unscented Kalman Filter (UKF) architecture. The prediction phase relies on the 6-DoF motion model detailed in the previous section. The update phase uses three consecutive LiDAR-based measurements to fuse and estimate the vehicle state.

The prediction phase manages the 6-DoF dynamic model described in the previous section to predict the system's state at time $(t+1)$. Along with the definition of the model, the model noise covariance matrix Q must be associated, as defined by the standard deviations evaluated above. The model noise covariance matrix is only defined in its main diagonal and is constant over time. Equation (14) represents the prediction phase of the filter.

$$\tilde{\mathbf{x}}(t+1) = A\hat{\mathbf{x}}(t) + Q, \tag{14}$$

where $\mathbf{x}$ is the 6-DoF state vector, as shown in Equation (15), and A matrix represents the developed dynamic model.

$$\mathbf{x}'(t) = \begin{bmatrix} x(t) \; y(t) \; z(t) \; \alpha(t) \; \theta(t) \; \psi(t) \end{bmatrix}. \tag{15}$$

In the filter update phase, the LiDAR odometry output is estimated. The estimated state vector, $\hat{\mathbf{x}}(t+1)$, is represented, in terms of the state variables, by Equation (16). The 6×6 matrix C is defined with the identity matrix, as the vectors $\mathbf{z}(t+1)$ and $\hat{\mathbf{x}}(t+1)$ contain the same measurement units. Finally, the matrix R is the covariance error matrix of the measurement, which is updated every odometry period in the measurement and fusion process, as explained in Section 5. The matrix R is only defined in its main diagonal, representing the uncertainty of each of the magnitudes measured in the process.

$$\mathbf{z}(t+1) = C\hat{\mathbf{x}}(t+1) + R. \tag{16}$$

LiDAR Sweep Correction

To use the LiDAR data in the update phase of the UKF, it is recommended to perform a so-called sweep correction of the raw data. The sweep correction phase is due to the nature of most LiDAR devices, which are composed of a series of laser emitters mounted on a spinning head (e.g., the Velodyne HDL-64E). The sweep correction process becomes crucial when the sensor is mounted on a moving vehicle, as the sensor spin requires a time span close to approximately 100 ms, as in the case of the Velodyne HDL-64E. The sweep correction process consists of assigning two poses for each sensor output and interpolates the poses with constant angular speed for all the LiDAR beams. These poses are commonly associated with the beginning and the end of the sweep. Thus, the initial pose is equal to the last filter estimation $\hat{\mathbf{x}}$ and the final pose is equal to the filter prediction $\tilde{\mathbf{x}}(t+1)$ to carry out the interpolation. The whole point cloud is corrected with the interpolated poses evaluated, solving the scene deformation issue when the LiDAR sensor is mounted on a moving platform. Figure 7a shows the key points on the sweep correction process.

Regarding the correction method, the authors in [30,31] proposed a point cloud correction procedure based on GPS data. The process requires synchronisation between each GPS and LiDAR output, a complex task when the GPS introduces small delays in its measurement. For this reason, in our case, the GPS data is replaced with the filter prediction to apply the sweep correction process. Figure 7b shows the same point cloud with and without sweep correction, captured in a roundabout with low angular speed vehicle movement. It can be seen that there is significant distortion concerning reality, as the difference of shapes between clouds is substantial, leading to errors of one meter in many of the scene elements. We can claim that the motion model accuracy is a determinant for the sweep correction process, as it improves the odometry results (as we depict in Section 7).

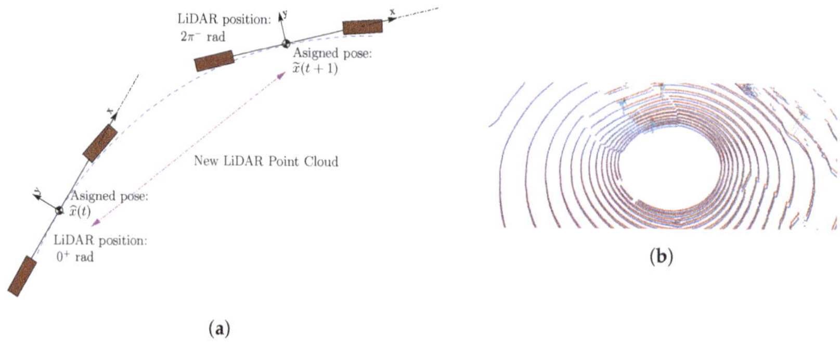

Figure 7. Sweep correction process in odometry: (**a**) the assignment of two poses to the point cloud when the vehicle is moving; and (**b**) raw (blue) and corrected measurements (red). An important difference between the measurement results of both point clouds is exposed, the correction being decisive for the result of the following stages.

5. Measurements Algorithms and Data Fusion

Three measurement methodologies based on LiDAR raw-data were developed to provide an accurate and robust algorithm. Two of them are based on ICP techniques, and another one relies on feature extraction and SVD. A 6-DoF measure $z(t)$ is the output of this process, after the fusion process is finished.

5.1. Multiplex General ICP

Using the ICP algorithm for the development of LiDAR odometry systems is very common, where the two most used versions are the Point-to-Point and Point-to-Plane schemes. Adaptations of both algorithms have been developed for our approach. For the first measurement system developed, we propose the use of the ICP point-to-point algorithm, which is based on aligning two partially overlapping point clouds to find the homogeneous transformation matrix (R, t) in order to align the two point clouds. The ICP used is based on minimising the cost function defined by the mean square error of the distances between points in both clouds, as expressed in Equation (17). The point cloud registration follows the criterion of the nearest neighbour distance between clouds.

$$\underset{R,T}{min}(error(R,T)) = \underset{R,T}{min}(\frac{1}{N_p}\sum_{1}^{N_p}\|p_i - (q_i R + T)\|), \tag{17}$$

where p_i represents the set of points that defines the cloud captured at a time instant $(t-1)$, q_i represents the set of points that define the cloud captured at a time instant t, N_p is the number of points considered in the minimisation process, R is the resulting rotation matrix, and T is the resulting translation matrix.

The ICP technique, as with many other gradient descent methods, can become stuck at a local minimum instead of the global minimum, causing measurement errors. The possibility of finding moving objects or a lack of features in the scene are some of the reasons why the ICP algorithm provides local minimum solutions. For this reason, an algorithm that computes the multiplex ICP algorithm for a set of distributed seeds was implemented. The selected seed, such as the ICP starting point, is evaluated with the Merwe Sigma Points method [32,33]. The error covariance matrix predicted by the filter $\widetilde{P}(t+1)$ and the predicted state vector $\widetilde{x}(t+1)$ are the input to assess the eight seeds needed. Figure 8 shows an example of seed distribution in the plane (x, y) for a time instant (t).

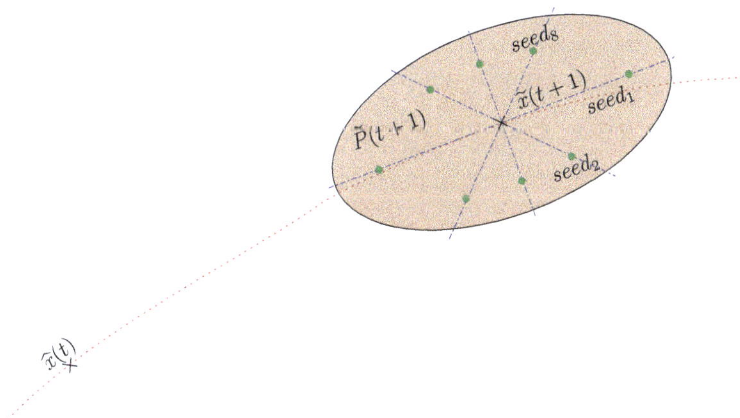

Figure 8. Sharing of Iterative Closest Points (ICP) initial conditions applying the Sigma Point techniques in the limits marked by $\widetilde{P}(t+1)$.

After evaluating the eight measures, the one with the best mean square error in the ICP process is selected. After the evaluation of two sequences of the KITTI data set, a decrease of error close to 9.5% was discerned, this being the determining reason why eight seeds were selected. However, the increase in computation time could be a disadvantage.

5.2. Normal Filtering ICP

For the design of a robust system, it is not enough to integrate only one measurement technique, as it may fail due to multiple factors. Therefore, a second measurement method based on ICP point-to-plane was developed to improve the robustness of the system, as it implies a lower computation time than the one above. In [34], the results with the point-to-plane method were more precise than those with the point-to-point method, improving the precision of the measurement. The cost function to be minimised in the point-to-plane process is as follows (18):

$$\min_{R,T}(error(R,T)) = \min_{R,T}(\frac{1}{N_p}\sum_{1}^{N_p}\left\|(p_i - (q_iR+T)) \cdot n_i^p\right\|), \tag{18}$$

where R and T are the rotation and translation matrices, respectively, N_p is the number of points used to optimize, p_i represents the source cloud, q_i represents the target cloud, and n_i^p represents the normal unit vector of a point in the target cloud. The point-to-plane technique is based on a weighting to register cloud points in the minimisation process, where $cos(\theta)$ from the vectorial product is the weight given in the process and θ is the angle between the unit normal vector n_i^p and the vector resulting from the operation $(p_i - (q_iR+T))$. Therefore, the smaller the angle θ is, the higher the contribution in the added term of this register point is. So, the normal unit vector n_i^p can be understood as rejecting or decreasing the impact over the added term of its register points when the alignment with the unitary vector is not right. The approach in this paper does not include all the points registered, as a filter process is carried out. The heading of the vehicle is the criteria to implement the filtering process. Thus, only those points that have a normal vector within the range $\widetilde{\psi}_v \pm \widetilde{\sigma}_{\psi\psi}rad$ are considered in the added term, where $\widetilde{\psi}_v$ represents the heading of the predicted vehicle and $\widetilde{\sigma}_{\psi\psi}$ represents the uncertainty predicted from the error covariance matrix. Equation (19) formulates the criteria applied in the minimisation to filter out points:

$$\begin{aligned} minJ(R,T) &= \min_{R,T}\left(\tfrac{1}{N_p}\textstyle\sum_1^{N_p}\left\|(p_i-(q_iR+T))\cdot n_i^p\right\|\right) \\ s.t. & \\ n_i^p > \widetilde{\psi}_v \pm \widetilde{\sigma}_{\psi\psi} & \\ n_i^p > \widetilde{\psi}_v \pm \widetilde{\sigma}_{\psi\psi}+\pi & \end{aligned} \tag{19}$$

Points that are not aligned with the longitudinal and transverse directions of the vehicle are eliminated from the process, improving the calculation time of this process as well as the accuracy of the measurement. Figure 9 represents an ICP iteration of the described technique, where the results achieved by RMSE are 20% better than if all the points of the cloud are considered.

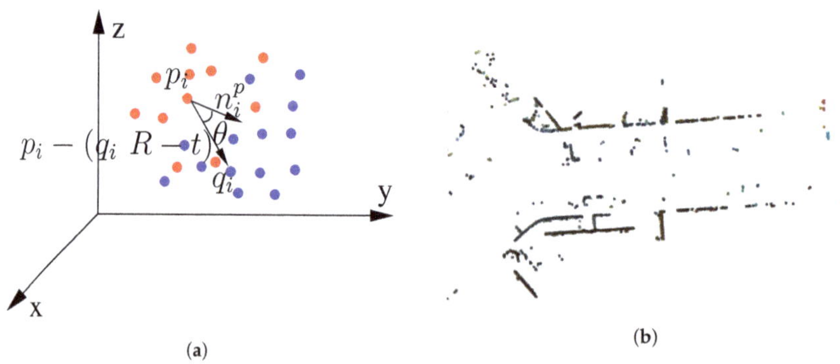

Figure 9. ICP process based on normals: (**a**) Graphical representation of the cost function with the normal unit vector n_i^p used to enter constraints in ICP; and (**b**) ICP output result applying constraints of normals. The figure shows the overlap of two consecutive clouds.

5.3. SVD Cornering Algorithm

The two previous systems of measurement are ICP-based techniques, where there is no known data association between the points of two consecutive point clouds. However, the third proposed algorithm uses synthetic points generated by the algorithm and the data association of the synthetic point between point clouds to evaluate the odometry step. An algorithm for extracting the characteristics within the point clouds is developed to assess the synthetic points. The corners built up with the intersection between planes are the features explored. The SVD algorithm uses the corners detected in consecutive instants to determine the odometry between point clouds. The new odometry complements the two previous measurements. The SVD algorithm is accurate and has low computational load, although the computation time increases in the detection and feature extraction steps.

5.3.1. Synthetic Point Evaluation

Plane Extraction

It is easy for humans to identify flat objects in an urban environment; for instance, building walls. However, identifying vertical planes in a point cloud with an algorithm is more complex. The algorithm identifies points that, at random heights, fit the same projection on the plane (x, y). Therefore, the number of repetitions that each beam of the LiDAR presents on the plane (x, y) is recorded. If the number of repetitions of the project exceeds the threshold of 20 counts, the points belong to a vertical plane. Figure 10 shows detected points that belong to vertical planes, although the planes in many cases are not segmented.

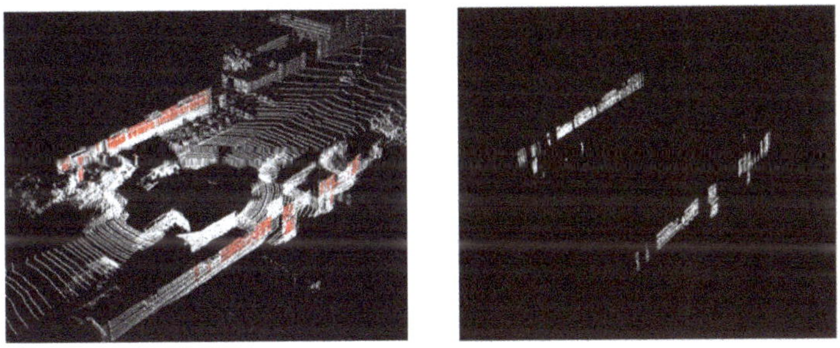

(**a**) (**b**)

Figure 10. Intermediate results of sequence 00, frame 482: (**a**) Input cloud to the plane detection algorithm; and (**b**) points detected on candidate planes.

Clustering

Clustering techniques are then used to group the previously selected points into sets of intersecting planes. Among those listed in the state-of-the-art, those that do not imply knowing the number of clusters to be segmented were considered valid, as it is not known a priori. Analysing the clustering results provided by the *Sklearn* library, DBSCAN was the one that obtained the best results, as it does not make mistakes when grouping points of the same plane in different clusters. In order to provide satisfactory results, the proposed configuration of the DBSCAN clustering algorithm sets the maximum distance between two neighbouring points (0.7) and the minimum number of samples between neighbours (50). The algorithm identifies solid structure corners, such as building walls, such that clusters associated to non-relevant structures are eliminated. For this purpose, clusters sized smaller than 300 points were filtered, eliminating noise produced by vegetation or pedestrians. Figure 11 represents the cluster segmentation of the point cloud depicted in Figure 10, where only the walls of buildings, street lights, or traffic signs are segmented as characteristic elements of the scene.

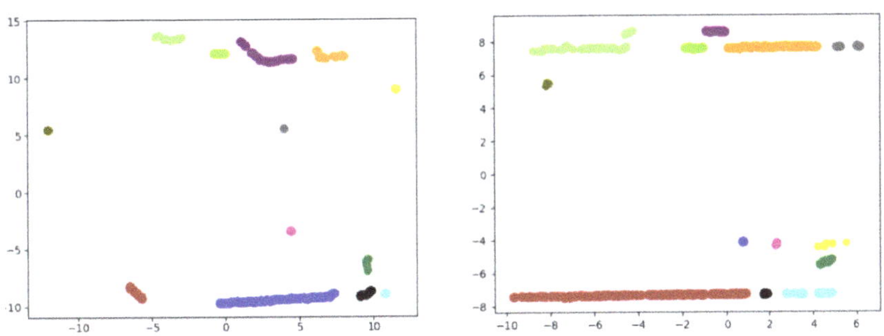

(**a**) (**b**)

Figure 11. Results of clustering, sequence 00 of the KITTI odometry data set: (**a**) Frame 0, and (**b**) Frame 482.

Corner Detection and Parameters Extraction

In addition, to eliminate straight walls, cylindrical points, or a variety of shapes that are not valid for the development of the algorithm, clusters that do not contain two vertical intersected planes are discarded, as shown in Figure 11. Thus, two intersecting planes are searched for in the cluster that satisfies the condition of forming an angle between both higher than 45° and less than 135°. Using the RANSAC algorithm on the complete set of points of the cluster, indicating that it selects a quarter of the total points and fixing the maximum number of iterations as 500 iterations, the algorithm returns the equation of a possible intersected plane in the cluster. Applying RANSAC again to the outlier points resulting from the first process and with the same configuration parameters, a second intersected plane in the cluster is achieved, as shown in Figure 12. If the angle formed between the two intersected planes fulfils the previous conditions, the intersection line of both planes is evaluated to obtain the synthetic points that define the evaluated corner.

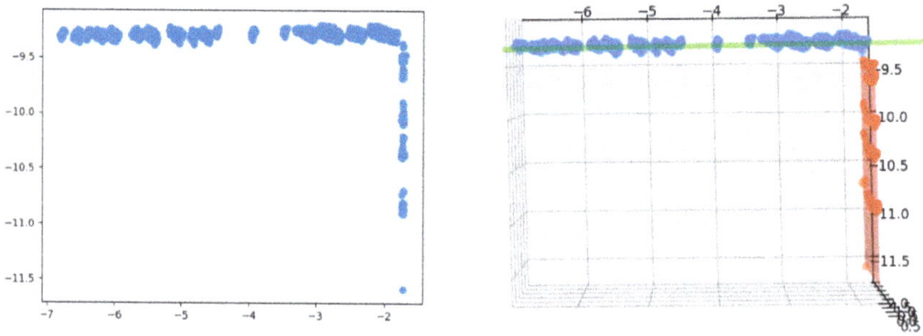

(a) (b)

Figure 12. Results of extraction of intersected planes: (**a**) Input data to cluster planes; and (**b**) detection results of two intersecting planes, represented in red and green.

Synthetic Points Evaluation

At this point, the objective is to generate three points that characterize the corners of the scene; these points are denoted as synthetic points. The synthetic points are obtained from the intersection line equation derived from the two intersecting planes. Figure 13a shows the criterion followed to evaluate three synthetic points for each of the detected corners. Two of the synthetic points, (M, J), belong to the intersection line and are located at a distance of 0.5 m. The third synthetic point, N, meets the criterion of being at 1 m of point separation from M with a value of $z = 0$. The process identifies, as the reference plane, the one that has the lowest longitudinal plane direction evaluated within the global co-ordinate system. Figure 13b shows the points (M, J, N) evaluated in two consecutive instants of time. In this situation, the SVD algorithm can be applied to assess the homogenous transform between two consecutive point clouds when the synthetic points data association is known.

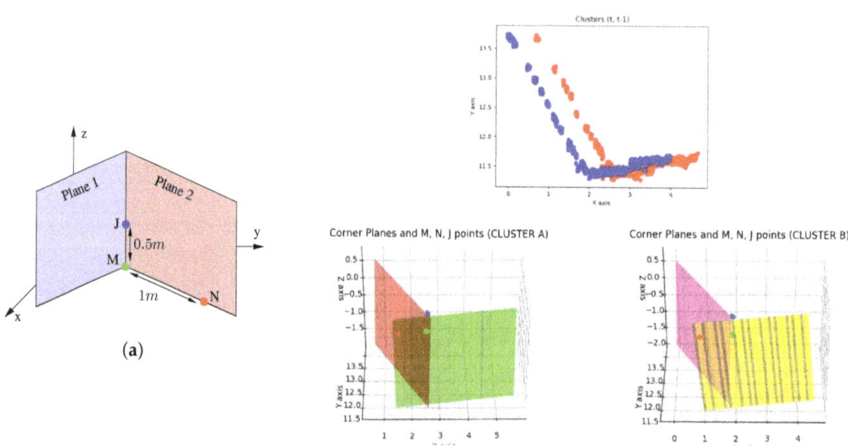

(a) (b)

Figure 13. Evaluation of synthetic points: (**a**) Nomenclature and position of calculated synthetic points; and (**b**) result of synthetic points detection in real clusters of two consecutive time instants.

5.3.2. SVD

Before applying the SVD, the registration of the extracted points of the corners between two consecutive instants need to be done. So, let us suppose that, for an instant t, there is a set of corners $X = x_1, x_2, x_3, ..., x_n$ where $x_1 = (M_1, N_1, J_1)$ and, for another instant $t+1$, there is another set of corners $Y = y_1, y_2, y_3, ...y_m$ where $y_1 = (M_1, N_1, J_1)$. Then, to register both sets, the Euclidean distance of points M is used. Only those corners that show a minimum distance less than 0.5 m are data associated. The non-data associated corners are removed.

Once the data association of synthetic corners is fulfilled, the objective is to find the homogeneous transformation between two consecutive scenes. Therefore, SVD minimises the distance error between synthetic points, first by eliminating the translation of both sets to exclude the unknown translation and then by solving the Procustes orthogonal problem to obtain the rotation matrix (R). Finally, it undoes the translation to obtain the translation matrix (T). Equation (20), described in more detail in [35], shows the mathematical expressions applied in the SVD algorithm to obtain the homogeneous transformation matrix between two sets of synthetic corners at consecutive time instants:

$$\begin{array}{c} X' = x_i - \mu_x = x_i' \\ Y' = y_i - \mu_y = y_i' \\ W = \sum_{i=1}^{N_p} x_i' y_i'^T \\ W = U \sum V^T \\ R = UV^T \\ t = \mu_x - R\mu_y \end{array} . \tag{20}$$

The SVD odometry measure z_{SVD} is fused with the other measurements, but the factor related to the uncertainty must be added to the SVD homogeneous transformation $\Delta\, Pose_{SVD}$. Therefore, Equation (21) defines the SVD measure added to the $\Delta\, Pose_{SVD}$, the UKF estimated state vector $\tilde{x}(t)$, and the uncertainty factor R_{SVD}. The uncertainty represents the noise covariance matrix of the SVD measurement and R_{SVD} is calculated with the RMSE returned by the RANSAC process applied within the method. The decision taken is a consequence of distinguishing a direct relationship between R_{SVD} and how well the intersection planes are fitted over the points of the cluster.

$$z_{SVD} = \tilde{x}(t) + \Delta\, Pose_{SVD} + R_{SVD}. \tag{21}$$

Figures 14 and 15 depict a successful scenario where SVD odometry is evaluated. The colour code used in the figure is: green (M points), blue (J points), and orange (N points).

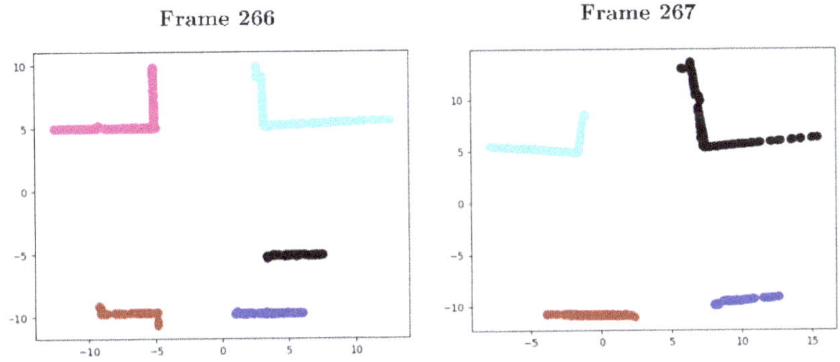

Figure 14. Odometry results with Singular Value Decomposition (SVD): Input cluster to extract synthetic points.

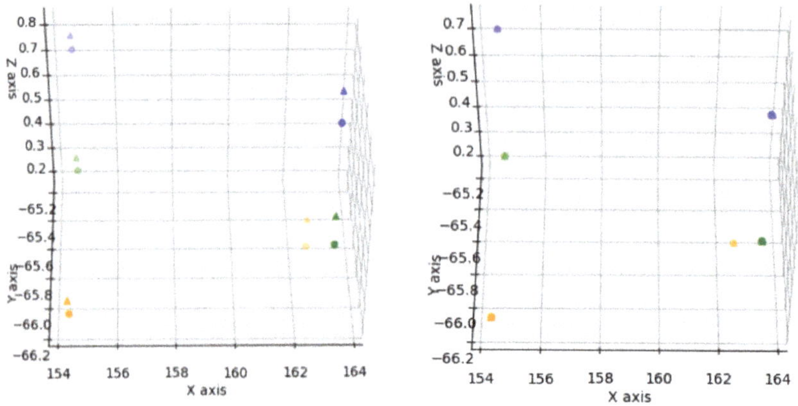

(**a**) (**b**)

Figure 15. Odometry results with SVD: (**a**) Representation of the synthetic points extracted from the previous clusters. A translation and rotation between them is shown; and (**b**) synthetic points are overlapped when applying the rotation and translation calculated by SVD.

5.4. Fusion Algorithm

An essential attribute in the design of a robust system is the redundancy. For the proposed work, three measurement techniques based on LiDAR were developed. Therefore, it is necessary to integrate sensor fusion techniques that allow for selecting and evaluating the optimal measurement from an input set to the solution. Figure 16a shows an architecture where the filter outputs—that is, the estimated state vectors—are fused. The main architecture characteristic is that multiplex filters have to be integrated into the solution. Figure 16b shows an architecture that fuses a set of measurements and then filters the fused measurement. For this architecture design, only blocks have to be designed, improving its simplicity. In this second case, all the measurements must represent the same magnitude to be measured. In [36], a system that merges the data from multiple sensors using the second approach was presented. The proposed fusion system implements this sensor fusion architecture, in which the resulting measurement vector comprises the 6-DoF of the vehicle.

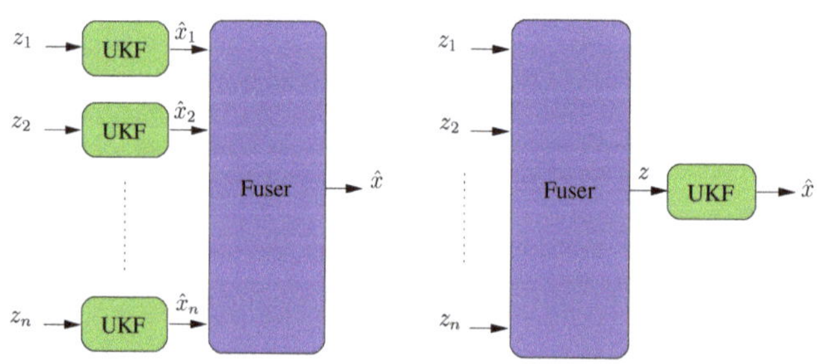

(**a**) (**b**)

Figure 16. Block diagram for two fusion philosophies: (**a**) Merging of the estimated state vector, which requires a filtering stage for each measure to be merged; and (**b**) merging of observations under a given criterion and subsequent filtering.

The proposed sensor fusion consists of assigning a weight to each of the measurements. The weights are evaluated considering the distance (x, y) between the filter prediction and the LiDAR-based measurements. Therefore, Equation (22) defines the weighting function. The assigned weight varies between 0 and 1 when the measurement is within the uncertainty ellipse. The assigned weight is 0 when the measurement is outside the uncertainty ellipse, as shown in Figure 17.

The predicted error covariance matrix $\widetilde{P}(t+1)$ defines the uncertainty ellipse. The weighted mean value is the fused measurement, as detailed in Equation (23). In the same way, the uncertainty associated with the fused measure is weighted with the partial measure weight. Thus, the sensor fusion output is a 6-DoF measure with an associated uncertainty matrix R.

$$\begin{cases} If \quad \frac{(z_x-\widetilde{x}_x(t+1))^2}{\sigma_{\widetilde{x}x}^2} + \frac{z_y-\widetilde{x}_y(t+1)}{\sigma_{\widetilde{y}y}^2} \leq 1 \;\Rightarrow\; w = \left|\sqrt{\frac{(z_x-\widetilde{x}_x(t+1))^2}{\sigma_{\widetilde{x}x}^2} + \frac{z_y-\widetilde{x}_y(t+1)}{\sigma_{\widetilde{y}y}^2}} - 1\right| \\ If \quad \frac{(z_x-\widetilde{x}_x(t+1))^2}{\sigma_{\widetilde{x}x}^2} + \frac{z_y-\widetilde{x}_y(t+1)}{\sigma_{\widetilde{y}y}^2} > 1 \;\Rightarrow\; w = 0 \end{cases} \tag{22}$$

$$z = \widetilde{x} + \frac{(z_1-\widetilde{x}(t+1))w_1 + (z_2-\widetilde{x}(t+1))w_2 + (z_3-\widetilde{x}(t+1))w_3}{w_1+w_2+w_3} \tag{23}$$

Fusing the set of available measurements provides the system with robustness and scalability. It is robust because, if any of the developed measurements fail, the system can continue to operate normally, and it is scalable as other measurement systems are easy to integrate using the weighting philosophy described above. Furthermore, the integrated measurement systems can be based on any of the available technologies, not only LiDAR. As the number of measurements increases, the result achieved should improve, considering the principles of Bayesian statistics.

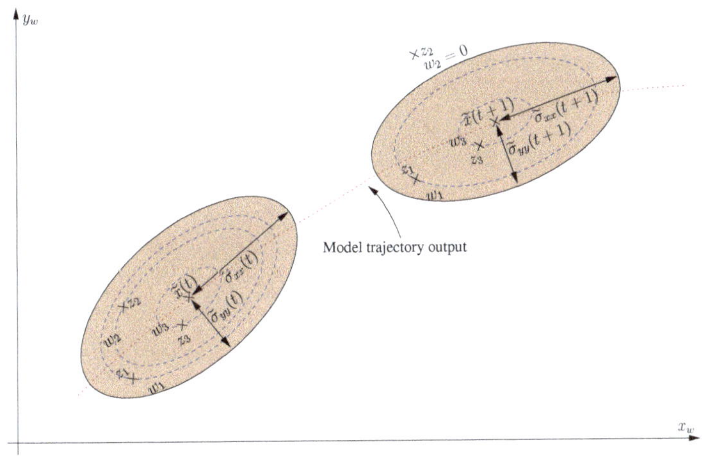

Figure 17. Representation of predicted ellipses of uncertainty and weight allocation to each measure to be applied in fusion.

6. Fail-Aware Odometry System

The estimated time window evaluated by the fail-aware indicator is recalculated for each instant of time, allowing the trajectory planner system to manage an emergency manoeuvre in the best way. In practice, most odometry systems do not implement this kind of indicator. Instead, our approach proposes the use of the evaluated heading error, as the heading error magnitude is critical for the localisation error. Thus, a small heading error at time t produces a huge localisation error at time $t+N$ if the vehicle has moved hundreds of meters away. For example, a heading error equal to 10^{-3} rad at t introduces a localisation error of 0.01 cm at $t+N$ if the vehicle moves only 100 m. This behaviour motivates us to use the heading error to develop the fail-aware indicator.

The estimated heading error has a significant dependence on the localisation accuracy. The developed fail-aware algorithm is composed of two parts: a function to evaluate the fail-aware indicator and a failure threshold, which is fixed as 0.001. This threshold value was chosen by using heuristic rules and analysing the system behaviour in sequences 00 and 03 of the KITTI odometry data set. We evaluated the fail-aware indicator (η) on each odometry execution period, in order to estimate the remaining time to overtake the fixed malfunction threshold. Equation (24) defines the fail-aware indicator, where $\sigma_{\psi\psi}$ is the estimated heading standard deviation and $\sigma_{\psi\psi}$ is identified as the variable most correlated with the localisation error; once again, regarding the error results in sequences 00 and 03.

For this reason, $\sigma_{\psi\psi}$ is useful to evaluate the fail-aware indicator. The second derivative of $\sigma_{\psi\psi}$ is used, representing the heading error acceleration, so how fast or slow this magnitude changes is used as a determinant to find the estimated time of reaching the malfunction threshold. If the acceleration of $\sigma_{\psi\psi}$ is low, the estimated time window is large and the trajectory planner has more time to perform an emergency manoeuvre. On the other hand, if the acceleration of $\sigma_{\psi\psi}$ is high, the estimated time window be decisive with respect to stopping the car safely in a short time.

The acceleration of $\sigma_{\psi\psi}$ can be positive or negative, but the main idea is to accumulate the absolute value for all the odometry interactions, in order to have an indicator that allows us to know the estimated time window. The limit time t_1 in the addition term represents when the LiDAR odometry system starts to work as a redundant system for localisation tasks. In this way, the speed η is calculated as the difference between two consecutive η values, in order to assess the time to reach the malfunction threshold. Figure 18 shows the behaviour of the fail-aware algorithm. In all the use-case studies, the Euclidean error $[x, y]$ is approximated as 0.6 m when the malfunction threshold is exceeded. The Euclidean XY error depicted in the image is calculated by comparing the LiDAR localisation and the available GT. The fail-aware algorithm provides a continuous diagnostic value of the LiDAR system, allowing for the development of more robust and safe autonomous vehicles.

$$\eta = \sum_{t=t1}^{\infty} \left\| \frac{d^2}{dt^2} \widehat{\sigma}_{\psi\psi} \right\| \tag{24}$$

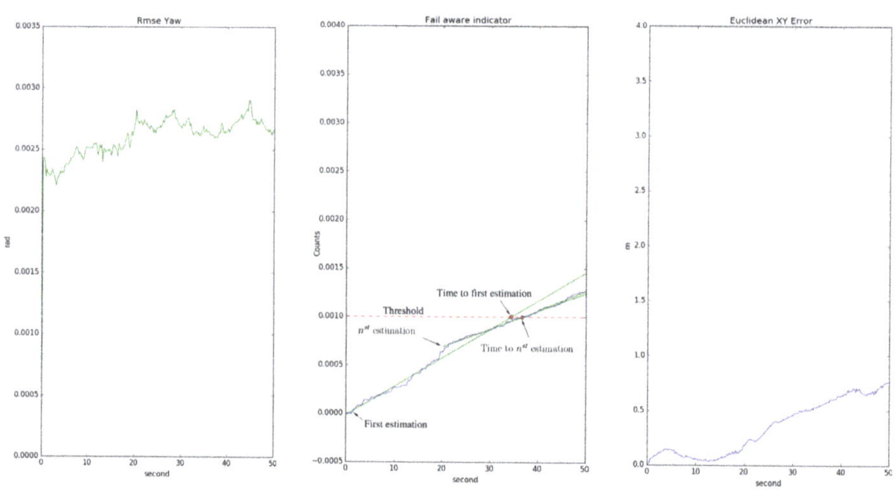

(a) (b) (c)

Figure 18. Fail-aware process. Sequence results 03. (**a**) Evolution of the signal standard deviation of $\widehat{\psi}$ estimated by the filter ($\sigma_{\psi\psi}$). (**b**) Representation of the failure threshold (red) and fail-aware indicator η (blue). The green lines represent the equation to evaluate the time window to reach the failure threshold. (**c**) Euclidean $[x, y]$ error compared with the ground truth (GT) of the data set.

7. Experimental Analysis

7.1. KITTI Odometry Data Set Evaluation

The presented algorithm was extensively tested. A total of approximately 50,000 point clouds from different environments and with a multitude of situations were processed, representing a total of 19.5 km processed. We defined four categories—urban, country, urban/country, and highway—to label each of the processed sequences. Table 2 lists the translation and rotation errors obtained in each sequence as a result of applying [37] for evaluation. Two use-cases, 6-DoF and 3-DoF, were evaluated to quantify the improvement introduced in the case of 6-DoF. The results show that the odometry system worked for different scenarios, without showing considerable differences in the results. However, sequence 01 (Highway) had considerable translation and rotation errors for the 6-DoF and 3-DoF cases, mainly because the road had a lower number of characteristics in the highway scenario. The average results for 6-DoF were 1.00% and 0.0039 deg/m in translation and rotation, respectively. In the case of 3-DoF, where the state vector $\hat{x}(t)$ was defined by the variables (x, y, ψ), the mean error values were 7.79% and 0.057 deg/m in translation and rotation, respectively. Figure 19 represents the results of processing sequence 00 in both cases.

Table 2. Numerical results when processing the sequences with 6-DoF or 3-DoF.

Sequence	Scene	6-DoF Error		3-DoF Error	
		Translational [%]	Angular [deg/m]	Translational [%]	Angular [deg/m]
00	Urban	1.28	0.0051	9.87	0.0793
01	Highway	2.36	0.0135	12.89	0.0462
02	Urban/Country	1.15	0.0028	4.42	0.0252
03	Country	0.93	0.0024	12.54	0.0864
04	Country	0.98	0.0033	1.34	0.0037
05	Urban	0.45	0.0018	10.01	0.0682
07	Urban	0.44	0.0034	3.39	0.0656
09	Urban/Country	0.64	0.0013	3.84	0.0219
10	Urban/Country	0.83	0.0017	12.29	0.0557

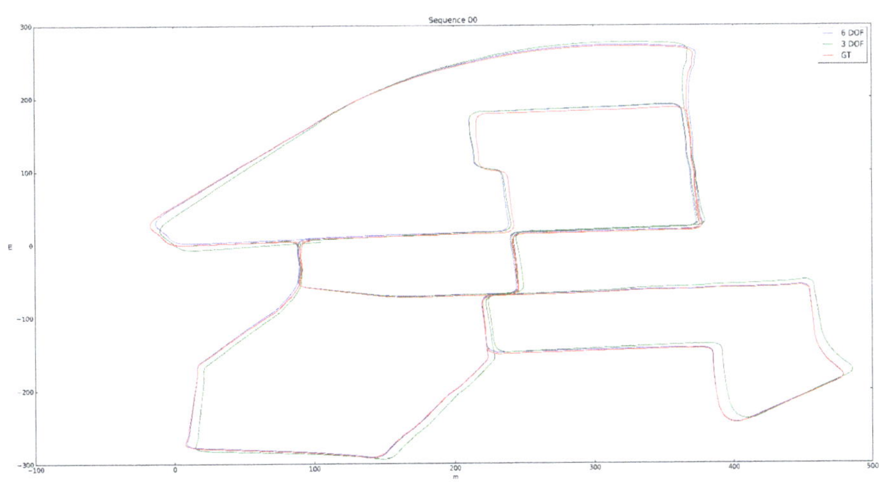

Figure 19. Visual results comparison using the 6-DoF and 3-DoF models in sequence 00.

On the other hand, the results analysed were evaluated with the integration of three or two measurements in the fusion system. The three-measurement fusion combined the three techniques described in the article, while the two-measurement fusion only combined the ICP-based techniques.

Table 3 shows the translation and rotation errors for each sequence in both cases. In the case of fusing only two measures, the average result was 1.61% and 0.0046 deg/m in translation and rotation, respectively. Therefore, the system with the three-measurement fusion improved the odometry behaviour by 62.3%, as compared to that with two-measurement fusion.

Table 3. Results of processed sequences with and without applying feature detection and SVD in the measurement fusion process.

Sequence	Scene	Fusion with 3 Measures		Fusion with 2 Measures	
		Translational [%]	Angular [deg/m]	Translational [%]	Angular [deg/m]
00	Urban	1.28	0.0051	1.31	0.0052
01	Highway	2.36	0.0135	7.08	0.0122
02	Urban/Country	1.15	0.0028	1.21	0.0030
03	Country	0.93	0.0024	0.97	0.0022
04	Country	0.98	0.0033	0.69	0.0031
05	Urban	0.45	0.0018	0.91	0.0052
07	Urban	0.44	0.0034	0.63	0.0022
09	Urban/Country	0.64	0.0013	0.93	0.0014
10	Urban/Country	0.83	0.0017	0.84	0.0017

One of the most well-known sequences in the KITTI odometry database is 00, as it has been analysed and referenced in many SLAM and odometry papers, with an approximate length of 3.8 km. Figure 20 shows the results of processing it, where the components (x, y, z, ψ) of the estimated state vector $\hat{x}$ are represented, as well as the 2D path followed. All plots overlap the ground truth (GT) information with the odometry results. The algorithm behaved properly visually, but ended the sequence with error in all its variables: $error_x = 3$ m, $error_y = 4$ m, $error_z = 0.6$ m, $error_\alpha = 0.02$ rad, $error_\theta = 0.002$ rad, and $error_\psi = 0.007$ rad.

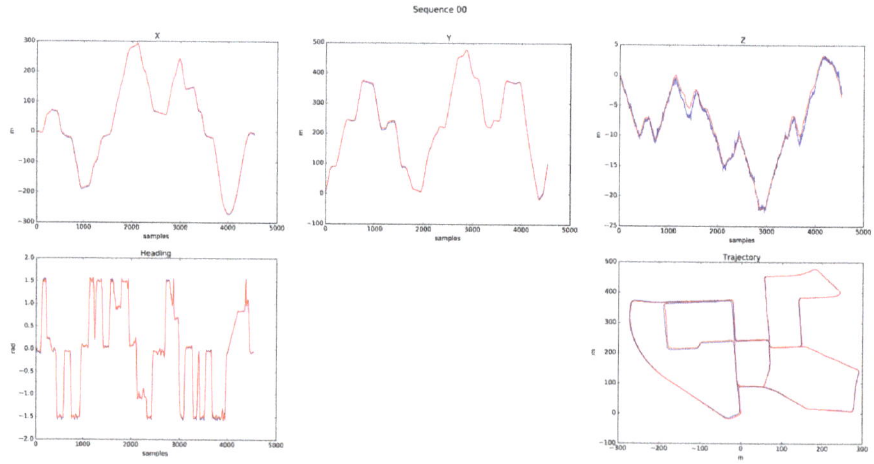

Figure 20. Sequence 00 results.

Figure 21 shows the behaviour of the pitch and roll angles, where (concerning the previous representations) the similarities with the GT are not as evident, due to the angle normalisation done between $\pm\pi$, besides putting in question whether the GT information was correct for the whole sequence.

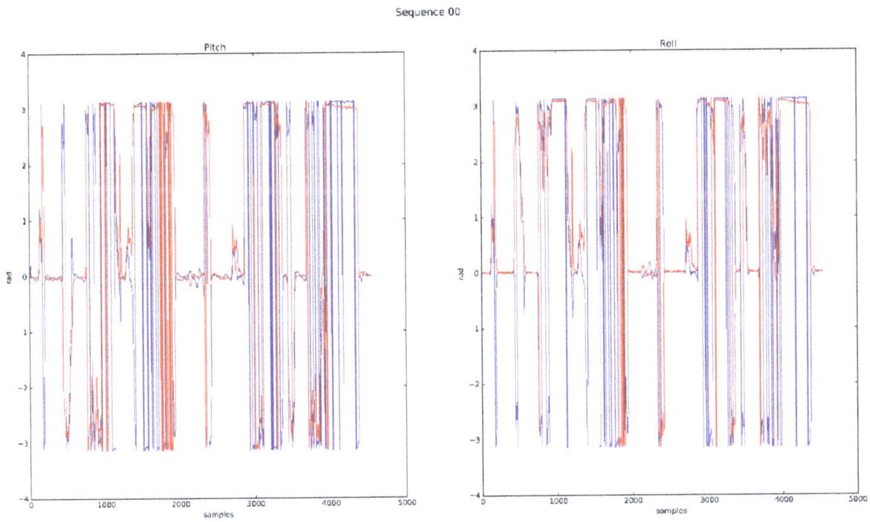

Figure 21. Pitch and roll results for sequence 00.

Figure 22 shows the path of the sequences available in the database. Sequences 06 and 08 were removed from the results, as the GT was not correct in both of these sequences. For all cases, a correct behaviour can be seen, except for sequence 01, which shows significant error concerning the GT. This error comes from the scenario in which the test was carried out, an open road without objects, where characteristics could not be extracted and which lacked relevant points to apply the ICP techniques correctly. Errors were caused when a local minimum was detected, such that the integrated odometry process made the vehicle trajectory drift and increased the error in its evaluation.

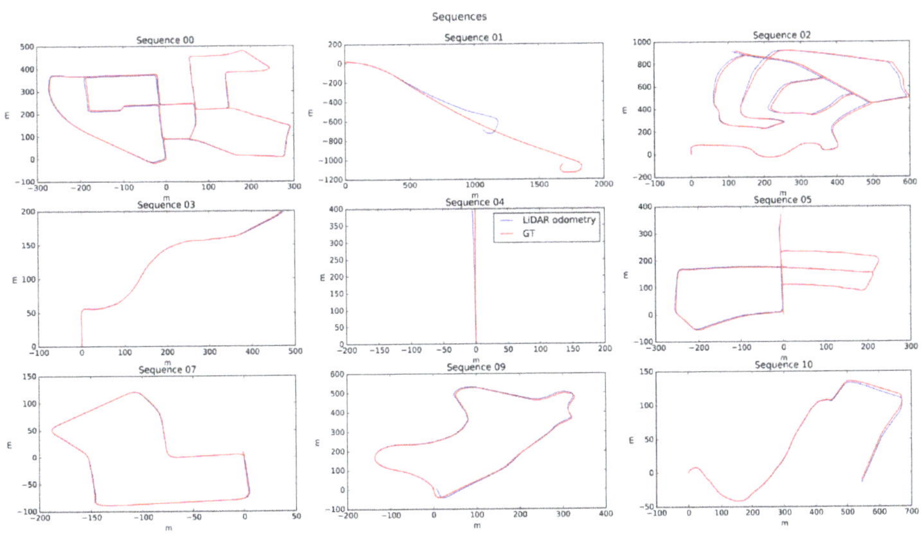

Figure 22. Sequence results.

The bad results shown in sequence 01 identify a system malfunction. The fault was detected when processing the first point cloud, and the fail-aware indicator showed a considerable difference from that in a sequence with acceptable results. Figure 23a shows the fail-aware indicator for sequence 01, as compared with the indicator for sequence 03 shown in Figure 23b The analysis shows that the estimated time to reach the incorrect operation threshold was lower in sequence 01 than in sequence 03, with values of 5 s and 40 s, respectively. Sequence 01 is characterised by a slow indicator change in its first seconds, estimating a failure time of approximately 10 s. However, after three seconds of operation, the indicator change increased considerably, reducing the failure time estimate to 5 s. The times listed were taken with regards to the beginning of the test; although, in a real scenario, these times are relative to the current instant. An estimated failure time of 5 s makes it practically impossible to carry out a safe stop manoeuvre. On the other hand, the speed of the indicator in sequence 03 is slow from the beginning of the test, and continues with a similar speed until reaching the failure threshold. As the indicator speed was slow, the system could operate with an error of less than 0.5 m for approximately 40 s, allowing the planning system to carry out a safe stop manoeuvre.

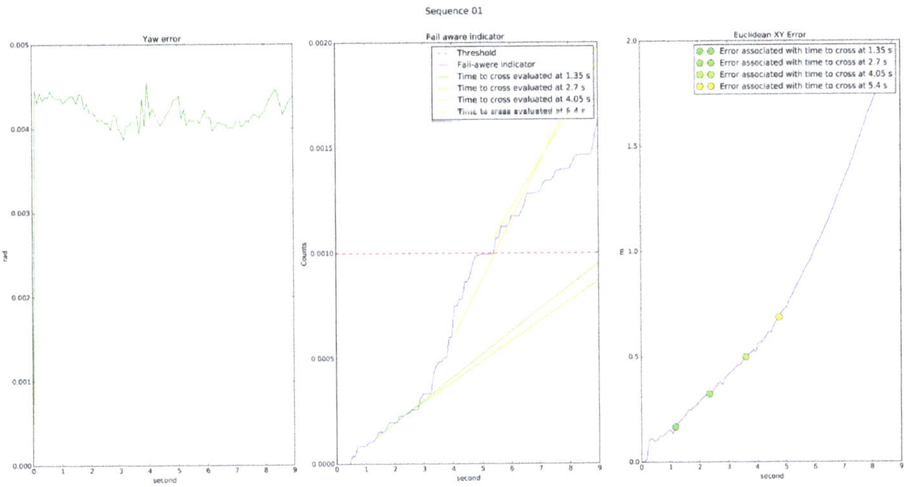

(**a**)

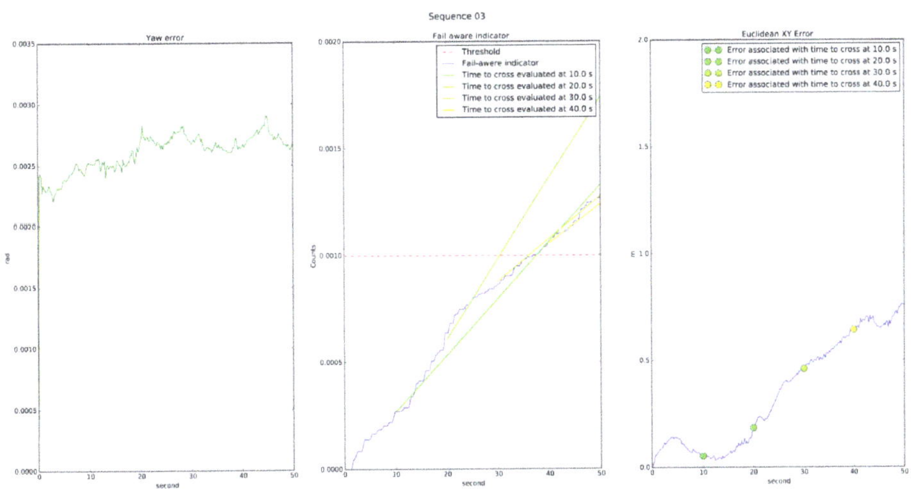

(**b**)

Figure 23. Dynamics of the fail-aware indicator: (**a**) Sequence 01 shows an estimated failure time of 5 s, which means that a high error is caused in a short time as a result of incorrect linear and angular localisation; (**b**) Sequence 03 shows an estimated failure time of 40 s.

On the other hand, Figure 24 shows the estimated height of all processed sequences. The good behaviour presented in the height estimation for six sequences should be noted. However, in sequences 01, 09, and 10, the estimated height was far from the GT. The poor odometry behaviour justifies the error in 01, but the errors in 09 and 10 are mainly due to the urban/country environment in which the sequences are developed, due to limited features to resolve the height estimate.

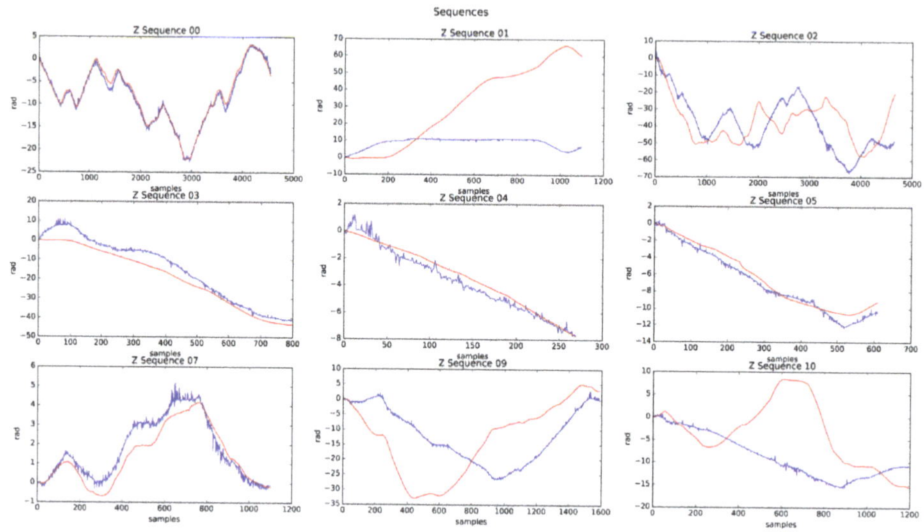

Figure 24. Height sequences results.

Finally, Figure 25 shows the estimated heading of each of the processed sequences. It is essential to highlight the excellent behaviour presented by the algorithm in practically all of them. However, the error at the end of 01 was 0.5 rad, much higher than that in the other sequences (i.e., close to 0.001 rad). This circumstance is essential to justify the excessive failure of 01.

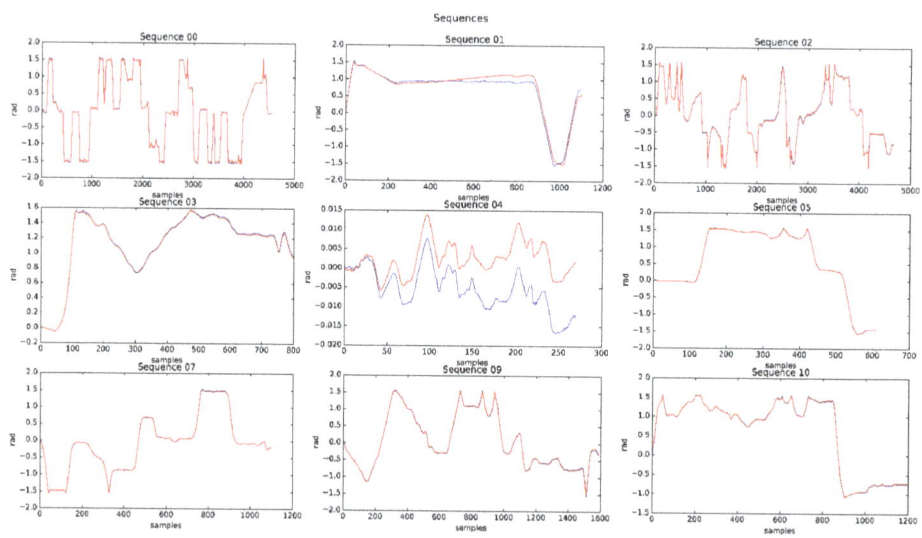

Figure 25. Heading sequences results.

7.2. Ranking Evaluation

In the set of listed algorithms in the KITTI odometry data set, there are a total of twenty-five pure LiDAR-based odometry algorithms, which we used to compare the results of the developed algorithm. Table 4 shows the first fifteen entries of the ranking where, between the LOAM and MDT-LO algorithms, the translation error is doubled. The implemented LiDAR odometry algorithm reached an average translation error of 1.00% and an average rotation error of 0.0039 deg/m; these errors were obtained when processing the totality of sequences for which the GT was available. The results achieved were encouraging, as they were within the first fifteen entries in a data set that is well-known at an international level by the scientific community. The ranked position was close to the 30th position, if all entries in the data set are considered. We believe these are quite acceptable results, considering those systems that have the objective of being redundant localisation systems in autonomous driving. In addition, the fail-aware functionality of the system adds a differentiating feature from the state-of-the-art. This feature allows the planning algorithms to compute safe trajectories even when the GPS localisation system fails, simply by using only our odometry method.

Table 4. Comparison of LIDAR-based methods ranked on KITTI data set.

Method	KITTI Ranking Position	Translational Error [%]	Angular Error [deg/m]	Fail-Aware
LOAM [11]	2	0.55	0.0013	
IMLS-SLAM [38]	4	0.69	0.0018	
MC2SLAM [26]	5	0.69	0.0016	
LIMO2-GP [24]	11	0.84	0.0022	
CAE-LO [25]	13	0.86	0.0025	
LIMO2	15	0.86	0.0022	
ICP-LO	16	0.87	0.0036	
CPFG [39]	17	0.87	0.0025	
PNDT LO [40]	22	0.89	0.0030	
LIMO	24	0.93	0.0026	
S4-SLAM	28	0.98	0.0044	
RIS	29	0.98	0.0026	
Ours	-	1.00	0.0039	x
KF-SLAM	30	1.00	0.0041	
S4OM	34	1.03	0.0053	
NDT-LO	35	1.05	0.0043	

8. Conclusions and Future Works

Although research and development into autonomous driving in recent decades has helped to achieve high SAE levels of automation, the presently existing control architectures are highly driver-dependent. Indeed, in the case of hardware failure, the driver must take control of the vehicle. However, to improve user acceptance of these systems, the vehicle itself should be able to solve the problem autonomously. This paper presents a LiDAR odometry system with an integrated fail-aware feature, which notifies high-level systems with the actual performance of our proposal. For instance, this allows a trajectory planner to plan a safe stop manoeuvre, guaranteeing the needed security in the environment.

In this paper, we presented a robust and scalable localisation system which, independently of the support or redundancy that it can offer to other systems, allows for fusion with any of the available localisation measures, improving its robustness in the measure. Thus, the fusion architecture presented reduces the undesirable consequences given in urban scenarios where the D-GPS measure may suffer from loss of accuracy due to satellite visibility.

Our proposal is based on a LiDAR-based odometry algorithm. An in-depth study of the topic was presented and it was pointed out that all existing odometry systems suffer from drift error in their process. To minimize and possibly eliminate this issue, the proposed system fuses different LiDAR-based localisation measurements using a UKF filter. The filter implementation takes into account 6-DoF dynamic models, which improve the correction process of the point cloud sweep, correcting the deformation of the scene when it is captured from a moving platform with high angular velocities. The model also allows us to estimate the vehicle roll and pitch variables, in order to reduce the measurement noise.

ICP techniques have been widely applied in LiDAR odometry systems, but these techniques have the disadvantage of being slow, despite their high precision. Therefore, the presented approach integrates multiple measurement stages to improve the accuracy of the final measurement. The first stage is based on multiple ICPs of lesser precision but with starting seeds distributed to improve its precision in the final measurement. The second stage is based on applying constraints to the ICP minimisation process, improving the accuracy and the computation time. Finally, the third stage is focused on the detection of vertical corner features above the point clouds, in order to apply SVD and estimate the homogenous transform between point clouds. However, the results of the third stage are conditioned to the type of scene, as shown in the highway results of Sequence 01.

The obtained linear and angular errors when processing the KITTI odometry data set were 1.00% and 0.0039 deg/m, respectively. These results are ranked within the first fifteen methods based only on LiDAR odometry. Furthermore, the proposed algorithm introduces a dynamic fail-aware indicator, a function of the standard error deviation associated with the estimation of the yaw vehicle angle.

As this work presents a fail-aware system based on LiDAR odometry, it could assist other systems of the vehicle (which is part of our planned future work) to decrease the linear and angular errors associated with localisation. For this purpose, a new localisation measure based on semantic segmentation and machine learning techniques should be added. On the other hand, building high-definition 3D maps is a booming topic, which may solve many of the problems that autonomous driving is prone to at present. Therefore, integrating a global D-GPS measurement into the developed system may eliminate the drift, allowing us to build high-definition 3D maps.

Author Contributions: Conceptualization, I.G.D. and M.R.; methodology, I.G.D.; software, I.G.D. and M.R.; validation, M.R., R.I.G., and C.S.M.; formal analysis, I.G.D.; investigation, I.G.D. and M.R.; resources, N.H.P.; writing—original draft preparation, I.G.D.; writing—review and editing, C.S.M., A.B., and N.H.P.; visualization, N.H.P.; supervision, D.F.L. All authors have read and agreed to the published version of the manuscript.

Funding: This work was supported in part by grant CCG2018/EXP-065 (UAH), in part by grant S2018/EMT-4362 SEGVAUTO 4.0 (CAM), in part by grant DPI2017-90035-R (Spanish Ministry of Science and Innovation), in part by grant 723021 (BRAVE, H2020), in part by the Electronic Component Systems for European Leadership Joint Undertaking through the European Union's Horizon 2020 Research and Innovation Program and Germany, Austria, Spain, Italy, Latvia, Belgium, The Netherlands, Sweden, Finland, Lithuania, Czech Republic, Romania, and Norway, under Grant 737469, in part by grant TRA2017-90620-REDT (Spanish Ministry of Science and Innovation) and in part by grant agreement Marie Skłodowska-Curie 754382.

Conflicts of Interest: The authors declare no conflict of interest.

Abbreviations

The following abbreviations are used in this manuscript:

GPS	Global Position System
DoF	Degrees Of Freedom
ICP	Iterative Closest Point
IMU	Inertial Measurement Unit
KITTI	Karsruhe Institute of Technology and Toyota Technological Institute
LiDAR	Laser imaging Detection And Ranging
LOAM	LiDAR Odometry And Mapping in real time
RANSAC	RANdom Sample Consensus
SLAM	Simultaneous Localisation And Mapping
SVD	Singular Value Decomposition
UKF	Unscented Kalman Filter

References

1. Hernández, N.; Daza, I.G.; Salinas, C.; Parra, I.; Alonso, J.; Llorca, D.F.; Sotelo, M.A. Intelligent Feature Selection Method for Accurate Laser-Based Mapping and Localisation in Self-Driving Cars. 2018. Available online: https://project.inria.fr/ppniv18/files/2018/10/paper10.pdf (accessed on 1 February 2020).

2. Parra, I.; Sotelo, M.A.; Llorca, D.F.; Ocaña, M. Robust visual odometry for vehicle localization in urban environments. *Robotica* **2010**, *28*, 441–452. [CrossRef]
3. Milanés, V.; Naranjo, J.E.; González, C.; Alonso, J.; de Pedro, T. Autonomous vehicle based in cooperative GPS and inertial systems. *Robotica* **2008**, *26*, 627–633. [CrossRef]
4. Hernández, N.; Corrales, H.; Parra, I.; Rentero, M., Llorca, D.F.; Sotelo, M.A. WiFi-based urban localisation using CNNs. In Proceedings of the 2019 IEEE Intelligent Transportation Systems Conference (ITSC), Auckland, New Zealand, 27–30 October 2019; pp. 1270–1275.
5. Parra Alonso, I.; Fernández Llorca, D.F.; Gavilan, M.; Álvarez Pardo, S.A.; Garcia Garrido, M.A.; Vlacic, L.; Sotelo, M.A. Accurate Global Localization Using Visual Odometry and Digital Maps on Urban Environments. *IEEE Trans. Intell. Transp. Syst.* **2012**, *13*, 1535–1545. [CrossRef]
6. Fraundorfer, F.; Scaramuzza, D. Visual Odometry: Part II: Matching, Robustness, Optimization, and Applications. *IEEE Robot. Autom. Mag.* **2012**, *19*, 78–90. [CrossRef]
7. Pavan, K.U.; Sahul, M.P.V.; Murthy, B.T.V. Implementation of stereo visual odometry estimation for ground vehicles. In Proceedings of the 2017 2nd IEEE International Conference on Recent Trends in Electronics, Information Communication Technology (RTEICT), Piscataway, NJ, USA, 19–20 May 2017; pp. 1173–1177.
8. Zhang, J.; Singh, S. Visual-lidar odometry and mapping: Low-drift, Robust, and Fast. In Proceedings of the IEEE International Conference on Robotics and Automation (ICRA 2015), Washington, DC, USA, 26–30 May 2015; pp. 2174–2181.
9. Rusinkiewicz, S.; Levoy, M. Efficient variants of the ICP algorithm. In Proceedings of the Third International Conference on 3-D Digital Imaging and Modeling, Washington, DC, USA, 28 May–1 June 2001; pp. 145–152.
10. Shi, X.; Liu, T.; Han, X. Improved Iterative Closest Point(ICP) 3D point cloud registration algorithm based on point cloud filtering and adaptive fireworks for coarse registration. *Int. J. Remote. Sens.* **2020**, *41*, 3197–3220. [CrossRef]
11. Zhang, J.; Singh, S. Low-Drift and Real-Time Lidar Odometry and Mapping. *Auton. Robot.* **2020**, *41*, 401–416. [CrossRef]
12. Low, K.L. Linear Least-Squares Optimization for Point-to-Plane ICP Surface Registration. 2004. Available online: http://www.cs.unc.edu/techreports/04-004.pdf (accessed on 10 December 2019)
13. Sun, S.; Song, W.; Tian, Y.; Fong, S. An ICP-Based Point Clouds Registration Method for Indoor Environment Modeling. In *Advanced Multimedia and Ubiquitous Engineering*; Park, J.J., Yang, L.T., Jeong, Y.S., Hao, F., Eds.; Springer: Singapore, 2020; pp. 339–344.
14. Besl, P.J.; McKay, N.D. Method for registration of 3-D shapes. In *Sensor Fusion IV: Control Paradigms and Data Structures*; Schenker, P.S., Ed.; International Society for Optics and Photonics, SPIE: Boston, MA, USA, 1992; Volume 1611, pp. 586–606. [CrossRef]
15. Myronenko, A.; Song, X. Point Set Registration: Coherent Point Drift. *IEEE Trans. Pattern Anal. Mach. Intell.* **2010**, *32*, 2262–2275. [CrossRef] [PubMed]
16. Fitzgibbon, A.W. Robust registration of 2D and 3D point sets. *Image Vis. Comput.* **2003**, *21*, 1145–1153. [CrossRef]
17. Stoyanov, T.; Magnusson, M.; Andreasson, H.; Lilienthal, A.J. Fast and accurate scan registration through minimization of the distance between compact 3D NDT representations. *Int. J. Robot. Res.* **2012**, *31*, 1377–1393. [CrossRef]
18. Ahuja, S.; Waslander, S.L. 3D Scan Registration Using Curvelet Features. In Proceedings of the 2014 Canadian Conference on Computer and Robot Vision, Montreal, QC, Canada, 6–9 May 2014; pp. 77–83.
19. Rusu, R.B.; Blodow, N.; Beetz, M. Fast Point Feature Histograms (FPFH) for 3D registration. In Proceedings of the 2009 IEEE International Conference on Robotics and Automation, Kobe, Japan, 12–17 May 2009; pp. 3212–3217.
20. Zhou, Q.Y.; Park, J.; Koltun, V. Fast Global Registration. In *Computer Vision—ECCV 2016*; Leibe, B., Matas, J., Sebe, N., Welling, M., Eds.; Springer International Publishing: Cham, UK, 2016; pp. 766–782.
21. Chen, X.; Milioto, A.; Palazzolo, E.; Giguère, P.; Behley, J.; Stachniss, C. SuMa++: Efficient LiDAR-based Semantic SLAM. In Proceedings of the 2019 IEEE/RSJ International Conference on Intelligent Robots and Systems (IROS), Macau, China, 3–8 November 2019; pp. 4530–4537. [CrossRef]
22. Leutenegger, S.; Lynen, S.; Bosse, M.; Siegwart, R.; Furgale, P. Keyframe-based visual–inertial odometry using nonlinear optimization. *Int. J. Robot. Res.* **2015**, *34*, 314–334. [CrossRef]

23. Polack, P.; Altché, F.; d'Andréa-Novel, B.; de La Fortelle, A. The kinematic bicycle model: A consistent model for planning feasible trajectories for autonomous vehicles? In Proceedings of the 2017 IEEE Intelligent Vehicles Symposium (IV), Los Angeles, CA, USA, 11–14 June 2017; pp. 812–818.
24. Graeter, J.; Wilczynski, A.; Lauer, M. LIMO: Lidar-Monocular Visual Odometry. In Proceedings of the 2018 IEEE/RSJ International Conference on Intelligent Robots and Systems (IROS), Madrid, Spain, 1–5 October 2018; pp. 7872–7879.
25. Yin, D.; Zhang, Q.; Liu, J.; Liang, X.; Wang, Y.; Maanpää, J.; Ma, H.; Hyyppä, J.; Chen, R. CAE-LO: LiDAR Odometry Leveraging Fully Unsupervised Convolutional Auto-Encoder for Interest Point Detection and Feature Description. *arXiv* **2020**, arXiv:2001.01354.
26. Neuhaus, F.; Koß, T.; Kohnen, R.; Paulus, D. MC2SLAM: Real-Time Inertial Lidar Odometry Using Two-Scan Motion Compensation. In *Pattern Recognition*; Brox, T., Bruhn, A., Fritz, M., Eds.; Springer International Publishing: Cham, UK, 2019; pp. 60–72.
27. Obialero, E. A Refined Vehicle Dynamic Model for Driving Simulators. 2013. Available online: http://publications.lib.chalmers.se/records/fulltext/179787/179787.pdf (accessed on 1 December 2019)
28. D'Alfonso, L.; Lucia, W.; Muraca, P.; Pugliese, P. Mobile robot localization via EKF and UKF: A comparison based on real data. *Robot. Auton. Syst.* **2015**, *74*, 122–127. [CrossRef]
29. Wan, E.A.; Van Der Merwe, R. The unscented Kalman filter for nonlinear estimation. In Proceedings of the IEEE 2000 Adaptive Systems for Signal Processing, Communications, and Control Symposium (Cat. No.00EX373), Lake Louise, AB, Canada, 4 October 2000; pp. 153–158.
30. Merriaux, P.; Dupuis, Y.; Boutteau, R.; Vasseur, P.; Savatier, X. LiDAR point clouds correction acquired from a moving car based on CAN-bus data. *arXiv* **2017**, arXiv:1706.05886.
31. Zhang, B.; Zhang, X.; Wei, B.; Qi, C. A Point Cloud Distortion Removing and Mapping Algorithm based on Lidar and IMU UKF Fusion. In Proceedings of the 2019 IEEE/ASME International Conference on Advanced Intelligent Mechatronics (AIM), Hong Kong, China, 8–12 July 2019; pp. 966–971. [CrossRef]
32. Van Der Merwe, R.; Wan, E.A. Sigma-Point Kalman Filters for Probabilistic Inference in Dynamic State-Space Models. Ph.D. Thesis, OGI School of Science Engineering at Oregon Health Science University, Portland, OR, USA, 2004.
33. Julier, S.J.; Uhlmann, J.K.; Durrant-Whyte, H.F. A new approach for filtering nonlinear systems. In Proceedings of 1995 American Control Conference—ACC'95, Seattle, WA, USA, 21–23 June 1995; Volume 3, pp. 1628–1632.
34. Chen, Y.; Medioni, G. Object modeling by registration of multiple range images. In Proceedings of the 1991 IEEE International Conference on Robotics and Automation, Sacramento, CA, USA, 9–11 April 1991; Volume 3, pp. 2724–2729. [CrossRef]
35. Wu, J.; Liu, M.; Zhou, Z.; Li, R. Fast Rigid 3D Registration Solution: A Simple Method Free of SVD and Eigen-Decomposition. *arXiv* **2018**, arXiv:1806.00627.
36. Gan, Q.; Harris, C.J. Comparison of two measurement fusion methods for Kalman-filter-based multisensor data fusion. *IEEE Trans. Aerosp. Electron. Syst.* **2001**, *37*, 273–279. [CrossRef]
37. Geiger, A.; Lenz, P.; Urtasun, R. Are we ready for Autonomous Driving? The KITTI Vision Benchmark Suite. In Proceedings of the Conference on Computer Vision and Pattern Recognition (CVPR), Providence, RI, USA, 18–20 June 2012.
38. Deschaud, J. IMLS-SLAM: Scan-to-Model Matching Based on 3D Data. In Proceedings of the 2018 IEEE International Conference on Robotics and Automation (ICRA), Brisbane, Australia, 21–25 May 2018; pp. 2480–2485.

39. Ji, K.; Chen, H.; Di, H.; Gong, J.; Xiong, G.; Qi, J.; Yi, T. CPFG-SLAM:a Robust Simultaneous Localization and Mapping based on LIDAR in Off-Road Environment. In Proceedings of the 2018 IEEE Intelligent Vehicles Symposium (IV), Changshu, China, 26–30 June 2018; pp. 650–655.
40. Hong, H.; Lee, B.H. Probabilistic normal distributions transform representation for accurate 3D point cloud registration. In Proceedings of the 2017 IEEE/RSJ International Conference on Intelligent Robots and Systems (IROS), Vancouver, BC, Canada, 24–28 September 2017; pp. 3333–3338.

MDPI
St. Alban-Anlage 66
4052 Basel
Switzerland
Tel. +41 61 683 77 34
Fax +41 61 302 89 18
www.mdpi.com

Sensors Editorial Office
E-mail: sensors@mdpi.com
www.mdpi.com/journal/sensors

www.ingramcontent.com/pod-product-compliance
Lightning Source LLC
Chambersburg PA
CBHW061445220926
51850CB00004B/19
* 9 7 8 3 0 3 9 4 3 4 0 2 2 *